HANDBOOK OF
SEPARATION PROCESS
TECHNOLOGY

HANDBOOK OF SEPARATION PROCESS TECHNOLOGY

Edited by

RONALD W. ROUSSEAU
Georgia Institute of Technology

A Wiley-Interscience Publication

JOHN WILEY & SONS

New York   **Chichester**   **Brisbane**   **Toronto**   **Singapore**

Library of Congress Cataloging in Publication Data:

Handbook of separation process technology.

"A Wiley-Interscience publication."
Includes index.
1. Separation (Technology)—Handbooks, manuals,
etc. I. Rousseau, Ronald W., 1943–

TP156.S45H34 1987   660.2′842   86-28134
ISBN 0-471-89558-X

Printed in the United States of America

10 9 8 7 6 5 4 3 2

In memory of my father, Ivy John,
and to my mother, Dorothy

Contributors

MICHAEL M. ABBOTT, Department of Chemical and Environmental Engineering, Rensselaer Polytechnic Institute, Troy, New York

K. P. ANANTHAPADMANABHAN, Henry Krumb School of Mines, Columbia University, New York, New York

RICHARD A. ANDERSON, Union Carbide Corporation, Tarrytown, New York

THOMAS W. CHAPMAN, Department of Chemical Engineering, University of Wisconsin, Madison, Wisconsin

REY T. CHERN, Department of Chemical Engineering, North Carolina State University, Raleigh, North Carolina

ANN N. CLARKE, AWARE Incorporated, Nashville, Tennessee

FRANCIS LOUIS DIRK CLOETE, Department of Chemical and Metallurgical Engineering, University of Stellenbosch, Stellenbosch, South Africa

WILLIAM EYKAMP, Koch Membrane Systems, Inc., Wilmington, Massachusetts

JAMES R. FAIR, Department of Chemical Engineering, The University of Texas, Austin, Texas

J. W. FRANKENFELD, Exxon Research and Engineering Company, Linden, New Jersey

LOUIS J. JACOBS, JR., Corporate Engineering Division, A. E. Staley Manufacturing Company, Decatur, Illinois

GEORGE E. KELLER II, Union Carbide Corporation, South Charleston, West Virginia

ROBERT M. KELLY, Department of Chemical Engineering, The Johns Hopkins University, Baltimore, Maryland

C. JUDSON KING, Department of Chemical Engineering, University of California, Berkeley, California

DONALD J. KIRWAN, Department of Chemical Engineering, University of Virginia, Charlottesville, Virginia

ELIAS KLEIN, Division of Nephrology, School of Medicine, University of Louisville, Louisville, Kentucky

ARTHUR L. KOHL, Rocketdyne Division, Rockwell International, Canoga Park, California

WILLIAM J. KOROS, Department of Chemical Engineering, The University of Texas, Austin, Texas

ROBERT E. LACEY, retired from Southern Research Institute, Birmingham, Alabama

NORMAN N. LI, Allied-Signal, Inc., Des Plaines, Illinois

CHARLES G. MOYERS, JR., Union Carbide Corporation, South Charleston, West Virginia

HAROLD R. NULL, Monsanto Company, St. Louis, Missouri

W. ROY PENNEY, Corporate Process Engineering, A. E. Staley Manufacturing Company, Decatur, Illinois

JOHN M. PRAUSNITZ, Department of Chemical Engineering, University of California, Berkeley, California

RONALD W. ROUSSEAU, School of Chemical Engineering, Georgia Institute of Technology, Atlanta, Georgia

HENRY G. SCHWARTZBERG, Department of Food Engineering, University of Massachusetts, Amherst, Massachusetts

A. H. P. SKELLAND, School of Chemical Engineering, Georgia Institute of Technology, Atlanta, Georgia

P. SOMASUNDARAN, Henry Krumb School of Mines, Columbia University, New York, New York

JONATHAN STEEN, Koch Membrane Systems, Inc., Wilmington, Massachusetts

MICHAEL STREAT, Department of Chemical Engineering and Chemical Technology, Imperial College, London, England

D. WILLIAM TEDDER, School of Chemical Engineering, Georgia Institute of Technology, Atlanta, Georgia

MILTON E. WADSWORTH, College of Mines and Mineral Industries, University of Utah, Salt Lake City, Utah

PHILLIP C. WANKAT, School of Chemical Engineering, Purdue University, West Lafayette, Indiana

RICHARD A. WARD, Division of Nephrology, School of Medicine, University of Louisville, Louisville, Kentucky

DAVID J. WILSON, Departments of Chemistry and Environmental Engineering, Vanderbilt University, Nashville, Tennessee

CARMEN M. YON, Union Carbide Corporation, Tarrytown, New York

Preface

Separation processes are central to the petroleum, chemical, petrochemical, pulp, pharmaceutical, mineral, and other industries. Major portions of capital and operating expenses of such industries are associated with one or more separation processes; consequently, the impact of separation process technology on corporate profitability is great in most of these industries. The growth of new industries, based on biotechnology or electronics, for example, requires the development of new separation techniques and application of historically successful technology in new environments.

The field of separation processes is broad, encompassing subject matter that ranges from phase-equilibrium thermodynamics to hardware design. It is the objective of the *Handbook of Separation Process Technology* to cover in a single volume the operations that constitute most of the industrially important separation processes. The philosophy that has guided the selection of topics and the formulation of instructions provided to authors is that the material presented be accessible to engineers and scientists with divergent backgrounds: it must be more than simply a collection of data and correlations—although those are included where appropriate—and there must be sufficient descriptive material for the reader to understand the principles on which an individual process is based, how the process is implemented, and where the process fits into the entire field of industrially important separation techniques.

In addition to the obvious factors that influence the selection, design, and operation of separation processes are several items that may be more obscure. These include the economics governing plant design, a renewed awareness of the impact of chemical processing on the environment, new products that require purity levels exceeding those typically encountered, a recognition of the hazards associated with production of dangerous materials, and a growing awareness of the importance of political and natural limitations on the availability of raw materials. Moreover, the methods by which processes of all types are conceived and designed have been altered substantially by the growth of computer hardware and software. Several of the chapters in this book address the influence of these factors on the individual separation processes.

This handbook begins with coverage of principles that intersect most separation processes: phase equilibria, mass transfer, and phase segregation. Separation processes cannot be understood without at least some acquaintance with the driving force for such processes—hence the discussion of phase equilibria—and with factors influencing the rate at which the processes occur—hence the discussion of mass transfer. Phase segregation often may be overlooked when considering processes that involve inter-phase mass transfer. However, as discussed by the authors of Chapter 3, the principles of separating phases cut across several separation techniques and have direct bearing on the performance of many multiphase processes. Part I of the handbook is concluded with Chapter 4, which discusses several general processing techniques. This chapter covers the principles of contacting phases in multistage or continuous-contacting cascades, introduces the effects of various process requirements on the synthesis and control of separation processes, and concludes with a brief examination of the needs associated with biological separations and processing thermally sensitive or high-purity materials.

Part II of this handbook contains descriptions of specific separation processes. It begins with four chapters on the most common operations: distillation, absorption and stripping, and extraction. Distillation, which is covered in Chapter 5, continues to be the workhorse of many industrial separations and it is likely to continue in that role. Absorption and stripping (Chapter 6) can be found in a number of industries, most notably in natural and synthetic gas processing but also in environmentally important scrubbing operations. Many general aspects of vapor–liquid and gas–liquid contacting are discussed in the chapters on distillation and absorption and stripping. The coverage of extraction is divided between Chapter 7, which deals with processes whose primary concern is the recovery of organic chemicals, and Chapter 8, whose focus is metals recovery. General features of extraction processes are included in Chapter 7, whereas Chapter 8 features the highly system-specific aspects associated with the chemistry involved in metals extraction.

Leaching deals with the removal of soluble material from an insoluble or partially soluble solid matrix. Chapters 9 and 10 cover the subject from the perspectives of metals and organic chemicals recovery, respectively. Chapter 11 discusses crystallization and demonstrates the relationships among crystallizer hardware, crystal nucleation and growth, and crystal size distribution. Chapters 12, 13, and 14 are devoted to processes that involve the concentration of a mixture constituent at a solid–fluid surface: Chapter 12 discusses the principles and commercially important implementations of adsorption, Chapter 13 focuses on ion exchange, and Chapter 14 examines the role of chromatography in making separations on a large scale.

One of the areas in separation process technology that seems to hold potential for recovery of dilute species from aqueous mixtures is the use of chemical reactions to form easily separable complexes. Such complexes are separated from the milieu of the original system and subsequently the reactions by which they were formed are reversed. Chapter 15 discusses the principles of processes based on this concept, and it provides guidelines in considering this approach for specific systems.

Chapters 16 and 17 examine the use of bubble and foam separation technology. Although the use of these techniques is not widespread, it is thought that the technology itself has the potential for implementation in accomplishing several important separations (e.g., protein separation). The two areas in which bubble and foam separations have been used are ore flotation, which is the primary emphasis of Chapter 16, and waste treatment, the focus of Chapter 17.

The use of membranes as a vehicle for accomplishing separations of many types has been envisioned for a long time, and several corporate entities have been spawned around this objective. Recent developments in the membrane industry have brought these notions to fruition, and the use of membrane-based separations is one of the most rapidly growing areas in process technology. Accordingly, a large portion of this handbook (Chapters 18–21) has been devoted to their coverage. The specific areas included in these chapters are reverse osmosis, ultrafiltration, liquid membranes, gas separations, dialysis, and electrodialysis.

The subject matter of this handbook is concluded by addressing the question of which separation process should be used in a given situation. The material in Chapter 22 suggests methods for screening separation processes based on matching physical properties of the system constituents with particular operations, and it examines the influence of scale of operation and design reliability on process selection. The chapter concludes with general observations regarding system synthesis, simulation, and design.

In choosing topics for inclusion in this handbook, the intent has been to provide a comprehensive coverage of those separation processes on which much of the chemical and related businesses are based. In addition, some separation processes were selected for coverage based on their potential industrial applications. A complete coverage of the latter category would have resulted in an overbalanced and exceedingly large volume; therefore, some separation techniques that will be developed into commercially successful operations undoubtedly have been overlooked.

The task of editing a volume of this scope has been educational, and it is hoped that the result will be judged a worthwhile contribution to the technical literature. By and large it has been a pleasure to work with the 35 authors whose chapters are included in this handbook and with numerous reviewers, potential authors who fell by the wayside, and secretaries who protected their bosses from my prodding phone calls. Thanks are extended to all authors, especially to those who completed their manuscripts on time and have been waiting patiently for appearance of their work in print.

I extend special thanks to the members of the Editorial Board, James R. Fair, George E. Keller, C. Judson King, and Harold R. Null; their counsel, particularly in the planning stage of this handbook, is acknowledged gratefully. Thurmond Poston, formerly Handbook Editor at John Wiley & Sons, is responsible for persuading me that this chore would be a worthwhile endeavor. Without his encouragement I would not have begun such a task. Upon Mr. Poston's retirement, James L. Smith, editor in the Wiley-Interscience Division, assumed the editorial responsibility, and he has carried on in grand style. Much gratitude is owed to both these men. Christina Mikulak is to be applauded for the care with which she performed the tedious job of copy editing. I would be remiss not to acknowledge the support of friends and colleagues in the Department of Chemical Engineering at North Carolina State University, where most of my work on this handbook took place.

Finally, I wish to thank Sandra for her help with the index and her companionship and encouragement throughout the preparation of this handbook.

RONALD W. ROUSSEAU

Atlanta, Georgia
February 1987

Contents

PART I

GENERAL PRINCIPLES

CHAPTER 1

Phase Equilibria

MICHAEL M. ABBOTT
Department of Chemical and Environmental Engineering
Rensselaer Polytechnic Institute, Troy, New York

JOHN M. PRAUSNITZ
Department of Chemical Engineering
University of California,
Berkeley, California

1.1 INTRODUCTION

Most of the common separation methods used in the chemical industry rely on a well-known observation: when a multicomponent two-phase system is given sufficient time to attain a stationary state called equilibrium, the composition of one phase is different from that of the other. It is this property of nature which enables separation of fluid mixtures by distillation, extraction, and other diffusional operations. For rational design of such operations it is necessary to have a quantitative description of how a component distributes itself between two contacting phases. Phase-equilibrium thermodynamics, summarized here, provides a framework for establishing that description.

If experimental phase-equilibrium measurements were simple, fast, and inexpensive, chemical engineers would have little need for phase-equilibrium thermodynamics because in that happy event all component-distribution data required for design would be obtained readily in the laboratory. Unfortunately, however, component-distribution data are not easily obtained because experimental studies require much patience and skill. As a result, required data are often not at hand but must be estimated using suitable physico-chemical models whose parameters are obtained from correlations or from limited experimental data.

It was Einstein who said that when God made the world, he was subtle but not malicious. The subtlety of nature is evident by our inability to construct models of mixtures which give directly to the chemical engineer the required information in the desired form: temperature, pressure, phase compositions. Nature, it seems, does not choose to reveal secrets in the everyday language of chemical process design but prefers to use an abstract language—thermodynamics.

To achieve a quantitative description of phase equilibria, thermodynamics provides a useful theoretical framework. By itself, thermodynamics cannot provide all the numerical information we desire but, when coupled with concepts from molecular physics and physical chemistry, it can efficiently organize limited experimental information toward helpful interpolation and extrapolation. Thermodynamics is not magic; it cannot produce something for nothing: some experimental information is always necessary. But when used with skill and courage, thermodynamics can squeeze the last drop out of a nearly dried-up lemon.

The brief survey presented here must necessarily begin with a discussion of thermodynamics as a language; most of Section 1.2 is concerned with the definition of thermodynamic terms such as chemical potential, fugacity, and activity. At the end of Section 1.2, the phase-equilibrium problem is clearly stated in several thermodynamic forms; each of these forms is particularly suited for a particular situation, as indicated in Sections 1.5, 1.6, and 1.7.

3

Section 1.3 discusses fugacities (through fugacity coefficients) in the vapor phase. Illustrative examples are given using equations of state.

Section 1.4 discusses fugacities (through activity coefficients) in the liquid phase. Illustrative examples are given using semiempirical models for liquid mixtures of nonelectrolytes.

Section 1.5 gives examples for vapor–liquid equilibria at ordinary pressures and for liquid–liquid equilibria. Section 1.6 discusses equilibria for systems containing a solid phase in addition to a liquid or gaseous phase, and Section 1.7 gives an introduction to methods for describing fluid-phase equilibria at high pressures.

This brief survey of applied phase-equilibrium thermodynamics can do no more than summarize the main ideas that constitute the present state of the art. Attention is restricted to relatively simple mixtures as encountered in the petroleum, natural gas, and petrochemical industries; unfortunately, limited space does not allow discussion of other important systems such as polymer mixtures, electrolyte solutions, metallic alloys, molten salts, refractories (such as ceramics), or aqueous solutions of biologically important solutes. However, it is not only lack of space that is responsible for these omissions because, at present, thermodynamic knowledge is severely limited for these more complex systems.

1.2 THERMODYNAMIC FRAMEWORK FOR PHASE EQUILIBRIA

1.2-1 Conventions and Definitions

Lowercase roman letters usually denote *molar* properties of a phase. Thus, g, h, s, and v are the molar Gibbs energy, molar enthalpy, molar entropy, and molar volume. When it is essential to distinguish between a molar property of a mixture and that of a pure component, we identify the pure-component property by a subscript. For example, h_i is the molar enthalpy of pure i. *Total* properties are usually designated by capital letters. Thus H is the total enthalpy of a mixture; it is related to the molar mixture enthalpy h by $H = nh$, where n is the total number of moles in the mixture.

Mole fraction is the conventional measure of composition. We use the generic symbol x_i to denote this quantity when no particular phase (solid, liquid, or gas) is implied. When referring to a *specific* phase, we use common notation, for example, x_i for liquid-phase mole fraction and y_i for the vapor-phase mole fraction. The dual usage of x_i should cause no confusion because it will be clear from the context whether an arbitrary phase or a liquid phase is under consideration.

The molar *residual function* m^R (or m_i^R) is the difference between molar property m (or m_i) of a real mixture (or pure substance i) and the value m' (or m_i') it would have were it an *ideal gas* at the same temperature (T), pressure (P), and composition:

$$m^R \equiv m - m' \tag{1.2-1}$$

$$m_i^R \equiv m_i - m_i' \tag{1.2-2}$$

The residual functions (e.g., g^R, h^R, and s^R) provide measures of the contributions of intermolecular forces to thermodynamic properties.

The molar *excess function* m^E is the difference between a molar mixture property m and the value m^{id} the mixture would have were it an *ideal solution* at the same temperature, pressure, and composition:

$$m^E \equiv m - m^{id} \tag{1.2-3}$$

Excess functions are related to the corresponding residual functions:

$$m^E = m^R - \sum_i x_i m_i^R, \tag{1.2-4}$$

Thus, the excess functions (e.g., g^E, h^E, and s^E) also reflect the contributions of intermolecular forces to mixture property m.

Partial molar property $\overline{m}_i$ corresponding to molar mixture property m is defined in the usual way:

$$\overline{m}_i \equiv \left(\frac{\partial M}{\partial n_i} \right)_{T,P,n_j} \equiv \left[\frac{\partial (nm)}{\partial n_i} \right]_{T,P,n_j} \tag{1.2-5}$$

where subscript n_j denotes constancy of all mole numbers except n_i. All $\overline{m}_i$ have the important feature that

$$m = \sum_i x_i \overline{m}_i \tag{1.2-6}$$

or

$$M = \sum_i n_i \overline{m}_i \qquad (1.2\text{-}7)$$

That is, a molar property of a mixture is the mole-fraction-weighted sum of its constituent partial molar properties. The partial molar property $\overline{m}_i$ of species i in solution becomes equal to molar property m_i of pure i in the appropriate limit:

$$\lim_{x_i \to 1} \overline{m}_i = m_i \qquad (1.2\text{-}8)$$

The *chemical potential* μ_i is identical to the partial molar Gibbs energy $\overline{g}_i$:

$$\mu_i = \overline{g}_i \equiv \left(\frac{\partial G}{\partial n_i}\right)_{T,P,n_j} = \left[\frac{\partial (ng)}{\partial n_i}\right]_{T,P,n_j} \qquad (1.2\text{-}9)$$

Thus, the μ_i, when multiplied by mole fractions, sum to the molar Gibbs energy of the mixture:

$$g = \sum_i x_i \mu_i \qquad (1.2\text{-}10)$$

or

$$G = \sum_i n_i \mu_i \qquad (1.2\text{-}11)$$

Partial molar properties play a central role in phase-equilibrium thermodynamics, and it is convenient to broaden their definition to include partial molar residual functions and partial molar excess functions. Hence, we define, analogous to Eq. (1.2-5),

$$\overline{m}_i^R \equiv \left(\frac{\partial M^R}{\partial n_i}\right)_{T,P,n_j} = \left[\frac{\partial (nm^R)}{\partial n_i}\right]_{T,P,n_j} \qquad (1.2\text{-}12)$$

and

$$\overline{m}_i^E \equiv \left(\frac{\partial M^E}{\partial n_i}\right)_{T,P,n_j} = \left[\frac{\partial (nm^E)}{\partial n_i}\right]_{T,P,n_j} \qquad (1.2\text{-}13)$$

1.2-2   Criteria for Phase Equilibria

Consider the situation shown in Fig. 1.2-1, where two phases α and β are brought into contact and allowed to interact until no changes are observed in their intensive properties. The condition where these properties assume stationary values is a state of *phase equilibrium*. It is characterized by temperature T and pressure P (both assumed uniform throughout the two-phase system) and by the sets of concentrations $\{z_i^\alpha\}$ and $\{z_i^\beta\}$, which may or may not be the sets of mole fractions $\{x_i^\alpha\}$ and $\{x_i^\beta\}$. The basic problem of phase equilibrium is this: given values for some of the intensive variables (T, P, and the concentrations), find values for the remaining ones.

The route to the solution of problems in chemical and phase equilibria is indirect; it derives from a formalism developed over a century ago by the American physicist J. W. Gibbs.[1] Let G be the total Gibbs energy of a closed, multiphase system of constant and uniform T and P. Equilibrium states are those for which G is a *minimum*, subject to material-balance constraints appropriate to the problem:

$$G_{T,P} = \text{minimum} \qquad (1.2\text{-}14)$$

Although Eq. (1.2-14) is sometimes used directly for solution of complex equilibrium problems, it is more often employed in equivalent algebraic forms which use explicitly the chemical potential or other related quantities. Consider a closed system containing π phases and N components. Introducing the chemical potential μ_i^p of each component i in each phase p and incorporating material-balance constraints, one obtains as necessary conditions to Eq. (1.2-14) a set of $N(\pi - 1)$ equations for phase equilibrium:

$$\mu_i^p = \mu_i^\pi \qquad (i = 1, 2, \ldots, N; \ p = \alpha, \beta, \ldots, \pi - 1) \qquad (1.2\text{-}15)$$

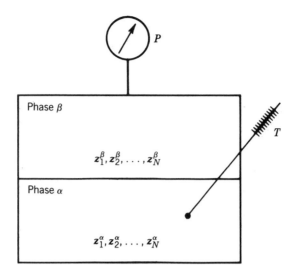

FIGURE 1.2-1 A multicomponent system in two-phase equilibrium.

Thus, temperature, pressure, and the chemical potential of each distributed component are uniform for a closed system in phase equilibrium. If the system contains chemically reactive species, then additional equations are required to characterize the equilibrium state.

Equation (1.2-15) is a basis for the formulation of phase-equilibrium problems. However, since the chemical potential has some practical and conceptual shortcomings, it is useful to replace μ_i with a related quantity, $\hat{f}_i$, the *fugacity*. Equation (1.2-15) is then replaced by the equivalent criterion for phase equilibrium,

$$\hat{f}_i^p = \hat{f}_i^\pi \qquad (i = 1, 2, \ldots, N; \, p = \alpha, \beta, \ldots, \pi - 1) \tag{1.2-16}$$

Equation (1.2-16) is the basis for all applications considered in this chapter. The major task is to represent the dependence of the fugacity on temperature, pressure, and concentration.

1.2-3 Behavior of the Fugacity

Table 1.2-1 summarizes important general thermodynamic formulas for the fugacity. Equations (1.2-17) and (1.2-18) *define* the fugacity $\hat{f}_i$ of a component in solution; Eqs. (1.2-19) and (1.2-20) similarly *define* the fugacity f of a mixture. For a component in an *ideal-gas* mixture, Eq. (1.2-18) implies that

$$\hat{f}_i = y_i P \quad \text{(ideal gas)} \tag{1.2-26}$$

which leads to the interpretation of a vapor-phase fugacity as a corrected partial pressure. Equations (1.2-21) and (1.2-22) are useful summarizing relationships, which provide by inspection general expressions for the temperature and pressure derivatives of the fugacities; note here the appearance of the residual enthalpy h^R. Equations (1.2-23) and (1.2-24) are partial-property relationships, and Eq. (1.2-25) is one form of the *Gibbs–Duhem equation*.

A pure substance i may be considered a special case of either a mixture or of a component in solution, in the limit as mole fraction x_i approaches unity. Thus, formulas for the fugacity f_i of pure i are recovered as special cases of Eqs. (1.2-21)–(1.2-22). In particular,

$$d \ln f_i \equiv \frac{dg_i}{RT} \quad \text{(constant } T) \tag{1.2-27}$$

$$\lim_{P \to 0} \left(\frac{f_i}{P} \right) \equiv 1 \tag{1.2-28}$$

$$d \ln f_i = -\frac{h_i^R}{RT^2} \, dT + \frac{v_i}{RT} \, dP \tag{1.2-29}$$

TABLE 1.2-1 Summary of Thermodynamic Relations for Fugacity

$$d \ln \hat{f}_i \equiv \frac{d\mu_i}{RT} \quad \text{(constant } T\text{)} \tag{1.2-17}$$

$$\lim_{P \to 0} \left(\frac{\hat{f}_i}{x_i P} \right) \equiv 1 \tag{1.2-18}$$

$$d \ln f \equiv \frac{dg}{RT} \quad \text{(constant } T, x\text{)} \tag{1.2-19}$$

$$\lim_{P \to 0} \left(\frac{f}{P} \right) \equiv 1 \tag{1.2-20}$$

$$d \ln \hat{f}_i = -\frac{\bar{h}_i^R}{RT^2} dT + \frac{\bar{v}_i}{RT} dP \quad \text{(constant } x\text{)} \tag{1.2-21}$$

$$d \ln f = -\frac{h^R}{RT^2} dT + \frac{v}{RT} dP \quad \text{(constant } x\text{)} \tag{1.2-22}$$

$$\ln \left(\frac{\hat{f}_i}{x_i} \right) = \left[\frac{\partial (n \ln f)}{\partial n_i} \right]_{T,P,n_j} \tag{1.2-23}$$

$$\ln f = \sum_i x_i \ln \left(\frac{\hat{f}_i}{x_i} \right) \tag{1.2-24}$$

$$\sum_i x_i \, d \ln \left(\frac{\hat{f}_i}{x_i} \right) = -\frac{h^R}{RT^2} dT + \frac{v}{RT} dP \tag{1.2-25}$$

The fugacity of a pure substance depends on T and P. *Absolute* values for f_i are computed from

$$f_i = P \exp \int_0^P (Z_i - 1) \frac{dP}{P} \quad \text{(constant } T\text{)} \tag{1.2-30}$$

which follows from Eqs. (1.2-28) and (1.2-29) upon introduction of the compressibility factor Z_i ($\equiv Pv_i/RT$). Use of Eq. (1.2-30) requires a *PVT* equation of state, valid from $P = 0$ to the physical state of interest at pressure P. *Relative* values of f_i are given by the *Poynting correction*:

$$\frac{f_i(P_2)}{f_i(P_1)} = \exp \int_{P_1}^{P_2} \frac{v_i}{RT} dP \quad \text{(constant } T\text{)} \tag{1.2-31}$$

which follows from Eq. (1.2-29). Equation (1.2-31) is most often used for calculation of the fugacity of a condensed phase, relative to the fugacity of the same phase at saturation pressure P_i^{sat}.

Suppose we require the absolute fugacity of a pure subcooled liquid at some pressure P and that available data include the vapor–liquid saturation pressure P_i^{sat}, an equation of state for the vapor phase, and molar volumes v_i^L for the liquid. Application of Eqs. (1.2-30) and (1.2-31), together with the criterion for pure-fluid vapor–liquid equilibrium,

$$f_i^L (P_i^{\text{sat}}) = f_i^V (P_i^{\text{sat}}) \tag{1.2-32}$$

gives the required result, namely,

$$f_i = P_i^{\text{sat}} \exp \left[\int_0^{P_i^{\text{sat}}} (Z_i^V - 1) \frac{dP}{P} + \int_{P_i^{\text{sat}}}^P \frac{v_i^L}{RT} dP \right] \tag{1.2-33}$$

Figure 1.2-2 shows the fugacity of nitrogen at 100 K, as computed from Eqs. (1.2-30), (1.2-32), and (1.2-33). Also shown are several commonly employed approximations. The dashed line $f_i^V = P$ is the ideal-gas approximation to the vapor fugacity; it is a special case of Eq. (1.2-26) and is a consequence of the definition, Eq. (1.2-28). Note that the ideal-gas approximation becomes *asymptotically* valid as P approaches zero.

The horizontal dashed line $f_i^L = P_i^{\text{sat}}$ is the approximation to f_i^L employed in Raoult's Law for vapor–

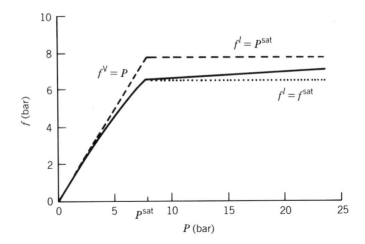

FIGURE 1.2-2 Pressure dependence of fugacity f of nitrogen at 100 K. Dashed and dotted lines represent approximations to real behavior.

liquid equilibrium. A much better approximation to f_i^L at moderate pressure is afforded by the horizontal dotted line, $f_i^L = f_i^{sat}$. This approximation involves neglect of the Poynting correction given by Eq. (1.2-31).

Since the molar volume of a condensed phase is frequently insensitive to pressure, Eq. (1.2-31) can often be approximated by

$$\frac{f_i(P_2)}{f_i(P_1)} \approx \exp\left[\frac{v_i(P_2 - P_1)}{RT}\right] \tag{1.2-34}$$

With v_i taken as the molar volume of the saturated liquid, relative fugacities computed from Eq. (1.2-34) for subcooled liquid nitrogen at 100 K produce results nearly identical to those given by the solid curve in Fig. 1.2-2.

The fugacity $\hat{f}_i$ of a component in solution depends on temperature, pressure, and composition. Figure 1.2-3 shows the variation $\hat{f}_i$ with x_i for acetone in two binary liquid mixtures (acetone–methanol and acetone–chloroform) at 1 bar and 50°C. Although they differ in detail, both $\hat{f}_i$ versus x_i curves have certain features in common. For example,

$$\lim_{x_i \to 0} \hat{f}_i = 0 \tag{1.2-35}$$

that is, the fugacity of a component in solution approaches zero as its concentration approaches zero. Moreover,

$$\lim_{x_i \to 1} \left(\frac{\hat{f}_i}{x_i}\right) = f_i \tag{1.2-36}$$

that is, the fugacity of a component in a nonelectrolyte solution asymptotically approaches the linear behavior represented by the dashed straight line

$$\hat{f}_i = x_i f_i \tag{1.2-37}$$

as its mole fraction approaches unity.

There is an analogous statement to Eq. (1.2-36) which applies to the limit of zero concentration in a binary mixture, namely,

$$\lim_{x_i \to 0} \left(\frac{\hat{f}_i}{x_i}\right) = \mathcal{H}_{i,j} \tag{1.2-38}$$

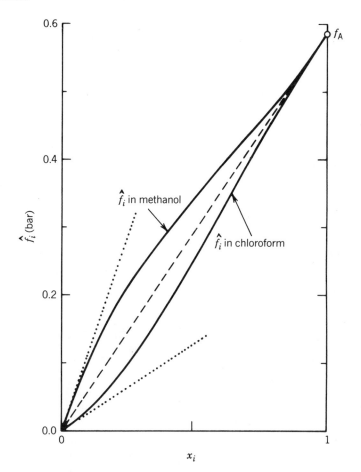

FIGURE 1.2-3 Composition dependence of fugacity $\hat{f}_i$ of acetone in two binary liquid mixtures at 50°C and 1 bar. Dashed and dotted lines represent approximations to real behavior.

where *Henry's constant* $\mathcal{K}_{i,j}$ is, for binary nonelectrolyte solutions, a positive definite number that depends on temperature and pressure. Unlike f_i in the analogous Eq. (1.2-36), the numerical value of $\mathcal{K}_{i,j}$ also depends on the identity of the other component j in the mixture; hence, the double subscript notation on $\mathcal{K}_{i,j}$. The dotted straight lines in Fig. 1.2-3 represent the equations

$$\hat{f}_i = x_i\,\mathcal{K}_{i,j} \tag{1.2-39}$$

which are given by construction as tangent lines drawn to the $\hat{f}_i$ versus x_i curves at $x_i = 0$. Henry's constants are then represented as intercepts of these tangent lines with the vertical axis $x_i = 1$.

Equations (1.2-35), (1.2-36), and (1.2-37) apply without modification to species i in a multicomponent mixture. However, Henry's constant, as defined by Eq. (1.2-38), can assume an infinity of values depending on the solvent composition. Thus, Henry's constant for a solute species in a multicomponent mixture is a function of temperature, pressure, and *composition*. The thermodynamic treatment of this topic is complex and is not considered in this chapter; the reader is referred to an article by Van Ness and Abbott.[2]

1.2-4 Normalized Fugacities

The group $x_i P$ appears as part of the definition of the component fugacity: see Eq. (1.2-18). It follows from this definition that $\hat{f}_i$ for a species in a *vapor* mixture is normally of the same order of magnitude as the partial pressure $y_i P$: see Eq. (1.2-26). Thus, it is convenient to introduce a normalized fugacity, called the *fugacity coefficient* $\hat{\phi}_i$, defined as the ratio of the component fugacity to the pressure–composition product:

$$\hat{\phi}_i \equiv \frac{\hat{f}_i}{x_i P} \tag{1.2-40}$$

Similarly, we write for a mixture that

$$\phi \equiv \frac{f}{P} \tag{1.2-41}$$

and for a pure component i that

$$\phi_i \equiv \frac{f_i}{P} \tag{1.2-42}$$

Fugacity coefficients are dimensionless; they are identically unity for ideal gases. For nonreacting *real* gases, their values approach unity as pressure approaches zero. Table 1.2-2 summarizes general thermodynamic relationships for the fugacity coefficients. Section 1.3 discusses the calculation of fugacity coefficients from $PVTx$ equations of state.

The composition dependence of the component fugacity $\hat{f}_i$ in condensed phases is conventionally represented through either of two normalized quantities called the *activity* and the *activity coefficient*. The motivation for the definitions of these quantities was provided by Fig. 1.2-3 and the accompanying discussion, where it was shown that for binary nonelectrolyte solutions the limiting $\hat{f}_i$ versus x_i behavior is a simple proportionality, given by Eq. (1.2-37) for $x_i \to 1$ and by Eq. (1.2-39) for $x_i \to 0$. Either of these limiting laws, when assumed to apply to *all* compositions at fixed temperature and pressure, can be used to define an *ideal solution*. We generalize this notion by writing

$$\hat{f}_i^{id} \equiv x_i f_i^{\circ} \quad \text{(constant } T, P) \tag{1.2-48}$$

where superscript id denotes ideal-solution behavior and f_i° is the *standard-state fugacity* of species i. If the ideal solution is defined so as to reproduce real behavior for $x_i \to 1$, then $f_i^{\circ} = f_i$, and Eq. (1.2-48) becomes

$$\hat{f}_i^{id} \text{ (RL)} \equiv x_i f_i \quad \text{(constant } T, P) \tag{1.2-49}$$

TABLE 1.2-2. Summary of Thermodynamic Relationships
for the Fugacity Coefficient

$\hat{\phi}_i \equiv \dfrac{\hat{f}_i}{x_i P}$	(1.2-40)
$\phi \equiv \dfrac{f}{P}$	(1.2-41)
$d \ln \hat{\phi}_i = -\dfrac{\bar{h}_i^R}{RT^2} dT + (\bar{Z}_i - 1)\dfrac{dP}{P} \quad \text{(constant } x)$	(1.2-43)
$d \ln \phi = -\dfrac{h^R}{RT^2} dT + (Z - 1)\dfrac{dP}{P} \quad \text{(constant } x)$	(1.2-44)
$\ln \hat{\phi}_i = \left[\dfrac{\partial(n \ln \phi)}{\partial n_i}\right]_{T,P,n_j}$	(1.2-45)
$\ln \phi = \sum_i x_i \ln \hat{\phi}_i$	(1.2-46)
$\sum x_i \, d \ln \hat{\phi}_i = -\dfrac{h^R}{RT^2} dT + (Z - 1)\dfrac{dP}{P}$	(1.2-47)

where RL indicates that we have chosen a *Raoult's-Law* standard state. If the ideal solution is defined so as to reproduce real behavior for $x_i \to 0$, then $f_i^\circ = \mathcal{K}_{i,j}$, and Eq. (1.2-48) becomes

$$\hat{f}_i^{\,id} \text{ (HL)} \equiv x_i \, \mathcal{K}_{i,j} \quad \text{(constant } T, P) \tag{1.2-50}$$

where HL denotes the choice of a *Henry's Law* standard state. In defining an ideal solution, it is not necessary that one use the same standard-state convention for all components in the mixture.

Equation (1.2-48) is the basis for the definitions of the activity $\hat{a}_i$ and the activity coefficient γ_i:

$$\hat{a}_i \equiv \frac{\hat{f}_i}{f_i^\circ} = \left(\frac{\hat{f}_i}{\hat{f}_i^{\,id}}\right) x_i \tag{1.2-51}$$

$$\gamma_i \equiv \frac{\hat{f}_i}{\hat{f}_i^{\,id}} = \frac{\hat{f}_i}{x_i f_i^\circ} \tag{1.2-52}$$

Clearly, $\hat{a}_i$ and γ_i are related:

$$\hat{a}_i = \gamma_i x_i \tag{1.2-53}$$

Hence, the name activity *coefficient* for γ_i. The activity and the activity coefficient, like the fugacity coefficient, are normalized fugacities. However, unlike the normalizing factor $x_i P$ in $\hat{\phi}_i$, the normalizing factors in $\hat{a}_i$ and γ_i contain property information, for pure i (when $f_i^\circ = f_i$) or for the mixture of which i is a component (when $f_i^\circ = \mathcal{K}_{i,j}$). Moreover, the numerical values of $\hat{a}_i$ and γ_i, unlike those of $\hat{\phi}_i$, are arbitrary to the extent that the choice of standard state is arbitrary.

Activities are identically equal to *mole fractions* for ideal solutions. For a *real* solution, $\hat{a}_i$ approaches x_i in an appropriate composition limit. Thus, for a Raoult's Law standard state,

$$\lim_{x_i \to 1} \hat{a}_i \text{ (RL)} = \lim_{x_i \to 1} \left(\frac{\hat{f}_i}{f_i}\right) = x_i \tag{1.2-54}$$

Similarly, for a Henry's Law standard state

$$\lim_{x_i \to 0} \hat{a}_i \text{ (HL)} = \lim_{x_i \to 0} \left(\frac{\hat{f}_i}{\mathcal{K}_{i,j}}\right) = x_i \tag{1.2-55}$$

Activity coefficients are identically *unity* for ideal solutions. For a real solution, the value of γ_i approaches unity in an appropriate composition limit. For a Raoult's Law standard state,

$$\lim_{x_i \to 1} \gamma_i \text{ (RL)} = \lim_{x_i \to 1} \left(\frac{\hat{f}_i}{x_i f_i}\right) = 1 \tag{1.2-56}$$

For a Henry's Law standard state,

$$\lim_{x_i \to 0} \gamma_i \text{ (HL)} = \lim_{x_i \to 0} \left(\frac{\hat{f}_i}{x_i \, \mathcal{K}_{i,j}}\right) = 1 \tag{1.2-57}$$

Table 1.2-3 summarizes general thermodynamic relationships for the activity coefficient. (One could construct a similar table for the activity, but in this chapter we favor use of the activity coefficient.) Section 1.4 discusses representation of activity coefficients through expressions for the molar excess Gibbs energy g^E.

1.2-5 Formulation of Phase-Equilibrium Problems

Consider the problem of equilibrium between two *N*-component phases α and β. By Eq. (1.2-16), the equilibrium criteria are

$$\hat{f}_i^\alpha = \hat{f}_i^\beta \quad (i = 1, 2, \ldots, N) \tag{1.2-62}$$

where uniformity of temperature and pressure are understood. As it stands, Eq. (1.2-62) displays explicitly none of the variables that figure in a phase-equilibrium calculation, that is, temperature, pressure, and concentration. Nor does its form suggest simplifications that might be made in the absence of complete data or under well-defined limiting conditions. Thus, while Eq. (1.2-62) is exact, it is not yet useful.

TABLE 1.2-3. Summary of Thermodynamic Relationships
for the Activity Coefficient

$$\gamma_i \equiv \frac{\hat{f}_i}{x_i f_i^\circ} \tag{1.2-52}$$

$$d \ln \gamma_i = -\frac{\bar{h}_i^E}{RT^2} dT + \frac{\bar{v}_i^E}{RT} dP \quad \text{(constant } x\text{)} \tag{1.2-58}$$

$$\ln \gamma_i = \left[\frac{\partial (ng^E/RT)}{\partial n_i} \right]_{T,P,n_j} \tag{1.2-59}$$

$$\frac{g^E}{RT} = \sum_i x_i \ln \gamma_i \tag{1.2-60}$$

$$\sum_i x_i \, d \ln \gamma_i = -\frac{h^E}{RT^2} dT + \frac{v^E}{RT} dP \tag{1.2-61}$$

The transformation of Eq. (1.2-62) from an abstract formulation to one appropriate for engineering calculations is accomplished by elimination of the component fugacities $\hat{f}_i$ in favor of the normalized auxiliary functions $\hat{\phi}_i$ and/or γ_i. For two-phase equilibrium, there are three general possibilities:

1. Introduce the activity coefficient for one phase (say α) and the fugacity coefficient for the other. Then, by Eqs. (1.2-52), (1.2-40), and (1.2-62), we obtain

$$x_i^\alpha \gamma_i^\alpha (f_i^\circ)^\alpha = x_i^\beta \hat{\phi}_i^\beta P \tag{1.2-63}$$

2. Introduce activity coefficients for both phases, obtaining

$$x_i^\alpha \gamma_i^\alpha (f_i^\circ)^\alpha = x_i^\beta \gamma_i^\beta (f_i^\circ)^\beta \tag{1.2-64}$$

3. Introduce fugacity coefficients for both phases, obtaining (since P is uniform)

$$x_i^\alpha \hat{\phi}_i^\alpha = x_i^\beta \hat{\phi}_i^\beta \tag{1.2-65}$$

For each of the formulations 1, 2, and 3, there are further choices one can make. For example, in Eq. (1.2-63) the choice of standard states for the activity coefficients has been left open.

Which of the above formulations one adopts for a particular problem is determined not only by the type of equilibrium (e.g., vapor–liquid, liquid–liquid, or solid–liquid) but also by the type and extent of thermodynamic data available for evaluation of the auxiliary functions. Representation and evaluation of the auxiliary functions is treated in the next two sections.

1.3 FUGACITY COEFFICIENTS

1.3-1 Fugacity Coefficients and the Equation of State

The route to a fugacity coefficient is through a $PVTx$ equation of state. By Eq. (1.2-44), we have for a mixture that

$$\ln \phi = \int_0^P (Z - 1) \frac{dP}{P} \quad \text{(constant } T, x\text{)} \tag{1.3-1}$$

and thus, as a special case, we obtain for pure component i that

$$\ln \phi_i = \int_0^P (Z_i - 1) \frac{dP}{P} \quad \text{(constant } T\text{)} \tag{1.3-2}$$

Determination of fugacity coefficients from these equations requires an expression for the compressibility factor as a function of temperature, pressure, and (for a mixture) composition. Such an expression, of functional form

$$Z = \mathcal{Z}\,(T, P, x)$$

is called a *volume-explicit* equation of state, because it can be solved to give the molar volume v as an algebraically explicit function of T, P, and x.

The analogous expression for $\ln \hat{\phi}_i$ follows from Eq. (1.2-43) or, equivalently, from Eq. (1.3-1) via the partial-property relationship Eq. (1.2-45). Thus,

$$\ln \hat{\phi}_i = \int_0^P (\bar{Z}_i - 1)\,\frac{dP}{P} \quad \text{(constant } T, x) \tag{1.3-3}$$

where $\bar{Z}_i$ is the partial molar compressibility factor:

$$\bar{Z}_i \equiv \left[\frac{\partial(nZ)}{\partial n_i}\right]_{T, P, n_j}$$

Determination of $\hat{\phi}_i$ therefore requires the same information as that required for the mixture ϕ. However, because of the differentiation required to find $\bar{Z}_i$ and hence $\hat{\phi}_i$, the *details* of the composition dependence of Z are crucial here. These details are conventionally expressed in the *mixing rules* for the equation-of-state parameters.

The above discussion presumes the availability of a volume-explicit equation of state. For applications to gases at moderate to high pressures or densities or to vapors *and* liquids, realistic equations of state are not volume explicit but are instead *pressure explicit*. That is, Z is expressed as a function of T, v, and x or, equivalently, of T, ρ (molar density $\equiv v^{-1}$), and x:

$$Z = \mathcal{Z}\,(T, \rho, x)$$

In this event, Eqs. (1.3-1), (1.3-2), and (1.3-3) are inappropriate; one uses instead the equivalent expressions

$$\ln \phi = \int_0^\rho (Z - 1)\,\frac{d\rho}{\rho} + Z - 1 - \ln Z \quad \text{(constant } T, x) \tag{1.3-4}$$

$$\ln \phi_i = \int_0^{\rho_i} (Z_i - 1)\,\frac{d\rho}{\rho} + Z_i - 1 - \ln Z_i \quad \text{(constant } T) \tag{1.3-5}$$

$$\ln \hat{\phi}_i = \int_0^\rho (\tilde{Z}_i - 1)\,\frac{d\rho}{\rho} - \ln Z \quad \text{(constant } T, x) \tag{1.3-6}$$

Here, quantity $\tilde{Z}_i$ is a partial molar compressibility factor evaluated at constant temperature and *total volume*:

$$\tilde{Z}_i \equiv \left[\frac{\partial(nZ)}{\partial n_i}\right]_{T, nv, n_j} \tag{1.3-7}$$

Again, the details of the composition dependence of the equation of state, as contained in $\tilde{Z}_i$, are crucial to the determination of accurate values for $\hat{\phi}_i$.

There is no known *PVTx* equation of state that is suitable for calculation of fugacity coefficients for all mixtures at all possible conditions of interest. The choice of an equation of state for an engineering calculation is therefore often made on an ad hoc basis. Guidelines are available, but they reflect the inevitable compromise between simplicity and accuracy. We treat in the remainder of this section three popular classes of equations of state commonly employed for practical calculations: the virial equations, used for gases at low to moderate densities; the cubic equations of state (exemplified by the Redlich–Kwong equations), used for dense gases and liquids; and equations inspired by the so-called "chemical theories," used for associating vapors and vapor mixtures.

1.3-2 Virial Equations of State

Virial equations of state are infinite-series representations of the gas-phase compressibility factor, with either molar density or pressure taken as the independent variable for expansion:

$$Z = 1 + B\rho + C\rho^2 + D\rho^3 + \ldots \tag{1.3-8}$$

$$Z = 1 + B'P + C'P^2 + D'P^3 + \ldots \tag{1.3-9}$$

Parameters B, C, D, ... are density-series virial coefficients, and B', C', D', ... are pressure-series virial coefficients. Virial coefficients depend only on temperature and composition; they are defined through the usual prescriptions for coefficients in a Taylor expansion. Thus, the *second* virial coefficients are given as

$$B = \frac{1}{1!}\left(\frac{\partial Z}{\partial \rho}\right)_{T,y;\rho=0}$$

$$B' = \frac{1}{1!}\left(\frac{\partial Z}{\partial P}\right)_{T,y;P=0}$$

Similarly, the *third* virial coefficients are defined as

$$C = \frac{1}{2!}\left(\frac{\partial^2 Z}{\partial \rho^2}\right)_{T,y;\rho=0}$$

$$C' = \frac{1}{2!}\left(\frac{\partial^2 Z}{\partial P^2}\right)_{T,y;P=0}$$

Higher virial coefficients are defined analogously as higher-order derivatives of Z, each of them evaluated at the state of zero density or zero pressure.

The pressure-series coefficients and density-series coefficients are related:

$$B' = \frac{B}{RT}$$

$$C' = \frac{C - B^2}{(RT)^2} \quad \text{and so on}$$

Thus, the virial expansion in pressure, Eq. (1.3-9), can be written in terms of density-series virial coefficients:

$$Z = 1 + \frac{BP}{RT} + \left[\frac{C - B^2}{(RT)^2}\right]P^2 + \cdots \tag{1.3-10}$$

This form is preferred to Eq. (1.3-9) because the density-series coefficients are the ones normally reported by experimentalists, and they are the ones for which correlations (for B and C) are available.

In practice, one must work with *truncations* of any infinite-series representation and, since virial coefficients beyond the third are rarely available, Eqs. (1.3-8) and (1.3-10) are normally truncated after two or three terms. For low pressures, the two-term truncation of Eq. (1.3-10) is sufficient:

$$Z = 1 + \frac{BP}{RT} \tag{1.3-11}$$

For more severe conditions, the three-term truncation of Eq. (1.3-8) is preferred:

$$Z = 1 + B\rho + C\rho^2 \tag{1.3-12}$$

Equation (1.3-11) should not be used for densities greater than about half the critical value, and Eq. (1.3-12) should not be used for densities exceeding about three-quarters of the critical value. Note that Eq. (1.3-11) can be considered either a volume-explicit or a pressure-explicit equation of state, whereas Eq. (1.3-12) is pressure explicit.

The great appeal of the virial equations derives from their interpretations in terms of molecular theory. Virial coefficients can be calculated from potential functions describing interactions among molecules. More importantly, statistical mechanics provides rigorous expressions for the composition dependence of the virial coefficients. Thus, the nth virial coefficient of a mixture is nth order in the mole fractions:

$$B = \sum_i \sum_j y_i y_j B_{ij} \tag{1.3-13}$$

$$C = \sum_i \sum_j \sum_k y_i y_j y_k C_{ijk} \quad \text{and so on} \tag{1.3-14}$$

The subscripted coefficients B_{ij}, C_{ijk}, ... depend only on T, and their numerical values are unaffected on permutation of the subscripts. Coefficients with identical subscripts (B_{11}, C_{222}, etc.) are properties of pure gases. Those with mixed subscripts ($B_{12} = B_{21}$, $C_{122} = C_{212}$, etc.) are mixture properties; they are called interaction virial coefficients or cross virial coefficients.

Expressions for fugacity coefficients follow from Eqs. (1.3-2), (1.3-3), and (1.3-11) or from Eqs. (1.3-5), (1.3-6), and (1.3-12). For applications at low pressures, we find for the two-term virial equation in pressure that

$$\ln \phi_i = \frac{B_{ii} P}{RT} \tag{1.3-15}$$

and

$$\ln \hat{\phi}_i = \frac{\bar{B}_i P}{RT} \tag{1.3-16}$$

Similarly, for conditions requiring the use of the three-term virial equation in density, we obtain

$$\ln \phi_i = 2 B_{ii} \rho + \tfrac{3}{2} C_{iii} \rho^2 - \ln Z \tag{1.3-17}$$

and

$$\ln \hat{\phi}_i = (B + \bar{B}_i)\rho + \tfrac{1}{2}(2C + \bar{C}_i)\rho^2 - \ln Z \tag{1.3-18}$$

In Eqs. (1.3-16) and (1.3-18), quantities $\bar{B}_i$ and $\bar{C}_i$ are partial molar virial coefficients, defined by

$$\bar{B}_i \equiv \left[\frac{\partial(nB)}{\partial n_i} \right]_{T,n_j}$$

$$\bar{C}_i \equiv \left[\frac{\partial(nC)}{\partial n_i} \right]_{T,n_j}$$

and determined from the mixing rules given by Eqs. (1.3-13) and (1.3-14). General expressions for $\bar{B}_i$ and $\bar{C}_i$ and summarized in Table 1.3-1; for components 1 and 2 in a binary mixture, they reduce to

$$\bar{B}_1 = B_{11} + \delta_{12} y_2^2 \tag{1.3-19a}$$

$$\bar{B}_2 = B_{22} + \delta_{12} y_1^2 \tag{1.3-19b}$$

TABLE 1.3-1 Expressions for the Partial Molar Virial Coefficients $\bar{B}_i$ and $\bar{C}_i$

$\bar{B}_i \equiv \left[\dfrac{\partial(nB)}{\partial n_i} \right]_{T,n_j}$	where
$= -B + 2 \sum_k y_k B_{ki}$	$B = \sum_k \sum_l y_k y_l B_{kl}$
$= \sum_k \sum_l y_k y_l (2B_{ki} - B_{kl})$	$= \sum_k y_k B_{kk} + \tfrac{1}{2} \sum_k \sum_l y_k y_l \delta_{kl}$
$= B_{ii} + \tfrac{1}{2} \sum_k \sum_l y_k y_l (2\delta_{ki} - \delta_{kl})$	and
	$\delta_{kl} \equiv 2 B_{kl} - B_{kk} - B_{ll}$
$\bar{C}_i \equiv \left[\dfrac{\partial(nC)}{\partial n_i} \right]_{T,n_j}$	where
$= -2C + 3 \sum_k \sum_l y_k y_l C_{kli}$	$C = \sum_k \sum_l \sum_m y_k y_l y_m C_{klm}$
$= \sum_k \sum_l \sum_m y_k y_l y_m (3 C_{kli} - 2 C_{klm})$	$= \sum_k y_k C_{kkk} + \tfrac{1}{3} \sum_k \sum_l \sum_m y_k y_l y_m \delta_{klm}$
$= C_{iii} + \tfrac{1}{3} \sum_k \sum_l \sum_m y_k y_l y_m (3\delta_{kli} - 2\delta_{klm})$	and
	$\delta_{klm} \equiv 3 C_{klm} - C_{kkk} - C_{lll} - C_{mmm}$

Source: Van Ness and Abbott.[1]

where

$$\delta_{ij} \equiv 2B_{ij} - B_{ii} - B_{jj} \tag{1.3-20}$$

and

$$\overline{C}_1 = C_{111} + y_2^3 \delta_{122} + y_1 y_2^2 (2\delta_{112} - \delta_{122}) \tag{1.3-21a}$$

$$\overline{C}_2 = C_{222} + y_1^3 \delta_{112} + y_1^2 y_2 (2\delta_{122} - \delta_{112}) \tag{1.3-21b}$$

where

$$\delta_{ijk} \equiv 3C_{ijk} - C_{iii} - C_{jjj} - C_{kkk} \tag{1.3-22}$$

For a binary gas mixture at low pressure, Eqs. (1.3-16) and (1.3-19) provide the following frequently used expressions for the fugacity coefficients:

$$\hat{\phi}_1 = \exp \frac{(B_{11} + \delta_{12} y_2^2)P}{RT} \tag{1.3-23a}$$

$$\hat{\phi}_2 = \exp \frac{(B_{22} + \delta_{12} y_1^2)P}{RT} \tag{1.3-23b}$$

Since $\delta_{12} = 2B_{12} - B_{11} - B_{22}$, the details of the composition dependence of $\hat{\phi}_1$ and $\hat{\phi}_2$ are directly influenced by the magnitude of the interaction coefficient B_{12}. The effect is illustrated in Fig. 1.3-1, which shows values of $\hat{\phi}_1$ versus y_1 computed from Eq. (1.3-23a) for a representative binary system for which the pure-component virial coefficients are $B_{11} = -1000$ cm^3/mol and $B_{22} = -2000$ cm^3/mol. The temperature is 300 K and the pressure is 1 bar; the curves correspond to different values of B_{12}, which range from -500 to -2500 cm^3/mol. All curves approach asymptotically the pure-component value $\phi_1 = 0.9607$

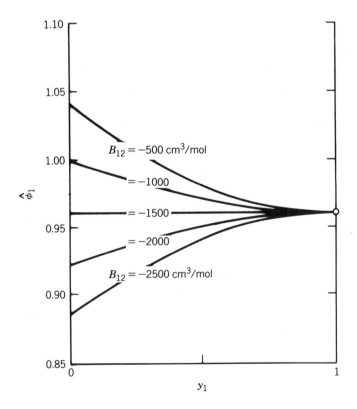

FIGURE 1.3-1 Composition dependence of fugacity coefficient $\hat{\phi}_1$ of component 1 in a binary gas mixture at 300 K and 1 bar. Curves correspond to different values of the interaction second virial coefficient B_{12}. (See text for discussion.)

as y_1 approaches unity, but the infinite-dilution behavior (as $y_1 \rightarrow 0$) varies from case to case. For the special case $B_{12} = -1500$ cm^3/mol, corresponding to $\delta_{12} = 0$, we see that $\hat{\phi}_1 = $ constant $= \phi_1$ for all y_1. This is ideal-solution behavior, which introduces a desirable simplification into the representation of vapor-phase fugacity coefficients. Unfortunately, most real gas mixtures are not ideal solutions, and the composition dependence of the $\hat{\phi}_i$ cannot generally be ignored.

Calculations with the virial equations require numerical values for the virial coefficients. Dymond and Smith[2] present an extensive compilation of experimental and recommended values, and new data appear frequently in the literature. Excellent corresponding-states correlations are available for B; the best are probably those of Hayden and O'Connell[3], and of Tsonopoulos.[4-6] A few correlations have also been proposed for C; the most recent are those of DeSantis and Grande[7] and of Orbey and Vera.[8] However, the data base for third virial coefficients is meager, and the correlations for C are not nearly as comprehensive or reliable as those for B.

1.3-3   Cubic Equations of State

Truncated virial equations are unsuitable for high-density applications; in particular, they are inappropriate for the liquid phase. For such applications, one must use more comprehensive but empirical equations of state. The simplest empirical equations of state are cubic in molar volume (or molar density). They may be represented by the general formula

$$P = \frac{RT}{v - b} - \frac{\theta(v - \eta)}{(v - b)(v^2 + \delta v + \epsilon)} \tag{1.3-24}$$

where quantities b, θ, δ, ϵ, and η are equation-of-state parameters, each of which may depend on temperature and composition.

Equation (1.3-24) is inspired by the van der Waals equation of state, to which it reduces under the assignments $\delta = \epsilon = 0$, $\eta = b = b(x)$, and $\theta = a(x)$:

$$P = \frac{RT}{v - b} - \frac{a}{v^2}$$

Given its simplicity, the van der Waals equation is remarkable for its ability to reproduce the qualitative features of real-fluid behavior. However, it is never used for engineering calculations; more flexible equations are required for quantitative work. These may be generated from Eq. (1.3-24) by incorporating other assignments for parameters θ, δ, ϵ, and η. It is customary to set $\eta = b$ and to express parameters δ and ϵ as specified multiples of b and b^2, respectively. By this procedure, one generates *two-parameter* variants of the van der Waals equation. Modern examples of cubic equations obtained in this way are the *Redlich–Kwong* equation of state,[9]

$$P = \frac{RT}{v - b} - \frac{\theta}{v^2 + bv} \tag{1.3-25}$$

and the *Peng–Robinson* equation of state,[10]

$$P = \frac{RT}{v - b} - \frac{\theta}{v^2 + 2bv + b^2}.$$

In these equations, parameter θ depends on composition and temperature, whereas parameter b is usually a function of composition only.

Of the two-parameter cubic equations, modifications of the Redlich–Kwong equation are among the most popular. In this chapter we consider only one cubic equation of state: Soave's[11] version of the Redlich–Kwong equation; its performance is typical of modern cubic equations. The Soave–Redlich–Kwong equation incorporates the following prescription for parameter θ:

$$\theta = \theta_c[1 + (0.480 + 1.574\omega - 0.176\omega^2)(1 - T_r^{1/2})]^2 \tag{1.3-26}$$

Here, ω is the acentric factor, and $T_r \equiv T/T_c$ is the reduced temperature. Parameter θ_c is related to the critical temperature and pressure,

$$\theta_c = 0.42748 \frac{R^2 T_c^2}{P_c} \tag{1.3-27}$$

as is parameter b:

$$b = 0.08664 \frac{RT_c}{P_c} \qquad (1.3\text{-}28)$$

Equations (1.3-27) and (1.3-28) follow from the classical critical constraints:

$$\left(\frac{\partial P}{\partial v}\right)_{T;cr} = \left(\frac{\partial^2 P}{\partial v^2}\right)_{T;cr} = 0$$

The expression for θ, Eq. (1.3-26), was obtained by forcing agreement of predicted with experimental vapor pressures of pure hydrocarbon liquids. (This procedure is essential if the equation of state is to be used for prediction or correlation of vapor–liquid equilibria.)

Equation (1.3-25) is explicit in pressure; it may be written in the alternative form

$$Z \equiv \frac{P}{\rho RT} = \frac{1}{1 - b\rho} - \frac{\theta}{RT} \frac{\rho}{1 + b\rho} \qquad (1.3\text{-}29)$$

where ρ is the molar density. Expressions for the fugacity coefficients then follow on application of Eqs. (1.3-4), (1.3-6), and (1.3-7). The results are

$$\ln \phi = Z - 1 - \ln (1 - b\rho)Z - \frac{\theta}{bRT} \ln (1 + b\rho) \qquad (1.3\text{-}30)$$

and

$$\ln \hat{\phi}_i = \ln \phi + \left(\frac{\bar{b}_i}{b} - 1\right)(Z - 1) + \frac{\theta}{bRT}\left(\frac{\bar{b}_i}{b} - \frac{\bar{\theta}_i}{\theta}\right) \ln (1 + b\rho) \qquad (1.3\text{-}31)$$

In Eqs. (1.3-30) and (1.3-31), all unsubscripted quantities refer to the *mixture*. Quantities $\bar{b}_i$ and $\bar{\theta}_i$ are partial molar equation-of-state parameters, defined by

$$\bar{b}_i \equiv \left[\frac{\partial(nb)}{\partial n_i}\right]_{T,n_j}$$

$$\bar{\theta}_i \equiv \left[\frac{\partial(n\theta)}{\partial n_i}\right]_{T,n_j}$$

Determination of $\bar{b}_i$ and $\bar{\theta}_i$ requires a set of mixing rules for parameters b and θ. The usual procedure is to assume that b and θ are quadratic in composition:

$$b = \sum_i \sum_j x_i x_j b_{ij} \qquad (1.3\text{-}32)$$

$$\theta = \sum_i \sum_j x_i x_j \theta_{ij} \qquad (1.3\text{-}33)$$

Here, x is a generic mole fraction and can refer to any phase. When subscripts i and j are identical in Eq. (1.3-32) or (1.3-33), the parameters refer to a pure component. When they are different, the parameters are called interaction parameters and these depend on the properties of the binary i-j mixture as indicated by the subscripts. To estimate these interaction parameters, we use *combining rules*, for example,

$$b_{ij} = \tfrac{1}{2}(b_{ii} + b_{jj})(1 - c_{ij}) \qquad (1.3\text{-}34)$$

$$\theta_{ij} = (\theta_{ii}\theta_{jj})^{1/2}(1 - k_{ij}) \qquad (1.3\text{-}35)$$

where c_{ij} and k_{ij} are empirical binary parameters, small compared to unity, that often are nearly independent of temperature over modest temperature ranges. Frequently, c_{ij} is set equal to zero, but it is almost always necessary to use for k_{ij} some number other than zero. With mixing rules given by Eqs. (1.3-32) and (1.3-33) and combining rules given by Eqs. (1.3-34) and (1.3-35), one finds the following expressions for $\bar{b}_i$ and $\bar{\theta}_i$ for components 1 and 2 in a binary mixture:

$$\bar{b}_1 = b_{11} - x_2^2(b_{11} + b_{22})c_{12} \tag{1.3-36a}$$

$$\bar{b}_2 = b_{22} - x_1^2(b_{11} + b_{22})c_{12} \tag{1.3-36b}$$

and

$$\bar{\theta}_1 = \theta_{11} - x_2^2[(\theta_{11}^{1/2} - \theta_{22}^{1/2})^2 + 2(\theta_{11}\theta_{22})^{1/2}k_{12}] \tag{1.3-37a}$$

$$\bar{\theta}_2 = \theta_{22} - x_1^2[(\theta_{11}^{1/2} - \theta_{22}^{1/2})^2 + 2(\theta_{11}\theta_{22})^{1/2}k_{12}] \tag{1.3-37b}$$

Calculation of fugacity coefficient $\hat{\phi}_i$ for component i in a binary mixture at specified temperature, pressure, and composition is straightforward but tedious and is best done with a computer. First, one finds the pure-component equation-of-state parameters from Eqs. (1.3-26), (1.3-27), and (1.3-28), and the interaction parameters from Eqs. (1.3-34) and (1.3-35). Application of the mixing rules, Eqs. (1.3-32) and (1.3-33), then given parameters b and θ for the mixture. Knowing these quantities, one determines the mixture ρ and Z from Eq. (1.3-29). Because the equation of state is cubic in molar density, an analytical solution for ρ (and hence Z) is possible; however, numerical techniques may often be just as fast. Given the mixture ρ and Z, one next finds the mixture ϕ from (1.3-30); these quantities, together with the $\bar{b}_i$ and $\bar{\theta}_i$ as given by Eqs. (1.3-36) and (1.3-37), finally permit calculation of the $\hat{\phi}_i$ from Eq. (1.3-31).

The behavior of Soave–Redlich–Kwong fugacity coefficients is best illustrated by numerical example. In Fig. 1.3-2 we show computed values of $\hat{\phi}_i$ for $i = H_2S$ in the H_2S–ethane system at 300 K. Two pressure levels are represented: 15 bar, for which states of superheated vapor are obtained at all compositions, and 50 bar, for which all states are subcooled liquids. In this example, interaction parameter c_{12} is set equal to zero; for each pressure level, the different curves correspond to different values of k_{12}, which varies from -0.20 to $+0.20$. For the vapor mixtures, behavior similar to that illustrated in Fig. 1.3-1 is observed: variations in k_{12} are reflected qualitatively in the shapes of the $\hat{\phi}_{H_2S}$ curves but, for these conditions, the quantitative effects are not large. The situation is dramatically different for the liquid mixtures. Here, small changes in k_{12} promote large changes in $\hat{\phi}_{H_2S}$; typically, the effect on $\hat{\phi}_i$ is greatest for mixtures dilute in component i. Analysis of vapor–liquid equilibrium data for this system shows that k_{12} is about 0.10 at 300 K; comparison of the curves in Fig. 1.3-2 illustrates the substantial effect of this apparently small

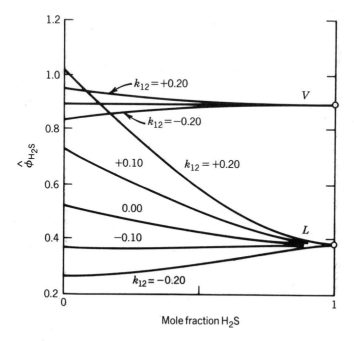

FIGURE 1.3-2 Composition dependence of fugacity coefficient of hydrogen sulfide in binary mixtures with ethane at 300 K. Curves labeled V are for superheated vapors at 15 bar; those labeled L are for subcooled liquids at 50 bar. All curves are computed from the Soave–Redlich–Kwong equation, with values of interaction parameter k_{12} as shown.

quantity. This example demonstrates an extremely important feature of applied equation-of-state thermo-dynamics: the implications of mixing rules and combining rules are seen most dramatically in fugacity calculations for dense phases. Application of an equation of state to vapor–liquid equilibrium calculations via formulation 3 of Section 1.2-5 therefore requires mixing rules of appropriate flexibility. Development and testing of such rules is a major area of research in chemical engineering thermodynamics.

1.3-4   Chemical Theories of Vapor-Phase Nonideality

It may happen that the nonideal behavior of gases results wholly or partly from stoichiometric effects attributable to the formation of extra chemical species. When this is the case, a "chemical theory" can be used to develop an equation of state from which fugacity coefficients may be determined. A general treatment of chemical theories is beyond the scope of this chapter; to illustrate the principles involved, we develop instead, by way of example, the procedure for treating strong dimerization in gases at low pressure.

For orientation, consider the following simple thought emperiment. A gas mixture, of total *apparent* number of moles n, is contained in a vessel of known total volume V which is submerged in a thermostated bath at known temperature T. A measurement of the equilibrium pressure of the gas permits calculation of the *apparent* compressibility factor Z:

$$Z = \frac{PV}{nRT} \tag{1.3-38}$$

Suppose now that the apparent number of moles n is not the correct value; that—for whatever reason—the *true* value is n'. Then the *true* compressibility factor Z' is

$$Z' = \frac{PV}{n'RT} \tag{1.3-39}$$

Now quantities T, P, and V are the same in Eqs. (1.3-38) and (1.3-39); they are values obtained by direct measurement or by calibration. Combination of the two equations thus produces the relation

$$Z = \frac{n'}{n} Z' \tag{1.3-40}$$

Equation (1.3-40) is one of the fundamental equations for the chemical theory of vapor-phase non-idealities. It asserts that the apparent, or *observed*, compressibility factor Z differs from the *true* value Z' because of differences between the apparent, or *assumed*, mole number n and the true value n'. In a chemical theory, such differences are assumed to obtain because of the occurrence of one or more chemical reactions. If the reactions are at *equilibrium*, then one finds the following relationship for the apparent fugacity coefficient $\hat{\phi}_i$:

$$\hat{\phi}_i = \frac{y_i'}{y_i} \hat{\phi}_i' \tag{1.3-41}$$

Equation (1.3-41) is the second fundamental equation for the chemical theory of vapor-phase nonidealities. As in Eq. (1.3-40), the primed quantities represent properties for the *true* mixture.

In applying Eqs. (1.3-40) and (1.3-41), one seeks expressions for the apparent quantities Z and $\hat{\phi}_i$. To do this, one must propose a reaction scheme: this provides relationships for n'/n and y_i'/y_i in terms of equilibrium conversions. One must also assume an expression for Z', which in turn implies an expression for the $\hat{\phi}_i'$. The true fugacity coefficients $\hat{\phi}_i'$, when incorporated into the criteria for chemical-reaction equilibrium for the true mixture, permit determination of the equilibrium conversions, and hence, finally, via Eqs. (1.3-40) and (1.3-41), expressions for Z and $\hat{\phi}_i$ as functions of T, P, and the set of apparent compositions $\{y_i\}$.

The simplest cases (the only ones considered here) obtain for pressures sufficiently low that the *true* mixtures can be considered ideal-gas mixtures. In this event, $Z' = 1$ and $\hat{\phi}_i' = 1$, and Eqs. (1.3-40) and (1.3-41) reduce to

$$Z = \frac{n'}{n} \tag{1.3-42}$$

and

$$\hat{\phi}_i = \frac{y_i'}{y_i} \tag{1.3-43}$$

Suppose that $n'/n > 1$, as would occur, for example, as the result of a dissociation reaction undergone by a nominally pure chemical species. Then, according to (1.3-42), the apparent compressibility factor is *greater* than unity. On the other hand, suppose that $n'/n < 1$, as would occur, for example, if a nominally pure substance underwent association. Then, by (1.3-42), the apparent compressibility factor is *less* than unity. In both cases—dissociation and association of a nominally pure substance at low pressure—the apparent fugacity coefficient is also different from the expected value of unity.

The simplest example of self-association in the vapor phase is *dimerization*, as exemplified by hydrogen bonding in carboxylic acids. Consider the dimerization of acetic acid:

$$2\left(CH_3-C\overset{O}{\underset{OH}{\lessgtr}}\right) \rightleftarrows CH_3-C\overset{O\,\cdots\,HO}{\underset{OH\,\cdots\,O}{\lessgtr}}C-CH_3$$

where the dots denote hydrogen bonds. Evidence for vapor-phase association of acetic acid is provided by the *PVT* data of MacDougall,[12] shown in Fig. 1.3-3 as a plot of Z versus P for a temperature of 40°C. Even though the pressure level is extremely low, the apparent compressibility factor is small (0.7 or less for $P > 0.005$ atm); at this pressure any normal vapor would exhibit a Z very close to unity.

If we write the acetic acid dimerization reaction as

$$2A \rightleftarrows A_2$$

then reaction stoichiometry provides the following material balance equations:

$$n'_A = n_A - 2\,\epsilon$$

$$n'_{A_2} = \epsilon$$

$$n' = n_A - \epsilon$$

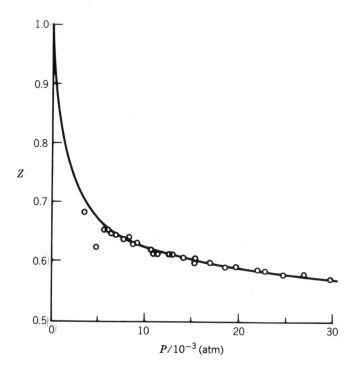

FIGURE 1.3-3 Compressibility factor Z for acetic acid vapor at 40°C. Circles are data; curve is computed from chemical theory, assuming dimerization, with $K = 380$.

Here n_A is the apparent number of moles of monomer, the primed quantities are true mole numbers, and ϵ is the number of moles of dimer formed. The material balance equations produce expressions for the true mole fractions:

$$y'_A = \frac{1 - 2\xi}{1 - \xi}$$

$$y'_{A_2} = \frac{\xi}{1 - \xi}$$

where ξ is a dimensionless extent of reaction:

$$\xi \equiv \frac{\epsilon}{n_A}$$

By Eq. (1.3-42), the apparent compressibility factor is

$$Z = \frac{n_A - \epsilon}{n_A} = 1 - \xi$$

and it remains to determine ξ. We do this by assuming that the true mixture is at chemical-reaction equilibrium:

$$\frac{y'_{A_2}}{(y'_A)^2} = KP$$

or

$$\frac{\xi(1 - \xi)}{(1 - 2\xi)^2} = KP$$

from which

$$\xi = \frac{1}{2}\left[1 - \left(\frac{1}{1 + 4KP}\right)^{1/2}\right] \qquad (1.3\text{-}44)$$

where K is the chemical-reaction equilibrium constant. Thus, we obtain finally the following expression for Z:

$$Z = \frac{1}{2}\left[1 + \left(\frac{1}{1 + 4KP}\right)^{1/2}\right] \qquad (1.3\text{-}45)$$

A test of the usefulness of the dimerization model is provided by the ability of Eq. (1.3-45) to represent MacDougall's volumetric data for acetic acid vapor. The solid line in Fig. 1.3-3, generated from Eq. (1.3-45) with $K = 380$, provides an excellent fit of the data; one concludes that the dimerization model is consistent with the observed PVT behavior at 40°C. The apparent fugacity coefficient for acetic acid vapor, found from Eq. (1.3-43) with $\hat{\phi}_i = \phi_A$ and $y_i = y_A = 1$, is given by

$$\phi_A = \frac{1}{2KP}[(1 + 4KP)^{1/2} - 1] \qquad (1.3\text{-}46)$$

and is plotted against pressure in Fig. 1.3-4. Significantly, ϕ_A is small (0.6 or less for $P > 0.005$ atm) at a pressure level where we would expect it to be very nearly unity.

So far we have only considered the behavior of apparently "pure" acetic acid vapor. Dimerization also occurs in vapor *mixtures* containing carboxylic acids. The effect on component fugacity coefficients is easily illustrated for binary vapor mixtures containing acetic acid and an inert substance I. For this example, the true mixture contains *three* species: monomer, dimer, and inert. A development similar to that just presented produces a similar expression for the dimensionless extent of reaction:

$$\xi = \frac{1 + 4KPy_A - [1 + 4y_A(2 - y_A)KP]^{1/2}}{2(1 + 4KP)} \qquad (1.3\text{-}47)$$

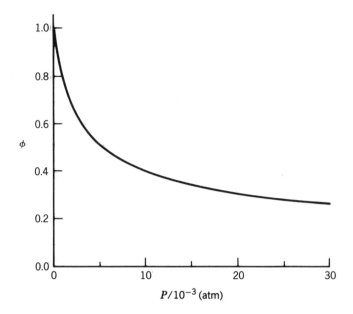

FIGURE 1.3-4 Fugacity coefficient ϕ of acetic acid vapor at 40°C. Curve is computed from chemical theory, assuming dimerization, with $K = 380$.

where now y_A is the apparent mole fraction of acetic acid in the mixture. The true mole fractions of monomer and inert are

$$y'_A = \frac{y_A - 2\xi}{1 - \xi} \qquad (1.3\text{-}48a)$$

$$y'_I = \frac{y_I}{1 - \xi} \qquad (1.3\text{-}48b)$$

and the apparent fugacity coefficients are

$$\hat{\phi}_A = \frac{y'_A}{y_A} \qquad (1.3\text{-}49a)$$

$$\hat{\phi}_I = \frac{y'_I}{y_I} = \frac{y'_I}{1 - y_A} \qquad (1.3\text{-}49b)$$

Equations (1.3-47)–(1.3-49), with $K = 380$, produce the $\hat{\phi}_i$ versus y_A curves shown in Fig. 1.3-5, for $t = 40°C$ and $P = 0.025$ atm. As expected, $\hat{\phi}_A$ differs significantly from unity. Perhaps surprisingly, so does $\hat{\phi}_I$. The behavior of $\hat{\phi}_I$, however, is conditioned by that of $\hat{\phi}_A$, because the two fugacity coefficients must satisfy the following form of the Gibbs–Duhem equation:

$$y_A \frac{d \ln \hat{\phi}_A}{dy_A} + y_I \frac{d \ln \hat{\phi}_I}{dy_A} = 0$$

Thus, chemical effects can influence the fugacity behavior of *all* the components in a mixture, even those that do not participate in the reactions.

In this example the pressure level was low enough so that the true mixtures could be treated as ideal-gas mixtures. Moreover, hydrogen bonding in carboxylic acid vapors is a "strong" phenomenon, which even at normal pressures can be expected to dominate the fugacity behavior. However, comprehensive treatments of the equation of state must allow for both chemical *and* physical effects. The second virial coefficient correlation of Hayden and O'Connell[3] and the modified van der Waals equation of Hu et al.[13] provide examples.

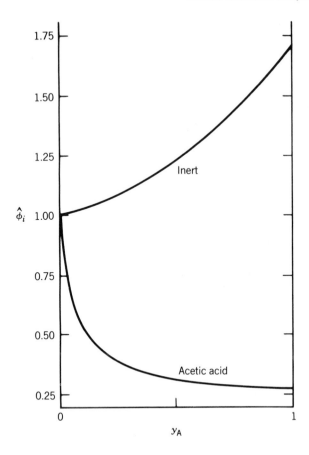

FIGURE 1.3-5 Composition dependence of fugacity coefficients at 40°C and 0.025 atm in binary gas mixture containing acetic acid and an inert component. Curves are computed from· chemical theory, assuming dimerization of acetic acid, with $K = 380$.

1.4 ACTIVITY COEFFICIENTS

1.4-1 Activity Coefficients and the Excess Gibbs Energy

The route to an activity coefficient is through an expression for the dimensionless excess Gibbs energy, g^E/RT, to which $\ln \gamma_i$ is related as a partial molar property:

$$\ln \gamma_i = \left[\frac{\partial(ng^E/RT)}{\partial n_i} \right]_{T,P,n_j} \tag{1.2-59}$$

For a binary mixture, $\ln \gamma_1$ and $\ln \gamma_2$ are conveniently expressed in terms of g^E and its mole-fraction derivative:

$$RT \ln \gamma_1 = g^E + x_2 \left(\frac{\partial g^E}{\partial x_1} \right)_{T,P} \tag{1.4-1a}$$

$$RT \ln \gamma_2 = g^E - x_1 \left(\frac{\partial g^E}{\partial x_1} \right)_{T,P} \tag{1.4-1b}$$

Hence, g^E is a *generating function* for the activity coefficient; given an expression for the composition dependence of g^E, expressions for the γ_i follow. Conversely, by (1.2-60), g^E/RT is the mole-fraction-weighted sum of the $\ln \gamma_i$; given values for the γ_i, values for g^E follow:

$$\frac{g^E}{RT} = \sum_i x_i \ln \gamma_i \qquad (1.2\text{-}60)$$

If g^E is known as a function of temperature, pressure, and composition, then the other excess functions can be derived from it. For example, the excess entropy is proportional to the temperature derivative of g^E:

$$s^E = -\left(\frac{\partial g^E}{\partial T}\right)_{P,x} \qquad (1.4\text{-}2)$$

Similarly, the excess volume ("volume change of mixing") is equal to the pressure derivative of g^E:

$$v^E = \left(\frac{\partial g^E}{\partial P}\right)_{T,x} \qquad (1.4\text{-}3)$$

The excess enthalpy ("heat of mixing") can be related either to g^E and s^E or, through Eq. (1.4-2), to the temperature derivative of g^E/RT:

$$h^E = g^E + Ts^E \qquad (1.4\text{-}4a)$$

$$h^E = -RT^2 \left[\frac{\partial(g^E/RT)}{\partial T}\right]_{P,x} \qquad (1.4\text{-}4b)$$

Equation (1.4-4b) is an example of a *Gibbs-Helmholtz equation*.

Equations (1.2-59), (1.2-60), and (1.4-1)–(1.4-4) are valid regardless of the standard-state conventions adopted for the components of the mixture. If Raoult's Law standard states are used for all components (the usual procedure), then

$$\lim_{x_i \to 1} \gamma_i = 1 \quad \text{(all } i\text{)}$$

and

$$\lim_{x_i \to 1} m^E = 0 \quad \text{(all } m^E; \text{ all } i\text{)}$$

In particular,

$$\lim_{x_i \to 1} g^E = 0 \quad \text{(all } i\text{)}$$

Thus, for a binary mixture, the excess functions are identically zero at the composition extremes; for any mixture, the activity coefficient of a component approaches unity as that component approaches purity.

The state of infinite solution is also of special interest. We *define* the activity coefficient of component i at infinite dilution by

$$\lim_{x_i \to 0} \gamma_i \equiv \gamma_i^\infty \qquad (1.4\text{-}5)$$

Usually, but not always, γ_i^∞ is the extreme (maximum or minimum) value assumed by γ_i for a component in a binary mixture. Hence, the γ_i^∞ are often used as measures of the magnitudes of nonidealities of binary liquid mixtures. Another measure is provided by g^E or g^E/RT for the equimolar mixture; for many binary solutions, this is near to the maximum (or minimum) value. For liquid solutions exhibiting positive deviations from ideal-solution behavior (activity coefficients greater than unity), a γ_i^∞ of about 5 or an equimolar g^E/RT of about 0.5 is considered "large."

For binary solutions, the auxiliary function

$$\mathcal{G} \equiv \frac{g^E}{x_1 x_2 RT} \qquad (1.4\text{-}6)$$

is convenient for displaying and smoothing experimental data. The activity coefficients are obtained from it via the expressions

$$\ln \gamma_1 = x_2^2 \left[\mathcal{G} + x_1 \left(\frac{\partial \mathcal{G}}{\partial x_1} \right)_{T,P} \right] \tag{1.4-7a}$$

$$\ln \gamma_2 = x_1^2 \left[\mathcal{G} - x_2 \left(\frac{\partial \mathcal{G}}{\partial x_1} \right)_{T,P} \right] \tag{1.4-7b}$$

which are analogues of Eqs. (1.4-1a) and (1.4-1b). In particular, the γ_i^∞ are simply related to the limiting values of $\mathcal{G}$:

$$\ln \gamma_1^\infty = \lim_{x_1 \to 0} \mathcal{G}$$

$$\ln \gamma_2^\infty = \lim_{x_1 \to 1} \mathcal{G}$$

The diversity of behavior exhibited by real liquid solutions is illustrated by Fig. 1.4-1, which shows plots of $\mathcal{G}$ and the $\ln \gamma_i$ versus mole fraction for six binary mixtures at 50°C and low pressures. Of the six mixtures, acetone–methanol shows the simplest behavior, with nearly symmetrical curves for the $\ln \gamma_i$ and an essentially horizontal straight line for $\mathcal{G}$. The magnitudes of the deviations from ideality are small: the γ_i^∞ are only about 1.9. Methyl acetate–1-hexene is an example of a simple nonsymmetrical system: the γ_i^∞ are again of modest size, but here $\mathcal{G}$ is described by a straight line of nonzero slope. Ethanol–chloroform is highly asymmetrical, and the γ_i exhibit interior extrema with respect to composition; here, the activity coefficient of chloroform shows a pronounced maximum at an ethanol mole fraction of about 0.85. Ethanol–n-heptane exhibits extremely large positive deviations from ideality, so large that the equimolar mixture is very close to a condition of instability with respect to liquid–liquid phasesplitting. Acetone–chloroform shows *negative* deviations from ideality, and benzene–hexafluorobenzene is a rare example of a system that shows negative *and* positive deviations: for low benzene concentrations g^E is negative, whereas for large concentrations it is positive.

One of the goals of applied solution thermodynamics is to develop expressions for g^E of minimal complexity but of sufficient flexibility to represent the various types of behavior illustrated by Fig. 1.4-1. It is desirable for such expressions to have a sound physicochemical basis, so that the numerical values of the parameters in the expressions are susceptible to correlation and estimation. Moreover, one would like to be able to reliably estimate g^E for multicomponent mixtures with parameter values determined from binary data only. Finally, a desirable expression for g^E would incorporate a "built-in" temperature dependence accurate enough to permit estimation of g^E at one temperature from parameters determined at another. By Eq. (1.4-4b), this requires that the expression for g^E imply realistic values for the excess enthalpy h^E.

No known model for g^E meets all the above criteria. As with $PVTx$ equations of state, the choice of an expression for g^E for an engineering calculation is frequently made on an ad hoc basis. We discuss in the next two sections two classes of expressions commonly employed for practical work: classical empirical expressions for g^E and more recent expressions for g^E based on the local-composition concept.

1.4-2 Empirical Expressions for g^E

The simplest procedure for generating an expression for g^E of a binary mixture is through a power-series expansion of the function $\mathcal{G}$ in mole fraction. For example, one can write

$$\mathcal{G} = (A_{21}x_1 + A_{12}x_2) - (D_{21}x_1 + D_{12}x_2)x_1x_2 + \cdots \tag{1.4-8}$$

where parameters $A_{12}, A_{21}, D_{12}, D_{21}, \ldots$ are functions of temperature and (in principle) pressure. However, for most liquid mixtures the pressure dependence of g^E is small; it is usually ignored or else incorporated through the excess volume via Eq. (1.4-3). Equation (1.4-8) is the *generalized Margules equation*.

The Margules equation is rarely used in a form containing more than four parameters. By Eq. (1.4-7), the four-parameter Margules equation gives the following expressions for the activity coefficients:

$$\ln \gamma_1 = x_2^2 [A_{12} + 2(A_{21} - A_{12})x_1 \\ + 2(D_{21}x_1 + D_{12}x_2)x_1 (x_1 - x_2) - D_{21}x_1^2] \tag{1.4-9a}$$

$$\ln \gamma_2 = x_1^2 [A_{21} + 2(A_{12} - A_{21})x_2 \\ + 2(D_{12}x_2 + D_{21}x_1)x_2 (x_2 - x_1) - D_{12}x_2^2] \tag{1.4-9b}$$

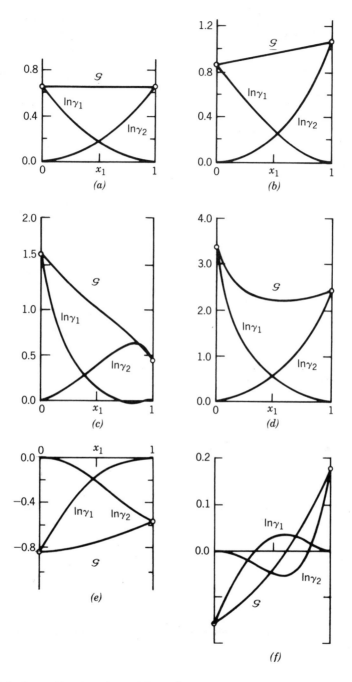

FIGURE 1.4-1 Composition dependence of $\mathcal{G}$ ($\equiv g^E/x_1x_2RT$) and of activity coefficients for six binary liquid mixtures at 50°C and low pressures: (*a*) acetone (1)–methanol(2); (*b*) methyl acetate(1)–1-hexene(2); (*c*) ethanol(1)–chloroform(2); (*d*) ethanol(1)–*n*-heptane(2); (*e*) acetone(1)–chloroform(2); and (*f*) benzene(1)–hexafluorobenzene(2).

Because they require four binary parameters, Eqs. (1.4-9a) and (1.4-9b) are infrequently used for process calculations. However, they are required for the precise representation of the γ_i for binary mixtures in which G exhibits a reversal in curvature, for example, the ethanol–chloroform system shown in Fig. 1.4-1.

Lower-order Margules equations follow from Eqs. (1.4-8) and (1.4-9) on appropriate assignments of the parameters. Thus, with $D_{12} = D_{21} = D$, we obtain the three-parameter Margules equation, for which g^E/RT and the activity coefficients are

$$\frac{g^E}{RT} = (A_{21}x_1 + A_{12}x_2)x_1x_2 - Dx_1^2x_2^2 \tag{1.4-10}$$

and

$$\ln \gamma_1 = x_2^2[A_{12} + 2(A_{21} - A_{12})x_1 + Dx_1(3x_1 - 2)] \tag{1.4-11a}$$

$$\ln \gamma_2 = x_1^2[A_{21} + 2(A_{12} - A_{21})x_2 + Dx_2(3x_2 - 2)] \tag{1.4-11b}$$

Equations (1.4-10) and (1.4-11) are appropriate for binary mixtures in which G exhibits modest curvature, such as the acetone–chloroform system shown in Fig. 1.4-1.

With $D_{12} = D_{21} = 0$, one obtains the two-parameter Margules equation, for which

$$\frac{g^E}{RT} = (A_{21}x_1 + A_{12}x_2)x_1x_2 \tag{1.4-12}$$

and

$$\ln \gamma_1 = x_2^2[A_{12} + 2(A_{21} - A_{12})x_1] \tag{1.4-13a}$$

$$\ln \gamma_2 = x_1^2[A_{21} + 2(A_{12} - A_{21})x_2] \tag{1.4-13b}$$

Here, G is linear with mole fraction: see the methyl acetate–1-hexene system in Fig. 1.4-1. Finally, if $D_{12} = D_{21} = 0$ and $A_{12} = A_{21} = A$, then Eq. (1.4-8) reduces to the one-parameter Margules equation, or *Porter equation*. Here,

$$\frac{g^E}{RT} = Ax_1x_2 \tag{1.4-14}$$

and

$$\ln \gamma_1 = Ax_2^2 \tag{1.4-15a}$$

$$\ln \gamma_2 = Ax_1^2 \tag{1.4-15b}$$

The Porter equation is the simplest realistic expression for g^E. It is appropriate for "symmetrical" binary mixtures showing small deviations from ideality, for example, the acetone–methanol system depicted in Fig. 1.4-1.

An alternative procedure to the series expansion of Eq. (1.4-8) is a power-series representation in mole fraction of the *reciprocal* of G. This expansion may be written

$$\frac{1}{G} = \frac{A'_{12}x_1 + A'_{21}x_2}{A'_{12}A'_{21}} - \left(\frac{D'_{12}x_1 + D'_{21}x_2}{D'_{12}D'_{21}}\right)x_1x_2 + \cdots \tag{1.4-16}$$

where again parameters $A'_{12}, A'_{21}, D'_{12}, D'_{21}, \ldots$ depend on temperature and (in principle) pressure. Equation (1.4-16) is a *generalized van Laar equation*. The van Laar equation is almost always applied in its two-parameter form, for which

$$\frac{g^E}{RT} = \frac{A'_{12}A'_{21}x_1x_2}{A'_{12}x_1 + A'_{21}x_2} \tag{1.4-17}$$

and

$$\ln \gamma_1 = \frac{A'_{12}(A'_{21})^2x_2^2}{(A'_{12}x_1 + A'_{21}x_2)^2} \tag{1.4-18a}$$

$$\ln \gamma_2 = \frac{A'_{21}(A'_{12})^2x_1^2}{(A'_{21}x_2 + A'_{12}x_1)^2} \tag{1.4-18b}$$

If $A'_{12} = A'_{21} = A$, then Eq. (1.4-17) reduces to the Porter equation.

The infinite-dilution activity coefficients given by the generalized Margules and van Laar equations are

$$\ln \gamma_1^\infty = A_{12}(\text{Margules}) = A'_{12}(\text{van Laar}) \qquad (1.4\text{-}19a)$$

$$\ln \gamma_2^\infty = A_{21}(\text{Margules}) = A'_{21}(\text{van Laar}) \qquad (1.4\text{-}19b)$$

These expressions justify the notation adopted for the parameters: subscript 12 denotes the state of infinite dilution of component 1 in component 2; similarly, subscript 21 denotes the state of infinite dilution of component 2 in component 1.

The Margules and van Laar equations have different correlating capabilities. These differences are most easily illustrated for the two-parameter expressions, Eqs. (1.4-12) and (1.4-17). Figure 1.4-2 shows plots of $\mathcal{G}$ and the $\ln \gamma_i$ generated from these equations for four cases. In each case, we have taken $\ln \gamma_2^\infty = 0.5$ and have fixed A_{21} and A'_{21} by Eq. (1.4-19b); the different cases represent different values of $\ln \gamma_1^\infty$, with A_{12} and A'_{12} given by Eq. (1.4-19a). For the smallest value of the ratio R_{12} ($\equiv A'_{12}/A'_{21} = A_{12}/A_{21}$), the generated curves differ only in minor detail; were this a data-fitting exercise, the two equations would show similar correlating abilities for a real mixture exhibiting modest deviations from ideality and modest asymmetry. However, differences in the behavior of the equations become apparent as R_{12} increases. With increasing R_{12}, the van Laar $\mathcal{G}$ exhibits increasingly greater curvature, and the Margules activity coefficients eventually (for $R_{12} > 2$) show interior extrema. Which (if either) of these two expressions one would choose for a highly asymmetrical mixture (e.g., as represented by Fig. 1.4-2d) would be determined by a

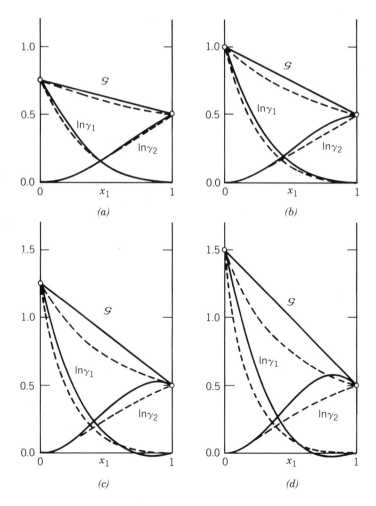

FIGURE 1.4-2 Composition dependence of $\mathcal{G}$ ($\equiv g^E/x_1 x_2 RT$) and of activity coefficients for binary mixtures, as represented by the two-parameter Margules equation (solid curves) and by the van Laar equation (dashed curves). For all cases, $\ln \gamma_1^\infty = 0.5$; different cases correspond to different values for $\ln \gamma_2^\infty$.

careful examination of the available data; both kinds of behavior illustrated in Fig. 1.4-2d are observed for real mixtures.

The two-parameter van Laar equation cannot represent maxima or minima in the activity coefficients, nor can it represent the mixed deviations from ideality exemplified by the benzene–hexafluorobenzene system of Fig. 1.4-1. However, it is superior to the Margules equation for some extremely nonideal systems such as alcohol–hydrocarbon mixtures, for example, the ethanol–n-heptane system of Fig. 1.4-1. For such mixtures the two-parameter Margules equation often incorrectly predicts liquid–liquid phase splitting. Higher-order Margules equations can sometimes be used for these systems, but at the expense of many additional parameters.

Multicomponent extensions of the Margules and van Laar equations are available. The simplest is the expression

$$g^E = \sum\sum_{i<j} g^E_{ij} \tag{1.4-20}$$

which asserts that g^E for a multicomponent mixture is just the sum of the g^E values for the constituent binary mixtures. Thus, for a ternary mixture containing components 1, 2, and 3

$$g^E = g^E_{12} + g^E_{13} + g^E_{23}$$

Although acceptable performance is sometimes obtained, neither Eq. (1.4-20) nor other formulations incorporating only binary parameters have proved *generally* reliable for estimating g^E and activity coefficients for moderately to highly nonideal multicomponent mixtures. This shortcoming derives ultimately from the inherent empiricism of Eqs. (1.4-8) and (1.4-16); these series expansions have little physical significance. Reasonable extrapolation from binary to multicomponent behavior requires models for g^E with a sound basis in molecular theory.

1.4-3 Local Composition Expressions for g^E

The inability of the classical expressions for g^E to provide adequate descriptions of multicomponent behavior has inspired the search for molecularly based models. The most popular class of models is based on the concept of *local composition*, which explicitly recognizes that the local environment of a molecule in solution with molecular species of other types is not the same as that provided by the overall composition of the solution. Local mole fractions are semiempirically related to overall mole fractions through Boltzmann factors incorporating characteristic energies of interaction; these terms appear in the parameters in the correlating expressions for g^E. Well-defined assumptions regarding the molecular interactions permit extension of binary local-composition expressions to the multicomponent case, without the introduction of additional parameters. Moreover, use of Boltzmann factors produces a built-in temperature dependence for the parameters which, while not generally precise, is sometimes acceptable for engineering applications.

The prototypical local-composition expression for g^E is the *Wilson equation*,[1] which for a binary mixture is

$$\frac{g^E}{RT} = -x_1 \ln (x_1 + \Lambda_{12}x_2) - x_2 \ln (x_2 + \Lambda_{21}x_1) \tag{1.4-21}$$

Parameter Λ_{ij} is given by

$$\Lambda_{ij} = \frac{v_j}{v_i} \exp\left(-\frac{\Delta\lambda_{ij}}{RT}\right) \tag{1.4-22}$$

where v_i and v_j are the molar volumes of the pure components and $\Delta\lambda_{ij}$ is a characteristic energy. Equation (1.4-21) yields the following expressions for the activity coefficients:

$$\ln \gamma_1 = -\ln (x_1 + \Lambda_{12}x_2) + x_2 \left(\frac{\Lambda_{12}}{x_1 + \Lambda_{12}x_2} - \frac{\Lambda_{21}}{x_2 + \Lambda_{21}x_1}\right) \tag{1.4-23a}$$

$$\ln \gamma_2 = -\ln (x_2 + \Lambda_{21}x_1) + x_1 \left(\frac{\Lambda_{21}}{x_2 + \Lambda_{21}x_1} - \frac{\Lambda_{12}}{x_1 + \Lambda_{12}x_2}\right) \tag{1.4-23b}$$

Infinite-dilution activity coefficients are related to parameters Λ_{12} and Λ_{21} via

$$\ln \gamma_1^\infty = 1 - \ln \Lambda_{12} - \Lambda_{21} \tag{1.4-24a}$$

$$\ln \gamma_2^\infty = 1 - \ln \Lambda_{21} - \Lambda_{12} \tag{1.4-24b}$$

Wilson's equation is particularly useful for homogeneous mixtures exhibiting large positive deviations from ideality, for example, alcohol–hydrocarbon systems. Like the van Laar equation, it cannot represent interior extrema in the activity coefficients; moreover, it is incapable of predicting liquid–liquid phase splitting. To remedy the latter shortcoming, the following three-parameter modification of Eq. (1.4-21) has been proposed:

$$\frac{g^E}{RT} = -C[x_1 \ln (x_1 + \Lambda_{12}x_2) + x_2 \ln (x_2 + \Lambda_{21}x_1)] \tag{1.4-25}$$

Here, C is an adjustable positive constant; when $C = 1$, the original Wilson equation is recovered.

A significant increase in flexibility over the Wilson equation is afforded by the *NRTL equation* of Renon and Prausnitz.[2] For a binary mixture, this three-parameter local-composition expression reads

$$\frac{g^E}{RT} = x_1 x_2 \left(\frac{\tau_{21}G_{21}}{x_1 + G_{21}x_2} + \frac{\tau_{12}G_{12}}{x_2 + G_{12}x_1} \right) \tag{1.4-26}$$

hence,

$$\ln \gamma_1 = x_2^2 \left[\tau_{21} \left(\frac{G_{21}}{x_1 + G_{21}x_2} \right)^2 + \frac{\tau_{12}G_{12}}{(x_2 + G_{12}x_1)^2} \right] \tag{1.4-27a}$$

$$\ln \gamma_2 = x_1^2 \left[\tau_{12} \left(\frac{G_{12}}{x_2 + G_{12}x_1} \right)^2 + \frac{\tau_{21}G_{21}}{(x_1 + G_{21}x_2)^2} \right] \tag{1.4-27b}$$

and

$$\ln \gamma_1^\infty = \tau_{21} + \tau_{12}G_{12} \tag{1.4-28a}$$

$$\ln \gamma_2^\infty = \tau_{12} + \tau_{21}G_{21} \tag{1.4-28b}$$

Parameters τ_{ij} and G_{ij} are related:

$$\tau_{ij} = \frac{\Delta g_{ij}}{RT} \tag{1.4-29}$$

and

$$G_{ij} = \exp (-\alpha_{ij}\tau_{ij}) \tag{1.4-30}$$

where Δg_{ij} is a characteristic energy and α_{ij} is a nonrandomness parameter. If $\alpha_{ij} = 0$, then $G_{ij} = 1$, and Eq. (1.4-26) reduces to the Porter equation. The NRTL equation is capable of reproducing most of the features of real-mixture behavior, but at the expense of a third parameter.

The UNIQUAC equation (Abrams and Prausnitz,[3] Anderson and Prausnitz,[4] Maurer and Prausnitz[5]) provides an example of a two-parameter local-composition equation of great flexibility. In its original form, it is written

$$g^E = g^E \text{ (combinatorial)} + g^E \text{ (residual)} \tag{1.4-31}$$

where, for a binary mixture,

$$\frac{g^E \text{ (combinatorial)}}{RT} = x_1 \ln \frac{\Phi_1}{x_1} + x_2 \ln \frac{\Phi_2}{x_2} + \left(\frac{z}{2}\right) \left(q_1 x_1 \ln \frac{\theta_1}{\Phi_1} + q_2 x_2 \ln \frac{\theta_2}{\Phi_2} \right) \tag{1.4-32}$$

$$\frac{g^E \text{ (residual)}}{RT} = -q_1 x_1 \ln (\theta_1 + \theta_2 \tau_{21}) - q_2 x_2 \ln (\theta_2 + \theta_1 \tau_{12}) \tag{1.4-33}$$

Here, z is a coordination number (usually set equal to 10), the Φ_i are molecular segment fractions, and the θ_i are molecular area fractions. Quantities Φ_i and θ_i are related to pure-component molecular-structure constants r_i and q_i by the prescriptions

$$\Phi_1 = \frac{x_1 r_1}{x_1 r_1 + x_2 r_2} \qquad \Phi_2 = \frac{x_2 r_2}{x_1 r_1 + x_2 r_2} \tag{1.4-34a}$$

and

$$\theta_1 = \frac{x_1 q_1}{x_1 q_1 + x_2 q_2} \qquad \theta_2 = \frac{x_2 q_2}{x_1 q_1 + x_2 q_2} \tag{1.4-34b}$$

Parameter τ_{ij} is given by

$$\tau_{ij} = \exp\left(-\frac{\Delta u_{ij}}{RT}\right) \tag{1.4-35}$$

where Δu_{ij} is a characteristic energy. Since structure constants r_i and q_i can be computed once and for all for each pure component i (see Prausnitz et al.[6] for a tabulation of values), the UNIQUAC equation is a two-parameter expression for g^E; characteristic energies Δu_{12} and Δu_{21} (and hence parameters τ_{12} and τ_{21}) must be found by analysis of binary mixture data, usually vapor–liquid equilibrium data. UNIQUAC activity coefficients for a binary mixture are given by

$$\ln \gamma_1 = \ln \frac{\Phi_1}{x_1} + \left(\frac{z}{2}\right) q_1 \ln \frac{\theta_1}{\Phi_1} + \Phi_2\left(l_1 - \frac{r_1}{r_2} l_2\right)$$

$$- q_1 \ln (\theta_1 + \theta_2 \tau_{12}) + \theta_2 q_1 \left(\frac{\tau_{21}}{\theta_1 + \theta_2 \tau_{21}} - \frac{\tau_{12}}{\theta_2 + \theta_1 \tau_{12}}\right) \tag{1.4-36a}$$

$$\ln \gamma_2 = \ln \frac{\Phi_2}{x_2} + \left(\frac{z}{2}\right) q_2 \ln \frac{\theta_2}{\Phi_2} + \Phi_1\left(l_2 - \frac{r_2}{r_1} l_1\right)$$

$$- q_2 \ln (\theta_2 + \theta_1 \tau_{21}) + \theta_1 q_2 \left(\frac{\tau_{12}}{\theta_2 + \theta_1 \tau_{12}} - \frac{\tau_{21}}{\theta_1 + \theta_2 \tau_{21}}\right) \tag{1.4-36b}$$

where

$$l_1 = \left(\frac{z}{2}\right)(r_1 - q_1) - (r_1 - 1) \tag{1.4-37a}$$

$$l_2 = \left(\frac{z}{2}\right)(r_2 - q_2) - (r_2 - 1) \tag{1.4-37b}$$

Hence,

$$\ln \gamma_1^\infty = \ln \frac{r_1}{r_2} + \frac{z}{2} q_1 \ln \frac{q_1 r_2}{q_2 r_1} + \left(l_1 - \frac{r_1}{r_2} l_2\right) + q_1 (1 - \ln \tau_{12} - \tau_{12}) \tag{1.4-38a}$$

$$\ln \gamma_2^\infty = \ln \frac{r_2}{r_1} + \frac{z}{2} q_2 \ln \frac{q_2 r_1}{q_1 r_2} + \left(l_2 - \frac{r_2}{r_1} l_1\right) + q_2 (1 - \ln \tau_{21} - \tau_{21}) \tag{1.4-38b}$$

Sometimes insufficient binary data are available for evaluation of parameters τ_{ij} in the UNIQUAC equation. In this event, parameters may be estimated by the UNIFAC correlation,[7] a group-contribution technique. The UNIFAC method is based on the notion that *functional groups* (e.g., methyl groups, hydroxyl groups), rather than entire molecules, are the key units of interaction in a mixture. In an exhaustive and continuing effort, Aa. Fredenslund, J. Gmehling, and their coworkers have evaluated and reduced available vapor–liquid equilibrium data to obtain interaction energy parameters for many of the functional groups present in substances of commercial importance. The UNIFAC procedure is treated in a monograph,[8] and updated tables of parameter values appear periodically in the open literature.[9-11] A special set of parameter values is available for liquid–liquid equilibrium calculations.[12]

Local-composition equations for g^E are readily extended to multicomponent mixtures. The multicomponent expressions contain parameters obtainable in principle from binary data only; they provide descriptions of g^E of acceptable accuracy for many engineering calculations of multicomponent vapor–liquid equilibria at subcritical conditions. Listed below are the multicomponent versions of the Wilson, NRTL, and UNIQUAC equations.

Multicomponent Wilson Equation

$$\frac{g^E}{RT} = -\sum_i x_i \ln \left(\sum_j \Lambda_{ij} x_j \right)$$

(1.4-39)

hence,

$$\ln \gamma_i = 1 - \ln \left(\sum_j \Lambda_{ij} x_j \right) - \sum_j \left(\frac{\Lambda_{ji} x_j}{\sum_k \Lambda_{jk} x_k} \right)$$

(1.4-40)

The Λ_{ij} are defined by Eq. (1.4–22).

Multicomponent NRTL Equation

$$\frac{g^E}{RT} = \sum_i x_i \left(\frac{\sum_j \tau_{ji} G_{ji} x_j}{\sum_j G_{ji} x_j} \right)$$

(1.4-41)

hence,

$$\ln \gamma_i = \frac{\sum_j \tau_{ji} G_{ji} x_j}{\sum_j G_{ji} x_j} + \sum_j \frac{G_{ij} x_j}{\sum_k G_{kj} x_k} \left(\tau_{ij} - \frac{\sum_k \tau_{kj} G_{kj} x_k}{\sum_k G_{kj} x_k} \right)$$

(1.4-42)

Quantities τ_{ij} and G_{ij} are defined by Eqs. (1.4-29) and (1.4-30).

Multicomponent UNIQUAC Equation

$$g^E = g^E \text{ (combinatorial)} + g^E \text{ (residual)}$$

(1.4-31)

$$\frac{g^E \text{ (combinatorial)}}{RT} = \sum_i x_i \ln \frac{\Phi_i}{x_i} + \frac{z}{2} \sum_i q_i x_i \ln \frac{\theta_i}{\Phi_i}$$

(1.4-43)

$$\frac{g^E \text{ (residual)}}{RT} = -\sum_i q_i x_i \ln \left(\sum_j \theta_j \tau_{ji} \right)$$

(1.4-44)

Here,

$$\Phi_i = \frac{r_i x_i}{\sum_j r_j x_j}$$

(1.4-45a)

$$\theta_i = \frac{q_i x_i}{\sum_j q_j x_j}$$

(1.4-45b)

where r_i and q_i are pure-component molecular-structure constants and parameter τ_{ij} is defined by Eq. (1.4-35). The activity coefficient γ_i is given by

$$\ln \gamma_i = \ln \frac{\Phi_i}{x_i} + \frac{z}{2} q_i \ln \frac{\theta_i}{\Phi_i} + l_i - \frac{\Phi_i}{x_i} \sum_j l_j x_j - q_i \ln \left(\sum_j \theta_j \tau_{ji} \right) + q_i - q_i \sum_j \left(\frac{\theta_j \tau_{ij}}{\sum_k \theta_k \tau_{kj}} \right)$$

(1.4-46)

where

$$l_i = \frac{z}{2} (r_i - q_i) - (r_i - 1)$$

(1.4-47)

1.5   VAPOR–LIQUID EQUILIBRIA AND LIQUID–LIQUID EQUILIBRIA

1.5-1   Subcritical Vapor–Liquid Equilibria

The essential features of vapor–liquid equilibrium (VLE) behavior are demonstrated by the simplest case: isothermal VLE of a binary system at a temperature below the critical temperatures of both pure components. For this case (''subcritical'' VLE), each pure component has a well-defined vapor–liquid saturation pressure P_i^{sat}, and VLE is possible for the full range of liquid and vapor compositions x_i and y_i. Figure 1.5-1 illustrates several types of behavior shown by such systems. In each case, the upper solid curve (''bubble curve'') represents states of saturated liquid; the lower solid curve (''dew curve'') represents states of saturated vapor.

Figure 1.5-1a is for a system that obeys *Raoult's Law*. The significant feature of a Raoult's Law system is the linearity of the isothermal bubble curve, expressed for a binary system as

$$P(\text{bubble}) = P_2^{sat} + x_1(P_1^{sat} - P_2^{sat}) \tag{1.5-1}$$

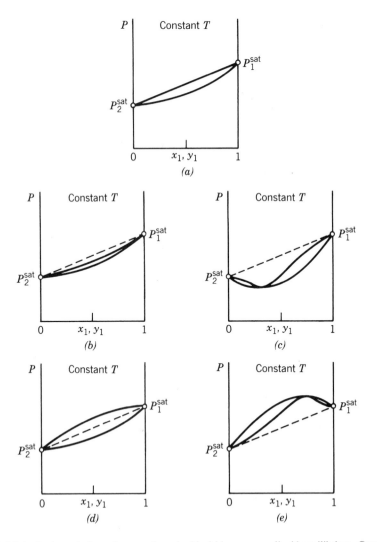

FIGURE 1.5-1   Isothermal phase diagrams for subcritical binary vapor–liquid equilibrium. Case (*a*) represents Raoult's-Law behavior. Cases (*b*) and (*c*) illustrate negative deviations from Raoult's Law; cases (*d*) and (*e*) illustrate positive deviations from Raoult's Law.

Although Raoult's Law is rarely obeyed by real mixtures, it serves as a useful standard against which real VLE behavior can be compared. The dashed lines in Figs. 1.5-1b–1.5-1e are the Raoult's Law bubble curves produced by the vapor pressure P_1^{sat} and P_2^{sat}.

Figures 1.5-1b and 1.5-1c illustrate *negative* deviations from Raoult's Law: the actual bubble curves lie below the Raoult's Law bubble curve. In Fig. 1.5-1b the deviations are moderate, but in Fig. 1.5-1c the deviations are so pronounced that the system exhibits a minimum-pressure (maximum-boiling) homogeneous azeotrope.

The systems of Figs. 1.5-1d and 1.5-1e show *positive* deviations from Raoult's Law, for which the true bubble curves lie above the Raoult's Law line. In Fig. 1.5-1d, the deviations are modest; in Fig. 1.5-1e they are large, and a maximum-pressure (minimum-boiling) homogeneous azeotrope occurs.

The goal of a subcritical VLE calculation is to quantitatively predict or correlate the various kinds of behavior illustrated by Fig. 1.5-1 or by its isobaric or multicomponent counterparts. The basis for the calculation is phase-equilibrium formulation 1 of Section 1.2-5, where liquid-phase fugacities are eliminated in favor of liquid-phase activity coefficients, and vapor-phase fugacities in favor of vapor-phase fugacity coefficients. Raoult's Law standard states are chosen for all components in the liquid phase; hence, $(f_i^{\circ})^L = f_i^L$, and Eq. (1.2-63) becomes

$$x_i \gamma_i^L f_i^L = y_i \hat{\phi}_i^V P \quad \text{(all } i\text{)} \tag{1.5-2}$$

where liquid-phase activity coefficients are defined by the Raoult's Law convention (see Section 1.2-4).

All quantities in Eq. (1.5-2) refer to the *actual* conditions of VLE. For applications it is convenient to eliminate the fugacity f_i^L in favor of the vapor pressure P_i^{sat} of pure i, and to refer the activity coefficients γ_i^L to a fixed reference pressure P^r. Equation (1.5-2) then becomes

$$x_i \gamma_i^r P_i^{\text{sat}} \phi_i^{\text{sat}} \frac{f_i}{f_i^{\text{sat}}} \frac{\gamma_i}{\gamma_i^r} = y_i \hat{\phi}_i P \quad \text{(all } i\text{)} \tag{1.5-3}$$

where we have deleted superscripts identifying a phase: here and henceforth in this subsection fugacities and activity coefficients are for the liquid phase, and fugacity coefficients are for the vapor phase. Quantities ϕ_i^{sat} and f_i^{sat} are evaluated for pure component i at the vapor pressure P_i^{sat}.

Equation (1.5-3) can be written in the equivalent form

$$x_i \gamma_i^r P_i^{\text{sat}} = y_i \Psi_i P \quad \text{(all } i\text{)} \tag{1.5-4}$$

where Ψ_i is a composite function defined by

$$\Psi_i = \frac{\gamma_i^r}{\gamma_i} \frac{f_i^{\text{sat}}}{f_i} \frac{\hat{\phi}_i}{\phi_i^{\text{sat}}} \tag{1.5-5}$$

Although Ψ_i is conveniently viewed as a correction factor of order unity, one or more of the ratios on the right-hand side of Eq. (1.5-5) may in particular applications differ appreciably from unity. Thermodynamics provides the following exact expressions for these ratios:

$$\frac{\gamma_i^r}{\gamma_i} = \exp \int_P^{P^r} \frac{\bar{v}_i^E}{RT} dP \tag{1.5-6}$$

$$\frac{f_i^{\text{sat}}}{f_i} = \exp \int_P^{P_i^{\text{sat}}} \frac{v_i}{RT} dP \tag{1.5-7}$$

$$\frac{\hat{\phi}_i}{\phi_i^{\text{sat}}} = \exp \left[\int_0^P (\bar{Z}_i - 1) \frac{dP}{P} - \int_0^{P_i^{\text{sat}}} (Z_i - 1) \frac{dP}{P} \right] \tag{1.5-8}$$

Equation (1.5-6) follows from Eq. (1.2-58); it represents the effect of pressure on the activity coefficient and requires liquid-phase excess molar volume data. Equation (1.5-7) is a Poynting correction [see Eq. (1.2-31) and the accompanying discussion], which requires volumetric data for pure liquid i. Equation (1.5-8) represents the contributions of vapor-phase nonidealities, which are represented by a *PVTx* equation of state. Note here that the effects of vapor-phase nonidealities enter through both $\hat{\phi}_i$ and ϕ_i^{sat}. Consistent description of subcritical VLE via Eq. (1.5-4) requires that $\hat{\phi}_i$ and ϕ_i^{sat} be evaluated in a consistent fashion.

Equation (1.5-4) is rarely used in its exact form; approximations are made reflecting the conditions and the nature of the system and also reflecting the nature and extent of available thermodynamic data. Vapor pressures P_i^{sat} are the single most important quantities in Eq. (1.5-4): they characterize the states of pure-fluid VLE (the "edges" of the phase diagrams); without them, even qualitative prediction of mixture VLE is impossible. Given good values for the P_i^{sat}, one introduces approximations by making statements about the ratios appearing on the right-hand side of Eq. (1.5-5). In roughly decreasing order of reasonableness, some common approximations are:

1. Assume that the activity coefficients are independent of pressure. By Eq. (1.5-6) this requires that $\bar{v}_i^E/RT$ be small.
2. Assume that liquid fugacities are independent of pressure. By Eq. (1.5-7), this requires that v_i/RT be small.
3. Assume that vapor-phase corrections are negligible. By Eq. (1.5-8), this occurs if both Z_i and $\bar{Z}_i$ are close to unity (the ideal-gas approximation, valid for sufficiently low pressures), but also if both Z_i and $\bar{Z}_i$ and P and P_i^{sat} are of comparable magnitudes. In the latter case, simplification results because of a fortuitous cancellation of effects.
4. Assume that the liquid phase is an ideal solution, that is, $\gamma_i = 1$ for all temperatures, pressures, and compositions.

Unless the reduced temperature of component i is high (say about 0.85 or greater), approximation 1 is nearly always reasonable. On the other hand, approximation 4 is rarely realistic: liquid-phase nonidealities are normally present and must be taken into account. If one invokes approximations 2, 3, and 4, then Raoult's Law is obtained:

$$x_i P_i^{sat} = y_i P \quad \text{(all } i\text{)} \tag{1.5-9}$$

Raoult's Law, however, is useful mainly as a standard of comparison. The simplest *realistic* simplification of Eq. (1.5-4) follows from approximations 1, 2, and 3:

$$x_i \gamma_i P_i^{sat} = y_i P \quad \text{(all } i\text{)} \tag{1.5-10}$$

Here, superscript r has been dropped from γ_i because the γ_i are assumed independent of pressure.

Equation (1.5-10), unlike Eq. (1.5-9), can reproduce all the qualitative features of subcritical VLE illustrated in Fig. 1.5-1. If we take (1.5-10) as representing actual behavior, then subtraction of (1.5-9) from (1.5-10) and summation over all components gives

$$P - P_{RL} = \sum_i x_i(\gamma_i - 1) \, P_i^{sat} \tag{1.5-11}$$

Equation (1.5-11) is an expression for the difference between the actual bubble pressure P given by Eq. (1.5-10) and the Raoult's Law bubble pressure P_{RL} given by Eq. (1.5-9). To the extent that the approximate Eq. (1.5-10) is valid, Eq. (1.5-11) asserts that deviations from Raoult's Law result from liquid-phase nonidealities: liquid-phase activity coefficients greater than unity promote positive deviations from Raoult's Law, and liquid-phase activity coefficients less than unity promote negative deviations from Raoult's Law.

Calculations at moderate pressure levels (say up to 5 atm) require the inclusion of at least some of the corrections represented by Eq. (1.5-5). A suitable formulation is developed as follows. First we ignore the effect of pressure on the activity coefficients; thus, we set $\gamma_i^r/\gamma_i = 1$ and write Eq. (1.5-4) as

$$x_i \gamma_i P_i^{sat} = y_i \Psi_i P \quad \text{(all } i\text{)} \tag{1.5-12}$$

where superscript r has been dropped from γ_i. Additionally, we assume that the liquid molar volume of pure i is independent of pressure and equal to its saturation value: $v_i = v_i^{sat}$. Equation (1.5-7) becomes

$$\frac{f_i^{sat}}{f_i} = \exp \frac{v_i^{sat} \, (P_i^{sat} - P)}{RT}$$

which is a special case of Eq. (1.2-34). Finally, we assume that the vapor phase is described by the two-term virial equation in pressure, Eq. (1.3-11). Then, by Eqs. (1.3-15) and (1.3-16) we have

$$\frac{\hat{\phi}_i}{\phi_i^{sat}} = \exp \frac{\bar{B}_i \, P - B_{ii} P_i^{sat}}{RT}$$

and the working expression for Ψ_i becomes

$$\Psi_i = \exp \frac{(B_{ii} - v_i^{sat}) \, (P - P_i^{sat}) + (\bar{B}_i - B_{ii})P}{RT} \tag{1.5-13}$$

where (see Table 1.3-1)

$$\overline{B}_i - B_{ii} = \frac{1}{2} \sum_j \sum_k y_j y_k (2\delta_{ji} - \delta_{jk}) \tag{1.5-14}$$

and

$$\delta_{ij} = 2B_{ij} - B_{ii} - B_{jj} \tag{1.5-15}$$

Equations (1.5-12)–(1.5-15) together constitute the most common formulation for predicting or correlating subcritical VLE at low to moderate pressures. When using the formulation for VLE *predictions*, one requires data or correlations for pure-component vapor pressures (e.g., Antoine equations), for the activity coefficients (e.g., the UNIQUAC equation or the UNIFAC correlation), for the second virial coefficients (e.g., one of the correlations referenced in Section 1.3-2), and for the molar volumes of the saturated liquid (e.g., the Rackett equation[1,2] for v_i^{sat}). The actual VLE calculations are iterative and require the use of a computer; details are given in the monograph by Prausnitz et al.[3]

In VLE *correlation* (data reduction), one again requires values for the P_i^{sat}, B_{ij}, and v_i^{sat}, but here the goal is to determine from VLE data best values for the parameters in an assumed expression for g^E: in effect, to find liquid-phase activity coefficients from VLE data. Data reduction procedures combine VLE calculations with nonlinear regression techniques and again require use of a computer; this topic is discussed by Van Ness and Abbott[4] and by Prausnitz et al.[3]

Figure 1.5-2 shows experimental and correlated binary VLE data for three dioxane–*n*-alkane systems at 80°C.[5,6] The pressure levels are modest (0.2–1.4 atm); liquid-phase nonidealities are sufficiently large to promote azeotropy in all three cases. Equations (1.5-12)–(1.5-15) were used for the data reduction, with experimental values for the P_i^{sat} and v_i^{sat}; virial coefficients were estimated from the correlation of Tsonopoulos.[7] Activity coefficients were assumed to be represented by the three-parameter Margules equation, and the products of the data reduction were sets of values for parameters A_{12}, A_{21}, and D in Eqs. (1.4-10) and (1.4-11). The parameters so determined produce the correlations of the data shown by the solid curves in Fig. 1.5-2. For all three systems, the data are represented to within their experimental uncertainty.

1.5-2 Supercritical Vapor–Liquid Equilibria

When the system temperature T is greater than the critical temperature T_{ci} of component i, then pure i cannot exist as a liquid. The procedure of Section 1.5-1, which incorporates the vapor–liquid saturation pressure P_i^{sat}, is therefore inappropriate for representing VLE for mixtures containing "supercritical" component i. Several methods are available for the quantitative description of such cases; the most powerful of them is that using an equation of state as discussed briefly in Section 1.7. Alternatively, one may use Eq. (1.5-2),

$$x_i \gamma_i^L f_i^L = y_i \hat{\phi}_i^v P \quad \text{(all } i)$$

incorporating for the supercritical components special correlations for f_i^L (or ϕ_i^L) for the *hypothetical* liquid state. Finally, one may employ a variant of the procedure of Section 1.5-1, in which Henry's Law standard states are adopted for the supercritical components in the liquid phase. We consider in this subsection the last of these techniques.

For simplicity, we restrict our discussion to binary systems containing a single supercritical component 1 (the "solute" species) and a single subcritical component 2 (the "solvent" species); rigorous extensions to more complicated situations (e.g., mixed solvents) are complex: see Van Ness and Abbott[8] for a discussion. The basis for the procedure is phase-equilibrium formulation 1 of Section 1.2-5. Subcritical component 2 is treated as in Section 1.5-1. However, we write for supercritical component 1 that

$$(f_1^\circ)^L = \mathcal{H}_{1,2}$$

where $\mathcal{H}_{1,2}$ is Henry's constant for 1 dissolved in 2. This choice of standard-state fugacity accommodates the fact that the full $\hat{f}_1^L$ versus x_1 curve (see Fig. 1.2-3) is undefined for component 1: $\mathcal{H}_{1,2}$ is an experimentally accessible quantity, whereas f_1^L is not. The equilibrium equation for component 1 is then

$$x_1 \gamma_1^* \mathcal{H}_{1,2} = y_1 \hat{\phi}_1 P \tag{1.5-16}$$

where the liquid-phase activity coefficient $\gamma_1^* \equiv \gamma_1(HL)$ is normalized by the Henry's Law convention (see Section 1.2-4).

All quantities in Eq. (1.5-16) refer to the *actual* conditions of VLE. For applications it is convenient

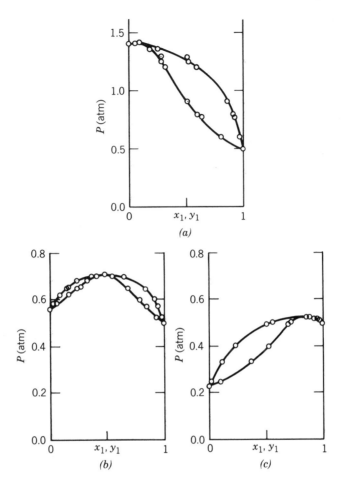

FIGURE 1.5-2 Correlation of subcritical vapor–liquid equilibria via Eqs. (1.5-12)–(1.5-15): (a) 1,4-dioxane(1)–n-hexane(2) at 80°C; (b) 1,4-dioxane(1)–n-heptane(2) at 80°C; and (c) 1,4-dioxane(1)–n-octane(2) at 80°C. Circles are data; curves are correlations.

to refer both γ_1^* and $\mathfrak{K}_{1,2}$ to a single reference pressure P^r, often taken as the vapor pressure P_2^{sat} of the solvent. Equation (1.5-16) then becomes

$$x_1(\gamma_1^*)^r \, \mathfrak{K}_{1,2}^r \exp \int_{P^r}^{P} \frac{\bar{v}_1}{RT} \, dP = y_1 \hat{\phi}_1 P \tag{1.5-17}$$

where $\bar{v}_1$ is the partial molar volume of the solute in the liquid phase. Note that the term containing $\bar{v}_1$ represents two contributions, the first,

$$\exp \int_{P^r}^{P} \frac{(\bar{v}_1^E)^*}{RT} \, dP,$$

representing the effect of P on the activity coefficient, and the second,

$$\exp \int_{P^r}^{P} \frac{\bar{v}_1^\infty}{RT} \, dP$$

representing the effect of P on Henry's constant. Quantity $\bar{v}_1^\infty$ is the partial volume at infinite dilution of the solute in the liquid; quantity $(\bar{v}_1^E)^* = \bar{v}_1 - \bar{v}_1^\infty$ is the partial molar excess volume of the solute.

Use of Eq. (1.5-17) requires an expression for a Henry's Law activity coefficient γ_1^*. Such expressions are usually developed from the conventional expressions for γ_1 based on Raoult's Law standard states. For a single-solute–single-solvent mixture the connection between the two is easily established. By Eqs. (1.2-52), (1.2-38), and (1.4-5), we have

$$\gamma_1 \equiv \gamma_1(\text{RL}) = \frac{\hat{f}_1}{x_1 f_1} \tag{1.5-18}$$

$$\lim_{x_1 \to 0} \left(\frac{\hat{f}_1}{x_1} \right) = \mathcal{K}_{1,2}$$

$$\lim_{x_1 \to 0} \gamma_1 = \gamma_1^\infty$$

Thus,

$$\mathcal{K}_{1,2} = \gamma_1^\infty f_1 \tag{1.5-19}$$

and we see that Henry's constant is directly related to the Raoult's Law infinite-dilution activity coefficient and to the fugacity of pure liquid 1. But, by definition,

$$\gamma_1^* \equiv \gamma_1(\text{HL}) = \frac{\hat{f}_1}{x_1 \mathcal{K}_{1,2}} \tag{1.5-20}$$

Combination of Eqs. (1.5-18), (1.5-19), and (1.5-20) gives

$$\gamma_1^* = \frac{\gamma_1}{\gamma_1^\infty}$$

or, equivalently,

$$\ln \gamma_1^* = \ln \gamma_1 - \ln \gamma_1^\infty \tag{1.5-21}$$

Equation (1.5-21) is a *general* relationship between Henry's Law and Raoult's Law activity coefficients for a component in a binary mixture. *Particular* expressions for γ_1^* are obtained from particular expressions for γ_1, which require a model for g^E for their determination: see Sections 1.4-1–1.4-3. To illustrate, suppose that g^E is described by the Porter equation, Eq. (1.4-14). Then, by Eqs. (1.5-21), (1.4-15a), and (1.4-19a), we find the one-parameter expression

$$\ln \gamma_1^* = A(x_2^2 - 1) \tag{1.5-22}$$

Similarly, for a binary mixture described by the two-parameter Margules equation for g^E, Eq. (1.4-12), we have

$$\ln \gamma_1^* = x_2^2[A_{12} + 2(A_{21} - A_{12})x_1] - A_{12}$$

Other expressions for γ_1^*, based on alternative models for the molar excess Gibbs energy, are easily developed. Note that all such expressions for γ_1^* satisfy the limiting condition

$$\lim_{x_1 \to 0} \gamma_1^* = 1$$

as required by Eq. (1.2-57).

Equilibrium equation (1.5-17) is rarely used in its exact form; just as with the analogous formulation for subcritical VLE, Eq. (1.5-4), approximations are introduced which reflect the conditions and the nature of the system. Some common approximations are:

1. Assume that activity coefficient γ_1^* is independent of pressure. This requires that $(\bar{v}^E)^*/RT$ be small. When this approximation is valid, then

$$\int_{P^r}^{P} \frac{\bar{v}_1}{RT}\, dP = \int_{P^r}^{P} \frac{\bar{v}_1^\infty}{RT}\, dP \tag{1.5-23}$$

2. Assume that *both* activity coefficient γ_1^* and Henry's constant $\mathcal{H}_{1,2}$ are independent of pressure. This requires that $\bar{v}_1/RT$ be small.

3. Assume that the vapor phase is an ideal-gas mixture, that is $\hat{\phi}_1 = 1$.

4. Assume that the liquid phase is an ideal solution, that is, $\gamma_1^* = 1$.

Approximation 1 is usually reasonable, provided the reduced temperature of the solvent is not too high; it is always reasonable if the solute mole fraction is sufficiently small, because

$$\bar{v}_1^\infty = \lim_{x_1 \to 0} \bar{v}_1$$

Thus, $\bar{v}_1^\infty$ is the key piece of volumetric information required for description of the liquid phase. Its measurement and correlation are discussed in a series of papers by Battino and coworkers.[9-11] If one invokes approximations 2, 3, and 4, then Henry's Law is obtained:

$$x_1 \, \mathcal{H}_{1,2} = y_1 P$$

Like Raoult's Law, Henry's Law is of limited use for accurate work. However, realistic versions of Eq. (1.5-17) suitable, for example, for gas-solubility calculations, are easily developed. If we adopt approximation 1 and also assume that $\bar{v}_1^\infty$ is independent of pressure, then Eqs. (1.5-17) and (1.5-23) yield

$$x_1 \gamma_1^* \, \mathcal{H}_{1,2}^r \, \exp \frac{\bar{v}_1^\infty (P - P^r)}{RT} = y_1 \hat{\phi}_1 P \qquad (1.5\text{-}24)$$

where superscript r has been dropped from γ_1^*. Provided that the reduced temperature of the solvent is not too high, Eq. (1.5-24) is satisfactory for calculation or correlation of gas-solubility data to quite high pressures. If g^E is represented by the Porter equation, then combination of Eqs. (1.5-24) and (1.5-22) gives on rearrangement the *Krichevsky–Ilinskaya equation*[12] for gas solubility:

$$\ln x_1 = \ln (y_1 \hat{\phi}_1 P) - \ln \mathcal{H}_{1,2}^r - A(x_2^2 - 1) - \frac{\bar{v}_1^\infty (P - P^r)}{RT} \qquad (1.5\text{-}25)$$

The Krichevsky–Ilinskaya equation is often used for precise correlation of high-quality gas-solubility data: see Prausnitz[13] and Van Ness and Abbott[4] for examples and discussion.

1.5-3 Liquid–Liquid Equilibria

Mutual solubilities of liquids vary greatly; at ambient temperature, water and ethyl alcohol are miscible in all proportions, water and benzene are only very slightly soluble in one another, while benzene and mercury show essentially no mutual solubility. For most liquid pairs, mutual solubility increases with rising temperature, but many exceptions are known; a few pairs are completely soluble in one another at low temperatures and at high temperatures with limited miscibility in between, while others (notably polymer-solvent systems) show complete miscibility only between lower and upper temperature limits.

Following a brief quantitative discussion of two-component liquid–liquid equilibria (LLE), we generalize the discussion to multicomponent systems with special attention to ternaries.

For a binary system containing two liquid phases at the same temperature and pressure, formulation 2 of Section 1.2-5 provides the following equations of equilibrium:

$$x_1^\alpha \gamma_1^\alpha = x_1^\beta \gamma_1^\beta \qquad (1.5\text{-}26a)$$

$$x_2^\alpha \gamma_2^\alpha = x_2^\beta \gamma_2^\beta \qquad (1.5\text{-}26b)$$

Here, subscripts 1 and 2 refer to the components, superscripts α and β refer to the equilibrated liquid phases, and the γ_i are activity coefficients. In these equations, for each component, the Raoult's Law standard-state convention is used for both phases.

Suppose that we have available an expression for g^E, the molar excess Gibbs energy, which holds for the entire composition range. We can find activity coefficients, as discussed in Section 1.4, from the relationship

$$RT \ln \gamma_i = \left[\frac{\partial (ng^E)}{\partial n_i} \right]_{T,P,n_j} \qquad (i = 1 \text{ or } 2, j = 2 \text{ or } 1) \qquad (1.5\text{-}27)$$

Since g^E is a function only of x at fixed temperature and pressure, Eqs. (1.5-26) and (1.5-27) are sufficient to find the equilibrium compositions x_1^α (or x_2^α) and x_1^β (or x_2^β). What is needed is an appropriate expression for g^E.

To illustrate, consider the simplest case where g^E is described by the Porter equation

$$\frac{g^E}{RT} = Ax_1x_2 \qquad (1.4\text{-}14)$$

with A a function of temperature only. The equations of equilibrium become

$$x_1^\alpha \exp [A(x_2^\alpha)^2] = x_1^\beta \exp [A(x_2^\beta)^2] \qquad (1.5\text{-}28a)$$

$$x_2^\alpha \exp [A(x_1^\alpha)^2] = x_1^\beta \exp [A(x_1^\beta)^2] \qquad (1.5\text{-}28b)$$

In addition, we have two conservation relationships:

$$x_1^\alpha + x_2^\alpha = 1 \qquad (1.5\text{-}29a)$$

$$x_1^\beta + x_2^\beta = 1 \qquad (1.5\text{-}29b)$$

The solution to these equations is discussed elsewhere;[4] here, we note only that, because of symmetry,

$$x_1^\alpha = 1 - x_1^\beta \qquad (1.5\text{-}30a)$$

$$x_2^\alpha = 1 - x_2^\beta \qquad (1.5\text{-}30b)$$

and therefore a single equation is sufficient to describe the equilibrium. It is

$$A(1 - 2x_1) = \ln \left(\frac{1 - x_1}{x_1}\right) \qquad (1.5\text{-}31)$$

This equation has the trivial solution $x_1 = \frac{1}{2}$ for all values of A. However, if $A > 2$, there are three real roots: $x_1 = \frac{1}{2}$, $x_1 = r$, and $x_1 = 1 - r$, where r lies between 0 and $\frac{1}{2}$. The latter two roots are the equilibrium compositions x_1^α and x_1^β. For example, if $A = 2.1972$, then $x_1^\alpha = 0.25$ and $x_1^\beta = 0.75$; when $A = 6.9206$, $x_1^\alpha = 0.001$ and $x_1^\beta = 0.999$.

Equation (1.5-31) gives realistic results only when $A > 2$, consistent with thermodynamic stability analysis. The condition for instability of a binary liquid phase is

$$\left(\frac{\partial^2 g}{\partial x_1^2}\right)_{T,P} < 0 \qquad (1.5\text{-}32)$$

where g is the molar Gibbs energy of the mixture:

$$g = x_1g_1 + x_2g_2 + RT(x_1 \ln x_1 + x_2 \ln x_2) + g^E \qquad (1.5\text{-}33)$$

When Eq. (1.4-14) is substituted into Eqs. (1.5-32) and (1.5-33), we find that a two-component liquid mixture is unstable only if $A > 2$. In other words, a single liquid phase described by the Porter equation splits into two liquid phases only if $A > 2$. As indicated also by Eq. (1.5-31), if $A \leqslant 2$, the two components are completely miscible.

The above discussion considers only the calculation of equilibrium compositions x_1^α and x_1^β at a single temperature. The temperature dependence of x_1^α and x_1^β, as represented by a liquid–liquid solubility diagram, is also of interest. This temperature dependence enters the equilibrium formulation through the temperature dependence of the parameters in the assumed expression for g^E; thermodynamically, it is determined by the behavior of the molar excess enthalpy h^E. If a binary liquid system is described by the Porter equation, then the T dependence of the single parameter A determines the shape of the solubility diagram. Figure 1.5-3 shows several types of solubility diagram, and the qualitative variations of A with T that give rise to them. *Quantitative* representation of a liquid–liquid solubility diagram is far more difficult than this figure suggests; it is one of the most exacting tests to which one can subject a model for the excess Gibbs energy.

For multicomponent mixtures, LLE calculations are complex. If a mixture contains N components and if we assume that there are two liquid phases (α and β) at equilibrium, we have N equilibrium equations of the form

$$x_i^\alpha \gamma_i^\alpha = x_i^\beta \gamma_i^\beta \qquad (i = 1, 2, \ldots, N) \qquad (1.5\text{-}34)$$

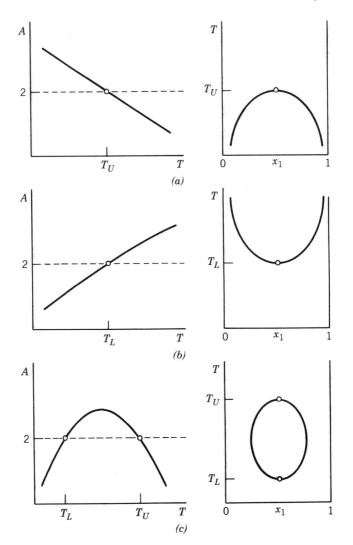

FIGURE 1.5-3 Binary liquid–liquid solubility diagrams implied by the Porter equation for g^E. T_U and T_L are upper and lower consolute temperatures; the T dependence of parameter A determines the shape of the solubility diagram.

and two conservation relationships:

$$\sum_i^N x_i^\alpha = 1 \qquad (1.5\text{-}35a)$$

$$\sum_i^N x_i^\beta = 1 \qquad (1.5\text{-}35b)$$

Activity coefficients are obtained by differentiation of a suitable multicomponent expression for g^E, for example, the NRTL equation or the UNIQUAC equation: see Section 1.4-3.

A general procedure for solving all simultaneous equations is given elsewhere;[3] the procedure is to solve a flash problem, as briefly outlined below.

Let F stand for the number of moles of a feed mixture that enters a separator operating at temperature T and pressure P; leaving this vessel we have two liquid phases at equilibrium. Let L^α stand for the number

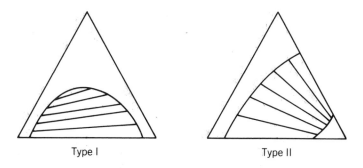

FIGURE 1.5-4 Two common types of ternary liquid–liquid phase diagram.

of moles of liquid phase α and L^β for the number of moles of liquid phase β. We then have N material balances of the form

$$z_i F = x_i^\alpha L^\alpha + x_i^\beta L^\beta \qquad (i = 1, 2, \ldots, N) \tag{1.5-36}$$

where z_i is the mole fraction of component i in the feed.

The number of unknowns is $2N + 2$; there are N unknown mole fractions in each phase and we do not know L^α and L^β. But we also have $2N + 2$ independent equations: phase equilibrium equation (1.5-34) and material balance equation (1.5-36), each taken N times, and the two conservation equations, Eqs. (1.5-35a) and (1.5-35b). In principle, therefore, the problem can be solved but the iterative numerical

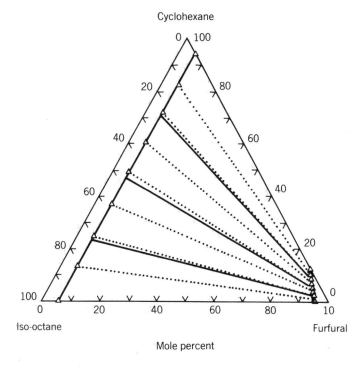

FIGURE 1.5-5 Predicted ternary liquid–liquid equilibria for a typical Type II system shows good agreement with experimental data, using parameters estimated from binary data only. Δ . . . , data at 25°C; —, predicted from UNIQUAC equation with binary data only.

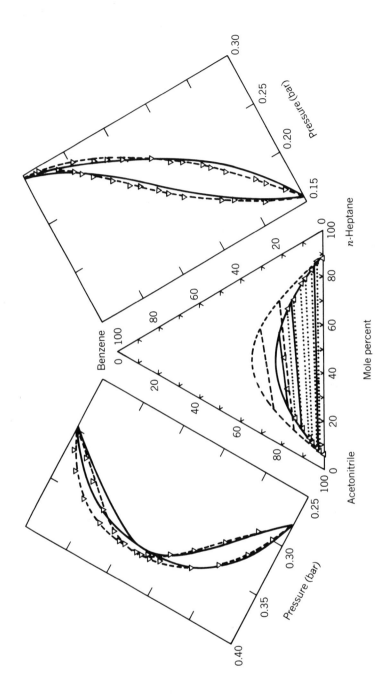

FIGURE 1.5-6 Representation of ternary Type I liquid–liquid equilibria via the UNIQUAC equation is improved by incorporating ternary tie-line data into binary-parameter estimation. Representation of binary VLE shows some loss of accuracy. △ · · · , data at 45°C; – – –, representation via binary data only; —, representation via binary data plus one tie line.

procedure for doing so is not necessarily simple. A computer program for performing these calculations is given by Prausnitz et al.[3]

We now consider the representation of ternary LLE. While other types of ternary phase diagram are possible, the two most common types are shown in Fig. 1.5-4.

To calculate phase equilibria for ternary mixtures of Type II, we require mutual solubility data for the two partially miscible binaries and vapor–liquid equilibrium data for the completely miscible binary. From such data we can obtain binary parameters as required in some expression for molar excess Gibbs energy g^E for the ternary.

For ternary mixtures of Type II, calculated equilibria are often in good agreement with experiment because the calculations are not highly sensitive to the binary parameters for the completely miscible pair.

For ternary mixtures of Type I, we require mutual-solubility data for the partially miscible pair and vapor–liquid equilibrium data for the two completely-miscible pairs. For ternary mixtures of Type I, calculated results are strongly sensitive to the choice of binary parameters obtained from binary vapor–liquid equilibria.

To illustrate, Fig. 1.5-5 compares calculated and experimental results reported by Anderson and Prausnitz[14] for the system cyclohexane–iso-octane–furfural. Calculations were performed using the UNIQUAC equation for g^E. Since the end points of the two connodal lines are fixed by binary mutual solubility data for cyclohexane–furfural and for iso-octane–furfural, the ternary calculations represent an interpolation. Calculated and experimental ternary results agree well.

Figure 1.5-6 compares calculated and experimental results reported by Anderson and Prausnitz[14] for the system benzene–acetonitrile–n-heptane. Also shown are calculated and observed vapor–liquid equilibria for the two completely miscible binaries. Figure 1.5-6 shows that calculated ternary results are sensitive to the choice of binary parameters in the UNIQUAC equation.

When binary vapor–liquid equilibrium data are reduced to yield binary parameters in UNIQUAC or in some other expression for g^E, it is usually not possible to obtain a unique set of binary parameters that are in some significant sense ''best'' for that binary. For a given binary mixture, binary parameters are almost always at least partially correlated so that, when experimental uncertainties are taken into account, there are many sets of binary parameters that can represent equally well the experimental data. When the goal of data reduction is limited to representing the binary data, it does not matter which of these many sets of binary parameters is used in the calculations. But when binary parameters are used to predict ternary liquid–liquid equilibria (Type I), calculated results depend strongly on which set of binary parameters is used.

Anderson and Prausnitz[14] found that it was necessary to use a few experimental ternary data to guide them in the choice of optimum binary parameters for representing ternary liquid–liquid equilibria. With such guidance, they were able to calculate good ternary liquid–liquid equilibria but at the cost of some inaccuracy in the calculated binary vapor–liquid equilibria.

A truly ''correct'' model for g^E for a ternary mixture should simultaneously give correct results for both vapor–liquid and liquid–liquid equilibria, using the same parameters. When the model for g^E contains only pure-component and binary parameters, this goal cannot as yet be achieved with generality.

When the purpose of a calculation is no more than to represent ternary liquid–liquid equilibria, the Anderson–Prausnitz method for choosing ''optimum'' binary parameters is useful. However, if the purpose of the calculation is to provide simultaneously liquid–liquid and vapor–liquid equilibria, a better method is to use the procedure described by Fuchs et al.[15] or that described by Cha and Prausnitz.[16]

1.6 FLUID–SOLID EQUILIBRIUM

1.6-1 Liquid–Solid Equilibria

Typical nonelectrolyte solids often dissolve appreciably in typical liquid solvents or compressed gases. These solubilities are usually strong functions of temperature and, when the solvent is supercritical or near its critical state, they are also strong functions of pressure.

When it is obvious that the solubility of a solid depends on intermolecular forces between solute and solvent, it is perhaps not as obvious that this solubility also depends on how far the system temperature is removed from the solid's melting point and on the solute's enthalpy of fusion. These dependencies become clear when we relate the standard-state fugacity of the solute in the solid phase to that in the liquid phase, as indicated below.

We consider a binary two-phase system at temperature T. One phase is a liquid and the other is a solid. Since the effect of pressure on condensed-phase properties is normally negligible at low or moderate pressures, we do not need to specify the pressure. Let component 1 be the liquid solvent and component 2 the solid solute.

The equations of equilibrium are

$$\hat{f}_1^L = \hat{f}_1^S \tag{1.6-1}$$

$$\hat{f}_2^L = \hat{f}_2^S \tag{1.6-2}$$

where $\hat{f}_i$ is component fugacity and where superscripts L and S indicates, respectively, liquid phase and solid phase.

We now make an important simplifying assumption: we assume that the solid phase consists only of component 2; that is, there is no solubility of solvent in the solid phase. (This is usually a good assumption whenever components 1 and 2 have different molecular structures because miscibility in the solid phase is highly sensitive to small details in molecular geometry.) In that event we need consider only Eq. (1.6-2) which we rewrite

$$\gamma_2 x_2 = \frac{f_2^S}{f_2^L} \tag{1.6-3}$$

where γ is the liquid-phase activity coefficient and x and is the liquid-phase mole fraction; x_2 is the solubility. The product $\gamma_2 x_2$ is the liquid phase activity $\hat{a}_2$ of component 2. Note that the standard-state fugacity for this activity is the fugacity of pure liquid 2 at system temperature. However, since system temperature T is lower than the melting temperature of pure liquid 2, this standard-state fugacity is hypothetical because pure liquid 2 cannot exist (at equilibrium) at T. Pure liquid 2 at T is a subcooled liquid. With special care, it is often possible to subcool a pure liquid below its melting point but regardless of whether or not this can be done in the laboratory, it is possible to relate the fugacity of a pure subcooled liquid to measurable equilibrium properties as briefly outlined in the next paragraphs.

Let

$$\Delta g^f = \text{Gibbs energy change of fusion} = RT \ln \frac{f^L}{f^S}$$

$$\Delta h^f = \text{enthalpy of fusion}$$

The effect of temperature of Δg^f is given by the Gibbs–Helmholtz equation

$$\frac{d(\Delta g^f/RT)}{dT} = -\frac{\Delta h^f}{RT^2} \tag{1.6-4}$$

The enthalpy of fusion depends on temperature according to

$$\frac{d\,\Delta h_f}{dT} = \Delta c \approx c_P^L - c_P^S \tag{1.6-5}$$

where c is the heat capacity along the saturation line; at modest pressures, this heat capacity is nearly identical to c_P, the heat capacity at constant pressure.

Substitution of Eq. (1.6-5) into Eq. (1.6-4), followed by integration, gives‡

$$\ln \frac{f_2^S}{f_2^L} = -\frac{\Delta h_2^f}{RT}\left(1 - \frac{T}{T_{t_2}}\right) + \frac{\Delta c_p}{r}\left(\frac{T_{t_2} - T}{T}\right) - \frac{\Delta c_P}{R}\ln\frac{T_{t_2}}{T} \tag{1.6-6}$$

where we have used the boundary condition that $\Delta g^f = 0$ at the triple-point temperature T_t; in Eq. (1.6-6), Δh^f referes to the enthalpy of fusion at T_t. However, for many typical cases, T_t is very close to T_m (the normal melting temperature). Therefore, for most practical purposes, in Eq. (1.6-6) we can substitute T_m for T_t and use for Δh^f the enthalpy of fusion at T_m.

In Eq. (1.6-6), the first term on the right-hand side is much more important than the remaining (correction) terms. To a reasonable approximation, these correction terms tend to cancel one another; whenever T is not very far removed from T_t, these correction terms can be neglected.

Substitution of Eq. (1.6-6) into Eq. (1.6-3) gives the activity of solute 2 in terms of measurable properties for pure solute 2. If we make one more (strong) assumption, namely, assume that the liquid phase is ideal (in the sense of Raoult's Law), we obtain the ideal solubility:

$$\ln x_2(\text{ideal}) = -\frac{\Delta h_2^f}{RT}\left(1 - \frac{T}{T_{m_2}}\right) + \frac{\Delta c_p}{R}\left(\frac{T_{m_2} - T}{T}\right) - \frac{\Delta c_p}{R}\ln\frac{T_{m_2}}{T} \tag{1.6-7}$$

Equation (1.6-7) shows the importance of melting point and enthalpy of fusion. When other effects are equal, the ideal solubility declines as the distance between system temperature and melting temperature

‡If there are solid–solid phase transitions in the range $T_t - T$, additional correction terms are required as discussed by Weimer and Prausnitz[1].

rises. The rate of this decline increases with rising entropy of fusion which is the ratio of enthalpy of fusion to melting temperature.

Equation (1.6-7) says that, at a given temperature, the ideal solubility of component 2 depends only on the properties of pure component 2, that is, the solubility is the same in all solvents. Furthermore, Eq. (1.6-7) says that the ideal solubility always rises with increasing temperature. These statements, especially the first, are not in agreement with experiment. To correct the ideal solubility, it is necessary to include the liquid-phase activity coefficient γ_2 which depends on temperature and composition and on the nature of components 1 and 2.

As in vapor–liquid equilibria and in liquid–liquid equilibria, experimental mixture data are required to find liquid–phase activity coefficients. For some solid–liquid systems, estimates for γ can be made using regular-solution theory (Preston and Prausnitz[2]) or UNIFAC (Gmehling et al.[3]), but for reliable work a few experimental measurements are necessary.

While Eq. (1.6-3) provides the conventional thermodynamic method for solid–liquid equilibria, an alternate technique is provided by using an equation of state to calculate the fugacity of the dissolved solute in the liquid phase. The equation of equilibrium is written

$$\hat{\phi}_2^L x_2 P = f_2^S \qquad (1.6\text{-}8)$$

where $\hat{\phi}_2^L$ is the fugacity coefficient of component 2 in the liquid phase at composition x_2 and system temperature; $\hat{\phi}_2^L$ can be calculated from a suitable equation of state as discussed briefly in Section 1.7. Total pressure P can be calculated by assuming that the liquid phase is at its vapor–liquid saturation condition. Pure-solid fugacity f_2^S is essentially equal to the vapor pressure of pure solid 2.

Equation (1.6-8) has been applied by Soave[4] to correlate solubilities of solid carbon dioxide in liquid hydrocarbons at low temperatures. Alternatively, Myers and Prausnitz,[5] Preston and Prausnitz,[2] and more recently Teller and Knapp[6] also correlated such solubilities but they used the conventional activity-coefficient method: Eqs. (1.6-3) and (1.6-6).

The equilibrium relationships discussed above can be used to generate solid–liquid phase diagrams where the freezing temperature and the melting temperature are given as functions of the composition. Although there are numerous intermediate cases, we can distinguish between two extremes: in one extreme, there is no mutual solubility in the solid phase and in the other extreme there is complete solubility in the solid phase, that is, all solids are mutually soluble in each other in all proportions.

For binary systems, Fig. 1.6-1, shows these two extremes. In Fig. 1.6-1a, there are two possible solid phases: one of these is pure solid 1 and the other is pure solid 2. The curve on the left gives the solubility of solid 2 in liquid 1; the curve on the right gives the solubility of solid 1 in liquid 2. These solubility curves intersect at the eutectic point. The curve on the left is the coexistence line for solid 2 and the liquid mixture; the curve on the right gives the coexistence line for solid 1 and the liquid mixture. At the eutectic point, all three phases are at equilibrium. For this type of system, there is no difference between freezing temperature and melting temperature.

In Fig. 1.6-1b there is only one possible solid phase because solids 1 and 2 are miscible here in all proportions. The top curve gives the freezing-point locus and the bottom curve gives the melting-point locus; the horizontal (tie) line drawn between these loci at any fixed temperature gives the equilibrium liquid-phase and solid-phase compositions.

For those binary mixtures where the solid phase is a solution, the equations of equilibrium are

$$\hat{f}_1^L = \hat{f}_1^S \quad \text{and} \quad \hat{f}_2^L = \hat{f}_2^S \qquad (1.6\text{-}9)$$

which can be rewritten

$$\gamma_1^L x_1 = \gamma_1^S z_1 \frac{f_1^S}{f_1^L} \quad \text{and} \quad \gamma_2^L x_2 = \gamma_2^S z_2 \frac{f_2^S}{f_2^L} \qquad (1.6\text{-}10)$$

where x and z are, respectively, mole fractions in the liquid phase and in the solid phase and where we now require activity coefficients γ^L and γ^S for *both* phases. The ratios of pure-component fugacities, for pure 1 and for pure 2, are obtained from Eq. (1.6-6). In addition, we have the necessary relationships $x_1 + x_2 = 1$ and $z_1 + z_2 = 1$.

Equations (1.6-10) are the general relationships for binary solid–liquid phase diagrams. They degenerate to the case discussed earlier (no mutual solubilities in the solid phases) on setting $\gamma_1^S z_1 = \gamma_2^S z_2 = 1$.

The two extreme-case phase diagrams shown in Fig. 1.6-1 indicate that the phase behavior of a solid–liquid mixture depends crucially on the extent to which the solids are mutually soluble. Thermodynamic calculations are relatively straightforward when it is reasonable to assume that all possible solid phases are pure, because in that event we need not be concerned with activity coefficients in the solid phase; at present, little is known about these activity coefficients.

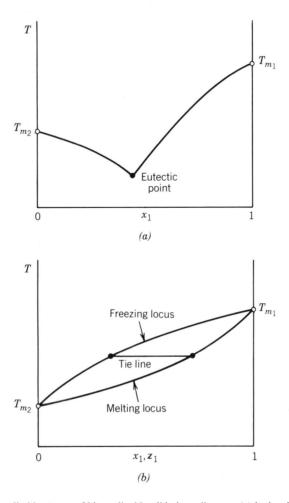

FIGURE 1.6-1 Two limiting types of binary liquid–solid phase diagram: (*a*) both solid phases are pure; (*b*) solids are miscible in all proportions.

Figure 1.6-1*a* indicates—as often observed—that the freezing point of a binary mixture is below that of either pure component. This feature of many typical solvent pairs can sometimes be applied to optimum design of a gas-absorption process: since the solubility of an absorbable gas usually rises with falling temperature, it may be useful to operate the absorber at as low a temperature as possible. If the absorber uses a mixed solvent, rather than a single solvent, the lowest possible operating temperature is likely to be appreciably below that for any of the single solvents in the solvent mixture.

1.6-2 Gas–Solid Equilibria

For binary two-phase equilibria where one phase is a solid and the other a gas, we employ the equilibrium criteria

$$\hat{f}_1^S = y_1 \hat{\phi}_1 P \tag{1.6-11}$$

and

$$\hat{f}_2^S = y_2 \, \hat{\phi}_2 P \tag{1.6-12}$$

where $\hat{\phi}_i$ is the gas-phase fugacity coefficient, y_i is the gas-phase mole fraction, and subscripts 1 and 2 refer, respectively, to the gas solvent and to the solid solute. If there is some solubility of component 1 in the solid phase, then both Eqs. (1.6-11) and (1.6-12) are required. An equation of state is used for $\hat{\phi}_1$ and $\hat{\phi}_2$, but the solid fugacities (particularly $\hat{f}_1^S$) require special treatment.

In the simplest case, the solid phase is pure 2 (no solubility of solvent in the solid), Eq. (1.6-11) does not apply, and in Eq. (1.6-12) we may set $\hat{f}_2^S$ equal to f_2^S, the fugacity of pure solid 2. The (single) equation for gas–solid equilibrium is then

$$f_2^S = y_2 \hat{\phi}_2 P \tag{1.6-13}$$

Fugacity coefficient $\hat{\phi}_2$ is evaluated from an appropriate equation of state by procedures described in Section 1.3. The fugacity of the solid is given by an analogue of Eq. (1.2-33), namely,

$$f_2^S = P_2^{\text{sat}} \exp\left[\int_0^{P_2^{\text{sat}}} (Z_2^V - 1)\, \frac{dP}{P} + \int_{P_2^{\text{sat}}}^P \frac{v_2^S}{RT}\, dP \right] \tag{1.6-14}$$

where P_2^{sat} is the vapor–solid saturation pressure of pure component 2 at system temperature T. (It is assumed here that T is less than T_{t_2}, the triple temperature of pure component 2.)

Equations (1.6-13) and (1.6-14) constitute an exact formulation of the equilibrium problem. The expression for f_2^S can be simplified if P_2^{sat} is sufficiently small and if v_2^S is assumed independent of pressure. In this event Eq. (1.6-14) becomes, approximately,

$$f_2^S = P_2^{\text{sat}} \exp \frac{v_2^S P}{RT} \tag{1.6-15}$$

Combination of Eq. (1.6-15) with Eq. (1.6-13) gives, on rearrangement, the following expression for the solubility y_2 of the solid in the gas phase:

$$y_2 = \frac{1}{\hat{\phi}_2} \left(\frac{P_2^{\text{sat}}}{P} \right) \exp \frac{v_2^S P}{RT} \tag{1.6-16}$$

Equation (1.6-16) can be simplified. If the system pressure is sufficiently low, then $\hat{\phi}_2 = 1$, the argument of the exponential is small, and one obtains the Raoult's Law solubility:

$$y_2(\text{RL}) = \frac{P_2^{\text{sat}}}{P} \tag{1.6-17}$$

If the pressure is high, but $\hat{\phi}_2 = 1$, we obtain the "ideal" solubility

$$y_2(\text{id}) = \frac{P_2^{\text{sat}}}{P} \exp \frac{v_2^S P}{RT} \tag{1.6-18}$$

so called because it incorporates the pressure effect on solid fugacity (Poynting correction) but assumes ideal-gas behavior for the gas phase.

We consider application of Eqs. (1.6-16)–(1.6-18) to the calculation of the solubility of naphthalene (component 2) in carbon dioxide (component 1) at 35 °C and at high pressures. Data for these conditions (Tsekhanskaya et al.[7]) are shown as open circles in Fig. 1.6-2. Particularly noteworthy is the dramatic enhancement in solubility—several orders of magnitude—that occurs with increasing pressure near the critical pressure of the solvent gas. The solubility enhancement, which obtains for temperatures slightly higher than the critical temperature of a solvent gas (the critical temperature of CO_2 is 31 °C), is the basis for certain "supercritical extraction" processes: see Paulaitis et al.[8,9] for discussions of this topic.

At 35 °C, the vapor–solid saturation pressure of pure naphthalene is 2.9×10^{-4} bar; Eq. (1.6-17) with this value of P_2^{sat} produces the bottom curve in Fig. 1.6-2. The high-pressure solubilities are underestimated by three to four orders of magnitude.

The second curve from the bottom is obtained from Eq. (1.6-18), which incorporates the Poynting correction for the solid fugacity. Some enhancement of solubility is predicted for sufficiently high pressures, but again the solubilities are grossly underestimated.

It is thus clear that the major contributor to the solubility enhancement is $\hat{\phi}_2$, the gas-phase fugacity coefficient of the dissolved solute. To estimate this quantity, we choose the Soave–Redlich–Kwong equation of state, Eq. (1.3-29):

$$Z = \frac{1}{1 - b\rho} - \frac{\theta}{RT} \frac{\rho}{1 + b\rho}$$

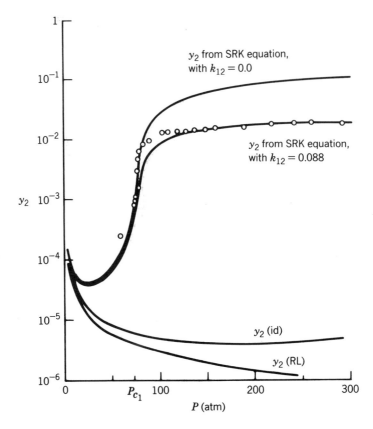

FIGURE 1.6-2   Solubility y_2 of naphthalene in gaseous carbon dioxide at 35°C and high pressures. Circles are data. Curves are computed as described in text. P_{c_1} is critical pressure of carbon dioxide.

Here, ρ is molar density, and b and θ are the equation-of-state parameters. Fugacity coefficient $\hat{\phi}_2$ is found from Eqs. (1.3-30) and (1.3-31),

$$\ln \hat{\phi}_2 = \frac{\bar{b}_2}{b}(Z - 1) + \frac{\theta}{bRT}\left(\frac{\bar{b}_2}{b} - \frac{\bar{\theta}_2}{\theta} - 1\right) \ln (1 + b\rho) - \ln (1 - b\rho)Z$$

and the following mixing rules and combining rules are assumed (see Section 1.3-3):

$$b = \sum_i \sum_j y_i y_j b_{ij}$$

$$\theta = \sum_i \sum_j y_i y_j \theta_{ij}$$

$$b_{ij} = \tfrac{1}{2}(b_{ii} + b_{jj})(1 - c_{ij})$$

$$\theta_{ij} = (\theta_{ii}\theta_{jj})^{1/2}(1 - k_{ij})$$

Partial molar parameters $\bar{b}_2$ and $\bar{\theta}_2$ are given by Eqs. (1.3-36b) and (1.3-37b):

$$\bar{b}_2 = b_{22} - y_1^2(b_{11} + b_{22})c_{12}$$

$$\bar{\theta}_2 = \theta_{22} - y_1^2[(\theta_{11}^{1/2} - \theta_{22}^{1/2})^2 + 2(\theta_{11}\theta_{22})^{1/2}k_{12}]$$

Numerical values for the pure-component parameters θ_{ii} and b_{ii} are determined from Eqs. (1.3-26)-(1.3-28):

$$\theta_{ii} = \theta_{ci}[1 + (0.480 + 1.574\omega_i - 0.176\omega_i^2)(1 - Tr_i^{1/2})]^2$$

$$\theta_{ci} = 0.42748 \frac{R^2 T_{ci}^2}{P_{ci}}$$

$$b_{ii} = 0.08864 \frac{RT_{ci}}{P_{ci}}$$

The above equations permit calculation of $\hat{\phi}_2$. To do the calculation, one requires values for the critical temperatures, critical pressures, and acentric factors of the pure components. Also needed are estimates for the binary interaction parameters c_{12} and k_{12}.

For the present calculations we have set $c_{12} = 0$. The two upper curves in Fig. 1.6-2 represent solubilities computed from Eq. (1.6-16) with the Soave–Redlich–Kwong equation and two different values of k_{12}. The topmost curve corresponds to $k_{12} = 0$: it is a purely *predictive* result and, although quantitatively poor for the higher pressures, it correctly reproduces the qualitative features of the solubility enhancement.

The second curve from the top corresponds to $k_{12} = 0.088$, a value obtained by forcing agreement between experimental and computed solubilities at higher pressures. This is a *correlative* result, which illustrates the extreme sensitivity of dense-phase fugacity predictions to the details of equation-of-state mixing rules and combining rules. Representation of solubilities in this case is nearly quantitative, the major discrepancies occurring in the near-critical region of the solvent gas. Similar results are obtained with other cubic equations of state, for example, the Peng–Robinson equation.

1.7 HIGH PRESSURE VAPOR–LIQUID EQUILIBRIA

In an equilibrated, multiphase mixture at temperature T and pressure P, for every component i, the fugacity $\hat{f}_i$ must be the same in all phases. Therefore, if a method is available for calculating $\hat{f}_i$ for phase as a function of temperature, pressure, and the phase's composition, then it is possible to calculate all equilibrium compositions at any desired temperature and pressure. When all phases are fluids, such a method is given by an equation of state.

An equation of state, applicable to all fluid phases, is particularly useful for phase-equilibrium calculations where a liquid phase and a vapor phase coexist at high pressures. At such conditions, conventional activity coefficients are not useful because, with rare exceptions, at least one of the mixture's components is supercritical; that is, the system temperature is above that component's critical temperature. In that event, one must employ special standard states for the activity coefficients of the supercritical components (see Section 1.5-2). That complication is avoided when all fugacities are calculated from an equation of state.

To fix ideas, consider a multicomponent, two-phase (liquid–vapor) mixture where the phases are denoted, respectively, by superscripts L and V. Pressure P and temperature T are the same in both phases. The condition of equilibrium is given by formulation 3 of Section 1.2-5:

$$x_i \hat{\phi}_i^L = y_i \hat{\phi}_i^V \quad \text{(all } i\text{)} \tag{1.7-1}$$

Here, $\hat{\phi}$ is the fugacity coefficient, x is the liquid-phase mole fraction, and y is the vapor-phase mole fraction. It follows that the familiar K factor is given by

$$K_i \equiv \frac{y_i}{x_i} = \frac{\hat{\phi}_i^L}{\hat{\phi}_i^V} \tag{1.7-2}$$

As discussed in Section 1.3-1, the fugacity coefficient is directly related to an integral obtained from the equation of state. For the liquid-phase fugacity coefficient, at constant temperature and composition,

$$\ln \hat{\phi}_i^L = \int_0^{\rho^L} (\bar{Z}_i - 1) \frac{d\rho}{\rho} - \ln Z^L \tag{1.7-3}$$

Here, ρ is molar density, and $\bar{Z}_i$ is a partial molar compressibility factor given by

$$\bar{Z}_i \equiv \left[\frac{\partial(nZ)}{\partial n_i} \right]_{T, nv, n_j} \tag{1.7-4}$$

where $nv \equiv n/\rho \equiv V$ is the total volume. An analogous equation holds for the fugacity coefficient in the vapor phase:

$$\ln \hat{\phi}_i^V = \int_0^{\rho^V} (\bar{Z}_i - 1) \frac{d\rho}{\rho} - \ln Z^V \tag{1.7-5}$$

To find $\hat{\phi}_i^L$ and $\hat{\phi}_i^V$ we need an equation of state of the form

$$Z \equiv \frac{P}{\rho RT} = \mathbb{Z}(T, \rho, n_1, n_2, \ldots) \qquad (1.7\text{-}6)$$

where n_1 is the number of moles of component 1, and so on. As indicated by the integral in Eq. (1.7-5), the equation of state must hold over the density range 0 to ρ^V; similarly, in Eq. (1.7-3) it must hold over the density range 0 to ρ^L. Furthermore, the equation of state must be valid not only for every pure component in the mixture but also for the mixture at any composition.

Before considering examples, it is important to recognize that, given a suitable equation of state, Eqs. (1.7-1)–(1.7-5) are sufficient to yield all equilibrium information (temperature, pressure, phase compositions) when the two-phase conditions are specified as dictated by the phase rule; for example, for a two-phase system, the number of degrees of freedom is equal to the number of components.

To illustrate, consider a two-phase mixture with N components. To specify that system, we must stipulate N independent variables. Suppose that we fix the total pressure P and the composition of the liquid phase, $x_1, x_2, \ldots, x_N$. We want to find the equilibrium temperature T and vapor-phase compositions y_1, $y_2, \ldots, y_N$. (This problem, frequently encountered in distillation calculations, is called the bubble-point T problem.)

The number of primary unknowns is $N + 1$. However, since we shall use Eqs. (1.7-3) and (1.7-5), there are in addition two secondary unknowns: ρ^L and ρ^V. Therefore, the total number of unknowns is $N + 3$. To determine these, we need $N + 3$ independent equations. These are Eq. (1.7-1), taken N times, once for each component; in addition, we have

$$\sum_1^N y_i = 1 \qquad (1.7\text{-}7)$$

and Eq. (1.7-6) taken twice, once for the vapor phase and once for the liquid phase. This adds to $N + 3$ independent equations.

In principle, a solution to the bubble-point T problem is possible because the number of unknowns is equal to the number of independent equations. However, since the equations are strongly nonlinear, simultaneous solution of $N + 3$ equations is not trivial, even if N is small. Fortunately, efficient and robust computer programs for such calculations have been prepared by several authors, including the authors of this article.

Numerous equations of state have been proposed and new ones appear regularly. For pure fluids, there are some highly accurate (essentially empirical) equations that can well represent pure-fluid thermodynamic properties, provided there is a large body of reliable experimental information available for determining the many (typically 15 or more) equation-of-state constants. Unfortunately, these equations of state are of little value for phase-equilibrium calculations in mixtures because it is not clear how to extend such equations to mixtures and because constants for the mixture are not easily determined unless there are many experimental data for the mixture.

For typical fluid mixtures of nonpolar (or slightly polar) fluids, good representation of phase equilibria can often be obtained using a simple equation of state of the van der Waals form; in such equations, the total pressure is given by the sum of two parts: the first part gives the pressure that the fluid would have if all molecules exhibit only repulsive forces (due to molecular size); the second part gives the (negative) contribution from attractive forces. A well-known example is the Soave–Redlich–Kwong (SRK) equation, discussed in Section 1.3-3 and used in Section 1.6-2 for calculation of fugacity coefficients in dense-gas mixtures. This equation of state, or modifications of it, is frequently used for calculation of vapor–liquid equilibria.

Figure 1.7-1 shows calculated phase equilibria for the binary system methane–propane; calculations were made with the SRK equation using $c_{12} = 0$ and $k_{12} = 0.029$. Agreement with experiment is good. A small change in k_{12} can significantly affect the results.

Calculations for a more complex mixture are shown in Fig. 1.7-2 for hydrogen sulfide–water. In this case $c_{12} = 0.08$ and $k_{12} = 0.163$. Good agreement with experiment is obtained at the temperature indicated but different binary parameters are required to obtain similar agreement at some other temperature. Furthermore, this calculation, using the SRK equation of state, neglects ionization of hydrogen sulfide in excess water and therefore equilibria calculated with the SRK equation are necessarily poor for very dilute aqueous solutions of hydrogen sulfide.

While Figs. 1.7-1 and 1.7-2 are for binary mixtures, the equation-of-state method for calculating vapor–liquid equilibria can be applied to mixtures with any number of components. When the quadratic mixing rules [Eqs. (1.3-32) and (1.3-33)] are used, only pure-component and binary parameters are required; these mixing rules therefore provide a powerful tool for "scale-up" in the sense that only binary mixture data are needed to calculate equilibria for a mixture containing more than two components. For example, in the ternary mixture containing components 1, 2 and 3, only binary constants k_{12}, k_{13}, k_{23} (and perhaps c_{12}, c_{13}

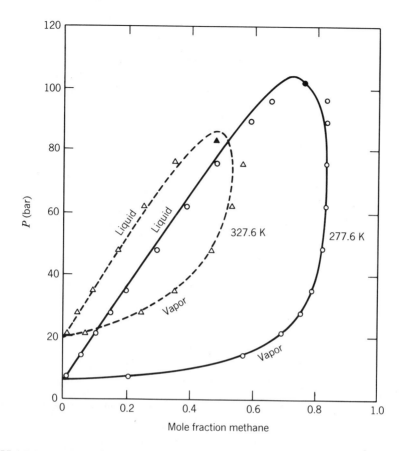

FIGURE 1.7-1 Isothermal VLE for methane–propane. $\triangle$, $\bigcirc$ are experimental data; $\blacktriangle$, $\bullet$ are observed critical points. Curves are computed via the Soave–Redlich–Kwong equation with $k_{12} = 0.029$.

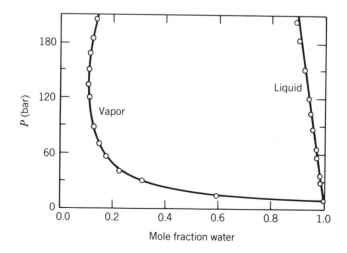

FIGURE 1.7-2 VLE for the hydrogen sulfide–water system at 171°C. Circles are data. Curves are computed via the Soave–Redlich–Kwong equation with $c_{12} = 0.08$ and $k_{12} = 0.163$ (Evelein et al.[7]).

53

and c_{23}) are required, in addition to pure-component constants. For relatively simple multicomponent mixtures, good results are often obtained when calculations are based exclusively on constants obtained from binary data.

An exhaustive study of high-pressure vapor–liquid equilibria has been reported by Knapp et al.,[1] who give not only a comprehensive literature survey but also compare calculated and observed results for many systems. In that study, several popular equations of state were used to perform the calculations but no one equation of state emerged as markedly superior to the others. All the equations of state used gave reasonably good results provided care is exercised in choosing the all-important binary constant k_{ij}. All the equations of state used gave poor results when mixtures were close to critical conditions.

Equations of state can also be used to calculate three-phase (vapor–liquid–liquid) equilibria at high pressures; the principles for doing so are the same as those used for calculating two-phase equilibria but the numerical techniques for solving the many simultaneous equations are now more complex.

Equation-of-state calculations are used routinely in the chemical and related industries for designing high-pressure flashes or distillation and absorption columns. For mixtures containing nonpolar (or slightly polar) fluids, such calculations are generally satisfactory but for mixtures containing some highly polar or hydrogen-bonded fluids, serious errors may result when the equation of state and, more importantly, when the mixing rules are those used for simple mixtures. Recent studies indicate that significant modifications are needed for complex mixtures (especially aqueous mixtures); one possible modification is to superimpose chemical equilibria (association and/or solvation) to take into account nonrandomness (preferential ordering of molecules) caused by strong polar forces or hydrogen bonding, as shown, for example, by Gmehling et al.[2] Another promising modification is to use density-dependent mixing rules (e.g., see Mollerup,[3] Whiting and Prausnitz,[4] and Mathias and Copeman[5]) where the equation-of-state parameters for the mixture depend not only on composition but also on density. Finally, new equation-of-state methods have been suggested by computer-simulation calculations (e.g., see Gubbins[6]). While these new methods are not as yet directly applicable to engineering-design calculations, they are of much use in providing guidance toward significant improvement of semitheoretical methods.

1.8 SUMMARY

The formulation of engineering problems in phase equilibria begins with a set of abstract equations relating the component fugacities $\hat{f}_i$ of each component i in a multiphase system. Translation of these equations into useful working relationships is done by definition (through the fugacity coefficient $\hat{\phi}_i$ and/or the activity coefficient γ_i) and by thermodynamic manipulation. Different problems require different translations; we have illustrated many of the standard procedures by examples in Sections 1.5, 1.6, and 1.7.

Numerical calculations of phase equilibria require thermodynamic data or correlations of data. For pure components, the requisite data may include saturation pressures (or temperatures), heat capacities, latent heats, and volumetric properties. For mixtures, one requires a $PVTx$ equation of state (for determination of $\hat{\phi}_i$), and/or an expression for the molar excess Gibbs energy g^E (for determination of γ_i). We have discussed in Sections 1.3 and 1.4 the correlating capabilities of selected equations of state and expressions for g^E, and the behavior of the fugacity coefficients and activity coefficients derived from them.

The coverage in this chapter has necessarily been selective and restricted in scope. The interested reader will find more thorough discussions of topics considered here (and others not considered here) in the general references cited for this section.[1-8] In addition, major data sources and bibliographies are available; a partial list is given by Van Ness and Abbott.[9]

NOTATION

Roman Letters

A	parameter in Porter equation: Eq. (1.4-14)
$A_{12}, A_{21},$ D_{12}, D_{21}	parameters in Margules equation: Eq. (1.4-8)
$A'_{12}, A'_{21},$ D'_{12}, D'_{21}	parameters in van Laar equation: Eq. (1.4-16)
a	parameter in van der Waals equation
$\hat{a}_i$	activity of component i
B, C, D	density-series virial coefficients: Eq. (1.3-8)
B', C', D'	pressure-series virial coefficients: Eq. (1.3-9)
B_{ij}, C_{ijk}	interaction virial coefficients
$\overline{B}_i, \overline{C}_i$	partial molar virial coefficients

b	parameter in cubic equation of state: Eq. (1.3-24)
$\bar{b}_i$	partial molar equation-of-state parameter
c_{ij}	binary interaction parameter: Eq. (1.3-34)
c_P	molar heat capacity at constant pressure
f	fugacity of a mixture
f_i	fugacity of pure component i
f_i°	fugacity of component i in a standard state
$\hat{f}_i$	fugacity of component i in solution
$\hat{f}_i^{id}$	fugacity of component i in an ideal solution
G	total Gibbs energy $= ng$
G_{ij}	parameter in NRTL equation: Eq. (1.4-30)
g	molar Gibbs energy
g^E	molar excess Gibbs energy
$\mathcal{G}$	normalized excess molar Gibbs energy: Eq. (1.4-6)
Δg_{ij}	characteristic energy in NRTL equation: Eq. (1.4-29)
Δg^f	molar Gibbs energy change of fusion
H	total enthalpy $= nh$
h	molar enthalpy
h^E	molar excess enthalpy ("heat of mixing")
h^R	molar residual enthalpy
$\bar{h}_i^E$	partial molar excess enthalpy
$\bar{h}_i^R$	partial molar residual enthalpy
$\mathcal{K}_{i,j}$	Henry's constant for component i in a binary i-j mixture: Eq. (1.2-38)
Δh^f	molar enthalpy change of fusion
K	chemical-reaction equilibrium constant: Section 1.3-4
K_i	K factor for VLE: Eq. (1.7-2)
k_{ij}	binary interaction parameter: Eq. (1.3-35)
l_i	parameter in UNIQUAC equation: Eq. (1.4-47)
M	total property of a solution $= nm$ (M can represent a variety of thermodynamic functions, including G, H, S, and V)
m	molar property of a solution (m can represent a variety of thermodynamic functions, including g, h, s, and v)
m^E	molar excess function: Eq. (1.2-3)
m^R	molar residual function: Eq. (1.2-1)
m^{id}	molar property of an ideal solution
m'	molar property of an ideal-gas mixture
$\bar{m}_i$	partial molar property: Eq. (1.2-5)
$\bar{m}_i^E$	partial molar excess function: Eq. (1.2-13)
$\bar{m}_i^R$	partial molar residual function: Eq. (1.2-12)
N	number of components in a mixture
n	number of moles
n_i	number of moles of component i
P	absolute pressure
P_c	critical pressure
P_i^{sat}	saturation pressure of pure component i
q_i, r_i	molecular-structure constants in UNIQUAC equation: Eqs. (1.4-45a) and (1.4-45b)
R	universal gas constant
s	molar entropy
s^E	molar excess entropy
T	absolute temperature
T_c	critical temperature
T_r	reduced temperature $\equiv T/T_c$

t	empirical temperature
Δu_{ij}	characteristic energy in UNIQUAC equation: Eq. (1.4-35)
V	total volume $= nv$
v	molar volume
v^E	molar excess volume ("volume change of mixing")
$\bar{v}_i$	partial molar volume
$\bar{v}_i^E$	partial molar excess volume
x_i	mole fraction of component i; also (in context), liquid-phase mole fraction of component i
y_i	vapor-phase mole fraction of component i
Z	compressibility factor $\equiv Pv/RT$
$\bar{Z}_i$	partial molar compressibility factor
$\tilde{Z}_i$	partial molar compressibility factor at constant T and V: Eq. (1.3-7)
z	coordination number
z_i	feed mole fraction of component i
z_i	concentration (arbitrary units) of component i; also (in context), solid-phase mole fraction of component i

Greek Letters

α_{ij}	nonrandomness parameter in NRTL equation: Eq. (1.4-30)
γ_i	activity coefficient of component i
$\delta, \epsilon, \eta, \theta$	parameters in cubic equation of state: Eq. (1.3-34)
$\delta_{ij}, \delta_{ijk}$	difference functions for virial coefficients: Eqs. (1.3-20) and (1.3-22)
θ_i	molecular area fraction in UNIQUAC equation: Eq. (1.4-45b)
$\bar{\theta}_i$	partial molar equation-of-state parameter
Λ_{ij}	parameter in Wilson equation: Eq. (1.4-22)
$\Delta\lambda_{ij}$	characteristic energy in Wilson equation: Eq. (1.4-22)
μ_i	chemical potential of component i
ξ	dimensionless extent of reaction: Section 1.3-4
π	number of phases in a multiphase system
ρ	molar density $= v^{-1}$
τ_{ij}	parameter in NRTL equation: Eq. (1.4-29); parameter in UNIQUAC equation: Eq. (1.4-35)
Φ_i	molecular segment fraction in UNIQUAC equation: Eq. (1.4-45a)
ϕ	fugacity coefficient of a mixture
ϕ_i	fugacity coefficient of pure component i
$\hat{\phi}_i$	fugacity coefficient of component i in solution
Ψ_i	composite correction term in equation for subcritical VLE: Eq. (1.5-5)
ω	acentric factor

Subscripts

c, cr	critical state
i, j, k	(arbitrary) components i, j, k
m	melting point
t	triple state

Superscripts

E	excess
id	ideal solution
L	liquid phase
p	(arbitrary) phase p
R	residual

r	reference state, at pressure P^r
S	solid phase
V	vapor phase
α, β	α, β phases
$^\circ$	standard state
$*$	denotes normalization with respect to Henry's Law convention: Eq. (1.5-16)
∞	infinite-dilution value

REFERENCES

Section 1.2

1.2-1   J. W. Gibbs, *The Scientific Papers*, Vol. I, Dover Publications, New York, 1961.

1.2-2   H. C. Van Ness and M. M. Abbott, *AIChE J.*, **25,** 645 (1979).

Section 1.3

1.3-1   H. C. Van Ness and M. M. Abbott, *Classical Thermodynamics of Nonelectrolyte Solutions: With Applications to Phase Equilibria*, McGraw-Hill, New York, 1982.

1.3-2   J. H. Dymond and E. B. Smith, *The Virial Coefficients of Pure Gases and Mixtures—A Critical Compilation*, Clarendon Press, Oxford, 1980.

1.3-3   J. G. Hayden and J. P. O'Connell, *Ind. Eng. Chem. Process Des. Dev.*, **14,** 209 (1975).

1.3-4   C. Tsonopoulos, *AIChE J.*, **20,** 263 (1974).

1.3-5   C. Tsonopoulos, *AIChE J.*, **21,** 827 (1975).

1.3-6   C. Tsonopoulos, *AIChE J.*, **24,** 1112 (1978).

1.3-7   R. DeSantis and B. Grande, *AIChE J.*, **25,** 931 (1979).

1.3-8   H. Orbey and J. H. Vera, *AIChE J.*, **29,** 107 (1983).

1.3-9   O. Redlich and J. N. S. Kwong, *Chem. Rev.*, **44,** 233 (1949).

1.3-10   D.-Y. Peng and D. B. Robinson, *Ind. Eng. Chem. Fundam.*, **15,** 59 (1976).

1.3-11   G. Soave, *Chem. Eng. Sci.*, **27,** 1197 (1972).

1.3-12   F. H. MacDougall, *J. Am. Chem. Soc.*, **58,** 2585 (1936).

1.3-13   Y. Hu, E. G. Azevedo, D. Luedecke, and J. M. Prausnitz, *Fluid Phase Equilibria*, **17,** 303 (1984).

Section 1.4

1.4-1   G. M. Wilson, *J. Am. Chem. Soc.*, **86,** 127 (1964).

1.4-2   H. Renon and J. M. Prausnitz, *AIChE J.*, **14,** 135 (1968).

1.4-3   D. S. Abrams and J. M. Prausnitz, *AIChE J.*, **21,** 116 (1975).

1.4-4   T. F. Anderson and J. M. Prausnitz, *Ind. Eng. Chem. Process Des. Dev.*, **17,** 552 (1978).

1.4-5   G. Maurer and J. M. Prausnitz, *Fluid Phase Equilibria*, **2,** 91 (1978).

1.4-6   J. M. Prausnitz, T. F. Anderson, E. A. Grens, C. A. Eckert, R. Hsieh, and J. P. O'Connell, *Computer Calculations for Multicomponent Vapor–Liquid and Liquid–Liquid Equilibria*, Prentice-Hall, Englewood Cliffs, NJ, 1980.

1.4-7   Aa. Fredenslund, R. L. Jones, and J. M. Prausnitz, *AIChE J.*, **21,** 1086 (1975).

1.4-8   Aa. Fredenslund, J. Gmehling, and P. Rasmussen, *Vapor–Liquid Equilibria using UNIFAC*, Elsevier, Amsterdam, 1977.

1.4-9   S. Skjold-Jørgensen, B. Kolbe, J. Gmehling, and P. Rasmussen, *Ind. Eng. Chem. Process Des. Dev.*, **18,** 714 (1979).

1.4-10   J. Gmehling, P. Rasmussen, and Aa. Fredenslund, *Ind. Eng. Chem. Process Des. Dev.*, **21,** 118 (1982).

1.4-11   E. A. Macedo, U. Weidlich, J. Gmehling, and P. Rasmussen, *Ind. Eng. Chem. Process Des. Dev.*, **22,** 676 (1983).

1.4-12. T. Magnussen, P. Rasmussen, and Aa. Fredenslund, *Ind. Eng. Chem. Process Des. Dev.*, **20,** 331 (1981).

Section 1.5

1.5-1   H. G. Rackett, *J. Chem. Eng. Data*, **15,** 514 (1970).

1.5-2   C. F. Spencer and R. P. Danner, *J. Chem. Eng. Data*, **17,** 236 (1972).

1.5-3   J. M. Prausnitz, T. F. Anderson, E. A. Grens, C. A. Eckert, R. Hsieh, and J. P. O'Connell, *Computer Calculations for Multicomponent Vapor–Liquid and Liquid–Liquid Equilibria*, Prentice-Hall, Englewood Cliffs, NJ, 1980.

1.5-4   H. C. Van Ness and M. M. Abbott, *Classical Thermodynamics of Nonelectrolyte Solutions: With Applications to Phase Equilibria*, McGraw-Hill, New York, 1982.

1.5-5   D. Tassios and M. Van Winkle, *J. Chem. Eng. Data*, **12,** 555 (1967).

1.5-6   H. V. Kehiaian, *Int. DATA Ser. Select. Data Mixtures Ser. A*, **1980,** 126, 129, 132 (1980).

1.5-7   C. Tsonopoulos, *AIChE J.*, **20,** 263 (1974).

1.5-8   H. C. Van Ness and M. M. Abbott, *AIChE J.*, **25,** 645 (1979).

1.5-9   R. Battino and H. C. Clever, *Chem. Rev.*, **66,** 395 (1966).

1.5-10   H. L. Clever and R. Battino, in M. R. J. Dack (Ed.), *Solutions and Solubilities*, Vol. 8, Part 1, p. 379, Wiley, New York, 1975.

1.5-11   J. C. Moore, R. Battino, T. R. Rettich, Y. P. Handa, and E. Wilhelm, *J. Chem. Eng. Data*, **27,** 22 (1982).

1.5-12   I. R. Krichevsky and A. A. Ilinskaya, *Acta Physicochim.*, *URSS*, **20,** 327 (1945).

1.5-13   J. M. Prausnitz, R. N. Lichtenthaler, and E. G. Azeredo, *Molecular Thermodynamics of Fluid-Phase Equilibria*, 2nd. ed., Prentice-Hall, Englewood Cliffs, NJ, 1986.

1.5-14   T. F. Anderson and J. M. Prausnitz, *Ind. Eng. Chem. Process Des. Dev.*, **17,** 561 (1978).

1.5-15   R. Fuchs, M. Gipser, and J. Gaube, *Fluid Phase Equilibria*, **14,** 325 (1983).

1.5-16   T.-H. Cha and J. M. Prausnitz, *Ind. Eng. Chem. Process Des. Dev.*, **24,** 551 (1986).

Section 1.6

1.6-1   R. Weimer and J. M. Prausnitz, *J. Chem. Phys.*, **42,** 3643 (1965).

1.6-2   G. T. Preston and J. M. Prausnitz, *Ind. Eng. Chem. Process Des. Dev.*, **9,** 264 (1970).

1.6-3   J. Gmehling, T. F. Anderson, and J. M. Prausnitz, *Ind. Eng. Chem. Fundam.*, **17,** 269 (1978).

1.6-4   G. Soave, *Chem. Eng. Sci.*, **34,** 225 (1979).

1.6-5   A. L. Myers and J. M. Prausnitz, *Ind. Eng. Chem. Fundam.*, **4,** 209 (1965).

1.6-6   M. Teller and H. Knapp, *Ber. Bunsges. Phys. Chem*, **87,** 532 (1983).

1.6-7   Y. V. Tsekhanskaya and M. B. Iomtev, and E. V. Mushkina, *Zh. Fiz. Khim.*, **38,** 2166 (1964).

1.6-8   M. E. Paulaitis, V. J. Krukonis, R. T. Kurnik, and R. C. Reid, *Rev. Chem. Eng.*, **1,** 179 (1983).

1.6-9   M. E. Paulaitis, J. M. L. Penninger, R. D. Gray, Jr., and P. Davidson (Eds.), *Chemical Engineering at Supercritical Fluid Conditions*, Ann Arbor Science Publishers, Ann Arbor, MI, 1983.

Section 1.7

1.7-1   H. Knapp, R. Doering, L. Oellrich, U. Ploecker, and J. M. Prausnitz, *Vapor–Liquid Equilibria for Mixtures of Low-Boiling Substances*, DECHEMA Chemistry Data Series, Vol. VI, Frankfurt/Main, 1982.

1.7-2   J. Gmehling, D. D. Liu, and J. M. Prausnitz, *Chem. Eng. Sci.*, **34,** 951 (1979).

1.7-3   J. Mollerup, *Fluid Phase Equilibria*, **15,** 189 (1983).

1.7-4   W. B. Whiting and J. M. Prausnitz, *Fluid Phase Equilibria*, **9,** 119 (1982).

1.7-5   P. M. Mathias and T. W. Copeman, *Fluid Phase Equilibria*, **13,** 91 (1983).

1.7.6   K. E. Gubbins, *Fluid Phase Equilibria*, **13,** 35 (1983).

1.7-7   K. A. Evelein, R. G. Moore, and R. A. Heidemann, *Ind. Eng. Chem. Process Des. Dev.*, **15,** 423 (1976).

Section 1.8 (General References)

1.8-1   M. B. King, *Phase Equilibria in Mixtures*, Pergamon Press, Oxford, 1969.

1.8-2   M. L. McGlashan, *Chemical Thermodynamics*, Academic Press, London, 1979.

1.8-3   M. Modell and R. C. Reid, *Thermodynamics and Its Applications*, 2nd ed., Prentice-Hall, Englewood Cliffs, NJ, 1983.

1.8-4   J. M. Prausnitz, R. N. Lichtenthaler, and E. G. Azeredo, *Molecular Thermodynamics of Fluid-Phase Equilibria*, 2nd ed., Prentice-Hall, Englewood Cliffs, NJ, 1986.

1.8-5   J. M. Prausnitz, T. F. Anderson, E. A. Grens, C. A. Eckert, R. Hsieh, and J. P. O'Connell, *Computer Calculations for Multicomponent Vapor–Liquid and Liquid–Liquid Equilibria*, Prentice-Hall, Englewood Cliffs, NJ, 1980.

1.8-6   I. Prigogine and R. Defay, *Chemical Thermodynamics*, Longmans, London, 1954.

1.8-7   R. C. Reid, J. M. Prausnitz, and T. K. Sherwood, *The Properties of Gases and Liquids*, 3rd ed., McGraw-Hill, New York, 1977.

1.8-8   J. S. Rowlinson and F. L. Swinton, *Liquids and Liquid Mixtures*, 3rd ed., Butterworth, London, 1982.

1.8-9   H. C. Van Ness and M. M. Abbott, *Classical Thermodynamics of Nonelectrolyte Solutions: With Applications to Phase Equilibria*, McGraw-Hill, New York, 1982.

CHAPTER 2

Mass Transfer Principles

DONALD J. KIRWAN
Department of Chemical Engineering
University of Virginia
Charlottesville, Virginia

2.1 INTRODUCTION

Mass transfer can be defined simply as the movement of any identifiable species from one spatial location to another. The mechanism of movement can be *macroscopic* as in the flow of a fluid in a pipe (convection) or in the mechanical transport of solids by a conveyor belt. In addition, the transport of a particular species may be the result of random *molecular* motion (molecular diffusion) or random *microscopic* fluid motion (eddy or turbulent diffusion) in the presence of a composition gradient within a phase. This chapter is concerned primarily with mass transfer owing to molecular or microscopic processes.

Chemical processes produce complex mixtures of compounds from various feedstocks. Proper operation of chemical reactors often requires that the feed contain only certain species in specified ratios. Thus, separation and purification of species from feedstocks, whether petroleum, coal, mineral ores, or biomass, must be accomplished. Similarly, a mixture leaving a reactor must be separated into purified products, by-products, unreacted feed, and waste materials. Separation processes are also of importance where no reaction is involved as in seawater desalination by reverse osmosis, crystallization, or evaporation; in the fractionation of crude petroleum; or in the drying of solids or devolatization of polymers where diffusion within a porous solid is of importance.

Since separation processes are based on the creation of composition differences within and between phases, a consideration of mass transfer principles is necessary to the analysis and design of such processes. It is the combination of mass transfer principles with phase-equilibrium relationships (Chapter 1), energy considerations, and system geometry or configuration that permits the proper description of the separation processes discussed in the remainder of this book. It is the goal of this chapter to provide quantitative descriptions of the rate of species transport within phases under the influence of composition differences. An understanding of the importance of physical properties, geometry, and hydrodynamics to mass transfer rates is sought.

2.1-1 Factors Influencing Mass Transfer

To set the stage for the subject matter of this chapter, consider a familiar mass transfer situation—sweetening a cup of coffee or tea. A series of experiments at our kitchen table can provide insight into a number of characteristics of mass transfer. It could be quickly determined that an equal amount of granulated sugar will dissolve more rapidly than a cube, that an increase in the amount of granulated sugar will increase the dissolution rate, that stirring greatly increases the rate, that an upper limit to the amount that can be dissolved is reached and this limit increases with temperature as does the rate. Careful sampling would also reveal a greater sugar concentration near the sugar–solution interface which would be very nearly the final saturation value that could be obtained.

The results of these simple experiments on mass transfer can be stated as follows:

1. Mass transfer occurs owing to a concentration gradient or difference within a phase.

2. The mass transfer rate between two phases is proportional to their interfacial area and not the volumes of the phases present.

3. Transfer owing to microscopic fluid motion or mixing is much more rapid than that due to molecular motion (diffusion).

4. Thermodynamics provides a limit to the concentration of a species within a phase and often governs the interfacial compositions during transfer.

5. Temperature has only a modest influence on mass transfer rates under a given concentration driving force.

This example also illustrates the use of the three basic concepts on which the analysis of more complex mass transfer problems is based; namely, *conservation laws*, *rate expressions*, and *equilibrium thermodynamics*. The conservation of mass principle was implicitly employed to relate a measured rate of accumulation of sugar in the solution or decrease in undissolved sugar to the mass transfer rate from the crystals. The dependence of the rate expression for mass transfer on various variables (area, stirring, concentration, etc.) was explored experimentally. Phase-equilibrium thermodynamics was involved in setting limits to the final sugar concentration in solution as well as providing the value of the sugar concentration in solution at the solution–crystal interface.

2.1-2 Scope of Chapter

The various sections of this chapter develop and distinguish the conservation laws and various rate expressions for mass transfer. The laws of conservation of mass, energy, and momentum, which are taken as universal principles, are formulated in both macroscopic and differential forms in Section 2.2.

To make use of the conservation laws it is necessary to have specific rate expressions or constitutive relationships connecting generation and transport rates to appropriate variables of the system of interest. In contrast to the universality of the conservation laws, both the form and the parameter values of rate expressions can be dependent on the specific material under consideration. Fourier's Law of heat conduction relating the heat flux to a temperature gradient, $q = -k \, dT/dz$, is applicable to both gases and metals yet the magnitude and the temperature dependence of the thermal conductivity differ greatly between gases and metals. Expressions for the dependence of the rate of a chemical reaction on the concentrations of participating species vary widely for different reactions. In complex geometries and flow patterns, useful expressions for the mass transfer flux are dependent on geometry, hydrodynamics, and physical properties in ways not completely predictable from theory. In any case, the conservation equations cannot be solved for any particular problem until appropriate rate expressions, whether obtained from theory or from experiment and correlation, are provided. In Section 2.3 theoretical foundations, correlations, and experimental measurements of molecular diffusion are treated along with problems of diffusion in static systems and in laminar flow fields. Theories, experimental observations, and engineering correlations for mass transfer in turbulent flow and in complex geometries and flow fields are presented in Section 2.4 with some examples illustrating their use in analyzing separation phenomena and processes. A further important use of examples in this chapter is to give the reader a "feel" for the order of magnitude of various mass transfer parameters and the length and time scales over which mass transfer effects are of importance.

2.2 CONSERVATION LAWS

Expressions of the conservation of mass, a particular chemical species, momentum, and energy are fundamental principles which are used in the analysis and design of any separation device. It is appropriate to formulate these laws first without specific rate expressions so that a clear distinction between conservation laws and rate expressions is made. Some of these laws contain a source or generation term, for example, for a particular chemical species, so that the particular quantity is not actually conserved. A conservation law for entropy can also be formulated which contributes to a useful framework for a generalized transport theory.[1,2] Such a discussion is beyond the scope of this chapter. The conservation expressions are first presented in their macroscopic forms, which are applicable to overall balances on energy, mass, and so on, within a system. However, such macroscopic formulations do not provide the information required to size equipment. Such analyses usually depend on a differential formulation of the conservation laws which permits consideration of spatial variations of composition, temperature, and so on within a system.

2.2-1 Macroscopic Formulation of Conservation Laws

In any particular region of space (Fig. 2.2-1) the macroscopic balances express the fact that the time rate of change of mass, species, momentum, or energy within the system is equal to the sum of the net flow across the boundaries of the region and the rate of generation within the region.

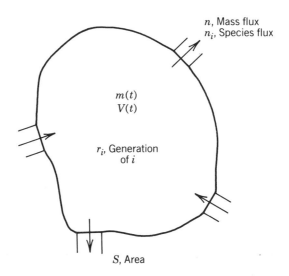

FIGURE 2.2-1 Mass balances on arbitrary volume of space.

Since, in non-nuclear events, total mass is neither created nor destroyed, the conservation law for total mass in a system having a number of discrete entry and exit points may be written

$$\frac{dm}{dt} = \sum_{\text{entries}} nS - \sum_{\text{exits}} nS \tag{2.2-1}$$

The total mass of the system is m, n is the total mass flux (mass flow per unit area) relative to the system boundary at any point, and S is the cross-sectional area normal to flow at that same location. The summations extend over all the mass entry and exit locations in the system. The mass flux at any point is equal to ρv, where ρ is the mass density and v is the velocity relative to the boundary at that point. Equation (2.2-1) can be applied equally well to a countercurrent gas absorption column or to a lake with input and output streams such as rainfall, evaporation, streams flowing to or from the lake, deposition of sediment on the lake bottom, or dissolution of minerals from the sides and bottom of the lake. The steady-state version of Eq. (2.2-1) ($dm/dt = 0$) is of use in chemical process analysis because it permits calculation of various flow rates once some have been specified.

The macroscopic version of the conservation law for a particular chemical species i that is produced by chemical reaction at a rate per unit volume‡ of r_i is written

$$\frac{dm_i}{dt} = \sum_{\text{entries}} n_i S - \sum_{\text{exits}} n_i S + \int_V r_i \, dV \tag{2.2-2}$$

Again it is assumed that there are a number of discrete entry and exit locations where species i crosses the system boundary with a flux n_i. The flux n_i is also equal to $\rho_i v_i$ where ρ_i is the mass concentration of i (mass of i/volume) and v_i is the velocity of species i relative to the system boundary. In a multicomponent mixture the velocities of each species may differ from one another and from the average velocity of the mixture. The mass average velocity of a mixture is defined by

$$v \equiv \frac{\sum_i \rho_i v_i}{\rho} \tag{2.2-3}$$

where the summation is over all species. The last term of Eq. (2.2-2) represents the total production (consumption) of species i owing to chemical reaction within the control volume. Since the total mass of

‡The term r_i would actually represent the sum of the rates of all reactions producing or consuming species i.

the system is the sum of the mass of each species, Eq. (2.2-1) can be obtained by summing Eq. (2.2-2) written for each species. The term $\Sigma_i \, r_i$ is identically zero since mass is neither created nor destroyed by chemical reaction.

Conservation of species can also be expressed on a molar basis as shown by Eq. (2.2-4):

$$\frac{d\overline{M_i}}{dt} = \underset{\text{entries}}{\Sigma} N_i S - \underset{\text{exits}}{\Sigma} N_i S + \int_V R_i \, dV \tag{2.2-4}$$

where moles of i in the system is $\overline{M_i}$. The molar flux of species i (moles i/time·area) relative to the system boundary at any location is $N_i \equiv C_i v_i$, where C_i is the molar concentration of species i. Summation of Eq. (2.2-4) written for all species would yield a conservation expression for total moles which is of limited use since a generation term representing net moles created or destroyed by reaction is present. A molar average velocity of a mixture v^* can be defined analogous to the mass average velocity

$$v^* \equiv \frac{\underset{i}{\Sigma} C_i v_i}{C} \equiv \frac{\underset{i}{\Sigma} N_i}{C} \tag{2.2-5}$$

The total molar flux $N = \Sigma_i N_i$ is Cv^*, where C is the total molar concentration $\Sigma_i C_i$.

The conservation of species equations, whether expressed in terms of mass [Eq. (2.2-2)] or moles [Eq. 2.2-4)], are entirely equivalent; the choice of which to use is made on the basis of convenience. The steady-state forms of these material balances are universally employed in the design and analysis of chemical processes including separation devices.[3,4] Many examples will be encountered in later chapters of this book. The primary purpose here is to set the stage for the differential balance formulation to follow shortly.

Since the macroscopic formulation of the momentum balance has little use in mass transfer theory it is not presented here. The differential form is of importance because many mass transfer problems involve the interaction of velocity and diffusion fields.

Similar to the mass conservation laws, the macroscopic formulation of the First Law of Thermodynamics—that is, energy is conserved—is applicable to a wide range of chemical processes and process elements. The energy conservation equation for a multicomponent mixture can be written in a bewildering array of equivalent forms.[5,6] For the volume of space shown in Fig. 2.2-1 containing i species, one form of the conservation equation for the energy (internal + kinetic + potential) content of the system is

$$\frac{d}{dt} \int_{V=1} \Sigma \, \rho_i (u_i + \tfrac{1}{2}|v_i|^2 + \Phi_i) \, dV = \underset{\text{entries}}{\Sigma} \left[\Sigma_i \, n_i (h_i + \tfrac{1}{2}|v_i|^2 + \Phi_i) \, S \right]$$
$$- \underset{\text{exits}}{\Sigma} \left[\Sigma_i \, n_i (h_i + \tfrac{1}{2}|v_i|^2 + \Phi_i) \, S \right] + W_S + qS + \dot{Q}_R \tag{2.2-6}$$

Equation (2.2-6) states that the change with time of the total energy content of a volume is due to the net difference between energy flowing into and out of the system with mass flow plus the shaft power being delivered to the system plus the rate of energy delivered to the system as heat (both by conduction and radiation). The partial mass enthalpy, specific kinetic energy, and specific potential energy of each species i in the flow terms are evaluated at the entry or exit point conditions. The potential energy term, a result of so-called body forces, can differ for different species, for example, the force of an applied electrical field on ions in solution, but most commonly it is only that resulting from gravity.

The macroscopic energy balances, with appropriate simplifying assumptions, are used in many different process systems. For example, under steady state conditions with unimportant mechanical energy and shaft work terms, it is the basis of the so-called process "heat" balance which may be applied (along with mass balances) to heat exchangers, chemical reactors, distillation columns, and so on to determine quantities such as heat duties or exit temperatures from a unit. Application of the macroscopic balances generally does not permit a determination of the composition or temperature profiles within a unit such as a packed column gas absorber or tubular reactor. A differential formulation of the conservation laws is therefore required.

2.2-2 Differential Formulation of Conservation Laws

We shall see that the conversion equations give rise to partial differential equations for the variation of velocity, density, concentration, and temperature as a function of position and time. Most practical problems require considerable simplification of the complete equations by appropriate assumptions. By far the most important part of an analysis is the proper choice of these assumptions based on physical reasoning. Most of this book is devoted to such analyses. This section derives formulations of balance equations and focuses on the physical meaning of the various terms, especially with application to mass transfer phenomena.

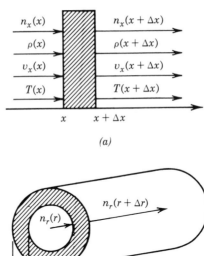

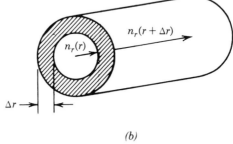

FIGURE 2.2-2   One-dimensional shell balances: (*a*) rectangular and (*b*) cylindrical coordinates.

Although general and rather elegant derivations of the differential balances are available,[6,7] it is more instructive to use the simple "shell" balance approach. Figure 2.2-2 depicts situations where changes in density, velocity, composition, and temperature occur continuously and in only one spatial direction, either x or r. The change with time of the total mass within the volume element of Fig. 2.2-2a, owing to the difference in mass flow into and out of the volume, is

$$\frac{\partial(\rho S\,\Delta x)}{\partial t} = Sn_x(x) - Sn_x(x + \Delta x)$$

where n_x denotes the mass flux in the x direction. In this rectangular coordinate system the cross-sectional area is independent of position x and as $\Delta x \rightarrow 0$

$$\frac{\partial \rho}{\partial t} = -\frac{\partial n_x}{\partial x} \qquad (2.2\text{-}7)$$

Equation (2.2-7), known as the one-dimensional continuity equation, expresses the fact that the density can only change at a particular location if there is a change in the mass flux with position. Conversely, if the density does not change with time, the mass flux is independent of position.

Equation (2.2-7) can be generalized readily to three dimensions as shown in Table 2.2-1. The mass flux in a particular coordinate direction (e.g., n_x) is simply equal to the product of the density and the component of the mass average velocity in that direction, ρv_x. These mass fluxes and velocities are with reference to a stationary or laboratory coordinate system.

This same shell balance can be applied in curvilinear coordinate systems‡ as illustrated in Fig. 2.2-2b. In this case the change of area with radial position must be taken into account, that is,

$$\frac{\partial \rho}{\partial t} = -\frac{1}{r}\frac{\partial(rn_r)}{\partial r}$$

The three dimensional analogue of this equation and that for spherical coordinates are also listed in Table 2.2-1. The continuity equation can also be more compactly expressed in vector or tensor notation.

‡It is preferable, however, to derive the equations in rectangular coordinates and then transform to another system.

TABLE 2.2-1. Differential Forms of the Conservation Equation for Total Mass

Rectangular Coordinates

$$\frac{\partial \rho}{\partial t} + \frac{\partial n_x}{\partial x} + \frac{\partial n_y}{\partial y} + \frac{\partial n_z}{\partial z} = 0$$

or

$$\frac{\partial \rho}{\partial t} + \frac{\partial(\rho v_x)}{\partial x} + \frac{\partial(\rho v_y)}{\partial y} + \frac{\partial(\rho v_z)}{\partial z} = 0$$

Cylindrical Coordinates

$$\frac{\partial \rho}{\partial t} + \frac{1}{r}\frac{\partial}{\partial r}(rn_r) + \frac{1}{r}\frac{\partial n_\theta}{\partial \theta} + \frac{\partial n_z}{\partial z} = 0$$

or

$$\frac{\partial \rho}{\partial t} + \frac{1}{r}\frac{\partial}{\partial r}(r\rho v_r) + \frac{1}{r}\frac{\partial}{\partial \theta}(\rho v_\theta) + \frac{\partial}{\partial z}(\rho v_z) = 0$$

Spherical Coordinates

$$\frac{\partial \rho}{\partial t} + \frac{1}{r^2}\frac{\partial}{\partial r}(r^2 n_r) + \frac{1}{r\sin\theta}\frac{\partial}{\partial \theta}(n_\theta \sin\theta) + \frac{1}{r\sin\theta}\frac{\partial n_\phi}{\partial \phi} = 0$$

or

$$\frac{\partial \rho}{\partial t} + \frac{1}{r^2}\frac{\partial}{\partial r}(r^2 \rho v_r) + \frac{1}{r\sin\theta}\frac{\partial}{\partial \theta}(\rho v_\theta \sin\theta) + \frac{1}{r\sin\theta}\frac{\partial}{\partial \phi}(\rho v_\phi) = 0$$

The continuity equation for total moles can also be written, but it must include a term representing the net production of moles by chemical reaction. In rectangular coordinates the equation would be

$$\frac{\partial C}{\partial t} = -\frac{\partial N_x}{\partial x} - \frac{\partial N_y}{\partial y} - \frac{\partial N_z}{\partial z} + \sum_i R_i \qquad (2.2\text{-}8)$$

where N_x, N_y, N_z are the components of the molar flux vector relative to stationary coordinates.

A shell balance applied to the mass of any particular species i would yield, in rectangular coordinates,

$$\frac{\partial \rho_i}{\partial t} + \frac{\partial n_{i,x}}{\partial x} + \frac{\partial n_{i,y}}{\partial y} + \frac{\partial n_{i,z}}{\partial z} = r_i \qquad (2.2\text{-}9)$$

The notation $n_{i,x}$ denotes the x component of the mass flux of species i relative to laboratory coordinates. Equation (2.2-9) expresses the fact that the mass concentration of a species will change with time at a particular point if there is a change in the flux with position at that point or if a chemical reaction is occurring there.

For analysis of mass transfer problems it is convenient to rewrite Eq. (2.2-9) in terms of a diffusive flux or diffusion velocity. The mass diffusive flux in the x direction ($j_{i,x}$) is defined relative to the mass average velocity in that direction:‡

$$j_{i,x} \equiv \rho_i(v_{i,x} - v_x) = n_{i,x} - \frac{\rho_i}{\rho} n_x \qquad (2.2\text{-}10)$$

The ratio ρ_i/ρ is the mass fraction of species i.

The conservation equation for species i can then be written in a form that explicitly recognizes transport by convection at the mass average velocity and by diffusive transport relative to the average velocity:

‡In three directions the diffusive flux vector is defined as the difference between the total flux of i and that carried at the mass average velocity.

$$\frac{\partial \rho_i}{\partial t} + \frac{\partial(\rho_i v_x)}{\partial x} + \frac{\partial(\rho_i v_y)}{\partial y} + \frac{\partial(\rho_i v_z)}{\partial z} + \frac{\partial j_{i,x}}{\partial x} + \frac{\partial j_{i,y}}{\partial y} + \frac{\partial j_{i,z}}{\partial z} = r_i \tag{2.2-11}$$

Equation (2.2-11) can also be written in the form of a *substantial* derivative of the mass fraction ($w_i = \rho_i/\rho$) by use of the continuity equation for total mass (Table 2.2-1)

$$\rho \frac{Dw_i}{Dt} + \frac{\partial j_{i,x}}{\partial x} + \frac{\partial j_{i,y}}{\partial y} + \frac{\partial j_{i,z}}{\partial z} = r_i \tag{2.2-12}$$

The substantial derivative can be interpreted as the variation of w_i with time as seen by an observer moving with the mass average velocity of the system:

$$\frac{Dw_i}{Dt} \equiv \frac{\partial w_i}{\partial t} + v_x \frac{\partial w_i}{\partial x} + v_z \frac{\partial w_i}{\partial y} + v_z \frac{\partial w_i}{\partial z} \tag{2.2-13}$$

The species conservation equations in cylindrical and spherical coordinates are shown in Table 2.2-2. In rectangular coordinates the corresponding conservation of species equations in *molar* units are

$$\frac{\partial C_i}{\partial t} + \frac{\partial N_{i,x}}{\partial x} + \frac{\partial N_{i,y}}{\partial y} + \frac{\partial N_{i,z}}{\partial z} = R_i \tag{2.2-14}$$

$$\frac{\partial C_i}{\partial t} + \frac{\partial(C_i v_x^*)}{\partial x} + \frac{\partial(C_i v_y^*)}{\partial y} + \frac{\partial(C_i v_z^*)}{\partial z} + \frac{\partial J_{i,x}}{\partial x} + \frac{\partial J_{i,y}}{\partial y} + \frac{\partial J_{i,z}}{\partial z} = R_i \tag{2.2-15}$$

The molar diffusion flux J_i is defined with respect to the *molar average* velocity,

$$J_i = C_i(v_i - v^*) = N_i - \frac{C_i}{C} N \tag{2.2-16}$$

were C_i/C is simply the mole fraction of i in the system. Other definitions of the diffusion flux exist in the literature[5] but they are not necessary to the analysis of mass transfer problems.

If we examine the continuity equations of total mass and any species i, it is quite clear that the equations cannot be solved for density and composition as a function of time and position until the velocity field is known. In addition, appropriate rate expressions (Sections 2.3 and 2.4) must be available for mass transport as well as for chemical reaction. These rates will depend on temperature so that the temperature field must

TABLE 2.2-2. Differential Forms of the Conservation Equation for a Species (Mass Units)[a]

Rectangular Coordinates

$$\frac{\partial \rho_i}{\partial t} + \frac{\partial n_{i,x}}{\partial x} + \frac{\partial n_{i,y}}{\partial y} + \frac{\partial n_{i,z}}{\partial z} = r_i$$

or

$$\rho \frac{\partial w_i}{\partial t} + \rho v_x \frac{\partial w}{\partial x} + \rho v_y \frac{\partial w_i}{\partial y} + \rho v_z \frac{\partial w_i}{\partial z} + \frac{\partial j_{i,x}}{\partial x} + \frac{\partial j_{i,y}}{\partial y} + \frac{\partial j_{i,z}}{\partial z} = r_i$$

Cylindrical Coordinates

$$\frac{\partial \rho_i}{\partial t} + \frac{1}{r}\frac{\partial}{\partial r}(rn_{i,r}) + \frac{1}{r}\frac{\partial n_{i,\theta}}{\partial \theta} + \frac{\partial n_{i,z}}{\partial z} = r_i$$

Spherical Coordinates

$$\frac{\partial \rho_i}{\partial t} + \frac{1}{r^2}\frac{\partial}{\partial r}(r^2 n_{i,r}) + \frac{1}{r \sin \theta}\frac{\partial}{\partial \theta}(n_{i,\theta} \sin \theta) + \frac{1}{r \sin \theta}\frac{\partial n_{i,\phi}}{\partial \phi} = r_i$$

[a] Equivalent forms in molar units could be written based on Eq. (2.2-14) and (2.2-15).

also be specified. Although many practical mass transfer problems are solved by independently specifying the velocity and temperature—for example, diffusion in a stagnant, isothermal system—the general problem requires that the velocity and temperature be determined simultaneously by using appropriate expressions of conservation of momentum and energy in differential form.

The differential linear momentum balance for a multicomponent mixture can be derived from a shell balance approach or from a more sophisticated analysis using continuum mechanics. Such formulations are tabulated in many texts, for example, that by Bird et al.,[5] and will not be presented here in any detail. By way of illustration the x component of the differential momentum balance in rectangular coordinates would be

$$\rho\,\frac{\partial v_x}{\partial t} + v_x\,\frac{\partial v_x}{\partial x} + v_y\,\frac{\partial v_y}{\partial y} + v_z\,\frac{\partial v_z}{\partial z} = -\frac{\partial p}{\partial x} - \left(\frac{\partial \bar{\tau}_{xx}}{\partial x} + \frac{\partial \bar{\tau}_{yx}}{\partial y} + \frac{\partial \bar{\tau}_{zx}}{\partial z}\right) + \rho g_x \qquad (2.2\text{-}17)$$

where p is the hydrostatic pressure, g_x is the x component of gravity, and $\bar{\tau}$ represents the stress tensor. It is important to note that v_x, v_y, and v_z in a multicomponent system are the components of the mass average velocity of the system. If the fluid is Newtonian, then the stress tensor, which may also be thought of as the momentum flux vector, is related to the velocity gradients in the system by Newton's Law of viscosity, and the well-known Navier–Stokes equations result. Equation (2.2-17) and its analogues are the starting point for the solution of problems in fluid mechanics and may be used simultaneously with the mass conservation equations to solve many problems in mass transfer.

The differential form of the energy balance for a multicomponent mixture can be written in a variety of forms.[5,6] It would contain terms representing heat conduction and radiation, body forces, viscous dissipation, reversible work, kinetic energy, and the substantial derivative of the enthalpy of the mixture. Its formulation is beyond the scope of this chapter. Certain simplified forms will be used in later chapters in problems such as simultaneous heat and mass transfer in air–water operations or thermal effects in gas absorbers.

As shown in the following examples, the conservation laws are employed in specific instances by making appropriate simplifying assumptions to reduce the general differential balances to a tractable form.

EXAMPLE 2.2-1 A DIFFUSION EXPERIMENT

Two pure fluids A and B are initially contained in vertical cylinders that are brought over one another at time zero (Fig. 2.2-3). Write down the appropriate equations to determine the composition profile at any later time if the fluids are (a) perfect gases and (b) liquids.

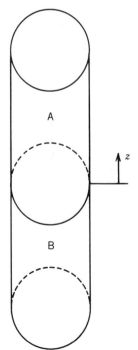

FIGURE 2.2-3 One-dimensional free diffusion geometry.

Solution: We make the following assumptions:

1. Variations of composition occur only in the z direction.
2. The system is at constant pressure and temperature.
3. No chemical reaction occurs.

The continuity and species equations in either mass or molar units may be used at the convenience of the analyst.

For *gases* it is convenient to use molar units because for an ideal gas mixture the molar concentration $C = p/RT$ is independent of composition. The one-dimensional continuity equation is

$$\frac{\partial C}{\partial t} + v_z^* \frac{\partial C}{\partial z} + C \frac{\partial v_z^*}{\partial z} = 0$$

Since C does not change, $\partial v_z^*/\partial z = 0$ and v_z^* is a constant. Furthermore, since there is no flux through the ends of the cylinder, $v_z^* = 0$.

The species balance for A reduces to

$$\frac{\partial C_A}{\partial t} = -\frac{\partial J_{A,z}}{\partial z}$$

This equation can be solved provided an appropriate rate expression is available for J_A, the diffusive flux relative to the molar average velocity.

If the conservation equations in mass units had been employed,

$$\frac{\partial \rho}{\partial t} + \frac{\partial}{\partial z} (\rho v_z) = 0 \tag{A}$$

$$\frac{\partial \rho_A}{\partial t} + \frac{\partial}{\partial z} (\rho_A v_z + j_{A,z}) = 0 \tag{B}$$

For an ideal gas mixture the mass density does depend on position since $\rho = Mp/RT$, where M is the mean molecular weight. Substitution for the density into Eqs. (A) and (B) would yield equations for v_z and ρ_A after an appropriate flux relationship for j_A was supplied. The z-component of the mass average velocity, v_z, is not zero.

For *liquids* the equations in mass units are convenient because liquid mass densities are generally weak functions of composition, so that the approximation $\partial \rho/\partial t = \partial \rho/\partial z = 0$ is made and $v_z = 0$ by Eq. (A). Equation (B) is then

$$\frac{\partial \rho_A}{\partial t} = -\frac{\partial j_{A,z}}{\partial z}$$

Solutions to the above equations will be given in Section 2.3. In this problem of free diffusion the continuity equation combined with thermodynamic relationships allows the specification of the velocity field.

EXAMPLE 2.2-2

Pure water can be obtained from brackish water by permeation through a reverse osmosis membrane. Consider the steady laminar flow of a salt solution in a thin channel between two walls composed of a membrane that rejects salt. Derive the governing equations for the salt distribution in the transverse direction for a given water permeation flux (see Fig. 2.2-4).

Solution. We make the following assumptions:

1. The system exists in an isothermal steady state.
2. The density is independent of composition because it is a liquid.
3. The effect of H_2O permeation through the walls on the velocity profile is neglected.

The continuity equation can be written

$$\rho \frac{\partial v_z}{\partial z} + \rho \frac{\partial v_y}{\partial y} = 0 \tag{A}$$

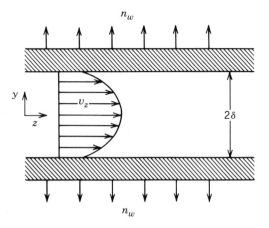

FIGURE 2.2-4 Laminar flow membrane channel.

The species balance for salt is

$$\frac{\partial n_{s,z}}{\partial z} + \frac{\partial n_{s,y}}{\partial y} = 0 \tag{B}$$

and the z-component of momentum balance is

$$v_z \frac{\partial v_z}{\partial z} + v_y \frac{\partial v_z}{\partial y} = -\frac{1}{\rho}\frac{\partial p}{\partial z} + \nu \frac{\partial^2 v_z}{\partial z^2} \tag{C}$$

To obtain the velocity profile assumption 3 is used, whereby we take $v_y = 0$ for low permeation rates.‡ From Eq. (A) $\partial v_z / \partial z = 0$ and Eq. (C) becomes

$$+\frac{1}{\rho}\frac{\partial p}{\partial z} = \nu \frac{\partial^2 v_z}{\partial y^2}$$

whose solution is the well-known parabolic velocity profile

$$v_z(y) = \tfrac{3}{2} v_{\mathrm{avg}} \left[1 - \left(\frac{y}{\delta}\right)^2 \right]$$

An approximate solution for the salt distribution under low permeation rates can be obtained by neglecting diffusion in the z direction with respect to convection ($n_{s,z} \approx \rho_s v_z$) and neglecting v_y. The species balance then becomes

$$v_z(y) \frac{\partial \rho_s}{\partial z} + \frac{\partial j_{s,y}}{\partial y} = 0$$

where $v_z(y)$ is known. This equation can be solved provided an appropriate rate expression for the diffusive flux is known as well as an expression relating the flux of water through the membrane to the salt concentration at the membrane surface and the pressure difference across the membrane. A discussion of membrane transport can be found in Chapter 10 while rate expressions for diffusion are the subject of Section 2.3.

The primary purpose of this section was to state the framework in which all mass transfer problems should be formulated, that is, the conservation equations for mass, species, momentum, and energy. As illustrated by the above examples complete solutions of problems in mass transfer require a knowledge of

‡Berman[8] has given the exact velocity profiles for the case of a uniform flux of water through the membrane.

rate expressions for species transport under various conditions. These rate expressions and their use are the subject of the remainder of this chapter.

2.3 MOLECULAR DIFFUSION

Ordinary diffusion can be defined as the transport of a particular species relative to an appropriate reference plane owing to the random motion of molecules in a region of space in which a composition gradient exists. Although the mechanisms by which the molecular motion occurs may vary greatly from those in a gas to those in a crystalline solid, the essential features of a random molecular motion in a composition gradient are the same, as will be seen in the simplified derivations discussed here.

Consider a perfect gas mixture at constant temperature and pressure (therefore the molar concentration C is constant and there is no bulk motion of the gas) in which a mole fraction gradient exists in one direction as shown in Fig. 2.3-1. Molecules within some distance $\pm a$, which is proportional to the mean free path of the gas (λ), can cross the $z = 0$ plane. The net molar flux from left to right across the $z = 0$ plane would be proportional to the product of the mean molecular speed $\bar{v}$, and the difference between the average concentration on the left and on the right.

$$\text{Net flux} \propto C\bar{v}\left[y_A\left(z = \frac{-a}{2}\right) - y_A\left(z = \frac{a}{2}\right) \right]$$

The difference in the average mole fractions can be approximated by $-a(dy_A/dz)_{z=0}$. The diffusive flux J_A is, therefore

$$J_A \propto -\lambda\bar{v}C\frac{dy_A}{dz} \quad \text{or} \quad -\lambda\bar{v}\frac{dC_A}{dz} \tag{2.3-1}$$

The molecular diffusivity (or diffusion coefficient) D, defined as the proportionality constant between the diffusive flux and the negative of the composition gradient, is therefore proportional to the product of the mean molecular speed and the mean distance between collisions. If the simple kinetic theory expressions for the mean molecular speed and the mean free path of like molecules are used, one finds a modest temperature and pressure dependence of the diffusivity.

$$D \propto \frac{T^{3/2}}{Pd_A^2 M_A^{1/2}} \tag{2.3-2}$$

where d_A is the collision diameter and M_A is the molecular weight of species A. Detailed kinetic theory analysis or the more exact results of Chapman and Cowling[1] taking into account intermolecular forces and

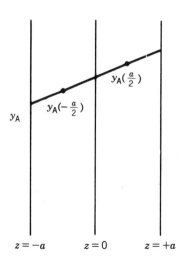

$z = -a$ $z = 0$ $z = +a$

FIGURE 2.3-1 Diffusion in a gas mixture owing to random molecular motion in a composition gradient.

differing sizes and masses of molecules are quoted in Section 2.3-2. The purpose here is simply to illustrate that a diffusive flux originates from the random motion of molecules in a composition gradient.

For diffusion in condensed phases whether solid or liquid, Eq. (2.3-1) could still hold except it must be recognized that not all molecules are free to move. Thus, the expression for the diffusive flux must be multiplied by a probability that the molecule is available to move or jump. This probability is generally related to an energy barrier which must be overcome. The nature of the barrier is dependent on the mechanism of the diffusive process, for example, the motion of atoms and vacancies in a metal crystal would be characterized by a different energy barrier than that for the movement of molecules in the liquid state. In any event, the diffusivity in the condensed phase would be given by an expression of the form[2] $\nu_0 b^2 \exp(-E/RT)$. In this expression the mean free path has been replaced by a characteristic intermolecular spacing b, and the mean speed by the product of a vibration frequency and this spacing. Again, the diffusion coefficient contains parameters characteristic of random molecular motion within a phase.

Appropriate definitions of diffusive fluxes and a phenomenological theory for the relationship of the fluxes to the driving forces or potentials for diffusion are presented first. Methods of predicting and correlating diffusion coefficients in gases, liquids, and solids are discussed in Section 2.3-2, along with some typical experimental values of diffusion coefficients. Section 2.3-3 describes a number of experimental methods for measuring diffusion coefficients along with the appropriate mathematical analysis for each technique. Finally, some sample problems of diffusion in stagnant phases and in laminar flow fields are considered.

2.3-1 Phenomenological Theory of Diffusion

A general theory of transport phenomena[3] suggests that the diffusive flux would be made up of terms associated with gradients in composition, temperature, pressure and other potential fields. Only ordinary diffusion, the diffusive motion of a species owing to a composition gradient, is considered in detail here. The *assumed* linear relationship between diffusive flux and concentration gradient is traditionally known as *Fick's First Law of Diffusion*. In one dimension,

$$J_{i,z} = -D \frac{dC_i}{dz} \qquad (2.3-3)$$

Equation (2.3-3) can be generalized to relate the diffusive flux vector to the gradient of composition in three coordinate directions. Cussler[4] has recently summarized the historical background to the development of Fick's Law. In fact, there have been a number of ambiguities or inconsistencies in the use of rate expressions for diffusion primarily arising from an unclear view of the definitions used in a particular analysis.

In Section 2.2 both the mass diffusive flux relative to the mass average velocity and the molar diffusive flux relative to the molar average velocity were defined:

$$j_i \equiv n_i - \rho_i v = n_i - w_i n \qquad (2.3-4)$$

$$J_i \equiv N_i - C_i v^* = N_i - x_i N \qquad (2.3-4a)$$

Other definitions of the diffusive flux such as the mass flux relative to the molar average velocity could be employed but they have no particular value. Equation (2.3-4) or (2.3-4a) can be used in conjunction with the definitions of v and v^* to show immediately that

$$\sum_i j_i = 0 \qquad (2.3-5)$$

and

$$\sum_i J_i = 0 \qquad (2.3-5a)$$

These relationships among the diffusive fluxes of various species provide restrictions on the diffusion coefficients appearing in the rate equations for the diffusive flux.

BINARY SYSTEMS

Consider a binary isotropic‡ system composed of species A and B at uniform temperature and pressure and ask what the appropriate rate expression for ordinary diffusion should be. The thermodynamics of irrever-

‡In crystal systems, anisotropy or variations of diffusivity with coordinate direction can occur. See Jost.[2]

sible processes suggests that the driving force for diffusion would be the isothermal, isobaric gradient in chemical potential, that is, the gradient in chemical potential owing to composition changes. It further postulates a linear relationship between flux and driving force. There is nothing to say whether this composition should be expressed in terms of a mass or mole fraction or a volume concentration; indeed, all are quite proper measures of composition.[5] In mass units, for one-dimensional diffusion, either Eq. (2.3-6) or (2.3-6a) can be used.

$$j_{A,z} \equiv -D_{AB}\, \rho\, \frac{dw_A}{dz} \qquad j_{B,z} \equiv -D_{BA}\, \rho\, \frac{dw_B}{dz} \tag{2.3-6}$$

$$j_{A,z} \equiv -D'_{AB}\, \frac{d\rho_A}{dz} \qquad j_{B,z} \equiv -D'_{BA}\, \frac{d\rho_B}{dz} \tag{2.3-6a}$$

The expressions are different, of course, unless the density is independent of position (composition).

If Eq. (2.3-6) is chosen as the definition, then it immediately follows by use of Eq. (2.3-5) that $D_{AB} = D_{BA}$; that is, there is a single mutual diffusion coefficient in a binary system. If Eq. (2.3-6a) is chosen, $D'_{BA} \neq D'_{AB}$, but there would be a relationship between them so that there still is only a single characteristic diffusivity for a binary system. The choice of Eq. (2.3-6) is preferred because of its simplicity although it certainly is not intrinsically better then Eq. (2.3-6a).

The analogous rate expressions for the molar diffusive flux relative to the molar average velocity can be used. The preferred definition relates the flux to the mole fraction gradient

$$J_{A,z} = -D_{AB}\, C\, \frac{dx_A}{dz} \tag{2.3-7}$$

It can be shown that D_{AB} in this definition is identical to that in Eq. (2.2-6). The particular choice of reference velocity and corresponding flux expression can be made for the convenience of the analyst as illustrated by the extension of the analysis carried out in Example 2.2-1 on the interdiffusion of two gases or liquids.

For the case of a perfect gas mixture whose molar concentration is independent of composition, the continuity and species balance equations reduce to the statements

$$v_z^* = 0$$

$$\frac{\partial C_A}{\partial t} = -\frac{\partial J_{A,z}}{\partial z}$$

Since C is constant $J_{A,z}$ can be written $-D_{AB}\, \partial C_A/\partial z$, and the final form of the species balance becomes

$$\frac{\partial C_A}{\partial t} = \frac{\partial (D_{AB}\, \partial C_A/\partial z)}{\partial z} \tag{2.3-8}$$

Equation (2.3-8) is known as *Fick's Second Law of Diffusion*, but recognize that it is the result of assuming constant molar concentration, determining $v^* = 0$ from the continuity equation, and substituting Fick's First Law in the species conservation equation. For a binary liquid solution whose mass density is approximately independent of composition, the species balance and continuity equation would also reduce to the form of Fick's Second Law

$$\frac{\partial \rho_A}{\partial t} = \frac{\partial (D_{AB}\, \partial \rho_A/\partial z)}{\partial z} \tag{2.3-9}$$

As illustrated in the following example, it is preferable to start with the conservation equations, rate expression, and thermodynamic relationships, rather than immediately applying Fick's Second Law to a diffusion problem.

EXAMPLE 2.3-1
Derive the governing equation for composition as a function of position and time for the case of the interdiffusion of two fluids in which the molar density of the mixture is *not* constant, but the partial molar volumes of each species are constant and equal to the pure component values, that is, a thermodynamically ideal solution.

Solution. The continuity equation and species balances in molar units are

$$\frac{\partial C}{\partial t} + \frac{\partial}{\partial z}(Cv^*) = 0 \tag{A}$$

$$\frac{\partial C_A}{\partial t} + \frac{\partial}{\partial z}(Cv^* + J_A) = 0 \tag{B}$$

By means of Eq. (A) the species balance can be written in the alternative form

$$C\frac{\partial x_A}{\partial t} + Cv^*\frac{\partial x_A}{\partial z} = -\frac{\partial J_A}{\partial z} \tag{C}$$

Application of the chain rule for the variation of the molar density allows the continuity equation to be written in the form

$$\frac{dC}{dx_A}\frac{\partial x_A}{\partial t} + v^*\frac{dC}{dx_A}\frac{\partial x_A}{\partial z} + C\frac{\partial v^*}{\partial z} = 0$$

Elimination of $\partial x_A/\partial t$ by Eq. (C) provides an expression for the velocity gradient

$$\frac{\partial v^*}{\partial z} = \frac{1}{C^2}\frac{dC}{dx_A}\frac{\partial J_A}{\partial z} \tag{D}$$

or

$$v^*(z, t) = -\int_{-\infty}^{z}\frac{d(1/C)}{dx_A}\frac{\partial J_A}{\partial z}\,dz + v^*(-\infty, t)$$

The experiment can usually be arranged so that $v^*(-\infty, t) = 0$. This general result for the average velocity in terms of the diffusive flux in a free diffusion experiment was published by Duda and Vrentas.[6]
The density is related to the partial molar volumes by

$$\frac{1}{C} = \bar{V}_A x_A + \bar{V}_B(1 - x_A) \qquad \frac{d(1/C)}{dx_A} = \bar{V}_A - \bar{V}_B$$

For our case V_A and V_B are equal to their pure-component values, V_A° and V_B°, so that the molar average velocity is

$$v^*(z, t) = -(\bar{V}_A^\circ - \bar{V}_B^\circ) J_A$$

Substitution for v^* in Eq. (B) and use of the relationship

$$J_A \equiv -D_{AB}C\frac{\partial x_A}{\partial z} = \frac{-D_{AB}\,\partial C_A/\partial z}{1 - C_A(\bar{V}_A^\circ - \bar{V}_B^\circ)}$$

yield after some algebra

$$\frac{\partial C_A}{\partial t} = \frac{\partial}{\partial z}\left(D_{AB}\frac{\partial C_A}{\partial z}\right)$$

Thus, Fick's Second Law can be rigorously applied for the case of the interdiffusion of ideal solutions. For nonideal solutions, the simultaneous solution of Eqs. (D) and (B) can be carried out if density versus composition data are available for the system.[6]

It should be emphasized that in the phenomenological theory the diffusivity is presumed independent of composition gradient but still may be strongly dependent on composition as well as temperature.

MULTICOMPONENT SYSYEMS

Obviously, diffusion in multicomponent systems is the rule rather than the exception in most systems of practical interest. Yet, unusual or significant effects occur only rarely in multicomponent systems. Such effects may be seen in systems that are thermodynamically highly nonideal, in membrane diffusion, when electrostatic effects occur, or when there are strong associations among particular species. Cussler has written an excellent book[4] on the experimental and theoretical aspects of multicomponent diffusion to which the reader is referred for an extensive discussion.

Early theories for multicomponent diffusion in gases were obtained from kinetic theory approaches and culminated in the Stefan–Maxwell equations[7,8] for dilute gas mixtures of constant molar density. In one dimension,

$$\frac{dy_i}{dz} = \sum \frac{1}{CD_{ij}} (y_i N_j - y_j N_i) \tag{2.3-10}$$

The D_{ij} are the binary diffusion coefficients for an $i - j$ mixture; thus, no additional information is required for computations of multicomponent diffusion in dilute gas mixtures although one might prefer a form in which the fluxes appeared explicitly. Generalization of the Stefan–Maxwell form to dense gases and to liquids has been suggested,[9] but in these cases there is no rigorous relationship to the binary diffusivities. Furthermore, the form of Eq. (2.3-10) in which the fluxes do not appear explicitly has little to recommend it.

A phenomenological theory of ordinary diffusion in multicomponent systems based on irreversible thermodynamics[3] suggests that the ordinary diffusive flux of any species is a linear function of all the independent composition gradients. In one dimension,

$$j_i = \sum_{k=1}^{c-1} \rho D_{ik} \frac{dw_k}{dz} \tag{2.3-11}$$

In a system containing c species there are $c - 1$ independent fluxes (since $\Sigma j_i = 0$) and $c - 1$ independent composition gradients. There would appear to be $(c - 1) \times (c - 1)$ diffusion coefficients, but use of the Onsager reciprocity relationships of irreversible thermodynamics shows that there are actually only $c(c - 1)/2$ independent coefficients. Because of the complexity of the relationships among the D_{ij} of Eq. (2.3-9) (requiring knowledge of the thermodynamic properties of the system), Cussler[4] suggests that as a practical matter the D_{ij} be treated as independent coefficients in many cases. The key aspect of Eq. (2.3-11) is the understanding that in a multicomponent system the diffusion rate of a particular species can be affected by the composition gradients of other species as well as by its own. The D_{ij}, of course, are composition dependent.

A striking example of the effects of a third component on the diffusive rate of a species is shown in Fig. 2.3-2 in which the acceleration of the flux of sodium sulfate in aqueous solution owing to the presence of acetone is shown. Some of the most dramatic examples of multicomponent effects in diffusion occur in both natural and synthetic membranes.

In many practical instances involving multicomponent diffusion the diffusive flux is simply modeled in terms of an effective diffusivity of i in the mixture (D_{im}) which relates the flux of i only to its own composition gradient.

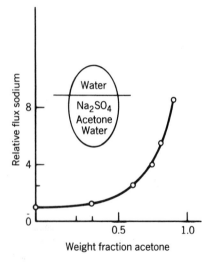

FIGURE 2.3-2 Enhancement of Na_2SO_4 flux in aqueous solution owing to acetone addition. Reprinted with permission from E. L. Cussler, *Multicomponent Diffusion*, Elsevier, Amsterdam, 1976.

$$j_{i,z} = -D_{im}\rho \frac{dw_i}{dz} \tag{2.3-12}$$

The quantity D_{im} would not only be a function of the composition of the mixture but, in fact, could be dependent on the gradients of various other species and therefore could depend on time and/or position during the diffusion process. Nevertheless, Eq. (2.3-12) is often employed with reasonable success unless some highly specific interactions exist. D_{im} should certainly be measured under the mixture composition of interest, yet it is often estimated as the binary diffusivity of i in the solvent.

2.3-2 Diffusion Coefficients in Gases, Liquids and Solids

This section provides some typical experimental values of interdiffusion coefficients in gases, in solutions of nonelectrolytes, electrolytes, and macromolecules, in solids, and for gases in porous solids. Theoretical and empirical correlations for predicting diffusivities will also be discussed for use in those cases where estimates must be made because experimental data are unavailable. A critical discussion of predicting diffusivities in a fluid phase can be found in the book by Reid et al.[10]

GASES

At ambient temperature and pressure, gas-phase diffusion coefficients are of the order of 10^{-4}–10^{-5} ft²/s (10^{-5}–10^{-6} m²/s). Table 2.3-1 presents some representative values for binary gas mixtures at 1 atm. Marrero and Mason[11] have provided an extensive review of experimental values of gas phase binary diffusivities.

Kinetic theory predicts the self-diffusion coefficient, that is, the diffusion coefficient for a gas mixture in which all molecules have identical molecular weights and collision diameters, to be[12]

$$D_{AA} \propto \frac{R^{3/2} T^{3/2}}{M_A^{1/2} d_A^2 p}$$

while that for mutual diffusion of A and B is

$$D_{AB} \propto \frac{R^{3/2} T^{3/2}[1/M_A + 1/M_B]^{1/2}}{[(d_A + d_B)/2]^2 p} \tag{2.3-13}$$

TABLE 2.3-1 Experimental Values of Binary Gaseous Diffusion Coefficients at 1 atm and Various Temperatures

Gas Pair	T (K)	$D_{AB} \times 10^5$ (m²/s)^a
Air–CO_2	317	1.77
Air–H_2O	313	2.88
Air–n-C_6H_{14}	328	0.93
Ar–O_2	293	2.0
CO_2–N_2	273	1.44
CO_2–N_2	298	1.66
CO_2–O_2	293	1.53
CO_2–CO	273	1.39
CO_2–H_2O	307	1.98
CO_2–H_2O	352	2.45
C_2H_4–H_2O	328	2.33
H_2–CH_4	298	7.26
H_2–NH_3	298	7.83
H_2–NH_4	473	18.6
H_2–C_6H_6	288	3.19
H_2–SO_2	473	12.3
N_2–NH_3	298	2.30
N_2–NH_3	358	3.28
N_2–CO	423	6.10
O_2–C_6H_6	311	1.01
O_2–H_2O	352	3.52

^{a}Diffusion coefficients in ft²/s are obtained by multiplying table entries by 10.76.

TABLE 2.3-2 Values of the Collision Integral for Diffusion

kT/ϵ_{AB}	Ω_D
0.30	2.662
0.50	2.066
0.75	1.667
1.00	1.439
1.50	1.198
2.00	1.075
3.00	0.9490
4.00	0.8836
5.0	0.8422
10	0.7424
20	0.6640
50	0.5756
100	0.5170

where d_i is the molecular diameter of i and M_i is its molecular weight. The pressure dependence of Eq. (2.3-13) appears to be correct for low pressures; but gas phase diffusivities vary somewhat more strongly with temperature than is indicated by Eq. (2.3-13).

By taking into account forces between molecules, Chapman and Enskog as described by Chapman and Cowling[1] derived the following expression for the binary diffusivity at low pressures:

$$D_{AB} = \frac{1.858 \times 10^{-7} \, T^{3/2} (1/M_A + 1/M_B)^{1/2}}{p\sigma_{AB}^2 \, \Omega_D} \qquad (2.3\text{-}14)$$

In Eq. (2.3-14) D_{AB} is in m^2/s, T, is in K, p is in atm, and σ_{AB} is in Å. The dimensionless collision integral for diffusion Ω_D is a function of the intermolecular potential chosen to represent the force field between molecules as well as of temperature.[1,13] The Lennard-Jones 6-12 potential is most commonly used for calculation of Ω_D and extensive tabulations as a function kT/ϵ_{AB} are available.[13,14] A small tabulation is included in Table 2.3-2, while Table 2.3-3 provides values of intermolecular diameters σ_i and energy parameters ϵ_i for some common gases. Empirical methods for estimating σ_i and ϵ_i from other physical property data are available.[10] The combining rules for the Lennard-Jones potential are $\sigma_{AB} = (\sigma_A + \sigma_B)/2$ and $\epsilon_{AB} = (\epsilon_A \epsilon_B)^{1/2}$. Equation (2.3-14) represents well the diffusivities of binary gas pairs that have approximately spherical force fields and are only slightly polar. Use of a different intermolecular potential function for polar gases has been suggested.[14] There also are no satisfactory theories and little data available for the diffusivities of gases at pressures greater than about 10 atm.

A number of empirical correlations for gaseous diffusion coefficients at low pressures based approximately on the ideal gas form have been proposed.[15-19] According to Reid et al.[10] the equation of Fuller, Schettler, and Giddings[19] appears to be the best, generally predicting diffusivities of nonpolar gases within 10%, although the temperature dependence may be underestimated.

TABLE 2.3-3 Lennard-Jones Parameters for Some Common Gases

Gas	$\sigma(Å)$	ϵ/k (K)
Air	3.617	97
CO	3.590	110
CO_2	3.996	190
CH_4	3.822	137
C_2H_4	4.232	205
C_3H_8	5.061	254
$n\text{-}C_6H_{14}$	5.909	413
C_6H_6	5.270	440
N_2	3.681	91.5
NO	3.470	119
O_2	3.433	113
SO_2	4.290	252

$$D_{AB} = \frac{10^{-7} \, T^{1.75}(1/M_A + 1/M_B)}{p[(\Sigma v)_A^{1/3} + (\Sigma v)_B^{1/3}]^2} \tag{2.3-15}$$

Equation (2.3-15) is a dimensional equation with D_{AB} in m²/s, T in K, and p in atm. This equation must make use of the diffusion volumes that are tabulated for simple molecules and computed from atomic volumes for more complex molecules. Table 2.3-4 provides some of these values. In summary, it may be said that gaseous diffusion coefficients at low or modest pressures can be readily estimated from either Eq. (2.3-14) or (2.3-15) if experimental data are not available.

EXAMPLE 2.3-2

Estimate the diffusivity of a CO_2–O_2 mixture at 20°C and 1 atm and compare it to the experimental value in Table 2.3-1. Repeat the calculation for 1500 K where the experimental value is 2.4×10^{-4} m²/s (2.6 $\times 10^{-3}$ ft²/s).

Solution. Using Eq. (2.3-14) and data from Table 2.3-3,

$$\sigma_{O_2-CO_2} = \frac{(3.433 + 3.996)}{2} = 3.7145$$

$$\frac{\epsilon_{O_2-CO_2}}{k} = [(113)(190)]^{1/2} = 146.5$$

At 293 K (from Table 2.3-2),

$$\frac{kT}{\epsilon} = 2.0 \qquad \Omega_D = 1.075$$

$$D_{O_2-CO_2} = 1.4 \times 10^{-5} \text{ m}^2/\text{s}$$

At 1500 K,

$$\frac{kT}{\epsilon} = 10.23 \qquad \Omega_D = 0.742$$

$$D_{O_2-CO_2} = 2.4 \times 10^{-4} \text{ m}^2/\text{s}$$

Using Eq. (2.3-15) and data from Table 2.3-4

$$(\Sigma v)_{CO_2} = 26.9 \qquad (\Sigma v)_{O_2} = 16.6$$

TABLE 2.3-4 Atomic Volumes for Use in Predicting Diffusion Coefficients by the Method of Fuller, et al.

Atomic and Structural Diffusion Volume Increments			
C	16.5	Cl	19.5
H	1.98	S	17.0
O	5.48	Aromatic ring	−20.2
N	5.69	Heterocyclic ring	−20.2
Diffusion Volumes for Simple Molecules			
H_2	7.07	CO_2	26.9
He	2.88	N_2O	35.9
N_2	17.9	NH_3	14.9
O_2	16.6	H_2O	12.7
Air	20.1	SF_6	69.7
Ar	16.1	Cl_2	37.7
CO	18.9	SO_2	41.1

Source. Reprinted with permission from E. N. Fuller, P. D. Schettler, and J. C. Giddings, *Ind. Eng. Chem.*, **58**, 18 (1966), Copyright 1966 American Chemical Society.

At 293 K,
$$D_{O_2-CO_2} = 1.5 \times 10^{-5}$$

At 1500 K,
$$D_{O_2-CO_2} = 2.6 \times 10^{-4} \text{ m}^2/\text{s}$$

LIQUIDS

Diffusion coefficients in liquids are of the order of 10^{-8} ft^2/s (10^{-9} m^2/s) unless the solution is highly viscous or the solute has a very high molecular weight. Table 2.3-5 presents a few experimental diffusion coefficients for liquids at room temperature in dilute solution. Ertl and Dullien,[20] Johnson and Babb,[21] and Himmelblau[22] provide extensive tabulations of diffusion coefficients in liquids. It is probably safe to say that most of the reported experimental diffusivities were computed based on Fick's Second Law without consideration of whether or not the system was thermodynamically ideal. Since the binary diffusion coefficient in liquids may vary strongly with composition, tabulations and predictive equations usually deal with the diffusivity of A at infinite dilution in B, D_{AB}°, and the diffusivity of B at infinite dilution in A, D_{BA}°. Separate consideration is then given to the variation of the diffusivity with composition.

Hydrodynamic theories for prediction of liquid-phase diffusion coefficients at infinite dilution are represented by the Stokes–Einstein equation[23] which views the diffusion process as the motion of a spherical solute molecule A through a continuum made up of solvent molecules B.

TABLE 2.3-5 Diffusion Coefficients in Solutions at Low Solute Concentrations Near Room Temperature

Solute	Solvent	T (°K)	$D_{AB}^0 \times 10^9$ (m^2/sec)
CH_3OH	H_2O	283	0.84
C_2H_5OH	H_2O	283	0.84
C_2H_4O	H_2O	293	1.16
Ethylene glycol	H_2O	293	1.06
Glycerol	H_2O	293	0.83
Ethylene	H_2O	298	1.87
Toluene	H_2O	296	6.19
O_2	H_2O	298	2.10
H_2	H_2O	293	5.0
CO_2	H_2O	298	1.92
NH_3	H_2O	285	1.64
H_2O	C_2H_5OH	298	1.24
H_2O	Ethylene glycol	298	0.18
H_2O	Glycerol	293	0.013
CH_3OH	Glycerol	294	0.0064
n-Hexyl alcohol	Glycerol	298	0.006
H_2O	C_2H_4O	298	4.56
Glycerol	C_2H_5OH	293	0.53
C_2H_4O	$CHCl_3$	298	2.35
C_6H_{12}	C_6H_6	298	2.09
C_2H_5OH	C_6H_6	288	2.25
CO_2	n-C_6H_{14}	298	6.0
C_3H_8	n-C_6H_{14}	298	4.87
CO_2	C_2H_5OH	290	3.20
O_2	C_2H_5OH	303	3.20
$C_2H_4O_2$	Ethyl acetate	293	2.18
H_2O	Ethyl acetate	298	3.20
Ribonuclease (MW = 13,683)	Aqueous buffer	293	0.119
Albumin (MW = 65,000)	Aqueous buffer	293	0.0594
Myosin (MW = 493,000)	Aqueous buffer	293	0.0116
Polystyrene (MW = 414,000)	Cyclohexane	308	0.022
Polystyrene (MW = 50,000)	Cyclohexanone	298	0.089
Polystyrene (MW = 180,000)	Cyclohexanone	298	0.05
Polystyrene (MW = 402,000)	Cyclohexanone	298	0.028
Polyacrylonitrile (MW = 127,000)	Dimethylformamide	298	0.10

$$D_{AB} = \frac{RT}{3\pi\mu_B \, d_A} \tag{2.3-16}$$

where μ_B is the viscosity of the solvent and d_A is the diameter of the solute molecule. The most widely used correlation for D_{AB} is that of Wilke and Chang[24] which represents an empirical modification of Eq. (2.3-16) to fit available experimental data:

$$D_{AB}^\circ = \frac{7.4 \times 10^{-12} \, (\phi \overline{M}_B)^{1/2} T}{\mu_B \, (\overline{V}_A^\circ)^{0.6}} \tag{2.3-17}$$

where $(\overline{V}_A^\circ)^{0.6}$ is the molar volume of pure A at its *normal* boiling point (cm³/mol) and ϕ is an association factor for the solvent. The constant in Eq. (2.3-17) is not dimensionless so the viscosity must be expressed in centipoise (cP) and the temperature in K to obtain D in units of m²/s. Wilke and Chang gave empirical values of ϕ ranging from 2.6 for water and 1.9 for methanol to 1.0 for nonassociated solvents. The temperature dependence of the diffusivity in Eq. (2.3-17) is primarily represented by the temperature dependence of the viscosity. Thus, the group $D_{AB}\mu_B/T$ may be assumed constant over modest temperature ranges. Over wider temperature ranges or when viscosity data are unavailable, the diffusivity is conveniently correlated by an equation of the Arrhenius form:

$$D_{AB} = D_0 \exp\left(\frac{-E_a}{RT}\right) \tag{2.3-18}$$

Figure 2.3-3 illustrates this correlation for a few systems. The activation energy E_a for a diffusion process is typically a few kilocalories per mole.

After comparing a number of correlations for liquid-phase diffusivities, Reid et al.[10] conclude that the Wilke–Chang correlation is to be preferred for estimating infinite-dilution coefficients of low-molecular-weight solutes in nonpolar and nonviscous solvents. For highly viscous solvents the Wilke–Chang corre-

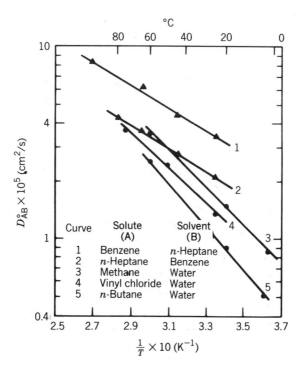

Curve	Solute (A)	Solvent (B)
1	Benzene	n-Heptane
2	n-Heptane	Benzene
3	Methane	Water
4	Vinyl chloride	Water
5	n-Butane	Water

FIGURE 2.3-3 Temperature dependence of liquid diffusion coefficients. Reprinted with permission from Ref. 10.

lation may underestimate the diffusivity by as much as an order of magnitude. The correlation of Gainer and Metzner,[25] based on Absolute Rate Theory, is quite successful at predicting diffusion coefficients in viscous liquids.

The application of the Wilke–Chang correlation to the diffusion of solutes that would be gases if they existed in a pure state at the temperature of the solution has yielded conflicting results. Wise and Houghton[26] and Shrier[27] found the diffusivities to be significantly underestimated while Himmelblau[22] found good agreement between the correlation and literature values. Witherspoon and Bonoli[28] found agreement between their measurements of the diffusivity of hydrocarbon gases in water and the Wilke–Chang equation. Akgerman and Gainer[29] presented a method based on Absolute Reaction Rate Theory to predict the diffusion coefficients of gases in solution.

Composition Dependence of D_{AB}. The diffusion coefficient of a binary solution is a function of composition. If the true driving force for ordinary diffusion is the isothermal, isobaric gradient of chemical potential, then the binary diffusivity can be written in the form

$$D_{AB} = \hat{D}_{AB} \left(\frac{1 + d \ln \gamma_A}{d \ln x_A} \right) \tag{2.3-19}$$

where γ_A is the activity coefficient at mole fraction x_A. This result, which suggests that the diffusivity can be corrected by a thermodynamic activity term, has been successful, in some cases, in accounting for much of the composition dependence of the diffusion coefficient (Fig. 2.3-4).

Vignes[30] correlated the composition dependence of binary diffusion coefficients in terms of their infinite-dilution values and this thermodynamic correction factor.

$$D_{AB} = (D_{AB}^{\circ})^{x_B} (D_{BA}^{\circ})^{x_A} \left(1 + \frac{d \ln \gamma_A}{d \ln x_A} \right) \tag{2.3-20}$$

Some representative curves from his paper are also shown in Fig. 2.3-4. The correlation appears to be quite successful unless the mixtures are strongly associating (e.g., CH_3Cl–acetone). Leffler and Cullinan[31] have provided a derivation of Eq. (2.3-20) based on Absolute Rate Theory while Gainer[32] has used Absolute Rate Theory to provide a different correlation for the composition dependence of binary diffusion coefficients.

Multicomponent Effects. Only limited experimental data are available for multicomponent diffusion in liquids. The binary correlations are sometimes employed for the case of a solute diffusing through a mixed solvent of uniform composition.[5,33] It is clear that thermodynamic nonidealities in multicomponent systems can cause significant effects. The reader is referred to Cussler's book[4] for a discussion of available experimental information on diffusion in multicomponent systems.

Ionic Solutions. Since salts are partially or completely dissociated into ions in aqueous solution, the diffusion coefficient of the salt must be related to those of its component ions. Comprehensive discussions of electrolytic solutions may be found in a number of references.[34–37] For a single salt in solution a *molecular* diffusivity can be assigned to the salt and is calculable at infinite dilution from the ionic conductances of the ions:[37]

$$D_{SW}^{\circ} = \frac{RT}{F} \frac{1/Z_+ + 1/Z_-}{1/\lambda_+ + 1/\lambda_-} \tag{2.3-21}$$

where $Z_{+,-}$ is the valence and $\lambda_{+,-}$ the ionic conductance of positive or negative ions. No adequate theory exists for the effect of increasing salt concentration on the diffusion coefficient, although it is generally found that the diffusion coefficient first decreases with increasing salt concentration and then increases. Diffusion of ions in mixed electrolyte solutions is a complex phenomenon involving transport owing to both composition and electrical potential gradients. Even without an applied potential the differences of mobility among the ions in solution give rise to an electrical potential. Mixed electrolytic solutions are in fact multicomponent solutions so that a number of diffusion coefficients are required to describe the system properly. Indeed, some interesting multicomponent effects are the result of electrostatic interactions.[4]

The diffusion coefficients of sparingly soluble gases in electrolytes appear to be influenced by the ionic strength of the solution (Table 2.3-6). Although no theoretical explanation is available, the diffusivities appear to decrease with increasing salt concentration. Since solubilities of such gases usually decrease with increasing salt concentration, one should be aware of these effects when dealing with the dispersion of a gas in an ionic solution.

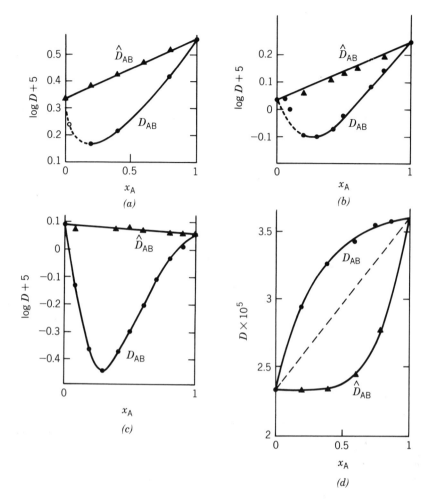

FIGURE 2.3-4 Composition dependence of binary diffusion coefficients in liquids. (in units of cm^2/s): (a) acetone (A)–CCl$_4$, (b) methanol (A)–H$_2$O (c) ethanol (A)–H$_2$O, (d) acetone (A)–CH$_3$Cl. Reprinted with permission from A. Vignes, *Ind. Eng. Chem. Fundam.*, **5**, 198 (1966). Copyright 1966 American Chemical Society.

Macromolecular Solutions. In these systems two diffusion phenomena could be of importance: the mutual diffusion coefficient of a macromolecule and its solvent and the influence of a macromolecule in a solution on the diffusion rate of smaller solutes. As can be seen in Table 2.3-5 the mutual diffusion coefficients of proteins in aqueous solution or of a synthetic polymer in a solvent are in the range of 10^{-10}–10^{-11} m^2/s at room temperature. Experimental methods and theoretical concepts dealing with the diffusion coefficients of macromolecules can be found in Tanford.[41] Representative data are those of Paul[42] and Osmers and Metzner.[43] Duda and Vrentas[44] have reviewed the theory and experiments dealing with the mutual diffusion coefficient of amorphous polymers and low molecular weight solvents. The diffusivities of macromolecules appear to vary with the square root of the molecular weight of the solute, although solute shape also has an effect.

The diffusivities of low molecular weight solutes in polymer solutions do not appear to follow a predictable pattern as a function of polymer concentration. It is clear that the results cannot be predicted with the Wilke–Chang equation by merely using the increased viscosity of the solution. Diffusivities have been reported to increase, decrease, or go through a maximum with increasing polymer concentration despite solution viscosity increases.[43-47] Li and Gainer[47] measured the diffusivities of four solutes in seven

TABLE 2.3-6   Diffusion Coefficients of Gases in Aqueous Electrolyte Solutions[a] at 25°C

Gas	Electrolyte	Concentration (mol/L)	$D_{AB}^0 \times 10^9$ (m^2/s)
O_2	None	—	2.10
O_2	NaCl	0.15	2.07
O_2	KCl	0.01	2.13
O_2	KCl	0.10	2.05
O_2	KCl	1.0	1.95
O_2	KCl	4.0	1.65
H_2	None	—	4.50
H_2	KCl	1	3.79
H_2	KCl	3	3.34
H_2	$MgSO_4$	0.05	3.90
H_2	$MgSO_4$	2.0	1.57
CH_4	None	—	1.50–3.00
CH_4	$MgSO_4$	0.5	1.88
CH_4	$MgSO_4$	2.0	0.73
CO_2	None	—	1.92
CO_2	NaCl	1.04	1.73
CO_2	NaCl	3.77	1.30
CO_2	$MgSO_4$	0.20	1.89
CO_2	$MgSO_4$	0.97	1.28

[a]Data from Ref. 38–40.

polymer solutions and developed a predictive theory that could account for both increases and decreases in diffusion coefficients. Navari et al.[48] examined the diffusion coefficients of small solutes in mixed protein solutions, for example, O_2 or glucose in solutions containing various amounts of albumin and gamma globin, and concluded that the ratio of diffusivity in the solution to that in the pure solvent was independent of the solute. They also observed some rather dramatic decreases in diffusivity at particular protein compositions. At the present time, it is advisable to obtain experimental data for the system of interest because an understanding of all the interactions that occur in such systems does not exist. Although these are multicomponent solutions, studies to date have only considered the situation where there is no gradient in polymer concentration.

EXAMPLE 2.3-3
It is desired to estimate the diffusivity of a thymol ($C_{10}H_{14}$)–salol ($C_{13}H_{10}O_3$) organic melt at 27°C. The measured viscosities of pure thymol and salol at 27°C are 11 and 29 cP, respectively.

Solution   Although the liquids are somewhat viscous, first estimate the infinite-dilution diffusivities using the Wilke–Chang correlation, Eq. (2.3-17). The molar volumes at the boiling point are estimated by Shroeder's method[10] as

$$\overline{V}_T^\circ = 189 \text{ cm}^3/\text{mol} \qquad M_T = 150$$
$$\overline{V}_S^\circ = 210 \text{ cm}^3/\text{mol} \qquad M_S = 214$$

It is not clear what the association factor ϕ should be for these mixtures. Aromatic mixtures have[10] $\phi \approx$ 0.7 while hydrogen bonded systems have $\phi \approx 1.0$. Use $\phi \approx 1.0$

$$D_{ST}^\circ = 9.9 \times 10^{-11} \text{ m}^2/\text{s}$$
$$D_{TS}^\circ = 6.9 \times 10^{-11} \text{ m}^2/\text{s}$$

A single experimental measurement[49] of the average diffusivity over the whole composition range gave a value of 7.4×10^{-11} m²/s. These solutions do form nearly ideal liquid mixtures so the diffusivity of a 50/50 mixture could be estimated from Vignes' correlation:

$$D_{TS} = (D_{ST}^\circ)^{0.5} (D_{TS}^\circ)^{0.5} = 8.3 \times 10^{-11} \text{ m}^2/\text{s}$$

The infinite dilution diffusivities can also be estimated using the Gainer–Metzner correlation[25] with the results: $D_{ST}^\circ = 8.0 \times 10^{-11}$ m²/s, $D_{TS}^\circ = 6.7 \times 10^{-11}$ m²/s and for the 50/50 mixture, $D_{ST} = 7.3 \times 10^{-11}$ m²/s.

Both methods give values in reasonable agreement with experiment although the Wilke–Chang method does have the adjustable parameter ϕ. The Gainer–Metzner correlation is to be preferred as the viscosity of a solution liquid increases.

SOLIDS

Diffusion rates of ions, atoms, or gases through crystalline solids often govern the rates of important solid-state reactions and phase changes such as the formation of oxide films. Extensive discussions of these phenomena are given by Barrer[50] and Jost,[2] including tabulations of diffusion coefficient data. Recent developments can be found in *Diffusion in Solids* by Nowick and Burton.[51] Since the solid phase can exist over a large temperature range, diffusivity values in solids can vary over an enormous range as can be seen from the representative data shown in Table 2.3-7. An Arrhenius form for diffusion in metals ($D_0 e^{-E_a/RT}$) has been shown to correlate diffusion data over many orders of magnitude of the diffusivity and hundreds of kelvins. Some representative data taken from Birchenall's text[52] for self-diffusion in various metals are shown in Fig. 2.3-5.

In metal alloys forming solid solutions over significant composition ranges, the binary diffusivity varies with composition. Figure 2.3-6 shows some typical data for diffusion in the Fe–Ni system at 1185°C. Vignes and Birchenall[54] have presented a correlation for the concentration dependence in binary alloys similar to Eq. (2.3-18) for liquids but they required an additional factor involving the melting temperature of the alloy to obtain a satisfactory correlation.

POROUS SOLIDS

The diffusion rates of fluids, particularly gases, through porous solid materials are of particular interest because of the common practice of using porous solids as catalysts and adsorbents. Diffusional transport rates within the pores may limit the rates by which such processes occur.

Satterfield's book, *Mass Transfer in Heterogeneous Catalysis,*[55] is an excellent source of information on diffusional rates in porous media. Transport in porous solids can occur by ordinary diffusion, by Knudsen diffusion, and by surface diffusion of adsorbed species. Ordinary diffusion simply represents the transport of material by diffusion in the gas phase within the pores. Since the whole cross-sectional area of the solid is not available for diffusion and the pores do not run straight through the material but consist of a randomly connected network, the effective diffusivity is given by

$$D_{AB,eff} = \frac{D_{AB}\epsilon}{\tau} \tag{2.3-22}$$

where D_{AB} is the gas phase diffusion coefficient, ϵ is the porosity of the solid, and τ is the tortuosity. The tortuosity factor, accounting for the random direction of pores, can vary widely but typical values are in the range of 1 to 10.[55]

When the pressure is lowered or the pore radius is small, the mean free path of a gas molecule becomes comparable to that of the pore radius. In this situation, called the Knudsen diffusion regime, collisions between gas molecules and the wall are as important as those among molecules. The Knudsen diffusion coefficient is given by the expression[55]

$$D_K = \frac{\epsilon}{\tau} \frac{4}{3} r_e \left[\frac{2 RT}{\pi M_A} \right]^{1/2} = \frac{\epsilon}{3\tau} \frac{8 \epsilon}{S_g \rho_b} \left(\frac{2 RT}{\pi M_A} \right)^{1/2} \tag{2.3-23}$$

where the effective pore radius is represented by $2\epsilon/\rho_b S_g$ and $\rho_b S_b$ is the total surface per volume of solid. The Knudsen diffusion coefficient is independent of pressure and only a weak function of temperature,

TABLE 2.3-7 Some Binary Diffusion Coefficients in Solids

System	Temperature (°K)	$D_{AB} \times 10^{12}$ (m²/sec)
H_2 in Fe	283	0.16
H_2 in Fe	373	12.4
He in Pyrex	293	0.0045
He in Pyrex	773	2
Al in Cu	293	1.3×10^{-26}
Al in Cu	1123	0.22
C in Fe	1073	1.5
C in Fe	1373	45
Ni (10%) in Fe	1293	0.0009
Ni (10%) in Fe	1568	0.010

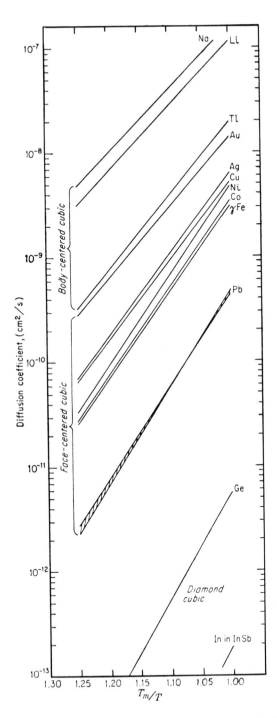

FIGURE 2.3-5 Temperature dependence of diffusion coefficients in metals. Reprinted with permission from Ref. 52.

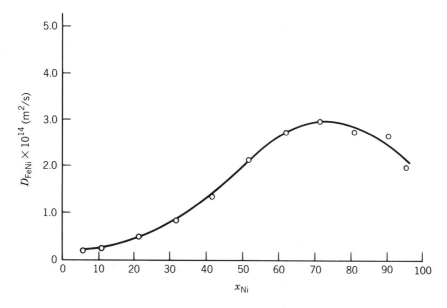

FIGURE 2.3-6 Composition dependence of diffusion coefficient in iron-nickel alloys at 1185°C.[53]

while the diffusion coefficient in a gas is inversely proportional to pressure and varies approximately as $T^{1.75}$. Whether ordinary or Knudsen diffusion governs the transport in the gas depends on the magnitudes of $D_{AB,eff}$ and D_K for a given situation. In the transition region where both mechanisms are of importance, it has been shown that an overall diffusion coefficient is given approximately by[55,56]

$$\frac{1}{D_{eff}} = \frac{1}{D_K} + \frac{1}{D_{AB,eff}} \tag{2.3-24}$$

Satterfield[55] summarizes experimental data illustrating the transition from one regime to the other as a function of pore size, pressure, and chemical composition of the gases involved.

A third transport mechanism operating in parallel with the above mechanisms in a porous solid occurs when the gaseous species adsorbs and migrates along the surface of the pores in the direction of decreasing surface concentration. For this to be a significant mechanism sufficient surface coverage must be achieved, yet the absorbed species cannot be tightly bound or little motion can occur. The total flux of a species owing to a concentration gradient in a porous solid can be represented by an expression of the form

$$J = -(D_{eff} + K D_S) \frac{d C_A}{dz} \tag{2.3-25}$$

if the relationship between adsorbed and gaseous concentrations can be represented by a linear isotherm, $C_{AS} = K C_A$. The surface diffusion coefficient (D_S) typically is in the range of 10^{-7}–10^{-11} m²/s for physically adsorbed species. Sladek[57] has developed an extensive correlation for surface diffusion coefficients of both physically and chemically adsorbed species.

Zeolites of controlled pore sizes of the order of a few angstroms have become popular catalysts and adsorbents because of their ability to discriminate effectively the shape and size differences among molecules. Weisz et al.[58] describe applications of these shape-selective catalysts in oxidation of normal paraffins and olefins in the presence of the branched isomers. Observed diffusion coefficients in these zeolites are often found to be quite small (10^{-14} m²/s) but precise interpretations of the measurements and mechanisms of transport in zeolites are not available as yet.

POLYMERS

Diffusive transport rates through a polymeric phase are very important to membrane separation processes such as reverse osmosis, ultrafiltration, and gas permeation. See Chapters 18–21 for discussion of membrane processes. Here, diffusion coefficients of solutes within a polymeric material are discussed.

The diffusive process in polymers is quite complex, depending on the degree of crystallinity and cross-linking of the polymer, the ability of the gas or liquid to swell the polymer, and specific interactions between solute and polymer. Transport in membranes is often characterized by the permeability coefficient—the product of the diffusivity and the solubility coefficient of the solute in the polymer. In this way the transport is described in terms of concentrations or partial pressures of the solute existing in the fluid phases adjacent to the polymer external surface. Thus, a high permeability may be the result of either a large solubility of the solute or a large diffusion coefficient within the polymeric phase.

Data and theory for diffusion coefficients in polymers are summarized in the book by Crank and Park[59] and the earlier books by Barrer[50] and Jost.[2] Michaels and Bixler[60,61] describe both solubility and diffusion coefficients for various nonswelling gases in a number of polymers, while Brubaker and Kammermeyer[62] report the permeabilities of a number of simple gases in a variety of commercial polymers. Diffusion coefficients for gases in polymers appear to exhibit on Arrhenius dependence on temperature. Some representative data for gases in polymers are given in Table 2.3-8.

For vapors that can interact with the polymer matrix, for example, hydrocarbons or chlorinated hydrocarbons in polyethylene, both the diffusivity and the solubility rapidly increase with increasing vapor concentration.[60,63,64] In a similar fashion, liquids that are able to swell the membrane have enhanced transport rates. Permeabilities for vapors that swell the membrane can easily be orders of magnitude greater than those for gases that do not chemically interact with the membrane.[65]

The removal of water from aqueous salt solutions by reverse osmosis, as in seawater desalination with cellulose acetate membranes or nylon hollow fibers, is believed to occur primarily by a diffusive transport mechanism for both water and solutes. On the other hand, in the use of membranes for the removal of water from aqueous solutions containing higher molecular weight solutes, such as the ultrafiltration of protein solutions, the solvent is believed transported by a viscous flow mechanism within the pores of the membrane and solute molecules are convected with the solvent in the larger pores.[66-69]

2.3-3 Experimental Methods for Measuring Molecular Diffusivity

In this section a number of methods are described for the experimental determination of molecular diffusion coefficients. The purpose is not only to acquaint the reader with some of these techniques but also to illustrate, by means of the associated analyses, the proper formulation and solution of the appropriate mathematical model for the particular experimental diffusion situation. In all cases, the governing equations follow from simplifying the conservation equations for total mass and particular species, using the flux expression and necessary thermodynamic relationships, and applying appropriate boundary conditions established from the physical situation. Further descriptions of experimental methods may be found in the books by Jost[2] and Cussler.[4] Many mathematical solutions of the diffusion equation are found in Crank.[70]

DIAPHRAGM CELL

The diaphragm cell technique,[2,4,71] primarily used for diffusion in liquids, is based on setting up a situation in which there is a steady flux of material as a result of a known composition gradient. The apparatus, shown schematically in Fig. 2.3-7, consists of two well-stirred chambers separated by a porous diaphragm of glass or other material. If the chambers are well mixed, the entire composition gradient occurs within the separating diaphragm. The compositions of the solutions to the reservoirs are measured as a function of time during the course of the experiment. The experiment only yields an average diffusion coefficient over the composition range employed.

TABLE 2.3-8 Some Diffusion Coefficients of Gases in Polymers at 25 °C

Gas	Polymer	$D_{AB} \times 10^{10}$ (m²/sec)
He	Polyethylene (amorphous)	20
O_2	Polyethylene (amorphous)	1.8
CH_4	Polyethylene (amorphous)	0.9
C_3H_8	Polyethylene (amorphous)	0.2
H_2	Butyl rubber	1.52
O_2	Butyl rubber	0.81
CO_2	Butyl rubber	0.058
H_2	Silicone rubber	67.0
O_2	Silicone rubber	17.0
O_2	Natural rubber	1.58
CO_2	Natural rubber	1.10
CH_4	Natural rubber	0.89

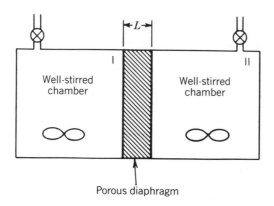

FIGURE 2.3-7 Diaphragm diffusion cell.

Within the diaphragm the conservation equations can be written in mass units for the binary solution in the pores as

$$\frac{\partial \rho}{\partial t} + \frac{\partial (n_A + n_B)}{\partial z} = 0 \tag{2.3-26}$$

$$\frac{\partial \rho_A}{\partial t} + \frac{\partial n_A}{\partial z} = 0 \tag{2.3-27}$$

The usual analysis assumes that a steady state applies if the reservoir volumes are much larger than the diaphragm volume. It is a ''pseudo'' steady state since the chamber concentrations change with time, but only slowly with respect to the time required to establish a concentration gradient across the diaphragm. Mills et al.[72] examined this assumption for the diaphragm cell. Equations (2.3-26) and (2.3-27) under steady-state conditions immediately yield that the total mass flux and the flux of A are independent of position z. The mass flux of A is given by

$$n_A = -D_{eff}\, \rho\, \frac{dw_A}{dz} + w_A(n_A + n_B) \tag{2.3-28}$$

The diffusion coefficient is written as $D_{eff}(D_{AB}\epsilon/\tau)$ because, in fact, diffusion occurs through a porous diaphragm whose porosity and tortuosity affect the observed flux based on the geometrical cross-sectional area. Integration of Eq. (2.3-28) under the assumption of constant fluid density relates the flux to the composition difference at any time:

$$n_A = D_{eff}\, \rho\, \frac{w_{A,I} - w_{A,II}}{L} \tag{2.3-29}$$

where L is the diaphragm thickness. The flux at any time can be obtained from measurement of the composition changes in the chambers. Alternatively, the expression for the flux in terms of the time derivative of the composition of the chambers can be substituted in Eq. (2.3-27) with the final result

$$\ln \frac{\Delta w_A}{\Delta w_A^\circ} = K D_{AB} t \tag{2.3-30}$$

where Δw_A is the mass fraction difference at any time, Δw_A° is the original difference, and K is an apparatus constant that contains a collection of only geometrical parameters for the apparatus.

The average diffusivity over the composition range is then obtained by plotting the logarithm of the measured composition differences versus time. The apparatus constant is determined by calibrating the cell with a solution of known diffusivity. Obviously, the pores in the diaphragm must be large enough not to affect the diffusing solutes yet not so large as to allow any bulk flow resulting from density differences. Care must also be taken that the porosity is not affected by deposits or adsorption of species during the calibration or the experiments.

A variation of the experimental technique to measure the diffusivities of slightly soluble gases, such as O_2 in solution, was employed by Tham et al.[73] and Gainer and coworkers.[29,48] One reservoir was maintained saturated with a diffusing solute gas by steady sparging with a gas stream while the other chamber was swept free of solute gas by sparging with helium. The steady flux across the cell was determined from gas flow rates and chromatographic measurements of the gas composition. This experiment established steady compositions on both sides of the diaphragm. In princple, this is a ternary diffusion system but in very dilute solutions and with an inert carrier gas (He), the measured diffusivity is the binary value.

STEFAN TUBE

A classical method for measuring the diffusion coefficient of the vapor of a volatile liquid in air or other gas (e.g., toluene in N_2) employs the Stefan tube shown in Fig. 2.3-8, a long tube of narrow diameter (to suppress convection) partially filled with a pure volatile liquid A and maintained in a constant-temperature bath. A gentle flow of air is sometimes established across the top of the tube to sweep away the vapor reaching the top of the tube. The fall of the liquid level with time is observed.

The conservation equations are conveniently employed in molar units for the analysis of gaseous diffusion in one direction.

$$\frac{\partial C}{\partial t} + \frac{\partial (N_A + N_B)}{\partial z} = 0 \tag{2.3-31}$$

$$\frac{\partial C_A}{\partial t} + \frac{\partial N_A}{\partial z} = 0 \tag{2.3-32}$$

The analysis of this problem is based on two assumptions: (1) the motion of the liquid boundary downward can be neglected in solving the diffusion equation in the vapor and (2) a pseudo-steady-state assumption can be made to calculate the composition profile in the vapor space.

Under these assumptions Eqs. (2.3-31) and (2.3-32) yield that the fluxes are independent of position in the tube. Furthermore, the net flux of B across the gas–liquid interface is zero, if B has a negligible solubility in the liquid. Therefore, N_B is zero everywhere in the tube.

The flux of A becomes

$$N_A = -\frac{D_{AB}C}{1 - y_A} \frac{dy_A}{dz} \tag{2.3-33}$$

Since N_A is independent of z, Eq. (2.3-33) may be integrated using the boundary conditions: $z = 0$, $y_A = 0$ and $z = L$, $y_A = p_A^*/p$ where p_A^* is the vapor pressure of pure A at the temperature of the experiment.

$$N_A = -\frac{D_{AB}C}{L} \ln \left(1 - \frac{p_A^*}{p}\right) \tag{2.3-34}$$

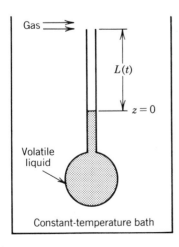

FIGURE 2.3-8   Stefan tube

This time-dependent flux can be related to the observed drop in the liquid level with time to obtain a working equation for the diffusivity:

$$L^2(t) = L_0^2 - D_{AB}\frac{2CM_A}{\rho^l}\ln\left(1 - \frac{p_A^*}{p}\right) \qquad (2.3\text{-}35)$$

D_{AB} is obtained from the slope of the experimental data plotted as L^2 versus t. The method, although limited to volatile liquids, is quite convenient since no chemical analyses need be conducted.

Loschmidt Cell

The Loschmidt cell was originally used for gases[2] but has also been adopted to measure average diffusivities in liquids.[74] As sketched in Fig. 2.3-9, the apparatus allows two long cylinders of fluid of different composition to be brought together so that at time zero, a step change in composition exists at $z = 0$. Diffusion is allowed to proceed for some time, the chambers are then separated, mixed, and the average composition in each measured. During the experiment great care must be exercised to prevent any effects of mixing or natural convection. The method only yields an average diffusion coefficient over the composition range if the diffusivity is composition dependent.

The experiment is actually one of free diffusion as described in Example 2.3-1. Under the assumptions that the solutions form thermodynamically ideal solutions, the combination of the total mass (mole) and species balances yields Fick's Second Law as the governing equation.

$$\frac{\partial C_A}{\partial t} = D_{AB}\frac{\partial^2 C_A}{\partial z^2} \qquad (2.3\text{-}36)$$

The experiment may be conducted and analyzed in two ways:[70] for short diffusion times no changes in composition at the ends of the two chambers have occurred so that the ends are effectively at $z = \pm\infty$, while for longer times composition changes may occur at $z = \pm L$ but the flux, of course, is zero at the ends.

The short time solution is obtained using the Boltzmann transformation on the independent variables $z/t^{1/2}$.

$$C_A = \tfrac{1}{2}(C_0 - C_1)\,\text{erf}\left(\frac{z}{2\sqrt{D_{AB}t}}\right) + \tfrac{1}{2}(C_0 + C_1)$$

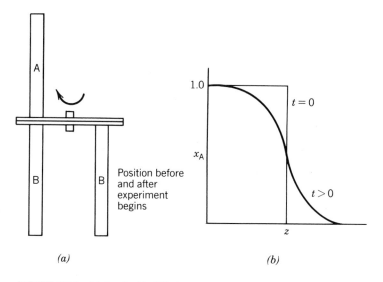

FIGURE 2.3-9 (a) Loschmidt diffusion apparatus, and (b) composition profiles

where C_0 and C_1 are the initial compositions in the two halves of the cell. The above expression is used to obtain the average composition in one half of the cell at a particular time from which the diffusivity can be calculated.

$$\overline{C}_A = C_0 - (C_0 - C_1) \frac{(D_{AB}t/\pi)^{1/2}}{L} \qquad (2.3\text{-}37)$$

For long diffusion times, where the concentration has reached the end of the chambers, the solution of Eq. (2.3-36) is best obtained by separation of variables. The result for the average concentration in the half cell is

$$\overline{C}_A = \frac{C_0 + C_1}{2} + \frac{4(C_0 - C_1)}{\pi^2} \sum_{n=0}^{\infty} \frac{1}{(2n+1)^2} \exp\left[-\frac{(2n+1)^2 \pi^2 D_{AB}t}{4L^2}\right] \qquad (2.3\text{-}38)$$

To obtain the diffusivity, a numerical solution is required in which enough terms of the infinite series are retained to provide the desired accuracy.

METHODS INVOLVING CONCENTRATION PROFILE MEASUREMENTS

If the diffusion coefficient is independent of composition as in gases or only an effective diffusion coefficient over a range of composition is desired, the above techniques are quite useful. If the variation of the diffusivity with composition is needed, then it is generally necessary to measure concentration profiles, although some investigators have used the Loschmidt cell method to measure average diffusivities in a series of experiments involving initial compositions that differ by only 10–20% in each experiment. For liquids, interferometric methods are commonly employed to determine the refractive index (hence the composition) variation with position. Cussler[4] critically reviews the various interferometers that can be employed. A microinterferometric method allowing use of very small samples, but having somewhat poorer accuracy, has also been used.[75,76] In all cases, the experiment performed is that of free diffusion in which two solutions of different composition are brought together at time zero and the concentration profile obtained at a series of later times. For metal systems a similar experiment is performed in which a diffusion couple consisting of two alloys of different composition are brought together and placed in a furnance at the desired temperature. After the desired time, the couple is removed and the composition profile determined most commonly by an electron microprobe, although careful sectioning and chemical analysis can be performed.

The analysis of these experiments follows a technique first introduced by Boltzmann.[2,77] Under the assumption of ideal solutions the diffusion equation with concentration dependent diffusivity is

$$\frac{\partial C_A}{\partial t} = \frac{\partial}{\partial z}\left[D_{AB} \frac{\partial C_A}{\partial z}\right] \qquad 2.3\text{-}39)$$

For short times, where diffusion effects have not reached the end of the cell, the Boltzmann transformation is used to yield an expression for the diffusivity at concentration C_A:

$$D_{AB}(C_A) = \frac{-\displaystyle\int_{C_1}^{C_A} z\, dC_A}{2t(dC_A/dz)} \qquad (2.3\text{-}40)$$

Figure 2.3-10 illustrates the integral and derivative that are to be computed from the experimental curve to calculate the diffusion coefficient. The $z = 0$ plane is known as the Matano interface[2,77,78] and must be located on the profile by applying the condition

$$\int_{C_0}^{C_1} z\, dC_A = 0 \qquad (2.3\text{-}41)$$

The $z = 0$ plane does not correspond to the average concentration $(C_0 + C_1)/2$ unless D_{AB} is independent of composition. Duda and Vrentas[6] have analyzed the more complex case of free diffusion, where there is a volume change of mixing, and have shown that neglecting volume changes may introduce significant errors in the diffusion coefficients calculated.

TRACER METHODS

Radioactive isotopes are often used in solid systems to determine either mutual or self-diffusion coefficients using the configurations sketched in Fig. 2.3-11. A known quantity q (in moles/area) of a tracer material

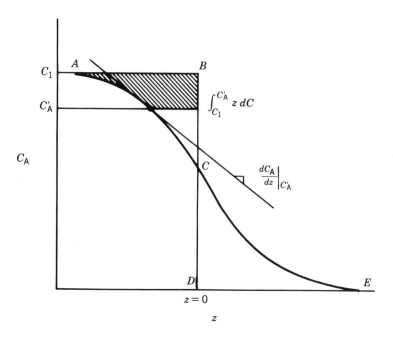

FIGURE 2.3-10 Matano–Boltzmann analysis of free-diffusion experiment. Matano plane ($z = 0$) Located by area ABC = area CDE.

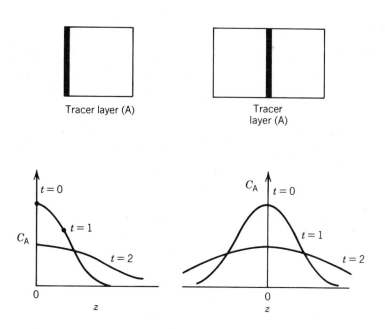

FIGURE 2.3-11 One- and two-sided tracer diffusion experiments and profiles.

is deposited as a thin layer on a surface or between two surfaces and the spread of this material is followed as a function of time. For example, a thin layer of radioactive iron can be deposited onto pure iron or onto another metal.

For the case of no volume change on mixing and constant diffusivity, the concentration profile of tracer A in the region $z > 0$ is governed by

$$\frac{\partial C_A}{\partial t} = D_{AB} \frac{\partial^2 C_A}{\partial z^2} \tag{2.3-42}$$

with auxiliary conditions:

1. $t = 0$, $C_A = 0$.
2. $z \to +\infty$, $C_A \to 0$.
3. $q = \int_0^\infty C_A dz$ (if the sandwich configuration were used, $q/2 = \int_0^\infty C_A dz$ since one-half the material would spread in each direction).

The solution of Eq. (2.3-42) can be obtained by the Laplace transform technique:

$$C_A(z, t) = \frac{q}{\sqrt{\pi D_{AB} t}} \exp\left(\frac{-z^2}{4 D_{AB} t}\right) \tag{2.3-43}$$

Plots of Eq. (2.3-43) for various times are also sketched in Fig. 2.3-11 and it can be seen to be one-half the Gaussian error distribution curve. If diffusion were allowed to occur in both the positive and negative z directions, the symmetrical profile would be described as

$$C_A(z, t) = \frac{q}{2\sqrt{\pi D_{AB} t}} \exp\left(\frac{-z^2}{4 D_{AB} t}\right) \tag{2.3-43a}$$

where q is the total amount deposited. Equation (2.3-43) is said to give the concentration distribution emanating from an instantaneous plane source at $z = 0$. The diffusion constant is determined from the slope of the line of the experimental data plotted as $\ln C_A$ versus z^2 for any particular diffusion time t. Jost[2] presents analogous solutions for the distributions emanating from line or point sources in cylindrical and spherical geometries.

2.3-4 Molecular Diffusion in Stagnant Phases and in Laminar Flow

In this section further illustration of the methods of solving diffusional mass transfer problems is provided by considering a number of classic problems that have been solved in the literature. The approach consists of combining the conservation equations and appropriate rate expressions. The results of examples such as these provide guidance for the consideration of more complex mass transfer situations such as in turbulent flow (Section 2.4) and in later chapters of this handbook.

STEADY TRANSPORT TO A SPHERE
Transport of solute from a fluid phase to a spherical or nearly spherical shape is important in a variety of separation operations such as liquid–liquid extraction, crystallization from solution, and ion exchange. The situation depicted in Fig. 2.3-12 assumes that there is no forced or natural convection in the fluid about the particle so that transport is governed entirely by molecular diffusion. A steady-state solution can be obtained for the case of a sphere of fixed radius with a constant concentration at the interface as well as in the bulk fluid. Such a model will be useful for crystallization from vapors and dilute solutions (slow-moving boundary) or for ion exchange with rapid irreversible reaction. Bankoff[79] has reviewed moving-boundary problems and Chapters 11 and 12 deal with adsorption and ion exchange.

A dilute solution of constant density and bulk composition w_A° is in contact with a sphere where the mass fraction is w_A^*. For dissolution or crystal growth w_A^* would be determined from solubility while for ion exchange it could be nearly zero at low saturations. The steady-state continuity and species balance equations, considering only radial variations, are

$$\frac{d}{dr}[r^2(n_A + n_B)] = 0 \tag{2.3-44}$$

$$\frac{d}{dr}[r^2 n_A] = 0 \tag{2.3-45}$$

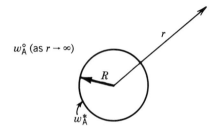

FIGURE 2.3-12 Transport between a sphere and a stagnant fluid.

where n_A and n_B are the mass fluxes of solute and solvent, respectively. The rate expression for the flux of A is

$$n_A = -D_{AB}\,\rho\,\frac{dw_A}{dr} + w_A(n_A + n_B) \tag{2.3-46}$$

Substitution of Eq. (2.3-46), integration, and use of the boundary conditions—$r = R$, $w_A = w_A^*$, $n_B = 0$ and $r \to \infty$, $w_A \to w_A^\circ$—result in the flux expressions

$$n_B = 0$$

$$n_A(r = R) = \frac{-D_{AB}\,\rho}{R}\ln\left(\frac{1 - w_A^*}{1 - w_A^\circ}\right) \tag{2.3-47}$$

Equation (2.3-47) can also be written in the form

$$n_A = \frac{(D_{AB}\,\rho/R)\,(w_A^* - w_A^\circ)}{[(1 - w_A^\circ) - (1 - w_A^*)]/\ln[(1 - w_A^\circ)/(1 - w_A^*)]}$$

The expression in the denominator is called the log mean of mass fraction of B, written $w_{B,lm}$. If w_A° and w_A^* are both small with respect to unity (dilute solution), $w_{B,lm} \approx 1.0$ and the flux is proportional to the composition difference,

$$n_A = \left(\frac{D_{AB}\,\rho}{R}\right)(w_A^* - w_A^\circ) \tag{2.3-48}$$

Equation (2.3-47), the pseudo-steady-state solution for the flux, could be used to predict the diffusion-controlled growth or dissolution rate of the crystal in a manner analogous to the Stefan problem solution. The result would indicate that the square of the particle radius varies linearly with time.

TRANSIENT DIFFUSION IN A STAGNANT PHASE
A number of problems such as the drying of porous solids of various shapes fall into this category. Solutions to such similar problems may be found in Crank, *Mathematics of Diffusion*,[70] and also in Carslaw and Jaeger, *Heat Conduction in Solids*,[80] since the unsteady-state diffusion equation often takes on the same form as that for heat conduction.

Consider the drying of a porous slab of material which is governed by the diffusion rate of vapor through the pores (Fig. 2.3-13). Assume that the slab is very large in two directions so that significant gradients only occur in the z direction. The initial concentration of vapor is C_A° within the solid and, at time zero, the slab is placed in contact with an atmosphere having a concentration C_A^*.

If the gas phase (e.g., water vapor and air) is considered to be ideal, the overall continuity equation in molar units reduces to a statement that the total molar flux ($N_A + N_B$) is independent of z. Since there is symmetry about the centerline of the slab, $N_A + N_B$ is zero at $z = 0$ and therefore is zero everywhere. The species continuity equation in this case reduces to

$$\frac{\partial C_A}{\partial t} = D_{eff}\,\frac{\partial^2 C_A}{\partial z^2} \tag{2.3-49}$$

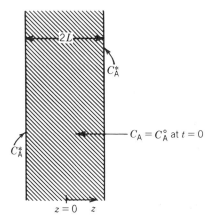

FIGURE 2.3-13   Transient diffusion in a porous slab.

where D_{eff} is written because the diffusion occurs in the pores of the solid. The appropriate boundary conditions are $z = 0$, $\partial C_A/\partial z = 0$; $z = L$, $C_A = C_A^*$; $t = 0$, $C_A = C_A^\circ$. The solution in dimensionless terms is readily found by separation of variables in terms of an infinite series.[70,78]

$$\Psi(\tau, z') = \frac{4}{\pi} \sum_{n=0}^{\infty} \frac{(-1)^n}{2n + 1} \cos\left[\frac{(2n + 1)\pi z'}{2}\right] \exp\left[-\left(\frac{(2n + 1)\pi}{2}\right)^2 \tau\right] \qquad (2.3\text{-}50)$$

where $\Psi = (C_A - C_A^*)/(C_A^\circ - C_A^*)$, $\tau = tD_{eff}/L^2$ and $z' = z/L$. The series solution is particularly convenient for long times because it converges rapidly in that instance. An alternative series suitable for short times obtained by Laplace transform techniques can be found in Crank.[70] The instantaneous drying rate is readily obtained from the above solution by computing $(-D_{eff} \partial C_A/\partial z)_L$. Solutions to the above problem and similar problems in spherical and cylindrical geometries were tabulated by Newman.[81] The solutions can also be obtained from the Gurney–Lurie heat conduction charts in Perry and Green[82] by replacing the dimensionless temperature with dimensionless concentration and the thermal diffusivity with D_{eff}.

A common alternative boundary condition to that of constant concentration at the outer surface is the flux boundary condition in which the flux of material out of the slab is taken as proportional to the difference in the concentration at the outer surface of the slab and that far into the surrounding medium:

$$-D_{eff} \left.\frac{\partial C_A}{\partial z}\right|_L \propto C_A(t, L) - C_A^\infty$$

Solutions in a variety of geometries with this boundary condition are also available.

Problems involving unsteady diffusion in more than one coordinate direction, such as a cylinder of finite length or a long slab of comparable width and depth dimensions, can usually be solved by separation of variables. For example, for a cylinder of radius R and length L the governing equation and boundary conditions for an ideal gas mixture and constant surface conditions would be

$$\frac{\partial C_A}{\partial t} = D_{eff} \left[\frac{\partial^2 C_A}{\partial z^2} + \frac{1}{r}\frac{\partial}{\partial r}\left(r\frac{\partial C_A}{\partial r}\right)\right]$$

with the following conditions:

1. $C_A(0, r, z) = C_A^*$.
2. $C_A(t, r, 0) = C_A^*$.
3. $C_A(t, r, L) = C_A^*$.
4. $C_A(t, R, z) = C_A^*$.
5. $\partial C_A/\partial z(t, 0, z) = 0$.

Again, Crank[70] and Carslaw and Jaeger[80] have tabulated many such solutions.

FILM THEORY OF MASS TRANSFER

In considering the transport of a species from a fluid in turbulent flow toward a solid surface, for example, an electrochemically active species to an electrode, Nernst assumed that the transport was governed by molecular diffusion through a stagnant film of fluid of thickness δ. This model, although having questionable physical relevance, is quite useful for correlating effects such as the influence of chemical reaction on mass transfer. A few simple examples of the use of film theory to describe mass transfer in the presence of chemical reaction are considered here.

Consider the situation depicted in Fig. 2.3-14 in which there is steady diffusional transport across a stagnant film of thickness δ in a binary or effectively binary system.

Case 1: No reaction. In the absence of chemical reaction, the steady-state one-dimensional species balances are

$$\frac{\partial N_A}{\partial z} = 0 \tag{2.3-51}$$

$$\frac{\partial N_B}{\partial z} = 0 \tag{2.3-51a}$$

so that N_A, N_B, and $N_A + N_B$ are independent of z. Substitution of $-D_{Am}C \, dx_A/dz + x_A(N_A + N_B)$ for N_A in Eq. (2.3-51) and integration using the boundary conditions $x_A = x_{A,1}$ at $z = 0$ and $x_A = x_{A,2}$ at $z = \delta$ yield the general solution,

$$N_A = \frac{N_A}{N_A + N_B} \frac{D_{AB}}{\delta} \ln \left(\frac{N_A/(N_A + N_B) - x_{A,2}}{N_A/(N_A + N_B) - x_{A,1}} \right) \tag{2.3-52}$$

D_{AB} represents the effective diffusivity of A in the mixture. Completion of the solution requires additional information on the flux ratio N_A/N_B.

The case of equimolar counterdiffusion ($N_A = -N_B$) is a good assumption in certain situations such as binary distillation of similar species. Equation (2.3-52) is indeterminate in this instance but Eq. (2.3-51) is readily solved.

$$N_A = \left(\frac{D_{AB}C}{\delta} \right) (x_{A,1} - x_{A,2}) = - \left(\frac{D_{AB}}{\delta} \right) (C_{A,2} - C_{A,1}) \tag{2.3-53}$$

The flux is linearly related to the composition difference and the diffusivity and inversely proportional to the hydrodynamically dependent film thickness. Equation (2.3-53) can also be employed in dilute solutions ($x_A \ll 1$) where the bulk flow term $x_A(N_A + N_B)$ is neglected with respect to the diffusive term.

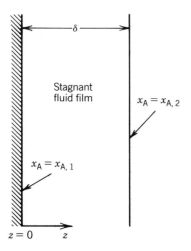

$z = 0$ z

FIGURE 2.3-14 Film theory geometry.

If the species B is stagnant, for example, a solvent that does not react at an electrode surface, then $N_B = 0$, and the flux is

$$N_A = \frac{D_{Am}C}{\delta} \ln \left(\frac{1 - x_{A,2}}{1 - x_{A,1}} \right)$$

$$N_A = -\frac{D_{Am}C}{\delta} \frac{x_{A,2} - x_{A,1}}{x_{B,lm}}$$

(2.3-54)

Case 2: Surface Reaction. Suppose A disappears by a first-order chemical reaction on the surface at $z = 0$ at a rate (mol/s · area) of $kC_{A,1}$. For the case of a dilute solution, where the bulk flow term can be neglected, Eq. (2.3-53) applies except that $C_{A,1}$ is not known but must be calculated by equating the flux of A at $z = 0$ to the negative of the reaction rate at the surface. Elimination of $C_{A,1}$ between these expressions allows the flux of A to be written

$$N_A = -\left[\frac{1}{\delta/D_{AB} + 1/k} \right] C_{A,2}$$

(2.3-55)

Equation (2.3-55) is in the form of a rate being governed by two resistances in series—diffusion and chemical reaction. If $1/k \ll \delta/D_{AB}$ (fast surface reaction), the rate is governed by diffusion, while if $1/k \gg \delta/D_{AB}$ (slow reaction rate), the rate is governed by chemical kinetics. This additivity of resistances is only obtained when linear expressions relate rates and driving forces and would not be obtained, for example, if the surface reaction kinetics were second order. More complex kinetic situations can be analyzed in a similar fashion where reaction stoichiometry at the surface provides information on the flux ratio of various species.

Case 3: Reaction Occurring Within Fluid Film. As a third example consider the situation when species A disappears by homogeneous reaction in the fluid film. Such a model has been used to predict the effect of chemical reaction on gas absorption rates (Chapter 6) or on carrier-facilitated transport in membranes (Chapter 19). For simplicity, assume the reaction to be first order and irreversible and the solution to be dilute so that bulk flow transport is negligible and the total molar concentration constant. The steady-state balance for A is obtained from simplification of Eq. (2.3-14):

$$\frac{dN_A}{dz} = kC_A$$

$$D_{Am} \frac{d^2 C_A}{dz^2} - kC_A = 0$$

(2.3-56)

For homogeneous reactions, the reaction kinetics enter into the balance equation while for heterogeneous reactions, the kinetics appear in the boundary conditions. The solution of Eq. (2.3-56) with the boundary conditions $C_A = C_{A,1}$ at $z = 0$ and $C_A = C_{A,2}$ at $z = \delta$ provides an expression for the flux of A into the film at $z = \delta$:

$$N_A = -D_{Am} \frac{dC_A}{dz}\bigg|_{\delta} = \frac{-\sqrt{kD_{Am}} \,(C_{A,2} - C_{A,1})}{\tanh \left(\delta\sqrt{k/D_{Am}} \right)}$$

(2.3-57)

Thus, for a film of thickness δ, the ratio of the flux with reaction to that without reaction is given by

$$\frac{N_A}{N_A^0} = \frac{\delta\sqrt{k/D_{Am}}}{\tanh \left(\delta\sqrt{k/D_{Am}} \right)}$$

(2.3-58)

and depends only on the parameter $\delta\sqrt{k/D_{Am}}$. As $k \to 0$, the ratio tends to unity, while for large values of the reaction rate constant, the transfer rate is enhanced owing to the steeper composition gradients near the gas-liquid interface. Figure 2.3-15 illustrates a few composition profiles for various values of $\delta\sqrt{k/D_{Am}}$.

DIFFUSION INTO A FLOWING LAMINAR FILM

Two situations involving transport of a species to a liquid in laminar flow can be modeled by the system in Fig. 2.3-16. One might represent absorption from a gas stream into liquid flowing over pieces of packing in a column. The other geometry could represent dissolution of a solid into a flowing liquid film or the transport from a membrane surface into a solution in laminar flow over the membrane. Both these simple systems illustrate the effect of velocity profiles or shear rates on mass transfer to a fluid.

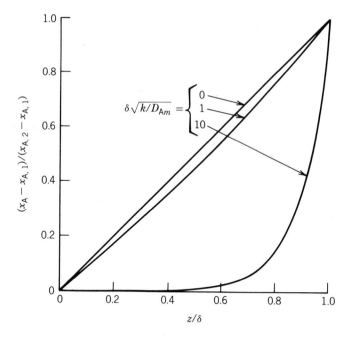

FIGURE 2.3-15 Composition profiles for diffusion and reaction in a fluid film

Consider the simple case in which a low rate of mass transfer normal to the surface does not influence the velocity profile ($v_y \approx 0$). The velocity profile $v_z(y)$ in the film is determined by solving the z component of the equation of motion for a Newtonian fluid in steady flow owing to the action of gravity. If the solid surface is inclined at an angle α with respect to the horizontal the momentum equation is

$$0 = \nu \frac{\partial^2 v_z}{\partial y^2} + g \sin \alpha$$

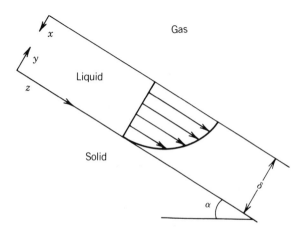

FIGURE 2.3-16 Mass transfer to a liquid in laminar flow on a solid surface.

The solution for the velocity profile is

$$v_z(y) = v_{max}\left[\frac{2y}{\delta} - \left(\frac{y}{\delta^2}\right)\right] \tag{2.3-59a}$$

or

$$v_z(x) = v_{max}\left[1 - \left(\frac{x}{\delta^2}\right)\right] \tag{2.3-59b}$$

where $v_{max} = g\delta^2 \sin \alpha/2\nu$ and the film thickness δ is $(3\nu Q/gW \sin \alpha)^{2/3}$. Q is the volumetric flow rate of the film and W is the film width. For a vertical surface the film becomes turbulent at a Reynolds number $(4Q/W\nu)$ of 1000–2000 but rippling or wave formation on the surface occurs at a Reynolds number of about 20.

Gas Absorption. For the situation of a pure gas A absorbing at the interface and diffusing into the moving liquid film under steady-state conditions, the material balance for A (in mass units) in the liquid film is

$$\frac{\partial n_{A,x}}{\partial x} + \frac{\partial n_{A,z}}{\partial z} = 0 \tag{2.3-60}$$

Substitution of the continuity equation for total mass

$$\frac{\partial(\rho v_x)}{\partial x} + \frac{\partial(\rho v_y)}{\partial y} = 0$$

into Eq. (2.3-60) yields

$$\frac{\partial j_{A,x}}{\partial x} + \rho v_x \frac{\partial w_A}{\partial x} + \frac{\partial j_{A,z}}{\partial z} + \rho v_z \frac{\partial w_A}{\partial z} = 0$$

This equation is usually further simplified by neglecting the convection term in the x direction and the diffusive term in the z direction. These simplifications appear reasonable for low mass transfer rates. The final equation to be solved under conditions of *constant* density and diffusivity is

$$D_{Am} \frac{\partial^2 \rho_A}{\partial x^2} = v_z(x) \frac{\partial \rho_A}{\partial x} \tag{2.3-61}$$

with the following boundary conditions:

1. $z = 0$, $\rho_A = \rho_A^\circ$
2. $x = 0$, $\rho_A = \rho_A^*$ (determined by the solubility of the gas in the liquid).
3. $x \to \infty$, $\rho_A = \rho_A^\circ$, where v_z is given by Equation (2.3-59b).

The third boundary condition, which is valid for short contact times, assumes that none of the diffusing solute has yet reached the wall so that the wall is effectively at infinity.

Johnstone and Pigford[83] obtained a series solution to the above problem using the complete velocity profile. For short contact times, the diffusing solute effectively encounters only the fluid moving at the surface velocity (v_{max}) since it has penetrated into the film only a short distance. Under these circumstances the solution for the composition profile is obtained in terms of the error function and the instantaneous flux into the liquid is given by

$$D_{Am} \rho \left.\frac{\partial w_A}{\partial x}\right|_{x=0} = (\rho_A^* - \rho_A^\circ)\left(\frac{D_{Am}}{\pi\Theta}\right)^{1/2}$$

The gas–liquid contact time z/v_{max} is denoted as Θ. The average flux over contact time $\Theta = L/v_{max}$ is

$$\bar{n}_A = \frac{1}{2}\left(\frac{D_{Am}}{\pi\Theta}\right)^{1/2}(\rho_A^* - \rho_A^\circ) \tag{2.3-62}$$

In this case of *transient* diffusion as compared to the steady-state film model, the mass transfer flux is proportional to the square root of the diffusivity rather than the first power.

The above expression can also apply to gas absorption in wetted-wall columns or into laminar liquid jets when there is only a small penetration of solute into the film or jet. In practice, Eq. (2.3-62) tends to underestimate the average flux because the liquid film tends to have ripples on the surface even under laminar flow conditions. This added interfacial area increases the total amount of transfer to the surface so that the observed average flux is greater than would be predicted based on a planar surface.[84,85]

Transfer at a Solid Surface. For the situation of mass transfer from a slightly soluble wall (e.g., benzoic acid in water) or from a membrane surface, there is a quite steep velocity gradient at the wall so it is not acceptable to approximate v_z as a constant. The mass conservation equation with constant density is more conveniently written in terms of the y coordinate as

$$D_{Am} \frac{\partial^2 \rho_A}{\partial y^2} = v_z(y) \frac{\partial \rho_A}{\partial z}$$

The solution to the corresponding heat transfer problem was obtained by Leveque[86] by assuming that the velocity profile near the wall was linear; that is, in Eq. (2.3-57a), $2y/\delta \gg y^2/\delta^2$ for small y/δ. Under this assumption the solution of the equation

$$\dot{\gamma} y \frac{\partial \rho_A}{\partial z} = D_{Am} \frac{\partial^2 \rho_A}{\partial y^2}$$

where $\dot{\gamma}$ is the velocity gradient at the wall, is conveniently carried out by the method of combination of variables, $y/z^{1/3}$. The average flux over a length L of surface is

$$\bar{n}_A = \frac{0.808 \, D_{Am}^{2/3}}{L^{1/3}} \gamma^{1/3}(\rho_A^* - \rho_A^\circ) \tag{2.3-63}$$

The factor 0.808 arises from the numerical computation of a definite integral in the composition profile expression. The above solution, which is valid only for short penetration distances where the velocity profile remains linear, predicts how the mass transfer rate is affected by the wall velocity gradient. Nusselt[87] provided a numerical solution of the above problem using the complete velocity profile expression, while Skelland[88] compares these results to experimental studies of mass transfer to laminar films. A number of similar problems are encountered when mass transfer in membrane systems is considered in Chapters 18–21.

BOUNDARY LAYER MASS TRANSFER

The concept of boundary layers in which velocity, temperature, or composition change rapidly with respect to position in the vicinity of a solid surface has provided useful solutions to a wide range of problems in hydrodynamics and heat and mass transfer. Schlichting[89] provides one of the most authoritative discussions of boundary layer theory in fluid mechanics and heat transfer. Skelland's text[88] has a particularly extensive discussion of mass transfer in boundary layers. By way of illustration consider the simplest case, that of developing velocity and composition profiles in a laminar boundary layer as a fluid approaches a thin plate (Fig. 2.3-17). Assume that the plate is made of a slightly soluble material A, so that the concentration in solution at the surface is constant and given by the equilibrium solubility. The distances over which velocity and composition change significantly are denoted by δ and δ_c, respectively.

By an order-of-magnitude analysis of the various terms in the equations of motion and species conservation equations, the governing equations, for a Newtonian fluid of constant density and viscosity, are.[88,89]

$$\frac{\partial v_x}{\partial x} + \frac{\partial v_y}{\partial y} = 0 \tag{2.3-64a}$$

$$v_x \frac{\partial v_x}{\partial x} + v_y \frac{\partial v_x}{\partial y} = \nu \frac{\partial^2 v_x}{\partial y^2} \tag{2.3-64b}$$

$$v_x \frac{\partial w_A}{\partial x} + v_y \frac{\partial w_Y}{\partial y} = D_{Am} \frac{\partial^2 w_A}{\partial y^2} \tag{2.3-64c}$$

Pohlhausen[90] solved the analogous heat transfer problem by a similarity transform and numerical solution of the ordinary differential equations. For low mass transfer rates and $Sc > 0.6$, the resulting expression for the local flux as a function of position from the leading edge is

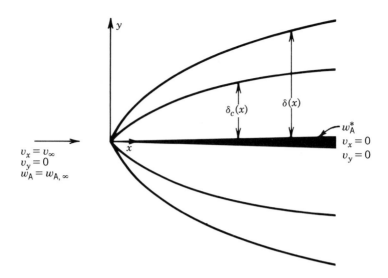

FIGURE 2.3-17 Hydrodynamic and composition boundary layers on a flat plate.

$$n_A = 0.332 \, \text{Re}_x^{1/2} \, \text{Sc}^{1/3} \left(\frac{D_{Am}}{x} \right) (\rho_A^* - \rho_A^\infty) \tag{2.3-65}$$

This equation applies only for laminar boundary layers that exist on a flat plate for $\text{Re}_x < 10^5 - 10^6$.

Of particular interest in Eq. (2.3-65), as well as in the earlier results of film theory and transient diffusion into a fluid, is the predicted dependence of the flux on geometry, hydrodynamics (Θ, $\dot{\gamma}$, or Re), physical properties (D_{Am} and Sc), and composition driving force. In all cases the flux varies linearly with the composition driving force but the dependence of flux on diffusivity ranges from the first to the one-half power. These simple models can be used to guide or interpret mass transfer rate observations in more complex situations.

Many additional studies have been conducted with the boundary layer model by taking into account the variation of physical properties with composition (or temperature) and by relaxing the assumption that $v_y = 0$ at $y = 0$ when mass transfer is occurring. Under conditions of high mass transfer rates one finds that mass transfer to the plate decreases the thickness of the mass transfer boundary layer while a mass flux away from the wall increases the boundary layer thickness[91,92]. The analogous problem of uniform flux at the plate has also been solved.[89] Skelland[88] describes a number of additional mass transfer boundary layer problems such as developing hydrodynamic and mass transfer profiles in the entrance region of parallel flat plates and round tubes.

2.4 MASS TRANSFER IN TURBULENT FLOW

A large number of practical mass transfer applications involve transport between phases in complex flow situations: the turbulent flow of a fluid in a turbular membrane, the counter current flow of gas and liquid in a packed column gas absorber, or the dispersion of droplets in an agitated vessel. In such situations our fundamental understanding, not only of the mass transport mechanisms but also of the fluid mechanics, is generally insufficient to obtain a theoretical solution. In these circumstances resort is made to conceptual models, experimental observations, and correlative methods to obtain estimates useful for engineering analysis and design.

The approach taken to describe mass transfer within a phase under these complex conditions is to employ a mass transfer coefficient that is defined as the ratio of the molar (or mass) flux of a species across a particular surface to the composition difference causing the transfer of mass. The coefficient for mass transfer between two locations 1 and 2 can be defined by any of the following equations:

$$N_A \equiv k_x(x_{A,1} - x_{A,2}) \equiv k_c(C_{A,1} - C_{A,2}) \equiv k_p(p_{A,1} - p_{A,2}) \tag{2.4-1a}$$

$$n_A \equiv k_W(w_{A,1} - w_{A,2}) \equiv k_\rho(\rho_{A,1} - \rho_{A,2}) \tag{2.4-1b}$$

The form involving partial pressures is only used for gases, while those involving mass concentrations or mass fractions are usually employed for liquids. When defined in this way the mass transfer coefficient includes effects of geometry, hydrodynamics, physical and transport properties of the fluid, and bulk flow contributions.

Although the definitions of mass transfer coefficients expressed in Eq. (2.4-1) are most commonly used, an alternative definition originally employed by Colburn and Drew[1] is useful under conditions of large convective flow in the direction of transport. The flux across a transfer surface at position 1 is given as

$$N_{A,1} - x_{A,1} \sum_i N_{i,1} \equiv k_x'(x_{A,1} - x_{A,2}) \tag{2.4-2}$$

Other driving force units and corresponding coefficients can be used in analogous definitions. Equation (2.4-2) explicitly allows for the bulk flow contribution to the molar flux relative to stationary coordinates. The coefficient k_x' or its analogues is thought to exhibit a somewhat less complex relationship to composition, flow conditions, and geometry. Further discussion of the relationships among those coefficients is provided in Section 2.4-1 in terms of particular models for turbulent mass transfer.

To provide some understanding of the origin and use of mass transfer coefficients some conceptual models for transfer under turbulent flow conditions are discussed. These models contain parameters whose values are dependent on understanding the hydrodynamics of a particular mass transfer situation. A brief description of theories of turbulence and of transport mechanisms in turbulent flows, which provide a basis for the analogies that have been developed to relate mass, heat, and momentum transfer under similar conditions, are presented. A number of the available engineering correlations for mass transfer coefficients in a variety of geometries and flow conditions are presented with some examples of their use. The development of such correlations, although guided by available theory, are founded on experimental observations.

2.4-1 Conceptual Models of Mass Transfer

The problem of predicting mass transfer rates between a turbulent fluid and a relevant surface can be illustrated by the data shown in Fig. 2.4-1. Fig. 2.4-1a shows the composition profiles that exist in a flat channel when water vapor is being transported from a water film on one wall to a $CaCl_2$ solution flowing down the other wall with turbulent flow of air in the channel. It is clear that at low Reynolds number significant composition gradients exist over the whole channel, while at higher Reynolds number most of the change in composition occurs near the bounding surfaces. If these observations are compared to those for transport in a turbulent liquid whose Schmidt number (ν/D_{AB}) is very much larger, it is seen that the composition is now confined to a very narrow region near the wall (Fig. 2.4-1b). Thus, it is clear that the

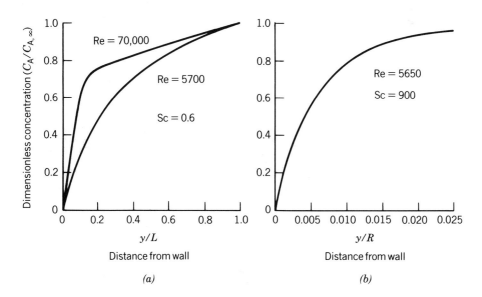

FIGURE 2.4-1 Effect of Reynolds and Schmidt numbers on mass transfer in turbulent flow. Adapted from Data of Refs. 2 and 3.

effects of both hydrodynamics and physical properties must be accounted for in describing turbulent mass transfer.

FILM THEORY

This earliest model for turbulent mass transport to a phase boundary, which is generally attributed to Nernst,[4] suggests that the entire resistance to mass transfer resides in a film of fixed thickness near the surface to which transfer occurs. Transport through the film is assumed to be governed by steady-state unidirectional molecular diffusion. Composition gradients are assumed to exist in the film only in the direction of transport. Such a model might be applied to transport from the wall of a conduit to a fluid in turbulent flow or to transport to a surface immersed in an agitated fluid. An examination of Fig. 2.4-1 would suggest that this model more closely approximates physical reality under conditions of high Reynolds and Schmidt numbers.

The solution of the film model is based on the steady-state species balances (mass or molar units) that were developed in Section 2.3 for a binary system.

$$N_A = \frac{N_A}{N_A + N_B} \frac{D_{AB}C}{\delta} \ln \left[\frac{N_A/(N_A + N_B) - x_{A,2}}{N_A/(N_A + N_B) - x_{A,1}} \right] \tag{2.4-3}$$

The influence of hydrodynamics and physical properties is contained within the fictitious film thickness δ; direct prediction of mass transfer rates is therefore not possible. The model is useful, however, in comparing two mass transfer situations for which δ is taken to be unchanged. For example, the effect of heterogeneous or homogeneous reaction on the transport rate can be predicted. Note that Eq. (2.4-3) could apply to the situation of species A diffusing through a mixture provided its transport could be characterized by an effective diffusivity of A in the mixture, D_{Am}, and N_B is interpreted as the sum of the fluxes of all other species.

The flux of A is not completely specified by Eq. (2.4-3), even if a film thickness were known, until a separate statement of the value of $N_A/(N_A + N_B)$ is provided. This auxiliary information might be obtained from solubility considerations—for example, the flux of air into water is zero during evaporation of H_2O— or possibly from the stoichiometry of a chemical reaction. Two common cases are often considered: equimolar counterdiffusion and the diffusion of a species A through an inert stagnant mixture (B).

For the case of $N_A = -N_B$, which is sometimes a good approximation during the distillation of a binary mixture, Eq. (2.4-3) reduces (via L'Hôpital's Rule) to

$$N_A = \frac{D_{AB}C}{\delta} (x_{A,1} - x_{A,2}) \tag{2.4-4}$$

The film theory expresses the mass transfer coefficients as

$$k_x = \frac{D_{AB}C}{\delta} \tag{2.4-5a}$$

or

$$k_c = \frac{D_{AB}}{\delta} \tag{2.4-5b}$$

The mass transfer coefficient k_x' is identical to k_x in this case since $N_A + N_B = 0$. The film theory suggests that the mass transfer coefficient is linearly related to the diffusivity—a result not generally supported by experimental observations of transport in turbulent fluids.

For the case of $N_B = 0$, Eq. (2.4-3) reduces to

$$N_A = \frac{D_{AB}C}{\delta} \ln \left[\frac{1 - x_{A,2}}{1 - x_{A,1}} \right] \tag{2.4-6}$$

and

$$k_x = \frac{D_{AB}C}{\delta} \frac{\ln \left[(1 - x_{A,1})/(1 - x_{A,2}) \right]}{(1 - x_{A,1}) - (1 - x_{A,2})} = \frac{D_{AB}C}{\delta} \frac{1}{x_{B,lm}} \tag{2.4-7a}$$

$$k_c = \frac{D_{AB}}{\delta} \frac{1}{C_{B,lm}} \tag{2.4-7b}$$

and

$$k_x' = \frac{k_x}{1 - x_{A,1}} \tag{2.4-7c}$$

Again, the mass transfer coefficient is predicted to be proportional to the diffusivity of A to the first power. If mass transfer is occurring in dilute solution, $x_B = 1 - x_A \approx 1.0$ and the mass transfer coefficient k_x is simply $D_{AB}C/\delta$ and $k'_x = k_x$ since the bulk flow term would be unimportant. The situation of A diffusing through stagnant B is not only a good approximation to many practical mass transfer situations but also represents the most common experimental situation used to measure mass transfer coefficients: for example, the sublimation of a pure solid or the evaporation of a pure liquid. For this reason, experimental data on mass transfer coefficients in concentrated gases or solutions are often correlated using the form suggested by Eq. (2.4-7a), that is, $k_x x_{B,lm}$, in the hope that this group would be less composition dependent. This form is justified for the film theory assumption of transport only by molecular diffusion, but it is not clear whether it can be justified for other models of turbulent transport. The investigation of Vivian and Behrmann[5] and of Yoshida and Hyodo[6] appear, however, to confirm this approach.

HIGH MASS TRANSFER FLUXES

The film theory model can be used to predict the effect of high mass transfer rates on the mass transfer coefficient[1] and a similar approach can also be employed to predict the influence of mass transfer on simultaneous heat transfer.[7,8] Since steady state, no axial velocity and composition gradients, and a fixed film thickness are assumed in the theory, only the influence of large convective flows in the direction of transport is predicted; allowance cannot be made for change in δ. The analysis is not restricted simply to bulk flow owing to molecular diffusion but can account for an additional imposed flow. For example, an inert gas might be forced through a porous wall from which water was evaporating into a turbulent stream.

Equation (2.4-3) is applicable regardless of the magnitude of the flux so it may be employed to demonstrate the effect of high mass flux. For any ratio of N_A/N_B as $N_A + N_B \to 0$, the mass transfer coefficients k_x and k'_x approach $D_{AB}C/\delta$. If $D_{AB}C/\delta$ is designated as k°_x, the coefficient under low transfer rates, an expression for the effect of total mass transfer flux on the ratio k'_x/k'_x can be derived.

$$\frac{k'_x}{k^\circ_x} = \frac{(N_A + N_B)/k^\circ_x}{\exp\left[(N_A + N_B)/k^\circ_x\right] - 1} \tag{2.4-8}$$

A plot of this equation is shown in Fig. 2.4-2 along with the predictions of some other theories to be discussed. For convection away from the phase boundary, $N_A + N_B > 0$ and the transfer coefficient is reduced, while for $N_A + N_B < 0$ the coefficient is enhanced. The total transfer flux of A is given by Eq. (2.4-2),

$$N_A = x_{A,1}(N_A + N_B) + k'_x(x_{A,1} - x_{A,2})$$

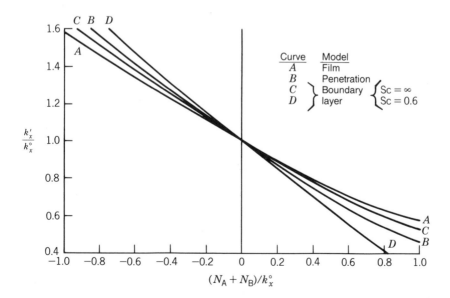

FIGURE 2.4-2 Variation of mass transfer coefficient with total flux. Reprinted with permission from R. B. Bird, W. E. Stewart, and E. N. Lightfoot, *Transport Phenomena*, p. 675, Wiley, New York, 1960.

so that the two effects of bulk flow on N_A tend to oppose one another. Extensive experimental investigations of this effect are not available although the work of Humphrey and van Ness[9] and Vivian and Behrman[5] support the use of Eq. (2.4-8).

A significant criticism of the above model is that the high mass transfer flux should influence the effective film thickness, yet no allowance has been made for this in the film theory model. A boundary layer model to be described shortly allows for the effect of the high flux on hydrodynamics.

EXAMPLE 2.4-1

Use the film theory approach to predict the effect of a simultaneous mass transfer flux on the heat transfer flux as in the condensation of a binary vapor on a cold surface. The bulk gas conditions are temperature T_2 and mole fraction $y_{A,2}$, while the conditions at the liquid surface are T_1 and $y_{A,1}$.

Solution. The film theory assumes steady state and that variations in concentration, temperature, and soon occur only in the z direction, which is taken to be normal to the transfer surface. The differential energy balance, written for negligible mechanical energy effects and the film theory assumptions, is

$$\frac{\partial}{\partial z}\left[\sum_i n_i h_i\right] + \frac{\partial q}{\partial z} = 0$$

where n_i and q are, respectively, the mass flux and heat conduction flux in the z direction and h_i is the partial mass enthalpy. The equation can also be written in molar units and integrated once to obtain

$$\sum_i N_i H_i + q = C_1$$

The heat conduction flux is given by $-k\,dT/dz$ and for ideal vapors partial molar enthalpies (H_i) depend only on temperature. Substitution and integration result in an expression for the ratio of the heat transfer coefficient with and without mass transfer:

$$\frac{h}{h°} = \frac{\sum (N_i C_{pi}/h°)}{\exp\left(\sum N_i C_{pi}/h°\right) - 1}$$

This equation, which is analogous in form to Eq. (2.4-8) for the effect of mass flux on mass transfer coefficient, is known as the Ackermann correction factor[7] for the effect of mass transfer on heat transfer. Note that the total flux of enthalpy to the vapor–liquid interface is the sum of the conductive flux and the enthalpy carried by the transferring molecules

$$q + \sum N_i H_i$$

Furthermore, if the molecules condense, the total enthalpy given up would include the latent heat. Sherwood et al.[8] describe how the above equations along with appropriate mass transfer expressions and phase equilibrium relationships are employed in the analysis of condensers and other processes involving simultaneous heat transfer and mass transfer.

TWO-FILM THEORY

In 1924 Lewis and Whitman[11] suggested that the film theory model could be applied to both the gas and liquid phases during gas absorption. This two-film theory has had extensive use in modeling steady-state transport between two phases. Transfer of species A occurring between a gas phase and liquid phase, each of which may be in turbulent flow, can be described by the individual rate expressions between the bulk of each phase and the interface,

$$N_A = k_y(y_{A,1} - y_{A,i}) \tag{2.4-9a}$$

$$N_A = k_x(x_{A,i} - x_{A,2}) \tag{2.4-9b}$$

where subscript i denotes an interfacial value. The usual assumption made is that interfacial equilibrium exists, that is, the kinetic process for transport or phase change at the interface is very rapid.‡ This appears to be satisfactory in gas–liquid and liquid–liquid systems unless surfactant materials accumulate at the interface and provide an additional resistance. If phase equilibrium does prevail, there is a relationship between the equilibrium compositions; for example, for a slightly soluble gas the compositions might be related by Henry's Law. In general, the equilibrium relationship would not be linear.

‡The concept of interfacial equilibrium is often not true for crystal growth. See Chapter 11.

Under the assumption of steady state and a linear equilibrium expression $y_{A,i} = mx_{A,i}$, the well-known resistance in series model can be derived.

$$N_A = \left[\frac{1}{1/k_x + 1/mk_y} \right] \left(\frac{y_{A,1}}{m} - x_{A,2} \right) \qquad (2.4\text{-}10a)$$

$$N_A = \left[\frac{1}{m/k_x + 1/k_y} \right] (y_{A,1} - mx_{A,2}) \qquad (2.4\text{-}10b)$$

The expressions in brackets are known as the overall coefficients for mass transfer, based on overall gas and liquid composition driving forces, and are analogous to overall heat transfer coefficients. The quantity $y_{A,1}/m$ can be interpreted as the liquid composition that would be in equilibrium with the bulk gas composition and $(y_{A,1}/m - x_{A,2})$ can be interpreted as the appropriate overall driving force.

The additivity of the individual resistances is dependent on the linearity of the flux expressions and of the equilibrium relationship. For nonlinear equilibrium relationships Eq. (2.4-10a) and (2.4-10b) can still be used provided m is recognized to be a function of the interfacial composition. Overall coefficients are often employed in the analysis of fluid–fluid mass transfer operations despite their complex dependence on the hydrodynamics, geometry and compositions of the two phases. In some instances the overall coefficients can be predicted from correlations for the individual coefficients for each phase provided the conditions in the apparatus are comparable to those for which the correlation was developed.

Equations (2.4-10a) and (2.4-10b) readily allow demonstration of the controlling resistance concept. For example, if $k_x \gg mk_y$, the transport rate reduces to

$$N_A = k_y(y_{A,1} - mx_{A,2})$$

Because of negligible liquid-side resistance the flux is governed by the gas phase transport. Note that $mx_{A,2} = y_{A,i}$, which implies that the interfacial gas-phase composition is in equilibrium with the *bulk* liquid composition. Even if the mass transfer coefficients in the two phases are comparable, Eqs. (2.4-10a) and (2.4-10b) show that phase equilibrium considerations can cause one or the other phase to "control." For a very soluble gas (m is small), the gas phase may control while for a sparingly soluble gas such as O_2 in water the liquid phase transport generally will govern.

The significant contribution of the two-film theory actually lies in the concept of the proper combination of rate processes occurring in series rather than the use of the film theory concept. This idea is often extended to mass transfer rates in combination with other kinetic processes.

Although the film theory is simple, conceptually useful, and successful in demonstrating effects of reaction on mass transfer[12] and the effects of mass flux on heat transfer rates,[8] it is based on physically unrealistic assumptions. Additional models that relax some of the film theory assumptions have therefore been developed.

PENETRATION THEORY
In 1935 Higbie[13] suggested that in mass transfer operations, such as gas absorption, the process consists of successive contacts between the fluid phases followed by mixing within the phases between contacts. Examples illustrated in Fig. 2.4-3 include the following. An absorbing liquid flows over a piece of packing

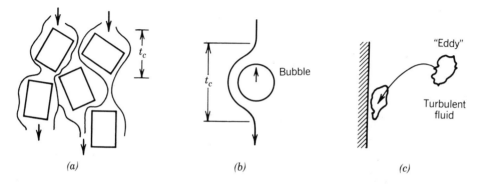

(a)   *(b)*   *(c)*

FIGURE 2.4-3 Penetration theory situations (*a*) flow of liquid over packing, (*b*) rise of bubble through a liquid, and (*c*) eddy residing at an interface.

in a column, it contacts the gas phase for some exposure time before becoming remixed. As a bubble rises through a liquid, the effective contact time of liquid flowing over the bubble is approximately the time required for the bubble to rise a distance equal to its diameter. Eddies in a turbulent fluid may reside at an interface for some exposure time prior to being mixed into the bulk stream. In all cases, mass transport occurs under transient conditions rather than steady state as assumed in the film theory.

In its simplest form, the penetration theory assumes that a fluid of initial composition x_A^o is brought into contact with an interface at a fixed composition $x_{A,1}$ for a time t. For short contact times the composition far from the interface ($z \rightarrow \infty$) remains at x_A^o. If bulk flow is neglected (dilute solution or low transfer rates), solution of the unsteady-state diffusion equation provides an expression for the average mass transfer flux and coefficient for a contact time Θ.

$$\overline{N}_A = 2 \left(\frac{D_{AB}}{\pi\Theta}\right)^{1/2} C(x_{A,1} - x_{A,2}) \tag{2.4-11}$$

$$k_c^o C = k_x^o = 2C\left(\frac{D_{AB}}{\pi\Theta}\right)^{1/2} \tag{2.4-12}$$

As in the film theory model the mass transfer coefficient depends on a parameter (the contact time) not readily predicted a priori, although in certain situations, such as a bubble rising through a liquid, reasonable estimates of the contact time might be made. In general, the exposure time is dependent on the hydrodynamics existing in the phase during transfer.

One of the most significant contributions of the penetration theory is the prediction that the mass transfer coefficient varies as $D_{AB}^{0.5}$. As will be seen in Section 2.4-3, experimental mass transfer coefficients generally are correlated with an exponent on D_{AB} ranging from $\frac{1}{2}$ to $\frac{2}{3}$. The penetration theory model has also been successfully used to predict the effect of simultaneous chemical reaction on mass transfer in gas absorption and in carrier-facilitated membranes.[12] Stewart[14] solved the penetration theory model by taking into account bulk flow at the interface. The results, as in the film theory case, are conveniently expressed as the dependence of the ratio k_x'/k_x^o on the dimensionless total flux, $(N_A + N_B)k_x^o$. This curve is also shown in Fig. 2.4-2 and generally predicts greater effects of convection than does the film theory.

EXAMPLE 2.4-2
The rise velocity in water of an air bubble 0.004 m in diameter is about 0.2 m/s. Estimate the liquid-side mass transfer coefficient k_c for oxygen transfer at 25°C.

Solution   Use Higbie'e Theory, Eq. (2.4-12), with Θ approximated as bubble diameter/rise velocity. The diffusivity is 2.1×10^{-9} m²/s (Table 2.3-6)

$$k_c^o = 1.8 \times 10^{-4} \text{ m/s} \; (5.9 \times 10^{-4} \text{ ft/s})$$

SURFACE RENEWAL THEORY
Further relaxation of the assumptions of the film and penetration theories was suggested by Danckwerts[15] who viewed the process as one of transient one-dimensional diffusion to packets or elements of fluid that reside at the phase interface for varying periods of time. Therefore, the model is that of the penetration theory with a distribution of contact times. The surface age distribution $\phi(t)$ is defined such that $\phi(t)dt$ is the fraction of surface that has resided at the interface for a time between t and $t + dt$. The mass transfer flux for the entire surface is obtained by integration of the instantaneous flux over all exposure times:

$$N_A = \int_0^\infty N_A(t)\phi(t)dt \tag{2.4-13}$$

where penetration theory provides an expression for $N_A(t)$.

Danckwerts assumed that the chance of a particular surface element being replaced is independent of its age, that is,

$$\phi(t) = \overline{S}e^{-\overline{S}t}$$

where $\overline{S}$ is the fractional rate of surface renewal (reciprocal of the average residence time of an element on the surface). The only known experimental confirmation of the above distribution is the gas absorption measurements of Lamb et al.[16] The surface renewal theory expression for the mass transfer coefficient in terms of the diffusivity and the surface renewal is

$$k_x^o = C(\overline{S}D_{AB})^{1/2} \tag{2.4-14}$$

Presumably, increasing the degree of turbulence of the fluid phase will increase the surface renewal rate $\overline{S}$, but no prediction of $\overline{S}$ from other characteristics of the hydrodynamics is possible at present.

EXTENSIONS OF FILM AND PENETRATION THEORIES
A number of extensions of these basic theories have been proposed in the literature. The Toor–Marchello[17] Film–Penetration Model views transport as being governed by a film of fixed thickness, but it allows unsteady transport to occur from eddies that penetrate from the bulk phase to the surface and reside there for varying contact times. A number of other variations on this theme of periodic penetration of eddies to the phase interface are available and all probably represent a more realistic view of the behavior of turbulent fluid in contact with a second phase. Nevertheless, these models all contain parameters (film thicknesses, renewal rates, etc.) that are not presently predictable from the fluid mechanics of the situation. All the models have a very useful conceptual role and often can allow the prediction of the influence of chemical reaction on the transport rate. Scriven[18] has reviewed many of these models and suggested some future directions in modeling, particularly taking into account the influence of convection and interfacial motion. Lightfoot and coworkers[19, 20] have proposed a surface-stretch model applicable to liquid–liquid mass transfer in situations where drops are oscillating.

BOUNDARY LAYER MODELS
The models discussed above, whether transient or steady state, do not permit variations in composition in more than one direction nor do they explicitly consider the effects of fluid velocity gradients. There are a number of situations, however, in which composition and velocity vary both parallel and normal to a transfer surface. Such situations often result when transport occurs from a solid surface to a fluid moving past the surface. Examples include transfer from the wall of a circular or flat-walled channel to a flowing fluid in a membrane separation device and dissolution or crystallization of solids in an agitated liquid. Although the bulk of the fluid stream may be in turbulent flow, a laminar boundary layer can be assumed in which the mass transfer resistance resides in a narrow region near the transfer surface where compositions and velocities are varying strongly with position. Numerical and in some cases anlytical results can be obtained which suggest how the mass transfer coefficient might vary with relevant hydrodynamics, physical property groups, and position.

The simplest situation, that of the transport from a flat plate at zero incidence to a flowing stream, was considered in Section 2.3. The solution to the problem of mass transfer with a constant surface composition based on Pohlhausen's heat transfer solution can be written in the form‡

$$k_\rho^\circ(x) \equiv \frac{n_A(x)}{\rho_{A,w} - \rho_{A,\infty}}$$

$$= \frac{D_{AB}}{x} 0.332 \, (\mathrm{Re}_x)^{1/2} \, (\mathrm{Sc})^{1/3}$$

The average mass transfer coefficient over a length L of the plate is given by

$$\overline{k}_\rho^\circ = \frac{D_{AB}}{L} (0.664) \, (\mathrm{Re}_L)^{1/2} \, (\mathrm{Sc})^{1/3} \tag{2.4-15}$$

and the average Sherwood number would be

$$\overline{\mathrm{Sh}} \equiv \frac{\overline{k}_\rho^\circ L}{D_{AB}} = 0.664 \, (\mathrm{Re}_L)^{1/2} \, (\mathrm{Sc})^{1/3} \tag{2.4-16}$$

This result suggests that the mass transfer coefficient varies as $D_{AB}^{2/3}$, a result intermediate to those of the film theory and the penetration theory. In fact, the $\frac{1}{3}$ power dependence of the Sherwood number on the Schmidt number is observed in a number of correlations for mass transfer in turbulent flow in conduits and for flow about submerged objects. (See Section 2.4-3.)

Many numerical and series solutions for the laminar boundary layer model of mass transfer are available for situations such as flow in conduits under conditions of fully developed or developing concentration or velocity profiles. Skelland[21] provides a particularly good summary of these results. The laminar boundary layer model has been extended to predict the effects of high mass transfer flux on the mass transfer coefficient from a flat plate. The results of this work are shown in Fig. 2.4-2 and, in contrast to the other theories, indicate a Schmidt number dependence of the correction factor.

‡Solution of boundary layer problems is usually most easily accomplished using mass units because the momentum equation contains the mass average velocity.

2.4-2 Turbulent Transport

Many texts and monographs are devoted to careful discussions of turbulent flow fields,[22-24] but only those aspects of turbulent flow that bear on the transport of particular species are discussed here. Davies' book[24] is a particularly good presentation of heat, mass, and momentum transport in turbulent fluids. The turbulent motion of a fluid is characterized by relatively rapid, irregular changes in velocity at any one point in the fluid and by limited correlation between the velocities at different spatial positions at any time. The random nature of the turbulent fluctuations imposed on the flow requires description using various statistical measures. Two characteristics of the flow field are the "intensity" and the "scale" of turbulence. The intensity refers to the magnitude of the velocity changes at any particular point with respect to the mean velocity; the scale of turbulence refers to the correlation between velocities measured some distance apart or to the correlation between velocities at the same point at two different times.

When the flow of a fluid becomes turbulent, momentum (and mass) transport is visualized to occur by the motion of eddies of various sizes and speeds continuously being formed, moving into other fluid elements and dissipating. The scale of turbulence referred to above can be roughly thought of as a measure of the size of the eddies, since velocities at two points located within the same eddy might be expected to have a high degree of correlation. Complete characterization of eddy motion, of course, is not possible so that eddies are best viewed as conceptual aids in understanding transport in turbulent flow.

EDDY VISCOSITY AND EDDY DIFFUSIVITY

In turbulent flow, momentum transfer is often characterized by an eddy viscosity E_V defined by analogy with Newton's Law of Viscosity. The time averaged momentum flux (shear stress) in the y direction owing to the gradient of the time averaged velocity in the z direction is given by

$$\bar{\tau}_{yz} = -(E_V + \nu)\,\rho\,\frac{d\bar{v}_z}{dy} \tag{2.4-17}$$

Equation (2.4-17) explicitly assumes that the contributions from the eddy viscosity and the kinematic viscosity are additive. While the kinematic viscosity, which characterizes molecular transport of momentum, is a physical property of the fluid, the eddy viscosity, which represents momentum transfer by eddy motion, depends on geometry, Reynolds number, and physical properties. In general, the eddy viscosity is very much larger than the kinematic viscosity except near a solid surface where turbulent flow is damped out. For example, Fig. 2.4-4 presents the calculated ratio of E_V/ν as a function of position and Reynolds number for turbulent flow in round tubes.[25] Such measurements are obtained by relating the time averaged shear stress to the observed friction losses and measuring the average velocities as a function of position to obtain the velocity gradient. Note that these data are not for the region very near the wall of the tube.

The eddy diffusion coefficient for mass transfer‡ is defined in a similar fashion to the eddy viscosity:

$$\bar{j}_{A,y} = -(E_D + D_{AB})\,\frac{d\bar{\rho}_A}{dy} \tag{2.4-18}$$

The transport mechanisms of molecular diffusion and mass carried by eddy motion are again assumed additive although the contribution of the molecular diffusivity term is quite small except in the region near a wall where eddy motion is limited. The eddy diffusivity is directly applicable to problems such as the dispersion of particles or species (pollutants) from a source into a homogeneously turbulent air stream in which there is little shear stress. The theories developed by Taylor,[26] which have been confirmed by a number of experimental investigations, can describe these phenomena. Of more interest in chemical engineering applications is mass transfer from a turbulent fluid to a surface or an interface. In this instance, turbulent motion may be damped out as the interface is approached and the contributions of both molecular and eddy diffusion processes must be considered. To accomplish this, some description of the velocity profile as the interface is approached must be available.

UNIVERSAL VELOCITY DISTRIBUTION

The Universal Velocity Distribution Law refers to the correlation of many sets of experimental measurements of the time averaged velocity profile near a wall during turbulent flow in smooth circular tubes or ducts.[27] These data are normally represented in terms of a dimensionless velocity u^+ and distance from the wall y^+:

$$u^+ = \frac{\bar{v}}{(\tau_W/\rho)^{1/2}} = \left(\frac{\bar{v}}{\bar{v}_{avg}}\right)\left(\frac{2}{f}\right)^{1/2} \tag{2.4-19}$$

$$y^+ = \left(\frac{y\bar{v}_{avg}}{\nu}\right)\left(\frac{f}{2}\right)^{1/2} \tag{2.4-20}$$

‡An eddy thermal diffusivity E_H can be defined by the equation $\bar{q}_y/\rho C_p = -(E_H + \alpha)\,dT/dy$.

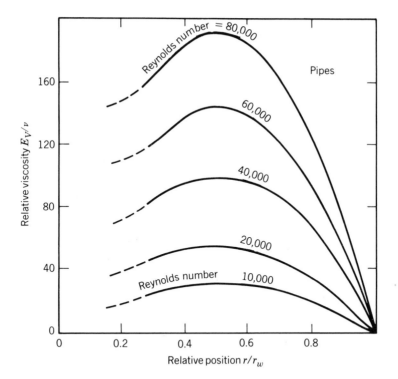

FIGURE 2.4-4 Ratio of eddy to molecular viscosity for the turbulent flow of gases and liquids in pipes. Reprinted with permission from T. K. Sherwood, R. L. Pigford, and C. R. Wilke, *Mass Transfer*, McGraw-Hill, New York, 1975.

where $\bar{v}$ is the time averaged velocity at a distance y from the wall, f is the Fanning friction factor, and $\bar{v}_{avg}$ is the average velocity in the conduit. Plotted in this way the data are well represented by a single curve up to Reynolds numbers of about 100,000, although it must be pointed out that data in the region $y^+ < 5$ are limited because of the very great difficulty in obtaining accurate measurements near the wall. Unfortunately, this is just the region where the dominant resistance to mass transfer occurs, at least for liquids.

Von Karman[28] first suggested that the profile might be represented by three functions:

$$\begin{array}{ll} u^+ = y^+ & y^+ < 5 \\ u^+ = -3.05 + 5.0 \ln y^+ & 5 < y^+ < 30 \\ u^+ = 5.5 + 2.5 \ln y^+ & 30 < y^+ \end{array} \qquad (2.4\text{-}21)$$

Although this representation is empirical, it reasonably represents the data and the analytical expressions are convenient for further mathematical manipulation. Furthermore, the statement that $u^+ = y^+$ near the wall corresponds to laminar flow where the molecular viscosity totally dominates. This has led to the concept that the flow field is made up of three regions: a laminar sublayer, a transition region, and a turbulent core. These regions correspond to systems in which molecular viscosity dominates, where molecular and eddy viscosity contributions are comparable, and where eddy transport dominates. However, experimental work using optical techniques suggests that even very close to the wall there persist significant velocity fluctuations which are not easily characterized.[29,30]

The universal velocity distribution or any other measured or assumed velocity profile can be used to obtain the eddy viscosity from Eq. (2.4-17). If various assumptions regarding the relationship between E_D and E_V and the relative importance of molecular and eddy transport processes are made, analogies between mass transfer and momentum transfer can be obtained. As more sophisticated information becomes available on the nature of the flows near a phase boundary, refined theories can be developed.

TRANSPORT ANALOGIES

A fruitful approach to characterizing mass transfer from turbulent fluids to surfaces is the comparison of momentum transfer and heat transfer under similar conditions.[31-34] These analogies are developed on two levels. By examining (in dimensionless form) the relevant conservation equations, flux expressions, and boundary conditions for a particular situation, one can demonstrate an expected analogy among the various transport coefficients. In other instances particular assumptions are made relating eddy diffusivities and eddy viscosities which allow the actual form of the analogy to be expressed.

The time averaged equations of mass, momentum, energy, and species conservation can be written in dimensionless form for a fluid in turbulent flow past a surface. If (1) radiant energy and chemical reaction are not present, (2) viscous dissipation is negligible, (3) physical properties are independent of temperature and composition, (4) the effect of mass transfer on velocity profiles is neglected, and (5) the boundary conditions are compatible, then dimensionless local heat and mass transfer coefficients can be shown to be described by equations of the form:

$$\mathrm{Nu}_x = h(\mathrm{Re}, \mathrm{Pr}) \tag{2.4-22a}$$

$$\mathrm{Sh}_x = g(\mathrm{Re}, \mathrm{Sc}) \tag{2.4-22b}$$

Furthermore, if the ratio E_D/E_V varies with Sc, Re, and distance from the wall in the same way as E_H/E_V varies with Re, Pr, and position, then a complete analogy exists between heat and mass transfer.

When the analogy holds, existing correlations for heat transfer can be used to obtain corresponding correlations for mass transfer by replacing Pr by Sc and vice versa. There are insufficient experimental data on eddy transport coefficients to establish the validity of the assumptions made to obtain the analogies. Early investigators often simply assumed $E_H = E_D = E_V$, which is certainly not generally true although the ratios of these quantities may not differ greatly from unity in many cases. The assumption of $E_H = E_D$ is perhaps somewhat more justified but can be in serious error if one compared fluids of greatly different Sc and Pr numbers, for example, air and a liquid metal. Nevertheless, as will be described shortly, heat and mass transfer analogies have been very useful and successful.

The vast majority of studies of turbulent transport have dealt with momentum transport, so that an analogy between heat or mass transfer and momentum transport is highly desirable. Rather more stringent assumptions are required to demonstrate that such an analogy is to be expected. Based on experimental observations such an analogy appears to hold under high Reynolds number conditions, but it is not as successful as those relating heat transfer and mass transfer.

A number of investigators have attempted to establish the form of the function $g(\mathrm{Re}, \mathrm{Sc})$ by simplifying the conservation equations and by making specific assumptions regarding the relationships among the eddy transport coefficients. Most of the analogies to date have dealt with a region relatively close to the transfer surface and have made the assumption that the average mass, heat, and momentum fluxes normal to the surface are independent of distance from the wall:

$$\frac{\partial \overline{j}_A}{\partial y} \simeq 0 \tag{2.4-23a}$$

$$\frac{\partial \overline{q}}{\partial y} \simeq 0 \tag{2.4-23b}$$

$$\frac{\partial \overline{\tau}}{\partial y} \simeq 0 \tag{2.4-23c}$$

This implies that the convective terms in the axial direction have been neglected in the species and energy equations and that the axial pressure drop and gravity terms have been neglected in the momentum equation. A number of the analogies between mass transfer and momentum transport can then be obtained by an analysis first described by Sherwood et al.[32]

The average mass flux and momentum flux *to* the wall are given by

$$\overline{\tau}_W = (E_V + \nu)\rho \frac{d\overline{v}_x}{dy} \tag{2.4-24a}$$

$$\overline{j}_{A,W} = (E_D + D_{Am}) \frac{d\overline{\rho}_A}{dy} \tag{2.4-24b}$$

where y is the distance *from* the wall. The above equations can be rearranged, put in dimensionless form, and integrated to yield

$$\left(\frac{2}{f}\right)^{1/2} = \int_0^{y_i^+} \frac{dy^+}{E_V/\nu + 1} \tag{2.4-25a}$$

$$\frac{\overline{v}_{\mathrm{avg}}}{k_\rho^\circ} = \left(\frac{2}{f}\right)^{1/2} \int_0^{y_0^+} \frac{dy^+}{E_D/\nu + 1} \tag{24.25b}$$

where y_1^+ is the dimensionless distance from the wall where the average composition and velocity are attained. To go further, specific assumptions about the eddy transport coefficients must be made.

The Reynolds analogy assumes that $E_V = E_D$ and neglects the molecular transport terms; that is, it assumes that the transport resistances reside wholly in the turbulent region. The result of combining Eqs. (2.4-25a) and (2.4-25b) is

$$\frac{\bar{v}_{avg}}{k_\rho^\circ} = \frac{2}{f}$$

or

$$\text{Sh} = \left(\frac{f}{2}\right)(\text{Re Sc})^{-1} \tag{2.4-26}$$

The quantity $k_\rho/\bar{v}_{avg}$ is known as the Stanton number. This analogy appears to hold only for Sc ≈ 1.0 and when only skin frictional effects are present.

The analysis can be improved by breaking the integrals in Eq. (2.4-25a) and (2.4-25b) into two parts corresponding to a turbulent core region (y_2^+ to y_1^+), where only eddy transport is important, and a wall region (0 to y_2^+) where both eddy and molecular transport are of importance. Furthermore, most authors assume $E_D \approx E_V$ with the result

$$\frac{1}{\text{St}} = \frac{2}{f} + \left(\frac{2}{f}\right)^{1/2} \int^{y_2^+} \left(\frac{1}{E_V/\nu + 1/\text{Sc}} - \frac{1}{E_V/\nu + 1}\right) dy^+$$

Evaluation of the integral requires that the eddy viscosity be available as a function of distance from the wall. For any particular assumed relationship for E_V, the resulting value of the integral is a function only of the Schmidt number.

Many different analogies have been proposed which make different assumptions about the relationship between E_V and position. Von Karman[28] employed the universal velocity profile with y_2^+ as 30 to obtain the relationship.

$$\frac{1}{\text{St}} = \frac{2}{f} + 5\left(\frac{2}{f}\right)^{1/2}\left[(\text{Sc} - 1) + \ln\left(\frac{1 + 5(\text{Sc} - 1)}{6}\right)\right] \tag{2.4-27}$$

Sherwood[35] modified Von Karman's analysis to retain the ratio E_D/E_V rather than assuming it to be unity. By using experimental results that indicated E_D/E_V to be about 1.6, he was able to correlate data for the evaporation of nine different liquids into air in a wetted-wall column. The Von Karman analogy is not as successful at high Schmidt numbers where most of the resistance is concentrated near the wall.

Both Diessler[36] and Lin et al.[3] abandoned the universal velocity profile in the region very near the wall and proposed essentially empirical expressions for E_V or E_D to account for some turbulence in this region. Both expressions provide reasonable predictions to quite high Schmidt numbers. Skelland[33] carefully compares these expressions with a relationship attributed to Gowariker and Garner who abandoned the concept of different layers or zones and proposed continuous expressions for E_V and E_D. Other workers[37-39] have proposed improvements in deriving the analogies by including more zones, entrance region effects, or other assumptions about the fluid mechanics near the wall. Further improvements will await experimental and theoretical advances in our understanding of the behavior of moving fluids very near a wall.

One of the most extensively used and successful analogies—the j factor of Chilton and Colburn[40]—was proposed as an empirical relationship based on available experimental data at that time:

$$j_D \equiv \frac{\text{Sh}}{\text{Re Sc}^{1/3}} = \frac{f}{2} \tag{2.4-28}$$

Since the j-factor correlation is based on experimental results, most of which were obtained under experimental conditions of a single species diffusing in a stagnant solvent or inert gas, the Sherwood number is best defined as

$$\text{Sh} \equiv \frac{k_\rho\, w_{\text{B,lm}} L}{D_{\text{A}m}} = \frac{k_C\, x_{\text{B,lm}} L}{D_{\text{A}m}} = \frac{k_x\, C\, x_{\text{B,lm}} L}{D_{\text{A}M}} \tag{2.4-29}$$

According to the film theory, the coefficients $k_\rho w_{\text{B,lm}}$ and so on represent the mass transfer coefficient at low mass transfer rates (k_ρ°) which should be less sensitive to the composition of the fluid. Vieth et al[41] later demonstrated that this analogy can be obtained by making a specific assumption relating E_D/ν to the distance from the wall and f. In any event, the j-factor correlations are probably the most successful approach to correlating mass transfer data with momentum and heat transfer results. The *j*-factor correlation adequately represents mass transfer data in tubes over a Schmidt number range of 0.6 to > 1000.

All the developments presented above can be used for heat transfer rather than mass transfer by replacing Sh by Nu and Sc by Pr. In fact, the more limited analogy relating heat transfer and mass transfer is usually more applicable than the relationships involving momentum transfer. Since heat transfer results are generally obtained under conditions of no mass transfer, the analogy is only expected to relate to the mass transfer coefficient under low mass transfer conditions. Also, variations of physical properties with temperature and composition must be neglected for the analogies to be expected. If the analogy is to be extended to momentum transfer, f must be interpreted as the "skin" friction factor; frictional losses owing to "form drag" about solid objects are not to be included.

2.4-3 Engineering Correlations

Although the analogy concept has been successful in predicting mass transfer coefficients in certain turbulent flow situations, for most mass transfer situations encountered in practice, correlations based primarily on experimental observations are required to obtain a reasonable estimate of a mass transfer coefficient. The use of such empirical or semiempirical correlations is dangerous if one attempts to extrapolate beyond the region where experimental support is available, yet such correlations represent the only practical approach when an engineering estimate is to be made. Only a few of the correlations available for common mass transfer situations are described here. Many correlations specific to particular equipment and separation processes will be found throughout this book. Extensive listings of correlations can be found in Perry and Green[42] and Treybal.[43]

FLAT PLATES

An analysis of the development of a *laminar* boundary layer on a flat plate provides the following expression for the average mass transfer coefficient over a distance L:

$$\text{Sh}_L = \frac{k_\rho^\circ L}{D_{Am}} = 0.664 \, \text{Re}_L^{1/2} \, \text{Sc}^{1/3} \tag{2.4-15}$$

The theory, as developed, applied only to low mass transfer rates, but corrections to the coefficient owing to finite fluxes at the plate are available from the curves in Fig. 2.4-2.

At Re_L of about 5×10^5, depending upon roughness or other disturbances, the boundary layer becomes turbulent provided the free-stream conditions are turbulent. If the turbulent boundary layer begins at the leading edge of the plate, the average Sherwood number could be obtained using the Chilton–Colburn analogy and an observed correlation for the turbulent friction factor on a flat plate, $f = 0.074 \, \text{Re}_x^{0.2}$:

$$j_D = \frac{f}{2} = 0.037 \, \text{Re}^{-0.2} \tag{2.4-30}$$

To account for the fact that for some distance (L_{CR}) from the leading edge the boundary layer is laminar, it is usually assumed that Eq. (2.4-30) is valid for the region beyond L_{CR}. An average Sherwood number over the entire plate of length L is estimated

$$\text{Sh} = 0.037 \, \text{Sc}^{1/3} \left[\text{Re}_L^{0.8} - \text{Re}_{cr}^{0.8} + \frac{0.664}{0.037} \, \text{Re}_{cr}^{1/2} \right] \tag{2.4-31}$$

where $\text{Re}_{cr} \equiv L_{CR} v_\infty / \nu$ is in the range of 3×10^5–3×10^6 depending on free-stream turbulence and roughness of the plate. Other authors[44-46] indicate that the Chilton–Colburn analogy can apply to the entire plate, so that $j_D = f/2$, where j_D and f are average values for the plate as a whole. If the active transfer surface does not begin at the leading edge of the plate, careful analysis is required to compute properly the average coefficient.

ROUND TUBES

Transfer from the wall of a tube to a fluid in turbulent flow is the situation most commonly considered for the development of the analogies. Sherwood, et al.[32] critically compared various analogies with both heat transfer data and mass transfer data at high Schmidt numbers (liquids) and low Schmidt numbers (gases) and concluded that the simple Chilton–Colburn analogy does about as well as any of the correlations in representing the results.

Gilliland and Sherwood[47] measured evaporation rates of various liquids into a turbulent air stream in a wetted-wall column for both cocurrent and countercurrent flows and proposed the equation

$$\text{Sh} \equiv \frac{k_\rho \, y_{B, \text{lm}} \, d}{D_{Am}} = 0.023 \, \text{Re}^{0.83} \, \text{Sc}^{0.44} \tag{2.4-32}$$

Because of limited variation in experimental values of the Schmidt number, some uncertainty is associated with its exponent. Also, the Reynolds number is based on gas velocity relative to the pipe wall rather than the liquid surface. The above correlation and data typically lie some 20% higher than the Chilton–Colburn prediction. Wetted-wall column studies by Jackson and Caegleske[48] and Johnstone and Pigford,[49] employing Re based on gas velocity relative to the liquid, generally are in reasonable agreement with the Chilton–Colburn analogy.

Further support for the Chilton–Colburn analogy, particularly for Sh $\propto$ Sc$^{1/3}$, is found from the data on dissolution of cast tubes of various soluble solids,[50–52] although Davies[53] suggests that dissolution data for high Sc are better correlated by

$$Sh = 0.01 \, Re^{0.91} \, Sc^{1/3} \tag{2.4-33}$$

He provides theoretical support for this form from the theory of Levich,[39] which recognizes that eddy diffusion can be neglected only in a very narrow region near the wall if high Schmidt numbers are involved.

It appears that for most engineering estimates of mass transfer to turbulent fluids in round tubes the equation

$$Sh = 0.023 \, Re^{0.8} \, Sc^{1/3} \tag{2.4-34}$$

can be used. This equation is in agreement with the Chilton–Colburn analogy since for smooth tubes a reasonable correlation of the Fanning friction factor is $f = 0.046 \, Re^{-0.2}$. For roughened tubes, protrusions from the wall can introduce additional eddies which can penetrate into the viscous sublayer. The effect of roughness on mass transfer is only expected at Sc > 1 where the resistance to mass transfer is primarily near the wall. Davies[53] obtains the result

$$Sh \propto f^{1/2} \, Re^{1.0} \, Sc^{0.5} \tag{2.4-35}$$

Since at high Reynolds numbers the friction factor is nearly independent of Re but increases with increasing roughness, mass transfer from a roughened pipe might exceed that from a smooth pipe by as much as a factor of 3.

EXAMPLE 2.4-1
Water is being evaporated into a countercurrent air stream at atmospheric pressure flowing at a rate of 2.2 lb$_m$ air/min (1 kg/min) in a vertical wetted-wall column 1.2 in. (0.03 m) in diameter. If the water flow rate is about 0.22 lb$_m$/min (0.1 kg/min), estimate the evaporative flux at a point in the column where the water temperature is 40°C and the mole fraction of H$_2$O in the bulk gas is 0.01.

Solution For the case of H$_2$O diffusing through stagnant air

$$N_{H_2O} = k_y \, (y^*_{H_2O} - y_{H_2O,b}) = k^\circ_y \ln \left[\frac{1 - y_{H_2O,b}}{1 - y^*_{H_2O}} \right]$$

At the water–air interface, equilibrium can be assumed so that

$$y^*_{H_2O} = \frac{p^*_{H_2O} \, (40°C)}{p} = 0.072$$

The mass transfer coefficient $k^\circ_y = k^\circ_c \, C$ can be estimated from the Chilton–Colburn correlation.

To use the correlation, the gas Reynolds number should be calculated with gas velocity relative to the liquid surface velocity. The liquid surface velocity for flow of the film down the wall is (Section 2.3)

$$v_L = \left(\frac{g}{\nu} \right) \frac{\delta^2}{2} = \frac{g}{2\nu} \left(\frac{3\nu Q}{g\pi D} \right)^{2/3} = 0.57 \, \text{ft/s}$$

which corresponds to a liquid Reynolds number of 105. At this Reynolds number some rippling of the liquid surface would be expected. Assume the air temperature is also approximately 40°C.

$$v_g = 70.4 \, \text{ft/s} \, (21.5 \, \text{m/s})$$

The gas Reynolds number relative to the liquid is

$$Re_g = 39,600$$

Equation (2.4-34) is used to estimate the mass transfer coefficient

$$Sh = 0.023 \, Re_g^{0.8} \, Sc^{1/3} = 90.5$$

$$k_y^\circ = \frac{Sh \, D_{H_2O\text{-air}} \, C}{d} = 6.82 \times 10^{-4} \text{ lb-mole/ft}^2 \cdot s \, (3.3 \times 10^{-3} \text{ kmol/m}^2 \cdot s)$$

Figure 2.4-2 can be used to verify that the effect of bulk flow on k_y is negligible.
The evaporative flux is

$$N_{H_2O} = k_y^\circ \ln \left(\frac{0.99}{0.928} \right) = 4.4 \times 10^{-5} \text{ lb-mole/ft}^2 \cdot s \, (2.1 \times 10^{-4} \text{ kmol/m}^2 \cdot s)$$

Complete analysis of such an evaporator would require taking into account variations in composition and temperature in the axial direction by integration of the species and energy equations for the gas and liquid phases as well as considering the energy flux to the surface.[8]

Transfer Between Fluid and Particles
Mass transfer from a fluid to solid particles, such as ion-exchange pellets, or crystals in suspension is of importance in many separation processes. In addition to this transfer step in the fluid phase, in many applications significant rate processes may be occurring within the particle or at its surface.

Single Spheres. As developed in Section 2.3, steady-state molecular diffusion to a single sphere in a stagnant fluid provides the asymptotic limit of

$$Sh = 2 \tag{2.4-36}$$

When flow occurs about a sphere the solution to this forced convection mass transfer problem is quite complex because of the complexity of the flow field.[54] At low flow rates (creeping flow) a laminar boundary layer exists about the sphere which separates from the surface at an angular position and moves toward the forward stagnation point as the flow rate increases. Wake formation occurs at the rear of the sphere. At still higher flow rates transition to a turbulent boundary layer occurs. Solutions to the problem of mass transfer during creeping flow about a sphere (Re < 1) have been developed by a number of authors[55-57] with the numerical solutions of Brian and Hales[55] being perhaps the most extensive. Their result is

$$Sh = [4.0 + 1.21(Re \, Sc)^{2/3}]^{1/2} \tag{2.4-37}$$

for Pe = Re Sc < 10,000 and Re < 1.
Very extensive experimental studies of mass transfer from a sphere to a fluid have been carried out at higher Re using methods such as dissolution of solids into a liquid,[55-59] evaporation from small rigid drops[60] or from wet porous spheres,[61] and sublimation of solid particles.[62,63] Use also has been made of the analogy to heat transfer by employing experimental observations on heat transfer to single spheres.[64] A large number of correlations have been proposed to represent these data—Skelland[33] lists 14 such correlations!
Generally, the correlations are of a form that assumes that the contributions of molecular diffusion and forced convection are additive; that is,

$$Sh = 2 + A \, Re^b \, Sc^{1/3}$$

For Re = $d_p \, v_\infty / \nu$ < 1, the exponent b is usually taken as $\frac{1}{3}$ while for Re > 1 the exponent is found to range from 0.5 to 0.62. The Frössling equation

$$Sh = 2 + 0.6 \, Re^{1/2} \, Sc^{1/3} \tag{2.4-38}$$

appears to represent the data about as well as any of the correlations for a wide range of Sc numbers and Re up to 1000. It should be noted that the mass transfer coefficient in these correlations is the average value for the sphere as a whole. In fact, the local coefficient varies from the front to the rear, being largest at the forward stagnation point.[65]
At low Reynolds numbers, natural convection effects can contribute to the mass transfer rate as well. Some authors[35,59] suggest equations of the form

$$Sh = 2 + Sh_{nc} + A \, Re^m \, Sc^{1/3}$$

where the natural convection contribution is of the form $B(Gr)^\alpha$ $(Sc)^\beta$. The Grashoff number Gr = $gd_p^3 \Delta\rho/\nu^2\rho$ can be looked on as a Reynolds number arising from natural convection. Typically, α and β range from $\frac{1}{3}$ to $\frac{1}{4}$ in different correlations and $B \approx 0.6$. The effects of natural convection can usually be neglected whenever the contribution to the Sherwood number from forced convection is equal to that from natural convection. It should be noted that the additivity of terms corresponding to molecular diffusion, natural convection, and forced convection has not been theoretically established.

Although heat transfer and mass transfer are quite analogous in these situations, none of the above equations for transfer to a sphere obey an analogy relationship between mass transfer and momentum transfer except at high Reynolds numbers where only the forced convection term is of importance. Even here only the skin friction contribution to momentum transfer can be used.

Additional effects on mass transfer to single spheres that have been considered are those of changing sphere diameter, significant radial fluxes, and the influence of free-stream turbulence on the mass transfer coefficient. Brian and Hales[55] examined the effect of changing sphere diameter and significant radial fluxes by numerically solving the governing mass transfer equation under creeping flow conditions. High mass transfer fluxes appear to be handled adequately by the film theory approach described previously. Some investigators[61,66] have examined the influence of the main-stream turbulent intensity (at a given particle Reynolds number) on the mass transfer coefficient between a sphere and fluid. Increases of as much as 40% have been found. Such effects may be particularly important in situations such as a suspension of particles in an agitated tank where the turbulent intensity varies greatly from the impeller region to other parts of the vessel.

Other Shapes. Rather limited experimental and theoretical work has been conducted on transport to immersed bodies of other than spherical shape. Pasternak and Gauvin,[67] who studied force convection heat transfer and mass transfer by evaporation of water from porous solids of 20 different shapes oriented in different directions relative to the flow field, proposed the correlation

$$j_D = j_H = 0.692 \, Re^{-0.49} \tag{2.4-39}$$

for $500 < Re < 5000$. The characteristic length dimension for use in this correlation was defined as the surface area of the body divided by the perimeter normal to the flow. The Skelland–Cornish[63] results on sublimation rates of oblate spheroids (a shape often assumed by drops) agreed with the correlation suggested by Eq. (2.4-39). The well-established correlation for heat or mass transfer to a long cylinder transverse to a flowing stream[68] is also in agreement with Eq. (2.4-39).

The above correlations hold in the region where molecular diffusion and natural convection effects are not of importance. For low relative velocities between fluid and particles, for example, for neutrally buoyant crystals in solution, the diffusion term can be important. Ohara and Reid[69] state that Eq. (2.4-36) may be employed with a constant term of 2 for cubes, $2\sqrt{6}$ for tetrahedrons, and $2\sqrt{2}$ for octahedrons.

Agitated Vessels. In common with many processes occurring in mixing vessels, mass transfer to a suspension of particles in an agitated vessel is a very important, yet not fully understood, situation. It is clear that reliable information is only obtained at stirring speeds sufficient to suspend completely the particles in the vessel. Suspended particles are acted on by at least two velocities—a settling velocity owing to a density difference between solid and fluid, and an effective velocity owing to turbulence in the tank. If proper methods for calculating and combining these velocities were known, the previous correlations could be used with the Reynolds number calculated using this resultant velocity.

Harriott[70] suggested that a minimum mass transfer coefficient could be calculated from the terminal velocity of the particle in the fluid. The calculation would be carried out by computing the terminal velocity of the sphere including correlations for the variation of the drag coefficient with Reynolds number. The minimum mass transfer coefficient is then calculated from Eq. (2.4-38). When these calculations are carried out, there is a range of particle sizes ($\sim$ 100–1000 μm) where the coefficient is nearly independent of size, while at higher and lower particle sizes the coefficient decreases with increasing size.[70] Harriott's experimental measurements of coefficients in agitated vessels gave values two to eight times this minimum. A reliable correction factor to the minimum coefficient was not developed although agitator power per unit volume raised to the $\frac{1}{3}$ power was found to correlate the results.

One of the more complete studies to date is that of Brian et al.,[71] who took into account effects of particle transpiration and size changes in examining all the experimental data. Based on the concept of Kolmogaroff's theory of isotropic turbulence applied to energy dissipation in an agitated vessel, they correlated mass transfer coefficients with the one-fourth power of the average energy dissipation per unit mass, E. In their studies they found the solid–liquid density difference to be of minor importance for density ratios between 0.8 and 1.25, while Harriott did find some effect in his studies. The data of Brian et al. were for both heat and mass transfer rates but covered only a limited range of particle sizes. Brian et al.[71] also suggested that the Sc number dependence is more complex than the $\frac{1}{3}$ power used in the correlation.

Levins and Glastonbury[72] provide perhaps the best correlating equation to date for neutrally buoyant particles in stirred vessels:

$$Sh = 2 + 0.47 \left(\frac{E^{1/3} d_p^{4/3}}{\nu} \right)^{0.62} Sc^{0.36} \left(\frac{D_S}{T} \right)^{0.17} \tag{2.4-40}$$

The geometrical ratio of impeller to tank diameter (D_S/T) is included in their correlation but it is not of great importance. They observed little effect of density difference on the mass transfer coefficient. The comparison of these correlations along with a number of others available in the literature was made by Armenante[73] and is shown in Fig. 2.4-5. The length of each line in the figure indicates the experimental range covered by the investigator.

The abscissa in Figure 2.4-5 ($E^{1/3} d_p^{4/3}/\nu$) can be interpreted as a particle Reynolds number (Re_p) and also as the ratio of particle size to minimum eddy size raised to the $\frac{4}{3}$ power. Only a few investigations have been conducted for $Re_p < 1$, that is, particle diameter smaller than eddy size, and some investigators have observed Sherwood numbers smaller than the limiting value of 2.[77,78] In a recent careful experimental study of mass transfer to such microparticles, Armenante[73] confirmed the limiting value of 2 for the Sherwood number. For the influence of turbulence (power input per unit mass) on transfer to microparticles he obtained the correlating equation

$$Sh = 2 + 0.52 \, Re_p^{0.52} \, Sc^{1/3} \tag{2.4-41}$$

EXAMPLE 2.4-4

In connectiion with a study of crystal growth kinetics it is desired to estimate the mass transfer effects for a single crystal of potassium alum ($KAl(SO_4)_2 \cdot 12 \, H_2O$) growing in a tube through which supersaturated solution flows at a velocity of 0.10 m/s. The solution at 25°C has an alum concentration of 150 kg/m³, while saturated solution at this temperature contains 140 kg/m³. The density of the solution is about 1100 kg/m³, the viscosity is about 1 mPa s, and the diffusivity of alum in solution is approximately 5×10^{-10} m²/s. The alum crystals whose density is about 1800 kg/m³) form regular octahedrals that might be approximated as spheres. Under these conditions a crystal of equivalent diameter of 3 mm is observed to grow at a linear rate dr/dt of 5×10^{-5} mm/s. Is it possible that this growth rate is mass transfer controlled?

Solution The mass flux of hydrate to the surface is

$$n_A = k_p^\circ \rho \ln \left(\frac{1 - \rho_A/\rho}{1 - \rho_A^*/\rho} \right)$$

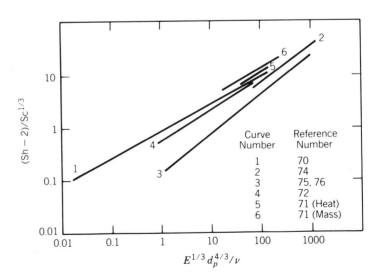

Curve Number	Reference Number
1	70
2	74
3	75, 76
4	72
5	71 (Heat)
6	71 (Mass)

FIGURE 2.4-5 Correlations of mass (heat) transfer to suspended particles in agitated systems.[73]

for the case of hydrate (A) moving through stagnant water in a relatively concentrated solution. The choice of mass units is preferred because mass density is approximately constant while the molar density would be less so.

Estimate the coefficient k_ρ°, from Eq. (2.4-38):

$$Re = 330$$

$$Sc = 1818$$

$$Sh = 2 + 0.6(330)^{1/2} (1818)^{1/3} = 134$$

$$k_\rho^\circ = 2.2 \times 10^{-5} \text{ m/s}$$

If the growth is mass transfer controlled, then the interfacial concentration is that corresponding to saturation and

$$n_A = (2.2 \times 10^{-5})(1100) \ln \left(\frac{1 - 150/1100}{1 - 140/1100} \right)$$

$$= -2.5 \times 10^{-4} \text{ kg/m}^2 \cdot \text{s} \quad (- \text{ means toward crystal})$$

The linear growth rate of the crystal corresponding to this flux would be

$$\frac{d}{dt} (\rho_c \tfrac{4}{3}\pi r^3) = n_A 4\pi r^2$$

$$\frac{dr}{dt} = 1.3 \times 10^{-7} \text{ m/s} \quad (1.3 \times 10^{-4} \text{ mm/s})$$

Since the mass transfer controlled rate is only about a factor of 2 greater than the observed rate, it is likely that mass transfer is affecting the observed values. Experiments need to be conducted at a series of solution flow rates to account properly for mass transfer effects and obtain the true interfacial growth kinetics.

Fixed and Fluidized Beds. Fixed beds of particles through which gases or liquids flow are extensively used in adsorption, ion exchange, chromatography, gas absorption and stripping, and heterogeneous catalytic reactions. Similarly, fluidized beds are employed for heterogeneous catalysis, for noncatalyzed heterogeneous reactions such as coal combustion and gasification, and for crystallization. In all these operations the rate of mass transfer between the fluid phase and the particles can be of vital importance in controlling the rate of the process and therefore the size or throughput for the processing unit.

Studies of the transport from a particle to a single-phase fluid in a fixed bed have usually been conducted by observing the sublimation of solid particles, the evaporation of a liquid film on the surface of the particle, or the dissolution of soluble particles. Sherwood et al.[68] review much of the available data for transport from particles to gases or liquids in fixed beds and suggest that the data can be represented by the equation

$$j_D = 1.17 \, Re_p^{-0.415} \tag{2.4-42}$$

for $10 < Re_p < 2500$. In this equation the Reynolds number is defined in terms of the diameter of a sphere having the equivalent area of the particle and the superficial velocity of the fluid through the bed. Since most of the data have been with beds having void fractions of about 0.4, the effect of voidage is not clear although a number of correlations have been proposed including voidage.[79-81]

Karabelas et al.[81] reviewed many of the published correlations for fixed-bed coefficients and proposed different correlations to be used depending on the flow regime; that is, at low Reynolds number the effects of molecular diffusion and natural convection must be considered. Kato et al.[82] reviewed mass transfer coefficients in fixed and fluid beds and observed considerable deviations from established correlations in both the literature and their own data for $Re < 10$. In some cases it appeared that the limiting Sherwood number could be less than 2 for gas–particle transfer. They suggested that for small Re and $Sc \approx 1$ the concentration boundary layers of the individual particles in a fixed bed would overlap considerably. They proposed two correlations for different flow regimes which also included a particle diameter to bed height term.

Kato et al.[82] and Davidson and Harrison[84] have reviewed available correlations for both heat and mass transfer coefficients in gas–solid fluidized beds. Some investigators have proposed that the same correlation can be applied[81-83] for both fixed and fluidized beds by proper choice of the fluid velocity term. Further discussions of mass transfer in fluidized beds can be found in Davidson and Harrison[84] and Kunii and Levenspiel.[85] Because of the difficulty of operating small laboratory fluidized beds under conditions comparable to those of a full-scale unit, the applicability of many of these studies may be questionable.

Gas- and liquid-side mass transfer coefficients in packed absorption columns with countercurrent flows exhibit complex dependencies on the gas and liquid rates and the column packing. In addition, the interfacial transfer area will be a function of the hydrodynamics and packing. Correlations for such situations, based on experimental data, are usually developed in terms of the height of a transfer unit (HTU) concept. Gas and liquid heights of transfer units are defined by the relationships,

$$H_L = \frac{\overline{L}}{k_x a} \qquad\qquad (2.4\text{-}43a)$$

$$H_G = \frac{\overline{G}}{k_y a} \qquad\qquad (2.4\text{-}43b)$$

where a is the interfacial area per unit column volume, $\overline{L}$ and $\overline{G}$ are the liquid and gas superficial velocities, and k_x and k_y are the liquid and gas individual mass transfer coefficients. These definitions arise naturally in separation column analysis and the height of a transfer unit varies less strongly with the flow rate of a phase than does the mass transfer coefficient. Specific correlations for HTUs[19] will be discussed in later chapters in this book. Some information is available in Perry and Green.[42]

The resistance-in-series model or two-film theory [Eq. 2.4-10] is often used in gas–liquid or liquid–liquid systems in terms of HTUs giving rise to the definition of overall heights of transfer units:

$$H_{OL} = H_L = \left(\frac{\overline{L}}{m\overline{G}}\right) H_G \qquad\qquad (2.4\text{-}44a)$$

$$H_{OG} = \left(\frac{m\overline{G}}{\overline{L}}\right) H_L + H_G \qquad\qquad (2.4\text{-}44b)$$

In many instances only the overall transfer unit heights are obtainable from column operating data.

One final mass transfer phenomenon relevant to fixed and fluidized beds is that of axial and radial dispersions. When a fluid flows through a packed (or in fact an empty bed) one finds that the concentration of a particular species is distributed radially and axially by a variety of mechanisms in addition to molecular diffusion. The origins of these effects are velocity gradients and mixing caused by the presence of particles. The most common models for these processes employ radial and axial dispersion coefficients defined in analogy to Fick's First Law to characterize the fluxes. Reviews of this important subject may be found in Levenspiel and Bischoff,[86] Gunn,[87] and Levenspiel.[88] While the effect of radial mixing is to help maintain a uniform transverse concentration profile, axial dispersion generally causes deleterious effects on the performance of separation columns.

FLUID–FLUID MASS TRANSFER
Transport rates between a fluid phase and a bubble or drop differ from those previously discussed in that the interfacial fluid velocity is not necessarily zero, circulation may occur within the dispersed phase, and interfacial motion and oscillations can occur.

Drops. Dispersion of a liquid phase into small droplets greatly enhances the interfacial area for mass transfer to a second fluid phase as in liquid–liquid extraction in sieve columns or mixer–settler systems and in spray columns for gas–liquid contacting. Mass transfer coefficients in both the continuous and the dispersed phases must be considered. Furthermore, mass transfer occurs as the drop goes through the stages of formation, acceleration to its terminal velocity, free fall at its terminal velocity, and coalescence. As might be expected in dealing with such complex phenomena, completely general theories and correlations are not yet available. Heertjes and DeNie[89] have reviewed research in both fluid mechanics and mass transfer while Kintner's review[90] deals primarily with drop mechanics in extraction. Skelland[91] and Treybal[92] list many of the proposed correlations for continuous- and dispersed-phase coefficients.

Mass transfer during drop formation can be quite significant. After formation the drop falls (or rises) through the continuous phase at its terminal velocity. Small drops (<2 mm), those in the presence of surfactants, or those for which the continuous-phase viscosity is much less than the drop viscosity behave as rigid spheres with little internal circulation. For this situation the continuous-phase coefficient can be obtained from correlations such as Eq. (2.4-38); indeed, much of the data for this correlation were obtained from evaporation rates of pure liquid drops in a gas. If no circulation is occurring within the drop, the mass transfer mechanism within the drop is that of transient molecular diffusion into a sphere for which solutions are readily available (see Section 2.3).

As a drop becomes larger, it begins to distort in shape, oscillate, and finally break into smaller drops. Circulation occurs within it. Skelland[91] provides a number of correlations for dispersed- and continuous-phase coefficients for various situations in the two phases, for example, a circulating drop with significant resistance in both phases, and an oscillating drop with resistance mainly within the drop. It is clear that much additional research is required to understand the interactions among the fluid mechanics and mass transfer processes. Further discussion relative to extraction processes may be found in Chapter 7.

   Bubbles.   Mass transfer between gas bubbles and a liquid phase is of importance in a variety of operations—gas–liquid reactions in agitated vessels, aerobic fermentations, and absorption or distillation in tray columns. As in liquid–liquid transfer, dispersion of a phase into small units greatly increases the available area for transfer. If the fractional holdup (volume gas/total volume) of the gas in a gas–liquid mixture is H_g, the interfacial area per unit volume for bubbles of diameter d_B is given by

$$a = \frac{6 H_g}{d_B} \qquad (2.4\text{-}45)$$

Increases in area therefore accompany decreases in bubble size. Because of the high diffusivity in the gas phase, the liquid-side mass transfer resistance almost always limits the mass transfer rate and is the coefficient of interest.

   The rise velocity of bubbles in water as a function of diameter is sketched in Fig. 2.4-6. For small bubble sizes, Stokes' Law is approximately obeyed; but, as the equivalent diameter increases, circulation reduces the relative velocity. At larger sizes the bubble shape becomes that of an oblate spheroid and then assumes that of a spherical cap (hemispherical). The presence of electrolytes or traces of surfactants in the solution tend to make the bubbles smaller and behave as rigid spheres so that in most practical applications only smaller bubble sizes are important.

   Calderbank and Moo Young[75,93] proposed two correlations for mass transfer coefficients to rising bubbles of diameters less than and greater than 2.5 mm, respectively.

$$Sh = 2 + 0.31 \, (Gr)^{1/3} \, (Sc)^{1/3} \quad \text{(small } d_B) \qquad (2.4\text{-}46)$$

$$Sh = 0.42 \, (Gr)^{1/3} \, (Sc)^{1/2} \quad \text{(larger } d_B) \qquad (2.4\text{-}47)$$

where $Gr = d_B^3 |\rho_g - \rho_l| g / \rho_l \nu^2$. They suggested that these correlations would also hold for swarms of bubbles in agitated tanks and in sieve columns. Furthermore, the small bubble-size correlation appeared to hold for literature data on natural-convection heat transfer and mass transfer to single spheres, growth of crystals in liquid fluidized beds, and dissolution of just-suspended solids in agitated vessels. Equation (2.4-46) can be thought to apply when the contribution of turbulence in the continuous phase is unimportant. This always occurs for a gas dispersion in agitated vessels because of the very large rise velocity. Similarly, for just-suspended particles the turbulence contribution is very small. As discussed earlier, at higher power input the contribution of turbulence to mass transfer to solid particles in agitated tanks becomes significant.

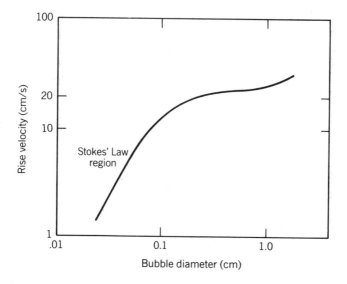

FIGURE 2.4-6   Typical rise velocity of a bubble in a liquid as a function of diameter.

Calderbank and Moo Young suggest that their small-bubble correlation applies to nearly rigid spherical bubbles or particles, whereas the large-bubble correlation applies when significant interfacial flow occurs.

For small particles the rise velocity is given by Stokes' Law ($d_B\Delta\rho/18$) so that the particle Reynolds number is simply ($Gr/18$). Equation (2.4-46) can therefore be put in the form

$$Sh = 2 + 0.31 \, (18)^{1/3} \, Re^{1/3} \, Sc^{1/3} \qquad (2.4\text{-}48)$$
$$Sh = 2 + 0.81 \, (Pe)^{1/3}$$

which is in reasonable agreement with the creeping flow solution for single spheres—Eq. (2.4-37). At higher Pe (>1) Eq. (2.4-46) can also be shown to be in approximate agreement with Eq. (2.4-38) if the drag coefficient for spheres is taken as inversely proportional to the square root of the Reynolds number. Thus, Eq. 2.4-46 seems to have reasonable theoretical and experimental support, although the effects of surfactants or other impurities and a non-Newtonian liquid phase are not taken into account. Note that Eq. (2.4-46) suggests that the mass transfer coefficient for very small bubbles is inversely proportional to their size and then becomes independent of size as the bubbles increase in size:

$$k_c^\circ = \frac{2 \, D_{Am}}{d_B} + 0.31 \left(\frac{|\Delta\rho| \, gD_{Am}^2}{\mu} \right)^{1/3} \qquad (2.4\text{-}49)$$

As bubble size increases above 2–3 mm, there is a transition regime and then the mass transfer coefficient is estimated by Eq. (2.4-47) and is again independent of bubble size. In this region of significant interfacial flow it is likely that the bubbles have assumed a spherical cap shape. The correlation is in agreement with Higbie's Theory which predicts $k_c \propto D^{1/2}$. Furthermore, Calderbank[93] has shown reasonable quantitative agreement between his correlation and available results for the rise velocity of spherical cap bubbles. More recent studies have examined the contribution to mass transfer that may occur at the rear of a spherical cap bubble.[94]

In the dispersion of gas bubbles for mass transfer purposes, the primary effect of agitator power or of gas velocity in sieve trays is on the interfacial area through the reduction of bubble size. The mass transfer coefficients themselves generally lie in a narrow range around 10^{-4} m/s, although the influences of contaminants and non-Newtonian fluids on the coefficient, the transition from "small" to "large" bubbles, and bubble–bubble interactions are not yet well characterized.

Interfacial Phenomena During Fluid–Fluid Mass Transfer. During mass transfer between two fluid phases certain interfacial processes often occur to affect the results that would be expected if transport were governed solely by the two bulk phases: interfacial turbulence and interfacial resistance.

It is often observed during mass transfer that significant interfacial motion or turbulence, unrelated to the bulk fluid hydrodynamics, occurs.[95–97] For example, the surface of a drop of water in contact with ethyl formate will undergo motion and oscillations,[98] or dissolving CO_2 into ethanolamine solutions will cause rapid turbulent motions in the liquid at the interface.[99] These effects accompany mass transfer and are generally believed to result from the transferring solute causing local gradients in interfacial tension, which results in surface flow. Some investigators have found that the effect is dependent on the direction of mass transfer. The phenomenon is often termed the Marangoni effect and usually serves to increase mass transfer rates. Davies and Rideal[100] suggest in fact, that the phenomenon generally accompanies mass transfer but may not be observed if the contact times between the two phases are small and a fresh surface is quickly supplied. Sternling and Scriven[101] provided an early theoretical analysis of the effect that is in reasonable agreement with observations, and many subsequent analyses have been conducted.[102, 103]

The presence of surfactants in a liquid that accumulate at the interface may provide an additional interfacial barrier to transport. Davies and Rideal[100] discuss many of these phenomena. It is well known that the presence of surfactants such as long chain alcohols will suppress evaporation rates from a quiescent water surface. In gas–liquid systems it appears that significant transport reduction only occurs when contact times are sufficiently long to allow the accumulation of surfactant at the interface. Thus, in packed gas absorption columns, where continual remixing of fluid surface elements occurs, little effect is expected. In sparged vessels the effect may be more significant with mass transfer coefficient reductions by a factor of 2 or more.[104–106] It is important to realize that a surfactant will likely also cause bubble size to decrease, resulting in significant increases in interfacial area so that the net effect of surfactant addition on mass transfer rates is not obvious.[107]

In liquid–liquid systems effects of surfactants are again quite complex: they serve to reduce circulation and depress interfacial turbulence, promote emulsification, and provide an additional barrier to mass transfer.[100, 108] At present, any particular system of interest should be investigated specifically. In practice, the effects are difficult to isolate because of the low concentration levels of contaminants that are effective and the many sources of contamination available in processing equipment.

2.5 NOTATION

Roman Letters

a	interfacial area per unit volume
b	intermolecular spacing
C	molar density
C_A	molar concentration of species A
C_{Pi}	partial molar specific heat of species i
D_{AB}	diffusivity in A-B system
$\hat{D}_{AB}$	activity-corrected binary diffusivity
D_{im}	diffusion coefficient of i in a mixture
D_K	Knudsen diffusion coefficient
d	diameter of tube
d_B	bubble diameter
d_P	particle diameter
E	energy dissipation rate per unit mass of fluid
E_a	activation energy
E_D	eddy diffusivity
E_V	eddy viscosity
f	Fanning friction factor
F	Faraday's constant
g	acceleration due to gravity
g_i	body force per unit mass on species i
$\overline{G}$	gas superficial velocity
G_m	mass velocity
Gr	Grashoff number
H_g	fractional holdup of gas in a volume
H_i	partial molar enthalpy
h	heat transfer coefficient
h_i	partial mass enthalpy
J_A	molar diffusive flux of species A
j_A	mass diffusive flux of species A
j_D	Colburn j factor for mass transfer
k	reaction rate constant
K	adsorption equilibrium constant
k_x	mass transfer coefficient, based on mole fraction driving force, relating mass flux relative to stationary coordinates. Other driving force units denoted by different subscripts
k_x'	mass transfer coefficient, based on mole fraction driving force, relating flux relative to bulk flow across transfer surface
$\overline{L}$	liquid superficial velocity
L	length parameter
m	Gas–liquid distribution coefficient
m_i	mass of species i
$\overline{M}_i$	moles of species i
M_i	molecular weight of species i
N	total molar flux relative to stationary coordinates
n	total mass flux relative to stationary coordinates
N_A	molar flux of A relative to stationary coordinates
n_A	mass flux of A relative to stationary coordinates
Pe	Peclet number
Pr	Prandtl number
p	pressure

p_A	partial pressure of A in gas phase
Q	volumetric flow rate
$\dot{Q}$	heating rate
$\dot{Q}_R$	heating rate per unit volume by radiation
q	heat conduction flux
R	gas constant or tube radius
R_i	molar production rate of i due to chemical reaction per unit volume
r_i	mass production rate of i due to chemical reaction per unit volume
Re	Reynolds number
S	surface area
$\bar{S}$	surface renewal rate
S_g	surface area per unit mass
Sc	Schmidt number
Sh	Sherwood number
St	Stanton number
t	time
T	temperature
u_i	partial mass internal energy
u^+	normalized velocity in Universal Velocity Distribution Law
$\bar{v}$	mean molecular speed
v	mass average velocity relative to stationary coordinates (v_x − x component of v)
v_i	velocity of species i relative to stationary coordinates
v^*	molar average velocity relative to stationary coordinates
V	volume
$\bar{V}_i$	partial molar volume of species i
$\bar{V}_i^\circ$	molar volume of pure i
W_S	shaft power
w_i	mass fraction of species i
x_i	mole fraction of i in a condensed phase
y_i	mole fraction of i in a gas phase
y^+	normalized distance from wall in Universal Velocity Distribution Law
$Z_{+,-}$	valence of an ion

Greek Letters

α	thermal diffusivity
α_i	stoichiometric coefficient
γ	velocity gradient
γ_i	activity coefficient of species i
δ	film or boundary layer thickness
ϵ	porosity
ϵ_i	energy parameter
Θ	contact time
λ	mean free path
$\lambda_{+,-}$	ionic conductance
μ	viscosity
ν	kinematic viscosity
ν_0	frequency
ρ	mass density
ρ_i	mass concentration of species i
ρ_b	mass density of a porous solid
ρ_c	mass density of a crystal

σ_{AB} collision diameter
τ tortuosity in a porous solid
$\bar{\bar{\tau}}$ stress tensor
Φ energy dissipation rate
$\phi(t)$ surface age distribution function
ϕ association factor in Wilke–Chang equation
Ω_D collision integral for diffusion

Subscripts

A, B, i species A, B, i
1, 2 locations 1, 2
W value at wall
x, y, z coordinate direction
∞ bulk value

For Mass Transfer Coefficients:

c molar concentration driving force
p partial pressure driving force
x mole fraction driving force
y mole fraction driving force (gas phase)
w mass fraction driving force
ρ mass concentration driving force

Superscripts

* equilibrium value

REFERENCES

Section 2.2

2.2-1 S. P. deGroot and P. Mazur, *Non-Equilibrium Thermodynamics*, North-Holland, Amsterdam, 1962.

2.2-2 R. B. Bird, C. F. Curtiss, and J. O. Hirschfelder, *Chem. Eng. Prog. Symp. Ser. No. 16*, **51,** 69 (1955).

2.2-3 R. M. Felder and R. W. Rousseau, *Elementary Principles of Chemical Processes*, 2nd ed., Wiley, New York, 1986.

2.2-4 R. H. Perry and D. W. Green (Eds.), *Perry's Chemical Engineers' Handbook*, 6th ed., McGraw-Hill, New York, 1984.

2.2-5 R. B. Bird, W. E. Stewart, and E. N. Lightfoot, *Transport Phenomena*, Wiley, New York, 1960.

2.2-6 J. C. Slattery, *Momentum, Energy, and Mass Transfer in Continua*, McGraw-Hill, New York, 1972.

2.2-7 C. Truesdell and R. A. Taupin, in S. Flugge (Ed.), *Handbuch der Physik*, Vol. 3, Springer-Verlag, Berlin, 1960.

2.2-8 A. S. Berman, *J. Appl. Phys.*, **24,** 1232 (1953)

Section 2.3

2.3-1 S. Chapman and T. G. Cowling, *Mathematical Theory of Non-Uniform Gases*, Cambridge University Press, New York, 1964.

2.3-2 W. Jost, *Diffusion in Solids, Liquids and Gases*, Academic Press, New York, 1960.

2.3-3   S. R. DeGroot and P. Mazur, *Non-Equilibrium Thermodynamics,* North Holland, Amsterdam, 1962.

2.3-4   E. L. Cussler, *Multicomponent Diffusion,* American Elsevier, New York, 1976.

2.3-5   T. K. Sherwood, R. L. Pigford, and C. R. Wilke, *Mass Transfer,* Chap. 2, McGraw-Hill, New York, 1975.

2.3-6   J. L. Duda and J. S. Vrentas, *Ind. Eng. Chem. Fundam.,* **4,** 301 (1965).

2.3-7   J. C. Maxwell, *Scientific Papers,* Vol. II, p. 629, Dover, New York, 1952.

2.3-8   C. F. Curtiss and J. O. Hirshfelder, *J. Chem. Phys.,* **17,** 550 (1949).

2.3-9   E. N. Lightfoot, E. L. Cussler, and R. L. Rettig, *AIChE J.,* **8,** 702 (1962).

2.3-10   R. C. Reid, J. M. Prausnitz, and T. K. Sherwood, *The Properties of Gases and Liquids,* 3rd ed., Chap. 11, McGraw-Hill, New York, 1977.

2.3-11   T. R. Marrero and E. A. Mason, *J. Phys. Chem. Ref. Data,* **1,** 3 (1972).

2.3-12   W. Kauzmann, *Kinetic Theory of Gases,* W. A. Benjamin, New York, 1966.

2.3-13   J. O. Hirshfelder, C. F. Curtiss, and R. B. Bird, *Molecular Theory of Gases and Liquids,* Wiley, New York, 1954.

2.3-14   R. S. Brokaw, *Ind. Eng. Chem. Process Des. Dev.* **8,** 240 (1969).

2.3-15   J. H. Arnold, *Ind. Eng. Chem.,* **22,** 1091 (1930).

2.3-16   C. R. Wilke and C. Y. Lee, *Ind. Eng. Chem.,* **47,** 1253 (1955).

2.3-17   J. C. Slattery and R. B. Bird, *AIChE J.,* **4,** 137 (1958).

2.3-18   N. H. Chen and D. P. Othmer, *J. Chem. Eng. Data,* **7,** 37 (1962).

2.3-19   E. N. Fuller, P. D. Schettler, and J. C. Giddings, *Ind. Eng. Chem.,* **58,** 18 (1966).

2.3-20   H. Ertl and F. A. Dullien, *AIChE J.,* **19,** 1215 (1973).

2.3-21   P. A. Johnson and A. L. Babb, *Chem. Rev.,* **56,** 387 (1956).

2.3-22   D. M. Himmelblau, *Chem. Rev.,* **64,** 527 (1964).

2.3-23   W. Sutherland, *Philos. Mag.,* **9,** 781 (1905).

2.3-24   C. R. Wilke and P. Chang, *AIChE J.,* **1,** 264, (1955).

2.3-25   J. L. Gainer and A. B. Metzner, *AIChE–Inst. Chem. Eng. Symp. Ser.,* **6,** 73 (1965).

2.3-26   D. L. Wise and A. Houghton, *Chem. Eng. Sci.,* **21,** 999 (1966).

2.3-27   A. L. Shrier, *Chem. Eng. Sci.,* **22,** 1391 (1967).

2.3-28   P. A. Witherspoon and L. Bonoli, *Ind. Eng. Chem. Fundam.,* **8,** 589 (1969).

2.3-29   A. Akgerman and J. L. Gainer, *Ind. Eng. Chem. Fundam.,* **9,** 84, 88 (1970).

2.3-30   A. Vignes, *Ind. Eng. Chem. Fundam.,* **5,** 189 (1966).

2.3-31   J. Leffler and H. Cullinan, *Ind. Eng. Chem. Fundam.,* **9,** 84, 88 (1970).

2.3-32   J. L. Gainer, *Ind. Eng. Chem. Fundam.,* **9,** 381 (1970).

2.3-33   R. B. Bird, W. E. Stewart, and E. N. Lightfoot, *Transport Phenomena,* Wiley, New York, 1960.

2.3-34   J. S. Neuman, *Electrochemical Systems,* Prentice-Hall, Englewood Cliffs, NJ, 1973.

2.3-35   R. P. Wendt, *J. Phys. Chem.,* **69,** 1227 (1965).

2.3-36   D. G. Miller, *J. Phys. Chem.,* **71,** 616 (1967).

2.3-37   J. S. Newman, in C. Tobias (Ed.), *Advances in Electrochemistry and Electrochemical Engineering,* Vol. 5, Interscience, New York, 1967.

2.3-38   K. E. Gubbins, K. K. Bhatia, and R. D. Walker, *AIChE J.,* **12,** 588 (1966).

2.3-39   G. A. Ratcliff and J. G. Holdcroft, *Ind. Eng. Chem. Fundam.,* **10,** 474 (1971).

2.3-40   T. Goldstick and I. Fatt, *Chem. Eng. Prog. Symp. Ser.,* **99,** No. 66, 101 (1970).

2.3-41   C. Tanford, *Physical Chemistry of Macromolecules,* Wiley, New York, 1963.

2.3-42   D. R. Paul, *Ind. Eng. Chem. Fundam.,* **6,** 217 (1967).

2.3-43   H. Osmers and A. B. Metzner, *Ind. Eng. Chem. Fundam.,* **11,** 161 (1972).

2.3-44   J. L. Duda and J. S. Vrentas, *AIChE J.,* **25,** 1 (1979).

2.3-45   G. Astarita, *Ind. Eng. Chem. Fundam.,* **4,** 236 (1965).

2.3-46   I. Zandi and C. D. Turner, *Chem. Eng. Sci.,* **25,** 517 (1979).

2.3-47   S. U. Li and J. L. Gainer, *Ind. Eng. Chem. Fundam.,* **7,** 433 (1968).

2.3-48   R. M. Navari, J. L. Gainer, and K. R. Hall, *AIChE J.,* **17,** 1028 (1971).

2.3-49   D. J. Kirwan, ''Rates of Crystallization from Pure and Binary Melts,'' Ph.D. Dissertation, University of Delaware, 1966.

2.3-50 R. M. Barrer, *Diffusion in and Through Solids,* 2nd ed., Cambridge University Press, London, 1952.

2.3-51 A. S. Nowick and J. Burton, *Diffusion in Solids,* Academic Press, New York, 1975.

2.3-52 C. E. Birchenall, *Physical Metallurgy,* Chapter 9, McGraw-Hill, New York, 1959.

2.3-53 D. J. Kirwan, ''Chemical Interdiffusion and the Kirkendall Effect,'' M.S. Thesis, University of Delaware, 1964.

2.3-54 A. Vignes and C. E. Birchenall, *Acta Meta.,* 1117 (1968).

2.3-55 C. N. Satterfield, *Mass Transfer in Heterogeneous Catalysis,* MIT Press, Cambridge, MA, 1970.

2.3-56 R. B. Evans, G. M. Watson, and E. A. Mason, *J. Chem. Phys., 33,* 2076 (1961).

2.3-57 K. J. Sladek, Sc.D. Thesis, MIT, 1967, cited in Ref. 55.

2.3-58 P. B. Weisz, V. J. Firlette, R. W. Maatman, and E. B. Mower, Jr., *J. Catal., 1,* 307 (1962).

2.3-59 J. Crank and G. S. Park, *Diffusion in Polymers,* Academic Press, New York, 1968.

2.3-60 A. S. Michaels and H. J. Bixler, in E. S. Perry (Ed.), *Progress in Separation and Purification,* Vol. 1, Chap. 5, Interscience, New York, 1968.

2.3-61 A. S. Michaels and H. J. Bixler, *J. Polymer Sci., 50,* 393, 413 (1961).

2.3-62 D. W. Brubaker and K. Kammermeyer, *AIChE J., 18,* 1015 (1972).

2.3-63 A. S. Michaels, R. F. Baddour, H. S. Bixler, and C. Y. Choo, *Ind. Eng. Chem. Process Des. Dev., 1,* 14 (1962).

2.3-64 C. E. Rogers, V. Stannett, and M. Szwarc. *J. Poly. Sci. 45,* 61 (1960).

2.3-65 D. R. Paul, *Sep. Purif. Methods, 4,* 33 (1976).

2.3-66 A. S. Michaels, *Chem. Eng. Progr., 64,* 31 (1968).

2.3-67 M. Sarblowski, *Sep. Sci. Tech., 17,* 381 (1982).

2.3-68 H. K. Lonsdale and H. Podall, *Reverse Osmosis Membrane Research,* Plenum Press, New York, 1972.

2.3-69 R. Sourirajan, *Reverse Osmosis,* Academic Press, New York, 1970.

2.3-70 J. Crank, *Mathematics of Diffusion,* Clarendon Press, Oxford, 1956.

2.3-71 R. H. Stokes, *J. Am. Chem. Soc., 72,* 2243 (1950).

2.3-72 R. Mills, L. A. Woolf, and R. O. Watts, *AIChE J., 14,* 671 (1968).

2.3-73 J. J. Tham, K. K. Bhatia, and K. E. Gubbins, *Chem. Eng. Sci., 22,* 309 (1967).

2.3-74 A. Vignes, *J. Chim. Phys., 58,* 991 (1960).

2.3-75 Y. Nishijima and G. Oster, *J. Polym. Sci., 19,* 337 (1956).

2.3-76 R. M. Secor, *AIChE J., 11,* 452 (1965).

2.3-77 L. Boltzmann, *Wied. Ann., 53,* 959 (1894).

2.3-78 C. Matano, *Jpn. J. Phys., 8,* 109 (1933).

2.3-79 G. Bankoff, in *Advances in Chemical Engineering,* Vol. 5, Academic Press, New York, 1964.

2.3-80 H. S. Carslaw and J. C. Jaeger, *Conduction of Heat in Solids,* 2nd ed., Oxford University Press, New York, 1959.

2.3-81 A. B. Newman, *Trans. AIChE, 27,* 310 (1931).

2.3-82 R. H. Perry and D. W. Green, *Perry's Chemical Engineers' Handbook,* 6th ed., Chap. 10, p. 11, McGraw-Hill, New York, 1984.

2.3-83 H. F. Johnstone and R. L. Pigford, *Trans. AIChE, 38,* 25 (1942).

2.3-84 C. Stirba and D. M. Hurt, *AIChE J., 1,* 178 (1955).

2.3-85 G. D. Fulford, in *Advances in Chemical Engineering,* Vol. 5, Academic Press, New York, 1964.

2.3-86 J. Leveque, *Ann. Mines, 13*(12)*,* 201, 305, 381 (1928).

2.3-87 N. Z. Nusselt, *Z. Ver. Dtsch Ing., 67,* 206 (1923).

2.3-88 A. H. P. Skelland, *Diffusional Mass Transfer,* Chap. 5, Wiley-Interscience, New York, 1974.

2.3-89 H. Schlichting, *Boundary Layer Theory,* 4th ed., McGraw-Hill, New York, 1960.

2.3-90 E. Pohlhausen, *Zamm, 1,* 115 (1921).

2.3-91 J. P. Hartnett and E. R. G. Eckert, *Trans. Am. Soc. Mech. Eng., 79,* 247 (1957).

2.3-92 W. E. Stewart, *AIChE J., 8,* 421 (1962).

Section 2.4

2.4-1 A. P. Colburn and T. B. Drew, *Trans. AIChE, 33,* 197, (1937).

2.4-2 T. K. Sherwood and B. B. Woertz, *Ind. Eng. Chem., 31,* 1034 (1939).

2.4-3   C. S. Lin, R. W. Moulton, and G. L. Putnam, *Ind. Eng. Chem.*, **45,** 636, 1377 (1953).

2.4-4   W. Nernst, *Z. Phys. Chem.*, **47,** 52 (1904).

2.4-5   J. E. Vivian and W. C. Behrmann, *AIChE J.*, **11,** 656 (1965).

2.4-6   T. Yoshida and T. Hyodo, *Ind. Eng. Chem. Process Des. Dev.*, **9,** 207 (1965).

2.4-7   G. Ackermann, *Forschungsheft*, **382,** 1 (1937).

2.4-8   T. K. Sherwood, R. L. Pigford, and C. R. Wilke, *Mass Transfer,* Chap. 7, McGraw-Hill, New York, 1975

2.4-9   D. W. Humphrey and H. C. van Ness, *AIChE J.* **3,** 283 (1957).

2.4-10   R. B. Bird, W. E. Stewart, and E. N. Lightfoot, *Transport Phenomena,* p. 675, Wiley, New York, 1960.

2.4-11   W. K. Lewis and W. G. Whitman, *Ind. Eng. Chem.*, **16,** 1215 (1924).

2.4-12   T. K. Sherwood, R. L. Pigford, and C. R. Wilke, *Mass Transfer,* Chap. 8, McGraw-Hill, New York, 1975.

2.4-13   R. Higbie, *Trans. AIChE,* **31,** 365 (1935).

2.4-14   W. E. Stewart, as quoted in R. B. Bird, W. E. Stewart, and E. N. Lightfoot, *Transport Phenomena,* pp. 669 ff, Wiley, New York, 1960.

2.4-15   P. V. Danckwerts, *Chem. Eng. Sci.*, **20,** 785 (1965).

2.4-16   W. B. Lamb, T. G. Springer, and R. L. Pigford, *Ind. Eng. Chem. Funda.*, **8,** 823 (1969).

2.4-17   H. Toor and J. M. Marchello, *AIChE J.*, **4,** 97 (1958).

2.4-18   L. E. Scriven, *Chem. Eng. Educ.*, **2,** 150 (Fall, 1968); 26 (Winter, 1969); 94 (Spring 1969).

2.4-19   W. E. Stewart, J. B. Angelo, and E. N. Lightfoot, *AIChE J.*, **16,** 771 (1970).

2.4-20   J. B. Angelo, E. N. Lightfoot, and D. W. Howard, *AIChE J.*, **12,** 751 (1966).

2.4-21   A. H. P. Skelland, *Diffusional Mass Transfer,* Chap. 5, Wiley, New York, 1974.

2.4-22   J. O. Hinze, *Turbulence,* McGraw-Hill, New York, 1959.

2.4-23   G. K. Batchelor, *The Theory of Homogeneous Turbulence,* Cambridge University Press, Cambridge, 1960.

2.4-24   J. T. Davies, *Turbulence Phenomena,* Academic Press, New York, 1972.

2.4-25   T. K. Sherwood, R. L. Pigford, and C. R. Wilke, *Mass Transfer,* p. 109, McGraw-Hill, New York, 1975.

2.4-26   G. I. Taylor, in G. K. Batchelor (Ed.), *Scientific Papers,* Vol. II, Cambridge Press, New York, 1959.

2.4-27   H. Schlichting, *Boundary Layer Theory,* Chap. XX, McGraw-Hill, New York, 1960.

2.4-28   T. Von Karman, *Trans. AIME,* **61,** 705 (1939).

2.4-29   T. K. Sherwood, K. A. Smith, and P. E. Fowles, *Chem. Eng. Sci.*, **23,** 1225 (1968).

2.4-30   A. T. Popovich and R. Hummel, *AIChE J.*, **13,** 854 (1967).

2.4-31   T. K. Sherwood, *Chem. Eng. Prog. Symp. Ser.*, **55,** (25), 71 (1959).

2.4-32   T. K. Sherwood, R. L. Pigford, and C. R. Wilke, *Mass Transfer,* Chap. 5, McGraw-Hill, New York, 1975.

2.4-33   A. H. P. Skelland, *Diffusional Mass Transfer,* Chap. 6, Wiley, New York, 1974.

2.4-34   R. B. Bird, W. E. Stewart, and E. N. Lightfoot, *Transport Phenomena,* Chaps. 5, 10, and 20, Wiley, New York, 1960.

2.4-35   T. Sherwood, *Trans. AIChE,* **36,** 817 (1940).

2.4-36   R. G. Diessler, *Natl. Adv. Comm. Aeronaut. Dept.*, 1210 (1955).

2.4-37   D. T. Wasan, C. L. Tien, and C. R. Wilke, *AIChE J.*, **9,** 567 (1963).

2.4-38   R. H. Notter and C. A. Sleicher, *Chem. Eng. Sci.*, **26,** 161 (1971).

2.4-39   V. Levich, *Physicochemical Hydrodynamics,* Prentice-Hall, Englewodd Cliffs, NJ, 1962.

2.4-40   T. H. Chilton and A. P. Colburn, *Ind. Eng. Chem.*, **26,** 1183 (1934).

2.4-41   W. R. Vieth, J. H. Porter, and T. K. Sherwood, *Ind. Eng. Chem. Funda.*, **2,** 1 (1963).

2.4-42   R. H. Perry and D. W. Green (Eds.), *Perry's Chemical Engineers' Handbook,* 6th ed., McGraw-Hill, New York, 1984).

2.4-43   R. E. Treybal, *Mass Transfer Operations,* McGraw-Hill, New York, 1979.

2.4-44   T. K. Sherwood and R. L. Pigford, *Absorption and Extraction,* 2nd ed., p. 66, McGraw-Hill, New York, 1952.

2.4-45   D. R. Davies and T. S. Walters, *Proc. Phys. Soc. (London),* **B65,** 640 (1952).

2.4-46   T. K. Sherwood and O. Tröss, *Trans. ASME J. Heat Transfer*, **82C,** 313 (1960).

2.4-47   E. R. Gilliland and T. K. Sherwood, *Ind. Eng. Chem.*, **26,** 516 (1934).

2.4-48   M. L. Jackson and N. H. Caegleske, *Ind. Eng. Chem.*, **42,** 1188 (1950).

2.4-49   H. F. Johnstone and R. L. Pigford, *Trans. AIChE*, **38,** 25 (1942).

2.4-50   W. H. Linton and T. K. Sherwood, *Chem. Eng. Prog.*, **46,** 258 (1950).

2.4-51   E. S. C. Meyerink and S. K. Friedlander, *Chem. Eng. Sci.*, **17,** 121 (1962).

2.4-52   P. Harriott and R. M. Hamilton, *Chem. Eng. Sci.*, **20,** 1073 (1965).

2.4-53   J. T. Davies, *Turbulence Phenomena*, Chap. 3, Academic Press, New York, 1972.

2.4-54   H. Brenner, *Low Reynolds Number Hydrodynamics*, Prentice-Hall, Englewood Cliffs, NJ, 1962.

2.4-55   P. L. T. Brian and H. B. Hales, *AIChE J.*, **15,** 419 (1969).

2.4-56   S. K. Friedlander, *AIChE J.*, **7,** 347 (1961).

2.4-57   V. Levich, *Physicochemical Hydrodynamics*, Prentice-Hall, Englewood Cliffs, NJ, 1962.

2.4-58   P. N. Rowe, K. T. Claxton and J. B. Lewis, *Trans. Inst. Chem. Eng. (London)*, **43,** 14 (1965).

2.4-59   R. L. Steinberger and R. E. Treybal, *AIChE J.*, **6,** 227 (1960).

2.4-60   W. E. Ranz and W. P. Marshall, Jr., *Chem. Eng. Prog.*, **48,** 141, 173 (1952).

2.4-61   R. A. Brown, K. Sato, and B. H. Sage, *Chem. Eng. Data Ser.*, **3,** 263 (1958).

2.4-62   N. Frössling, *Beitr. Geophys.*, **52,** 170 (1938).

2.4-63   A. H. P. Skelland and A. P. H. Cornish, *AIChE J.* **9,** 73 (1963).

2.4-64   S. Evnochides and G. Thodos, *AIChE J.*, **5,** 178 (1959).

2.4-65   K. Lee and H. Barrow, *Int. J. Heat Mass Transfer*, **8,** 403 (1965).

2.4-66   D. S. Maisel and T. K. Sherwood, *Chem. Eng. Prog.*, **46,** 131, 172 (1950).

2.4-67   I. S. Pasternak and W. H. Gauvin, *AIChE J.*, **7,** 254 (1961).

2.4-68   T. K. Sherwood, R. L. Pigford, and C. R. Wilkie, *Mass Transfer*, Chap. 6, McGraw-Hill, New York, 1975.

2.4-69   M. Ohara and R. C. Reid, *Modeling Crystal Growth Rates from Solution*, Chap. 3, Prentice-Hall, Englewood Cliffs, NJ 1973.

2.4-70   P. Harriott, *AIChE J.*, **8,** 93 (1962).

2.4-71   P. L. T. Brian, H. B. Hales, and T. K. Sherwood, *AIChE J.*, **15,** 727 (1969).

2.4-72   D. M. Levins and J. R. Glastonbury, *Trans. Inst. Chem. Eng.*, **50,** 132 (1972).

2.4-73   P. Armenante, "Mass Transfer to Microparticles in Agitated Systems," Ph.D. Dissertation, University of Virginia, August 1983.

2.4-74   D. N. Miller, *Ind. Eng. Chem. Process Des. Dev.*, **10,** 365 (1971).

2.4-75   P. H. Calderbank and M. Moo-Young, *Chem. Eng. Sci.*, **16,** 49 (1961).

2.4-76   A. W. Hixson and S. W. Baum, *Ind. Eng. Chem.*, **33,** 478, 1433 (1941).

2.4-77   S. Nagata, *Mixing*, Chap. 6, Wiley, New York, 1975.

2.4-78   J. Garside and S. J. Jančić, *AIChE J.*, **22,** 887 (1976).

2.4-79   J. T. L. McConnache and G. Thodos, *AIChE J.*, **9,** 60 (1963).

2.4-80   R. E. Riccetti and G. Thodos, *AIChE J.*, **7,** 442 (1961).

2.4-81   A. J. Karabelas, T. N. Wegner, and T. J. Hanratty, *Chem. Eng. Sci.*, **26,** 1581 (1971).

2.4-82   K. Kato, H. Kubota, and C. Y. Wen, *Chem. Eng. Prog. Symp. Ser.*, **66**(105), 100 (1970).

2.4-83   K. Kato and C. Y. Wen, *Chem. Eng. Prog. Symp. Ser.*, **66**(105), 100 (1970).

2.4-84   J. F. Davidson and D. Harrison, *Fluidised Solids*, Cambridge University Press, London, 1963.

2.4-85   D. Kunii and O. Levenspiel, *Fluidization Engineering*, Wiley, New York, 1968.

2.4-86   O. Levenspiel and K. G. Bishoff, in T. B. Drew, J. W. Hoopes, and T. Vermeulen (Eds.), *Advances in Chemical Engineering*, Vol. 4, Academic Press, New York, 1963.

2.4-87   D. J. Gunn, *Chem. Eng. (London)*, p. CE 153 (June 1968).

2.4-88   O. Levenspiel, *Chemical Reactor Omnibook*, Chaps. 61–68, Oregon State University Press, Corvallis, 1979.

2.4-89   P. M. Heertjes and L. H. DeNie, in C. Hanson (Ed.), *Recent Advances in Liquid–Liquid Extraction*, Chap. 10, Pergamon, New York, 1971.

2.4-90   R. C. Kintner, *Advances in Chemical Engineering*, Vol. 4, pp. 52–92, Academic Press, New York, 1963.

2.4-91   A. H. P. Skelland, *Diffusional Mass Transfer*, Chap. 8, Wiley, New York, 1974.

2.4-92   R. Treybal, *Mass-Transfer Operations,* 3rd ed., Chap. 10, McGraw-Hill, New York, 1980.

2.4-93   P. N. Calderbank, in V. Uhl and J. Grey (Eds.), *Mixing,* Vol. 2, Chap. 6, Academic Press, New York, 1967.

2.4-94   J. H. C. Coppus, ''The Structure of the Wake Behind Spherical Cap Bubbles and Its Relation to the Mass Transfer Mechanism,'' Doctoral Dissertation, Technische Hogeschool Eindhoven, 1977.

2.4-95   T. K. Sherwood and J. Wei, *Ind. Eng. Chem.,* **49,** 1030 (1957).

2.4-96   F. H. Garner, C. W. Nutt, and M. F. Mohtodi, *Nature,* **175,** 603 (1955).

2.4-97   A. Ovell and J. W. Westwater, *AIChE J.,* **8,** 350 (1962).

2.4-98   L. J. Austin, W. E. Ying, and H. Sawistowski, *Chem. Eng. Sci.,* **21,** 1109 (1966).

2.4-99   J. C. Berg and C. R. Movig, *Chem. Eng. Sci.,* **24,** 937 (1969).

2.4-100   J. T. Davies and E. K. Rideal, *Interfacial Phenomena,* Chap. 7, Academic Press, New York, 1961.

2.4-101   C. Sternling and L. E. Scriven, *Chem. Eng. Sci.,* **5,** 514 (1959).

2.4-102   P. L. T. Brian and K. A. Smith, *AIChE J.,* **18,** 231 (1972).

2.4-103   K. H. Wang, V. Ludviksson, and E. A. Lightfoot, *AIChE J.,* **17,** 1402 (1971).

2.4-104   F. Goodridge and D. J. Bricknell, *Trans. Inst. Chem. Eng. (London),* **40,** 54 (1962).

2.4-105   S. Aiba, A. E. Humphrey, and N. F. Millis, *Biochemical Engineering,* 2nd ed., Academic Press, New York, 1973.

2.4-106   S. Aiba and K. Toda, *J. Gen. Microbiol.,* **9,** 443 (1963).

2.4-107   W. W. Eckenfelder, Jr. and E. L. Barnhart, *AIChE J.,* **7,** 631 (1967).

2.4-108   J. T. Davies, *AIChE J.,* **18,** 169 (1972).

CHAPTER 3

Phase Segregation

LOUIS J. JACOBS, JR.
Director, Corporate Engineering Division
A. E. Staley Manufacturing Company
Decatur, Illinois

W. ROY PENNEY
Director, Corporate Process Engineering
A. E. Staley Manufacturing Company
Decatur, Illinois

3.1 BASIC MECHANISMS AND ANALOGIES

This chapter covers separations that are strictly a segregation of well-defined phase materials, that is, gas, solid, or liquid. Excluded are separations involving equilibrium considerations or phase changes of any of the components.

Separations of this type are always performed following some other contacting operation, often an equilibrium separation, such as, entrainment separation following distillation, filtration after crystallization, decantation after liquid–liquid extraction, and collection of solids from air exiting a dryer. Sometimes, the phase segregation is performed by the equilibrium separator hardware, but most frequently a separate piece of equipment is used.

These operations will be grouped by the phases that are to be separated: gas–liquid, liquid–liquid, solid–gas, and solid–liquid. The fundamental mechanisms in each group are similar. There are four basic mechanisms that contribute to each group of separations:

1. Gravity.
2. Centrifugal force.
3. Impaction (or interception).
4. Electromotive force.

Within each separation group there are variations and differences of effects, but most of the methods and equipment fall into one of these four categories.

Gravity separations depend essentially on the density differences of the gas, solid, or liquids present in the mix. The particle size of the dispersed phase and the properties of the continuous phase are also factors with the separation motivated by the acceleration of gravity. The simplest representation of this involves the assumption of a rigid spherical particle dispersed in a fluid with the terminal or free-settling velocity represented by

$$U_t = \left[\frac{4 g d_p (\rho_p - \rho)}{3 \rho C} \right]^{1/2} \tag{3.1-1}$$

129

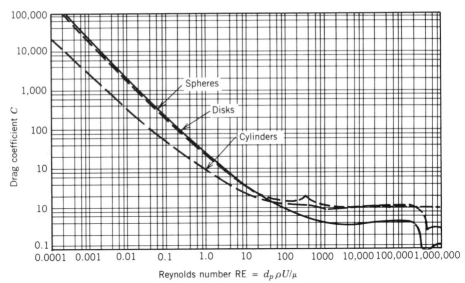

FIGURE 3.1-1 Drag coefficient versus Reynolds number for various particle shapes. Reprinted with permission from Lapple and Shepherd, *Ind. Eng. Chem.*, **32**, 605 (1940), © American Chemical Society.

where g is the local acceleration of gravity, d_p is the particle diameter, ρ_p is the particle density, ρ is the density of the continuous phase, and C is the drag coefficient. Any consistent set of units can be used in Eq. (3.1-1). The drag coefficient is a function of the particle shape and the Reynolds number, Re $=$ $d_p \rho U/\mu$, where U is the relative velocity between the fluid and the particle, and μ is the continuous fluid viscosity. Correlations have been developed by Lappel and Shepherd[1] and by many others for the relationship of the drag coefficient to the Reynolds number for various types of dispersed- and continuous-phase systems. A common correlation of the drag coefficient to Re is that of Lappel and Shepherd (Fig. 3.1-1). Such relationships serve as the basis for design of many types of gravity separator. They represent only the simplest approach since factors such as particle geometrical variations, nonrigid particles, and hindrance effects of other particles complicate an actual design. At Re $<$ 0.3, Stokes' Law applies with $C = 24/\text{Re}$ and the terminal velocity equation becomes

$$U_t = \frac{g d_p^2 (\rho_p - \rho)}{18\,\mu} \qquad (3.1-2)$$

Separations or segregations using centrifugal force are very similiar in principle to those for gravity separation. In this situation, the acceleration of gravity in Eq. (3.1-2) is replaced by the centrifugal acceleration $\omega^2 r$, where ω is the rate of rotation and r is the radial distance from the center of rotation. For high-speed mechanical separators, the force driving the separation could be several orders of magnitude greater than that of gravity. Geometrical effects, dispersed-phase concentrations, and other factors will complicate any actual calculations, but the above functionality will hold and is useful to understand the phenomenon.

Impaction or impingement is another mechanism common to the various groups of separations. Fluid carrying a discrete dispersed phase, that is, solid or liquid, impacts or impinges on a body. This body could be another solid or liquid particle, a plate, a fiber, and so on. The fluid will deflect around the body while the particle having greater inertia will impact the body allowing the opportunity for separation. Figure 3.1-2 shows an example of this phenomenon. Impingement is estimated by calculating a target efficiency $\eta_t = X/D_b$, where D_b is the body face dimension and X is the dimension approaching the body in which all dispersed particles of a specified size will impact the body. The target efficiency has been correlated for various body shapes as in Fig. 3.1-3 using a separation number $N_s = U_t V_0/g D_b$, where U_t is the terminal settling velocity of the particle assuming Stokes' Law applies, V_0 is the bulk fluid velocity, g is the acceleration of gravity, and D_b is the characteristic body dimension. Since η_t and N_s are both dimensionless, any consistent set of units may be used. This calculation assumes the body is in an infinite fluid. The presence of other target bodies, as in a rain of spray droplets or a fiber bed, would allow a higher collection efficiency.

A subset of the impaction mechanism is that of filtration for solid dispersed-phase particles. A screen or cloth may be used in which the space between components of the filter media is smaller than the particle

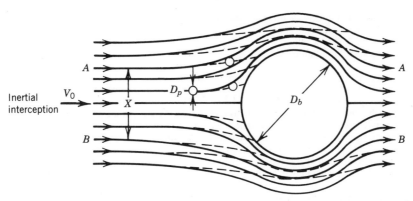

FIGURE 3.1-2 Mechanism of inertial interception. Adapted from *Perry's Chemical Engineers' Handbook*, 6th ed., 1984, Courtesy of McGraw-Hill Book Co.

size and the particle is sieved out of the fluid. Frequently, however, initial filter operation involves the basic impaction mechanism to remove particles even smaller than the filter media opening. Once solids build up on the media, the depth of solids ("the cake") serves as the impaction or sieving media, not the filter cloth or screen.

Another subset of the impaction mechanism involves collection of very small particles, less than 0.3 μm, via contact with target bodies by Brownian movement. Brownian movement is a random diffusion of these small particles; since they are so small they have little momentum. In dense fields of target bodies, these particles diffuse randomly into contact with a body and are separated from the bulk fluid.

Electromotive or electrically induced charge separations are the fourth basic segregation mechanism common to the various phase combinations. The most common example of this mechanism is the gas–solid electrostatic precipitator. An electrical field is also used commercially for gas–liquid separations and for removing liquid water from organic liquids and solids from organic liquids. In most cases, the mixture is passed through an electrical field which charges the particles. Fixed bodies such as plates or grids with an opposite electrical charge are used to attract and collect the particles. The basic separation equations involve an efficiency calculation:

$$n = 1 - e^{-V_m A/Q} \tag{3.1-3}$$

where n = collection efficiency (<1)

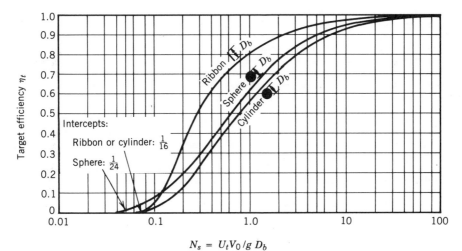

$$N_s = U_t V_0 / g\, D_b$$

FIGURE 3.1-3 Target efficiency of various shapes for particles in a Stokes' Law condition. From Langmuir and Blodgett, U.S. Army Air Forces Tech. Rep. 5418, 1946, U.S. Dept. of Commerce, Office of Technical Service PB27565.

V_m = particle velocity toward collection electrode
A = collector electrode area
Q = bulk fluid flow rate
e = natural logarithm base

The V_m term is directly related to particle diameter and the electrical field strength of each electrode, and inversely proportional to the fluid viscosity. This term is difficult to calculate for performance predictions, but it is typically between 0.1 and 0.7 ft/s (0.03 and 0.21 m/s) for solid-in-gas systems.

Many of the commercial phase segregations involve a combination of the mechanisms discussed, either in the same piece of equipment or in multiple equipment items in series.

In addition to common separation mechanisms for the various combinations of phase segregation, the approach to solving problems also is similar. In each case, the quantity of each phase must be known, as well as the density of each phase, the viscosity of the continuous phase, the dispersed-phase particle size, and the desired specification for the quality of the finished separation. In most cases, this involves sampling and experimental work.

Selection of a specific separation technique or scheme requires the above information plus comparison of capital cost, operating cost, space available for equipment installation, and material of construction requirements. If the desired separation quality is obtainable by several approaches, the last four factors become trade-off parameters.

3.2 GAS–LIQUID SEGREGATION

3.2-1 General

Gas–liquid phase segregation is typically required following some other separation process, such as distillation, absorption, evaporation, gas–liquid reactions, and condensation. Before separation techniques or equipment can be selected, the parameters of the separation must be defined. Information on volume of gas or liquid, volume ratio of the phases, and dispersed-phase particle size—that is, drop or bubble size—should be known or estimated. For existing operations, measurements are a possibility, while for new facilities analogies from data on other processes could be used. Laboratory or pilot plant tests may be considered, but it is difficult to maintain all fluid dynamic properties constant while changing scale and wall effects may be significant on the small scale.

The most common of the gas–liquid separations are those involving liquid dispersed in gas such as entrainment from distillation trays, evaporators, or condensers. The usual objective is to prevent environmental emissions, downstream process contamination, or to conserve valuable material. The ratio of liquid to gas volumes is usually small although the weight ratio may be very high. Drop size is a key to selecting the most effective separation technique, but size is difficult to measure. It must also be measured at the point where the separation is to be made owing to coalescence of collected particles. Sample trains consisting of a series of cyclone or various impaction collectors (Fig. 3.2-1) are available and can give an indication of liquid quantities in various particle size cuts, although they are best used for solid particle cut evaluation.

Sampling must be done in an isokinetic manner to assure a representative sample. Isokinetic means that the sample must be collected in-line with the bulk gas flow so that the velocity in the sample tube equals the bulk flow velocity (Fig. 3.2-2). This sampling approach works in a dispersed two-phase flow regime, but it is difficult in either two-phase gas–liquid systems such as annular or stratified flow regimes. Sampling is also more difficult in horizontal flow than in vertical flow because drops larger than 5 μm tend to a nonuniform distribution in the stream, owing to gravity. Small particles (less than 3–5 μm) tend to remain uniformly dispersed owing to Brownian movement. Samples should be taken at more than one position across a duct.

For new gas–liquid systems or where measurements of particle size are too difficult to obtain, predictions from existing data can be used as analogies. Four sources of such information are spray nozzle correlations and data on gas bubbling through a liquid, vapor condensation, and two-phase flow in pipes. A good review of methods for predicting particle size can be found in Perry and Green.[1] Table 3.2-1 indicates approximate size ranges for each of these operations if detailed correlations are not available.

Methods for estimating quantities of entrainment from gas–liquid contacting devices are also scarce. Distillation tray correlations by Fair[2] are useful for tray-contacting operations (Figs. 3.2-3 and 3.2-4). $C_{SB, Flood}$ is obtained from Fig. 3.2-3; the flooding velocity is calculated by

$$V_f = C_{SB, \, Flood} \left(\frac{\sigma}{20}\right)^{0.2} \left(\frac{\rho_L - \rho_G}{\rho_G}\right)^{0.5} \tag{3.2-1}$$

where V_f = velocity (ft/s).

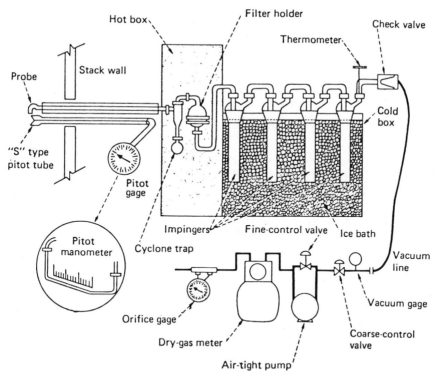

FIGURE 3.2-1 Typical vapor stream sampling train configuration. Excerpted by special permission from *Chemical Engineering*, August 5, 1985, © McGraw-Hill, Inc., New York, NY 10020.

σ = liquid surface tension (dynes/cm)
ρ_L = liquid density (lb$_m$/ft^3)
ρ_G = gas or vapor density (lb/ft^3)
W_L, W_G = mass flow rates for liquid and vapor (gas) in Figs. 3.2-3 and 3.2-4

The percent of flood is calculated by $(V_f/V_{\text{actual}}) \times 100$. The entrainment ratio of liquid/gas is obtained from the y axis in Fig. 3.2-4. These correlations can also be used to approximate entrainment from bubbling liquids, as in evaporators, by setting W_L/W_G equal to 1 and using the sieve tray correlations in Fig. 3.2-4. Below 30% flood the amount of entrainment is almost inversely proportioned to the (vapor rate)^3 down to a liquid/gas ratio of 0.0001. Below this value, liquid/gas entrainment varies directly with the gas rate indicating a change in drop formation mechanism.

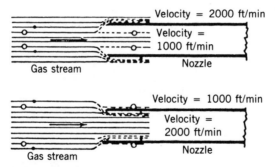

FIGURE 3.2-2 Illustration of the need for isokinetic sampling.

TABLE 3.2-1 Particle Size for Various Processes

Conventional liquid spray nozzles	100–5000 μm
Two-fluid atomizing spray nozzles	1–100 μm
Gas bubbling through liquid	20–1000 μm
Condensation processes with fogging	0.1–30 μm
Annular two-phase pipe flow	10–2000 μm

3.2-2 Gravity Segregation

Gravity separation occurs by merely reducing the velocity of a stream so that terminal particle settling or rise velocities due to gravity exceed the velocity of the bulk flow. The terminal particle or bubble velocities can be estimated by the methods described in Section 3.1.

This approach is usually not practical for drops less than 100 μm, since for most situations large cross-sectional areas are needed to reduce the bulk velocity below the terminal settling velocities of such drops. An air stream with a 10 ft/s vertical velocity will entrain water drops less than about 700 μm, while a bulk velocity of 1.0 ft/s will entrain drops of about 100 μm or less. Gravity techniques can be used to remove large quantities of the larger-size dispersed-phase material prior to some other segregation technique.

Horizontal bulk flow configurations can be more effective than vertical flow by providing the dispersed material a shorter path to travel to be separated. Figure 3.2-5 shows a configuration where the minimum size particle that is removed at 100% efficiency is calculated by

$$d_{p,\min} = \left[\frac{18\, V_0 h \mu}{L g (\rho_L - \rho_G)} \right]^{1/2} \tag{3.2-2}$$

This equation neglects the effect of stream turbulence. If the bulk stream flow has Re > 2100, then the drop settling velocity should exceed the velocity of turbulent eddies, which can be estimated by $V_e = V_0 (f/2)^{1/2}$, where f is the friction factor determined from a Reynolds number correlation.

The above discussion holds for both liquid drops in gas or gas bubbles in liquid. It is especially difficult to remove small gas bubbles from liquids because of the high bulk fluid viscosity. A useful technique for gas removal involves use of thin liquid films or use of parallel plates in the bulk stream to minimize the vertical distance for a bubble to travel before coalescing with other gas volumes.

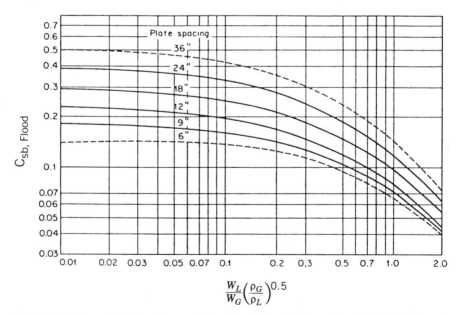

FIGURE 3.2-3 Vapor–liquid flooding correlation. From Fair, *Pet./Chem. Eng.*, **33**(10), 45 (September 1961).

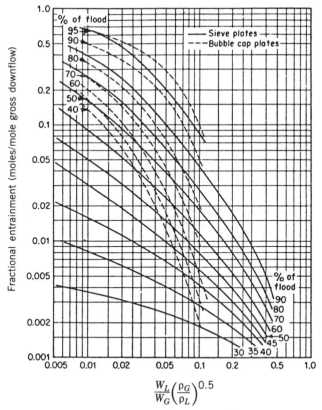

$$\frac{W_L}{W_G}\left(\frac{\rho_G}{\rho_L}\right)^{0.5}$$

FIGURE 3.2-4 Entrainment correlation for vapor–liquid contacting. From Fair, *Pet./Chem. Eng.*, **33**(10), 45 (September 1961).

3.2-3 Centrifugal Separators

The most common type of centrifugal separator for segregation is the cyclone. Many designs are used for gas–liquid and gas–solid separations. A design is shown in Fig. 3.5-2 which is common for gas–solid separations. It has also been used for gas–liquid separation but problems are often encountered. These problems are apparently due to re-entrainment of liquid from the cyclone wall or to liquid running down the outside of the outlet stack which is then carried out with the exit gas. A much improved design for gas–liquid cyclone or tangential entry separator is shown in Fig. 3.2-6. This design is based on information by Ludwig[3] and Penney and Jacobs.[4] The key differences over conventional cyclone design are the following:

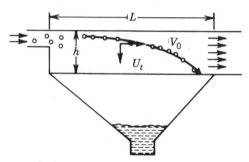

FIGURE 3.2-5 Horizontal gravity-settler model.

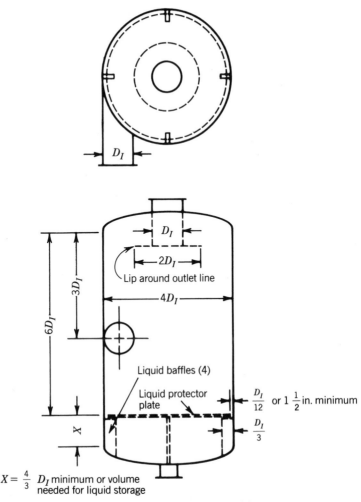

$$X = \frac{4}{3} \, D_I \text{ minimum or volume needed for liquid storage}$$

FIGURE 3.2-6   Recommended vapor–liquid tangential separator geometry.

1. The cone bottom is replaced with a dished or flat head.
2. Liquid in the bottom head is isolated from the rest of the separator volume by a horizontal plate with an open annular space at the vertical vessel walls to allow the separated liquid to pass into the bottom of the vessel.
3. The gas–liquid inlet is located well below the top of the vessel.
4. A lip is attached to the entrance of the outlet stack to cause liquid running down the outside of the outlet pipe to drop through the separator away from the high-velocity outlet gas vortex.

Figure 3.2-7 illustrates the potential problem with liquid re-entrainment if some of these measures are not applied. The horizontal liquid protector plate, which separates accumulated liquid from swirling gases is probably best supported by vessel wall baffles. This prevents rotation of the collected liquid and re-entrainment.

As the inlet velocity is increased from very low values, the separation efficiency rapidly increases to a maximum in the 80–120 ft/s range. At higher velocities, the liquid on the wall can be re-entrained by gas shear into the rotating gas vortex, and the separation efficiency drops significantly. The design limitation should be on the basis of gas shear which is proportional to $\rho_G V_I^2$, where ρ_G is the gas density (in lb_m/ft^3) and V_I is the inlet velocity (in ft/s); $\rho_G V_I^2$ should be less than 1000.

For gas–liquid separators designed as in Fig. 3.2-6, the separation can be estimated using a common cyclone equation:

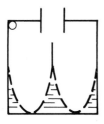

FIGURE 3.2-7 Potential problems of vapor-liquid tangential separators. From Stern et al., *Cyclone Dust Collectors*, American Petroleum Institute, New York, 1955.

$$d_{pc} = \left[\frac{9 \, \mu D_I}{2 \pi N \, V_I \, (\rho_L - \rho_G)} \right]^{1/2} \qquad (3.2\text{-}3)$$

where d_{pc} is the drop cut diameter for 50% removal, N is the number of vapor spirals in the separator body, D_I is the inlet nozzle diameter (in ft), and V_I is the inlet velocity (in ft/s). Table 3.2-2 provides an indication of variation of N with inlet velocity. Figure 3.2-8 indicates prediction of removal efficiency for other particle sizes. From Eq. (3.2-3), it can be observed that larger flow rates require larger diameters which increase the size of the drops effectively removed. Table 3.2-3 indicates the effect of separator size on drop collection. For very large flows, multiple units allow more efficient separation, although careful piping design is required to assure equal distribution to each unit.

Pressure drops through tangential entry separators are not well documented, but they are generally in the range of 8 velocity heads; that is, $\Delta p = 8 \, (\rho_G V_I^2 / 2 g_c)$

Several commercial separators are available using centrifugal principles. All are similar, in that stationary internal vanes direct the flow into a centrifugal motion which causes denser liquid drops to contact a surface where they coalesce and are separated by draining. These commercial devices are either in-line or inside vessels near the vapor outlet (Figs. 3.2-9 and 3.2-10). These devices are limited to smaller fractions of liquid than the tangential separators described above. They are, however, about 99% efficient in removing 10 μm drops. Sizing and pressure drop estimation are based on vendor's literature. Common application is in steam lines or evaporator heads to improve the "quality" of steam generated.

3.2-4 Impaction and Impingement

This category involves a wide range of styles and devices, all based on some obstruction to the bulk vapor flow causing a directional change in the flow. The entrained liquid particles having greater momentum impinge or impact on a surface, coalesce with other liquid material, and separate via gravity.

Several manufacturers supply units with "wavy plates" which force several direction changes in series (Fig. 3.2-11). Channel spacing may vary from a fraction of an inch to 6–8 in. Devices with four bends are common with channel spacing of 2–3 in. The best separation efficiency is obtained in the range of 7–20 ft/s bulk velocity with a pressure drop in the range of 0.05–0.15 in. H_2O. Vendors should be consulted for actual performance data. These devices are limited to about 5% liquid by weight, and they can handle some entrained solids with the liquid flushing away collected solids. Separation efficiencies of 90% are possible for particles down to about 10 μm.

Some of the units have gutters (Fig. 3.2-11) on the vanes to trap separated liquid and force it to drain. Other designs such as the Petersen Candle Separator (Fig. 3.2-12) have close channel spacing down to 0.025–0.03 in., with a high turning angle which allow 80–90% separation of drops down to 2–3 μm. This is accomplished with a much higher pressure drop, 12–20 in. H_2O.

Knitted wire mesh pads are a common impingement device for removing liquid entrainment from vapor or gas streams. The Otto H. York Company pioneered this concept with the Demister®. Several vendors are now active in the field.

The wire or fiber pad is generally 4–12 in. in depth with 6 in. being the most common size. Wire diameter ranges from 0.006 to 0.15 in. and the density of the mesh varies from 5 to 13.5 lb_m/ft^3 for the

TABLE 3.2-2 Number of Tangential Separator Spirals as a Function of Inlet Vapor Velocity

V_I (ft/s)	N
30	3
65	4
100	5

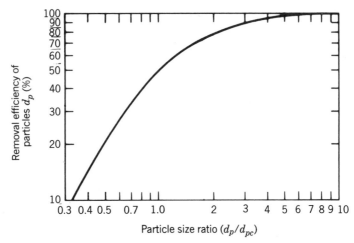

FIGURE 3.2-8   Adjustment of separation efficiency for various particle sizes for tangential separators.

common 316 stainless steel material. The wire provides a target surface on which the liquid drops impinge as the mixture passes through. The larger, heavier drops have enough momentum that they are unable to follow the gas through the many layers of wire. As drops impact the wires, they coalesce with liquid already held on the wire until the mass of retained liquid grows large enough to drain and drop off the mesh pad. The pads are generally placed horizontally with vertical vapor flow up through the pad. Figure 3.2-13 shows several configurations.

As with other separators, there is an optimum vapor velocity through the pad. The allowable velocity is calculated by

$$V_a = K \left(\frac{\rho_L - \rho_G}{\rho_G} \right)^{1/2} \tag{3.2-4}$$

K is a design constant based on experience which varies with disengaging height above the source of entrainment, system pressure, liquid loading, and liquid viscosity. The most common recommendation of the vendors is $K = 0.35$. A disengaging height below about 10 in. should have lower K values: $K = 0.12$ for 3 in. or $K = 0.22$ for 6 in. For low-pressure applications the value of K should be reduced; at 2 in. Hg abs, $K = 0.2$ or at 16 in. Hg abs, $K = 0.27$ is recommended. Most vendors agree that for pressure surges a design velocity of a new separator should be $V_D = 0.75\,V_a$. The separator diameter is calculated by $D = (4Q/\pi V_D)^{1/2}$. The calculated diameter is that available for the vapor flow and does not include the portion of the pad diameter that is ineffective because of support rings or bars.

Good separation efficiency is obtained from $0.3V_a$ to V_a. Efficient removal of smaller particles increases with velocity until re-entrainment occurs at velocities above V_a. Mesh pads are effective in drop removal down to 5–10 μm. Some solids may be handled, but the lower-density mesh and liquid sprays above the pad are recommended to prevent pluggage.

Pressure drop across a mesh pad is usually less than 1 in. H$_2$O and is of little concern except in vacuum operations or systems involving large flows with fans or blowers at near atmospheric pressure. Pressure drop is estimated by $\Delta p = CV_G^2 \rho_G$, where Δp is the pressure drop (in in. H$_2$O), V_G is the gas velocity (in ft/s), and ρ_G is the vapor density (in lb$_m$/ft^3; $C = 0.2$ for a standard-density 4 in. thick pad, and $C = 0.12$ for a low-density 6 in. thick pad. For situations where pressure drop is critical, methods presented

TABLE 3.2-3   Separator Diameter Effect on Drop Size Removed

Separator Diameter (ft)	Drop Diameter 50% Removed	Drop Diameter 90% Removed
4	10	30
8	15	40
12	19	50

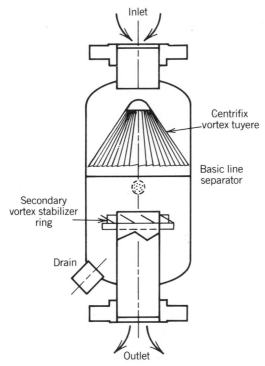

FIGURE 3.2-9 Commercial in-line centrifugal vapor–liquid separator. Courtesy of the Centrifix Corp.

by York and Poppele[5] should be used which include physical property, liquid loading, and mesh density effects.

Pads are usually made in sections to fit through vessel access paths with hold down bars above and below the mesh. These sections may be wired together and then wired to an annular vessel support ring.

Mesh material is usually 316 stainless steel wire although monel, nickel, copper, titanium, alloy 20, teflon, polyethylene, or other plastics are also available.

A wide variety of other packing materials such as rings, saddles, and Tellerettes are used to eliminate entrainment. Correlations for flow and pressure drop through these materials can be found in Chapter 6. Particles strike the packing and, if the surface is wet, adhere and coalesce. Drops as small as 3 μm can be separated with 90% efficiency with vapor velocities of up to 30 ft/s. With high liquid loading, the packing may be adequately wetted by the entrainment. If not, irrigation of the packed bed will be needed. Vapor flow can be either vertical or horizontal with countercurrent or cocurrent liquid irrigation in the vertical case or crossflow liquid design in the horizontal case. Horizontal crossflow has advantages of smaller total liquid rates to obtain the same wetting and more effective flushing of solids. Detailed packing design information is available from vendors. Unirrigated efficiency is less than mesh pads and frequently a packed section is followed by a mesh pad to catch final entrainment.

Tightly packed fiber bed mist eliminators are another option for impingement separations. These devices were developed by the Monsanto Company for sulfuric acid plant mist applications, but they are now also available from various companies. Their main advantage is to allow efficient removal of particles down to the 0.1–1 μm range. These very small particles are not separated by having greater momentum than the vapor or gas, but by the random Brownian movement or random diffusion. The particles diffuse to the fiber surface in the very densely packed beds. Brownian movement actually increases as particle size decreases, improving the separation of very small particles.

Various configurations are used, but a cylindrical bed as in Fig. 3.2-14 is the most common. The typical size ranges from 8.5 in. in diameter and 24 in. long to 24 in. in diameter and 120 in. long with multiple parallel units used for high gas flows.

Since the bed is very dense, the pressure drop will be very high unless low velocity and large cross-sectional flow areas are used. The units can be sized for a specific pressure drop since this quantity is not a significant factor in the efficiency obtained. Pressure drops of 2–20 in. H$_2$O are typical with bed velocities in the 0.25–0.7 ft/s range.

While these devices are unsurpassed for particles less than 3 μm, they are very expensive when compared

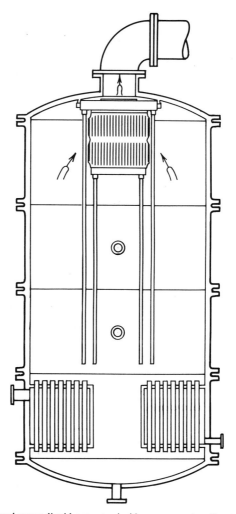

FIGURE 3.2-10 Centrifugal vapor–liquid separator inside an evaporator. Courtesy of the Centrifix Corp.

to mesh pads and packing. Information on particle size should be obtained to be sure removal below 3 μm is required. Devices such as packing or mesh pads upstream of the fiber bed may be desirable to limit the liquid loading on the fiber bed, thereby minimizing pressure drop and possible re-entrainment.

Other techniques using an impingement or impaction principle involve collision of drops with larger drops of the same or different liquids. One way of facilitating this is to use spray nozzles to create a large population of liquid drops that intercept or contact the drops to be removed. Spray nozzles can be selected by style and pressure drop to obtain the desired droplet size that maximizes removal of entrained material and is large enough to be collected easily by gravity. Figure 3.1-3 is a plot of collection efficiency of a single drop. The V_0 term is the relative velocity between the spray particle and the bulk gas flow. With target efficiency η_T, the overall efficiency of the spray can be estimated by Eq. (3.2-5):

$$E_0 = 1 - \exp\left(\frac{-3\eta_T L Q_L}{2 d_c Q_G}\right) \tag{3.2-5}$$

where L is the effective spray length (in ft) and d_c is the effective spray drop diameter (in ft). For multiple stages of sprays, as in Fig. 3.2-15, the separation efficiency is estimated by Eq. (3.2-6):

$$E_m = 1 - (1 - E_0)^m \tag{3.2-6}$$

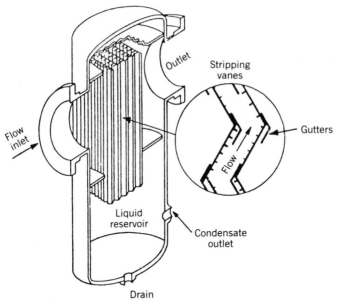

FIGURE 3.2-11 Wavy plate impingement separator. Courtesy of the Peerless Manufacturing Co.

How Petersen Separator works

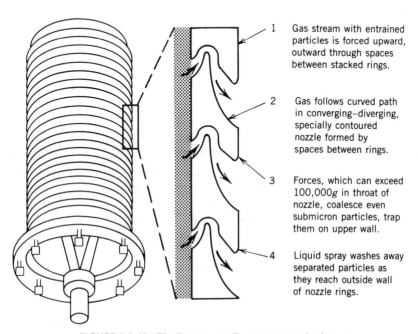

1 Gas stream with entrained particles is forced upward, outward through spaces between stacked rings.

2 Gas follows curved path in converging–diverging, specially contoured nozzle formed by spaces between rings.

3 Forces, which can exceed 100,000g in throat of nozzle, coalesce even submicron particles, trap them on upper wall.

4 Liquid spray washes away separated particles as they reach outside wall of nozzle rings.

FIGURE 3.2-12 The Peterson candle separator mechanism.

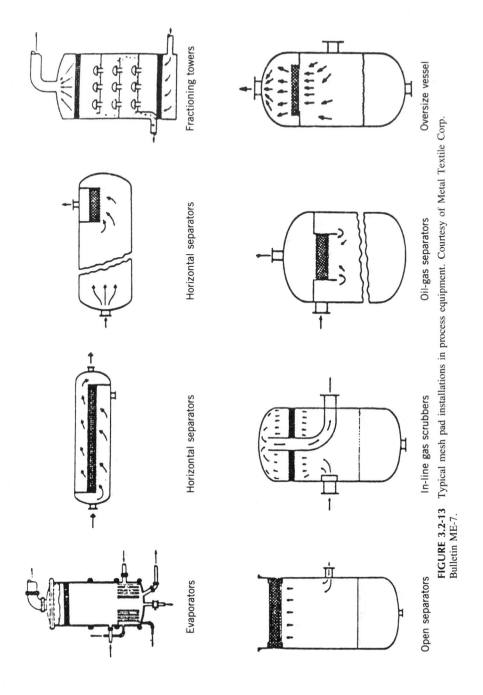

FIGURE 3.2-13 Typical mesh pad installations in process equipment. Courtesy of Metal Textile Corp. Bulletin ME-7.

Fractioning towers

Oversize vessel

Horizontal separators

Oil-gas separators

Horizontal separators

In-line gas scrubbers

Evaporators

Open separators

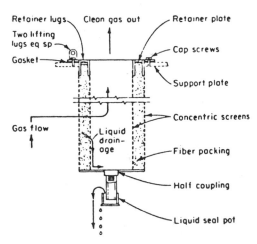

FIGURE 3.2-14 Fiber bed mist eliminator configuration. Courtesy Monsanto Envirochem Co.

where m is the number of stages of sprays. Desirable drop sizes of sprays are in the 200–800 μm range.

Numerous commercial designs are available using sprays. Open towers with sprays can remove particles down to about 10 μm with about 0.5–1.0 in. H_2O pressure drop. Spray systems have the advantage of handling solid entrainment as well as liquid, and they operate with a low irrigation rate, about 0.134 ft³ liquid/ft³ vapor.

Venturi scrubbers are an enhancement of the spray approach. A contacting liquid is added to the entrainment-contaminated gas stream just upstream of, or at the throat of, a venturi (Fig. 3.2-16). The turbulence of high-velocity gas in the venturi causes break-up of the introduced liquid into small drops that intercept and coalesce with the small, entrained particles. Collection efficiency greater than 90% of particles

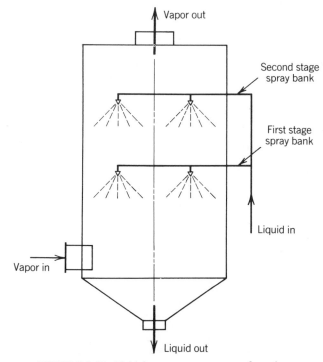

FIGURE 3.2-15 Multiple-stage spray tower configuration.

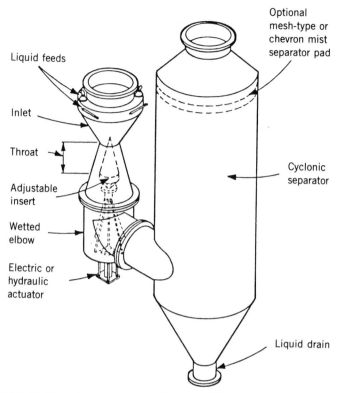

FIGURE 3.2-16   Venturi separator system. Courtesy of *Plant Engineering*, Sept. 30, 1982.

down to 0.5 μm is possible with a venturi, but with a high pressure drop of about 20–40 in. H$_2$O. A venturi can handle liquid and solid entrainment and can also allow for gas absorption by using appropriate contacting liquids. The scrubbing liquid is generally recirculated with purge and make-up added as needed.

Venturi scrubbers are available in various configurations from several vendors. All the designs have another device following the venturi, such as a cyclone or mesh pad to separate the liquid–gas mixture leaving the venturi.

Venturi scrubbers are costly to operate because of the high pressure drop associated with their use, particularly for large gas flows. Figure 3.2-17 indicates some of the trade-offs of pressure drop, liquid/gas ratio, and entrainment size removal.[6] Various correlations are given in the literature, particularly by Calvert[6] and a good review is found in Perry and Green,[1] but vendors should be consulted for actual design data.

3.2-5   Electromotive Devices

Electrostatic precipitators are most often used for the removal of solids from large gas volumes, but they work equally well for submicron- to low-micron-range liquid entrainment. At dc voltages of 25–100 kV, a corona discharge occurs close to the negative electrode, which is usually a wire or grid. Gas is ionized and migrates to a grounded collector electrode, which is usually a vertical plate. Particles in the gas are charged and drawn to the collector plate where their charge is neutralized and they are collected. Liquid removal is less of a problem than solid removal since collected liquid coalesces and drains readily. Gas usually flows horizontally past the electrodes at velocities in the 2–8 ft/s range to allow particles time to migrate to the collector without significant turbulence. In some designs, vertical cylinders are used as the collector with the charging electrode being a concentric wire with gas flow up through the cylinder.

Pressure drop in these devices is only about 1 in. H$_2$O which is a big advantage for very small particle removal in comparison to fiber bed or venturi separators. The low gas velocity and low pressure drop, however, require a unit with a large volume that has high capital cost.

Theoretical calculations are not recommended for design, but Eq. (3.1-3) gives an indication of the effect of several variables. Increasing collection efficiency from 90 to 99% requires doubling the collector area.

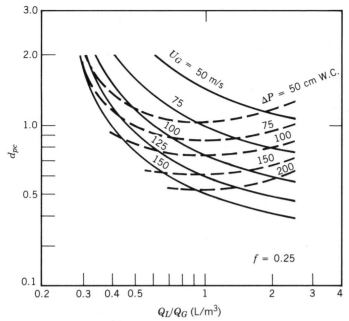

FIGURE 3.2-17 Venturi scrubber parameter trade-offs. Reproduced with permission from S. Calvert, *J. Air Pollut. Control Assoc.*, **24**, 929 (Oct. 1974).

3.2-6 Selection Approach

Selection of the optimal gas–liquid separation technique or equipment configuration depends on the following factors:

1. Primary objective of the separation (clean gas or clear liquid).
2. Concentration of liquid and vapor (or gas).
3. Size of drops (or bubbles if the liquid phase is continuous).
4. Separation efficiency required.
5. Presence of any solids with liquid entrainment.
6. Material of construction requirements.
7. Capital and operating costs.

These factors are listed in the order in which they are normally considered in making a selection. Figure 3.2-18 presents a logic diagram to aid in selection.

The first decision in the equipment-selection process is determining whether the objective of the separation is a good quality liquid or gas. A clean gas, free of entrainment, is the more common separation goal, especially in situations where gas or vapor is to be vented to the environment. The option seeking a good quality liquid effluent is a much less common problem. Typical situations involving the separation of gas from liquid include the removal of bubbles or foam to measure liquid properties for process control or final liquid product packaging.

The concentration of one phase in the other dictates the equipment configuration. If a clean gas stream is the primary objective, a stream with a high liquid loading, greater than 10% by volume of liquid, will need some technique to remove the gross liquid prior to a final clean-up device. A gravity-separator vessel or tangential-entry vessel can provide removal of most of the liquid with a mesh pad, wavy vane, or fiber bed following to provide the desired clean-up.

The particle size of liquid drops or gas bubbles to be removed significantly influences the selection. Figure 3.2-19 indicates the working range of the common gas–liquid separators as well as liquid particle sizes formed in several common processes.

Efficiency of separation from the stream is a major factor in the economics of separation. Clean water in gas from an evaporator or dryer may be much less critical than removal of sulfuric acid mist from an acid plant vent. Therefore, separation specifications and complexity of equipment can vary greatly.

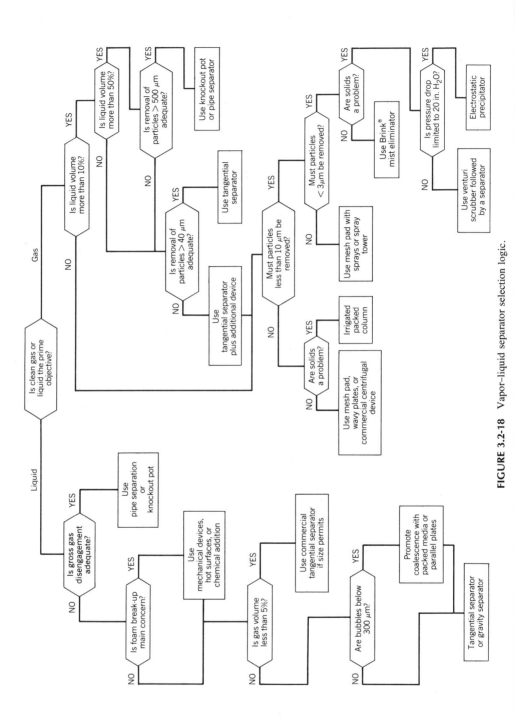

FIGURE 3.2-18 Vapor–liquid separator selection logic.

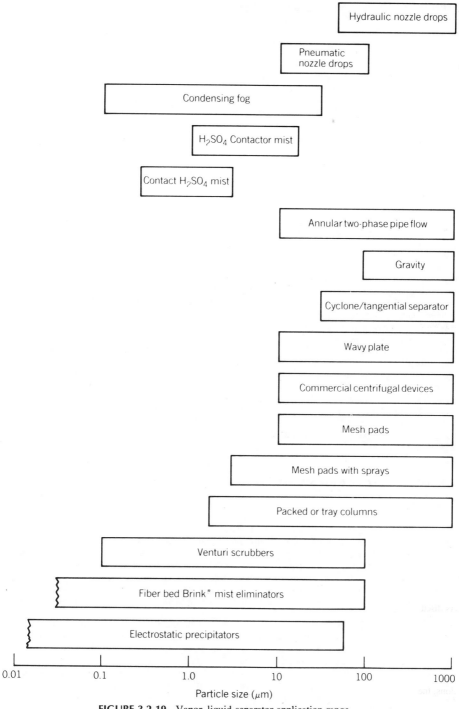

FIGURE 3.2-19 Vapor–liquid separator application range.

TABLE 3.2-4 Relative Costs of Gas–Liquid Separation

	Capital	Operating
LOW	Open pipes	Open pipes
	Mesh pads	Gravity chambers
	Centrifugal vanes	Mesh pads
	Mesh pad with vessel	Wavy plates
	Wavy plates	Spray columns
	Tangential entry separators	Packed sections
	Packed sections	Centrifugal turning vanes
	Spray towers	Electrostatic precipitators
	Venturi with separator	Fiber beds
	Gravity settling chamber	Tangential entry separators
	Fiber bed	Venturi scrubbers
HIGH	Electrostatic precipitator	

The presence of solids, even in small quantities, can be an extreme nuisance factor in a system design. Solids can readily foul packed columns, mesh pads, or fiber beds unless prescrubbing or significant irrigation is used to flush contact surfaces.

Material of construction is not a big factor in equipment selection because most equipment is available in a wide variety of metals, glasses, or plastics. In some situations where plastics are available, significant cost savings may be possible compared to use of some of the exotic metals.

Capital and operating costs are always the key factor in an equipment selection. They become a factor, however, only after multiple approaches have been identified as viable solutions to the defined problems. It is very difficult to generalize cost information for the amount of material to be handled, the separation requirements, and the materials of construction needed. Table 3.2-4 indicates relative comparisons of the general equipment types with regard to capital and operating costs. Operating costs should consider pressure drop through the equipment if the loss must be compensated with fan or blower energy, liquid handling such as pumping costs, chemical usage for cleaning, and maintenance. For situations where a process is vented from some significant pressure to atmosphere, the pressure drop through a separation may not be a factor if the loss is available.

The gravity settling chamber shows up as a high capital cost technique, if it is used separately to approach a high degree of separation efficiency. A much smaller gravity chamber to remove gross entrainment plus a follow-up device such as a mesh pad or centrifugal vane device could make a very economical separator.

3.3 IMMISCIBLE LIQUID SEGREGATION

3.3-1 General

As discussed in Section 3.1, the principles of all phase segregations are very similar regardless of whether the individual phases are gas, liquid, or solid. However, separation of immiscible liquid phases is distinctive compared to the others in that the important density difference between the phases is usually very small, sometimes less than 0.1 g/cm^3. Liquid–gas, liquid–solid, or gas–solid separations typically have phase density differences of 1.0 g/cm^3 or more. Segregation of phases with a small density difference requires large equipment, application of large forces, or unique equipment geometries. As with gas–liquid separations, the common mechanisms are gravity, centrifugal force, impaction and interception, and electromotive effects.

Immiscible liquid phases are formed because of chemical effects, namely, the mutual solubilities of the two phases. The design for liquid–liquid separations is affected therefore by changes in temperature, pressure, presence of contaminants such as surfactants, and stream mixing effects. In this section, however, we will not consider any solubility factors, only the effect of physical forces.

The presence of two liquid phases may be the result of an extraction, washing, or reaction operation. The two phases may have been contacted in pipes, static mixers, agitated vessels, or at the impeller of a pump. The type of contacting, the level of shear forces applied, the concentration of the phases, and the time of contacting all influence the difficulty of the separation process.

To define the immiscible-liquid-segregation problem, the drop size range of the dispersed phase must be known or approximated. Weinstein[1] and Middleman[2] present drop correlations for pipeline contacting and static mixers while Coulaloglou and Tavlarides[3] present correlations of liquid in liquid drop sizes for agitated vessels. All use the Sauter mean drop size definitions and involve use of the Weber number, a dimensionless group relating inertial forces and surface-tension forces as in Eqs. (3.3-1) and (3.3-2):

$$We = \frac{DV^2 \rho_C}{\sigma} \quad \text{(for pipes)} \tag{3.3-1}$$

$$We = \frac{N^2 D_I^3 \rho_C}{\sigma} \quad \text{(for agitated vessels or pump contacting)} \tag{3.3-2}$$

where D is the pipe diameter, V is the average velocity in the pipe, ρ_C is the continuous-phase density, σ is the interfacial tension, N is the impeller rotational speed, and D_I is the impeller diameter. Figure 3.3-1 gives correlations for open pipes or Kenics static mixers, where μ_D is the dispersed-phase viscosity, μ_C is the continuous-phase viscosity and d_{32} is the Sauter mean drop size. A design drop size of about 0.8 d_{32} is reasonable. For agitated vessels or pumps, Eq. (3.3-3) gives an estimate of drop size:

$$\frac{d_{32}}{D_I} = 0.081 \ (1 + 4.47 \ \phi) \ (We)^{0.6} \tag{3.3-3}$$

where d_{32} is the Sauter mean drop size in the same units as D_I and ϕ is the weight fraction of the dispersed phase. The drop size range may be from 0.5 d_{32} to 2.0 d_{32}.[3]

Part of the drop size prediction requires knowing which phase is dispersed and which is continuous. Selker and Sleicher[4] provide a useful correlation to predict which is the dispersed phase based on phase volume ratios and density and viscosity of each phase. This information may be approximated by Eq. (3.3-4),

$$X = \frac{Q_L}{Q_H} \left(\frac{\rho_L \ \mu_H}{\rho_H \ \mu_L} \right)^{0.3} \tag{3.3-4}$$

where Q_L is the volume of the light phase and Q_H the volume of the heavy phase is consistent units. The following guidelines are suggested:

X	Result
<0.3	Light phase always dispersed
0.3–0.5	Light phase probably dispersed
0.5–2.0	Phase inversion possible; design for the worst case
2.0–3.3	Heavy phase probably dispersed
>3.3	Heavy phase always dispersed

In situations where either phase could be dispersed, the phase being added is normally the dispersed phase; that is, in an agitated vessel one phase is being mixed and a second phase is added. As concentrations, temperature, and physical properties change, phase inversion may occur.

3.3-2 Gravity Separation

A wide variety of designs are used for gravity segregation in immiscible-liquid separators: horizontal or vertical vessels, troughs (API separators), and vessels with various internal configurations or parallel plates.

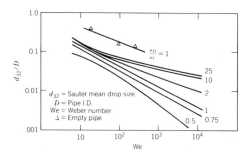

FIGURE 3.3-1 Drop size correlation for pipes and static mixes. Reprinted with permission from S. Middleman, *IEC Chem Process Des Dev.*, **13**, 78 (1974), © American Chemical Society.

The vessels are often referred to as decanters and may operate as a batch or a continuous operation. The emphasis here will be on continuous designs. The basic design approach is as follows:

Calculate settling velocity for drops.

Pick a preliminary size based on overflow rate.

Estimate the time for drop coalescence.

Specify feed and outlet geometries.

Drop settling velocity is estimated from Stokes' Law using Newton's basic drag equation:

$$U_t = \frac{gd^2\,(\rho_D - \rho_C)}{18\,\mu_C} \tag{3.3-5}$$

where d is the drop size, g is the acceleration of gravity ρ_D and ρ_C are the dispersed- and continuous-phase densities, and μ_C is the continuous-phase viscosity. A number of assumptions for derivation of Eq. (3.3-5) are often violated in the design of continuous equipment:

1. The continuous phase is a quiescent fluid.
2. The drop is a rigid sphere.
3. The drop moves in laminar flow (i.e., $\mathrm{Re} = Ud\,\rho_C/\mu_C \leq 10$).
4. The drop is large enough so that Brownian movement can be ignored.
5. The drop is not hindered by other drops or by vessel surfaces.

Assumption 1 is probably the most significant; however, the difficulty in obtaining or specifying a drop diameter overshadows these other factors. Most drop sizes calculated by typical correlations as in Fig. 3.3-1 are above 300 μm. However, because of the limitations of the assumptions of the design equation, a 150 μm ($d = 0.0005$ ft) design drop diameter is recommended.

The overflow rate in a continuous gravity separator is defined as Q_C/A_I, where Q_C is the continuous-phase flow rate and A_I is the interfacial area between the coalesced phases. The continuous phase must flow vertically from the entry elevation to the outlet. Drops must move countercurrently to this velocity to reach the interface and coalesce with other dispersed-phase material. Ideally, the continuous phase moves in a uniform plug flow with a velocity equal to the overflow rate. The minimum or design drop size should have a settling velocity greater than the overflow rate, $Q_C/A_I < U_t$. This provides an approach to set the separator size based on interface area. The approach is made less accurate because of deviations from plug flow, but with a conservative design drop size, reasonable estimates can be obtained.

The performance of gravity separators is dependent on two parts: (1) the movement of dispersed-phase drops to the interface and (2) the coalescing of the dispersed-phase drops at the interface. Either part could be the controlling factor. The coalescence is dependent on the purity of the phases and the interfacial tension. There is no simple equation to predict the time required for coalescence; which may range from a fraction of a second to 2–3 min. The time gets shorter as the density difference between the phases increases, the continuous-phase viscosity decreases, the viscosity inside the drop decreases, and the interfacial tension increases.

Drops waiting to cross the interface often form a deep band of dispersed drops. The depth of this dispersion band, H_D, is correlated by $H_D = (Q_C/A_I)^n$. Values of n range from 2.5 to 7 for different chemical systems. If H_D becomes a significant fraction of the separator depth, a slight increase in feed rate could cause the separator to become filled with the dispersion band, thereby eliminating the desired separation. Roughly half the volume of the dispersion band is occupied by dispersed-phase drops. Residence time in the dispersion band is then $\Theta = \frac{1}{2}(H_D A_I/Q_D)$. A good design practice is to keep $H_D < 10\%$ of the separator height and Θ greater than 2–5 min.

The above conditions assume two relatively pure liquids. The presence of a surface-active agent or fine dispersed solids can interfere with the coalescing process and result in a stable emulsion. Many liquid–liquid separators form a stable emulsion at the interface called a ''rag layer'' because of these agents and may require draw-off nozzles near the interface to prevent accumulation. The ''rag layer'' is like foam in liquid–gas systems and is typically stabilized by very fine solids. If the rag is drawn off it may be de-emulsified or broken by filtration, heating, chemical addition, or reversing the phase that is dispersed.

Both the overflow rate and the residence time in dispersion band are decreased by increasing the interfacial area. Accordingly, long horizontal vessels provide the most desirable geometry (Fig. 3.3-2). This configuration is good for vessels up to about a 12 ft (3.7 m) diameter and for pressure vessel requirements. For much larger volume applications, either a large horizontal tank with center feed and peripheral discharge, such as a water clarifier, or a large horizontal trough such as an API separator is a better selection.

In horizontal decanters, the continuous phase flows perpendicular to the drops. This causes turbulence which interferes with the settling process. If the cross-flow were fully laminar, settling would not be a

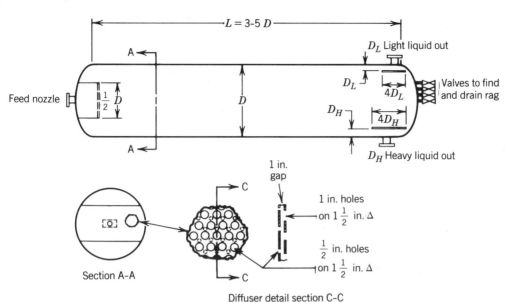

Diffuser detail section C-C

FIGURE 3.3-2 Recommended horizontal decanter configuration.

problem; however, it is usually impractical to operate with laminar flow. The degree of turbulence is expressed by the Reynolds number Re $= VD_h\, \rho_C/\mu_C$, where V is the continuous-phase crossflow velocity and D_h is the hydraulic diameter of the continuous-phase layer (D_h equals four times the flow area divided by the perimeter of the flow channel including the interface). Experience indicates that the following guidelines hold for decanters:

Reynolds Number	Effect
<5,000	Little problem
5,000–20,000	Some hindrance
20,000–50,000	Major problem may exist
>50,000	Expect poor separation

It is good practice to proportion the decanter so that the velocities of both phases are similar, normally with a ratio less than 2:1. Some velocity difference, however, seems to be desirable to aid coalescence.

The proper introduction to feed to a gravity separation is the key to its performance. For optimum design, the feed should be (1) introduced uniformly across the active cross-section of the decanter and (2) done in a way to leave no residual jetting or turbulence. An uncontrolled mixture inlet without proper diffusion will enter as a strong jet, shooting across the separator and causing significant turbulence and mixing. A practical approach is to limit the inlet velocity head at the nozzle to $\rho V^2 = 250$ lb$_m$/ft·s^2. For water, this corresponds to about 2 ft/s. In addition, forcing a direction change on the inlet flow can reduce the jet effect. A very simple approach is to use a plate just inside the inlet at least two times the nozzle diameter and located one-half nozzle diameter in from the inlet. This deflects the jet and significantly reduces the velocity.

For critical designs or retrofitting a marginal existing separator, more elaborate designs are required. To force good distribution and to calm inlet velocity effects, diffusers having a peak velocity of no more than 2–5 times the separator mean velocity are recommended. Successful designs have two closely spaced, perforated or slotted, parallel plates. The first plate takes a high pressure drop with open flow area 3–10% of the separator cross-sectional area. The second plate has its open area off-set from the first plate to dissipate velocity. The second plate should have an area of 20–50% of the separator cross section. The spacing of the plates is about the dimension of the plate perforations or slot width.

High outlet velocity of the heavy and light phases can also adversely affect separation if a vortex is formed and nonseparated dispersion band material is drawn out in the vortex. The exit flow of each separated phase should be drawn from the separator at velocities not more than 10 times the average velocity in the separator. The ideal outlet consists of overflow and underflow weirs extending across the separator. If standard nozzles are used at the top or bottom of a horizontal vessel, vortex breaker plates four times the nozzle diameter inside the vessel can prevent vortexing.

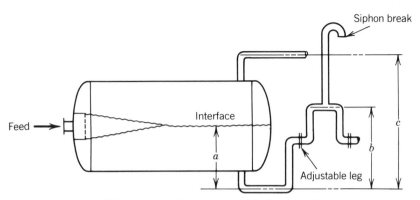

FIGURE 3.3-3   Simple decanter interface control.

For some critical designs, small-scale dynamic models are useful in evaluating performance. Care must be taken to assure that the same flow regime exists on both scales, that is, laminar or turbulent.

All gravity separators require good control of the liquid–liquid interface for consistent performance. For liquid-filled systems, level control instruments or interface detectors are used. A simpler, less costly method is to use a weir or underflow leg to fix the interface. The weirs or legs are made adjustable and are set on start-up. Positioning is estimated by doing a force balance [Eq. (3.3-6)] based on the a, b, and c dimensions indicated in Fig. 3.3-3:

$$b = a + \frac{\rho_L}{\rho_H} (c - a) \qquad\qquad (3.3\text{-}6)$$

Flow friction losses are assumed negligible and ρ_L and ρ_H are the light and heavy densities, respectively.

Batch-operated gravity separations avoid the complications of turbulence in continuous separators. They can be designed based on Stokes' Law for dilute dispersions, where the separation time is that for a dispersed-phase drop to go from the farthest point to the interface: $\Theta = H/U_t$ where Θ is separation time, H is the height from the farthest point to the interface, and U_t is the Stokes' Law settling velocity. For systems where coalescence is the limiting mechanism, the ratio of the volume to cross-sectional area should be proportional to separation time and is the key parameter. Small-scale batch tests can be used to confirm sizing or separation time.

Batch settlers are best used for separation following batch contacting of two phases where both operations can be done in the same vessel. The difficult part is making a clean cut between phases when draining or pumping out. This is normally done visually with a sight glass or with a conductivity probe. Prevention of vortexing in the exit line is important to prevent premature entrainment of the light phase into the heavy phase. Use of a vortex breaker over a bottom outlet nozzle, a slow exit velocity, or taking an intermediate cut and recycling are alternate approaches.

Since all gravity settling depends on the distance a drop must travel to reach an interface, or the interface area available for coalescence, various designs having parallel plates have been promoted in recent years. The parallel plates create individual flow channels and, in effect, are many separators in parallel. Plate spacing may range from about 0.75 in. (1.9 cm) to several inches. The maximum distance a dispersed-phase drop must travel to reach an interface is the distance between the plates, rather than several feet in a large open gravity-settling vessel. The parallel plates also maximize the interfacial area to aid coalescence. Separations keyed to overflow rate Q/A can be performed with much greater interface area in a much smaller vessel. The parallel plates also address the problem of turbulence affecting gravity settling. Since the separator Reynolds number is a function of hydraulic diameter, the presence of the plates greatly reduces the hydraulic diameter and allows laminar flow (Re < 2100) to be obtained in a reasonably sized vessel with 0.75–1.5 in. (1.9–3.8 cm) plate spacing.

Many variations of parallel-plate separators are now available; some have flat plates, some use wavy plates, some use parallel chevrons or tubes. Most of the plates are placed on an angle to allow the light and heavy phases to disengage as they pass through the plate section. New commercial designs use a crossflow approach, while some of the earlier designs required the collected dispersed phase to flow back countercurrent to the bulk flow. Some disengagement volume is needed after the bulk flow exits the plate sections, but at this point, the dispersed phase has coalesced into large drops which readily separate by gravity as they exit the plate surfaces. Units requiring countercurrent flow are not recommended for large volumes of dispersed phase. Parallel plates added to ineffective existing gravity separator vessels can greatly improve separation quality or increase capacity.

3.3-3 Centrifugal Separators

Cyclones for liquid–liquid segregations have been tried experimentally but are not in use commercially owing to the relatively small density differences and the high shear flow in the device which would break down liquid drops.

Mechanical centrifuges, however, are used commercially for immiscible-liquid separations. Centrifuges work on the basic Stokes' Law functionality by increasing the settling velocity many times over that of gravity by the application of centrifugal force. This is referred to as the number of G's developed by the machine, $G = r(2\pi N)^2/g$, where r is the centrifuge bowl radius, N is the number of revolutions of the bowl per second, and g is the acceleration due to gravity.

The simplest liquid–liquid centrifuge is a tubular device of small diameter (< 8 in.) and very high speed. It is fed at one end with the separated liquid phases removed at the opposite end through tubes of different radii. Maximum capacity is about 0.045 ft³/s (1.3 L/s).

The most common liquid–liquid centrifuge is a disk type design (Fig. 3.3-4). Forces up to 9000 times gravity are typical. The inclined disks in the centrifuge enhance settling similar to the application of parallel plates in gravity settlers. The lower-density liquid phase moves to the top of the channel between disks and toward the center of the machine. The higher-density liquid phase moves toward the outside of the bowl but also exits at the top in a separate channel. Some solids can be handled with either periodic shutdown and clean-out or an automatic solids ejection by either parting the bowl momentarily or opening nozzles to eject the solids.

A minimum disk spacing must be maintained to prevent choking of the channels. Capacities of disk centrifuges range from 0.011 to 1.11 ft³/s (0.3 to 30 L/s). The specific gravity difference of the pure components should be at least 0.01 and the drop size of the dispersed phase should be at least 1 μm. Some reported results indicate removals down to 0–5 ppm on feed streams of 15% dispersed phase.

3.3-4 Interception

Mixtures of immiscible liquids can be separated by an interception method analogous to mesh pads or fiber beds in gas–liquid segregation. The mixture is passed through a dense bed of fibers or wire mesh and the dispersed drops contact with media surface. Most systems are designed with media materials that are preferentially wetted by the dispersed drops. The dispersed material is held by the media until enough material coalesces and large globules disengage from the media. The large globules, now of significant size, are readily separated by gravity. These devices are often referred to as coalescers.

Metal wires are generally wetted by aqueous phases, while organic polymeric materials are usually wetted by organic chemicals. Glass or treated glass fibers can be tailored for either aqueous or organic phase wetting. Fiber size and media pore size are decreased with the need to separate smaller drops. Particles down to 1–5 μm have been removed successfully. Pressure drop typically ranges from < 1 to 20 lb$_f$/in.²

For very efficient operations, two-stage combinations of coalescing elements are used. The second stage may be a continuous-phase wetted media, very densely packed to prevent entry of dispersed drops (Fig. 3.3-5).

Coalescers are good for clean liquid–liquid systems, but systems containing solids will rapidly plug the

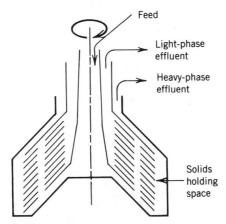

FIGURE 3.3-4 Liquid–liquid disk centrifuge configuration. Figure reproduced with the permission of McGraw-Hill Book Co. from *Perry's Chemical Engineers' Handbook*, 6th ed., 1984.

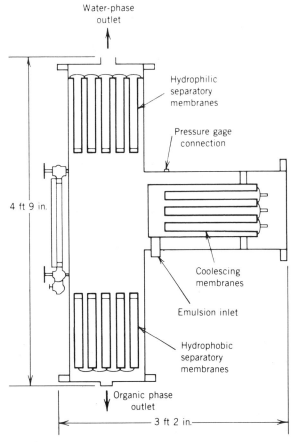

FIGURE 3.3-5 Liquid–liquid coalescer configuration. Courtesy of Osmonics, Inc., Minnetonka, MN.

media unless a prefiltration step is used. When significant solids are present, a packed bed similar to a deep bed filter may be used with periodic backflush.

The performance of coalescers cannot be predicted theoretically and such systems require experimental work to determine sizing, using the same media, flow rate per area of media, and dispersion preparation method. Tests need to be run over sufficient time to assure that the media is fully loaded with the dispersed phase and that steady-state operation has been obtained. Periods up to 2 h may be needed to assure complete loading. A material balance of dispersed phase in and out of the system is a good check.

Another impaction technique is that of flotation. Gas bubbles are used to intercept dispersed liquid drops and float them out of the bulk phase. Three methods are used to form the bubbles: (1) mechanical dispersion, (2) electrolytic gas formation, and (3) dissolved air. Mechanically dispersed systems involve either agitators or gas spargers to form the bubbles. Bubble size is usually large and not efficient in contacting small liquid drops. Electrolytic gas uses a direct current between electrodes to create small oxygen and hydrogen bubbles from water solutions. Dissolved air flotation is the most common method; in this procedure, liquid is saturated with gas by various techniques followed by a sudden pressure reduction. Bubbles 30–100 μm are evolved with optimum separation occurring with 50–60 μm bubbles.[5] Gas requirements of 0.01–0.03 lb_m air/lb_m dispersed phase are typical. Oil refinery applications of flotation to remove oil from wastewater are used, after gross oil separation by gravity settlers. Separation efficiencies of greater than 90% are obtained with 50–250 ppm oil concentration feed.

Membrane separators are used for immiscible liquid segregation. These devices are extensively covered elsewhere in this book. For actual segregation, ultrafiltration or microfiltration membranes are another variation of an impaction or interception mechanism. Ultrafiltration membranes have discrete pores about 0.001–0.02 μm in diameter which allow the continuous phase to pass while trapping the much larger individual drops, which are usually greater than 0.1 μm. Flow velocity across the membrane surface resuspends trapped dispersed-phase material to prevent blinding of the pores. Pressure drop requirements for ultrafiltration are under 100 $lb_f/in.^2$. Ideally, the membrane material is not wetted by the dispersed phase. These devices are good for obtaining a clarified continuous phase, for example, removing dilute oil from water. The dispersed phase, however, can only be concentrated, not truly separated, since a strong continuous-phase velocity is needed across the surface to prevent fouling. An example is feeding a 2–3% oil in water mimxture with a clear water stream passing through the membrane containing about 0.0005% oil, and an exit bulk flow of about 20–30% oil. The exit bulk flow could then be separated by more conventional settling devices.

3.3-5 Electroseparators

Electrical charging methods may be used to remove aqueous materials and solids from organic fluids as in crude oil desalting. Dispersed materials passed between electrodes are charged with either dipoles or net charges such that they are attracted either to each other or to an electrode surface where they coalesce with other dispersed-phase materials. The coalesced materials are readily separated by gravity settling. Some commercial units handle up to 125,000 bbl/day of crude oil.

3.3-6 Selection of Liquid–Liquid Segregation Devices

The following factors need to be considered in selection of the optimal device or combination of separation equipment items:

   Concentration of feed
   Particle size of dispersed phase
   Effluent quality required
   Capital cost
   Operating costs
   Space for equipment
   Material of construction

These factors are considered in the logic diagram shown as Fig. 3.3-6.

   Systems with greater than 30% concentration of dispersed phase are difficult to handle in anything but large-volume gravity settlers. Coalescence will be a dominant factor due to the high frequency of dispersed-phase drop interaction. Coalescence will be greatly affected by the physical properties of the fluids, especially the interfacial tension. Fluid pairs with interfacial tensions greater than 10 dynes/cm should readily coalesce, while those <1 dynes/cm will readily emulsify. Concentrations below 30% disperse material can be handled by disk centrifuges also. Below 15%, concentration plate settlers can be added because settling, not coalescence, is likely to be the dominant mechanism. Below 5% concentration, all equipment is applicable and selection must be based on other criteria.

   Dispersions may be considered in two categories—primary dispersions having drops 25 μm and larger and secondary dispersions of particles below 25 μm. Some secondary dispersions or haze are formed as a result of the coalescing of primary drops. Secondary dispersions require techniques other than gravity settling such as fiber coalescers, flotation, or centrifugation to obtain good separation efficiency.

   Emulsions are another category of dispersion with particles generally less than 1 μm and with low interfacial tension. Dense media coalescers are sometimes effective as are membranes. Chemical or heat treatment may be needed to break the emulsion by changing the interfacial tension followed by conventional techniques. For water dispersed in an organic emulsion, the electroseparators are a possible choice. Figure 3.3-7 gives an indication of the range of application of the various devices as a function of particle size.

   The required effluent quality is a prime factor in the final separator system configuration. Combinations of two or three separator devices or techniques may be needed. Gravity settlers can do a gross separation of drops down to about 150 μm. Centrifuges, fiber bed coalescers, or membranes would be required for very high efficiencies of particles down to 1 μm. To clear a mixture of 30% dispersed-phase concentration to only a few parts per million residual would require gravity settling followed by some polishing device.

   Capital costs are difficult to generalize owing to the wide range of feed properties, output specifications, and material of construction requirements. Table 3.3-1 is a relative positioning of the various separation techniques.

   Operating costs of the various devices vary from essentially no cost for gravity chambers with weir outlet controls, to significant electrical and maintenance costs for high-speed centrifuges. Some periodic costs for clean-out of solids accumulations may be required with most of the devices. Relative positioning of the devices by operating cost is essentially in the order of capital cost shown in Table 3.3-1.

   Space available for equipment is frequently a key factor, especially when retrofitting existing operations. For locations with considerable real estate available, large gravity decanters, API separators, or circular clarifier vessels are applicable. For space-limited situations, use of parallel plates can greatly reduce the separator area and volume, sometimes by a factor of 10–20. High-speed disk centrifuges require the smallest

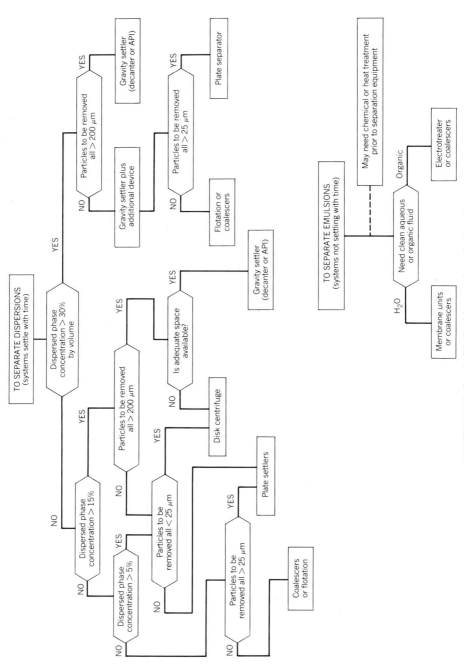

FIGURE 3.3-6 Immiscible liquid separation selection chart.

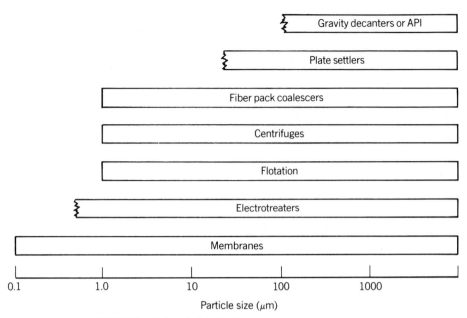

FIGURE 3.3-7 Immiscible liquid separator particle size range.

area and volume of the liquid separators for large flow applications, but they are the devices with highest purchase and operating costs.

Material of construction should not pose a major problem since most devices are available in a wide range of metals and plastics. If lower-cost metals or reinforced plastics are not acceptable and exotic metals are needed, a smaller size device such as the centrifuge may be more attractive, compromising the typical capital cost comparison of Table 3.3-1.

3.4 LIQUID–SOLID SEGREGATION

3.4-1 Gravity Settlers

TYPE I SEDIMENTATION

The settling of dilute slurries where there is little particle-to-particle interaction is commonly referred to as Type I settling. All particles settle independently; consequently, if a particle size distribution is known and the settling rate of individual particles is known, then a settler can be designed. The design method for solid–liquid gravity settlers is the same as the previously described design method for gas–liquid gravity

TABLE 3.3-1 Relative Capital Costs of Liquid–Liquid Separators

Low capital
  Decanter vessels
  Open-trough separators (API type)
  Plate separator devices

Moderate capital
  Coalescers
  Flotation systems

High capital
  Electroseparators
  Membranes
  Centrifuges

settlers (see Section 3.2). If the terminal velocity (U_t) of the smallest particle to be separated is known or can be calculated, then the overflow area (A_t) can be calculated from the equation

$$A_I = \frac{Q_C}{U_t} \tag{3.4-1}$$

where Q_C is the volumetric flow rate of the liquid.

As for gas–liquid and gas–solid systems, the liquid depth must be sufficiently great to minimize the suspending effects of turbulent liquid flow. As discussed in Section 3.2, the particles will not be resuspended provided the flow velocity is less than about 20 times the terminal velocity. To be conservative the flow velocity should be 10 times the particle terminal velocity. Thus, the flow area (A_F) is given by

$$A_F = \frac{Q_C}{U_F} = \frac{Q_C}{10\,U_t} \tag{3.4-2}$$

If the height of the basin is one-third the width,

$$H = \frac{W}{3} = \left(\frac{Q_C}{30\,U_t}\right)^{1/2} \tag{3.4-3}$$

For an example of 10 μm particles of specific gravity = 3, the terminal velocity is 0.033 ft/s. To separate these particles and all larger particles from a water stream of 100 ft^3/min (833 gal/min), the basin dimensions are calculated as follows:

$$A_I = \frac{Q_C}{U_t}$$

$$= \frac{100}{(60)(0.033)} = 50 \text{ ft}^2$$

$$H = \frac{W}{3} = \left[\frac{100}{(60)(30)(0.033)}\right]^{1/2} = 1.3 \text{ ft}$$

$$W = (1.3)(3) = 3.9 \text{ ft}$$

$$L = \frac{A_I}{W} = \frac{50}{3.9} = 13 \text{ ft}$$

In actual practice the effects of turbulence and other nonuniform flow characteristics would necessitate making the basin at least 19 ft long.

TYPE II SEDIMENTATION

Slurries exhibiting Type II behavior are sufficiently thick and flocculent that the solids tend to settle as a mass, giving a rather sharp line of demarcation between the clear liquid overflow and the settling solids. For such systems the design is normally controlled by the thickening capability of the basin, although the basin design must be adequate to provide sufficient overflow area to clarify the liquid overflow.

A settler for Type II slurries is normally referred to as a thickener. They are sometimes constructed as a rectangular basin; however, most often they are of circular cross-section. In the rectangular basin, solids are normally removed by a traveling syphon that moves longitudinally back and forth along the basin. In the circular design a raking mechanism is used to convey the settled solids slowly to the center of the basin where, as with the rectangular basin, a syphon is used for their removal. Figure 3.4-1 depicts a circular thickener. The feed is introduced below the liquid surface and above the sludge blanket. Density differences between the feed, clarified liquid, and settled solids cause the feed to spread laterally or radially from the feed point, producing the effect of feeding the basin uniformly across the area just above the sludge layer.

The design methods most commonly used for Type II settlers and thickeners are discussed by Fitch,[1] Talmadge and Fitch,[2] Cremer and Davies,[3] Metcalf and Eddy,[4] and Clark et al.[5] The methods rely on the taking of experimental data and on empirical analysis of the data to obtain a design. For this reason the recommended approach is best presented by considering a specific example. The settling data (Fig. 3.4-2) for a water–cement slurry, presented by Cremer and Davies, will be used. The important independent design parameters are as follows:

W_s, mass flow rate of solid	100,000 lb$_m$/h
ρ_s, true density of solid	190 lb$_m$/ft^3
Q_s, volumetric flow rate of solid	525 ft^3/h

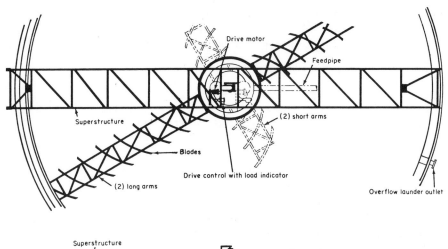

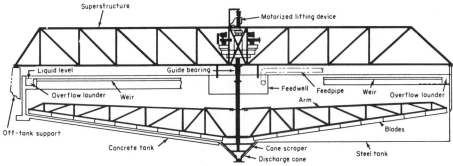

FIGURE 3.4-1 Clarifier/thickener configuration. Courtesy of Eimco Process Equipment Co.

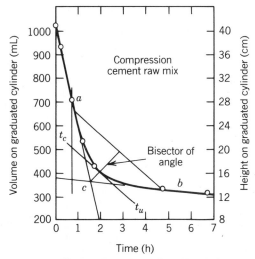

FIGURE 3.4-2 Settling test data utilization. Reproduced from H. W. Cremer and T. Davis, *Chemical Engineering Practice*, Vol. 3, Chap. 9, Academic Press, New York, 1957.

Q_L, volumetric flow rate of water	2,625 ft³/h
Q_t, total volumetric flow to basin	3,150 ft³/h
C_0, concentration of solid in feed	32 lb$_m$/ft³

The dependent parameters are as follows:

X_w, volume fraction of water in underflow	0.33
ρ_u, average density of underflow	147 lb$_m$/ft³
Q_0, water overflow rate = [2625 − (525/0.667)(0.333)]	2363 ft³/h

Clark et al.[5] present the method for determining the surface area large enough to prevent the rate of liquid rise from exceeding the velocity of subsidence of the sludge interface:

$$A_s = \frac{Q_0}{V_s} \qquad (3.4\text{-}4)$$

The subsidence velocity (V_s) is obtained by measuring the slope of the plot of interface height versus time (Fig. 3.4-2). For the cement slurry $V_s = 40/2 = 20$ cm/h $\Rightarrow 0.66$ ft/h. Then

$$A_s = \frac{Q_0}{V_s} = \frac{2363}{0.66} = 3580 \text{ ft}^2$$

Next, the area required for thickening in the bottom of the basin is determined by the method explained by Fitch, where H_0 is the initial height of the slurry in the graduated test cylinder:

$$G_t = \frac{C_0 H_0}{t_u} \qquad (3.4\text{-}5)$$

The underflow time (t_u) is determined graphically as shown in Fig. 3.4-2. For this example $t_u = 3.2$ h.

$$G_t = \frac{(32 \text{ lb}_m/\text{ft}^3)(40 \text{ cm})}{(3.2 \text{ h})(30.5 \text{ cm/ft})} = 13.1 \text{ lb}_m/\text{h} \cdot \text{ft}$$

$$A_t = \frac{W_s}{G_t} = \frac{100,000}{13.1} = 7625 \text{ ft}^2$$

It is recommended that this value be multiplied by a safety factory of 1.33; thus,

$$A_t = (7625)(1.33) = 10,000 \text{ ft}^2$$

The diameter of the circular basin would be 113 ft. As in most thickeners the area is determined by the requirement to thicken the sludge and not by the requirement to clarify the overflow. Note that the thickener can be sized from one settling test. In practice, one should conduct a settling test in at least two different cylinder heights.

As explained by Fitch[1] the depth of the sludge layer needs to be calculated. From Fig. 3.4-2 note that the time in the compression layer is about $7 - 2 = 5$ h. The volumetric flow of the slurry out of the basin bottom is equal to the total flow to the basin minus the water overflow, that is $(3150 - 2363) = 787$ ft³/h. Knowing the flow area (10,000 ft²) and the residence time (5 h), the depth of the compression layer can be computed:

$$H_c = \frac{t Q_u}{A_t} = \frac{(5)(787)}{10,000} = 0.4 \text{ ft}$$

As Fitch[1] indicates, the accepted procedure is to use a minimum depth of 3 ft for the sludge layer. This then makes the method conservative because more compression will be realized in a 3 ft layer than in a 0.4 ft layer.

The total depth is not well defined; however, in this case a total depth of 6–9 ft would be adequate.

Gravity settlers can be reduced significantly in volume by using multiple parallel plates spaced about 1 in. apart. The plates are inclined to the vertical so that the solids will slide off the plates. The minimum area requirement can be computed for the parallel plate designs by the previously described calculation procedure for a gravity settling basin. The actual vertical distance is calculated from the geometry and the residence time is calculated from the flow rate and volume of the plate stack. In practice, turbulence normally limits the performance; consequently, a manufacturer should be consulted for design help.

3.4-2 Centrifugal Separators

HYDROCYCLONES

The liquid–solid hydrocyclone, shown schematically in Fig. 3.4-3, functions like a gas–solid cyclone. The hydrocyclone is also known as a hydroclone. The primary independent parameters that influence the ability of a hydrocyclone to make a separation are size and geometry of the hydrocyclone, particle size and geometry, solids loading, inlet velocity, split between overflow and underflow, density differential, and liquid viscosity. A reasonable estimate of the particle cut diameter (50% in underflow and overflow) (d_{50}) is given by the following dimensionless relationship, developed initially by Bradley:[6]

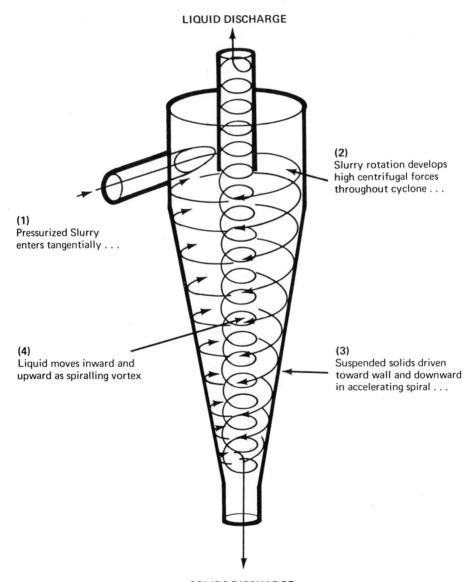

OPERATION OF THE HYDROCYCLONE

LIQUID DISCHARGE

(2)
Slurry rotation develops
high centrifugal forces
throughout cyclone . . .

(1)
Pressurized Slurry
enters tangentially . . .

(4)
Liquid moves inward and
upward as spiralling vortex

(3)
Suspended solids driven
toward wall and downward
in accelerating spiral . . .

SOLIDS DISCHARGE

FIGURE 3.4-3 Operation of the hydrocyclone. From R. W. Day, *Chem. Eng. Prog.*, **69**(9), 67 (1973). Reproduced by permission of the American Institute of Chemical Engineers.

$$\frac{d_{50}}{D_i} = 2.2 \left[\frac{18\pi\mu D_i(1 - R_f)}{16\ Q(\rho_s - \rho_L)}\right]^{0.5} \left(\frac{2.3\ D_o}{D_c}\right)^{0.8} \left(\frac{D_i}{L}\right)^{0.5} \tag{3.4-6}$$

where D_c, D_i, and D_o are the body, inlet, and overflow diameters, respectively, μ is the liquid viscosity, R_f is the fractional underflow rate, Q is the volumetric feed rate, ρ_s and ρ_L are solid and liquid densities, and L is the cyclone length. This equation predicts reasonable dependencies for the primary independent variables except for the cyclone diameter. The cut diameter is more nearly proportional to D_i rather than $D_i^{1.5}$.

To estimate the performance of hydrocyclones, Day[7] has published Fig. 3.4-4. This chart can be used to estimate the performance of hydrocyclones of various sizes and it can be used to predict the effect of specific gravity difference and liquid viscosity. Figure 3.4-5 allows prediction of capacities at various flows and pressure drops as a function of hydrocyclone size.

Equation (3.4-6) in conjunction with Figs. 3.4-4 and 3.4-5 will allow preliminary sizing of a hydrocyclone. Sometimes this information in conjunction with vendor consultation will be adequate to determine the final sizing of a hydrocyclone, provided the solids are readily characterized and the solids concentration in the feed is steady. For applications where the solids vary widely in size, where the solids are far from spherical, where clarity of the overflow is critical, and where solids concentration in the underflow is crucial, a test program should be undertaken.

Hydrocyclones are applicable where extreme clarity of the overflow is not needed and where the separated solids are not expected to be very dry. Hydrocyclones are extremely useful where a liquid stream contains few relatively large solids that must be separated. For this situation a catch pot will suffice to capture the solids and then they can be discharged intermittently.

Discharge of the solids is a special problem with hydrocyclones. Under circumstances of a constant

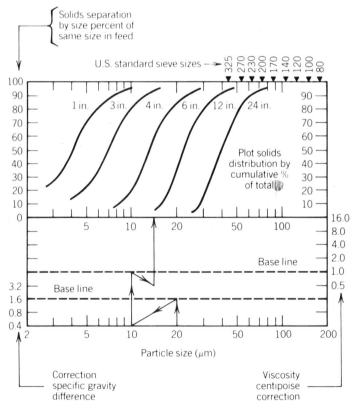

FIGURE 3.4-4 Hydrocyclone separation performance for various sized cyclones. This chart is for guidance engineering purposes only. Final sizing of cyclones must be determined by tests on the expected slurry or prior experience in similar situations. From R. W. Day, *Chem. Eng. Prog.*, **69**(9), 67 (1973). Reproduced by permission of the American Institute of Chemical Engineers.

CYCLONE CAPACITY NOMOGRAPH

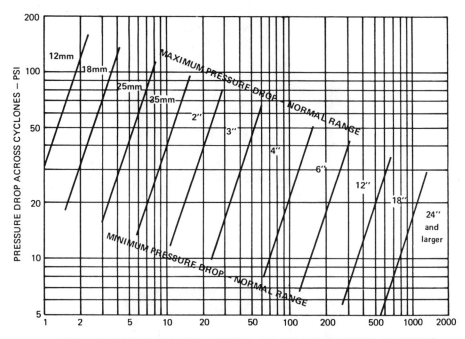

SINGLE HYDROCYCLONE FEED CAPACITY – U.S. GALLONS OF WATER PER MINUTE

FIGURE 3.4-5 Hydrocyclone capacity ranges versus size. From R. W. Day, *Chem. Eng. Prog.*, **69**(9), 67 (1973). Reproduced by permission of the American Institute of Chemical Engineers.

feed composition, the installation of a cone of given size will result in steady discharge of solids at a given liquid composition. However, if solids in the feed fluctuate, then the underflow will fluctuate. The conditions necessary to obtain proper discharge of underflow without plugging the hydrocyclone or getting too high a flow out the underflow must be determined experimentally. Experiments are normally necessary to determine the proper size of cone extension in order to obtain the proper split between underflow and overflow. The control of the split is a unique problem that requires careful consideration of the design. A flow control scheme can be devised to give a constant fraction of the total flow as overflow so that the volumetric flow of underflow is automatically controlled. Also, a control valve can be placed in the underflow line.

A good discussion of the hydrocyclone applications and design is given by Poole and Doyle[8] and they list 193 references covering selection, operation, and design.

CENTRIFUGES
Perry's Chemical Engineers' Handbook[9] and manufacturers' literature[10,11] are excellent sources of technical information about the selection and design of centrifuges.

Centrifuges are of two general types: (1) sedimentation and (2) filtration. The sedimentation machines apply centrifugal forces to enhance the sedimentation velocity of the solids toward a rotating, collecting solid surface; the filtration machines apply centrifugal forces to force the solid phase to collect first as a cake on a filter cloth or screen and subsequently to force the liquid through the cake and the screen. Conceptually, the sedimentation centrifuges operate analogously to the gravity settler except the gravitational force is enhanced from 1,000 to 20,000 times by the centrifugal acceleration. The filtration machines operate analogously to pressure or vacuum filters except the force causing the liquid to flow through the cake is a result of centrifugal forces rather than pressure forces.

Sedimentation Centrifuges. Various sedimentation centrifuges are illustrated schematically in Figs. 3.4-6–3.4-9. The machines differ primarily in the manner in which the *solids* are discharged. In the *solids-retaining* solid-bowl centrifuge, either conical disks or cylindrical inserts are features that increase the separation area. In both types, the solids must be discharged manually by periodic opening of the bowl. In the *solids-ejecting* solid-bowl centrifuge, the solids collect at the periphery of the bowl and periodically

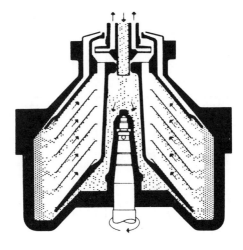

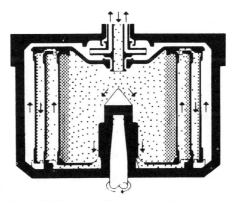

FIGURE 3.4-6 Solids-retaining solid-bowl centrifuges: (*a*) bowl with conical disks and (*b*) bowl with cylinder-insert solids retainers. Courtesy of Alfa-Laval Inc.

the bottom of the bowl is lowered very rapidly and the collected solids are ejected. In the *nozzle* solid-bowl centrifuge, the solids are discharged continuously through small nozzles. Washing can be accomplished on these machines by injecting wash liquid between the disk stack and the discharge nozzles. By recycling solids, the solids concentration in the nozzle effluent can be maintained at a reasonable level somewhat independent of the solids content of the feed. The *decanter* solid-bowl centrifuge has a helical auger conveyor operating within a rotating solid bowl. The bowl has a straight cylindrical portion and a cylindrical–conical portion ("beach"). The solids are centrifuged to the outer wall of the cylindrical section where they are conveyed out of the liquid pool up the "beach," where they drain, and out the solids discharge port. The depth of the liquid pool is controlled by the radial location of liquid discharge ports.

Filtration Centrifuges. The most common machines are the basket (Fig. 3.4-10), the peeler (Fig. 3.4-11), and the pusher (Fig. 3.4-12). All the machines consist of a perforated basket with a cloth or wire mesh inset as a filtration media to retain the solids. As with the sedimentation centrifuges, the major differences are with the mechanics of solids discharge. The *basket* centrifuge is operated intermittently as the slurry is fed, spun dry, washed, spun dry, and discharged. The solids can be discharged manually or automatically by a plough. The machine must be stopped for manual discharge and it must be operated at very slow speed for plough discharge. Unlike the basket machine, the *peeler* and *pusher* both operate at a constant speed. The *peeler* operates with intermittent feed and solids discharge. The *pusher* operates with continuous feed and intermittent, but very frequent, solids discharge. Figure 3.4-12 shows a two-stage pusher machine. Multistage baskets are used to minimize buckling of the column of cake. An inner screen

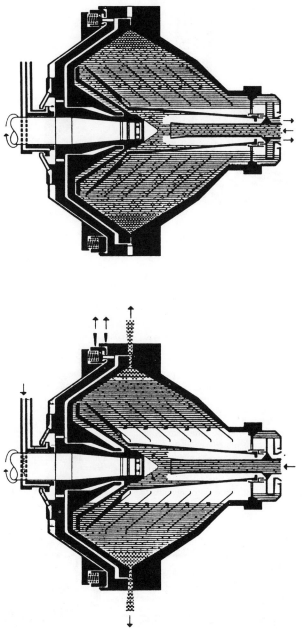

FIGURE 3.4-7 Solids-ejecting solid-bowl centrifuges: (*a*) the bowl is closed hydraulically and (*b*) the bowl is opened hydraulically during the partial discharge of solids only. Courtesy of Alfa-Laval Inc.

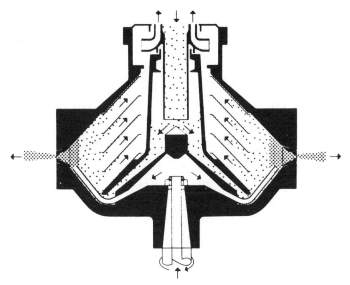

FIGURE 3.4-8 Nozzle solid-bowl centrifuge. The peripheral nozzles are for solids discharge. Solids are washed before discharge. Courtesy of Alfa-Laval Inc.

is attached to the pusher plate, and as the pusher plate oscillates back and forth the solids are pushed off the inner screen onto the outer screen and also from the outer screen into the discharge chute.

Ranges of Applicability. The suitability of a particular machine will depend strongly on the solid and liquid characteristics (liquid viscosity, density difference between liquid and solid, particle size, particle shape, percent solids in feed, and solids flowability) and on the required clarification of the liquid and the required dryness of the solids. The solid-bowl machines are ideally suited for applications requiring liquid clarification; the order of preference is solids-retaining, nozzle, solids-ejecting, and decanter. In all these machines the solids will normally contain more than 50 vol.% liquid. The filtering machines will produce the dryest cakes. Cake liquid contents below 10 wt.% are common. However, the filtering machines will normally not produce a solids-free filtrate. Sometimes a sedimentation and a filtering centrifuge are used together to obtain the driest solids and the clearest filtrate.

Particle size and percent solids in the feed are also key parameters that affect equipment selection. Figure 3.4-13 indicates the suitable choices for given ranges of these parameters. The filtering centrifuges will not normally handle small particles as well as sedimenting centrifuges because of problems with filter media blinding.

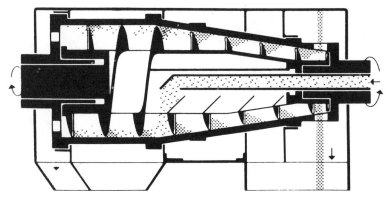

FIGURE 3.4-9 Decanter solid-bowl centrifuge with cylindrical–conical rotor ad open liquid outlet. Courtesy of Alfa-Laval Inc.

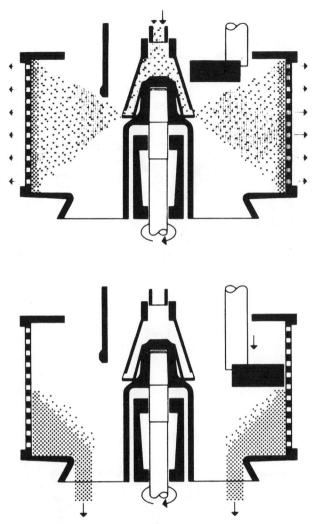

FIGURE 3.4-10 Filtering vertical basket centrifuges: (*a*) feed and separation phase and (*b*) solids-discharging phase.

Capacities. Ultimate capacities and machine operability must be determined experimentally; however, typical ranges for various machines can be given. Table 3.4-1 gives capacity-related data for various types and sizes of sedimentation machine. Note that a tubular machine is included but not discussed here. It is simply a single rotating tube used to clarify liquids at low flow rates.

The capacities of filtering machines are even more variable than the capacities of solid-bowl machines; consequently, it is even more difficult to give an estimate of capacity. However, the data presented in Table 3.4-2 give approximate capacities as a function of machine size for a two-stage pusher. The capacities for the peeler would be somewhat less and for a basket less still. The best source of information regarding capacities are the manufacturers of the various machines.

3.4-3 Impaction (Filtration)

In liquid–solid systems, the impaction or interception mechanism is referred to as filtration. Filtration can be classified in several ways: vacuum, pressure, batch, continuous, cake, or depth. Filtration is generally considered to be a separation involving passage of liquid through a porous medium that retains most of the particulates contained in the liquid. The porous medium may be a wire screen, cloth, paper, or a bed of

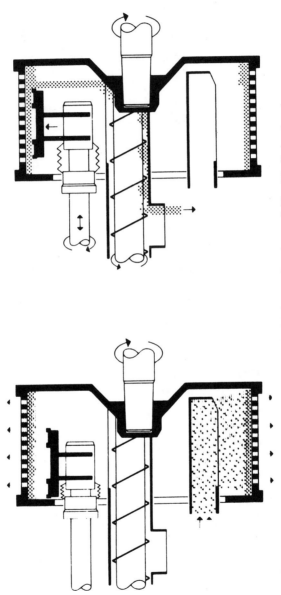

FIGURE 3.4-11 Peeler centrifuge: (*a*) feed and separation phase and (*b*) solids-discharging phase.

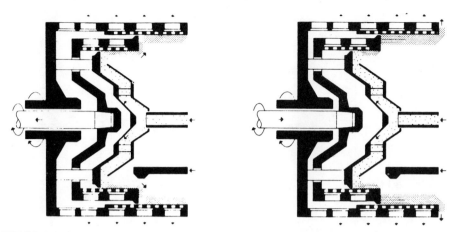

FIGURE 3.4-12 Pusher centrifuge: two-stage model with washing device; (*a*) backward stroke and (*b*) forward stroke.

solids. The medium in most filtrations has only a minor effect; as solids are trapped by the medium they begin to pile up on top of each other, forming a cake of solids that actually becomes the filter medium. This is called *cake filtration*. In the other common filtration mechanism, called *depth filtration*, the solids are trapped in the pores or body of the medium.

Filtration is facilitated by establishing flow through the medium either by gravity, by applying pressure upstream of the filter medium, or by applying vacuum downstream of the filter medium. Filtration can be thought of in terms of the common engineering relationship: rate is proportional to driving force divided by resistance. For filtration, the resistance is equal to the resistance through the cake plus the resistance through the filter medium:

$$\frac{Q_L}{A} = \frac{\Delta P}{\mu(aW/A + R)} \tag{3.4-4}$$

where Q_L is liquid flow rate, A is the filter area, ΔP is the pressure drop across the cake and medium, μ is the liquid viscosity, a is the specific cake resistance, W is the mass of accumulated dry cake solids corresponding to Q_L, and R is the resistance of the medium.

Particle size (μm)	0.1	1	10	100	1000	10000	100000
Pusher							
Peeler							
Basket							
Decanter							
Nozzle							
Solids-ejecting							
Solids-retaining							

Solids (%)	0	10	20	30	40	50	60	70	80	90	100
Pusher											
Peeler											
Basket											
Decanter											
Nozzle											
Solids-ejecting											
Solids-retaining											

FIGURE 3.4-13 Centrifuge range of applications. Courtesy of Alfa-Laval Inc.

TABLE 3.4-1 Specifications and Performance Characteristics of Typical Sedimentation Centrifuges

Type	Bowl Diameter (in.)	Speed (rev/min)	Maximum Centrifugal Force (× gravity)	Throughput Liquid (gal/min)	Throughput Solids (tons/h)	Typical Motor Size (hp)
Tubular	$1\frac{3}{4}$	50,000[a]	62,400	0.05-0.25	—	a
	$4\frac{1}{8}$	15,000	13,200	0.1-10	—	2
	5	15,000	15,900	0.2-20	—	3
Disk	7	12,000	14,300	0.1-10	—	$\frac{1}{2}$
	13	7,500	10,400	5-50	—	6
	24	4,000	5,500	20-200	—	$7\frac{1}{2}$
Nozzle discharge	10	10,000	14,200	10-40	0.1-1	20
	16	6,250	8,900	25-150	0.4-4	40
	27	4,200	6,750	40-400	1-11	125
	30	3,300	4,600	40-400	1-11	125
Helical conveyor	6	8,000	5,500	To 20	0.03-0.25	5
	14	4,000	3,180	To 75	0.5-1.5	20
	18	3,500	3,130	To 50	0.5-1.5	15
	25	3,000	3,190	To 250	2.5-12	150
	32	1,800	1,470	To 250	3-10	60
	40	1,600	1,450	To 375	10-18	100
	54	1,000	770	To 750	20-60	150
Knife discharge	20	1,800	920	b	1.0[c]	20
	36	1,200	740	b	4.1[c]	30
	68	900	780	b	20.5[c]	40

Source: R. H. Perry and D. Green, *Perry's Chemical Engineers' Handbook*, 6th ed., McGraw-Hill, New York, 1984.
[a]Turbine drive, 100 lb_m/h (45 kg/h) of stream at 40 $lb/in.^2$ gauge (372 kPa) or equivalent compressed air.
[b]Widely variable
[c]Maximum volume of solids that the bowl can contain (in ft^3).

Note: To convert inches to millimeters, multiply by 25.4; to convert revolutions per minute to radians per second, multiply by 0.105; to convert gallons per minute to liters per second, multiply by 0.063; to convert tons per hour to kilograms per second, multiply by 0.253; and to convert horsepower to kilowatts, multiply by 0.746.

TABLE 3.4-2 Size and Typical Approximate Capacities for Two-Screen Pusher Centrifuges

Type	SHS	252	352	452	602	802	1002	1202
Nominal basket	mm	250	350	450	600	800	1000	1200
Capacity up to	ton/h	2.5	4.5	8	15	25	40	60
Drive capacity	kW	7.7	10.5	22.5	33	45	75	85
Length	mm	1500	1500	1900	2050	2250	2900	2900
Width	mm	850	950	1100	1820	2100	2500	2700
Height	mm	800	850	1050	1400	1570	2020	2220
Weight	kg	900	1000	1800	3000	3800	8200	9800

Source: Courtesy of TEMA, Inc.

Sizing of filtration equipment is impossible from a theoretical approach because of the varying nature of the solid particles. Almost all applications are determined with the aid of test work. Typical test arrangements are a vacuum leaf test, a buchner funnel, and a pressure bomb method.

A leaf test consists of a 10–12 cm diameter frame to which is attached a filter medium such as a screen or cloth. This frame is connected by tubing to a filtrate collection reservoir which is evacuated by a vacuum source. The leaf assembly is placed into a well-mixed vessel containing the slurry to be separated. Time to form various depths of filter cake and the amount of liquid passed are observed versus time and the level of vacuum applied. Potential for drying the cake is observed by removing the leaf assembly from the slurry and letting air be drawn through the cake to allow drying. A slight alternative to this approach is the use of a small Buchner funnel with an appropriate filter medium. The slurry is poured into the funnel and the time of filtration and resulting cake dryness observed. These tests can be used to determine sizing parameters for various vacuum-motivated filtrations such as nutsche filters, rotary drum filters, pan filters, and belt filters.

A bomb test consists of a 6–25 cm^2 leaf covered with appropriate filter medium and placed in a small pressure vessel containing enough slurry to form a desired amount of cake on the leaf. The quantity of filtrate liquid flowing through the cake is measured as a function of time under various pressures and the depth of cake formed is observed. This type of test can provide information for sizing pressure leaf filters or plate and frame filters. These tests are best conducted by filter vendors who have experience in translating the results to full-scale equipment.

Filtration equipment will be briefly reviewed for the following categories: batch cake filters, continuous cake filters, and clarifying filters.

BATCH CAKE FILTERS

The *nutsche filter* is one of the simplest designs, consisting of a vertical tank with a false bottom that is either perforated or porous and that may be covered with a cloth or screen. Slurry is fed into the vessel and separation occurs by using gravity, vacuum, or gas pressure to motivate the liquid through the media. Washing of accumulated cake is possible. This type of filter is very similar to the laboratory Buchner funnel. Most nutsche filters are self-manufactured, although the Rosenmund filter is a commercial unit of 1–10 m^2 designed for closed operation with automated cake discharge.

The *horizontal plate filter* is a pressure filter with a number of horizontal circular drainage plates in a stack in a cylindrical shell. A filter cloth or paper is placed on each plate. Filter aid may be applied if necessary. Filtration is continued until the cake capacity of the unit is reached or the filtration rate becomes too slow owing to cake resistance. Pressure drop across the filter medium is generally designed for 50 lb$_f$/in.^{2}. Cake may be washed or air-blown prior to manual removal. This type of filter is flexible and easily cleaned or sterilized, but it has a high labor requirement.

The *filter press* was one of the most common filters in the past and is still used today. The plate-and-frame design consists of a series of rectangular plates with filter medium, usually cloth, attached to both sides with a hollow frame placed between each plate in an alternating manner to allow accumulation of cake. The plates and frames are usually hung vertically from two parallel horizontal bars or tubes and pressed together mechanically; with new larger designs a hydraulic ram is often used. Slurry is fed to the frame sections, with filtrate passing through the medium and exiting from the plate sections. Nozzles for feed and filtrate discharge are connected to manifolds with flexible tubing so that the press may be opened and spread apart for cake removal. Various nozzle configurations are used depending on the type of slurry and washing requirements. Plates and frames are constructed of metals, plastics, or wood with some designed for operating pressures up to 100 lb$_f$/in.^{2}.

The filter press has the advantages of simplicity, low capital cost, flexibility, ease of cleaning and media replacement, and moderate pressures. Disadvantages include high labor input to handle each plate and frame for cake discharge, frequent cloth replacement due to the handling and scraping of the cloth, and housekeeping problems that leaks or solid discharge. Some of these disadvantages are being addressed by automated mechanical systems for press opening and cake discharge. Operator exposure to process materials during the opening of the filter at each cycle limits the application to nontoxic, nonflammable materials.

Various *tubular filters* exist in which the cake is formed either on the inside or outside surface of horizontal or vertical tubes. Manually operated tubular units often use paper or cloth liners in a tubular screen with slurry fed into the tubes and operated until the tube is full and significant pressure drop occurs. Solids are then manually removed. These devices have the advantages of easily replaced amd inexpensive disposable filter medium, and flexibility of operation. Disadvantages are high labor requirements and the tendency of solids to accumulate in the tubesheet channel. Vertical designs are common for automated operation with solids collected on the outsides of tubes covered with a screen or cloth medium. Some designs for higher pressure are now using a sintered metal tubular form. Most designs use tubes hung vertically from a tubesheet. Slurry is fed to the body of the chamber containing the tubes, with filtrate passing through the medium up the inside of the tube and out through the top head above the tubesheet. As pressure drop across the medium increases to the desired limit, the filtrate discharge is closed. With same designs, a vapor space in the top head is allowed to compress. Blowback of liquid motivated by the expansion of compressed gas in the head when opening the bottom discharge valve causes sufficient force

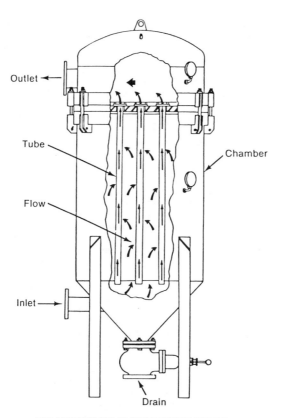

Outlet

Tube

Chamber

Flow

Inlet

Drain

TOP-OUTLET TUBULAR FILTER "HYDRA-SHOCK"

FIGURE 3.4-14 Tubular filter. Courtesy of the Industrial Filter and Pump Manufacturing Co.

to displace cake from the outside surface of the tubes. An example of this design is shown in Fig. 3.4-14. This system has the advantage of being a totally closed operation, which allows handling of toxic materials and low labor requirements. The blowback fluid requirements limit the ability to obtain maximum cake dryness.

Pressure leaf filters consist of flat filter elements supported inside a pressure vessel. The leaves may be circular, arc-sided, or rectangular, and they have filter medium on both faces of a durable screen frame. The vessel axis and the leaf orientation can be any combination of vertical or horizontal. Slurry is fed to the vessel body and cake accumulates on the surface of the leaves. Filtrate passes into the leaf assembly and exits via a manifold. Cake is allowed to build to a limited thickness to allow for washing or ease of discharge. Cake discharge is facilitated by sluicing with liquid sprays, vibration of the leaves, or rotation of the leaves against a knife, wire, or brush. If wet cake is allowed, a high-pressure liquid sluicing is used. With rotation of the leaves a center axis manifold and leaf support are used. Some units open the filter for cake discharge by separating the leaf assembly from the shell. Some horizontal vessels with vertical leaves use an internal conveyor to obtain dry cake removal and closed operation. Centrifugal-discharge designs consist of horizontal disks with filter elements on one side attached to a vertical hollow shaft. This filter is appropriate for closed operation on toxic materials. It is also flexible for ease of washing of cake, drying of cake with a gas purge, and a relatively dry discharge of cake due to high-speed centrifugal discharge by rotation of the vertical shaft. Some designs use a brush to aid in cake removal during the rotation for cake discharge.

Continuous Cake Filters
Continuous cake filters are used where cake formation is rapid, solid concentration in the slurry is greater than 1%, particle size is greater than 100 μm, and slurry flow rate is greater than about 0.083 L/s (0.017–0.03 gal/s). Liquid viscosity is usually below 100 cP to allow good liquid flow through the cake as it forms. Some applications can compromise some of these parameters if filter aid is used and a pure cake is not the desired product.

Rotary drum filters are the most common of the continuous cake filters. Designs may be either pressure or vacuum motivated with a horizontal axis drum and a filter medium placed on the cylindrical portion of the drum supported by a backup screen or grid. Sizes range from 4 to 1000 ft^2 and material of construction may be either metals or plastics. The filter operates by slowly rotating the drum in a trough of slurry with the drum from 10 to 35% submerged. An oscillating rake or agitator may be needed in the trough for rapid settling slurries. Cake forms on the outside of the drum while in the trough, and filtrate passes through the cake and is collected by a manifold system beneath the surface of the drum. As the drum rotates out of the slurry pool, air or gas is drawn through the cake, displacing the filtrate. A wash may be applied with sprays to the cake surface followed by further drying. An air or gas blowback can be applied to dislodge the cake prior to the drum rotating back through the slurry trough.

Filtrate and washes collected by the manifold system exit via a rotary valve placed in one of the drum trunnions. Drum rotation is usually in the 0.1–10 rev/min range. Various cake discharge designs are used. Scraper discharge uses a blade to deflect the cake just prior to the drum re-entry into the pool, usually assisted by an air blowback. A taut wire may be used in place of a blade. String discharge consists of many parallel endless strings or wires spaced about 1.25 cm apart. The strings separate tangentially from the drum surface, lifting the cake from the medium. The strings are separated from the cake and return to the drum surface by two small rollers. Some designs involve removal of the entire medium from the drum surface, passing it over rollers that discharge the cake and then return to the drum, often with medium washing sprays applied. A roll discharge involves a small diameter roll in close proximity to the drum and rotating in the opposite direction at a speed slightly higher than the drum. The cake on the drum adheres to the roller and is separated from the drum. A blade or wire may be used to remove the cake from the roller.

A variation of the rotary drum filter is the continuous precoat filter. This type is used for clarification of dilute slurries containing 50–5000 ppm of solids where only very thin unacceptable cakes are formed on other filters. Design is similar to the standard rotary drum with a blade discharge. Prior to feeding slurry, a precoat layer of filter aid or other solids are applied 75–125 mm thick. Slurry is then fed and solids are trapped in the outer layer of the precoat. A progressively advancing blade cuts the solids plus some of the precoat on each pass. When the blade advances to the minimum point, the operation must be suspended and the precoat re-established.

A *disk filter* is a vacuum filter consisting of a series of vertical disks attached to a horizontal axis shaft rotating in a pool of slurry similar to the rotary drum filter. The disks consist of several sectors of metal or plastic covered on both sides with a filter cloth. Filtrate exits from the sectors via a nipple into the center shaft. The disks are usually 40–45% submerged in the slurry trough. Wash may be applied with sprays, but this is unusual. As the sectors rotate to the discharge point, the vacuum is cut off and an air blowback helps dislodge the cake, with scraper blades directing the cake into discharge chutes positioned between the disks. This filter provides a large area with minimum floor space and has the lowest cost per unit area, and it is best used for high-volume dewatering applications.

Horizontal vacuum filters are found in two basic designs, rotary circular and belt filters. They provide flexibility in cake thickness, washing configuration, and drying time. They are effective for heavy dense solids, but they are expensive, require a large floor space, and are difficult to enclose for hazardous applications.

The rotary design is a circular pan with slurry fed across the radius of the pan and the cake discharge located about 300° around the circle using a screw discharge. Sizes range to 81 ft (25 m) diameter and the most common application is on free-draining inorganic salt slurries. A modification of the rotary horizontal design is the tilting pan design. The two designs are similar in concept except that the tilting pan is divided into sectors. Each sector has its own sides and is pivoted on a radial axis to allow it to be inverted for cake discharge, usually assisted by an air blast.

The horizontal belt filter consists of a slotted or perforated elastomer belt driven as a conveyor belt supporting a filter fabric belt. Both belts are supported by a deck which is sectioned to form vacuum chambers to collect filtrate and multiple wash zones. Slurry is fed via an overflow weir or a fantail chute at one end. Wash liquid is fed with either a weir or spray arrangement. Cake is removed from the belt as it passes over the pulleys at the end opposite the feed point. The filter medium belt is separated from the elastomer belt as it returns to the feed and, allowing for wash spraying of the filter medium if needed. Belt speeds up to 1.6 ft/s (0.5 m/s) are typical, with some units having variable-speed drives. Cake thickness of 100–150 mm is possible for some fast-draining solids. Advantages are complete cake removal and effective filter media washing. A disadvantage is that at least half of the filter area is idle on the return to the feed point.

CLARIFYING FILTERS

These filters are used on slurries with small amounts of solids, usually less than 0.1%, and generally do not form any visible cake. Solids to be removed are usually very small particles that may be trapped on the surface of the filter medium or within the medium. This type of filter is usually used in a polishing application where excellent quality liquids are needed as in food or beverage, pharmaceutical, and electronic processing operations. The most common clarifying filters are disk and plate presses, cartridge filters, precoat pressure filters, deep bed filters, and membrane filters.

Disk or plate filters are very similar to horizontal plate pressure filters used in cake filtration. A paper or cloth filter medium may be attached to the horizontal screen and serve as the filter medium to trap solids on the surface. Flow rates to 3 gal/min and operating pressures to 50 $lb_f/in.^2$ are typical. Filter presses similar to the cake filter applications are used with good quality media to trap solids on the media surface for clarifying applications. These filters are often used to provide complete particle removal for a specified particle size cut. Newer filter media such as the Zeta Plus from AMF Cuno Division are composed of cellulose and inorganic filter aids which have a positive charge and provide an electrokinetic attraction to hold solids that are usually negatively charged. This provides both mechanical straining and electrokinetic adsorption. Other new plate-type media for clarification include composite nonwoven media combining activated carbon and cellulose fibers. These filters are not attractive for hazardous solvent applications because the units must be opened for media replacement.

Cartridge filters are probably the most common of the clarifying filters. They come in a wide variety of configurations—horizontal, vertical, single cartridge, and multiple cartridge—and in a wide variety of cartridge designs—wound fiber, resin-impregnated papers of pleated design, porous metals, and ceramics. Cartridge elements come with a wide variety of particle removal ratings, some down to 1 μm. The lower the rating, the higher the pressure drop and the lower the filtrate capacity. For large-capacity applications, parallel units of multiple cartridges are used to provide the desired capacity. The cartridges are operated until a solids build-up causes the pressure drop across the elements to increase to an undesirable level at which time the cartridges are replaced. Most cartridges of fiber and paper are disposable, but some of the metal and ceramic units may be made reusable by off-line chemical treatment or incineration of organic contaminants. Cartridges have the advantage of flexibility: cartridges with different ratings or materials may be inserted to adapt to changed conditions or improved filtration performance requirements. These filters have the disadvantages of very limited solids capacity (usually slurries of 0.01% solids or less) and the exposure of operating personnel to the process of cartridge replacement. Some recent constructions using bonded metal fibers are designed with parallel units that can be either chemically cleaned or back-flushed without opening the system. These units can operate at up to 480°C and at pressures up to 5000 $lb_f/in.^2$ These designs are particularly good for filtering polymer melts prior to forming operations.

Precoat pressure filters are of the same design as the cake filters described above. The filters are precoated with filter aid to provide a medium on which they can then form an effective surface for clarifying applications. These applications provide filtration by removing the solid contaminants in the depth of the media, in this case in the filter aid. Additional filter aid may be fed with the slurry for systems where the solids are sticky or slimy and tend to blind the surface. This is called a "body feed." Initial precoat of filter aid to prevent bleed-through of fine solid particles is important. Filter-aid material, while inexpensive on a unit weight basis, can add significant cost to the process and cause problems in solids disposal. Techniques have been developed to reuse filter aid for several cycles. This is done by backwashing filter aid and separated solids from the outside of tubular filter elements, and then obtaining a uniform mix and redepositing the mix as a new precoat. This may be possible until the separated solids build up in the mix to a system-dependent limiting point.

Deep bed filters are a vast contrast to the types discussed above in that liquid volumes handled can be extremely large. These filters consist of a bed of solid material, typically sand or carbon, at least 3 ft deep in a vertical vessel up to several feet in diameter. The vessel may be either a large open unit made of concrete, as in water treatment plants, or a process vessel that could operate under pressure. The bed sometimes may be composed of layers of various sized particles of different materials, such as both sand and carbon. Various designs are used which feed slurry from above, below, or in the middle of the bed. The bottom feed or takeoff points may be accomplished with perforated pipes or a false perforated vessel bottom. Upflow designs have a greater solids handling capacity as larger particles in the feed are removed by the coarser solids of the bed, with the finest solids in the stream removed by the finer grained particles near the top of the bed. Downflow designs tend to load rapidly the fine layer of solids at the top of the bed, and they require more frequent bed flushing to maintain flow capacity. The bed is backflushed with liquid and sometimes with an air sparge. The contaminant solids flushed from the bed must be removed from the backflush stream by a more conventional filter. Since a backflush is required the filtrate flow must be interrupted or parallel units used.

Deep bed filters are used for liquid clarification on large liquid flows where the solids are less than 1000 ppm. Filtration rates to 0.027 $ft^3/s \cdot ft^2$ are possible.

A deep bed system that avoids batch backwashing to clean the filter is the Dyna Sand™ filter. The bed design is continuously cleaned and regenerated by the internal recycling of solids from the bottom of the bed to the top through an air-lift pipe and a sand washer. This filter allows a constant pressure drop across the bed and avoids the need for a parallel filter to allow backwashing.

While membranes are discussed extensively elsewhere in this book, some membrane separators function as a filter sieve to clarify liquids. Solids as low as 0.001 μm can be removed. Membranes may be in the form of hollow fibers, spiral wound sheets, or membrane disks.

3.4-4 Electromotive Separators

Electromotive separation for solid–liquid processes is not commercially well developed. Particles suspended or slurried in water become electronically charged. Solids such as clays are generally positively charged while materials such as coal, proteins, and other organic solids are typically negatively charged. Use of electrical fields can cause particles to agglomerate, and thereby improve settling, or to be attracted or repelled from surfaces. Experimental applications have been in the areas of enhanced settling in sludge ponds, electronically augmented filtration, and membrane dewatering processes. This should be a promising development area.

3.4-5 Selection Approach

The size and handling characteristics of the settled solids have the largest impact on separator selection. Also, the desired clarity of the liquid and the dryness of the solids are key selection criteria. For biological sludges, settling basins are often used because of the inexpensive open construction. For small-diameter particles (1–10 μm) solid-bowl centrifuges are most appropriate. For applications requiring dry cakes of solids which are larger than about 10 μm in size, either a filtering centrifuge or a vacuum or pressure filter are the logical choices. For applications requiring extreme clarity of the liquid a solid-bowl centrifuge is the logical choice. For applications requiring low clarification of the liquid and not very dry solids, hydro-clones should be considered. For applications requiring the driest solids, a filtering centrifuge or a filter are the logical choices. For almost every application more than one type of equipment will suffice. The selection procedure involves a careful study of the characteristics of the solids, the needed process require-ment, usually a test program and, normally, contact with several equipment manufacturers.

3.5 GAS–SOLID SEGREGATION

3.5-1 General

Gas–solid phase segregation is required following solids processing operations such as drying, calcining, grinding, and solids air conveying. Before attempting to design a separation system, one should determine pertinent characteristics of the solids. The shape and size distributions are key parameters, as is the solids density. The ''stickiness'' of the solids, as indicated by the angle of repose and the general flow charac-teristics of the solids must be determined. The solids loading will also affect equipment selection and design.

3.5-2 Gravity Settling Chambers

The gravity settler is the simplest of all the gas–solid separation equipment. Refer to Sections 3.2-2 and 3.3-2 for discussions of gravity separators used for liquid–gas and liquid–liquid phase segregation. In its simplest form, this device is an empty chamber with a well-designed inlet diffuser to introduce the gas–solid stream to the separator while producing a minimum level of turbulence. (Refer to Section 3.3-2 for guidelines about inlet diffuser design.) With negligible suspending action owing to turbulence, the design of the gravity settler is very simple: the residence time of the gas in the chamber must be sufficient to allow the smallest solid particle to settle from the top to the bottom of the chamber. For a rectangular chamber, the required horizontal area A_I is given by

$$A_I = LW = \frac{Q_C}{U_t} \tag{3.5-1}$$

and the height of the chamber H is given by

$$H = \frac{Q_C}{U_F W} \tag{3.5-2}$$

where L and W are the length and width of the separator, Q_C is the volumetric flow rate, and U_t is the particle terminal velocity of the smallest particle to be separated, and U_F is the horizontal flow velocity of the gas stream.

The flow velocity has to be sufficiently low so that the turbulence in the flowing stream will not suspend the settled particles. According to Dryden[1] the settled particles will not be resuspended provided the velocity of the turbulent eddies (refer to Section 3.2) does not exceed the particle terminal velocity. On this basis, the flow velocity can be about 20 times the terminal velocity; however, it is reasonable to use only about 10 times the terminal velocity. On this basis,

$$H = \frac{Q_C}{10 U_t W} \tag{3.5-3}$$

The terminal velocities of solid particles must be calculated. This can be done by using the drag

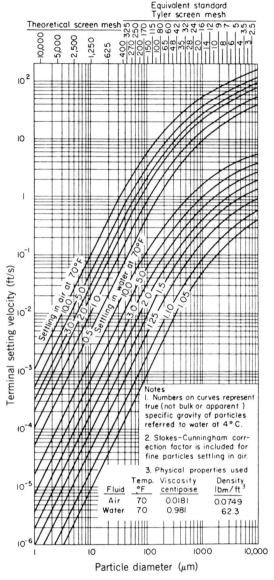

FIGURE 3.5-1 Terminal velocities of spherical particles of different densities settling in air and water at 70°F under the action of gravity. From Lapple et al., *Fluid and Particle Mechanics*, University of Delaware, Newark, 1951.

coefficient curves for falling spheres (Fig. 3.1-1) and Eq. (3.1-1); however, Fig. 3.5-1 will normally be sufficient for design purposes for particles falling in air.

An example is now considered: 100 μm particles of specific gravity = 2.58 (U_t = 1.9 ft/s) are to be separated from a 10,000 ft³/min ambient air stream in a separator with $H = W$.

From Eq. (3.5-3),

$$H = W = \left[\frac{10,000}{(60)(10)(1.9)} \right]^{1/2} = 3.0 \text{ ft}$$

From Eq. (3.5-1)

$$L = Q_c/WU_t = \frac{10,000}{(60)(3)(1.9)} = 29.2 \text{ ft}$$

and from Eq. (3.5-2)

$$U_F = \frac{Q_C}{HW} = \frac{10,000}{(3.0)^2} = 1100 \text{ ft/min} = 18.5 \text{ ft/s}$$

Particles larger than about 50 μm can often be successfully separated in an oversized transfer line. However, very fine particles require excessively large chambers; for example, if the particles in this example are reduced to 20 μm, the chamber size is increased to 14 ft × 14 ft × 140 ft.

3.5-3 Centrifugal Separators (Cyclones)

Typical geometries for gas–solid cyclones are indicated in Fig. 3.5-2. In the cyclone, the gas–solid stream flows in a helical path; consequently, centrifugal action results in a force on the solid particles which moves them toward the wall of the cyclone. Conceptually, the cyclone separates solids in the same manner as in a gravity settling chamber with the gravitational force being replaced with a centrifugal force that can vary frm 50 to 1000 times the normal gravitational force.

The cyclone has an inlet velocity limitation due to settled solids being re-entrained from the cyclone wall. The design method presented here allows calculation of the optimum inlet velocity.

Depending on the size of the cyclone, particles from 2 to 20 μm can be economically separated. With large cyclones (3–10 ft diameter) a reasonable lower limit for successful separation is 10–20 μm. For banks of smaller cyclones (4–10 in. diameter), the reasonable lower limit is 2–5 μm. Figure 3.5-3 (for air at 100°F with a solid particle density of 2.58 g/cm^3) gives a good qualitative idea of the range of applicability of cyclone separators.

The Leith and Licht method[6-8] for predicting cyclone performance is presented here because it is considered to be the most accurate and comprehensive. However, the method is too long and tedious for hand use in its most comprehensive form; consequently, an abbreviated version is presented. As the step-by-step procedure is described, the simplifications and compromises will be noted. Only a rating procedure is given, and to use it for design, an iterative procedure is required. The recommended rating procedure is as follows:

1. Select a geometry from Fig. 3.5-2. For design purposes the Stairmand geometry is a good one. For rating an existing cyclone, select the Fig. 3.5-2 geometry that comes closest to the existing equipment. The value of C, which is needed to calculate the grade efficiency, is given for the various geometries in the table in Fig. 3.5-2.

2. Calculate the optimum value of the cyclone inlet velocity V_I by Eq. (3.5-4).

$$V_I = 17 \, W \left[\frac{(b/D_c)^{1.2}}{1 - b/D_c} \right] D_c^{0.2} \tag{3.5-4}$$

where b and D_c are in ft and

$$W = \frac{4g\mu(\rho_p - \rho_g)}{3 \, \rho_g^2} \tag{3.5-5}$$

g is the gravitational acceleration, μ is the gas viscosity, ρ_p is the density of the solid particle, and ρ_g is the gas density.

4. Calculate the parameter n using the equation

$$n = 1 - \left[1 - \frac{(12 \, D_c)^{0.14}}{2.5} \right] \left[\frac{°F + 460}{530} \right]^{0.3} \tag{3.5-6}$$

5. Calculate the product of ψ and C:

$$C\psi = C \left(\frac{\rho_p d_p^2 V_I}{18 \, \mu D_c} \right)(n + 1) \tag{3.5-7}$$

Values of C are presented in Fig. 3.5-2. Note that C appears on both sides of the equation.

6. Calculate the fractional collection efficiency for the particle diameter d_p:

$$\eta = 1 - \exp\left[-2(C\psi)^{1/(2n+2)} \right]$$

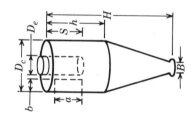

	Nomenclature	High Efficiency		General Purpose		
		Stairmand[2]	Swift[3]	Lapple[4]	Swift[3]	Peterson and Whitby[5]
D_c	body diameter	1.0	1.0	1.0	1.0	1.0
a	inlet height	0.5	0.44	0.5	0.5	0.583
b	inlet width	0.2	0.21	0.25	0.25	0.208
S	outlet length	0.5	0.5	0.625	0.6	0.583
D_e	outlet diameter	0.5	0.4	0.5	0.5	0.5
h	cylinder height	1.5	1.4	2.0	1.75	1.333
H	overall height	4.0	3.9	4.0	3.75	3.17
B	dust outlet diameter	0.375	0.4	0.25	0.4	0.5
l	natural length	2.48	2.04	2.30	2.30	1.8
G $8 K_c/K_b^2$		551.3	699.2	402.9	381.8	324.8
N_H $16\ ab/D_e^2$		6.40	9.24	8.0	8.0	7.76
G/N_H		86.14	75.67	50.36	47.7	41.86
C $8 K_c/K_a K_b$		55.1	64.6	54.4	47.7	39.4

FIGURE 3.5-2 Design configurations for tangential entry solid–gas cyclones. Excerpted by special permission from *Chemical Engineering*, Nov. 7, 1977, pp. 80–88. © by McGraw-Hill, Inc., New York, NY 10020.

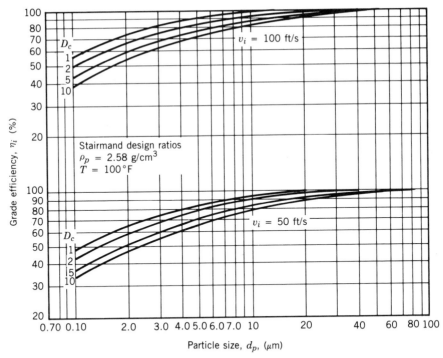

FIGURE 3.5-3 Grade efficiency of Stairmand cyclone correlated with particle size, inlet velocity, and cyclone diameter. Excerpted by special permission from *Chemical Engineering*, Nov. 7, 1977, pp. 80–88, © by McGraw-Hill, Inc., New York, NY 10020.

7. Repeat the procedure for as many size fractions as needed to generate the grade efficiency curve for this particular geometry.

As Leith and Licht[6] and Koch and Licht[7] indicate, this design method does not include dust loading[9] as a parameter. Higher dust loading improves efficiency; however, the quantitative effect cannot be predicted. Also, a purge flow of 10% of the gas out the bottom of the cyclone can increase overall collection efficiency by as much as 20–28%, according to Crocker et al.[9]

The pressure drop through a cyclone can be calculated by

$$\Delta P \text{ (in } H_2O) = 0.003\, \rho_g V_I^2 N_H \qquad (3.5\text{-}8)$$

where N_H is given in Fig. 3.5-2. and V_T, is in ft sec^{-1}.

3.5-4 Impaction

WET SCRUBBERS

Seive-plate columns, packed columns, fibrous beds, spray nozzles, venturis, and many other specialized designs are used. Calvert[10] gives a good discussion of the various types with their advantages and disadvantages and an approximate cut diameter chart (Fig. 3.5-4). The most widely used scrubbers are the so-called venturi types. Figure 3.5-5 presents a schematic of a particular commercial design that functions by atomizing the liquid with a high-velocity gas flow that shears a liquid film into droplets. Approximate sizing information is given here for this type of scrubber. The operating parameters of significant economic importance are:

1. The gas-phase pressure drop required to separate a given particle size.
2. The amount of liquid required per unit of gas flow.
3. The size (and capital) cost of the equipment.

The scrubber system consists of the scrubber, a liquid recirculating system, and sometimes a mist

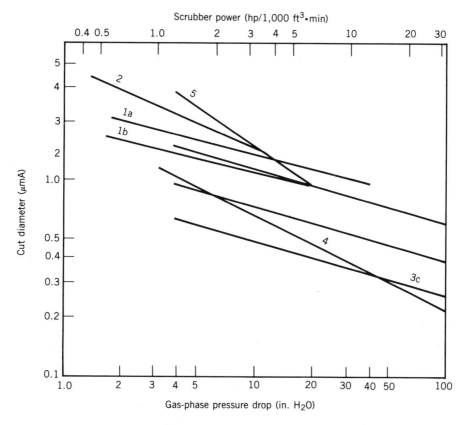

Scrubber power (hp/1,000 ft³•min)

Cut diameter (μmA)

Gas-phase pressure drop (in. H₂O)

1a. Sieve-plate column with foam density of 0.4 g/cm³ and 0.2-in. hole diameter. the number of plates does not affect the relationship much. (Experimental data and mathematical model.)

1b. Same as 1a except 0.125-in. hole diameter.

2. Packed column with 1-in. rings or saddles. Packing depth does not affect the relationship much (Experimental data and mathematical model.)

3a. Fibrous packed bed with 0.012-in. diameter fiber—any depth. (Experimental data and mathematical model.)

3b. Same as 3a except 0.004-in. diameter fibers.

3c. Same as 3a except 0.002-in. diameter fibers.

4. Gas-atomized spray. (Experimental data from large venturis, orifices, and rod-type units, plus mathematical model.)

5. Mobile bed with 1 to 3 stages of fluidized hollow plastic spheres. (Experimental data from pilot-plant and large-scale power-plant scrubbers.)

FIGURE 3.5-4 Particle cut versus pressure drop or scrubber power for various separators. Excerpted by special permission from *Chemical Engineering*, Aug. 29, 1977, p. 54, © by McGraw-Hill, Inc., New York, NY 10020.

eliminator on the outlet of the scrubber. Scrubbers normally use the cyclonic action of the gas flow to separate all but a small fraction of the atomized liquid from the gas stream. The required gas-phase pressure drop to separate a particular particle size can be determined from Fig. 3.5-4. For the scrubber of Fig. 3.5-5, the liquid requirement per unit of gas flow can be estimated by Fig. 3.5-6. To determine the scrubber diameter for use with Fig. 3.5-6, a superficial velocity of about 15 ft/s (900 ft/min) can be used.

As an example, removal of 50% of 0.3 μm particles from a stream flowing at a rate of 100,000 ft³/min requires a scrubber about 12 ft in diameter (set by superficial velocity) by 48 ft tall, with a gas pressure drop of 50 in. H₂O and a scrubber power requirement of 1500 hp (Fig. 3.5-4), and a liquid requirement of 700 gal/min (Fig. 3.5-6).

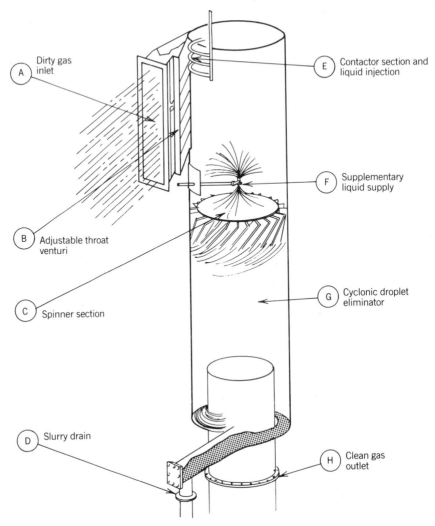

FIGURE 3.5-5 Configuration of Fisher–Klosterman high-energy venturi scrubber. Courtesy of Fisher–Klosterman Inc.

It could well be that other scrubber designs might be more economical. To determine the economic optimum is beyond the scope of this chapter; for the necessary information refer to Calvert et al.[11]

FILTERS

Typical fractional collection efficiencies for gas–solid filters are indicated in Fig. 3.5-4. Below a particle size of 2–5 μm, where even banks of small-diameter cyclones are ineffective, filters are often the most economical equipment. A filter can often be selected with a minimum of experimental work. The availability of standard equipment to handle capacities up to 100,000–200,000 ft^3/min is very good. Consequently, filters are normally the equipment that is given first consideration after a determination has been made that neither gravity settlers nor cyclones are economical.

Types of Filters. Two broad general classes of filters are commonly used: (1) depth filters, which normally employ inertial impaction and Brownian movement as collection mechanisms and which are often discarded or cleaned with a liquid when they become dirty; and (2) fabric (or bag) filters, which build a bed of solids that acts as the filter medium and which is normally cleaned mechanically at frequent intervals (similar to liquid–solid cake filters).

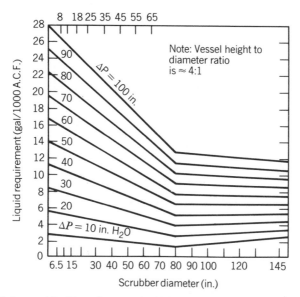

FIGURE 3.5-6 Estimate of liquid requirements for high-energy venturi scrubber. Courtesy of Fisher–Klosterman Inc.

Depth Filters. These are either fiber bed or pleated-paper media that are built into standard-size rectangular panels. The panels can be easily inserted into a suitable frame. Tables 3.5-2 and 3.5-3 give the characteristics and pertinent operational parameters for the filter types as classified by the American Society of Heating, Refrigerating and Air-Conditioning Engineers (Types I, II, and III) and for nuclear facilities (Type HEPA—high efficiency particulate air). The depth filters are normally used only for gas streams containing small amounts of solids. The service life of depth filters depends on the dust loading of the gas. Table 3.5-4 presents typical dust loading for various environments. Viscous filters are wetted with a tacky liquid (e.g., mineral oil or adhesives) to retain the dust. Dry filters use finer fibers and have much smaller pores than viscous media. Dry filters are normally deeper than viscous filters and have higher flow resistances. The filtration velocities are normally lower than those used with viscous filters. In general, viscous-type filters will average 2–5 months with average city air, whereas for dry types 1–2 months is typical. The service life is inversely related to the dust loading.

Fabric Filters. Fabric filters normally use either woven or felt type cloth in the common bag-filter configuration. The fabric is made into bags 6–10 in. in diameter and 6–40 ft long. The air flow can be in either direction through the bag wall. The dust is periodically removed from the bag either by a shaking mechanism or by a reverse flow pulse-jet. Means are provided in either style so that only a fraction of the total number of bags are being cleaned at any one time.

The collection efficiency of bag filters for various dust particle sizes is given approximately in Fig. 3.5-13. These collection efficiencies are realized for bags in good condition with minimal holes. The wearing of bag fabric is a severe problem and means should be provided to check the filter performance to detect bag failures before they become catastrophic.

For units cleaned by mechanical shaking, cleaning may take up to 3 min, while for air-pulsed units cleaning can be done in 2–10 s.

Bag filters are normally designed for a 4–8 ft/min face velocity at pressure drops of 1–6 in. H_2O. The normal mode of operation is to remove the solids buildup based on pressure drop across the filter. For typical dust loadings of normal chemical process gas streams, the cleaning frequency will be from 1 to 15 min. The pressure drop through the bed of solids can be calculated[9] by assuming that the properties of the gas and solids are known. These calculations are normally not necessary because once it appears that a bag filter is the separation technique of choice, a reputable vendor can provide specific information about costs and operating procedures.

The air pulse for a bag filter is normally 90 $lb_f/in.^2$. Filters are normally supplied in standard units up to 10,000 ft² in area with larger units often assembled from standard units of 10,000 ft² each.

For the pulsed filter, the solids removal may be very effective; consequently, there will be no solids layer initially to act as a depth filter. Felt fabrics must be used to provide for effective filtration until the solids layer builds to a level where particulate removal is very effective.

TABLE 3.5-2 Comparative Air-Filter Characteristics

	Unit Filters				Automatic Filters
	Viscous Type		Dry Type		
	Cleanable	Throwaway	Throwaway	Cleanable	
Dust capacity	1. Well adapted for heavy dust loads (up to 2 grains/1000 ft³) due to high dust capacity		1. Well adapted to light or moderate dust loads of less than 1 grain/1000 ft³		1. Well adapted for heavy dust loads (>2 grains/1000 ft³) since it is serviced automatically
Filter size		1. Common size of unit filter is 20 × 20 in. face area handling 800 ft³ min at rated capacity 2. Face velocity is generally 300–400 ft/min for all types			1. Automatic viscous units supplied to handle 1000 ft³/min and over 2. Face velocity is 350–750 ft/min
Air velocity	1. Rated velocity is 300–400 ft/min through the filter medium 2. Entrainment of oil may occur at very high velocities		1. Rated velocity is 10–50 ft/min through the medium. (Some dry glass types run as high as 300 ft/min.) 2. Higher velocities may result in rupture of filter medium		1. Rated velocity is 350–750 ft/min through the filter medium for viscous types. For dry types, it is 10–50 ft/min
Resistance		1. Resistance ranges from 0.05–0.30 in. when clean to 0.4–0.5 in. when dirty 2. When the resistance exceeds a given value, the cells should be replaced or reconditioned 3. Cycling cells in large installations will serve to maintain a nearly constant resistance 4. Excessive pressure drops resulting from high dust loading may result in rupture of filter medium			1. Resistance runs about 0.3–0.4 in. water
Efficiency	4. High resistance due to excessive dust loading results in channeling and poor efficiency		1. In general, give higher efficiency than viscous type, particularly on fine particles 2. Efficiency increases with increased dust load and decreases with increased velocity		
	1. Commercial makes are found in a variety of efficiencies, these depending roughly on filter resistance for similar types of medium 2. Efficiency decreases with increased dust load and increases with increased velocity up to certain limits				

Operating cycle	1. Well adapted for short-period operations (less than 10 h/day) due to relatively low investment cost 2. Operating cycle is 1–2 months for general "average" industrial air conditioning	2. Operating cycle is 2–4 weeks for general "average" industrial air conditioning	1. Well adapted for continuous operation
Method of cleaning	1. Washed with steam, hot water, or solvents and given fresh oil coating	1. Filter cell replaced. Life may in some cases be lengthened by shaking or vacuum cleaning, but this is not often successful	1. Vacuum cleaned, blown with compressed air, or dry cleaned	1. Automatic. Filter may clog in time and cleaning by blowing with compressed air may be necessary
Space requirement	1. Well adapted for low headroom requirements 2. Form of banks can be chosen to fit any shaped space 3. Space should be allowed for a man to remove filter cells for cleaning or replacement 4. Requires space for washing, reoiling, and draining tanks	4. Requires space for mechanical loader in some cases	1. Have a high headroom requirement 2. Take up less floor space than other types
Type of filter medium	1. Crimped, split, or woven metal, glass fibers, wood shavings, hair—all oil coated	1. Cellulose pulp, felt, cotton gauze, spun glass 2. Dry medium cannot stand direct wetting. Oil-impregnated mediums are available to resist humidity and prevent fluff entrainment	1. Metals screens, packing, or baffling. One type uses cellulose pulp
Character of dust	1. Not well suited for linty materials 2. Well adapted for make-up air and granular materials	1. Not well suited for handling oily dusts 2. Well adapted for linty material 3. Better adapted for fine dust than other types	1. Not suited for linty material if of viscous types
Temperature limitations	1. All metal types may be used up as high as 250°F if suitable oil or grease is used. Those using cellulosic materials are limited to 180°F	1. Limited to 180°F, except for glass types which may be used up to 700°F if suitable frames and gaskets are used	1. Viscous may be used up to 250°F if suitable oil is used. Dry type limited to 180°F

Source: R. H. Perry and D. W. Green, *Perry's Chemical Engineers' Handbook*, 6th ed., McGraw-Hill, New York, 1984.

TABLE 3.5-3 Classification of Common Air Filters

Group	Efficiency	Filter type	Removal Efficiency (%) for Particle Size of				Air-flow Capacity (ft^3/min·ft^2 of frontal area)	Resistance, in Water	
			0.3 μm	1.0 μm	5.0 μm	10.0 μm		Clean Filter	Used Filter
I	Low	Viscous impingement, panel type	0–2	10–30	40–70	90–98	300–500	0.05–0.1	0.3–0.5
II	Moderate	Extended medium, dry type	10–40	40–70	85–95	98–99	250–750	0.1–0.5	0.5–1.0
III	High	Extended medium, dry type	45–85	75–99	99–99.9	99.9	250–750	0.20–0.5	0.6–1.4
HEPA	Extreme	Extended medium, dry type	99.97 min	99.99	100	100	125–250	0.5–1.0	0.5–2

Source: Burchsted et al., *Nuclear Air Cleaning Handbook*, ERDA 76-21, Oak Ridge, TN, 1976.

TABLE 3.5-4   Average Atmospheric-Dust Concentrations
(1 grain/1000 ft^3 = 2.3 mg/m^3)

Location	Dust Concentration (grain/1000 ft^3)
Rural and suburban districts	0.02–0.2
Metropolitan districts	0.04–0.4
Industrial districts	0.1–2.0
Ordinary factories or workrooms	0.2–4.0
Excessive dusty factors or mines	4.0–4.00

Source: Heating Ventilating Air Conditioning Guide, American Society of Heating, Refrigerating and Air-Conditioning Engineers, New York, 1960, p. 77.

3.5-5   Electrostatic Precipitators

Figures 3.5-7 and 3.5-8 illustrate the important features of an electrostatic precipitator. The typical precipitator consists of collector plates arranged parallel to the gas flow. Midway between the collector plates are placed discharge electrodes with a high negative, pulsating voltage. Typical voltages are 70–100 kV peak which is sufficient to cause gas ionization (corona formation) around the discharge electrode. The applied voltage must be controlled to produce corona formation without excessive sparkover (i.e., arcing). As the particles pass through the electrostatic field created by the discharge electrode, they become charged and the charge attracts them to the collector plates, where they migrate and are collected. The collected dust is shaken loose by mechanical plate rappers, falls into a hopper, and is discharged.

Thus, for an electrostatic precipitator to function properly, the following features are essential:

1. Ionized gas around an electrode.
2. Charging of solid particles as they pass through the ionized gas.
3. Gas retention to permit particle migration into a collector plate.
4. Removal of collected particles.

In a single-stage precipitator, the discharge electrode is negatively charged and the collector plates are the positive electrodes in the same electrical circuit. The single stage has the advantage of simplicity; however, it does produce more ozone than a positive discharge two-stage system (Fig. 3.5-9) and lower voltages must be used for collection with the single-stage unit. A two-stage unit has the advantage of being able to use different voltages on the ionization and collecting stages. Normal industrial practice is to use single-stage units for most process applications (where the small amount of ozone produced does not have significant detrimental effects) and to use two-stage units for air-conditioning applications. Precipitators on home heating and cooling systems are all two-stage units.

For electrostatic precipitators to be economically applied, great care must be exercised during both design and operation, because many variables have a very important effect on the optimum operation. An appreciation of the complexity of the electrostatic precipitator system can be realized by considering typical operational problems and their solutions.

1. Dust re-entrainment. As the dust collects on the plates or is removed from the plates by rapping, it can be re-entrained into the gas stream. The gas velocity, the geometry of the collector plate, the resistivity of the dust, electrical arcing, and gas flow maldistribution can all contribute to excessive re-entrainment. Good gas flow distribution, optimum sparkover frequency, effective collector plate geometry, proper dust resistivity, proper rapping force and frequency, and optimum gas velocity are necessary to obtain the best operation. With low dust resistivity (below 100 ohm · cm), the dust readily loses its charge to the collector plate and is sheared off into the gas stream. When the resistivity is too high ($2-5 \times 10^{10}$ ohm · cm), excessive voltage drops occur across the dust layer and corona discharge can occur (in the dust layer) causing the dust to explode from the collecting surface; for example, carbon black has too low a resistivity and low-sulfur coal fly ash often has too high a resistivity.

2. The precipitator can have excessive arcing. This is controlled in modern precipitators with electrical circuits that give the optimum voltage to produce 50–100 sparks/min.

3. When the dust resistivity is too low and excessive re-entrainment occurs, the precipitator will normally cause agglomeration of the particles so that they can be captured in a downstream cyclone.

4. When dust resistivity is too high, the gas can often be conditioned to lower the resistivity. Temperature and gas composition are the variables that give the most control. Proper humidification of the gas and chemical conditioning (NH_3, SO_3, etc.) are often effective. The resistivity of fly ash as a function of

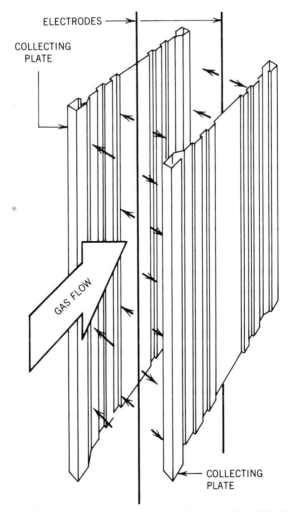

FIGURE 3.5-7 Electrostatic precipitator mechanism. Courtesy of the MikroPul Corp.

temperature and coal sulfur content is given in Fig. 3.5-10; Fig. 3.5-11 shows the effects of temperature and moisture. Figure 3.5-12 presents experimental data for the performance of electrostatic precipitators removing fly ash.

Electrostatic precipitators are often applied for very high-volume applications where the capital expenditures are very large. For such applications, considerable funds can be spent to develop the optimum system. Extensive testing in conjunction with an established reputable manufacturer is recommended. To design *and* build an electrostatic precipitator are beyond the capability of almost all users. Thus, unlike some of the other equipment covered here, such as cyclones, it is not appropriate to present sufficient information for design. The performance data presented in Table 3.5-5 give an approximate idea of the sizes of equipment needed for typical applications. Note that a key design parameter, the average drift, is tabulated. More in-depth treatment of the subject requires mathematical relationships to calculate the average drift and other pertinent design parameters, and equipment manufacturers can supply estimates of design parameters and capital and operating costs.

3.5-6 Selection Procedure

Figure 3.5-13 gives an indication of the collection efficiency of various dust collectors. This figure in conjunction with Fig. 3.5-14 will give a reasonable basis for the preliminary selection of the appropriate

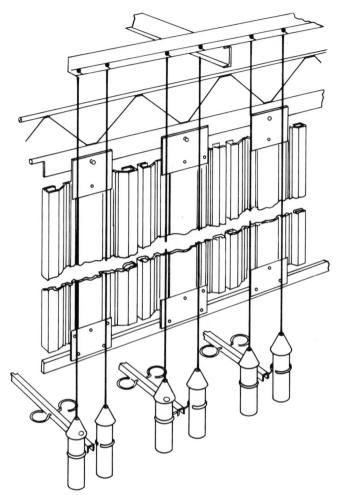

FIGURE 3.5-8 Electrostatic precipitator hardware configuration. Courtesy of the MikroPul Corp.

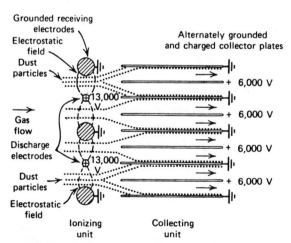

FIGURE 3.5-9 Operating principle of two-stage electrostatic precipitation. Courtesy of McGraw-Hill Book Co., *Perry's Chemical Engineers' Handbook*, 6th ed., 1984.

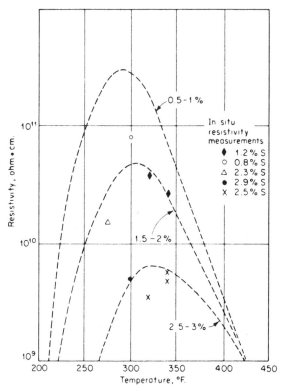

FIGURE 3.5-10 Trends in resistivity of fly ash with variation in flue-gas temperature and coal sulfur content. From Oglesby and Nichols, *A Manual of Electrostatic Precipitator Technology, Part II*, Southern Research Institute, Birmingham, AL, 1970.

equipment for a given solid–gas separation application. There are often other independent parameters and qualitative considerations that affect equipment selection; the most important of those will be discussed.

GRAVITY SETTLERS

These will economically collect dusts with particle sizes exceeding 50 μm; however, for heavy dust loadings and for sticky dusts, the problems of removing dust from the settler can be intolerable. These devices are

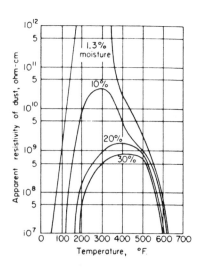

FIGURE 3.5-11 Apparent resistivity of fume from an open hearth furnace as a function of temperature and varying moisture content (volume). Reprinted with permission from Sproull and Nakada, *Ind. Eng. Chem.*, **43**, 1355 (1951). Copyright by American Chemical Society.

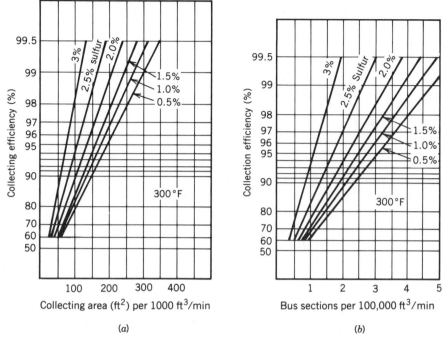

FIGURE 3.5-12 Design curves for electrostatic precipitators for fly ash. Collection efficiency for various levels of percent sulfur in coal versus specific collecting surface, and bus sections per 100,000 ft³/min. From Ramsdell, Design Criteria for Precipitators for Modern Central Power Plants, American Power Conference, Chicago, 1968.

ideally suited for applications with low loadings and large particles. For applications with high loadings, a low-efficiency cyclone may be the best choice for particles larger than 50 μm because of the ease of solids removal. Similarly, a wet scrubber may be the best choice for a sticky solid.

CYCLONES

The range of applicability indicated by Fig. 3.5-14 is certainly appropriate for nonsticky, free-flowing solids. For solids that tend to plug a cyclone, a wet scrubber may again be the best choice. Also, for solids that would infrequently plug a cyclone, the consequences of plugging banks of multiclones are much more severe than the plugging of a single large-diameter cyclone.

Table 3.5-5   Performance Data on Typical Single-Stage Electrical Precipitator Installations

Type of Precipitator	Type of Dust	Gas Volume (ft/min)	Average Gas Velocity (ft/s)	Collecting Electrode Area (ft²)	Overall Collection Efficiency (%)	Average Particle Migration Velocity (ft/s)
Rod curtain	Smelter fume	180,000	6	44,400	85	0.13
Tulip type	Gypsum from kiln	25,000	3.5	3,800	99.7	0.64
Perforated plate	Fly ash	108,000	6	10,900	91	0.40
Rod curtain	Cement	204,000	9.5	26,000	91	0.31

Source: Research-Cottrell, Inc. To convert cubic feet per minute to cubic meters per second, multiply by 0.00047; to convert feet per second to meters per second, multiply by 0.3048; and to convert square feet to square meters, multiply by 0.0929.

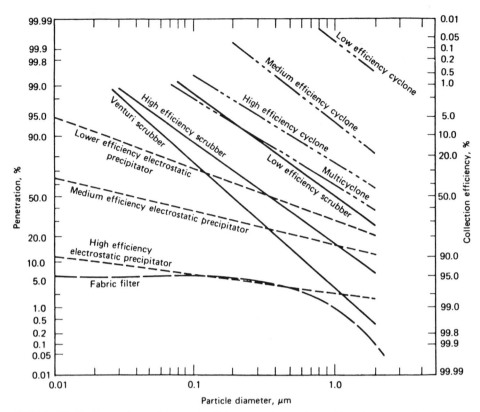

FIGURE 3.5-13 Penetration and fractional efficiency for fine particles with various separation devices. Excerpted by special permission from *Chemical Engineering*, June 18, 1973, p. 107, © by McGraw-Hill, Inc., New York, NY 10020.

FABRIC FILTERS

For low-flow applications ($<$ 10,000 ft^3/min) the fabric filter is often the equipment of choice even with relatively large particulates. This is because the exact size distribution of solids is often not known, and the collection efficiency of fabric filters is very good. The expedient of quickly and confidently selecting equipment that will do the job is often the cost-effective alternative. For small-diameter, nonsticky, free-flowing particles with gas flows below about 50,000 ft^3/min, the fabric filter is invariably the most cost-effective equipment. Furthermore, even though the operating costs for a fabric filter will be significantly higher than for an electrostatic precipitator, fabric filters are often used for small particles and with gas flows above 50,000 ft^3/min. This is because (1) the capital cost of a fabric filter is lower than for an electrostatic precipitator, (2) a fabric filter is much easier to design that an electrostatic precipitator, and (3) the fabric filter is more easily operated than an electrostatic precipitator.

WET SCRUBBERS

Wet scrubbers are ideally suited for applications where the recovered solid needs to be dissolved or slurried in a liquid for return to the process, or where the agglomerated solids are difficult to handle. Outside these limits, other equipment should be carefully considered in lieu of wet scrubbers.

ELECTROSTATIC PRECIPITATORS

These devices are ideally suited for large-volume applications with small particles where their operating cost advantage more than offsets their first-cost and operational complexity disadvantages. They also are at a disadvantage compared to other equipment because the experimental program leading to a design is often extensive, especially on applications where vendors have little or no experience. The lead time to install a precipitator is longer than for other equipment. However, in applications where vendors have adequate experience to specify a demonstrated design, they are much more competitive, even at gas flow below 50,000 ft^3/min.

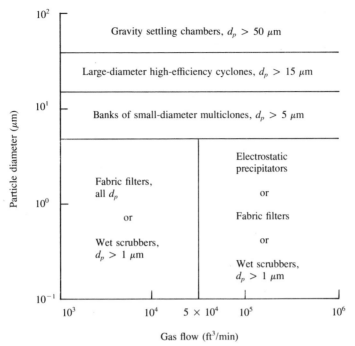

FIGURE 3.5-14 Selection guide for dust collection equipment based on particle diameter and capacity.

NOTATION

Roman Letters

a	specific cake resistance
A	area
A_F	flow area
A_I	interfacial area
A_s	surface area
A_t	area for thickening
C	drag coefficient (dimensionless)
d_{32}	Sauter mean drop diameter
d_C	effective spray drop diameter
d_p	particle diameter
d_{pc}	particle cut diameter 50% removal
D_b	body face dimension
D_c	diameter of cyclone barrel
D_h	hydraulic diameter
D_I	inlet diameter, or agitator impeller diameter
E_m	multistage efficiency (dimensionless)
E_0	overall separation efficiency (dimensionless)
f	friction factor (dimensionless)
g	local acceleration of gravity
G	number of times gravity for centrifugal separation
h	height
H	height of separator
H_C	height of compression layer

H_D depth of dispersion band
H_0 height of slurry in graduated cylinder at start of gravity settling test
K mesh pad design factor (dimensionless)
L length
n collection efficiency (dimensionless)
Re Reynolds number (dimensionless)
N rotation rate (rev/s)
Q bulk volumetric flow rate
Q_C continuous-phase volumetric flow rate
Q_G vapor volumetric flow rate
Q_L liquid volumetric flow rate
Q_s volume flow rate of solids
r radial distance from center of rotation
R filter medium resistance
R_f fractional underflow in hydroclone
t time in compression layer
U relative velocity between fluid and particle
U_F flow velocity
U_t terminal settling velocity
V velocity
V_a allowable velocity
V_D design velocity
V_e velocity of turbulent eddies
V_f flooding velocity
V_I inlet velocity
V_m particle velocity toward electrode
V_0 bulk fluid velocity
W_e Weber number (dimensionless)
W_G mass flow rate, vapor
W_L mass flow rate, liquid
W_s mass flow rate, solid
x distance

Greek Letters

Δp pressure drop
μ viscosity
μ_C continuous-phase viscosity
μ_D dispersed-phase viscosity
μ_H viscosity heavy liquid phase
μ_L viscosity light liquid phase
η_t target efficiency (dimensionless)
ρ fluid density
ρ_C continuous-phase density
ρ_G gas (vapor) density
ρ_H density heavy liquid phase
ρ_L liquid density or density light liquid phase
ρ_p particle density
ρ_s true density of solid
σ surface tension
Θ residence time in dispersion band
ω rotation rate (rad/s)

REFERENCES

Section 3.1

3.1-1   C. E. Lappel and C. B. Shepherd, Calculation of Particle Trajectories, *Ind. Eng. Chem.*, **32,** 605 (1940).

Section 3.2

3.2-1   R. H. Perry and D. W. Green, *Perry's Chemical Engineers' Handbook*, 6th ed., Sec. 18, McGraw-Hill, New York, 1984.

3.2-2   J. R. Fair, *Pet/Chem. Eng.*, **33**(10), 45 (Sept. 1969).

3.2-3   E. E. Ludwig, *Applied Process Design for Chemical and Petrochemical Plants*, Vol. I, Gulf Publishing, Houston, 1964.

3.2-4   W. R. Penney and L. J. Jacobs, "Design of Tangential Entry Gas–Liquid Separators," St. Louis AIChE Symposium, 1973.

3.2-5   O. H. York and E. W. Poppele, Wire Mesh Mist Eliminators, *Chem. Eng. Prog.*, **59,** 6 (June 1963).

3.2-6   S. Calvert, Engineering Design of Fine Particle Scrubbers, *J. Air Pollut. Control Assoc.*, **24,** 929 (1974).

Section 3.3

3.3-1   B. Weinstein, "Drop Size and Distribution of Water Dispersions Generated by Static Mixer Elements," AIChE National Meeting, Houston, TX, March 1975.

3.3-2   S. Middleman, Drop Size Distribution Produced by Turbulent Flow of Immiscible Fluids Through a Static Mixer, *IEC Chem. Process Des. Dev.*, **13**(1), 78 (1974).

3.3-3   C. A. Coulaloglou and L. L. Tavlarides, Drop Size Distribution and Coalescence Frequencies of Liquid–Liquid Dispersions in Flow Vessel, *AIChE J.*, **22,** 289 (1976).

3.3-4   A. H. Selker and C. A. Sleicher, Factors Effecting Which Phase Will Disperse When Immiscible Liquids are Stirred Together, *Can. J. Chem. Eng.*, **43,** 298 (Dec. 1965).

3.3-5   E. A. Cassell, K. M. Kaufman, and E. Matijevic, The Effects of Bubble Size on Microflotation, *Water Res.*, **9,** 1017 (1975).

Section 3.4

3.4-1   B. Fitch, Batch Tests Predict Thickener Performance, *Chem. Eng.*, **78**(19) 83 (1971).

3.4-2   W. P. Tallmadge and E. B. Fitch, Determining Thickener Unit Areas, *Ind. Eng. Chem.*, **47,** 38 (1955).

3.4-3   H. W. Cremer and T. Davies, *Chemical Engineering Practice*, Vol. 3, Chap. 9, Academic Press, New York, 1957.

3.4-4   Metcalf and Eddy, Inc. *Wastewater Engineering*, Chap. 8, McGraw-Hill, New York, 1972.

3.4-5   J. W. Clark, W. Viessman, and M. J. Hammer, *Water Supply and Pollution Control*, Chap. 9, International Textbook, Scranton, PA, 1971.

3.4-6   D. Bradley, A Theoretical Study of the Hydraulic Cyclone, *Ind. Chem. Manuf.*, **34,** 473 (1958).

3.4-7   R. W. Day, The Hydroclone in Process and Pollution Control, *Chem. Eng. Prog.*, **69**(9), 67 (1973).

3.4-8   J. B. Poole and D. Doyle, *Solid–Liquid Separation*, Chemical Publishing, New York, 1958.

3.4-9   R. H. Perry and D. W. Green, *Perry's Chemical Engineers' Handbook*, 6th ed., Chap. 19, McGraw-Hill, New York, 1984.

3.4-10 H. Hemfort, *Separators: Centrifuges for Clarification, Separation and Extraction Processes*, Bulletin No. 7394/681, Westfalia Separator AG, 1979.

3.4-11 *Centrifuges for the Chemical and Process Industries*, Bulletin No. IB 40200EZ, Alfa-Laval.

Section 3.5

3.5-1   H. L. Dryden (Ed.), *Fluid Mechanics and Statistical Methods in Engineering*, pp. 41–55, University of Pennsylvania Press, Philadelphia, 1941.

3.5-2   C. J. Stairmand, *Trans. Inst. Chem. Eng.*, **29,** 356 (1951).

3.5-3   P. Swift, *Steam Heating Eng.*, **38,** 453 (1969).

3.5-4   C. E. Lapple, in J. A. Danielson (Ed.), *Air Pollution Engineering Manual*, p. 95, U.S. Department of Health, Education, and Welfare, Public Health Publication No., 999-AP-40, 1967.

3.5-5   C. M. Peterson and K. T. Whitby, *ASHRAE J.*, **7**(5), 42 (1965).

3.5-6   D. Leith and W. Licht, The Collection Efficiency of Cyclone Type Particle Collectors—A New Theoretical Approach, *AIChE Symp. Ser. 120*, **68,** 196–206 (1972).

3.5-7   W. H. Koch and W. Licht, New Design Approach Boosts Cyclone Efficiency, *Chem. Eng.*, 80–88 (Nov. 7, 1977).

3.5-8   Y. M. Shah and R. T. Price, Calculator Program Solves Cyclone Efficiency Equations, *Chem. Eng.*, 99–102 (Aug. 28, 1978).

3.5-9   B. B. Crocker, R. A. Novak, and N. H. Sholle, Air Pollution Control Methods, in *Kirk–Othmer Encyclopedia of Chemical Technology*, Vol. 1, 3rd ed., pp. 649–716, Wiley, New York, 1978. (Note: This publication is an excellent source of references.)

3.5-10 S. Calvert, How to Choose a Particulate Scrubber, *Chem. Eng.*, 54 (Aug. 29, 1977).

3.5-11 S. Calvert, J. Goldschmidt, D. Leith, and D. Mehta, "Scrubber Handbook," and PB213-017, National Technical Information Service, Springfield, VA, 1972, #PB213-016.

3.5-12 A. E. Vandergrift, L. J. Shannon, and P. G. Gorman, Controlling Fine Particles, *Chem. Eng.* (*Deskbook Issue*), 107 (June 18, 1973).

CHAPTER 4

General Processing Considerations

ROBERT M. KELLY
Department of Chemical Engineering
The Johns Hopkins University, Baltimore, Maryland

In the design and operation of separation processes, there are several factors to be considered that are important regardless of the particular separation methodology. These general processing considerations include contacting modes, effective integration of the separation unit or units into the overall process, development and implementation of a process control strategy, and potential difficulties that arise from particular situations.

The purpose of this chapter is to provide some insight into several areas of importance to the designer or practitioner of separation processes.

4.1 METHODS OF OPERATION

Separation processes are based on the mass transfer of one or more components from one phase to another due to the existence of a driving force arising from differences in chemical potentials. In practice, this requires the intimate contact of the phases for some period of time, during which equilibrium is approached as mass transfer proceeds. A *stage* is the unit in which the contacting occurs and where the phases are separated physically. A *single-stage process* is one in which the contacting is done once and, if equilibrium actually is achieved, further classification as an *ideal* or *theoretical stage* is appropriate. Because of economic considerations, the contacting time in a particular stage may not be long enough for equilibrium to be reached; the fractional approach to equilibrium is a measure of the *mass transfer efficiency* of the process.

Consider the following *single-stage cocurrent contacting process*:

$$y_0, A_0 \longrightarrow \boxed{\begin{array}{c} \text{Single-} \\ \text{stage} \\ \text{process} \end{array}} \begin{array}{c} \longrightarrow y_1, A_1 \\ \\ \longrightarrow x_1, B_1 \end{array}$$

$$x_0, B_0 \longrightarrow$$

A steady-state material balance can be used to track the separation taking place in the process. As a simplification, only the transfer of a single component will be considered. Taking into account only the bulk-averaged compositions of the flows into and out of the stage, we can make the following balances:

Total Balance

$$A_0 + B_0 = A_1 + B_1 \tag{4.1-1}$$

Component Balance

$$A_0 y_0 + B_0 x_0 = A_1 y_1 + B_1 x_1 \tag{4.1-2}$$

where A and B refer to the bulk mass or molar flow rates entering and leaving the stage and x and y are the mass or mole fractions of the transferring component in A and B.

The separation taking place in the *single-stage process* is represented best on an *operating diagram* which can be drawn from phase equilibrium information available for the system along with the material balance data. In this example, and for any case where the inlet streams to the process are mixed together, the contacting pattern is said to be *cocurrent*. An operating diagram for the steady-state mass transfer of a single component between two phases is shown in Fig. 4.1-1. This diagram shows the transfer of a component from phase A to phase B. For transfer in the other direction, the equilibrium line remains the same, but the operating line would be plotted below the equilibrium relationship from the case shown above. As the end of the operating line, representing the exit stream from the process, approaches the equilibrium curve, the single-stage process approaches an ideal or theoretical stage. If the operating line reaches the equilibrium curve, the single-stage device is a theoretical stage.

The shape of the equilibrium curve on the operating diagram arises from the phase equilibria of the system which must reflect changes in temperature, pressure, ionic strength, and so on that occur in the single-stage process. The shape of the operating line reflects changes in the quantity of material in streams A and B as mass is transferred from one phase to another. For reasons that will become apparent shortly, it is often desirable, for calculational purposes, to work with linear or nearly linear equilibrium relationships and operating lines.

To reduce the degree of curvature of the operating line, the use of mole or mass ratios often is preferred to mole or mass fractions. This conversion is accomplished by taking the ratio of the number of moles or mass of a transferring species to the number of moles or mass of those components that do not transfer:

$$A_0 y_0 = \frac{A_s y_0}{1 - y_0} = A_s Y_0 \tag{4.1-3}$$

where A_s is the mass or molar flow rate of the nontransferring portion of A and

$$Y_0 = \frac{\text{moles of transferring component}}{\text{moles of nontransferring component}}$$

A similar expression can be written for stream B. A component balance can be used to derive an operating line expression in terms of mole or mass ratios:

$$A_s(Y_0 - Y_1) = B_s(X_1 - X_0) \tag{4.1-4}$$

or

$$-A_s(Y_1 - Y_0) = B_s(X_1 - X_0) \tag{4.1-5}$$

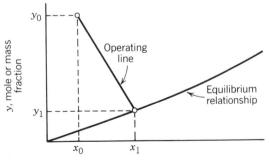

FIGURE 4.1-1   Operating diagram: single-stage device.

The operating line is therefore a straight line passing through points (X_0, Y_0) and (X_1, Y_1) with a slope of $-B_s/A_s$. All points in the single-stage device must satisfy the material balance and lie on the operating line, although their position on the operating line is a function of the conditions and related phenomena of the separation process of interest.

Because, at best, the exiting streams from a single-stage device will be in equilibrium, it is usually desirable to use multiple stages for a given separation. When multiple stages are used, some thought must be given to the pattern of contacting of the two phases. *Cocurrent* contacting, where the inlet stream from one phase is mixed with the inlet stream of another phase, can provide, at best, the equivalent of only one theoretical stage no matter how many actual stages are used. Nonetheless, multiple-stage cocurrent contacting is sometimes used to promote better mixing or heat transfer.

To maximize driving forces throughout a particular system, *countercurrent* contacting is frequently used. In this contacting mode, the inlet stream for one phase is mixed with the outlet stream of the other phase. For the two-stage countercurrent contacting system,

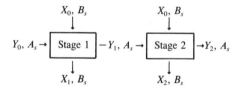

the following material balance can be made, using mole or mass ratios:

Overall Component Balance

$$A_s Y_0 + B_s X_3 = A_s Y_2 + B_s X_1 \tag{4.1-6}$$

The operating line for countercurrent contacting is shown in Fig. 4.1-2. Here, the diagram shows the transfer from phase A to phase B. If transfer were in the opposite direction, the operating line would be located below the equilibrium line. In general, for a given number of stages, countercurrent contacting yields the highest mass transfer efficiency for interphase mass transfer where at least one stream is not a pure component. (This is not to be confused with tray efficiencies which usually represent the fractional approach to equilibrium on a given stage.) The reason for this is that the average mass transfer driving force across the device is greater than would be the case with cocurrent contacting.

Crosscurrent contacting, which is intermediate in mass transfer efficiency to cocurrent and countercurrent contacting, is shown below for a two-stage cascade:

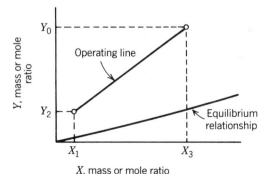

Although the phase B feed to both stages is the same, this need not be the case; the feed rate to each stage can be different as can the composition of that feed stream. To draw the operating diagram, the following balances can be made:

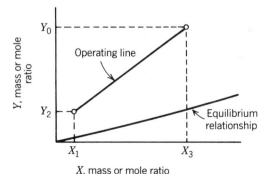

X, mass or mole ratio

FIGURE 4.1-2 Operating diagram: countercurrent contacting.

Stage 1

$$A_s(Y_0 - Y_1) = B_s(X_1 - X_0) \tag{4.1-7}$$

Stage 2

$$A_s(Y_1 - Y_2) = B_s(X_2 - X_0) \tag{4.1-8}$$

As can be seen, the balance for each stage is similar to that for a cocurrent process. However, the stages are coupled. For this case, the operating diagram is presented in Fig. 4.1-3. Crosscurrent contacting is not used as commonly as cocurrent and countercurrent contacting but may be found in extraction, leaching, and drying operations.

EXAMPLE 4.1-1   COMPARISON OF CONTACTING MODES
To illustrate best the relative efficiencies of the various contacting modes, consider the following example. Suppose we have two discrete stages that can mix and separate phases, and that the stages can be connected cocurrently, crosscurrently, or countercurrently for gas–liquid contacting. Find the contacting pattern that will give the maximum removal of a single transferring component from the gas phase if a fixed amount of solvent is to be used. We have the following data:

Inlet liquid flow, $B = 20$ units/time.

Inlet liquid composition = 0% transferring component.

Inlet gas flow, $A = 20$ units/time.

Inlet gas composition = 30 mol % transferring component.

Only one component transfers between gas and liquid.

Process is isothermal and isobaric.

Each stage can be considered to be an ideal stage.

Equilibrium data in terms of mole ratios: $Y = 0.408X$, where X and Y are the ratios of the transferring component to the nontransferring component.

Solution.   Consider first a single stage:

$$A_s, Y_0 \rightarrow \boxed{} \rightarrow A_s, Y_1$$
$$B_s, X_0 \rightarrow \phantom{\boxed{xxxx}} \rightarrow B_s, X_1$$

The inlet gas mole ratio is $Y_0 = 0.3/(1 - 0.3) = 0.429$. The inlet liquid mole ratio is $X_0 = 0.0$. The ratio of flows of nontransferring components is

$$\frac{B_s}{A_s} = \frac{20}{20 - 0.3(20)}$$

$$= \frac{20}{14}$$

$$= 1.429$$

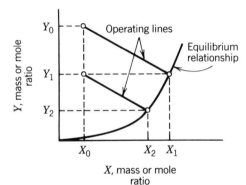

FIGURE 4.1-3   Operating diagram: crosscurrent contacting.

The outlet mole ratios, X_1 and Y_1, can be found through graphical construction on the operating diagram by constructing a line of slope $-B_s/A_s$ from (X_0, Y_0) to the equilibrium line since this is a theoretical stage.

Since the inlet mole ratio $X_0 = 0.0$, an analytical expression can also be developed to find X_1 and Y_1. By material balance,

$$A_s Y_0 + B_s X_0 = A_s Y_1 + B_s X_1$$

But, $X_1 = Y_1/0.408$ from the equilibrium relationship. Noting that $X_0 = 0.0$ and rearranging, we obtain

$$Y_1 = \frac{Y_0}{1 + B_s/(0.408A_s)} = \frac{Y_0}{1 + K}$$

where K is the absorption factor, $K = B_s/mA_s$, and m is the slope of equilibrium line. By either method,

$$Y_1 = 0.095 \quad \text{and} \quad X_1 = 0.233$$

Note that K is used to represent the absorption factor in this and later examples. However, the symbol for this quantity is often A.

1. *Cocurrent Contacting.* As stated previously, cocurrent contacting, even for a cascade of stages, yields at best one theoretical stage. This can be demonstrated as follows. Two cocurrent stages in series are shown below:

$$
\begin{array}{lll}
A_s, Y_0 \rightarrow & \boxed{\text{Stage 1}} & - A_s, Y_1 \rightarrow \boxed{\text{Stage 2}} \rightarrow A_s, Y_2 \\
B_s, X_0 \rightarrow & & - B_s, X_1 \rightarrow \phantom{\text{Stage 2}} \rightarrow B_s, X_2
\end{array}
$$

In Fig. 4.1-4a, it is evident that the streams leaving the first stage are in equilibrium. The addition of a second cocurrent stage will not result in any further mass transfer. Using a material balance around stage 2, given the exiting conditions from stage 1, the following analytical expression to determine Y can be developed:

$$Y_2 = \frac{Y_1 + (B_s/A_s)X_1}{1 + K}$$

With $Y_1 = 0.095$ and $X_1 = 0.233$, the outlet gas composition remains at 0.095. Thus, the addition of another cocurrent stage does nothing to enhance the transfer of solute as will be the case no matter how many stages are added.

2. *Countercurrent Contacting.* The two stages are now arranged countercurrently:

$$
\begin{array}{lll}
Y_0, A_s \longrightarrow & \boxed{\text{Stage 1}} & - Y_1, A_s \rightarrow \boxed{\text{Stage 2}} \rightarrow Y_2, A_s \\
\leftarrow X_1, B_s - & & \leftarrow X_2, B_s - \phantom{\text{Stage 2}} \leftarrow X_3, B_s
\end{array}
$$

A solute balance around stage 1 yields the following result:

$$Y_1 = \frac{Y_0 + (B_s/A_s)X_2}{1 + K}$$

A balance around stage 2 gives a similar result:

$$Y_2 = \frac{Y_1 + (B_s/A_s)X_3}{1 + K}$$

Recognizing that $X_3 = 0.0$ and that $(B_s/A_s)X_2$ is the same as KY_2, the following expression can be derived:[1]

$$Y_2 = \frac{Y_0}{1 + K + K^2}$$

This means that $Y_2 = 0.026$, which is much less than that obtained from the cocurrent case. This result is shown graphically in Fig. 4.1-4b. For the special case in which the inlet liquid contains no transferring

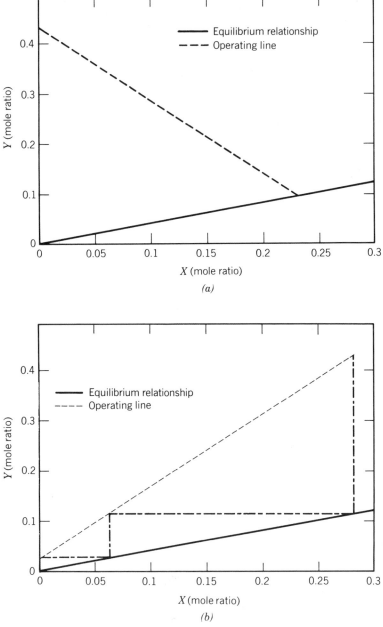

FIGURE 4.1-4 (*a*) Cocurrent contacting, (*b*) countercurrent contacting (two stages), and (*c*) crosscurrent contacting (two stages).

component, the above result can be generalized for any number of stages to

$$Y_N = \frac{Y_0}{\displaystyle\sum_{n=0}^{N} K^n}$$

which converges to $Y_N = 0$ for $K \geq 1$ and to $(1 - K)Y_0$ for $K < 1$ as $N \to \infty$.

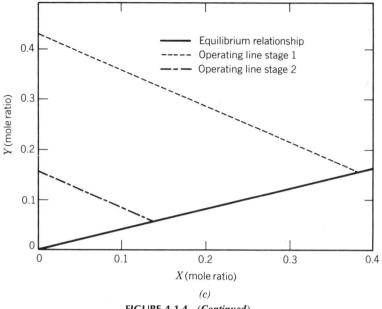

FIGURE 4.1-4   (*Continued*)

3. *Crosscurrent Contacting.*   The following arrangement is used for crosscurrent contacting:

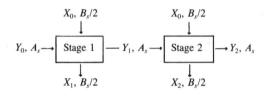

Using a material balance around stage 1, with $X_0 = 0.0$, the following expression for Y_1 can be developed:

$$Y_1 = \frac{Y_0}{1 + K/2}$$

Similarly, for stage 2, with $X_0 = 0.0$,

$$Y_2 = \frac{Y_1}{1 + K/2}$$

Combining the two expressions, Y_2 can be found as a function of Y_0:

$$Y_2 = \frac{Y_0}{(1 + K/2)^2}$$

Thus, Y_2 thus can be calculated to be 0.057, which is a value intermediate to cocurrent and countercurrent contacting. This result is illustrated in Fig. 4.1-4c. The above result can be generalized for any number of stages:[1]

$$Y_N = \frac{Y_0}{(1 + K/N)^N}$$

which approaches $Y_0/\exp(K)$ as $N \to \infty$.

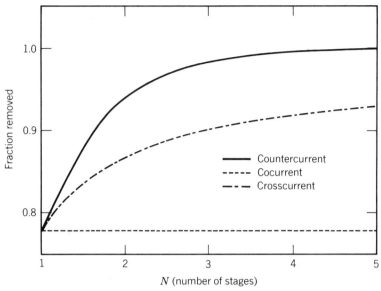

FIGURE 4.1-5 Comparison of contacting modes.

SUMMARY

In our example, the lowest value of Y_2 that can be attained with crosscurrent contacting with the inlet liquid stream split equally among N stages is 0.013. Theoretically, Y_2 can be reduced to zero with countercurrent contacting in our example. These results are plotted in Fig. 4.1-5. Thus, this simple example illustrates the relative effectiveness of the three classic contacting modes, which are listed in Table 4.1-1.

4.2 PROCESS SYNTHESIS

After one has obtained the necessary data on physical properties, transport properties, phase equilibria, and reaction kinetics, sorted out the flow of material through the system, and exploited any particular charac-

TABLE 4.1-1 Comparison of Contacting Modes

$$\text{Fraction removed for } N \text{ stages} = 1 - \frac{Y_N}{Y_0}$$

$$= 1 - \frac{1}{1 + K} \quad \text{cocurrent}$$

$$= 1 - \frac{1}{\displaystyle\sum_{n=0}^{N} K^n} \quad \text{countercurrent}$$

$$= 1 - \frac{1}{(1 + K/N)^N} \quad \text{crosscurrent}$$

	Fraction Removed for a Given K		
N	Cocurrent	Crosscurrent	Countercurrent
1	0.778	0.778	0.778
2	0.778	0.868	0.940
3	0.778	0.902	0.983
4	0.778	0.919	0.995
5	0.778	0.930	0.999
$N \to \infty$	0.778	0.970	1.000

teristics for a given separation, perhaps the hardest task remains: For a multicomponent system, how should the choice and sequence of separation tasks be made to maximize efficiency and minimize costs? Process synthesis, despite its critical importance in the design and analysis of separation processes, has only recently emerged as a distinct area of study and research. The pioneering work of Rudd in the late 1960s has led to a high degree of interest in this subject by those involved in plant design and/or the improvement of existing processes. This interest in process synthesis has been fueled by numerous examples of how systematic and informed development of process flowsheets can lead to substantial economic benefits.

Process synthesis represents the inventive aspect of process design. The analytical aspect of design, which is emphasized in undergraduate engineering school curricula, is relatively well established and includes the sizing and specification of the separator and its operating conditions. Sophisticated computer algorithms, which are constantly being improved, can now often be relied on to execute design calculations for individual separators. However, the systematic sequencing of separators (often along with other units such as reactors and heat exchangers) in the best possible arrangement is a complex problem that has not been solved completely except for a few simplified systems.

Process synthesis can be broken down into two broad tasks: (1) selection of the separation technique or techniques to be used for a particular system and (2) the arrangement or sequencing of separators and auxiliary equipment (e.g., heat exchangers) to provide an optimal process from both economic and engineering perspectives. Unfortunately, neither of these objectives can be satisfied independently. For a particular multicomponent separation, there might be several ways in which the products could be recovered; each process to be considered might be composed of a variety of separation techniques arranged in several different configurations. The optimum design must arise from careful consideration of all feasible alternatives, which is usually a complex undertaking.

To illustrate the complexity that must be dealt with in developing an optimal process, Thompson and King[1] made the following calculation. If a process stream containing N components is to be separated into N pure-component products, using M different separation methods, the number of possible sequences R can be determined as follows:

$$R = \frac{[2(N - 1)]!}{N!(N - 1)!} M^{N-1} \qquad (4.2\text{-}1)$$

The assumption made in this calculation is that each separator will yield two product streams from one feed stream and each component can exit in only one of these streams. This ignores, for example, distillation columns with sidestream removal. Nonetheless, the combinatorial problem that arises can be monumental for even apparently simple synthesis problems:

Components	Separation Methods	Possible Sequences
3	1	2
3	2	8
4	1	5
4	2	40
5	1	14
5	2	224
5	3	1,134
10	3	95,698,746

It should be apparent that as the number of components and possible separation methods increase, the development of an optimal design becomes nearly impossible without some systematic method to discard alternatives that are not feasible.

4.2-1 Separation Process Selection

Before one can begin addressing the problem of sequencing separators, the choice of a particular separation method or methods must be made. Of course, decisions are based on both technical and economic merits and it is not uncommon for conflicts to arise. For instance, although the technical feasibility of a given separation method might be attractive, it may not be compatible with the expected product value.

The wide variety of possible choices of separation processes has been catalogued by King.[2] Also, a detailed discussion of procedures for selecting a separation process is provided by Null in Chapter 22. Here, some comments on separation process selection are included to provide some perspective for later discussion.

SEPARATION METHODS
In general, separation methods fall into one of three categories:

Energy-related separations

Mass-related separations

Transport-related separations

The most common example of the use of an energy-related separating agent is distillation where heat is used to produce two phases of differing composition. In general, the use of an energy-related separating agent also results in a relatively low amount of energy consumption compared to the use of other separating agents. On the other hand, the separation potential (e.g., relative volatilities in distillation) tends to be the smallest for cases in which an energy-related separating agent is used. Crystallization is another example in which an energy-related separation agent is used.

Mass separation agents are employed in processes where an additional phase or phases are added to the system to effect the separation. Extraction and gas absorption are two well-known examples. Because mixing of homogeneous materials is an irreversible process, energy is needed to separate products from the mixture at constant temperature and pressure. This is one reason why mass separating agents usually are associated with lower thermodynamic efficiencies than energy-related separating agents.

Transport-related separations are usually the least energy-efficient but will often produce the highest separation potential. The contacted phases are not allowed to equilibrate and hence a relatively large driving force is maintained across the separator. Advantage is taken of the fact that components will migrate from phase to phase at different rates governed by the relevant transport phenomena. Many diffusional processes fall into this category, including membrane separations.

It should be pointed out that combinations of the above three separation types are frequently found. Extractive and azeotropic distillation, in which a mass separating agent is added to enhance or make feasible a particular separation, are two examples. A recent invention in which distillation and centrifugation are combined to improve the separation potential of ordinary distillation is another example.[3]

STAGING

As was shown in a preceding section, the method of contacting phases can be a crucial consideration in designing an efficient separator. When at least one phase contains more than one component, countercurrent contacting will usually improve the efficiency of the separation. The concept of staging thus becomes an important consideration in choosing a separator. Recall that staging requires the mixing and subsequent separation of phases. (Note that for continuous-contacting devices, such as packed columns, effective staging occurs but not in the discrete manner that is associated with tray columns, for instance. The concept of the height equivalent to a theoretical stage, which is frequently used for systems such as chromatographic separations, is an attempt to make the connection between the efficiencies of the two types of contacting device.) For separation methods such as distillation, where countercurrency and staging can be accomplished within the same vessel, high separation efficiencies can be achieved economically. Membrane processes, however, are often difficult to stage although they generally do have large driving forces for single-stage devices. Therefore, in selecting a separation method, consideration must be given to the ease with which the separator of choice can be used in a multistage configuration.

OTHER CONSIDERATIONS

There are numerous other factors that must be weighed in choosing a separation method. These include the sensitivity of products to operating conditions, the amount of material that must be processed, the value of the product, environmental factors, safety problems, corrosion problems related to materials of construction, and special regulations that may apply (e.g., in the food and drug industries).

Another important but often overlooked consideration is the capability to design accurately and control the separator. It is often the case that the most attractive technical option is the most difficult to design. This is especially true for new technologies but examples can be found of systems that are difficult to design and have been used for 50 years. Precise design usually results in a better functioning process and more attractive economics. Also, where transport-related separations are concerned, overdesign can lead to a failure to accomplish processing objectives. Finally, it is always easier to control a process whose design is well conceived and understood.

Additional discussion of processing considerations will follow later in the chapter.

4.2-2 Sequencing Separation Processes

Once the possible separation methods have been established, it becomes necessary to assemble them into a process flowsheet. Motard and Westerberg[4] cite three problems that arise at this point:

1. How can it be ensured that all alternatives are considered but that infeasible options are ruled out?
2. Can an evaluation process be developed that balances speed and accuracy?
3. In an evolutionary sense, what strategy can be adopted that continually moves toward better alternatives?

Despite a fair amount of effort to find a process synthesis methodology that can handle the above problems,

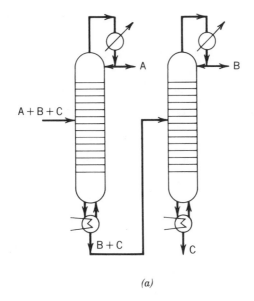

(a)

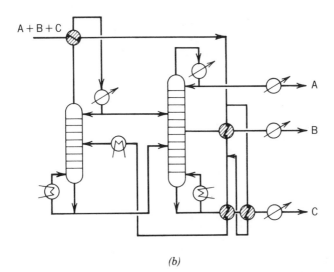

(b)

FIGURE 4.2-1 Separation of feed stream into pure components A, B, and C through (*a*) ordinary distillation and (*b*) a more complex arrangement that includes thermal coupling and sidestream feed and removal.

no techniques currently exist that are universally applicable. Most of the work reported in the literature deals with *simple* (a single feed stream is divided into two product streams) or *sharp* (each entering component exits in only one product stream) separations. In considering only these types of separations, many economically attractive alternatives are neglected; for instance, multiple feed and removal streams from a distillation column may reduce the number or size of columns needed for a particular separation. Also, effective heat exchange may improve the thermodynamic efficiency of the system. (See Fig. 4.2-1.) Nonetheless, the complexity associated with sequencing *simple*, *sharp* separators still represents a considerble challenge as evidenced by the previously cited calculation of Thompson and King.[1]

PROCESS SYNTHESIS METHODOLOGIES
Although there are numerous approaches to synthesizing an optimal process for a particular separation or separations, they usually can be categorized as heuristic, evolutionary, or algorithmic.

In practice, most methodologies used today are hybrid techniques based on features of at least two of these methods. However, before discussing how the methodologies are combined for process synthesis, each will be considered separately.

Heuristic Approaches.   To facilitate the development of flowsheets and subsequently improve an existing flowsheet, rules of thumb have been suggested by those experienced with the design and sequencing of separators. These rules of thumb, or heuristics, are somewhat empirical but useful guidelines for choosing and sequencing separations. It must be stressed that the heuristics are not absolutes and must be applied carefully if they are to serve a purpose. Also, the most commonly used heuristics can be contradictory, which can lead to two or more sequences being suggested. Nevertheless, even with the advent of digital computing for computer-aided process synthesis, heuristics almost always are factored into sequencing and selection decisions. This is because their usage reduces an overwhelmingly complex combinatorial problem into a more tractable one.

Process synthesis heuristics fall into several categories. Many of these were proposed originally for multicomponent distillation but have since been applied to other types of separation. Some heuristics have been verified through calculation and experimentation while others are described best as common sense. In any case, they serve as effective guidelines in the selection and sequencing of separation processes.

Presented below is a list of those heuristics most commonly cited in separation process synthesis. This is certainly not a complete list but a compilation of several proposed in the literature. If anything, what is needed is a more detailed list of heuristics that have been tested through calculation and experimentation. As more emphasis is placed on process synthesis as a vital part of process design, a more comprehensive and useful set of heuristics will evolve.

General Heuristics.   Below are listed a set of heuristics that generally are applicable to separation process synthesis. A summary of these heuristics appears in Table 4.2-1.

1. *Select the Separation Methods First.*   Rudd et al.[5] point out that for a given process the most difficult task involves separating components from a mixture. In the section on selection of separation techniques, many of the factors that must be taken into account during the separation task selection were pointed out. The most important consideration involves evaluating the basis for separation which requires examining data on physical and chemical properties. Ranked lists are then used to evaluate possible separation methods. For instance, if components A, B, C, D, and E are to be separated, their relative volatilities and solubilities in a particular solvent may be pertinent. The following ranked lists, in descending order, might result:

Normal Boiling Point	Solubility in Solvent
A	C
B	E
C	A
D	B
E	D

Depending on the objectives of the process, distillation or extraction (or, perhaps, a combination of both) might be used to affect the separation. If component C is the only desired product, an extraction scheme might be preferred while distillation might be used if only A is desired. For multiple products, the separation problem becomes more complex but can be approached by examining the possible splits in the ranked lists.

When using ranked lists as a basis for separation process selection, avoid separation factors close to one and draw on experience to weigh theoretical potential with practical expectations. A more detailed discussion of the selection of a separation process is given in Chapter 22.

TABLE 4.2-1   Summary of General Heuristics

1. Select the separation methods first.
2. Always attempt to reduce the separation load.
3. Remove corrosive and unstable materials early.
4. Separate the most plentiful components early.
5. Save the most difficult separations for last.
6. Separations with high recovery fractions should be done last.
7. Move toward sequences with the smallest number of products.
8. Avoid adding foreign species to the separation sequence.
9. If used, immediately recover a mass separating agent.
10. Do not use another mass separating agent to recover the original one.
11. Avoid extreme operating conditions.

2. *Always Attempt to Reduce the Separation Load.* Rudd et al.[5] assume that operations to mix and divide process streams are relatively inexpensive as long as no component separation is required. It is usually better to blend two streams of similar composition and separate the species of interest than to treat each stream independently.

3. *Remove Corrosive and Unstable Materials Early.* The early separation of corrosive and unstable materials is desirable. The early removal of corrosive materials probably will increase the lifetime of downstream processing units or allow them to be constructed of less expensive materials. Unstable species that are sensitive to temperature, shear, and so on should be removed early, as should potentially toxic components. Components that might undergo unfavorable side reactions also should be removed early.

4. *Separate the Most Plentiful Components Early.* Components comprising a large fraction of the feed should be removed first. Clearly, the separation load will depend on the amount of material to be processed. By reducing the amount of this material early in the sequence, processing costs can be cut.

5. *Save the Most Difficult Separations for Last.* When the differences in properties (boiling point, density, etc.) of the components to be separated are not far apart, separation of those components is done best in the absence of other nonkey components. In the case of gas absorption, as the solubility of a particular component in a solvent decreases, the number of trays or height of packing required for the separation increases. Also, as the liquid and gas inventories in the column go up, the diameter of the column goes up. Thus, as the amount of material to be processed is reduced, so too is the cost of the separation. The same argument is equally valid when a particularly difficult separation requires specialized equipment.

6. *Separations with High Recovery Fractions Should Be Done Last.* This heuristic follows 5 and is justified by the same arguments.

7. *Move Toward Sequences with the Smallest Number of Products.* Usually, following this strategy will result in the need for the minimum number of separators.

8. *Avoid Adding Foreign Species to the Separation Sequence.* It is best to avoid adding mass separating agents, such as in azeotropic distillation and leaching, because these agents subsequently must be recovered, often creating a separation problem. This is one reason why energy separating agents often are preferred.

9. *If Used, Immediately Recover a Mass Separating Agent.* The mass-related separating agent should be recovered in a subsequent step so that it does not interfere with downstream processing. This is important because mass separating agents usually are found in large quantities and often are chemically different from the original feed stream. Of course, if the mass separating agent facilitates downstream separations (which is rare), it should be left in the system.

10. *Do Not Use Another Mass Separating Agent to Recover the Original One.* It is obvious that this might complicate the original separation problem and negate any benefit associated with adding a mass separating agent in the first place.

11. *Avoid Extreme Operating Conditions.* This heuristic is aimed primarily at conditions of extreme temperature and pressure. If excursions in temperature or pressure are necessary, it is better to aim high rather than low. For the case of temperature, heating costs are usually significantly lower than refrigeration costs, often by a factor of 5 or more. Also, pressure equipment is easier to build and maintain than vacuum equipment.

Heuristics for Multicomponent Distillation Sequences. The origin of many heuristics can be traced to multicomponent distillation problems and, with the possible exception of heat exchange network synthesis, most process synthesis study has been in this area. This is not surprising since distillation is the backbone of the chemical industry. Heuristics for multicomponent distillation are summarized in Table 4.2-2 and discussed below.

1. *Favor Distillation.* Because of the vast experience associated with the design and operation of distillation columns, this separation technique almost always is considered for fluid–fluid separations. Because energy is used as the separating agent, thermodynamic efficiencies can be high and the addition of foreign components to the system is avoided. (Azeotropic and extractive distillation are two notable exceptions.) Also, staging is accomplished easily within the column. Even though energy costs are constantly rising (distillation is said to account for 3% of the energy consumption in the United States), possible alternatives to distillation such as adsorption, chromatography, and membrane processes are technologically

---

TABLE 4.2-2 Summary of Heuristics for Multicomponent Distillation

1. Favor distillation.
2. For distillation, favor sequences that remove components one-by-one as overhead product.
3. As a rule, favor splits that yield equal-sized parts.
4. Favor sequences that recover the most valuable species as a distillate product.

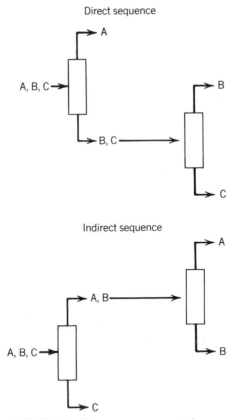

FIGURE 4.2-2   Three-component separation.

young and are only beginning to be tried at scales necessary to be competitive. Because of the years of operating experience and the vast amount of study and research put into distillation, it is known to be a reliable and, therefore, desirable separation method.

2. *For Distillation, Favor Sequences that Remove Components One-by-One as Overhead Product.*   This advice is based on the need to condense or remove heat from overhead streams and add heat to bottom streams. Adding nonkey components to the overhead stream increases the cooling load on the overhead condenser and the heating load on the reboiler for a simple column. Removing components one-by-one in order of decreasing volatility also minimizes the vapor flow in the column supporting a direct sequence.

For example, if A, B, and C are components to be separated and their boiling points increase as A → B → C, there are two possibilities for sequences for a simple, sharp separation as shown in Fig. 4.2-2.

This heuristic favors the direct sequence because it removes the products one-by-one as distillates and therefore minimizes vapor flow in the column. Lockhart[6] points out, however, that the direct sequence is not optimal when the least-volatile component is the primary constituent of the feed stream. In this case, the indirect sequence is preferred.

3. *As a Rule, Favor Splits that Yield Equal-Sized Parts.*   Splitting the feed stream into equal-sized parts leads to a better thermodynamic efficiency in a simple distillation column. This is true because the column traffic in the sections above and below the feed will have better balance.

4. *Favor Sequences that Recover the Most Valuable Species as a Distillate Product.*   By forcing the product away from the bulk stream, higher purities generally will result. Recovery as a distillate product is desirable because it avoids the higher temperatures present in the reboiler which could lead to degradation.

Discussion.   The use of heuristics for sequencing multicomponent distillation systems, as well as for separation processes in general, must be used with some discretion on the part of the design engineer. For certain separation problems, some heuristics will be more pertinent than others and for this reason the systematic application of heuristics is difficult. Since heuristics also usually apply to simple, sharp separations, more complex schemes that may be advantageous often are ignored. For example, Petlyuk et al.[7] and Stupin and Lockhart[8] showed that more complex separation schemes can be an improvement to the direct and indirect sequences normally considered. For distillation, they pointed out that the use of intermediate condensers and reboilers can lead to a system that more closely approaches a thermodynamically reversible distillation.

Attempts to apply heuristics systematically to sequencing and selecting separation processes will be discussed in a later section.

Evolutionary Methods. The objective of all evolutionary methods for selecting and sequencing separation processes is a systematic approach to an optimal system. These methods all seek to improve the existing flowsheet. Stephanopoulos[9] describes the elements of an evolutionary strategy as follows:

1. Generation of the initial separation sequence.
2. Identification of evolutionary rules.
3. Determination of the evolutionary strategy.

Generation of the initial separation sequence is a critical step in finding an optimal process. It is not surprising then that the heuristics discussed in the previous section often are used in the generation of the initial flowsheet. In addition to these heuristics, several workers have suggested additional rules to be followed or have suggested simply a prescribed order in which existing heuristics be applied (Seader and Westerberg,[10] Nath and Motard,[11] Westerberg,[12] Nadgir and Liu[13]). Often, the best way to generate an initial flowsheet is to examine similar processes (found in practice or in the literature) that will lead to a feasible initial sequence. Carefully selected initial sequences will be those that are closest to the optimum, and, conversely, poor initial choices may lead to failure in finding the optimal or nearly optimal sequence.

Identification of evolutionary rules establishes criteria to evaluate possible changes in the initial sequence. The best set of rules will not only help to identify feasible alternatives but also ensure consideration of all promising sequences. Major changes at each iteration should be avoided so that alternatives worth consideration are not passed over. Examples of rules developed for evolutionary methods can be found in the literature (Stephanopoulos and Westerberg[14]). Often the rules involve picking the separation method, characterizing its effect on the sequence, moving the separator earlier in the process, and finally evaluating alternative separation methods at that point in the sequence. As changes are made, downstream product distributions will vary and existing separators in the flowsheet must be altered to achieve the desired product distribution.

The most important step in evolutionary methods is the *determination of the evolutionary strategy.* Perhaps the most difficult part of any design process is developing a strategy that directly leads to improvements in existing flowsheets. Many strategies can be characterized as either *depth-first* or *breadth-first,* both of which are tree-search or network methods. *Depth-first* strategies are aimed at generating a sequence that appears to be an optimum through the repeated application of one or more evolutionary rules. Additional criteria are applied to determine if an optimum in fact has been reached and, if necessary, to suggest an improvement. In this strategy, the optimum will arise through repeated modification of an existing detailed flowsheet.

Breadth-first strategies involve the generation of a number of possible sequences that may arise from a single change in an existing flowsheet. All are evaluated and the most promising is selected as a basis for generating another set of flowsheets. Heuristics may be applied through a prescribed weighting procedure to help identify the best choices. The optimum is near when the next generation of flowsheets offer little, if any, improvement over the one from which they were created.

Effective evolutionary methods are important for systematic process synthesis. They usually contain either heuristic or algorithmic elements and can be used effectively with computer-aided design packages.

Algorithmic Methods. The use of algorithmic approaches to process synthesis is, in theory, the only certain way to develop an optimal process. This is true because every possible process sequence can be considered rigorously. Unfortunately, because of the nature of the combinatorial problem that arises, even for relatively simple situations, the use of strictly algorithmic approaches is often unrealistic. However, when they are used in conjunction with heuristic and evolutionary strategies, they represent the best hope for rational process synthesis.

As the space of feasible alternatives is increased, the difficulty in finding an optimum process sequence also increases. Factored into most algorithmic approaches is some method to reduce the space of alternatives. In this case, heuristics often are employed in some systematic fashion to rule out options. *Decomposition methods* also are used to attempt to break the problem down into subproblems that can be handled by existing technology. *Bounding methods,* which often are programmed on a computer, can be used to provide some structure to the problem and to reduce the set of alternatives. The idea here is to set some criteria with which to evaluate options at each level of the problem so that only the better alternatives are explored. While bounding methods can be used to find directions to take on each level of the sequence, there is always the chance that the optimum may not be found; setting the bounds in itself can be a critical step.

Decomposition and bounding methods are ways to reduce the problem of balancing speed and accuracy which often plagues strictly algorithmic approaches to process synthesis. It should be noted that some or all of the above approaches can be combined to reduce the space of alternatives that must be considered by any algorithmic method.

4.2-3   Energy Integration in Process Synthesis

The motivation behind many efforts to sequence separation processes optimally is to reduce energy consumption. Examination of the lists of process heuristics points out this orientation. Unfortunately, addition of energy integration considerations to the often overwhelmingly complex sequencing problem creates an even more difficult situation. This is mainly the result of the fact that the sequencing and energy integration tasks cannot be handled separately. For example, the balance between reflux ratio and the number of stages for columns in a distillation sequence must take into account energy costs. These energy costs are influenced by the degree of heat integration that might be achieved with other parts of the process. Despite the complexity of the problem, effective energy integration for multicomponent separations can produce significant cost reductions and always must be considered if an optimal process is to be developed.

MINIMUM WORK
The concepts of *minimum work* and *reversibility* often are applied in the energy analysis of single-stage or multistage separations. From a thermodynamic standpoint, energy efficiency will be best for those processes that approach *reversibility* or minimize the generation of entropy. In a conceptual framework, a process that is thermodynamically reversible can be used as a standard of comparison for evaluating the energy efficiency of a real process. When energy separating agents are used, such as for simple distillation and crystallization, the process is potentially reversible because energy is added as heat. When mass separating agents are used, such as in absorption and extraction, the resulting process is partially reversible because of the inherent irreversibility of homogeneous mixing operations. Separations based on transport phenomena are rate governed and generally irreversible, as would be the case for membrane separations.

To a large extent, *minimum work* for a single separation unit is a function of the separation method used and arises from the degree of irreversibility associated with a given technique. In a sense, *minimum work* can be viewed as a measure of the difficulty of a particular separation. Thus, the actual work for the separation can be compared to the minimum work to provide a measure of energy efficiency. The use of this analysis for separations in which an energy separating agent is involved requires some mechanism for converting heat into work (net work consumption). One possibility for this conversion is through a reversible heat engine. The result of a calculation of this sort is to determine the efficiency of the process; efficiencies of less than 100% are the result of irreversibilities or lost work.

In a practical sense, this type of analysis can be put to use in energy integration schemes. For example, when heat is to be exchanged, it should be done in as thermodynamically reversible a way as possible; that is, exchange heat between streams that are close in temperature.

When sequences of separators are considered, the energy analysis can no longer be applied only to single separators. The potential for heat exchange will have a bearing on the sequencing as well as on the selection of separation tasks. Thus, the minimum work for the separation sequence should be considered when a search for an optimum is done.

For more information, the reader is referred to the works of Linnhoff,[15] De Nevers,[16] Sussman,[17] and De Nevers and Seader.[18] Also, the texts of King[2] and Henley and Seader[19] contain some discussion of thermodynamic analysis based on minimum work of separation and net work consumption.

HEURISTICS FOR REDUCED ENERGY CONSUMPTION IN SEPARATIONS
King[2] presents a list of heuristics to be considered in designing an energy-efficient process. Several of these heuristics remind the designer to select options that reduce the degree of irreversibility of the process and are related to those heuristics cited previously. For instance, he advises designers "to avoid the mixing of streams of dissimilar composition and/or temperature," and "to endeavor to use the full temperature differences between heat sources and sinks efficiently." He also suggests that the most-reversible separation method be used; energy separating agents are preferred over mass separating agents, and equilibration processes are preferred over rate-governed processes. King[2] demonstrates the application of these heuristics with two examples.

SYSTEMATIC APPROACHES TO ENERGY INTEGRATION
As might be expected, several workers have attempted to include energy integration considerations in their approaches for optimal process synthesis. Most, if not all, of their efforts involve heat integration for multicomponent distillation sequencing. In this case, overhead vapor streams and bottom products can serve as heat sources while feed and vapor recycle streams from the reboiler serve as heat sinks. The problem here is to find a suitable matching of hot and cold streams to minimize energy consumption within with the usual constraints inherent in finding an optimal sequence. As can be imagined, the considerable amount of work that has gone into optimization of heat exchanger networks is useful in addressing this problem.

Rathore et al.[20] suggested a technique using the list processing approach of Hendry and Hughes[21] along with an energy match matrix that specified all feasible energy exchanges between hot and cold streams. The energy matrix approach followed the branch and bound heat integration technique of Lee et al.[22] Heuristics were applied to rule out unlikely sequences and then the proper sequence was developed using dynamic programming methods. Although they did not consider vapor recompression, they did find that heat integration opportunities were best when columns in a distillation train were operated at different pressures. Their approach, however, proved to be too cumbersome for general usage.

Freshwater and Zigou[23] examined several four- and five-component systems with and without heat integration. They noted that without heat integration a direct sequence was usually best. However, if heat integration were considered, the optimal sequence was often distinctly different. Their work suggests the importance of heat integration in distillation column sequencing.

Siirola[24] more recently discussed an energy integration scheme, based on the use of heuristics, that makes feasible the simultaneous consideration of sequencing and heat integration. A bounding strategy is used to rule out unattractive options to eliminate the need for complete optimization. Using as an example a four-component separation into pure products by distillation, he compared heat integration schemes to the simple, direct sequence without heat integration or pressure optimization. The results of his study show that if sequencing and heat integration are considered simultaneously, a significant improvement can be achieved over the cases where they are considered separately. Unfortunately, although this method required only limited computer time, it still required the generation of all possible schemes of all design variables before an optimum was apparent. This becomes less feasible as the number of components and potential separation methods is increased. Although it was not useful in his example, Siirola[24] suggests using conventional process synthesis strategies of decomposition and bounding whenever possible. He also illustrates the potential value of thermodynamic availability analysis in setting a lower bound on the cost of the separation sequence. If a lower thermodynamic bound differs appreciably from the apparent optimum, this may indicate that additional heat integration is desirable through use of additional levels of utilities and heat pumping.

Sophos et al.[25] employed an optimal, heat-integrated distillation sequence. By first identifying favorable unintegrated sequences, they are able to limit the problem. They assume that the integrated optimum is a member of the set developed in the first screening. These sequences are then looked at for heat integration opportunities given whatever constraints are appropriate.

Thermodynamic analysis is the basis of the efforts by Umeda et al.[26] and Naka et al.[27] The former work applies the use of a heat availability diagram to the unintegrated schemes to help to identify optimal heat-integrated approaches and follows techniques used for heat exchanger network synthesis. The latter effort attempts to expand on the work of Umeda et al.[26] to include heat sources and sinks from other processes and uses in developing an optimal arrangement.

There are many efforts now underway to develop procedures for incorporating heat integration into process synthesis methodology; several of the references for this section should be examined for more information.

It should be apparent that the degree of sophistication for heat integration in separation sequences is not very high. This is not surprising given the nature of the problem. Advances should come as process synthesis methods in general are improved and as applications from the relatively well-developed heat exchange network design are implemented.

4.2-4 Examples of Process Synthesis

Many examples of methods to develop optimum process arrangements can be found in the literature. More recently, with the increased usage of digital computers to evaluate sophisticated algorithms, more complex sequencing problems are being addressed. Nevertheless, one must be somewhat familiar with the concepts applied earlier in process synthesis to understand and use the newer methodologies. These concepts are illustrated best by examining several proposed synthesis strategies; these will provide some insight into the types of problem that one encounters in process synthesis. Several will be presented here for illustrative purposes.

METHOD OF HENDRY AND HUGHES
The approach used here is to treat separators as list-processing operations using dynamic programming and ranked lists to develop the optimum sequence. The procedure, as originally described by Hendry and Hughes[21] and later summarized by Hendry et al.,[28] can be described as follows:

1. Given an initial mixture of A, B, C, and D that must be separated into individual products, the components are ordered into ranked lists based on one or more physical and/or chemical properties. Following Hendry et al.,[28] consider two properties that result in the following ordering:

Property 1:   A B C D

Property 2:   B A C D

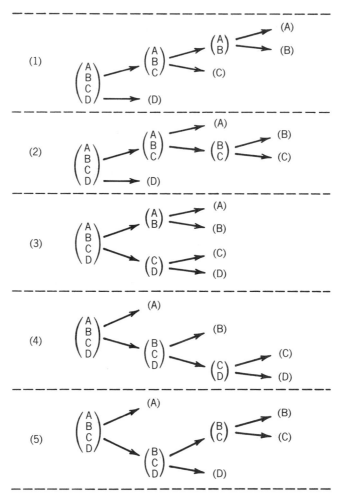

FIGURE 4.2-3 Five possible sequences for separating a four-component feed stream into pure components.

2. Based on these ranked lists, generate all possible subgroups. Each component must appear in at least one list. Since only simple, sharp separations are considered, splits must be between adjacent components in the ranked lists. The optimum process must be composed from these subgroups:

Binaries: A,B B,C C,D A,C

Ternaries: A,B,C B,C,D A,C,D

3. Working backward through the network, determine the least expensive way of separating each binary. This may be done for a variety of feed conditions.

4. Once this is done, evaluate the cost of each three-component separation from the original mixture, which may involve considering several feed conditions. Add this cost to the previously determined binary separation cost and select the sequence with the lowest combined cost for the separation.

Figure 4.2-3 shows all five possible sequences for the separation of the four components into pure products. Each sequence involves the use of three separators. If only one separation method were used and only simple, sharp separations were considered, these five sequences would represent all possibilities. However, with two separation methods, using the formula [Eq. (4.2-1)] of Thompson and King,[1] one finds that there are 40 possible sequences. The combination of list processing and dynamic programming used by Hendry and Hughes[21] can reduce this combinatorial problem into a manageable one. Shown below is one of the possible sequences which is arrived at through the use of two separation methods:

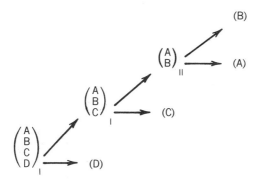

Hendry and Hughes[21] applied this approach in detail to an *n*-butylene purification system involving six components and two separation techniques (distillation and extractive distillation). Unfortunately, because all unique subgroups must be evaluated, computational requirements become excessive for large problems. Nevertheless, their work represents a relatively simple but powerful algorithmic method for addressing the combinatorial problem encountered in process synthesis and helps to initiate the development of other process synthesis techniques.

METHOD OF THOMPSON AND KING
An approach combining heuristic and evolutionary elements, this method[1] of process synthesis attempts to minimize the formation of extra products while also considering separator costs as flowsheets are generated. Process flowsheets are developed and optimized as follows:

1. Select a feasible product set for the process with the aim of minimizing the generation of extra products that would have to be mixed downstream to form the required products.

2. Combining heuristics with cost information, flowsheets that will produce the set of products chosen in step 1 are generated. A cost factor that incorporates the number of stages and the type of separator is applied; in the first iteration, no weighting is given to the type of separator so that a wide range of options might be considered. Beginning with the feed stream, the first separation technique is chosen to be consistent with the desired product set and based on the heuristic that the cheapest option is to be used.

3. The effluent streams from the first separator are examined and, if necessary, subjected to further separation. The sequence is developed, guided by the cost factor and the heuristic to choose the cheapest of all candidate separations next. The flowsheet is complete when all final effluent streams contain only one of the products of the predetermined set.

4. With the completed flowsheet, a detailed design and cost estimate of the units of the sequence are determined and the weighting factors for particular separators are then redefined.

5. The sequencing process is then repeated with the more accurate cost information until no further improvements are possible. If the apparent optimal sequence contains mass separating agent processes, the initial requirement of a minimum number of products is relaxed and the procedure is repeated using energy separating agents instead.

Thompson and King[1] point out that their method serves as a general strategy for process synthesis and more specific rules should be incorporated for a particular situation. It can be used to reduce the need for large amounts of computer time and will generally produce a set of near-optimal solutions perhaps containing the optimum. It may not be appropriate for highly complex and unusual separations problems, which is true of all existing systematic sequencing procedures.

METHODS BASED ON THE ORDERING OF HEURISTICS
Intuitively, one might imagine that the heuristics presented earlier are not of equal merit for all situations. Indeed, many of the process synthesis methods used today rely on the ordering of these heuristics either to generate an initial sequence that can be improved through an evolutionary strategy or to arrive at a final optimal sequence.

Seader and Westerberg[10] used a set of six heuristics along with several evolutionary rules to construct an optimal sequence for the separation problems presented by Hendry and Hughes[21] and Thompson and King.[1] They noted that their technique resulted in a much more rapid and concise treatment of the problem than earlier methods.

Nath and Motard[29] presented a method that extends the techniques used by Seader and Westerberg.[10] They used eight heuristics along with five evolutionary rules to develop the best sequence. Also included is a strategy to refine the best flowsheets. Like Seader and Westerberg, they use heuristics to generate an initial feasible sequence and evolutionary rules to improve on the initial configuration. They emphasize the importance of a good initial sequence since this will speed up the solution of the problem and favor global over local optima. Their method is shown to work for several previously presented sequencing problems.

More recently, Nadgir and Liu[13] suggest that an ordered heuristic approach to sequencing based on seven previously described heuristics will produce initial sequences that are very close to the final optimal sequences of Seader and Westerberg[10] and Nath and Motard.[29] They also make use of an auxiliary sequencing parameter called the coefficient of ease of separation (CES). They categorize heuristics in the following way: method heuristics; design heuristics; species heuristics; and composition heuristics following the procedure suggested by Tedder.[30] These are applied along with the CES parameter to generate initial sequences for previously presented problems that are close to the final sequences arrived at by the above heuristic–evolutionary approaches. Nadgir and Liu[13] also claim to have avoided the complicated mathematics that are associated with more sophisticated optimization methods.

Apparently the above methods are all successful in providing an optimal or near-optimal sequence for a well-defined separations problem. The utility of these approaches to problems that include separations beyond ordinary and extractive distillation is unclear.

METHOD OF TEDDER AND RUDD

Most of the work in developing process synthesis methodology centers on the use of simple, sharp separators and ignores the use of more complex separator arrangements. In distillation, where most of the process synthesis efforts have been focused, thermal coupling and sidestream removal are largely ignored despite the advantages this represents in many situations. Given current computational limitations, the only way that more complex configurations can be examined is by defining a relatively limited space of alternatives and by examining the options carefully. This was the approach used by Tedder and Rudd.[31–33]

In a comprehensive study of the distillation of ternary feeds of light hydrocarbons, Tedder and Rudd[31–33] considered eight distillation systems including complex arrangements with recycle streams and sidestream removal. Using a parametric approach, they were able to evaluate the eight designs and draw conclusions about the relative merits of each design. The basis used for comparison was the feed composition and a rating system they called the ease of separation index (ESI). They then analyzed their data using an equilateral triangle diagram showing the range of optimality for each of the eight configurations studied. For a given ESI, the spaces for which each design is optimal can be determined once the feed composition is known. In the diagram, each component is represented as one of the vertices of the equilateral triangle. While this limits the technique to ternary systems, the authors suggested that this approach can be extended to larger systems through the generation of pseudoternaries with the most difficult ternary separation taken last. While their technique does not account for complexities that factor into the final cost estimates for these systems, Tedder and Rudd[31–33] present a useful technique for reducing the set of alternatives for these complex systems. This type of approach, although not algorithmic in the strictest sense, nonetheless provides a mechanism for quantitative evaluation of a restricted set of process options.

COMPUTER-AIDED PROCESS SYNTHESIS METHODOLOGIES

Often the overall objective in process synthesis is to arrive at an optimal design for a complete chemical plant. This includes consideration of chemical synthesis, process control, energy efficiency, as well as separation process sequencing. Ultimately, a systematic methodology requiring only the definition of process objectives is desirable. While the use of artificial intelligence or knowledge engineering might someday make this notion a reality, this will not be the case in the near future. The work that has been done in this area takes one of two approaches: (1) no initial process structure is assumed or (2) the process is developed by considering all possible alternatives (integrated or structural parameter approach).

Several procedures have been suggested for systematic process synthesis in which no initial process structure is present. Most of these are heuristic–evolutionary methods but they often contain algorithmic elements as well. Siirola and Rudd[34] and Powers[35] developed a heuristics-based method for the computer-aided synthesis of process designs. The program, called AIDES (Adaptive Initial Design Synthesizer), used the repeated application of specific design heuristics to the original feed stream and its subunits until the final product set was achieved. Certain heuristics were applied to select the appropriate separation methods first and then others were used to determine the sequencing. The concept of ranked lists was used along with a weighting procedure based on heuristics for each split from each ranked list. The best process is the one with the most favorable score that is both chemically and physically realistic. AIDES is an attempt to apply design heuristics systematically for process synthesis but relies heavily on the heuristics employed and their relative weighting in the sequencing and selection process. It has been tested on several processes and has been successful in generating alternate flowsheets; the optimum, however, may not be within the set of alternatives. Nonetheless, it was one of the first reasonable attempts to develop a systematic process synthesis methodology and provided the basis on which more sophisticated methods could be developed.

There have been several other proposed procedures for systematic generation of optimal process flowsheets without an initial process structure. Of note is one developed by Mahalec and Motard[36] and incorporated into a computer program they called BALTAZAR. Like AIDES, it can be used for the synthesis of an entire chemical plant but takes a more sophisticated approach to processing objectives and constraints. It is based on a systematic resolution of conflicts between a set of processing rules and the processing objectives. The procedure is based on a sequential depth-first approach which employs several structural rules that help to produce feasible flowsheets. An objective function that incorporates heuristic arguments into the flowsheet evaluation process helps to identify promising options. Evolutionary rules also are used to improve the process structure. BALTAZAR has been used to generate alternate flowsheets for several industrial processes including one previously analyzed by AIDES.

Nishida et al.[37] provide an excellent comparison of AIDES and BALTAZAR.

More recent efforts to develop computer-aided process synthesis methodologies can be characterized as either *sequential modular, simultaneous modular,* or *equation oriented.* Sequential-modular approaches are best for steady-state simulation where process inputs are defined and process parameters are available. The best feature of sequential-modular approaches is that they are flowsheet oriented, but they are not as flexible as the other methodologies in performing design and optimization tasks.[38]

Equation-oriented approaches are based on sets of equations that are written for the units in a particular flowsheet.[38] Unlike the sequential-modular systems, which often contain the necessary information for a variety of process units, equation-oriented synthesizers require the practitioner to develop the model equations. These are solved through iterative techniques with standard numerical methods.

Simultaneous-modular methods attempt to combine the features of the sequential-modular and equation-oriented approaches to make use of models developed for the former and to include the flexibility of the latter.[39] For more information, the reader is referred to the recent review by Biegler.[39]

There are several reviews that provide additional information about computer-aided process synthesis.[37,40] A brief discussion of one particular methodology is presented here.

Westerberg and coworkers[41,42] have described an equation-oriented process synthesis methodology that is being developed at Carnegie–Mellon University called ASCEND-II. It is capable of performing both simulation and optimization calculations. Westerberg describes the following guidelines, characteristic of an evolutionary approach, that were inherent in the development of ASCEND-II and should be considered in its use:

1. *Evolve from Simple to Complex.* It is more important at the outset of a design not to lose perspective by attempting to be overly quantitative. The purpose of initial calculations is to provide some qualitative understanding of the problem.

2. *Use a Depth-First Approach.* The point here is to generate an initial flowsheet that later can be optimized and to avoid backtracking. A great deal is learned on completing an initial feasible design even if it is far from optimal.

3. *Develop Approximate Criteria.* Final design criteria cannot be applied without an initial design in hand. Therefore, use approximate criteria, such as heuristics, to generate alternate flowsheets that can be subjected to more stringent standards.

4. *Use Top-Down/Bottom-Up Design Strategies Alternately.* Approaches that scope out options in light of overall process design objectives should be used alternately with approaches that evaluate these objectives at the local level for feasibility.

5. *Be Optimistic in Generating Design Alternatives.* Most initial design concepts will not work. It therefore pays to be creative in the initial phases of design. It is unlikely that the optimum sequence will be generated immediately. Creativity, on the other hand, may produce an unexpectedly effective design.

Westerberg[42] comments that ASCEND-II offers the following elements to the design engineer: simulation, design, dynamics, and optimization. It is a tool for developing interactive computer models of varying complexities for the process under development; these can be used to improve the design. Westerberg[42] notes that the flexibility of ASCEND-II in helping the design engineer to evaluate options interactively is an advantage of this system over traditional flowsheeting systems.

Others have reported computer-aided process synthesis methodologies that are algorithmic approaches. The so-called *integrated or systems approaches* to process synthesis depend heavily on optimization theory and linear and nonlinear programming techniques in finding an optimal processing arrangement. The determination of the best processing sequence is performed as a multiobjective optimization problem with consideration given to the entire space of possible alternatives. In practice, this can be done by combining all possible flowsheets into one flowsheet and creating some mechanism for dealing with interconnections between the elements of the process. While the systems or structural parameter approach theoretically will yield the optimum, the solutions to the resulting large-scale nonlinear programming problems are often not available. Also, because many direct optimization approaches are based on the representation of discrete variables by continuous variables, the apparent solution may turn out to be nonoptimal due to numerical difficulties. Nevertheless, as computer-aided process synthesis develops and becomes more useful for design, the systems approach necessarily will become more widely employed as a framework with which to attack the problem. For more information on integrated approaches to overall process synthesis, the excellent reviews of this subject by Umeda[40] and Takama et al.[43] should be examined.

4.2-5 Future of Process Synthesis

Several of the review papers referenced at the end of this chapter point out the future needs and directions of research in process synthesis. Some of the recommendations for areas for further work in separation processes are summarized below:

1. Synthesis of separation schemes for processes other than distillation.
2. Heat integration, multiple feed and withdrawal, partial splits, and medium recovery fractions in distillation column sequencing.
3. Use of second law analysis in considering the thermodynamic efficiency of separation schemes.
4. Consideration of nonideal systems and the effect of short-cut or approximate design methods on process synthesis approaches.
5. Integration of process control strategies into the sequencing problem including the consideration of the dynamic behavior of the system.
6. Improvements in methods for developing optimal flowsheets for the overall process, including a more sophisticated approach to incorporating separations into the flowsheet.

Without a doubt, the continued development of process synthesis methodologies has had and will have a significant impact on the design of chemical processes. There is little chance, however, that any systematic procedure or computer-based technique will supplant the need for experienced design engineers in the near future. The most promising developments will involve interactive computer programs that can make the best use of the power of the computer and the experience of the design engineer for both flowsheet simulation (steady-state and dynamic cases) and optimization.

4.3 CONTROL OF SEPARATION PROCESSES

A detailed look at the use of control in separation processes requires a much more comprehensive treatment than is possible here. The literature contains many well-written books and articles dealing with control in general and distillation column control in particular. Several references in each case are included at the end of this chapter.[1-9]

It would be useful perhaps to provide some perspective as to how the nature of process control is evolving, particularly in the context of design and synthesis of separation processes.

In a classic sense, process control usually implies a single-input–single-output relationship that can be monitored and regulated to ensure that a given system will operate as designed. For a given unit operation or separation, control objectives are specified, following which controlled, manipulated, and measured variables are selected. These are synthesized into a control loop. While this methodology is still a functional one in the chemical industry, it nonetheless is changing rapidly. The proliferation of digital computers and the incentive to design and operate processes and entire chemical plants optimally have led to an expansion of the role that process control plays in the chemical industry.

In most instances, the standard PID (proportional-integral-derivative) controller still is used for the control of single-input–single-output systems. However, individual control loops may be placed in the hierarchy of a distributed control system. By doing so, the control of many single-input–single-output systems can be orchestrated better, and several more sophisticated control methodologies can be considered including cascade control and split-range control. Distributed control also provides a framework for considering multiple-input–multiple-output systems and the interactions that may be inherent in these situations.

Most of the more recent developments in control are linked to improvements in computing at all levels. Digital computers have been developed to the point where the elements of a chemical process can be treated more realistically as the complex, highly nonlinear, and multi-interactive systems that they are. This has led to a more sophisticated use of dynamic modeling and simulation in process control. As computing costs have come down and computing capabilities have expanded, control strategy is being developed in a more global sense using simulation to consider the interaction of process elements and sometimes different processes. As a result, the focus in process control has expanded from the tuning and performance of controllers for individual unit operations to the broader context of controlling and coordinating the operation of an entire chemical process or plant.

A distributed control system may involve the use of microcomputers at the local level and the use of more powerful machines to coordinate overall plant control objectives. In this context, it has been suggested that process control might be described better as process management.[10,11] Local control of the operation of individual separators is still important but, with the use of distributed control, reliability is maintained (microcomputers are dedicated to particular process units) while overall technical and economic objectives are pursued (mainframe computers can perform complex on-line/off-line optimization). The advantage to distributed control is that it makes effective use of current technology and provides a framework within which control and optimization developments can be implemented. These developments probably will include better simulation and optimization routines that will help to assess the current state of the process plant and to suggest improvements.

There is no doubt that, in many instances, relatively simple approaches are still sufficient to accomplish particular control objectives. However, for example, in the case of separation sequences that involve complex flow arrangements and thermal coupling, more advanced process control strategies will be necessary. As process synthesis methodologies improve, control strategy evaluation and process design and analysis will be done simultaneously. Efforts to develop such a procedure have been reported by Fisher et al.[12] among others.

4.3-1 Process Modeling and Simulation

In the perspective of process synthesis, process control should be viewed not as a separate element in process design and optimization but rather as a component of a coordinated approach. Therefore, the design and sequencing of separation processes must consider the relationship that process control will have to the final process structure. This can be done only through process modeling and simulation.

The most important aspect in the control of separation processes or, for that matter, any chemical process is understanding the process itself. Previously, without the availability of digital computers, overdesign and the neglect of process dynamics were practical necessities. Today, however, through process modeling and simulation with digital computers, a better understanding of the process is possible before any hardware is fabricated.[13,14] From a control standpoint, this enables the process engineer to investigate the characteristics of the system over a wide range of operating conditions and to obtain a better parametric understanding of the process. This, of course, provides a better basis for selecting appropriate control strategies.

With the continued increase in sophistication of process modeling and simulation, it is becoming apparent that control of chemical processes often requires a unique application of the foundations of control theory. Early work in chemical process control borrowed heavily from other disciplines and practice severely lagged behind theory. Chemical engineers often looked deeper and deeper into sophisticated control theory for answers to problems that arose in process control. It now seems, after useful experience with process modeling and simulation, that less-sophisticated applications of control theory often are adequate for even complex problems. The key to successful process control has been found in an increased understanding of the process rather than sophisticated applications of control theory.

Thus, process modeling and simulation has become a key element in process control. Complex computer programs for these purposes have evolved and become well-accepted tools for chemical engineers. These simulation packages can be used to design new processes as well as to modify and optimize existing ones. Usually, they are comprised of sophisticated physicochemical estimation routines along with process models for a variety of unit operations, reactors, and auxiliary equipment. Typical simulators might include over 1000 subroutines totaling 150,000 lines of FORTRAN computer code.[15-18] Their overall function is to coordinate material and energy balance information for a variety of processing arrangements and conditions. The utility and accuracy of these simulation programs are a function of how well conceived the models of the components of the process are as well as how effective the necessary convergence techniques are implemented.

The first process simulators developed for general use were for steady-state operation. Process dynamics usually were ignored to avoid excessive computation times and computational difficulties. While steady-state information is useful for control strategy evaluation, process control systems are designed best with some knowledge of process dynamics. As nonlinearities and interactions become more pronounced, information from the time-dependent behavior of a process or processes becomes crucial. There is currently a great deal of effort focused on upgrading process simulators so that they can handle unsteady-state operation. Several of these programs are able to model process dynamics for specific situations.

Because of the computational complexities associated with dynamic process simulation for multiunit processes, there is still much to be done before simulators of this type become available for general application. Another problem complicating their development is that process models for even individual separation units are usually for steady-state cases; this is the result of both incomplete understanding of the chemical and physical principles involved and computational difficulties. This is one of the main reasons why process control considerations are difficult to incorporate into chemical process simulation and synthesis and why on-line plant optimization is still far away in most instances.

Process modeling and simulation are nevertheless extremely important tools in the design and evaluation of process control strategies for separation processes. There is a strong need, however, for better process models for a variety of separations as well as process data with which to confirm these models. Confidence in complex process models, especially those that can be used to study process dynamics, can come only from experimental verification of these models. This will require more sophisticated process sensors than those commonly available for temperature, pressure, pH, and differential pressure. Direct, reliable measurement of stream composition, viscosity, turbidity, conductivity, and so on is important not only for process model verification but also for actual process control applications. Other probes, which could be used to provide a better estimate of the state of the system, are needed to contribute to the understanding of the process in the same time frame as that of changes occurring in the process.

There are still many obstacles to overcome before the use of process modeling, simulation, and control reaches the potential that many think it holds. One of the most interesting possibilities is the development of systems capable of performing on-line optimization functions. Although on-line plant optimization is still in the future, optimization of subprocesses is already a possibility.[19] This is true for both continuous and batch processes,[20-23] and there are now many examples of how process modeling and simulation have enabled process control strategy and design to move beyond strictly performance considerations.

4.3-2 Process Synthesis of Control Structures

The objective in the control of separation processes has evolved from the control of the components of a process that already has been designed and often constructed to an integrated plan involving simulation, design, and control. Chemical process synthesis approaches that include control structures have been suggested only recently and are not well developed.

There have been several attempts at control structure development within process synthesis methodology. Govind and Powers[24-26] started with steady-state flowsheets and used cause-and-effect diagrams to represent control logic for a particular process. Dynamics were modeled by a first-order lag and deadtime in series. A set of heuristics were used to rule out unfavorable alternatives before any detailed simulation was carried out to determine the best options. Govind and Powers, noting that their work was the first attempt to use non-numerical problem-solving techniques for control structure synthesis, discussed the manner in which their approach could be used to translate steady-state process flowsheets into piping and instrumentation diagrams.

Morari and coworkers[27-29] describe a procedure for control structure development based on the multilayer–multiechelon approach of hierarchical control theory, which is free of heuristics. The key to the procedure is the effective application of decomposition to produce manageable subsystems of the problem. Examples are provided to show how structural controllability and observability can lead to a control system that is consistent with processing objectives.

Umeda[30] reviews some of his own work, as well as that of his colleagues, which relates to control system synthesis, instrumentation, and alarm system design. To handle simpler systems, he notes that the development of control heuristics to rule out unfavorable pairings of controlled and manipulated variables might be beneficial. However, even for highly interactive systems with complex control objectives, reasonably simple procedures that can be used by process engineers are desirable even if they are useful only as design tools. Umeda also points out that one should not rule out the rearrangement of process structures to suit control structure synthesis, if necessary.

For additional information on the synthesis of process control structures, the reader is referred to the reviews of Umeda,[30] Nishida et al.,[31] Morari,[32] and Stephanopoulos[33] and the ongoing work of Fisher et al.[34]

4.4 SPECIAL PROBLEMS

Many situations exist in which conventional approaches to design and operation of separation processes must be modified or changed altogether to meet processing objectives. Usually, these cases involve systems that are characterized by certain constraints that limit the potential range of operating conditions or scope of separation alternatives. Sometimes, even completely new separation methods must be developed.

Some examples of situations that require special processing consideration are discussed below.

4.4-1 Biological Separations

To be able to bring the advances in biotechnology to the marketplace, the products of many biotransformations somehow must be recovered from the complex reaction media in which they are formed. Because of the nature of biological products, conventional separation methods (e.g., distillation) often are not appropriate, and completely new approaches often have been developed.[1-7]

The recovery of most biological products begins with the isolation of cells or proteins from fermentation broths. These broths can be viscous solutions which must be treated carefully because of the damaging effects of heat and mechanical stresses on their contents. If the desired product is extracellular, it must be removed from a complex mixture of similar chemicals which often contains significant amounts of biomass. Another important consideration is that the concentration of the product in the mixture may be less than a fraction of a gram per liter. If the product is intracellular, it must first be released from the cell, which may involve harvesting and disrupting the cells, and then recovered from the resulting debris.

The approach taken to recover biological products depends on the nature of the product. Biological products can be classified as either high-value–low-volume or low-value–high-volume. In the case of the former, many of the techniques used in analytical applications have been adapted to larger scale to handle product purification. For low-value–high-volume products, such as ethanol, more conventional procedures can be used, but these often require some modification.

For any intracellular product, the cells first must be harvested from the fermentation broth and lysed, or broken open, to release their contents. The liquid fraction, which contains the product, must then be separated from the cells. Cellular debris can be removed by either filtration or centrifugation, or a combination of both. Crossflow filtration, using microporous media or ultrafiltration membranes, has been shown to be extremely effective for this step, but problems with membrane blockage have been reported.

For extracellular products, one is faced with the problem of removing the liquid phase, which contains the product, from the biomass. This can be accomplished using techniques similar to those used for removing cellular debris. It should be noted that there are emerging genetic engineering techniques that induce microorganisms to secrete normally intracellular products into the fermentation broth to facilitate recovery. Also, for biotransformations that do not involve microbial growth, immobilization techniques can be implemented which allow the use of a biocatalyst in much the same way that conventional heterogeneous catalysts are used. Cells or enzymes are attached or entrapped in a solid support and are contacted with a solution containing the reactants. This mode of processing greatly simplifies the downstream processing steps.

After harvesting and, if necessary, cell lysis, one is still left with a very dilute mixture of proteins or other biological compounds which is subjected usually to some form of chromatography. High performance liquid chromatography (HPLC), which may include normal-phase, reverse-phase, ion-exchange, and gel-permeation columns, has been the focus of many efforts to scale analytical procedures to commercial applications. Packing stability and liquid distribution and collection often present scale-up problems. Solutions to these problems as well as better approaches to packing selectivity probably will come in the not too distant future. Affinity chromatography, based on the unique attraction of one molecule for another (i.e., antigen–antibody interaction), represents a potential solution to the problem of recovering trace amounts of a high-value product from dilute liquid solutions. The compound of interest is removed from solution by complexing or reacting with another compound immobilized on the column packing. Of course, each application of this technique requires the development of the particular chemistry.

Continuous electrophoretic separations also have been proposed for the recovery of biological material from dilute solution. Under an electric field, and often in a pH gradient as well, proteins will migrate to their isoelectric point or point of zero net charge.[8] As a laboratory technique, electrophoresis has long been a valuable tool. Commercial scale applications, which have been limited because of problems with remixing of the separated fractions, are seen to be possible with the development of a new system by Harwell for the United Kingdom's Atomic Energy Authority.[1] This system avoids remixing by processing under laminar flow using a rotating outer electrode.

While it is often economically justifiable to implement sophisticated separation schemes for high-value-low-volume biological products, this is not the case for commodity chemicals, such as solvents and organic acids. Distillation still is used to recover ethanol from fermentation broths or from the product stream of an immobilized cell reactor. This represents a fairly energy-intensive undertaking since the distilled solution is usually 90% (weight) water. Molecular sieves, which avoid the water–ethanol azeotrope, and supercritical extraction with carbon dioxide have been used in an attempt to improve both energy and recovery efficiencies. The recovery of organic acids through the use of membrane technology, including electrodialysis, also has been tried to avoid distillation.

Indeed, biological separations present some unique problems for separations engineers. It is likely that this will lead to concentrated efforts to develop new techniques that may be used in place of or along with more conventional separation processes.

4.4-2 Thermally Sensitive Materials

In addition to biological products, there are many chemical compounds that will degrade or change if, during recovery or concentration from solution, processing temperatures are increased to any great extent. This is particularly true in the foods industry and in polymer processing. Fruit juices, sugars, and gelatins in addition to many plastics and resins are good examples of materials that are sensitive to thermal processing. They present particularly difficult separations problems because, in addition to their thermal sensitivity, they often are present in highly viscous solutions which makes large amounts of heat transfer in short residence times particularly difficult.

In the concentration of fruit juices,[9,10] it is important to keep temperatures low and processing times as short as possible. Operation under vacuum is common. Because the concentrate contains solid particulates, the slurry must be moved quickly across any heat transfer surface to avoid the loss of solids due to surface adhesion. Also, because many vitamins are affected adversely by thermal processing, short residence times at elevated temperatures are crucial.

Most polymers are dissolved in either aqueous or organic solvent at some point in their processing. The concentration of rubber latex from emulsion polymerization solutions and the recovery of plastics and resins from batch polymerization reactors are two examples. In addition to problems of polymer degradation due to exposure to high temperatures for any significant period of time, there exists the additional problem of molecular weight distributions. In many cases, holding polymer solutions at elevated temperatures will lead to additional polymerization of any residual monomer and hence to a product that might not meet specifications. Many polymer recovery processes must be operated carefully to avoid degradation as well as to eliminate product variability arising from any temperature excursions.

For the concentration of viscous, heat-sensitive materials, thin-, agitated- or wiped-film evaporators can be used.[11] Among the materials that have been processed through these systems are foods, fruit juices, plastics and resins, pharmaceuticals, vitamins, and radioactive wastes. The heat transfer surfaces in these evaporators are either wiped by a rotating blade or held at a small, fixed distance from the rotating blade to reduce surface fouling and to promote high heat transfer rates. The vapors leaving the heat transfer surface can be recovered by internal or external condensers. These evaporators, which often are operated under vacuum, are designed to promote short residence times to reduce the adverse effects of thermal processing as well as to facilitate the treatment of viscous materials. By adjusting the heat transfer surface temperature, the feed rate, and the vacuum level, processing conditions can be adapted for a wide variety of applications. Another advantage of these types of evaporator is that they can be operated continuously and are scaled and instrumented easily.

4.4-3   High-Purity Materials

There are many instances in which the objective of a separation process is to recover a given compound at a very high level of purity. When this is the case, it is sometimes necessary to take a different approach to the particular separation. For example, the ethanol–water azeotrope precludes the recovery of pure ethanol using ordinary distillation. For higher purities, azeotropic distillation, solvent extraction, or the use of molecular sieves can be used. Clearly, the necessity to recover ethanol in concentrations above the azeotrope requires an alternative separations strategy.

The need for high purity in a separations process is common in many industries: semiconductor manufacture, pharmaceuticals processing, and the foods industry, as well as in many cases of more-conventional chemical processing. It is also very important in separation processes that are oriented to cleaning gas, liquid, and solid streams for environmental purposes. The low concentrations required of many environmentally significant compounds prior to discharge from a chemical plant have created a need for a new class of separation methods and have focused attention on many techniques that often have been ignored. Adsorption, ultrafiltration, electrostatic precipitation, reverse osmosis, and electrodialysis are just a few examples of separation processes in which there has been an increased level of interest partly because of their potential in environmental applications.

4.4-4   Summary

It is important in the design of a separation process to be cognizant of the limitations presented by the available alternatives. As the required level of purification increases, the technical and economic feasibilities of a given separation method may be reduced to the point at which some alternative must be identified. Failure to recognize this in the design stages will lead to unsatisfactory processing results.

REFERENCES

Section 4.1

4.1-1   Ernest J. Henley and J. D. Seader, *Equilibrium-Stage Separation Operations in Chemical Engineering*, Wiley, New York, 1981.

Section 4.2

4.2-1   R. W. Thompson and C. J. King, "Systematic Synthesis of Separation Schemes," *AIChE J.*, **18,** 941 (1972).

4.2-2   C. J. King, *Separation Processes*, 2nd ed., McGraw-Hill, New York, 1980.

4.2-3   H. Short, "New Mass-Transfer Find Is a Matter of Gravity," *Chem. Eng.*, **90**(4), 23 (1983).

4.2-4   R. L. Motard and A. W. Westerberg, *Process Synthesis*, AIChE Advanced Seminar Lecture Notes, AIChE, New York, 1978.

4.2-5   D. F. Rudd, G. J. Powers, and J. J. Siirola, *Process Synthesis*, Prentice-Hall, Englewood Cliffs, 1973.

4.2-6   F. J. Lockhart, "Multi-column Distillation of Gasoline," *Pet. Refiner*, **26,** 104 (1947).

4.2-7   F. B. Petlyuk, V. M. Platonov, and D. M. Slavinskii, "Thermodynamically Optimal Method for Separating Multicomponent Mixtures," *Int. Chem. Eng.*, **5,** 555 (1965).

4.2-8   W. J. Stupin and F. J. Lockhart, "Thermally Coupled Distillation—A Case History," *Chem. Eng. Prog.*, **68,** 71 (1972).

4.2-9   G. Stephanopoulos, "Optimization of Nonconvex Systems and the Synthesis of Optimum Process Flowsheets," Ph.D. Thesis, University of Florida, 1974.

4.2-10 J. D. Seader and A. W. Westerberg, "A Combined Heuristic and Evolutionary Strategy for Synthesis of Simple Separation Sequences," *AIChE J.*, **23,** 951 (1977).

4.2-11 R. Nath and R. L. Motard, "Evolutionary Synthesis of Separation Processes," 85th National Meeting of AIChE, Philadelphia, PA, June 1978.

4.2-12 A. W. Westerberg, "A Review of Process Synthesis," *ACS Symp. Ser.*, **124** (1980).

4.2-13 V. M. Nadgir and Y. A. Liu, "Studies in Chemical Process Design and Synthesis: Part V: A Simple Heuristic Method for Systematic Synthesis of Initial Sequences for Multicomponent Separations," *AIChE J.*, **29**, 926 (1983).

4.2-14 G. Stephanopoulos and A. W. Westerberg, "Studies in Process Synthesis—II. Evolutionary Synthesis of Optimal Process Flowsheets," *Chem. Eng. Sci.*, **31**, 195 (1976).

4.2-15 B. Linnhoff, Entropy in Practical Process Design, in *Foundations of Computer-Aided Chemical Process Design*, Vol. II, Engineering Foundation and AIChE, New York, 1981.

4.2-16 N. De Nevers, Two Fundamental Approaches to Second Law Analysis, in *Foundations of Computer-Aided Chemical Process Design*, Vol. II, Engineering Foundation and AIChE, New York, 1981.

4.2-17 M. V. Sussman, *Availability (Energy) Analysis—A Self-Instruction Manual*, M. V. Sussman, Publisher, Tufts University, Medford, MA, 1980.

4.2-18 N. De Nevers and J. D. Seader, "Mechanical Lost Work, Thermodynamic Lost Work, and Thermodynamic Efficiencies of Processes," AIChE National Meeting, Houston, TX, April 1979.

4.2-19 E. J. Henley and J. D. Seader, *Equilibrium-Stage Separation Operations in Chemical Engineering*, Wiley, New York, 1981.

4.2-20 R. N. S. Rathore, K. A. van Wormer, and G. J. Powers, "Synthesis Strategies for Multicomponent Separation Systems with Energy Integration," *AIChE J.*, **220**, 491 (1974).

4.2-21 J. E. Hendry and R. R. Hughes, "Generating Separation Process Flowsheets," *Chem. Eng. Prog.*, **68**, 71 (1972).

4.2-22 K. F. Lee, A. H. Masso, and D. F. Rudd, "Branch and Bound Synthesis of Integrated Process Designs," *Ind. Eng. Chem. Fundam.* **9**, 48 (1970).

4.2-23 D. C. Freshwater and E. Zigou, "Reducing Energy Requirements in Unit Operations," *Chem. Eng. J.*, **11**, 215 (1976).

4.2-24 J. J. Siirola, Energy Integration in Separation Processes, in *Foundations of Computer-Aided Process Design*, Vol. II, Engineering Foundation and AIChE, New York, 1981.

4.2-25 A. Sophos, G. Stephanopoulos, and M. Morari, "Synthesis of Optimum Distillation Sequences with Heat Integration Schemes," 71st Annual AIChE Meeting, Miami, FL, November 1978.

4.2-26 T. Umeda, K. Niida, and K. Shiroko, "A Thermodynamic Approach to Heat Integration in Distillation Systems," *AIChE J.*, **25**, 423 (1979).

4.2-27 Y. Naka, M. Terashita, and T. Takamatsu, "A Thermodynamic Approach to Multicomponent Distillation System Synthesis," *AIChE J.*, **28**, 812 (1982).

4.2-28 J. E. Hendry, D. F. Rudd, and J. D. Seader, "Synthesis in the Design of Chemical Processes," *AIChE J.*, **19**, 1 (1973).

4.2-29 R. Nath and R. L. Motard, "Evolutionary Synthesis of Separation Processes," *AIChE J.*, **27**, 578 (1981).

4.2-30 D. W. Tedder, "The Heuristic Synthesis and Topology of Optimal Distillation Networks," Ph.D. Dissertation, University of Wisconsin, Madison, WI, 1975.

4.2-31 D. W. Tedder and D. F. Rudd, "Parametric Studies in Industrial Distillation: Part I. Design Comparisons," *AIChE J.*, **24**, 303 (1978).

4.2-32 D. W. Tedder and D. F. Rudd, "Parametric Studies in Industrial Distillation: Part II. Heuristic Optimization," *AIChE J.*, **24**, 316 (1978).

4.2-33 D. W. Tedder and D. F. Rudd, "Parametric Studies in Industrial Distillation: Part III. Design Methods and Their Evaluation," *AIChE J.*, **24**, 323 (1978).

4.2-34 J. J. Siirola and D. F. Rudd, "Computer-Aided Synthesis of Chemical Process Designs," *Ind. Eng. Chem. Fundam.* **10**, 353 (1971).

4.2-35 G. J. Powers, "Recognizing Patterns in the Synthesis of Chemical Processing Systems," Ph.D. Dissertation, University of Wisconsin, Madison, WI 1971.

4.2-36 V. Mahalec and R. L. Motard, "Procedures for the Initial Design of Chemical Processing Systems," *Comput. Chem. Eng.*, **1**, 57 (1977).

4.2-37 N. Nishida, G. Stephanopoulos, and A. W. Westerberg, "A Review of Process Synthesis," *AIChE J.*, **27**, 321 (1981).

4.2-38 J. D. Perkins, Equation-Oriented Flowsheeting, in *Proceedings of the Second International Conference on Foundations of Computer-Aided Design*, A. W. Westerberg and H. H. Chien (Eds.), CACHE Publications, Ann Arbor, MI, 1984.

4.2-39 L. T. Biegler, Simultaneous Modular Simulation and Optimization, in *Proceedings of the Second International Conference on Foundations of Computer-Aided Design*, A. W. Westerberg and H. H. Chien (Eds.), CACHE Publications, Ann Arbor, MI, 1984.

4.2-40 T. Umeda, "Computer-Aided Process Synthesis," *Comput. Chem. Eng.*, **7**, 279 (1983).

4.2-41 M. H. Locke, S. Kuru, P. A. Clark, and A. W. Westerberg, "ASCEND-II: An Advanced System for Chemical Engineering Design," 11th Annual Pittsburgh Conference on Modeling and Simulation, University of Pittsburgh, May 1–2, 1980.

4.2-42 A. W. Westerberg, "Design Research: Both Theory and Strategy," *Chem. Eng. Educ.*, **16**, 12–16, 62–66 (1982).

4.2-43 N. Takama, T. Kuriyama, K. Niida, A. Kinoshita, K. Shiroko, and T. Umeda, "Optimal Design of a Processing System," *Chem. Eng. Process.*, **78**(9), 83 (1982).

Section 4.3

4.3-1 P. S. Buckley, *Techniques of Process Control*, Wiley, New York, 1964.

4.3-2 D. R. Coughanowr and L. B. Koppel, *Process Systems Analysis and Control*, McGraw-Hill, New York, 1965.

4.3-3 W. L. Luyben, *Process Modeling, Simulation and Control for Chemical Engineers*, McGraw-Hill, New York, 1973.

4.3-4 W. H. Ray, *Advanced Process Control*, McGraw-Hill, New York, 1981.

4.3-5 F. G. Shinskey, *Distillation Control*, McGraw-Hill, New York, 1977.

4.3-6 F. G. Shinskey, *Process Control Systems*, 2nd ed., McGraw-Hill, New York, 1979.

4.3-7 C. L. Smith, *Digital Computer Process Control*, Intext Educational Publishers, Scranton, PA, 1972.

4.3-8 G. Stephanopoulos, *Chemical Process Control*, Prentice-Hall, Englewood Cliffs, NJ, 1984.

4.3-9 T. W. Weber, *An Introduction to Process Dynamics and Control*, Wiley, New York, 1973.

4.3-10 J. Haggin, "Process Control No Longer Separate from Simulation, Design," *Chem. Eng. News*, **62**(14), 7 (1984).

4.3-11 J. Haggin, "Process Control on Way to Becoming Process Management," *Chem. Eng. News*, **62**(21), 7 (1984).

4.3-12 W. R. Fisher, M. F. Doherty, and J. M. Douglas, "An Evolutionary and Hierarchical Procedure for Optimization of Preliminary Process Designs," presented at the Annual Meeting of AIChE, paper 104c, San Francisco, CA, November 1984.

4.3-13 R. G. E. Franks, *Modeling and Simulation in Chemical Engineering*, Wiley, New York, 1972.

4.3-14 C. D. Holland and A. I. Liapsis, *Computer Methods for Solving Dynamic Separation Problems*, McGraw-Hill, New York, 1983.

4.3-15 J. A. Liles, "Computer Aids for Process Engineering," *Chem. Eng. Prog.*, **79**(6), 43 (June 1983).

4.3-16 S. I. Proctor, "The FLOWTRAN Simulation System," *Chem. Eng. Prog.*, **79**(6), 49 (June 1983).

4.3-17 S. S. Grover, A. M. Peiser, M. K. Sood, M. T. Tayyabkhan, and G. M. Weber, "Bench-Marking Problems for Simulation," *Chem. Eng. Prog.*, **79**(6), 54 (June 1983).

4.3-18 T. B. Challand, "Computerized Optimization of Complete Process Flowsheets," *Chem. Eng. Prog.*, **79**(6), 65 (June 1983).

4.3-19 D. E. Haskins, "Restraints on Entire Plant Optimization," *Chem. Eng. Prog.*, **79**(6), 39 (June 1983).

4.3-20 J. Haggin, "Faster, Smaller Integrated Sensors in Offing for Process Control," *Chem. Eng. News*, **62**(23), 7 (1984).

4.3-21 G. Severns, "Planning Control Methods for Batch Processes," *Chem. Eng.*, **90**, 69 (April 18, 1983).

4.3-22 R. M. Felder, "Simulation—A Tool for Optimizing Batch-Process Production," *Chem. Eng.*, **90**(8), 79 (1983).

4.3-23 R. M. Felder, P. M. Kester, and J. M. McConney, "Simulation/Optimization of a Specialties Plant," *Chem. Eng. Prog.*, **79**(6), 84 (1983).

4.3-24 R. Govind and G. J. Powers, "Control System Synthesis Strategies," presented at the 82nd AIChE National Meeting, Atlantic City, NJ, 1976.

4.3-25 R. Govind, "Control System Synthesis Strategies," Ph.D. Dissertation, Carnegie–Mellon University, Pittsburgh, PA 1978.

4.3-26 R. Govind and G. J. Powers, "Control System Synthesis Strategies," *AIChE J.*, **28**, 60 (1982).

4.3-27 M. Morari, Y. Arkun, and G. Stephanopoulos, "Studies in the Synthesis of Control Structures for Chemical Processes. Part I: Formulation of the Problem. Process Decomposition and the Classification of the Control Tasks. Analysis of the Optimizing Control Structures," *AIChE J.*, **26**, 220 (1980).

4.3-28   M. Morari and G. Stephanopoulos, "Studies in the Synthesis of Control Structures for Chemical Processes. Part II: Structural Aspects and the Synthesis of Alternative Feasible Control Schemes," *AIChE J.*, **26**, 232 (1980).

4.3-29   M. Morari and G. Stephanopoulos, "Studies in the Synthesis of Control Structures for Chemical Processes. Part III: Optimal Selection of Secondary Measurements Within the Framework of State Estimation in the Presence of Persistent Unknown Disturbances," *AIChE J.*, **26**, 247 (1980).

4.3-30   T. Umeda, "Computer-Aided Process Synthesis," *Comput. Chem. Eng.*, **7**, 279 (1983).

4.3-31   N. Nishida, G. Stephanopoulos, and A. W. Westerberg, "A Review of Process Synthesis," *AIChE J.*, **27**, 321 (1981).

4.3-32   M. Morari, Integrated Plant Control: A Solution at Hand or a Research Topic for the Next Decade?, in *Chemical Process Control II*, T. F. Edgar and D. E. Seborg (Eds.), Proceedings of the Engineering Foundation Conference, Sea Island, GA, 1981, AIChE, New York, 1982.

4.3-33   G. Stephanopoulos, "Synthesis of Control Systems for Chemical Plants," Proceedings of the International Symposium on Process Systems Engineering, Kyoto, Japan, 1982.

4.3-34   W. R. Fisher, M. F. Doherty, and J. M. Douglas, "Synthesis of Steady-State Control Structures for Complete Chemical Plants Based on Process Economics," presented at the Annual Meeting of AIChE, paper 82b, San Francisco, CA, November 1984.

Section 4.4

4.4-1   Gordon M. Graff, "Gene-Splicing Methods Move from Lab to Plant," *Chem. Eng.*, **90**(12), 22 (1983).

4.4-2   T. J. O'Sullivan, A. C. Epstein, S. R. Korchin, and N. C. Beaton, "Applications of Ultrafiltration in Biotechnology," *Chem. Eng. Prog.*, **80**(1), 68 (1984).

4.4-3   B. Atkinson and P. Sainter, "Downstream Biological Process Engineering," *The Chem. Engineer*, 410 (November 1982).

4.4-4   P. Hawtin, "Downstream Processing in Biochemical Technology," *The Chem. Engineer*, 11–13 (January 1982).

4.4-5   S. J. Hochhauser, "Bringing Biotechnology to Market," *High Technology*, 55–60 (February 1983).

4.4-6   T. K. Ng, R. M. Busche, C. C. McDonald, and R. W. F. Hardy, "Production of Feedstock Chemicals," *Science*, **219**, 733 (1983).

4.4-7   E. L. Gaden, Jr., "Production Methods in Industrial Microbiology," *Sci. Am.*, **245**(3), 180 (1981).

4.4-8   M. Bier, O. A. Palusinski, R. A. Mosher, and D. A. Saville, "Electrophoresis: Mathematical Modeling and Computer Simulation," *Science*, **219**, 1281 (1983).

4.4-9   S. E. Charm, *The Fundamentals of Food Engineering*, 2nd ed., Avi Publishing, Westport, CT, 1971.

4.4-10   C. J. Geankoplis, *Transport Processes and Unit Operations*, 2nd ed., Allyn and Bacon, Newton, MA, 1983, pp. 503–504.

4.4-11   D. B. Arlidge, "Wiped-Film Evaporators as Pilot Plants," *Chem. Eng. Prog.*, **79**(8), 35 (1983).

PART II

INDIVIDUAL SEPARATION PROCESSES

CHAPTER 5

Distillation

JAMES R. FAIR
Department of Chemical Engineering
The University of Texas
Austin, Texas

5.1 INTRODUCTION

5.1-1 Definitions

Distillation is a method of separation based on the difference in composition between a liquid mixture and the vapor formed from it. The composition difference is due to differing effective vapor pressures, or *volatilities*, of the components of the liquid. When such a difference does not exist, as at an azeotropic point, separation by distillation is not possible. The most elementary form of the method is *simple distillation* in which the liquid mixture is brought to boiling and the vapor formed is separated and condensed to form a product; if the process is continuous, it is called *flash distillation* or an *equilibrium flash*, and if the feed mixture is available as an isolated batch of material, the process is a form of *batch distillation* and the compositions of the collected vapor and residual liquid are thus time dependent.

The term *fractional distillation* (which may be contracted to "fractionation") originally was applied to the collection of separate fractions of condensed vapor, each fraction being segregated. Currently, the term is applied to distillation separations in general, where an effort is made to separate an original mixture into several components by means of distillation. When the vapors are enriched by contact with counterflowing liquid *reflux*, the process often is called *rectification*. When operated with a continuous feed of liquid mixture and continuous removal of product fractions, the process is *continuous distillation*. When steam is added to the vapors to reduce the partial pressures of the components to be separated, the term *steam distillation* is used; if such a process is altered to eliminate the steam, *dry distillation* ("conventional distillation") results.

Most distillations conducted commercially operate continuously, with a more volatile fraction recovered as *distillate* and a less volatile fraction remaining as *residue* or *bottoms*. If a portion of the distillate is condensed and returned to the process to enrich the vapors, the liquid is called *reflux*. The apparatus in which the enrichment occurs is usually a vertical, cylindrical vessel called a *still* or *distillation column*. This apparatus normally contains internal devices for effecting vapor–liquid contact; the devices may be *trays* or *packings*.

As stated above, a separation by distillation involves differing volatilities of the components to be separated. If the volatility difference is so small that an adequate separation cannot be made, it may be possible to increase the volatility difference by the addition of an extraneous material that can be separated later; this process is known as *extractive distillation* if the added material is relatively nonvolatile and leaves the column with the residue. If the added material forms an azeotrope with one or more of the components of the mixture and in so doing enhances the separability of the original mixture, *azeotropic distillation* results. In this last-named mode, the extraneous material, or azeotropic agent, may leave the column in the distillate (low-boiling azeotrope) or in the residue (high-boiling azeotrope).

It is clear that the difference in volatility of the various components of a liquid mixture is a key to the successful application of distillation. This difference can be related to the thermodynamic equilibrium that can exist between the liquid and vapor mixtures under conditions that can be associated with the distillation at hand. The *phase equilibrium* relationships are embodied in the general area of solution thermodynamics and can be measured or, in some cases, predicted from the properties of the pure materials involved. The resulting equilibrium compositions often are referred to as *vapor–liquid equilibrium data*, shortened to *vapor–liquid equilibria* and abbreviated simply as VLE. There are occasional instances when a second immiscible liquid phase is involved, with compositions of the three phases at thermodynamic equilibrium known simply as *vapor–liquid–liquid equilibria*, or VLLE.

5.1-2   Areas of Application

Distillation is the most widely used separation method to be found in the chemical and petroleum processing industries. It is normally the least expensive of possible methods for separating a given mixture and in many cases is the only feasible method. It has the disadvantage of requiring energy in the form of heat to produce the necessary vaporization, and this can represent a significant cost. It also subjects the mixture components to the temperatures of vaporization, and this can be detrimental to heat-sensitive materials (although operation under vacuum, to reduce temperature levels, is quite common).

By definition, distillation involves the liquid phase and if it is to be applied to normally gaseous mixtures, temperatures must be lowered by the use of refrigeration. The common example of distilling a normally gaseous mixture is the cryogenic fractional distillation of air to produce high-purity streams of oxygen, nitrogen, and argon.

Distillation is applied at small scales on a batch or continuous basis to separate and purify specialty chemicals. At the other extreme it is the basic method for separating commodity chemicals on a tonnage basis, a familiar example being the distillation of crude petroleum to obtain a number of hydrocarbon fractions.

The reader should recognize that many of the principles involved in distillation separations apply also to absorption and stripping separations. Indeed, these unit operations all involve the contacting of liquids and gases (vapors) and utilize similar equipment when operated at the commercial scale. The reader may wish to consult Chapter 6 for detailed information on absorption and stripping.

5.1-3   General Approach to Solving Distillation Problems

Problems associated with separations of any type may be approached from *design* or *rating* points of view. In the design approach, a separation is specified and the necessary equipment is designed such that the specification can be met. In the rating approach, the equipment is specified (it may be existing equipment) and the degree of separation is rated. For any distillation problem, whether rating or design, there is a sequence of steps that must be taken, and these are described below.

1. *Define the Mixture.* While this first step might seem obvious, it often is not considered properly. For precise separations, *all* components of the mixture must be identified. Furthermore, possible future additions to the mixture must be considered when new equipment is being designed. The simple textbook two-component mixtures are not often encountered commercially, and unanticipated components can play havoc with many distillations. In some instances it may not be possible to identify individual components of the mixture, as in the handling of ''boiling-range'' materials; in such cases it is necessary to identify pseudocomponents that can serve the computational purposes of the design or rating procedures.

2. *Establish Separation Criteria.* These criteria include product purities as well as component recoveries. For example, a light hydrocarbon distillation column might be designed to produce an overhead product containing a minimum ethylene mole fraction of 99.5 mol %, and at the same time to have the requirement of recovering at least 95% of the ethylene that enters with the feed. To use this example further, the separation criteria to be established should recognize whether future markets might require, say, 99.7 mol % purity and whether some safety factor should be added to these purity requirements in carrying out the design of the distillation system.

3. *Obtain Property Data.* These can include transport, physical, and thermochemical data as needed for computations. Importantly, they include the necessary vapor–liquid equilibrium (VLE) data, measured or predicted for the ranges of composition, temperature, and pressure to be encountered in the computations. As will be noted later, the reliability of the VLE can seriously influence many distillation designs.

4. *Select a Model for Computing Stages or Transfer Units.* The difficulty of making a given separation is computed in terms of *equilibrium stages*, if the process is to be carried out on a stagewise basis, or *transfer units*, if the process is to be conducted as a differential vapor–liquid contacting operation. Distillation columns with plates or trays handle the stagewise operations, and columns with packings or other special devices handle the differential operations. The models in question can range from empirical and nonfundamental approaches to the rigorous stagewise methods that require computers to handle situations involving many mixture components and many theoretical contacts. In some cases, an approximate

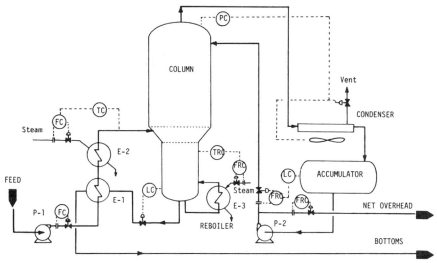

FIGURE 5.1-1 Flow diagram of typical continuous distillation system.

model may suffice in the earlier stages of design or rating, to be followed by a more exact model for the final analysis.

5. *Calculate Required Stages or Transfer Units.* After the model is selected, the number of theoretical stages or transfer units is computed. This is an index of the difficulty of the separation and is dependent on the amount of reflux that is used. It is in this step that the familiar stages/reflux relationship is developed, with the final combination of these two parameters dependent on economics.

6. *Size the Distillation Column.* This includes the hydraulic analysis to establish operating ranges, pressure drop, and mass transfer efficiency. The result is a set of dimensions, including column diameter and height, number of actual trays (or height of packed bed), details of internal devices, and profiles of temperature and pressure.

7. *Complete the System Design.* A typical distillation system is shown in Fig. 5.1-1. It is clear that the column is only a part of the system, which also includes heat exchangers, vessels, instruments, and piping that must be dealt with by the chemical process engineer. However, in the present chapter, the column and its characteristics are the principal thrust; the other components of the system represent individual technologies (such as process control or process heat transfer) that are best covered separately.

This sequence of steps has been discussed from the design approach, that is, with a needed separation leading to a final equipment specification. Close scrutiny will show that the sequence applies also to rating situations. For example, if an existing column is to be used for a new service, steps 1–5 will not change. In step 6, the accommodation of the existing column for the needed stages and vapor/liquid flows will be tested by trial-and-error until either a fit is made or modifications of the existing equipment will be defined.

5.1-4 Contents of this Chapter

This chapter covers all important aspects of distillation column rating or design. It follows generally the sequence of the steps noted immediately above. The common case for discussion is conventional distillation operated in the continuous mode. Exceptions to this mode, such as extractive distillation or batch distillation, will be handled separately. Much of the material in Sections 5.7–5.10, dealing with hydraulics and mass transfer in columns, can also be used in the design and analysis of absorption columns. Perusal of Chapter 6 will show the reader how the connection may be made.

5.2 PHASE EQUILIBRIUM

5.2-1 Thermodynamic Relationships

At thermodynamic equilibrium, the vapor and liquid phases (Fig. 5.2-1) show a distribution of mixture components between the phases that is determined by the *relative volatility* between pairs of components. This separation factor has its equivalent term in other methods of separation that are based on the equilibrium

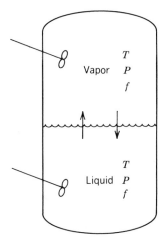

FIGURE 5.2-1 Equilibrium between vapor and liquid. The conditions for equilibrium are $T^V = T^L$ and $P^V = P^L$. For a given T and P, $f^V = f^L$ and $f_i^V = f_i^L$.

concept. For components i and j their relative volatility (often called the "alpha value") is defined as

$$\alpha_{ij} = \frac{K_i}{K_j} \tag{5.2-1}$$

Each component of the mixture has a distinct *vapor–liquid equilibrium ratio* or "*K* value":

$$K_i \equiv \frac{y_i^*}{x_i} \tag{5.2-2}$$

where y_i^* is a vapor mole fraction in equilibrium with liquid mole fraction x_i.

Thermodynamic relationships described in Section 1.5-1 of Chapter 1 may be used to develop the following general form for the K value, which uses Raoult's law as the basis of the liquid reference state:

$$K_i = \frac{\gamma_i \phi_i^{\text{sat}} p_i^{\text{sat}} \exp\left(\dfrac{1}{RT} \displaystyle\int_{p_i^{\text{sat}}}^{P} v_i^L \, dP\right)}{\hat{\phi}_i P} \tag{5.2-3}$$

where γ_i is the activity coefficient describing the deviation of the liquid phase from ideal solution behavior, $\hat{\phi}_i$ is the fugacity coefficient describing the deviation of the vapor mixture from ideal behavior, and ϕ_i^{sat} is the fugacity coefficient of pure vapor i at the system temperature and a pressure equal to the vapor pressure of i at that temperature. As described in Chapter 1, both γ_i and $\hat{\phi}_i$ can be determined from equations of state in some instances, but it is more common to evaluate $\hat{\phi}_i$ from an equation of state and γ_i from a model of the liquid solution. The exponential term in Eq. (5.2-3) is known as the Poynting correction.

Perhaps the most important term in Eq. (5.2-3) is the liquid-phase activity coefficient, and methods for its prediction have been developed in many forms and by many workers. For binary systems the Van Laar [Eq. (1.4-18)], Wilson [Eq. (1.4-23)], NRTL [Eq. (1.4-27)], and UNIQUAC [Eq. (1.4-36)] relationships are useful for predicting liquid-phase nonidealities, but they require some experimental data. When no data are available, and an approximate nonideality correction will suffice, the UNIFAC approach [Eq. (1.4-31)], which utilizes functional group contributions, may be used. For special cases involving regular solutions (no excess entropy of mixing), the Scatchard–Hildebrand method[1] provides liquid-phase activity coefficients based on easily obtained pure-component properties.

The vapor phase at low pressure normally is considered to be an ideal gas and under such circumstances the fugacity coefficient $\hat{\phi}_i$ is unity. For more careful determinations of the K value, however, values of $\hat{\phi}_i$ may be estimated from the following relationship:

$$\ln \hat{\phi}_i = \frac{1}{RT} \int_0^P \left(\bar{v}_i - \frac{RT}{P}\right) dP \tag{5.2-4}$$

For an ideal vapor solution, Eq. (5.2-4) can be simplified to the expression

$$\ln \hat{\phi}_i = \frac{1}{RT} \int_0^P \left(v_i - \frac{RT}{P} \right) dP \tag{5.2-5}$$

Equation (5.2-3) provides a rigorous thermodynamic basis for the prediction of the vapor–liquid equilibrium ratio. Sometimes it can be simplified, as the following special cases demonstrate:

Liquid Incompressible. If this is the case, the Poynting correction (PC) becomes

$$PC = \exp \left(\frac{v_i^L (P - p_i^{sat})}{RT} \right) \tag{5.2-6}$$

Poynting Correction Negligble. This is usually the case at pressures less than 20 atm and temperatures greater than 273 K. The resulting equation is

$$K_i = \frac{\gamma_i \phi_i^{sat} p_i^{sat}}{\hat{\phi}_i P} \tag{5.2-7}$$

Vapor Solution Ideal. For this case $\hat{\phi}_i = \phi_i$, which is usually the case at pressures less than 20 atm and temperatures greater than 273 K. Equation (5.2-3) is modified further then to give

$$K_i = \frac{\gamma_i \phi_i^{sat} p_i^{sat}}{\phi_i P} \tag{5.2-8}$$

Vapor Obeys the Ideal Gas Law. This is a familiar and frequent situation and one that can be checked easily through estimation of the compressibility factor. At pressures below about 2 atm and temperatures above 273 K, it is very probable that the ideal gas law is obeyed and Eq. (5.2-3) further reduces to

$$K_i = \frac{\gamma_i p_i^{sat}}{P} \tag{5.2-9}$$

Since Raoult's law is $p_i = x_i p_i^{sat}$, the liquid-phase activity coefficient in Eq. (5.2-9) is a "Raoult's law correction factor" that takes into account liquid-phase nonideality. Since most distillations are carried out at relatively low pressure and moderate-to-high temperature, Eq. (5.2-9) is the most generally used relationship in distillation system analysis and design.

Liquid Solution Ideal. This is the ultimate reduction in the K value equation and is represented by Raoult's law. it is likely to apply when very similar molecules form the binary or multicomponent solution. Accordingly, Eq. (5.2-3) then reduces to the simplest possible K value relationship:

$$K_i = \frac{p_i^{sat}}{P} \tag{5.2-10}$$

or the simple ratio of vapor pressure to total pressure. Vapor pressures for many substances have been tabulated and usually are correlated well by the Antoine equation:

$$\ln p_i^{sat} = A_i - \frac{B_i}{C_i + T} \tag{5.2-11}$$

Values of the Antoine constants for various materials may be found in many reference works; a particularly useful listing of constants may be found in the book by Reid, Prausnitz, and Sherwood.[2]
As pointed out in Chapter 1, several of the correction factors for Eq. (5.2-3) may be obtained from equations of state, when there is a significant pressure effect on the equilibrium mixture.

5.2-2 Binary Systems

For a system containing only two components (a *binary system*), relative volatility is

$$\alpha_{12} = \frac{K_1}{K_2} = \frac{y_1^*(1 - x_1)}{x_1(1 - y_1^*)} \tag{5.2-12}$$

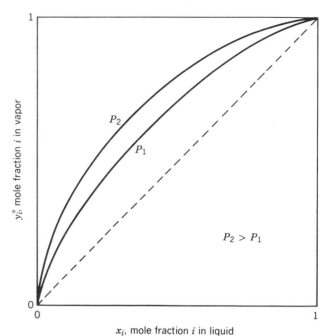

FIGURE 5.2-2   Typical y^*-x equilibrium diagram for a binary system.

By convention, component 1 is the more volatile. Equation (5.2-12) can be rearranged to

$$y_1^* = \frac{\alpha_{12}x_1}{1 + (\alpha_{12} - 1)x_1} \tag{5.2-13}$$

which may be plotted as the familiar y^*-x curve shown in Fig. 5.2-2. The value of α_{12} might not be constant across the entire range, and thus Eq. (5.2-12) strictly applies to a given value of the liquid mole fraction x_1, at which composition the slope is

$$m_{12} = \frac{dy_1^*}{dx_1} = \frac{\alpha_{12}}{[1 + (\alpha_{12} - 1)x_1]^2} \tag{5.2-14}$$

This slope is used primarily in mass transfer calculations, to be discussed in Sections 5.9 and 5.10.

Alternative equilibrium diagrams are temperature–composition (T-x) and pressure–composition (P-x) as shown in Figs. 5.2-3 and 5.2-4. While not as useful in distillation as the y^*-x diagram, they present clearly the concepts of the bubble point, dew point, and composition changes during simple vaporization and condensation. For example, a liquid of composition A may be heated at constánt pressure to point B, the *bubble point*, at which the initial vapor is formed (Fig. 5.2-3). This vapor has the composition C, significantly richer in the light component than is the initial liquid. As vaporization proceeds, the vapor composition moves from C to D and the liquid composition from B to E. When the last remaining liquid disappears, the *dew point* has been reached and with continued heating the mixture moves to point F, in the superheated vapor region. The process could be reversed, starting with vapor at point F and finally ending with subcooled liquid at point A.

For an isothermal process in which pressure is changed, the subcooled liquid at point A (Fig. 5.2-4) may be moved to its bubble point B by lowering the pressure. With continued pressure lowering, the vapor composition moves from C to D and the liquid composition from B to E; point D is the dew point. Further lowering of pressure produces the superheated vapor at point F.

Figure 5.2-5 shows a variety of T-x and y^*-x diagrams that represent different types of binary systems. It should be clear that the simplified diagrams in Figs. 5.2-3 and 5.2-4 do not always prevail, and yet they represent a fairly large body of binary equilibrium compositions. The important point is that the system at hand must be characterized carefully with respect to vapor–liquid equilibria. The application of nonideal correlating relationships such as the Van Laar equation is, however, much simpler for binary systems than for multicomponent systems.

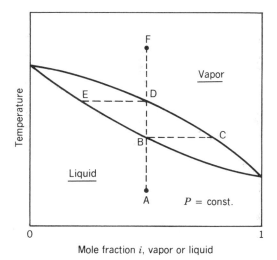

FIGURE 5.2-3   Temperature–composition equilibrium diagram for a binary system.

5.2-3 Multicomponent Systems

Equilibrium data for multicomponent systems may be measured, but because of the ranges of temperature and pressure likely to be encountered in the distillation, some sort of model must still be available for extending such experimental data. The best approach is to start with the binary data for all possible binary pairs—a chore not too difficult for, say, ternary or quarternary systems, but one that is quite challenging for mixtures with more than four components. A six-component mixture, for example, would require evaluation of 15 binary pairs.

The various correlating equations (Wilson, Van Laar, etc.) have multicomponent forms, but the most versatile method for combining binary pair data is that of Chien and Null.[3] This method permits use of the best-fit correlating method for each binary pair followed by combination of the binary data to give multicomponent equilibrium distributions.

5.2-4 Sources of Equilibrium Data

A very large amount of effort has gone into the measurement and correlation of vapor–liquid equilibria, and the published effort has been referenced by Hala and coworkers in several bibliographic treatises.[4-6]

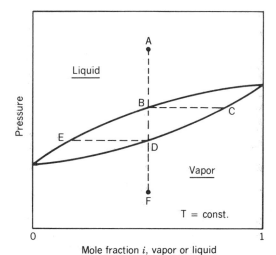

FIGURE 5.2-4   Pressure–composition equilibrium diagram for a binary system.

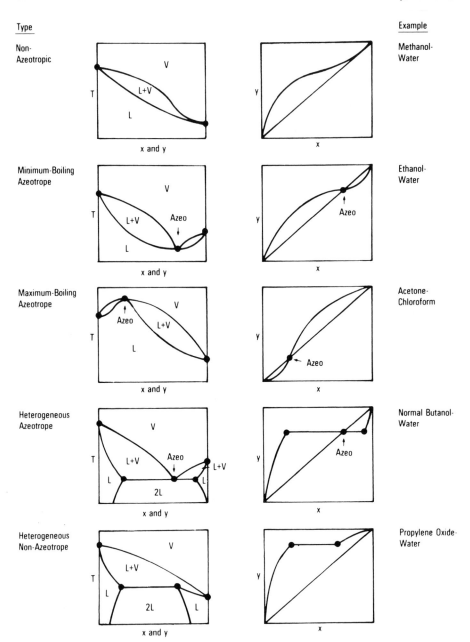

FIGURE 5.2-5 Common types of equilibrium diagrams for binary systems.

These works direct the reader to the source publications but do not in general report scientific data. An older work that does report data is by Chu et al.;[7] newer and more reliable data compilations have been published by Hirata et al.[8] and by Gmehling et al.[9] The latter is a continuing series and includes liquid–liquid equilibria; an example page of data is shown in Fig. 5.2-6. Both Refs. 8 and 9 include parameters for the more popular models for correlating and predicting liquid-phase activity coefficients. It is important to note that the data of Ref. 9 are available for computer retrieval.[10]

For hydrocarbon systems, many approximate K value data have been presented in the forms of graphs and nomograms. An oft-cited nomogram is that of Hadden and Grayson,[11] which includes also hydrogen,

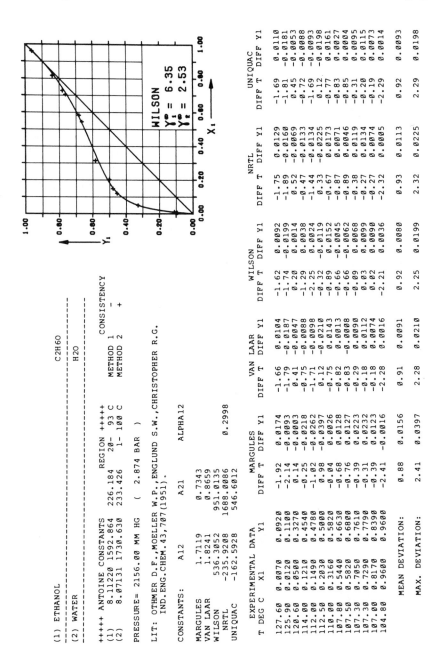

WILSON
$Y_1^\infty = 6.35$
$Y_2^\infty = 2.53$

(1) ETHANOL   C2H6O
(2) WATER   H2O

```
++++ ANTOINE CONSTANTS              REGION ++++    CONSISTENCY
(1)  8.11220 1592.864  226.184   20-  93 C   METHOD 1   -
(2)  8.07131 1730.630  233.426    1- 100 C   METHOD 2   +
```

PRESSURE= 2156.00 MM HG (2.874 BAR)

LIT: OTHMER D.F.,MOELLER W.P.,ENGLUND S.W.,CHRISTOPHER R.G.
     IND.ENG.CHEM.43,707(1951).

CONSTANTS: A12   A21   ALPHA12

	A12	A21	ALPHA12
MARGULES	1.7119	0.7343	
VAN LAAR	1.8241	0.8659	
WILSON	536.3052	951.0135	
NRTL	-235.9208	1688.0086	0.2998
UNIQUAC	-162.5928	546.6012	

EXPERIMENTAL DATA			MARGULES		VAN LAAR		WILSON		NRTL		UNIQUAC	
T DEG C	X1	Y1	DIFF T	DIFF Y1	DIFF T	DIFF Y1	DIFF T	DIFF Y1	DIFF T	DIFF Y1	DIFF T	DIFF Y1
127.60	0.0070	0.0920	-1.92	0.0174	-1.66	0.0104	-1.62	0.0092	-1.75	0.0129	-1.69	0.0110
125.90	0.0120	0.1100	-2.14	-0.0093	-1.79	-0.0187	-1.74	-0.0199	-1.89	-0.0160	-1.81	-0.0181
120.60	0.0500	0.3270	0.14	-0.0003	0.41	-0.0047	0.20	0.0014	0.52	-0.0069	0.45	-0.0053
114.00	0.1210	0.4540	-0.25	-0.0218	-0.75	-0.0088	-1.29	0.0038	-0.47	-0.0133	-0.72	-0.0088
112.00	0.1490	0.4780	-1.02	-0.0262	-1.71	-0.0098	-2.25	0.0024	-1.44	-0.0134	-1.69	-0.0093
112.50	0.2030	0.5000	0.98	-0.0397	0.12	-0.0210	-0.32	-0.0119	-0.33	-0.0225	-0.12	-0.0198
110.00	0.3160	0.5820	-0.04	0.0026	-0.75	0.0143	-0.89	0.0152	-0.67	0.0173	-0.77	0.0161
107.80	0.5440	0.6630	-0.68	-0.0128	-0.82	0.0013	-0.66	-0.0045	-0.87	-0.0071	-0.83	0.0027
107.50	0.5820	0.6800	-0.76	0.0127	-0.83	-0.0008	-0.66	-0.0062	-0.89	-0.0046	-0.85	-0.0004
107.50	0.7050	0.7610	-0.39	0.0223	-0.29	0.0090	-0.09	-0.0068	-0.38	0.0119	-0.31	0.0095
107.30	0.7290	0.7790	-0.31	-0.0232	-0.18	0.0112	0.03	0.0099	-0.27	0.0134	-0.20	0.0115
107.00	0.8170	0.8390	-0.39	0.0123	-0.18	0.0074	0.02	0.0090	-0.27	-0.0074	-0.19	0.0073
104.80	0.9600	0.9600	-2.41	-0.0016	-2.28	0.0016	-2.21	0.0036	-2.32	-0.0005	-2.29	0.0014
MEAN DEVIATION:			0.88	0.0156	0.91	0.0091	0.92	0.0080	0.93	0.0113	0.92	0.0093
MAX. DEVIATION:			2.41	0.0397	2.28	0.0210	2.25	0.0199	2.32	0.0225	2.29	0.0198

FIGURE 5.2-6 Vapor-liquid equilibrium data for ethanol–water at 1 atm. Example page from Ref. 9.

hydrogen sulfide, carbon dioxide, and water under special conditions. Perhaps the most reliable graphical presentation of hydrocarbon K values is that of the Natural Gas Processors Association.[12]

A listing of azeotropic data has been published by Horsley.[13] General source data are included in the volume by Reid et al.[2] More information on sources of equilibrium data will be found in Chapter 1. In general, a thorough search of the literature should be made before investing in laboratory measurements, but the searcher should beware of faulty experimental data that are not thermodynamically consistent.

5.3 EQUILIBRIUM STAGES

5.3-1 Basic Relationships

Separation calculations for distillation are based on the concept of a process that involves contacting of vapor and liquid in one or more equilibrium stages. In this process one or more feed streams enter a stage, and one or more streams leave the stage. Energy may be added to or withdrawn from the stage. Importantly, thermodynamic equilibrium is required to exist on the stage, and this concept has been discussed in the previous section (e.g., Fig. 5.2-1).

Mass flows to and from a generalized vapor–liquid contacting stage are shown in Fig. 5.3-1. The indication of "interstage flows" is in the context of a cascade of stages as in a distillation column, but the figure applies to other contexts as well. Figure 5.3-2 shows the equivalent energy flows to and from the stage. A *material balance* across the stage is

$$F_n z_{i,n} + L_{n+1} x_{i,n+1} + V_{n-1} y_{i,n-1} = (V_n + V'_n)y_{i,n} + (L_n + L'_n)x_{i,n} \qquad (5.3\text{-}1)$$

There must be *equilibrium* on the stage:

$$K_{i,n} \equiv \frac{y_{i,n}}{x_{i,n}} \qquad (5.3\text{-}2)$$

where $K_{i,n}$ is the vapor–liquid equilibrium ratio for component i. In general,

$$K_i = f(T, P, x_i, x_j, \ldots .) \qquad (5.3\text{-}3)$$

Also, the *summation of mole fractions* must equal unity:

$$\sum_i x_{i,n} = \sum_i y_{i,n} = \sum_i z_{i,n} = 1.0 \qquad (5.3\text{-}4)$$

Finally, an *enthalpy balance* must prevail on the stage:

$$Q_n + F_n H_{f,n} + L_{n+1} h_{n+1} + V_{n-1} H_{n-1} = (V_n + V'_n)H_n + (L_n + L'_n)h_n \qquad (5.3\text{-}5)$$

where, in general,

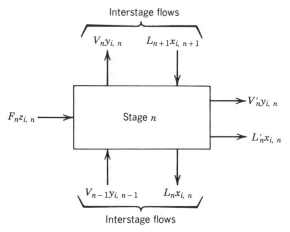

FIGURE 5.3-1 Component mass flows on a contacting stage.

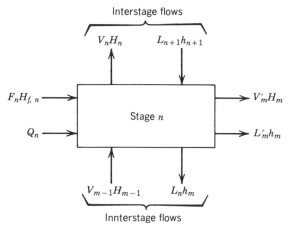

FIGURE 5.3-2 Stream energy flows on a contacting stage.

$$H = f(T, P, y_i, y_j, \ldots) = \text{vapor enthalpy} \tag{5.3-6}$$

$$h = f(T, P, x_i, x_j, \ldots) = \text{liquid enthalpy} \tag{5.3-7}$$

These four relationships, the "M-E-S-H equations," are the critical ones for equilibrium stage calculations in distillation, and they are sufficient if the pressure effect in Eqs. (5.3-3), (5.3-6), and (5.3-7) can be determined deductively. Otherwise, momentum balance equations must be added.

5.3-2 Equilibrium Flash

The stage shown in Figs. 5.3-1 and 5.3-2 may be simplified for the case of a single feed and two products, one vapor and one liquid, as shown in Fig. 5.3-3. When operated on a steady-state continuous basis, this arrangement is a simple takeover distillation, or equilibrium flash separation. If there is a net increase in the total number of moles in the liquid, the arrangement can be called equilibrium condensation. The MESH equations can be used in a straightforward way to obtain material and energy balances. The separation may be isothermal, with heat added or removed, or it may be adiabatic. It often operates with a liquid feed and a reduction of pressure in the contacting zone, hence the common use of the term *equilibrium flash*.

For the general case of a multicomponent system, the total number of moles of vapor produced by the flash is

$$V_n = \sum_i (V_n y_{i,n}) = \sum_i \left(\frac{F_n z_{i,n}}{L_n/V_n K_{i,n} + 1} \right) \tag{5.3-8}$$

Alternatively, the total number of moles of liquid produced is

$$L_n = \sum_i (L_n x_{i,n}) = \sum_i \left(\frac{F_n z_{i,n}}{1 + V_n K_{i,n}/L_n} \right) \tag{5.3-9}$$

These equations involve iterative solutions, usually based on the initial assumption of the L/V or V/L ratios. It should be noted that the terms $L_n/V_n K_n$ and $V_n K_n/L_n$ are used elsewhere in this book as the absorption factor A_i and the stripping factor S_i, respectively. While Eqs. (5.3-8) and (5.3-9) are basic to the equilibrium flash process, one should not overlook the implied use of an energy balance to obtain the temperature and pressure conditions of the flash.

As noted, both vapor and liquid products are withdrawn from the flash chamber of Fig. 5.3-3. This means that the flash conditions must lie between the dew point and the bubble point of the feed mixture and that the conditions of the Gibbs phase rule must be met.

5.3-3 Multistage Distillation

For a multiple-stage cascade, and for a feed that enters in the midregion of the cascade, the representation of a multistage distillation column emerges, as shown in Fig. 5.3-4. In this column, vapor stream V_1 is

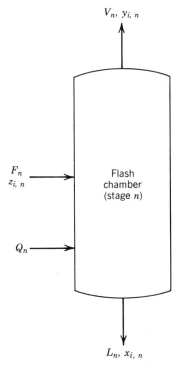

$$V_n, y_{i,n}$$

$$F_n$$
$$z_{i,n}$$

Flash
chamber
(stage n)

$$Q_n$$

$$L_n, x_{i,n}$$

FIGURE 5.3-3 Single-stage equilibrium flash.

generated at the base by a boiling process and this vapor passes up through the column, its composition and mass changing as the distillation process occurs. The vapor is completely or partially condensed at the top of the column at stage N. All or part of the condensed vapor is returned to the column as *reflux* stream L_N, and this sets up a counterflow of vapor and liquid, to meet the context of the "interstage flows" in Figs. 5.3-1 and 5.3-2. The requirement of equilibrium on each stage establishes a composition profile for the column with the result of material transport between the phases, and the model assumes that the transport is so rapid that there is no departure from equilibrium at any point in the column.

5.3-4 Binary Distillation

ANALYTICAL METHOD
The simplest case of multistage distillation involves only two components. Figure 5.3-5 is an adaptation of Figure 5.3-4 for the binary case, and it shows two envelopes that are amenable to energy and material balancing. For the top envelope a mass balance on the lighter (more volatile) component is

$$y_n = \frac{L_{n+1}}{V_n} x_{n+1} + \frac{D}{V_n} x_D \tag{5.3-10}$$

This equation provides a relationship between the composition of the vapor leaving a stage and the composition of the liquid entering the stage from above. The compositions of the vapor and liquid leaving the tray are related by the equilibrium requirement

$$y^* = \frac{\alpha x}{1 + (\alpha - 1)x} \tag{5.3-11}$$

or

$$x^* = \frac{y}{\alpha - (\alpha - 1)y} \tag{5.3-12}$$

For the upper envelope, an energy balance is

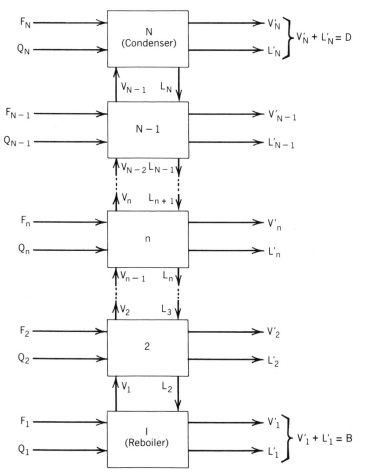

FIGURE 5.3-4 Cascaded stages in a distillation column.

$$V_nH_n = L_{n+1}h_{n+1} + Dh_D + Q_d \qquad (5.3\text{-}13)$$

Thus, with Eqs. (5.3-10)–(5.3-13), plus the requirement that on each stage the liquid and vapor mole fractions must each equal unity, the number of theoretical stages for a given separation may be determined analytically. One may start from the top condition and work down the column, taking into account the addition or removal of material or energy at any of the stages, or one may start from the bottom and work up toward the top condition.

As an example of the procedure, consider the case of starting at the top of the column. Streams D and L_{N+1} and their compositions are determined by the separation specification and the assigned reflux ratio. This then provides the flow and composition of the vapor leaving the top stage N. Equation (5.3-12) gives the composition of the liquid on the top tray, x_N. The temperature of the top tray may be obtained from a dew point calculation on the vapor composition y_N.

For tray $N - 1$, Eq. (5.3-10) gives

$$y_{N-1} = \frac{L_N}{V_{N-1}}x_N + \frac{D}{V_{N-1}}x \qquad (5.3\text{-}14)$$

Recognizing that $V_{N-1} = D + L_N$, the value of y_{N-1} is obtained. In turn, a dew point calculation provides the value of x_{N-1}. An energy balance confirms the dew point temperature. And thus, calculations proceed down the column.

Calculations from the bottom toward the top are analogous, with a mass balance across the bottom envelope (Fig. 5.3-5) providing the material balance equation,

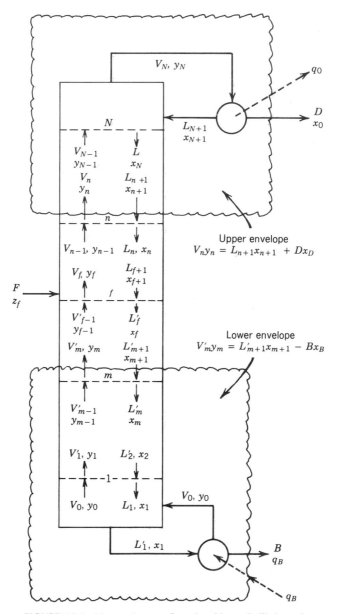

FIGURE 5.3-5 Mass and energy flows in a binary distillation column.

$$y_m = \frac{L'_{m+1}}{V'_m} x_{m+1} - \frac{B}{V'_m} x_B \qquad (5.3\text{-}15)$$

and Eq. (5.3-11) providing the equilibrium vapor composition. The terms B and x_B are obtained from the separation specification, and the rate V_0 also is specified in connection with the reflux ratio. The composition y_0 is obtained from a bubble point calculation on x_B, which also gives the temperature of V_0. The material balance $L_1 = V_0 + B$ thus gives the flow and composition of the liquid from stage 1. A bubble point on this composition yields the temperature of stage 1 and its vapor composition. Material balance $L_2 = V_1 + B$ must be checked by heat balance to arrive at specific values of L_2 and V_1. In like fashion, calculations are continued up the column.

These stagewise calculations are easily programmed for computer solution. If the calculations are made by hand, it may not be necessary to run an energy balance at every stage, unless there are large differences between the molal latent heats of vaporization of the mixture components. At any rate, if the equilibrium and enthalpy data are at hand, the computations proceed quite rapidly, even if done by hand.

McCABE–THIELE GRAPHICAL METHOD

This method is based on an interpretation of Fig. 5.3-5 and Eq. (5.3-10) (upper envelope) and (5.3-15) (lower envelope). It was first published in 1925[1] and is described in some detail in most unit operations textbooks (e.g., see Ref. 4). It is by far the best known and most used graphical method for determining theoretical stage requirements for binary systems.

In Eq. (5.3-10), if the vapor molar flow rate V_n and the liquid molar flow rate L_{n+1} are constant throughout the upper envelope, then the equation represents a straight line with slope L_{n+1}/V_n when y_n is plotted against x_{n+1}. The line is known as the *upper operating line* and is the locus of points coupling the vapor and liquid compositions of streams passing each other (see Fig. 5.3-5).

In a like fashion, a mass balance for the lower envelope of Fig. 5.3-5 produces Eq. (5.3-15), and if the vapor and liquid molar flow rates are constant, the equation is a straight line on a plot of y_m versus x_{m+1} and is the *lower operating line*. At the feed stage, reference to Fig. 5.3-1 shows a mass balance as follows:

$$F + V_{f-1} + L_{f+1} = V_f + L_f \qquad (5.3-16)$$

If the upper and lower envelopes are extended to the feed stage, then there is an intersection of the operating lines:

$$y_n = y_m = y_f \qquad (5.3-17)$$

$$x_{n+1} = x_{m+1} = x_f \qquad (5.3-18)$$

The thermal condition of the feed stream introduces the need for an energy balance at the feed stage. The feed may be a subcooled liquid, a saturated liquid, a mixture of liquid and vapor, a saturated vapor, or a superheated vapor. A special term may be defined that can account for the thermal condition of the feed:

$$q = \frac{\text{heat required to convert 1 mole of feed to saturated vapor}}{\text{molar latent heat of the feed}} \qquad (5.3-19)$$

(for feed at its boiling point, $q = 1$; for a saturated vapor feed, $q = 0$).

With this q term, an energy balance across the feed stage yields

$$qF + L_{f+1} = L_f \qquad (5.3-20)$$

$$V_f = V_{f-1} + (1 - q)F \qquad (5.3-21)$$

Combining these equations with the upper and lower operating line equations and the material balance across the feed stage gives

$$y = \frac{q}{q-1}x - \frac{x}{q-1} \qquad (5.3-22)$$

This "q line equation" also represents a straight line on a y-x plot, with slope $= q/(q-1)$.

The operating lines, the q line, and the equilibrium curve [see Eq. (5.2-13)] are all y-x functions where, by convention, the mole fractions refer to the lighter, or more volatile, component of the binary pair. It was on this basis that McCabe and Thiele[1] developed their graphical approach to binary distillation stage determination and their approach is summarized in Fig. 5.3-6.

In the figure, the upper and lower operating lines are shown intersecting at the feed point with each other and with the q line. The location of the lines depends on their slopes and on the compositions of the distillate (overhead) and residue (bottoms) products, x_D and x_B. These compositions usually are specified on the basis of process needs for purity and recovery. The slope of the upper operating line is determined by the *reflux ratio* to be used. This ratio is

$$\text{Reflux ratio} = R = \frac{\text{moles of liquid returned to the column}}{\text{moles of product distillate}} = \frac{L_{N+1}}{D} \qquad (5.3-23)$$

Thus, the slope of the upper operating line is

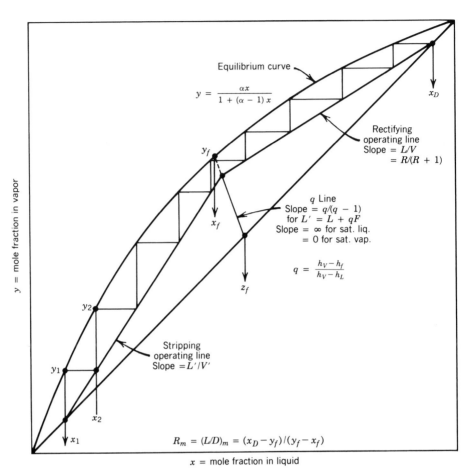

Equilibrium curve

$$y = \frac{\alpha x}{1 + (\alpha - 1)\, x}$$

x_D

Rectifying
operating line
Slope = L/V
     = $R/(R + 1)$

y_f

q Line
Slope = $q/(q - 1)$
for $L' = L + qF$
Slope = ∞ for sat. liq.
     = 0 for sat. vap.

x_f

$$q = \frac{h_V - h_f}{h_V - h_L}$$

z_f

y_2

Stripping
operating line
Slope = L'/V'

y_1

x_2

x_1 $R_m = (L/D)_m = (x_D - y_f)/(y_f - x_f)$

x = mole fraction in liquid

y = mole fraction in vapor

FIGURE 5.3-6   Summary of McCabe–Thiele graphical method for binary distillation.

$$\frac{L}{V} = \frac{R}{R + 1} = \frac{L_{n+1}}{V_n} \tag{5.3-24}$$

and is always less than 1. The slope of the lower operating line is L'_{m+1}/V'_m and is always greater than 1. In Fig. 5.3-6, which represents an arbitrary example, the theoretical stages are stepped off according to the equations of the equilibrium line and operating lines, and a total of nine theoretical stages are shown. It should be clear that a different number of stages would result if the slopes of the operating lines were changed, that is, if the reflux ratio was varied. This leads to the concept of the stages versus reflux curves, shown in Fig. 5.3-7. Each curve is the locus of points for a given separation. At their extremes, the curves become asymptotic to *minimum stages* and *minimum reflux* values.

Minimum stages are stepped off when the operating lines coincide with the diagonal line and have a slope of unity; from Eq. (5.3-24) it can be seen that at this condition of minimum stages there can be no net product distillate takeoff. This is the condition of *total reflux* and is an important condition for starting up columns and for conducting efficiency tests on columns.

Again, with reference to Fig. 5.3-6, as the point of intersection of the operating lines is moved closer to the equilibrium curve (reflux ratio being decreased), more and more stages would be stepped off, and when the lines and curve intersect, a "pinch" results with an infinite number of stages being stepped off, in theory. This is the condition of *minimum reflux* and is associated with an infinite number of stages, as indicated in Fig. 5.3-7. It is clear that the design reflux ratio for a column must lie between an infinite value (minimum stages) and a minimum value (infinite stages), and thus these limiting parameters assume great importance in the analysis and design of distillation equipment. They will be taken up in greater detail later.

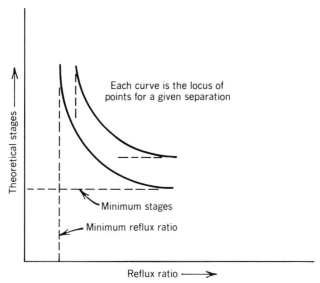

FIGURE 5.3-7 Stages–reflux relationship.

The McCabe–Thiele graphical method is useful for preliminary designs and for general orientation of the effects of process variables. It includes one basic assumption; that there is no change in the molar flow of liquid and vapor in each section (the *rectifying section* above the feed and the *stripping section* below the feed). This assumption implies that the molar latent heat of vaporization of each component is the same, and for many practical cases this is essentially true. It is possible to remove this restriction by using heat balances and changing the slopes of the operating lines (giving, in effect, curved operating lines), but this added embellishment is rarely justified.

The McCabe–Thiele method also may be used for cases of multiple feeds and of product withdrawals from intermediate stages. As mentioned earlier, standard texts on unit operations should be consulted for additional details.

Steps in the use of the McCabe–Thiele method for design are:

1. Draw the equilibrium curve, based on information from sources such as those given in Chapter 1 or in Section 5.2 of the present chapter.

2. Locate the feed, distillate product, and bottoms product compositions on the diagonal line.

3. Draw the q line, based on the condition of feed, starting with the feed composition on the diagonal. The line is vertical for a saturated liquid (liquid at its boiling point) and is horizontal for a saturated vapor feed.

4. Draw the upper operating line, starting with the distillate composition on the diagonal and using a slope determined from the selected reflux ratio.

5. Draw the lower operating line, starting with the bottoms composition on the diagonal and connecting with the point of intersection of the q line and the upper operating line.

6. Step off the theoretical stages as indicated in Fig. 5.3-6, changing operating lines at the feed stage.

PONCHON–SAVARIT GRAPHICAL METHOD

An alternative graphical method for handling binary mixtures is that of Ponchon[2] and Savarit,[3] and while more cumbersome to use than McCabe–Thiele, it allows for variations in the molar latent heat of vaporization and thus removes the principal assumption of the McCabe–Thiele method. As basic information it requires not only a y^*-x equilibrium relationship but also data on enthalpy of vaporization as a function of composition, and, except for a few mixtures, such data are not readily available.

The Ponchon–Savarit method is summarized in Fig. 5.3-8. The method involves an enthalpy–concentration diagram, and the enthalpies of the saturated liquid and vapor are first plotted on the diagram. Next, the equilibrium tie lines are added, based on phase equilibria. Compositions of feed, distillate, and bottoms are then located on the diagram (in the example shown the feed is mixed vapor–liquid and the distillate and bottoms are saturated liquids). A reflux ratio is chosen, and the enthalpy of the reflux is located as the top difference point, Δ_D. (The reflux ratio is equal numerically to the vertical distance from the difference point to the value of y_N, divided by the vertical distance from y_N to x_D.)

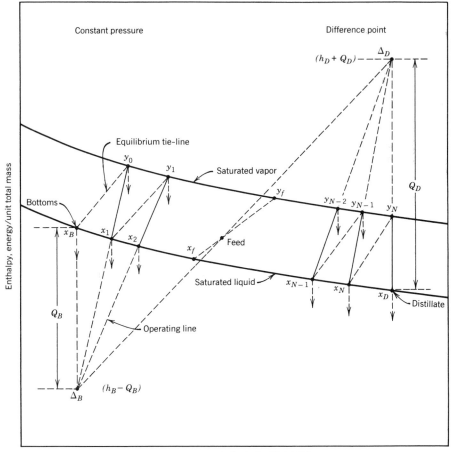

FIGURE 5.3-8 Summary of Ponchon–Sayarit graphical method for binary distillation.

A series of operating lines are drawn from the top difference point Δ_D to connect the compositions of streams passing each other (e.g., x_N and y_{N-1}), the equilibrium values (e.g., x_N and y_N) being connected by tie lines. In this fashion, stages are stepped off from the top of the column down to the feed stage, at which point a shift is made to the bottom operating line. This latter line is based on a bottom difference point that is colinear with the feed point and the top difference point, and also with a composition equal to the bottoms composition.

The lower part of the column is covered by stepping off stages in a fashion similar to that in the upper part of the column, and the final count of theoretical stages is then determined. The Ponchon–Savarit method may be used for many situations more complex than the simple one just described: mixed vapor–liquid distillate product, side draw streams, multiple feeds, and so on. Standard unit operations textbooks should be consulted for more details on this method. As mentioned, it suffers from a need for enthalpy-concentration data, but even a crude approximation based on linear variation of enthalpy with concentration can be better than the McCabe–Thiele approach if there is a very large difference in the latent heats of vaporization of the two components being distilled.

5.3-5 Minimum Stages

When a distillation column is operated under conditions of total reflux, all the vapor passing up from the top stage is condensed and returned to the column as reflux. Similarly, all the liquid flowing down from the bottom stage is vaporized and returned to the column. This is a closed system, with no feed and no products, but a concentration gradient prevails in the column and mass transfer between the phases takes place. Since there is no distillate product, the reflux ratio is said to be infinite ("total reflux"). Thus, in Fig. 5.3-7 the stages–reflux curve shows the asymptotic approach to a minimum stage value at very high values of reflux ratio.

Operation of a column at total reflux is important in two ways: it is a convenient startup condition that enables a column to be lined out at steady state before feed is processed, and in experimental work it is a simple and yet effective means for obtaining mass transfer information. The number of stages at total reflux is also important in design calculations in that it represents a lower limit to the required stages and it also represents a parameter used in short cut estimates of stage requirements (to be discussed in Section 5.3-7).

If graphical methods are used for binary systems, the minimum number of stages may be estimated directly. For McCabe–Thiele, the upper and lower operating lines coincide with the diagonal line at total reflux, and stages are stepped off as shown in Fig. 5.3-9. For Ponchon–Savarit, the difference points are at positive and negative infinity values on the enthalpy–concentration diagram, and thus the operating lines are vertical. Stages are stepped off as shown in Fig. 5.3-10.

Minimum stages may be determined analytically by means of the Fenske relationship.[5] For a binary mixture of i (lighter component) and j, and with the recognition that at total reflux and at any stage n, $x_{i,n} = y_{i,n-1}$ and $x_{j,n} = y_{j,n-1}$, it can be shown that

$$\left(\frac{y_i}{y_j}\right)_N = \alpha_N \alpha_{N-1} \cdots \alpha_2 \alpha_1 \left(\frac{x_i}{x_j}\right)_1 \tag{5.3-25}$$

since

$$(\bar{\alpha})^N = \alpha_N \alpha_{N-1} \cdots \alpha_2 \alpha_1 \tag{5.3-26}$$

Equation (5.3-25) becomes

$$\left(\frac{y_i}{y_j}\right)_N = \bar{\alpha}^N \left(\frac{x_i}{x_j}\right)_1 \tag{5.3-27}$$

A convenient final form of the Fenske equation is

$$N_{\min} = \frac{\ln(y_i/y_j)_N (x_j/x_i)_1}{\ln \bar{\alpha}} \tag{5.3-28}$$

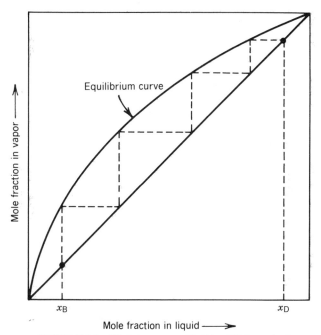

FIGURE 5.3-9 Minimum stages, McCabe–Thiele method.

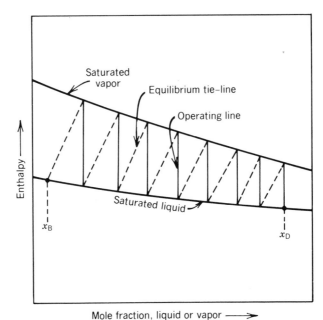

FIGURE 5.3-10   Minimum stages, Ponchon–Savarit method.

Some comments on the use of the Fenske equation are:

1. It can be applied over any part of the column.
2. The average value of the relative volatility, $\bar{\alpha}$, is subject to interpretation. Strictly, it is

$$\bar{\alpha} = (\alpha_N \alpha_{N-1} \cdots \alpha_2 \alpha_1)^{1/N} \tag{5.3-29}$$

Normally, it can be simplified to a geometric average of the volatility values at the top and bottom of the column:

$$\bar{\alpha} = (\alpha_{top}\alpha_{bottom})^{1/2} \tag{5.3-30}$$

If there is wide variation of relative volatility between top and bottom, an additional value can be used in the average:

$$\bar{\alpha} = (\alpha_{top}\alpha_{feed}\alpha_{bottom})^{1/3} \tag{5.3-31}$$

3. The value of N includes the reboiler as an equilibrium stage.
4. It can be applied to multicomponent mixtures, as will be discussed below.

5.3-6   Minimum Reflux Ratio

Minimum stages at total reflux represents one limit of operation of a column. The opposite limit is represented by an infinite number of stages at a theoretically minimum reflux ratio. This is shown in Fig. 5.3-7 as the asymptotic value of reflux ratio at a very high number of theoretical stages. This lower limit of reflux ratio represents a design parameter for shortcut estimates of stage requirements (to be discussed in Section 5.3-7) and it also guides the designer in assigning an operating reflux ratio as some multiple of the minimum reflux ratio. It does not represent a feasible operating condition, as does the minimum stages case.

   If graphical methods are used for binary systems, the minimum reflux ratio may be estimated directly. For McCabe–Thiele, it can be determined from the minimum slope of the upper operating line, which can be found graphically and often represents coincidence with the intersection of the q line with the equilibrium curve, as shown in Fig. 5.3-11. Unusual curvature of the equilibrium relationship can limit the minimum reflux ratio to a higher value as also shown in Fig. 5.3-11. In stepping off stages, starting at the top, a

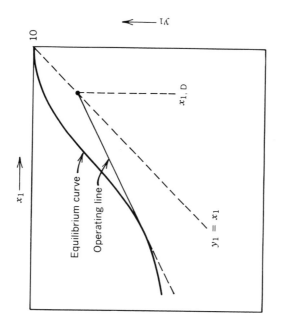

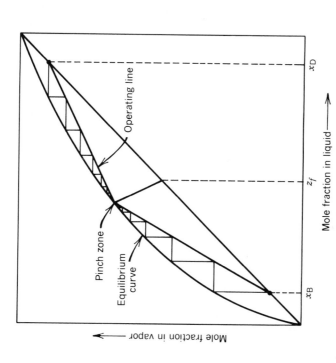

FIGURE 5.3-11 Minimum reflux by McCabe–Thiele method: (*a*) normal equilibrium curve and (*b*) equilibrium curve with concavity.

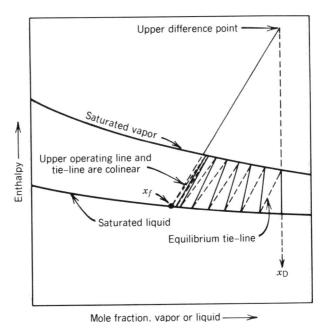

FIGURE 5.3-12 Minimum reflux by Ponchon-Savarit method.

point would be reached where no further progress could be made; this represents a "pinch zone" where there is no change of composition from tray to tray. Stepping off trays from the bottom would result also in a "pinch zone," very near the one obtained from the top–down direction. With reference to Fig. 5.3-11, the minimum reflux ratio for the case of a saturated liquid feed is

$$R_{min} = \frac{y_D - y_f}{(x_D - x_f) + (y_f - y_D)} \tag{5.3-32}$$

For the Ponchon-Savarit method, a pinch occurs when an operating line coincides with a tie line, thus preventing further stepping of theoretical stages. This is shown in Fig. 5.3-12, and normally the pinch occurs in the vicinity of the feed stage.

An analytical approach for the estimation of minimum reflux ratio has been published by Underwood[6] and is useful for multicomponent as well as binary systems. There are three basic assumptions made by Underwood:

1. Constant value of relative volatility
2. Constant molal flow of vapor and liquid in the column
3. Pinch occurs at the feed stage

The resulting equations of Underwood are

$$\frac{\alpha_i x_{i,f}}{\alpha_i - \phi} + \frac{\alpha_j x_{j,f}}{\alpha_j - \phi} = 1 - q \tag{5.3-33}$$

where q is defined as in Eq. (5.3-19) (e.g., $q = 0$ for saturated vapor feed and $q = 1$ for saturated liquid feed) and ϕ is a root to be obtained from the equation and which has a value between α_i and α_j, the latter being assigned a value of unity.

Equation (5.3-34) is then used to obtain the minimum reflux ratio:

$$R_{min} = \frac{\alpha_i x_{i,D}}{\alpha_i - \phi} + \frac{\alpha_i - x_{j,D}}{\alpha_j - \phi} + 1 \tag{5.3-34}$$

A more rigorous method for estimating the minimum reflux ratio has been published by Chien.[7]

5.3-7 Multicomponent Distillation

When the feed mixture contains more than two components, the separation is generally termed *multicomponent*, even though a three-component mixture is also known as a ternary, a four-component mixture as a quarternary, and so on. The key point is that when there are more than two components, the simple procedures such as McCabe–Thiele cannot be used with reliability and it is necessary to use analytical rather than graphical approaches. However, the same principles are used, and the MESH equations are applied at each theoretical stage.

KEY COMPONENTS

For a multicomponent mixture to be split into two streams (distillate and bottoms) by distillation, it is common to specify the separation in terms of two "key components" of the mixture. The *light key* will have a specified maximum limit in the bottoms product and the *heavy key* will have a specified maximum limit in the distillate product. Normally, the keys are adjacent to each other in the ranking of the mixture components according to relative volatility but this is not always the case, and *distributed components* may have volatilities intermediate to those of the keys.

MINIMUM STAGES

For multicomponent systems, an approximate value of the minimum number of stages (at total reflux) may be obtained from the Fenske relationship [Eq. (5.3-28)]. In the use of this relationship for multicomponent mixtures, the mole fractions and the relative volatility refer to the light and heavy keys only. However, values for the nonkey components may be inserted in the equation to determine their distribution *after* the number of minimum stages has been determined through the use of the key components. For a more rigorous approach to the determination of minimum stages, see the paper by Chien.[7]

MINIMUM REFLUX RATIO

For multicomponent systems, the Underwood method[6] may be used for estimating minimum reflux ratio. Its limitations should be recognized (see Section 5.3-6). The general equations are

$$\sum_i \frac{\alpha_i x_{if}}{\alpha_i - \phi} = 1 - q \tag{5.3-35}$$

$$\sum_i \frac{\alpha_i (x_{id})}{\alpha_i - \phi} = R_{min} + 1 \tag{5.3-36}$$

In Eq. (5.3-35), the value of the root ϕ which lies between the values α_{LK} and α_{HK} must be used.

For a more rigorous approach to the determination of minimum reflux ratio for multicomponent systems, see the paper by Chien.[8]

DESIGN APPROACH

This approach, simply stated, starts with a given separation and reflux ratio and determines the number of theoretical stages required to make the separation. It follows the general approach given in Section 5.3-4 for the analytical handling of binary system stage determinations. The pioneer paper on this approach was published in 1932 by Lewis and Matheson[9] and involves stage-by-stage calculations starting at the bottom and working up the column, or starting at the top and working down the column, or working from both ends and matching the feed composition at the feed stage. The difficulty with the method is that until the work is complete, the distribution of components between the distillate and the bottoms will not be known; thus the procedure is iterative and when many components are coupled with a number of stages, it is only feasible to handle the method by computer.

Techniques for speeding up a converged computer solution are given by Holland.[10] The input–output information for the Lewis–Matheson method is listed in Table 5.3-1.

RATING APPROACH

This approach starts with an assumed number of stages and reflux ratio and provides as output the separation that can be made. In other words, it *rates* a given column instead of designing it. Since this approach has convergence advantages, it is the one normally used for computer solutions to multicomponent, multitray distillation systems.

The rating approach usually is referenced as the Thiele–Geddes method, based on the original reference in 1933.[11] The technique is to assume a temperature profile throughout the column (number of stages being given) and, by starting at either end and working along the profile, arrive at a composition at the other end. If the outcome of such an iteration is a split that does not check the material balance requirements, the profile is varied through successive iterations. The experienced designer can bracket the desired answer with relatively few passes through the computer. Table 5.3-1 shows typical input–output characteristics of the Thiele–Geddes method.

**TABLE 5.3-1 Input–Output for Rigorous
Stagewise Calculations**

Input	Output
Rating Method—Thiele–Geddes	
Number of stages	Distillate composition
Feed stage number	Bottoms composition
Feed rate	
Feed composition	
Feed enthalpy	
Reflux ratio	
Distillate to feed ratio	
Pressure	
Design Method—Lewis–Matheson	
Distillate composition	Number of stages
Bottoms composition	Feed stage
Feed rate	Reflux ratio
Feed composition	Distillate rate
Feed enthalpy	
Design/minimum reflux ratio	
Optimum feed stage	
Pressure	

The Thiele–Geddes approach is summarized as follows. Consider a stage n in a cascade, the stage being intermediate between the feed and either end of the cascade and not complicated by a side draw or by extraneous heat addition or removal. A mass balance for component i on the tray is

$$v_{n-1,i} + \ell_{n+1,i} - v_{n,i} - \ell_{n,i} = 0 \tag{5.3-37}$$

For phase equilibria on the stage,

$$y_{n,i} = K_{n,i} x_{n,i} \tag{5.3-38}$$

or

$$\frac{v_{n,i}}{V_n} = K_{n,i} \frac{\ell_{n,i}}{L_n}$$

$$\ell_{n,i} = \frac{L_n}{V_n K_{n,i}} v_{n,i} = A_{n,i} v_{n,i} \tag{5.3-39}$$

where $A_{n,i}$ is the absorption factor on the stage. Combining the mass balance and phase equilibrium relationships, we obtain

$$v_{n-1,i} + A_{n+1,i} v_{n+1,i} - v_{n,i} - A_{n,i} v_{n,i} = 0$$

and

$$A_{n+1,i} v_{n+1,i} - (1 + A_{n,i}) v_{n,i} + v_{n-1,i} = 0 \tag{5.3-40}$$

At the top stage N,

$$A_{N+1,i} v_{N+1,i} - (1 + A_{N,i}) v_{N,i} + v_{N-1,i} = 0 \tag{5.3-41}$$

At the bottom stage 1,

$$A_2 v_{2,i} - (1 + A_{1,i}) v_{1,i} + v_{0,i} = 0 \tag{5.3-42}$$

At the feed stage, by material balance,

$$A_{f+1,i} v_{f+1,i} - (1 + A_{f,i}) v_{f,i} + v_{f-1,i} = z_{f,i} F \tag{5.3-43}$$

Thus, there is an equation for each component for each stage. This set of equations may be solved simultaneously by matrix methods, there being a tridiagonal matrix, N by N ($N + 1$ by $N + 1$ including the reboiler or $N + 2$ by $N + 2$ if there is also a partial condenser), shown in Fig. 5.3-13. This approach to the simultaneous (not stagewise) solution of the MESH equations was introduced by Amundson and Pontinen.[12]

There is a large literature dealing with design and rating methods for handling multicomponent, multistage calculations rigorously. The methods of solution were discussed in some depth by Friday and Smith[13] and by Wang and Henke,[14] among others. A more recent review was provided by Wang and Wang.[15] It is clear that the computational procedures for the design approach tend to be numerically unstable and are difficult to apply to complex columns with multiple feeds and multiple side draws. Procedures used for the rating approach are not susceptible to important round-off errors and can be used more easily for complex columns. Computer software for rigorous stagewise calculations may be accessed through commercial organizations such as Chemshare Corporation, Simulation Sciences, Inc., and Aspen Technology.

APPROXIMATE (SHORTCUT) METHODS

It is often sufficient, and certainly expedient, to use approximate methods for estimating the stage requirements for a multicomponent separation. These methods proceed according to the following sequence:

1. Specify the separation, based on the two key components.
2. Calculate the minimum number of stages, using the Fenske relationship (Section 5.3-7).
3. Determine the distribution of nonkey components and the ratio of rectifying to stripping stages, both by the Fenske relationship.
4. Calculate the minimum reflux ratio, using the Underwood method (Section 5.3-7).
5. For a selected operating reflux ratio, employ the results from steps 2 and 4 in an empirical correlation to obtain the required number of theoretical stages.

For step 3, some clarification is necessary. The Fenske equation may be used for any two pairs of components. Forcing their distributions to give a minimum number of stages equal to that obtained for the keys is a reasonable basis for estimating distribution:

$$N_{min} = \frac{\ln\,[(y_{LK}/y_{HK})_N(x_{HK}/x_{LK})_1]}{\ln\,\overline{\alpha}_{LK/HK}} = \frac{\ln\,[(y_{LLK}/y_{HK})_N(x_{HK}/x_{LLK})_1]}{\ln\,\overline{\alpha}_{LLK/HK}}$$

$$= \frac{\ln\,[(y_{LK}/y_{HHK})_N(x_{HHK}/x_{LK})_1]}{\ln\,\overline{\alpha}_{LK/HHK}} \qquad (5.3\text{-}44)$$

where LLK = component lighter than the light key
HHK = component heavier than the heavy key

Also,

$$N_{min,\,rect} = \frac{\ln\,[(y_{LK}/y_{HK})_N(x_{HK}/x_{LK})_f]}{\ln\,\overline{\alpha}_{LK/HK}} \qquad (5.3\text{-}45)$$

and

$$N_{min,\,strip} = \frac{\ln\,[(y_{LK}/y_{HK})_f(x_{HK}/x_{LK})_1]}{\ln\,\overline{\alpha}_{LK/HK}} \qquad (5.3\text{-}46)$$

It will be found that the sum of the stages from Eqs. (5.3-45) and (5.3-46) equals the number of stages computed by Eq. (5.3-44).

For step 5, the Gilliland correlation[16] is most often used. It was developed originally as a graphical fit to a number of stagewise calculations that were carried out by hand and is shown in Fig. 5.3-14. A fit of the curve has been provided by Eduljee:[17]

$$\frac{N_t - N_{min}}{N_t + 1} = 0.75 - 0.75\left(\frac{R - R_{min}}{R + 1}\right)^{0.5668} \qquad (5.3\text{-}47)$$

An alternative to the Gilliland method has been provided by Erbar and Maddox[18] and is shown in Fig. 5.3-15. This correlation is based on more extensive stagewise calculations, using rigorous computer solutions, and should give slightly better results than the method of Gilliland.

FIGURE 5.3-13 Tridiagonal matrix formulation of multistage distillation material balances. A similar matrix is required for each component.

Stage labels (top to bottom): Top stage (N), Stage $N-1$, Stage $N-2$, ..., Stage $f+1$, Feed stage (f), Stage $f-1$, ..., Stage 2, Stage 1, Reboiler (stage 0).

Matrix coefficient rows:

$$
\begin{aligned}
&A_{N+1} \quad -(1+A_N) \quad 1 \\
&A_N \quad -(1+A_{N-1}) \quad 1 \\
&A_{N-1} \quad -(1+A_{N-2}) \\
&\vdots \\
&A_{f+2} \quad -(1+A_{f+1}) \quad 1 \\
&A_{f+1} \quad -(1+A_f) \quad 1 \\
&A_f \quad -(1+A_{f-1}) \\
&\vdots \\
&A_3 \quad -(1+A_2) \quad 1 \\
&A_2 \quad -(1+A_1) \quad 1 \\
&A_1 \quad -(1+A_0)
\end{aligned}
$$

Unknown vector:
$$v_{N+1},\; v_N,\; v_{N-1},\; \ldots,\; v_{f+2},\; v_{f+1},\; v_f,\; \ldots,\; v_3,\; v_2,\; v_1$$

Right-hand side vector:
$$0,\; 0,\; 0,\; \ldots,\; 0,\; -z_f F,\; 0,\; \ldots,\; 0,\; 0,\; 0$$

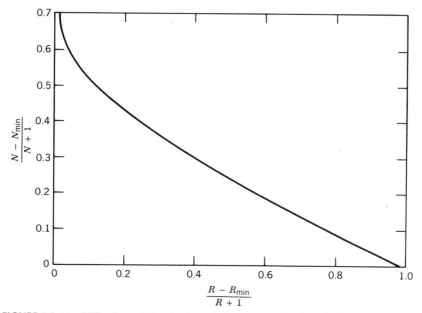

FIGURE 5.3-14 Gilliland correlation for theoretical stages as a function of reflux ratio (Ref. 16).

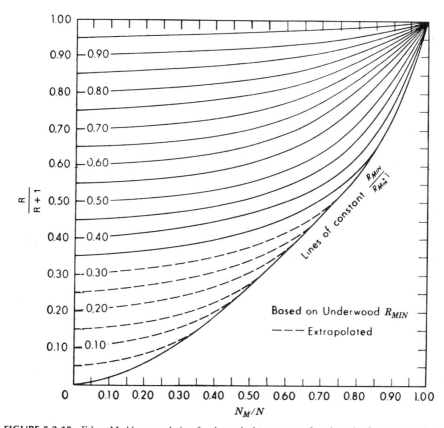

FIGURE 5.3-15 Erbar–Maddox correlation for theoretical stages as a function of reflux ratio (Ref. 18).

FEED STAGE LOCATION

The optimum feed stage location is the one that gives the fewest numbers of stages for the required separation. On this basis, Gilliland[19] developed criteria for evaluating the optimum feed location. He dealt with liquid, liquid–vapor, and vapor feed cases. For a liquid feed, the criterion is

$$\left(\frac{x_{LK}}{x_{HK}}\right)_f \leq \left(\frac{x_{LK}}{x_{HK}}\right)_{feed} \leq \left(\frac{x_{LK}}{x_{HK}}\right)_{f+1} \tag{5.3-48}$$

where f is the feed stage and $f + 1$ the stage above the feed. This equation states that the ratio of key components in the feed should be matched closely with the ratio of keys in the downflow to the feed tray and in the downflow from the feed stage.

In practice, an effort is made to match as closely as possible the composition of the feed and that of the appropriate stream (vapor or liquid) leaving the feed stage. Alternative feed locations usually are provided to allow for variations in column operation as well as inaccuracies in composition profile calculations.

5.3-8 Example Design

A deethanizing column is to be designed to recover an ethane concentrate from a feed stream containing six light hydrocarbons. Operating conditions and product specifications are given in Table 5.3-2. Determine the required number of theoretical and actual stages when the operating reflux ratio is 1.25 times the minimum. Use both shortcut and rigorous computational approaches. (*Note:* This problem is taken from the paper by Fair and Bolles.[20])

The Shortcut Method.   The shortcut calculation is carried out as follows: First, a preliminary material balance is set up on the assumption that no methane appears in the bottoms and no propane or heavier components appear in the distillate. The balance, in lb-mole/h is as follows:

	Feed	Distillate	Bottoms
Methane	5.0	5.00	—
Ethane (LK)	35.0	31.89	3.11
Propylene (HK)	15.0	0.95	14.05
Propane	20.0	—	20.00
i-Butane	10.0	—	10.00
n-Butane	15.0	—	15.00
	100.0	37.84	62.16

TABLE 5.3-2 Operating Conditions and Product Specifications for a Six-Component Deethanizing Column

Feed Stream Components	Mole %	α
Methane	5.0	7.356
Ethane	35.0	2.091
Propylene	15.0	1.000
Propane	20.0	0.901
i-Butane	10.0	0.507
n-Butane	15.0	0.408

Design Separation

Concentration of propylene in distillate, 2.5 mol %
Concentration of ethane in bottoms, 5.0 mol %

Operating Conditions

Condenser pressure, 400 psia (27.2 atm)
Top-plate pressure, 401 psia (27.3 atm)
Pressure drop per plate, 0.1 psi
Feed condition, liquid at bubble point
Distillate condition, liquid at bubble point
Reflux subcooling, none
Murphree plate efficiency, 75%

Minimum stages are then calculated by the Fenske equation (5.3-28) using mole ratios of the keys in place of mole-fraction ratios:

$$N_M = \frac{\log\ [(31.89/0.95)\ (14.05/3.11)]}{\log 2.091} = 6.79$$

Stages for the rectifying section (including the feed stage) are calculated by means of the same equation:

$$N_{M,R} = \frac{\log\ [(31.89/0.95)\ (15.0/35.0)]}{\log 2.091} = 3.60$$

Calculating in a similar fashion the stages below the feed (including the reboiler), $N_{M,S} = 3.19$.

Minimum reflux ratio is calculated by the Underwood equations. First, Eq. (5.3-35) is used to obtain a value of the root ϕ, noting that for a liquid feed at its boiling point $q = 1.0$:

$$\sum_i \frac{(\alpha_i x_{i,F})}{(\alpha_i - \phi)} = 0 = \frac{(7.356 \times 0.05)}{(7.356 - \phi)} + \frac{(2.091 \times 0.35)}{(2.091 - \phi)} + \frac{(1.000 \times 0.15)}{(1.000 - \phi)}$$

$$+ \frac{(0.901 \times 0.20)}{(0.901 - \phi)} + \frac{(0.507 \times 0.10)}{(0.507 - \phi)} + \frac{(0.408 \times 0.15)}{(0.408 - \phi)}$$

From which, $\phi = 1.325$. Then, substituting in Eq. (5.3-36),

$$\sum_i \frac{\alpha_i x_{i,D}}{(\alpha_i - \phi)} = R_M + 1 = \frac{(7.356 \times 0.132)}{7.356 - 1.325} + \frac{(2.091 \times 0.843)}{(2.091 - 1.325)} + \frac{(1.000 \times 0.025)}{(1.000 - 1.325)}$$

From which, $R_M = 1.378$.

For an operating reflux ratio of 1.722 (1.25 times the minimum), the Erbar–Maddox correlation is used to find the stage requirement. With reference to Fig. 5.3-15,

$$\frac{R}{R + 1} = \frac{1.722}{2.722} = 0.63$$

$$\frac{R_M}{R_M + 1} = \frac{1.378}{2.378} = 0.58$$

from which, $N_M/N = 0.47$. Accordingly, $N = N_M/0.47 = 6.79/0.47 = 14.5$.

The results of the calculations are summarized in Table 5.3-3. At the operating reflux of 1.722, the column requires 13.5 theoretical plates or 18 actual plates, assuming an overall column efficiency of 75% and a reboiler efficiency of 100%. A final step in the calculation would be to adjust the material balance

TABLE 5.3-3   Results of Shortcut Hand Calculations
for the Six-Component Deethanizer

Minimum Parameters		
Minimum reflux ratio, R_M		1.378
Minimum theoretical stages		
Feed and above		3.6
Below feed, including reboiler		3.2
		6.8
Results at Reflux Ratio (min. × 1.25)		
Operating reflux ratio, R		1.722
	Theoretical	Actual
Overall column efficiency, %	100	75
Plates		
Feed and above	7.7	10
Below feed, including reboiler	6.8	9
	14.5	19

for propane and heavier materials that would be found in the distillate. The Fenske equation is used, along with the appropriate relative volatility.

Rigorous Calculation. The problem was then solved using a rigorous plate-to-plate rating-type computer program. The results are shown in Table 5.3-4.

The same vapor–liquid equilibrium ratio (K) charts were used for the rigorous solution as for the shortcut. For the rigorous, however, values of K_i were combined with total pressure P, and the $K_i P$ product was treated as "effective vapor pressure" in the Antoine equation. The rigorous program was run in a trial-and-error fashion, with constant reflux ratio of 1.722 (from the shortcut) and with iterations of D/F ratio, total plates, and feed plate location.

The number of actual plates in the column turned out to be 20. Table 5.3-4 shows stage-by-stage flows, temperatures, pressure, and compositions. It also shows condenser and reboiler duties. The final material balance, in lb-mole/h, is:

	Feed	Distillate	Bottoms
Methane	5.0	5.00	—
Ethane	35.0	31.99	3.01
Propylene	15.0	0.95	14.05
Propane	20.0	0.56	19.44
i-Butane	10.0	—	10.00
n-Butane	15.0	—	15.00
	100.0	38.50	61.50

After the theoretical or actual stage requirements have been calculated, the final step is that of specifying the optimum distillation column (fractionator); that is, the proper combination of column height, column diameter, and contacting internals must be chosen.

5.4 SPECIFICATION OF VARIABLES

5.4-1 Separation

There are four common ways of expressing the required separation to be made by the distillation column: purity of one or more products, allowable impurities in one or more products, recovery of one or more feed components, and measurable properties of one or more of the products. Examples of these are as follows:

Purities	99.6 mol % minimum purity of light key in distillate product
Impurities	0.3 mol % maximum content of heavy key in distillate product
Recovery	95% minimum of light key in feed, to be contained in distillate product
Properties	Distillate product to have a Reid vapor pressure of 45 lb$_f$/in.^{2} minimum

These may be in combination and not limited to one of the keys. However, one must be careful not to overspecify or underspecify the separation, as discussed below.

5.4-2 Problem Specification

To execute the design of a distillation column, $C + 6$ variables must be specified, where C is the number of components in the feed mixture. (This assumes that reflux is saturated.) Of these, $C + 2$ are always specified:

Feed composition	$C - 1$
Feed rate	1
Feed enthalpy	1
Pressure	1
	$C + 2$

There are many possible combinations of the other four variables. Some typical groupings are as follows:

TABLE 5.3-4 Computer Results of a Rigorous Plate-to-Plate Design Calculation for a Six-Component Deethanizer

	Plate	Temperature	Pressure	Liquid Flow	Vapor Flow	Plate Liquid Compositions (mol %)					
						Methane	Ethane	Propylene	Propane	i-Butane	n-Butane
Condenser	21	-9.68	400.00	66.30	0.0	12.9869	83.0946	2.4644	1.4441	0.0075	0.0025
	20	38.80	401.00	70.34	104.80	3.4141	88.1861	5.1180	3.2485	0.0244	0.0089
	19	50.71	401.10	69.60	108.84	1.6045	84.4861	8.2317	5.5954	0.0586	0.0237
	18	57.48	401.20	67.88	108.10	1.2649	78.2023	11.7862	8.5638	0.1260	0.0568
	17	64.01	401.30	66.02	106.38	1.2080	70.9312	15.4686	12.0137	0.2513	0.1272
	16	70.77	401.40	64.33	104.52	1.2050	63.6003	18.8250	15.6297	0.4708	0.2692
	15	77.35	401.50	62.92	102.83	1.2113	56.9325	21.4482	19.0331	0.8336	0.5414
	14	83.42	401.60	61.76	101.42	1.2178	51.3311	23.1064	21.9000	1.4047	1.0399
	13	88.92	401.70	60.69	100.26	1.2230	46.8458	23.7421	24.0044	2.2662	1.9185
	12	94.14	401.80	59.57	99.19	1.2271	43.2805	23.3906	25.1818	3.5116	3.4083
	11	99.62	401.90	58.05	98.07	1.2303	40.3218	22.1141	25.2863	5.2221	5.8253
Feed tray	10	106.40	402.00	162.96	96.55	1.4896	37.3924	18.5154	22.7923	8.2519	11.5584
	9	113.34	402.10	165.03	101.46	0.5988	36.1336	19.5079	23.7906	8.3456	11.6234
	8	118.46	402.20	166.03	103.53	0.2379	33.8948	20.6364	24.9772	8.4937	11.7599
	7	123.30	402.30	166.69	104.53	0.0938	30.9776	21.9061	26.3750	8.6939	11.9536
	6	128.47	402.40	167.30	105.19	0.0368	27.5497	23.2732	27.9693	8.9576	12.2134
	5	134.16	402.50	168.00	105.80	0.0143	23.7632	24.6409	29.6978	9.3116	12.5722
	4	140.38	402.60	168.82	106.50	0.0055	19.7934	25.8566	31.4384	9.8037	13.1024
	3	147.12	402.70	169.70	107.32	0.0021	15.8328	26.7129	32.9951	10.5127	13.9443
	2	154.44	402.80	170.54	108.20	0.0008	12.0668	26.9503	34.0792	11.5546	15.3483
	1	162.68	402.90	168.89	109.04	0.0003	8.6417	26.2505	34.2761	13.0846	17.7468
Reboiler	0	176.86	403.00	61.50	107.39	0.0001	4.8920	22.8475	31.6163	16.2554	24.3887

TABLE 5.3.4. (*Continued*)

	Top Liquid	Bottom Liquid	Feed
Reflux ratio	1.722		Condenser duty	4.96345D 05
Reflux rate	66.3		Reboiler duty	5.76678D 05
Plate Number	21	0	10
Flow rate	38.50	61.50	100.00
Temperature	−9.68	176.86	90.50
Pressure	400.00	403.00	402.00
Enthalpy	−938.77	3908.51	1238.98
Composition			
Methane	12.9869	0.0001	5.0000
Ethane	83.0946	4.8920	35.0000
Propylene	2.4644	22.8475	15.0000
Propane	1.4441	31.6163	20.0000
i-Butane	0.0075	16.2554	10.0000
n-Butane	0.0025	24.3887	15.0000
Percent Recovery			
Methane	99.9990	0.0010	
Ethane	91.4041	8.5959	
Propylene	6.3252	93.6748	
Propane	2.7799	97.2201	
i-Butane	0.0290	99.9710	
n-Butane	0.0063	99.9937	

1. For "rating" approach (Thiele–Geddes, Table 5.3-1)
   - (a) Number of stages
   - (b) Feed stage number
   - (c) Reflux ratio
   - (d) Distillate to feed ratio
2. Alternative for "rating" approach
   - (a) Rectifying stages ⎫ or ⎧ Number of stages
   - (b) Stripping stages ⎭ ⎩ Feed stage
   - (c) Fractional recovery of light key in distillate
   - (d) Concentration of light key in distillate
3. For "design" approach (Lewis–Matheson, Table 5.3-1)
   - (a) Distillate composition
   - (b) Bottoms composition
   - (c) Design/minimum reflux ratio
   - (d) Feed location (usually at optimum)
4. Alternative for "design" approach
   - (a) Composition of light key in distillate
   - (b) Composition of light key in bottoms
   - (c) Reflux ratio
   - (d) Feed location (usually at optimum)

It should be clear that the above groupings are examples, and other possibilities exist. Alternative 4 commonly is used for approximate calculations. Care must be taken not to overspecify the distillation separation problem.

5.5 SPECIAL DISTILLATIONS

5.5-1 Azeotropic Distillation

AZEOTROPES

Many liquid mixtures exhibit azeotropes at intermediate concentrations such that the liquid and its equilibrium vapor have the same composition. No separation of this concentration is possible by partial vaporization. A binary mixture may have a *minimum boiling azeotrope*, where the boiling temperature of the azeotrope is less than that of the pure components, or a *maximum boiling azeotrope*, where the boiling temperature is higher than that of the pure components. About 90% of the known azeotropes are of the minimum variety.

The presence of an azeotrope is one indication that a mixture is not ideal, that it has deviations from Raoult's law (see Chapter 1 and Section 5.2 of the present chapter). Close-boiling mixtures are more likely to exhibit azeotropism than wide-boiling mixtures; when there is more than 30°C boiling point difference, it is quite unlikely that an azeotrope will be present. Thus, the combination of close-boiling and nonideality is one that can lead to the presence of an azeotrope.

Figure 5.2-5 shows equilibrium diagrams for binary systems containing maximum boiling (acetone–chloroform) and minimum boiling (ethanol–water) azeotropes. It also shows an example of an azeotrope which, when condensed, forms two liquid phases (n-butanol–water); this is called a *heterogeneous azeotrope*.

Most investigations of azeotropes have dealt with binary or ternary mixtures. The binary pairs of a multicomponent mixture may separately form azeotropes, but these are submerged by the possible azeotropes involving the full mixture. It is sometimes possible to utilize binary pair azeotrope information in estimating the role that a multicomponent azeotrope will play in the distillation separation. As an example, the ethanol–water system exhibits a minimum boiling azeotrope. At atmospheric pressure, the boiling points are

Ethanol–water azeotrope	78.2°C
Ethanol	78.4°C
Water	100°C

The azeotrope distills overhead and its water content (4.0 vol %) is supplied fully by the usual content of water in the feed. If benzene is added to the azeotrope and the mixture is fed to another column, an expanded range of possible boiling points (at 1 atm) is found:

Ethanol–benzene–water azeotrope	64.9°C
Ethanol–benzene azeotrope	68.2°C
Benzene–water azeotrope	69.3°C
Ethanol–water azeotrope	78.2°C
Ethanol	78.4°C
Benzene	80.1°C
Water	100°C

Now for limited amounts of water and benzene, the low-boiling ternary azeotrope dominates and becomes the distillate product. With careful control of the amount of benzene added, all the water and the benzene are taken overhead. This particular system will be used as an example later in the present section.

An extensive tabulation of azeotropes has been compiled by Horsley.[1] An older compilation is that of Lecat.[2] Certain nonideal vapor–liquid equilibrium models are useful for predicting azeotropic behavior of binary systems; in particular, the model of Renon and Prausnitz[3] is useful in this regard because it can handle the two liquid phases associated with heterogeneous azeotropes. The Horsley book also contains guidelines for the prediction of azeotropes.

AZEOTROPIC DISTILLATION

This term usually is applied to cases where an extraneous material, called an *entrainer*, is added to a mixture to make a distillation separation feasible. In this way, a problem involving a close-boiling mixture is made tractable by the addition of an entrainer that will azeotrope with one of the components to give, in effect, a respectable relative volatility between the nonazeotroping component and the azeotrope (treated as a pseudocomponent). Typically, the azeotrope has the higher volatility and becomes the distillate product.

A difficulty is that the extraneous material must itself be separated from the product and, if valuable, recycled back to the process. This leads to the desirability of employing a heterogeneous azeotrope that through phase separation can provide the entrainer recovery that is needed. Other entrainer separation methods include solvent extraction and distillation under a pressure such that the azeotropic composition is different or that the azeotrope is absent. Figure 5.5-1 shows a flow diagram for the example of separating water from ethanol, using benzene as the entrainer. A low-boiling ternary azeotrope is formed, as indicated above, and is taken overhead. Note that when the azeotrope vapor is condensed, two liquid layers result (heterogeneous azeotrope). The water layer contains residual ethanol and benzene and is taken to a stripping column for recovery of the entrainer. Figure 5.5-2 shows the distillation path on the ternary phase diagram.[4]

In some cases, a mixture is separated on the basis of its contained azeotropes (i.e., no extraneous material is added) and this may also be called azeotropic distillation. An example, shown in Fig. 5.5-3, is the azeotropic drying of toluene.[5] The water–toluene azeotrope is taken overhead, leaving the desired dry toluene as bottoms.

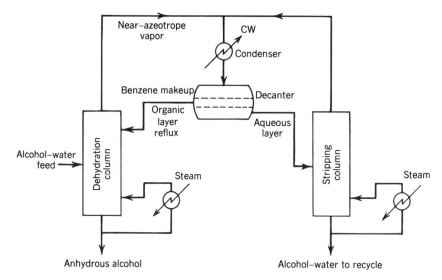

FIGURE 5.5-1 Dehydration of alcohol by azeotropic distillation, using benzene as entrainer to form a low-boiling ethanol–water–benzene azeotrope.

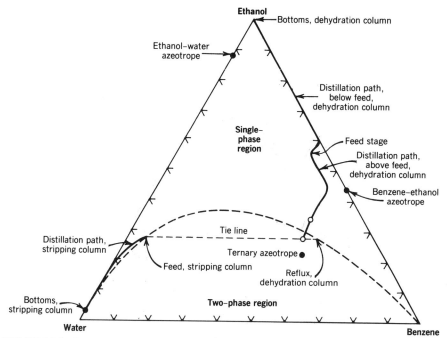

FIGURE 5.5-2 Ternary phase diagram for ethanol–benzene–water, showing distillation paths for the columns shown in Fig. 5.5-1. Composition in mole percent or mole fraction. (Based on Ref. 4.)

AZEOTROPIC DISTILLATION CALCULATIONS

Methods for determining equilibrium stages for azeotropic distillation are the same as those discussed in Section 5.3. As indicated in Fig. 5.5-3, the distillation path may cross over into the two liquid phase region, and one must be cautious about the design of contacting equipment when this condition is present. Research has shown that with good aeration, conventional trays may be used when two liquid phases are present.[6]

Hoffman[7] has discussed in detail methods for designing azeotropic distillation systems. Black et al.[8] provide insight on the use of computers for making azeotropic distillation calculations. Gerster[9] and Black and Ditsler[10] show comparisons between azeotropic and extractive distillation approaches for handling the same problem.

SELECTION OF ENTRAINER

Guidelines for entrainer selection have been provided by Gerster[9] and Berg.[11] The latter author has listed the following desirable properties of an entrainer, for hydrocarbon separations:

1. Should boil within a limited range (0–30°C) of the hydrocarbon to be separated.
2. Should form, on mixing with the hydrocarbon, a large positive deviation from Raoult's law to give a minimum azeotrope with one or more of the hydrocarbon types in the mixture.
3. Should be soluble in the hydrocarbon.
4. Should be easily separable from the azeotrope.
5. Should be inexpensive and readily obtainable.
6. Should be stable at the temperatures of the distillation.
7. Should be nonreactive with the hydrocarbons being separated and with the materials of construction of the equipment.

5.5-2 Extractive Distillation

As in the case of azeotropic distillation, an extraneous material is added to the mixture in the case of extractive distillation. However, the function of this material, called an *extractive agent* or simply a *solvent*, is different. It is designed to enhance the relative volatility of the key components by influencing the nonideality of the mixture. The solvent is normally a high-boiling material and in some cases may form

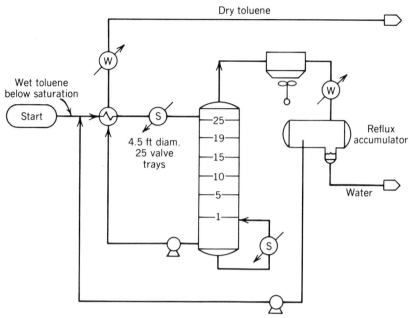

FIGURE 5.5-3   Azeotropic drying of toluene, with residual water content of less than 5 ppmw (Ref. 5).

loose chemical bonds with one of the keys, with the result of a suppressed effective vapor pressure; this is discussed in connection with solvent selection by Berg[11] and by Ewell et al.[12]

As in the case of azeotropic distillation, there is the problem of separating the solvent from the product, but for extractive distillation the problem is much simpler, since the solvent is normally quite high boiling, beyond the range of azeotrope formation, and simple distillation usually is sufficient to separate the solvent for reuse.

EXTRACTIVE DISTILLATION CALCULATIONS

Determination of stage requirements for extractive distillation follows the approaches discussed in Section 5.3. It is necessary to have vapor–liquid equilibrium data, and these are measured conventionally. A representative relation between solvent/nonsolvent ratio and relative volatility is shown in Fig. 5.5-4, taken from the papers of Gerster[9] and Drickamer et al.[13] From the figure it is clear that various combinations of solvent content and stage requirements are possible, and the optimum solvent ratio must be determined. Figure 5.5-4 deals with the separation of toluene from methylcyclohexane using phenol as the solvent. The "natural" volatility of the binary mixture is shown as the bottom line (no phenol present) and the enhancement of this volatility by phenol addition is significant. A flow diagram of this separation process is shown in Fig. 5.5-5.

The solvent, being a low boiler, must be added high in the column to ensure volatility enhancement throughout the column. Additional stages are used above the solvent feed point to remove any entrained or volatilized solvent from the vapors leading to the distillate product.

In many cases it is possible to carry out stage determinations on a solvent-free basis, using a graphical approach such as McCabe–Thiele. An example of this approach is shown in Fig. 5.5-6.[14] Reflux ratio and boilup specifications are made as if there were no solvent present. If the feed stream causes a large change in the solvent/nonsolvent ratio at the feed stage, then separate equilibrium curves are needed for the rectifying and stripping sections.

Computations for extractive distillation are covered in detail by Hoffman[7] and Smith.[15] Algorithms for rigorous stage-to-stage energy and material balances (such as Thiele–Geddes) may be used if the multicomponent equilibria are established. Process design considerations are discussed by Kumar et al.[16] Case studies of extractive distillation separations are provided by Black and Ditsler,[10] Gerster,[17] Hafslund,[18] and Bannister and Buck.[19] Stage efficiencies are characteristically low, because of the heavy solvents involved, but are amenable to mass transfer modeling[16] and simulation in small-scale Oldershaw equipment (see Section 5.9).

SOLVENT SELECTION

Approaches to the selection of an extractive distillation solvent are discussed by Berg,[11] Ewell et al.[12] and Tassios.[20] In general, selection criteria follow those for azeotropic entrainers:

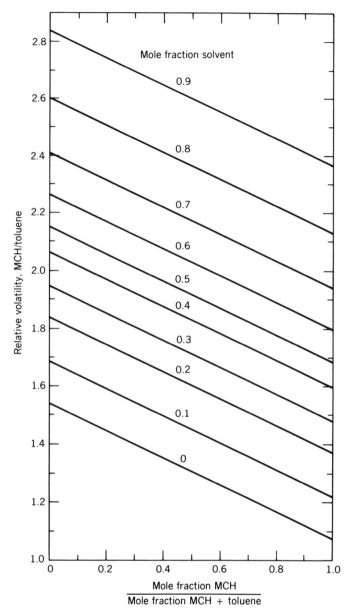

FIGURE 5.5-4 Influence of solvent/nonsolvent ratio on relative volatility of methylcyclohexane (MCH) to toluene. The solvent is phenol. (Taken from Refs. 8 and 13.)

1. Should enhance significantly the natural relative volatility of the key components.
2. Should not require an excessive ratio of solvent to nonsolvent (because of cost of handling in the column and in auxiliary equipment).
3. Should be miscible with the system to be separated.
4. Should be easily separable from the bottoms product (for the usual case of a heavy solvent).
5. Should be inexpensive and readily available.
6. Should be stable at the temperatures of the distillation and solvent separation.
7. Should be nonreactive with the materials of construction of the equipment.

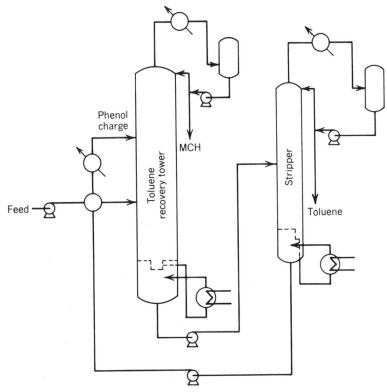

FIGURE 5.5-5 Flow diagram for the extractive distillation separation of toluene from methylcyclohexane (MCH), using phenol as the solvent.

5.5-3 Steam Distillation

For mixtures with very high boiling points it may be desirable to add an inert gas to the mixture to reduce the partial pressures of the mixture constituents and thus enable boiling at a lower temperature. The most convenient inert gas is steam, which can be condensed and does not need handling by the vacuum-producing device. In the typical case, a high-boiling organic mixture, such as crude tall oil, is to be distilled under vacuum. Steam is added at the reboiler, as shown in Fig. 5.5-7,[21] does not condense in the column, and is removed in an aftercondenser that is located upstream of the vacuum-producing steam jet ejector.

The amount of steam needed depends on the vapor pressure of the boiling mixture, the partial pressure being determined from a corrected form of Raoult's law:

$$p_i = \gamma_i x_i p_i^* \tag{5.5-1}$$

$$P = \Sigma p_i + p_{\text{steam}} \tag{5.5-2}$$

$$\frac{\text{moles steam}}{\text{moles of mixture in vapor from reboiler}} = \frac{p_{\text{steam}}}{\Sigma p_i} = \frac{P - \Sigma p_i}{\Sigma p_i} \tag{5.5-3}$$

Rigorous stage calculations can be made for steam distillation columns, using methods described in Section 5.3. It is also possible to make approximate calculations on a steam-free basis, as if the total pressure in the column were equal to the total vapor pressure of the material being distilled.

5.5-4 Complex Mixture Distillation

This type of distillation involves feeds and products that are not easily described in terms of identifiable components. Instead, the feed and product streams are characterized in terms of their boiling ranges, measurements that can be made in simple takeover stills of the type specified by the American Society for

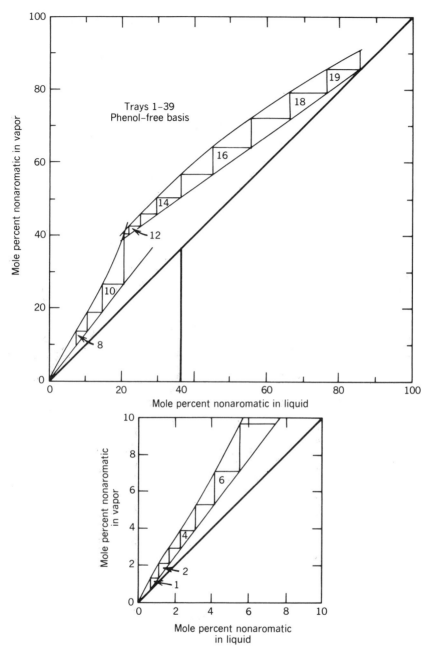

FIGURE 5.5-6 McCabe–Thiele diagram for toluene recovery column, plotted on a solvent-free basis (Ref. 14).

Testing Materials (ASTM) or in more elaborate, refluxed true boiling point (TBP) stills. Complex mixtures have been handled successfully for years in the petroleum industry, using empirical methods such as those of Packie[22] and Houghland et al.[23] Such methods have been summarized and converted to stepwise computational procedures by Van Winkle;[24] they may be regarded as approximate methods.

For more rigorous work with boiling range materials, the pseudocomponent method should be used. The true boiling point curve for the mixture (feed or product) is broken up into a series of steps as shown in Fig. 5.5-8, each step representing a pseudocomponent with boiling point as indicated. Next, the material

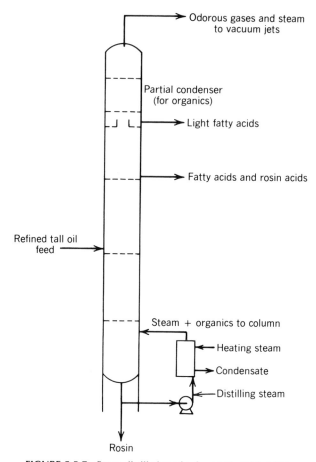

FIGURE 5.5-7 Steam distillation of refined tall oil (Ref. 21).

representing each step is characterized for properties, using homolog-series approaches such as that of Dreisbach[25] for vapor pressure and enthalpy.

Next, rigorous stage calculation models may be applied to determine the separation as a function of reflux ratio. Finally, TBP curves are developed from the pseudocomponent contents of the overhead and bottoms products. The rigorous method is now preferred, especially since rigorous stage computation software is so readily available for computer use.

5.5-5 Batch Distillation

Although most commercial distillations are run continuously, there are certain applications where batch distillation is the method of choice. Such applications include the following:

1. Semiworks operations producing interim amounts of product in equipment that is used for multiple separations.
2. Distillations of specialty chemicals where contamination can be a problem. The batch equipment can be cleaned or sterilized between batch runs.
3. Operations involving wide swings in feed compositions and product specifications, where batch operating conditions can be adjusted to meet the varying needs.
4. Laboratory distillations where separability is being investigated without concern over the scaleup to commercial continuous operations.

Batch distillations are generally more expensive than their continuous counterparts, in terms of cost per unit of product. Close supervision and/or computer control are required, the equipment is more complex

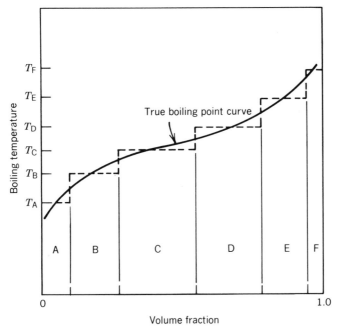

FIGURE 5.5-8 Use of pseudocomponents for complex mixture distillation.

if several products are to be recovered, and total throughput is limited by the needs for changing operating modes and for recharging the system with feed material.

RAYLEIGH DISTILLATION

The simplest batch distillation method is the straight takeover approach, without reflux, often called a *Rayleigh distillation*. This is the traditional "moonshine still" technique, in which a batch of material is charged to a kettle (stillpot), brought to boiling, and the vapor condensed as it passes over from the stillpot. A diagram of this type of operation is shown in Fig. 5.5-9.

The analytical method for handling a Rayleigh distillation may be summarized as follows. Let $S =$ total moles in stillpot with light key mole fraction x. A material balance for a differential amount of vaporization is

$$Sx_s = (S - ds)(x_s - dx_s) + x_D dS$$

(amount before vaporization) = (remaining amount) + (vaporized amount)   (5.5-4)

Rearranging and integrating between initial stillpot moles S_0 and final stillpot moles S_t, we obtain

$$\ln\frac{S_0}{S_t} = -\int_{x_{s0}}^{x_{st}} \frac{dx_s}{x_D - x_s} \quad \text{(Rayleigh equation)} \tag{5.5-5}$$

This equation may be evaluated by graphical integration. An analytical solution is possible, if the vapor is assumed to be in equilibrium with the liquid at any time and if the relative volatility is constant. Since the vapor is fully condensed, $x_D = y^*$ and by Eq. (5.2-13),

$$x_s = \frac{x_D}{\alpha - x_D(\alpha - 1)} \tag{5.5-6}$$

and

$$dx_s = \frac{\alpha\, dx_D}{[\alpha - x_D(\alpha - 1)]^2} \tag{5.5-7}$$

Substituting in Eq. (5.5-5) and rearranging, we obtain

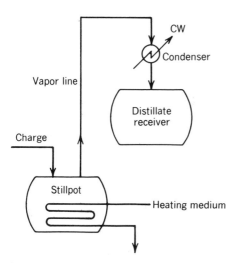

FIGURE 5.5-9 Rayleigh, or simple takeover, batch distillation.

$$\ln \frac{S_0}{S_t} = \frac{1}{\alpha - 1}\left(\ln \frac{x_{s0}}{x_{st}} + \alpha \ln \frac{1 - x_{st}}{1 - x_{s0}}\right) \tag{5.5-8}$$

An alternative form, in terms of distillate compositions, is given as follows:

$$\ln \frac{S_0}{S_t} = \frac{1}{\alpha - 1}\left(\ln \frac{x_{D0}}{x_{Dt}} + \alpha \ln \frac{1 - x_{Dt}}{1 - x_{D0}} + (\alpha - 1)\ln \frac{\alpha - x_{D0}(\alpha - 1)}{\alpha - x_{Dt}(\alpha - 1)}\right) \tag{5.5-9}$$

where x_D values are related to x_s values by Eq. (5.5-6). At any time during the distillation a component material balance gives

$$S_0 x_{s0} = S_t x_{st} + D_t x_{Dt} \tag{5.5-10}$$

and the composite distillate composition is

$$\bar{x}_D = \bar{y}^* = \frac{S_0 x_{s0} - S_t x_{st}}{S_0 - S_t} \tag{5.5-11}$$

EXAMPLE 5.5-1

100 moles of a 50–50 mixture of A (light key) and B (heavy key) are to be batch distilled in a simple takeover process. The relative volatility between A and B is 2.0. The distillation is to proceed until the mole fraction of A in the stillpot is 0.2. (a) How many total moles remain in the pot at the end of the distillation? (b) What are the initial, final, and composite mole fractions of A in the overhead receiver?

Solution.   (a) By Eq. (5.5-7),

$$\ln \frac{S_0}{S_t} = \frac{1}{1}\ln \frac{0.5}{0.2} + 2\ln \frac{1 - 0.2}{1 - 0.5}$$

$$\frac{S_0}{S_t} = 6.40 \qquad S_t = \frac{100}{6.40} = 15.6 \text{ moles left}$$

By Eq. (5.5-6),

$$x_{D0} = \frac{2(0.5)}{1 + 1(0.5)} = 0.667 \quad \text{(mole fraction A at start)}$$

$$x_{Dt} = \frac{2(0.2)}{1 + 1(0.2)} = 0.333 \quad \text{(mole fraction A at end)}$$

By Eq. (5.5-10),

$$\bar{x}_D = \frac{100(0.5) - 15.6(0.2)}{100 - 15.6} = 0.555 \quad \text{(composite mole fraction A at end)}$$

BATCH DISTILLATION WITH REFLUX

The foregoing equations apply to simple takeover batch distillation, with no reflux. If a column is installed above the stillpot and provision is made to return a portion of the condensed overhead vapor to the column as reflux, it is possible to obtain a much better separation. As an extreme case, near-total reflux is used to get the best possible separation, without regard to the length of time required for the revaporization of the reflux. Under such a condition, the Fenske relationship [Eq. (5.3-28)] may be used to relate the stillpot vapor composition to that of the vapor passing over to the condenser:

$$\alpha^N = \frac{x_D}{1 - x_D} \frac{x_s}{1 - x_s} \tag{5.5-12}$$

Combining with Eqs. (5.5-5) and (5.5-6), we obtain

$$-\ln \frac{S_0}{S_t} = \ln \frac{S_t}{S_0} = \frac{1}{\alpha^N - 1} \ln \frac{x_{st}(1 - x_{s0})}{x_{s0}(1 - x_{st})} + \ln \frac{1 - x_{s0}}{1 - x_{st}} \tag{5.5-13}$$

EXAMPLE 5.5-2

For the batch distillation of the mixture of Example 5.5-1, a column with five theoretical stages is added above the stillpot and operation is maintained at near-total reflux. After half of the original mixture has distilled over: (a) What is the composition of A in the stillpot? (b) What is the composite composition of the distillate?

Solution.   (a) For this case, Eq. (5.5-13) is used, with $N = 6$ (column stages plus stillpot):

$$-\ln \frac{S_0}{S_t} = -\ln \frac{100}{50} = \frac{1}{2^6 - 1} \ln \frac{x_{st}(1 - 0.50)}{0.50(1 - x_{st})} + \ln \frac{(1 - 0.50)}{(1 - x_{st})}$$

Solving for x_{st} we obtain $x_{st} = 0.044$.

(b) The composite composition of the distillate is

$$\bar{x}_D = \frac{100(0.50) - 50(0.044)}{100 - 50} = 0.956$$

Note that the combination of contacting stages plus high reflux enables high recovery of A at a high purity.

Batch distillation columns may be operated under either of two reflux policies: *constant reflux* with varying distillate composition, and *variable reflux*, with constant distillate composition. These policies have the usual limitations of total reflux and minimum stages.

If a batch distillation is to be run at constant reflux, then it follows that the distillate product composition must change with time. This is illustrated in Fig. 5.5-10, which shows a McCabe–Thiele diagram with constant slope of the upper operating line and movement of distillate composition down along the diagonal. If stage efficiency is constant, then the stillpot composition changes accordingly. In the example shown, a total of three equilibrium stages are provided; as the distillate composition moves from x_{D0} to x_{Dt}, the stillpot composition moves from x_{s0} to x_{st}, and so on.

The operating policy for constant reflux operation would be to allow movement of the distillate purity until the composite purity, from time 0 to time t, meets specification. It might be desirable at this point to shift to variable reflux as described below.

If a batch distillation is carried out at variable reflux ratio, it is possible to maintain a constant purity of distillate and to do this until blocked by material balance limitations. A McCabe–Thiele diagram illustrating this method of operation is shown in Fig. 5.5-11. The value of x_D is kept constant, and the slope of the operating line is varied in accordance with the overall material balance. The figure shows the conditions at time 0 and time t. Note that the three equilibrium stages include the stillpot.

Figures 5.5-10 and 5.5-11 are intended to be illustrative and are useful for approximations and for binary systems. Further information on the use of the McCabe–Thiele method in this connection may be found in the early work of Bogart[26] and in several other sources.

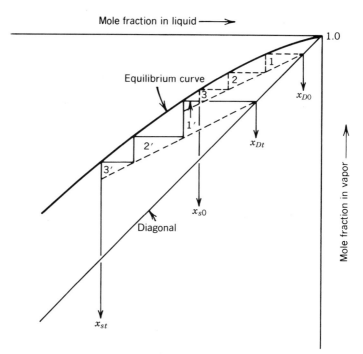

FIGURE 5.5-10 Batch distillation with constant reflux ratio (three theoretical stages).

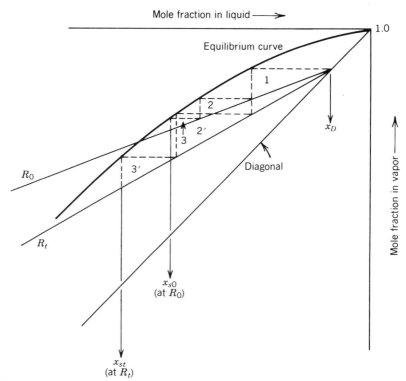

FIGURE 5.5-11 Batch distillation with constant distillate composition and varying reflux (three theoretical stages).

EQUILIBRATION TIME

The complicating factor in batch distillation is the transient nature of the process. For a new design, the engineer must be concerned with optimizing the total cycle, which includes charging and equilibration, product takeoff at constant or varying rate, and cleanout.

Equilibration time is of interest in continuous system startups as well as in batch distillation. Clearly, the amount of liquid holdup in the condenser, column, and stillpot is a major factor in the time requirement. For a stage n (see Fig. 5.3-1), the basic material balance equation is

$$J_n\left(\frac{dx_{i,n}}{dt}\right) = L_{n+1}x_{i,n+1} + V_{n-1}Y_{i,n-1} - L_nx_{i,n} - V_nY_{i,n} \qquad (5.5\text{-}14)$$

where J_n is the molar holdup of liquid on the stage. Jackson and Pigford,[27] starting with this basic equation and considering total reflux operation (normally used during equilibration), developed a generalized correlation for estimating equilibration time when tray-type columns are used.

Their results are shown in Fig. 5.5-12, for the case of a 90% approach to steady-state operation. The ordinate value includes the time required to fill the trays with their steady-state content of low-boiling

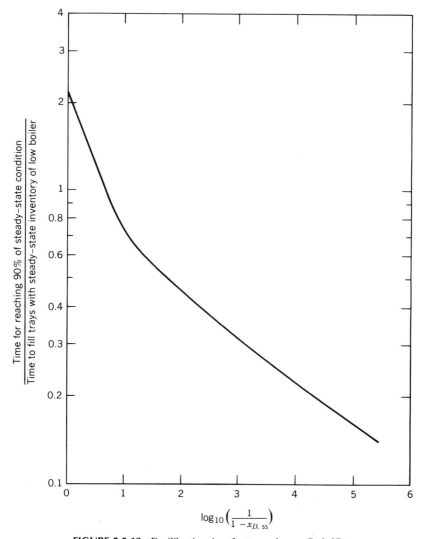

FIGURE 5.5-12 Equilibration time for tray columns (Ref. 27).

material. It is here where tray holdup enters into the picture, and estimates of holdup may be made based on methods given in Section 5.7. The abscissa term includes the steady-state content of low boiler in the stillpot. For packed columns, estimates of holdup may be made, based on methods given in Section 5.8.

For a more rigorous approach to the evaluation of equilibration time, the book by Holland and Liapis[28] should be consulted.

OPERATING POLICY
Batch distillations may be carried out under several different policies. These include constant distillate purity, constant reflux ratio, intermediate cuts that are recycled, and equilibration between cuts. Frequently, a good policy is to take over each cut at constant reflux ratio, take over intermediate cuts between each pair of products, and not equilibrate between cuts. This is shown graphically in Fig. 5.5-13.

CYCLE TIME
The total cycle for batch distillation includes charging of the stillpot, equilibration of the system, period of product takeover, and shutdown/cleanout. Time for pumping in the charge is determined by straight-forward methods. Equilibration time has been discussed above. The time for taking over product is a function of the column capacity at the operating reflux, and the approach to the maximum capacity (flooding) of the column that is chosen. Bogart[26] suggests the following equation for approximate determination of distillation time when product composition is kept constant (i.e., when reflux ratio is varied):

$$t = \frac{S_c(\bar{x}_D - x_{s\,0})}{U_s} \int_{x_{st}}^{x_{s0}} \frac{dx_s}{(1 - L/V)(\bar{x}_D - x_s)^2} \tag{5.5-15}$$

In the use of this equation, values of x are selected and the integration performed graphically or analytically. The value of the superficial vapor velocity, U_s, is determined from column capacity correlations given in Section 5.7 and 5.8.

RIGOROUS METHODS FOR MODELING BATCH DISTILLATION
It is clear that for a multicomponent system with varying holdup and the composition transients inherent in batch processing, rigorous models become quite complex. The calculations require a large digital computer, and the underlying models are discussed by Holland and Liapis,[28] Meadows,[29] Distefano,[30] and Boston et al.[31] The last-named authors developed what is probably the most comprehensive computer program for batch distillation, BATCHFRAK, and this program is available through arrangements with Aspen Technology, Inc., Cambridge, MA.

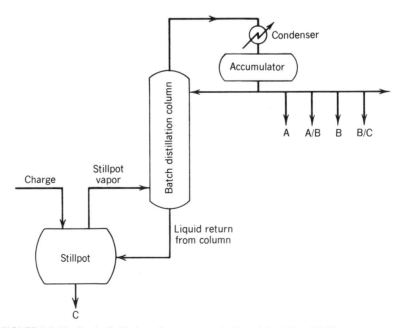

FIGURE 5.5-13 Batch distillation of components A, B, and C. A/B and B/C cuts go to recycle.

5.6 TRANSFER UNITS

Computationally, distillation columns are treated as staged devices, even though in some designs they function as countercurrent contactors such as packed columns. Accordingly, stage concepts as detailed in Section 5.3 may be applied to packed columns as long as the physical contacting mechanisms are not under study. This approach has led to the concept of the "height equivalent to a theoretical plate," discussed later in Section 5.10. When it is desirable to consider the true countercurrent contacting mode, the concept of the *transfer unit* should be used.

Consider a differential height of a packed distillation column of *unit cross section* as shown in Fig. 5.6-1. For component i,

$$L\,dx_i + x_i\,dL = V\,dy_i + y_i\,dV \qquad (5.6\text{-}1)$$

For equimolar counterdiffusion (approximately the case for distillation), $dV = dL$, and for net transfer of i from bulk vapor to the liquid interface,

$$V\,dy_i = (k_V a)(y_i - y_{i,\,int})\,dZ \qquad (5.6\text{-}2)$$

$$\frac{dy}{y_i - y_{i,\,int}} = \frac{k_V a\,dZ}{V} = \frac{1}{H_V}\,dZ \qquad (5.6\text{-}3)$$

Note that this expression applies to the vapor phase only (transfer *to* the interface from the bulk vapor). For overall transfer from the bulk vapor to the bulk liquid,

$$\frac{dy_i}{y_i - y_i^*} = \frac{K_{OV}a\,dZ}{V} = \frac{1}{H_{OV}}\,dZ \qquad (5.6\text{-}4)$$

where H_V is defined as the *height of a vapor-phase transfer unit* and H_{OV} is defined as the *height of an overall vapor-phase transfer unit*, using vapor-phase concentration units. (Similar expressions may be derived for heights of liquid-phase transfer units, but they are not often used in distillation calculations.) It should be noted that in Eq. (5.6-3), the term $y_{i,\,int}$ is the vapor mole fraction at the interface, whereas in Eq. (5.6-4), y_i^* is the vapor mole fraction that would be in equilibrium with the bulk liquid mole fraction. Integration of Eqs. (5.6-3) and (5.6-4) gives

$$\int_1^2 \frac{k_V a\,dZ}{V} = \int_1^2 \frac{1}{H_V}\,dZ = \int_1^2 \frac{dy_i}{y_i - y_{i,\,int}} = N_V \qquad (5.6\text{-}5)$$

$$\int_1^2 \frac{K_{OV}a\,dZ}{V} = \int_1^2 \frac{1}{H_{OV}}\,dZ = \int_1^2 \frac{dy_i}{y_i - y_i^*} = N_{OV} \qquad (5.6\text{-}6)$$

The right sides of Eqs. (5.6-5) and (5.6-6) define the *number of vapor-phase transfer units* and the *number of overall vapor-phase transfer units* (vapor concentration basis), respectively. Thus, the required height

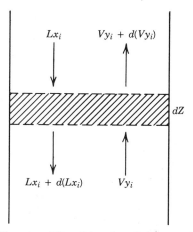

FIGURE 5.6-1 Flows in a differential section of a countercurrent contactor.

of packing in a column is obtained from

$$Z = N_V H_V = N_{OV} H_{OV} \tag{5.6-7}$$

where

$$H_V = \frac{V}{k_V a} \tag{5.6-8}$$

and

$$H_{OV} = \frac{V}{K_{OV} a} \tag{5.6-9}$$

evaluated under conditions such that the terms can be brought outside the integral signs of Eq. (5.6-5) and (5.6-6).

For a binary distillation system of components A and B, $N_{V,A} = N_{V,B}$ and $N_{OV,A} = N_{OV,B}$. For multicomponent systems, it is possible for each component to have a different value of the transfer unit. For a discussion of the multicomponent problem, see Krishnamurthy and Taylor.[1] The usual practice is to deal with the multicomponent mixture as if a staged column were to be used and then convert from theoretical stages to transfer units by the relationships

$$N_{OV} = N_t \frac{\ln \lambda}{\lambda - 1} \tag{5.6-10}$$

where N_t = the computed number of theoretical stages
λ = the ratio of slopes of the equilibrium and operating lines

Equation (5.6-10) will be discussed in more detail in Section 5.10.

5.7   TRAY-TYPE DISTILLATION COLUMNS

5.7-1   General Description and Types

The typical distillation column has dimensions sufficient for handling the required flows of vapor and liquid and for making the desired separation. The vessel contains internal devices that are designed to promote intimate contacting of the vapor and liquid. The amounts and properties of the streams usually follow from the equilibrium stage or transfer unit calculations (Sections 5.3 and 5.6).

The internal devices may be grouped into two general categories: *tray-type* and *packing-type*. The former provides a stagewise contacting mode whereas the latter provides a countercurrent mode. In the present section the tray-type devices will be considered.

An enormous amount of work has gone into the study of contacting devices, and the literature on their performance characteristics is extensive. Most of the work dealing with larger equipment has been based on plant observations, but a significant portion of it has been based on controlled experiments in commercial-scale equipment by Fractionation Research, Inc. (FRI). The methods for analysis and design of distillation columns, presented in this section, are based on a combination of fundamental research papers, results released by FRI, and many reported plant tests.

General types of tray columns are shown in Fig. 5.7-1. The crossflow tray column is the most prominent in industry, with the trays containing vapor dispersers in the form of bubble caps, liftable valves, or simple round perforations. Liquid flows down the column, from tray to tray, via connecting downcomer channels and thus intermittently comes in contact with the upflowing vapor. The counterflow tray ("dualflow tray") column has no downcomers; liquid and vapor use the same openings, which normally are round perforations but which in some cases are in the form of slots. Clearly, for such a device to operate stably, the hydrodynamics must be controlled carefully. The baffle tray column contains simple baffles, or shower decks, over which the liquid flows in a turbulent fashion, contacting the vapor during the fall from one baffle to the next. Views of typical tray-type devices are shown in Fig. 5.7-2.

When one recognizes that in addition to trays, a number of different packing materials may be used for contacting devices, the problem of the selection of the optimum device becomes apparent. Table 5.7-1 presents several criteria that should be considered both for new designs and for the analysis of existing equipment. The listings are generally self-explanatory, but special mention should be made of the "design background" criterion. The engineer must have reasonable confidence that the selected device will behave in the manner expected; this gives some advantage to the better-known devices but may exclude newer as yet untested devices that could have strong potential for cost savings. Such devices are often of a proprietary nature, with the owners persuading the users on the basis of undocumented experience in ill-defined installations. Designers must be wary of claims not based on hard results in known services.

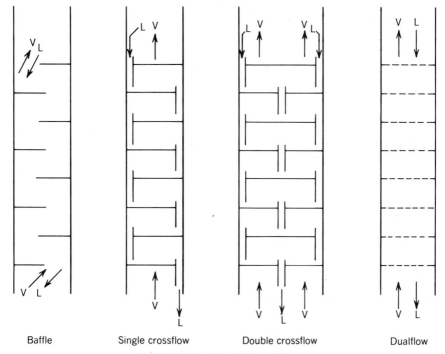

Baffle	Single crossflow	Double crossflow	Dualflow

FIGURE 5.7-1 Types of tray columns.

The contacting devices, or internals, of the distillation column may be classified as shown in Table 5.7-2. The categories represent convenient bases for presentation of design methods and as such are followed in this chapter of the handbook. For many years the crossflow tray dominated the chemical and petroleum processing industries, with the other devices being used only for special services. In recent years the packings have become prominent and today tend to be dominant both for new designs and for the retrofitting of existing columns to improve their efficiency and capacity.

5.7-2 Properties of Tray Froths and Sprays

The sieve tray will be selected as the most common crossflow device, and detailed attention to its performance characteristics will be given. The other crossflow devices will then be considered as modifications of the sieve tray. A schematic diagram of this basic sieve, or perforated, tray is shown in Fig. 5.7-3. Allocations of cross-sectional area are shown in Fig. 5.7-4.

Liquid enters from the downcomer at the left, flows through the zone aerated by the upflowing vapor, and departs into the downcomer at the right. A two-phase mixture exists on the tray and it may be either liquid-continuous (a bubbly froth) or vapor-continuous (a spray), or some combination of the two. The objective of the designer is to determine the mass transfer efficiency and pressure drop brought about by this contacting action; resultant typical performance profiles are shown in Fig. 5.7-5.

It is convenient to define a *contacting unit* as the space between trays, including the downcomer(s) associated with one of the trays (Fig. 5.7-6). Within the unit three zones can be defined. Zone A is immediately above the perforations and is liquid continuous. Zone C is in the downcomer and is both liquid- and vapor-continuous. Zone B is at the vapor outlet end of the unit and is vapor-continuous. The locations of the boundaries shown in Fig. 5.7-6 are intended to be illustrative only; even if clear demarcation between zones were possible, the boundaries would change with many design and operating parameters.

For a so-called "well-behaved" sieve tray, Zone A comprises a froth (bubbly, or aerated, mixture of vapor and liquid) with observable height. Liquid droplets are projected or carried into Zone B, and some of them may be entrained from that zone to the tray above. There is also droplet movement into Zone C, in addition to normal movement of froth over the outlet weir. For many designs an attempt is made to have Zone A predominate in the mass transfer process; the well-behaved sieve tray operates in the froth contacting mode if at all possible.

Under some circumstances, usually at high volumetric ratios of vapor to liquid flow (as in vacuum fractionation), Zone A inverts to a vapor-continuous spray, with Zone B representing an extension of the

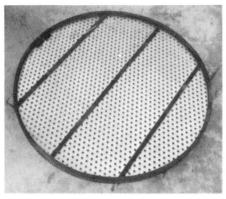

Dualflow tray (Fractionation Research, Inc.).

Typical valve tray (Koch Engineering Co.).

FIGURE 5.7-2   Typical tray-type devices.

Crossflow sieve tray (Fractionation Research, Inc.).

spray but with a higher fraction of vapor than for Zone A. When this happens, the liquid tends to move into the tray as a turbulent cloud of spray droplets. The region of inversion from froth to spray (and vice versa) can be predicted and this will be discussed later.

FROTH CONTACTING

When Zone A is expected to be largely froth, contacting efficiency can be predicted through the use of the following equations. The volume of the froth zone may be defined as

$$Q_f = Z_f A_a \tag{5.7-1}$$

and the average porosity of the froth is

$$\epsilon = \frac{Z_f - h_L}{Z_f} = 1 - \frac{h_L}{Z_f} \tag{5.7-2}$$

The average density of the froth is

$$\rho_f = \frac{h_L \rho_L + (Z_f - h_L)\rho_V}{Z_f} \tag{5.7-3}$$

The relative froth density (dimensionless) is

$$\frac{\rho_f}{\rho_L} = \phi_f = \frac{h_L}{Z_f} + \epsilon \frac{\rho_V}{\rho_L} \tag{5.7-4}$$

TABLE 5.7-1 Criteria for the Selection of Distillation Column Contacting Devices

	General
Vapor-handling capacity	The device must permit reasonable volumetric flow of vapor without excessive entrainment of liquid or, at the maximum vapor rate, flooding.
Liquid-handling capacity	There must be channels for liquid flow that will be nonconstrictive, otherwise the column will flood due to excessive liquid backup.
Flexibility	The device should allow for variations in vapor and liquid flow, to accommodate those periods when demand for production fluctuates.
Pressure drop	For situations when pressure drop can be costly, for example, in vacuum distillations of heat-sensitive materials, the device should maximize the ratio of efficiency to pressure drop.
Cost	The device should not be excessively complex, and therefore costly to manufacture. However, one must consider the total cost of the system; an expensive device might permit a smaller column, lower cost auxiliary equipment, and so on.
Design background	The designer should work with a device in which he or she has confidence and an understanding of the physical principles by which the device will operate.

Special

1. Possible fouling should be considered; some devices resist fouling better than others.
2. Potential corrosion problems place limitations of the type of material and the techniques for fabricating the device to be used.
3. If foaming is expected, some devices can provide a built-in foam-breaking capability.

TABLE 5.7-2 Classification of Contacting Devices

Crossflow trays	Bubble-cap trays Sieve (perforated) trays Valve trays Slotted sieve trays
Counterflow trays	Round openings (''Dualflow'') Rectangular openings (''Turbogrid'')
Baffle trays	Segmental baffles Disk-and-donut baffles Shower decks with perforations
Random packings	Raschig rings, plain and slotted, metal and ceramic Other ring-type packings Berl saddles Intalox saddles, ceramic and metal Other saddle-type packings Specialty random packings
Ordered packings	Corrugated metal and plastic gauze Corrugated sheet metal Mesh-structured Grid-arranged

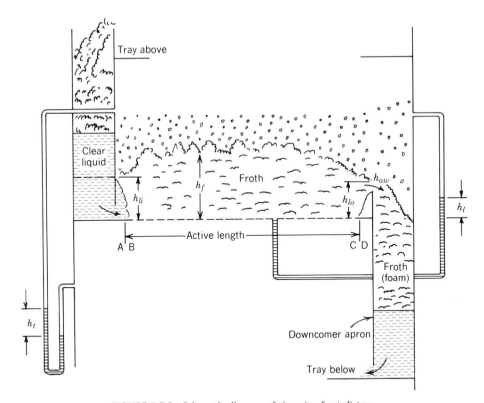

FIGURE 5.7-3 Schematic diagram of sieve (perforated) tray.

When $\rho_L \gg \rho_V$,

$$\phi_f \sim \frac{h_L}{Z_f} \quad \text{and} \quad 1 - \phi_f \sim \epsilon \tag{5.7-5}$$

An average residence time of vapor flowing through the froth may be estimated as

$$\bar{t}_V = \frac{\epsilon Z_f A_a}{Q_V} = \epsilon \frac{Z_f}{U_a} \frac{(1 - \phi_f)h_L}{\phi_f U_a} \tag{5.7-6}$$

In a similar fashion, average liquid residence time may be estimated:

$$\bar{t}_L = \frac{(1 - \epsilon)Z_f A_a}{q} = \frac{(1 - \epsilon)Z_f}{U_a} \tag{5.7-7}$$

Equations (5.7-6) and (5.7-7) involve froth height Z_f. Estimation of this height may be made by the use of Fig. 5.7-7, which is adapted from work sponsored by the American Institute of Chemical Engineers (AIChE).[1] The more recently published approach of Bennett et al.[2] permits avoidance of this parameter by use of Eqs. (5.7-8) and (5.7-9):

$$\bar{t}_V = \frac{(1 - \phi_e)h_L}{12 \phi_e U_a} \tag{5.7-8}$$

$$\bar{t}_L = \frac{h_L A_a}{q} \tag{5.7-9}$$

which imply that

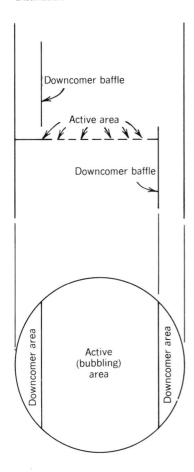

FIGURE 5.7-4 Allocations of cross-sectional area, crossflow tray: A_N = net area = $A_A + A_D$ = active area + one downcomer area; A_T = total tower cross-sectional area = $A_A + 2A_D$.

$$Z_f = \frac{h_L}{\phi_e} \qquad (5.7\text{-}10)$$

Values of the terms in Eqs. (5.7-8)–(5.7-10) may be obtained from Eqs. (5.7-25)–(5.7-28), presented later in the discussion of pressure drop through sieve trays.

SPRAY CONTACTING
As mentioned, under certain high ratios of the flow rates of vapor and liquid, a vapor-continuous region can exist in the contacting zone. While there is not a great deal of information on the characteristics of the resulting spray, at least with the implications regarding mass transfer, there are methods for predicting whether a froth is likely *not* to exist. One method, published by Hofhuis and Zuiderweg,[3] will be shown in connection with the prediction of flooding (Fig. 5.7-10). Another useful method for predicting whether a froth might exist is given by Loon et al.[4] and is shown graphically in Fig. 5.7-8. This chart indicates that the following conditions favor the spray regime:

High vapor rate
Low liquid rate
Low hole area (i.e., high hole velocity)
Large hole size

5.7-3 Vapor Capacity

The usual approach for a new design is first to determine on a tentative basis the required diameter of the column. This diameter often is controlled by the amount of vapor to be handled, although in some cases it may be more a function of liquid flow. After the tentative diameter is calculated, based on an assumed

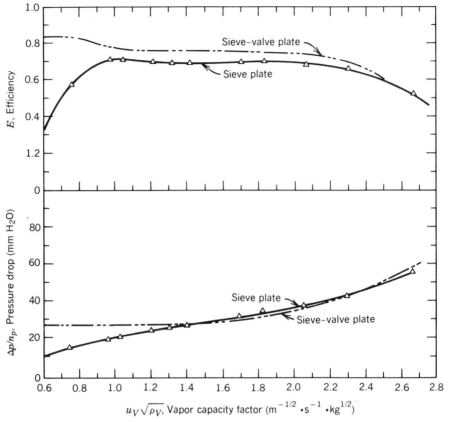

FIGURE 5.7-5 Typical performance profiles for a sieve tray and a combination sieve and valve tray: ethylbenzene-styrene system at 0.13 atm (Ref. 19).

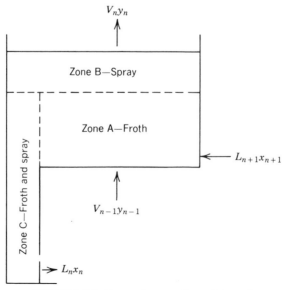

FIGURE 5.7-6 Zones of a contacting unit (tray n).

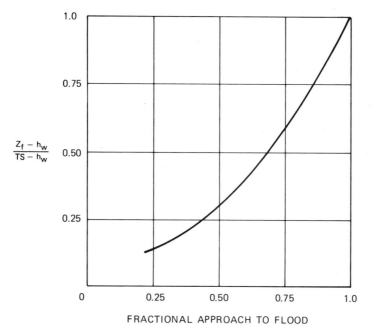

$$\frac{Z_f - h_w}{TS - h_w}$$

FRACTIONAL APPROACH TO FLOOD

FIGURE 5.7-7 Chart for estimating height of froth on sieve trays.

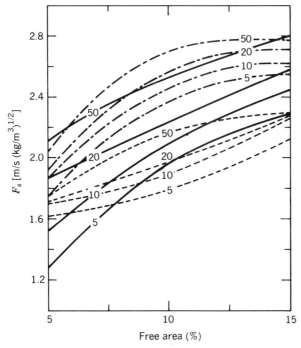

F_s [m/s (kg/m^3)$^{1/2}$]

Free area (%)

FIGURE 5.7-8 Chart for estimating whether froth contacting can be expected. When operating point falls below the appropriate curve, the froth regime prevails (Ref. 4). —————: $\frac{1}{4}$ in. (0.635 cm) holes; —: $\frac{1}{2}$ in. (1.27 cm) holes; —‑‑: $\frac{3}{4}$ in. (1.91 cm) holes. Values on curves are liquid loadings in (m^3/h)/m weir.

geometry of the contacting device, the capacity for liquid is checked and any needed adjustments are then made. Thus, initial attention is given to vapor capacity.

5.7-4 Maximum Vapor Flow

It is clear that if the vapor rate through the tray is excessively high, liquid droplets ("liquid entrainment") will be carried to the tray above, either passing through its perforations or coalescing into larger drops that can fall downward. Also, the very high vapor rate is accompanied by a high pressure drop. It has been observed that when the *flood point* is approached, operation of the tray becomes unstable, mass transfer efficiency drops to a very low value (see Fig. 5.7-5), liquid builds up in the downcomers, and the maximum vapor loading has been reached. This flood point thus represents an upper operating limit and it is convenient for the designer to locate it in context with the other operating parameters (see Fig. 5.7-9).

If a droplet of liquid is considered suspended above the two-phase mixture, with the drag force of the vapor being exactly counterbalanced by the force of gravity, the vapor velocity for suspension may be determined from the equation

$$U_N = \left(\frac{4\ gD_d}{3\ C_d}\right)^{1/2} \left(\frac{\rho_L - \rho_V}{\rho_V}\right)^{1/2} \tag{5.7-11}$$

where
U_N = velocity of approach to the tray (based on net area A_N in Fig. 5.7-4)
D_d = diameter of the droplet
C_d = drag coefficient (dimensionless)
ρ_V, ρ_L = vapor and liquid densities, respectively

The first term on the right-hand side of Eq. (5.7-11) has a range of values depending on the droplet size distribution above the two-phase mixture. For practical design purposes it is termed the capacity parameter C_{SB}, after the work of Souders and Brown.[5] For a condition of maximum vapor capacity ("entrainment flood capacity"),

$$U_{Nf} = C_{SBf} \left(\frac{\rho_L - \rho_V}{\rho_V}\right)^{1/2} \tag{5.7-12}$$

Equation (5.7-12) is the basis for most correlations used to predict maximum allowable vapor velocity. The correlations are based on observed flood conditions in operating columns as evidenced by a sharp drop in efficiency (Fig. 5.7-5), a sharp rise in pressure drop, or a liquid loading condition giving difficulty in maintaining operating stability. A discussion of entrainment flood measurement has been given by Silvey and Keller.[6]

Of the correlations for C_{SBf}, the one that has best stood the test of time is shown in Fig. 5.7-10.[7] The abscissa term, a *flow parameter*, represents a ratio of liquid to vapor kinetic energies. It also indicates

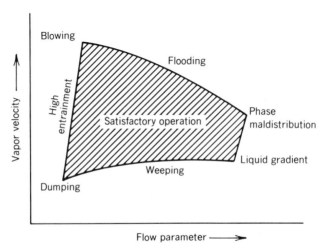

FIGURE 5.7-9 Generalized performance diagram, crossflow trays.

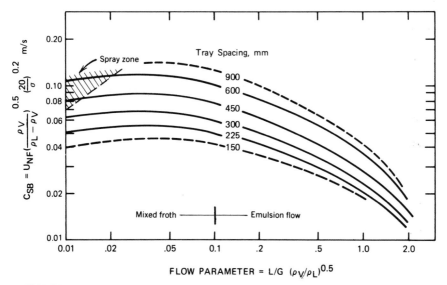

FIGURE 5.7-10 Flooding correlation for crossflow trays (sieve, valve, and bubble-cap trays).

zones where spray or froth might predominate.[3] For flow parameter values higher than about 0.1, a froth (two-phase bubbly mixture) is usually dominant.

Figure 5.7-10 may be used for design purposes, for sieve, bubble-cap, or valve trays, to obtain the maximum velocity:

$$U_{Nf} = C_{SBf} \left(\frac{\rho_L - \rho_V}{\rho_V} \right)^{0.5} \left(\frac{\sigma}{20} \right)^{0.2} \tag{5.7-13}$$

with the following restrictions:

1. Low-foaming to nonfoaming system.
2. Weir height less than 15% of tray spacing.
3. Hole diameter 12.7 mm (0.5 in.) or less (sieve trays)
4. Hole or riser area 10% or more of the active, or bubbling, area (Fig. 5.7-4). Smaller hole areas tend to produce jetting because of the high hole velocities[8] and require correction:

A_h/A_a	U_{Nf}/U_{Nf} from Chart
0.10	1.00
0.08	0.90
0.06	0.80

Figure 5.7-10 has been found to represent flood data for all crossflow trays. Since the correlation first appeared in 1961, a significant number of large-scale flood tests have been reported, many of them in the 1.2 m column of FRI. Analysis has shown that the correlation is conservative,[9] and one can use 90% of the predicted flood values as suitable operating levels for design.

5.7-5 Liquid Entrainment

Reference has been made to liquid entrainment that occurs in increasing amounts as the flood point is approached. The entrained liquid is recycled back to the tray above, negating the effect of countercurrent contacting and decreasing tray efficiency. The recirculation due to entrainment is shown in Fig. 5.7-11.

The deentraining device of Fig. 5.7-11 may not be needed if the "contaminated" overhead vapor meets distillate specifications. Entrainment data for sieve and bubble-cap trays have been correlated by Fair and coworkers[7,10] as shown in Fig. 5.7-12. The sieve tray data are for trays with small (less than 7 mm) diameter. Visual data of FRI, released as a movie,[11] show that under distillation conditions 3 mm holes entrain significantly less than 12.7 mm holes, with hole areas and gross vapor rates being equal.

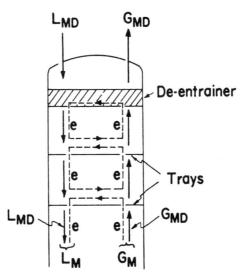

FIGURE 5.7-11 Entrainment recirculation in column. L_{MD} and G_{MD} are liquid and vapor molar flows on a dry basis.

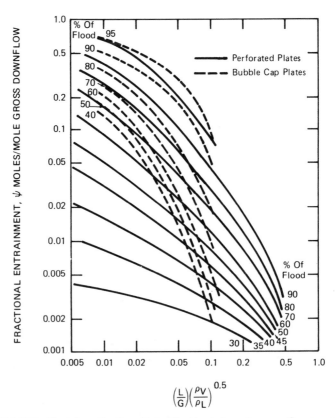

FIGURE 5.7-12 Chart for estimating effect of liquid entrainment on crossflow tray efficiency.

The parameter in the chart is the ratio of entrained liquid to gross downflow:

$$\psi = \frac{e}{L_M} = \frac{e}{L_{MD} + e} \tag{5.7-14}$$

The terms are clear from Fig. 5.7-11. The ψ parameter represents a fraction of the total liquid entering that is entrained upward. It may be used to correct a "dry" efficiency as follows:

$$\frac{E_a}{E_{mv}} = \frac{1}{1 + E_{mv}\,[\psi/(1 - \psi)]} \tag{5.7-15}$$

where E_{mv} is the dry Murphree tray efficiency, to be discussed in Section 5.9. Equation (5.7-15) is based on the early work of Colburn.[12]

Figure 5.7-12 appears to be useful for estimating valve tray entrainment, although the available data are scarce.[9] Absolute values of entrainment obtained from the figure may not be very accurate, but the predicted effect of the entrainment on efficiency appears quite reliable.

EXAMPLE 5.7-1

A distillation column is separating a methanol–water mixture and produces a 98.0 mol % distillate product. The reflux ratio is 1.0, the top of the column operates at 1.0 atm, and the condensing temperature of the overhead product liquid is 65°C (149°F). For the top condition, the vapor density is 1.22 kg/m³, the liquid density is 751 kg/m³, and the surface tension is 20 mN/m (dynes/cm). The net distillate rate is 13,500 kg/h (29,768 lb$_m$/h). The column dimensions are as follows:

Tray Type: Sieve, Single Crossflow, Segmental Downcomers

Column diameter	1.98 m	6.5 ft
Tray spacing	0.61 m	24 in.
Weir height	50 mm	2 in.
Weir length	1.43 m	4.7 ft
Hole diameter	6 mm	$\frac{1}{4}$ in.
Downcomer		
clearance	25 mm	1 in.
Cross-sectional area	3.08 m²	33.2 ft²
Downcomer area	0.31 m²	3.32 ft²
Net area	2.77 m²	29.9 ft²
Active area	2.46 m²	26.6 ft²
Hole area (12%		
open)	0.295 m²	3.18 ft²
Tray metal thickness	2.6 mm	0.10 in.

Calculate the following: (a) approach to flooding and (b) efficiency discount due to entrainment, based on a dry Murphree efficiency of 0.75 (75%). The calculations are to be based on the top plate conditions.

Solution

(a) Vapor mass rate = distillate + reflux = 13,500 + 13,500 = 27,000 kg/h

Vapor volumetric rate = $Q = \dfrac{27,000}{3600 \times 1.22} = 6.148$ m³/s

Net area vapor velocity = $U_N = \dfrac{Q}{A_N} = \dfrac{6.148}{2.77} = 2.22$ m/s

Souders–Brown coefficient = $C_{SB} = U_N \sqrt{\dfrac{\rho_V}{\rho_L - \rho_V}} = 2.22 \sqrt{\dfrac{1.22}{751 - 1.22}} = 0.090$ m/s

Flow parameter = $\dfrac{L}{G}\sqrt{\dfrac{\rho_V}{\rho_L}} = \dfrac{13,500}{27,000}\sqrt{\dfrac{1.22}{751}} = 0.020$

From Fig. 5.7-12, at 610 mm tray spacing, $C_{SBf} = 0.12$ m/s

$$\% \text{ flood} = \frac{0.090}{0.12} \times 100 = 75\%$$

(b) For 75% flood and a flow parameter of 0.020, Fig. 5.7-12 gives $\psi = 0.018$. From Eq. (5.7-15) and
for $E_{mv} = 0.75$ (dry basis),

$$\frac{E_a}{E_{mv}} = \frac{1}{1 + 0.75(0.18/0.82)} = 0.859$$

Thus, the dry efficiency is discounted by about 86%, to a value of $E_a = 0.859(0.75) = 0.644$, or 64%.

5.7-6 Weeping

Whereas bubble-cap trays have a built-in seal against liquid draining through the tray during operation,
this is not the case for sieve trays and valve trays. The sieve tray is especially prone to this drainage, which
in relatively small amounts is called *weeping* and in large amounts, *dumping*. Liquid draining through the
holes causes some short-circuiting, but at the same time provides surface for mass transfer; small amounts
of weeping appear not to be detrimental to the operation and performance of a tray. Thus, it is the dumping
that is normally to be avoided.

Reference to Fig. 5.7-5 shows that at low vapor rates there is a decline in efficiency, and this must be
taken into account in determining the "turndown ratio" (ratio of maximum allowable rate to minimum
allowable rate) of the tray device. It has been found that a reasonable prediction of the minimum rate can
be made with the use of Fig. 5.7-13. This chart was developed earlier by Fair[13] on the basis of visual tests
and was later confirmed by Zanelli and Del Bianco.[14] In practice, it predicts the "point of minimum
turndown" as denoted by Fig. 5.7-5. The pressure loss in forming a gas bubble may be estimated from
the dimensional relationship

$$h_\sigma = \frac{390\,\sigma}{\rho_L d_h} \tag{5.7-16}$$

5.7-7 Pressure Drop

The pressure loss experienced by the vapor in flowing through a tray is that which might be measured by
a manometer as shown in Fig. 5.7-3. It may be assumed that this is the sum of contributions by the
dispersers (holes, caps, valves) and by the head of two-phase mixture:

$$h_t = h_d + h_L' \tag{5.7-17}$$

The term h_L' is treated as a residual and is not necessarily equal to the equivalent head of liquid h_L on the
tray. It also takes into account the fact that the orifice (disperser) has changed characteristics when it is
wetted.

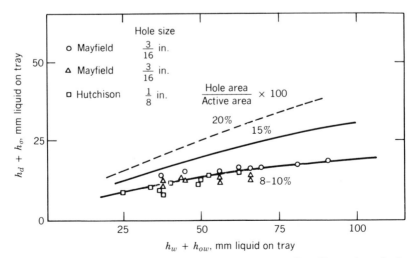

FIGURE 5.7-13 Chart for estimating the weep point of a sieve tray. (*Note:* If operating point lies above
the appropriate curve, weeping is not expected.)

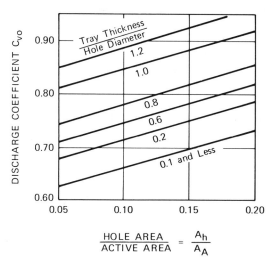

FIGURE 5.7-14 Discharge coefficient for sieve trays.

For sieve trays, the simplified orifice equation is used to estimate the pressure loss for flow through the holes:

$$h_d = \frac{50.8}{C_v^2} \frac{\rho_V}{\rho_L} U_h^2 \qquad (5.7\text{-}18)$$

This is the so-called "dry drop." The discharge coefficient C_v is obtained from Fig. 5.7-14, from the work of Leibson et al.[15] The plot was developed originally for small-hole sieve trays (7 mm diameter and less). It includes the corrections for velocity of approach and for sharpness of the orifice.

For pressure loss through the two-phase mixture, the residual term is estimated from the relationship

$$h_L' = \beta(h_W + h_{ow}) \qquad (5.7\text{-}19)$$

where β is an aeration factor, originally utilized by Hutchinson et al.[16] for sieve tray experiments. This factor has been correlated as shown in Fig. 5.7-15[9,17] and leads to the final equation for estimating pressure loss across a sieve tray:

$$h_t = \frac{50.8}{C_v^2} \frac{\rho_v}{\rho_L} U_h^2 + \beta(h_W + h_{ow}) \qquad (5.7\text{-}20)$$

The curves of Fig. 5.7-15 may be represented by the equation

$$\beta = 0.19 \log_{10} L_W - 0.62 \log_{10} F_{vh} + 1.679 \qquad (5.7\text{-}21)$$

where $L_W = q/l_W$ in m³/s·m. The value of the weir crest h_{ow} in Eqs. (5.7-19) and (5.7-20) is calculated by the classical Francis weir equation:

$$h_{ow} = 664 \, L_W^{2/3} \qquad (5.7\text{-}22)$$

even though the two-phase mixture actually flows over the weir (unless "calming zones" (unperforated sections) are used upstream of the weir. If one wishes to consider that froth actually flows over the weir,

$$\frac{h_{ow}}{\phi_f} = 664 \left(\frac{L_W}{\phi_f}\right)^{2/3} \qquad (5.7\text{-}23)$$

or

$$h_{ow} = 664 \, \phi_f^{1/3} \, L_W^{2/3} \qquad (5.7\text{-}24)$$

In the two foregoing equations, ϕ_f is a relative froth density, ρ_f/ρ_L.

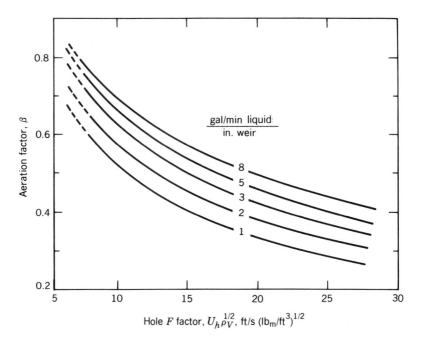

FIGURE 5.7-15 Aeration factor for sieve trays. (*Note:* ft/s $(lb_m/ft^3)^{1/2} \times 1.22 = $ m/s $(kg/m^3)^{1/2}$ and [(gal/min)/in. weir] $\times$ 0.00249 = (m^3/s)/m weir.

A more exact analysis of the hydraulic parameters is possible, and for this the paper by Bennett et al.* should be consulted. In that work, an effective froth density is defined as follows:

$$\phi_e = \exp\left[-12.55\ (C'_{SB})^{0.91}\right] \qquad (5.7-25)$$

where $C'_{SB} = C_{SB}(A_N/A_a)$ is a capacity parameter based on the active area instead of the net area. Liquid head above the perforations is given by

$$h_L = \phi_e\left[h_W + 15{,}330\ C'\left(\frac{q}{\phi_e}\right)^{2/3}\right] \qquad (5.7-26)$$

with the correction factor C' obtained from

$$C' = 0.0327 + 0.0286\ \exp\ (-0.1378\ h_W) \qquad (5.7-27)$$

Finally, the residual drop is obtained from Eqs. (5.7-16) and (5.7-27):

$$h'_L = h_o + h_L \qquad (5.7-28)$$

and this is added to the dry tray drop to obtain the total tray pressure drop. The model of Bennett and coworkers correlates a large bank of data with an average error of -0.8% and a mean absolute error of 6.0%. The reliability of the aeration factor approach is about 10% mean absolute error. It should be noted that Bennett and coworkers suggest the use of the Leibson plot (Fig. 5.7-14) to obtain dry tray pressure loss.

For valve trays, design manuals of the vendors should be used; these manuals are readily available. The approach used is the same as the aeration factor approach described above, except that constant values of this factor are used. For bubble-cap trays, the "dry drop" is more complex because of the unique geometry and mode of flow. For example, the total dry drop can be the sum of losses for flow through the riser, through the annulus, and through the slots. The 1963 work by Bolles[18] is undoubtedly the best currently available method for determining the hydraulics of bubble-cap trays. Since these trays are rarely used for new designs, and since no advances in the technology have been made since 1963, the Bolles paper should be consulted when it is necessary to analyze the performance of existing bubble-cap installations.

5.7-8 Liquid Handling Capacity

The downflowing liquid is transported from a tray to the tray below by means of conduits called *downcomers*, and it is evident that if the downcomer is not sufficiently large to handle the required liquid load, the pressure drop associated with liquid flow will serve as a constriction and a point of flow rate limitation. In fact, downcomers usually serve to bottleneck operations of high-pressure fractionators and absorbers. They must be sized such that they do not fill completely under the highest flow rates expected for the column. As will be shown, the vapor flow rate contributes toward the liquid capacity limitation.

Figure 5.7-16 shows a diagram of three trays with downcomers, on which is superimposed the liquid backup in one of the downcomers. This backup may be calculated from a pressure balance:

$$h_{dc} = h_t + h'_L + h_{da} + \Delta \tag{5.7-29}$$

The segments of the buildup are (a) the equivalent clear liquid head on the tray h'_L, (b) any hydraulic gradient Δ caused by resistance to liquid flow across the tray, which usually is not significant for sieve trays, (c) liquid head equivalent to pressure loss due to flow under the downcomer apron, h_{da}, and (d) total pressure loss across the tray above, necessarily included to maintain the dynamic pressure balance between point A (just above the floor of tray 3) and point B in the vapor space above tray 2.

Segments a and b are covered by the tray pressure drop calculations discussed in Section 5.7-6. For segment b, only bubble-cap trays require an evaluation, and methods for this are detailed in Ref. 18. For segment c, the head loss for flow under the downcomer may be estimated from the following empirical expression:

$$h_{da} = 165.2 \, (U_{da})^2 \tag{5.7-30}$$

where U_{da} is in m/s and h_{da} is in mm of liquid. Equation (5.7-30) applies to flow under a simple downcomer apron, with no recessed discharge area or tray inlet weir (see Fig. 5.7-17 for liquid inflow arrangements). If an inlet weir is used, the value of h_{da} from Eq. (5.7-30) should be doubled.

This so-called "clearance under the downcomer," represented by the area A (see Fig. 5.7-4), bears special mention, since improperly fitted tray sections can lead to inadequate clearance at one or more points in the column. Thus, the clearance is a dimension that should be checked very carefully during tray installation.

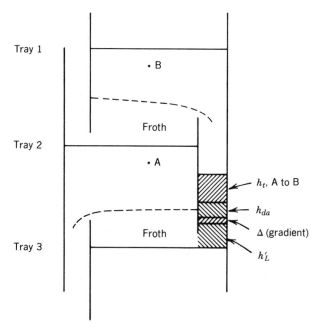

FIGURE 5.7-16 Components of downcomer backup.

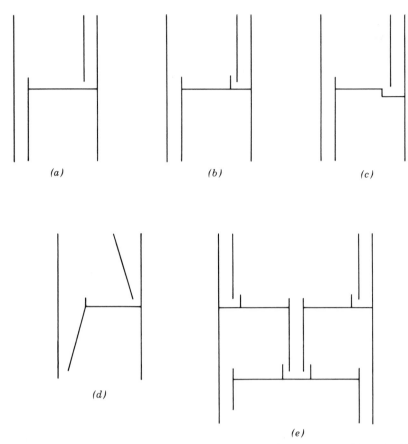

FIGURE 5.7-17 Possible downcomer arrangements: (*a*) vertical baffle, no recess, no inlet weir (this type is quite common); (*b*) vertical baffle, no recess, inlet weir; (*c*) vertical baffle, recess, no inlet weir; (*d*) sloped baffle, no recess, no inlet weir; and (*e*) double crossflow, vertical baffles, no recesses, inlet weirs.

Returning to Fig. 5.7-16, we see that the backup relationship [Eq. (5.7-29)] is based on clear liquid heads. Actually, the fluid in the downcomer contains a large amount of entrapped vapor and can be represented as a froth with an average density of $\overline{\phi}_{dc}$. Accordingly, the *actual* downcomer backup is

$$h'_{dc} = \frac{h_{dc}}{\overline{\phi}_{dc}} \tag{5.7-31}$$

The value of $\overline{\phi}_{dc}$ in Eq. (5.7-31) is obtained as follows. If the flow into the downcomer is a tray froth of density ϕ_f, then $\phi_f = \phi_{dc}$, and if the flow out of the downcomer is clear liquid ($\phi_{dc} \sim 1.0$), then the average value is some mean of these two froth densities. If there is very rapid disengagement of vapor from liquid in the downcomer, the average may be as high as 0.8 or 0.9. On the other hand, for slow disengagement, as in fractionators operating near the critical point of the tray mixture, the poor buoyancy of vapor bubbles can lead to an average downcomer froth density of as low as 0.2 or 0.3. Designers often use an average value of $\overline{\phi}_{dc} = 0.5$, but this value should be considered carefully on the basis of the system as well as the geometry of the downcomer.

It is possible for vapor to be entrained downward with the liquid, and this "reverse entrainment" has been studied by Hoek and Zuiderweg,[20] using Fractionation Research, Inc. data taken at pressures of 20 and 27 atm and a system operating close to its critical point. The downward entrainment was found to affect significantly the overall efficiency of the column. A maximum superficial velocity in the downcomer should be about 0.12 m/s, based on clear liquid and the smallest cross section of the downcomer.

The above discussion of downcomer sizing has dealt with flow rates to minimize downward entrainment. From another point of view, the downcomer must have enough volume to allow collapse of any stable

TABLE 5.7-3 Downflow Capacity Discount Factors, Foaming Systems

Nonfoaming systems	1.00
Moderate foaming, as in oil absorbers and amine and glycol regenerators	0.85
Heavy foaming, as in amine and glycol absorbers	0.73
Severe foaming, as in methyl ethyl ketone units	0.60
Foam-stable systems, as in caustic regenerators	0.15

foam that might develop from the tray aeration. This is handled by "system discount factors," developed empirically over the years for mixtures known to exhibit foaming tendencies. Typical factors are shown in Table 5.7-3.[21] The discount factor should be multiplied by the maximum allowable velocity of 0.12 m/s, mentioned above.

In summary, the downcomer can limit column capacity when liquid flow rates are high, as in absorbers and pressure fractionators. Two viewpoints are used (and these are not necessarily independent of each other): height of froth buildup in the downcomer, obtained from a pressure balance, and residence time in the downcomer, obtained from an entrainment velocity limitation. When the downcomer backs up liquid, the vapor entrains more liquid, and a flooding condition can be approached.

5.7-9 Miscellaneous Comments

The foregoing material relates directly to a "standard" single crossflow tray with vertical downcomer baffles as shown in Fig. 5.7-3. Departures from this design can be handled as follows:

1.   For multiple crossflow trays, divide the liquid flow according to the number of liquid streams (e.g., two streams for a double crossflow tray), and then use the various hydraulic equations as given. Note that for a double crossflow tray the center weir is approximately equal to the column diameter, and for side-to-center liquid movement there is diverging flow with the possibility of stagnant zones if the column size is quite large (4 m or larger). For center-to-side flow, stagnation is less likely.

2.   For sloped downcomers, the critical liquid velocity is at the bottom, insofar as final disengagement of vapor is concerned. The total volume of the filled portion of the downcomer can be used in estimating residence time. For downcomers with bottom recesses, where the liquid must make an extra turn before entering the tray, the pressure loss under the downcomer may be estimated as twice that calculated from Eq. (5.7-30). This rule of thumb applies also to the case where an inlet weir is used to distribute the liquid after it has flowed under the downcomer baffle.

3.   When it is necessary to decrease the hole area of an existing sieve tray, small blanking strips can be used; these are metal pieces that can be fastened directly to the plate and that block the vapor flow through the holes that are covered by the strips. Care should be taken to distribute these strips throughout the tray.

4.   Because of changing vapor and liquid flows throughout the column, it may be desirable to vary the parameters such as hole area and downcomer area. For the former, a fixed design plus variable blanking strip arrangements is often feasible. Variations in downcomer area usually are limited to cases with very wide-ranging liquid flow rates.

5.   For columns with wide variations in total pressure (vacuum columns where the bottom pressure may be twice or more the pressure at the top of the column), it may be desirable to vary the tray spacing, progressing from a larger value at the top to a smaller value at the bottom. Economics usually mitigate against this.

6.   The various hydraulic parameters should be checked at various locations in the column. As a minimum this should be done at four points for a simple, single-feed column: top tray, tray above the feed tray, tray below the feed tray, and bottom tray.

7.   Care should be taken in feeding mixed vapor–liquid streams to the column. The reboiler return is of this character, and its flow should be directed away from the bottom seal pan so as not to hinder liquid flow from that pan (see Fig. 5.7-17). Various baffling arrangements are possible for separating the liquid and vapor for a mixed-phase feed stream.

EXAMPLE 5.7-2
For the problem of Example 5.7-1, calculate the following: (a) pressure drop across the top tray, (b) turndown ratio, and (c) downcomer backup for the tray below the top tray.

Solution.   (a) Vapor flow F-factor is obtained as follows:

$$\text{Based on active area} = F_{va} = \frac{Q}{A_a}\, \rho_V^{1/2} = \frac{6.148}{2.46}\,(1.22)^{1/2} = 2.76\ \frac{\text{m}}{\text{s}}\left(\frac{\text{kg}}{\text{m}^3}\right)^{1/2}$$

$$\text{Based on hole area} \quad = F_{vh} = F_{va}\left(\frac{A_a}{A_h}\right) = 2.76\left(\frac{2.42}{0.295}\right) = 23.0\,\frac{m}{s}\left(\frac{kg}{m^3}\right)^{1/2}$$

The discharge coefficient (Fig. 5.7-14), for a tray thickness/hole diameter of $2.6/6 = 0.43$ and $A_h/A_a = 0.295/2.46 = 0.12$, is $C_v = 0.75$. The weir rate is

$$L_w = \frac{q}{l_w} = \frac{13,500}{3600(751)(1.43)} = 0.00349 \text{ m}^3/\text{s}\cdot\text{m}$$

The aeration factor [Eq. (5.7-21)] is

$$\beta = 0.19\log_{10}(0.00349) - 0.62\log_{10}(23) + 1.679 = 0.37$$

For the weir crest [Eq. (5.7-22)]

$$h_{ow} = 664(0.00349)^{2/3} = 15.3 \text{ mm liquid}$$

For the dry tray drop [Eq. (5.7-18)]

$$h_d = \frac{50.8(1.22)(6.148/0.295)^2}{(0.75)^2(751)} = 63.7 \text{ mm liquid}$$

The total tray pressure drop [Eq. (5.7-20)] is given by

$$h_t = h_d + \beta(h_w + h_{ow}) = 63.7 + 0.37(50 + 15.3) = 87.9 \text{ mm liquid}$$

(b) Estimation of the turndown ratio utilizes Fig. 5.7-13 and a trial-and-error procedure. The pressure loss for bubble formation [Eq. (5.7-16)] is given by

$$h_\sigma = \frac{390(20)}{751(6)} = 1.73 \text{ mm liquid}$$

If we assume 50% of design vapor and liquid rates,

$$\text{Dry drop} = 63.7(0.25) = 15.9 \text{ mm liquid}$$
$$\text{Weir crest} = 9.62 \text{ mm liquid}$$

Then,

$$h_\sigma + h_d = 1.7 + 15.9 = 17.6 \text{ mm liquid}$$
$$h_w + h_{ow} = 50 + 9.62 = 59.6 \text{ mm liquid}$$

These values are on the curve of Fig. 5.7-13. Thus, our assumption is correct;

$$\text{Turndown ratio} = \frac{100}{50} = 2$$

(c) For this part,

$$\text{Downcomer discharge area} = 0.025(1.43) = 0.0358 \text{ m}^2$$
$$\text{Velocity under downcomer} = U_{da} = \frac{0.0050}{0.0358} = 0.140 \text{ m/s}$$

The pressure loss for flow under downcomer [Eq. (5.7-30) is

$$h_{da} = 165.2\,(U_{da})^2 = 165.2(0.140)^2 = 3.24 \text{ mm liquid}$$
$$\text{Backup} = h_{dc} = h_t + h_L' + h_{da} + \Delta \qquad \Delta \sim 0$$
$$h_{dc} = 87.9 + 0.37(50 + 15.3) + 3.24 + 0 = 115.3 \text{ mm liquid}$$

Based on tray spacing,

$$\text{Fractional backup} = \frac{115.3}{610} = 0.189 \quad (18.9\% \text{ of tray spacing})$$

5.8 PACKED-TYPE DISTILLATION COLUMNS

5.8-1 Types of Packing Materials

Packings for distillation columns come in many types, shapes, and sizes. Many of them are nonproprietary and are provided by more than one supplier, but most of the newer and more important devices are proprietary, patent-protected, and obtained from a single source. As noted in Table 5.7-2, the packing elements may be placed in the column in a random fashion, as by dumping. Alternatively, they may be stacked as individual elements or they may be fashioned as rigid meshes, grids, or multiple plates and inserted carefully into the column. This latter type of column internal is known as *structured packing*; the former type is known as *random packing*. A diagram of a "composite" packed column, taken from the paper by Chen,[1] is shown in Fig. 5.8-1.

RANDOM PACKINGS

Traditionally, packed columns have contained random packings. Descriptions of packings in 1934 ranged from jackchain and ceramic rings to carpet tacks and birdshot. Through the years various handy materials were used to provide contact surface for vapor and liquid. Emerging from early experimentation were the fairly standard Raschig rings and Berl saddles in the 1930s and 1940s. In more recent years, two particular random packings have been quite popular—the slotted ring (Pall ring, Flexiring, Ballast ring) and the modified saddle (Intalox saddle, Flexisaddle). These packings have become available in ceramic, plastic, or metal materials. For convenience in the present work, these packings will be called Pall rings and Intalox saddles.

Newer random packings include the Metal Intalox saddle (Norton Company), called IMTP, the Nutter ring (Nutter Engineering Co.), and the Cascade Miniring (Glitsch, Inc.), called CMR. The IMTP and Nutter packings are available only in metal, whereas the CMR is available in both metal and plastic. These newer packings have received extensive field and laboratory testing, and the vendors should be consulted for a review of such testing.

Views of various random packings, useful particularly for distillation service, are shown in Fig. 5.8-2. Properties of several of these packings are shown in Table 5.8-1. It is important to note that the listed specific surface area of the packing (the amount of surface of a single element times the number of elements that can be packed in a given volume) is not necessarily indicative of the mass transfer capability of the packing.

The packings in Fig. 5.8-2 may be separated into two types according to their resistance to the flows of liquid and vapor. The older packings require that the fluids flow *around* them, thus causing pressure loss both by form drag and by skin friction. The Raschig ring, the Berl saddle, and the ceramic Intalox saddle are representative of this type. The newer packings permit fluids to flow *through* them, with greatly reduced form drag. The distinction between these types is significant, the "through-flow" concept having led to the so-called "high efficiency packings" that produce increased mass transfer while minimizing pressure drop.

STRUCTURED PACKINGS

An early form of structured packing was introduced by Stedman[2] who cut pieces of metal gauze and placed them horizontally in small-scale laboratory and pilot plant columns. This packing was found to be quite efficient for mass transfer[3,4] but was regarded as practical only for small columns. In later times, expanded metal was used to fabricate structured elements for larger columns,[5] and knitted wire mesh was developed as a high-efficiency packing,[6] but this too was considered impractical and uneconomical for larger columns. It was not until the 1960s that a structured packing was developed that could be considered a proven and practical material for commercial distillation columns; this was the packing known as Sulzer BX (Sulzer Brothers, Winterthur, Switzerland) and was fabricated from metal gauze. The first test data of significance were published by Billet in 1969,[7] following the granting of a United States patent to Huber.[8] Representative data of Billet, for the ethylbenzene–styrene system, are compared against sieve trays and Pall rings (from Ref. 9) in Fig. 5.8-3. It is clear that the Sulzer BX packing gives a superior combination of high mass transfer efficiency and low pressure drop.

A disadvantage of the Sulzer packing was its high cost of fabrication; in time more economical methods of manufacture plus the substitution of sheet metal for gauze led to a new group of structured packings: Mellapak (Sulzer Brothers), Flexipac (Koch Engineering), Gempak (Glitsch, Inc.), and Montz B (Julius

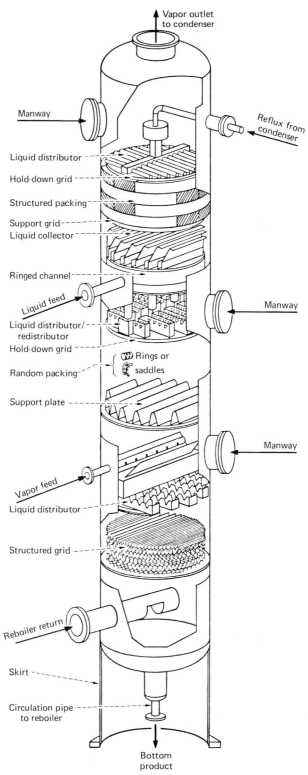

FIGURE 5.8-1 Composite packed column (Ref. 1).

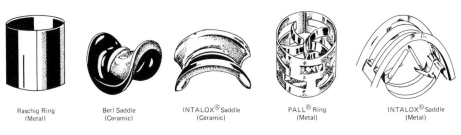

Raschig Ring Berl Saddle INTALOX® Saddle PALL® Ring INTALOX® Saddle
(Metal)   (Ceramic)   (Ceramic)   (Metal)   (Metal)

FIGURE 5.8-2 Representative random packings.

Montz). These structured packings come in various dimensions and are available in a number of materials of construction (typically plastic as well as metals of various types).

Views of several different structured packings are shown in Fig. 5.8-4; careful inspection will show that the packing elements are made from sheets of corrugated sheet metal or gauze and that the sheets are perforated. Furthermore, the sheet metal has been given a special surface treatment to aid in the spreading of the liquid film and thus to emulate the gauze surface, which by capillary action promotes liquid spreading.

The elements of structured packing are installed in layers in the column, with care being taken that the adjoining elements of adjacent layers are oriented such that the liquid flow direction is reversed and that gas is redistributed. For the simple pilot plant element shown in Fig. 5.8-4, adjacent layers are usually rotated by 90°.

Properties of representative packings are shown in Table 5.8-2. More details on typical geometries will be given when the topics of pressure drop and mass transfer efficiency are addressed.

5.8-2 Packed Column Hydraulics

LIQUID DISTRIBUTION

It has always been recognized that for a packed bed to exhibit good performance it must be fed with a uniform and well-distributed liquid flow. The early work of Baker et al.[10] led to the conclusion that for randomly packed beds, unless the column diameter is at least eight times the nominal diameter of the packing pieces, there will be excessive liquid flow along the walls, with resulting loss of efficiency due to channeling. Through the years this limitation has been a guideline and has hampered "scaledown" studies in pilot plant equipment, it being necessary to use at least a 10 cm (4 in.) diameter pilot column if data on minimum size packings (12 mm) were to be scaled up to some commercial size.

Fortunately for the designers, the traditional "bluff body" packings such as Raschig rings and Berl saddles have had some capability to correct any maldistribution of liquid in the center portions of the bed, especially when the column diameter to packing element diameter was very large. Silvey and Keller[11] reported on tests with 3.8 cm (1.5 in.) ceramic Raschig rings in a 1.2 m (4 ft) column and showed how the distribution of the liquid improved as it moved down the packed bed. Albright[12] has provided a method for estimating the "characteristic distribution" of liquid in randomly packed beds, based on a theoretical model involving the diversion of flow by the packing pieces. A number of researchers have found that for bluff packings an initial liquid distribution of at least 40 pour points per square meter of column cross section (4 points/ft²) is generally satisfactory.

For the higher efficiency packings, those having the "through-flow" characteristic mentioned earlier, the packed bed appears to offer little correction of a poor initial liquid distribution. This matter is discussed by Chen,[1] who also provides an excellent coverage of packed column internals in general, and by Kunesh et al.[13], who report on studies at Fractionation Research, Inc. The consensus of designers and researchers is that for the high-efficiency packings there should be at least 100 pour points per square meter (about 10 points/ft²).

Types of liquid distributors are shown in Fig. 5.8-5. The orifice/riser device can be developed in a variety of forms, and the risers are often of a rectangular cross section. The orifices must be distributed properly over the cross section and must offer enough resistance to liquid flow that an adequate head of liquid can be maintained over them. This can lead to difficulties: small holes can be easily plugged, and out-of-levelness of the distributor can lead to dry sections, especially when liquid rate is considerably below the design value.

For orifice-type distributors, the head of liquid required to produce a given flow through the orifice can be calculated from a dimensional form of the orifice equation:

$$h_{\text{dist}} = 100 \ U_h^2 \tag{5.8-1}$$

where h_{dist} is in millimeters of liquid and U_h is the velocity of liquid through the hole in meters per second.

The perforated pipe distributor (Fig. 5.8-5) is capable of providing many pour points and can be

TABLE 5.8-1 Properties of Random Packings

Packing Type	Nominal Size (mm)	Elements (per m³)	Bed Weight (kg/m³)	Surface Area (m²/m³)	ε Void Fraction	F_p Packing Factor (m⁻¹)	$F_p \epsilon^2$	Vendors[a]
Intalox saddles (ceramic)	13	730,000	720	625	0.78	660	402	Norton, Koch, Glitsch
	25	84,000	705	255	0.77	320	190	
	38	25,000	670	195	0.80	170	108	
	50	9,400	670	118	0.79	130	81	
	75	1,870	590	92	0.80	70	45	
Intalox saddles (metal)	25	168,400	350	n.a.[b]	0.97	135	127	Norton
	40	50,100	230	n.a.	0.97	82	77	
	50	14,700	181	n.a.	0.98	52	50	
	70	4,630	149	n.a.	0.98	43	41	
Pall rings (Ballast rings, Flexirings) (metal)	16	49,600	480	205	0.92	230	195	Norton, Glitsch, Koch
	25	13,000	415	130	0.94	157	139	
	38	6,040	385	115	0.95	92	93	
	50	1,170	270	92	0.96	66	61	
	90				0.97	53	50	
Raschig rings (ceramic)	13	378,000	880	370	0.64	2000	819	Norton, Glitsch, Koch, others
	25	47,700	670	190	0.74	510	279	
	38	13,500	740	120	0.68	310	143	
	50	5,800	660	92	0.74	215	118	
	75	1,700	590	62	0.75	120	68	
Berl saddles (ceramic)	13	590,000	865	465	0.62	790	304	Koch, others
	25	77,000	720	250	0.68	360	166	
	38	22,800	640	150	0.71	215	108	
	50	8,800	625	105	0.72	150	78	
Intalox saddles (plastic)	25	55,800	76	206	0.91	105	87	Norton, Glitsch, Koch
	50	7,760	64	108	0.93	69	60	
	75	1,520	60	88	0.94	50	44	
Pall rings (plastic)	16	213,700	116	341	0.87	310	325	Norton, Glitsch, Koch
	25	50,150	88	207	0.90	170	138	
	50	6,360	72	100	0.92	82	69	

[a]Identification of vendors: Norton Company, Akron, OH; Glitsch, Inc., Dallas, TX; Koch Engineering Co., Wichita, KS.
[b]n.a.—not available.

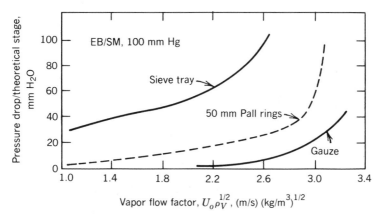

FIGURE 5.8-3 Pressure drop/efficiency comparisons, trays and packings: ethylbenzene–styrene system at 100 nm Hg pressure (Refs. 8 and 9).

equipped with a central reservoir to allow variations in liquid head as flow rates change (and as the holes become partly fouled), and it also has the advantage of offering very little resistance to the flow of vapor. However, the need for resistance to liquid flow limits the total hole area (as in the case of the orifice/riser unit) that can be used. Also, for complex lateral piping (containing the holes), careful fluid flow calculations must be made to ensure that all holes discharge equally. For each hole, Eq. (5.8-1) may be applied.

The trough-type distributor can accommodate to wide swings in liquid rate, since the V-notch orifices can function at various heads. A disadvantage is that a uniform and extensive distribution of pour points may be difficult to achieve. Also, the streams from the V-notch weirs tend to be coarse and with attendant splashing, and the troughs offer some resistance to vapor flow. However, the trough distributor is more resistant to fouling and indeed can collect significant amounts of solids before losing its ability to distribute the liquid. Thus, troughs are often used when distribution is not overly critical and when dirty liquids are used. If the trough openings are in the form of V-notches, the following equation is suitable for estimating flow as a function of liquid head over the bottom of the notch:

$$h = 851 \left(\frac{q}{\tan \theta/2} \right)^{0.4}$$

(5.8-2)

where θ is the notch angle and q is in $m^2/s \cdot$notch.

The effect of distributor type on packing mass transfer efficiency is shown in Fig. 5.8-6, taken from the presentation by Kunesh et al.[13] The "tubed drip pan" is an orifice riser distributor with tubes extending toward the packing from each orifice. The system studied was cyclohexane–n-heptane. The term HETP, to be discussed in Section 5.10, refers to "height equivalent to a theoretical plate" and is a measure of reciprocal efficiency of the packed bed. The tubed drip pan shows a 50% higher mass transfer efficiency than the notched trough unit when a through-flow packing such as slotted Raschig rings is used.

Sprays also may be used for liquid distribution. When more than one nozzle is needed, overlapping (or underlapping) of spray patterns is inevitable. Care must be taken to select a full cone spray nozzle that does not have too fine a mean drop size (to prevent entrainment of liquid by the rising vapor) and which has a good pattern of spray over the needed cross section. Nozzles are available for handling liquids containing suspended solids, and thus spray distribution is a possible solution to the need to distribute a dirty liquid.

LIQUID REDISTRIBUTION

As shown in Fig. 5.8-1, when a stream is added or withdrawn from the side of the column, it is necessary to collect the liquid from the bed above and redistribute to the next bed below. Considering the bed from which the liquid is to be collected, a support plate (see below) is used for the packing support, and liquid flowing from this plate is collected in a device that resembles the orifice/riser distributor, except that instead of the orifices a sump or weir box is used to channel the liquid to a distributor serving the bed below. In the instance of a liquid feed being admitted between these beds, the liquid is fed to the lower distributor and mixed with the liquid from the collector above. This arrangement is diagrammed in Fig. 5.8-7. For a vapor feed, or a feed that is a mixture of vapor and liquid, a special baffling arrangement can be used to separate the two phases.

Another aspect of liquid redistribution involves the handling of liquid that has migrated to the walls

FIGURE 5.8-4 Structured packing (Flexipac) of the sheetmetal type: (*a*) fabricated to fit a pilot-scale column and (*b*) arranged to fit through the manways of a larger-diameter column (Koch Engineering Co.).

within the packed bed itself. This can be handled with wall wiping devices, if the bed contains random packing; such devices are simple ring-type affairs that slope downward (Fig. 5.8-8) and that must be placed in the bed during the packing process. Wall wipers are generally not needed if the column diameter is more than 10 times the packing piece diameter. Silvey and Keller[11] found that for Raschig rings and a trough-type distributor, no redistribution was needed for bed heights up to about 10 m (33 ft). A good general rule is to limit the height of randomly packed beds to about 7 m (23 ft).

VAPOR DISTRIBUTION
While it is apparent that vapor (or gas) should be distributed uniformly to the packing at the bed support, there is usually the assumption that if the initial distribution is not good, pressure drop in the bed will make the necessary correction. While this assumption is valid for the traditional bluff body packings, it does not apply to the high-efficiency through-flow random packings or to the modern structured packings. The inherent low pressure drop of such packings may not be sufficient to correct vapor maldistribution.

There are no quantitative guidelines for determining how "good" the distribution should be. The bed support may offer little help in ironing out maldistribution. Reasonable care should be used to disperse

TALE 5.8-2 Properties of Structured Packings

Name	Material	Nominal Size	Surface Area (m^2/m^3)	Void Fraction	Packing Factor (m^{-1})	Vendor
Flexipac	Metal	1	558	0.91	108	(1)
		2	250	0.93	72	
		3	135	0.96	52	
		4	69	0.98	30	
Gempak	Metal	4a	525	—	—	(2)
		4a	394	—	—	
		2a	262	—	—	
		1a	131	—	—	
Sulzer	Metal gauze	CY	700	0.85	—	(3)
		BX	500	0.90	—	
		AX	250	0.95	—	
Kerapak	Ceramic	BX	450	0.75	—	(3)
Munters	Plastic	6,560	400	—	193	(4)
		12,060	220	—	89	
		17,060	148	—	56	
		25,060	98	—	33	
Rombopak	Metal	—	755	—	—	(5)
Montz	Metal	—	—	—	—	(6)
Goodloe	Metal knit	—	—	—	—	(2)

Vendors: (1) Koch Engineering Co., Wichita, KS.
        (2) Glitsch, Inc., Dallas, TX.
        (3) Sulzer Brothers Ltd., Winterthur, Switzerland.
        (4) Munters Corp., Fort Myers, FL.
        (5) Kuhni, Ltd., Basle, Switzerland.
        (6) Julius Montz Co., Hilden, West Germany.

side-entering vapor (e.g., the vapor–liquid return from a reboiler) across the column cross section. For critical cases, one or two crossflow trays may be used to improve the distribution, if the pressure drop they consume can be tolerated.

LIQUID HOLDUP

The void space in a packed bed is occupied by vapor and liquid during operation. As might be expected for countercurrent flow, higher vapor rates restrict liquid downflow and increase liquid holdup. In the extreme case, a sufficient amount of holdup can represent an incipient flooding condition. Figure 5.8-9 shows typical holdup and interfacial areas for 1 in. ceramic Raschig rings. At low liquid rates there is little influence of vapor rate on holdup or interfacial area, the latter approximating the specific surface area of the packing itself. At higher rates of liquid and vapor flows, there is clear interaction between the streams and a buildup of liquid takes place. This happens in the so-called "loading zone."

The holdup shown in Fig. 5.8-9 is given as a fraction of the total bed volume. It is a dynamic, or *total holdup*, and is given the notation H_t. If the vapor and liquid flows were to be stopped, and the bed then allowed to drain, a certain amount of liquid would be retained in the interstices of the bed; this is the *static holdup*, H_s. The difference between these holdup values is the *operating holdup*, H_o. Thus,

$$H_t = H_o + H_s \tag{5.8-3}$$

The effective void fraction under operating conditions is

$$\epsilon' = \epsilon - H_t \tag{5.8-4}$$

For random packings, Shulman et al.[14] proposed the following relationship for the static holdup:

$$H_s = \frac{2.79\ (C_1\mu_L C_2\sigma C_3)}{\rho_L\ 0.37} \tag{5.8-5}$$

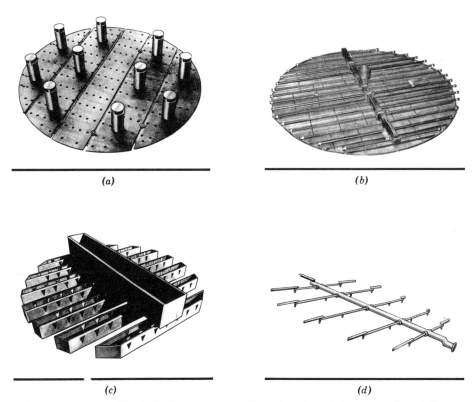

(a)   (b)

(c)   (d)

FIGURE 5.8-5 Types of liquid distributors: (a) orifice/riser; (b) perforated pipe; (c) trough; and (d) spray nozzle.

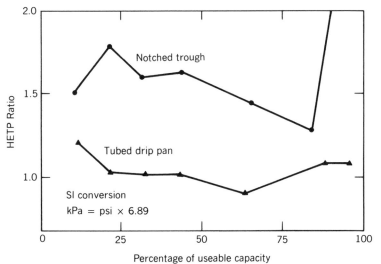

FIGURE 5.8-6 Effect of liquid distributor type on packing efficiency (Ref. 13) for 1 in. Pall rings, 24 psia, cyclohexane-*n*-heptane.

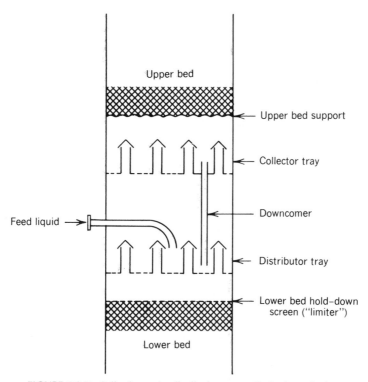

FIGURE 5.8-7 Collection and redistribution at a packed-column feed tray.

FIGURE 5.8-8 Wall-wiper liquid redistributor.

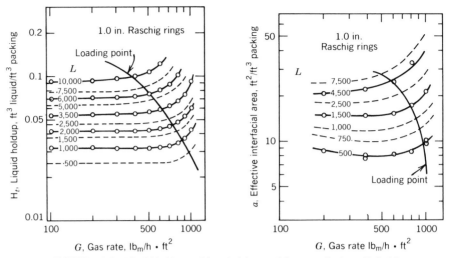

FIGURE 5.8-9 Liquid holdup and interfacial area, 1 in. ceramic rings (Ref. 14).

Values of the constants in Eq. (5.8-5) are as follows:

	C_1	C_2	C_3
25 mm carbon Raschig rings	0.086	0.02	0.23
25 mm ceramic Raschig rings	0.00092	0.02	0.99
25 mm ceramic Berl saddles	0.0055	0.04	0.55

The packings tested by Shulman and coworkers are of the traditional bluff body type. One would expect holdup values to be lower for the through-flow type packings. Vendors usually have available holdup values based on air–water tests.

Operating holdup may be estimated from the dimensionless equation of Buchanan:[15]

$$H_o = 2.2 \left(\frac{\mu_L U_L}{g \rho_L d_p^2} \right) + 1.8 \left(\frac{U_L^2}{g d_p} \right)^{1/2} \tag{5.8-6}$$

The first group on the right is a film number and the second is the Froude number. More will be said about these dimensionless parameters when structured packings are discussed. Representative operating holdup data for a structured packing material are given in Fig. 5.8-10.

LOADING/FLOODING

As mentioned above, when the vapor rate is very high, downflowing liquid tends to be held up by the drag of the vapor. This reduces the net available cross section for vapor flow and causes the pressure drop to increase. If vapor rate continues to increase, liquid will be carried overhead and the column will reach a state of incipient flooding. The same situation can arise if there is a large increase in liquid rate, when vapor rate is kept constant, and excessive rates of liquid also can lead to a state of incipient flooding.

These effects of liquid and vapor flow rates are exhibited in Fig. 5.8-11, based on the studies of Sarchet.[17] At low liquid rates, there is essentially vapor flow through a dry bed; that is, liquid occupies a relatively small portion of the available cross section for vapor flow. The effect on pressure drop of higher liquid rates is evident; the pressure drop curves bend upward, as a result of the greatly increased liquid holdup. At the very high rates of vapor and liquid flows, *flooding* occurs, and this represents a maximum possible operating condition. As for tray-type devices, it is convenient to relate certain aspects of packed bed operation to this maximum condition, even though it does not represent a practical operating condition.

A discussion of the techniques for measuring the flood condition has been given by Silvey and Keller;[18] as one considers making such a measurement he or she soon realizes that the flood point determination is influenced by the measurement approach and that it is really not a "point" but a fairly narrow range of values. Even so, methods for making an estimate of the flood point are useful and are outlined in the following section dealing with pressure drop.

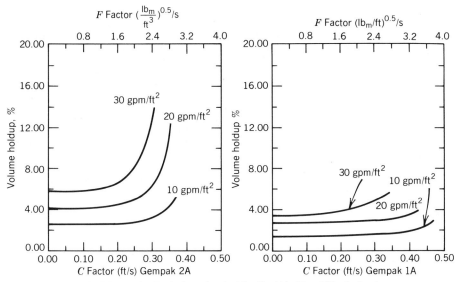

FIGURE 5.8-10 Typical vendor data for liquid holdup (Glitsch, Inc.).

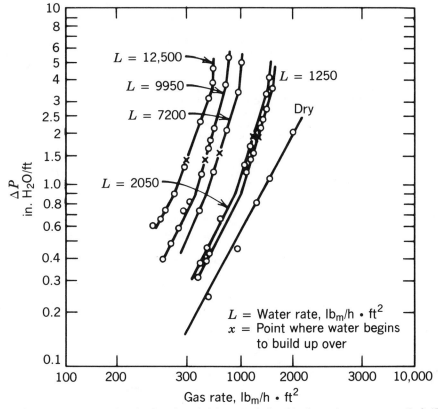

FIGURE 5.8-11 Pressure drop for flow through 1-in. ceramic Raschig rings, air–water system (Ref. 17).

INTERFACIAL AREA

The amount of interfacial area that is available in the packed column is of vital importance in determining the mass transfer efficiency of the packed column. However, it is rarely known with any degree of accuracy. It bears some relation to the specific area of the packing, that is, the amount of dry packing surface provided by the packing elements. But even this parameter is not really known with accuracy in the case of the random packings, since an exact count of elements contained in the column is not feasible for commercial scale columns, and even for controlled pilot-scale units the elements may nest or otherwise block out area that can then serve no purpose for mass transfer.

Figure 5.8-9 shows effective interfacial areas that were deduced by Shulman et al.[14] from experiments on the sublimation of naphthalene Raschig rings. Many other studies have been made in an attempt to relate effective area to specific packing surface, as a function of liquid and vapor flow rates. Generally, it is thought that the effective area reaches the specific area near the onset of flooding, but Fig. 5.8-9 shows that this may not be the case. It also should be noted that the total area available for mass transfer includes film surface ripples, entrained liquid within the bed, and vapor bubbling through pockets of liquid held up in the bed. Another important point: relative specific surface of different packings might indicate relative mass transfer efficiencies, but this is often not the case.

In the present state of development, at least for the random packings, it is prudent to deal with an "effective mass transfer surface", as used by Bravo and Fair,[16] that can be deduced by semiempirical means, or simply to couple the area term with the mass transfer coefficient to form a volumetric-type coefficient.

Pressure Drop

RANDOM PACKINGS

It is clear from experimental data such as those shown in Fig. 5.8-11 that the loss of pressure experienced by the vapor as it flows through the packing is a function of flow rates, system properties, and packing characteristics. The resistance offered by the bed is a combination of form drag and skin friction, and thus the shape of the packing elements must be considered. While some efforts have been made to correlate pressure drop in randomly packed beds by means of a dry bed friction factor that is modified according to vapor flow, the efforts have not been successful for handling the many different possible packing shapes and sizes. (Such an approach has been more successful for the structured packings and will be described in the next subsection.) As a result, it has been necessary to use empirical approaches for estimating pressure drop in beds containing the random packings. These approaches are based on the observation that at flooding, or incipient flooding, most beds exhibit a pressure drop of about 20 mbar/m of packed height (2.5 in. water/ft).

In working with a modification of an early flooding correlation of Sherwood et al.,[19] Eckert[20] developed a graphical method for pressure drop estimation, shown in Fig. 5.8-12. The Eckert diagram originally was designed to provide both flooding and pressure drop information. The effects of packing type and size are forced to merge through the use of a *packing factor* that is presumed to be a constant for a given packing type and size. Representative values of the factor are given in Table 5.8-1, and vendors of packings not listed can supply values of the packing factor for their products. This approach to the estimation of pressure drop is quite simple and rapid. It gives approximate values and is recommended for applications when the pressure drop estimation is not of critical importance.

A more fundamental approach to pressure drop prediction is being pursued by current researchers. The usual first step is to consider the dry pressure drop only, utilizing a conventional Fanning or Darcy type relationship:

$$\frac{\Delta P_0}{Z} = \frac{f\rho_V U_i^2}{d_p g_c} = \frac{\text{pressure drop}}{\text{height of bed}} \tag{5.8-7}$$

Any consistent set of units may be used. After dry pressure drop is determined, the influence of irrigation liquid is introduced:

$$\frac{\Delta P}{Z} = \frac{\Delta P_0}{Z}\left(\frac{1}{1 - KH_t}\right)^5 \tag{5.8-8}$$

In this equation K is a constant, characteristic of packing type and size. Pressure drops through beds of random packings have been correlated by these equations,[21-23] but it is not yet possible to point toward a particular piece of work that gives reliable results for the full range of possible packings.

An adaptation of the basic theory has been presented by Billet and Mackowiak[23] and is stated to be valid for loadings up to about 65% of the flood point. Liquid holdup is obtained from

$$H_t = C_h \, \text{Fr}_L^{1/3} \tag{5.8-9}$$

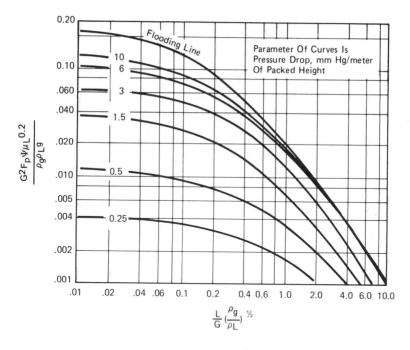

L   = Liquid rate, kg/s m²
G   = Gas rate, kg/s m²
ρ_L = Liquid density, kg/m³
ρ_g = Gas density, kg/m³
F_p = Packing factor, m⁻¹
μ_L = Viscosity of liquid, mPa s
Ψ = Ratio, (density of water)/(density of liquid)
g   = Gravitational constant, 9.81 m/s²

FIGURE 5.8-12 Eckert method for estimating packed column flooding and pressure drop (Ref. 20). (*Note:* mm Hg/m × 0.163 = in. H₂O/ft.)

where

$$Fr_L = \text{Froude number} = \frac{U_L^2}{\epsilon^2 g d_p} \tag{5.8-10}$$

$$\frac{\Delta P_0}{Z} = \text{dry bed drop} = \frac{f(1 - \epsilon)}{\epsilon^3} \frac{F_v^2}{g d_p K'} \tag{5.8-11}$$

In this expression d_p is the effective packing size $[= 6(1 - \epsilon)/a_p]$ and K' is the Ergun–Brauer wall factor:

$$\frac{1}{K'} = 1 + \frac{2}{3(1 - \epsilon)} \frac{d_p}{d_t} \tag{5.8-12}$$

The friction factor in Eq. (5.8-11) is taken as constant at 2.45 for vapor Reynolds numbers of 2100 or higher. For lower Reynolds numbers,

$$f = 10 \, Re_V^{-0.18} \tag{5.8-13}$$

$$Re_V = \frac{d_p U_V \rho_V}{(1 - \epsilon)\mu_V} \tag{5.8-14}$$

Finally, the ratio of irrigated to dry pressure drop is obtained:

TABLE 5.8-3 Constants for Billet Pressure Drop Equations

Packing	Material	f	ϕ	χ	C_h
Pall rings	Metal	2.45	0.81	0.525	1.0
	Ceramic	1.95	0.92	0.64	1.15
	Plastic	2.45	0.83	0.57	0.75
NSW rings*	Plastic	1.025	1.0	0.61	0.75
Bialecki rings	Metal	2.45	0.79	0.596	1.0

*Also known as Nor-Pac packing, product of NSW Co., Roanoke, VA.

$$\frac{\Delta P/Z}{\Delta P_0/Z} = 1 - \left(\frac{H_t}{2\epsilon\phi\chi^{5/3}}\right)^{-5} \tag{5.8-15}$$

where ϕ = geometric packing factor
  χ = aperture correction factor
  ϵ = packing void fraction

Values of the several constants, each determined experimentally for a particular packing type (but for a range of sizes), are given in Table 5.8-3.

A final approach to the problem of estimating pressure drop for random packings should not be overlooked: test data from the packing vendors. These data usually are based on air–water and are taken in the vendor laboratories. They are presented in charts such as that shown in Fig. 5.8-13. In the example shown, pressure drop is given as a function of gas mass velocity. For gases other than air, the abscissa should be multiplied by the square root of the ratio, air density to the density of the gas under consideration:

$$\text{Abscissa} = G\left(\frac{\rho_{\text{air}}}{\rho_{\text{gas}}}\right)^{0.5}$$

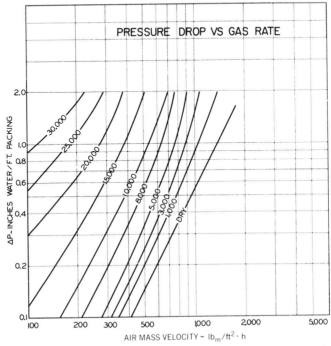

FIGURE 5.8-13 Typical vendor data on random packing pressure drop. One-inch ceramic Raschig rings, air–water system (Norton Co.).

EXAMPLE 5.8-1

For the conditions of Example 5.7-1, that is, the top zone of a methanol–water fractionator, and for the case of 50 mm Pall rings replacing the sieve trays, estimate the pressure drop in the top 0.5 m of packing, using (a) the Eckert graphical method and (b) the Billet analytical method.

Solution. (a) From Example 5.7-1, the value of the flow parameter is

$$\frac{L}{G}\frac{\rho_V}{\rho_L} = 0.5\frac{1.22}{751} = 0.020$$

For the ordinate scale parameter of the Eckert chart (Fig. 5.8-12),

$$G = \frac{27,000}{3600(3.08)} = 2.44 \text{ kg/s·m}^2$$

$$F_p = 66 \quad \text{(Table 5.8-1)}$$

$$\Psi = \frac{1000}{751} = 1.33$$

$$\mu_L = 0.30 \text{ cP} = 0.30 \text{ mPa·s}$$

$$\rho_V = 1.22 \text{ kg/m}^3$$

$$\rho_L = 751 \text{ kg/m}^3$$

$$g = 9.81 \text{ m/s}^2$$

$$\frac{G^2 F_p \Psi \mu_L^{0.2}}{\rho_V \rho_L g} = \frac{(2.44)^2(66)(1.33)(0.30)^{0.2}}{(1.22)(751)(9.81)} = 0.0455$$

From Fig. 5.8-12,

$$\text{Pressure drop} = 2.0 \text{ mm Hg/m} \quad (0.33 \text{ in. } H_2O/\text{ft})$$

Thus, for a packing height of 0.5 m,

$$\Delta P = 1.0 \text{ mm Hg}$$

(b) Using the Billet method,

$$U_L = \frac{13,500}{(3600)(751)(3.08)} = 0.00162 \text{ m/s}$$

$$\epsilon = 0.96 \quad \text{(Table 5.8-1)}$$

$$a_p = 115 \text{ m}^{-1} \quad \text{(Table 5.8-1)}$$

$$d_p = \frac{6(1 - \epsilon)}{a_p} = \frac{6(0.04)}{115} = 0.00209 \text{ m}$$

$$U_V = \frac{27,000}{(3600)(1.22)(3.08)} = 2.00 \text{ m/s}$$

$$F_v = U_V(\rho_V)^{1/2} = 2.00(1.22)^{1/2} = 2.21 \text{ m/s (kg/m}^3)^{1/2}$$

$$Fr_L = \frac{U_L^2}{\epsilon^2 g d_p} = \frac{(0.00162)^2}{(0.96)^2(9.81)(0.00209)} = 1.39(10^{-4})$$

$$K' = 0.98 \quad \text{[by Eq. (5.8-12)]}$$

$$Re_V = \frac{d_p U_V \rho_V}{(1 - \epsilon)\mu_V} = \frac{(0.00209)(2.0)(1.22)}{(0.04)(9.3)(10^{-6})} = 13,700$$

Since Re_V is greater than 2100, $f = 2.45$. By Eq. (5.8-11),

$$\frac{\Delta P_0}{Z} = \frac{f(1 - \epsilon)F_v^2}{\epsilon^3 g d_p K'}$$

$$= \frac{2.45(0.04)(2.21)^2}{(0.96)^3(9.81)(0.00209)(0.98)} = 26.9 \text{ kg/m}^2 \cdot \text{m}$$

$$= 1.98 \text{ mm Hg/m} \quad \text{(dry bed drop)}$$

$$H_t = C_h \text{Fr}_L^{1/3} = 1.0[1.39(10^{-4})]^{1/3} = 0.052$$

Finally, by Eq. (5.8-15),

$$\frac{\Delta P}{Z} = \frac{\Delta P_0}{Z} \left[1 - \frac{0.052}{2(0.96)(0.81)(0.525)^{5/3}} \right]^{-5} = 1.98(1.674)$$

$$= 3.31 \text{ mm Hg/m} \quad \text{or} \quad 1.66 \text{ mm Hg per 0.5 m of bed}$$

STRUCTURED PACKINGS

The pressure drop through structured packings may be estimated according to the method of Bravo et al.[24] as outlined below. The method applies to conditions below the loading point.

Equation (5.8-6) may be adapted to allow for vapor flow through the channels of the packing:

$$\Delta P_0 = \frac{f \rho_V U_{ve}^2}{d_{eq} g_c} \tag{5.8-16}$$

The effective gas velocity inside the flow channel takes into account the slope of the corrugations, as proposed in an earlier paper by Bravo et al.:[24]

$$U_{ve} = \frac{U_{vs}}{\epsilon \sin \theta} \tag{5.8-17}$$

where θ is the angle of inclination of the flow channel from the horizontal. The equivalent diameter of the flow channel d_{eq} is taken as the side of a corrugation or crimp of the packing (see Table 5.8-4).

The friction factor for Eq. (5.8-16) is obtained from the general relationship

$$f = C_1 + \frac{C_2}{\text{Re}_V} \tag{5.8-18}$$

where C_1 and C_2 are constants for a particular packing type and size. The Reynolds group of Eq. (5.8-18) is defined as

$$\text{Re}_V = \frac{d_{eq} U_{ve} \rho_V}{\mu_V} \tag{5.8-19}$$

This friction factor is applicable to gas flow only and incorporates both turbulent and laminar contributions. Since the shapes of the flow channels of the structured packings are geometrically similar, one would expect that a single set of constants for Eq. (5.8-18) would cover all packing sizes.

When the packing is irrigated, the influence of the presence of liquid may be related to the operating holdup, using a form alternate to that given earlier as Fig. 5.8-6. Bemer and Kallis[22] concluded that the following simple form is adequate:

$$H_t = A' \left(\frac{U_L^2}{d_{eq} g} \right)^\alpha = A' \text{Fr}^\alpha \tag{5.8-20}$$

where A' and α are constants for the packing type.

Equation (5.8-8) is then adapted for the total pressure drop:

$$\Delta P = \Delta P_0 \left(\frac{1}{1 - C_3 \text{Fr}^\alpha} \right)^5 \tag{5.8-21}$$

where

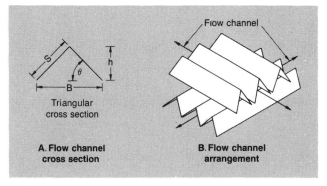

FIGURE 5.8-14 Geometric factors for estimating structured packing pressure drop (Ref. 24).

$$C_3 Fr^\alpha = KH_t = KA'Fr^\alpha \tag{5.8-22}$$

Figure 5.8-14 shows the flow channel geometry factors. Bravo et al.[24] found that for the corrugated-type structured packings only the constant C_3 was variable; values of the other constants in Eqs. (5.8-18) and (5.8-22) are $C_1 = 0.171$, $C_2 = 92.7$, and $\alpha = 0.5$. Thus, the final working equation is

$$\Delta P = \left(0.171 + \frac{92.7}{Re_V}\right)\left(\frac{\rho_V U_{ve}^2}{d_{eq}g_c}\right)\left(\frac{1}{1 - C_3 Fr^{0.5}}\right)^5 \tag{5.8-23}$$

Values of the constant C_3 are given in Table 5.8-4.

EXAMPLE 5.8-2

In the preceding example, the use of 50 mm Pall rings was considered for the methanol–water fractionator of Example 5.7-1. For the case of a 12.5 mm crimp height structured packing of the sheet metal type (Flexipac 2, Gempak 2A), estimate the pressure drop in the top 0.5 m of the packed bed.

Solution. For the No. 2 structured packing and from Table 5.8-4,

$$d_{eq} = 0.018 \text{ m}$$
$$\theta = 45°$$
$$C_3 = 3.08$$
$$\epsilon = 0.93 \quad \text{(Table 5.8-2)}$$

TABLE 5.8-4 Pressure Drop Parameters for Structured Packings

Packing		Angle* (degrees)	Equivalent Diameter (mm)	Constant C_3
Flexipac	1	45	9	3.38
	2	45	18	3.08
	3	45	36	4.50
	4	45	72	7.26
Gempak	1A	45	36	4.50
	2A	45	18	3.08
	3A	45	13.5	3.87
	4A	45	9	3.38
Sulzer	BX	60	9	3.38

*Angle of channel from horizontal.

From Example 5.8-1,

$$Re_V = 13,700$$

$$Fr_L = \frac{U_L^2}{d_{eq}g} = \frac{(0.00162)^2}{(0.018)(9.81)} = 1.49(10)^{-5}$$

$$U_{ve} = \frac{U_{vs}}{\epsilon \sin \theta} = \frac{2.00}{0.93(0.707)} = 3.04 \text{ m/s}$$

Finally,

$$\frac{\Delta P}{Z} = 0.171 + \frac{92.7}{13,700} \frac{(1.22)(3.04)^2}{0.018} \left(\frac{1}{1 - 3.08[1.45(10^{-5})]}\right)^5$$

$$\frac{\Delta P}{Z} = 111.5 \text{ Pa/m} = 0.836 \text{ mm Hg/m}$$

or, for 0.5 m of bed, $\Delta P = 0.42$ mm Hg.

5.8-4 Vapor Capacity

Generalized correlations for flood point prediction have been based on the early work of Sherwood et al.[19] Modifications of this work have been made from time to time, with that of Eckert,[20] mentioned earlier in connection with pressure drop estimation, being representative of those now in use. The flood line is designated as the top curve of Fig. 5.8-12. Thus, as for pressure drop, the Eckert chart enables a quick estimate of the maximum allowable vapor flow through the packed bed in question.

The ordinate group of Fig. 5.8-12 is a form of the capacity factor used in Fig. 5.7-10 for estimating flooding in tray columns. It is possible to replace the group with the much simpler Souders–Brown coefficient and obtain quite good correlations of flooding data,[25] and a generalized chart on this basis is shown in Fig. 5.8-15. Parameters for the chart are shown in Table 5.8-5. This graphical approach to flooding is recommended over the one of Eckert, and it can be used for the structured packings as well as the random packings.

5.9 MASS TRANSFER IN TRAY COLUMNS

5.9-1 Efficiency Definitions and Approaches to Prediction

In Section 5.3 methods were developed for the determination of the number of theoretical stages, or contacts, needed for a given separation. For a tray column, the designer must specify the *actual* number of trays (stages, contacts) to be installed, and each of these actual stages must operate efficiently enough to give an equivalent theoretical stage, or portion of an equivalent theoretical stage.

Several different efficiency terms will be used in this section, and the one ultimately needed for design is *overall column efficiency*:

$$E_{oc} = \frac{N_t}{N_a} \tag{5.9-1}$$

where the computed number of theoretical stages is N_t and the actual number of column trays is N_a. Note that a reboiler or a partial condenser would not be included in N_t: one is concerned here only with the stages in the column itself. Note further that the number of trays actually specified for the column may be greater than N_a, to allow for future contingencies, column control problems, and so on.

The efficiency of a distillation column depends on three sets of design parameters:

1. The system being processed, its composition and properties.
2. Flow conditions and degree of loading.
3. The geometry of the contacting device and how it influences the intimate interaction of vapor and liquid.

It is clear that each of these parameters is not completely independent of the others and that the designer may not have much control over all of them. Still, it is convenient to consider efficiency from these three viewpoints and there are optimum approaches to combining them.

Since conditions vary throughout the column, it is more fundamental to deal with a tray efficiency.

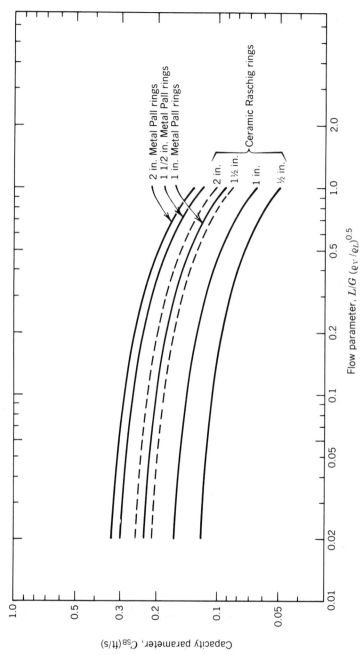

FIGURE 5.8-15 Design chart for packed column flooding. See Table 5.8-5 for factors of other packings.

313

TABLE 5.8-5 Relative Flooding Capacities for Packed Columns Using 50 mm Pall Rings as the Basis

Pall Rings—Metal		Intalox Saddles—Ceramic		Koch Sulzer BX	
50 mm	1.00*	50	0.89	BX	1.00
38	0.91	38	0.75		
25	0.70	25	0.60	Koch Flexipac	
12	0.65	12	0.40	No. 1	0.69
				No. 2	1.08
Raschig Rings—Metal		Berl Saddles—Ceramic		No. 3	1.35
50	0.79	50	0.84		
38	0.71	38	0.70	Tellerettes	
25	0.66	25	0.54		1.0
12	0.55	12	0.37		
				Nor-Pak Plastic	
Raschig Rings—Ceramic		Intalox Saddles—Metal		No. 25	1.0
50	0.78	No. 25	0.88	No. 35	1.20
38	0.65	No. 40	0.98		
25	0.50	No. 50	1.10		
12	0.37	No. 70	1.24		

*Value is ratio of C_{SB} for packing type and size to the C_{SB} for 50 mm Pall rings for the same value of flow parameter.

Figure 5.9-1 shows the vapor and liquid streams entering and leaving the nth tray in the column, and the efficiency of the vapor–liquid contact may be related to the approach to phase equilibrium between the two exit streams:

$$E_{mv} = \frac{y_{n-1} - y_n}{y_{n-1} - y_n^*} \tag{5.9-2}$$

where $y_n^* = K_{VL}x_n$ (K_{VL} being the vapor–liquid equilibrium ratio discussed in Section 5.2) and the compositions refer to one of the components in the mixture (often to the lighter component of a binary pair or to the "light key" of the mixture). The graphical equivalent of Eq. (5.9-2) is shown in Fig. 5.9-1, where actual trays have been stepped off on a McCabe–Thiele-type diagram.

Equation (5.9-2) defines the *Murphree vapor efficiency*; an equivalent expression could be given in terms of liquid concentrations, but the vapor concentration basis is the one most used, and it is equivalent to its counterpart dealing with the liquid. It should be noted that the value of y_n^* is based on the *exit concentration* of the liquid and is silent on the idea that liquid concentration can vary from point to point of the tray; when liquid concentration gradients are severe, as in the case of very long liquid flow paths, it is possible for the Murphree efficiency to have a value greater than 1.0.

A more fundamental efficiency expression deals with some point on the tray:

$$E_{ov} = \left[\frac{y_{n-1} - y_n}{y_{n-1} - y_n^*} \right]_{\text{point}} \tag{5.9-3}$$

This is the *local efficiency* or *point efficiency*, and because a liquid concentration gradient is not involved, it cannot have a value greater than 1.0. Thus, it is a more fundamental concept, but it suffers in application since concentration profiles of operating trays are difficult to predict. As in the case of the Murphree tray efficiency, it is possible to use a liquid-phase point efficiency.

There are four general methods for predicting the efficiency of a commercial distillation column:

1. Comparison with a very similar installation for which performance data are available.
2. Use of an empirical or statistical efficiency model.
3. Direct scaleup from carefully designed laboratory or pilot plant experiments.
4. Use of a theoretical or semitheoretical mass transfer model.

In practice, more than one of these methods will likely be used. For example, available data from a similar installation may be used for orientation and a theoretical model may then be used to extrapolate to the new conditions.

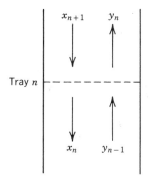

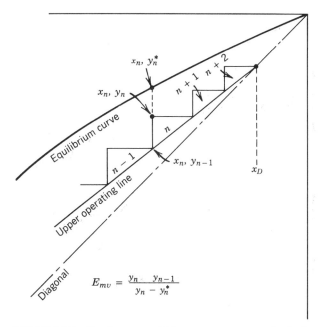

FIGURE 5.9-1 Murphree stage efficiency, vapor concentration basis.

EFFICIENCY FROM PERFORMANCE DATA

There are many cases where an existing fractionation system is to be duplicated as part of a production expansion. If careful measurements are available on the existing system (e.g., with good closures on material and energy balances), then the new system can be designed with some confidence. But caution is advised here; subtle differences between the new and the old can require very careful analysis and the exercise of predictive models.

EFFICIENCY FROM EMPIRICAL METHODS

A number of empirical approaches have been proposed for making quick and approximate predictions of overall column efficiency. The one shown in Fig. 5.9-2, from O'Connell,[1] represents such an approach and partially takes into account property variations. Its use should be restricted to preliminary designs.

EFFICIENCY FROM LABORATORY DATA

It is possible to predict large-scale tray column efficiencies from measurements in laboratory columns as small as 25 mm (1 in.) in diameter. Development work at Monsanto Company showed that the use of a

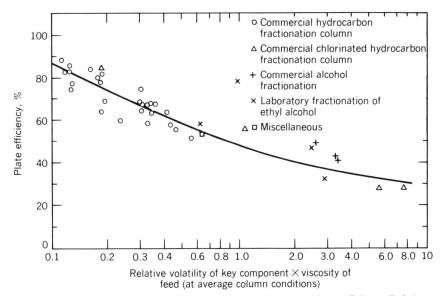

FIGURE 5.9-2   O'Connell method for the estimation of overall column efficiency (Ref. 1).

special laboratory sieve tray column, the Oldershaw column, produces separations very close to those that can be accomplished in the large equipment. A description of the underlying research is given by Fair et al.[2] A view of a glass Oldershaw tray is shown in Fig. 5.9-3. Typical comparisons between Oldershaw and commercial-scale efficiencies are shown in Fig. 5.9-4.

The Oldershaw testing provides values of the point efficiency. It is still necessary to correct this efficiency to the Murphree and overall efficiencies, and approaches for this will be given later. For the moment it is important to consider the potential importance of the Oldershaw scaleup method: a specific definition of the system may not be necessary; the matter of vapor–liquid equilibria can be bypassed (in effect, the Oldershaw column serves as a multiple-stage equilibrium still) and efficiency modeling can be minimized or even eliminated.

The procedure for using the Oldershaw scaleup approach is given in Ref. 2 and may be summarized as follows:

1. Arbitrarily select a number of Oldershaw stages and set up the column in the laboratory. Oldershaw columns and their auxiliaries are available from laboratory supply houses and are available in a number of actual stages.

2. For the system to be studied, run the column and establish the upper operating limit (flood condition). Take data at a selected approach to flood, using total reflux or finite reflux. At steady conditions, take overhead and bottom samples.

3. If the separation is satisfactory, the commercial column will require no more than the number of stages used in the laboratory. Depending on the degree of liquid mixing expected, the large column

FIGURE 5.9-3   Glass Oldershaw tray, 50 mm diameter.

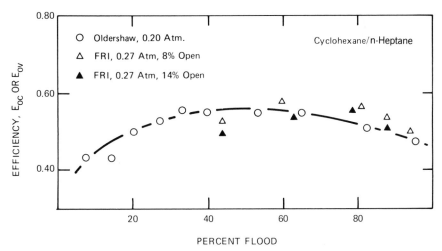

FIGURE 5.9-4 Comparison of Oldershaw column (25 mm diameter) efficiency with that of a test column at Fractionation Research, Inc. (FRI) of 1.2 m diameter (Ref. 2).

may require fewer stages. Note that comparisons must be made at the same approach to the flood point.

Unfortunately, a method such as this is not available for packed columns; more on this will be given in Section 5.10.

EFFICIENCY FROM MASS TRANSFER MODELS

Prediction of efficiency based on mass transfer theory and character of the two-phase contacting zone (Section 5.7-2) is a goal with parallels in other types of unit operations equipment. The development of suitable models for efficiency has been underway for a number of years, since the pioneering paper of Geddes in 1946.[3] From a practical point of view the first model suitable for design emerged from the AIChE Distillation Efficiency in 1958.[4] As mentioned earlier, mass transfer models are useful not only for new designs but for studying effects of changed operating conditions in existing designs.

5.9-2 Phase Transfer Rates

GENERAL EXPRESSIONS

Mass transfer on a distillation tray proceeds according to the general flux relationship:

$$N_A = k_V(y_A - y_{Ai}) = k_L(x_{Ai} - x_A)$$

$$= K_{OV}(y_A - y_A^*) = k_{OL}(x_A^* - x_A) \tag{5.9-4}$$

where N_A = flux of species A across a unit area of interface

y_A = mole fraction of A in bulk vapor

y_{Ai} = mole fraction of A in vapor at the interface

y_A^* = mole fraction of A in vapor that would be in equilibrium with mole fraction of A in bulk liquid, x_A

x_{Ai} = mole fraction of A in liquid at the interface

x_A^* = mole fraction of A in liquid that would be in equilibrium with mole fraction of A in bulk vapor, y_A

Equation (5.9-4) may be written in a similar fashion for net transfer from liquid to vapor, and indeed in distillation there is usually a nearly equal number of total moles transferred in each direction. The mass transfer coefficients in Eq. (5.9-4) may be combined as follows:

$$K_{OV} = \frac{1}{1/k_V + m/k_L} \tag{5.9-5}$$

and

$$K_{OL} = \frac{1}{1/k_L + 1/mk_V} \tag{5.9-6}$$

In these equations, m is the slope of the equilibrium line. For a binary system, and from Section 5.2, this slope is

$$m = \frac{dy_A}{dx_A} = \frac{\alpha_{AB}}{[1 + (\alpha_{AB} - 1)x_A]^2} \tag{5.9-7}$$

where α_{AB} is the relative volatility and A is the lighter of the two components in the mixture. It can be shown that for a multicomponent system the pseudobinary value of m is equal to the equilibrium ratio K_{VL} for the lighter of the pseudocomponents.

The mass transfer coefficients in Eq. (5.9-4) may be evaluated by various means. According to the film model, and for the example of vapor-phase transfer,

$$k_V = \frac{D^V}{\Delta Z} \tag{5.9-8}$$

where ΔZ is the thickness of the laminar-type film adjacent to the interface and D^V is the vapor-phase diffusion coefficient. For the penetration model,

$$k_V = 2\sqrt{\frac{D^V}{\pi\Theta}} \tag{5.9-9}$$

where Θ is the time of exposure at the interface for an element of fluid that has the propensity to transfer mass across the interface. Both film and penetration (or, alternatively, surface renewal) models are used and differ primarily in the exponent on the diffusion coefficient. The weight of experimental evidence supports the one-half power dependence, but neither the renewal time nor the film thickness is a parameter easily determined. Thus, models deal primarily with the mass transfer coefficients.

Even though tray columns are normally crossflow devices (as opposed to counterflow devices such as packed columns), certain aspects of counterflow theory are used with trays. This involves the use of transfer units as measures of the difficulty of the separation. For a simple binary system, and for distillation, the number of transfer units may be obtained from

$$N_V = \int \frac{dy}{y_i - y} \tag{5.9-10}$$

$$N_L = \int \frac{dx}{x_i - x} \tag{5.9-11}$$

$$N_{OV} = \int \frac{dy}{y^* - y} \tag{5.9-12}$$

with heights of transfer units obtained from

$$H_V = \frac{G_M}{k_V a_i} \tag{5.9-13}$$

$$H_L = \frac{L_M}{k_L a_i} \tag{5.9-14}$$

$$H_{OV} = \frac{G_M}{K_{OV} a_i} \tag{5.9-15}$$

Considering vapor-phase transfer as an example, and recognizing that $Z = N_V H_V$, the number of transfer units may be expressed as

$$N_V = \frac{k_V a_i M Z}{\rho_V U_V} \tag{5.9-16}$$

In the context of a vapor passing up through the froth on a tray, this expression may be modified to

$$N_V = \frac{k_V a_i \epsilon Z_f}{U_a} = k_V a_i \bar{t}_V \qquad (5.9\text{-}17)$$

Note here that Eq. (5.7-6) has been used. Thus, the height of a vapor-phase transfer unit is $H_V = U_a/k_V a_i \epsilon$. In a similar fashion, and with the aid of Eq. (5.7-7), the number of liquid-phase transfer units may be obtained:

$$N_L = \frac{k_L a_i (1 - \epsilon) Z_f}{\mu_A} = k_L a_i \bar{t}_L \qquad (5.9\text{-}18)$$

On the assumption that the transfer resistances of the phases are additive,

$$\frac{1}{N_{OV}} = \frac{1}{N_V} + \frac{\lambda}{N_L} \qquad (5.9\text{-}19)$$

where λ is the ratio of slopes of the equilibrium and operating lines; that is,

$$\lambda = \frac{m}{L/V} \qquad (5.9\text{-}20)$$

Thus, the slope of the operating line is obtained from equilibrium stage calculations and the slope of the equilibrium line is obtained as discussed above.

When N_{OV} is evaluated at a point on the tray, the basic expression for point efficiency is obtained:

$$E_{ov} = 1 - \exp(-N_{OV}) \qquad (5.9\text{-}21)$$

LIQUID-PHASE MASS TRANSFER
Equation (5.9-18) is used to obtain values of N_L. The residence time for liquid in the froth or spray is obtained from Eq. (5.7-9). For sieve trays, the experimental correlation of Foss and Gerster,[5] included in Ref. 4, is recommended:

$$k_L a_i = (3.875 \times 10^8 \, D^L)(0.40 \, F_{va} + 0.17) \qquad (5.9\text{-}22)$$

For bubble-cap trays,

$$k_L a_i = (4.127 \times 10^8 \, D^L)(0.21 \, F_{va} + 0.15) \qquad (5.9\text{-}23)$$

In these equations, D^L is the liquid-phase molecular diffusion coefficient, in m^2/s, and the volumetric coefficient $k_L a_i$ has units of reciprocal seconds.

VAPOR-PHASE MASS TRANSFER
In distillation the vapor-phase resistance to mass transfer often controls and thus the estimation of vapor-phase transfer units tends to be a critical issue. Equation (5.9-17) is used to evaluate the number of transfer units, and residence time is obtained from Eq. (5.7-6) or (5.7-8). The volumetric mass transfer coefficient is obtained from the expression of Chan and Fair:[6]

$$k_V a_i = \frac{D^V (1030F - 867F^2)}{(h'_L)^{1/2}} \qquad (5.9\text{-}24)$$

In this expression, the volumetric mass transfer coefficient is in reciprocal seconds, the molecular diffusion coefficient in m^2/s, and the liquid holdup h'_L in meters. The term F refers to fractional approach to flooding.

For bubble-cap trays, the expression developed by the AIChE[4] should be used to obtain the number of transfer units directly:

$$N_V = \frac{0.776 + 0.0045 \, h_w - 0.238 \, F_{VA} + 0.0712 \, W}{Sc_V^{1/2}} \qquad (5.9\text{-}25)$$

where h_w = weir height, mm
$\quad F_{VA}$ = F factor through active portion of tray, $m/s(kg/m^3)^{1/2}$
$\quad W$ = liquid flow rate, $m^3/s \cdot m$, of width of flow path on the tray
$\quad Sc_V$ = vapor-phase Schmidt number, dimensionless

5.9-3 Local Efficiency

Once the vapor- and liquid-phase transfer units have been determined, they are combined to obtain the number of overall transfer units (vapor concentration basis) by Eq. (5.9-19). Finally, the local, or point, efficiency is computed:

$$E_{ov} = 1 - \exp(-N_{OV}) \tag{5.9-21}$$

The next step in the process is to correct this point efficiency for vapor and liquid mixing effects.

5.9-4 Liquid and Vapor Mixing

The concentration gradients in the liquid and vapor on the tray are analyzed to determine whether the point efficiency needs correction to give the Murphree efficiency. If the phases are well mixed, and no concentration gradients exist,

$$E_{mv} = E_{ov} \quad \text{(complete mixing)} \tag{5.9-26}$$

If the liquid moves across the tray in plug flow, and the vapor rising through it becomes completely mixed before entering the tray above,

$$E_{mv} = \frac{1}{\lambda} (e^{\lambda E_{ov}} - 1) \tag{5.9-27}$$

Thus, it is necessary to consider tray mixing effects to arrive at a value of E_{mv} for design. Entrainment effects also must be considered, as discussed in Section 5.7-4. A chart showing the overall development of a design sequence is shown in Fig. 5.9-5.

For trays with relatively narrow spacing, the vapor may not be able to mix completely before leaving the tray. For the case of plug flow of liquid and nonmixing of vapor, and with liquid alternating in flow direction on successive trays, representative effects of vapor mixing are shown in Table 5.9-1.[7] A more complete view of mixing effects may be seen in Fig. 5.9-6.[8]

The partial liquid mixing case is of considerable interest in design. Various models for representing partial mixing have been proposed: for most instances the *eddy diffusion model* appears appropriate. This model uses a dimensionless Peclet number:

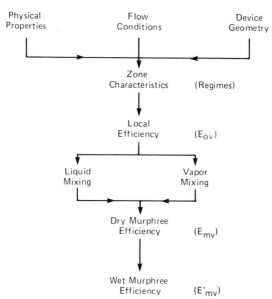

FIGURE 5.9-5 Sequence of steps for evaluating the overall tray efficiency or for deducing the point efficiency.

TABLE 5.9-1   Effect of Vapor Mixing on Plug Flow Efficiency

	Plug E_{mv}, $\lambda = 1$	
E_{ov}	Vapor Mixed	Vapor Unmixed
0.60	0.82	0.81
0.70	1.02	0.98
0.80	1.22	1.16

$$Pe = \frac{w'^2}{D_E t_L} \tag{5.9-28}$$

where w' is the length of travel across the tray. A high value of the Peclet number indicates a close approach to plug flow. As the Peclet number approaches zero, complete mixing is approached. The relationship between E_{mv}, E_{ov}, and Pe is shown in Fig. 5.9-7.

A simpler approach uses the mixing pool model of Gautreaux and O'Connell.[9] These workers visualized a crossflow tray as being comprised of a number of well-mixed stages, or mixing pools, in series. Their general expression for mixing is

$$E_{mv} = \frac{1}{\lambda} [(1 + \lambda E_{ov}/s)^s - 1] \tag{5.9-29}$$

where s, the number of pools, was estimated by Gautreaux and O'Connell to be some function of length of liquid path. It is more convenient to estimate s as a function of Peclet number:

$$s = \frac{Pe + 2}{2} \tag{5.9-30}$$

which is simpler than using Fig. 5.9-7.

It still remains that an estimate of the eddy diffusion coefficient must be made to evaluate the Peclet number. For sieve trays the work of Barker and Self[10] is appropriate:

$$D_E = 0.00668 \, U_a^{1.44} + 0.922(10^{-4})h_L' - 0.00562 \tag{5.9-31}$$

For bubble-cap trays, the AIChE work[4] provides the following:

$$(D_E)^{0.5} = 0.00378 + 0.0171U_a + 3.689 \frac{q}{W} + 0.116h_w \tag{5.9-32}$$

Equations (5.9-31) and (5.9-32) are based on unidirectional flow, the expected situation for the rectangular simulators that were used in the experiments. For large-diameter trays, there can be serious retrograde flow effects, with resulting loss of efficiency due to stagnant zones. The work of Porter et al.[11] and Bell[12] elucidates the retrograde flow phenomenon, but the efficiency loss appears to be important only for column diameters of 5 m or larger, especially for relatively short segmental weirs.

5.9-5   Overall Column Efficiency

As indicated in Fig. 5.9-6, overall column efficiency is obtained from the Murphree efficiency E_{mv} (or from the Murphree efficiency corrected for entrainment, E_a). The conversion is by the relationships

$$E_{oc} = \frac{\log [1 + E_{mv}(\lambda - 1)]}{\log \lambda}$$
$$E_{mv} = \frac{\lambda^{E_{oc}} - 1}{\lambda - 1} \tag{5.9-33}$$

5.9-5   Multicomponent Systems

The foregoing material is based on studies of binary systems. For such systems the efficiency for each component is the same. When more than two components are present, it is possible for each component to have a different efficiency. Examples of efficiency distribution in multicomponent systems are given in

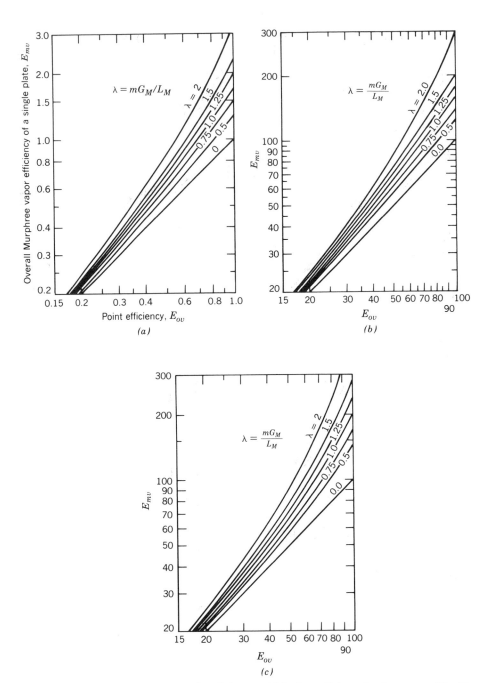

FIGURE 5.9-6 Relationship between point efficiency and Murphree efficiency for three cases of liquid-vapor flow combinations. Liquid in plug flow in all cases (Ref. 8). (a) Relationship of point efficiency to Murphree plate efficiency. Case 1: Vapor mixed under each plate. (b) Relationship of point efficiency to Murphree plate efficiency. Case 2: Vapor not mixed; flow direction of liquid reversed on successive plates. (c) Relationship of point efficiency to Murphree plate efficiency. Case 3: Vapor not mixed; liquid flow in same direction on successive plates.

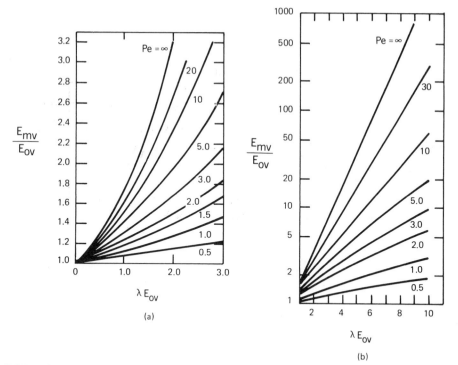

FIGURE 5.9-7 Relationship between point and Murphree efficiencies for partial mixing of liquid and complete mixing of vapor (Ref. 4).

the experimental results of Young and Weber,[13] Krishna et al.,[14] and Chan and Fair.[15] For ideal systems the spread in efficiency is normally small.

The basic theory for multicomponent system efficiency determination has been given by Toor and Burchard[16] and by Krishna et al.[14] In a series of experiments, confirmed by theory, Chan and Fair determined that a good approximation of multicomponent efficiency may be obtained by considering the dominant pair of components at the location in the column under consideration. This permits use of a pseudobinary approach, undoubtedly realistic for most commercial designs.

5.10 MASS TRANSFER IN PACKED COLUMNS

5.10-1 Efficiency Definitions and Approaches to Prediction

For packed columns, which are classified as *counterflow columns* (most trays are crossflow devices), mass transfer is related to transfer rates in counterflow vapor–liquid contacting. The most-used criterion of efficiency is the *height equivalent to a theoretical plate* (HETP) that was first introduced by Peters:[1]

$$\text{HETP} = \frac{\text{Height of packed zone}}{\text{Number of theoretical stages achieved in zone}} \qquad (5.10\text{-}1)$$

In using this expression to determine the required height of packing, one would calculate the number of ideal stages (Section 5.3) and obtain a value of HETP (likely from a packing vendor); the height of packing would then follow directly.

Alternately, the height of packing may be obtained by a more fundamental approach, the use of transfer units:

$$Z = H_V N_V = H_L N_L = H_{OV} N_{OV} = H_{OL} N_{OL} \qquad (5.10\text{-}2)$$

where H_V, H_L, H_{OV}, and H_{OL} are known as "heights of a transfer unit" (HTU) based on vapor-phase, liquid-phase, and combined-phase concentration driving forces, respectively. In the present work, a vapor-phase concentration basis will be adopted, and thus the term H_{OL} will not be used. The equivalent "numbers of transfer units," N_V, N_L, and N_{OV}, will also be used.

As is the case for a tray column, the efficiency depends on the system properties, the flow conditions, and the geometry of the device. Because of the many possible variations in the geometry of the packed bed, the last-named parameter assumes great importance in the estimation of packed column efficiency.

Methods for predicting efficiency also parallel those for tray columns: comparison against a similar installation, use of empirical methods, direct scaleup from laboratory or pilot plant, and use of theoretically derived models. Approaches by vendors of packing usually center on comparisons with similar installations (the so-called "vendor experience") and empirical approximations. Direct scaleup from small column studies is difficult with packed columns because of the unknown effects of geometrical factors and the variations of liquid distribution that are required for practical reasons. Theoretical or semitheoretical models are difficult to validate because of the flow effects on interfacial area. It may be concluded that there is no very good way to predict packed column efficiency, at least for the random type packings.

The approach used in the present work is that of semitheoretical models for mass transfer prediction. The reader should remember, however, that the packing vendors should be consulted as early as possible. These people have a great deal of experience with their own packings and, in addition, have done a great deal of their own test work. The models can then be used in connection with the vendor information to arrive at a final design.

5.10-2   Vendor Information

Vendors can provide values of HETP on the basis of their experience. They also conduct controlled tests on packings, using as a basis the absorption of carbon dioxide from air into a dilute solution of sodium hydroxide. The results of such tests normally are given in terms of an overall volumetric mass transfer coefficient, $K_{OV}a$. A discussion of the test method has been provided by Eckert et al.[2] Because a reaction occurs in the liquid phase, and because the rate of the reaction is a function of the degree of conversion of the hydroxide to the carbonate, it is difficult to construe the test results in terms of distillation needs. Also, the process involves significant mass transfer resistance in both the gas and liquid phases. As a result, such test values should be used with caution and, in general, considered only as a guide to the relative efficiencies of different packings.

Some vendors also have conducted their own tests under distillation conditions and can make such test data available. This is particularly the case for the vendors of structured packings. Standard distillation test mixtures are used, with operation usually under total reflux conditions. The use of these data, in connection with a mass transfer model, enables the most reliable approach to the estimation of the efficiency that a packed bed will produce.

5.10-3   Mass Transfer Models

For distillation, the height of a transfer unit is defined as

$$H_V = \frac{G_M}{k_V a_i} \tag{5.10-3}$$

$$H_L = \frac{L_M}{k_L a_i} \tag{5.10-4}$$

$$H_{OV} = \frac{G_M}{K_{OV} a_i} \tag{5.10-5}$$

A combination of these units leads to

$$H_{OV} = H_V + \lambda H_L \tag{5.10-6}$$

This approach enables one to consider separately the vapor- and liquid-phase mass transfer characteristics of the system and the packing.

RANDOM PACKINGS

Several methods have been proposed for predicting values of H_V and H_L as functions of system, flow conditions, and packing type. These include the methods of Cornell et al.,[3] Onda et al.,[4] Bravo and Fair,[5] and Bolles and Fair.[6,7] The last-named has had the broadest validation and will be discussed here.

The Bolles–Fair method may be summarized by the following equations:

$$H_V = \frac{\psi Sc_V^{0.5} D_T^{n_1}}{L_M M_L f_1 f_2 f_3} \tag{5.10-7}$$

$$H_L = \phi C_F \text{Sc}_L^{0.5} \tag{5.10-8}$$

$$f_1 = \left(\frac{\mu_L}{2.42}\right)^{0.16} \tag{5.10-9}$$

$$f_2 = \left(\frac{1000}{\rho_L}\right)^{0.16} \tag{5.10-10}$$

$$f_3 = \left(\frac{72}{\sigma}\right)^{0.8} \tag{5.10-11}$$

Values of exponent n_1 are 1.24 and 1.11 for ring-type and saddle-type packings, respectively. Values of m_1 are 0.6 and 0.5 for ring-type and saddle-type packings, respectively. Parameter C_F accounts for the effect of gas flow rate on the liquid-phase mass transfer coefficient, as the flood point is approached (below the loading point, there is relatively little effect of gas flow rate on the liquid-phase mass transfer process); suitable values of C_F are

50% flood	1.0
60% flood	0.90
80% flood	0.60

The D_T term (tower diameter) has an upper limit correction if there is good liquid distribution. The upper limit is for $D_T = 0.6$ m (2.0 ft). Thus, $D_T^{n_1} = 2.36$ maximum for rings, 2.16 for saddles.

Values of the parameters ψ and ϕ are given in Figs. 5.10-1 and 5.10-2. It is clear that the variety of packing types is limited, but one may use the Berl saddle curves for ceramic Intalox saddles (*not* for metal Intalox saddles).

It should be noted that the smallest size packings in the Bolles–Fair correlation are nominally 12 mm (0.5 in.) in diameter. If the general rule of 8:1 column:packing diameter is to be maintained, then a minimum column size of 100 mm (4 in.) is indicated. This is a crucial point in the design of pilot facilities and the development of scaleup parameters. It is not yet possible to go to the tiny packings for laboratory tests (using, say, column diameters of 25–50 mm) and still obtain reliable scaleup data.

One difficulty that arises in the application of the above equations is that of handling multicomponent mixtures. For such systems (and also for binaries) it may be convenient to compute the number of theoretical stages by conventional approaches and then convert to transfer units:

$$N_{OV} = N_t \frac{\ln \lambda}{\lambda - 1} \tag{5.10-12}$$

which applies to cases (e.g., individual column sections) where the equilibrium relationship may be considered linear. When that relationship and the operating line are both straight and parallel, $N_{OV} = N_t$.

Another point regarding the Bolles–Fair correlation: the height correction term to be used is that for a single bed having its own liquid distribution (or redistribution).

The Bravo–Fair model[5] for random packings involves the prediction of an effective interfacial area. The same data bank published by Bolles and Fair was used in conjunction with the individual-phase mass transfer correlation of Onda et al.[4] to deduce values of an effective area for transfer. The final equation for area is

$$a_e = 0.310 a_p \left(\frac{\sigma^{0.5}}{Z^{0.4}}\right) (\text{Ca}_L \text{Re}_V)^{0.392} \tag{5.10-13}$$

The dimensionless capillary number for the liquid, used in Eq. (5.10-13), is defined as follows:

$$\text{Ca}_L = \frac{L\mu_L}{\rho_L \sigma g_c} \tag{5.10-14}$$

where μ_L = liquid viscosity
$\quad L$ = liquid superficial mass velocity
$\quad \rho_L$ = liquid density
$\quad g_c$ = conversion factor (not needed for SI units)

The dimensionless Reynolds number for the vapor is

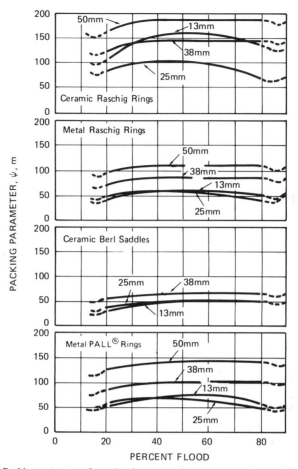

FIGURE 5.10-1 Packing parameters for estimating vapor-phase mass transfer, random packings (Ref. 7).

$$\text{Re}_V = \frac{6G}{a_p \mu_V} \tag{5.10-15}$$

This model was found to give a slightly better fit of the distillation data than did the Bolles–Fair model. It is somewhat more cumbersome to use, however, because of the need to evaluate individual-phase transfer coefficients.

STRUCTURED PACKINGS

The channels formed by the structured packing elements are presumed to pass liquid and vapor in counterflow contacting. Since the geometry is well defined and the packing surface area is known precisely, it has been found possible to model the mass transfer process along lines of wetted wall theory. The following material is based on recent work by Bravo et al.[8] at the University of Texas and has been developed specifically for structured packings of the gauze type, the principal dimensions of which are listed in Table 5.10-1.

Using Fig. 5.8-14 as a guide, the equivalent diameter of a channel is defined as

$$d_{eq} = Bh\left(\frac{1}{B + 2S} + \frac{1}{2S}\right) \tag{5.10-16}$$

where B = base of the triangle (channel cross section)
$\quad\ \ h$ = height of the triangle ("crimp height")
$\quad\ \ S$ = corrugation spacing (channel side)

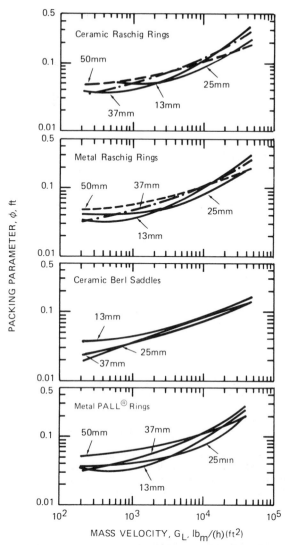

FIGURE 5.10-2 Packing parameters for estimating liquid-phase mass transfer, random packings (Ref. 7). (*Note:* The value of ϕ from the chart is in feet. $(lb_m/h \cdot ft^2) \times 0.00136 = kg/s \cdot m^2$.)

TABLE 5.10-1 Geometric Information, Sulzer BX Packing

Crimp height, h (mm)	6.4
Channel base, B (mm)	12.7
Channel side, S (mm)	8.9
Hydraulic radius, r_h (mm)	1.8
Equivalent diameter, d_{eq} (mm)	7.2
Packing surface, a_p (m²/m³)	492
Void fraction, ϵ	0.90
Channel flow angle from horizontal (degrees)	60

The effective velocity of vapor through the channel is

$$U_{ve} = \frac{U_{vs}}{\epsilon \sin \theta}$$

(5.10-17)

The effective liquid velocity is based on the falling film relationship for laminar flow:

$$U_{Le} = \left(\frac{3\Gamma}{2\rho_L}\right)\left(\frac{\rho_L^2 g}{3\mu_L \Gamma}\right)^{0.333}$$

(5.10-18)

where $\Gamma = L/PA_t$, kg/s · m perimeter
$\quad A_t$ = tower cross-sectional area, m^2
$\quad P$ = available perimeter, m/m^2, tower cross section
$\quad\quad = 1/2[(4S + 2B)/Bh + 4S/8h]$

For the vapor-phase mass transfer coefficient, the early work of Johnstone and Pigford[9] is used in a slightly modified form:

$$\text{Sh}_V = 0.0338 \text{Re}_V^{0.8}\text{Sc}_V^{0.333}$$

(5.10-19)

where Sh_V = Sherwood number for the vapor = $k_V d_{eq}/D^V$
$\quad \text{Re}_V$ = Reynolds number for the vapor, defined for the present geometry as $(d_{eq}\rho_V/\mu_V)(U_{ve} + U_{Le})$
$\quad \text{Sc}_V$ = Schmidt number for the vapor = $\mu_V/\rho_V D^V$

For the liquid phase, the penetration theory is used, with one exposure being the residence time for liquid flow between corrugation changes:

$$k = 2\left(\frac{D^L U_{Le}}{\pi S}\right)^{0.5}$$

(5.10-20)

Finally, an analysis of the experimental data on the gauze-type structured packing showed that the total surface was fully wetted. This was attributed in part to the capillary action that takes place with the finely woven wire gauze.

Thus, heights of transfer units can be calculated for the structured packing case, using information from the above relationships. The method was checked against 132 data points and found to have an average deviation of 14.6%.

Methods for handling the sheet metal structured packings have not yet been published. However, it seems clear that such packings do not have the benefit of capillary action and thus may not be completely wetted. General experience shows that the same approach may be used but that the transfer area must be discounted to about 60% of the specific surface area of the packing. Under high loading conditions, the amount of discount is likely to be much less.

5.11 DISTILLATION COLUMN CONTROL

The control of distillation systems is a subject that has been treated extensively in the literature. At least four books on distillation control have been published in recent years.[1-4] It is appropriate that in the present work only a few general guidelines on control be presented.

The column variables of concern, in a simple one-feed two-product column with pressure controlled separately, may be categorized as follows:

1. Controlled Variables
   (a) Throughput
   (b) Distillate composition
   (c) Bottoms composition
   (d) Bottoms sump level
   (e) Reflux accumulator level
2. Manipulated Variables
   (a) Feed flow
   (b) Distillate flow

(c) Reflux flow

(d) Heat medium flow

(e) Bottoms flow

For this situation there are generally four schemes that might be considered, as shown in Fig. 5.11-1, taken from the paper by Tolliver and McCune.[5] These schemes are for *composition control* and are described as follows:

Scheme 1. This scheme directly adjusts the column material balance by manipulation of the distillate flow. The main advantage of this scheme is that it has the least interaction with the energy balance. In terms of a McCabe–Thiele diagram, this means that the slopes of the column operating lines can be held constant in spite of energy balance upsets. This independence from energy balance upsets is achieved by the scheme's ability to maintain a constant internal reflux even for variations in external reflux subcooling. When the temperature of the external reflux varies, the external reflux adjustment to maintain accumulator level offsets temporary internal reflux variations. If the accumulator level loop responds rapidly, the disturbance will not propagate down the column, and the column's overall material balance remains undisturbed.

The principal disadvantage of scheme 1 is that it sometimes responds slowly to feed rate upsets in

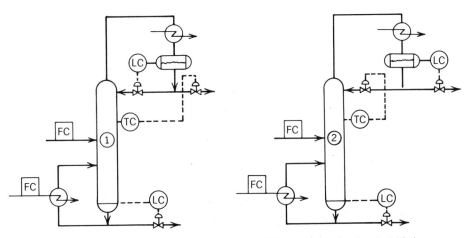

Scheme 1: Direct material balance   Scheme 2: Indirect material balance

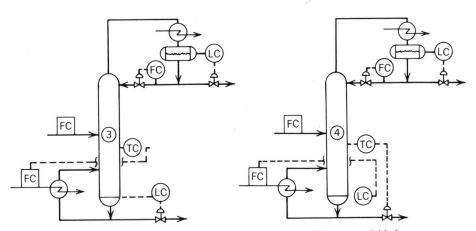

Scheme 3: Indirect material balance   Scheme 4: Direct material balance

FIGURE 5.11-1 Some basic control schemes for composition control, distillation columns.

stripping columns where the column objectives strongly emphasize control of bottom product composition. Also, in columns with relatively large accumulator holdup, often found in pilot plant columns, the level control (within the composition loop) can be too slow for effective composition control. Scheme 1 should be selected whenever the distillate flow is one of the smaller flows in the column. If column dynamics favor other schemes, but the guidelines do not, then additional feed-forward controls may be applied to this scheme to improve the dynamics.

Scheme 2. This scheme indirectly adjusts the material balance through the two level control loops. This arrangement has the advantage of reducing the ratio of effective dead time to total lag time within the composition loop. It has the disadvantage of allowing greater interaction between the material and energy balances because internal reflux is not held constant. This scheme should be considered when the reflux is smaller than other flows in the column and when the reflux to distillate ratio is 0.8 or less. It should also be considered for applications where reducing the ratio of effective dead time to total lag time in the composition loop is a significant and necessary consideration.

Scheme 3. This scheme is a common arrangement in the chemical industry, used primarily on stripping columns. This arrangement indirectly adjusts the material balance in the same manner as scheme 2. It has the advantage of offering the minimum effective dead time to total lag time in the composition loop. It has the disadvantage of having the greatest interaction between the material and energy balances. Because the reboiler is this interaction, scheme 3 is *not* recommended in columns where the reboiler boilup to column feed ratio is 2.0 or greater.

Scheme 4. This scheme directly manipulates the material balance by adjusting the bottom product flow. This scheme has the advantage of little interaction between the material and energy balances, similar to scheme 1. However, the sump level loop can significantly increase the ratio of effective dead time to total lag time in the composition loop. Furthermore, the sump level loop is often subject to an inverse response, which can make the loop difficult to implement or even impossible to control. Because of these difficulties, this arrangement is recommended only when the bottom product flow is significantly smaller than other flows in the column and is less than 20% of the boilup from the reboiler.

The authors point out that the schemes shown in Fig. 5.11-1 are conceptual arrangements and the reader is cautioned against inferring any control hardware requirements from such simplified diagrams.

In Fig. 5.11-1, composition of a flowing stream in the column is inferred from its temperature. This is the most common inferential measurement of composition; others include density, refractive index, viscosity and thermal conductivity. In some cases, especially for fractionating close-boiling mixtures, direct composition measurement by stream analyzer may be desired. The most common analyzer used is the gas chromatograph, although other techniques such as mass spectrometry may be used in some instances. The location of the point of measurement is determined by the gradient of composition or by column dynamic considerations.

For *column pressure control* there are three general approaches: vent bleed (to atmosphere or to vacuum system), hot vapor bypass, and flooded condenser. These approaches are illustrated in Fig. 5.11-2.[5] For an atmospheric column, the vent approach is quite simple. The vapor bypass represents a temperature blending method. Partial flooding of the condenser surface adjusts the heat transfer capability of the condenser. The schemes are generally self-explanatory.

Distillation control schemes may be analyzed either on a steady-state (sensitivity analysis) or on a dynamic basis. The latter requires a dynamic model that takes into account the dynamic response of the column and the control loops. An example of a dynamic model is described by McCune and Gallier,[6] but it should be apparent from the material presented earlier that the holdup characteristics of distillation column devices can vary widely, and such variation should be accommodated by the model. The development of the new high-efficiency packings has caused a new look at the system dynamics when the liquid holdup in the column is quite low, and thus the existing models for trays may not be adjustable to application to packings. The use of a tray-type dynamic model is described in the article by Gallier and McCune;[7] no work to date has been reported for packed column dynamic models.

Steady-state analysis is based on the determination of a new set of operating conditions that would be required by an imposed upset of the system. In this case there is the constraint of a fixed number of trays or stages in the column, and the assumption is often possible that there would be no significant change in efficiency caused by the upset. Thus, in a simplistic example, one could modify a McCabe–Thiele plot to accommodate a changed feed composition, by trial-and-error, finding a new set of operating lines that would not change the required theoretical stages. The steady-state approach, called "sensitivity analysis," determines the new set of operating conditions but says nothing about how long it takes to reach the new state—nor does it say anything about the gyrations that might occur during the period of transition. Still, it should always be carried out for new column designs.

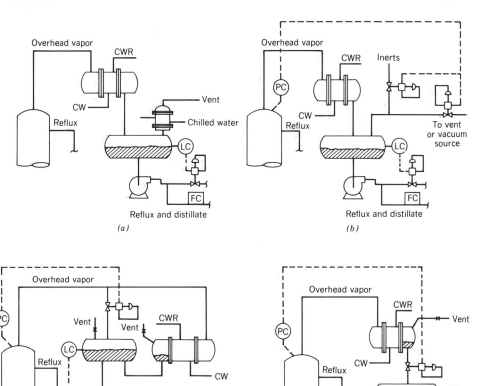

FIGURE 5.11-2 Schemes for control of column pressure: (*a*) pressure control for an atmospheric column (vent bleed to atmosphere); (*b*) split range valves in a block and bleed arrangement (vent bleed to vacuum); (*c*) hot vapor bypass pressure control; and (*d*) flooded condenser pressure control.

NOTATION

a	interfacial area, m^2/m^3
a_e	effective interfacial area, m^2/m^3
a_p	specific surface area of packing, m^2/m^3
A	absorption factor, L/VK
A	cross-sectional area, m^2
A_a	active or bubbling area
A_d	downcomer area
A_h	hole area
A_n	net area
A_T	total tower area
A_i	constant in Antoine equation
B	bottoms flow rate, kmol/s
B_i	constant in Antoine equation
C	number of components

C'	constant in Eq. (5.7-26)
C_d	drag coefficient, dimensionless
C_i	constant in Antoine equation
C_{SB}	capacity (Souders–Brown) coefficient based on net area, m/s
C_{SBf}	capacity coefficient at flood, m/s
C'_{SB}	capacity coefficient based on active area, m/s
C_v	discharge (orifice) coefficient, dimensionless
Ca	capillary number, dimensionless
CW	cooling water
d_{eq}	equivalent diameter of packing, mm
d_h	hole diameter, mm
d_p	diameter of random packing element, mm
D_d	diameter of liquid droplet, mm
D^L	liquid diffusion coefficient, m²/s
D^V	vapor diffusion coefficient, m²/s
D_E	eddy diffusion coefficient, m²/s
E	efficiency, fractional
E_a	efficiency, corrected for entrainment
E_{mv}	Murphree vapor efficiency
E_{oc}	overall column efficiency
E_{ov}	point (local) efficiency, vapor basis
e	exponential
e	entrainment rate, moles/time
f	friction factor, dimensionless
F_v	vapor flow F factor, $U_V \rho^{1/2}$, m/s (kg/m³)$^{1/2}$
F_{va}	F factor through active area
F_{vh}	F factor through hole area
F_{vs}	F factor through total tower cross-sectional area
F_n	molar flow rate of feed to stage n, kmol/s
F_p	packing factor, m
Fr	Froude number, dimensionless
g	gravitational constant, 9.81 m/s²
g_c	conversion factor, 32.2 lb$_m$/lb$_f$ (ft/s²); not used with SI system
G	vapor mass velocity, kg/s·m²
G_M	molar velocity of vapor, kmol/s·m²
h	liquid enthalpy, J/kg
h	head, mm liquid (pressure loss)
h_d	pressure loss through holes
h_{da}	pressure loss under downcomer apron
h_{dist}	pressure loss for flow through distributor
h'_L	residual pressure loss
h_t	total pressure drop across tray
h_{dc}	height of liquid in downcomer, mm
h_L	equivalent height of liquid holdup on tray, mm liquid
h_{ow}	crest of liquid over weir, mm
h_w	weir height, mm
H	enthalpy of vapor, J/kg
H	height of a transfer unit (HTU), m
H_L	height of a liquid-phase transfer unit
H_{OL}	height of an overall liquid transfer unit
H_{OV}	height of an overall vapor transfer unit
H_V	height of a vapor transfer unit

H	liquid holdup, volume fraction
H_o	operating holdup
H_s	static holdup
H_t	total holdup
HETP	height equivalent to a theoretical plate, m
J	liquid holdup on tray, kg·moles
k	individual-phase mass transfer coefficient, m/s or kmol/(s·m^2·mole fraction)
k_V	vapor-phase coefficient
k_L	liquid-phase coefficient
K	overall mass transfer coefficient, m/s or kmol/(s·m^2·mole fraction)
K_{OV}	overall coefficient based on vapor concentration
K_{OL}	overall coefficient based on liquid concentrations
K	vapor–liquid equilibrium ratio
ℓ	component molar flow rate, kmol/s
l_w	weir length, mm or m
L	molar liquid rate, kmol/s
L'	molar liquid rate in stripping section, kmol/s
L_M	molar liquid velocity, kmol/s·m^2
L_W	liquid rate, m^3/s·m weir
m	slope of the equilibrium line
m	stage designation, stripping section
n	stage designation, rectifying section (or general)
N	top stage
N	number of transfer units
N_L	number of liquid transfer units
N_{OL}	number of overall transfer units, liquid concentration basis
N_{OV}	number of overall transfer units, vapor concentration basis
N_V	number of vapor transfer units
N_A	molar flux of species A, kmol/s·m
N_a	number of actual trays
N_t	number of theoretical stages
$N_{\min}$	minimum number of theoretical stages
p^{sat}	vapor pressure, mm Hg
p	partial pressure, mm Hg
P	total pressure, mm Hg
PC	Poynting correction [Eq. (5.2-3)]
Pe	Peclet number, dimensionless
q	heat ratio [Eq. (5.3-19)]
q	volumetric flow rate of liquid, m^3/s
Q	heat flow, J/s
Q_B	flow to reboiler
Q_D	flow from condenser
Q_n	flow to or from stage n
Q_f	volume of froth, m^3
Q	volumetric flow rate of vapor, m^3/s
R	gas constant
R	reflux ratio
$R_{\min}$	minimum reflux ratio (also R_m)
Re	Reynolds number, dimensionless
s	number of mixing stages on tray
S	number of moles in stillpot
Sc	Schmidt number, dimensionless
Sh	Sherwood number dimensionless

t	time, s
T	temperature, K
TS	tray spacing, mm
u	liquid velocity, m/s
u_{da}	velocity under downcomer apron, m/s
U	vapor velocity, m/s
U_a	velocity through active area
U_h	velocity through hole area
U_n	velocity through net area
U_{nf}	flooding velocity through net area
U_0	superficial velocity (also U_s)
U_L	liquid velocity, m/s
U_{ve}	effective vapor velocity, m/s
U_{vs}	superficial vapor velocity in packed column, m/s
v	molar volume, m³/kmol
$\bar{v}$	partial molar volume, m³/kmol
v	component vapor flow rate, kmol/s
V	molar flow rate of vapor, kmol/s
V'_n	rate of sidestream vapor withdrawal from tray n
V'	molar flow rate of liquid in stripping section, moles/time
W	liquid flow rate across tray [Eq. (5.9-25)], m³/s·m
w'	length of liquid flow path, m
x	mole fraction in liquid
y	mole fraction in vapor
$y*$	equilibrium mole fraction in vapor
z	mole fraction in feed
Z	height of packing, m
Z_f	froth height on tray, m

Subscripts

a	active area basis
f	feed
HK	heavy key
i	component i
j	component j
L	liquid
LK	light key
n	stage n
o	operating
s	stillpot
ss	steady state
t	time t
V	vapor
0	initial time (time 0)
1, 2	components 1 and 2
1, 2	stages 1 and 2

Superscripts

L	liquid
V	vapor

Greek Letters

α	relative volatility
$\bar{\alpha}$	average relative volatility
β	aeration factor
γ	activity coefficient
Γ	liquid flow rate, based on length, $m^3/s \cdot m$
Δ	hydraulic gradient, mm liquid
Δ	difference point, Ponchon–Savarit method
Δp	pressure drop, mm liquid
Δp_0	pressure drop through dry bed of packing
ϵ	void fraction or froth porosity
θ	angle of inclination of channel in structured packing
θ	notch angle, degrees
Θ	exposure time, s
λ	ratio of slope of equilibrium line to that of operating line
μ	viscosity, $kg/m \cdot s$
π	$3.1416 \ldots \ldots$
ρ	density, kg/m
σ	surface tension, mN/m (dynes/cm)
ϕ	parameter for packed column mass transfer, Eq. (5.10-8)
ϕ	geometric packing factor, Eq. (5.8-16)
ϕ_e	effective froth density, dimensionless
ϕ_i	fugacity coefficient, dimensionless
$\hat{\phi}_i$	fugacity coefficient for i in mixture
ϕ_i^{sat}	fugacity coefficient for pure i, at system temperature and pressure equal to the vapor pressure of i
ϕ_f	froth density, dimensionless
χ	aperature correction factor, Eq. (5.8-16)
ψ	parameter for packed column mass transfer, Eq. (5.10-7)
Ψ	ratio, density of liquid to density of water

REFERENCES

Section 5.2

5.2-1 J. H. Hildebrand, J. M. Prausnitz, and R. L. Scott, *Regular and Related Solutions*, Van Nostrand Reinhold, New York, 1970.

5.2-2 R. C. Reid, J. M. Prausnitz, and T. K. Sherwood, *The Properties of Gases and Liquids*, 3rd ed., McGraw-Hill, New York, 1977.

5.2-3 H. H. Y. Chien and H. R. Null, *AIChE J.*, **18,** 1177 (1972).

5.2-4 E. Hala, J. Pick, V. Fried, and O. Vilim, *Vapor–Liquid Equilibrium*, 2nd ed., Pergamon, Oxford, 1967.

5.2-5 E. Hala, I. Wichterle, J. Polak, and T. Boublik, *Vapor–Liquid Equilibrium at Normal Pressures*, Pergamon, Oxford, 1968.

5.2-6 I. Wichterle, J. Linek, and E. Hala, *Vapor–Liquid Equilibrium Data Bibliography*, Elsevier, Amsterdam, 1975.

5.2-7 J. C. Chu, S. L. Wang, S. L. Levy, and R. Paul, *Vapor–Liquid Equilibrium Data*, J. B. Edwards, Ann Arbor, 1956.

5.2-8 M. Hirata, S. Ohe, and K. Nagahama, *Computer Aided Data Book of Vapor–Liquid Equilibria*, Elsevier, Amsterdam, 1975.

5.2-9 J. Gmehling, U. Onken, and W. Arlt, *Vapor–Liquid Equilibrium Collection* (continuing series), DECHEMA, Frankfurt, 1979.

5.2-10 Chemshare Corp., Houston, Texas.

5.2-11 S. T. Hadden and H. G. Grayson, *Petrol. Refiner*, **40**(9), 207 (1961).

5.2-12 Natural Gas Processors Suppliers Assn., *Engineering Data Book*, 9th ed., Tulsa, 1972.

5.2-13   L. H. Horsley, *Azeotropic Data—III*, American Chemical Society Advances in Chemistry Series 116, Washington, DC, 1973.

Section 5.3

5.3-1   W. L. McCabe and E. W. Thiele, *Ind. Eng. Chem.*, **17**, 605 (1925).

5.3-2   M. Ponchon, *Tech. Mod.*, **13**, 20, 55 (1921).

5.3-3   R. Savarit, *Arts Metiers*, 65, 142, 178, 241, 266, 307 (1922).

5.3-4   W. L. McCabe and J. C. Smith, *Unit Operations of Chemical Engineering*, 3rd ed., McGraw-Hill, New York, 1976.

5.3-5   M. R. Fenske, *Ind. Eng. Chem.*, **24**, 482 (1932).

5.3-6   A. J. V. Underwood, *Chem. Eng. Prog.*, **44**, 603 (1948).

5.3-7   H. H. Chien, *Chem. Eng. Sci.*, **28**, 1967 (1973).

5.3-8   H. H. Chien, *AIChE J.*, **24**, 606 (1978).

5.3-9   W. K. Lewis and G. L. Matheson, *Ind. Eng. Chem.*, **24**, 494 (1932).

5.3-10   C. D. Holland, *Fundamentals of Multicomponent Distillation*, McGraw-Hill, New York, 1981.

5.3-11   E. W. Thiele and R. L. Geddes, *Ind. Eng. Chem.*, **25**, 290 (1933).

5.3-12   N. R. Amundson and A. J. Pontinen, *Ind. Eng. Chem.*, **50**, 730 (1958).

5.3-13   J. R. Friday and B. D. Smith, *AIChE J.*, **10**, 698 (1964).

5.3-14   J. C. Wang and G. E. Henke, *Hydrocarbon Proc.*, **45**(8), 155 (1966).

5.3-15   J. C. Wang and Y. L. Wang, A Review on the Modeling and Simulation of Multistaged Separation Processes, in *Foundations of Computer-Aided Chemical Process Design*, Vol. 2, R. S. H. Mah and W. D. Seider (Eds.), American Institute of Chemical Engineers, New York, 1981.

5.3-16   E. R. Gilliland, *Ind. Eng. Chem.*, **32**, 1220 (1940).

5.3-17   H. E. Eduljee, *Hydrocarbon Proc.*, **54**(9), 120 (1975).

5.3-18   J. H. Erbar and R. N. Maddox, *Petrol. Ref.*, **40**(5), 183 (1961).

5.3-19   E. R. Gilliland, *Ind. Eng. Chem.*, **32**, 918 (1940).

5.3-20   J. R. Fair and W. L. Bolles, *Chem. Eng.*, **75**(9), 156 (Apr. 22, 1968).

Section 5.5

5.5-1   L. H. Horsley, *Azeotropic Data—III*, American Chemical Society Advances in Chemistry No. 116, Washington, DC, 1973.

5.5-2   M. Lecat, *Tables azeotropiques*, l'Auteur, Brussels, 1949.

5.5-3   H. Renon and J. M. Prausnitz, *AIChE J.*, **14**, 135 (1968).

5.5-4   C. S. Robinson and E. R. Gilliland, *Elements of Fractional Distillation*, 4th ed., McGraw-Hill, New York, 1950.

5.5-5   B. T. Brown, H. A. Clay, and J. M. Miles, *Chem. Eng. Prog.*, **66**(8), 54 (1970).

5.5-6   B. K. Kruelski, C. C. Herron, and J. R. Fair, "Hydrodynamics and Mass Transfer on Three-Phase Distillation Trays," presented at Denver AIChE Meeting, Aug. 1983.

5.5-7   E. J. Hoffman, *Azeotropic and Extractive Distillation*, Interscience, New York, 1964.

5.5-8   C. Black, R. A. Golding, and D. E. Ditsler, Azeotropic Distillation Results from Automatic Computer Calculations, in *Extractive and Azeotropic Distillation*, Chap. 5, American Chemical Society Advances in Chemistry No. 115, Washington, DC, 1972.

5.5-9   J. A. Gerster, *Chem. Eng. Prog.*, **65**(9), 43 (1969).

5.5-10   C. Black and D. E. Ditsler, Dehydration of Aqueous Ethanol Mixtures by Extractive Distillation, in *Extractive and Azeotropic Distillation*, Chap. 1, American Chemical Society Advances in Chemistry No. 115, Washington, DC, 1972.

5.5-11   L. Berg, *Chem. Eng. Prog.*, **65**(9), 52 (1969).

5.5-12   R. H. Ewell, J. M. Harrison, and C. Berg, *Ind. Eng. Chem.*, **36**, 871 (1944).

5.5-13   H. G. Drickamer, G. G. Brown, and R. R. White, *Trans. AIChE*, **41**, 555 (1945).

5.5-14   H. G. Drickamer and H. H. Hummel, *Trans. AIChE*, **41**, 607 (1945).

5.5-15   B. D. Smith, *Design of Equilibrium Stage Processes*, McGraw-Hill, New York, 1963.

5.5-16   R. Kumar, J. M. Prausnitz, and C. J. King, Process Design Considerations for Extractive Distillation: Separation of Propylene–Propane, in *Extractive and Azeotropic Distillation*, Chap. 2, American Chemical Society Advances in Chemistry No. 115, Washington, DC, 1972.

5.5-17 J. A. Gerster, T. Mizushina, T. N. Marks, and A. W. Catanach, *AIChE J.*, **1,** 536 (1955).

5.5-18 E. R. Hafslund, *Chem. Eng. Prog.*, **65**(9), 58 (1969).

5.5-19 R. R. Bannister and E. Buck, *Chem. Eng. Prog.*, **65**(9), 65 (1969).

5.5-20 D. P. Tassios, Rapid Screening of Extractive Distillation Solvents. Predictive and Experimental Techniques, in *Extractive and Azeotropic Distillation*, Chap. 4, American Chemical Society Advances in Chemistry Series No. 115, Washington, DC, 1972.

5.5-21 J. Drew and M. Probst (Eds.), *Tall Oil*, Pulp Chemicals Assn., New York, 1981.

5.5-22 J. W. Packie, *Trans. AIChE*, **37,** 51 (1941).

5.5-23 G. S. Houghland, E. J. Lemieux and W. C. Schreiner, *Oil Gas J.*, **52,** 198 (July 26, 1954).

5.5-24 M. W. Van Winkle, *Distillation*, McGraw-Hill, New York, 1967.

5.5-25 R. R. Dreisbach, *Pressure–Volume–Temperature Relationships of Organic Compounds*, Handbook Publishers, Cleveland, 1952.

5.5-26 M. J. P. Bogart, *Trans. AIChE*, **33,** 139 (1937).

5.5-27 R. F. Jackson and R. L. Pigford, *Ind. Eng. Chem.*, **48,** 1020 (1932).

5.5-28 C. D. Holland and A. I. Liapis, *Computer Methods for Solving Dynamic Separation Problems*, McGraw-Hill, New York, 1983.

5.5-29 E. L. Meadows, *Chem. Eng. Prog. Symp. Ser. No. 46*, **59,** 48–55 (1963).

5.5-30 G. P. Distefano, *AIChE J.*, **14,** 190 (1968).

5.5-31 J. F. Boston, H. I. Britt, S. Jirapongphan, and V. B. Shah, An Advanced System for the Simulation of Batch Distillation Operations, in *Foundations of Computer-Aided Chemical Process Design*, Vol. 2, American Institute of Chemical Engineers, New York, 1981.

Section 5.6

5.6-1 R. Krishnamurthy and R. Taylor, *Ind. Eng. Chem. Proc. Des. Dev.*, **24,** 513 (1985).

Section 5.7

5.7-1 American Institute of Chemical Engineers, *Bubble-Tray Design Manual*, New York, 1958.

5.7-2 D. L. Bennett, R. Agrawal, and P. J. Cook, *AIChE J.*, **29,** 434 (1983).

5.7-3 P. A. M. Hofhuis and F. J. Zuiderweg, *Int. Chem. Eng. Symp. Ser.*, **56,** 2.2/1 (1979).

5.7-4 R. E. Loon, W. V. Pincewski, and C. J. D. Fell, *Trans. Inst. Chem. Eng.*, **51,** 374 (1974).

5.7-5 M. Souders and G. G. Brown, *Ind. Eng. Chem.*, **26,** 98 (1934).

5.7-6 F. C. Silvey and G. J. Keller, *Chem. Eng. Prog.*, **62**(1), 69 (1966).

5.7-7 J. R. Fair, *Petro/Chem. Eng.*, **33**(10), 45 (1961).

5.7-8 E. Kirschbaum, *Distillier-und Rektifiziertechnik*, 4th ed., Springer Verlag, Berlin, 1969.

5.7-9 J. R. Fair, *Distillation in Practice,* course notes for Today course, American Institute of Chemical Engineers, New York, 1981.

5.7-10 J. R. Fair and R. L. Matthews, *Petrol. Ref.*, **37**(4), 153 (1958).

5.7-11 T. Yanagi, "Performance of Trays in Low Liquid Rate Fractionation," motion picture shown at AIChE Miami meeting, Nov. 1978 (film available from Fractionation Research, Inc.)

5.7-12 A. P. Colburn, *Ind. Eng. Chem.*, **28,** 526 (1936).

5.7-13 J. R. Fair, Chap. 15 in B. D. Smith (Ed.), *Design of Equilibrium Stage Processes*, McGraw-Hill, New York, 1963.

5.7-14 S. Zanelli and R. Del Bianco, *Chem. Eng. J.*, **6,** 181 (1973).

5.7-15 I. Leibson, R. E. Kelley, and L. A. Bullington, *Petrol. Ref.*, **36**(2), 127 (1957).

5.7-16 M. H. Hutchinson, A. G. Buron and B. P. Miller, Paper presented at AIChE meeting, Los Angeles, May 1949.

5.7-17 W. L. Bolles and J. R. Fair, *Distillation*, Vol. 16 of *Encyclopedia of Chemical Processing and Design*, J. J. McKetta (Ed.), Marcel Dekker, New York, 1982.

5.7-18 W. L. Bolles, Chap. 14 in B. D. Smith (Ed.), *Design of Equilibrium Stage Processes*, McGraw-Hill, New York, 1963.

5.7-19 R. Billet, S. Conrad, and C. M. Grubb, *Int. Chem. Eng. Symp. Ser. No. 32*, **5,** 111 (1969).

5.7-20 P. J. Hoek and F. J. Zuiderweg, *AIChE J.*, **28,** 535 (1982).

5.7-21 Glitsch, Inc., *Ballast Tray Design Manual, Bulletin 4900*, Dallas, Texas, 1961.

Section 5.8

5.8-1   G. K. Chen, *Chem. Eng.*, **91**(5), 40 (Mar. 5, 1984).
5.8-2   D. F. Stedman, *Trans. AIChE*, **33**, 153 (1937).
5.8-3   L. B. Bragg, *Trans. AIChE*, **37**, 19 (1941).
5.8-4   A. J. Hayter, *Ind. Chemist*, 59 (Feb. 1952).
5.8-5   R. C. Scofield, *Chem. Eng. Prog.*, **46**, 405 (1950).
5.8-6   L. B. Bragg, *Ind. Eng. Chem.*, **49**, 1062 (1957).
5.8-7   R. Billet, *Int. Chem. Eng. Symp. Ser. No. 32*, **4**, 42 (1969).
5.8-8   M. Huber, U.S. Patent 3,285,587 (Nov. 15, 1966).
5.8-9   J. R. Fair, *Chem. Eng. Prog.*, **66**(3), 45 (1970).
5.8-10   T. Baker, T. H. Chilton, and H. C. Vernon, *Trans. AIChE*, **31**, 296 (1935).
5.8-11   F. C. Silvey and G. J. Keller, *Int. Chem. Eng. Symp. Ser. No. 32*, **4**, 18 (1969).
5.8-12   M. A. Albright, *Hydrocarbon Proc.*, **63**(9), 173 (1984).
5.8-13   J. G. Kunesh, L. L. Lahm, and T. Yanagi, "Liquid Distribution Studies in Packed Beds", presented at Chicago AIChE Meeting, Nov. 1985.
5.8-14   H. L. Shulman, C. F. Ullrich, and N. Wells, *AIChE J.*, **1**, 247, 253 (1955).
5.8-15   J. E. Buchanan, *Ind. Eng. Chem. Fundam.*, **6**, 400 (1967).
5.8-16   J. L. Bravo and J. R. Fair, *Ind. Eng. Chem. Proc. Des. Dev.*, **21**, 162 (1982).
5.8-17   B. R. Sarchet, *Trans. AIChE*, **38**, 283 (1942).
5.8-18   F. C. Silvey and G. J. Keller, *Chem. Eng. Prog.*, **62**(1), 69 (1966).
5.8-19   T. K. Sherwood, G. H. Shipley, and F. A. L. Holloway, *Ind. Eng. Chem.*, **30**, 765 (1938).
5.8-20   J. S. Eckert, *Chem. Eng. Prog.*, **66**(3), 39 (1970).
5.8-21   J. E. Buchanan, *Ind. Eng. Chem. Fundam.*, **8**, 502 (1969).
5.8-22   G. G. Bemer and G. A. J. Kallis, *Trans. Inst. Chem. Eng.*, **56**, 200 (1978).
5.8-23   R. Billet and J. Mackowiak, *Fette-Seifen-Anstrichmittel*, **86**, 349 (1984).
5.8-24   J. L. Bravo, J. A. Rocha, and J. R. Fair, *Hydrocarbon Proc.*, **65**(3), 45 (1986).
5.8-25   W. L. Bolles and J. R. Fair, *Distillation in Practice*, Notes for American Institute of Chemical Engineers course, New York, 1984.

Section 5.9

5.9-1   H. E. O'Connell, *Trans AIChE*, **42**, 741 (1946).
5.9-2   J. R. Fair, W. L. Bolles, and H. R. Null, *Ind. Eng. Chem. Proc. Des. Dev.*, **22**, 53 (1983).
5.9-3   R. L. Geddes, *Trans. AIChE*, **42**, 79 (1946).
5.9-4   AIChE [American Institute of Chemical Engineers], *Bubble Tray Design Manual*, New York, 1958.
5.9-5   A. S. Foss and J. A. Gerster, *Chem. Eng. Prog.*, **52**, 28 (1956).
5.9-6   H. Chan and J. R. Fair, *Ind. Eng. Chem. Proc. Des. Dev.*, **23**, 814 (1984).
5.9-7   D. A. Diener, *Ind. Eng. Chem. Proc. Des. Dev.*, **6**, 499 (1967).
5.9-8   W. K. Lewis, Jr., *Ind. Eng. Chem.*, **28**, 399 (1936).
5.9-9   M. F. Gautreaux and H. E. O'Connell, *Chem. Eng. Prog.*, **51**, 232 (1955).
5.9-10   P. E. Barker and M. F. Self, *Chem. Eng. Sci.*, **17**, 541 (1962).
5.9-11   K. E. Porter, M. J. Lockett, and C. T. Lim, *Trans. Inst. Chem. Eng.*, **50**, 91 (1972).
5.9-12   R. J. Bell, *AIChE J.*, **18**, 498 (1972).
5.9-13   G. G. Young and J. H. Weber, *Ind. Eng. Chem. Proc. Des. Dev.*, **11**, 440 (1972).
5.9-14   R. Krishna, H. F. Martinez, R. Sreedhar, and G. L. Standart, *Trans. Inst. Chem. Eng.*, **55**, 178 (1977).
5.9-15   H. Chan and J. R. Fair, *Ind. Eng. Chem. Proc. Des. Dev.*, **23**, 820 (1984).
5.9-16   H. L. Toor and J. K. Burchard, *AIChE J.*, **6**, 202 (1960).

Section 5.10

5.10-1   W. A. Peters, *J. Ind. Eng. Chem.*, **14**, 476 (1922).
5.10-2   J. S. Eckert, E. H. Foote, L. R. Rollison, and L. F. Walter, *Ind. Eng. Chem.*, **59**(2), 41 (1967).

5.10-3   D. Cornell, W. G. Knapp, and J. R. Fair, *Chem. Eng. Prog.*, **56**(7), 68 (1960).

5.10-4   K. Onda, H. Takeuchi, and Y. Okumoto, *J. Chem. Eng. Japan*, **1,** 56 (1968).

5.10-5   J. L. Bravo and J. R. Fair, *Ind. Eng. Chem. Proc. Des. Dev.*, **21,** 162 (1982).

5.10-6   W. L. Bolles and J. R. Fair, *Int. Chem. Eng. Symp. Ser. No. 56*, 3.3/35 (1979).

5.10-7   W. L. Bolles and J. R. Fair, *Chem. Eng.*, **89**(14), 109 (July 12, 1982).

5.10-8   J. L. Bravo, J. A. Rocha, and J. R. Fair, *Hydrocarbon Proc.*, **64**(1), 91 (1985).

5.10-9   H. F. Johnstone and R. L. Pigford, *Trans. AIChE*, **38,** 25 (1942).

Section 5.11

5.11-1   A. E. Nisenfeld and R. C. Seeman, *Distillation Columns*, Instrument Society of America (ISA), Research Triangle Park, NC, 1981.

5.11-2   F. G. Shinskey, *Distillation Control*, 2nd ed., McGraw-Hill, New York, 1984.

5.11-3   P. B. Deshpande, *Distillation Dynamics and Control*, Instrument Society of America (ISA), Research Triangle Park, NC, 1984.

5.11-4   P. S. Buckley, W. L. Luyben, and J. P. Shunta, *Design of Distillation Control Systems*, Instrument Society of America (ISA), Research Triangle Park, NC, 1985.

5.11-5   T. L. Tolliver and L. C. McCune *ISA Transactions*, **17**(3), 3 (1978).

5.11-6   L. C. McCune and P. W. Gallier, *ISA Transactions*, **12**(3), 193 (1973).

5.11-7   P. W. Gallier and L. C. McCune, *Chem. Eng. Prog.*, **70**(9), 71 (1974).

CHAPTER 6

Absorption and Stripping

ARTHUR L. KOHL
Program Manager, Rocketdyne Division
Rockwell International
Canoga Park, California

6.1 BASIC CONCEPTS

6.1-1 Definitions

The term absorption as used in this chapter refers to the transfer of one or more components of a gas phase to a liquid phase in which it is soluble. Stripping is exactly the reverse, the transfer of a component from a liquid phase in which it is dissolved to a gas phase. The same basic principles apply to both operations; however, for convenience, the discussions that follow refer primarily to the absorption case.

Technically, the liquid used in a gas absorption process is referred to as the absorbent and the component absorbed is called the absorbate. As used in this text, these terms are considered synonymous with solvent and solute, respectively. In practical usage, the absorbent often is designated as the lean solution or the rich solution depending on whether it is entering or leaving the absorber.

The operation of absorption can be categorized on the basis of the nature of the interaction between absorbent and absorbate into the following three general types:

1. *Physical Solution.* In this case, the component being absorbed is more soluble in the liquid absorbent than are the other gases with which it is mixed but does not react chemically with the absorbent. As a result, the equilibrium concentration in the liquid phase is primarily a function of partial pressure in the gas phase. One example of this type of absorption operation is the recovery of light hydrocarbons with oil. This type of system has been the subject of a great many studies and is analyzed quite readily. It becomes more complicated when many components are involved and attempts are made to include subtle interactions and heat effects in the analysis.

2. *Reversible Reaction.* This type of absorption is characterized by the occurrence of a chemical reaction between the gaseous component being absorbed and a component in the liquid phase to form a compound that exerts a significant vapor pressure of the absorbed component. An example is the absorption of carbon dioxide into a monoethanolamine solution. This type of system is quite difficult to analyze because the vapor–liquid equilibrium curve is not linear and the rate of absorption may be affected by chemical reaction rates.

3. *Irreversible Reaction.* In this case, a reaction occurs between the component being absorbed and a component in the liquid phase which is essentially irreversible. An example of this type is the absorption of ammonia by sulfuric acid solution. In this example, the compound formed, ammonium sulfate, exerts a negligible vapor pressure of ammonia under the conditions of the absorption operation. Another example is the absorption of H_2S by strongly oxidizing solutions which completely change the chemical nature of sulfur compounds in the liquid phase. Engineering analysis of systems involving irreversible reactions is

TABLE 6.1-1. Typical Applications of Absorption for Product Recovery

Product	Process	Absorbent
Acetylene	High Temperature Cracking of Hydrocarbons	Dimethylformamide
Acrylonitrile	Sohio process	Water
Ammonia	Phosam process	Ammonium phosphate in water
Butadiene	Houdry process	Oil
Carbon dioxide	Recovery from flue gas	Monoethanolamine in water
Ethanol	Hydration of ethylene	Water
Formaldehyde	Partial oxidation	Water
Hydrochloric acid	Synthesis from hydrogen and chlorine	Water
Light hydrocarbons	Oil absorption	Oil
Sodium carbonate	Solvay process (CO_2 absorption)	Ammoniated brine
Sulfur dioxide	Recovery from smelter gas	Dimethylaniline
Sulfuric acid	Contact process (SO_3 absorption)	Sulfuric acid
Urea	Synthesis (CO_2 and NH_3 absorption)	Ammonium carbamate solution

simplified somewhat by the absence of a vapor pressure over the absorbent; however, it can become quite complicated if the irreversible reaction is not instantaneous or involves several steps.

The operation of absorption is considered to include cases where the liquid phase contains suspended particles of reactive solid. An example is the removal of sulfur dioxide from flue gas with an aqueous slurry of lime or limestone. However, the transfer of a component from a gas phase to a dry solid is not included as an absorption operation. Normally, such a process is referred to as "adsorption" when the gas is held on the surface of the solid and "chemisorption" when the gas combines chemically with the dry solid material.

6.1-2 Applications

Generally, commercial applications of absorption can be described as product recovery or gas purification depending, respectively, on whether the absorbed or the unabsorbed portion of the feed gas has the greater value. Obviously, this is not a rigorous classification. Typical product recovery applications are listed in Table 6.1-1 which is based on processes described by Shreve and Brink.[1] The absorption of SO_3 in water or dilute sulfuric acid to make concentrated sulfuric acid is probably the most widely used chemical recovery application of absorption. The recovery of natural gas liquids by absorption in oil is also quite important but very few new plants of this type are being built because the process is being replaced largely by low-temperature fractionation systems employing turboexpanders for energy recovery and refrigeration. The production of sodium carbonate by the Solvay process, which involves the absorption of CO_2 in ammoniated brine, is also almost obsolete. On the other hand, a number of new absorption processes have been introduced recently for gas purification. The most notable example is the use of lime/limestone slurries for the absorption of SO_2. The basic technology of this process was established in 1931, but the process was not applied commercially to a significant extent until the 1970s. Literally hundreds of plants based on the lime/limestone process for flue gas desulfurization are now in operation, primarily in Japan and the United States.

Examples of absorption processes used for gas purification are listed in Table 6.1-2, which is based on processes described by Kohl and Riesenfeld.[2] The table lists only a small fraction of the solvents used or proposed for H_2S and CO_2 removal. The removal of these acid gases has widespread application for treating fuel and synthesis gas streams and is being improved continuously by the development of new solvents, process configurations, and design techniques. Two relatively new solvents, for example, are methyldi-ethanolamine (MDEA) and diglycolamine (DGA). These materials, which are not listed in Table 6.1-2, recently gained commercial acceptance because of their unique properties compared to the more commonly used alkanolamines. MDEA has shown the capability of absorbing H_2S very efficiently while allowing a significant portion of the CO_2 to pass through the system unabsorbed. DGA is nonselective, analogous to monoethanolamine (MEA) in performance, but has a much lower vapor pressure than MEA.

In some applications of absorption, the liquid absorbent is used on a once-through basis as, for example, the absorption of traces of H_2S from carbon dioxide by potassium permanganate solution. More typically, the absorbent is regenerated and recycled to the absorber. Usually, regeneration is accomplished by a stripping operation in which pressure reduction, temperature elevation, and a stripping gas are used alone or in combination to enhance the removal of absorbed components from the solvent. Important examples of processes that use stripping for absorbent regeneration are the recovery of liquid hydrocarbons by oil absorption and the removal of the acid gases, carbon dioxide, and hydrogen sulfide from gas streams with

TABLE 6.1-2. Typical Applications of Absorption for Gas Purification

Impurity	Process	Absorbent
Ammonia	Indirect process (for coke oven gas)	Water
Carbon dioxide and hydrogen sulfide	Ethanolamine	Mono- or diethanolamine in water
Carbon dioxide and hydrogen sulfide	Water wash	Water (at elevated pressure)
Carbon dioxide and hydrogen sulfide	Estasolvan	Tributylphosphate
Carbon dioxide and hydrogen sulfide	Benfield	Potassium carbonate and activator in water
Carbon dioxide and hydrogen sulfide	Fluor solvent	Propylene carbonate
Carbon dioxide and hydrogen sulfide	Giammarco-vetrocoke	Potassium arsenite in water
Carbon dioxide and hydrogen sulfide	Purisol	N-Methyl-2 pyrrolidone
Carbon dioxide and hydrogen sulfide	Rectisol	Methanol
Carbon dioxide and hydrogen sulfide	Selexol	Dimethyl ether of polyethylene glycol
Carbon dioxide and hydrogen sulfide	Sulfinol	Diisopropanolamine, sulfolane, and water
Carbon monoxide	Copper ammonium salt	Cuprous ammonium carbonate and formate in water
Hydrogen chloride	Water wash	Water
Hydrogen fluoride	Water wash	Water
Hydrogen sulfide	Ferrox	Ferric hydroxide, sodium carbonate, and water
Hydrogen sulfide	Stretford	Sodium carbonate, sodium vanadate, and anthraquinone disulfonic acid
Hydrogen sulfide	Thylox	Sodium thioarsenate in water
Naphthalene	Oil wash	Oil
Sulfur dioxide	Aqueous carbonate process (ACP)	Sodium carbonate solution (in spray dryer)
Sulfur dioxide	ASARCO	Dimethylaniline
Sulfur dioxide	Double alkali process	Sodium sulfite
Sulfur dioxide	Wellman–Lord	Sodium sulfite
Sulfur dioxide	Lime/limestone scrubbing	Slurry of lime or limestone in water
Sulfur dioxide	Dry lime	Lime slurry (in spray dryer)
Water	Glycol dehydration	Di- or triethylene glycol
Water	Kathabar	Lithium chloride in water

mildly alkaline solutions. More complex techniques for absorbent regeneration are not uncommon, particularly in gas purification applications. Examples are the dual alkali process for sulfur dioxide removal in which the absorbed components are removed from the solution by precipitating them as insoluble calcium salts, and the Stretford process for hydrogen sulfide removal in which the impurity is absorbed as sulfide and removed from solution by oxidation to elemental sulfur. Although stripping is used characteristically as an adjunct to absorption to regenerate the absorbent, it is used occasionally as a means of removing dissolved gases from a liquid stream. The deaeration of boiler feedwater is an example of such an application.

6.1-3 Selection of Equipment

The application of absorption as a practical process requires the generation of an extensive area of liquid surface in contact with a gas phase under conditions favoring mass transfer. This can be accomplished by three basic techniques: (1) breaking up the gas into small bubbles within a continuous liquid phase (e.g., tray columns), (2) dividing the liquid stream into numerous thin films that flow through a continuous gas phase (e.g., packed columns), and (3) dispersing the liquid as a multitude of discrete droplets within a continuous gas phase (e.g., spray contactors). A great variety of equipment designs have been developed for the performance of absorption and stripping operations. The three most commonly used are the examples given above, that is, tray columns, packed columns, and spray contactors, although numerous other types are used for special applications. Tray and packed columns used for absorption are essentially identical to those used for distillation, and a detailed discussion covering the application and design of these types of gas/liquid contactor is included in Chapter 5. Spray contactors, which are seldom used for distillation, are discussed in more detail in a subsequent section of this chapter. A simplified selection guide for evaluating the most commonly used types of contactor for gas absorption service is given in Table 6.1-3.

Tray columns (also called plate columns) are particularly well suited for large installations; clean, noncorrosive, nonfoaming liquids; and low-to-medium liquid flow rate applications. They also are preferred when internal cooling is required in the column. With appropriate tray design, cooling coils can be installed on individual trays, or alternatively, liquid can be removed from the column at one tray, cooled, and returned to another tray. Tray columns also are advantageous for separations that require a large number of transfer units because they are not subject to channeling of vapor and liquid streams which can cause problems in tall packed columns.

Two types of tray are listed in Table 6.1-3: perforated (also called sieve) trays, which are used widely because of their simplicity and low cost, and bubble-cap trays. Bubble-cap columns were formerly very popular for general absorption and stripping applications but they now are used primarily for cases of very low liquid rate or high turndown requirement where the bubble-cap design assures that a liquid level will be maintained on each tray. Another type of tray, which is not shown in the table but is growing in popularity, is the valve tray which represents a modification to the perforated tray and which permits operation over a wider range of flow rates. Several proprietary types of valve tray are available such as the Glitsch ballast tray[3] and the Koch Flexitray.[4]

In general, tray columns are constructed with downcomers to provide a separate flow path for the liquid stream from tray to tray. However, in some designs such as the Turbogrid, Dualflow, Ripple Tray, Kittle

TABLE 6.1-3. Selection Guide for Absorbers and Strippers

	Tray Columns		Packed Columns (Random)	Spray Contactors
	Perforated	Bubble Cap		
Low liquid rate	D	A	C	D
Medium liquid rate	A	C	B	C
High liquid rate	B	C	A	A
Difficult separation (many stages)	A	B	A	D
Easy separation (one stage)	C	C	B	A
Foaming system	B	C	A	C
Corrosive fluids	B	C	A	A
Solids present	B	D	C	A
Low ΔP	C	D	B	A
High turndown ratio	C	A	B	D
Versatility (can be changed)	C	C	A	D
Multiple feed & drawoff points	B	A	C	C

Key: A, best selection; B, usually suitable; C, evaluate before specifying; D, generally not applicable.

Tray, and Marcy Tray (see Oliver[5] p. 310 for description), the liquid flows from one tray to the next one below as free falling drops or streams distributed across the entire column cross section. These trays have not proved as popular as trays equipped with downcomers. Their main disadvantage is a lack of flexibility. Tray holdup and operating characteristics are strongly dependent on vapor, liquid, and gas flow rates.

Packed columns are gaining favor for a wide range of applications because of the development of new packings which provide higher capacity and better performance as well as the development of better design techniques for packed columns. The packed column listing in Table 6.1-3 refers to random packing, which is the type most commonly used in large absorption columns. However, it should be noted that some of the newer very high efficiency and very low pressure drop packings involve packing elements that are stacked individually within the column in a special ordered arrangement.

Packed columns are preferable to tray columns for small installations, corrosive service, liquids with a tendency to foam, very high liquid-to-gas ratios, and low pressure drop applications. In addition, packed columns offer greater flexibility because the packing can be changed with relative ease to modify column operating characteristics, if necessary. Commercially available random packings, which offer better overall performance than the older Raschig rings and Berl saddles, include Pall rings and Intalox saddles.[6] Examples of very high efficiency, structurally ordered packing elements include Sulzer packing[7] and Goodloe packing.[8]

Spray contactors are used almost entirely for applications where pressure drop is critical, such as flue gas scrubbing. They are also useful for slurries that might plug packing or trays. Generally, they are not suitable for applications requiring several stages of contact or a close approach to equilibrium. The listing in Table 6.1-3 refers to spray contactors rather than spray columns since many spray devices do not have the configuration generally attributed to columns. In spray contactors, the contact area for mass transfer is related directly to the volume of liquid sprayed into the unit. It is therefore common practice to recycle the absorbent to increase mass transfer. Types of equipment classified as spray contactors include countercurrent spray columns, venturi scrubbers, cyclone scrubbers, showerdeck towers, and spray dryers. The use of spray dryers for gas absorption is a relatively recent innovation that is gaining favor for flue gas desulfurization. In this type of apparatus, the feed gas must be sufficiently hot and dry to evaporate all the water from the absorbent solution (or slurry) which is sprayed into the unit as a fine mist. The gas to be absorbed (normally SO_2) is transferred from the gas phase to the droplets at the same time that water is transferred by evaporation in the opposite direction. The absorbent must contain a component that reacts with the absorbate to form a compound which can be dried to the solid form without decomposing. The final products of a spray dryer–absorber system are a purified gas that is more humid and cooler than the feed gas but normally not saturated with water and a fine powder containing the product of the absorption reaction.

Bubble columns, which are basically simple vessels filled with liquid through which a gas is bubbled, are used more frequently as reactors than as absorbers. However, even when used as a reactor, the gas that is bubbled through the continuous liquid phase is normally one of the reactants and must be absorbed before it can react. Advantages of bubble columns relative to other contactors include: low maintenance due to the absence of moving parts; high heat transfer rates per unit volume of reaction; ability to handle slurries without plugging or erosion; and long residence time for the liquid, which permits slow reactions to proceed. A key disadvantage compared to tray and packed columns is the occurrence of backmixing which prevents true countercurrent operation. A very comprehensive review of design parameter estimation for bubble column reactors has been presented by Shah et al.[9]

6.1-4 General Design Approach

The initial step in the design of the absorption system is selection of the absorbent and overall process to be employed. There is no simple analytical method for accomplishing this step. In most cases, the process requirements can be met by more than one solvent, and the only satisfactory approach is an economic evaluation which may involve the complete but preliminary design and cost estimate for more than one alternative. It should be noted that a thorough evaluation also may require the consideration of other separation processes such as adsorption, chemisorption, membrane permeation, or low-temperature condensation.

Once an absorbent has been selected, the design of the absorber will require the procurement of basic physical property data on the absorbent–absorbate system. In addition to the standard physical properties such as density, viscosity, surface tension, and heat capacity, specific data will be required on vapor–liquid equilibrium and the heat of solution or reaction.

The second step is selection of the type of contactor to be used. This may be accomplished on the basis of system requirements and experience factors such as those listed in Table 6.1-3 or, again, an economic comparison may be required to evaluate several alternatives. This step includes at least a preliminary selection of the specific tray, packing, or spray contactor configuration to be employed.

Following the above decisions, it is necessary to perform computations to establish material and heat balances around the contactor; define the mass transfer requirements; determine the height of packing or number of trays needed to meet the desired separation requirements; and calculate the contactor size required to accommodate the liquid and gas flow rates with the selected column internals. The computations are not performed necessarily in the above sequence nor are the steps completely independent. For example, heat effects can affect equilibrium significantly within the contactor which in turn can affect the required flow rates and contactor dimensions.

6.1-5 Equilibrium Data

The most important physical property data required for the design of absorbers and strippers are gas–liquid equilibria. Since equilibrium represents the limiting condition for any gas–liquid contact, such data are needed to define the maximum gas purity and rich solution concentration attainable in absorbers and the maximum lean solution purity attainable in strippers. Equilibrium data also are needed to establish the mass transfer driving force, which can be defined simply as the difference between the actual and equilibrium conditions at any point in a contactor.

Equilibrium data are presented in a variety of ways. Frequently, the solubilities of gases in liquids in which they are sparingly soluble are given in terms of Henry's Law constant H. Henry's Law states simply that the solubility of a gas in a liquid is directly proportional to its partial pressure in the gas phase; that is,

$$p_A = Hx_A \qquad (6.1\text{-}1)$$

where p_A is the partial pressure of A in the gas phase, and x_A is the concentration of A in the liquid phase. The constant H has the units of pressure per composition. Typical units are atmospheres for p_A and mole fraction for x_A. H is dependent on temperature but relatively independent of system pressure at moderate pressure levels for systems where Henry's Law applies. Henry's Law constants for many gases and solvents are given in the critical tables.[10] Examples of Henry's Law constants for a number of common gases in pure water are given in Fig. 6.1-1. An example of the applicability of Henry's Law to a nonaqueous solvent is illustrated by Fig. 6.1-2. The linearity of the solubility curves for several gases in the solvent used in the Selexol process (the dimethyl ether of polyethylene glycol) shows that Henry's Law applies to these systems over the range of partial pressures reported.

Occasionally, vapor–liquid equilibrium can be correlated by Raoult's Law:

$$p_A = P_A x_A \qquad (6.1\text{-}2)$$

where p_A is the partial pressure of A in the gas phase, P_A is the vapor pressure of pure compound A, and x_A is the mole fraction of A in the liquid phase. Raoult's Law states that a component exerts a vapor pressure over a liquid equivalent to its vapor pressure as a pure material times its mole fraction in the liquid. This relationship implies an ideal solution with no interaction between the dissolved material and the solvent and is seldom directly applicable to systems involved in absorption or stripping operations.

Usually, hydrocarbon equilibrium data are presented in the form of equilibrium constants:

$$K = \frac{y_A}{x_A} \qquad (6.1\text{-}3)$$

where y_A is the mole fraction of A in the gas phase and x_A is the mole fraction of A in the liquid phase. Although, ideally, K values can be derived from pure-component vapor pressures using Raoult's Law, in fact, K values vary with total system pressure, temperature, and composition. Fortunately, extensive charts and correlations have been developed for predicting K values for many components, particularly those associated with the natural gas and oil refining industries.

Gas–liquid equilibrium data are available in the literature in one form or another for most gases and solvents commonly encountered in absorption and stripping operations. *Perry's Chemical Engineers' Handbook*[13] is a readily available source of data for many commonly used systems. Hildebrand and Scott[14] also present solubility data for a wide range of gases. Kohl and Riesenfeld[2] present equilibrium charts and data for a large number of solvents and gases that are important in gas purification processes. Hydrocarbon equilibrium data are available from several sources. The Gas Processors Suppliers' Association *Engineering Data Book*[15] is particularly valuable with regard to hydrocarbons normally encountered in natural gas processing. Extensive hydrocarbon data also are given by Edmister.[16]

Experimental equilibrium data have been determined for most absorption and stripping operations of commercial significance. However, since many commercial processes are considered to be proprietary, the experimental data are not always published in the open literature. In the development of new processes the procurement of experimental equilibrium data is usually the first step. For example, Zawacki et al.,[17] in an attempt to develop the ideal solvent for cleaning gas from coal hydrogasification, screened more than 100 solvents and selected 26 for detailed study. Experimental equilibrium data were obtained on these solvents at 1100 psia (7.58×10^3 kPa) and 80°F (26.7°C). The results, which are presented in Table 6.1-4 in terms of equilibrium constants, are of interest because they illustrate the solubilities of several gases frequently of interest in gas absorption processes in a wide range of physical solvents. The authors

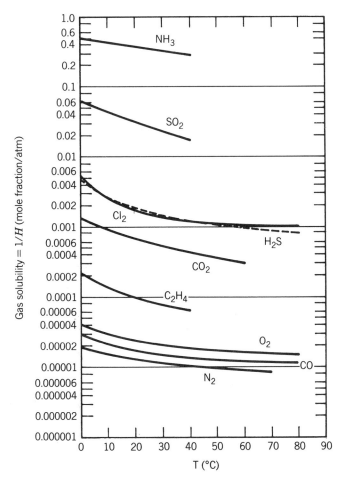

FIGURE 6.1-1 Solubilities of various gases in water expressed as the reciprocal of the Henry's Law constant. Valid for pressures of 5 atm (500 kPa) or less. (From Schmidt and List,[11] copyright 1962. Reprinted with permission of Prentice-Hall, Inc., Englewood Cliffs, NJ.)

finally selected the dimethyl ether of tetraethylene glycol and *N*-formyl morpholine for detailed economic analysis.

Generally, adequate computational procedures are not available for predicting equilibrium data for gas–liquid systems based entirely on the physicochemical properties of the individual components. However, correlations have been proposed which can provide approximate values for simple systems. Osborne and Markovic,[18] for example, present a nomograph that relates the solubilities of common gases in organic liquids to the liquid surface tension and molar volume.

Excellent correlations have been developed for interpolating and extrapolating a minimum amount of experimental data to cover conditions outside the range of experimentation. Oliver[5] presents a review of correlations available to 1966. The correlations generally relate to systems involving physical solubility without chemical reaction and, more particularly, to hydrocarbon and related oil industry species. A correlation, which is intended to cover a wide range of hydrocarbons and hydrogen, has been developed by Chao and Seader.[19] Unfortunately, the correlation does not apply to polar or reacting materials. Prausnitz and Chueh[20] have developed a procedure for high-pressure systems and give complete computer program listings. Parameters are provided for most natural gas components.

Systems in which the absorbed gas reacts with the solvent to produce a compound that exhibits a significant vapor pressure are quite difficult to correlate. Kent and Eisenberg[21] have devised a workable approach for the case of H_2S and CO_2 in alkanolamines. Predicted vapor pressures generated by this correlation compare quite favorably with experimentally determined data. The Kent–Eisenberg model is an expansion of work by Danckwerts and McNeil[22] and is based on the development of correlations for a

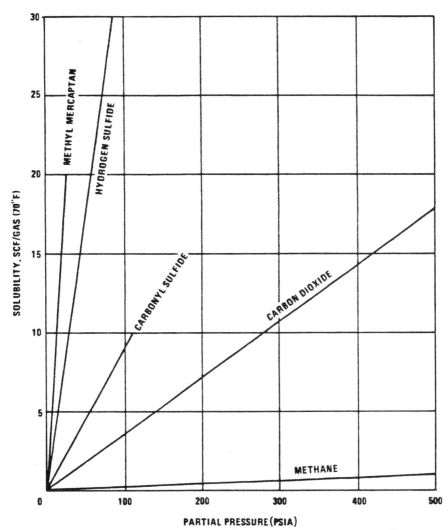

FIGURE 6.1-2 Solubility of gases in Selexol solvent. (Data of Sweny and Valentine,[12] excerpted by special permission from *Chemical Engineering*, September 7, 1970. Copyright 1970 by McGraw-Hill, Inc., New York.) SI conversions: 1 psia $= 6.895 \times 10^3$ N/m²; 1 scf/gal $= 7.48$ m³/m³; 70°F $= 21.2$°C.

series of nine equilibrium equations. These equations describe (1) the equilibrium between the alkanolamine molecules and ions in solution, (2) the equilibrium between acid gas in the gas phase and dissolved acid gas molecules (as a Henry's Law constant), and (3) the equilibria between H_2S, CO_2, H_2O, and the ions formed from these species in aqueous solution. Kent and Eisenberg were able to obtain a correlation by using literature data for items (2) and (3) and then determining values for (1) by forcing a fit with experimentally available vapor–liquid equilibrium data. The model was found to match available experimental data extremely well as indicated by Fig. 6.1-3, which is for the case of H_2S absorption in 15.3 wt.% MEA. Since this approach has a theoretically sound basis, it should be applicable to other systems involving reversible reactions in aqueous solution.

A comprehensive treatment of the thermodynamics of acid gas absorption by reactive solvents is presented by Astarita et al.[23] These authors discuss the theoretical approach to modeling such systems and provide equilibrium data on a number of commercially important solutes and solvents. Figures 6.1-4 and 6.1-5 show comparative equilibrium curves for CO_2 in four different alkaline solvents at conditions representing absorption and stripping, respectively.

Although vapor–liquid equilibrium data represent the most critical data required for the design of absorbers and strippers, other physical property data also are needed. Properties are required for both the

TABLE 6.1-4 Summary of Experimental Vapor–Liquid Equilibrium Data for Physical Solvents

Physical Solvents	Molecular Weight	Density (g/mL)	K_{CO_2}	Selectivity				
				K_{H_2S}/K_{CO_2}	$K_{C_2H_6}/K_{CO_2}$	K_{CH_4}/K_{CO_2}	K_{CO}	K_{H_2}
Acetone	58.08	0.790	1.29	0.33	1.81	7.48	18.8	45.2
4-Butyrolactone	86.09	1.129	1.71	0.23	3.57	14.00	117.0	102.0
Cellosolve® Acetate	132.16	0.975	0.85	0.26	1.98	1.68	25.0	34.3
Cyclohexanone	98.15	0.947	1.28	0.30	1.56	7.91	28.4	46.0
Diethylacetamide	115.18	0.925	1.02	0.15	1.40	6.63	27.9	38.6
Diethyl Carbitol®	162.23	0.908	0.80	0.23	1.36	7.37	—	—
Diethyl sulfate	154.18	1.180	1.00	0.30	2.20	7.68	26.7	41.6
Dimethylacetamide	87.12	0.943	1.17	0.16	1.70	8.41	41.4	52.0
Dimethyl ether of diethylene glycol	134.18	0.945	0.87	0.21	1.41	7.33	21.6	42.0
Dimethyl ether of ethylene glycol	90.12	0.868	0.92	0.28	1.14	6.95	12.2	31.8
Dimethyl ether of tetraethylene glycol	222.28	1.013	0.76	0.18	2.00	8.65	25.0	50.0
Dimethylformamide	73.09	0.949	1.17	0.14	2.96	11.60	86.9	68.0
1,4-Dioxane	88.11	1.036	0.89	0.29	1.70	9.80	57.7	55.1
Ethyl acetate	88.11	0.901	0.97	0.35	1.55	2.65	21.5	33.1
Ethyl acetoacetate	158.20	0.986	1.00	0.28	2.39	8.32	25.8	42.2
Ethyl morpholine	115.18	0.914	0.89	0.24	1.25	7.02	35.7	33.3
N-Formyl morpholine	115.13	1.153	1.42	0.18	4.20	17.50	>220	124.0
Isopropyl acetate	102.13	0.874	0.86	0.34	1.37	5.63	11.6	27.0
Malononitrile	66.06	1.049	8.90	0.36	10.00	17.00	254.0	208.0
Methanol	32.04	0.792	2.64	0.34	2.24	5.93	27.7	76.3
Methyl Cellosolve acetate	118.14	1.007	0.86	0.28	2.12	8.61	53.9	40.2
Methyl cyanoacetate	99.09	0.734	1.49	0.34	3.45	13.70	—	—
Methyl ethyl ketone	72.10	0.806	1.05	0.33	2.00	6.25	—	31.6
Methyl morpholine	101.15	0.921	0.95	0.21	1.34	7.28	26.8	39.4
Methyl pyrrolidone	99.13	1.026	1.24	0.12	2.55	11.80	107.0	59.1
Propylene carbonate	102.09	1.205	1.58	0.29	3.34	13.20	48.6	77.2

Source: Data from Zawaki et al.[17] Reproduced with permission from *Hydrocarbon Processing*, April 1971.

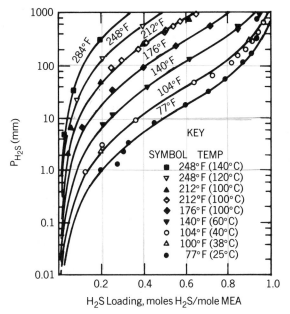

FIGURE 6.1-3 Comparison of calculated curves for H$_2$S equilibrium partial pressure over 15.3 wt.% MEA with experimental data points. (From Kent and Eisenberg,[21] reproduced with permission from *Hydrocarbon Processing*, February 1976.) SI conversion: 1 mm Hg = 1.33×10^2 N/m^2.

gaseous and liquid components involved and for various concentrations of absorbed gas in the liquid. Data that are required generally for the complete design of absorption and stripping systems include heats of solution or reaction, heat of vaporization of the absorbent, heat capacities, densities, viscosities, surface tensions, and diffusivities. The presentation of data or correlations for predicting values of these properties is beyond the scope of this chapter. In some cases, the best source of such data is the manufacturer of the absorbent, particularly if it is an organic solvent or reactant such as the alkanolamines. Kohl and Riesenfeld[2] present physical property data for many absorbents used in commercial gas purification processes. *Perry's Chemical Engineers' Handbook*[13] is a good source of general physical property data and also provides information relative to the estimation of data for cases where experimental values are not available. *The Properties of Gases and Liquids* by Reid et al.[24] is a comprehensive text on the subject.

6.1-6 Material and Heat Balances

A countercurrent absorption tower is illustrated in Fig. 6.1-6. The tower may contain either trays or packing. For purposes of this preliminary discussion, it is assumed that a single solute is being absorbed. Since the molar flow rates of the liquid and gas streams, L_M and G_M, respectively, vary over the length of the tower due to the gain of material by the liquid and loss from the gas, it is convenient to base the material balance on flow parameters that can be considered constant. The flow rates used are usually the solute-free liquid, designated L'_M, and the solute-free (inert) gas, G'_M; however, other flow parameters may be used. The Kremser[25] and Souders and Brown[26] design equations, for example, are based on the lean solvent rate and the rich gas rate.

Use of the solute-free flow rates requires a redefinition of concentration terms from mole fractions x and y to mole ratios X and Y, where the mole ratios are defined by the equations

$$X = \frac{x}{1 - x} \tag{6.1-4}$$

and

$$Y = \frac{y}{1 - y} \tag{6.1-5}$$

A material balance around the top portion of the tower then becomes

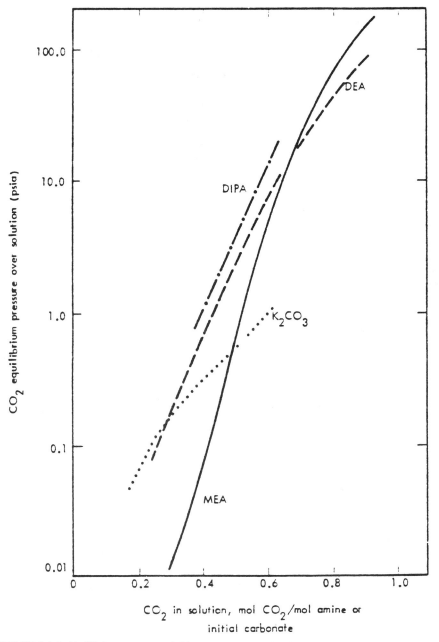

FIGURE 6.1-4 Equilibrium pressures of CO_2 over aqueous aminoalcohol and potassium carbonate solutions at absorber top conditions (40°C, 104°F). (From Astarita et al.,[23] copyright 1983. Reproduced with permission of John Wiley & Sons, New York.) SI conversion: 1 psia = 6.895 × 10³ N/m².

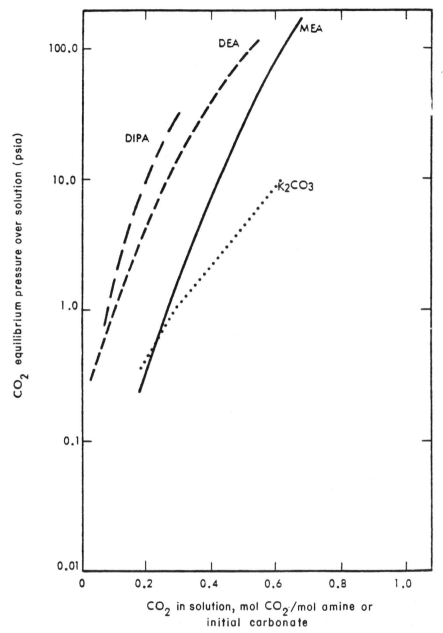

FIGURE 6.1-5 Equilibrium pressures of CO_2 over aqueous aminoalcohol and potassium carbonate solutions at regenerator conditions (100°C, 212°F). (From Astarita et al.,[23] copyright 1983. Reproduced with permission of John Wiley & Sons, New York.) SI conversion: 1 psia = 6.895×10^3 N/m².

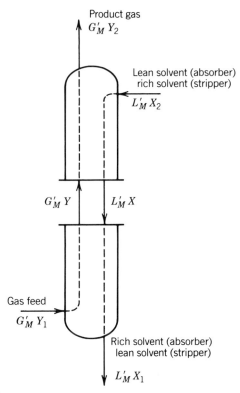

FIGURE 6.1-6   Material balance diagram for countercurrent contactor.

$$G'_M (Y - Y_2) = L'_M (X - X_2) \tag{6.1-6}$$

where G'_M = solute-free gas flow rate, lb-mole/h · ft² (or kmol/h · m²)
$\quad\quad L'_M$ = solute-free liquid flow rate, lb-mole/h · ft² (or kmol/h · m²)
$\quad\quad x, y$ = mole fraction solute in liquid and gas phases, respectively
$\quad\quad X, Y$ = mole ratio of solute in liquid and gas phases, respectively

Rearranging Eq. (6.1-6) gives

$$Y = \frac{L'_M}{G'_M} (X - X_2) + Y_2 \tag{6.1-7}$$

which is a straight line with a slope of L'_M/G'_M. This is referred to as the operating line. It can be plotted readily on rectangular coordinate paper by locating the X, Y values corresponding to the top and bottom of the column, that is, X_2, Y_2 (top) and X_1, Y_1 (bottom) as shown in Fig. 6.1-7a.

Frequently, gases and liquids involved in gas absorption operations are quite dilute. When this is the case, Y is approximately equal to y and X is approximately equal to x. Also, the total molar mass flow rates L_M and G_M are approximately equal to the corresponding solute-free flow rates L'_M and G'_M so that the following approximate equation can be used:

$$y = \frac{L_M}{G_M} (x - x_2) + y_2 \tag{6.1-8}$$

This greatly simplifies subsequent calculations since vapor–liquid equilibrium relationships usually are given in terms of mole fractions rather than mole ratios.

The equilibrium curve may be plotted on the same chart as the operating line using X and Y coordinates as shown in Fig. 6.1-7 (or mole fractions for the case of a dilute system). Typically, in the design of an absorber, the equilibrium curve is known and can be plotted. Also known are the flow rate of the feed gas

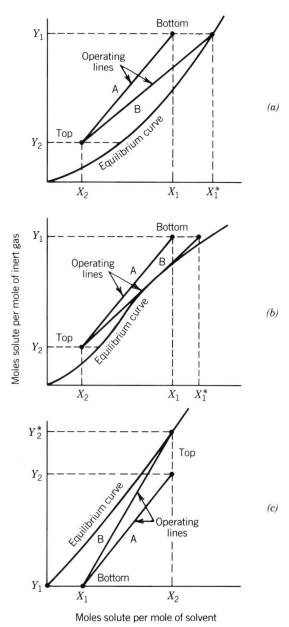

FIGURE 6.1-7 Operating line–equilibrium curve diagrams: (*a*) absorption, (*b*) absorption, and (*c*) stripping.

G'_M; the mole ratio of solute in the feed gas Y_1; the mole ratio of solute (if any) in the lean solvent X_2; and the required mole ratio of solute in the product gas Y_2. Usually, it is desired to estimate the flow rate of liquid required to accomplish the separation and ultimately the dimensions of the tower.

Referring to Fig. 6.1-7*a*, two of the possible operating lines have been drawn, together with the equilibrium curve. Since the operating line represents actual gas and liquid compositions over the length of the tower, the distance from the operating line to the equilibrium curve represents the driving force for absorption. If the operating line touches the equilibrium curve at any point, no driving force exists and an infinitely tall column would be required. Solute mole ratios in the liquid and gas at the top of the column are known, and this fixes one end of the operating line at X_2, Y_2. The other end of the operating line must be at the ordinate Y_1, but its slope L_M'/G_M' may be varied. For line B, the slope of the operating line has been set so that it touches the equilibrium curve at Y_1, X_1^*, where X_1^* represents the equilibrium mole ratio of solute in the solvent at Y_1. This would require an infinitely tall column and represents a limiting condition, that is, the minimum liquid/gas ratio for the desired separation. The actual liquid/gas ratio selected must be somewhat higher than the minimum indicated by operating line B. Typically, L/G ratios 20–100% higher than the minimum are used. The lower values require less solvent but a taller column, while high L/G ratios require a high liquid flow rate with a short column. An economic analysis is necessary to define the optimum L/G with precision. Operating line A represents typical column operating conditions.

Although usually the minimum liquid/gas ratio can be calculated based on the assumption of equilibrium at the bottom of the column where both the gas and liquid have the highest concentration of solute, this is not always the case. Occasionally, the equilibrium curve may be shaped in a manner that causes it to touch the operating line at some point near the middle of the column. Such a case is illustrated in Fig. 6.1-7b with operating line B again representing the minimum L/G ratio. This type of operating line–equilibrium curve relationship may be caused by heat effects that alter the equilibrium conditions within the column.

The graphical analysis of a stripping column is shown in Fig. 6.1-7c. In this case, the mole ratio of solute in the rich solution fed to the top of the stripper, X_2, and the required stripped solution mole ratio, X_1, are usually known. The mole ratio of solute in the stripping gas, Y_1 (often 0), is also known, and it is necessary to determine the minimum amount of stripping gas (i.e., the maximum L/G) that will accomplish the required stripping. Thus, operating line B is drawn from X_1, Y_1 so that it contacts the equilibrium curve and extends to X_2. The point of intersection with X_2 is at the gas composition Y_2^* which represents maximum concentration of solute that can be attained in the stripping gas. An actual operating line, A, is then selected to give a lower value for Y_2 and a reasonable driving force over the length of the column.

The occurrence of heat effects may require that the equilibrium curve be modified to represent more accurately the conditions at all points in the contactor. The major heat effect during absorption is the heat of solution (or reaction) of the solute in the solvent. The heat of solution is normally exothermic and results in an increase in temperature of the solvent at the point where absorption occurs. The final distribution of the heat released between the liquid and gas streams is determined largely by the relative magnitudes of the overall heat capacities of the two streams: $L_M C_q$ and $G_M C_p$, where L_M is the flow rate of the liquid, G_M is the flow rate of the gas, C_q is the heat capacity of the liquid, and C_p is the heat capacity of the gas, all in consistent units and measured at the same point in the column. If $L_M C_q$ is considerably higher than $G_M C_p$ at the top of the column, the entering liquid will remove heat from the rising gas by heat exchange so that the gas leaves the column at approximately the temperature of the feed solvent. In this case, all the heat released by absorption reactions will be carried out of the column as an increase in temperature of the rich solvent. The entering gas stream will be heated by the high-temperature rich solvent near the bottom of the tower causing some of the thermal energy to be carried back into the tower and creating a temperature bulge as illustrated in Fig. 6.1-8. The temperature profile shown in Fig. 6.1-8 is for an actual tray tower used to absorb CO_2 and H_2S from natural gas using a solution of monoethanolamine, diethylene glycol, and water. In this example, which is from Kohl and Riesenfeld,[2] $L_M C_q/G_M C_p = 2.5$; therefore, at the top of the tower, the outlet gas and inlet solution are essentially the same temperature [107°F (41.7°C)], while at the bottom, the two streams have widely different temperatures [rich solution 175°F (79.4°C), feed gas 90°F (32.2°C)].

Similarly, if the heat capacity of the gas stream is significantly higher than that of the solvent, most of the heat of reaction will be carried out of the column with the gas. Such a case is illustrated in Fig. 6.1-9 which represents an absorber treating natural gas containing a relatively low level of acid gas. As a result, a low solvent flow rate is required and $L_M C_q/G_M C_p \simeq 0.2$. The solvent is cooled by the gas as it flows down through the tower and leaves at a temperature very close to that of the inlet gas. All the heat evolved in the column is carried out as sensible heat in the product gas.

If $L_M C_q$ and $G_M C_p$ are approximately equal and heat is released in the absorber, both gas and liquid products will leave the tower at temperatures in excess of the entering streams. In this case, the distribution of heat between the liquid and gas streams will be determined by the location in the tower where heat release occurs and a detailed analysis is required to determine the temperature profile.

Overall heat balances with approximations for the distribution of the heat released are useful for correcting equilibrium curves and establishing operating lines for preliminary design of absorption or stripping columns. However, more sophisticated calculation methods are required for the final design of such towers and particularly for cases involving multicomponent or high heat of reaction systems. Such techniques are discussed in subsequent sections.

6.2 MULTISTAGE CONTACTORS

6.2-1 Number of Trays: Graphical Approach

In an ideal stage, the gas and liquid are brought into intimate contact, attain equilibrium, and then are separated. In most tray columns, where the liquid flows horizontally across each tray while the gas flows

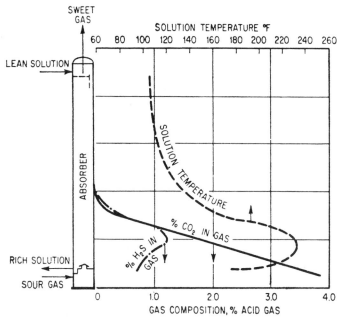

FIGURE 6.1-8 Temperature and composition profiles for absorber employing a reactive solvent to purify a gas containing high concentrations of H_2S and CO_2. (From Kohl and Riesenfeld,[2] copyright 1985 by Gulf Publishing Co., Houston, TX. All rights reserved; used with permission.)

vertically through it, the individual trays are not ideal stages for two reasons: (1) complete equilibrium is not attained and (2) the gas contacts liquids of different compositions at various points on the tray. Nevertheless, it is convenient to consider the trays as potentially ideal stages and to define a theoretical or ideal tray as one where the average composition of the gas leaving the tray is in equilibrium with the average composition of the liquid discharged from the tray. Tray efficiencies then can be defined in terms of how closely the actual average gas or liquid composition approaches the ideal case. The Murphree vapor-phase tray efficiency for tray n is defined as

$$E_{MV} = \frac{y_n - y_{n+1}}{y_n^* - y_{n+1}} \qquad (6.2\text{-}1)$$

where
y_n = average mole fraction of solute in gas leaving tray n
y_{n+1} = average mole fraction of solute in gas entering tray n (leaving the tray below)
y_n^* = mole fraction of solute in gas in equilibrium with liquid leaving tray n

A similar efficiency can be defined for the liquid phase E_{ML}; however, the vapor-phase efficiency is used more widely in column design calculations.

If a countercurrent absorber or stripper could be designed in which all the trays are ideal stages, the composition of the gas and liquid leaving each tray would be represented by a point on the equilibrium line. This permits the required number of theoretical trays to be determined by a simple graphical procedure. The technique is illustrated by a practical example shown in Fig. 6.2-1 which represents the absorber of a natural gas dehydration plant employing triethylene glycol as the absorbent. It should be noted that the trays are numbered from the top tray downward, and the entering and leaving streams have different subscripts than used in the previous overall tower material balance discussions. The nomenclature for tray column calculations is shown in Fig. 6.2-2.

Any consistent set of units may be used to define the concentrations of solute in the liquid and gas phases; however, the same units must be used for both the equilibrium and operating lines. For purposes of the illustrated example, the concentration units commonly used by the U.S. natural gas industry are employed. The feed gas is assumed to be saturated with water at 90°F (32.2°C) and 500 psia (3.447 × 10^3 k/Pa^2) (i.e., it contains 80 lb$_m$ $H_2O/10^6$ scf of gas). It is desired to produce a gas containing no more than 10 lb$_m$ $H_2O/10^6$ scf [dew point 28°F (−2.2°C)] using triethylene glycol absorbent which has been regenerated to a concentration of 98.5% (0.0152 lb$_m$ H_2O/lb_m glycol) in a distillation column. To fix the

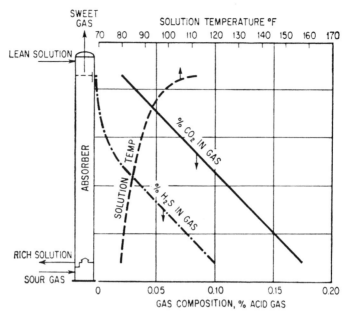

FIGURE 6.1-9   Temperature and composition profiles for absorber employing a reactive solvent to purify a gas containing low concentrations of H_2S and CO_2. (From Kohl and Riesenfeld,[2] copyright 1985 by Gulf Publishing Co., Houston, TX. All rights reserved; used with permission.)

top of the operating line, it is necessary to select a desired rich solution concentration (or operating line slope). For this case, a glycol circulation rate of 4 gal/lb_m of water absorbed is selected based on industrial practice. This results in a water concentration of the discharged glycol of 0.0428 lb_m water/lb_m glycol. The operating line is then drawn between the points (0.0152, 10) and (0.0428, 80) and the equilibrium curve for the water–triethylene glycol system at 90°F (32.2°C) and 500 psia (3.447 $\times$ 10^3 kPa^2) is plotted on the chart.

Starting at the top point on the operating line (representing conditions at the bottom of the absorber), we draw a vertical line until it intersects the equilibrium curve. This point, marked 2*, represents a theoretical tray at the bottom of the tower with the liquid leaving it (and therefore leaving the tower) in equilibrium with the gas rising from the tray. A horizontal line is then drawn from point 2* until it intersects the operating line and a second vertical line is drawn from this intersection back down to the equilibrium curve. It intersects the equilibrium curve at point 1* representing another theoretical tray. Since the equilibrium gas leaving this tray is dehydrated more completely than necessary, the diagram indicates that less than two theoretical trays are necessary.

Experience with absorbers of this type has indicated that actual trays have a Murphree vapor-phase efficiency on the order of 40%. This can be taken into account by drawing a pseudoequilibrium curve 40% of the distance (vertically) from the operating line to the equilibrium curve and stepping off the trays as before. Results indicate that six actual trays will be required. Heat effects, of course, must be considered in establishing the final position of the equilibrium curve but, in this case, the effects are negligible because the large volume of gas carries away the heat of reaction with only a very small temperature increase.

An identical process can be employed to determine the number of theoretical (and actual) trays for stripping towers, except that the operating line must be below the equilibrium curve for stripping to occur, and the highest point on the operating line represents the top of the stripper. The reason for this reversal of the operating line is obvious when a typical absorber–stripper combination system is considered. In such a system the rich solution from the bottom of the absorber becomes the feed to the top of the stripping tower while the stripped solution from the bottom of the stripping tower is pumped to the top of the absorber as lean feed solvent.

6.2-2   Number of Trays: Analytical Approach

The graphical technique for estimating the number of trays illustrated in Fig. 6.2-1 can be used even if the operating and equilibrium lines are curved. For the special case where both are straight, an analytical solution has been derived by Kremser[1] and improved by Souders and Brown.[2] Their equation is based on

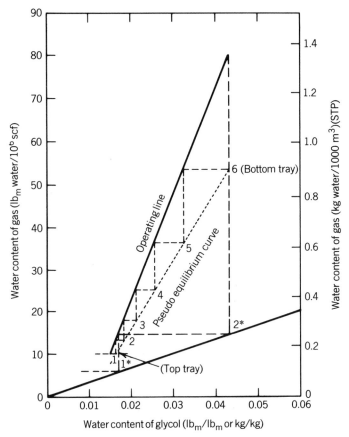

FIGURE 6.2-1 Graphical determination of number of trays for gas dehydration column employing triethylene glycol.

the use of the absorption factor A which is defined as

$$A = \frac{L_M}{mG_M} \tag{6.2-2}$$

where A = absorption factor
  L_M = liquid flow rate
  G_M = vapor flow rate
  m = slope of the equilibrium curve

The absorption factor normally varies somewhat from tray to tray because of slight changes in all three variables. For purposes of characterizing the overall tower, Kremser and Souders and Brown recommend defining A in terms of the lean solution flow rate and the solute-rich gas rate; that is,

$$A = \frac{L_{M0}}{mG_{M(N+1)}} \tag{6.2-3}$$

The fractional absorption of any component by an absorber of N plates is given by the following equation, often referred to as the Kremser equation:

$$\frac{Y_{N+1} - Y_1}{Y_{N+1} - Y_0^*} = \frac{A^{N+1} - A}{A^{N+1} - 1} \tag{6.2-4}$$

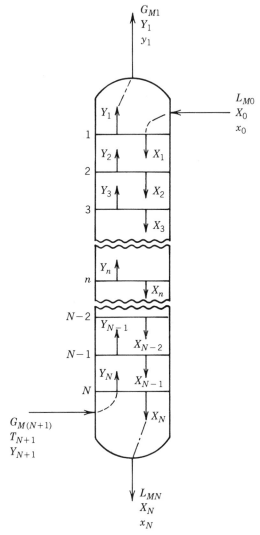

FIGURE 6.2-2 Nomenclature for tray absorption columns.

where Y_0^* is the value of Y for the component which would be in equilibrium with the value of X_0; that is, $Y_0^* = mX_0$. In the above equation, Y is expressed in terms of the moles of one component per mole of entering rich gas.

For a stripper of N plates, the Kremser equation becomes

$$\frac{X_0 - X_N}{X_0 - X_{N+1}^*} = \frac{S^{N+1} - S}{S^{N+1} - 1}$$

(6.2-5)

where S is $1/A$ and X_{N+1}^* is the value of X for the component which would be in equilibrium with the value of Y_{N+1}; that is, $X_{N+1}^* = mY_{N+1}$.

For the design of columns in which the absorption requirements are specified but the number of trays is unknown, the Kremser equation can be converted to the form

$$N = \frac{\log\left[\frac{Y_{N+1} - mX_0}{Y_1 - mX_0}\left(1 - \frac{1}{A}\right) + \frac{1}{A}\right]}{\log A}$$

(6.2-6)

For the special case of $A = 1$, the equation simplifies to

$$N = \frac{Y_{N+1} - Y_1}{Y_1 - mX_0} \tag{6.2-7}$$

In the above equations, $m = Y/X$ at equilibrium. The equations also may be used with mole fractions for dilute gas and liquid streams. In this case, x and y are used instead of X and Y and m is equal to the conventional equilibrium constant, $m = K = y/x$ at equilibrium. For the case of liquid-phase concentrations as commonly used for stripping towers, Eq. (6.2-5) becomes

$$N = \frac{\log\left[\frac{X_0 - Y_{N+1}/m}{X_N - Y_{N+1}/m}(1 - A) + A\right]}{\log (1/A)} \tag{6.2-8}$$

For the special case of $A = 1$ (or $S = 1$), the number of theoretical trays for the liquid phase becomes

$$N = \frac{X_0 - X_N}{X_N - mY_{N+1}} \tag{6.2-9}$$

Equations (6.2-6) and (6.2-8) are plotted in Fig. 6.2-3, which can be used for the quick estimation of the number of theoretical trays required for absorbers or strippers handling dilute gas and liquid mixtures where the absorption (or stripping) factor is relatively constant over the column.

6.2-3 Multicomponent, Nonisothermal Systems

Generally, techniques for designing absorbers and strippers for multicomponent systems have been aimed at the recovery of light hydrocarbons from natural or refinery gas streams using tray towers. Numerous shortcut calculation methods have been developed to obviate the need for tedious tray-by-tray calculations; however, the importance of these techniques has declined somewhat because of (1) the advent of small powerful computers and (2) the use of low-temperature separation instead of oil absorption to recover light hydrocarbons. As a result, the calculational techniques for multicomponent hydrocarbon absorbers are treated rather briefly in this section.

An excellent review of hydrocarbon absorber design is presented by Diab and Maddox.[4] The simplest approach uses the design equations developed by Kremser and Souders and Brown and discussed in the preceding section for each individual component in the system. The application of this approach to multicomponent absorption is illustrated by this example from the GPSA *Engineering Data Book.*[5]

It is desired to recover 75% of the propane (called the key component) from a gas stream of the composition given in Table 6.2-1 using oil absorption. The absorber is equivalent to six theoretical trays and operates at 1000 psig (7 × 10³ kPa). It is assumed that the entering lean oil is stripped completely of rich gas components and that the absorber operates at a constant temperature of 104°F (40°C). What oil circulation rate is required, and what will be the composition of the residue gas leaving the absorber?

1. Using appropriate equilibrium charts, obtain the K values for each component at 104°F (40°C) and 1000 psig (7 × 10³ kPa). Values obtained from the GPSA *Engineering Data Book* are listed in Column K of the table.

2. For 75% absorption of propane, solve Eq. (6.2-4) for A. To simplify the calculation, the chart of Fig. 6.2-4 is provided. The chart is entered at the ordinate value of 0.75 [since $(Y_{N+1} - Y_1)/(Y_{N+1} - 0) = 0.75$ for propane] and a line is drawn horizontally to the curve representing $N = 6$. The point of intersection marks the average absorption factor for the tower, in this case 0.8.

3. The absorption factor and K values are then used in Eq. (6.2-3) to calculate the lean oil rate. For 100 moles of rich gas, $L_{MO} = 0.8 \times 0.37 \times 100 = 29.6$ moles of lean oil.

4. Using the oil rate calculated in Step 3 and K values from the table, Eq. (6.2-3) is used to calculate the absorption factor for the remaining components. For example, for methane $A = 29.6/3.25 \times 100 = 0.091$.

5. Using the absorption factor values determined in Step 4, read values for the factor $(Y_{N+1} - Y_1)/(Y_{N+1} - Y_0^*)$ referred to as B in the table.

6. Calculate the mole ratio of all components in the residue gas. For example, for methane $(0.906 - Y_1)/(0.906 - 0) = 0.091$; $Y_1 = 0.8236$.

7. Calculate the moles of each component in the rich oil, l. For 100 mol of rich gas (or 29.6 mol of lean oil), $l = 100(Y_{N+1} - Y_1 + Y_0)$. For methane, this is $100 (0.906 - 0.8236 + 0) = 8.24$.

8. Check the assumed column temperature [104°F (40°C) in the example] by a heat balance around the absorber. If the calculated absorption performance results in a significantly different temperature than

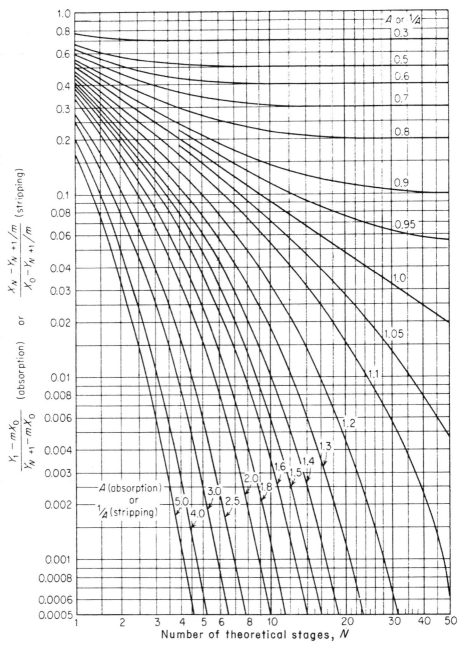

FIGURE 6.2-3 Number of theoretical stages for countercurrent columns with Henry's Law equilibrium and constant absorption and stripping factors. (From Treybal,[3] copyright 1980. Used with permission of McGraw-Hill Book Company, New York.)

TABLE 6.2-1 Example of Light Hydrocarbon Absorption Problem Solved by Kremser Equation

	Rich Gas (mol %)	K	A	B	Y_1	l	Lean Gas (mol %)
C_1	90.6	3.25	0.091	0.091	0.8236	8.24	95.67
C_2	4.3	0.90	0.329	0.329	0.0289	1.41	3.36
C_3	3.2	0.37	0.80	0.75	0.0080	2.40	0.93
iC_4	0.5	0.21	1.41	0.96	0.0002	0.48	0.02
nC_4	1.0	0.17	1.74	0.985	0.00015	0.985	0.02
C_6	0.4	0.035	8.46	1.0	—	0.40	0
Totals	100.0				0.86085	13.915	100.00

where $K = y/x$ at average column conditions
A = average absorption factor as defined for Kremser equations
$B = (Y_{N+1} - Y_1)/(Y_{N+1} - Y_0^*)$
Y_1 = moles of component in the product gas per mole of rich gas
l = moles of component in the rich oil per 100 moles of rich gas

Source: Gas Processors Suppliers' Association, *Engineering Data Book*, 9th ed., Tulsa, OK, 1972.

originally assumed, it is necessary to revise the temperature assumption and repeat the calculations until reasonable agreement is attained. Design basis data and results for this simplified multicomponent absorption calculation for the 104°F (40°C) case are given in Table 6.2-1.

An improvement over the Kremser and Souders and Brown approach was suggested by Edmister[6] who developed the following equation to define an effective absorption factor for the tower:

$$A_e = [A_N(A_1 + 1) + 0.25]^{1/2} - 0.5 \qquad (6.2\text{-}10)$$

The Edmister approach requires knowledge of conditions on the top and bottom trays and the method of Horton and Franklin[7] is suggested for estimating these conditions.

Douglas[8] analyzed the Kremser equation to determine the limits of its applicability. He presents several absorber design equations representing both simplified shortcut techniques and more sophisticated correlations which cover absorbers and in which the equilibrium curve changes shape and temperature effects occur. For very approximate analyses of cases involving pure solvents and high solute recoveries as well as the assumptions implicit in the Kremser equation, Douglas suggests that the number of theoretical trays can be estimated by the use of the simple equation

$$N = 6 \log \left(\frac{Y_{N+1}}{Y_1} \right) - 2 \qquad (6.2\text{-}11)$$

This is based on the use of the rule-of-thumb value of $A = 1.4$.

The Kremser equation and its various modified forms consider only the number of theoretical plates, not the number of actual plates in absorbers and strippers. For cases where the Murphree stage efficiency is known and relatively constant over the length of the tower, a correlation proposed by Nguyen[8] can be used to estimate the total required number of actual plates N_T as follows:

$$N_T = \frac{\ln \left[(y_{N+1} + \alpha)/(y_n + \alpha) \right]}{\ln \beta_v} \qquad (6.2\text{-}12)$$

where N_T is the total number of actual plates required and β_v is a modified absorption factor defined by the equation

$$\beta_v = \frac{1}{1 + E_{MV}(mG_M/L_M - 1)} \qquad (6.2\text{-}13)$$

and α is defined as follows:

$$\alpha = \frac{y_1 - (L_M/mG_m)(mx_0 + C)}{L_M/mG_M - 1} \qquad (6.2\text{-}14)$$

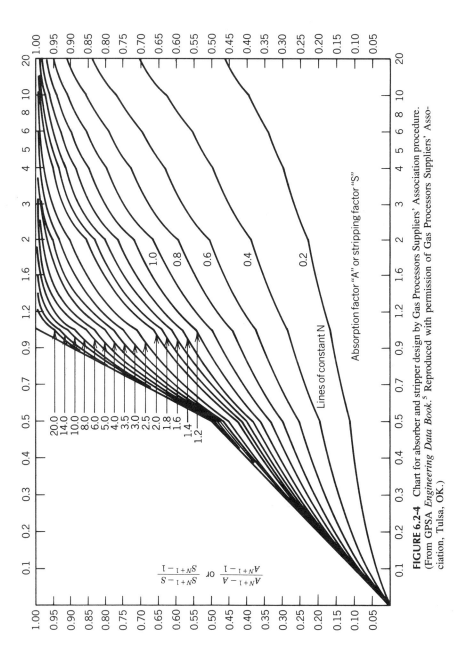

FIGURE 6.2-4 Chart for absorber and stripper design by Gas Processors Suppliers' Association procedure. (From GPSA *Engineering Data Book*.[5] Reproduced with permission of Gas Processors Suppliers' Association, Tulsa, OK.)

In the above expression, C is a constant in the equation defining the gas–liquid equilibrium. It is assumed that a linear relationship exists between x and y of the form $y^* = mx + C$, where y^* is the equilibrium concentration in the gas, m is the slope of the equilibrium curve, and C is the y intercept. Nguyen gives the following example of the use of these equations.

It is desired to desorb a solute from a solution by contacting the liquid with an insoluble gas. The solution enters the top of the stripper at 20 mol % of solute A and leaves at 2 mol %. The gas entering the bottom of the tower is free of solute. The slope of the operating line is 1.1 and the equilibrium line is defined by the equation $y = x + 0.02$. Calculate the number of actual plates required if the Murphree efficiency $E_{MV} = 0.80$ (80%).

The values given in the equation are $y_{N+1} = 0$, $x_N = 0.02$, $x_0 = 0.20$, $m = 1$, $C = 0.02$, and $L/G = 1.1$. y_1 can be calculated on the basis of a material balance around the tower: $y_1 = (L_M/G_M)(x_0 - x_N) + y_{N+1} = 1.1 (0.20 - 0.02) + 0 = 0.198$.

The modified absorption factor is calculated from Eq. (6.2-13):

$$\beta_v = \frac{1}{1 + 0.8 (1/1.1 - 1)} = 1.08$$

The factor α is calculated from Eq. (6.2-14):

$$\alpha = \frac{0.198 - 1.1 (0.20 + 0.02)}{1.1 - 1} = -0.44$$

The number of actual plates required is calculated from Eq. (6.2-12):

$$N_T = \frac{\ln [(0 - 044)/(0.198 - 0.44)]}{\ln 1.08} = 8.0$$

When $E_{MV} = 1$, $\beta_v = L_M/mG_M$ and the above equations can be solved for the number of theoretical plates, N. In this case,

$$N = \frac{\ln [(0 - 044)/(0.198 - 0.44)]}{\ln 1.10} = 6.3$$

and the overall column efficiency is

$$\frac{N}{N_T} = \frac{6.3}{8.0} = 0.79 \ (79\%)$$

The most precise methods for calculating the performance of absorbers and strippers are based on tray-to-tray heat and material balances, and these usually are accomplished by computer. One such method was proposed by Sujata.[10] This method makes use of an overall and individual component material balance across each tray which can be adjusted to account for a feed or drawoff stream at the tray. An initially assumed temperature profile is employed for the first iteration and a set of n simultaneous heat balance equations must be solved to provide a correction to the temperature profile and flow rates. Recalculation of the material balance equations and corrections to the temperature profile lead to a converging solution for the entire tower.

Diab and Maddox[4] have compared solutions to a typical light hydrocarbon absorption problem using the methods of Kremser, Edmister, and Sujata. The problem involved the design of a column to recover 70% of the propane from a specified light hydrocarbon gas stream at 1000 psia (6.9 × 10³ kPa) in an absorber equivalent to six theoretical trays. Calculations for the Edmister effective absorption factor technique were performed using a computer program developed by Erbar and Maddox.[11] Two cases were run for the Edmister and Sujata approaches to compare the results with a narrow-boiling-range absorption oil and a wide-boiling-range oil. The boiling range of the oil frequently becomes wider during operation owing to the retention of a variety of high-molecular-weight hydrocarbons absorbed from the gas. The results are summarized in Table 6.2-2 and show that the rigorous tray-by-tray calculation method predicts considerably higher propane absorption than the Edmister method, which in turn predicts more than the Kremser approach (for the same oil). The results also show that changing the oil to a wider boiling range mixture reduces the predicted recovery significantly. For purposes of the example, it was assumed that both the lean oil and rich gas were chilled to 0°F (−18°C) before entering the column. A single-column temperature is assumed for the Kremser approach while the other two approaches provide more detailed temperature data. As the data in the table indicate, the different approaches predict significantly different temperature profiles for the tower.

Proprietary computer programs are used by many companies to handle tray-to-tray design calculations

TABLE 6.2-2 Comparison of Tray Column Design Techniques for a Multicomponent Hydrocarbon Absorption Problem

Calculation Technique	Kremser	Edmister		Sujata	
Absorption oil[a]	A	A	B	A	B
Pressure (psia)	1000	1000	1000	1000	1000
Theoretical trays	6	6	6	6	6
Feed gas rate (mol/h)	1846	1846	1846	1846	1846
Column temperatures (°F)					
Rich gas	0	0	0	0	0
Lean oil	0	0	0	0	0
Product gas	40	26	26	31	39
Rich oil	40	4	87	13	77
Propane absorbed (%)	70	84	68	90	88

[a] Oil A is a narrow-boiling-range oil fraction typical of an absorber design using a "pure component" characterization. Oil B is a wide-boiling-range oil fraction representing an actual absorption oil after extended use with a rich gas containing some heavy (C_6 +) components.

Source: Diab and Maddox.[4]

for absorbers, strippers, and fractionation columns. A wide variety of computer programs also have been made available to others on a reasonable-fee basis. The ChemShare Corporation,[12] for example, offers a comprehensive program that handles all equilibrium, energy, entropy, and material balance calculations for absorbers and strippers. The columns can have up to 8 feed streams, 8 product streams, 8 side heaters, 8 intercoolers, and 300 theoretical trays. Several equilibrium value options are available and correlations for thermodynamic and physical property data are updated regularly. It is claimed that the program involves a rigorous calculation procedure with special techniques to provide rapid convergence.

6.3 DIFFERENTIAL CONTACTORS

6.3-1 Mass Transfer Coefficients

To determine the required size of an absorption or stripping unit, it is necessary to know not only the equilibrium solubility of the solute in the solvent and the material balance around the column but also the rate at which solute is transferred from one phase to the other within the tower. This rate directly affects the volume of packing needed in a packed tower, the degree of dispersion required in a spray contactor, and (somewhat less directly) the number of trays required in a tray tower. The last effect occurs as a result of the influence of mass transfer rate on tray efficiency which is discussed in a later section. Because of its direct effect on packed tower design and the importance of this type of contactor in absorption, this discussion of mass transfer is aimed primarily at the packed tower case. A more detailed review of mass transfer theory is given in Chapter 2.

Many models have been proposed to explain and correlate mass transfer. The most widely accepted are the film theory first proposed by Whitman,[1] the penetration model suggested by Higbie,[2] the surface renewal theory proposed by Danckwerts[3] as an improvement to the penetration theory, and the film-penetration theory of Toor and Marchello.[4] Although the film theory is admittedly an inexact representation of conditions at the gas–liquid interface, it has proved to be an effective correlation tool and is used most widely for the design of absorption and stripping equipment. The film theory is illustrated diagrammatically by Fig. 6.3-1. It is based on the premise that the gas and liquid are in equilibrium at the interface and thin films separate the interface from the main bodies of the two phases. The main bodies of the liquid and gas phases are assumed to be well mixed, while little or no fluid motion occurs within the films and the process of molecular diffusion becomes the primary mechanism of mass transfer. Two absorption coefficients are defined then as k_L, the quantity of material transferred through the liquid film per unit time, per unit area, per unit of driving force in the liquid, and k_G, the quantity transferred through the gas film per unit time, per unit area, per unit of gas-phase driving force. A material balance across the interface yields the following simple relationship:

$$N_A = k_G(p - p_i) = k_L(c_i - c) \tag{6.3-1}$$

where N_A = quantity of component A transferred per unit time, per unit area
    p = partial pressure of A in the main body of gas
    p_i = partial pressure of A in the gas at the interface

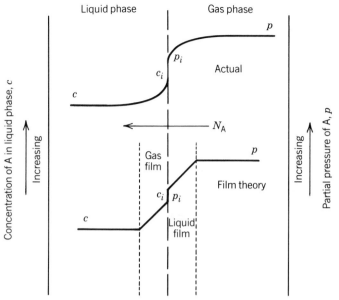

FIGURE 6.3-1 Diagram of two-film concept. c_i and p_i represent equilibrium conditions at the interface.

c = concentration of A in the main body of liquid
c_i = concentration of A in the liquid at the interface

The application of this equation to design requires information on concentrations at the interface which is seldom known. As a result, absorption data frequently are correlated in terms of overall coefficients. These are based on the total driving force from the main body of the gas to the main body of the liquid. The overall coefficients K_G and K_L are defined by the following relationship:

$$N_A = K_G(p - p^*) = K_L(c^* - c) \tag{6.3-2}$$

where p^* = partial pressure of A in equilibrium with a solution having the composition of the main body of liquid
c^* = concentration in a solution in equilibrium with the main body of gas

If a suitable equilibrium relationship exists for relating gas-phase partial pressures and liquid-phase concentrations, the overall coefficients can be expressed in terms of the individual film coefficients. For the case where Henry's Law applies, the following relationships hold:

$$\frac{1}{K_G} = \frac{1}{k_G} + \frac{H}{k_L} \tag{6.3-3}$$

and

$$\frac{1}{K_L} = \frac{1}{Hk_G} + \frac{1}{k_L} \tag{6.3-4}$$

When the solute is very soluble, the Henry's Law constant H is low which makes the term H/k_L much smaller than $1/k_G$ so that $1/K_G \simeq 1/k_G$. In such a case, the gas film represents the controlling resistance, and mass transfer data can be correlated best in terms of K_G. The reverse is true with low-solubility gases; the liquid film is controlling and K_L is the preferred overall coefficient.

Actually, neither K_G nor K_L can be used alone for practical design problems because the effective interfacial area per unit volume of contactor, a, cannot be determined readily. As a result, the products K_Ga or K_La are used. These combined terms represent volume coefficients, that is, the quantity transferred per unit time, per unit of contactor volume, per unit of driving force. Then Eqs. (6.3-3) and (6.3-4) can be written

$$\frac{1}{K_G a} = \frac{1}{k_G a} + \frac{H}{k_L a} \tag{6.3-5}$$

$$\frac{1}{K_L a} = \frac{1}{H k_G a} + \frac{1}{k_L a} \tag{6.3-6}$$

According to King[5] the following criteria must be met for Eqs. (6.3-5) and (6.3-6) to be valid:

1. H must be constant; if it is not, the value of the slope of the equilibrium curve at the properly defined value of liquid-phase concentration must be employed.
2. There must be no significant resistance present other than those represented by $k_G a$ and $k_L a$.
3. The hydrodynamic conditions for the case in which the resistances are to be combined must be the same as for the measurements of the individual phase resistances. Similarly, the solute diffusivities must be the same. These factors usually are taken into account through correlations for $k_G a$ and $k_L a$.
4. The mass transfer resistances of the two phases must not interact; that is, the magnitude of k_L must not depend on the magnitude of k_G or vice versa.
5. The ratio $H k_G / k_L$ must be constant at all points of the interface.

In spite of their limitations, the addition of resistances equations [Eqs. (6.3-5) and (6.3-6)] represent the best available basis for establishing overall mass transfer coefficients which can be used for the design of commercial equipment.

6.3-2 Transfer Units

The height of packing required to perform a specified amount of mass transfer can be related to the mass transfer coefficient and a material balance for the absorbed component at any point in a tower. Referring to Fig. 6.3-2, which represents a packed absorption tower, a material balance within a differential height of packing, dz, yields

$$-d(G_M y) = -G_M' \frac{dy}{(1-y)^2} = k_G a P \frac{y - y_i}{(1-y)_{iM}} dZ \tag{6.3-7}$$

where
G_M = molar mass velocity of the gas
G_M' = molar mass velocity of the solute-free gas
k_G = gas phase mass transfer coefficient
a = effective interfacial area per unit volume of packing
P = absolute pressure
y = mole fraction of solute in the main gas stream
y_i = mole fraction of solute in the gas at the interface
$(1 - y)_{iM}$ = logarithmic mean of the mole fraction of inert gas in the main gas stream and in the gas at the interface
Z = height of packing

The required height of packing to change the gas composition from y_1 to y_2 is

$$Z = G_M' \int_{y_2}^{y_1} \frac{(1-y)_{iM}\, dy}{k_G a P\, (1-y)^2 (y - y_i)} \tag{6.3-8}$$

Since k_G is roughly proportional to the gas velocity G_M, the ratio $G_M'/k_G a P\, (1 - y)$ is approximately constant over the column and can be removed from the integral, yielding

$$Z = \frac{G_M'}{k_G a P\, (1 - y)} \int_{y_2}^{y_1} \frac{(1-y)_{iM}\, dy}{(1-y)(y - y_i)} \tag{6.3-9}$$

The expression in front of the integral has the units of length and is called the height of a transfer unit (HTU) or, for this case, the height of a gas-phase transfer unit, H_G, which is defined as

$$H_G = \frac{G_M}{k_G a P} \tag{6.3-10}$$

Similarly, an overall gas-phase transfer unit is defined as

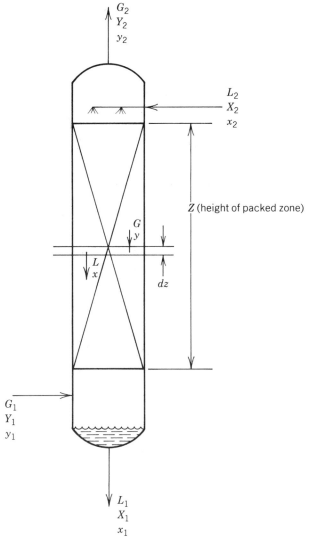

FIGURE 6.3-2 Diagram of packed absorber showing terminal streams and nomenclature.

$$H_{OG} = \frac{G_M}{K_G a P} \tag{6.3-11}$$

The integral in Eq. (6.3-9) is dimensionless and indicates how many transfer units are required to accomplish the separation. It is called the number of gas-phase transfer units, N_G. Thus,

$$N_G = \int_{y_2}^{y_1} \frac{(1 - y)_{iM} \, dy}{(1 - y)(y - y_i)} \tag{6.3-12}$$

and for the overall gas-phase transfer unit,

$$N_{OG} = \int_{y_2}^{y_1} \frac{(1 - y)_{iM} \, dy}{(1 - y)(y - y^*)} \tag{6.3-13}$$

where y^* is the value of y in equilibrium with the main body of the liquid.

Similar equations can be developed in terms of liquid concentrations. The results for the overall liquid case are

$$H_{OL} = \frac{L_M}{K_{OL}a\rho_M}$$ (6.3-14)

and

$$N_{OL} = \int_{x_2}^{x_1} \frac{(1 - x)_{iM}\, dx}{(1 - x)(x^* - x)}$$ (6.3-15)

where x^* is the value of x in equilibrium with the main body of the gas.

In all cases, when consistent expressions are used, the same total height is calculated, that is, $Z = H_G N_G$; $Z = H_{OG} N_{OG}$; $Z = H_L N_L$; and $Z = H_{OL} N_{OL}$.

Identical relationships hold for stripping columns. The driving forces, $y - y^*$ and $x^* - x$, become negative, but since $x_2 > x_1$ and $y_2 > y_1$ the result is a positive value for Z.

Equations based on overall mass transfer resistance are used more readily for design because they do not require knowledge of interfacial concentrations. Where data are available for H_G and H_L, the height of an overall gas-phase transfer unit H_{OG} can be calculated by use of the equation

$$H_{OG} = \frac{(1 - y)_{iM}}{(1 - y)_{*M}} H_G + \frac{mG_M}{L_M} \frac{(1 - x)_{iM}}{(1 - y)_{*M}} H_L$$ (6.3-16)

where $(1 - y)_{iM}$ = the logarithmic mean of the mole fraction of inert gas in the main gas stream and in the gas at the interface

$(1 - y)_{*M}$ = the logarithmic mean of the mole fraction of inert gas in the main gas stream and in the gas at equilibrium with the bulk of the liquid

6.3-3 Simplified Equations for Transfer Units

Simplified forms of the above equations have been developed to permit them to be used more widely in design calculations. The simplifications discussed below are based on the assumption that the following conditions hold:

1. The equilibrium curve is linear over the range of concentrations encountered (and therefore overall coefficients can be used).
2. The partial pressure of the inert gas is essentially constant over the length of the column.
3. The solute concentrations in the liquid and gas phases are sufficiently low that mole ratio and mole fraction values are approximately equal. With these assumptions, the tower height can be estimated by the use of one of the following equations:

$$Z = \frac{G_M}{K_G aP} \int_{y_2}^{y_1} \frac{dy}{y - y^*}$$ (6.3-17)

$$Z = \frac{L_M}{\rho_M K_L a} \int_{x_2}^{x_1} \frac{dx}{x^* - x}$$ (6.3-18)

The calculations can be based on the volume coefficients of mass transfer ($K_G a$ and $K_L a$) or on HTU values, where H_{OG} and H_{OL} are defined by Eqs. (6.3-11) and (6.3-14). When it is necessary to estimate H_{OG} from values of H_G and H_L, the above assumptions permit a simplification of Eq. (6.3-16) to

$$H_{OG} = H_G + \frac{mG_M}{L_M} H_L$$ (6.3-19)

Whether the mass transfer coefficient or HTU approach is used for design, the value of the integral, which is equal to the number of transfer units required, must be estimated. Since the equilibrium curve and the operating line are assumed to be linear over the composition range of the column, it is theoretically correct to use a logarithmic mean driving force. The integral then reduces to

$$N_{OG} = \frac{y_1 - y_2}{(y - y^*)_M}$$ (6.3-20)

where $(y - y^*)_M$ is the logarithmic mean of the overall driving force $(y - y^*)$ at the top and at the bottom of the tower.

The equation is simple to use but requires a computation of the exit liquid composition and y^* at both ends of the column. This can be avoided by use of an equation developed by Colburn[6] which incorporates the equilibrium relationship $y^* = mx$ and the material balance expression $L_M(x_1 - x) = G_M(y - y_1)$ to eliminate the need for y^*. This equation is for absorption:

$$N_{OG} = \frac{\ln\left[\left(1 - \frac{mG_M}{L_M}\right)\left(\frac{y_1 - mx_2}{y_2 - mx_2}\right) + \frac{mG_M}{L_M}\right]}{1 - mG_M/L_M} \tag{6.3-21}$$

and for stripping (based on N_{OL})

$$N_{OL} = \frac{\ln\left[\left(1 - \frac{L_M}{mG_M}\right)\left(\frac{x_2 - y_1/m}{x_1 - y_1/m}\right) + \frac{L_M}{mG_M}\right]}{1 - L_M/mG_M} \tag{6.3-22}$$

In these expressions the term L_M/mG_M is called the absorption factor A, and its reciprocal is called the stripping factor S. A graphical solution to Eqs. (6.3-21) and (6.3-22) is presented in Fig. 6.3-3.

6.3-4 Graphical Determination of Transfer Units

The number of transfer units can be calculated by a graphical technique developed by Baker.[8] The procedure resembles that used for the determination of the number of stages in a tray column in that it involves a plot of the operating line and the equilibrium curve on a diagram with liquid and gas compositions as the coordinates. The technique is based on Eq. (6.3-20) which indicates that one overall gas transfer unit results when the change in gas composition equals the average overall driving force causing the change.

Referring to Fig. 6.3-4, a line is drawn vertically halfway between the operating and equilibrium lines as shown by the dashed line. Starting at point A on the operating line, a line is drawn horizontally toward the equilibrium line and extending to point C such that $AB = AC$. A vertical line is then drawn from point C to the operating line to the point marked D. The step ACD represents one overall gas transfer unit. In a similar manner, additional transfer units are stepped off from the exit gas to the inlet gas composition to yield the total number of transfer units, N_{OG}, required. If N_{OL} is desired, the dashed line would be drawn halfway horizontally between the operating and equilibrium lines, and the steps would be constructed so that their vertical lines are divided into equal segments by the dashed line.

6.3-5 Correction for Curved Operating and Equilibrium Lines

Equations (6.3-21) and (6.3-22) cannot be used for systems in which the operating and equilibrium lines are curved because the absorption factor will vary from point to point in the tower. However, an approximate solution can be determined by finding effective average values for m and for L_M/G_M and using these in the absorption factor term. Sherwood et al.[9] found approximate effective average values based on an evaluation of numerous hypothetical absorber designs. It was found that the L_M/G_M ratio could be correlated best as a function of values of the ratio at each end of the tower, the ratio of the mole fraction of solute in the inlet gas to its mole fraction in the outlet gas, and the fractional approach to equilibrium at the bottom of the tower. The resulting correlation, which is applicable only for cases in which $L_M/G_M > 1$, is reproduced in Fig. 6.3-5. The correlation for the effective average slope is shown in Fig. 6.3-6.

In the recommended design procedure, values for $(L_M/G_M)_{av}$ and $\overline{m}$ $(= M_{av})$ from Figs. 6.3-5 and 6.3-6 are used in Eq. (6.3-21) to give an approximate value of N_{OG}. Since the equation for N_{OG} is based on the assumption of very dilute solutions, a further correction can be made to account for errors introduced by this assumption. For cases where $(1 - y)_{*M}$ can be represented adequately by an arithmetic mean, Wiegand[11] has shown that a correction ΔN_{OG} should be added to the N_{OG} calculated by Eq. (6.3-21) as follows:

$$\Delta N_{OG} = \tfrac{1}{2} \ln\left(\frac{1 - y_2}{1 - y_1}\right) \tag{6.3-23}$$

The final design equation then becomes

$$N_{OG}\text{(corrected)} = \frac{\ln\left[\left(1 - \frac{1}{A_{av}}\right)\left(\frac{y_1 - m_{av}X_2}{y_2 - m_{av}X_2}\right) + \frac{1}{A_{av}}\right]}{1 - 1/A_{av}} + \frac{1}{2}\ln\left(\frac{1 - y_2}{1 - y_1}\right) \tag{6.3-24}$$

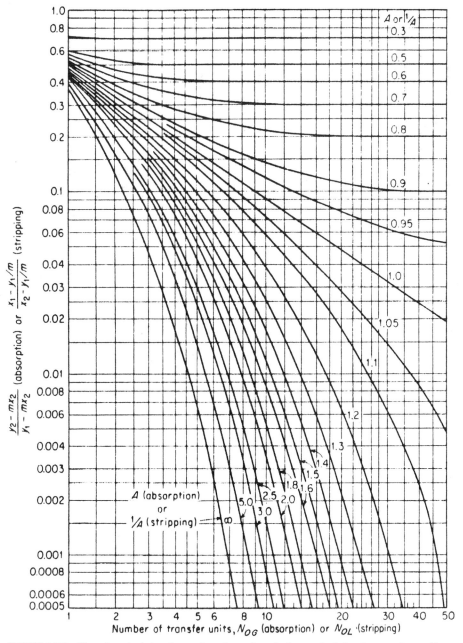

FIGURE 6.3-3 Chart for calculating the number of transfer units in a packed absorber or stripper. (From Treybal,[7] copyright 1980. Used with permission of McGraw-Hill Book Company, New York.)

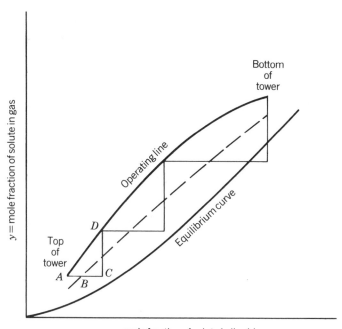

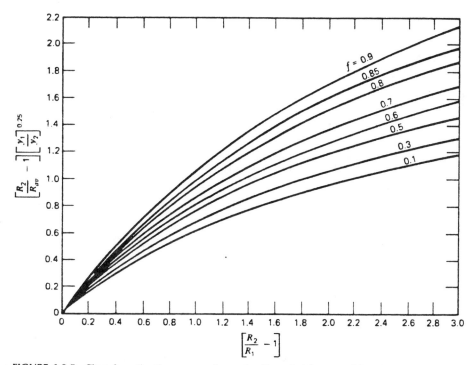

FIGURE 6.3-4 Graphical determination of the number of transfer units.

FIGURE 6.3-5 Chart for estimating average flow ratio: $R_1 = L_M/G_M$ at gas inlet; $R_2 = L_M/G_M$ at gas outlet; $y_1 =$ mole fraction in inlet gas; $y_2 =$ mole fraction in outlet gas; $R_{av} =$ effective average L_M/G_M; $f = y_1^*/y_1 =$ fractional approach to equilibrium. (From Wilke and von Stockar,[10] copyright 1978. Used with permission of John Wiley & Sons, New York.)

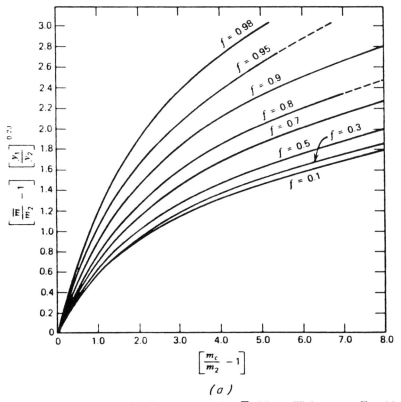

FIGURE 6.3-6 Correlation of the effective average slope $\overline{m}$ of the equilibrium curve. Chart (a) equilibrium curve concave upward; chart (b) equilibrium curve concave downward. (From Wilke and von Stockar,[10] copyright 1978. Used with permission of John Wiley & Sons, New York.)

6.3-6 HETP Concept

To relate packed tower performance to trays, packing performance is defined occasionally in terms of the height equivalent to a theoretical plate (HETP). The HTU concept is theoretically more correct for packed towers, in which mass transfer is accomplished by a differential action rather than a series of discrete stages. However, some data still are presented as HETPs. When the operating and equilibrium lines are straight, the two concepts can be related as follows:

$$\frac{H_{OG}}{\text{HETP}} = \frac{(mG_M/L_M) - 1}{\ln(mG_M/L_M)} \tag{6.3-25}$$

For the special case where the equilibrium and operating lines are parallel, that is, $mG_M/L_M = 1$, HETP and HTU values are equal. The height of the packing zone, Z, may be estimated using either HTU or HETP concepts; that is,

$$Z = (H_{OG})(N_{OG}) = (\text{HETP})(N) \tag{6.3-26}$$

where N_{OG} is the number of overall gas units and N is the number of theoretical plates.

6.3-7 Heat Effects in Packed Towers

Absorbers and strippers are very seldom isothermal. In the case of absorbers, the heat of solution or reaction of the solute with the solvent tends to increase the temperature, while partial evaporation of the solvent provides some cooling. In the case of strippers, the decomposition of solvent–solute compounds and the release of solute from the solvent are usually endothermic, while heat often is provided by the condensation of stripping vapor within the tower. In both cases, heat is transferred between the gas and liquid streams

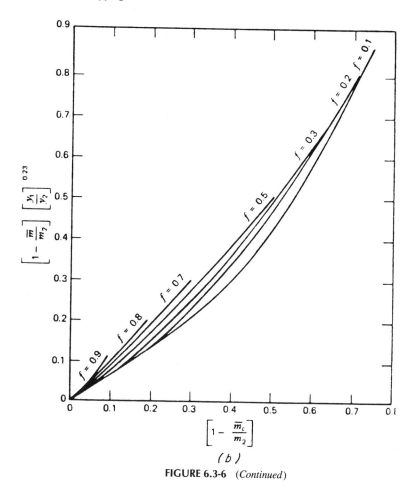

The y-axis is labeled $\left[1 - \dfrac{\overline{m}}{m_2}\right]\left[\dfrac{y_1}{y_2}\right]^{0.23}$ and the x-axis is labeled $\left[1 - \dfrac{\overline{m}_c}{m_2}\right]$. Curves are labeled $f = 0.1$, $f = 0.2$, $f = 0.3$, $f = 0.5$, $f = 0.7$, $f = 0.8$, $f = 0.9$.

(b)

FIGURE 6.3-6 (*Continued*)

and from both streams to the shell of the tower or to cooling coils. The resulting temperature changes within the tower affect performance by altering equilibrium relationships and physical properties.

The simplest approach to the problem is to assume that tower operation is isothermal at a temperature that is estimated on the basis of the temperatures of the feed streams; however, this approach is valid for relatively few systems (such as those involving the physical solution of low-solubility solutes). A somewhat more accurate approach is to assume that all the heat of reaction appears as an increase in temperature of the liquid stream. This "adiabatic" procedure requires relating the temperature increase of the liquid to the increase in concentration of solute in the liquid by a simple enthalpy balance and then adjusting the equilibrium line on an x-y diagram so that it corresponds to the estimated temperature at several selected increments of liquid composition.

Neither of the above two simplifying assumptions yields satisfactory results for the large number of absorption problems in which heat effects are significant, and considerable effort has gone into the development of more precise representations of nonisothermal operations. Because of the complexity of the problem, such solutions have required the use of computer models.

von Stockar and Wilke[12] developed a computer algorithm with very stable convergence behavior based on simulating the startup procedure of the column dynamically. The procedure requires that the differential unsteady-state mass and enthalpy balances be formulated and integrated with respect to time up to the steady state. To simulate the accumulation rates at various locations in the tower, it is necessary to divide the packed zone into an arbitrary number of segments. A large number of segments provides more accuracy but requires more computer time. The recommended procedure is to start with a small number of segments (e.g., 25) then refine the initially calculated tower profile by using a larger number. The computer algorithms used for the von Stockar and Wilke procedure are shown in Figs. 6.3-7 and 6.3-8.

As indicated in Fig. 6.3-7, the computations are done at a single (estimated) column height h_T. If this height does not give the required recovery, it is adjusted iteratively. When the required recovery is obtained, the value of h_T is checked by numerical integration of the rigorous expression for column height using the calculated composition profile. This procedure results in convergence to the true solution as the number of segments is increased.

A somewhat similar approach has been proposed by Feintuch and Treybal[13] for the design of complex multicomponent adiabatic systems. This work is an extension of a previously developed method of estimating heat effects for simple three-component systems and takes into account the mass and heat transfer resistances of both the liquid and gas phases.[14]

The proposed method for multicomponent systems involves the use of a computer program to calculate heat and mass balances rigorously for a preselected number of packing increments, reiterating the calculation until satisfactory convergence is obtained. Details of the physical property correlations used in the program, as well as a summary of the equations used, are given in Feintuch's thesis.[15] The overall program requires 20 calculation steps per iteration and includes a main program together with a set of 15 subroutines.

Feintuch and Treybal[13] also compared shortcut calculation methods, based on Eq. (6.3-21), against the rigorous computer program for a four-component system. Two shortcut methods were used—a simple approach and an Edmister-type approach. In the simple approach, G and L values are used which correspond to the conditions of the inlet gas and liquid streams, and the equilibrium constants used are based on the temperature of the inlet liquid. In the Edmister-type approach, G, L, and m are evaluated separately at the top and bottom of the tower. The Edmister equation [Eq. (6.2-10)] is used, except that A_1 corresponds to the absorption factor at the top of the packing and A_N to the factor at the bottom of the packing instead of absorption factors for the top and bottom trays, respectively.

The results of the comparison, with regard to the required packing height, are given in Table 6.3-1. Four cases are presented covering recoveries of the key component, propane, ranging from 10 to 90%. The simple shortcut method is seen to give a reasonable estimate of packing height for all except the 90% propane recovery case.

It is concluded that the simple shortcut method will provide reasonably close agreement to a rigorously calculated solution of packing height and outlet composition for problems that do not involve a high recovery. For a high-recovery case (approximately 90%), the packing height can be estimated by an Edmister-type approach. However, neither shortcut approach provides a good evaluation of product compositions, and a rigorous calculation procedure is necessary to provide these data when high recovery is required.

A shortcut method capable of handling nonisothermal systems, which produce a temperature bulge in the tower, has been developed by von Stockar and Wilke.[16] A review of the method also is given by the same authors in the *Encyclopedia of Chemical Technology*;[10] however, the original paper presents a more detailed discussion and includes an illustrative example. The method is based on the empirical correlation of rigorously calculated results for a selected set of over 90 hypothetical design cases. The study was limited to solute concentrations in the gas below 15 mol % and to recoveries ranging from 90 to 99%. Water was considered to be the most important solvent, although the effects of solvent properties were incorporated into the correlation to a limited extent. Atmospheric pressure was assumed in all calculations; however, no significant effect of pressure would be expected below 10 atm.

The method is based on the development of a mathematical expression to correlate the temperature profile; using this temperature profile to establish a revised equilibrium curve; breaking the equilibrium curve into two segments; and finally using the average slopes of the two segments in conjunction with the Colburn equation [Eq. (6.3-21)] to determine the necessary number of transfer units.

The following specific steps are required to estimate the packing height by the von Stockar–Wilke shortcut procedure.

1. Estimate the temperature of the product gas by use of the following semiempirical equations:

$$T_{G,2} = T_{L,2} + \left(\frac{dT_L}{dx_A}\right)_2 \left(\frac{G_M}{L_M}\right)_2 \left(\frac{H_{GQ}}{H_{OG,A}}\right)(y_{A2} - y_{A2}^*) \qquad (6.3\text{-}27)$$

$$\left(\frac{dT_L}{dx_A}\right)_2 \approx \frac{L_{M2}H_{OS} - G_{m2}H_V m_{B2}}{L_{M2}C_{q2} - G_{M2}\,C_{p2} - G_{M2}\,H_V\,(1 - x_{A2})\,(dm_B/dT_L)_2} \qquad (6.3\text{-}28)$$

where
T_G = gas temperature (°C)
T_L = liquid temperature (°C)
$H_{GQ} = G_M C_P/k_G a$ = the height of a gas-phase heat transfer unit (m)
$H_{OG,A} = H_{OG}$ for the solute (m)
H_{OS} = integral heat of solution for solute (kJ/mol)
H_V = latent heat of pure solvent, kJ/mol
$m_{B2} = y_B^*/x_B$ at the top of the tower

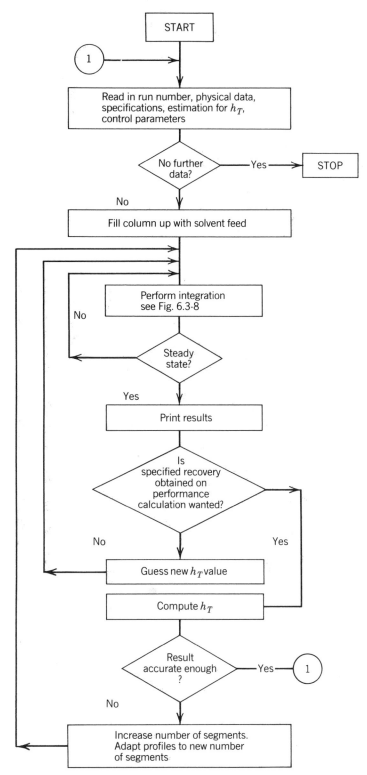

FIGURE 6.3-7 Computer flow diagram for packed absorber design by the method of von Stockar and Wilke.[12] (Reprinted with permission from *Industrial Engineering Fundamentals*, Vol. 16, No. 1, copyright 1977 by American Chemical Society.)

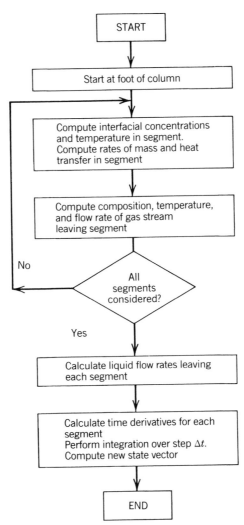

FIGURE 6.3-8 Algorithm for integrating unsteady-state mass and enthalpy balances. The state vector is an array of numbers representing the liquid compositions and enthalpies in all the segments. (From von Stockar and Wilke.[12] Reprinted with permission from *Industrial Engineering Fundamentals*, Vol. 16, No. 1, copyright 1977 by American Chemical Society.)

TABLE 6.3-1 Comparison of Calculation Methods for Four-Component Adiabatic Absorption System

	Case			
	1	2	3	4
Recovery of C_3(%)	10	50	60	90
Calculated packing height (ft)				
Method A	0.507	0.554	0.636	0.423
Method B	0.471	0.540	0.665	0.750
Method C	0.539	0.480	0.555	0.446

Note: Method A = rigorous computer calculation.
  Method B = simple approach.
  Method C = Edmister-type approach.

Source: Feintuch and Treybal.[13]

$(dm_B/dT_L)_2$ = temperature coefficient of m_B at the top of the tower
C_{p2} = mean molar heat capacity of gas at top of column (kJ/mol·K)
C_{q2} = mean molar heat capacity of liquid at top of column (kJ/mol·K)

The temperature of the product liquid can then be estimated by an enthalpy balance around the column:

$$T_{L1} = T_{L2} + \left[\left(\frac{G_M}{L_M} \right)_{av} \frac{C_p}{C_q} (T_{G1} - T_{G2}) + H_V (y_{B1} - y_{B2}) \right] + \frac{H_{OS}}{C_q} (x_{A1} - x_{A2}) \qquad (6.3\text{-}29)$$

where C_p = mean molar heat capacity of the gas averaged over the tower (kJ/mol·K)
C_q = mean molar heat capacity of the liquid averaged over the tower (kJ/mol·K)

2. Estimate the maximum temperature ΔT_{max} of the convex portion of the liquid temperature profile by the use of Fig. 6.3-9. The nomenclature for this figure is as follows:

$\Delta Y = y_{A1} - y_{A2}$
m_A = slope of equilibrium line for solute at liquid feed temperature
m_B = slope of equilibrium line for solvent, $y_B^*/(1-x_A)$ at liquid feed temperature
y_B/y_B^* = fraction saturation of solvent in feed gas
$H_{OG,B} = H_{OG}$ for solvent
$H_{OG} = H_G + (mG_M/L_M)H_L$ with m evaluated at the temperature of the liquid feed

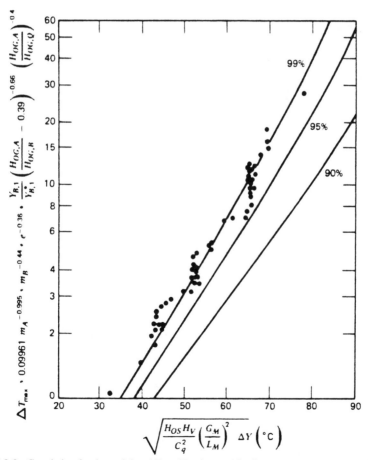

FIGURE 6.3-9 Correlation for determining ΔT_{max}. Numbers on the lines represent recoveries; the points are for the 99% recovery case. (From Wilke and von Stockar,[10] copyright 1978. Used with permission of John Wiley & Sons, New York.)

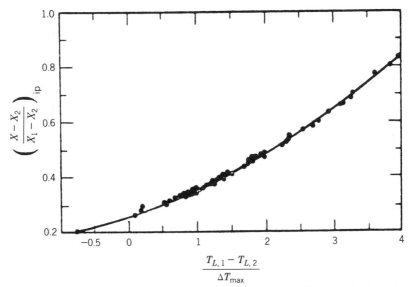

FIGURE 6.3-10 Correlation for determining the liquid concentration at which the inflection point of the nonisothermal equilibrium occurs. (From Wilke and von Stockar,[10] copyright 1978. Used with permission of John Wiley & Sons, New York.)

3. Determine the inflection point by the use of Fig. 6.3-10. In this figure $(x_A - x_{A2})(x_{A1} - x_{A2})_{ip}$ corresponds to the inflection point and represents the normalized concentration at which the highest temperature occurs. The temperature at the inflection point is calculated by use of the following empirical equation which defines the liquid temperature at any point in the column:

$$T_L = T_{L2} + (T_{L1} - T_{L2}) X_N + 74.34(X_N^{1.074} - X_N^{1.114})\Delta T_{max} \qquad (6.3\text{-}30)$$

where

$$X_N = \frac{x_A - x_{A2}}{x_{A1} - x_{A2}}$$

Once the temperature and liquid composition are known for the inflection point, the equilibrium gas composition, y_{ip}^*, can be determined from the appropriate equilibrium relationship. These calculations can be repeated at other liquid concentrations to establish the entire temperature profile for the column along with the temperature-corrected equilibrium curve as a function of liquid concentration. The required column height can be determined by conventional graphical integration. As an alternative to graphical integration, von Stockar and Wilke propose an analytical approach as defined in the next step.

4. Determine the effective average slopes for the two sections of the equilibrium curve by the use of Figs. 6.3-11 and 6.3-12. The various slopes used in these figures ($\overline{m}$, m_c, m_2, and m_{ip}) are defined in Fig. 6.3-13. It is not necessary to calculate the entire column profile for the analytical approach; however, it is necessary to calculate y^* at the inflection point, at the bottom and top of the column, and at one other point with a value of x_A slightly higher than the inflection point to permit m_{ip} to be estimated. The gas composition at the inflection point, y_{ip}, can be calculated on the basis of the liquid composition by assuming a straight operating line. The values of $\overline{m}$ for each section are used in Eq. (6.3-21) to determine the necessary number of transfer units. The overall height of a transfer unit is evaluated using the effective average slopes and Eq. (6.3-19). The total number of transfer units and the total height of packing required are obtained finally by adding the respective values for the two sections together.

von Stockar and Wilke state that no serious temperature bulge will occur if $(T_{L1} - T_{L2})/T_{max} \geq 4.3$, in which case the simple model of adiabatic gas absorption may be used to determine the required packing height. Figure 6.3-11, which provides a correlation of the effective average slope for the dilute (top) portion of the absorber, may be used for the entire absorber and the predicted $\overline{m}$ used in conjunction with Eqs. (6.3-21) and (6.3-19) to estimate the number of transfer units required and the total height of packing.

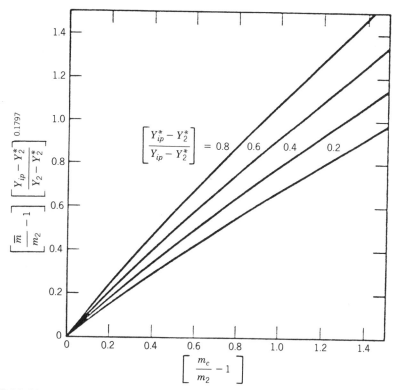

FIGURE 6.3-11 Correlation for determining the effective average slope of the equilibrium line in the portion of the absorber with a concave-upward equilibrium line (dilute region). (From Wilke and von Stockar,[10] copyright 1978. Used with permission of John Wiley & Sons, New York.)

6.4 PREDICTING CONTACTOR PERFORMANCE

6.4-1 Physical Absorption in Packed Towers

Numerous experimental and theoretical studies have been made relative to mass transfer in packed towers. The results usually are presented in terms of the mass transfer coefficients, K_{OG} or K_{OL}, HTUs, or HETPs. In general, the preferred basis for design is experimental data on the specific solute–solvent system and packing type, and such data are available for a large number of cases. *Perry's Chemical Engineers' Handbook*,[1] Section 18, is a good source of specific mass transfer data. Such data are also available from the manufacturers of packing and from specialized texts, such as Kohl and Riesenfeld[2] and Astarita et al.[3]

Reliable correlations for predicting the mass transfer performance of any absorption system and tower packing from basic physical data are not yet available. However, this situation is improving steadily. The preferred approach is to develop correlations for the individual heights of transfer units, H_G and H_L, then to combine these by Eq. (6.3-19) to determine H_{OG}, the height of an overall transfer unit. Of course, experimental data or correlations expressed in terms of mass transfer coefficients can be converted to heights of transfer units by equations such as (6.4-1) and (6.4-2):

$$H_G = \frac{G}{k_G a \, PM_G} \tag{6.4-1}$$

$$H_L = \frac{L}{k_L a \, \rho_L} \tag{6.4-2}$$

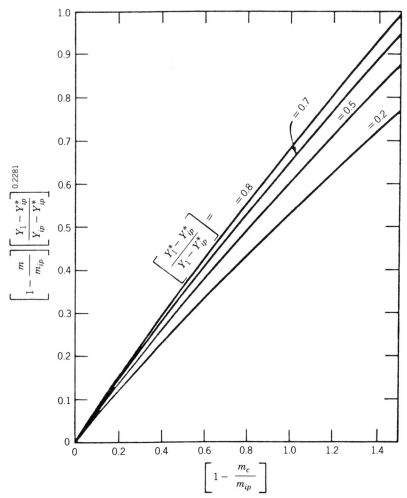

FIGURE 6.3-12 Correlation for determining the effective average slope of the equilibrium line in the portion of the absorber with a concave-downward equilibrium line (concentrated region). (From Wilke and von Stockar,[10] copyright 1978. Used with permission of John Wiley & Sons, New York.)

where a = the effective interfacial area for mass transfer (ft^2/ft^3 or m^2/m^3)
M_g = molecular weight of the gas
G = mass flow rate of gas (lb$_m$/s·ft^2 or kg/s·m^2)
L = mass flow rate of liquid (lb$_m$/s·ft^2 or kg/s·m^2)
p = pressure (atm or Pa)
k_G = gas-phase mass transfer coefficient (lb mole/s·ft^2 · atm or kmol/N·s)
k_L = liquid-phase mass transfer coefficient (ft/s or m/s)
ρ_L = density of liquid (lb$_m$/ft^3 or kg/m^3)

See also Eq. (6.3-10) and the accompanying discussion. Generalized correlations, based on the individual transfer units, have been proposed by Cornell et al.,[4] Onda et al.,[5] Bolles and Fair,[6] and Bravo and Fair.[7] The correlations have covered such common packings as Raschig rings, Berl saddles, Pall rings, and related configurations. The key element in the development of generalized correlations has been the definition of a, the effective interfacial area. This value is equal for the gas and liquid phases, since it represents the interfacial area for mass transfer between the two phases. It includes not only the wetted area of the packing but also any area provided by droplets, gas bubbles within the liquid stream, ripples on the liquid surface, and liquid film on the walls of the tower. In the Onda model, the individual mass transfer coefficients are given by the following dimensionally consistent equations:

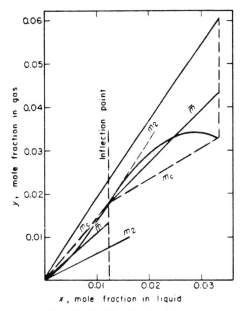

FIGURE 6.3-13   Diagram of curved equilibrium line showing inflection point and effective average slopes for both sections. (From Wilke and von Stockar,[10] copyright 1978. Used with permission of John Wiley & Sons, New York.)

$$k_G \left(\frac{RT}{a_p D_G} \right) = 5.23 \left(\frac{G}{a_p \mu_G} \right)^{0.7} (Sc_G^{1/3}) (a_p d_p)^{-2} \tag{6.4-3}$$

$$k_L \left(\frac{\rho L}{g \mu_L} \right)^{1/3} = 0.0051 \left(\frac{L}{a_w \mu_L} \right)^{2/3} (Sc_L^{-1/2})(a_p d_p)^{0.4} \tag{6.4-4}$$

In this model, a is assumed to be equal to a_w, the wetted area of the packing, and is calculated by the equation

$$a = a_w = a_p \left\{ 1 - \exp \left[-1.45 \, Re_L^{-0.1} \, Fr_L^{-0.05} \, We_L^{0.2} \left(\frac{\sigma}{\sigma_c} \right)^{-0.75} \right] \right\} \tag{6.4-5}$$

where d_p = effective packing diameter (the diameter of a sphere with equal surface area)
  a_p = dry outside surface area of packing
  a_w = wetted area of packing
  σ_c = a critical surface tension = 61 dyn/cm for ceramic packing, 75 dyn/cm for steel packing, and 33 dyn/cm for polyethylene packing

$$Re_L = L/a_p \mu_L \tag{6.4-6}$$

$$Fr_L = a_p L^2 / g \rho_L^2 \tag{6.4-7}$$

$$We_L = L^2 / a_p \, \rho_L \tag{6.4-8}$$

$$Sc_L = \frac{\mu_L}{\rho_L D_L} \tag{6.4-9}$$

$$Sc_G = \frac{\mu_G}{\rho_G D_G} \tag{6.4-10}$$

Bolles and Fair[6] evaluated a large data base of published tests on the performance of columns containing random packing and developed a mass transfer model that appears to be superior to those previously published. Their model, which actually represents an improved version of the earlier Cornell–Knapp–Fair[4] approach, is general with regard to fluid properties and flow rates but requires two specific inputs: (1) prior validation for the specific packing type and size and (2) a consistent method for predicting flooding conditions for the system and packing under study. The Bravo–Fair model[7] represents an attempt to avoid the limitations of the Bolles–Fair approach and to develop a completely general design model. Although the Bravo–Fair approach is more fundamental and ultimately may provide the basis for an excellent design correlation for absorbers and strippers, it was developed and tested primarily for distillation and therefore will not be described in this chapter.

In the Bolles–Fair model, the height of a gas-phase transfer unit is given by

$$H_G = \psi \, \frac{(d'_c)^m \, (Z_p/10)^{1/3}}{[3600 \, L(\mu_L/\mu_W)^{0.16} \, (\rho_L/\rho_W)^{-1.25} \, (\sigma_L/\sigma_W)^{-0.8}]^n} \, Sc_G^{1/2} \tag{6.4-11}$$

where ψ = packing parameter for vapor-phase mass transfer (ft)
   d'_c = diameter of column (ft) (adjusted; the lesser of the actual diameter, d_c, or 2)
   Z_p = height of each packed bed (ft)
   μ = viscosity (lb$_m$/ft·s or N·s/m^2)
   ρ = density (lb$_m$/ft^3 or kg/m^3)
   σ = surface tension (lb$_m$/s^2 or dyn/cm)
   Sc_G = Schmidt number for gas = $\mu_G/\rho_G D_G$
   D_G = diffusion coefficient for key component (ft^2/s or m^2/s)
   m = 1.24 for rings; 1.11 for saddles
   n = 0.6 for rings; 0.5 for saddles

and the height of a liquid-phase transfer unit is given by

$$H_L = \phi \, C_{fL} \left(\frac{Z_p}{10}\right)^{0.15} Sc_L^{1/2} \tag{6.4-12}$$

where ϕ = packing factor for liquid-phase mass transfer (ft)
   C_{fL} = coefficient for effect of approach to flood point on liquid-phase mass transfer
   Sc_L = Schmidt number for liquid = $\mu_L/\rho_L D_L$

The subscripts in both of the above equations are L = liquid, G = gas, W = water.

Graphical representations of packing parameters ψ and ϕ are given in Figs. 6.4-1 and 6.4-2. The vapor-load coefficient, which accounts for the effect of approach to flood point on H_L, is plotted as a function of the flood ratio F_r in Fig. 6.4-3. As used in Figs. 6.4-1, 6.4-2, and 6.4-3, L is the mass velocity of the liquid (lb$_m$/h·ft^2) and F_r is the flood ratio at constant $L/G = U_{Gs}/U_{Gsf}$, where U_{Gs} is the velocity (ft/s) of vapor based on superficial area and U_{Gsf} is the velocity (ft/s) of vapor based on superficial area at flood point (from Eckert model[8]).

Since Eqs. (6.4-11) and (6.4-12) are not dimensionally consistent, they cannot be used directly in SI units. However, SI units may be used in all dimensionless numbers and ratios. The final values for H_G or H_L will be in feet, which then of course can be converted to meters. In using the Bolles–Fair model for tower design, the individual heights of transfer units are estimated separately by Eqs. (6.4-11) and (6.4-12), then combined to yield the overall HTU by Eq. (6.3-19). The height equivalent to a theoretical plate, HETP, can be calculated, if desired, by Eq. (6.3-25).

Bolles and Fair tested their correlation against a data bank comprising 545 observations from 13 original sources covering column diameters from 0.82 to 4.0 ft (0.25 to 1.2 m), packing sizes from 0.6 to 3.0 in. (0.015 to 0.076 m), pressures from 0.97 to 315 psia (6.7 to 2.17 × 10^3 kPa), and a wide range of distillation, absorption, and stripping regimes. It was found to fit the data better than previously available models and to have a safety factor of 1.7 (based on a 95% probability of success).

Generalized correlations are not available for predicting absorption coefficients for ordered packings; however, test data have been published which can be used to evaluate specific types. A comparison of a sieve tray column, a packed column, and a column packed with Sulzer BX gauze packing is given in Fig. 6.4-4.[9] In this figure, F_s (the gas load factor) = $U_G \, \rho_G^{1/2}$, and ΔP/TP refers to the pressure drop per theoretical plate. HETP for the Sulzer packing is indicated to be less than half that of the other two types. Very-high-efficiency ordered packings, such as Goodloe and Sulzer, which are based on mesh or gauze materials, generally are employed for difficult distillation separations, particularly under vacuum conditions, and more open, low-cost structures are preferred for commercial absorption applications.

Ordered packings, which appear suitable for general-purpose absorption and stripping towers, have been developed based on the use of perforated sheetmetal instead of wire mesh or gauze. These materials exhibit properties intermediate between the very-high-efficiency mesh packings and random packings. One such material is Mellapak developed by Sulzer Brothers, Ltd.[10] The Mellapak packing has the same geometrical structure as Sulzer gauze packing. It is made up of elements that occupy the entire cross section

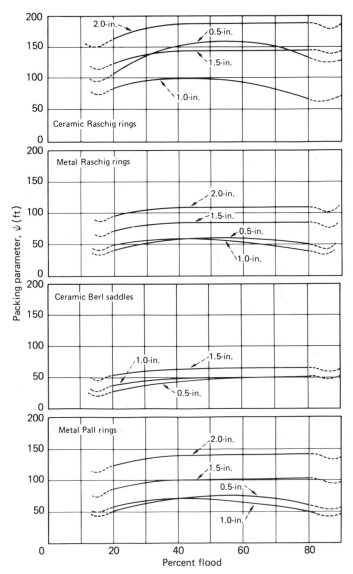

FIGURE 6.4-1 Packing parameters for the Bolles–Fair gas-phase mass transfer model. (From Bolles and Fair,[6] excerpted by special permission from *Chemical Engineering*, July 12, 1982. Copyright 1982 by McGraw-Hill, New York.)

of the tower and that are fabricated from perforated, corrugated sheetmetal or plastic sheets. The individual sheets are aligned vertically in the tower and are corrugated at an angle. Adjacent sheets have opposite corrugation angles to produce intersecting flow passages and enhance the mixing of gas and liquid flowing through the tower. There are several types of Mellapak differing in both the hydraulic diameter of the channels and the angle of inclination of the flow channels. Absorption data for Type 250Y are presented in Fig. 6.4-5.[9] This type has a surface area of 250 m^2/m^3, a corrugation depth of approximately 12 mm, and an angle of inclination of 45° from the vertical. The data were obtained in an 8-in. (0.2 m) diameter absorber operating at atmospheric pressure. Fresh water was fed to the top of the tower and dry air containing either SO_2 or NH_3 was fed to the bottom. N_{OG} values were calculated using Eq. (6.3-20) and HTU (H_{OG}) values were determined from the relationship $H_{OG} = Z/N_{OG}$. Pressure drop data also are included in the figure. Data for 2 in. (0.05 m) Pall rings obtained in the same manner are included for comparison.

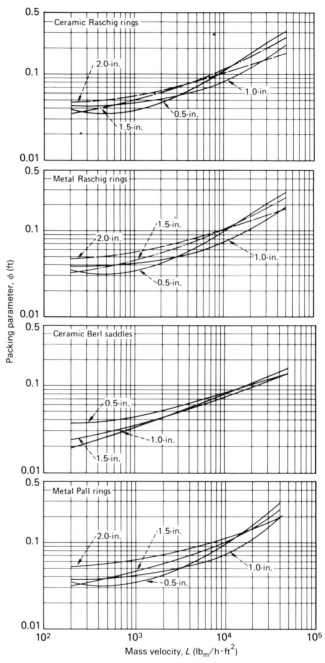

FIGURE 6.4-2 Packing parameters for the Bolles–Fair liquid-phase mass transfer model. (From Bolles and Fair,[6] excerpted by special permission from *Chemical Engineering*, July 12, 1982. Copyright 1982 by McGraw-Hill, New York.)

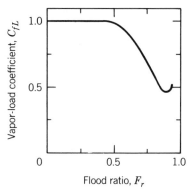

FIGURE 6.4-3 Vapor load coefficient for the Bolles–Fair liquid-phase mass transfer model. (From Bolles and Fair,[6] excerpted by special permission from *Chemical Engineering*, July 12, 1982. Copyright 1982 by McGraw-Hill, New York.)

6.4-2 Absorption in Spray Contactors

Spray contactors are particularly important for the absorption of impurities from large volumes of flue gas where low pressure drop is of key importance. They are used where materials in the liquid phase (e.g., particles of limestone) or in the gas phase (e.g., droplets of tar) may cause plugging of packing or trays. Other important applications of spray contactors (which are outside the scope of this discussion) include particulate removal and hot gas quenching. When used for absorption, spray devices are not applicable to difficult separations and generally are limited to about four transfer units even with countercurrent spray column designs. The low efficiency of spray columns is believed to be due to entrainment of droplets in the gas and backmixing of the gas induced by the sprays.

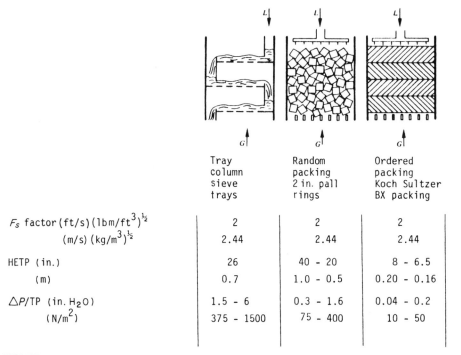

	Tray column sieve trays	Random packing 2 in. pall rings	Ordered packing Koch Sultzer BX packing
F_s factor $(ft/s)(lbm/ft^3)^{\frac{1}{2}}$	2	2	2
$(m/s)(kg/m^3)^{\frac{1}{2}}$	2.44	2.44	2.44
HETP (in.)	26	40 – 20	8 – 6.5
(m)	0.7	1.0 – 0.5	0.20 – 0.16
$\Delta P/TP$ (in. H_2O)	1.5 – 6	0.3 – 1.6	0.04 – 0.2
(N/m^2)	375 – 1500	75 – 400	10 – 50

FIGURE 6.4-4 Simplified comparison of sieve trays, random packing, and high-efficiency ordered packing. (From Meier et al.,[9] reprinted with permission from *Chemical Engineering Progress*, November 1977.)

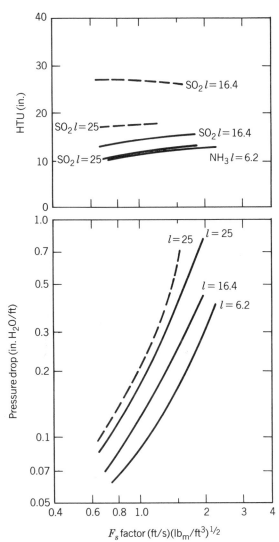

FIGURE 6.4-5 Comparison of the performance of Mellapak Type 250Y ordered packing with 2 in. Pall rings for absorption of SO_2 (and NH_3) in water. l = liquid flow rate, gpm/ft². (From Meier et al.,[9] reprinted with permission from *Chemical Engineering Progress*, November 1977.) SI conversions: 1 gpm/ft² = 6.78 $\times 10^{-4}$ m³/(s·m²); 1 in. = 0.0254 m; 1 in. H_2O/ft = 816 (N/m²)/m; 1 (ft/s)(lb$_m$/ft³)$^{1/2}$ = 1.22 (m/s)(kg/m³)$^{1/2}$.

Many different types of spray device have been developed and no generalized correlations are available for predicting performance. They can be categorized into two basic types: (1) preformed spray, which includes countercurrent, cocurrent, and crosscurrent spray towers, spray dryers, cyclonic spray devices, and injector venturis; and (2) gas atomized spray, which consists primarily of venturi scrubbers and related designs.

It is not unusual for exhaust gas scrubber systems to incorporate more than one type of spray contactor. In fact, combinations with trays and packed sections are used extensively. Figure 6.4-6 shows diagrams for four typical scrubber systems used for removing HF and SiF_4 from phosphoric acid plants.[11] The following systems are illustrated.

1. *Venturi Cyclonic.*—This system consists of a venturi followed by a cyclonic spray scrubber. A low-pressure pump circulates scrubber liquid to the venturi where it becomes atomized and thoroughly mixed with the gas by the extremely high velocity in the throat. Gas from the venturi enters the cyclonic scrubber

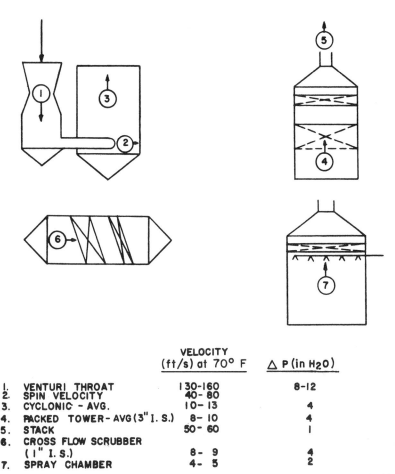

		VELOCITY (ft/s) at 70° F	$\triangle$ P (in H_2O)
1.	VENTURI THROAT	130-160	8-12
2.	SPIN VELOCITY	40-80	
3.	CYCLONIC - AVG.	10-13	4
4.	PACKED TOWER- AVG (3" I. S.)	8-10	4
5.	STACK	50-60	1
6.	CROSS FLOW SCRUBBER (1" I. S.)	8-9	4
7.	SPRAY CHAMBER	4-5	2

FIGURE 6.4-6 Typical atmospheric pressure scrubber designs showing approximate gas velocities and pressure drops in key zones. (From Hansen and Danos,[11] reprinted with permission from *Chemical Engineering Progress*, March 1982.) SI conversions: 1 ft/s (fps) = 0.305 m/s; 1 in. H_2O = 249 N/m^2.

tangentially with a high spin velocity as indicated. Typically, additional liquor is sprayed into the gas by high-pressure spray nozzles [e.g., 60 psig (413 kPa)] aimed radially inward or outward within the cyclone.

2. *Packed Tower.*—A conventional countercurrent packed tower is shown for comparison. When used for flue gas scrubbing, a short section of low-pressure-drop packing (e.g., 3 in. Intalox saddles) is employed with a mist eliminator above the liquid feed point.

3. *Crosscurrent Packed Scrubber.*—Typically, this type of scrubber includes a spray chamber followed by one or more thin beds of packing with the final bed designed as a mist eliminator. For efficient performance, high-pressure sprays are required in the spray section and efficient coverage of the packing with scrubber liquid must be maintained, although low-pressure sprays can be used for packing irrigation.

4. *Countercurrent Spray Chamber.*—This is the simplest design and the least subject to plugging or erosion problems due to dust in the gas stream. However, as indicated by the low gas velocity range given in Fig. 6.4-6, spray chambers tend to be larger than the other designs, and since high pressure sprays plus a large volume of liquid are required, energy requirements are not insignificant.

For many spray scrubber applications, a large excess of water is used to absorb a very soluble gas (such as HF, SiF_4, HCl, and NH_3) from a dilute gas stream. In such cases, the equilibrium concentration in the gas phase y^* can be neglected and Eq. (6.3-13) can be simplified to

$$N_{OG} = \int_{y_2}^{Y_1} \frac{dy}{y} = \ln \frac{y_1}{y_2}$$ (6.4-13)

Since the absorption efficiency E is related directly to y_1 and y_2, $E = [(y_1 - y_2)/y_1] \times 100$, the number of transfer units can be estimated based on the required efficiency from the following equation:

$$N_{OG} = \ln \frac{1}{1 - E/100} \tag{6.4-14}$$

In accordance with this equation, an absorption efficiency of 95% requires three transfer units, while 99% requires about five transfer units.

Spray contactor performance cannot be predicted by analogy to packed or tray systems because of fundamental differences in the contact mechanism, particularly in regard a, the effective area for mass transfer. In spray contactors, a is not a simple function of contactor height or volume but is related more closely to the number and diameter of liquid droplets in contact with the gas at any time. Since in many spray devices these values are determined largely by the liquid flow rate and the pressure drop across the spray nozzles, it is not surprising that attempts have been made to correlate spray system performance with power consumed by the contactor. Such a correlation was proposed originally by Lunde.[12] Data on several types of spray contactor correlated in this manner are presented in Fig. 6.4-7. On the basis of the data used in this plot, it appears that the spray tower is the most energy efficient system and the venturi scrubber is the least efficient with spray cyclonic scrubbers and crosscurrent contactors falling in between. Efficiency with regard to the scrubber volume required is in approximately the reverse order; that is, the spray chamber requires the greatest volume. The energy requirement for spray towers is divided between the gas and liquid pumping requirements, while for both packed towers and venturi scrubbers, most of the energy is required to overcome the gas-phase pressure drop through the apparatus. Although not shown in the chart, grid packed towers are reported to have an even lower power requirement than spray towers, primarily because the liquid is distributed without the requirement for high-pressure-drop spray nozzles. Venturi ejectors have a relatively high power requirement, in the same range as gas atomized venturis; however, in their case, all the power must be supplied to the liquid which provides the energy to move the gas through the device.

Attempts have been made to develop design equations for preformed spray contactors on the basis of a more fundamental approach by considering mass transfer to individual drops. However, this requires a characterization of the drop size distribution, time of flight, and other factors. The approach is useful when test data are not available but cannot be expected to provide accurate results for commercial equipment because the effects of drop agglomeration, entrainment, wall contact, gas recirculation, drop velocity changes, and other factors are very difficult to model analytically.

Approximate drop size distributions from spray nozzles usually can be obtained from the manufacturers. Drop diameters are typically in the range from 10 to 600 μm. Hobbler[13] suggests that a 600-μm drop size be assumed as a conservative value for approximate calculations.

The gas-film mass transfer coefficient to a drop can be estimated on the basis of the following semiempirical equation:[14]

$$\frac{k_G R T d}{D_G} = 2 + 0.552 \, Re^{1/2} Sc^{1/3} \tag{6.4-15}$$

where k_G = gas-film mass transfer coefficient (mol/cm·s·atm)
R = ideal gas constant (82.06 atm·cm^3/mol·K)

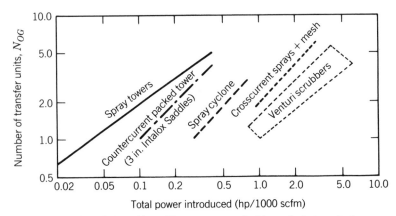

FIGURE 6.4-7 Correlation of N_{OG} with total horsepower required by typical atmospheric pressure scrubbers. SI conversions: 1 hp/1000 scf = 27.8 W/m^3 at STP.

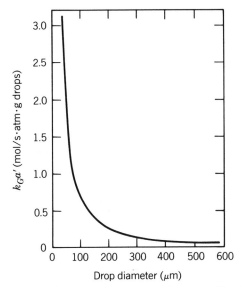

FIGURE 6.4-8 Gas-film mass transfer rate per gram of water for drops falling through air. (From Calvert et al.[15])

T = absolute temperature (K)
d = diameter (cm)
D_G = diffusivity of the solute in the gas (cm²/s)
Re = Reynolds number for the gas flowing past the drop
Sc = Schmidt number for the gas

The velocity to be used in Re is the relative velocity between the drop and the gas stream.

Calvert et al.[15] have used the above equation to calculate $k_G a'$ for water drops falling through air at their terminal velocity, where a' is equal to the surface area per unit weight of water. The result, which is useful for crosscurrent contactors where droplets may fall at their terminal velocity, is shown in Fig. 6.4-8. For purposes of this plot, a solute diffusivity of 0.1 cm²/s in air and a temperature of 25°C were assumed. Terminal settling velocities and Reynolds numbers for water drops in air are given in Fig. 6.4-9 from the same source. Calvert and coworkers suggest the following equations for estimating the liquid-side mass transfer coefficient:

$$k_L = 2\left(\frac{D_L}{\pi\Theta}\right)^{1/2} \quad \begin{array}{l}\text{for short contact times} \\ \text{(penetration theory)}\end{array} \qquad (6.4\text{-}16)$$

$$k_L = 10\frac{D_L}{d_d} \quad \text{for long contact times} \qquad (6.4\text{-}17)$$

where k_L = liquid-film mass transfer coefficient (cm/s)
D_L = diffusivity of the solute in the liquid (cm²/s)
Θ = time of exposure of the drop to the gas stream (s)
d_d = the drop diameter (cm)

Values of θ for water drops falling through air at their terminal velocity are shown in Fig. 6.4-10 for several heights as a function of drop diameter. Calvert et al.[15] have calculated $k_L a'$ using the above equations together with calculated values for a', the surface area per unit weight of water, to produce Fig. 6.4-11. A representative value of diffusivity, D_L, of 2×10^{-5} cm²/s at 25°C was used for the calculations on which the plot is based.

Spray dryer absorbers represent a special case of preformed spray contactor. They are finding increasing use in SO_2 absorption because of their low-gas side pressure drop, freedom from wet scrubber problems of corrosion and plugging, production of spent absorbent in a dry form, and production of a purified gas that is not completely saturated with water. The design of this type of unit is complicated by dual requirements: a dry flowable solid must be produced and the desired SO_2 absorption efficiency must be attained. The two do not necessarily go together. Highly efficient drying may not result in good SO_2 absorption

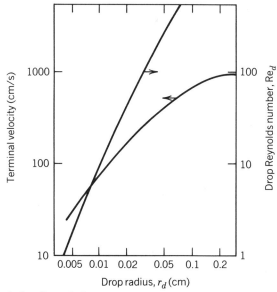

FIGURE 6.4-9 Terminal settling velocity and Reynolds number for water drops in air at 20°C and 1 atm. (From Calvert et al.[15])

efficiency. In fact, the most efficient SO_2 absorption is attained when the droplets remain moist until they are very close to the dryer wall at which point they should be dry enough to flow out of the dryer without sticking but should retain some moisture to aid in the further absorption of SO_2 in the particle collection apparatus.

The basic concepts of spray dryer design with regard to the drying requirements have been described adequately by Gauvin and Katta,[16, 17] Masters,[18] and others and are not described here. The absorption of gases in spray dryers has not yet been modeled adequately analytically although numerous plant descriptions have appeared in the literature. A qualitative model of the phenomena occurring in a spray dryer employing lime slurry to absorb SO_2 has been described by Getler et al.[19] They studied a unit that used a rotary

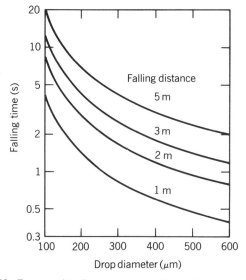

Fig. 6.4-10 Exposure time for water drops falling in air. (From Calvert et al.[15])

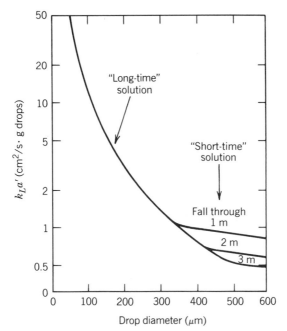

FIGURE 6.4-11 Liquid-film mass transfer rate per gram of water for drops falling in air. (From Calvert et al.[15])

atomizer to create droplets in the 50–8 μm mean diameter range and concluded that the operation proceeds in two phases. During phase I, the droplets leaving the atomizer wheel decelerate to velocities below 30 ft/s in less than 0.1 s. The liquid represents the continuous phase in each droplet during this period and the liquid, gas, and solid are mixed intimately. Chemical reactions in the liquid phase are very rapid and the rate-controlling step is either gas-phase diffusion or dissolution of the solid lime particles. Rapid evaporation occurs until the solid particles touch each other, restricting the remaining moisture paths.

During phase II, evaporation of water continues and SO_2 is absorbed by the moist particle which is composed of a number of the elementary solid particles. The rate of absorption is determined by the diffusion of SO_2 through the interstices between elementary particles and through the layer of spent reactant ($CaSO_3 \cdot 0.5\ H_2O$) which builds up on each elementary particle. Although a quantitative model for spray dryer absorption has not been developed, it has been observed that SO_2 absorption efficiency increases with increased lime/SO_2 ratio, reduced drop size, and reduced approach to water saturation of the outlet gas. A plot of absorption efficiency as a function of lime/SO_2 mole ratio for a typical spray dryer SO_2 absorption system is given in Fig. 6.4-12.[20]

6.4-3 Tray Efficiency

The number of actual trays required in a tower to attain the performance of a calculated number of theoretical stages is determined by the tray efficiency. Several types of tray efficiency have been proposed; however, the two most widely used are the Murphree vapor efficiency E_{MV}, defined in Section 6.2 [Eq. (6.2-1)], which refers to individual trays, and the overall column efficiency E_O, which is defined simply as

$$E_O = \frac{N}{N_T} \qquad (6.4\text{-}18)$$

where N is the total number of equilibrium stages and N_T is the total number of actual stages.

A third type of tray efficiency, the Murphree point efficiency, is used sometimes in attempts to correlate efficiency with fundamental mechanisms occurring on the tray. The point efficiency E_{OG}, is defined as follows:

$$E_{OG} = \frac{y_{out} - y_{in}}{y_{out}^* - y_{in}} \qquad (6.4\text{-}19)$$

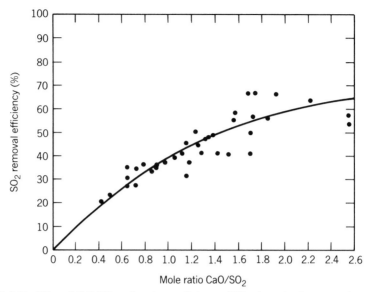

FIGURE 6.4-12   Effect of CaO/SO$_2$ mole ratio on efficiency of SO$_2$ absorption in a spray dryer. Inlet SO$_2$ concentration 740–1640 ppm. (From Gehri et al.[20])

where y_{out} and y_{in} refer to the mole fraction of solute in the gas leaving and entering the tray at a specific point on the tray, and y_{out}^* is the mole fraction of solute gas in equilibrium with the liquid on the tray at the specific point.

Numerous attempts have been made to develop useful correlations for predicting the efficiency of tray towers. The most comprehensive study was carried out under sponsorship of the American Institute of Chemical Engineers (AIChE) and resulted in the publication of the AIChE bubble tray design manual.[21] The AIChE prediction method, which is aimed primarily at bubble-cap trays, involves first estimating the number of gas-phase and liquid-phase transfer units, N_G and N_L, for the specific conditions on the tray, combining these to obtain the number of overall gas transfer units N_{OG}, and predicting the point efficiency by the relationship

$$E_{OG} = 1 - e^{-N_{OG}} \qquad (6.4\text{-}20)$$

A value for the Murphree vapor-phase efficiency E_{MV} is then calculated for the overall tray by adjusting the point efficiency E_{OG} to take into account the effects of liquid mixing and entrainment. A summary of the AIChE design procedure for bubble-cap trays and an example of its application are given by King.[22]

A very approximate estimate of the overall efficiency of bubble-cap columns used for absorption can be obtained from Fig. 6.4-13. This simple correlation was proposed by O'Connell.[23]

Properly designed perforated trays (also called sieve trays) are generally somewhat more efficient than bubble-cap trays. A comprehensive review of sieve tray performance is given by Zuiderweg.[24] He proposes a fundamental but highly simplified approach for predicting E_{MV} for perforated trays which includes the following steps:

1. Estimate the gas-phase mass transfer coefficient by the equation

$$k_G = \frac{0.065}{\rho_G^2} \qquad (6.4\text{-}21)$$

where k_G = gas-phase mass transfer coefficient (m/s)
ρ_G = the gas density (kg/m^3)

2. Estimate the liquid-phase mass transfer coefficient by the equation

$$k_L = \frac{2.6 \times 10^{-5}}{\mu_L^{0.25}} \qquad (6.4\text{-}22)$$

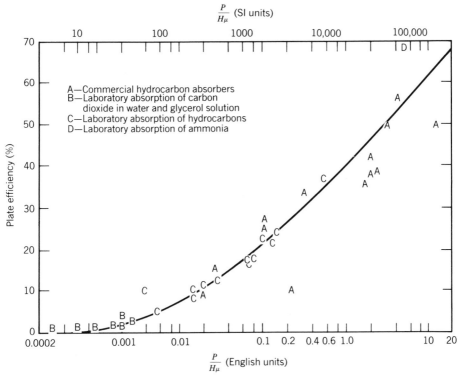

FIGURE 6.4-13 Correlation of overall tray efficiency for absorption columns. H = Henry's Law coefficient, atm/(lb-mole/ft^3) or (N/m^2)/(mol/m^3), P = total pressure (atm or N/m^2), and μ = viscosity (cP or N·s/m^2. (From O'Connell,[23] reprinted with permission from *Trans. AIChE*.)

where k_L = liquid-phase mass transfer coefficient (m/s)
        μ_L = viscosity of the liquid (N·s/m^2)

3. Calculate the overall gas mass transfer coefficient

$$K_G = \frac{k_G k_L}{k_L + mk_G} \tag{6.4-23}$$

where m is the distribution coefficient,

$$m = K\frac{\rho_G M_L}{\rho_L M_G} \tag{6.4-24}$$

and K = equilibrium constant expressed in mole fraction
    ρ = density (kg/m^3)
    M = molecular weight (g/mol)

4. Determine the point efficiency

$$E_{OG} = 1 - e^{-(aK_G)/u_G} \tag{6.4-25}$$

where a = effective interfacial area per unit of bubbling area
    u_G = vapor velocity in the bubbling area (m/s)

The value of a is a function of the flow regime on the tray. The tray is said to operate in the spray regime when the liquid is dispersed almost completely into small droplets by the action of the gas jets. This occurs at high gas velocities and low liquid loads. In the spray regime

$$a = \frac{40}{F^{0.3}} \left(\frac{F_{ba}^2 h_L FP}{\sigma}\right)^{0.37} \tag{6.4-26}$$

The tray operates in the "emulsion flow regime" when transport across the tray and over the weir is mainly by continuous flow of liquid containing emulsified vapor. This regime exists at high liquid loads and relatively low gas rates and is probably the dominant regime for high-pressure gas absorbers. For the emulsion flow regime and for mixed regimes

$$a = \frac{43}{F^{0.3}} \left(\frac{F_{ba}^2 h_L FP}{\sigma}\right)^{0.53} \tag{6.4-27}$$

The operating regime can be determined by the parameter FP/bh_L. When this parameter is below about 3–4, the tray will operate in the spray or mixed regime. When it is above this range, the tray will operate in the emulsion regime. In the above equations,

F = fraction hole area per unit bubbling area
F_{ba} = vapor F factor on bubbling area (m/s) $(kg/m^3)^{0.5}$
F_{ba} = $u_G \rho_G^{0.5}$
FP = flow parameter = $(u_L/u_G)(\rho_L/\rho_G)^{0.5}$ $\qquad$ (6.4-28)
σ = surface tension (N/m)
h_L = liquid holdup (m) = $0.6 H_w^{0.5} p^{0.25} b^{-0.25}(FP)^{0.25}$ $\qquad$ (6.4-29)
H_w = weir height (m)
p = pitch of holes in perforated area (m)
b = weir length per unit bubbling area (m^{-1})

5. Determine the Murphree vapor-phase efficiency

$$E_{MV} = \frac{1}{S}\left(e^{SE_{OG}} - 1\right) \tag{6.4-30}$$

where S is the stripping factor, mG_M/L_M.

In the development of the above series of equations, Zuiderweg has used the work of many prior investigators and relied heavily on the data recently released by Fractionation Research, Inc. (FRI), as reported by Sakata and Yanagi[25,26] on the performance of two types of commercial perforated tray. Zuiderweg also presents correlations that define the flow regime transitions, pressure drop, entrainment, column capacity, and other operating parameters of sieve trays.

6.4-4 Absorption with Chemical Reactions

Many commercial absorption processes involve a chemical reaction between the solute and the solvent. The occurrence of a reaction affects not only gas–liquid equilibrium relationships but also the rate of mass transfer. Since the reaction occurs in the solvent, only the liquid mass transfer rate is affected. Normally, the effect is an increase in the liquid mass transfer coefficient k_L. The development of correlations for predicting the degree of enhancement for various types of chemical reaction and system configuration has been the subject of numerous studies. Comprehensive discussions of the theory of mass transfer with chemical reaction are presented in recent books by Astarita,[27] Danckwerts,[28] and Astarita et al.[3]

In the presence of chemical reaction, the mass transfer coefficient k_L is affected by reaction kinetics as well as by all factors affecting physical mass transfer in the area of the interface. As a result of the complexities of the combined effects, little progress has been made toward the prediction of k_L from first principles. An approach that has proved successful is based on the use of the ratio of the actual mass transfer coefficient k_L to the mass transfer coefficient that would be experienced under the same circumstances if no reaction occurred, designated k_L°. This ratio is called the enhancement factor E and is defined as follows:

$$E = \frac{k_L}{k_L^\circ} \tag{0.4-31}$$

When the chemical reaction is very slow, the absorbed component A will diffuse into the bulk of the liquid before reaction occurs. As a result of the reaction, the concentration of A in the bulk of the liquid is kept low, and the driving force for transfer from the interface remains higher than it would be in the absence of chemical reaction. In this case, called "the slow reaction regime," $k_L = k_L^\circ$ and $E = 1$. There is no enhancement effect, and the only effect of the chemical reaction is its effect on the driving force.

At the other extreme, when the chemical reaction is extremely fast, molecules of the solute react with molecules of the reactant whenever both are present at the same point in the liquid. Chemical equilibrium therefore exists everywhere in the liquid phase and further increases in the reaction rate would have no effect. In this situation, designated the "instantaneous reaction regime," the rate of mass transfer is independent of chemical kinetics and dependent only on factors affecting the physical transfer of reactants and reaction products. The enhancement factor can be very large in this regime, particularly when the concentration of reactant in the liquid is high. According to Astarita et al.,[3] the values of E for the instantaneous reaction regime can be on the order of 10^2–10^4.

The broad range of conditions between the slow reaction regime in which $E = 1$ and the instantaneous reaction regime in which E is very large but independent of reaction rate is classified generally as the "fast reaction regime." In this regime, the mass transfer coefficient k_L is a function of the reaction rate. Both k_L and k_L^o are affected also by fluid mechanics, but, fortunately, their ratio E has been found to be relatively independent of these factors.

The enhancement of mass transfer due to chemical reaction depends on the order of the reaction as well as its rate. Order is defined as the sum of all the exponents to which the concentrations in the rate equation are raised. In elementary reactions, this number is equal to the number of molecules involved in the reaction; however, this is only true if the correct reaction path has been assumed. Danckwerts[28] presents a review of many cases of importance in gas absorption operations. He compares the results of using the film model and the Higbie and Danckwerts surface-renewal models and concludes that, in general, the predictions based on the three models are quite similar. Mass transfer rate equations for a few of the cases encountered in a gas absorption operation are summarized in the following paragraphs, which are based primarily on discussions presented by Danckwerts.[28]

First-Order Irreversible Reaction

$$A \xrightarrow{k_1} P$$

The following expression for this case was first derived by Hatta[29] based on the film model:

$$N_A = k_L^o \left(c_{Ai} - \frac{c_{Ab}}{\cosh \sqrt{M}} \right) \frac{\sqrt{M}}{\tanh \sqrt{M}} \tag{6.4-32}$$

where N_A = rate of absorption per unit area (mol/cm$^2 \cdot$s)
k_1 = first-order reaction rate constant (s^{-1})
c_{Ai} = concentration of A at interface (mol/cm^3)
c_{Ab} = concentration of A in the bulk liquid (mol/cm^3)
M = chemical reaction parameter = $D_A k_1/(k_L^o)^2$ (for first-order reactions)
D_A = diffusivity of A in the liquid (cm^2/s)

When the factor $\sqrt{M}$ is much greater than 1, $\tanh \sqrt{M}$ is equal to 1 and the solute reacts before reaching the bulk of the liquid. In this case, $C_{Ab} = 0$ and Eq. (6.4-32) becomes

$$N_A = c_{Ai}\sqrt{D_A k_1} \tag{6.4-33}$$

and, since $N_A = k_L^o c_{Ai}$ if no reaction occurs,

$$E = \frac{\sqrt{D_A k_1}}{k_L^o} = \sqrt{M} \tag{6.4-34}$$

Instantaneous Irreversible Reaction

$$A + bB \rightarrow cP$$

$$N_A = k_L^o\, c_{Ai} \left(1 + \frac{D_B c_{Bb}}{b D_A c_{Ai}} \right) \tag{6.4-35}$$

$$E_i = 1 + \frac{D_B c_{Bb}}{b D_A c_{Ai}} \tag{6.4-36}$$

where c_{Bb} = concentration of B in the bulk liquid (mol/cm^3)
D_B = diffusivity of B in the liquid (cm^2/s)
E_i = enhancement factor for instantaneous reaction
b = number of moles of B reacting with 1 mol of A

For the case when the concentration of A at the interface, c_{Ai}, is much smaller than the concentration of B in the bulk of the liquid, c_{Bb}, Eq. (6.4-35) reduces to

$$N_A = k_L^\circ \frac{D_B c_{Bb}}{b D_A} \tag{6.4-37}$$

In this case, the rate of absorption is independent of c_{Ai} and is controlled by the rate at which B can diffuse to the surface.

IRREVERSIBLE SECOND-ORDER REACTION

$$A + bB \xrightarrow{k_2} cP$$

The pioneering work on this case was done by van Krevelen and Hoftijzer in 1948 based on the film model.[30] They computed an approximate solution to the mass transfer chemical reaction equations for the case of $c_{Ab} = 0$ and showed that the results could be represented with an accuracy of about 10% by the equation

$$E = \frac{[M (E_i - E)/(E_i - 1)]^{1/2}}{\tanh [M (E_i - E)/(E_i - 1)]^{1/2}} \tag{6.4-38}$$

where M, the chemical reaction parameter, is

$$M = \frac{D_A k_2 c_{Bb}}{(k_L^\circ)^2} \quad \text{(for second-order reactions)}$$

and

$$E_i = \left(1 + \frac{D_B c_{Bb}}{b D_A c_{Ai}}\right)$$

$$E = \frac{N_A}{k_L^\circ c_{Ai}}$$

A graphical representation of these equations is given in Fig. 6.4-14. van Krevelen and Hoftijzer originally developed their correlation only for irreversible second-order reactions (first-order in each reactant) and for equal diffusivities of the two reactants. Danckwerts[28] pointed out that the results also are applicable to the case where D_A is not equal to D_B. Decoursey[31] developed an approximate solution for absorption with irreversible second-order reaction based on the Danckwerts surface-renewal model. The resulting expression, which is somewhat easier to use than the van Krevelen–Hoftijzer approach, is

$$E = \frac{M}{2 (E_i - 1)} + \left(\frac{M^2}{4 (E_i - 1)^2} + \frac{E_i M}{(E_i - 1)} + 1\right)^{1/2} \tag{6.4-39}$$

where E_i = enhancement factor for instantaneous irreversible reaction
$M = D_A k_2 c_{Bb}/(k_L^\circ)^2$

Matheron and Sandall[32] solved the equations describing gas absorption accompanied by a second-order chemical reaction numerically and found the results to be in good agreement with the approximate analytical formula of Decoursey. They also calculated a range of values for E_i and presented a table showing the value of this parameter for various ratios of c_{Bb}/c_{Ai} and D_B/D_A. They proposed that, for large values of E_i, E_i can be estimated by the equation

$$E_i = \sqrt{\frac{D_A}{D_B}} + \frac{c_{Bb}}{b c_{Ai}} \sqrt{\frac{D_B}{D_A}} \tag{6.4-40}$$

When $D_A = D_B$, E_i has the same form for the film and surface-renewal models [i.e., Eq. (6.4-36) with $D_A = D_B$; $E_i = 1 + c_{Bb}/b c_{Ai}$].

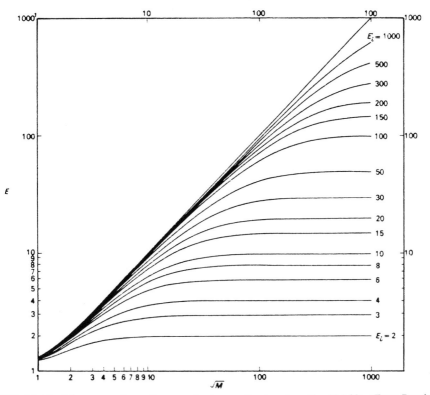

FIGURE 6.4-14 Enhancement factor for second-order reaction based on Eq. (6.4-38). (From Danckwerts,[28] copyright 1970. Reprinted with permission of McGraw-Hill Book Company, New York.)

REVERSIBLE FIRST-ORDER REACTION

$$A \underset{k_{-1}}{\overset{k_1}{\rightleftharpoons}} P$$

The forward rate constant for the reaction is k_1, and the rate constant for the reverse reaction is k_{-1}. For the case of equal diffusivities of A and P, the rate of mass transfer is given by the following equation proposed by Danckwerts and Kennedy:[33]

$$N_A = \frac{k_L^\circ (c_{Ai} - c_{Ab})(1 + K)}{1 + \{K \tanh [D_A k_1(1 + K)/(k_L^\circ)^2 K]^{1/2}/[D_A k_1(1 + K)/(k_L^\circ)^2 K]^{1/2}\}} \qquad (6.4\text{-}41)$$

where

$$K = (p/c_A)_{\text{equil}} = k_1/k_{-1}$$

As $K \to \infty$, $c_{Ab} \to 0$ and Eq. (6.4-41) reduces to

$$N_A = k_L^\circ c_{Ai} \frac{[D_A k_1/(k_L^\circ)^2]^{1/2}}{\tanh [D_A k_1/(k_L^\circ)^2]^{1/2}} \qquad (6.4\text{-}42)$$

which is the same as the expression for a first-order irreversible reaction [Eq. (6.4-32) with $c_{Ab} = 0$].

INSTANTANEOUS REVERSIBLE REACTION

$$A + bB \rightleftharpoons cP$$

A method of solution for several types of reaction which fall into this category has been developed by Olander.[34] Danckwerts[28] has generalized his conclusions to yield the equation

$$N_A = k_L^o \left[\left(c_{Ai} + \frac{D_P c_{Pi}}{D_A c} \right) - \left(c_{Ab} + \frac{D_P c_{Pb}}{D_A c} \right) \right]$$ (6.4-43)

When the diffusivities of A and P are equal, the equation simplifies to

$$N_A = k_L^o \left[\left(c_{Ai} + \frac{c_{Pi}}{c} \right) - \left(c_{Ab} + \frac{c_{Pb}}{c} \right) \right]$$ (6.4-44)

or

$$N_A = k_L^o (c_{Ai}^* - c_{Ab}^*)$$ (6.4-45)

where c_{Ai}^* = the total concentration of A at the interface, including both reacted and unreacted forms

c_{Ab}^* = the total concentration of A in the bulk, including both reacted and unreacted forms

c = the number of molecules of P formed in the chemical reaction equation

In the above equations, c_{Ai}, c_{Ab}, and c_{Pb} are known, but c_{Pi} is not. According to Danckwerts,[28] c_{Pi} is, in fact, the concentration of P which would be obtained by saturating a liquid of the bulk concentration with the gas, the resulting concentration of unreacted gas being c_{Ai}. The driving force $(c_{Ai}^* - c_{Ab}^*)$ is the total amount of A in both reacted and unreacted forms which can be taken up by the liquid having the initial bulk concentration of c_{Ab}, c_{Bb}, and c_{Pb} if it is saturated with A at a sufficient pressure to give a concentration c_{Ai} of unreacted A in equilibrium with the products of reaction.

For the limiting case of a reversible reaction where the reaction rate approaches infinity and the diffusivities of all species are assumed to be equal,

$$E_i = \frac{c_{Ai}^* - c_{Ab}^*}{c_{Ai} - c_{Ab}}$$ (6.4-46)

APPLICATIONS

The equations covering mass transfer with chemical reaction are not yet widely used in the design of absorbers and strippers. More commonly, designs are based on experimental data obtained with the same chemical system and similar equipment. However, examples have been worked out for a number of commercially important cases, and these are described in the recent literature.

A methodology for predicting the performance of an isothermal packed tower used for chemical absorption has been developed by Joshi et al.[35] The study was based on the absorption of CO_2 in a hot aqueous solution of potassium carbonate; however, the general approach is applicable to other chemical absorption processes.

For the case of mass transfer with chemical reaction, the local flux of component A per unit volume of packing becomes

$$N_A a = k_G a P (y - y_i) = \frac{E k_L^o a P}{H} (y_i - y^*)$$ (6.4-47)

where N_A = quantity of component A transferred per unit time per unit area

a = interfacial area per unit volume of packing

k_G = individual gas-phase mass transfer coefficient

P = total pressure

y = mole fraction of A in bulk of the gas

y_i = mole fraction of A in the gas at the interface

y^* = mole fraction of A in the gas corresponding to equilibrium with the bulk liquid composition

k_L^o = individual liquid-phase mass transfer coefficient in the absence of chemical reaction

H = Henry's Law constant

E = mass transfer enhancement factor = k_L/k_L^o

A local overall gas-phase coefficient for mass transfer with chemical reaction may be defined by the equation

$$N_A a = K_G a P (y - y^*)$$ (6.4-48)

From Eqs. (6.4-47) and (6.4-48) and the definition of the overall physical mass transfer coefficient, it can be shown that

$$K_G a = \frac{1 + B}{1 + B/E} K_G^\circ a \qquad (6.4\text{-}49)$$

where $B = k_G H/k_L^\circ$

K_G° = overall physical mass transfer coefficient

Combining the above equations with a mass balance for a differential section of the column and integrating, we obtain

$$Z = \frac{G'}{K_G^\circ a P} \int_{y_1}^{y_2} \frac{dy}{(1 - y)^2 (y - y^*) (1 + B)/(1 + B/E)} \qquad (6.4\text{-}50)$$

Equation (6.4-50) is a general design equation for calculating packing height. The equation bears a superficial resemblance to Eq. (6.3-9) for physical absorption. $G'/K_G^\circ a P$ is the height of a transfer unit for physical absorption and the integral can be considered to be the number of transfer units. However, Eq. (6.4-50) differs from the physical absorption case in that the value of the integral is a function of the individual mass transfer coefficients k_G and k_L and of the enhancement factor E. As a result, the use of a logarithmic mean driving force as in Eq. (6.3-20) is not even approximately valid for absorption with chemical reaction.

It should be noted that much of the absorption and stripping data available in the literature for systems involving chemical reaction have been correlated using logarithmic mean driving forces over the length of the column. The result is that the reported values of $K_G a$ are observed to vary widely with partial pressure of the reactant gas. This correlation technique has the advantage of simplicity, but great care must be used in applying it to conditions outside the range of experimental data because it is theoretically incorrect and may lead to significant errors.

The derivation of the above design equation [Eq. (6.4-50)] is presented in more detail by Joshi et al.[35] and Astarita et al.[3] Joshi et al. point out that its use to predict the performance of a packed tower has the following requirements:

1. Overall vapor–liquid equilibrium data, y^* versus x.
2. Physical solubility of the solute in the liquid, H.
3. Mass transfer enhancement factor E.
4. Individual mass transfer coefficients k_G and k_L° (and also the interfacial area per unit volume of packing, a).

They describe the development of the required information for the system CO_2–K_2CO_3–H_2O and its use to predict absorber performance. They conclude that an equation of the form proposed by Decoursey[31] [Eq. (6.4-39)] for a bimolecular irreversible reaction provides a satisfactory approximation of rate enhancement for this system. Pilot plant data originally published by Benson et al.[36] are compared to values predicted by the proposed methodology. The correlations of Onda et al.[5] [Eqs. (6.4-3) and (6.4-4)] are used to predict values for $k_L^\circ a$ for the Benson–Field–Jimeson pilot plant.[36] The final results show good agreement, indicating that the approach has promise for the design of commercial absorbers.

The simultaneous absorption of two gases that react with the solvent at different rates has been studied by Ouwerkerk.[37] The specific system which he selected for analysis was the selective absorption of H_2S in the presence of CO_2 into amine solutions. This operation is a feature of several commercially important gas purification processes. Bench scale experiments were conducted to collect the necessary physico-chemical data. An absorption rate equation was developed for H_2S based on the assumption of instantaneous reaction. For CO_2 it was found that the rate of absorption into diisopropanolamine (DIPA) solution at low CO_2 partial pressures can best be correlated on the basis of a fast pseudo-first-order reaction. A computer program was developed which took into account the competition between H_2S and CO_2 when absorbed simultaneously, and the computer predictions were verified by experiments in a pilot scale absorber. Finally, the methodology was employed successfully to design a large commercial plant absorber.

Stripping in the presence of chemical reaction has only recently been considered in detail. The most comprehensive treatment appears in the papers by Astarita and Savage[38] and by Savage et al.[39] In the first paper, it is concluded that if the reaction rate is first order with regard to the solute, the theory of chemical absorption can also be used for chemical desorption. Specifically, the absorption theory is directly applicable at reaction rates up to the fast reaction regime. At higher rates, the mathematical procedure is the same for absorption and stripping; however, the results are different because of different ranges in the parameters involved. In the case of desorption, it is always necessary to consider the reversibility of the reactions.

In the second paper, absorption and desorption rate data for the system CO_2–hot potassium carbonate solution are interpreted on the basis of the film theory model described in the first paper. The agreement is very satisfactory. Both absorption and desorption were found to take place under fast reaction conditions with occasional occurrence in the fast-to-instantaneous transition region at low values of k_L^o.

A comprehensive study of the stripping of carbon dioxide from monoethanolamine solutions in a packed column is described by Weiland et al.[40] These investigators developed a design approach that uses only fundamental physicochemical data and tested the approach against 173 experiments on the mass transfer performance of a pilot scale stripping column. Predicted mass transfer coefficients agreed with observed values within $\pm 25\%$.

The design approach is based on the film model in combination with an enhancement factor to account for the effect of liquid-phase chemical reaction. Because of the high temperature involved in stripping, it was assumed that reaction rates would be high enough to permit the equations for a reversible instantaneous reaction to be used. For column design, the approach involves selecting a small incremental packing height; calculating mass transfer, gas and liquid flow rates, compositions, and heat balances in this section; using the results to evaluate the next lower section; and continuing section-to-section calculations until conditions corresponding to the desired column bottom conditions are reached. The sum of the incremental heights represents the required total column height. The results indicate that stripping columns can be designed from first principles and also provide useful insight with regard to the effect of liquid rate, stripping steam rate, operating pressure, CO_2 to amine ratio, and amine concentration on the stripping of CO_2 from monoethanolamine solutions in a packed column.

NOTATION

A	absorption factor
a	effective interfacial area per packed volume (m^2/m^3 or ft^2/ft^3) (or per unit of bubbling area of sieve plate)
a_p	surface area of dry packing (m^2/m^3 or ft^2/ft^3)
a_W	interfacial area of wetted packing (m^2/m^3 or ft^2/ft^3)
b	weir length per unit bubbling area of sieve tray, (m/m^2 or ft/ft^2)
C_p	specific heat of gas (kJ/mol·K or Btu/lb-mole·°F)
C_q	specific heat of liquid, same units as C_p
c	concentration of solute in liquid, various units (typically mol/m^3 or lb-mole/ft^3)
c_i	concentration of solute in liquid at interface, same units as c
D	diffusion coefficient (m^2/s or ft^2/h)
d_c	diameter of column (m or ft)
d_d	diameter of drops (m or ft)
d_p	diameter of packing (m or ft)
E	enhancement factor for absorption with chemical reaction
E_i	enhancement factor for instantaneous reaction
E_{MV}	Murphree vapor tray efficiency
E_O	overall column efficiency
F	fraction hole area in bubbling area of sieve plate
G	mass flow rate of gas (kg/h·m^2 or lb_m/h·ft^2
G_M	gas flow rate (mol/h·m^2 or lb-mole/h·ft^2
G_M'	flow rate of inert gas, same units as G_M
H	Henry's Law constant (typically kPa or atm)
H_G	height of a gas-phase transfer unit (m or ft)
H_L	height of a liquid-phase transfer unit (m or ft)
H_{OG}	height of an overall gas-phase transfer unit (m or ft)
H_{OL}	height of an overall liquid-phase transfer unit (m or ft)
H_{GQ}	height of a gas-phase heat transfer unit (m or ft)
H_{OS}	integral heat of solution for the solute (kJ/mol or Btu/lb-mole)
H_v	latent heat of pure solvent, same units as H_{OS}
H_W	height of weir on sieve tray (m or ft)
HETP	height equivalent to a theoretical plate (m or ft)
h_L	liquid holdup on tray (m or ft)
h_T	total height of packing (von Stockar–Wilke correlation)

K	equilibrium constant, y/x at equilibrium
K_G	overall gas-phase mass transfer coefficient (m/s, mol/N·s, or lb-mole/h·ft²·atm)
K_L	overall liquid-phase mass transfer coefficient (m³/s·m² or ft/h)
k_G	gas-phase mass transfer coefficient (m/s, mol/N·s, or lb-mole/h·ft²·atm)
k_L	liquid-phase mass transfer coefficient (m³/s·m² or ft/h)
k_L^o	liquid-phase mass transfer coefficient experienced if no chemical reaction occurred, same units as k_L
k_1	reaction rate for first order reaction (s⁻¹)
k_2	reaction rate for second order reaction (m³/mol·s or ft³/lb-mole·h)
L	mass flow rate of liquid (kg/h·m² or lb_m/h·ft²)
L_M	molar flow rate of liquid (mol/h·m² or lb-mole/h·ft²)
L_M'	flow rate of solute-free solvent, same units as L_M
m	slope of equilibrium line
M	molecular weight
M	chemical reaction parameter
N	number of theoretical stages in a column
N_A	flux of solute A through interface (mol/s·m² or lb-mole/h·ft²)
N_T	number of actual trays in a column
N_G	number of gas-phase transfer units
N_{OG}	number of overall gas-phase transfer units
N_L	number of liquid-phase transfer units
N_{OL}	number of overall liquid-phase transfer units
P	pressure (kPa, atm, or psi)
p	pitch of holes in sieve tray (m or ft)
P_A	partial pressure of A (kPa, atm, or psi)
p_i	partial pressure of solute at the interface (kPa atm, or psi)
S	stripping factor
T	temperature (K or °R)
u	vapor velocity in bubbling area of sieve tray
x	mole fraction in liquid
X	mole ratio in liquid $= x_A/(1 - x_A)$
y	mole fraction in gas
Y	mole ratio in gas $= y_A/(1 - y_A)$ (also, moles per mole of rich gas in Kremser equations)
Z	packing height (m or ft)
μ	viscosity (Pa·s, cP, or lb_m·ft·s)
ρ	density (g/cm³ or lb_m/ft³)
σ	surface tension (N/m, dyn/cm, or lb_m/s²)

Subscripts

av	average
A	component A (solute)
B	component B (usually the solvent)
G	gas
L	liquid
i	at gas–liquid interface (or instantaneous)
M	molar (or logarithmic mean)
N	bottom tray of column with N trays
n	from nth tray
p	packing
W	water
0	liquid feed to tray columns
1	top tray of tray column (or bottom of column for overall heat and material balances)
2	second tray from top in tray column (or top of column for overall heat and material balances)

REFERENCES

Section 6.1

6.1-1   R. N. Shreve and J. A. Brink, *Chemical Process Industries*, 4th ed., McGraw-Hill, New York, 1977.

6.1-2   A. L. Kohl and F. C. Riesenfeld, *Gas Purification*, 4th ed., Gulf Publishing, Houston, TX, 1985.

6.1-3   Glitsch, Inc., Bulletin H-4900, 3rd ed., *Ballast Tray Design Manual*, Dallas, TX, 1980.

6.1-4   Koch Engineering Company, Inc., Bulletin KT-5, *Koch Flexitrays*, Wichita, KS, 1968.

6.1-5   E. D. Oliver, *Diffusional Separation Processes*, Wiley, New York, 1966.

6.1-6   The Norton Co., Bulletin DC-11, *Design Information for Packed Towers*, Akron, OH, 1977.

6.1-7   Koch Engineering Company, Inc., Bulletin KS-1, *The Koch Sulzer Packing*, Wichita, KS, 1978.

6.1-8   Glitsch, Inc., Bulletin 520A, *Goodloe*, Dallas, TX, 1981.

6.1-9   Y. T. Shah, B. G. Kelkar, S. P. Godbole, and W. D. Deckever, *AIChE J.*, **28**, 353 (1982).

6.1-10   *International Critical Tables*, McGraw-Hill, New York, Vol. III, 1928; Vol IV, 1929; Vol. V, 1929.

6.1-11   A. X. Schmidt and H. L. List, *Material and Energy Balances*, Prentice-Hall, Englewood Cliffs, NJ, 1972.

6.1-12   J. W. Sweny and J. P. Valentine, *Chem. Eng.*, **77**, 54 (Sept. 7, 1970).

6.1-13   R. H. Perry and D. W. Green (Eds.), *Perry's Chemical Engineer's Handbook*, 6th ed., McGraw-Hill, New York, 1984.

6.1-14   J. Hildebrand and R. L. Scott, *Regular Solutions*, Prentice-Hall, Englewood Cliffs, NJ, 1962.

6.1-15   Gas Processors Suppliers' Association, *Engineering Data Book*, 9th ed., Tulsa, OK, 1972.

6.1-16   W. C. Edmister, *Applied Hydrocarbon Thermodynamics*, Gulf Publishing, Houston, TX, 1961.

6.1-17   T. S. Zawaki, D. A. Duncan, and R. A. Macriss, *Hydrocarbon Processing*, **60**(4), 143 (1981).

6.1-18   O. R. Osborne and P. L. Markovic, *Chem. Eng.*, **76**, 105 (Aug. 25, 1969).

6.1-19   K. C. Chao and J. D. Seader, *AIChE J.*, **7**(4), 598 (1961).

6.1-20   J. M. Prausnitz and P. L. Chueh, *Computer Calculations for High Pressure Vapor–Liquid Equilibrium*, Prentice-Hall, Englewood Cliffs, NJ, 1968.

6.1-21   P. L. Kent and B. Eisenberg, *Hydrocarbon Processing*, **55**(2), 87 (1976).

6.1-22   P. V. Danckwerts and K. M. McNeil, *Trans. Inst. Chem. Eng.*, **45**, T32–T38 (1967).

6.1-23   G. Astarita, D. W. Savage, and A. Bisio, *Gas Treating with Chemical Solvents*, Wiley, New York, 1983.

6.1-24   R. C. Reid, J. M. Prausnitz, and T. K. Sherwood, *The Properties of Gases and Liquids*, 3rd ed., McGraw-Hill, New York, 1977.

6.1-25   A. Kremser, *Natl. Petrol News*, **22**(21), 48 (1930).

6.1-26   M. Souders and G. G. Brown, *Ind. Eng. Chem.*, **24**, 519 (1932); *Oil & Gas J.*, **31**(5), 34 (1932).

Section 6.2

6.2-1   A. Kremser, *Natl. Petrol News*, **22**(21), 48 (1930).

6.2-2   M. Souders and G. G. Brown, *Ind. Eng. Chem.*, **24**, 519 (1931); *Oil & Gas J.*, **31**(5), 34 (1932).

6.2-3   R. E. Treybal, *Mass Transfer Operations*, 3rd ed., McGraw-Hill, New York, 1980.

6.2-4   S. Diab and R. N. Maddox, *Chem. Eng.*, **89**, 42 (Dec. 27, 1982).

6.2-5   Gas Processors Suppliers' Association, *Engineering Data Book*, 9th ed., Tulsa, OK, 1972.

6.2-6   W. C. Edmister, *Ind. Eng. Chem.*, **35**, 837 (1943).

6.2-7   G. Horton and W. B. Franklin, *Ind. Eng. Chem.*, **32**, 1384 (1940).

6.2-8   J. M. Douglas, *Chem. Eng.*, **86**, 135 (Aug. 13, 1979).

6.2-9   H. X. Nguyen, *Chem. Eng.*, **86**, 113 (Apr. 9, 1979).

6.2-10   A. D. Sujata, *Hydrocarbon Processing*, **40**(12), 137 (1961).

6.2-11   J. H. Erbar and R. H. Maddox, *Computer Programs for Natural Gas Processing*, University of Oklahoma Press, Norman, 1982.

6.2-12   ChemShare Corporation, 1900 Lummus Tower, Houston, TX, private communication, 1983.

Section 6.3

6.3-1   W. G. Whitman, *Chem. Met. Eng.*, **24**, 147 (1923).

6.3-2   R. Higbie, *Trans. Am. Inst. Chem. Eng.*, **31**, 365 (1935).

6.3-3   P. V. Danckwerts, *Ind. Eng. Chem.*, **43**, 1460 (1951).

6.3-4   H. L. Toor and J. M. Marchello, *AIChE J.*, **4**, 97 (1958).

6.3-5   C. J. King, *Separation Processes*, McGraw-Hill, New York, 1971.

6.3-6   A. P. Colburn, *Trans. Am. Inst. Chem. Eng.*, **35**, 211 (1939).

6.3-7   R. E. Treybal, *Mass Transfer Operations*, McGraw-Hill, New York, 1968.

6.3-8   T. C. Baker, *Ind. Eng. Chem.*, **27**, 977 (1935).

6.3-9   T. K. Sherwood, R. L. Pigford, and C. R. Wilke, *Mass Transfer*, McGraw-Hill, New York, 1975.

6.3-10 C. R. Wilke and U. von Stockar, Absorption, in *Encyclopedia of Chemical Technology*, Vol. 1, 3rd eds., Kirk-Othmer (Eds.), Wiley, New York, 1978.

6.3-11 J. H. Wiegand, *Trans. Am. Inst. Chem. Eng.*, **36**, 679 (1940).

6.3-12 U. von Stockar and C. R. Wilke, *Ind. Eng. Chem. Fundam.*, **16**, 88 (1977).

6.3-13 H. M. Feintuch and R. E. Treybal, *Ind. Eng. Chem. Process Des. Dev.*, **17**, 505 (1978).

6.3-14 R. E. Treybal, *Ind. Eng. Chem.*, **61**(7), 36 (1969).

6.3-15 H. M. Feintuch, Ph.D. thesis, New York University, New York, 1973.

6.3-16 U. von Stockar and C. R. Wilke, *Ind. Eng. Chem. Fundam.*, **16**, 94 (1977).

Section 6.4

6.4-1   R. H. Perry and D. W. Green (Eds.), *Perry's Chemical Engineers' Handbook*, 6th ed., McGraw-Hill, New York, 1984.

6.4-2   A. L. Kohl and F. C. Riesenfeld, *Gas Purification*, 4th ed., Gulf Publishing, Houston, TX, 1985.

6.4-3   G. Astarita, D. W. Savage, and A. Basio, *Gas Treating with Chemical Solvents*, Wiley, New York, 1983.

6.4-4   D. Cornell, W. G. Knapp, and J. R. Fair, *Chem. Eng. Progr.*, **56**, 68 (1960).

6.4-5   K. Onda, H. Takeuchi, and Y. Okumoto, *J. Chem. Eng. Jpn.*, **1**, 56 (1968).

6.4-6   W. L. Bolles and J. R. Fair, *Chem. Eng.*, **89**, 109 (July 12, 1982).

6.4-7   J. L. Bravo and J. R. Fair, *Ind. Eng. Chem. Process Des. Dev.*, **21**, 162 (1982).

6.4-8   J. S. Eckert, *Chem. Eng. Progr.*, **66**, 39 (1970).

6.4-9   W. Meier, W. D. Stoecker, and B. Weinstein, *Chem. Eng. Progr.*, **73**, 71 (1977).

6.4-10 W. Meier, R. Hunkeler, and W. D. Stoecker, ''Performance of New Regular Tower Packing/ Mellapak,'' presented at 3rd International Symposium on Distillation, London, England, April 3–6, 1979.

6.4-11 A. O. Hansen and R. J. Danos, *Chem. Eng. Progr.*, **78**, 40 (1982).

6.4-12 K. E. Lunde, *Ind. Eng. Chem.*, **50**, 243 (1953).

6.4-13 T. Hobbler, *Mass Transfer and Absorbers*, Pergamon, New York, 1966.

6.4-14 T. K. Sherwood and R. L. Pigford, *Absorption and Extraction*, McGraw-Hill, New York, 1952.

6.4-15 S. Calvert, J. Goldschmid, D. Leith, and D. Mehta, *Wet Scrubber System Study*, Vol. 1, *Scrubber Handbook*, prepared for Environmental Protection Agency, NTIS No. PB-213-106, 1972.

6.4-16 W. H. Gauvin and S. Katta, *AIChE J.*, **22**, 713 (1976).

6.4-17 S. Katta and W. H. Gauvin, *AIChE J.*, **21**, 143 (1975).

6.4-18 K. Masters, *Spray Drying, An Introduction to Principles, Operating Practices, and Applications*, 2nd ed., Halsted, New York, 1976.

6.4-19 J. L. Getler, H. L. Shelton, and D. A. Furlong, *J. Air Pollut. Assoc.*, **29**, 1270 (1979).

6.4-20 D. C. Gehri, R. L. Adams, and J. H. Phelan, *U.S. Patent 4,197,278*, April 8, 1980.

6.4-21 American Institute of Chemical Engineers, *Bubble Tray Design Manual*, AIChE, New York, 1958.

6.4-22 C. J. King, *Separation Processes*, McGraw-Hill, New York, 1971.

6.4-23 H. E. O'Connell, *Trans. Am. Inst. Chem. Eng.*, **42**, 741 (1946).

6.4-24 F. J. Zuiderweg, *Chem. Eng. Sci.*, Review Article No. 9, **37**, 1441 (1982).

6.4-25 M. Sakata and T. Yanagi, *AIChE Symp. Ser.*, **56**, 121 (1979).

6.4-26 T. Yanagi and M. Sakata, 90th National AIChE Meeting, Symp. 44, Houston, TX, April 1981.

6.4-27 G. Astarita, *Mass Transfer with Chemical Reaction*, Elsevier, Amsterdam, 1967.

6.4-28 P. V. Danckwerts, *Gas–Liquid Reactions*, McGraw-Hill, New York, 1970.

6.4-29 S. Hatta, *Technol. Rept. (Tohoku Imp. Univ.*), **10,** 119 (1932).

6.4-30 D. W. van Krevelen and P. J. Hoftijzer, *Chem. Eng. Progr.*, **44,** 529 (1948).

6.4-31 W. J. Decoursey, *Chem. Eng. Sci.*, **29,** 1867 (1974).

6.4-32 E. R. Matheron and O. C. Sandall, *AIChE J.*, **24,** 552 (1978).

6.4-33 P. V. Danckwerts and A. M. Kennedy, *Trans. Inst. Chem. Eng.*, **32,** 549 (1954).

6.4-34 D. R. Olander, *AIChE J.*, **6,** 233 (1960).

6.4-35 S. V. Joshi, G. Astarita, and D. W. Savage, "Transport with Chemical Reactions," *AIChE Symp. Ser.*, No. 202, **77,** 63 (1981).

6.4-36 H. E. Benson, J. H. Field, and R. M. Jimeson, *Chem. Eng. Prog.*, **50,** 356 (1954).

6.4-37 C. Ouwerkerk, *Hydrocarbon Processing*, **57,** 89 (1978).

6.4-38 G. Astarita and D. W. Savage, *Chem. Eng. Sci.*, **35,** 649 (1980).

6.4-39 D. W. Savage, G. Astarita, and S. Joshi, *Chem. Eng. Sci.*, **35,** 1513 (1980).

6.4-40 R. H. Weiland, M. Rawal, and R. G. Rice, *AIChE J.*, **28,** 963 (1982).

CHAPTER 7

Extraction—Organic Chemicals Processing

A. H. P. SKELLAND
D. WILLIAM TEDDER
School of Chemical Engineering
The Georgia Institute of Technology
Atlanta, Georgia

The increasing diversity in the applications of liquid extraction has led to a correspondingly diverse proliferation of extraction devices that continue to be developed. This chapter focuses on those fundamental principles of diffusion, mass transfer, phase equilibrium, and solvent selection that provide a unifying basis for the entire operation. Design procedures for both stagewise and differential contactors also receive consideration, including packed and perforated plate columns and mixer–settlers. Some mechanically aided columns are discussed and an attempt is made to compare the performance of various equipment designs.

7.1 DIFFUSION AND MASS TRANSFER

First, we define concentration and flux.

7.1-1 Flux Relationships

Let the moles and mass of component A per unit volume of mixture be c_A and ρ_A, respectively. Then the mole fraction of A is c_A/c or x_A, and the mass fraction is ρ_A/ρ or w_A.

Consider a nonuniform fluid mixture with n components that is experiencing bulk motion. The statistical mean velocity of component i in the x direction with respect to stationary coordinates is u_i. The molal average velocity of the mixture in the x direction is defined as

$$U = \frac{1}{c} \sum_{i=1}^{n} c_i u_i \tag{7.1-1}$$

and the mass average velocity in the x direction is

$$u = \frac{1}{\rho} \sum_{i=1}^{n} \rho_i u_i \tag{7.1-2}$$

The corresponding definitions of molal fluxes in the x direction for component i are as follows:

Relative to Stationary Coordinates

$$N_{ix} = c_i u_i \tag{7.1-3}$$

Relative to the Mass Average Velocity

$$I_{ix} = c_i(u_i - u) \tag{7.1-4}$$

Relative to the Molal Average Velocity

$$J_{ix} = c_i(u_i - U) \tag{7.1-5}$$

Analogous expressions of course may be written for the mass fluxes. Relationships between the various fluxes are obtained in the following manner.

To relate molal fluxes I_{ix} and N_{ix}, consider Eqs. (7.1-1), (7.1-3), and (7.1-4):

$$I_{ix} = c_i u_i - c_i u$$

$$= N_{ix} - \frac{c_i}{\rho} \sum_{i=1}^{n} \rho_i u_i = N_{ix} - \frac{w_i}{M_i} \sum_{i=1}^{n} n_{ix} \tag{7.1-6}$$

and, for a binary mixture,

$$I_{Ax} = N_{Ax} - w_A\left(N_{Ax} + \frac{M_B}{M_A} N_{Bx}\right) \tag{7.1-7}$$

Also,

$$\sum_{i=1}^{n} I_{ix} = c(U - u) \quad \text{and} \quad \sum_{i=1}^{n} N_{ix} = cU \tag{7.1-8}$$

To relate molal fluxes J_{ix} and N_{ix}, consider Eqs. (7.1-2), (7.1-3), and (7.1-8):

$$J_{ix} = c_i u_i - c_i U$$

$$= N_{ix} - \frac{c_i}{c} \sum_{i=1}^{n} c_i u_i = N_{ix} - x_i \sum_{i=1}^{n} N_{ix} \tag{7.1-9}$$

and, for a binary mixture,

$$J_{Ax} = N_{Ax} - x_A(N_{Ax} + N_{Bx}) \tag{7.1-10}$$

Also,

$$\sum_{i=1}^{n} J_{ix} = 0 \tag{7.1-11}$$

7.1-2 Steady-State Equimolal Counterdiffusion and Unimolal Unidirectional Diffusion

The kinetic theory of gases is much more developed than that of liquids. Consequently, diffusional relationships for liquids are written largely by analogy with those for gases. Relationships for the latter are therefore developed first.

When the composition of a fluid mixture varies from one point to another, each component has a tendency to flow in the direction that will reduce the local differences in concentration. If the bulk fluid is either stationary or in laminar flow in a direction normal to the concentration gradient, the mass transfer reducing the concentration difference occurs by a process of molecular diffusion. This mechanism is characterized by random movement of individual molecules.

We confine our attention here to nonreacting systems of two components A and B, for which *Fick's First Law* of molecular diffusion may be written for steady one-dimensional transfer with constant c as

$$J_{Az} = -D_{AB} \frac{dc_A}{dz} \quad \text{and} \quad J_{Bz} = -D_{BA} \frac{dc_B}{dz} \tag{7.1-12}$$

where J_{Az} and J_{Bz} are molal fluxes of A and B relative to the molal average velocity of the whole mixture, the latter with respect to stationary coordinates; z is distance in the direction of diffusion; c_A and c_B are molar concentrations of A and B; and D_{AB} and D_{BA} are molecular diffusivities of A in B and of B in A, respectively. More generally,

$$J_{Az} = -cD_{AB} \frac{dx_A}{dz} \quad \text{and} \quad J_{Bz} = -cD_{BA} \frac{dx_B}{dz} \tag{7.1-13}$$

for which constancy of c in the z direction is not required (see Ref. 1, pp. 10–11).

Now for a perfect gas, $c_A = p_A/RT$ and $c_B = p_B/RT$ so that Eqs. (7.1-12) become

$$J_{Az} = -\frac{D_{AB}}{RT} \frac{dp_A}{dz} \quad \text{and} \quad J_{Bz} = -\frac{D_{BA}}{RT} \frac{dp_B}{dz} \tag{7.1-14}$$

Consider first the general case in which a steady total or bulk flow is imposed on the fluid mixture in the direction in which component A is diffusing. The magnitude of this molal flux of the whole mixture relative to stationary coordinates will be $N_{Az} + N_{Bz}$. The fluxes of components A and B relative to stationary coordinates are now the resultants of two vectors, namely, the flux caused by the bulk flow and the flux caused by molecular diffusion. Whereas these two vectors are in the same direction for component A, they are clearly in opposite directions for component B. The total flux of component A relative to stationary coordinates is then the sum of that resulting from bulk flow and that due to molecular diffusion as follows for a gaseous mixture:

$$N_{Az} = (N_{Az} + N_{Bz}) \frac{p_A}{P} - \frac{D_{AB}}{RT} \frac{dp_A}{dz} \tag{7.1-15}$$

This relationship is clearly another expression of Eq. (7.1-10) given earlier. Then, assuming constant D_{AB},

$$\frac{D_{AB}}{RT} \int_{p_{A1}}^{p_{A2}} \frac{dp_A}{N_{Az} - [(N_{Az} + N_{Bz})/P] \, p_A} = -\int_{z_1}^{z_2} dz$$

Integrating for constant N_{Az} and N_{Bz}, we obtain

$$N_{Az} = \frac{D_{AB}P}{RTz} \left(\frac{1}{1+\gamma} \right) \ln \left[\frac{1 - (1+\gamma) \, p_{A2}/P}{1 - (1+\gamma) \, p_{A1}/P} \right] \tag{7.1-16}$$

where $\gamma = N_{Bz}/N_{Az}$.

Equation (7.1-16) reduces to two special cases of molecular diffusion which are customarily considered. In *equimolal counterdiffusion*, component A diffuses through component B, which is diffusing at the same molal rate as A relative to stationary coordinates, but in the opposite direction. This process is often approximated in the distillation of a binary system. In *unimolal unidirectional diffusion*, only one molecular species—component A—diffuses through component B, which is motionless relative to stationary coordinates. This type of transfer is approximated frequently in the operations of gas absorption, liquid–liquid extraction, and adsorption.

STEADY-STATE EQUIMOLAL COUNTERDIFFUSION IN GASES

In this case the total molal flux with respect to stationary coordinates is zero, so that $N_{Az} = -N_{Bz}$. Then from Eqs. (7.1-14) and (7.1-15),

$$N_{Az} = J_{Az} = -N_{Bz} = -J_{Bz} \tag{7.1-17}$$

but $p_A + p_B = P = \text{constant}$, therefore,

$$\frac{dp_A}{dz} = -\frac{dp_B}{dz} \tag{7.1-18}$$

From Eqs. (7.1-14) and (7.1-17),

$$D_{AB} = D_{BA} = D \tag{7.1-19}$$

Application of L'Hôpital's rule to Eq. (7.1-16) for $\gamma = -1$ gives

$$N_{Az} = \frac{D}{RTz}(p_{A1} - p_{A2}) \tag{7.1-20}$$

where z is $z_2 - z_1$; p_{A1} and p_{A2} are partial pressures of A at z_1 and z_2, respectively.

Equations (7.1-14), (7.1-17), and (7.1-20) demonstrate that the partial-pressure distribution is linear in the case of steady-state equimolal counterdiffusion.

STEADY-STATE UNIMOLAL UNIDIRECTIONAL DIFFUSION IN GASES

In this case, the flux of component B in one direction resulting from bulk flow is equal to the flux of B in the opposite direction because of molecular diffusion. Component B is therefore motionless relative to stationary coordinates and N_{Bz} equals zero. Setting γ equal to zero in Eq. (7.1-16) and recalling that $P - p_A = p_B$,

$$N_{Az} = \frac{DP}{RTz} \ln \frac{p_{B2}}{p_{B1}} \tag{7.1-21}$$

which may be written

$$N_{Az} = \frac{DP}{RTz}\left(\frac{p_{B2} - p_{B1}}{p_{BLM}}\right) = \frac{D}{RTz}\left(\frac{P}{p_{BLM}}\right)(p_{A1} - p_{A2}) \tag{7.1-22}$$

where

$$p_{BLM} = \frac{p_{B2} - p_{B1}}{\ln (p_{B2}/p_{B1})}$$

The increase in transfer—by the factor P/p_{BLM}—resulting from bulk flow in the direction of diffusion of A is indicated by a comparison between Eqs. (7.1-20) and (7.1-22).

Equation (7.1-21) demonstrates that the partial-pressure distribution is nonlinear in the case of steady-state unidirectional diffusion.

7.1-3 Molecular Diffusion in Liquids

In the absence of a fully developed kinetic theory for liquids, the relationships for molecular diffusion are usually assumed to parallel those for gases, although diffusivities are often more substantially dependent on concentration of the diffusing components. In the case of equimolal counterdiffusion, the expression analogous to Eq. (7.1-20) is

$$N_{Az} = \frac{D}{z}(c_{A1} - c_{A2}) \tag{7.1-23}$$

If $c_A + c_B = c$, then $c_A = x_A c$ and $c_B = x_B c$, where x_A and x_B are mole fractions of A and B, respectively, and

$$N_{Az} = \frac{Dc}{z}(x_{A1} - x_{A2}) \tag{7.1-24}$$

For unimolal unidirectional diffusion, the liquid-phase analog of Eq. (7.1-22) is

$$N_{Az} = \frac{D}{z}\left(\frac{c}{c_{BLM}}\right)(c_{A1} - c_{A2}) = \frac{Dc}{z}\frac{(x_{A1} - x_{A2})}{x_{BLM}} \tag{7.1-25}$$

where

$$c_{BLM} = \frac{c_{B2} - c_{B1}}{\ln (c_{B2}/c_{B1})} \quad \text{and} \quad x_{BLM} = \frac{x_{B2} - x_{B1}}{\ln (x_{B2}/x_{B1})}$$

In addition to variation in D, the total molal concentration c also varies and a mean value of $(c_1 + c_2)/2$ is used when variations are not excessive.

7.1-4   Unsteady-State Diffusion in a Sphere

Many forms of extraction involve transfer between two liquid phases, one of which is dispersed as droplets in the other. Various attempts at theoretical analysis have assumed that the droplets may be regarded as spheres, in which mass transfer occurs by unsteady-state molecular diffusion. The following assumptions are made:

1. The concentration of solute (component A) is uniform at c_{A0} throughout the sphere at the start of diffusion ($t = 0$).

2. The resistance to transfer in the medium surrounding the sphere is negligible, so that the surface concentration of the sphere is constant at c_A^*, in equilibrium with the entire continuous phase—the latter having constant composition.

3. Diffusion is radial, there being no variation in concentration with angular position, and physical properties are constant.

The origin of coordinates is at the center of the sphere and the concentration at the spherical surface of radius r is c_A at time t. At the same instant, the concentration at the spherical surface of radius $r + dr$ is $c_A + dc_A$. A control volume is defined between the two surfaces at r and $r + dr$. The rate of flow of solute into the control volume is

$$-D(4\pi r^2) \frac{\partial c_A}{\partial r}$$

and the rate of flow out of the control volume is

$$-D[4\pi(r + dr)^2] \left[\frac{\partial c_A}{\partial r} + d\left(\frac{\partial c_A}{\partial r}\right) \right]$$

The difference between these two expressions, neglecting second- and third-order differentials, gives the rate of accumulation of solute in the control volume. Equating this to the rate of solute accumulation expressed as

$$(4\pi r^2 \, dr) \frac{\partial c_A}{\partial t}$$

leads to

$$\frac{\partial c_A}{\partial t} = D\left(\frac{\partial^2 c_A}{\partial r^2} + \frac{2}{r} \frac{\partial c_A}{\partial r} \right) \tag{7.1-26}$$

The boundary conditions follow from the initial assumptions as

$$c_A(r, 0) = c_{A0}$$

$$c_A(r_s, t) = c_A^*$$

$$\lim_{r \to 0} c_A(r, t) = \text{bounded}$$

where r_s is the radius of the sphere. The solution to Eq. (7.1-26) for the local $c_A(r, t)$ is then (Ref. 1, pp. 21–28)

$$c_A = c_A^* + \frac{2r_s}{\pi} (c_{A0} - c_A^*) \sum_{n=1}^{\infty} \frac{(-1)^{n+1}}{n} \frac{1}{r} \sin\left(\frac{n\pi r}{r_s}\right) \exp\left(\frac{-Dn^2\pi^2 t}{r_s^2}\right) \tag{7.1-27}$$

The fractional extraction from the sphere at time t is as follows, where $\bar{c}_A$ is the average concentration throughout the sphere at t (Ref. 1, pp. 21–28):

$$\frac{c_{A0} - \bar{c}_A}{c_{A0} - c_A^*} = 1 - \frac{6}{\pi^2} \sum_{n=1}^{\infty} \frac{1}{n^2} \exp\left(\frac{-Dn^2\pi^2 t}{r_s^2}\right) \tag{7.1-28}$$

For ready solution of numerical problems Newman[2] gives a graphical representation of Eq. (7.1-28) in the form of a plot of $1 - (c_A - \bar{c}_A)/(c_{A0} - c_A^*)$ against the dimensionless quantity Dt/r_s^2.

An individual coefficient of mass transfer for the disperse phase during free rise (or fall), k_{dr}, may now be formulated. A balance on component A diffusing into a stagnant, spherical droplet rising between points 1 and 2 during time dt is

$$k_{dr}\,\pi d_p^2\,(c_A^* - \bar{c}_A)\,dt = \frac{\pi d_p^3}{6}\,d\bar{c}_A \tag{7.1-29}$$

Integrating for constant c_A^* (at its average value) between locations 1 and 2,

$$k_{dr} = -\frac{d_p}{6t}\ln\left(\frac{c_A^* - \bar{c}_{A1}}{c_A^* - \bar{c}_{A1}} - \frac{\bar{c}_{A2} - \bar{c}_{A1}}{c_A^* - \bar{c}_{A1}}\right) = -\frac{d_p}{6t}\ln(1 - E_f) \tag{7.1-30}$$

where d_p is the droplet diameter and E_f is the fractional extraction in time t. Combining Eqs. (7.1-28) and (7.1-30), we obtain

$$k_{dr} = -\frac{d_p}{6t}\ln\left[\frac{6}{\pi^2}\sum_{n=1}^{\infty}\frac{1}{n^2}\exp\left(\frac{-D_d n^2 \pi^2 t}{(d_p/2)^2}\right)\right] \tag{7.1-31}$$

The k_{dr} here is for use with a driving force expressed as Δc_A. To obtain a k_{dr} suitable for use with Δy_A, the right-hand side of Eq. (7.1-31) must be multiplied by $(\rho/M)_{av}$ for the disperse phase between 1 and 2.

Equation (7.1-31) is used in the design of extraction equipment such as spray columns, perforated plate columns, and mixer–settlers.

7.1-5 Molecular Diffusivities in Liquids

The theory of molecular diffusion has been the subject of extensive investigation because of its close relationship to the kinetic theory of gases. Detailed reviews are available[3-8] and a valuable critical comparison of the various correlations that have been presented for the prediction of diffusivities, including electrolytes and nonelectrolytes under a variety of conditions, is provided by Reid et al.[9] Treybal (Ref. 10, pp. 150–165) reviews procedures for predicting diffusivities in liquids. Skelland (Ref. 1, Chap. 3) surveys the relationships available for estimating diffusivities for electrolytes and nonelectrolytes at low and high concentrations.

It is expected that the reader will consult the above references for a detailed understanding; the following classified relationships are presented only as an outline treatment.

NONELECTROLYTES—DILUTE SOLUTIONS
For dilute diffusion in organic solvents Lusis and Ratcliff[11] give the following relationship:

$$\frac{D_{AB}\mu_B}{T} = 8.52\,(10^{-8})\,V_{bB}^{-1/3}\left[1.40\left(\frac{V_{bB}}{V_{bA}}\right)^{1/3} + \frac{V_{bB}}{V_{bA}}\right] \tag{7.1-32}$$

where D_{AB} is in cm^2/s; μ_B is viscosity of the solvent B in cP; T is in K; V_{bA} and V_{bB} are molal volumes of the solute and solvent at their normal boiling temperatures, in cm^3/mol. For water as solute it is usually necessary to evaluate V_{bA} as for four moles of water, that is, assuming water to diffuse as a tetramer. Lusis and Ratcliff[11] discuss problems arising from strong solute–solvent interaction and in the diffusion of long straight-chain hydrocarbon molecules.

For diffusion in dilute aqueous solutions the following equation is provided by Hayduk and Laudie:[12]

$$D_{AB} = \frac{13.26 \times 10^{-5}}{\mu_{wt}^{1.4}\,V_{bA}^{0.589}} \tag{7.1-33}$$

where μ_{wt} is the viscosity of water at $T(K)$, in cP.

NONELECTROLYTES—CONCENTRATED SOLUTIONS
Leffler and Cullinan[13] give

$$(D_A\mu_{AB})_{conc} = (D_{AB}^\circ\mu_B)^{x_B}\,(D_{BA}^\circ\mu_A)^{x_A}\left(1 + \frac{d\ln\gamma_A}{d\ln x_A}\right) \tag{7.1-34}$$

where D_{AB}° and D_{BA}° are diffusivities of very dilute A in B and B in A, respectively, in cm^2/s; μ_A, μ_B, and μ_{AB} are viscosities of A, B, and the mixture of A and B, in cP; x_A and x_B are mole fractions of solute A and solvent B; and γ_A is the activity coefficient of A.

ELECTROLYTES—DILUTE SOLUTIONS
For strong (fully dissociated) electrolytes at infinite dilution, Nernst's[14] equation is

$$D_A^\circ = 8.931 \ (10^{-10}) \ T\left(\frac{l_+^\circ \ l_-^\circ}{l_+^\circ + l_-^\circ}\right)\left(\frac{z_+ + z_-}{z_+ \ z_-}\right) \tag{7.1-35}$$

where D_A° is diffusivity, in cm^2/s; l_+° and l_-° are the anionic and cationic conductances at infinite dilution, in mho/equivalent; $l_+^\circ + l_-^\circ$ is the electrolyte conductance at infinite dilution; z_+ and z_- are absolute values of the cation and anion valences; and T is the absolute temperature, in K.

ELECTROLYTES—CONCENTRATED SOLUTIONS
Gordon[15] gives

$$(D_A)_{conc} = D_A^\circ \left(1 + \frac{m\partial \ln \gamma\pm}{\partial m}\right)\frac{1}{c_B' \overline{V}_B}\left(\frac{\mu_B}{\mu_{AB}}\right) \tag{7.1-36}$$

where D_A° is calculated from Nernst's equation (7.1-35); m is molality; c_B' is the number of moles of water per cm^3 of solution; $\overline{V}_B$ is the partial molal volume of water in solution, in cm^3/mol; μ_B and μ_{AB} are the viscosities of water and of the solution; and $\gamma\pm$ is the mean ionic activity coefficient based on molality. Skelland (Ref. 1, Chap. 3) illustrates the use of all these relationships.

A comprehensive tabulation of experimental diffusivities for nonelectrolytes is given by Johnson and Babb.[16] Similar data for electrolytes are provided by Harned and Owen[17] and Robinson and Stokes.[18]

7.1-6 Mass Transfer Coefficients

Consider the distribution of a solute such as component A between two immiscible liquid phases in contact with each other. Under conditions of dynamic equilibrium the rate of transfer of A from the first to the second phase is equal to the rate at which A is transferred in the reverse direction. The equilibrium relationship may be represented by a plot such as Fig. 7.1-1 over a range of compositions of each phase.

It is assumed that local equilibrium prevails at the interface between phases, where the compositions are Y_A^* and X_A^*. If transfer of A takes place from the lighter liquid to the heavier liquid, the individual coefficients k_Y and k_X for the light and heavy phases, respectively, are defined as follows:

$$N_A A = k_Y A(Y_A - Y_A^*) = k_X A(X_A^* - X_A) \tag{7.1-37}$$

A is the area of the interface, where the flux is N_A, and Y_A and X_A are concentrations of component A in the bulk of the light- and heavy-liquid phases. Equation (7.1-37) shows that

$$\frac{Y_A - Y_A^*}{X_A - X_A^*} = -\frac{k_X}{k_Y} \tag{7.1-38}$$

This relationship is plotted in Fig. 7.1-2 on the assumption of interfacial equilibrium.

Interfacial concentrations (X_A^*, Y_A^*) are often unknown at a given location within a two-phase system and it is then more convenient to use overall coefficients K_Y and K_X, defined in terms of overall concentration differences or "driving forces," as shown below.

$$N_A A = K_Y A(Y_A - Y_{AL}) = K_X A(X_{AG} - X_A) \tag{7.1-39}$$

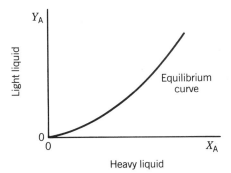

FIGURE 7.1-1   Equilibrium distribution of component A between two conjugate phases.

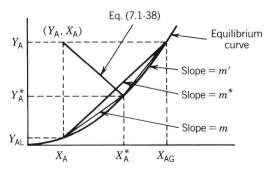

FIGURE 7.1-2 Individual and overall concentration "driving forces" for a two-phase system with distributed component A. Bulk concentrations of the two phases are X_A and Y_A, respectively, and interfacial equilibrium is assumed.

where Y_{AL} is the light liquid concentration that would be in equilibrium with the existing heavy liquid concentration; X_{AG} is the heavy liquid concentration that would be in equilibrium with the existing light liquid concentration. The location of these quantities is shown in Fig. 7.1-2. The relationships between the individual coefficients of Eq. (7.1-37) and the overall coefficients of Eq. (7.1-39) can be shown to be the following (Ref. 1, pp. 91–93).

$$\frac{1}{K_Y} = \frac{1}{k_Y} + \frac{m}{k_X}, \qquad \frac{1}{K_X} = \frac{1}{k_X} + \frac{1}{m'k_Y}, \qquad \frac{1}{K_X} = \frac{1}{m''K_Y} \qquad (7.1\text{-}40)$$

where the first two relationships express the overall resistance to transfer (in terms of the light and heavy liquids, respectively) as equaling the sum of the individual (or single-phase) resistances to mass transfer. The final expression relates the two forms of overall resistance to mass transfer, $1/K_X$ and $1/K_Y$ (note that constancy of K over a given concentration range requires constant k_X, k_Y, and m).

The interfacial area A is often unknown in many types of mass transfer equipment. In such cases mass transfer rates frequently are based on unit volume of the equipment, instead of unit interfacial area. The rate equations (7.1-37) and (7.1-39) are then modified to the following form:

$$N_A A = k_Y a(Y_A - Y_A^*)_m V_0 = k_X a(X_A^* - X_A)_m V_0$$
$$= K_Y a(Y_A - Y_{AL})_m V_0 = K_X a(X_{AG} - X_A)_m V_0 \qquad (7.1\text{-}41)$$

where subscript m denotes a suitable mean, V_0 is the contacting volume of the equipment, a is the interfacial area per unit volume, and the combined quantities $k_X a$, $k_Y a$, $K_X a$, and $K_Y a$ are called volumetric or capacity coefficients. Dividing Eq. (7.1-40) by a, we obtain

$$\frac{1}{K_Y a} = \frac{1}{k_Y a} + \frac{m}{k_X a}, \qquad \frac{1}{K_X a} = \frac{1}{k_X a} + \frac{1}{m'k_Y a}, \qquad \frac{1}{K_X a} = \frac{1}{m''K_Y a} \qquad (7.1\text{-}42)$$

7.1-7 The Two-Film Theory

This theory, developed by Lewis[19] and Whitman,[20] supposes that motion in the two phases dies out near the interface and the entire resistance to transfer is considered as being contained in two fictitious films on either side of the interface, in which transfer occurs by purely molecular diffusion. It is postulated that local equilibrium prevails at the interface and that the concentration gradients are established so rapidly in the films compared to the total time of contact that steady-state diffusion may be assumed.

Individual or single-phase mass transfer coefficients (k) are defined customarily as follows for a mechanism involving bulk molal flow:

Light Phase

$$N_A = k_y \Delta y_A = k_c \Delta c_A \qquad (7.1\text{-}43)$$

Heavy Phase

$$N_A = k_c \Delta c_A = k_x \Delta x_A = k_c \left(\frac{\rho}{M}\right)_{av} \Delta x_A \tag{7.1-44}$$

In the important case of unimolal unidirectional diffusion, Eqs. (7.1-25) and (7.1-43) show the following relationship:

$$k_y = \frac{D_G c}{z_{fG}(1 - y_A)_{LM}} = \frac{D_G c}{z_{fG} y_{BLM}} \tag{7.1-45}$$

where D_G is the solute diffusivity in the light phase, which has a fictitious film thickness z_{fG}; k_y is the individual light-phase coefficient of mass transfer for use with Δy_A.

Similarly, for the heavy liquid phase, from Eqs. (7.1-25) and (7.1-44),

$$k_x = \frac{D_L c}{z_{fL}(1 - x_A)_{LM}} = \frac{D_L c}{z_{fL} x_{BLM}} \tag{7.1-46}$$

where D_L is the solute diffusivity in the heavy phase, which has a fictitious film thickness z_{fL}; k_x is the individual heavy-phase coefficient of mass transfer for use with Δx_A.

Equations (7.1-45) and (7.1-46) show that the two-film theory predicts that the mass transfer coefficient is directly proportional to the molecular diffusivity to the power unity. The complexity of flow normally prevents evaluation of z_f, but it will decrease with increasing turbulence.

7.1-8 Other Theories

In the general relationship

$$k \propto D^n \tag{7.1-47}$$

the value of n was seen to be unity according to the two-film theory. In contrast to this, a variety of theories have been developed on the basis of some form of surface-renewal mechanism, which often show $n = \frac{1}{2}$, but which include $0 \leq n \leq 1$.

The work of Higbie[21] in 1935 provided the foundation for the *penetration theory*, which supposes that turbulence transports eddies from the bulk of the phase to the interface, where they remain for a short but constant time before being displaced back into the interior of the phase to be mixed with the bulk fluid. Solute is assumed to "penetrate" into a given eddy during its stay at the interface by a process of unsteady-state molecular diffusion, in accordance with Fick's Second Law and appropriate boundary conditions:

$$\frac{\partial c_A}{\partial t} = D \frac{\partial^2 c_A}{\partial z^2} \tag{7.1-48}$$

The solution yields $k = 2 (D/\pi t_e)^{1/2}$, where t_e is the (constant) time of exposure of the eddy at the interface.

Danckwerts'[22] *theory of penetration with random surface renewal* modifies this picture by proposing an "infinite" range of ages for elements of the surface. The probability of an element of surface being replaced by a fresh eddy is considered to be independent of the age of that element. Danckwerts[22] introduced this modification by defining a surface age distribution function, $\phi(t)$, such that the fraction of surface with ages between t and $t + dt$ is $\phi(t)dt$. If the probability of replacement of a surface element is independent of its age, Danckwerts showed that $k = (sD)^{1/2}$, where s is the fractional rate of surface renewal.

The *film–penetration theory*, presented by Toor and Marchello,[23] represents a combination of the three earlier theories reviewed above. The entire transfer resistance is considered to lie in a laminar surface layer of thickness z_L, where c_A is uniform at $c_{A\infty}$ for all z greater than z_L. Surface renewal occurs by eddies that penetrate the surface from the bulk of the phase. Thus, transfer through young elements of surface obeys the penetration theory ($k \propto D^{1/2}$), transfer through old elements follows the film theory ($k \propto D$), and transfer through elements of intermediate age is characterized by both mechanisms.

In a series of papers between 1949 and 1954 Kishinevskii and coworkers[24-27] proposed a surface-renewal mechanism which, in contrast to the theories described above, postulates that transfer into an eddy at the interface occurs predominantly by convective mass flow and not by molecular diffusion. The authors also dispute the suggestion that the probability of replacement of a surface element is independent of its age.

King[28] has proposed another general model for turbulent liquid-phase mass transfer to and from a free gas–liquid interface. The model requires the evaluation of three parameters and involves concepts of surface renewals in which surface tension exerts a damping effect on the smaller eddies. Allowance is made for a continuous eddy diffusivity profile near the free interface, thereby avoiding the postulate of a "film" or discontinuity in transport properties as required by the film–penetration theory.

Harriott[29] presents a model that incorporates the penetration mechanism, but which includes thin, laminar, interfacial films that are stabilized by interfacial tension and are therefore excluded from intermittent mixture with the bulk fluid. The penetration theory has been extended by Angelo et al.[30] to allow for "stretching" surfaces, such as those that occur, for example, in large, oscillating droplets. Ruckenstein[31] has further refined the penetration theory to accommodate the effects of velocity distributions within the eddies during solute penetration.

Skelland (Ref. 1, p. 106) gives the results of a theory of penetration with surface renewal in which it is assumed that the fractional rate of surface renewal s is related to the age of the surface through the constants ξ and n as $s = \xi t^n$, where $(\xi, n + 1) > 0$. A theory of penetration with periodically varying rates of surface renewal has been developed by Skelland and Lee[32] for two-phase dispersions in turbine-agitated vessels. This attempts to allow for the varying degrees of turbulence encountered by bubbles or drops as they circulate through all regions of the vessel.

Comparisons between experimental observation and the predictions of the film theory on the one hand and various forms of surface-renewal theory on the other are reviewed briefly in Ref. 33 (Chap. 5). It is interesting that, although the evidence generally appears to favor the surface-renewal mechanisms, the two-film theory contributes to the design of complex processes in a manner that continues to be very useful. An example of this will be given later in the formulation of transfer unit relationships for packed column design.

Although application of some of the theories outlined above is hampered by lack of knowledge of some essential components (such as z_{fG}, z_{fL}, t_e, s, or z_L, as appropriate), they are nevertheless valuable in indicating limits on n between 0 and 1 in Eq. (7.1-47). Empirical correlations of data which show an exponent on D outside this range should probably be rejected.

7.2 EQUILIBRIUM CONSIDERATIONS

Both liquid–liquid and vapor–liquid equilibria are important for the recovery of organic chemicals by solvent extraction. The former equilibria determine the required liquid–liquid contacting equipment sizes and operating conditions while the latter usually determine similar requirements for the solvent regeneration and product recovery system. Although other flowsheet configurations are possible, it is common to follow the liquid–liquid extraction step by solvent and product recovery to minimize the solvent inventory in the process and the sizes of subsequent equipment. Also, the chosen solvent, or mass separating agent (MSA), generally will affect the vapor–liquid equilibria of the solute–diluent binary as well as determine the liquid–liquid equilibrium properties. Hence, a solvent that is added to effect the solute–diluent separation via formation of a liquid–liquid mixture may be useful in a subsequent azeotropic or extractive distillation step to achieve further solute–diluent separation.

Clearly, the accurate determination and correlation of these equilibrium properties are of paramount importance in the rational design of extraction operations. In general, it is desirable to use computer-aided design techniques for the modeling of such a system. The more attractive correlation techniques for multicomponent systems include the UNIQUAC and UNIFAC methods, which are described in Chapter 1 of this handbook and are not discussed here. These techniques are suitable for computer-based computations.

On the other hand, it is often useful to complete simple hand calculations, especially during the initial screening of alternative solvents. When experimental data are available, the distribution coefficients for both diluent and solute can be correlated effectively using equations with the form:

$$\ln D_A = b_0 + b_1 W_A + b_2 V_M + b_3 W_D + b_4/T \qquad (7.2\text{-}1)$$

where D_A is the distribution coefficient of A, W_A and W_D are weight fractions, V_M is a volume fraction, T is temperature, and the b's are empirical parameters.

Similarly, it is often useful to correlate solute and diluent vapor–liquid equilibria over the solvent with an equation of the form:

$$\ln K_A = b_0 + b_1 X_A + b_2 X_B + b_3/T \qquad (7.2\text{-}2)$$

where K_A is a vapor–liquid distribution coefficient, X_A and X_B are mole fractions, and the b's are again empirical parameters. The liquid–liquid distribution coefficients usually are expressed as weight fraction ratios, while the vapor–liquid equilibria are expressed more often in terms of the mole fraction ratios.

Although these equations are quite useful with experimental data in the range of low solute concentrations, Eq. (7.2-1) is limited by the fact that it cannot be used to predict the mutual solubility curve. Consequently, it can lead to extrapolation beyond the two-phase region and therefore its use must be constrained.

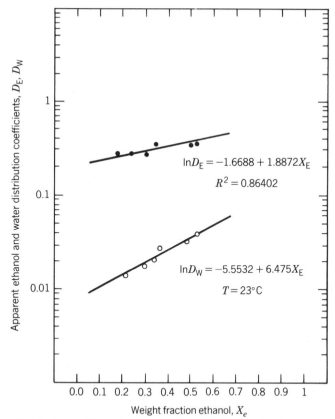

FIGURE 7.2-1   Distribution coefficient for ethanol and water using dimethyl heptanone as the solvent.[1]

On the other hand, the simplicity of Eqs. (7.2-1) and (7.2-2) and their accuracy over the intended range favor their use for preliminary calculations. They permit consideration of the effects of inextractable materials (e.g., salts or dissolved solids) on the liquid–liquid equilibrium and the temperature with the introduction of relatively few adjustable parameters. In addition, they can be expanded into a polynomial form to account for situations where the logarithm of the apparent distribution coefficient is not a linear function of one of the important variables. Several examples are given in Fig. 7.2-1, which is taken from Tawfik.[1]

The typical forms of Type 1 and Type 2 liquid–liquid systems are shown in Figs. 7.2-2[1] and 7.2-3[2]. Also illustrated in Fig. 7.2-2 is the effect of temperature on the mutual solubility curve, which may be so pronounced in some cases as to cause transition from one type of system to the other.

During the extraction of organic species, it may be desirable to modify the solvent. An inert paraffinic compound or mixture may be blended with a suitable modifier (e.g., a species that hydrogen bonds) to enhance the solvent properties. Such properties might include viscosity, density, surface tension, or attraction for the solute. In these cases, the mutual solubility curve may appear as in Fig. 7.2-4 when the solvent mixture is plotted at one vertex. Reasons for solvent blending may include improved solvent selectivity, interfacial tension, reduced solvent phase viscosity, and increased density differences between the two phases. A solvent that forms stable emulsions when mixed with the diluent phase, for example, may be suitable for use when it is modified with a suitable inert paraffinic material.

Other representations of equilibrium data are in common use. For example, a Hand[3] plot is often attractive as well as distribution diagrams and the use of other coordinate systems (e.g., rectangular or Janecke coordinates). Since the use of these systems has been described in detail elsewhere,[4–6] they simply are mentioned here.

7.3   STAGEWISE AND DIFFERENTIAL CONTACTING CALCULATION METHODS

Sections 7.3–7.5 deal with mass transfer between two liquid phases in countercurrent flow through packed and plate columns and in agitated vessels. In such cases the details of velocity distribution in the two phases

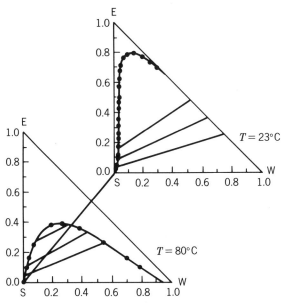

FIGURE 7.2-2 Effect of temperature on the mutual solubility curve of ethanol–water–20% tridecyl alcohol in Norpar 12.[1]

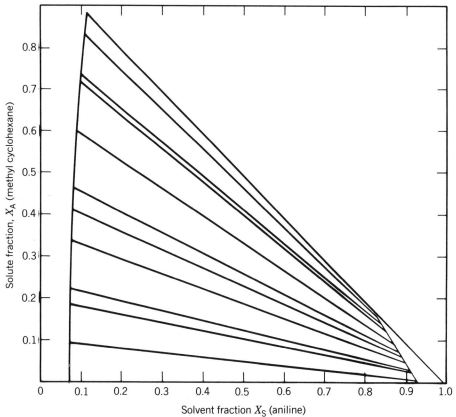

FIGURE 7.2-3 Equilibrium data for the system n-heptane, methylcyclohexane, and aniline at 25°C. (Data from Ref. 2).

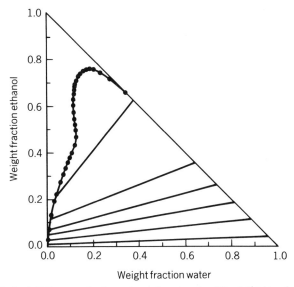

FIGURE 7.2-4 Mutual solubility curve for the system: ethanol–water–50 vol. % tridecyl alcohol in Norpar-12.[1]

are unknown, so that the convective contribution to transfer cannot be formulated quantitatively. Therefore, one must resort to empirical correlations of mass transfer rates with operating conditions, physical properties, and system geometry in the manner to be described.

Continuous-contact or packed columns are usually filled with Raschig rings, Pall rings, Berl saddles, Lessing rings, or other types of packing to promote intimate contact between the two phases. Continuous contact is maintained between the two countercurrent streams throughout the equipment, which necessitates a treatment based on a differential section of packing.

In contrast, stagewise contactors feature intermittent, rather than continuous, contact between the phases. The stages often take the form of plates or trays of varied design, arranged vertically above each other in a column. Alternatively, stagewise contactors can be constructed from a series of mixer–settlers. The two phases enter a stage from opposite directions in countercurrent flow, mix together to facilitate transfer, then separate and leave the stage. When the two phases leave in a state of equilibrium, the stage is said to be an "ideal" or "theoretical" one. This concept has been extended to packed columns by defining the *height* (of packing) *equivalent* to a *theoretical stage* (HETS), such that the streams leaving this section are in equilibrium.

7.3-1 The Operating Line

The relationship giving the composition of the two phases passing one another at a given location in the system is obtained by material balance and is called the operating line. Consider the column sketched in Fig. 7.3-1. Note carefully the locations of sections 1 and 2, which are directionally the same as that used by some authors and the reverse of that used by others.

A balance on component A over the differential volume $S\, dH$ gives

$$d(Gy_A) = d(Lx_A) \tag{7.3-1}$$

Integrating between section 1 and any section within the column leads to

$$y_A = \frac{L}{G} x_A + \frac{G_1 y_{A1} - L_1 x_{A1}}{G} \tag{7.3-2}$$

This is the equation for the operating line; it is valid for either packed or stagewise contactors and it relates y_A to x_A at any section within the column. In the general case, L and G may vary with location, giving a curved operating line. Equation (7.3-2) is effectively linear, however, in those cases of very dilute streams for which composition changes due to mass transfer have a negligible effect on L and G. Graphical representation of the operating line is also readily possible when phases G and L each consist only of inert

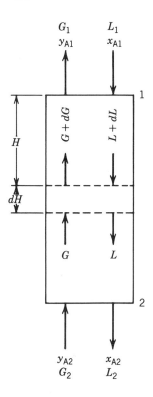

FIGURE 7.3-1 Terminology for a continuous column. A, total interfacial area; G, L, flow rates of phases G and L (moles/time); H, column height; S, cross-sectional area of empty column; x_A, y_A, concentration of component A in phases L and G (mole fraction). G is the raffinate phase and L is the extract phase.

(nontransferring) material and component A. Thus, if G' and L' are flow rates of nontransferring components in phases G and L (in moles/time), then

$$G' = G(1 - y_A) = G_1 (1 - y_{A1}) = G_2(1 - y_{A2}) \qquad (7.3\text{-}3)$$

$$L' = L(1 - x_A) = L_1 (1 - x_{A1}) = L_2(1 - x_{A2}) \qquad (7.3\text{-}4)$$

Substituting for L, L_1, G, and G_1 in Eq. (7.3-2) and rearranging,

$$\frac{y_A}{1 - y_A} = C_1 + C_2 \frac{x_A}{1 - x_A} \qquad (7.3\text{-}5)$$

where

$$C_1 = \frac{y_{A1}}{1 - y_{A1}} - C_2 \frac{x_{A1}}{1 - x_{A1}} \qquad (7.3\text{-}6)$$

$$C_2 = \frac{L'}{G'} \qquad (7.3\text{-}7)$$

The quantities x_{A1}, x_{A2}, y_{A1}, y_{A2}, G_2, and L_1 are normally known or can be calculated, so that L' and G' can be determined from Eqs. (7.3-3) and (7.3-4). The operating line is plotted readily from Eq. (7.3-5), with the terminal points (x_{A1}, y_{A1}) and (x_{A2}, y_{A2}).

When conditions required for the two simple cases just described are not met, the operating curves may be located on the x_A-y_A diagram by appropriate use of the triangular diagram, as described below.

7.3-2 Triangular Diagrams and the Line Ratio Principle

Ternary systems are often represented on equilateral or right triangular coordinates, as sketched in Fig. 7.3-2a.

Any point on one of the sides of the triangle represents a binary mixture of the components at each end of that side. For example, mixture J contains (line length CJ/line length AC) 100% of A, with the remainder being C.

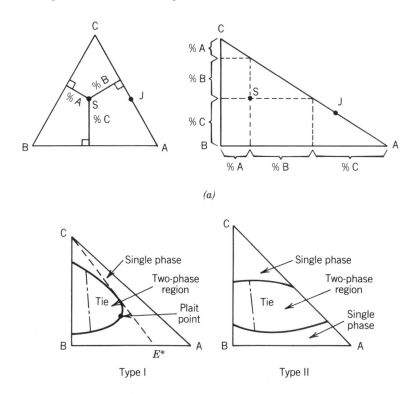

FIGURE 7.3-2 Triangular diagrams: (*a*) mixture location and (*b*) phase equilibria.

A point inside the triangle, such as S, represents a ternary mixture. The perpendicular distance of the point S from a given side of the equilateral triangle gives the percentage (or fraction) of the component represented by the apex opposite that side. The vertical height of the equilateral triangle represents 100% (or 1.0). In the right triangle—which need not be isosceles—the composition of a ternary mixture S is identified conveniently in terms of the scales of axes AB and BC. A point outside the triangle represents an imaginary mixture, of negative composition.

The right-triangular diagram is more convenient since it may be constructed readily to any scale, with enlargement of any particular region of interest. Accordingly, this representation will be used here.

Liquid extraction exhibits at least two common forms on the triangular diagram. Figure 7.3-2*b* shows both type I and type II systems, the former being characterized by the plait point. In type I systems, the tie lines linking equilibrium phases shrink to a point at the plait point, so that the two conjugate phases become identical.

When reflux, which is discussed later in more detail, is used in a type II system the feed is theoretically separable into pure A and B after solvent removal. This contrasts with type I systems for which, even with the use of reflux, a feed mixture is theoretically separable only into pure B at one end of the unit and a *mixture* of A and B at the other end—after removal of solvent. Thus, the most concentrated extract obtainable is E*.

Any point in the two-phase region of Fig. 7.3-2*b* represents an overall mixture that separates into two phases linked, at equilibrium, by a tie line.

Consider the steady flow of two ternary streams into the mixer shown in Fig. 7.3-3. The streams enter at the rates of M and N mol/h and a third stream leaves at P mol/h. In representing this process on the triangular diagram it can be shown that, when streams M and N are mixed, the resultant stream P lies *on* the line MN such that

$$\frac{\text{moles (or mass) of } N/\text{hour}}{\text{moles (or mass) of } M/\text{hour}} = \frac{N}{M} = \frac{\text{(length of) line } MP}{\text{(length of) line } PN}$$

Since the triangle used in Fig. 7.3-3 is neither equilateral nor isosceles, it follows that graphical addition of streams and the line ratio principle are independent of the shape of the triangle.

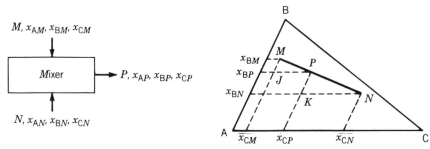

FIGURE 7.3-3 Mixing of streams and the line ratio principle.

7.3-3 Use of Triangular Diagrams to Locate Operating Curves on Distribution Diagrams

Whereas the determination of the number of theoretical stages required for a prescribed separation can usually be performed by appropriate construction on the triangular diagram, the number of transfer units (used to establish the necessary height of a packed column) is evaluated most readily via the distribution $(x_A\text{-}y_A)$ diagram. The latter may also be applied to stagewise evaluation using the usual McCabe–Thiele-type stepwise construction procedure. To facilitate both stagewise and packed column design, therefore, the location of the operating and equilibrium curves on the distribution diagram are outlined, via appropriate construction on the triangular diagram. This procedure is generally useful in cases where the simplified conditions for location of the operating line, given earlier, do not apply.

EXTRACTION WITHOUT REFLUX

Consider the countercurrent extraction of solute A from solution in B using pure solvent C. The operation is performed in the column shown in Fig. 7.3-1, which may be either continuous-contact (packed) or stagewise.

To obtain the operating curve on the $x_A\text{-}y_A$ distribution diagram, the terminal streams, raffinate G_1, solvent L_1, and feed G_2, are located on the triangular diagram of Fig. 7.3-4. This diagram shows the solubility or binodal curve for a type I system.

A material balance around the column in Fig. 7.3-1 gives

$$G_2 + L_1 = G_1 + L_2 = \Sigma \tag{7.3-8}$$

so that from the line ratio principle,

$$\frac{G_2}{G_2 + L_1} = \frac{\text{length of line } L_1\Sigma}{\text{length of line } L_1G_2} \tag{7.3-9}$$

This enables the location of Σ on Fig. 7.3-4; the point L_2—the extract product—then lies at the intersection of the binodal curve and the extended line $G_1\Sigma$. Next, consider the net flow of material in the G direction in Fig. 7.3-1: $G_1 - L_1 = G - L = G_2 - L_2 = \Delta$. The lines L_2G_2 and L_1G_1 are extended to intersect at the difference point Δ in accordance with the graphical addition (or subtraction) of streams principle demonstrated in Fig. 7.3-3.

The fixed location of Δ for the entire column is readily demonstrated by successive balances on any two of the three components involved. Random lines from Δ intersect the binodal curve at w_{AB} on the B-rich side and at w_{AC} on the C-rich side. These intersections, when converted to mole fractions, give points (x_A, y_A) on the operating curve of Fig. 7.3-5. (Note: w_{AB} = mass fraction of A in the B-rich phase.) It is also acceptable to construct a W_{AB} versus W_{AC} diagram and perform calculations similar to those that will be described for the x-y diagram.

The equilibrium curve is plotted in Fig. 7.3-5 from the terminal points of each experimentally determined tie line on the triangular diagram. The use of Fig. 7.3-5 in determining *either* the number of theoretical stages needed (for stagewise contact) *or* the number of transfer units needed (for packed column contact) will be discussed shortly.

Tie lines, of course, cannot cross in the two-phase region within the binodal curve of Fig. 7.3-4. Furthermore, a line from Δ must not coincide with a tie line in the region between lines $L_1\Delta$ and $L_2\Delta$, since this would cause contact (pinch) between the operating and equilibrium curves of Fig. 7.3-5, signifying zero driving force for the transfer of A. Such coincidence occurs in the use of a *minimum solvent-to-feed ratio*. The condition is avoided by using solvent in excess of the minimum solvent-to-feed ratio for the prescribed separation, which is established as follows with reference to Fig. 7.3-6.

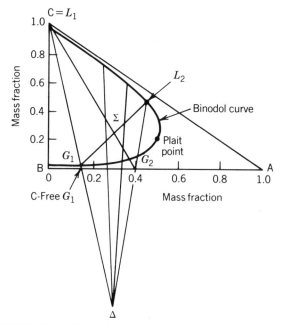

FIGURE 7.3-4 Triangular diagram for extraction without reflux.

All tie lines to the left of—and including—the one that extends through G_2 are extrapolated to intersect the extended line L_1G_1. The farthest intersection (if below the triangle) or the nearest (if above the triangle) gives Δ_m corresponding to the minimum solvent-to-feed ratio. The line $\Delta_m G_2$ then locates L_{2m} at its intersection with the binodal curve. Σ_m is found at the intersection of lines L_1G_2 and G_1L_{2m}, so that

$$L_{1,\text{min}} = (\text{feed rate } G_2)\frac{\text{line length } G_2\Sigma_m}{\text{line length } L_1\Sigma_m}$$

An operable solvent flow rate will exceed $L_{1,\text{min}}$.

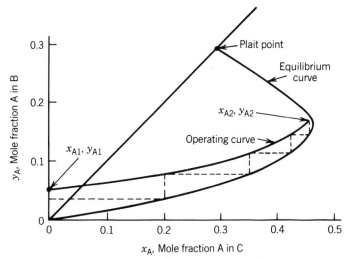

FIGURE 7.3-5 Distribution diagram for extraction without reflux, as in Fig. 7.3-1, showing both the operating and equilibrium curves.

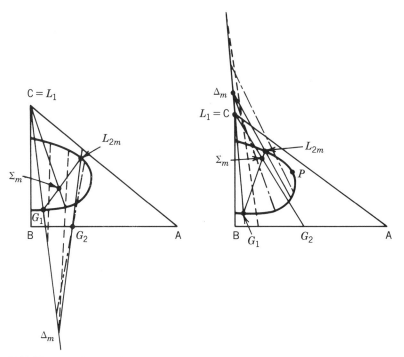

FIGURE 7.3-6 Construction to determine the minimum solvent-to-feed ratio, $L_{1,\min}/G_2$.

EXTRACTION WITH REFLUX

Reflux can be employed in countercurrent columns or in a series of mixer–settlers in liquid–liquid extraction. This allows the solute-rich stream leaving the unit to attain a higher concentration of solute than that which would be in equilibrium with the feed. In considering the separation of a binary mixture it must be realized, however, that reflux does not enhance the degree of separation between components of the feed unless they both transfer significantly into the phase that extracts the solute. (The proportions in which the two components transfer must be different from that in the feed for the separation to be feasible.)

The operation is shown in flowsheet form on the left side of Fig. 7.3-7, while the right side of the same figure identifies some of the terminology to be used. The C remover may be a distillation column, an evaporator, or another liquid extraction unit. It withdraws sufficient C from phase L_1 to convert it to a G phase.

MATERIAL BALANCES IN FIG. 7.3-7

A balance around the C remover results in

$$L_1 - G_0 = L_0 + G_p = \Delta_1 \tag{7.3-10}$$

A similar balance around the C remover plus any portion of the column above the feed plane gives

$$L_1 - G_0 = L - G = \Delta_1 \tag{7.3-11}$$

Converting Eq. (7.3-11) into component B and C balances,

$$L_1(x_B)_{L_1} - G_0(y_B)_{G_0} = L(x_B) - G(y_B) = \Delta_1(x_B)_{\Delta_1} \tag{7.3-12}$$

$$L_1(x_C)_{L_1} - G_0(y_C)_{G_0} = L(x_C) - G(y_C) = \Delta_1(x_C)_{\Delta_1} \tag{7.3-13}$$

Equations (7.3-11)–(7.3-13) demonstrate that the differences between adjacent streams in the column section above the feed plane are constant in amount and composition. These differences therefore may be represented by the single point Δ_1 on the triangular diagram. The "difference point" Δ_1 is located in Fig. 7.3-8 from the following development in terms of the reflux ratio, G_0/G_p.

Streams G_0, G_p, and G_0 in Fig. 7.3-7 are identical in composition and are represented by a common

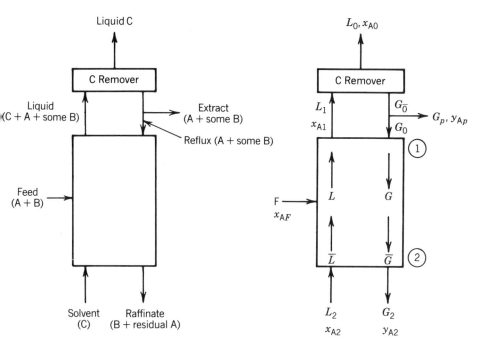

A = Solute
B = Diluent solvent
C = Extracting solvent

Liquid extraction: L = Extracting liquid, G = Raffinate liquid

Solute transfer is from G to L throughout

FIGURE 7.3-7   Liquid–liquid extraction with extract reflux G_0.

point in Fig. 7.3-8. The compositions of L_0 and G_p typically are specified and Eq. (7.3-10) shows that the points representing streams L_0, G_p, G_0, L_1, and Δ_1 all lie on the same straight line because of the graphical addition of streams and the line ratio principle. Then from Eq. (7.3-10),

$$\frac{G_p}{\Delta_1} = \frac{\text{line } L_0\Delta_1}{\text{line } L_0G_p} \quad \text{and} \quad \frac{G_0}{\Delta_1} = \frac{\text{line } L_1\Delta_1}{\text{line } L_1G_0}$$

The reflux ratio typically is specified and since

$$\frac{G_0}{G_p} = \frac{G_0}{\Delta_1} \cdot \frac{\Delta_1}{G_p} = \frac{\text{line } L_1\Delta_1}{\text{line } L_1G_0} \cdot \frac{\text{line } L_0G_p}{\text{line } L_0\Delta_1} \tag{7.3-14}$$

Δ_1 can be located from the readily measurable line lengths L_1G_0, L_0G_p, and L_0L_1 or from the projections of these lines onto the appropriate axes of the graph.

A material balance over the lower portion of the column below the feed plane in Fig. 7.3-7 shows that

$$\overline{G} - \overline{L} = G_2 - L_2 = \Delta_2 \tag{7.3-15}$$

This equation may be converted into component B and C balances analogous to Eqs. (7.3-12) and (7.3-13), to demonstrate that the difference between adjacent streams in the column section beneath the feed is constant in amount and composition. This quantity can therefore be represented by a second single difference point Δ_2 on the triangular diagram. Location of Δ_2 requires the following material balance about the feed plane:

$$F + \overline{L} + G = L + \overline{G}$$
$$F = L - G + \overline{G} - \overline{L}$$
$$= \Delta_1 + \Delta_2 \tag{7.3-16}$$

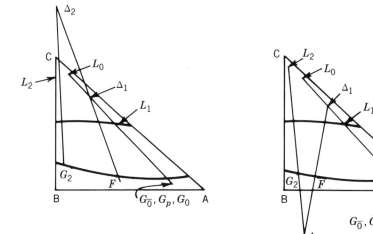

FIGURE 7.3-8   Countercurrent extraction with reflux as in Fig. 7.3-7, showing location of Δ_1 and Δ_2.

The problem specifications will enable points F, G_2, and L_2 to be located in Fig. 7.3-8. Equations (7.3-15) and (7.3-16) show that Δ_2 lies at the intersection of the extended lines L_2G_2 and Δ_1F, in accordance with the graphical addition of streams principle. It is clear that whether Δ_2 lies above or below the triangle merely depends on the relative positions of points L_2, G_2, F, and Δ_1. Thus, when Δ_2 lies above the diagram, Eq. (7.3-16) [together with Eq. (7.3-15)] shows that $\Delta_1 - F = -\Delta_2$. If L_0 or L_2 contains no A or B the corresponding point coincides with apex C.

The operating line on the distribution diagram (y_A versus x_A) may now be obtained from the triangular diagram of Fig. 7.3-8 in the following manner. Corresponding x_A and y_A values for the two phases at a given column cross-section are obtained from the intersection of random lines from the appropriate difference point (Δ_1 or Δ_2) with the upper and lower boundaries, respectively, of the two-phase region. Where reflux is used, as in Fig. 7.3-7, two difference points are required and a discontinuity will occur between the two operating curves representing the two sections of the column. The discontinuity will appear at the point on the distribution diagram corresponding to the intersection of the line $F\Delta_1\Delta_2$ with the upper and lower boundaries of the two-phase region on the triangular diagram. The procedure is sketched in Fig. 7.3-9.

The equilibrium curve on the distribution diagram may be obtained from the values of y_A and x_A read from the terminals of tie lines on the triangular diagram. One such point (L^*, G^*) is shown on the equilibrium curve in Fig. 7.3-9. The distribution diagram is now ready for the evaluation of *either* the number of theoretical stages *or* the NTU, as described later.

As noted earlier, reflux will only be of value in increasing the degree of separation between the feed components when the transfer of A into the C-rich phase is accompanied by significant transfer of B. Consequently, the single-phase region near the C apex of Fig. 7.3-8 will always be present to a significant extent in such cases. (In other words, the upper boundary of the two-phase region will not coincide with line AC when reflux will enhance separation.) The construction is unchanged if the lower boundary of the two-phase region is indistinguishable from the AB axis of the triangle.

EVALUATION OF TERMINAL STREAM FLOW RATES IN FIG. 7.3-7

In the general case, streams F, L_2, L_0, G_p, and G_2 will each contain all three components, A, B, and C, and the composition of these five streams will be specified. Also given will be the flow rate of F and the reflux ratio G_0/G_p to be used. It is then necessary to estimate the flow rates of streams G_p, G_2, L_2, and L_0, the latter determining the size of the C remover needed for the operation. Estimation of these flow rates is performed conveniently on the triangular diagram.

An overall material balance around the extraction unit in Fig. 7.3-7 gives

$$F + L_2 = G_2 + G_p + L_0 = G_2 + \Delta_1 = \Sigma \qquad (7.3\text{-}17)$$

This enables location of Σ and other relevant points in Fig. 7.3-10. Application of the line ratio principle then allows evaluation of both L_2 and Σ, since

$$L_2 = F\left(\frac{\text{line } \Sigma F}{\text{line } \Sigma L_2}\right)$$

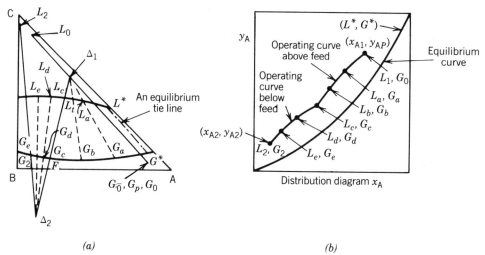

(a) (b)

FIGURE 7.3-9 Construction of the operating lines on the distribution diagram via random lines from Δ_1 and Δ_2 on the triangular diagram. The operation portrayed is that in Fig. 7.3-7.

Next, let

$$\frac{G_2}{\Delta_1} = N_1 \quad \text{so } \Delta_1(1 + N_1) = \Sigma$$

where $N_1 = (\text{line } \Sigma\Delta_1)/(\text{line } G_2\Sigma)$ is known from Fig. 7.3-10; hence, we can evaluate Δ_1 and G_2. From Eq. (7.3-10),

$$L_0 + G_p = \Delta_1 = \text{known}$$

$$\frac{L_0}{G_p} = \frac{\text{line } \Delta_1 G_p}{\text{line } \Delta_1 L_0} = N_2 = \text{known}$$

$$L_p = N_2 G_p \quad \therefore \ G_p(1 + N_2) = \Delta_1$$

hence, we can evaluate G_p and L_0.

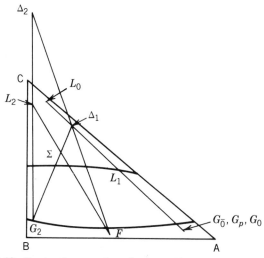

FIGURE 7.3-10 Construction to evaluate flow rates of terminal streams in Fig. 7.3-7.

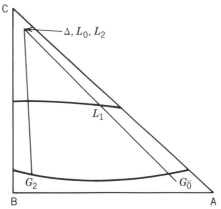

FIGURE 7.3-11   Operation at total reflux in Fig. 7.3-7.

TOTAL REFLUX IN FIG. 7.3-7

A limiting condition in column operation is that of total reflux, in which the flow rates of feed and product streams G_2 and G_p are all zero and $L_2 = L_0 = \Delta$ in amount and composition. The result on triangular coordinates is shown in Fig. 7.3-11. The intersection of random lines from Δ with the upper and lower bounds of the two-phase region would provide points on the single operating curve on an x_A-y_A diagram. No discontinuity would appear in the curve since $F = 0$.

In the special case of L_2 and L_0 consisting of pure C, these points coincide with apex C. This condition corresponds to the conventional definition of total reflux. However, there is an infinite number of operating lines corresponding to total reflux, depending on the particular identical composition of L_2 and L_0 and the consequent location of Δ.

MINIMUM REFLUX RATIO IN FIG. 7.3-7

Suppose that, on the distribution diagram of Fig. 7.3-9, the operating curve intersects or becomes tangent to the equilibrium curve at some point between the terminals of the column. At such a pinch point the two adjacent phases have attained a state of equilibrium, and the driving force causing mass transfer ($y_A - y_{AL}$) has become zero. This, however, requires an infinitely tall column: borrowing from Eq. (7.3-30) developed later in this section, the NTU$_{OG}$ needed to reach this condition is infinite. Referring to the triangular diagram of Fig. 7.3-9, the condition just described evidently corresponds to coincidence between a tie line and a random line from Δ_1 or Δ_2. This amounts to location of Δ_1 or Δ_2 such that some tie line above or below the feed will extrapolate through the difference point appropriate to that section of the column. The situation must be avoided as follows to ensure a column of finite dimensions.

Equation (7.3-14) shows that Δ_1 approaches L_0 on the diagram as the reflux ratio increases. Thus the minimum reflux ratio corresponds to Δ_1 located at the closest intersection to L_0 of all extended tie lines in the column section above the feed with line L_0G_p. ("Above the feed" is in the sense of the right-hand side of Fig. 7.3-7; this could be the lower part of the column in a liquid extraction operation in which the extract phase is more dense than the raffinate phase.)

The point Δ_{1m} located in this manner is then used to find Δ_{2m} at the intersection of lines $F\Delta_{1m}$ and G_2L_2 extended, as shown in Fig. 7.3-12.

If, however, any extended tie lines in the column section below the feed intersect extended line G_2L_2 between Δ_{2m} and L_2, then the intersection closest to L_2 finally determines Δ_{2m}. In this case, the intersection of the line $\Delta_{2m}F$ with line L_0G_p finally determines Δ_{1m}, at a point farther from L_1 than found earlier by extension of tie lines between G_p and F.

The minimum reflux ratio is then obtainable from Eq. (7.3-14) and Fig. 7.3-12 as

$$\left(\frac{G_0}{G_p}\right)_{\min} = \frac{\text{line } L_1\Delta_{1m}}{\text{line } L_1G_0} \cdot \frac{\text{line } L_0G_p}{\text{line } L_0\Delta_{1m}} \tag{7.3-18}$$

The corresponding minimum ratio of solvent to feed is found from the relationship

$$\left(\frac{L_2}{F}\right)_{\min} = \frac{\text{line } \Sigma_m F}{\text{line } \Sigma_m L_2} \tag{7.3-19}$$

This follows from Eq. (7.3-17).

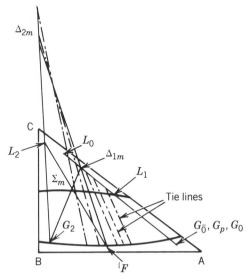

FIGURE 7.3-12 Construction to evaluate the minimum reflux ratio and the minimum solvent-to-feed ratio in Fig. 7.3-7.

7.3-4 Stagewise Contact Calculations

Suppose that the extraction in Figs. 7.3-1 and 7.3-5 is occurring in an extraction unit with stagewise contacting. If the feed stage, labeled "2" in Fig. 7.3-1, is symbolized by a horizontal line, the streams entering and leaving it may be sketched as in Fig. 7.3-13a. If the stage is ideal, y'_{A2} and x_{A2} are in equilibrium and therefore constitute a point on the equilibrium curve in Fig. 7.3-13b.

Furthermore, as explained when deriving the operating line earlier, the points (x_{A2}, y_{A2}) and (x'_{A2}, y'_{A2}) lie on the operating line. Thus, the feed stage (2) may be represented by the shaded "step" between the operating and equilibrium curves of the distribution diagram in Fig. 7.3-13b.

The use of these principles in solving mass balances and equilibrium relationships in sequence leads to a determination of the number of ideal stages required for a specified separation. An example illustrating the graphical solution of component mass balances and equilibrium relationships is given by the broken line construction in Fig. 7.3-5, which shows the need for approximately 3.6 ideal stages. The procedure for determination of the stage requirements in an extraction unit with reflux is entirely analogous to that just described. Considering the x-y diagram in Fig. 7.3-9, the ideal stage at which the stagewise calculations move from one operating line to the other (at the point L_c, G_c) is the stage at which the feed is to be introduced.

7.3-5 Differential Contact in Packed Columns

The continuous contact between phases provided by packed columns requires a differential treatment for a horizontal "slice" of the column, followed by integration over the column height.

TRANSFER UNITS IN EXTRACTION

The transfer of component A from one phase to the other is not accompanied by significant transfer in the reverse direction, so that L and G are not constant between sections 1 and 2 of Fig. 7.3-1. The rate equations (7.1-41) are written in differential form in terms of x_A and y_A as follows:

$$\frac{d(N_A A)}{S} = k_y a(y_A - y_A^*)dH = k_x a(x_A^* - x_A)dH$$

$$= K_y a(y_A - y_{AL})dH = K_x a(x_{AG} - x_A)dH \qquad (7.3-20)$$

where $dV_0 = S\, dH$. Now

$$d(N_A A) = d(Gy_A) = -d(Lx_A) \qquad (7.3-21)$$

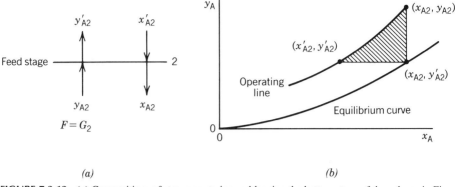

(a)   *(b)*

FIGURE 7.3-13 *(a)* Compositions of streams entering and leaving the bottom stage of the column in Fig. 7.3-1 *(b)* Representation of this stage—assuming it to be ideal—on the distribution diagram.

If it is assumed either that the solute (A) is the only component being transferred or that the solute transfer is accompanied by an equimolal countertransfer of the respective solvents (non-A) between phases, then

$$dG = d(Gy_A) = G \, dy_A + y_A \, dG$$

$$dG = d(Gy_A) = \frac{G \, dy_A}{1 - y_A} \tag{7.3-22}$$

and

$$dL = d(Lx_A) = \frac{L \, dx_A}{1 - x_A} \tag{7.3-23}$$

Combining Eqs. (7.3-20)–(7.3-23) and integrating,

$$\int_0^H dH = \int_{y_{A1}}^{y_{A2}} \frac{G}{Sk_y a} \frac{dy_A}{(1 - y_A)(y_A - y_A^*)}$$

$$= \int_{x_{A1}}^{x_{A2}} \frac{L}{Sk_x a} \frac{dx_A}{(1 - x_A)(x_A^* - x_A)}$$

$$= \int_{y_{A1}}^{y_{A2}} \frac{G}{SK_y a} \frac{dy_A}{(1 - y_A)(y_A - y_{AL})}$$

$$= \int_{x_{A1}}^{x_{A2}} \frac{L}{SK_x a} \frac{dx_A}{(1 - x_A)(x_{AG} - x_A)} \tag{7.3-24}$$

If the process is interpreted in terms of the two-film theory, Eqs. (7.1-45) and (7.1-46) show that the mass transfer coefficients are dependent on the concentration of component A because of the term $(1 - y_A)_{LM}$ or $(1 - x_A)_{LM}$ in the denominator of the right-hand side of each equation. Accordingly, from Eqs. (7.1-45) and (7.1-46), if D_G and D_L do not vary, the quantities $k_y(1 - y_A)_{i,LM}$, $k_x(1 - x_A)_{i,LM}$, $K_y(1 - y_A)_{0,LM}$, and $K_x(1 - x_A)_{0,LM}$ should be independent of concentration (assuming constant m in the case of the overall mass transfer coefficients), where

$$k_y(1 - y_A)_{i,LM} = \frac{k_y(y_A - y_A^*)}{\ln \left[(1 - y_A^*)/(1 - y_A) \right]} \tag{7.3-25}$$

$$k_x(1 - x_A)_{i,LM} = \frac{k_x(x_A^* - x_A)}{\ln \left[(1 - x_A)/(1 - x_A^*) \right]} \tag{7.3-26}$$

$$K_y(1 - y_A)_{0,LM} = \frac{K_y(y_A - y_{AL})}{\ln \left[(1 - y_{AL})/(1 - y_A) \right]} \tag{7.3-27}$$

$$K_x(1 - x_A)_{0,LM} = \frac{K_x(x_{AG} - x_A)}{\ln \left[(1 - x_A)/(1 - x_{AG}) \right]} \tag{7.3-28}$$

Equation (7.3-24) may now be multiplied and divided throughout by either $(1 - y_A)_{LM}$ or $(1 - x_A)_{LM}$, to obtain

$$
\begin{aligned}
H &= \int_{y_{A1}}^{y_{A2}} \frac{G/S}{k_y a(1 - y_A)_{i,LM}} \cdot \frac{(1 - y_A)_{i,LM} \, dy_A}{(1 - y_A)(y_A - y_A^*)} \\
&= \int_{x_{A1}}^{x_{A2}} \frac{L/S}{k_x a(1 - x_A)_{i,LM}} \cdot \frac{(1 - x_A)_{i,LM} \, dx_A}{(1 - x_A)(x_A^* - x_A)} \\
&= \int_{y_{A1}}^{y_{A2}} \frac{G/S}{K_y a(1 - y_A)_{0,LM}} \cdot \frac{(1 - y_A)_{0,LM} \, dy_A}{(1 - y_A)(y_A - y_{AL})} \\
&= \int_{x_{A1}}^{x_{A2}} \frac{L/S}{K_x a(1 - x_A)_{0,LM}} \cdot \frac{(1 - x_A)_{0,LM} \, dx_A}{(1 - x_A)(x_{AG} - x_A)}
\end{aligned}
\tag{7.3-29}
$$

Various experimental correlations for mass transfer coefficients show that k_y or k_x is proportional to $G^{0.8}$ or $L^{0.8}$ and this relationship—at least in terms of mass velocities—has been extended to include the capacity coefficients $k_y a$ and $k_x a$.[1,2] It is therefore customary to consider the capacity coefficients as varying roughly with the first power of flow rate of the corresponding phase between sections 1 and 2. The considerations following Eq. (7.3-24) to the present point provide some justification for regarding quantities such as $G/k_y a(1 - y_A)_{i,LM}$ as constant for a particular situation—generally at an average of the values at sections 1 and 2. These quantities are therefore removed from the integral signs in Eq. (7.3-29) to give

$$
\begin{aligned}
H &= \left[\frac{G/S}{k_y a(1 - y_A)_{i,LM}} \right]_{av} \int_{y_{A1}}^{y_{A2}} \frac{(1 - y_A)_{i,LM} \, dy_A}{(1 - y_A)(y_A - y_A^*)} \\
&= \left[\frac{L/S}{k_x a(1 - x_A)_{i,LM}} \right]_{av} \int_{x_{A1}}^{x_{A2}} \frac{(1 - x_A)_{i,LM} \, dx_A}{(1 - x_A)(x_A^* - x_A)} \\
&= \left[\frac{G/S}{K_y a(1 - y_A)_{0,LM}} \right]_{av} \int_{y_{A1}}^{y_{A2}} \frac{(1 - y_A)_{0,LM} \, dy_A}{(1 - y_A)(y_A - y_{AL})} \\
&= \left[\frac{L/S}{K_x a(1 - x_A)_{0,LM}} \right]_{av} \int_{x_{A1}}^{x_{A2}} \frac{(1 - x_A)_{0,LM} \, dx_A}{(1 - x_A)(x_{AG} - x_A)}
\end{aligned}
\tag{7.3-30}
$$

If the correction factors $(1 - y_A)_{LM}/(1 - y_A)$ or $(1 - x_A)_{LM}/(1 - x_A)$ are neglected for conceptual purposes, then the integrals represent the total change in composition of a given phase between sections 1 and 2 divided by the available driving force causing the transfer. Each integral is therefore a measure of the difficulty of separation and has been defined by Chilton and Colburn[3] as the *number of transfer units* (NTU). Clearly, the ratio $H/$NTU may be called the *height of a transfer unit* (HTU) and is given by the quantity outside each integral in Eq. (7.3-30). The latter relationship shows that there is an individual and an overall NTU expression for each of the G and L phases, combined with *corresponding* HTU expressions as follows:

$$
H = (HTU)_G(NTU)_G = (HTU)_L(NTU)_L = (HTU)_{OG}(NTU)_{OG}
$$
$$
= (HTU)_{OL}(NTU)_{OL}
\tag{7.3-31}
$$

By setting the NTU equal to unity, and with the simplification noted below Eq. (7.3-30), it is evident from that equation that the HTU is the column height necessary to effect a change in phase composition equal to the average driving force in the region under consideration. The evaluation of NTU and HTU to determine the column height needed to obtain a specified separation will be considered shortly.

It may be noted that, although the rate equations (7.3-20) could be integrated to obtain H using mass transfer coefficients, the procedure in terms of transfer units is preferable. This is because the coefficients are strongly dependent on flow rate and composition and therefore would vary with position. In contrast, the HTU has been shown to be less dependent on both flow rate and composition changes in a given application, and this greater stability renders it more suitable for design.

APPROXIMATE EXPRESSIONS FOR NTU IN UNIMOLAL UNIDIRECTIONAL DIFFUSION
In many cases the evaluation of the integrals in Eq. (7.3-30) is facilitated by the use of the arithmetic mean in place of the logarithmic mean $(1 - y_A)_{LM}$ or $(1 - x_A)_{LM}$, incurring only a small error.[5] Thus, in the case of $(NTU)_G$,

$$
(1 - y_A)_{i,LM} \doteq \frac{(1 - y_A^*) + (1 - y_A)}{2}
\tag{7.3-32}
$$

Insertion in the first integral of Eq. (7.3-30) leads to

$$(\text{NTU})_G = \int_{y_{A1}}^{y_{A2}} \frac{dy_A}{y_A - y_A^*} + \tfrac{1}{2} \ln \frac{1 - y_{A1}}{1 - y_{A2}} \tag{7.3-33}$$

The acceptability of Eq. (7.3-32) must be considered in any given case. Application of this approximation to the remaining three integrals in Eq. (7.3-30) results in

$$(\text{NTU})_L = \int_{x_{A1}}^{x_{A2}} \frac{dx_A}{x_A^* - x_A} + \tfrac{1}{2} \ln \frac{1 - x_{A2}}{1 - x_{A1}} \tag{7.3-34}$$

$$(\text{NTU})_{OG} = \int_{y_{A1}}^{y_{A2}} \frac{dy_A}{y_A - y_{AL}} + \tfrac{1}{2} \ln \frac{1 - y_{A1}}{1 - y_{A2}} \tag{7.3-35}$$

$$(\text{NTU})_{OL} = \int_{x_{A1}}^{x_{A2}} \frac{dx_A}{x_{AG} - x_A} + \tfrac{1}{2} \ln \frac{1 - x_{A2}}{1 - x_{A1}} \tag{7.3-36}$$

The reader should perhaps be cautioned regarding the incorrect forms of Eqs. (7.3-34) and (7.3-36) which are often are seen in the literature, where the logarithmic terms are erroneously inverted.

Evaluation of the NTU

The integrals in Eqs. (7.3-30) and (7.3-33)–(7.3-36) are usually evaluated numerically, for example, by graphical integration. Information for this procedure is obtained from the equilibrium curve–operating line plot on (x_A, y_A) coordinates, as sketched in Fig. 7.3-14 for unimolal unidirectional diffusion (liquid extraction).

It is often inconvenient to determine interfacial compositions (x_A^*, y_A^*) corresponding to each point on the operating line, so that overall NTU values are frequently determined in preference to the individual ones. In evaluating the number of G-phase transfer units, values of $1 - y_A$ and $y_A - y_{AL}$ are readily obtained to enable evaluation of the integrand for a series of y_A values between y_{A1} and y_{A2}. (The quantity $y_A - y_A L$ is evidently the vertical distance between the operating line and the equilibrium curve at a given y_A.) The quantity $(1 - y_A)_{o_{LM}}$ for use in NTU_{OG} is defined by Eq. (7.3-27). A plot such as that sketched in Fig. 7.13-15 is then prepared and the NTU_{OG} is given by the area under the curve between the limits of integration.

It should be noted that, in the case of operation with reflux, the change in L and G below the feed plate necessitates the determination of separate NTU values for the column segments above and below the feed—namely, above and below point (L_c, G_c) in Fig. 7.3-9b.

The evaluation of $(\text{NTU})_{OL}$ may be performed in an analogous manner, noting that $x_{AG} - x_A$ is the horizontal distance between the operating line and the equilibrium curve at a given x_A.

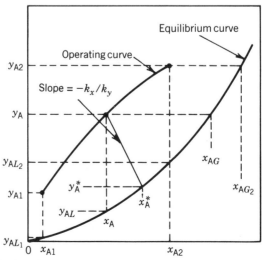

FIGURE 7.3-14 Evaluation of components in the expressions for NTU. The operation is shown in Fig. 7.3-1.

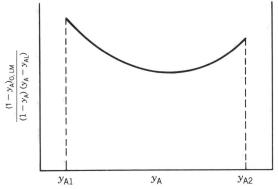

FIGURE 7.3-15 Graphical integration for $(NTU)_{OG}$ in liquid extraction.

The selection of an NTU expression for computation is arbitrary, but normally it is considered appropriate to use the relationships corresponding to the phase offering the greater resistance to mass transfer.

RELATIONSHIP BETWEEN OVERALL AND INDIVIDUAL HTUs
In the case of unimolal unidirectional diffusion, the HTU expressions in Eq. (7.3-30) may be rearranged to give

$$\frac{1}{K_y} = (HTU)_{OG} \frac{a(1 - y_A)_{0,LM}}{G/S} \qquad (7.3\text{-}37)$$

$$\frac{1}{k_y} = (HTU)_G \frac{a(1 - y_A)_{i,LM}}{G/S} \qquad (7.3\text{-}38)$$

$$\frac{1}{k_x} = (HTU)_L \frac{a(1 - x_A)_{i,LM}}{L/S} \qquad (7.3\text{-}39)$$

Inserting these expressions in the first part of Eq. (7.1-40) and multiplying throughout by $G/a(1 - y_A)_{0,LM}S$, we obtain

$$(HTU)_{OG} = (HTU)_G \frac{(1 - y_A)_{i,LM}}{(1 - y_A)_{0,LM}} + \frac{mG}{L} (HTU)_L \frac{(1 - x_A)_{i,LM}}{(1 - y_A)_{0,LM}} \qquad (7.3\text{-}40)$$

A parallel development using the second part of Eq. (7.1-40) will yield

$$(HTU)_{OL} = (HTU)_L \frac{(1 - x_A)_{i,LM}}{(1 - x_A)_{0,LM}} + \frac{L}{m'G} (HTU)_G \frac{(1 - y_A)_{i,LM}}{(1 - x_A)_{0,LM}} \qquad (7.3\text{-}41)$$

When the controlling resistance to transfer lies in the G phase,

$$(1 - y_A)_{i,LM} \doteq (1 - y_A)_{0,LM}$$

If the solutions are also dilute,

$$(1 - x_A)_{i,LM} \doteq (1 - y_A)_{0,LM}$$

Similarly, when the L phase is controlling,

$$(1 - x_A)_{i,LM} \doteq (1 - x_A)_{0,LM}$$

If the solutions are also dilute,

$$(1 - y_A)_{i,LM} \doteq (1 - x_A)_{0,LM}$$

Equations (7.3-40) and (7.3-41) reduce to obviously simple forms in such cases.

Packed column design is often effected by the use of HTU measurements made in a pilot plant in which the system, packing, and flow rates are the same as those to be used on the full scale.

Alternatively, interfacial areas per unit volume, a, and individual mass transfer coefficients are estimated from some existing correlations, leading to formulation of the overall HTU via Eqs. (7.3-37)–(7.3-41). These procedures are described at length by Skelland and Culp (Ref. 4, Chap. 6), who also outlines the correction methods available to allow for axial mixing. The latter results in an increase in the number of transfer units required to achieve a given separation compared to the number expected on the assumption of plug flow of both phases. This assumption is often invalidated by convective mixing currents, channeling effects, recirculation, and entrainment behind the disperse-phase drops.

7.4 STAGEWISE CONTACT IN PERFORATED PLATE COLUMNS

Two partially miscible liquid phases may be contacted for extraction purposes in a discontinuous or stage-wise manner in columns where the stages take the form of perforated plates, fitted with downcomers to facilitate flow of the continuous phase. Bubble-cap plates have been found ineffective in extraction, because of the lower density difference, lower interfacial tension, and higher viscosity of the disperse phase, compared to gas–liquid systems.

Skelland and co-workers[1-6] have developed a procedure for the design of perforated plate extraction columns. This eliminates the need for experimentally measured stage efficiencies, which are usually costly and troublesome to obtain. Additionally, the validity of such efficiencies in scaled-up application is frequently uncertain. Currently, the procedure involves use of rate equations for mass transfer during drop formation either at the perforations or at the end of jets issuing from the perforations, during free rise or fall of the drops, and during coalescence beneath each plate, to locate a pseudoequilibrium curve. The latter is employed instead of the actual equilibrium curve on the $x_A - y_A$ distribution diagram in a stepwise construction between the operating and pseudoequilibrium curves to obtain the number of real stages needed for a given separation.

When flow rates of the disperse phase are low, drop formation and detachment occur at the perforations on each plate. At higher flow rates, however, drops form at the tips of jets emerging from the perforations. Either form of operation is accommodated by the design procedure, but the jetting mode is the more desirable, since throughputs are higher and plate efficiency increases up to 2.5-fold.[7]

The original presentation of the design procedure, in FORTRAN IV computer language and for nonjetting conditions,[1,3] has been both revised and expanded to incorporate jetting operation. The computer printout gives the column diameter needed, the cross-sectional area of the downcomers, the number of perforations per plate, and the number of actual plates needed to achieve a specified separation. Good agreement was obtained with published measurements.[6] The structure of the design procedure will now be outlined.

In contrast to Section 7.3, where the two phases were symbolized by G and L without regard to which was the disperse phase, it is now convenient to identify the continuous and disperse phases as C and D, respectively, denoting the flow rates of each phase through the column cross section in moles/time.

Figure 7.4-1 shows the nth stage of a perforated plate extraction column, where transfer is from the continuous phase to the disperse phase. Drop formation is taking place under jetting conditions and the agitation resulting from motion of the droplets ensures constancy of y_{An}^* for a given stage. The mass transfer rate in stage n can be written in terms of the disperse phase (subscript d) as

$$q = K_{dj}A_j(y_{An}^* - y_{Aj})_{LM} + K_{df}A_f(y_{An}^* - y_{Af})_{LM}$$
$$+ K_{dr}A_r(y_{An}^* - y_{Ar})_{LM} + K_{dc}A_c(y_{An}^* - y_{Ac})_{LM} \qquad (7.4-1)$$

This is the sum of the mass transfer rates to the jets (subscript j), during drop formation at the tip of each jet (subscript f), during free rise (subscript r), and during coalescence (subscript c) beneath plate n + 1. But

$$(y_{An}^* - y_{Aj})_{LM} \doteq (y_{An}^* - y_{Af})_{LM} \doteq (y_{An}^* - y_{An}) \qquad (7.4-2)$$

$$(y_{An}^* - y_{Ac})_{LM} \doteq (y_{An}^* - y_{An+1}) \qquad (7.4-3)$$

and if the variation in D over stage n is slight,

$$(y_{An}^* - y_{Ar})_{LM} \doteq \frac{(y_{An}^* - y_{An}) - (y_{An}^* - y_{An+1})}{\ln\left[(y_{An}^* - y_{An})/(y_{An}^* - y_{An+1})\right]} \qquad (7.4-4)$$

Insertion in Eq. (7.4-1) results in

$$q = (K_{dj}A_j + K_{df}A_f)(y_{An}^* - y_{An})$$
$$+ K_{dr}A_r\left(\frac{y_{An+1} - y_{An}}{\ln\left[(y_{An}^* - y_{An})/(y_{An}^* - y_{An+1})\right]}\right) + K_{dc}A_c(y_{An}^* - y_{An+1}) \qquad (7.4-5)$$

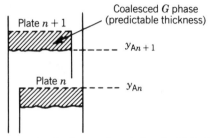

FIGURE 7.4-1   Plates $n + 1$ and n in a perforated plate liquid–liquid extraction column.

Next, the assumption is made either that solute transfer is accompanied by equimolal countertransfer of solvents between phases or that only solute (A) is transferred. Then

$$q = D_{n+1} y_{An+1} - D_n y_{An} \tag{7.4-6}$$

If the inlet for D is at section 2 of the column, a material balance on non-A gives

$$D_n = \frac{D_2(1 - y_{A2})}{1 - y_{An}} \tag{7.4-7}$$

$$D_{n+1} = \frac{D_2(1 - y_{A2})}{1 - y_{An+1}} \tag{7.4-8}$$

A trial-and-error process will yield y_{An+1} corresponding to a given pair of y_{An} and y_{An}^* values in the following way, if A_j, A_f, A_r, A_c, K_{dj}, K_{df}, K_{dr}, and K_{dc} are all predictable:

1. A value of y_{An+1} is assumed, corresponding to a given pair of y_{An} and y_{An}^* values in Fig. 7.4-2.
2. D_n and D_{n+1} corresponding to y_{An} and the assumed y_{An+1} are calculated next from Eqs. (7.4-7) and (7.4-8).
3. Values of q are computed from Eqs. (7.4-5) and (7.4-6).

The value assumed for y_{An+1} is correct when these two estimates of q coincide. This enables construction of the pseudoequilibrium curve, which is then used with the operating curve to step off the number of real plates needed to accomplish the desired change in D-phase composition from y_{A2} to y_{A1}.

The original references[1-6] give details of the methods of estimation of A_j, A_f, A_r, A_c, K_{dj}, K_{df}, K_{dr}, and K_{dc}. An outline is given here, assuming that there are n_0 nozzles or perforations on each plate:

$$A_j \doteq n_0 \pi \left(\frac{d_n + d_{jc}}{2} \right) L_j \tag{7.4-9}$$

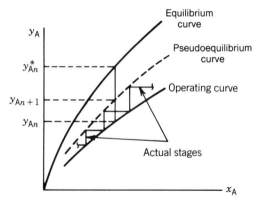

FIGURE 7.4-2   Location of the pseudoequilibrium curve and determination of actual stages.

where d_{jc} is the diameter of the contracted jet at breakup, d_n is the nozzle or perforation diameter, and L_j is the length of the jet. The latter may be predicted with reasonable success.[5,6] Also,

$$A_f = n_0 \pi \, d_p^2 \tag{7.4-10}$$

in which d_p is the predictable diameter[5,6] of the formed drops; A_f is for use with coefficients based on the drop surface at detachment. Next,

$$A_r = \frac{(A_0 - A_D)(H_p - h_c)\,\phi_d}{\text{volume per drop}} \quad \text{(surface area per drop)} \tag{7.4-11}$$

where the coalesced layer thickness h_c and the disperse-phase holdup ϕ_d are predictable.[3] Finally,

$$A_c = A_0 - A_D \tag{7.4-12}$$

The overall mass transfer coefficients based on the disperse phase during jetting, drop formation, free rise, and coalescence are assembled from the corresponding individual coefficients for the disperse and continuous phases, in accordance with Eq. (7.1-40), as

$$\frac{1}{k_{dj,f,r,\text{ or }c}} = \frac{1}{k_{dj,f,r,\text{ or }c}} + \frac{m}{k_{cj,f,r,\text{ or }c}} \tag{7.4-13}$$

During free rise or fall, drops may be stagnant, internally circulating due to the drag of the surrounding continuous phase, or oscillating, depending on drop size, physical properties, and the presence of trace amounts of surface-active contaminants.

The correlations currently preferred for predicting the individual mass transfer coefficients in Eq. (7.4-13) are listed in Table 7.4-1 for systems nominally free from surface-active contamination. Criteria are available[3] for detecting whether drops of a given system are stagnant, circulating, or oscillating.

Reductions in mass transfer rates due to the presence of trace amounts of surface-active contaminants may be substantial. These effects have been measured[8] for each of the two phases during drop formation, free fall, and coalescence and, although correlation was not achieved, at least those existing relationships that came closest to the data in each case were identified. These observations were systematized by Skelland and Chadha,[9] who also developed criteria for selection of the disperse phase in spray and plate extraction columns both in the presence and the absence of surface-active contamination.

7.5 STAGEWISE CONTACT IN MIXER–SETTLERS

Mixer–settlers have been used extensively in liquid–liquid extraction; nevertheless, they have received little theoretical or experimental attention until recently. This was perhaps due to the fact that increased stage efficiency was often attainable merely by increasing the agitator speed, without concern for the corresponding increase in energy consumption. Recent escalation of energy costs has rendered this solution less appealing, so that a closer understanding of the process is needed.

7.5-1 Mixers

Hydrodynamic aspects of the operation include the minimum impeller speed needed to ensure complete dispersion of one liquid phase in the other, and the size of droplets produced under a given set of conditions, preferably during the process of mass transfer.

Skelland and Seksaria[1] and Skelland and Ramsay[2] showed that in some cases an impeller speed of 1000 rpm is insufficient to ensure complete dispersion. Skelland and Ramsay[2] combined their own measurements with those from two other published sources to obtain the following correlation for the minimum impeller speed needed to ensure complete dispersion in a batch unit:

$$(N_{\text{Fr}})_{\min} = C^2 \left(\frac{T}{D}\right)^2 \phi^{0.106}(N_{\text{Ga}}N_{\text{Bo}})^{-0.084} \tag{7.5-1}$$

C and α are tabulated in their paper for flat- and curved-blade turbines, disk turbines, three-bladed propellers, and pitched-blade turbines as a function of position in the vessel. The average absolute deviation between Eq. (7.5-1) and experimental values was 12.7%, and the lowest value of $N_{\min}$ was provided by the six-flat-blade turbine located centrally in the vessel.

Skelland and Lee[3] showed that gross uniformity of dispersion was obtained with values of N that exceeded $N_{\min}$ by an average of about 8%. Furthermore, it was found by Skelland and Moeti[4] that Eq.

TABLE 7.4-1. Relationships Used to Estimate Coefficients in Eqs. (7.4-5) and (7.4-13)

Coefficient	Equation Number	Equation	Reference
k_{dj}	(7.4-14)	$k_{dj} = 2\left(\dfrac{\rho}{M}\right)_{av}\left(\dfrac{D_d}{\pi t_e}\right)^{1/2} \; ; \quad t_e = \dfrac{L_j}{U_j}$	5
k_{df} (based on A_f)	(7.4-15)	$k_{df} = 0.0432\,\dfrac{d_p}{t_f}\left(\dfrac{\rho}{M}\right)_{av}\left(\dfrac{u_0^2}{d_p g}\right)^{0.089}\left(\dfrac{d_p^2}{t_f D_d}\right)^{-0.334}\left(\dfrac{\mu_d}{\sqrt{\rho_d d_p\,\sigma g_c}}\right)^{-0.601}$	10
k_{dr} ($E > 0.5$), stagnant	(7.4-16)	$k_{dr} = \dfrac{-d_p}{6t}\left(\dfrac{\rho}{M}\right)_{av}\ln\left[\dfrac{6}{\pi^2}\sum_{n=1}^{\infty}\dfrac{1}{n^2}\exp\left(\dfrac{-D_d n^2 \pi^2 t}{(d_p/2)^2}\right)\right]$	2
k_{dr} ($E < 0.5$), stagnant	(7.4-17)	$k_{dr} = \dfrac{-d_p}{6t}\left(\dfrac{\rho}{M}\right)_{av}\ln\left(1 - \dfrac{\pi D_d^{1/2} t^{1/2}}{d_p/2}\right)$	11, 12
k_{dr}, circulating	(7.4-18)	$k_{dr} = 31.4\,\dfrac{D_d}{d_p}\left(\dfrac{\rho}{M}\right)_{av}\left(\dfrac{4 D_d t}{d_p^2}\right)^{-0.34}\left(\dfrac{\mu_d}{\rho_d D_d}\right)^{-0.125}\left(\dfrac{d_p u_s^2 \rho_c}{\sigma g_c}\right)^{0.37}$	13
k_{dr}, oscillating	(7.4-19)	$k_{dr} = 0.32\,\dfrac{D_d}{d_p}\left(\dfrac{\rho}{M}\right)_{av}\left(\dfrac{4 D_d t}{d_p^2}\right)^{-0.14}\left(\dfrac{d_p u_s \rho_c}{\mu_c}\right)^{0.68}\left(\dfrac{\sigma^3 g_c^3 \rho_c^2}{g \mu_c^4 \Delta_p}\right)^{0.10}$	13
k_{dc}	(7.4-20)	$k_{dc} = 0.173\,\dfrac{d_p}{t_f}\left(\dfrac{\rho}{M}\right)_{av}\left(\dfrac{\mu_d}{\rho_d D_d}\right)^{-1.115}\left(\dfrac{\Delta\rho g d_p^2}{\sigma g_c}\right)^{1.302}\left(\dfrac{u_s^2 t_f}{D_d}\right)^{0.146}$	10
k_{cj}	(7.4-21)	$k_{cj} = 2\left(\dfrac{\rho}{M}\right)_{av}\left(\dfrac{D_c}{\pi t_e}\right)^{1/2} \; ; \; t_e = \dfrac{L_j}{U_j}$	6
k_{cf} (based on A_f)	(7.4-22)	$k_{cf} = 0.386\left(\dfrac{\rho}{M}\right)_{av}\left(\dfrac{D_c}{t_f}\right)^{0.5}\left(\dfrac{\rho_c \sigma g_c}{\Delta\rho g t_f \mu_c}\right)^{0.407}\left(\dfrac{g t_f^2}{d_p}\right)^{0.148}$	3

TABLE 7.4-1 (*Continued*)

Coefficient	Equation Number	Equation	Reference
k_{cr}, stagnant	(7.4-23)	$k_{cr} = 0.74 \dfrac{D_c}{d_p}\left(\dfrac{\rho}{M}\right)_{av}\left(\dfrac{d_p u_s \rho_c}{\mu_c}\right)^{1/2}\left(\dfrac{\mu_c}{\rho_c D_c}\right)^{1/3}$	14
k_{cr}, circulating	(7.4-24)	$k_{cr} = 0.725\left(\dfrac{\rho}{M}\right)_{av}\left(\dfrac{d_p u_s \rho_c}{\mu_c}\right)^{-0.43}\left(\dfrac{\mu_c}{\rho_c D_c}\right)^{-0.58} u_s\,(1-\phi_d)$	15, 16
k_{cr}, oscillating	(7.4-25)	$k_{cr} = \dfrac{D_c}{d_p}\left(\dfrac{\rho}{M}\right)_{av}\left[50 + 0.0085\left(\dfrac{d_p u_s \rho_c}{\mu_c}\right)^{1.0}\left(\dfrac{\mu_c}{\rho_c D_c}\right)^{0.7}\right]$	17
k_{cc}	(7.4-26)	$k_{cc} = 5.959\,(10^{-4})\left(\dfrac{\rho}{M}\right)_{av}\left(\dfrac{D_c}{t_f}\right)^{0.5}\left(\dfrac{\rho_c u_s^3}{g\mu_c}\right)^{0.332}\left(\dfrac{d_p^2 \rho_c \rho_d u_s^3}{\mu_d \sigma g_c}\right)^{0.525}$	3

(7.5-1) gave good correlation of N_{min} in the presence of various surface-active agents in the aqueous phase, when the diminished value of interfacial tension σ due to the presence of the surfactant is used in the correlation. Thus the expression may be adaptable to contaminated industrial systems.

Coulaloglou and Tavlarides[5] compiled a tabulation of published correlations of drop size in agitated liquid systems, but all the expressions were for the period after steady state had been attained with regard to drop size and mass transfer. In contrast, the correlation by Skelland and Lee[6] is for the Sauter mean droplet diameter $\bar{d}_{32}$ when about 50% of the possible mass transfer has occurred—a significantly different condition.[6] For agitation with a centrally located six-flat-blade turbine with radial baffles they obtained

$$\frac{\bar{d}_{32}}{d_i} = 6.713(10^{-4})\phi^{0.188} \left(\frac{d_i}{T}\right)^{-1.034} (N_{Re})^{-0.558} (N_{Oh})^{-1.025} \tag{7.5-2}$$

The average absolute deviation between Eq. (7.5-2) and data from five systems was 10.9%.

Mass transfer rates in agitated liquid–liquid systems were measured by Shindler and Treybal[7] and Keey and Glen,[8] but correlation over a range of physical properties was prevented by the confinement of their work to single systems.

There are indications[9] that it is often preferable to make the phase offering the controlling resistance to mass transfer continuous. However, some systems may exhibit significant transfer resistance in both phases; alternatively, in some cases the controlling resistance may lie unavoidably in the disperse phase. In such instances, the disperse-phase mass transfer coefficient could be estimated, in principle, from Eq. (7.1-31)—at least in those systems for which the drops may reasonably be assumed to be internally stagnant. This would usually be the case for small drops, high interfacial tension, and in the presence of trace amounts of surface-active impurities. One difficulty, however, lies in the estimation of t, the lifetime of an individual drop within the vessel. This is because t should be regarded as the average time between the formation of a given drop by shearing influences in the vicinity of the impeller, and the coalescence of that drop with others in less-agitated regions of the vessel. The phenomenon appears related to the disperse-phase holdup and the interfacial tension of the system[7,8] but has not been studied to date. Equation (7.1-31) shows that k_d decreases with increasing t, so that a conservative (low) estimate of k_d might be made using a t equal to the average residence time of a given element of disperse phase in the vessel.

Continuous-phase mass transfer coefficients in five liquid–liquid systems agitated by six-flat-blade turbines, centrally located and with radial baffling, were measured and correlated by Skelland and Lee[6] by the expression

$$\frac{k_c}{(ND)^{1/2}} = 2.932(10^{-7})\phi^{-0.508} \left(\frac{d_i}{T}\right)^{0.548} (N_{Re})^{1.371} \tag{7.5-3}$$

for $T = H$; the average absolute deviation from the data was 23.8%.

Since the interfacial area per unit volume, a, is related to the Sauter mean drop diameter as $a = 6\phi/\bar{d}_{32}$, the continuous-phase capacity coefficient for a six-flat-blade turbine, centrally mounted and with radial baffling, is

$$k_c a = 2.621(10^{-3})\frac{(ND)^{1/2}}{d_i} \phi^{0.304} \left(\frac{d_i}{T}\right)^{1.582} (N_{Re})^{1.929} (N_{Oh})^{1.025} \tag{7.5-4}$$

Then for a fixed transfer rate of solute A, q_A(kg/s), in a baffled and agitated vessel of volume V,

$$V = \frac{q_A}{k_c a \Delta \rho_A} \tag{7.5-5}$$

Since vessel size decreases while power consumption increases with increasing impeller speed for a specified q_A, Skelland and Lee[6] were able to obtain optimum expressions for impeller speed, impeller diameter, vessel volume, and power consumption at which total costs are a minimum in a continuous-phase controlled system as follows:

$$N_{opt} = 0.894 \left(\frac{C_1}{C_5}\right)^{0.6469} \left(\frac{q_A}{C_2 \Delta \rho_A}\right)^{-0.3873} \lambda_1^{-1.7136} \tag{7.5-6}$$

$$d_{i,opt} = 1.1009 \left(\frac{C_1}{C_5}\right)^{-0.294} \left(\frac{q_A}{C_2 \Delta \rho_A}\right)^{0.363} \lambda_1^{1.044} \tag{7.5-7}$$

$$V_{opt} = 1.0479 \left(\frac{C_1}{C_5}\right)^{-0.882} \left(\frac{q_A}{C_2 \Delta \rho_A}\right)^{1.0893} \lambda_1^{0.132} \tag{7.5-8}$$

$$P_{opt} = 1.1554 \frac{C_5}{C_3} \left(\frac{C_1}{C_5}\right)^{0.4707} \left(\frac{q_A}{C_2 \Delta \rho_A}\right)^{0.6536} \lambda_1^{0.0792} \qquad (7.5\text{-}9)$$

where C_1, C_2, C_3, and C_5 are constants involving physical properties or cost factors for the case at hand.[6]

Scale-up relationships to obtain equal mass transfer rates per unit volume on two different scales of operation are also given in the original reference.[6]

The concentration driving force $\Delta \rho_A$, appearing in Eqs. (7.5-5)–(7.5-9), is evaluated by assuming that the continuous-phase concentration is uniform throughout the vessel at its effluent value, but that the disperse-phase concentration changes from its inlet to outlet value during the process. This results in

$$\Delta \rho_A = \frac{(\rho^*_{A,C_i} - \rho_{A,C_o}) - (\rho^*_{A,C_o} - \rho_{A,C_o})}{\ln\left[(\rho^*_{A,C_i} - \rho_{A,C_o})/(\rho^*_{A,C_o} - \rho_{A,C_o})\right]} \qquad (7.5\text{-}10)$$

where ρ^*_{A,C_i} and ρ^*_{A,C_o} are the concentrations of solute A in the continuous phase that would be in equilibrium with the solute concentrations in the adjacent disperse phase at their dual inlet and their dual outlet locations, respectively, in a continuously operated mixing vessel. The point is treated in further detail by Skelland and Culp.[10]

7.5-2 Settlers

The two-phase mixture leaving the mixer passes to a settler or decanter, the design of which is performed usually by highly empirical means. Some primary references to settler design are Refs. 10–16. Fundamental work includes a variety of studies on various aspects of the coalescence of a single drop at a plane surface and of two drops into one. One application has been to the case of lower flow rates of dispersion into the horizontal settling vessel, where, instead of covering the entire interface, the dispersion band may assume the form of a "coalescence wedge." The maximum height of the wedge is near the inlet, tapering to single-droplet thickness toward the outlet end of the vessel as coalescence proceeds along its length.[15] Jeffreys and Davies[16] have analyzed this phenomenon in terms of measured "coalescence times"—a statistically distributed quantity requiring a large number of measurements on the particular phases in question. Some success was achieved in predicting the horizontal length of the coalescence wedge, but the approach has not yet emerged as an established procedure for the design of settlers.

One must therefore resort to methods based on flow rates resulting in an acceptable height of the dispersion band[11] or on calculations of the rate of settling of individual drops of disperse phase through the (quiescent) continuous phase. The latter design approach is described in detail by Skelland and Culp[10] and by Hooper and Jacobs.[12]

7.6 MECHANICALLY AGITATED COLUMNS

A number of countercurrent liquid–liquid extraction devices that provide mechanically agitated internals are available. These devices can be divided into two general catagories: (1) columns that contain rotating components and (2) columns that provide reciprocating, or vibrating, internals.

Mechanically agitated columns are useful because they provide one or more additional degrees of freedom which may be adjusted to enhance contactor performance. For example, the frequency of rotation may be adjusted in columns with rotating internals. For reciprocating columns, both the reciprocation frequency and the vertical stroke length are adjustable. To a degree, these variables may be used to adjust the average droplet size in a liquid–liquid system and, in many cases, these features enable operation with reduced HTU or HETS values compared to contacting devices, such as packed beds, that have stationary internals.

In addition, mechanically agitated columns may facilitate a greater volumetric throughput of the two phases compared to an equivalent, stationary column. This attribute is due primarily to the reduced void fraction which the mechanically agitated internals occupy compared to a stationary bed. For example, reciprocating column components typically occupy only about 20% or less of the column cross-sectional area. This feature also tends to make mechanically agitated columns less susceptible to plugging when one of the liquid phases contains significant quantities of undissolved solids or the liquid–liquid system has a tendency to form interfacial debris.

7.6-1 Rotary-Agitated Columns

These contactors provide components that rotate around the vertical axis of the column (see Fig. 7.6-1). Generally, a rotational drive is provided at the top of the contactor. The rotational frequency can be adjusted to increase the interfacial area and the rate of mass transfer between the two phases. To reduce vertical mixing, baffles are often inserted within the column and interspersed between the rotating internals. A number of designs are commercially available.

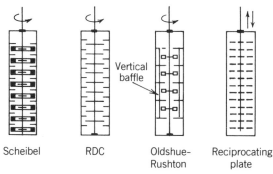

FIGURE 7.6-1 Typical types of mechanically agitated columns. (Reprinted from Ref. 45 by permission of McGraw-Hill.)

YORK–SCHEIBEL COLUMN

There are several versions of this column. The earliest model was introduced[1,2] around 1948 and it was the first to enjoy wide commercial application. It provides alternating compartments to aid dispersion with impellers and coalescence with a wire mesh (about 97% void space). Capacity and mass transfer[3-5] data have been developed for columns with diameters from 25 to 300 mm and in three different liquid–liquid systems. The reported column capacity depends on the system properties but varies from about 14,000 to 24,000 L/h · m² (or 350–600 gal/h · ft²). The Murphree-stage efficiency for a 12 in. diameter column can be correlated[4] as

$$\frac{E_{MD}}{1 - E_{MD}} = 1.45 \times 10^{-11} \left(\frac{H_c}{d_i}\right) (Nd_i)^4 \left(\frac{\delta a'}{\delta c_d}\right) \left(\frac{\Delta\rho'}{\sigma'}\right)^{1.5} \tag{7.6-1}$$

The effect of column diameter on the HETS varies with the square root[5] of the column diameter:

$$\frac{(HETS)_2}{(HETS)_1} = \left(\frac{D_2}{D_1}\right)^{0.5} \tag{7.6-2}$$

Several studies of the drop size distribution[6,7] and axial mixing[8] have also been reported for the York–Scheibel column. The more recent designs[9,10] are reported to give even higher efficiencies. Commercial columns with diameters up to 2.59 m (8.5 ft) have been operated successfully.

ROTATING-DISK CONTACTOR(RDC)

Reman[11] developed this device in 1951. It consists of a series of stator rings which form compartments that have a rotating disk, driven by a central rod, located in the center of each compartment. This device is in widespread use, especially in the petrochemical industry where it has been used for propane deasphalting, aromatic–aliphatic separations using sulfolane, the recovery of caprolactam, and in furfural and sulfur dioxide extraction. It has also been widely studied[12-18] and much is known about scale-up.

Oldshue–Rushton Column

This column also has been available[19] since the 1950s and has been used for a variety of systems. It uses turbine impellers and compartments separated by horizontal stator-ring baffles. The use of columns as large as 2.7 m (9 ft) in diameter[20] have been reported in use.

Most of the research publications[20-28] are based on studies with smaller diameter columns. Bibaud and Treybal[22] give the following correlations for the axial-dispersion efficiencies.
For the Continuous Phase

$$\frac{E_c\psi_c}{V_c H_c} = -0.14 + 0.0268 \left(\frac{d_i N\psi_c}{V_c}\right) \tag{7.6-3}$$

For the Dispersed Phase:

$$\frac{d_i^2 N}{E_d} = 3.93 \times 10^{-8} \left(\frac{d_i^3 N^2 \rho_c}{\sigma g_c}\right)^{1.54} \left(\frac{\rho_c}{\Delta\rho}\right)^{4.18} \left(\frac{d_i^2 N\rho_c}{\mu_c}\right)^{0.61} \tag{7.6-4}$$

ASYMMETRIC ROTATING DISK(ARD)

This column is a variation of the RDC in which the rotating disks are mounted on a vertical shaft that is off center in the column. This design configuration achieves a higher separation efficiency, owing to reduced axial mixing, but at the expense of reduced capacity. It was developed originally in Czechoslovakia by Misek and Marek[29] and extensive research has been completed[30] using this device. It has been used for the extraction of a variety of petrochemicals and columns up to 2.4 m (8 ft) in diameter are used commercially.

The ARD contactor is reported to exhibit satisfactory performance for a wide range of liquid–liquid density differences, interfacial tensions, and solvent-to-feed ratios. Considerable research has been completed on large-scale columns including studies on columns with diameters up to 3.9 m (13 ft). The typical column throughput is on the order of 10,000–30,000 L/h · m² (250–750 gal/h · ft²), depending on the characteristics of the liquid–liquid system. Detailed scale-up procedures[31] also have been developed.

KUHNI CONTACTOR

This unit[32] is similar to the York–Scheibel column except that the turbine impeller is shrouded to promote radial discharge within the mixing compartments. Extensive studies on droplet size, holdup, and back-mixing characteristics have been reported[33–35] and columns with diameters up to 4.95 m (16.5 ft) have been constructed.

GRAESSER RAINING-BUCKET CONTACTOR

This unit differs from the other contactors described previously in that it is a horizontal design. It is also unique in that the rotating buckets[36] disperse both liquid phases in turn. It is particularly useful for systems where the settling time is relatively large because it generates a large droplet size distribution and has been used in the coal-tar industry. Units with diameters of 1.8 m (6 ft) have been constructed which have a reported throughput of about 8600 L/h·m² (210 gal/h·ft²).

7.6-2 Reciprocating-Plate Column

These contactors are vertical columns that promote mixing through the reciprocation of the internals. For a particular column, it is possible to adjust the frequency of reciprocation and the stroke length to optimize the unit efficiency. The original concept of pulsing the internals generally is attributed to Van Dijck[37] in 1935, but working units were not developed until the mid-1950's. Compared to rotating internals, this column design often may have a larger specific throughput. However, for columns with a very large diameter it may suffer from vibrational problems during operation. Small laboratory units (i.e., 1 and 2 in. diameter Karr columns) require stabilization at the top of the column as well as the bottom and make considerable noise during operation.

KARR CONTACTOR

One of the more widely studied and used columns is that designed by Karr and Lo.[38] The column is simple in construction and exhibits a high throughput and stage efficiency (i.e., low HETS) and a great degree of flexibility. As already noted, small laboratory units also have some vibrational problems and require stabilization. This type of design has been widely studied[38–44] and simple scale-up procedures have been reported.[41,44] For scale-up from experimental data in a column with diameter D_1,

$$\frac{(\text{HETS})_2}{(\text{HETS})_1} = \left(\frac{D_2}{D_1}\right)^{0.38}$$

The reciprocating speed at which the larger-diameter column should be operated is given by

$$\frac{(\text{SPM})_2}{(\text{SPM})_1} = \left(\frac{D_1}{D_2}\right)^{0.14}$$

Although these equations are approximate, they have been used successfully to scale-up from laboratory columns with diameters from 2.5–7.5 cm (1–3 in.) to 0.9 m (3 ft).

More recently, Tawfik[45] has developed a generalized correlation for the Karr column based on experimental data obtained at the Georgia Institute of Technology and literature data. He claims the fit is within ±17%.

$$\frac{\text{HETS}}{D} = 1.03 \left(\frac{\rho_c u_T^2}{D \, \Delta\rho \, g}\right)^{0.075} \left(\frac{H_c u_T}{AFt_m}\right)^{1.3} \left(\frac{\sigma}{D\rho_c u_T^2}\right)^{0.625}$$

This equation was based on data from columns ranging from 1 in. to 3 ft in diameter and is confined to low-viscosity fluids. Other variable ranges relating to this correlation are: $1 < AF < 10$ cm/s, $200 < u_T < 2000$ gal/h·ft^2, $10 \leqslant \sigma \leqslant 40$ dyn/cm, and $0.055 \leqslant (1 - \psi_c) \leqslant 0.075$.

A variety of other contactors are available. These are described in more detail elsewhere.[46-48]

7.7 PERFORMANCE AND EFFICIENCY OF SELECTED CONTACTORS

As is apparent from Section 7.6, a wide variety of liquid–liquid contacting devices exist. Consequently, the design engineer is faced with the problem of selecting the most appropriate device for a particular application. In addition, it is usually necessary to estimate an overall extraction efficiency for the operation. Of course, it is preferable to obtain experimental data for a given application. However, in the absence of such data, it is often possible to obtain preliminary efficiency estimates from correlations which are available in the literature.

7.7-1 General Considerations

In many cases, qualitative arguments can be used to arrive at a logical choice for a contactor. It is suggested that these arguments be tabulated (i.e., either pro or con) for a particular case and then used to select the device that appears to have the fewest arguments against it. Of course, it is possible to have overriding considerations (e.g., limited head space).

CENTRIFUGAL CONTACTORS

Since these devices are relatively small, they are favored whenever the installation space is limited. They are generally very compact and can be arranged into banks requiring little head space. Centrifugal contactors are also attractive for circumstances where the solute is particularly valuable (e.g., pharmaceuticals) or the requirements for product cleanliness are strict. If the phase break time (defined in Section 7.8-1) is large, a centrifugal contactor may be chosen for that reason. In the processing of radioactive materials, these contactors facilitate a short residence time and therefore reduce the solvent degradation time compared to alternative devices. Short contact times may also be required if the solute undergoes degradation due to solvent exposure. Finally, centrifugal contactors have the advantages of achieving steady state in a relatively short period and require low inventories of the liquid–liquid phases. These factors make them attractive for batch, or campaign, operations where frequent startup and shutdown are required or where strict inventory control is needed.

There are several disadvantages to high-speed centrifugal contactors. Although they can be efficiently operated over a range of feed-to-solvent flow ratios that is narrower than for alternative devices, they may form stable emulsions in circumstances where other devices, such as the Karr column, would not. This problem may be particularly severe during the treatment of waste streams containing dissolved silica even in concentrations as low as 5 ppm. In addition, centrifugal contactors are subject to plugging and so are not recommended whenever undissolved solids are present in the feed or solvent stream. Finally, they usually require significantly more maintenance than alternative contacting equipment.

MIXER-SETTLERS

Mixer–settlers have the advantage of being relatively insensitive to interfacial debris and undissolved solids. They are also relatively inexpensive and can be scaled-up easily for large-volume applications. Consequently, they are employed frequently in mining where the feed streams are large and interfacial debris is common.

They do require extensive floor space, however, and this can be a disadvantage if many stages are needed. Also, mixer–settlers generally have large residence times and therefore require considerable time to achieve steady state and large liquid–liquid inventories. Another disadvantage of mixer–settlers is that most commercial units are difficult to seal. If the main purpose of the extraction step is to dehydrate an organic material, this feature can preclude their use.

MECHANICALLY AGITATED COLUMNS

For the dehydration of organic materials, mechanically aided columns have many advantages. They can be sealed easily and, compared to packed columns, offer higher volumetric throughputs and good efficiencies for many applications. Because of their larger holdups, they require more time to achieve steady state than centrifugal contactors, but they may be started up and shut down more efficiently than mixer–settlers. On the other hand, they do not retain their concentration profiles because the disperse phase will coalesce at one end of the column during shutdown. This feature is a disadvantage compared to either centrifugal contactors or mixer–settlers. On the other hand, mechanically agitated columns may require less space if head room is available and they are relatively insensitive to undissolved solids and the formation of stable emulsions because they can be operated to form larger droplets. Where reduced shear action is required, they also may be preferred and they are insensitive to silica contamination in the feed stream.

7.7-2 Efficiency and HTU Correlations

In the case of mechanically aided or pulsed columns, the choice of the continuous phase is a key consideration since the columns can be operated with the interface either at the top or the bottom of the column. For spray columns, it is usually more efficient to operate the column with mass transfer from the continuous to the disperse phase[1-3] and the available HTU correlations are more reliable on that basis. To some degree, the same arguments apply to mechanically aided and pulsed columns. In any event, the continuous phase is often chosen so as to maximize the interfacial area per unit length of column. If the solvent-to-raffinate flow ratio is greater than 1, this will occur when the raffinate phase is continuous.

Several correlations are available for estimating HTU values in pulsed columns. Smoot et al.[4,5] studied the system methyl–isobutyl ketone–acetic acid–water and arrived at the following expression for the overall HTU based on the continuous (aqueous) phase:

$$ \text{HTU} = 504 H_c \left(fA_p \frac{d_0 \rho_d}{\mu_d} \right)^{-0.4} \left(\frac{V_c}{fA_p} \right)^{0.43} \left(\frac{V_c}{V_d} \right)^{0.56} \left(\frac{d_0}{H_c} \right)^{0.62} \tag{7.7-1} $$

Logsdail and Thornton[6] correlated the performance of a pulsed plate column with several solvents to obtain the following general correlation for the overall HTU based on the continuous phase:

$$ \text{HTU} = K \left(\frac{\mu_c g}{V_0^3 (1-x)^3 \rho_c} \right)^{2m/3} \left(\frac{\Delta \rho}{\rho_c} \right)^{2(m-1)/3} \left(\frac{V_c^3 \rho_c}{g \mu_c x^3} \right)^{1/3} \left(\frac{V_d}{V_c} \right)^{0.5} \left(\frac{\mu_c^2}{g \rho_c^2} \right)^{1/3} \exp \left(\frac{D}{2} \right) \tag{7.7-2} $$

where K and m are to be evaluated from tests in a laboratory column. In general, the correlation technique is a function of the type of contactor. Various investigators have used a variety of techniques which are described in more detail elsewhere.[7]

For mixer–settlers and centrifugal contactors, it is useful to define an overall stage efficiency relative to the Kremser equations. For ethanol extraction from dilute aqueous mixtures, Tawfik[8] obtained the following empirical, overall efficiency correlation for miniature mixer–settlers:

$$ \eta = 1 - \exp (G) \tag{7.7-3} $$

$$ G = -0.38 (1000 N_{\text{Di}})^{-0.12} (E)^{0.78} \left(\frac{N_{\text{St}}}{100} \right)^{0.9} (N_{\text{Fr}})^{-1.24} \tag{7.7-4} $$

The dispersion number N_{Di} is an approximation technique that embodies the hydrodynamic features and physical properties of the two-phase system. Its use is described in more detail by Leonard et al.[9] Although it is more accurate to correlate efficiencies in terms of phase viscosities, interfacial tensions, and density differences, the dispersion number is a useful first approximation.

Similarly, for an eight-stage high-speed miniature centrifugal contactor (Argonne design, see Leonard et al.[9]), Tawfik[8] found the following expression for ethanol extraction:

$$ \eta = 1 - \exp (G) \tag{7.7-3} $$

$$ G = -0.35 (1000 N_{\text{Di}})^{-1.14} (1 + E)^{-0.82} \left(\frac{N_{\text{St}} N_{\text{Fr}}}{1000} \right)^{0.26} \tag{7.7-5} $$

Since these efficiency correlations are based on limited systems (ethanol extraction only) and in only one size contactor (16-stage Savannah River mini-mixer-settlers or 2 cm Argonne high-speed centrifugal contactor), their applicability to other systems and/or equipment is unknown.

Godfrey and Slater[10] suggest the following empirical relationships for mixer–settler design. For batch data, the efficiency may be expressed as the approach to equilibrium:

$$ \eta_B = \frac{C_0 - C_t}{C_0 - C_e} \tag{7.7-6} $$

Empirically, it is often true that

$$ 1 - \eta_B = \exp (-k t_B) \tag{7.7-7} $$

This enables evaluation of k, which may then be used to relate η_c to t_c in a continuous mixer of the same dimensions. Thus,

TABLE 7.7-1 Characteristic Data of the Extractors Studied

Make	Diameter (mm)	Active Length (m)	Relative Free Cross-Section (%)	Aperture Diameter (mm)	Cell Length (cm)
Graesser	100	1.0	—	—	2.5
PSE	50	3.0	20	2	10
RZE	150	2.0	10.8	—	7.5
Karr	50	2.6	53	9	2.5
PFK	70	3.0	—	—	—
FK	70	3.0	—	—	—
SE	75	2.0	3	2 and 3	15, 25, 45, 60
MS	100	1.2	—	—	—
Kuhni	150	2.0	11.8	—	—
RDC	70	1.7	—	—	—

Source: Excerpted by special permission from *Chemical Engineering*, 76 (Sept. 17, 1984). Copyright © 1984 by McGraw-Hill, New York.

$$\eta_c = \frac{kt_c}{1 + kt_c} \qquad (7.7\text{-}8)$$

Since much of the literature describes different contactors separating different liquid–liquid systems and the evaluation criteria for the various devices have not always been the same, some confusion exists as to whether or not one device is clearly superior to another. At least one study was completed by Stichlmair[11] in which 10 different countercurrent devices were evaluated on a comparative basis for the system toluene–acetone–water. Table 7.7-1 shows the characteristics of the contactors that were examined. Figure 7.7-1 summarizes the results obtained in terms of two key design parameters: the apparent number of theoretical stages/unit length versus the contactor capacity (in m³/h · m²). The mixer-settler results are for a single stage and indicate that the system was operating near 100% efficiency. As can be seen, the Karr column and the static perforated plate column give the highest capacities for this liquid–liquid system, but the Karr column exhibited significantly lower HETP values. The performance of the other mechanically agitated columns was near that of the Karr column, but slightly lower in capacity.

Although of comparative interest, this study is limited by the fact that only a single liquid–liquid system was examined. Also, the ternary mixture was free from impurities and the equipment diameters were relatively small (50–450 mm). Hence, significant differences which are not apparent from this study may result from differences in scale-up efficiencies.

From Fig. 7.7-1, the required column diameter and height for a given task are a function of the equipment design. Similarly, the FOB equipment costs also depend on the equipment type. Humphrey et

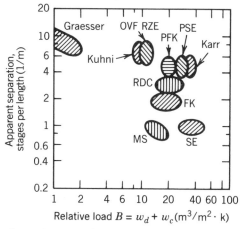

FIGURE 7.7-1 Comparative performance of several contactors for the system toluene–acetone–water. (From Ref. 11 with permission.)

TABLE 7.7-2 Approximate FOB Costs of Selected Extraction Equipment

Type	Material of Construction	Cost[a] Equation ($1000)	Range
Rotating disk	Carbon steel	$34 (ZD^{1.5}/10)^{0.61}$	$0.5 \le ZD^{1.5} \le 75$
Oldshue–Rushton	304 Stainless steel	$57 (ZD/4)^{0.85}$	$0.1 \le ZD \le 200$
		$117 (ZD^{1.5}/10)^{0.66}$	$0.5 \le ZD^{1.5} \le 55$
Pulsed plate	316 Stainless steel	$130 (ZD^{1.5}/10)^{0.81}$	$0.5 \le ZD^{1.5} \le 100$
Reciprocating plate	316 Stainless steel	$130 (ZD^{1.5}/10)^{0.75}$	$0.5 \le ZD^{1.5} \le 100$
Centrifugal contactor	316 Stainless steel	$66 (Q/2.2)^{0.25}$	$0.03 \le Q \le 2.2$
	Alloy 20	$109 (Q/2.2)^{0.48}$	$1.9 \le Q \le 20$
Mixer-settler	Carbon steel	$96 (Q/10)^{0.22}$	$1 \le Q \le 10$
		$96 (Q/10)^{0.60}$	$10 \le Q \le 100$

[a]July 1984 costs, M&S = 781.7. Z = height (m); Q = aqueous feed rate (L/s); D = diameter (m).
Source: From Ref. 12, with permission from McGraw-Hill.

al.[12] summarize the relative FOB costs (shown as Table 7.7-2) for several extraction devices based on information from Woods.[13] Cost evaluation therefore requires that the equipment type be chosen and sized from available correlations that are specific to that device.

7.8 SOLVENT AND PROCESS SELECTION

Liquid–liquid extraction (LLE) is a versatile unit operation which may be used for many applications. Some examples include the recovery of a solute from a raffinate phase, treatment of a wash stream for secondary solute recovery, and the decontamination of wastewaters. In many applications, the solvent selection is a design variable and this choice remains largely a matter of trial and error. Although predictive techniques are available, there is really no substitute for a strong experimental program in screening solvents.

The problem of solvent selection is relatively complex and a thorough treatment requires considerable information. In addition to basic liquid–liquid equilibrium data, knowledge of the phase densities, viscosities, and the liquid–liquid interfacial tension is also important. Moreover, the economics of LLE systems are often dominated by the solvent regeneration costs. If, for example, solvent regeneration is to be accomplished by extractive or azeotropic distillation, then vapor–liquid equilibrium data for the ternary system must also be available. Insofar as the most interesting LLE systems are often those which are least ideal, the generation of a physical property data base to complete cost analysis is usually a significant problem.

There are a number of qualitative considerations, however, that can be used to guide the screening process without detailed knowledge of the physical properties of the multicomponent system. The solvent cost, for example, is often of paramount importance. In general, this factor becomes more important whenever the solute value is relatively low or it is available in the feed stream in low concentrations. Another important factor in this regard is the solvent solubility in the raffinate phase; in general, this decreases with increasing solvent molecular weight and often can be used to guide preliminary screening.

Other solvent properties such as freezing and boiling points, toxicity, corrosiveness, and flammability can also be used to guide the preliminary solvent selection since this information is usually available from other literature data or vendors. In addition, solvent reactivity or decomposition properties should be considered.

7.8-1 Batch Shakeout Tests

A simple experiment in which a proposed solvent is mixed with the feed mixture can be used to guide solvent selection even if the equilibrium phases are not analyzed to determine concentrations. If laboratory facilities are available, batch shakeout tests are often the most attractive way to screen solvents and quickly identify a few suitable candidates for further study.

It is useful in such experiments to define a primary and secondary break time. The primary break time is that time required to form a clearly defined interface between the two phases. The secondary break time is that period needed for the quiescent liquid–liquid mixture to achieve clarity in each of the bulk phases after rapid agitation. Generally, the secondary break time is significantly greater than the primary break time.

If, for example, 5 mL of each phase are placed in a sample vial and vigorously agitated for 10–20 s, then the primary break time should be on the order of 1–2 min, although Treybal[1] has suggested this time could be as great as 5–10 min. However, it is usually more advantageous if the primary break time is less than 1 min.

Leonard et al.[2] have defined a dispersion number as

$$N_{\mathrm{Di}} = \frac{1}{t_B} \left(\frac{\Delta z}{g} \right)^{1/2} \tag{7.8-1}$$

They explain how N_{Di} can be generalized for use in conjunction with high-speed centrifugal contactor design. Although t_B is a function of how the phases are agitated (i.e., of the droplet size distribution) and Δz, it is not a particularly strong function of these variables for many LLE systems. Moreover, simple laboratory screening tests can be completed with reasonable, comparative accuracy simply by standardizing the sample vials, sample aliquot sizes, and the agitation technique.

Liquid–liquid systems in which t_B is short and clear phases are immediately formed are not necessarily attractive since this result probably indicates that little or no solute transfer occurred. On the other hand, systems that disengage more slowly and result in hazy phases are not necessarily bad because they often suggest that at least one species has transferred. If the phase disengagement at room temperature is unacceptable, batch shakeout tests can be conducted using a warm bath.

In addition to noting the primary and secondary break times, any change in the location of the interface after shakeout should also be noted. Such measurements are completed easily if standardized sample vials with flat bottoms are employed in the tests. Other changes, such as the transfer of color from one phase to another, are indicators of mass transfer which can be used to screen solvents without undue effort. In this manner, a large number of candidate solvents can be examined to identify a few systems for more detailed analysis as described below.

7.8-2 Solvent Selection and Modification

The predictive techniques that are now available are useful for solvent screening, but generally they are not accurate enough for design work. These methods usually estimate activity coefficients at the required temperatures and compositions. This information can then be used to compute the equilibrium-phase compositions. Hildebrand and Scott[3,4] developed a method based on regular solution theory which uses the pure-component latent heats of vaporization. Pierotti et al.[5] present a technique that correlates the infinite dilution activity coefficients with the molecular structure. Francis[6,7] identified a way to estimate solvent selectivity based on the critical solution temperatures. More recently, several organizations and investigators[8,9] have begun to develop data banks for liquid–liquid equilibrium data and to correlate these data using the UNIQUAC and NTRL equations (see Chapter 1).

An excellent review of these correlation techniques and their use is presented by Ashton et al.[10] for unreactive systems. Sharma[11] describes extraction with chemical reaction in some detail.

Although the chemistry of organic recovery by liquid–liquid extraction is complex, it is possible to devise reasonable strategies for solvent screening. In all cases, there must be some attractive force that causes the solute to transfer into the solvent phase. Hence, it is necessary to understand the molecular structure of the solute and to postulate solvent characteristics that might be used to effect transfer. Solutes that include basic moieties, for example, will be attracted to acidic extractants. Hydrogen bonding and dipole–dipole interactions also can be exploited. Therefore, an understanding of the chemistry underlying the process is essential to develop a good intuition for solvent candidates.

Reversible, chemical complexation can often be used to effect mass transfer (see Chapter 15). Munson and King[12] provide a review of factors that influence the extraction of ethanol from aqueous solutions. Much of this information can be applied by analogy to other systems. Synergism, which results from blending two or more solvents, can be exploited in some instances. For example, Wardell and King[13] identified a synergism for the mixed solvent, trioctylamine and chloroform, when it is used to extract acetic acid from water. The rationale used to explain this synergism is that the trioctylamine–acetic acid adduct is more extractable into chloroform than acetic acid alone.

Another unpublished synergism has been identified by Tedder for the extraction of lignin from acidified black liquors as they are generated in the Kraft pulping process. The solvent consists of a 1 : 1 by volume mixture of acetone and 2-ethylhexanol. Neither of these solvents alone is effective for lignin extraction from black liquor; acetone by itself is miscible and 2-ethylhexanol alone does not extract lignin. It is postulated that in this case the acetone solvates the acidified lignin moieties by displacing hydration waters and that the resulting dehydrated lignin–acetone adduct is then able to transfer into the 2-ethylhexanol phase. For large, polyfunctional molecules in water, such possibilities should not be overlooked. In the case of lignin extraction, the results are dramatic. Upon acidification in the presence of the mixed solvent, the aqueous, black liquor phase becomes water white while the organic phase develops a ruby red color.

In some cases, a usable solvent may be obtained by blending a viscous, but otherwise suitable, extractant and a diluent. Typical extractants include species such as tri-n-butyl phosphate or tridecyl alcohol which are slow to disengage when used neat. However, these species can be blended with diluents such as kerosene or diisopropylbenzene to form useful blended solvents. To an extent, the solvent selectivity and capacity also can be adjusted in this fashion.

When blended solvents are used, they may form third phases. In these cases, the solute–extractant adduct separates from the diluent to form the third liquid phase. This result is usually troublesome, but it often can be overcome in one of several ways. In some cases, third-phase formation may be avoided by controlling the solvent loading. Although this technique can sometimes be accomplished in the lab, it usually results in operational difficulties for large-scale operations.

Another approach to avoid third-phase formation consists of either using another diluent or adding a modifier (a third component) to the solvent mixture. In the former case, a more polar diluent (e.g., an aromatic blend) may solve the problem. In the latter case, the modifier should consist of a species that has an affinity for both the adduct which forms the third phase and the diluent. Compounds such as tridecyl alcohol and tri-n-butyl phosphate have been used as modifiers.

7.8-3 Solvent Characterization

The two basic parameters that must be estimated for design work are the solute distribution coefficient and the solvent selectivity. At equilibrium, the activity of each component in the B phase is equal to its activity in the C phase; that is,

$$a_{AB} = a_{AC} \tag{7.8-2}$$

$$a_{BB} = a_{BC} \tag{7.8-3}$$

$$a_{CB} = a_{CC} \qquad (7.8\text{-}4)$$

If the distribution coefficients are defined in terms of mole fractions, then

$$K_A = \frac{x_{AC}}{x_{AB}} = \frac{\gamma_{AB}}{\gamma_{AC}} \qquad (7.8\text{-}5)$$

$$K_B = \frac{x_{BC}}{x_{BB}} = \frac{\gamma_{BB}}{\gamma_{BC}} \qquad (7.8\text{-}6)$$

$$K_C = \frac{x_{CC}}{x_{CB}} = \frac{\gamma_{CB}}{\gamma_{CC}} \qquad (7.8\text{-}7)$$

where the first subscript refers to the species and the second refers to the phase. Also, A refers to the solute, B refers to the raffinate phase, and C refers to the solvent. With these definitions (i.e., on a mole fraction basis) the distribution coefficients are equal to the activity coefficient ratios as shown.

Solvent selectivity is then defined as

$$\beta_{AB} = \frac{K_A}{K_B} = \frac{\gamma_{AB}\gamma_{BC}}{\gamma_{AC}\gamma_{BB}} \qquad (7.8\text{-}8)$$

For the extraction of many organic species, it is observed that solvents that yield higher distribution coefficients also yield lower selectivities. In addition, it is observed frequently that as the solvent loading is increased (e.g., due to higher solute concentrations in the raffinate phase), the solvent selectivity decreases. Furthermore, the addition of inextractable species D_{DB} into a feed mixture may increase both the solute distribution coefficient and the solvent selectivity. Hence,

$$K_A = f_1 (x_{AC}, x_{DB}, T) \qquad (7.8\text{-}9)$$

$$\beta_{AB} = f_2 (x_{AC}, x_{DB}, T) \qquad (7.8\text{-}10)$$

The effect of other species besides A and B on the liquid–liquid equilibria is sometimes overlooked. It is, however, a good reason to use actual feed material during solvent characterization experiments. In addition to affecting the distribution coefficients and solvent selectivity, inextractable (or extractable) species may also increase the primary break time (perhaps causing stable emulsion formation) or decrease it significantly.

It is useful to compute the material solubility curve for a liquid–liquid system prior to generating detailed tie line information since the former data are often easier to obtain and can be used to justify simplifying assumptions for the tie line analysis.

Typically, this determination is carried out by starting with either pure solvent or feed material (either of which may be treated as pseudocomponents even if they are mixtures) and then titrating the curve by subsequent additions of the solute and solvent or the solute and feed mixture (without the solute).

Figure 7.8-1 illustrates this process. For example, if one begins with pure B and titrates into it pure species C, then the resulting mixture composition moves from the B apex in Fig. 7.8-1 toward the C apex. This titration is carried out until turbidity results in the mixture (point 1). Subsequently, species A is added until clarity in the mixture is achieved and the composition is at point 2 in Fig. 7.8-1. Continuation of this process leads to points 3, 4, and 5 as indicated. Subsequently, the other side of the solubility curve may be generated by starting with an aliquot of pure C and then titrating into it successive amounts of B and A until turbidity is achieved and removed, respectively.

If the mutual solubility curve is such that the solvent concentrations in the equilibrated raffinate phase are negligible in the region of interest (i.e., the curve nearly touches the A-B edge in Fig. 7.8-1), then the tie line determinations can be simplified. Consider, for example, the case where the phase analyses are carried out by gas chromatography (GC). It is often true that GC peaks for species A and B are obtained readily, but not for the solvent C. In this situation, a reference peak can be added to each phase and estimates of both phase compositions can be determined. This approach requires calibration curves for both species A and B.

On the other hand, the mutual solubility curve can be used together with a single calibration curve which indicates the relative weights of A and B in the chromatographed sample. (Typically, calibration curves are obtained by plotting standard weight ratios versus peak area ratios.) In this case, the sample must be prepared so that known weights (M_{AO}, M_{BO}, and M_{CO}) of A, B, and C are added to the sample vial. After equilibration, injection of samples from each phase yields the A to B weight fraction ratios. In the solvent phase,

$$r_C = \frac{W_{BC}}{W_{AC}} \qquad (7.8\text{-}11)$$

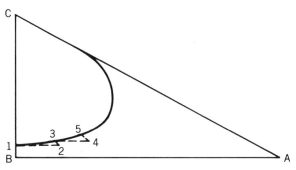

FIGURE 7.8-1   Typical mutual solubility curve with points illustrating titration path.

and in the raffinate phase,

$$r_B = \frac{W_{AB}}{W_{BB}} \tag{7.8-12}$$

Since the solvent concentration in the aqueous phase is negligible,

$$W_{CB} \simeq 0 \tag{7.8-13}$$

and hence

$$W_{BB} \simeq \frac{1}{r_B + 1} \tag{7.8-14}$$

Combining the overall material balances around A and B in the sample yields an estimate of the equilibrated raffinate weight:

$$M_B \simeq \frac{r_C M_{A0} - M_{B0}}{r_C W_{AB} - W_{BB}} \tag{7.8-15}$$

since the total sample weight M_T is known, the solvent weight fraction can be computed as

$$W_{CC} \simeq \frac{M_{C0}}{M_T - M_B} \tag{7.8-16}$$

and the weight fraction of solute in the organic phase is

$$W_{AC} \simeq \frac{1 - W_{CC}}{1 + r_C} \tag{7.8-17}$$

Reference peak spiking must be used whenever the solvent concentrations in the equilibrated raffinate phase cannot be neglected. Although useful in many instances, the above approximations are less likely to be valid when the solute concentration or the equilibration temperature is increased.

For design work, it is useful to correlate equilibrium data in terms of D_i, the weight fraction distribution coefficients. An empirical correlation technique that is simple to use is given by a reduced form of Eq. (7.2-1):

$$\ln D_i = a + b W_{AB} + c V_M + \frac{d}{T} \quad i = A, B \tag{7.8-18}$$

Although useful, it should be noted that the weight fraction distribution coefficient usually is not equal to the mole fraction ratio which is used to define activity coefficients (e.g., for use in predictive equations) and/or thermodynamic properties for the equilibrated phases. On the other hand, Eq. (7.8-18) can be expanded to include higher-order terms when required by the experimental results. The best results for design purposes are obtained when the relevant variables are perturbed using factorial designs[14] for the equilibration experiments in the anticipated LLE operating region.

7.8-4 Process Selection and Examples

The economics of solute recovery or raffinate decontamination by LLE are frequently dominated by the costs of solvent regeneration. Consequently, the design engineer must consider how the solvent is to be regenerated during the solvent selection process since the extract properties largely determine the options. For example, in the recovery of low molecular weight species such as ethanol or acetic acid from water, either extractive or azeotropic distillation may be employed for solvent regeneration. In the former case, a high-boiling solvent is required while in the latter the solvent is low boiling.

When LLE is considered for solute recovery, it should always be compared economically to fractional distillation. Generally, recovery systems based on LLE will be more complex than equivalent distillation systems and they may require more columns. On the other hand, LLE systems may require less energy and lower-pressure steam utilities. As already mentioned, solvents that exhibit higher selectivities usually have lower solute distribution coefficients. Consequently, an economic interaction exists between the LLE unit operations. Solvents that yield higher loadings also require more energy to regenerate since the raffinate impurity, species B, generally is recovered as a distillate. On the other hand, more selective solvents may require less energy to regenerate, but they also result in higher solvent recycle rates and therefore higher equipment costs.

Volatile organics are often recovered from solvent extracts by either azeotropic or extractive distillation. A key consideration in process synthesis is the volatility of the solvent C relative to the solute A and the diluent B. Typically, the normal boiling temperatures of species A and B, T_A and T_B, will be close to each other and the binary pair may even form an azeotrope with boiling temperature T_{AB}—hence the need to consider using extraction to separate A and B. Figures 7.8-2–7.8-5 illustrate several flowsheets that may be used in various situations. The primary differences among these flowsheets are the methods by which the solvent is regenerated. In the discussion that follows, it is helpful to keep in mind that the solute A is typically an organic compound and that it is being extracted most often from an aqueous diluent B by an organic solvent C. From the nature of these compounds, it is more common for C to form an azeotrope with B, having a boiling temperature T_{BC}, than it is to form an azeotrope with A or for there to be a ternary azeotrope ABC, especially if $T_A > T_B$. As shown in the figures, the binary azeotrope obtained in solvent regeneration is often heterogeneous.

Figure 7.8-2 illustrates a process that is suggested for situations in which the following order of boiling points exists:

$$T_A, T_B > T_C > T_{BC} \tag{7.8-19}$$

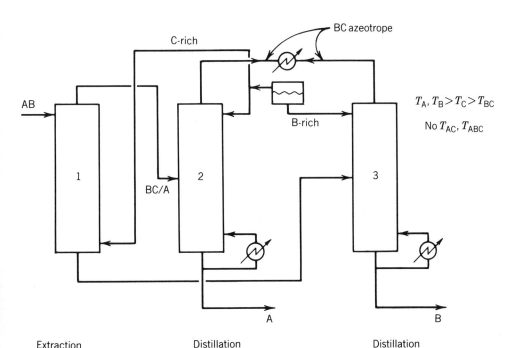

FIGURE 7.8-2 Liquid–liquid extraction and azeotropic distillation recovery system using a more volatile solvent. Column 2 reflux is provided by solvent vaporization.

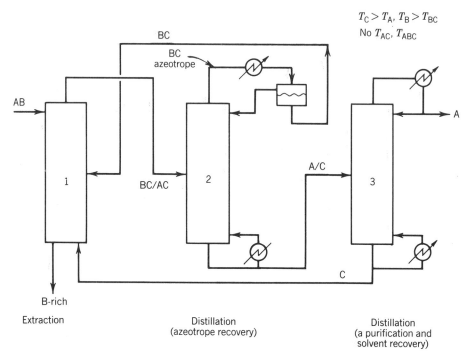

$T_C > T_A, T_B > T_{BC}$
No T_{AC}, T_{ABC}

BC

BC
azeotrope

AB

A

BC/AC

A/C

C

B-rich

| Extraction | Distillation
(azeotrope recovery) | Distillation
(a purification and
solvent recovery) |

FIGURE 7.8-3 Liquid–liquid extraction and azeotropic distillation recovery using a less volatile solvent. Column 2 reflux is provided by solvent vaporization.

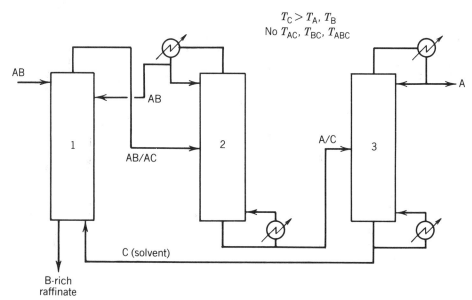

$T_C > T_A, T_B$
No T_{AC}, T_{BC}, T_{ABC}

AB

A

AB

AB/AC

A/C

C (solvent)

B-rich
raffinate

FIGURE 7.8-4 Liquid–liquid extraction and extractive distillation recovery using a less volatile solvent. Column 2 reflux is provided without vaporization of solvent.

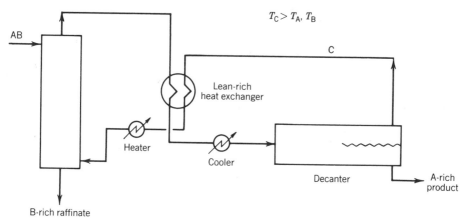

FIGURE 7.8-5 Liquid–liquid extraction and temperature swing recovery of solvent and product. Regeneration is accomplished by exploiting the temperature dependence of the mutual solubility curve.

As is indicated, column 2, the azeotropic column, is performing the split between the BC azeotrope and A. Also, as shown in Fig. 7.8-2, the volatile solvent is condensed to provide reflux for both columns 2 and 3, with the latter column performing the split between B and C. Hence, essentially 100% of the solvent C is volatilized during recycle through columns 2 and 3.

However, if the solvent is less volatile than either A or B but still forms an azeotrope with B, then often

$$T_C > T_A, \qquad T_B > T_{BC} \tag{7.8-20}$$

Even if $T_A > T_B$ it is more common for the BC azeotrope to occur than for the AC or ABC azeotropes to exist since species A is preferentially soluble in C.

In such cases, Fig. 7.8-3 represents a useful process flowsheet. Under these circumstances, column 2 performs the BC/AC split (i.e., it is used to remove the BC azeotrope). Column 3 is used to perform the A/C split.

Comparison of Figs. 7.8-2 and 7.8-3 reveals that while virtually all of species C is vaporized in Fig. 7.8-2, in the latter case only a portion of the solvent C is vaporized during regeneration. Also, while species A may be either more or less volatile than B as a pure component, it is common for B to be more volatile than A in the presence of the solvent. For example, if species A and B are nearly ideal in the vapor phase, which is in equilibrium with the extract at a given temperature, then

$$K_{AV} = \frac{y_A}{x_{AC}} = \frac{\gamma_{AC} P_A}{\pi_V} = \frac{\gamma_{AB} P_A}{\pi_V K_A} \tag{7.8-21}$$

$$K_{BV} = \frac{y_B}{x_{BC}} = \frac{\gamma_{BC} P_B}{\pi_V} = \frac{\gamma_{BB} P_B}{\pi_V K_B} \tag{7.8-22}$$

$$\alpha_{BA} = \frac{K_{BV}}{K_{AV}} = \left(\frac{\gamma_{BB} P_B}{\gamma_{AB} P_A}\right) \beta_{AB} \tag{7.8-23}$$

From Eq. (7.8-23), the relative volatility of species B to A in the vapor phase is directly proportional to the solvent selectivity [Eq. (7.8-8)]. Hence, solvents that are highly selective will tend to result in extractive distillation configurations where the volatility of B is greater than that of species A. Also, the extractive distillation step is enhanced whenever

$$T_A > T_B \tag{7.8-24}$$

The flowsheet in Fig. 7.8-4 is illustrative of situations in which the solvent C is much less volatile than either A or B, so that

$$T_C \gg T_A, T_B \tag{7.8-25}$$

and neither AC nor BC azeotropes are formed. In this process, extractive distillation is used to aid with the B/AC split, and simple distillation is used to achieve the A/C split in column 3. Note that regenerated solvent is fed a few stages from the top of column 2 to maintain reasonable concentrations of C in the sections of the column where extractive distillation is occurring. In those sections

$$\alpha_{BA} \gg 1 \qquad\qquad (7.8\text{-}26)$$

Since species C is a very high boiler, its latent heat of vaporization is also usually greater than that of either A or B and the technique that is suggested in Fig 7.8-4 minimizes the amount of species C which is vaporized.

As shown in Fig. 7.8-4, the B-rich overhead from column 2 is returned to extraction column 1. This is done because of the favorable economics associated with using extraction to complete the separation of A and B. Were this not the case, pure B might be recovered directly from the column 2 overhead stream, and recycle to column 1 would not be favored.

Comparison of Figs. 7.8-2 to 7.8-4 suggests that while virtually all of the solvent is vaporized in Fig. 7.8-2, little or none of species C is vaporized during solvent regeneration in Fig. 7.8-4. In Fig. 7.8-3, intermediate amounts of species C are vaporized. Since the solute loadings are frequently low, the conditions in Fig. 7.8-4 usually result in reduced energy requirements, if all other factors are approximately equal.

In the above-mentioned flowsheets, both species A and B are volatilized while varying amounts of species C are vaporized. Another approach is possible as illustrated in Fig. 7.8-5 in which a temperature swing between the LLE unit and the solvent regeneration step (a decantation) is used to provide solute recovery. In this configuration, favorable changes in the liquid–liquid mutual solubility curve are used to perform the separation.

To illustrate the thermodynamic behavior required by the process in Fig. 7.8-5, consider Fig. 7.8-6. If the LLE step is carried out at temperature T_1 to achieve an extract with composition 1, then cooling this hot extract to temperature T_2 will result in two liquid phases with the compositions at points 2 and 3, which are connected by the equilibrium tie line (points 2 and 3 at T_2), that also intersects the hot extract at composition 1. The relative amounts of raffinate to solvent phase at temperature T_2 are given by the inverse lever rule (i.e., the length from point 1 to 2 divided by the length from point 1 to 3).

An additional requirement for the liquid–liquid system in Fig. 7.8-5 is that the solute distribution coefficient must increase with temperature. Since, from the van't Hoff relationship,

$$\frac{\partial \ln K}{\partial T} = \frac{\Delta H_e}{RT^2} \qquad\qquad (7.8\text{-}27)$$

it follows that the excess heat of equilibration must be greater than zero. Therefore, ΔH_e must be endothermic and the tie line at T_1 (between compositions 1 and 4) is to the left of that at T_2 as drawn in Fig. 7.8-6.

The potential advantages of this last scheme for solvent regeneration are reduced energy requirements and simplicity of operation. The disadvantages are potential high solvent losses to the raffinate phase, low solute recovery, and reduced selectivity. However, it is a potential regeneration technique that often is overlooked.

The recovery of organic solutes (see Fig. 7.8-7) with higher molecular weights may be carried out by reversible chemical complexation. Distillation is seldom attractive in these cases due to the low volatility of the solute and thermal decomposition. Typically, acid–base equilibria may be used in the desired separation. For example, the solute may extract under alkaline conditions in column 1, but strip under acidic conditions in column 2. Usually, the solute A, which is recovered in the extract phase from column 2, will be more dilute than it was in the column 1 feed, but it will be separated from the B diluent as desired. Hence, the column 2 extract is often subjected to subsequent concentration (e.g., evaporation and/ or crystallization) to obtain the solute in the desired form.

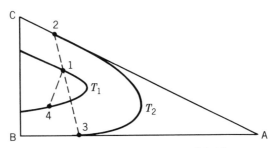

FIGURE 7.8-6   Use of temperature swing to accomplish solvent regeneration.

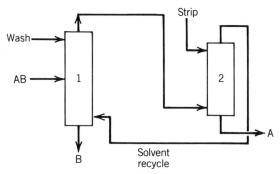

FIGURE 7.8-7 Conceptual flowsheet for solute recovery by reversible chemical complexation.

7.8-5 Acetic Acid Extraction

Acetic acid is commonly recovered from dilute aqueous mixtures using liquid–liquid extraction. A typical flowsheet is shown in Fig. 7.8-8 where liquid–liquid extraction is combined with azeotropic distillation for solvent regeneration. A variety of solvents may be used including acetates, ethers, alcohols, ketones, and chlorinated hydrocarbons.[15-19] In addition, several excellent reviews[20-22] on acid recovery by liquid–liquid extraction are available. Recently, Siebenhofer and Marr[23] presented data on the extraction of several carboxylic acids (formic, acetic, propionic, and butyric) using tertiary amines. Kawano and Kusano[24] present data on acetic acid extraction rates using long-chain alkylamines.

7.8-6 Aromatics Extraction

Liquid–liquid extraction provides one of the most important commercial processes for performing group separations between aromatic and aliphatic species in the petrochemical industry. One of the earliest proven processes, the Udex[25] system, is summarized in Fig. 7.8-9. It is based on the use of diethylene glycol and water, a polar solvent, which selectively extracts the more polar aromatic compounds from the nonpolar aliphatic species. A variety of solvents have been used including diethylene and triethylene glycols,[26] N-methyl pyrrolidine,[27] sulfolane,[28] and other polar solvents. The basic process configurations are similar to those already described, but in some cases a dual solvent system[29] may be used to remove asphaltic materials simultaneously. More recent studies[30,31] give further consideration to alternative solvents for this application.

7.8-7 Pharmaceuticals Extraction

One of the most important applications of LLE in the pharmaceuticals industry is in the recovery of penicillins. Souders et al.[32] describe the early industrial process. Ridgeway and Thorpe[33] give an excellent

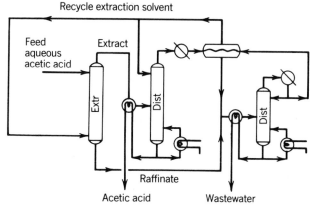

FIGURE 7.8-8 Acetic acid recovery by solvent extraction and azeotropic distillation. (From Ref. 22 with permission.)

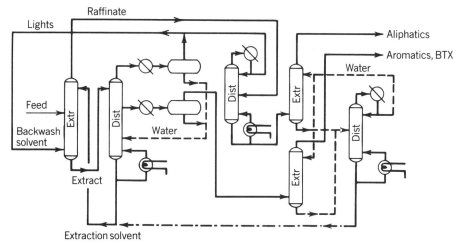

FIGURE 7.8-9 Udex process flowsheet for aromatics extraction. (From Ref. 22 with permission.)

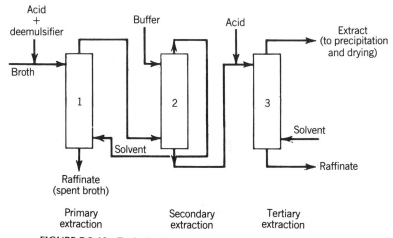

FIGURE 7.8-10 Typical solvent extraction process for penicillin recovery.

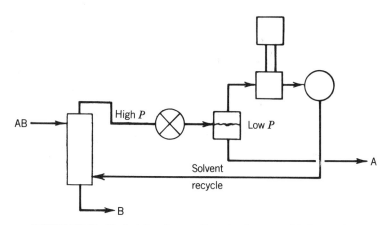

FIGURE 7.8-11 Typical flowsheet configuration for supercritical extraction.

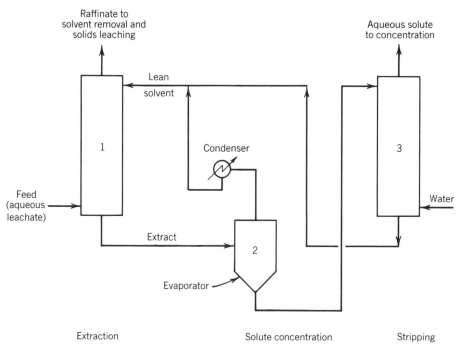

Raffinate to
solvent removal and
solids leaching

Aqueous solute
to concentration

Lean
solvent

1

Condenser

3

Feed
(aqueous
leachate)

Water

Extract

2

Evaporator

Extraction   Solute concentration   Stripping

FIGURE 7.8-12 Typical liquid–liquid extraction system in the food industry. The solvent is never in direct contact with the solid (food).

review of penicillin recovery as well as the recovery of antibiotics and other pharmaceuticals. Reschke and Schugerl[34,35] describe the reactive extraction of penicillin as a function of the type of amine used. A typical penicillin recovery flowsheet in shown is Fig. 7.8-10.

Because of the high molecular weight and reactivities of many pharmaceuticals, they are typically transferred to and from the solvent phase by reversible complexation (e.g., pH adjustment). In another interesting LLE application, Kula et al.[36] describe the use of polyethylene glycol 400 for the purification of enzymes.

7.8-8 Food Industry

An excellent review of LLE applications in the food industry is given by Hamm.[37] Since solvent toxicity is a major consideration, supercritical solvents, such as CO_2, which are nontoxic and LLE from an aqueous leachate are two popular means of dealing with this problem. Figure 7.8-11 is a conceptual supercritical extraction flowsheet. Because T_C is much lower than either T_A or T_B, the solvent (e.g., CO_2) simply is flashed from the solute by throttling (an isenthalpic process). Then the vapors are compressed (an isentropic process) and cooled (an isobaric step) to complete the solvent recycle. Usually, costs are determined by the compressor requirements. Several potential applications of supercritical or near-critical solvents are discussed in more detail elsewhere.[38–42]

Even if atmospheric extraction is employed, applications in the food industry tend to favor the use of volatile, low-molecular weight solvents that are distilled easily from the solutes. Typically, the extract is produced, concentrated, and then backwashed with water to recover the solute. Figure. 7.8-12 illustrates a typical configuration. In this case, water is used to leach the solute from the solid food to form an aqueous leachate feed to the extraction unit. Then the aqueous leachate is contacted with a suitable solvent (which may be toxic) that extracts the solute. The extract is concentrated by solvent evaporation (perhaps 90% of the solvent is volatilized), and the concentrated solvent may be contacted again with water to remove the solute from the concentrate. The lean solvent is combined with the condensed solvent from the evaporator and recycled to LLE contactor 1. In this fashion, the food is never in direct contact with the toxic solvent and the solute is recovered in an aqueous stream. In order for such a LLE process to be efficient, the solvent should possess (1) a high volatility, (2) a low heat of vaporization, (3) a low aqueous solubility, and (4) a high solute solubility.

NOTATION

Section 7.1

A	interfacial area
a	interfacial area per unit volume
c	total molar concentration (moles/volume)
c_A, c_{A*}	molar concentration of component A; equilibrium value at the interface (moles/volume)
$\bar{c}_A$; $\bar{c}_{A1}$, $\bar{c}_{A2}$	average concentration of component A throughout phase; at locations 1 and 2 (moles/volume)
c_{A0}	uniform concentration of component A throughout at $t = 0$ (moles/volume)
c_{A1}, c_{A2}	concentrations of component A at locations 1 and 2 (moles/volume)
c_B; c_{B1}, c_{B2}	concentration of component B; at locations 1 and 2 (moles/volume)
c_i	concentration of component i (moles/volume)
D; D_{AB}, D_{BA}	volumetric molecular diffusivity; of A in B and B in A [(length)^2/time]
D_A°	volumetric molecular diffusivity of A in very dilute solution [(length)^2/time]
D_d, D_G, D_L	volumetric molecular diffusivity of solute in the disperse phase, in the light phase, and in the heavy phase [(length)^2/time]
I_{ix}	molal flux of component i relative to the mass average velocity, both in the x direction (moles/area·time)
J_{Az}, J_{Bz}	molal flux of component A relative to the molal average velocity, both in the z direction (mole/area·time)
J_{ix}	molal flux of component i relative to the molal average velocity, both in the x direction (mole/area·time)
k_{dr}	individual coefficient of mass transfer for the disperse phase during free rise or fall (length/time)
K_X, K_Y	overall mass transfer coefficients based on ΔX and ΔY (mole/area·time) (unit of ΔX or ΔY)
k_X, k_Y	individual mass transfer coefficients based on ΔX and ΔY (mole/area·time) (unit of ΔX or ΔY)
k_c, k_x, k_y	individual mass transfer coefficients based on Δc_A, Δx_A, and Δy_A, defined in Eqs. (7.1-43) and (7.1-44)
M_A, M_B, M_i	molecular weights of components A, B, and i
m	slope of equilibrium curve, see also Fig. 7.1-2
m', m''	defined in Fig. 7.1-2
N_{Az}, N_{Bz}	molal fluxes of A and B in the z direction relative to stationary coordinates (mole/area·time)
N_{ix}	molal flux of component i in the x direction relative to stationary coordinates (mole/area·time)
P	total pressure (force/area)
p_A; p_{A1}, p_{A2}	partial pressure of component A; at locations 1 and 2 (at z_1 and z_2) (force/area)
p_B; p_{B1}, p_{B2}	partial pressure of component B; at locations 1 and 2 (at z_1 and z_2) (force/area)
p_{BLM}	log mean of p_{B1} and p_{B2} (force/area)
R	gas constant (e.g., 1545 ft·lb$_f$/lb-mole·°R)
r, r_s	radius and radius of a sphere
s	fractional rate of surface renewal [(time)$^{-1}$]
T	absolute temperature
t	time or time of free fall of a single drop
t_e	time of exposure
U	molal average velocity in the x direction, Eq. (7.1-1) (length/time)
u	mass average velocity in the x direction, Eq. (7.1-2) (length/time)
u_i	statistical mean velocity of component i in the x direction with respect to stationary coordinates (length/time)
V_{bA}, V_{bB}	molal volumes of components A and B at their normal boiling temperatures (volume/mole)

V_0	contacting volume
w_A	mass fraction of component A
X_A	composition of the heavy phase (any convenient units)
X_A^*	local equilibrium concentration in the heavy phase at the interface (any convenient units)
X_{AG}	heavy-phase concentration that would be in equilibrium with the existing light-phase concentration (any convenient units)
$x_A; x_{A1}, x_{A2}$	mole fraction of component A (in the heavy phase); at locations z_1 and z_2
$x_B; x_{B1}, x_{B2}$	mole fraction of component B; at locations z_1 and z_2
x_{BLM}	log mean of x_{B1} and x_{B2}
Y_A	composition of the light phase (any convenient units)
Y_A^*	local equilibrium concentration in the light phase at the interface (any convenient units)
Y_{AL}	light-phase concentration that would be in equilibrium with the existing heavy-phase concentration (any convenient units)
y_A	mole fraction of component A in the light phase
$z; z_1, z_2$	distance in the direction of diffusion; at locations 1 and 2 along z
z_{fG}, z_{fL}	thickness of fictitious light- and heavy-phase films
z_L	thickness of laminar surface layer in the film–penetration theory
γ	N_{Bz}/N_{Az}
γ_A	activity coefficient of component A
μ_A, μ_{AB}, μ_B	viscosity of A, of the mixture A + B, and of B (cP)
μ_{wT}	viscosity of water at $T(K)$ (cP)
ρ, ρ_A, ρ_i	total density, mass of component A per unit volume, and mass of component i per unit volume (mass/volume)
$(\rho/M)_{av}$	mean value for the phase under consideration (M = molecular weight)
$\phi(t)$	surface age distribution function

Section 7.2

b_i	adjustable parameters, determined by experiment
D_A	solute distribution coefficient (weight fraction of solute in the solvent divided by the solute weight fraction in the raffinate phase)
T	temperature in absolute units
V_M	volume fraction of modifier in the solvent phase
W_A, W_D	weight fractions of solute A and inextractable dissolved species D (e.g., salting agent D) in the raffinate phase
X_A, X_B	mole fractions of solute A and impurity B in the solvent phase

Section 7.3

A	solute
A	area or interfacial area
a	interfacial area per unit volume
B	diluent
C	extracting solvent
F	feed (moles/time)
$G, \overline{G}$	flow rate of raffinate-phase G, also above feed and below feed (moles/time)
G'	flow rate of nontransferring component in G (moles/time)
$G_0, G_{\overline{0}}, G_p$	defined in Fig. 7.3-7 (moles/time)
G_1, G_2	flow rate of phase G at sections 1 and 2 (moles/time)
$G_{\overline{2}}, G_{\overline{2}}$	defined in Fig. 7.3-18 (moles/time)
H	column height
$(HTU)_G, (HTU)_L$	heights of individual G- and L-phase transfer units (length)
$(HTU)_{OG}, (HTU)_{OL}$	heights of overall G- and L-phase transfer units (length)
K_x, K_y, k_x, k_y	overall and individual mass transfer coefficients based on Δx_A and Δy_A (mole/area·time)

$L, \bar{L}$ | flow rate of extract phase L, also above feed and below feed (moles/time)

L' | flow rate of nontransferring component in L (moles/time)

L_0 | flow rate of phase leaving C remover (not $G_{\bar{0}}$) (moles/time)

L_1, L_2 | flow rate of phase L at sections 1 and 2 (moles/time)

M | flow rate of stream M (moles/time)

m, m' | slope of equilibrium curve, see Fig. 7.1-2

N | flow rate of stream N (moles/time)

N_A | molal flux of A relative to stationary coordinates (mole/area·time)

$(NTU)_G, (NTU)_L$ | number of individual G- and L-phase transfer units

$(NTU)_{OG}, (NTU)_{OL}$ | number of overall G- and L-phase transfer units

P | flow rate of stream P (moles/time)

S | cross-sectional area of empty column

w_{AB}, w_{AC} | mass fraction of A in the B- and C-rich phases, respectively

$x_A, x_{A0}, x_{A1}, x_{A2}$ | mole fraction of A in stream L, L_0, L_1, and L_2

x_A^* | local equilibrium concentration of A in the L phase at the interface (mole fraction)

x_{AG} | L-phase concentration that would be in equilibrium with the existing G-phase concentration (mole fraction)

x_{AM}, x_{AN}, x_{AP} | mole fraction of A in streams M, N, and P

x_{BM}, x_{BN}, x_{BP} | mole fraction of B in streams M, N, and P

x_{CM}, x_{CN}, x_{CP} | mole fraction of C in streams M, N, and P

y_A, y_{A1}, y_{A2} | mole fraction of A in stream G, G_1, and G_2

y_A^* | local equilibrium concentration of A in the G phase at the interface (mole fraction)

y_{AL} | G-phase concentration that would be in equilibrium with the existing L-phase concentration (mole fraction)

y_{AP} | defined in Fig. 7.3-7 (mole fraction)

y_B, y_C | mole fraction of components B and C in the G phase

z_{AF} | mole fraction of A in the feed

$\Delta; \Delta_1, \Delta_2$ | difference between flow rates of adjacent streams at any horizontal plane in a column or between consecutive stages; for the portion between section 1 and the feed and for the portion between section 2 and the feed (moles/time)

Σ, Σ_m | sum as defined in Eqs. (7.3-8) and (7.3-17) and corresponding to conditions of minimum reflux ratio (moles/time)

Section 7.4

A_D | cross-sectional area of downcomer

A_j, A_f, A_r, A_c | total interfacial area between two consecutive plates for the respective stages of jetting, drop formation, free rise or fall, and coalescence

A_0 | cross-sectional area of column

D_c, D_d | molecular diffusivity of solute in the continuous and disperse phases [(length)^2/time]

D_n, D_{n+1}, D_2 | flow rates of disperse phase through the column at plate n, at plate $n+1$, and entering the column (moles/time)

d_{jc} | jet diameter at breakup

d_n | perforation or nozzle diameter

d_p | drop diameter (assumed spherical)

g | acceleration due to gravity [length/(time)^2]

g_c | conversion factor (mass·acceleration/force)

h_c | thickness of the coalesced layer

H_p | height between consecutive plates

$K_{dj}, K_{df}, K_{dr}, K_{dc}$ | overall mass transfer coefficients based on the disperse phase during jetting, droplet formation, free rise or fall, and coalescence (moles/area·time)

$k_{cj}, k_{cf}, k_{cr}, k_{cc}$ | individual continuous phase mass transfer coefficients during jetting, droplet formation, free rise or fall, and coalescence (moles/area·time)

$k_{dj}, k_{df}, k_{dr}, k_{dc}$ | individual disperse-phase mass transfer coefficients during jetting, droplet formation, free rise or fall, and coalescence (moles/area·time)

L_j	jet length
M	molecular weight
n_0	number of perforations or nozzles per plate
q	rate of mass transfer (mass/time)
t	time of free rise or fall of a single drop
t_e	time of exposure
t_f	time of formation of a single drop
u_j	jet velocity (length/time)
u_0	velocity through perforation or nozzle (length/time)
u_s	slip velocity (length/time)
y_{Aj}, y_{Af}, y_{Ar}, y_{Ac}	mole fraction of component A in disperse phase during jetting, droplet formation, free rise or fall, and coalescence
y_{An}, y_{An+1}	mole fractions of A in the disperse phase on plates n and $n + 1$
y_{An}^*	disperse-phase concentration that would be in equilibrium with existing continuous-phase concentration in stage n (mole fraction)
y_{A2}	mole fraction of A in the disperse phase entering the column
μ_c, μ_d	viscosity of continuous and disperse phases (mass/length·time)
ρ; ρ_c, ρ_d	density; of continuous and disperse phases (mass/volume)
$\Delta\rho$	$|\rho_c - \rho_d|$
σ	interfacial tension (force/length)
ϕ_d	volume fraction of disperse phase

Section 7.5

a	interfacial area per unit volume (1/length)
C	constant in Eq. (7.5-1), tabulated in Ref. 2 for various impeller types and locations
C_1, C_2, C_3, C_5	constants in Eqs. (7.5-5)–(7.5-9), involving physical properties or cost factors as shown in Ref. 6
D	molecular diffusivity [(length)^2/time]
d_i	impeller diameter
$\bar{d}_{32}$	Sauter mean drop diameter
g	acceleration due to gravity
k_c	continuous-phase mass transfer coefficient (length/time)
N	impeller speed (revolutions/time)
N_{Bo}	Bond number, $d_i^2 g\Delta\rho/\sigma$
N_{Fr}	Froude number, $d_i^2 N^2 \rho_m/g\Delta\rho$
N_{Ga}	Galileo number, $d_i^3 \rho_m g\Delta\rho/\mu_m^2$
N_{Oh}	Ohnesorge number, $\mu_c/(\rho_c d_i \sigma)^{1/2}$
N_{Re}	impeller Reynolds number, $d_i^2 N\rho_c/\mu_c$
P	power consumption [(length·force/time)]
q_A	mass transfer rate of solute A (mass/time)
T	vessel diameter
t	lifetime of an individual drop within the vessel
V	filled volume of vessel
α	a constant corresponding to C above
λ_1	d_i/T
μ_c, μ_d	continuous- and disperse-phase viscosities [force·time/(length)^2]
μ_m	$\dfrac{\mu_c}{1-\phi}\left(1 + \dfrac{1.5\,\mu_d\phi}{\mu_d + \mu_c}\right)$
ρ_c, ρ_d	continuous- and disperse-phase densities (mass/volume)
ρ_m	$\phi\rho_d + (1 - \phi)\rho_c$
ρ_{A,C_i}^*, ρ_{A,C_o}^*,	defined below Eq. (7.5-10)
ρ_{A,C_o}	mass concentration of A in continuous phase leaving vessel (mass of A/volume)
$\Delta\rho$	$|\rho_c - \rho_d|$

$\Delta\rho_A$	concentration driving force
σ	interfacial tension (force/length)
ϕ	volume fraction of dispersed phase

SUBSCRIPTS
min minimum
opt optimum

Section 7.6

A	reciprocation amplitude (ft)
a'	activity in dispersed phase
c_d	concentration of dispersed phase (kg·mol/m³)
D	column diameter (ft)
D_1, D_2	diameters of columns 1 and 2
d_i	rotor or impeller diameter (ft)
E_c, E_d	axial dispersion efficiencies for the continuous and disperse phases
E_{MD}	Murphree stage efficiency
F	reciprocation frequency (s⁻¹)
g	acceleration due to gravity (ft/s²)
g_c	conversion factor (4.18×10^8 lb$_m$·ft/lb$_f$·h²)
H_c	plate spacing or compartment height (ft)
(HETS)$_1$, (HETS)$_2$	height equivalent to a theoretical stage for columns 1 and 2 (ft)
N	rotational speed (rev/h)
SPM$_1$, SPM$_2$	strokes per minute in columns 1 and 2
t_m	mean plate thickness (ft)
u_T	total volumetric flux (ft/s)
V_c	velocity of continuous phase (ft/h)
δ	differential operator
μ_c	continuous-phase viscosity (lb$_m$/ft·h)
ρ_c, ρ_d	continuous- and disperse-phase densities (lb$_m$/ft³)
$\Delta\rho$	positive phase density difference (lb$_m$/ft³)
$\Delta\rho'$	positive phase density difference (g/mL)
σ	interfacial tension (lb$_m$/h² or lb$_m$/s²)
σ'	interfacial tension (dyn/cm)
ψ_c	volume fraction of continuous phase in the contactor active zone

Section 7.7

A	volumetric flow rate of raffinate phase
A_p	pulse amplitude (stroke)
C_e, C_0, C_t	Equilibrium, initial, and after-time-t solute concentrations
D	column diameter
d_0	plate hole diameter
d_p	plate diameter
E	extraction factor
f	pulse frequency
g	acceleration due to gravity
H_c	axial distance between adjacent plates
HTU	overall height of a transfer unit
K	a constant
k	batch data correlation coefficient
L	characteristic mixer-paddle length
N_{Di}	dispersion number, $(1/t_B)\,(\Delta z/g)^{1/2}$
N_{Fr}	the Froude number, $L\omega^2/g$

N_{St}	the Strohal number, $V\omega/(O + A)$
O	volumetric flow rate of solvent phase
Q	aqueous feed flow rate (L/s)
t_B	primary break time
t_c	residence time per stage during continuous mixer–settler operation
V	stage holdup
V_c	velocity of continuous phase
V_d	velocity of disperse phase
$\overline{V}_0$	mean droplet velocity
x	fractional holdup of the disperse phase
Z	column height (m)
Δz	dispersion zone height
η	overall fractional efficiency relative to the Kremser equation
η_B	batch efficiency, or approach to equilibrium, as defined by Eq. (7.7-6)
η_c	estimated approach to equilibrium in a continuous device
μ_c	viscosity of continuous phase
ρ_c, ρ_d	densities of continuous and disperse phases
$\Delta\rho$	density difference of phases
ω	rotational frequency

Section 7.8

a, b, c, d	experimental parameters
a_{ij}	activity of species i in phase j
D_i	W_{iB}/W_{iC}, weight fraction ratio distribution coefficient for species i
g	gravitational acceleration
K_A, K_B, K_C	mole fraction distribution ratio for species A, B, and C
M_{i0}	initial weight of species i in sample
N_{Di}	dispersion number $[(1/t_B)(\Delta Z/g)^{1/2}]$
P_i	pure-component vapor pressure of species i
r_j	weight fraction ratio [Eq. (7.8-12)]
T	absolute temperature
T_A, T_B, T_C	boiling temperature of pure species A, B, and C
T_{AC}, T_{BC}, T_{AB}	boiling temperatures of binary azeotropes AC, BC, and AB
T_{ABC}	boiling temperature of ABC azeotrope
t_B	primary break time
V_m	volume (or weight) fraction modifier (or extractant) in blended solvents
W_{ij}	weight fraction of species i in phase j
x_{ij}	mole fraction of species i in phase j
y_i	mole fraction of species i in vapor phase at equilibrium with the liquid–liquid mixture
ΔH_e	excess heat of equilibration
Δz	height of liquid–liquid mixture in the sample bottle
α_{BA}	relative volatility of species B to A
β_{AB}	solvent selectivity [see Eq. (7.8-8)] of species A to B
γ_{ij}	activity of species i in phase j
π_V	total pressure in equilibrium with liquid–liquid system

REFERENCES

Section 7.1

7.1-1   A. H. P. Skelland, *Diffusional Mass-Transfer*, Wiley-Interscience, New York, 1974.

7.1-2   A. B. Newman, *Trans. AIChE*, **27**, 310 (1931).

7.1-3   J. Crank, *The Mathematics of Diffusion*, Clarendon Press, Oxford, England, 1956.

7.1-4   R. B. Bird, W. E. Stewart, and E. N. Lightfoot, *Transport Phenomena*, Wiley, New York, 1960.

7.1-5   W. Jost, *Diffusion in Solids, Liquids, Gases*, Academic Press, New York, 1952.

7.1-6   R. B. Bird, in T. B. Drew and J. W. Hoopes, Jr. (Eds.), *Advances in Chemical Engineering*, Vol. 1, pp. 155–239, Academic Press, New York, 1956.

7.1-7   J. O. Hirschfelder, C. F. Curtiss, and R. B. Bird, *Molecular Theory of Gases and Liquids*, Wiley, New York, 1954.

7.1-8   R. M. Barrer, *Diffusion in and Through Solids*, Cambridge University Press, New York, 1941.

7.1-9   R. C. Reid, J. M. Prausnitz, and T. K. Sherwood, *The Properties of Gases and Liquids*, 3rd ed., Chap. 11, McGraw-Hill, New York, 1977.

7.1-10 R. E. Treybal, *Liquid Extraction*, 2nd ed., pp. 150–165, McGraw-Hill, New York, 1963.

7.1-11 M. A. Lusis and G. A. Ratcliff, *Can. J. Chem. Eng.*, **46,** 385 (1968).

7.1-12 W. Hayduk and H. Laudie, *AIChE J*, **20,** 611 (1974).

7.1-13 J. Leffler and H. T. Cullinan, *Ind. Eng. Chem. Fundam.*, **9,** 84–88 (1970).

7.1-14 W. Nernst, *Z. Phys. Chem.*, **2,** 613 (1888).

7.1-15 A. R. Gordon, *J. Chem. Phys.*, **5,** 522 (1937).

7.1-16 P. A. Johnson and A. L. Babb, *Chem. Rev.*, **56,** 387 (1956).

7.1-17 H. S. Harned and B. B. Owen, *The Physical Chemistry of Electrolytic Solutions*, ACS Monograph 95, American Chemical Society, Washington, DC, 1950.

7.1-18 R. A. Robinson and R. H. Stokes, *Electrolyte Solutions*, 2nd ed., Academic Press, New York, 1959.

7.1-19 W. K. Lewis, *Ind. Eng. Chem.*, **8,** 825 (1916).

7.1-20 W. G. Whitman, *Chem. Met. Eng.*, **29,** 147 (1923).

7.1-21 R. Higbie, *Trans. AIChE*, **31,** 365 (1935).

7.1-22 P. V. Danckwerts, *Ind. Eng. Chem.*, **43,** 1460 (1951).

7.1-23 H. L. Toor and J. M. Marchello, *AIChE J.*, **4,** 97 (1958).

7.1-24 M. Kishinevskii and A. V. Pamfilov, *J. Appl. Chem. USSR*, **22,** 118 (1949).

7.1-25 M. Kishinevskii, *J. Appl. Chem. USSR*, **24,** 542 (1951).

7.1-26 M. Kishinevskii and M. A. Keraivarenko, *J. Appl. Chem. USSR*, **24,** 413 (1951); *Novik*, **26,** 673 (1953).

7.1-27 M. Kishinevskii, *J. Appl. Chem. USSR*, **27,** 359 (1954).

7.1-28 C. J. King, *Ind. Eng. Chem. Fundam.*, **5,** 1 (1966).

7.1-29 P. Harriott, *Chem. Eng. Sci.*, **17,** 149 (1962).

7.1-30 J. B. Angelo, E. N. Lightfoot, and D. W. Howard, *AIChE J.*, **12,** 751 (1966).

7.1-31 E. Ruckenstein, *Chem. Eng. Sci.*, **23,** 363 (1968).

7.1-32 A. H. P. Skelland and J. M. Lee, *AIChE J.*, **27,** 99 (1981).

7.1-33 T. K. Sherwood, R. L. Pigford, and C. R. Wilke, *Mass Transfer*, Chap. 5, McGraw-Hill, New York, 1975.

Section 7.2

7.2-1   W. V. Tawfik, "Optimization of Fuel Grade Ethanol Recovery Systems Using Solvent Extraction," Ph.D. Thesis, Chemical Engineering, Georgia Institute of Technology, Atlanta, GA, 1986.

7.2-2   J. O. Maloney and A. E. Shubert, *Trans. AIChE*, **36,** 741 (1940).

7.2-3   D. B. Hand, *J. Phys. Chem.*, **34,** 1961 (1930).

7.2-4   R. H. Perry and D. W. Green, *Perry's Chemical Engineers' Handbook*, 6th ed., Chap. 15, McGraw-Hill, New York, 1984.

7.2-5   E. J. Hoffman, *Azeotropic and Extractive Distillation*, Interscience, New York, 1964.

7.2-6   B. D. Smith, *Design of Equilibrium Stage Operations*, McGraw-Hill, New York, 1963.

Section 7.3

7.3-1   G. G. Brown et al., *Unit Operations*, pp. 529–530, Wiley, New York, 1950.

7.3-2   W. L. Badger and J. T. Banchero, *Introduction to Chemical Engineering*, pp. 446–449, McGraw-Hill, New York, 1955.

7.3-3   T. H. Chilton and A. P. Colburn, *Ind. Eng. Chem.*, **27,** 255 (1935).

7.3-4   A. H. P. Skelland and G. L. Culp, *Extraction—Principles of Design*, AIChE Today Series, AIChE, New York, 1982.

7.3-5   J. H. Wiegand, *Trans. AIChE*, **36,** 679 (1940).

Section 7.4

7.4-1   A. H. P. Skelland, *Diffusional Mass Transfer*, Chap. 8., Wiley-Interscience, New York, 1974.

7.4-2   A. H. P. Skelland and A. R. H. Cornish, *Can. J. Chem. Eng.*, **43,** 302 (1965).

7.4-3   A. H. P. Skelland and W. L. Conger, *Ind. Eng. Chem. Process Des. Dev.*, **12,** 448 (1973).

7.4-4   A. H. P. Skelland and A. V. Shah, *Ind. Eng. Chem. Process Des. Dev.*, **14,** 379 (1975).

7.4-5   A. H. P. Skelland and Y-F Huang, *AIChE J.*, **23,** 701 (1977).

7.4-6   A. H. P. Skelland and Y-F Huang, *AIChE J.*, **25,** 80 (1979).

7.4-7   F. W. Mayfield and W. L. Church, *Ind. Eng. Chem.*, **44,** 2253 (1952).

7.4-8   A. H. P. Skelland and C. L. Caenepeel, *AIChE J.*, **18,** 1154 (1972).

7.4-9   A. H. P. Skelland and N. Chadha, *Ind. Eng. Chem. Process Des. Dev.*, **20,** 232 (1981).

7.4-10 A. H. P. Skelland and S. S. Minhas, *AIChE J.*, **17,** 1316 (1971).

7.4-11 T. Vermeulen, *Ind. Eng. Chem.*, **45,** 1664 (1953).

7.4-12 A. I. Johnson, A. E. Hamielec, D. Ward, and A. Golding, *Can. J. Chem. Eng.*, **36,** 221 (1958).

7.4-13 A. H. P. Skelland and R. M. Wellek, *AIChE J.*, **10,** 491–496, 789 (1964).

7.4-14 A. H. P. Skelland and A. R. H. Cornish, *AIChE J.*, **9,** 73 (1963).

7.4-15 C. L. Ruby and J. C. Elgin, *Chem. Eng. Prog. Symp. Ser.*, **51**(16), 17 (1955).

7.4-16 R. E. Treybal, *Liquid Extraction*, 2nd ed., p. 480, McGraw-Hill, New York, 1963.

7.4-17 F. H. Garner and M. Tayeban, *An. Real Soc. Esp. Fis. Quim. Ser. B*, **LVI(B)**, 479 (1960).

Section 7.5

7.5-1   A. H. P. Skelland and R. Seksaria, *Ind. Eng. Chem. Process Des. Dev.*, **17,** 56 (1978).

7.5-2   A. H. P. Skelland and G. G. Ramsay, *Ind. Eng. Chem. Process Des. Dev.*, in press.

7.5-3   A. H. P. Skelland and J. M. Lee, *Ind. Eng. Chem. Process Des. Dev.*, **17,** 473 (1978).

7.5-4   A. H. P. Skelland and L. Moeti, *AIChE* meeting in San Francisco, Nov. 1984.

7.5-5   C. A. Coulaloglou and L. L. Tavlarides, *AIChE J.*, **22,** 289 (1976).

7.5-6   A. H. P. Skelland and J. M. Lee, *AIChE J.*, **27,** 99 (1981); **28,** 1043 (1982).

7.5-7   H. D. Shindler and R. E. Treybal, *AIChE J.*, **14,** 790 (1968).

7.5-8   R. B. Keey and J. B. Glen, *AIChE J.*, **15,** 942 (1969).

7.5-9   A. H. P. Skelland and N. Chadha, *Ind. Eng. Chem. Process Des. Dev.*, **20,** 232 (1981).

7.5-10 A. H. P. Skelland and G. L. Culp, *Extraction—Principles of Design*, AIChE Today Series, AIChE, New York, 1982.

7.5-11 J. Mizrahi and E. Barnea, *Process Eng.*, 60–65 (Jan. 1973).

7.5-12 W. B. Hooper and L. L. Jacobs, Jr., in P. A. Schweitzer (Ed.), *Handbook of Separation Techniques for Chemical Engineers*, pp. 1-343–1-358, McGraw-Hill, New York, 1979.

7.5-13 R. E. Treybal, *Liquid Extraction*, 2nd ed., pp. 440–457, McGraw-Hill, New York, 1963.

7.5-14 R. E. Treybal, *Mass Transfer Operations*, 3rd ed., pp. 527–529, McGraw-Hill, New York, 1980.

7.5-15 R. E. Treybal, in R. H. Perry and C. H. Chilton (Eds.), *Chemical Engineers' Handbook*, 5th ed., Vol. 21, pp. 11–13, McGraw-Hill, New York, 1973.

7.5-16 G. V. Jeffreys and G. A. Davies, in C. Hanson (Ed.), *Recent Advances in Liquid–Liquid Extraction*, Chap. 14, Pergamon Press, New York, 1971.

Section 7.6

7.6-1   E. G. Scheibel, *Chem. Eng. Prog.*, **44,** 681 (1948).

7.6-2   E. G. Scheibel, U.S. Patent No. 2,493,265 (1950).

7.6-3   E. G. Scheibel and A. E. Karr, *Ind. Eng. Chem.*, **42,** 1048 (1950).

7.6-4   A. E. Karr and E. G. Scheibel, *Chem. Eng. Prog. Symp. Ser.*, **10,** 73 (1954).

7.6-5   E. G. Scheibel, *AIChE J.*, **2,** 74 (1956).

7.6-6   J. R. Honekamp and L. E. Burkhart, *Ind. Eng. Chem. Process Des. Dev.*, **1**(3), 176 (1962).

7.6-7   G. V. Jeffreys, G. A. Davies, and H. B. Piper, *Proc. Int. Solv. Ext. Conf.*, 680 (1971).

7.6-8   N. I. Gel'perin et al., *Theor. Found. Chem. Eng.*, **1**, 552 (1967).

7.6-9   E. G. Scheibel, U.S. Patent No. 2,856,362 (1958).

7.6-10   E. G. Scheibel, U.S. Patent No. 3,389,970 (1968).

7.6-11   G. H. Reman, *Proceedings of the 3rd World Petroleum Congress*, Sec. III, p. 121, Wiley, New York, 1951.

7.6-12   J. O. Hinze, *AIChE J.*, **1**, 289 (1955).

7.6-13   S. Stemerding, E. C. Lumb, and J. Lips, *Chem. Eng. Tech.*, **35**, 844 (1963).

7.6-14   F. P. Stainthorp and N. Sudall, *Trans. Inst. Chem. Eng.*, **42**, 198 (1964).

7.6-15   C. P. Strand, R. Olney, and G. H. Ackerman, *AIChE J.*, **8**, 252 (1962).

7.6-16   L. Lu, Z. Fan, and C. Y. Chen, *Proc. Int. Solv. Ext. Conf.*, 98 (1983).

7.6-17   F. G. Zhu, X. D. Ni, and Y. F. Su, *Proc. Int. Solv. Ext. Conf.*, 143 (1983).

7.6-18   W. Y. Fei and M. J. Slater, *Proc. Int. Solv. Ext. Conf.*, 174 (1983).

7.6-19   J. Y. Oldshue and J. H. Rushton, *Chem. Eng. Prog.*, **48**, 297 (1952).

7.6-20   Mixer/Extraction Column, *Chemmunique* (June 1967).

7.6-21   J. Y. Oldshue, *Biotech. Bioeng.*, **8**, 3 (1966).

7.6-22   R. Bibaud and R. Treybal, *AIChE J.*, **12**, 472 (1966).

7.6-23   H. F. Haug, *AIChE J.*, **17**, 585 (1971).

7.6-24   J. Ingham, *Trans. Inst. Chem. Eng.*, **50**, 372 (1972).

7.6-25   J. Dykstra, B. H. Thompson, and R. J. Clouse, *Ind. Eng. Chem.*, **50**, 161 (1958).

7.6-26   R. A. Gustison, R. E. Treybal, and R. C. Capps, *Chem. Eng. Prog. Symp. Ser.*, **58**(39), 8 (1962).

7.6-27   E. B. Gutoff, *AIChE J.*, **11**, 712 (1965).

7.6-28   T. Miyauchi, H. Mitsutake, and I. Harase, *AIChE J.*, **12**, 508 (1966).

7.6-29   T. Misek and J. Marek, *Br. Chem. Eng.*, **15**, 202 (1970).

7.6-30   B. Seidlova and T. Misek, *Proc. Int. Solv. Ext. Conf.*, **3**, 2365 (1974).

7.6-31   T. Misek and J. Marek, in T. C. Lo, M. H. I. Baird, and C. Hanson (Eds.), *Handbook of Solvent Extraction*, p. 407, Wiley-Interscience, New York, 1983.

7.6-32   A. Mogli and U. Buhlmann, in T. C. Lo, M. H. I. Baird, and C. Hanson (Eds.), *Handbook of Solvent Extraction*, p. 441, Wiley-Interscience, New York, 1983.

7.6-33   U. Buhlmann, J. C. Godfrey, and J. Breysse, *Proc. Int. Solv. Ext. Conf.*, 28 (1983).

7.6-34   D. Melzner, A. Mohrmann, W. Halwachs, and K. Schugerl, *Proc. Int. Solv. Ext. Conf.*, 64 (1983).

7.6-35   G. Vatai and A. Tolic, *Proc. Int. Solv. Ext. Conf.*, 100 (1983).

7.6-36   J. Coleby, in T. C. Lo, M. H. I. Baird, and C. Hanson (Eds.), *Handbook of Solvent Extraction*, p. 449, Wiley-Interscience, New York, 1983.

7.6-37   W. J. D. Van Dijck, U.S. Patent 2,011,186 (1935).

7.6-38   A. E. Karr and T. C. Lo, *Proc. Int. Solv. Ext. Conf.*, **1**, 299 (1971).

7.6-39   T. C. Lo and A. E. Karr, *Ind. Eng. Chem. Process Des. Dev.*, **11**, 495 (1972).

7.6-40   A. E. Karr and T. C. Lo, *Proc. Int. Solv. Ext. Conf.*, 355 (1977).

7.6-41   T. C. Lo and J. Prochazka, in T. C. Lo, M. H. I. Baird, and C. Hanson (Eds.), *Handbook of Solvent Extraction*, p. 373, Wiley-Interscience, New York, 1983.

7.6-42   N. V. R. Rao, N. S. Srinivas, and Y. B. G. Varma, *Proc. Int. Solv. Ext. Conf.*, 102 (1983).

7.6-43   A. Bensalem, L. Steiner, and S. Hartland, *Proc. Int. Solv. Ext. Conf.*, 130 (1983).

7.6-44   A. E. Karr, *Proc. Int. Solv. Ext. Conf.*, 151 (1983).

7.6-45   W. V. Tawfik, "Optimization of Fuel Grade Ethanol Recovery Systems Using Solvent Extraction," Ph.D. Thesis, Chemical Engineering, Georgia Institute of Technology, Atlanta, GA, 1986.

7.6-46   T. C. Lo, *Handbook of Separation Techniques for Chemical Engineers*, McGraw-Hill, New York, 1979.

7.6-47   M. H. I. Baird, in T. C. Lo, M. H. I. Baird, and C. Hanson (Eds.), *Handbook of Solvent Extraction*, p. 453, Wiley-Interscience, New York, 1983.

7.6-48   J. L. Humphrey, J. Antonio Rocha, and J. R. Fair, *Chem. Eng.*, 76 (Sept. 17, 1984).

Section 7.7

7.7-1   L. Steiner, M. Horvath, and S. Hartland, *Proc. Int. Solv. Ext. Conf.*, 366, (1977).

7.7-2   A. R. Smith, J. E. Caswell, P. P. Larson, and S. D. Cavers, *Can. J. Chem. Eng.*, **41**, 150 (1963).

7.7-3 E. Muller and H. M. Stonner, in T. C. Lo, M. H. I. Baird, and C. Hanson (Eds.), *Handbook of Solvent Extraction*, p. 311, Wiley-Interscience, New York, 1983.

7.7-4 L. D. Smoot, B. W. Mar, and A. L. Babb, *Ind. Eng. Chem.*, **51**, 1005 (1959).

7.7-5 L. D. Smoot and A. L. Babb, *Ind. Eng. Chem. Fundam.*, **1**, 93 (1962).

7.7-6 D. H. Logsdail and J. D. Thornton, *Trans. Inst. Chem. Eng.*, **35**, 331 (1957).

7.7-7 T. C. Lo, M. H. I. Baird, and C. Hanson (Eds.), *Handbook of Solvent Extraction*, Wiley-Interscience, New York, 1983.

7.7-8 W. Y. Tawfik, "Efficiency of Ethanol Extraction from Aqueous Mixtures," M.S. Thesis, School of Chemical Engineering, Georgia Institute of Technology, Atlanta, 1982.

7.7-9 R. A. Leonard, G. J. Bernstein, R. H. Pelto, and A. A. Ziegler, *AIChE J.*, **27**, 495 (May 1981).

7.7-10 J. C. Godfrey and M. J. Slater, in T. C. Lo, M. H. I. Baird, and C. Hanson (Eds.), *Handbook of Solvent Extraction*, p. 275, Wiley-Interscience, New York, 1983.

7.7-11 J. Stichlmair, *Chem. Ing. Tech.*, **52**, 3 (1980).

7.7-12 J. L. Humphrey, J. A. Rocha, and J. R. Fair, *Chem. Eng.*, **76** (Sept. 17, 1984).

7.7-13 D. R. Woods, in T. C. Lo, M. H. I. Baird, and C. Hanson (Eds.), *Handbook of Solvent Extraction*, p. 919, Wiley-Interscience, New York, 1983.

Section 7.8

7.8-1 R. E. Treybal, *Liquid Extraction*, McGraw-Hill, New York, 1963.

7.8-2 R. A. Leonard, G. J. Bernstein, R. H. Pelto, and A. A. Ziegler, *AIChE J.*, **27**, 495 (1981).

7.8-3 J. H. Hildebrand and R. L. Scott, *Solubility of Non-Electrolytes*, 3rd ed., Dover, New York, 1963.

7.8-4 J. H. Hildebrand and R. L. Scott, *Regular Solutions*, Prentice-Hall, Englewood Cliffs, NJ, 1962.

7.8-5 G. J. Pierotti, C. H. Deal, and E. L. Den, *Ind. Eng. Chem.*, **51**, 95 (Jan. 1959).

7.8-6 A. W. Francis, *Ind. Eng. Chem.*, **36**, 764 (1944).

7.8-7 A. W. Francis, *Critical Solution Temperatures*, American Chemical Society, Washington, DC, 1961.

7.8-8 J. M. Sorensen, T. Magnussen, P. Rasmussen, and A. Fredenslund, *Fluid Phase Equil.*, **3**, 47 (1979).

7.8-9 J. Sorensen and W. Aret, *Liquid–Liquid Equilibrium Data Collection*, DECHEMA Chemistry Data Series, Frankfurt, Part 1, 1979; Parts 2 and 5, 1980.

7.8-10 N. F. Ashton, C. McDermott, and A. Brench, Chemistry of Extraction of Nonreacting Solutes, in T. C. Lo, M. H. I. Baird, and C. Hanson (Eds.), *Handbook of Solvent Extraction*, Wiley-Interscience, New York, 1983.

7.8-11 M. M. Sharma, Extraction with Reaction, in T. C. Lo, M. H. I. Baird, and C. Hanson (Eds.), *Handbook of Solvent Extraction*, Wiley-Interscience, New York, 1983.

7.8-12 C. L. Munson and C. J. King, *Ind. Eng. Chem. Proc. Des. Dev.*, **23**, 109 (1984).

7.8-13 J. M. Wardell and C. J. King, *J. Chem. Eng. Data*, **23**, 144 (1978).

7.8-14 J. D. Murphy, Jr., *Chem. Eng.*, **84**, 180 (June 6, 1977).

7.8-15 P. Eaglesfield, B. K. Kelly, and J. F. Short, *Ind. Chem.*, **29**, 147, 243 (1953).

7.8-16 K. S. McMahan, Acetic Acid, in J. J. McKetta and W. A. Cunningham (Eds.), *Encyclopedia of Chemical Processing and Design*, Vol. 1, pp. 216–240, Marcel Dekker, New York, 1976.

7.8-17 W. V. Brown, *Chem. Eng. Prog.*, **59**, 65–68 (1963).

7.8-18 D. G. Weaver and W. A. Briggs, Jr., *Ind. Eng. Chem.*, **53**, 773 (1961).

7.8-19 E. Lloyd-Jones, *Chem. Ind.*, 1590 (1967).

7.8-20 J. Coleby, Industrial Organic Process, in C. Hanson (Ed.), *Recent Advances in Liquid/Liquid Extraction*, pp. 118–119, Pergamon Press, New York, 1971.

7.8-21 C. J. King, Acetic Acid Extraction, in J. C. Lo, M. H. I. Baird, and C. Hanson (Eds.), *Handbook of Solvent Extraction*, pp. 567–573, Wiley-Interscience, New York, 1983.

7.8-22 L. A. Robbins, Liquid–Liquid Extraction, in P. A. Schweitzer (Ed.), *Handbook of Separation Techniques for Chemical Engineers*, pp. 1-255, 1-282, McGraw-Hill, New York, 1979.

7.8-23 M. Siebenhofer and R. Marr, *Proc. Int. Solv. Ext. Conf.*, 219–220 (1983).

7.8-24 Y. Kawano and K. Kusano, *Proc. Int. Solv. Ext. Conf.*, 311–312 (1983).

7.8-25 H. W. Grate, *Chem. Eng. Prog.*, **54**, 43 (1958).

7.8-26 J. A. Vidueira, Union Carbide Tetra Process, in T. C. Lo, M. H. I. Baird, and C. Hanson (Eds.), *Handbook of Solvent Extraction*, pp. 531–539, Wiley-Interscience, New York, 1983.

7.8-27 E. Muller, NMP (Anosolvan) Process for BTX Separation, in T. C. Lo, M. H. I. Baird, and C. Hanson (Eds.), *Handbook of Solvent Extraction*, pp. 523–529, Wiley-Interscience, New York, 1983.

7.8-28 W. C. G. Kosters, Sulfolane Extraction Processes, in T. C. Lo, M. H. I. Baird, and C. Hanson (Eds.), *Handbook of Solvent Extraction*, p. 541, Wiley-Interscience, New York, 1983.

7.8-29 B. M. Sankey and D. A. Gudelis, Lube Oil Extraction, in T. C. Lo, M. H. I. Baird, and C. Hanson (Eds.), *Handbook of Solvent Extraction*, p. 549, Wiley-Interscience, New York, 1983.

7.8-30 M. C. Annesini, L. Marrelli, R. DeSantis, and I. Kikic, *Proc. Int. Solv. Ext. Conf.*, 217 (1984).

7.8-31 R. Loutaty and C. Yacono, *Proc. Int. Solv. Ext. Conf.*, 215 (1984).

7.8-32 M. Souders, G. J. Pierotti, and C. L. Dunn, *Chem. Eng. Prog. Symp. Ser.*, **66,** Chap. 5 (1970).

7.8-33 K. Ridgeway and E. E. Thorpe, Use of Solvent Extraction in Pharmaceuticals Manufacturing Processes, in J. C. Lo, M. H. I. Baird, and C. Hanson (Eds.), *Handbook of Solvent Extraction*, pp. 583–591, Wiley-Interscience, New York, 1983.

7.8-34 M. Reschke and K. Schugerl, *Chem. Eng. J.*, **28,** B1 (1984).

7.8-35 M. Reschke and K. Schugerl, *Chem. Eng. J.*, **28,** B11 (1984).

7.8-36 M. R. Kula, K. H. Krover, and H. Hustedt, Purification of Enzymes by Liquid–Liquid Extraction, in A. Fiechter (Ed.), *Advances in Biochemical Engineering*, Vol. 24, pp. 73–118, Springer-Verlag, New York.

7.8-37 W. Hamm, Liquid–Liquid Extraction in the Food Industry, in T. C. Lo, M. H. I. Baird, and C. Hanson (Eds.), *Handbook of Solvent Extraction*, pp. 593–603, Wiley-Interscience, New York, 1983.

7.8-38 E. Lark, G. Bunzenberger, and R. Man, *Proc. Int. Solv. Ext. Conf.*, 519–520 (1984).

7.8-39 G. Brunner, *Proc. Int. Solv. Ext. Conf.*, 521 (1984).

7.8-40 H. H. Delert and U. Lutz, *Proc. Int. Solv. Ext. Conf.*, 522 (1984).

7.8-41 T. R. Bott, M. B. King, and D. M. Kassim, *Proc. Int. Solv. Ext. Conf.*, 556–557 (1984).

7.8-42 G. Brunner and S. Peter, *Proc. Int. Solv. Ext. Conf.*, 558–559 (1984).

CHAPTER 8

Extraction—Metals Processing

THOMAS W. CHAPMAN
Department of Chemical Engineering
University of Wisconsin
Madison, Wisconsin

8.1 INTRODUCTION

Although electrolytes normally do not exhibit significant solubility in nonpolar solvents, many metal ions can react with a wide variety of organic compounds to form species that are soluble in organic solvents. Such solubility, which depends on a chemical reaction, provides a basis for separating and concentrating metals that are present as ions in aqueous solution.

As with many separation processes, solvent extraction of metals was first developed as a tool of the analytical chemist. Chemical reagents and solvents are known for isolating virtually every metallic element of the Periodic Table.[1-4] In the 1940s and 1950s some of this basic analytical chemistry was used to develop continuous processes for separating nuclear[5] and rare-earth[6] elements. Subsequently, the availability of inexpensive or especially effective reagents led to the establishment of large-scale processes for extraction of copper, zinc, uranium, and other metals from hydrometallurgical leach liquors.[7-10] The flowsheet of two proposed processes, based on solvent extraction, for separating the constituents of deep-sea manganese nodules are shown in Figs. 8.1-1 and 8.1-2.[11]

The solvent extraction of metals is similar in concept to ion exchange, which is treated in Chapter 13, and provides specific examples of separation by chemical complexation, discussed in Chapter 15. Technological implementation is based on the concepts of ordinary solute extraction, presented in Chapter 7. The choice of solvent extraction over alternative separation schemes is a complex one, but it clearly depends on the availability of an effective extraction agent. A reagent that provides a very high distribution coefficient from dilute solution usually can be found, but reagent selectivity, stability, solubility, kinetics, and cost are all important factors. The ability to reverse the extraction reaction to recover the metal and regenerate the reagent is crucial. In favor of extraction is the relative ease of handling liquid–liquid systems, compared with the solid–liquid operations required in most of the alternative processes.

An extensive literature has developed that offers many choices of chemical systems, and progress is being made in refining the engineering of metal extraction processes. Relevant publications appear in the analytical chemistry literature, the *Journal of Inorganic and Nuclear Chemistry, Separation Science and Technology*, metallurgical journals and conferences, *Hydrometallurgy*, and most recently in the standard chemical engineering periodicals, such as *Industrial and Engineering Chemistry*. Although publications in this field have been widely dispersed, a focal point has been the *Proceedings* of the triennial *International Solvent Extraction Conference*. The *Handbook of Solvent Extraction*[12] provides a number of useful chapters, and Ritcey and Ashbrook[13] discuss many practical aspects of metal extraction processes.

8.2 EXTRACTION CHEMISTRY AND REAGENTS

The extraction of a metal ion from an aqueous solution into an organic solvent is accomplished by the chemical formation of an uncharged species that is soluble in the organic phase. Because metal salts usually

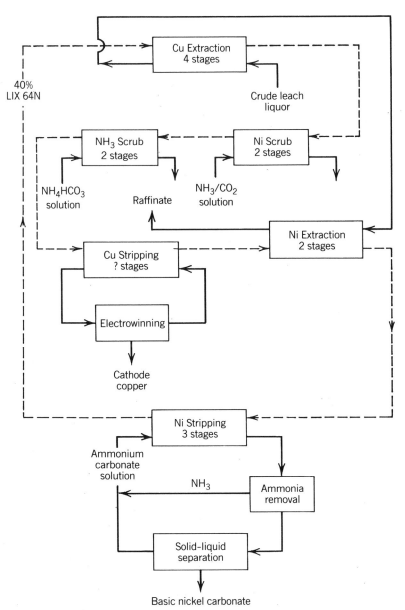

FIGURE 8.1-1 Solvent extraction portion of the Kennecott copper–nickel carbonate process for deep-sea manganese nodules. Adapted from U.S. Patent 3,907,966 in Ref. 11.

are not soluble in organic solvents, the process requires the introduction of a reagent, or *extractant*, that will combine with the metal ion to form an organic-soluble species. For economic and environmental reasons, the extractant chosen to separate a metal must be practically insoluble in water. Thus, the extractant sometimes can constitute the organic-phase solvent of the process. In most cases, however, an inert organic carrier, or *diluent*, is used as the solvent of the organic phase. Additional organic components, called *modifiers*, also are employed. The proper use of diluents and modifiers can prevent formation of very viscous or solid organic phases, can modify extraction kinetics as well as equilibria, can reduce the aqueous solubility of the extractant, and allows flexibility in the choice of phase ratios in contacting equipment.

The extractants that are used to form organic-soluble metal species from aqueous metal salts usually are grouped into three classes according to the type of reaction that occurs. *Solvating extractants* compete with water in coordination bonds around a metal ion and carry neutral salts into the organic phase. *Cation*

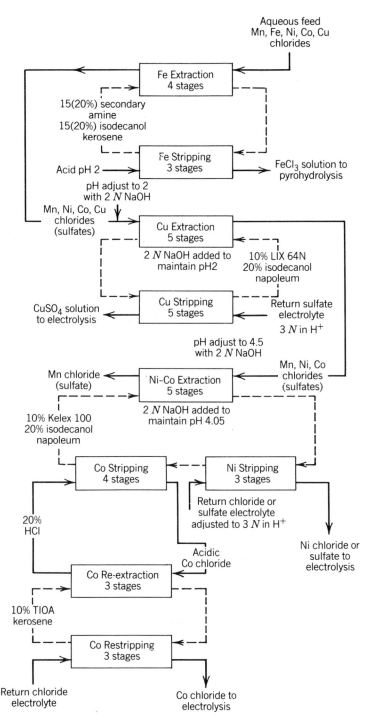

FIGURE 8.1-2 Deepsea Ventures solvent extraction flowsheet for separation of metals in a manganese nodule leach liquor. From Ref. 11, with permission.

TABLE 8.2-1 Examples of Some Common Metal Extractants[a]

Extractant	Trade Name	Chemical Structure	Molecular Weight
Tributyl phosphate	TPB	$\begin{array}{c} R\,O \\ \quad\diagdown \\ R\,O\!-\!P\!=\!O \\ \quad\diagup \\ R\,O \end{array}$ $R\ [=]\ CH_3(CH_2)_2CH_2-$	266
Trioctyl phosphine oxide	TOPO	$\begin{array}{c} R \\ \diagdown \\ R\!-\!P\!=\!O \\ \diagup \\ R \end{array}$ $R\ [=]\ CH_3(CH_2)_6CH_2-$	386
Di-2-ethylhexyl phosphoric acid	DEHPA	$\begin{array}{c} R\text{-}O \quad\ \ O \\ \diagdown\ \diagup \\ P \\ \diagup\ \diagdown \\ R\text{-}O \quad OH \end{array}$ $R\ [=]\ CH_3(CH_2)_3\ CHCH_2- \atop \qquad\qquad\ \ |\atop \qquad\qquad C_2H_5$	322
Versatic acids	Versatic	$\begin{array}{c} R_1 \quad CH_3 \\ \diagdown\ \diagup \\ C \\ \diagup\ \diagdown \\ R_2 \quad COOH \end{array}$	~200

Naphthenic acids	R₁, R₂ [=] alkyl C₄-C₅ groups 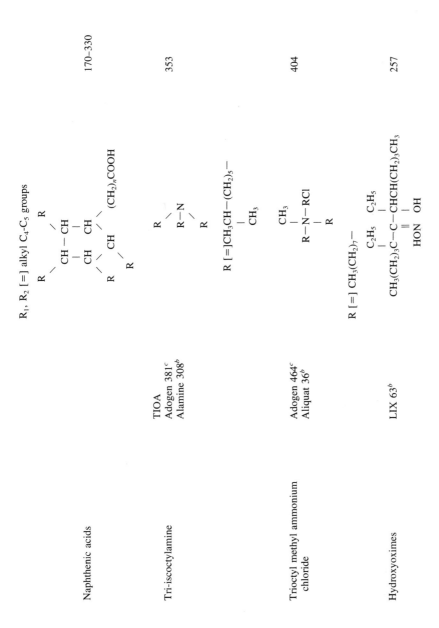	170–330
Tri-iscoctylamine TIOA Adogen 381[c] Alamine 308[b]		353
Trioctyl methyl ammonium chloride Adogen 464[c] Aliquat 36[b]		404
Hydroxyoximes LIX 63[b]		257

$R_1, R_2 \; [=]$ alkyl C_4-C_5 groups

$R \; [=] CH_3CH-(CH_2)_5-$, with CH_3

$R \; [=] CH_3(CH_2)_7-$

TABLE 8.2-1 (*Continued*)

Extractant	Trade Name	Chemical Structure	Molecular Weight
Hydroxybenzohexone oximes	LIX 65N[b]	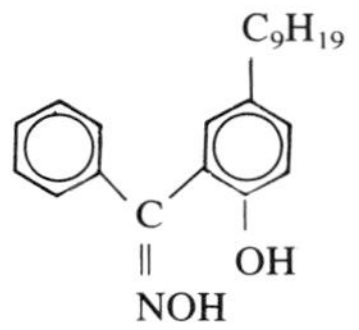 C_9H_{19}, OH, $\overset{\parallel}{C}$, NOH	339
Substituted 8-hydroxyquinoline	LIX 64N[b] Kelex 100[c]	LIX 65N + LIX 63 (1%) CH(CH_2)_2CH(CH_2)_3CH_3, OH, CH_3, C_2H_5	311

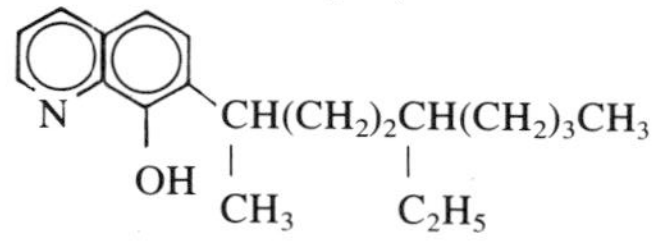

[a] Ritcey and Ashbrook[1] and Flett et al. give more complete lists.
[b] Henkel Corp.
[c] Sherex Corp.

472

exchangers include organic acids as well as specific chelating extractants. *Anion exchangers* extract metals that form anionic complexes. Synergistic *mixtures* of extractants are sometimes found to be effective in promoting a particular separation. The chemical structures of some common extractants are shown in Table 8.2-1.

8.2-1 Solvating Extractants

Extraction of metals by solvating extractants occurs formally by the reaction

$$MX_n + p\overline{S} \rightleftharpoons \overline{MX_nS_p} \tag{8.2-1}$$

where MX_n is an uncharged salt coming out of the aqueous phase and S is the extractant. Organic-phase species are denoted by an overline. Inorganic acid can also be extracted by a reaction of the form

$$q\,HX + \overline{S} \rightleftharpoons \overline{(HX)_q S} \tag{8.2-2}$$

The most common solvating extractants are alkyl phosphate esters, such as tributyl phosphate (TBP) and trioctyl phosphine oxide (TOPO), but thiophosphate esters, phosphine sulfides, ketones, alcohols, and esters also have been used.

The common features of these solvating extractants is that they contain both electron-rich oxygen or sulfur atoms that can coordinate metal cations and multiple alkyl chains that make them hydrophobic. It is the electron-rich atoms that can bind protons to extract acids. Generally, those reagents possessing a carbon-bound oxygen tend to extract water with the metal salt, whereas the phosphorus-based reagents do not.

Because the bonds formed in the solvation-extraction reactions are not highly specific, the stoichiometric coefficients p and q, as well as the amount of coextracted water, are not definite and may vary with extraction conditions. Increased polarity of the functional group in the extractant, as is obtained in eliminating the ester linkages in trioctyl phosphate to obtain TOPO, increases the strength of the solvation, but it also tends to increase the aqueous solubility of the reagent. Solvating extractants can be used undiluted, but usually they are mixed with an inexpensive diluent such as kerosene.

The solvation extraction of a metal, according to the form of reaction (8.2-1), depends on the formation in the aqueous phase of an uncharged ion pair or complex MX_n to some significant level of activity. Therefore, the extent of the extraction reaction will depend strongly on the aqueous solution chemistry as complexation, association, and hydrolysis affect the distribution of metal species. Similarly, the reversal of reaction (8.2-1) to strip the metal from the organic phase usually depends on altering the aqueous-phase chemistry. This may be accomplished by stripping with an aqueous phase low in the concentration of the ligand X or by using a competing aqueous-phase complexing agent. Temperature may have some effect on the distribution equilibrium, and precipitation or redox reactions might be used to reverse the extraction reaction.

Table 8.2-2 summarizes some specific applications of solvating extractants. More extensive reviews are given in Refs. 1–5.

8.2-2 Liquid Cation Exchangers

Solvent extraction of metals can be accomplished by use of organic-phase extractants that function as ion exchangers. For metal cations extractants possessing an acidic functionality are appropriate. There are two types of cationic, or acidic, ion-exchange extractant available. The same functional groups that are used in cation-exchange resins, namely, carboxylic, phosphoric, and sulfonic acids, can be incorporated in a soluble organic molecule to obtain a liquid ion exchanger. Structures of moderate molecular weight (around 300) are used such that the extractant is soluble in a diluent like kerosene but exhibits a low aqueous solubility. Common examples of these *acidic extractants* include di-2-ethylhexyl phosphoric acid (DEHPA) and alkyl monocarboxylic acids with 10–20 carbons in various branched (e.g., Versatic acids) or cyclic (e.g., naphthenic acid) structures.

The other type of cation-exchange extractant available is a class of *chelating extractants*. Chelating extractants contain two functional groups that form bidentate complexes with metal ions. Although there are many chelating reagents used in analytical chemistry to achieve highly selective complexing of a wide variety of metals, only a few have been developed for large-scale commercial use. The notable example is the family of LIX reagents developed especially for copper extraction by General Mills Inc., now Henkel Corp., which are hydroxyoximes and possess an acidic hydrogen. Similar reagents based on the oxime group have been developed by several other chemical companies. Other acidic chelating extractants, which act as cation exchangers, include other hydroxyoximes, such as SME259 of Shell Chemical Company and the P5000 reagents of Acorga Limited, and substituted 8-hydroxyquinolines, such as the Kelex reagents of Sherex Chemical Company.

With both simple and chelating acidic extractants the metal extraction reaction generally follows the form

TABLE 8.2-2 Hydrometallurgical Applications of Some Common Solvating Extractants

Extractant	Application
Tributyl phosphate (TBP)	U extraction from nitric acid
	Zr separation from Hf in nitric acid
	FeCl$_3$ separation from Cu and Co in HCl solution
Methyl isobutyl ketone (MIBK)	Extraction of Nb and Ta from HF solutions, followed by selective stripping
Trioctyl phosphine oxide (TOPO)	Extraction of U from phosphoric acid

$$M^{+n} + n\overline{HR} \rightleftharpoons \overline{MR_n} + nH^+ \qquad (8.2\text{-}3)$$

such that the reaction is strongly pH dependent and stripping occurs into strong acid.

Table 8.2-3 summarizes the applications of some acidic extractants. Ritcey and Ashbrook[6] provide a historical account of the evolution of the commercial chelating extractants.

8.2-3 Liquid Anion Exchangers

In some electrolytic solutions, especially those strong in halides, many metal cations are complexed by anions to the extent that they exist primarily as neutral ion pairs or anionic species. The formation of anionic complexes may retard metal extraction by solvation or cation exchange, but it can be exploited by use of *anion-exchanging extractants*.

The most important liquid anion exchangers for commercial use are quaternary ammonium salts and primary, secondary, and tertiary amines, which become anion exchangers by combining with mineral acids to form alkyl ammonium salts. In the former case, such a metal extraction reaction might be written

$$MX_n^{m-n} + \overline{R_4NX} \rightleftharpoons \overline{R_4NMX_{m+1}} + (n - m)X^- \qquad (8.2\text{-}4)$$

where m is the charge number of the metal cation M^{+m} and X^- is a univalent anion. Because aqueous complexes of various coordination numbers n coexist in an equilibrium distribution, the overall extraction reaction may be viewed as either anion exchange, with $n > m$, or an association reaction with $n = m$. In either case, the stability of the organic-soluble species depends on the tendency of the metal ion to form complexes with the anion.

With other amines, such as tri-isooctylamine (TIOA), an overall metal extraction reaction might be written

$$MX_m + 2H^+ + 2X^- + 2\overline{R_3N} \rightleftharpoons \overline{MX_{m+2}(HNR_3)_2} \qquad (8.2\text{-}5)$$

where HX and MX$_m$ both combine with the amine in an apparent addition reaction. Alternatively, this reaction could be written, and can occur, in two separate steps,

TABLE 8.2-3 Hydrometallurgical Applications of Some Common Cation-Exchanging Extractants

Extractant	Applications
Versatic 911 (carboxylic acid)	Separation of rare earths in nitrate solution
	Separation of Co from Ni in sulfate solution
Naphthenic acids (carboxylic acids)	Extraction of Cu, Ni, and Co
Di-2-ethylhexyl phosphoric acid (DEHPA)	Extraction of rare earths, V, Be, Al, Co, Ni, Zn
	U extraction from sulfuric acid
LIX 64N	Selective extraction of Cu from acidic solutions
	Extraction of Cu and Ni from alkaline ammonia solution
Kelex 100	Selective extraction of Cu from acidic solutions
	Extraction of Cu, Ni, Co, and Zn from alkaline ammonia solution

$$H^+ + X^- + \overline{R_3N} \rightleftharpoons \overline{R_3NHX} \qquad (8.2\text{-}6)$$

amine salt formation, followed by

$$MX_m + 2\,\overline{R_3NHX} \rightleftharpoons \overline{MX_{m+2}(HNR_3)_2} \qquad (8.2\text{-}7)$$

an addition reaction between the ion pair and the amine salt. If attention were focused on an anionic metal complex MX_n^{m-n} rather than the uncharged ion pair MX_m, an alternative to reaction (8.2-7) would be an anion-exchange reaction analogous to reaction (8.2-4),

$$MX_n^{m-n} + 2\,\overline{R_3NHX} \rightleftharpoons \overline{MX_{m+2}(HNR_3)_2} + (n-m)\,X^- \qquad (8.2\text{-}8)$$

Different amine extraction systems exhibit different stoichiometries from that shown here, and a number of different organic species may form in any one system, but the common feature of such extractions is that they depend on combining metal cations, anions, and amine into uncharged organic-soluble molecules. The extraction may be reversed by suppressing the activity of any of the product's constituents. For example, a metal extracted with a tertiary amine can usually be stripped by an aqueous phase that has either a sufficiently high pH to reverse the amine salt formation or sufficiently low coordinating-anion concentration to suppress formation of the complex.

For quantitative process calculations it is important to know what chemical species form in a given extraction system. The stoichiometry of the overall reactions affects the process material balance. Also, the number of different species formed determines the number of degrees of freedom required to specify the thermodynamic state of a system. On the other hand, the detailed mechanism of a reaction, for example, whether amine extraction occurs by addition [reaction (8.2-5)] or ion exchange [reaction (8.2-7)], is of no direct concern unless slow reaction kinetics become an issue.

Table 8.2-4 summarizes applications of amines for metal extraction. The viability of amine extraction depends directly on the identity of the anions present in the aqueous phase. Amine extraction has been discussed by Schmidt.[7]

8.2-4 Diluents

Although the solvent extraction of a metal occurs by reaction with an extractant that is essentially insoluble in water, it is seldom practical to use pure extractant as the solvent phase. The pure extractants are usually very viscous and may have specific gravities near 1 so that both mixing and separating phases may be difficult. They are intrinsically surface active and tend to form emulsions with water. To avoid these problems and to provide flexibility with respect to phase ratios and the hydraulics of liquid–liquid contacting, one uses a diluent, which is often the major component in the organic phase.

Inexpensive hydrocarbon solvents are chosen as diluents in large-scale processes. Table 8.2-5 lists some examples. These solvents are chosen for their low solubilities in water and high flash points as well as some less obvious chemical properties.

The diluent in a solvent extraction system does not serve simply as an inert carrier for the extractant and its metal compounds. It has been found repeatedly that the choice of diluent can affect significantly the performance of various extractants, presumably through both chemical and physical interactions with the solute species. For example, Murray and Bouboulis[8] reported the effects of systematically varying the aromatic content of the diluent on a copper extraction process, and Akiba and Freiser[9] demonstrated solvent effects on system chemistry. Ritcey and Ashbrook[10] review the various solvent properties that influence extractant performance.

From the combination of a hydrocarbon solvent as a diluent with extracted metal species, which often contain some water molecules or other polar groups, there is a tendency in many systems for a third phase

TABLE 8.2-4 Hydrometallurgical Applications of Some Common Anion-Exchanging Extractants

Extractant	Applications
Tertiary amines (TIOA)	Separation of Co and Cu from Ni in acidic chloride solutions
	Extraction of Zn and Hg from acidic chloride solutions
	Extraction of U, V, W, and Mo from acidic sulfate solutions
	Extraction of Re from acidic nitrate solution
Quaternary amines	Extraction of Mo from carbonate solution
	Extraction of Na$_2$CrO$_4$
	Extraction of V from caustic solution

TABLE 8.2-5 Some Commercial Solvents Used as Diluents in Metal Extraction Processes

	Composition (%)		
	Paraffins	Naphthenes	Aromatics
Amsco odorless mineral Spirits[a]	85	15	0
Chevron ion-exchange solvent[b]	52.3	33.3	14.4
CycloSol 63[c]	1.5		98.5
Escaid 100[d]	80		20
Escaid 110[d]	99.7		0.3
Escaid 350 (formerly Solvesso 150)[d]	3.0	0	97.0
Kermac 470B (formerly Napoleum 470)[e]	48.6	39.7	11.7
MSB210[c]	97.5		2.5
Shell 140[c]	45	49	6.0
Shellsol R[c]	17.5		82.5

[a] Union Oil Co.
[b] Chevron Oil.
[c] Shell.
[d] Esso or Exxon.
[e] Kerr McGee.
  Source: Reference 2.

to form when the extraction proceeds. This presents serious problems in fluid handling and is avoided by the addition of *modifiers*. The most common modifiers are alcohols, phenols, and TBP, which act as cosolvents by virtue of their polar groups. Modifier concentrations are usually 5–10% by volume of the organic phase, a level comparable to that of the extractant in many cases. Like the diluent, the modifier can also interact with the solute species and affect the extraction reaction in various ways.[10]

8.2-5 Competing and Supporting Reactions

In the outline of common extraction mechanisms given above, the phenomenon of metal extraction is viewed as a chemical reaction occurring across a liquid–liquid phase boundary. It is clear, however, that more than one reaction occurs in almost all extraction systems. In the aqueous solutions the metal ions often interact with anions or complexing agents or water (by hydrolysis) to form a variety of species. Such reactions may promote the extraction of the metal, as in the case of anionic complexes with amines, or they may hinder it and promote stripping, as with precipitation of hydroxides or sulfides or with ammonia complexing that competes with cation exchange. In the organic phase there are parallel interactions, such as association of extractant molecules into dimers and even polymers or micelles. Between phases other reactions may occur simultaneously. Water, acids, or other metals may be extracted by interaction with the extractant, and multiple species of organic-soluble metal may also form. Introduction of modifiers as well as diluents increases the number of possible interactions and the complexity of these systems.

In spite of the large number of experimental studies of various extraction systems reported in the literature, the range of variables available in these highly interacting, multicomponent systems leaves many regions unexplored. Therefore, the development of a solvent-extraction process requires extensive experimentation. The general behavior of many available reagents is known, but the effects of interactions in a specific system must be determined experimentally.

It must also be recognized that one is seldom working with pure chemicals in these extraction systems. Not only are the aqueous feed streams usually complex mixtures, but the commercial extractants, diluents, and modifiers are themselves mixtures and contain unknown impurities. Therefore, each system of interest must be characterized by experimental investigation.

To attain a quantitative understanding of a particular metal extraction system, one must gather as much information as possible about the chemical species forming in the system. Molecular weights and states of aggregation should be determined for both extractant and extracted metal species. Maximum metal loading should be measured as an indication of the extraction-reaction stoichiometry. Degree of water coextraction should be determined, and the tendency of other species such as acids to extract should be examined. Although extensive analytical work is required, such information is needed to construct meaningful material balances as well as to formulate equilibrium and rate analyses.

8.3 PHASE EQUILIBRIA

Often the equilibrium extent of extraction of a metal is described in terms of the phase distribution coefficient,

$$D = \frac{\text{(concentration of metal in organic phase)}}{\text{(concentration of metal in aqueous phase)}}$$

Much of the published information about phase equilibria is presented in terms of D. Although the distribution coefficient depends on many variables, because of the multicomponent and reactive nature of these systems, many studies report D for a trace amount of metal in fixed compositions of aqueous and organic solutions, usually at ambient temperature. Such data are independent of metal concentration.

Early reagent-screening experiments are summarized by Marcus and Kertes[1] in a form such as shown in Figs. 8.3-1 and 8.3-2. Figure 8.3-1 presents low-concentration distribution coefficient values for various metals in a 5% solution of TOPO in toluene for several concentrations of hydrochloric acid in the aqueous phase. Figure 8.3-2 shows a similar presentation of distribution coefficients from sulfuric acid solutions into undiluted TBP.

As an example of the use of such information, suppose that one wished to separate zinc and copper present in an acidic chloride solution. Figure 8.3-1 indicates that the distribution coefficients of the two metals with 5% TOPO in toluene are on the order of 100 and 0.01, respectively, for an aqueous phase containing 1.0 M HCl. Therefore, this extractant should be effective in separating zinc from copper if the aqueous chloride level is not too high. Furthermore, the distribution coefficient of zinc falls to 0.01 in 12 M HCl so that a concentrated hydrochloric acid solution would be a candidate for accomplishing the stripping of zinc from the loaded organic phase. This information suggests that some experimental work with TOPO and the zinc–copper solution would be justified to determine the phase equilibria more completely.

Trace-level distribution coefficients are often plotted as functions of acid or anion concentration for a given extractant and presented as an overlay of the Periodic Table to indicate the efficacy and selectivity of the reagent. Figures 8.3-3 and 8.3-4 summarize extraction of metals by some tertiary and quaternary amines from hydrochloric acid solutions. The ordinate of each graph is log D, which ranges from 10^{-2} to 10^4, and the abscissa is HCl concentration, ranging from 0 to 12 M. The high distribution coefficients of ferric chloride with amines shown in Figs. 8.3-3 and 8.3-4 indicate the basis for using amine extraction as the first step of the process shown in Fig. 8.1-2. Numerous charts of this form are available for various extraction systems.[2–4]

Another way to present tracer-level metal equilibrium data is to plot the fraction of the metal extracted into the organic phase as a function of some composition variable, often pH. Implicit in such a presentation is a phase volume ratio of unity. Figure 8.3-5 shows the percent extraction of several metal cations by naphthenic acid as a function of pH. Such data indicate the potential for separating certain metals with a particular reagent; they also show the possibility of stripping from the loaded organic by pH control. For example, Fig. 8.3-5 suggests that cupric and ferric ions can both be extracted by naphthenic acid if the pH of the aqueous phase at equilibrium is held above 4 by buffering or addition of base. On the other hand, only ferric ion should be extracted if the pH is adjusted to 2.5. With regard to stripping, addition of acid to lower the pH to 1.0 should strip all metal from the organic phase. Alternatively, the curve for copper in ammonium nitrate suggests that copper can be stripped at high pH in the presence of ammonia.

Although representative values of the distribution coefficients of various metals with different extractants are useful in guiding the selection of a reagent for a particular separation process, the information provided by low-concentration D values is incomplete. For example, Fig. 8.3-3 indicates that the distribution coefficient of mercuric ion between hydrochloric acid and a solution of tri-isooctylamine in xylene (0.14 M) is on the order of 100. Figure 8.3-6 shows the distribution coefficient of mercuric ion from 4.63 M NaCl solution at pH = 1 and indicates that the value can be considerably higher and that it varies with the level of mercury in solution. At high concentrations of metal, extractant is consumed by the extraction reaction, and the distribution coefficient falls as the finite stoichiometric loading of the organic phase is approached.

A further complication in the equilibrium is shown in Fig. 8.3-7, which shows the mercury distribution coefficient as a function of amine concentration for a fixed level of mercury. At low amine concentrations the distribution coefficient increases with amine level in the way that is expected for reaction (8.2-7) or (8.2-8), but above a certain amine level the distribution coefficient becomes constant. This behavior indicates that some separate phenomenon, such as limited solubility or association of the amine–hydrochloride salt, keeps its activity constant with increasing concentration. This is a common experimental observation in such systems.

Although distribution coefficients in metal extraction systems are highly variable, they are thermodynamic quantities that can be measured and correlated as functions of the state of the system. Because of the multicomponent nature of metal extraction systems, one must take care to identify a proper set of independent variables in constructing a correlation of phase equilibrium data. Another consideration in treating equilibrium extraction data is that one desires a systematic method for treating the effects of

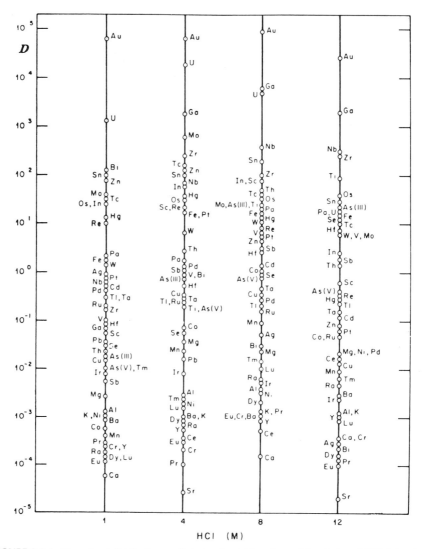

FIGURE 8.3-1 Trace-level distribution coefficients of metal ions into a 5% solution of TOPO in toluene as functions of aqueous HCl concentration. From Ref. 1, with permission.

additional species, such as other metals, modifiers, and so on, on the equilibrium distribution of a particular metal. A *chemical-reaction model* of extraction equilibria provides an appropriate framework for correlating data because the large, but reversible, distribution coefficients of most interesting systems are the result of reversible chemical reactions.

For a cation-exchanging extractant HR, which extracts a divalent metal ion M^{2+} according to reaction (8.2-3), a chemical-reaction model of the phase equilibrium would be the mass-action equilibrium expression

$$\frac{[\overline{MR_2}]\,[H^+]^2}{[M^{2+}][\overline{HR}]^2} = K \qquad (8.3-1)$$

where the quantities in brackets are activities and the equilibrium constant K for a specific metal–extractant system depends on temperature and the identity of the diluent. If activities were identical to analytical

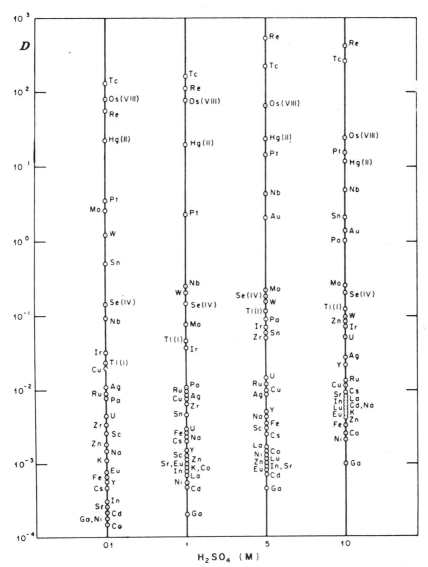

FIGURE 8.3-2 Trace-level distribution coefficients of metal ions into undiluted TBP as functions of aqueous sulfuric acid concentration. From Ref. 1, with permission.

concentrations, knowledge of K in Eq. (8.3-1) would allow calculation of the equilibrium metal distribution, or $D = (\overline{MR_2})/(M^{2+})$, for any given experimental conditions. For example, for the case of a trace amount of metal, equal phase volumes, and an extractant concentration fixed at 0.1 M, Eq. (8.3-1) with constant K yields the metal distribution curves shown in Fig. 8.3-8, which may be compared with Fig. 8.3-5. A value of $K = 1.0$ represents an extraction equilibrium that provides complete extraction at pH > 2 and stripping at pH < 0.

To account for nonidealities in solution one can use activity coefficients or simply replace activities with species concentrations in Eq. (8.3-1) to obtain an empirical mass-action equilibrium quotient that is concentration dependent. Bauer and Chapman[5] treated data for copper extraction by Kelex 100 by a chemical model, both with and without explicit activity coefficient terms, and were successful in extrapolating from results with 5% Kelex to obtain an accurate copper distribution curve for 20% Kelex. Hoh and Bautista[6] used a similar approach to correlate equilibrium data for copper with LIX 65N and with Kelex 100. The concentration-based equilibrium constant at 25°C was found to be on the order of 0.03 for LIX 65N in

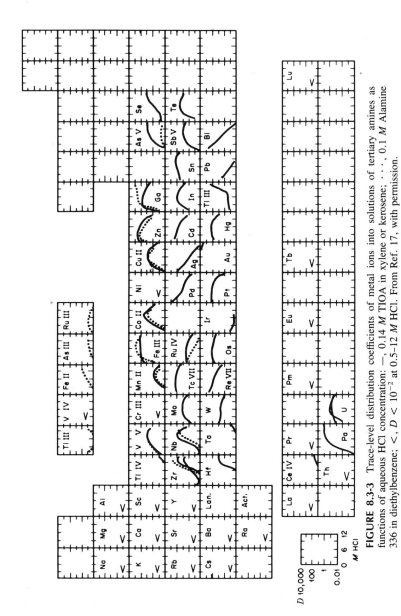

FIGURE 8.3-3 Trace-level distribution coefficients of metal ions into solutions of tertiary amines as functions of aqueous HCl concentration: —, 0.14 M TIOA in xylene or kerosene; · · ·, 0.1 M Alamine 336 in diethylbenzene; <, $D < 10^{-2}$ at 0.5–12 M HCl. From Ref. 17, with permission.

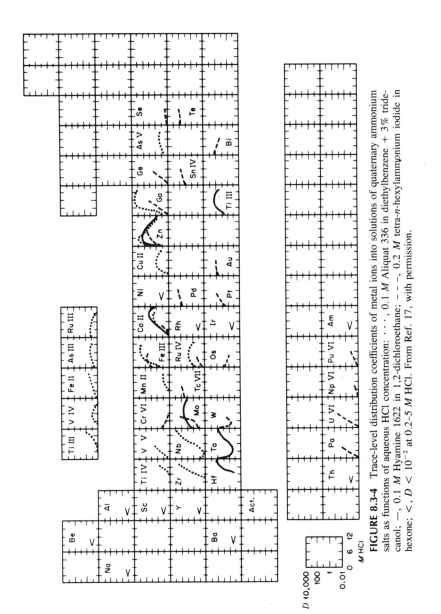

FIGURE 8.3-4 Trace-level distribution coefficients of metal ions into solutions of quaternary ammonium salts as functions of aqueous HCl concentration: · · · , 0.1 M Aliquat 336 in diethylbenzene + 3% tridecanol; —, 0.1 M Hyamine 1622 in 1,2-dichloroethane; – – –, 0.2 M tetra-n-hexylammonium iodide in hexone; <, $D < 10^{-2}$ at 0.2–5 M HCl. From Ref. 17, with permission.

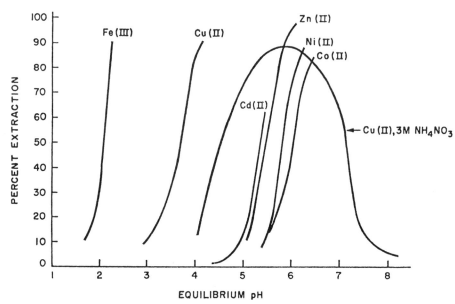

FIGURE 8.3-5 Fractional amount of extraction of trace levels of some metal cations from aqueous solution into organic solutions of naphthenic acid as a function of pH. From Ref. 18, with permission.

toluene,[6] 5 for Kelex 100 in toluene,[6] and 90 for Kelex 100 in xylene[5]. From trace-level measurements Akiba and Freiser[7] found the equilibrium constant for copper extraction by LIX 65N to range from $10^{-0.2}$ to $10^{1.3}$ in various solvents.

Virtually no metal extraction system involves only one reaction. In aqueous metal salt solutions there occur homogeneous association and complexation reactions that make true species concentrations differ from the analytical concentrations that are measured and that enter into material balances. In sulfate solutions bisulfate formation occurs at low pH, and metal–sulfate ion pairing may take place. In chloride and ammonia solutions metal complexation is common. Dimerization and other association reactions often occur in the organic phases. And two-phase extraction reactions of acids or other species may take place simultaneously with metal extraction or stripping. The equilibria of these reactions must be modeled concurrently with the

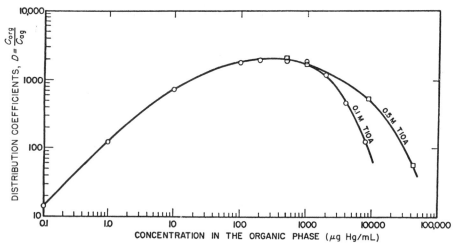

FIGURE 8.3-6 Distribution coefficient of mercuric ion from 4.63 M NaCl solution at 22°C and pH = 1 as a function of organic-phase mercury concentration. The organic phase is TIOA in xylene at 0.1 M and 0.5 M as indicated. From Ref. 13, with permission.

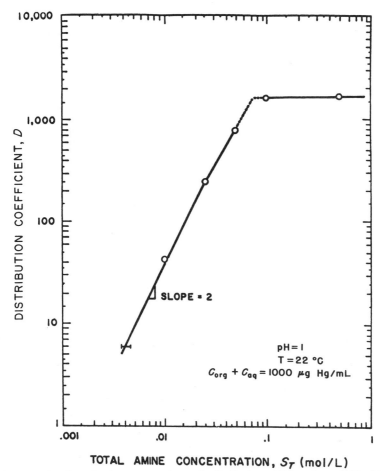

FIGURE 8.3-7   The effect of amine concentration on the distribution coefficient shown in Fig. 8.3-6. From Ref. 13, with permission.

metal-extraction reaction if the true reactive species concentrations are affected by them. If such reactions are overlooked in an equilibrium model, system behavior will appear to be highly nonideal, and extreme values of empirical activity coefficients will be required.

The relatively large values found for the extraction equilibrium constant of copper with Kelex 100 (5 and 90) indicate that stripping of copper from this reagent should be difficult. It is found, however, that copper does strip readily into sulfuric acid solutions because Kelex 100 reacts with sulfuric acid in preference to copper. Fitting the extraction of sulfuric acid by Kelex 100 by a chemical-reaction equilibrium constant,

$$\frac{[\overline{H_2SO_4 \cdot 2HR}]}{[H^+]^2[SO_4^{2-}][\overline{HR}]^2} = K_A = 1.8 \ M^{-4} \tag{8.3-2}$$

at 50°C, and solving the two equilibria simultaneously describes adequately the equilibrium distribution of copper under stripping conditions.[5] This is shown in Fig. 8.3-9.

The acidic copper-chelating extractants also can extract other metals from alkaline ammonia solutions, which are used in some hydrometallurgical processes to exclude iron. DeRuiter[8] modeled the extraction of copper(II) by LIX 64N by considering the following reversible reactions: complexation of aqueous copper by four and by five ammonia ligands, the ion-exchange extraction of the free copper cation, hydrolysis of ammonia to ammonium, and extraction of ammonia by both physical solubility and by chemical association with LIX. Not only does this chemical-reaction model fit the copper-extraction equilibrium, but the equilibrium constant for extraction of free copper cations, found to be 0.02, is in good agreement with that determined in acidic sulfate solutions.

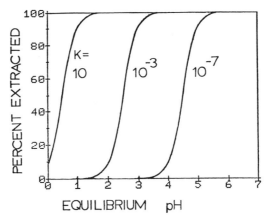

FIGURE 8.3-8 Fractional amount of extraction of trace levels of metal cations from aqueous solution into organic solutions of an acidic extractant as a function of pH. Computed from Eq. (8.3-1) for several values of the equilibrium constant K.

Nickel is extracted from alkaline ammonia solutions by LIX 64N, but its equilibrium constant is five orders of magnitude smaller than that of copper.[8] Also, it forms ammonia complexes more extensively than copper, which enhances the selectivity of LIX 64N for copper over nickel. Under conditions of simultaneous extraction the parameters of the chemical-reaction model for each individual metal are able to predict the two-metal equilibria provided that the equilibrium distribution of species in the aqueous phase is taken into account.

It appears that solutions of the common chelating agents behave ideally in the sense that the organic-phase species in these systems usually exhibit activities that are proportional to their concentrations. Unfortunately, such behavior is the exception rather than the rule. Figure 8.3-7 illustrated nonideal behavior of TIOA in the extraction of mercuric chloride. Organic-phase nonideality is commonly encountered with solvating extractants and most cation exchangers as well as amines. As with the aqueous phase most of the nonideality can be accounted for by identifying association reactions in the organic phase and modeling the chemical equilibrium among species in solution.

An example of nonideal organic-phase behavior arises with di-2-ethylhexyl phosphoric acid (DEHPA) in the extraction of metal cations. Equilibrium data of Troyer[9] for extraction of copper in the presence of nickel from sulfate solution into xylene solutions of DEHPA are shown in Fig. 8.3-10. Although the organic-phase copper concentration would be expected to rise in proportion to that in the aqueous phase,

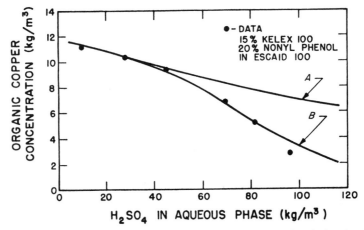

FIGURE 8.3-9 Comparison of chemical-equilibrium extraction models with stripping data for copper in Kelex 100. Curve A follows Eq. (8.3-1). Curve B is obtained by solving Eq. (8.3-1) simultaneously with Eq. (8.3-2). From Ref. 5, with permission.

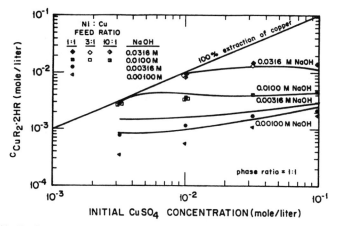

FIGURE 8.3-10 Predicted and observed organic copper concentrations in equilibrium with 0.1 M DEHPA in xylene for different amounts of base added and several copper-to-nickel feed ratios. Curves are predicted by the model for the indicated initial NaOH concentrations. From Ref. 8.4-10, with permission.

the data indicate that the former is rather insensitive to the latter. Detailed data analysis indicates that the activity of free extractant is nearly independent of its analytical concentration, and this is caused by reversible polymerization of DEHPA. The equilibrium data can be modeled by accounting for the following reactions: complete dimerization of DEHPA, formation of an organic metal species of the form $\overline{MR_2 \cdot 2HR}$, metal hydrolysis to form a monohydroxy complex, formation of a metal–sulfate ion pair, formation of bisulfate, extraction of sodium to form $\overline{NaR \cdot 3HR}$, and association of the extractant dimers to form a hierarchy of oligomers and polymers. The equilibrium of each of these reactions is governed by a mass-action equilibrium expression. Many aqueous-phase stability constants are reported in the literature,[10] and the sodium-extraction equilibrium constant can be measured separately. Thus, metal-extraction equilibrium data can be fit with only two additional parameters, the metal-extraction-reaction equilibrium constant K and a parameter characterizing the degree of polymerization. The latter parameter must relate the concentration of the reactive extractant species to the total analytical concentration of extractant in solution.

The curves in Fig. 8.3-9 have been computed from such a chemical-reaction model.[9] For an extraction reaction of the form

$$M^{2+} + 2\,(\overline{HR})_2 \rightleftharpoons \overline{MR_2 \cdot 2HR} + 2H^+ \tag{8.3-3}$$

the equilibrium constant is found to be 0.148 M^{-1} for copper and $1.65 \times 10^{-4}\ M^{-1}$ for nickel. These values indicate that copper is extracted at a lower pH than is nickel, starting at 2 or 3 compared with 3 or 4. If enough base is used, the extractant can be loaded with either metal, but when both metals are present, and the amount of DEHPA or base is limited, the extractant is quite selective for copper. The individual and simultaneous extractions of both metals as well as stripping are described adequately by the chemical-reaction model.

Amine-extraction equilibria can also be modeled by chemical-reaction equilibrium constants. Figure 8.3-3 indicates that cations such as iron(III), zinc, cobalt(II) and copper(II) exhibit high distribution coefficients with chloride solutions, whereas nickel, iron(II), and manganese are not extracted to any great extent. The basis for the differences in distribution coefficients lies mainly in the tendency for the former group of cations to form chloride complexes. Stability constants for these complexes are available in the literature,[11] and they can be used to develop quantitative phase-equilibrium models.

For TIOA with hydrochloric acid the concentration-based equilibrium constant for salt formation[12] according to reaction (8.2-6) is $1.51 \times 10^4\ M^{-2}$, and the equilibrium constant for amine-hydrochloride salt dimerization[13] is 8.0 M^{-1}. Combination of these parameters and the ion-complex stability constants with experimental metal-distribution data allows determination of the equilibrium constants for reactions (8.2-5) or (8.2-7). This completes the description of the amine–metal extraction-phase equilibria. For cobalt(II) in acidic sodium chloride solutions the equilibrium constant[14] for reaction (8.2-7) with TIOA is $2.0 \times 10^4\ M^{-2}$, and that for copper(II) is 370 M^{-2}. The corresponding value for zinc[15] is $7.5 \times 10^4\ M^{-2}$ In spite of these relative values, the order of selectivity of TIOA for extraction of the metals is Zn > Cu > Co because of the relative extent of chloride complex formation. For the same reason, zinc stripping is difficult in this system, and copper has a tendency to be reduced to cuprous, which also complexes and extracts extensively.

Unfortunately, few of the published studies of extraction equilibria have provided complete quantitative models that are useful for extrapolation of data or for predicting multiple metal distribution equilibria from single metal data. The chemical-reaction equilibrium formulation provides a framework for constructing such models. One of the drawbacks of purely empirical correlations of distribution coefficients is that pH has often been chosen as an independent variable. Such a choice is suggested by the form of Figs. 8.3-5 and 8.3-8. Although pH is readily measured and controlled on a laboratory scale, it is really a dependent variable, which is determined by mass balances and simultaneous reaction equilibria. An appropriate phase-equilibrium model should be able to predict equilibrium pH, at least within a moderate activity coefficient correction, concurrently with other species concentrations.

Workers at Oak Ridge National Laboratory have announced the establishment of a Separations Science Data Base, which should be useful in seeking equilibrium data for specific extraction systems.[16]

8.4 EXTRACTION KINETICS

Because the metal-extraction separation process involves chemical reactions, rates may be slow compared with ordinary liquid extraction. Although slow kinetics have important process design implications, more attention has been focused on mechanistic interpretations than on quantitative phenomenonological characterization. Cox and Flett[1] have reviewed some chemical aspects of extraction kinetics.

One issue in the study of extraction kinetics is the question of whether the metal-extraction reaction, or its rate-determining step, occurs heterogeneously or homogeneously in the aqueous phase.[2] For practical purposes, the locale of the reaction is usually irrelevant because the kinetics of most systems can be fit by a heterogeneous model. In practical extraction systems, the aqueous solubility of the extractant is necessarily very low; organic-to-aqueous distribution coefficients of some common extractants have been determined to be 10^3 and higher.[3] A small homogeneous reaction rate constant combined with very low extractant concentrations in the aqueous phase would provide an extraction rate so slow as to make the process uninteresting. On the other hand, if the homogeneous rate constant is large enough to provide finite extraction rates, the Hatta number[4] for the reaction between metal ion and aqueous extractant will normally be large. The Hatta number is a dimensionless ratio of the homogeneous reaction rate constant to the film mass transfer coefficient. According to the theory of mass transfer with homogeneous reaction, a Hatta number greater than 3 causes the reaction to go to completion within the mass transfer boundary layer. This condition requires only that the pseudo-first-order rate constant be greater than about 10 s^{-1}, whereas reported values of rate constants in "slow" extraction systems have generally been found to be much larger.[5]

Furthermore, for a large ratio of bulk metal ion concentration to the interfacial aqueous extractant concentration, the reaction zone is located very close to the liquid–liquid interface, and the extraction rate per unit area of interface becomes independent of the aqueous-film mass transfer coefficient as well as the system volume. Under these conditions a truly heterogeneous reaction and a homogeneous process are indistinguishable in the sense that the rate of each will be a definite function of reactant concentrations in the vicinity of the interface. In either case, the functional form of the kinetics and its parameters must be determined experimentally.

The critical question regarding metal-extraction (or stripping) kinetics is whether the chemical reactions that accomplish the phase transfer of the metal are fast or slow relative to the prevailing mass transfer rates. Mass transfer rates may be characterized conveniently in terms of a two-film model, with film mass transfer coefficients being typically on the order of 10^{-2} cm/s in liquid–liquid systems. (This value corresponds to an effective stagnant-film thickness of 10^{-3} cm; a homogeneous reaction zone might lie within 10^{-5} cm of the interface for a bulk metal concentration 100 times the aqueous extractant solubility.) Within this context maximum, or mass-transfer-limited, extraction rates, obtained when the extraction reaction maintains equilibrium at the interface, can be computed.

For a simple cation-exchange extraction, such as reaction (8.2-3), concentration profiles near the interface are represented by Fig. 8.4-1. The fluxes at the interface of all four species are related by the reaction stoichiometry. The interfacial concentrations depend on the bulk-solution concentrations, the interfacial metal flux, and the respective mass transfer resistances. Requiring the interfacial concentrations to be in equilibrium according to Eq. (8.3-1) yields the following equation[6] for the metal-extraction rate:

$$K = \frac{[C_{\text{MR}_2, \infty} + (N_{\text{M0}}/k_{\text{MR}_2})]\,[C_{\text{H}^+, \infty} + (2N_{\text{M0}}/k_{\text{H}^+})]^2}{[C_{\text{M}^{2+}, \infty} - (N_{\text{M0}}/k_{\text{M}^{2+}})]\,[C_{\text{HR}, \infty} - (2N_{\text{M0}}/k_{\text{HR}})]^2} \tag{8.4-1}$$

in which the interfacial metal flux N_{M0} is an implicit function of bulk concentrations, film mass transfer coefficients k_i, and the equilibrium constant. Figure 8.4-2 is a plot of Eq. (8.4-1) for several combinations of bulk-phase concentrations; the ordinate is the ratio of the metal flux to its value for $(M^{2+}) = 0$ at the interface and the abscissa represents deviation from equilibrium of the bulk concentrations. Figure 8.4-2 demonstrates that the extraction rate is affected by transport of all reactive species because all transport rates must be considered in estimating interfacial concentrations.

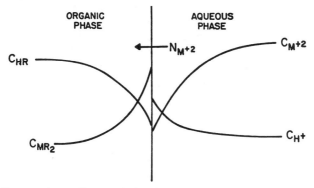

FIGURE 8.4-1   Concentration profiles near the interface during metal extraction by an acidic extractant.

If slow heterogeneous kinetics affect the metal-extraction rate, an appropriate kinetics model should be expressed in terms of interfacial species concentrations, which are obtained from bulk values by correcting for mass transfer resistances. Figure 8.4-3 combines an empirical kinetics model[7] for copper extraction by Kelex 100 with a two-film mass transfer model to indicate the relative importance of mass transfer and kinetics in different regions of composition. For this system, the reaction kinetics become fast at higher pH and high extractant and metal concentrations. As the flux approaches the mass-transfer-limited flux, given by Eq. (8.4-1), the interfacial concentrations approach the corresponding equilibrium values. At lower pH and low extractant level, the kinetics are considerably slower. Under these conditions the interfacial concentrations maintain their bulk values.

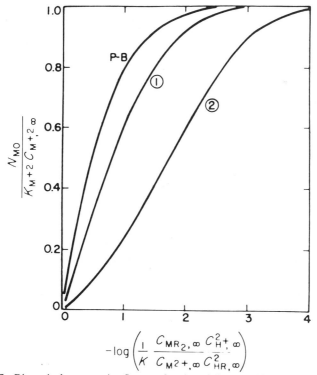

FIGURE 8.4-2   Dimensionless extraction flux as a function of distance of bulk-phase concentrations from equilibrium as computed from Eq. (8.4-1). Curves 1 and 2 represent different bulk concentration ratios. Curve P-B represents a pseudobinary calculation where gradients in H^+ and HR are neglected. From Ref. 6, with permission.

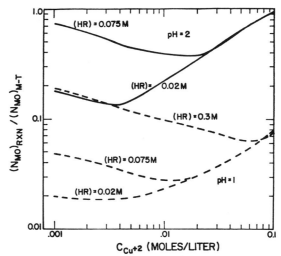

FIGURE 8.4-3 The rate of extraction of copper from sulfate solution by Kelex 100 in xylene. The experimental flux is presented relative to the mass-transfer-limited metal flux for $k_{M^{2+}} = 10^{-2}$ cm/s. From Ref. 6, with permission.

There is considerable confusion in the literature about extraction kinetics, and little information directly applicable to process calculations is available. Part of the problem is that many rate studies have been done under conditions that were not well characterized with respect to mass transfer effects or interfacial area so that quantitative heterogeneous rate models cannot be derived. Another problem is irreproducibility caused by unknown impurity levels in extractant solutions. Kinetics are much more sensitive to impurities, additives, and changes in diluent than are equilibrium properties. Finally, it is clear that interfacial phenomena have a strong, and variable, effect on extraction kinetics. Most extractants are surface-active agents as are many modifiers and impurities. Interactions with the surface can affect extraction kinetics directly by blocking sites for heterogeneous reaction steps or providing an interfacial transport resistance, or indirectly by affecting hydrodynamic conditions. It is difficult to control these effects to obtain reproducible and meaningful data.

A pragmatic approach to extraction kinetics is to measure extraction rates under mass transfer conditions that are similar to those expected in processing equipment and that are characterized well enough to allow estimation of interfacial concentrations. With such measurements one can determine whether the kinetics of a particular system are fast or slow compared to mass transfer, and if they are slow, the rate can be correlated with interfacial conditions. Several methods are used to measure extraction kinetics; these include measurements with (1) well-stirred vessels, (2) single drops or jets, and (3) the Lewis cell.

Rate measurements with well-stirred vessels may be continuous or batch. An example of the latter is simply a stirred flask. The former include the AKUFVE apparatus[8] and bench-scale models of mixer–settler contactors.[9] Because the interfacial area and mass transfer conditions in these systems are not known, they generally do not yield intrinsic heterogeneous rate data for metal-extraction systems. Nevertheless, they are useful for screening relative rates and identifying cases where slow kinetics may be a problem.

The advantage of single-drop and liquid-jet experiments is that the interfacial area is known. Also, the hydrodynamics may be simple enough to allow calculation of film mass transfer coefficients. With rising or falling drops, however, adsorption of surfactants can alter the hydrodynamics and possibly interfere with the extraction as well. A growing-drop experiment,[10] analogous to the dropping mercury electrode of polarography, offers the advantages of continuous generation of fresh interface as well as known mass transfer conditions. Laminar-jet measurements[11] offer similar advantages except that surfactants tend to build up on the downstream end of the interface and make the surface rigid.

The Lewis cell[12] is a stirred cell with a planar interface between the two liquid phases. Although the hydrodynamics are complex and surfactants can build up on the interface, it is possible to calibrate the mass transfer coefficients empirically. Also, it appears that sufficiently intense stirring can disrupt stagnant surface layers.[13]

Although not a great amount of systematic kinetics data are yet available for metal-extraction systems, a few cases have been studied in detail. It is well established that the acidic chelating extractants generally react so slowly compared to ordinary mass transfer rates that special precautions must be taken in process design to provide adequate contacting time. According to the commercial importance of the copper–LIX extraction systems, much of the kinetics work to date has been aimed at this system.[14–16] All the methods

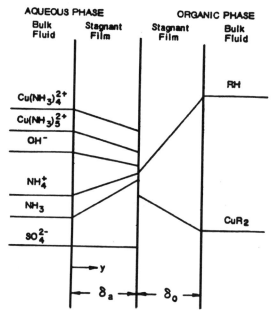

FIGURE 8.4-4 Schematic drawing of the species profiles near the interface during the extraction of copper from ammonia solution by an acidic extractant.

listed above have been used, but differences in experimental conditions and mechanistic interpretations make quantitative comparison of the results difficult. With alkaline solutions containing ammonia the extraction of copper by LIX 64N becomes mass transfer limited, but extraction of nickel is controlled partly by slow kinetics.[17]

From other experiments in which the effects of mass transfer have been analyzed, it appears that the following systems are mass transfer limited: uranyl nitrate extraction by TBP,[18] copper extraction by sodium-loaded DEHPA,[19] and extraction of zinc and copper(II) chlorides by TIOA.[20] Zinc extraction by dithizone in carbon tetrachloride is mass transfer limited at high zinc concentrations but kinetically controlled at low zinc levels.[21] Ferric ion extractions are reputed to be slow because of its sluggish ligand-exchange kinetics.[22] Extraction of ferric chloride by TIOA, for example, is controlled by a slow heterogeneous reaction.[23]

In mass-transfer-controlled systems in which extensive complexing or association takes place in the bulk phases, a proper mass transfer model must account for transport of all species. Otherwise, the transport model will not be consistent with a chemical model of phase equilibrium. For example, Fig. 8.4-4 indicates schematically the species concentration profiles established during the extraction of copper from ammonia–ammonium sulfate solution by a chelating agent such as LIX. In most such cases the reversible homogeneous reactions, like copper complexation by ammonia, will be fast and locally equilibrated. The method of Olander[24] can be applied in this case to compute individual species profiles and concentrations at the interface for use in an equilibrium or rate equation. This has been done in the rate analyses of several of the chloride and ammonia systems cited above.[25]

Stripping rates are as important for process viability as extraction rates, but less attention has been paid to their determination. In principle, stripping rates should be related through microscopic reversibility to extraction rates, but far from equilibrium the complete kinetics model is difficult to discern.

An interesting aspect of kinetics is the question of how species interact when multiple extraction reactions occur. Most extractants are not perfectly selective and are capable of extracting several species from a multicomponent aqueous solution. Complete equilibrium models should indicate extraction selectivity for a given system at equilibrium, but selectivity may vary considerably in a nonequilibrium process because of rate differences. In fact, iron(III) can be extracted by some of the copper-chelating extractants, but their effective selectivity for copper is based on the very slow kinetics of iron extraction.

Even in mass-transfer-limited processes, excursions in selectivity can be observed at finite contact times. This is predicted by rate models as simple as Eq. (8.4-1) for two metals with different equilibrium constant values.[6,26] The phenomenon involves initially fast coextraction followed by crowding out of the less preferred metal during competition for extractant. This has been observed during simultaneous extraction of copper and zinc chlorides by TIOA in a growing-drop experiment[10,25] and in extraction of uranyl nitrate and nitric acid by TBP in a Lewis cell,[27] as shown in Fig. 8.4-5.

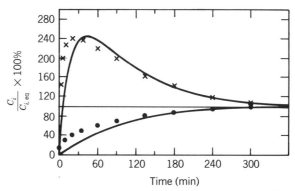

FIGURE 8.4-5 Measured and calculated organic-phase concentrations of nitric acid (×) and uranium (○) during coextraction by TBP in a Lewis cell. Concentrations are expressed as a percentage of their equilibrium values. From Ref. 28, with permission.

8.5 CONTACTING EQUIPMENT AND DESIGN CALCULATIONS

The processing equipment used to conduct solvent extraction of metals is the same as that used in conventional liquid–liquid extraction.[1-2] The most common choices have been mixer–settlers, columns with agitated internals, and static mixers. Some advantages and disadvantages of several classes of equipment are summarized in Table 8.5-1. Many of the practical aspects of equipment selection are discussed by Pratt and Hanson[3] and by Ritcey and Ashbrook.[4]

Because detailed models of the chemical effects on phase equilibrium and rates in metal-extraction systems have not been available, standard chemical engineering design procedures have not been applied very extensively in equipment sizing and comparison. Although McCabe–Thiele-type calculations are often suggested for designing equilibrium-stage operations, the results are only approximately correct because the locus of the metal-distribution equilibrium curve changes from stage to stage according to changes in pH and free extractant concentration. The problem is analogous to that encountered with multicomponent distillation or nonisothermal gas absorption in the sense that the state of the system cannot be represented on a two-dimensional plot. Multiple conservation and rate equations must be solved simultaneously to simulate process paths.

If the chemistry of an extraction system is well characterized and the equilibria among species and between phases are modeled adequately, equilibrium-stage calculations can be done numerically by solving simultaneously the algebraic equations that describe the equilibria and mass balances in each stage. Such an approach has been demonstrated,[5] and Table 8.5-2 shows calculated and experimental interstage concentrations during multistage extraction of copper by Kelex. In such calculations it becomes clear that analytical reagent concentrations in feed streams must be used as independent process variables and that pH is a dependent variable along with other individual species concentrations.

Although equilibrium-stage calculations are useful for preliminary design of staged contacting, characterization of equipment efficiency for a particular application requires experimental study. With mixer-settlers it is common to vary experimentally the number of real stages used as well as the residence time in each stage. Scale-up then is based on these parameters as well as power input per unit volume and superficial velocity in the settlers.[3] By this approach large-scale copper–LIX processes have been designed with three mixers for extraction and two for stripping.[6,7] The slow kinetics of the reaction are accommodated by a 2 min residence time in the mixers. Differential contactors can be characterized similarly in terms of transfer units if the equilibria and rates are sufficiently well known.[8]

An interesting comparative study of six different contactors is summarized in Fig. 8.5-1.[9] An alkaline feed stream containing copper, nickel, zinc, and ammonia was contacted with a Kelex 100 solution in each of the pilot-scale contactors. At a pH above 7 the reaction kinetics of Kelex are probably fast, but it loses some of its intrinsic selectivity for copper because of the tendency of nickel and zinc to be extracted. Although the various contactors could be made to operate at comparable levels with respect to specific throughput and copper-extraction efficiency, they exhibited quite different selectivities. Detailed liquid–liquid reactor models would help to identify and predict such variations in efficiency and selectivity.

With both staged equipment and differential contactors, availability of adequate phase-equilibrium models and rate expressions would allow application of existing correlations and simulation algorithms. For example, knowledge of metal-extraction kinetics in terms of interfacial species concentrations could be combined with correlations of film mass transfer coefficients in a particular type of equipment to obtain the interfacial flux as a function of bulk concentrations. Correlations or separate measurements of interfacial area and an estimate of dispersion characteristics would allow calculation of extraction performance as a

TABLE 8.5-1. Some Advantages and Disadvantages of Several Classes of Liquid–Liquid Contactors

	Mixer–Settlers	Nonagitated Differential	Agitated Differential	Centrifugal
Advantages	Good contacting of phases	Low initial cost	Good dispersion	Can handle low gravity difference
	Can handle wide range of flow ratios (with recycle)	Low operating cost	Reasonable cost	Low holdup volume
	Low headroom	Simplest construction	Many stages possible	Short holdup time
	High efficiency		Relatively easy scale-up	Small space requirement
	Many stages			Small inventory of solvent
	Reliable scale-up			
	Low cost			
	Low maintenance			
Disadvantages	Large holdup	Limited throughput with small gravity difference	Limited throughput with small gravity difference	High initial cost
	High power costs	Cannot handle wide flow ratio	Cannot handle emulsifying systems	High operating cost
	High solvent inventory	High headroom	Cannot handle high flow ratio	High maintenance
	Large floor space	Sometimes low efficiency	Will not always handle emulsifying systems, except perhaps pulse column	Limited number of stages in single unit, although some units have 20 stages
	Interstage pumping may be necessary	Difficult scale-up		

Source: From Ref. 4, with permission.

491

TABLE 8.5-2. Comparison of Experimental and Calculated Stream Concentrations for Copper Extraction by Kelex 100 in a Three-Stage Countercurrent Cascade of Mixer–Setters

	Aqueous Copper (g/L)		Organic Copper (g/L)		Aqueous H_2SO_4 (g/L)		Free Reagent (M)
	Data	Model	Data	Model	Data	Model	Model
Aqueous feed	33.6	33.6	—	—	13.5	13.5	—
1st stage	17.7	2.9	8.63	8.67	53	60.5	0.229
2nd stage	9.55	0.11	4.83	1.19	74.5	64.8	0.462
3rd stage	0.27	0.046	2.75	0.51	89	65.0	0.484
Organic feed	—	—	0.5	0.5	—	—	0.484

Source: From Ref. 5, with permission.

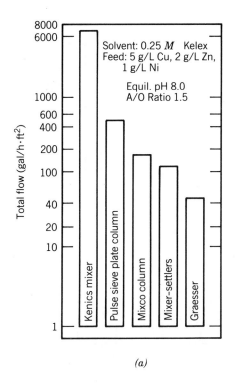

(a)

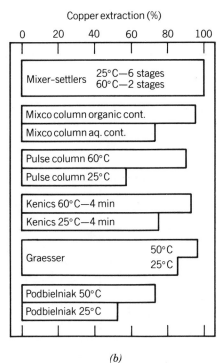

(b)

FIGURE 8.5-1 Comparisons of the operating characteristics of several contacting devices for the separation of copper from nickel and zinc by extraction from ammonia (pH 8) solution with Kelex. (*a*) Comparison of specific throughput at flooding. (*b*) Comparison of copper extraction efficiency. (*c*) Comparison of Cu–Zn selectivity. (*d*) Comparison of Cu–Ni selectivity. From Ref. 9, with permission.

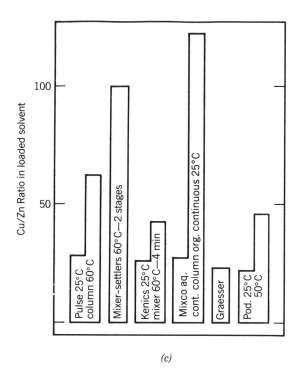

(c)

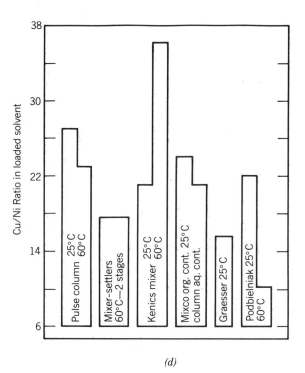

(d)

FIGURE 8.5-1 (*Continued*)

function of equipment size. A strategy for developing such contactor models has been outlined,[10] and the approach has been demonstrated with a reaction-controlled[11] and mass-transfer-controlled[12] extractions in a Kenics static mixer.

8.6   PROCESS DESIGN AND ENGINEERING

Process flowsheets for separations of numerous metals have been published.[1-5] Synthesis and design of such processes for a given feed stream require consideration of the following factors:

1. *Choice of Extractant.*   For separation of a particular metal, an effective extractant must be chosen that has the capability of providing high distribution coefficients and high organic-phase loadings. This choice depends on the chemistry of the metal and the composition of the aqueous feed solution. For example, a chelating extractant may be appropriate for extracting copper from a weakly acidic sulfate solution, but an amine may be required for a solution that has a high level of hydrochloric acid.

A practical extractant must also be subject to regeneration. The extraction must be reversible such that stripping is possible with minimal consumption of reagents. Stripping should yield a purified and concentrated form of the metal product.

With feed streams that contain several metals, selectivity may be a major concern. To obtain selectivity one may seek a different extractant for each metal that is highly specific by virtue of its equilibrium or kinetics properties, or one may use a single extractant for several metals and design the contacting process to achieve effective fractionation by exploiting quantitative rather than qualitative differences in metal chemistry. For example, in the process[6] represented by Fig. 8.6-1 TBP is chosen to extract iron from a chloride feed stream containing copper and cobalt. Then copper and cobalt chlorides are both extracted into an amine solution. The separation of cobalt from copper is accomplished by controlling the water-to-organic phase ratios in two separate stripping circuits. Similar strategies are represented by the flowsheets shown in Figs. 8.1-1 and 8.1-2. Process models for multimetal fractionations with a single extractant have been presented[7] for an idealized process chemistry.

2. *Choice of Diluent, Modifiers, and Extractant Concentration.*   The formulation and composition of the organic-phase solution play a major role in determining both the chemical and physical performance characteristics of an extraction system. The use of modifiers and mixed extractants alters both equilibrium and kinetics properties and can have a strong influence on interfacial phenomena. It is particularly important to formulate a system that minimizes organic solubility in the aqueous phase, promotes phase disengagement, and prevents emulsion formation. Problems in any of these areas can quickly render a process uneconomical or infeasible.

3. *Choice of Process Configuration.*   The sequence of processing steps and flowsheet configuration must be selected concurrently with the choice of extraction chemistry. Some of the established heuristics of separation process synthesis[8] may be helpful here. For example, in multimetal fractionation it is probably better to try selective stripping from a single organic stream than to do multiple selective-extraction oper-

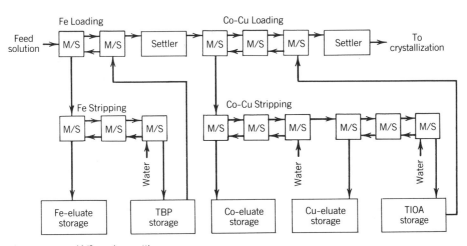

M/S = mixer-settler

FIGURE 8.6-1   Solvent extraction circuit of the Falconbridge Nikkelvert A/S matte leach plant for separation of iron, copper, and cobalt. From Ref. 6, with permission.

ations. The former approach, if possible, would minimize handling of the total feed stream volume and presumably reduce solvent losses and reagent consumption.

4. *Choice of Contacting Equipment.* Although each extraction and stripping operation in a metals-separation process is normally operated in a countercurrent mode, either stagewise or in differential contact, the large distribution coefficients provided by the chemical-reaction basis of the separations suggest that a large number of stages, or transfer units, is not required. On the other hand, slow kinetics may restrict efficiency factors considerably so that high interfacial areas and residence times may be required. These conditions have favored the use of mixer–settlers. Where kinetics are fast, differential contactors may be preferable because of smaller power requirements and solvent inventory requirements.

While efficiency is a factor in equipment selection, mechanical considerations often provide the determining criteria. One must always be sure to minimize solvent losses, and concern about entrainment and emulsion formation can dictate the mode of operation if not the choice of contactor. For example, power input for mixing may be limited or the less viscous phase chosen to be continuous to ensure good phase disengagement.

Many of these and other practical aspects of process design and operation, such as appropriate materials of construction, prevention of "crud" formation, effluent treatment methods, and typical process costs, are discussed by Ritcey and Ashbrook.[9]

If any given extraction system can be characterized with respect to the dominant species appearing and reactions occurring, experimental equilibrium and rate data can be correlated in terms of the species concentrations. Such information can be combined with material balances and available engineering correlations for liquid–liquid systems to generate process models useful for design, optimization, and control.

8.7 SUMMARY

Organic reagents are available that can extract metal ions into an organic solvent by virtue of liquid–liquid ion exchange or association reactions. Such reactions are reversible and provide a basis for effective metal-separation processes. Large distribution coefficients are observed, but they vary with system composition.

Phase equilibria can be modeled in terms of equilibrium constants for the relevant reactions. Because of low mutual solubilities of the phases, extraction reactions appear to be heterogeneous. Some reactions exhibit slow chemical kinetics, with the reaction step constituting a resistance to extraction in series with intraphase mass transfer.

Large-scale solvent extraction of metals is conducted in equipment that is similar to that used for conventional liquid extraction. Nonlinearities in the multicomponent, reactive phase equilibria and possibly slow kinetics complicate design calculations, but existing correlations and methods for treating mass transfer, interfacial phenomena, and dispersion can be used if the effects of process chemistry are properly characterized.

An extensive technology and a large body of experience have developed from both laboratory studies and commercial process operations. A variety of extractants, equipment, and processes is available for application to a wide range of metals-separation problems.

REFERENCES

Section 8.1

8.1-1   Y. Marcus and A. S. Kertes, *Ion Exchange and Solvent Extraction of Metal Complexes*, Wiley, New York, 1969.

8.1-2   H. Freiser, *Crit. Rev. Anal. Chem.*, **1**, 47 (1970).

8.1-3   J. Stary, *The Solvent Extraction of Metal Chelates*, Pergamon, Oxford, 1964.

8.1-4   A. K. De, S. M. Khopkar, and R. A. Chalmers, *Solvent Extraction of Metals*, Van Nostrand, New York, 1970.

8.1-5   H. A. C. McKay, T. V. Healy, I. L. Jenkins, and A. Nyalor (Eds.), *Solvent Extraction Chemistry of Metals*, MacMillan, London, 1965.

8.1-6   J. Korkisch, *Modern Methods for the Separation of Rarer Metal Ions*, Pergamon, Oxford, 1969.

8.1-7   J. F. C. Fisher and C. W. Notebaart, Commercial Processes for Copper, in T. C. Lo, M. H. I. Baird, and C. Hanson (Eds.), *Handbook of Solvent Extraction*, Chap. 25.1, Wiley-Interscience, New York, 1983.

8.1-8   G. Thorsen, Commercial Processes for Cadmium and Zinc, in T. C. Lo, M. H. I. Baird, and C. Hanson (Eds.), *Handbook of Solvent Extraction*, Chap. 25.5, Wiley-Interscience, New York, 1983.

8.1-9   P. J. D. Lloyd, Commercial Processes for Uranium from Ore, in T. C. Lo, M. H. I. Baird, and

C. Hanson (Eds.), *Handbook of Solvent Extraction*, Chap. 25.11, Wiley-Interscience, New York, 1983.

8.1-10 G. M. Ritcey and A. W. Ashbrook, *Solvent Extraction, Principles and Applications to Process Metallurgy*, Part II, Chaps. 4–6, Elsevier, Amsterdam, 1979.

8.1-11 A. J. Monhemius, The Extractive Metallurgy of Deep-Sea Manganese Nodules, in R. Burkin (Ed.), *Topics in Non-ferrous Extractive Metallurgy*, Blackwell Scientific Publications for the Society of Chemical Industry, Oxford, 1980.

8.1-12 T. C. Lo, M. H. I. Baird, and C. Hanson (Eds.), *Handbook of Solvent Extraction*, Wiley-Interscience, New York, 1983.

8.1-13 G. M. Ritcey and A. W. Ashbrook, *Solvent Extraction, Principles and Applications to Process Metallurgy*, Parts I and II, Elsevier, Amsterdam, 1979.

Section 8.2

8.2-1 G. M. Ritcey and A. W. Ashbrook, *Solvent Extraction, Principles and Applications to Process Metallurgy*, Elsevier, Amsterdam, 1979.

8.2-2 D. S. Flett, J. Melling, and M. Cox, Commercial Solvent Systems for Inorganic Processes, in T. C. Lo, M. H. I. Baird, and C. Hanson (Eds.), *Handbook of Solvent Extraction*, Chap. 24, Wiley-Interscience, New York, 1983.

8.2-3 *Proceedings of the International Solvent Extraction Conference, The Hague 1971*, Society of Chemical Industry, London, 1971.

8.2-4 *Proceedings of the International Solvent Extraction Conference, Lyons 1974*, Society of Chemical Industry, London, 1974.

8.2-5 *Proceedings of the International Solvent Extraction Conference, Toronto 1977*, Canadian Institute of Mining and Metallurgy, Montreal, 1979.

8.2-6 G. M. Ritcey and A. W. Ashbrook, *Solvent Extraction, Principles and Applications to Process Metallurgy*, Part I, Chap. 3, Elsevier, Amsterdam, 1979.

8.2-7 V. S. Schmidt, *Amine Extraction*, Israel Program for Scientific Translation, Keter Press, Jerusalem, 1971.

8.2-8 K. J. Murray and C. J. Bouboulis, *Eng. Mining J.*, **174**, 74 (1973).

8.2-9 K. Akiba and H. Freiser, *Anal. Chim. Acta*, **136**, 329 (1982).

8.2-10 G. M. Ritcey and A. W. Ashbrook, *Solvent Extraction, Principles and Applications to Process Metallurgy*, Part I, Chap. 4, Elsevier, Amsterdam, 1979.

Section 8.3

8.3-1 Y. Marcus and A. S. Kertes, *Ion Exchange and Solvent Extraction of Metal Complexes*, Appendix F, Wiley, New York, 1969.

8.3-2 F. G. Seeley and D. J. Crouse, *J. Chem. Eng. Data*, **11**, 424 (1966).

8.3-3 F. G. Seeley and D. J. Crouse, *J. Chem. Eng. Data*, **16**, 393 (1971).

8.3-4 Japan Atomic Energy Research Institute, *Data of Inorganic Solvent Extraction*, JAERI 1047 (1963), JAERI 1062 (1964), JAERI 1106 (1966).

8.3-5 G. L. Bauer and T. W. Chapman, *Met. Trans.*, **7B**, 519 (1976).

8.3-6 Y. C. Hoh and R. G. Bautista, *Met. Trans.*, **9B**, 69 (1978).

8.3-7 K. Akiba and H. Freiser, *Anal. Chim. Acta*, **136**, 329 (1982).

8.3-8 R. A. DeRuiter, "Copper and Nickel Extraction from Ammoniacal Solution by LIX 64N," M.S. Thesis, University of Wisconsin, Madison, 1981.

8.3-9 S. D. Troyer, "Liquid–Liquid Extraction of Copper and Nickel with Di-(2-Ethylhexyl) Phosphoric Acid," M.S. Thesis, University of Wisconsin, Madison, 1975.

8.3-10 L. G. Sillen and A. E. Martell, *Stability Constants of Metal Ion Complexes*, Special Publication No. 17 (1962), and *Supplement No. 1*, Special Publication No. 25 (1971), The Chemical Society, Burlington House, London.

8.3-11 J. J. Christensen and R. M. Izatt, *Handbook of Metal Ligand Heats*, Marcel Dekker, New York, 1983.

8.3-12 T. Kojima, H. Fukutomi, and H. Kakihana, *Bull. Chem. Soc. Jpn.*, **42**, 875 (1969).

8.3-13 R. Caban and T. W. Chapman, *AIChE J.*, **18**, 904 (1972).

8.3-14 S. W. Tse, "Mass Transfer Rates in the Solvent Extraction of Metal Chlorides by Tri-isooctylamine," M.S. Thesis, University of Wisconsin, Madison, 1978.

8.3-15   M. W. Thiel, "Separation of Copper and Zinc by Liquid–Liquid Extraction with Tri-isooctylamine," M.S. Thesis, University of Wisconsin, Madison, 1974.

8.3-16   W. J. McDowell, D. C. Michelson, B. A. Moyer, and C. F. Coleman, "A Data Base for Solvent Extraction Chemistry," presented at International Solvent Extraction Conference, Denver, 1983.

8.3-17   C. F. Coleman, *Amines as Extractants-Survey of the Descriptive and Fundamental Extraction Chemistry*, Report ORNL-3516, Oak Ridge National Laboratory, Oak Ridge, TN, 1963. See also *Nucl. Sci. Tech.*, **17**, 274 (1963) and *Prog. Nucl. Energy, Ser. III, Process Chemistry, Vol. 4*, C. E. Stevenson, et al., eds., Pergamon, Oxford, pp. 233–285, 1970.

8.3-18   G. M. Ritcey and A. W. Ashbrook, *Solvent Extraction, Principles and Applications to Process Metallurgy*, Part I, p. 112, Elsevier, Amsterdam, 1979.

Section 8.4

8.4-1   M. Cox and D. S. Flett, Metal Extractant Chemistry, in T. C. Lo, M. H. I. Baird, and C. Hanson (Eds.), *Handbook of Solvent Extraction*, Chap. 2.2, Wiley-Interscience, New York, 1983.

8.4-2   T. Kojima and T. Miyauchi, *Ind. Eng. Chem. Fundam.*, **20**, 14 (1981).

8.4-3   K. Akiba and H. Freiser, *Anal. Chim. Acta*, **136**, 329 (1982).

8.4-4   G. F. Froment and K. B. Bischoff, *Chemical Reactor Analysis and Design*, p. 310, Wiley, New York, 1979.

8.4-5   T. Kojima and T. Miyauchi, *Ind. Eng. Chem. Fundam.*, **21**, 220 (1982).

8.4-6   T. W. Chapman, R. Caban, and M. E. Tunison, *AIChE Symp. Ser.*, **71** (152), 128 (1976).

8.4-7   G. L. Bauer, "Solvent Extraction of Copper: Kinetics and Equilibrium Studies," Ph.D. Thesis, University of Wisconsin, Madison, 1974.

8.4-8   J. Rydberg, H. Reinhardt, and J. O. Liljensin, in J. A. Marinsky and Y. Marcus (Eds.), *Ions Exchange and Solvent Extraction*, Vol. 3, p. 111, Marcel Dekker, New York, 1973.

8.4-9   M. J. Slater, G. M. Ritcey, and R. F. Pilgrim, in *Proceedings of the International Solvent Extraction Conference, Lyon*, Vol. 1, p. 107, Society of Chemical Industry, London, 1974.

8.4-10   T. W. Chapman, Solvent Extraction of Metals—Metal Transfer Rates and Contactor Design, in R. G. Bautista (Ed.), *Hydrometallurgical Process Fundamentals*, Plenum, New York, 1984.

8.4-11   R. W. Freeman and L. L. Tavlarides, *Chem. Eng. Sci.*, **35**, 559 (1980).

8.4-12   J. B. Lewis, *Chem. Eng. Sci.*, **3**, 248 (1954).

8.4-13   W. Nitsch and J. G. Kahni, *Ger. Chem. Eng.*, **3**, 96 (1980).

8.4-14   D. S. Flett, D. N. Okuhara, and D. R. Spink, *J. Inorg. Nucl. Chem.*, **35**, 2471 (1973).

8.4-15   R. L. Atwood, D. N. Thatcher, and J. D. Miller, *Met. Trans.*, **6B**, 465 (1975).

8.4-16   R. J. Whewell, M. A. Hughes, and C. Hanson, *J. Inorg. Nucl. Chem.*, **38**, 2071 (1976).

8.4-17   R. A. DeRuiter, "Copper and Nickel Extraction from Ammoniacal Solution by LIX64N," M.S. Thesis, University of Wisconsin, Madison, 1981.

8.4-18   W. Nitsch and U. Schuster, *Separation Sci. Tech.*, **18**, 1509 (1983).

8.4-19   C. H. Hales, "The Effect of Bulk Concentrations on Liquid–Liquid Extraction of Copper with Sodium-Loaded Di-(2-Ethylhexyl)Phosphoric Acid," M.S. Thesis, University of Wisconsin, Madison, 1977.

8.4-20   S. W. Tse, "Mass Transfer Rates in the Solvent Extraction of Metal Chlorides by Tri-isooctylamine," M.S. Thesis, University of Wisconsin, Madison, 1978.

8.4-21   W. Nitsch and B. Kruis, *J. Inorg. Nucl. Chem.*, **40**, 857 (1978).

8.4-22   J. W. Roddy, C. F. Coleman, and S. Arai, *J. Inorg. Nucl. Chem.*, **33**, 1099 (1971).

8.4-23   J. C. S. Cassa and A. J. Monhemius, Kinetics of Extraction of Iron(III) from Chloride Solutions by Trioctylamine, in R. G. Bautista (Ed.), *Hydrometallurgical Process Fundamentals*, p. 327, Plenum, New York, 1984.

8.4-24   D. R. Olander, *AIChE J.*, **6**, 233 (1960).

8.4-25   S. W. Tse, E. S. Vargas, and T. W. Chapman, Mass Transfer and Reaction Rates in the Solvent Extraction of Metals, in *Hydrometallurgical Recovery of Metals from Ores, Concentrates, and Secondary Sources*, Chap. 19, SME–AIME, New York, 1981.

8.4-26   A. Altway, M. E. Tunison, and T. W. Chapman, Effect of Contactor Type on Metal Separation by Solvent Extraction, in *Hydrometallurgical Recovery of Metals from Ores, Concentrates, and Secondary Sources*, Chap. 25, SME–AIME, New York, 1981.

8.4-27   W. Nitsch, *Faraday Discuss. Chem. Soc.*, **77**, paper 77/8 (1984).

8.4-28   W. Nitsch and A. Van Schoor, *Chem. Eng. Sci.*, **38**, 1947 (1983).

Section 8.5

8.5-1 R. E. Treybal, *Liquid Extraction*, 2nd ed., McGraw-Hill, New York, 1963.

8.5-2 G. S. Laddha and T. E. Degalessan, *Transport Phenomena in Liquid Extraction*, McGraw-Hill, New York, 1978.

8.5-3 H. P. C. Pratt and C. Hanson, Selection, Pilot Testing, and Scale-Up of Commercial Extractors, in T. C. Lo, M. H. I. Baird, and C. Hanson (Eds.), *Handbook of Solvent Extraction*, Chap. 16, Wiley-Interscience, New York, 1983.

8.5-4 G. M. Ritcey and A. W. Ashbrook, *Solvent Extraction, Principles and Applications to Process Metallurgy*, Elsevier, Amsterdam, 1979.

8.5-5 T. W. Chapman, Equilibrium Stage Calculations for the Solvent Extraction of Metals, in *Fundamental Aspects of Hydrometallurgy*, AIChE Symposium Series, Vol. 74, No. 173, p. 120, AIChE, New York, 1978.

8.5-6 G. M. Ritcey and A. W. Ashbrook, *Solvent Extraction, Principles and Applications to Process Metallurgy*, Part II, pp. 198–220, Elsevier, Amsterdam, 1979.

8.5-7 K. L. Power, in *Proceedings of the International Solvent Extraction Conference, The Hague, 1971*, p. 1409, Society of Chemical Industry, London, 1971.

8.5-8 A. Altway, M. E. Tunison, and T. W. Chapman, Effect of Contactor Type on Metal Separation by Solvent Extraction, in *Hydrometallurgical Recovery of Metals from Ores, Concentrates, and Secondary Sources*, Chap. 25, SME–AIME, New York, 1981.

8.5-9 G. M. Ritcey and A. W. Ashbrook, *Solvent Extraction, Principles and Applications to Process Metallurgy*, Part II, p. 125, Elsevier, Amsterdam, 1979.

8.5-10 T. W. Chapman, Solvent Extraction of Metals—Metal Transfer Rates and Contactor Design, in R. G. Bautista (Ed.), *Hydrometallurgical Processes Fundamentals*, Plenum, New York, 1984.

8.5-11 M. E. Tunison and T. W. Chapman, Characterization of the Kenics Mixers as a Liquid–Liquid Extractive Reactor, in *Fundamental Aspects of Hydrometallurgical Processes*, AIChE Symposium Series, Vol. 74, No. 173, p. 112, AIChE, New York, 1978.

8.5-12 S. W. Tse, "Mass Transfer Rates in the Solvent Extraction of Metal Chlorides by Tri-isooctylamine," M.S. Thesis, University of Wisconsin, Madison, 1978.

Section 8.6

8.6-1 G. M. Ritcey and A. W. Ashbrook, *Solvent Extraction, Principles and Applications to Process Metallurgy*, Part II, Elsevier, Amsterdam, 1979.

8.6-2 T. C. Lo, M. H. I. Baird, and C. Hanson (Eds.), *Handbook of Solvent Extraction*, Wiley-Interscience, New York, 1983.

8.6-3 *Proceedings of the International Solvent Extraction Conference, The Hague, 1971*, Society of Chemical Industry, London, 1971.

8.6-4 *Proceedings of the International Solvent Extraction Conference, Lyons 1974*, Society of Chemical Industry, London, 1974.

8.6-5 *Proceedings of the International Solvent Extraction Conference, Toronto 1977*, Canadian Institute of Mining and Metallurgy, Montreal, 1979.

8.6-6 P. G. Thornhill, E. Wigstol, and G. Van Veert, *J. Metals*, **23**, 13 (1971).

8.6-7 F. J. Brana-Mulero, "Studies on the Solvent Extraction of Metals: Process Synthesis Strategies and Mathematical Consistency of the Models," Ph.D. Thesis, University of Wisconsin, Madison, 1980.

8.6-8 N. Nishida, G. Stephanopoulos, and A. W. Westerberg, *AIChE J.*, **27**, 320 (1981).

8.6-9 G. M. Ritcey and A. W. Ashbrook, *Solvent Extraction, Principles and Applications to Process Metallurgy*, Elsevier, Amsterdam, 1979.

CHAPTER 9

Leaching—Metals Applications

MILTON E. WADSWORTH
College of Mines and Mineral Industries
University of Utah
Salt Lake City, Utah

9.1 INTRODUCTION

Hydrometallurgy is a relatively recent technology. Modern, large-scale hydrometallurgical processing plants depend on electrical power for transport and agitation of large tonnages of ore slurries and solutions. Pumps, agitators, autoclaves, and air compressors of the design needed for hydrometallurgical processing are historically recent developments. Reduction of dissolved metals electrochemically, as in electrorefining and electrowinning, has broadened the application of hydrometallurgy for separation and recovery. Prior to the availability of electrical energy, reduction from solution was possible using chemical reductants, but rarely on a large scale. A noteworthy example is the reduction of copper contained in acidic mine waters using metallic iron. Contact reduction of one metal on another, more electronegative metal is commonly called *cementation*. This practice was carried out as early as the sixteenth century in Europe[1,2] and is common practice today for recovery of copper from acidic leach liquors.

Leaching refers to the process in which solid ores or concentrates are contacted by an aqueous lixiviant or leach liquor capable of dissolving all or part of the ore or concentrate. The loaded or pregnant liquor is then subjected to a separation process for purification or recovery or both of desired metals or metal compounds. The dissolution processes may involve oxidation and reduction or may be simple dissolution, without exchange and reduction or may be simple dissolution, without exchange of charge between the solid and the lixiviant. In the former, the lixiviant must contain a suitable oxidant or reductant for the dissolution process to occur. Since leaching includes various solution treatment steps, these steps must be considered briefly to understand the advent of and expanding role of leaching as a separation process.

9.1-1 Historical

It is quite apparent that reduction of metal in aqueous solutions had to wait for the development of electrical energy on a commercial scale. Copper was almost a unique exception. Hydrometallurgy therefore followed the developments of Michael Faraday[3] who in 1831 provided the basis that led to the dynamo and to electrical machinery. Electrical energy was available on a large scale some 40 years later. The first electrolytic reduction plant for copper[4] and was constructed near Swansea in 1869.

The solvent action of cyanide for gold and silver was noted as early as 1783 and cyanidation became a commercial process following patents by MacArthur and Forrest in 1887 and 1889.[5] Reduction of metals from aqueous solutions using hydrogen and other reductant gases was an important development.[6-8] Nickel and cobalt have been produced commercially by hydrogen reduction since 1954. The chemistry of gaseous reduction was understood prior to this time. Patents issued between 1906 and 1941 covered the use of SO_2, CO, and H_2 as gaseous reductants in pressurized vessels. The detailed work of the Ipatievs[6] between 1909 and 1931 provided the major body of work for high-temperature, high-pressure reduction of metals to the metallic or the oxide state in aqueous solutions. Thus, by the early 1900s three methods for the direct reduction of metals from aqueous solutions were available to the metallurgical industry: (1) cementation (contact reduction), (2) electrolytic reduction, and (3) gaseous reduction.

In the period 1946–1955[6] researchers at Sherritt Gordon Mines, Ltd, in Canada and Chemical Construction Corporation in the United States developed commercial processes for leaching of nickel and cobalt sulfides and arsenides. Coworkers[7] at the University of British Columbia published an important series of papers beginning in the 1950s on pressure leaching of sulfides. These advances coupled with research on gaseous reduction led to the development of several commercial processes.[7,8] Nonoxidative low-pressure leaching of bauxite was developed by Bayer[8,9] in 1892 and is currently the worldwide method for the production of high-purity alumina feed for the Hall Herault electrolytic cell for aluminum metal production.

An important step in many hydrometallurgical processing schemes involves the upgrading of solutions prior to electrowinning or crystallization. The advent of ion exchange and solvent extraction represented important developments for both recovery and upgrading of dilute process streams. Cation exchange on zeolites was first noted in 1876 by Lemberg[10] and a significant advance resulted from the work of Adams and Holmes in 1935 when they developed synthetic ion-exchange resins. Anion exchange became possible and available on a large scale with increased opportunity for metals extraction. Early attempts in 1945–1950 to recover uranium by cation exchange were not successful. In 1948, however, H. E. Bross successfully extracted uranium on anion exchangers. The first ion-exchange plant for uranium recovery was built in South Africa in 1950.[10] Similarly, solvent extraction has early origins since it has been used routinely for many organic chemical separations. In the 1950s Oak Ridge researchers developed solvent extraction techniques for recovery of uranium. The first solvent extraction process for uranium was installed in 1955 at the Kerr–McGee plant in Shiprock, New Mexico.[11] The LIX reagents developed by General Mills were an important development for processing low-grade copper streams. The first (LIX) plant coupling dump leaching, solvent extraction, and electrowinning was the Ranchers Bluebird Mine in Miami, Arizona, constructed in 1968.[11,12]

Today, hydrometallurgy is well established as the principal method for extraction of many important industrial metals. Hydrometallurgy for the direct treatment of base metal sulfide concentrates, as a widely used technology, must yet prove itself. The roast-leach electrowinning of zinc is a noteworthy exception and is evolving as standard practice in the zinc industry worldwide. Relatively recent developments by way of jarosite and iron oxide hydrolysis and precipitation processes have improved recovery and helped secure zinc hydrometallurgy as standard in the industry.[13]

There has been a tendency to place hydrometallurgy strictly on a competitive basis with smelting for the treatment of base metal sulfides. Too many overly optimistic and, in some cases, false claims have been made of the relative virtues of hydrometallurgy compared to smelting. Modern smelting plants can meet the SO_2 ambient air requirements by the installation of double contact-acid plants. If hydrometallurgy is to be a competitive technology it must in fact be so in the marketplace with a clear and decisive edge. For this reason it would appear that hydrometallurgy should not be viewed as a competitive technology but supplemental to and in some cases compatible with smelting. In fact, lower energy versions of hydrometallurgical processing of base metal sulfides often involve combined smelting and hydrometallurgical applications.[14] Hydrometallurgy applied to concentrates involves high capital costs, multiple steps, and in general is more energy intensive than is pyrometallurgy.[14,15] Arguments that hydrometallurgy is nonpolluting compared to smelting are not persuasive. Effluents may carry heavy metals and must be carefully treated or impounded. The main throwaway product in the case of hydrometallurgy is gangue plus a mixture of iron oxides and jarosites and other metal salts as compared with slag from a smelter which is easily stored and relatively nonreactive. Sulfur in smelting is emitted into the atmosphere at accepted levels and goes to sulfuric acid. In hydrometallurgical processes sulfur goes to sulfate salts or to elemental sulfur. Consequently, there is no clear advantage of hydrometallurgy over smelting when best technology is compared relative to effluent discharge.

Hydrometallurgy has an increasing role to play in the future. It is unique in its application to low-grade ores which cannot be beneficiated economically. It has distinct potential for in situ extraction of complex sulfide ores and concentrates. It should be viewed as an alternate technology having high chemical specificity. It may be concluded that, with only a relatively short history, hydrometallurgy has become an inportant part of separation technology. There exists a sound basis in theory, materials of construction, and in design capability for its implementation.

9.1-2   Scope of Leaching Practice

Figure 9.1-1 illustrates the general flow of ores and concentrates in leaching practice. Ore usually is transported from the mineral deposit for leaching. In some special cases leaching may occur in situ but these examples are few and will be treated separately. Normally, the ore is transported and treated by methods that depend on the value of the ore. It is useful to consider three ore types: low-grade ore, direct-leaching ore, and high-grade ore, the latter being suitable for beneficiation. The major example of the leaching of a low-grade ore is in copper hydrometallurgy commonly practiced in the western United States where copper is placed on vast leach dumps and is treated by recycling leach solutions. These dumps contain both sulfides and oxides of such low grade that they are not suitable for beneficiation. Leaching of

HYDROMETALLURGY-LEACHING

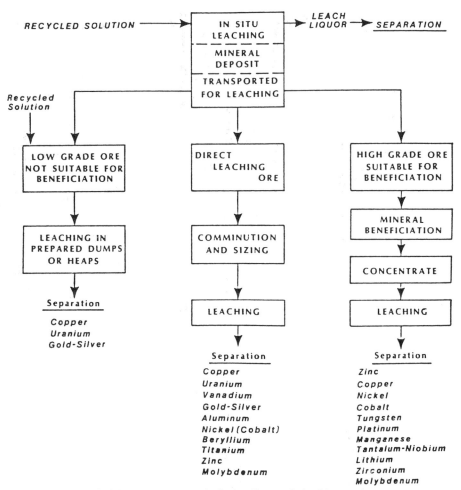

FIGURE 9.1-1 Various paths for handling ore in leaching practice.

dumps extends into years and in the best leaching operation achieves approximately a 15–20% ongoing or steady-state recovery of the new material content being placed on the dump. Heap leaching usually refers to the leaching of oxides and is carried out on prepared leaching pads, with leach cycles of days to months. Uranium, gold, and silver are also leached, though in relatively minor tonnages, by heap-leaching technology. The second type of ore (direct-leaching ore) is one which has sufficient value that it can be crushed or ground to appropriate sizes for vat leaching or agitation leaching. Examples are copper, uranium, vanadium, gold, silver, aluminum, nickel, cobalt, and beryllium. Pretreatment by oxidative, reductive, or salt roasting may be necessary to render the material suitable for leaching. High-grade ore is of sufficient value that it may be beneficiated to increase recovery and produce a concentrate. The concentrates produced may be pretreated by roasting prior to leaching.

Each process produces a leach liquor which must be treated by some separation process for the recovery of metal values as metals or as salts requiring further treatment. The various leaching processes produce either a dilute process stream or a concentrated liquor. Low-grade ores invariably produce dilute process streams. For example, copper dump-leaching effluents usually contain 0.5–1 g per liter of copper. In the case of direct-leaching ore, the leach liquor may be either a dilute process stream requiring upgrading or a concentrated stream requiring no additional upgrading. The leaching of concentrates produces a high-grade liquor that can be sent directly to metal recovery by reduction or precipitation. Figure 9.1-2 illustrates

SEPARATIONS

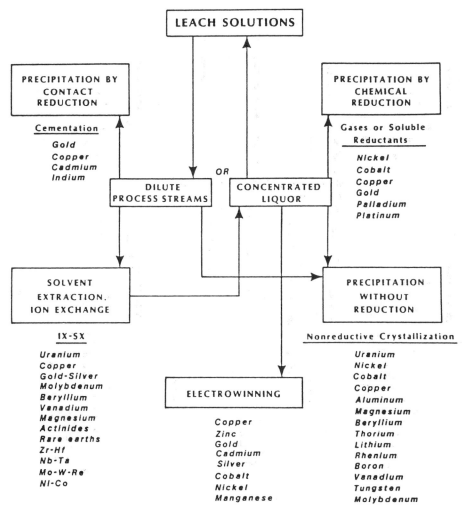

FIGURE 9.1-2 Methods employed for separating metals from dilute and concentrated process streams.

the various methods for treating both dilute process streams and concentrated liquors. In the case of dilute process streams, metal values may be recovered by either precipitation on other metals through contact reduction or cementation. Examples are recovery of copper by precipitation on iron and recovery of gold by precipitation on zinc. An alternate path is to upgrade the solution to higher concentration levels by means of solvent extraction or ion exchange. Important examples are uranium and copper and recently the adsorption of gold and silver on activated carbon. The stripped organic extractant or ion exchanger produces a concentrated liquor. The concentrated liquor may then be suitable for electrowinning directly. Metals such as copper, zinc, and gold fall into this category. Often, however, it is necessary to recover the concentrated metal values from solution either by gaseous or chemical reduction to the metallic state or to precipitated reduced oxides. Recovery may occur by nonreductive precipitation. Metals following these various paths are indicated in Fig. 9.1-2.

Many hydrometallurgical processes or process steps are used to upgrade concentrates, process recycled scrap metal, or purify aqueous process steams. Examples are (1) the leaching of molybdenite concentrate to remove impurities;[16] (2) leaching of tungsten carbide and molybdenum scrap;[17] (3) removal of copper impurities in nickel anolyte by cementation on metallic nickel;[18] and (4) various methods for treating nuclear fuel elements.

TABLE 9.2-1 1978 Copper Production: Statistics for Western U.S. Copper Operations with Significant Leach Ouput

Company	Copper Production (tons)		Percent of Total Copper Produced by Leaching
	Concentrating	Leaching	
Kennecott	287,200	88,200	23.5
Phelps–Dodge	283,600	34,550	10.8
Duval (Pennzoil)	112,300	10,200	8.3
Anamax Mining	67,200	35,810	34.8
Asarco	85,780	10,720	11.1
Cities Service	72,800	7,750	9.6
Cyprus Mines	61,600	12,150	16.5
Inspiration Consol.	20,700	18,000	46.5
	991,180	217,380	18.0

Source: Reference 1.

9.2 LEACHING PRACTICE

A few examples of leaching practice are presented with the aim of illustrating types of leaching processes rather than an extensive treatise of current practice. The general outline of ore types presented earlier will be followed.

9.2-1 Leaching of Low-Grade Ores

Table 9.2-1 presents the copper production by major copper producers in the western United States for the year 1978.[1] The tonnage produced by dump leaching and the leaching of oxide ores was 18% of total year tonnage. An estimated two-thirds of this, or approximately 12% of the total, may be attributed to dump leaching of low-grade predominantly sulfide-waste materials.

In copper dump leaching practice, waste rock (usually below 0.2% copper) is placed on dumps by truck or rail haulage. These dumps vary greatly in size and shape. Depths extend from a few tens of feet to as much as 1200 ft. It is generally recognized that good aeration is required as well as good permeability. Consequently, the stepped surfaces in the dump are usually ripped to provide needed permeability. As-mined ore, newly placed on a dump, will have permeabilities of approximately 1000 darcys.* Weathering of intrusive material can cause dramatic changes in porosity. Weathering plus deposition of salts can also cause significant changes in permeability with time. The impact of such induced weathering is an important consideration in assessing expected recoveries, since leaching may extend for years. The porosity of newly dumped rock will be in the range of 35–40% percent, but the weight of haulage trucks can cause compaction of as much as 10 ft in 100 ft. Porosities of as low as 25% may result from compaction and weathering.

Figure 9.2-1 illustrates a dump cross section according to Whiting,[2] showing important hydrological and structural characteristics.

1. *Channeling.* Channeling of each liquor occurs as a result of compaction and salt precipitation. Fluid flow down channels essentially bypasses regions of the dump. It is enhanced if solution application is by surface ponding.

2. *Seep or Blowout.* Compacted zones may cause entrapment of solutions, resulting in the formation of a perched water table within the dump. The buildup of hydrostatic pressure can cause surface seepage and even expulsion of solid material from the dump wall.

3. *Stratification.* During the dumping of waste rock, the coarse material travels further down the slope or the dump than the fine material, causing stratification. Without appropriate ripping, stratification may affect the flow pattern within the dump.

4. *Sorption.* Solutions bearing dissolved metals may pass through regions containing nongravitational water at lower concentrations. Trapping of metal values can occur by inward diffusion into the pore structure of the rocks and into stagnant aqueous regions.

5. *Aeration.* Aeration is best near the face of the dump, providing optimal conditions of temperature and bacterial activity. Aeration by convection through the dump is an essential part of the leaching mechanism. In regions of high oxidation potential, iron is oxidized by bacterial activity to the ferric state resulting in the precipitation of ferric oxides and jarosites.

*Darcy refers to hydraulic conductivity or permeability. Permeability in cm/s is converted to darcys by multiplying by 1045.

TYPICAL WASTE DUMP SECTION

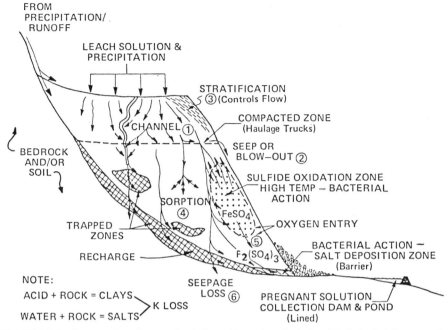

FIGURE 9.2-1 Cross section of copper leach dump illustrating physical and hydrological features. (Reproduced from Whiting.[2])

6. *Seepage Losses.* Seepage can occur through the foundation of the dump, although the formation of salts and the presence of fines often keeps this at a surprisingly low level. Runoff waters may also recharge or dilute the percolating liquors through the foundation of the dump.

In practice, dump leaching depends on sequence of processes. Within the dump three conditions are essential for leaching to occur and continue. These are:

Effective air circulation

Good bacterial activity

Uniform solution contact with the particle

The major unknowns in dump leaching for any given dump are:

A knowledge of air circulation relative to the dump configuration

The hydrology in terms of channeling and bypass

The effect of fines and precipitated salts

The effect of weathering as a function of time

Figure 9.2-2 illustrates the general flow of solution to the dump, to a holding basin, and to copper extraction. Copper extraction is achieved either by cementation on detinned scrap iron as indicated or by solvent extraction using one of the LIX reagents for selective removal of cupric ion from sulfate leach liquors. The general trend is toward solvent extraction due to the high costs of iron scrap. Schlitt[1] has indicated that the operating costs for solvent extraction are less than those for cementation, although the capital costs may be higher. Also, solvent extraction–electrowinning produces a marketable copper cathode.

Following extraction, solutions are recycled or enter a containment pond where some aeration occurs. It should be noted, however, that the iron balance for the greater part is achieved by precipitation of iron salts throughout the dump itself, as hydrated oxides or jarosites. In general, intermittent leaching with alternate leach and rest cycles is preferred to continuous leaching.[3] This practice conserves energy consumed in pumping and is effective since pore leaching continues during the rest period, under conditions of good aeration, building up dissolved metal values in the liquid phase. Continuous leaching without the rest cycle,

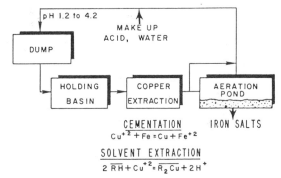

FIGURE 9.2-2 Flow of leach liquors in copper dump leaching.

because of the large volume of water, extracts large quantities of heat from the dump (up to one-half of the exothermic heat of reaction), adversely affecting leaching rates.

Of special importance is the method of solution application.[2,3] The general methods employed are:

1. Pond irrigation.
2. Trickle.
3. Low-pressure multiple sprays.
4. High-pressure single spray.
5. Well injection.

The trend in dump leaching is toward trickle leaching or sprays. In ponding, channeling can cause excessive dilution with loss of control over effluent quality. Trickle leaching is carried out by using a network of perforated PVC pipe. This system provides a more gradual application of solutions and more uniform air and solution access to the dump. Spraying, using low-pressure multiple sprays or high-pressure single sprays, also provides uniform coverage. Spray systems may suffer excessive solution loss by evaporation in areas having high evaporation rates. Both trickle and spray leaching suffer in some areas where excessive ice formation may occur during winter months. In such cases more than one method of solution application may be needed. The last method is injection down wells. This is the method used in uranium solution mining. Wells are drilled on a grid pattern and lined with perforated pipe. Solution flow is controlled by combined down-well and up-well pumping through a flooded formation. Percolation leaching using this method for copper recovery suffers because uniform coverage is difficult. Solutions move generally downward under free-flow conditions, requiring a close network of injection wells for adequate coverage. This method has been used by Anaconda in Butte, Montana, where ice formation is a serious problem.

Jackson and Ream[3] recently reviewed the results of an extensive field test study at Kennecott's Bingham Mine. A comparison was made between trickle and spray leaching. In general, spray leaching resulted in effluent solutions containing somewhat high concentrations of copper, illustrating the importance of uniform coverage with minimal channeling. In typical dump-leaching practice, the solution application rate differs from the irrigation rate, since the latter includes the rest portion of the total rest cycle.[3] The application rate is the leach solution flow rate per unit area of surface to which it is being applied, typically $\frac{1}{4}$ to $\frac{1}{5}$ of the total area available for application. On the average, application rates observed to give the best results vary from less than 5.56×10^{-6} m^3/m$^2 \cdot$s (0.5 gal/ft$^2 \cdot$h) for sprinklers, to 2.22×10^{-6} m^3/m$^2 \cdot$s (2 gal/ft$^2 \cdot$h) for trickle leaching. For pond leaching, the application rate may be as high as 5.56×10^{-5} m^3/m$^2 \cdot$s (5 gal/ft$^2 \cdot$h). The irrigation rate is the total dump-system solution flow rate divided by the total surface area available for solution application.

It is uneconomical to remove salts from leach liquors before recycle. Usual practice is to add make-up water as needed and in some cases sulfuric acid. The solutions will build up in metal and sulfate concentration until salt deposition occurs. Iron is soon in equilibrium with a variety of jarosite salts, depending on the general chemistry of the dump system. Aluminum and magnesium can increase to very high values, for example, 5–10 g/L, and sulfate concentrations may approach 1 M. In general, a successfully operating sulfide leach dump is capable of generating acid internally; that is, a dynamic buffering effect, balancing acid-producing and acid-consuming reactions, must produce pH values and solution oxidation potentials conducive to the promotion of bacterial activity and solubilization of copper and iron in solution. Acid additions to influent leach liquors do not alter the balance of acid-consuming and acid-producing reactions of such massive systems but serve to prevent precipitation of iron salts in the upper strata of the dump, preserving adequate permeability for uniform solution penetration.

The steady-state generation of ferric iron, in all of its soluble forms, Fe(III), and hydrogen ions is of primary importance in an acid dump-leaching system. At 30°C the solubility of oxygen in pure water is 2.3×10^{-4} mol/L. In the high-ionic-strength liquors produced in recycled leach liquors, the solubility is considerably less. As may be shown, the oxygen oxidation of sulfide minerals is kinetically less important than oxidation by complex ferric ions in solution, present in much greater concentration. Oxygen is essential, serving to oxidize ferrous iron to ferric and to provide conditions for the growth of chemoautotrophic bacteria.

Normally, the oxidation of ferrous iron to ferric is slow. The bacterium *Thiobacillus ferrooxidans*, an aerobic chemoautotroph deriving its energy from the oxidation of ferrous iron, greatly accelerates the oxidation of ferrous iron according to the reaction

$$0.5O_2 + 2H^+ + 2Fe^{2+} \overset{bac}{=} 2Fe^{3+} + H_2O \qquad (9.2\text{-}1)$$

Oxygen is essential to the metabolic cycle of the bacteria. The bacterium *Thiobacillus thiooxidans* is also an aerobic chemoautotroph deriving its energy from the oxidation of elemental sulfur, thiosulfate, or sulfide as contained in heavy metal sulfides. Sulfur oxidation produces sulfuric acid in place, an essential feature in maintaining open porosity in dump leaching. Pyrite is a strong acid producer, supplying ferrous iron which is subsequently oxidized to the ferric state in the presence of *T. ferrooxidans*. Pyrite also greatly influences the net consumption of oxygen in the system which often may be as high as 7–20 moles of oxygen per mole of Cu^{2+} produced.

Ferric iron in solution exists in several forms. The more important forms are $Fe(SO)_2^-$, $FeSO_4^+$, $FE_2(OH)_2^{2+}$, and $FeOH^{2+}$. The sulfate complexes are greatly favored over the hydroxyl complexes.[4] To illustrate the importance of iron complex formation, a typical leach solution will be considered in equilibrium with precipitated hydrogen jarosite, having an approximate free-sulfate activity of 0.02. The equilibrium is represented by

$$(Fe)_3(SO_4)_2(OH)_5 \cdot 2H_2O + 5H^+ = 3Fe^{3+} + 2SO_4^{2-} + 7H_2O \qquad (9.2\text{-}2)$$

for which $\log K = -2.7$. The iron sulfate complex equilibria are

$$Fe^{3+} + SO_4^{2-} = FeSO_4^+ \qquad (9.2\text{-}3)$$

and

$$Fe^{3+} + 2SO_4^{2-} = Fe(SO_4)_2^- \qquad (9.2\text{-}4)$$

with $\log K$ values of 4.15 and 5.4, respectively. Accordingly, at pH = 2.3, the ferric ion activity would be approximately 8.6×10^{-5}. Using activity coefficients of approximately 0.7, the corresponding ferric sulfate complex concentrations of $FeSO_4^+$ and $Fe(SO_4)_2^-$, respectively, would be 0.035 and 0.012 M. This corresponds to a total maximum Fe(III) concentration of approximately 2.6 g/L at this pH. Under conditions of dump-leaching, ferric ion complexes are present with activities much greater than the activity of dissolved oxygen and are kinetically more important than oxygen in the mineral oxidation. In the total dump-leaching system, the oxidation sequence is:

1. Aeration by convection, promoting bacterial activity.
2. Oxidation of ferrous iron to ferric.
3. Ferric iron dissolution of sulfides with metal release and acid generation.

The dissolution reactions typically of importance in copper dump leaching are:

Chalcopyrite (CuFeS₂) Oxidation

$$CuFeS_2 + 4Fe(III) = Cu^{2+} + 5Fe(II) + 2S^0 \qquad (9.2\text{-}5)$$

Chalcocite (Cu₂S) Oxidation

$$Cu_2S + 2Fe(III) = CuS + Cu^{2+} + 2Fe^{2+} \qquad (9.2\text{-}6)$$

Covellite (CuS) Oxidation

$$CuS + 2Fe(III) = Cu^{2+} + 2Fe^{2+} + S^0 \qquad (9.2\text{-}7)$$

Bornite (Cu₅FeS₄) Oxidation

$$Cu_5FeS_4 + 12Fe(III) = 5Cu^{2+} + 13Fe^{2+} + 4S^0 \qquad (9.2\text{-}8)$$

Sulfur Oxidation

$$S^0 + H_2O + 1.5O_2 \overset{\text{bac}}{=} H_2SO_4 \tag{9.2-9}$$

Pyrite (FeS$_2$) Oxidation

$$FeS_2 + 14Fe(III) + 8H_2O = 15Fe^{2+} + 2H_2SO_4 + 12H^+ \tag{9.2-10}$$

Pyrite is a strong acid producer. The bacterial oxidation of Fe(II) to Fe(III) [Eq. (9.2-1)] is acid consuming. When coupled with pyrite oxidation, the net result is the formation of 1 mole of H_2SO_4 per mole of pyrite oxidized to sulfate. This is possible only in the presence of bacteria under conditions of good aeration. Hydrolysis reactions forming precipitated salts such as Eq. (9.2-2) are also acid producing.

Many gangue mineral reactions are acid consuming, resulting in a dynamic steady-state pH suitable to support bacteria and maintain open porosity. Important acid consuming reactions are:

Malachite

$$CuCO_3 \cdot Cu(OH)_2 + 4H^+ = 2Cu^{2+} + CO_2 + 3H_2O \tag{9.2-11}$$

Azurite

$$2CuCO_3 \cdot Cu(OH)_2 + 6H^+ = 3Cu^{2+} + 2CO_2 + 4H_2O \tag{9.2-12}$$

Chrysocolla

$$CuO \cdot SiO_2 \cdot 2H_2O + 2H^+ = Cu^{2+} + SiO_2 + 3H_2O \tag{9.2-13}$$

Calcite

$$CaCO_3 + 2H^+ + SO_4^{2-} + H_2O = CaSO_4 \cdot 2H_2O + CO_2 \tag{9.2-14}$$

Chlorites

$$H_2Mg_2Si_2O_9 \cdot yH_4Mg_2Al_2Si_2O_9 + (6 + 6y)H^+$$
$$= (2 + 2y)SiO_2 + (3 + 2y)Mg^{2+} + 2y\,Al^{3+} + (5 + 5y)H_2O \tag{9.2-15}$$

Biotite

$$H_2K(Mg)_{3x}(Fe)_{6-3x}Al(SiO_4)_3 + 10H^+$$
$$= K^+ + 3xMg^{2+} + (6 - 3x)Fe^{2+} + Al^{3+} + 3SiO_2 + 6H_2O \tag{9.2-16}$$

K-Feldspar Alteration to K-Mica

$$3KAlSi_3O_8 + 2H^+ + 12H_2O = KAl_3Si_3O_{10}(OH)_2 + 2K^+ + 6H_4SiO_4 \tag{9.2-17}$$

K-Mica Alteration to Kaolinite

$$2KAl_3Si_3O_{10}(OH)_2 + 2H^+ = 3Al_2Si_2O_5(OH)_4 + 2K^+ \tag{9.2-18}$$

Kaolinite

$$Al_2Si_2O_5(OH) + 6H^+ = 3Al^{3+} + H_2O + 2H_4SiO_4 \tag{9.2-19}$$

Expected recovery from massive dumps is not known. Steady-state daily recovery, the fraction of the daily tonnage of copper added to a dump which is recovered by leaching, generally falls in the range of 15–20%. For example, at Kennecott, Bingham, Utah, 250,000 tons of ore (0.17% copper) goes to the waste dump daily. This ore contains approximately 850,000 lb_m of copper of which 150,000 lb_m or 18% is recovered daily.

U.S. Bureau of Mines researchers pioneered the commerical implementation of heap leaching applied to low-grade gold–silver ores and have demonstrated that agglomeration by the use of Portland cement, added in amounts of up to 10 lb_m per ton of feed, can improve the percolation rate of leach liquors significantly.[5,6] Pilot-plant studies have demonstrated improved uniform percolation rates with increased metal recovery. Also, Bureau of Mines researchers[7] have promoted earlier technology[8] for gold recovery from heap-leaching solutions by adsorption of gold cyanide complexes on activated carbon. This technology was first applied commercially by Smokey Valley Mining Company in 1977.[9]

9.2-2 Direct Leaching Ore

Direct-leaching ores are those having sufficient value that they can be sized and subjected to leaching in stirred reactors (pachuca or agitation) or by percolation (vat leaching). Important examples of direct-leaching ores are bauxites treated by the Bayer process, gold ores suitable for slime leaching, nickel laterites for nickel and cobalt recovery, and acid and base leaching of uranium ores. All these ores will experience expanding use of hydrometallurgy in the future.

Gold leaching has experienced some important recent developments[7,10] and is an excellent example of oxidation leaching. An important development in gold leaching has been the reintroduction of older technology in the form of the carbon-in-pulp process.[11,12]

The following steps occur in the carbon-in-pulp process:

Oxidative Leaching

$$2Au + 4CN^- + O_2 + 2H_2O = 2Au(CN)_2^- + 2OH^- + H_2O_2 \qquad (9.2\text{-}20)$$

Adsorption (Ion Exchange)

$$C \cdot OH + Au(CN)_2^- = C \cdot Au(CN)_2 + OH^- \qquad (9.2\text{-}21)$$

Hot Caustic Stripping

$$C \cdot Au(CN)_2^- + OH^- = C \cdot OH + Au(CN)_2^- \qquad (9.2\text{-}2)$$

Electrolysis

$$e^- + Au(CN)_2^- = Au^0 + 2CN^- \qquad (9.2\text{-}23)$$

The leaching step, Eq. (9.2-20), produces gold cyanide anion in solution which is adsorbed on an ion-exchange site on the surface of activated carbon as indicated in Eq. (9.2-21). The surface-active site on the activated carbon ($C \cdot OH$) is depicted here as an ion-exchange site. The actual mechanism is still uncertain. Hot caustic stripping, Eq. (9.2-22), produces a high-grade gold stream suitable for recovery of gold by electrolysis, Eq. (9.2-23). The flowsheet of the Homestake Mining Company carbon-in-pulp process is illustrated in Fig. 9.2-3.[11] The process permits the countercurrent flow of activated carbon to each liquor. Following stripping, the carbon must be reactivated by a thermal treatment. The need to treat more difficult gold ores has resulted in the development of innovative new methods of separation.[12]

The Bayer process has similarly experienced significant change in recent years. It is an excellent example of large-scale, nonoxidative pressure leaching. Changes in Bayer process technology have resulted mainly from the need to treat more refractory ores. Bauxite ore types are important since the difficulty of digestion depends on the form of the hydrated aluminum oxide present. Aluminum in bauxite usually exists in one of three forms:[13]

Gibbsite ($\gamma Al(OH)_3$)
Boehmite ($\gamma AlOOH$)
Diaspore ($\alpha AlOOH$)

and in addition contains iron to variable degrees. Bauxite ores high in gibbsite are less costly to treat because of the relative ease of dissolution of gibbsite. Bauxites containing the monohydrated minerals in large quantity are more difficult to leach, require higher temperatures and pressure for complete digestion, and are more costly to treat. Consequently, bauxites may be classified as low cost or high cost. Although the majority of the bauxite processed is low cost (high gibbsite), this type is estimated to constitute only 12% of total bauxite reserves.[13]

The cost of capital equipment for a Bayer processing plant is related to the amount of monohydrated alumina present (boehmite or diaspore) and to the amount of iron oxide present. The first relates to the refractory nature of ore and the second to the amount of material that must be handled and separated. Hill and Robson[14] have estimated capital cost ratios for Bayer plants treating various types of bauxite ore. Capital costs of a bauxite containing 50% of the alumina as the monohydrate are estimated to be 40% higher than for a typical low-cost bauxite.

HOMESTAKE MINING CO.
LEAD, SOUTH DAKOTA

CARBON-IN-PULP, CYANIDATION OF SLIME
2350 TPD (MAX) - 0.08 TO 0.10 OZ Au PER TON PER TON

SLIME THICKENER UNDERFLOW

2 - 18' ⌀ X 20' PROPELLOR TYPE. SERIES
CONDITIONING AGITATORS

2 LBS/TON
MILK OF LIME - AIR

6 - 30' ⌀ X 22' AND 1 - 35' ⌀ X 18' DORR TYPE
AIR LIFT DISSOLUTION AGITATORS IN SERIES

1 LB/TON
NACN & AIR

4 - 18' ⌀ X 16' AIR LIFT TYPE
ADSORPTION AGITATORS IN SERIES WITH 2' X 4' VIBRATING
24 MESH SCREENS FOR COUNTER FLOW OF CARBON AS FOLLOWS

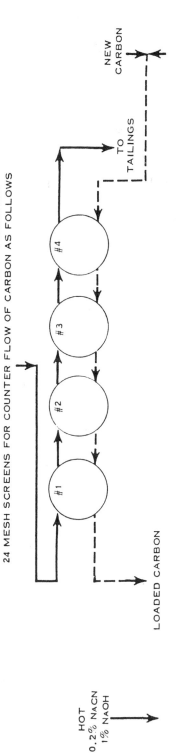

#1 #2 #3 #4

NEW CARBON

TO TAILINGS

LOADED CARBON

HOT
0.2% NACN
1% NAOH

PULP FLOW ———
CARBON FLOW – – –
SOLN FLOW –·–·–

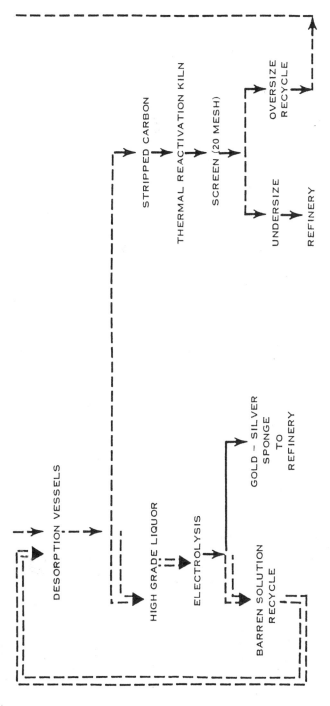

FIGURE 9.2-3 Carbon-in-pulp process for recovery of gold at Homestake Mining Company. (Reproduced from McQuiston and Shoemaker.[11])

The Bayer process involves the following steps:

1. Nonoxidative dissolution of hydrated aluminum minerals.
2. Solid–liquid separation to remove ferric oxide, silica, and silicates.
3. Cooling, seeding, and precipitation of gibbsite, $Al(OH)_3$.
4. Calcining of gibbsite to produce pure Al_2O_3.

The chemical reactions involved in dissolution are

$$Al(OH)_3 + NaOH = Al(OH)_4^- + Na^+ \qquad (9.2\text{-}24)$$

$$AlOOH + NaOH + H_2O = Al(OH)_4^- + Na^+ \qquad (9.2\text{-}25)$$

Ores containing essentially pure gibbsite may be digested at temperatures of 150°C while boehmitic ores require temperatures in the range of 230–250°C. These temperatures reflect the differences in the solubilities of the various hydrated aluminum minerals. Figure 9.2-4 illustrates a typical Bayer process flowsheet.

Nickel laterites represent an important example of direct-leaching ores. In 1924 Caron obtained a U.S. Patent for a hydrometallurgical process for treating nickel laterites.[15] The process involves the use of ammoniacal ammonium carbonate leach solutions for the dissolution of an iron–nickel alloy produced in a reductive roast pretreatment step. The leaching takes place at atmospheric pressure with oxygenation by air. This process became the Nicaro process, first used in Cuba in 1944.[16] An alternate process was also introduced in Cuba at Moa Bay. It involved sulfuric-acid pressure leaching of contained NiO and is suitable for low-acid-consuming ores. Power and Geiger[17] reviewed the application of the ammoniacal ammonium carbonate (AAC) leaching process worldwide. Of 16 plants listed, 8 used hydrometallurgical processing. One of these is the Moa Bay pressure-leaching process and the other 7 use the AAC process based on the original technology developed by Caron. In the Caron process nickel carbonate is precipitated. In various modifications, the carbonate may be reduced by coke to a highly metallized sinter or may be simply calcined to NiO. A modification proposed by Sherritt Gordon[17] produces metallic nickel by autoclave hydrogen reduction and is the basis for the Marinduque Surigao Nickel Refinery.[18] Recently, Amax[19] introduced a combined high-pressure and atmospheric-pressure acid-leach process capable of treating both limonitic and garnierite-type ores. Dissolved nickel and cobalt are precipitated as sulfides. Hydrometallurgy applied to the leaching of nickel laterites undoubtedly will continue to be an important method for the treatment of nickel laterites.

The AAC process consists of the following steps:[17]

Gaseous Reduction

$$NiO + H_2 \rightarrow Ni^0 + H_2O \qquad (9.2\text{-}26)$$

$$3Fe_2O_3 + H_2 \rightarrow 2Fe_3O_4 + H_2O \qquad (9.2\text{-}27)$$

where the Ni^0 is an alloy containing reduced iron.

Ammoniacal Oxidizing Leach

$$Ni(Fe) + O_2 + 8NH_3 + 3CO_2 + H_2O \rightarrow Ni(NH_3)_6^{2+} + Fe^{2+} + 2NH_4^+ + 3CO_3^{2-} \qquad (9.2\text{-}28)$$

Iron Precipitation

$$4Fe^{2+} + O_2 + 2H_2O + 8OH^- \rightarrow 4Fe(OH)_3 \qquad (9.2\text{-}29)$$

Ammonia Removal and Nickel Precipitation

$$Ni(NH_3)_6^{2+} + 2CO_2 + 2OH^- \xrightarrow[\text{strip}]{\text{steam}} Ni(HCO_3)_2 \downarrow + 6NH_3 \uparrow \qquad (9.2\text{-}30)$$

Calcining

$$Ni(HCO_3)_2 \xrightarrow{\text{roast}} NiO + CO_2 \uparrow + H_2O \uparrow \qquad (9.2\text{-}31)$$

Reduction

$$NiO + C(coke) = Ni^0 + CO \qquad (9.2\text{-}32)$$

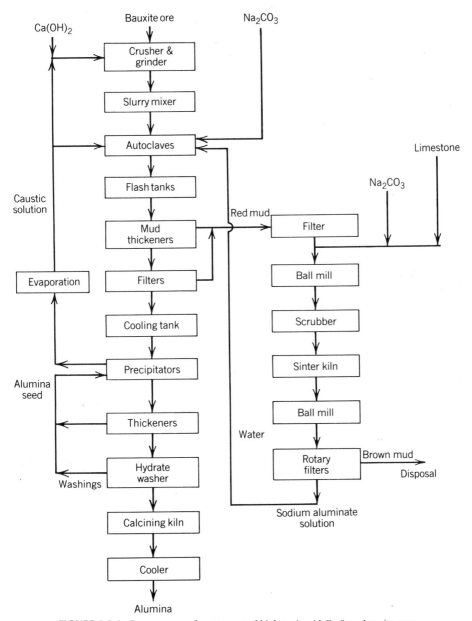

FIGURE 9.2-4 Bayer process for recovery of high-purity Al_2O_3 from bauxite ores.

In the process, cobalt is removed by precipitation as a sulfide containing a high Ni:CO ratio (Ni:Co >
500). Figure 9.2-5 illustrates the flowsheet of the Townsville[17] process.

Uranium slurry leaching is a standard technology for the treatment of uranium ores. Uranium may be
leached in acidic solutions or in basic solutions since the uranyl ion will form complexes with both sulfate
and carbonate ions. Uranium in the reduced U(IV) state, as in pitchblende and uraninite, must be oxidized
to the U(VI) state, producing the divalent uranyl ion in solution which will form sulfate and carbonate
complex anions. The majority of the uranium leaching plants use acidic leach solutions and may be operated
at atmospheric pressure. Chemical oxidants such as sodium chlorate or manganese dioxide are added to
oxidize U(IV) to U(VI). These oxidants couple effectively with the Fe(III)/Fe(II) naturally present is most
ores. Carbonate leaching is suitable for high-acid-consuming ores, but the reaction is slow at ambient

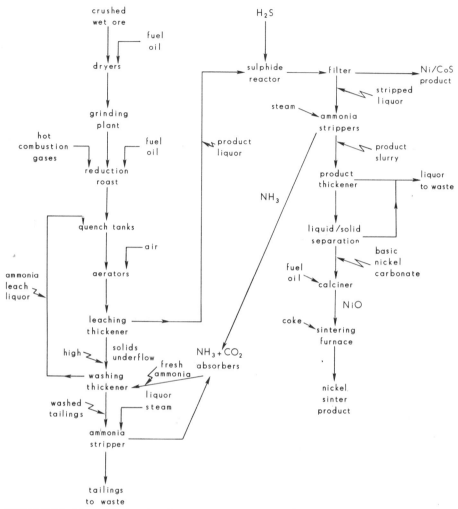

FIGURE 9.2-5 Townsville flowsheet for recovery of nickel from laterite ores. (Reproduced from Power and Geiger.[17])

pressure. Pressure leaching is commonly used with oxygen as the lixiviant. The chemistry and technology for uranium treatment are now some 30 years old and have experienced a variety of important changes in the intervening period. Most noteworthy were the application of anion-exchange resins and the introduction by Oak Ridge researchers of a solvent extraction process for the upgrading of uranium leach liquors. The ores containing vanadium usually are leached on the acid side if high recoveries of vanadium are desired. In general, the process involves either acidic or basic carbonate dissolution, upgrading of solutions by solvent extraction or ion exchange, stripping, and precipitation of uranium salts.

Acidic and basic dissolution of uraninite (UO_2) is depicted below as an example of U(IV) leaching. Similar reactions may be written for pitchblende (U_3O_8). Uraninite and pitchblende are the two most important uranium minerals. The leaching of oxidized [U(VI)] uranium minerals such as carnotite, $K_2O \cdot 2UO_3 \cdot V_2O_5 \cdot 3H_2O$, is readily achieved in both acid and basic circuits since oxidation is not required. Oxidation-leaching processes are rate limited by the oxidation step.

Acid Leaching (MnO_2–Fe(II/III) Coupled Oxidant)

$$UO_2 + 3HSO_4^- + 2Fe^{3+} = UO_2(SO_4)_3^{2-} + 3H^+ + 2Fe^{2+} \tag{9.2-33}$$

$$2Fe^{2+} + 2MnO_2 + 4H^+ = Mn^{2+} + 2Fe^{3+} + 2H_2O \tag{9.2-34}$$

Carbonate (Oxygen) Leaching

$$UO_2 + 0.5O_2 + 2HCO_3^- + CO_3^{2-} = UO_2(CO_3)_3^{4-} + H_2O \qquad (9.2\text{-}35)$$

In the carbonate, basic leaching process, both carbonate and bicarbonate ions are required to buffer the solution and prevent precipitation of uranium salts. In both acidic and basic leaching, complex anions are formed. The solutions may therefore be upgraded by anion exchange or anion solvent extraction (SX) processes. The upgraded solution is normally treated by selective precipitation of uranium by pH adjustment.

URANIUM PRECIPITATION:
Uranium is recovered by pH adjustment between 5 and 6, causing precipitation of a mixture of uranium salts including diuranates, hydrated oxides, and basic uranyl sulfate.[20] The ammonium hydroxide precipitation reaction of the diuranate salt is

$$2UO_2SO_4 + 6NH_4OH = (NH_4)_2U_2O_7 \downarrow + 2(NH_4)_2SO_4 + 3H_2O \qquad (9.2\text{-}36)$$

Uranium may also be precipitated as a peroxide using hydrogen peroxide

$$UO_2SO_4 + H_2O_2 + 2H_2O = UO_4 \cdot 2H_2O + H_2SO_4 \qquad (9.2\text{-}37)$$

Figure 9.2-6 illustrates the features of an acid leach circuit.[20]

Table 9.2-2 summarizes the various methods used for the treatment of direct leaching ores. Typical minerals are listed, and those that can be leached by nonoxidative dissolution (NOX) are indicated. Those requiring oxidants (OX) are also indicated. In some cases pretreatment is required, as in the case of reduction roasting of nickel laterites or vanadium ore which may be subjected to an oxidative salt roast prior to nonoxidative dissolution. Also indicated in Table 9.2-2 are the typical lixiviants used in the process, the methods used for metal recovery, and the form of the final product produced.

9.2-3   High-Grade Ores

High-grade ores are those which may be economically beneficiated to produce a concentrate prior to leaching. Examples of such concentrates are iron–nickel sulfide in the case of the Sherritt Gordon Process, roasted zinc sulfide concentrates, standard in the roast-leach–electrowinning practice for zinc recovery, zinc sulfides in the case of the new Sherritt Gordon Cominco pressure leaching process,[21] and flotation and gravity concentrates for gold recovery by cyanidation. Recovery of gold from concentrates and roast-leach

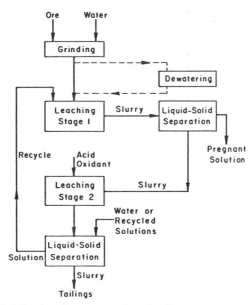

FIGURE 9.2-6   Simplified flowsheet for recovery of uranium by acid leaching. (Reproduced from Merritt.[20])

TABLE 9.2-2 Direct Leaching Ores, Indicating General Methods of Treatment

Mineral	Lixiviant	Method of Metal Recovery	Final Product
Oxidized Copper			
$CuCO_3 \cdot Cu(OH)_2$, malachite	NOX acid	Electrowinning	Metallic copper
$2CuCO_3 \cdot Cu(OH)_2$, azurite	NOX acid	Electrowinning	Metallic copper
$CuSiO_3 \cdot 2H_2O$, chrysocolla	NOX acid	Electrowinning	Metallic copper
Aluminum Oxides			
$Al(OH)_3$, AlOOH, bauxite	NOX caustic	Precipitation and roast	Al_2O_3
Uranium Oxides			
UO_2, uraninite	OX acid, base	Precipitation	Oxidized salts
U_3O_8, pitchblende	OX acid, base	Precipitation	Oxidized salts
$K_2(UO_2)_2(VO_4)_2 \cdot (1-3)H_2O$, carnotite	NOX acid	Precipitation	Oxidized salts
$Ca(UO_2)_2(VO_4)_2 \cdot (5-8)H_2O$, tyuyamunite	NOX acid	Precipitation	Oxidized salts
$U(SiO_4)_{1-x}(OH)_{4x}$, coffinite	NOX acid, base	Precipitation	Oxidized salts
$Ca(UO_2)_2(SiO_3)_2(OH)_2$, uranophane	NOX acid, base	Precipitation	Oxidized salts
Nickel Laterite			
$(FeNi)O(OH) \cdot nH_2O$, limonitic laterite	RR–OX or NOX AAC acid	Precipitation or H_2 reduction precipitation	NiO, metallic nickel, sulfides
NiO substitution, nickeliferous silicates	RED–OX AAC	Precipitation	NiO, metallic nickel
Gold-Silver			
Free leaching ores, metallic Au, Ag	OX caustic	Precipitation, electrowinning	Metallic Au, Ag
Vanadium			
Uranium ores and vanadiferous clays	SR–NOX water	Precipitation	Oxide

NOX = Nonoxidative dissolution.
OX = Oxidative dissolution.
AAC = Ammoniacal ammonium carbonate.
RR = Reductive roast.
SR = Salt roast.

electrowinning applied to zinc represent commercial processes used throughout the world. The application of hydrometallurgy to the treatment of conventional base metal sulfide concentrates has not emerged as a promising new technology. Even if various smelter sulfide intermediates are included as feed materials, the total number of plants treating such materials remains few in number.

The Sherritt Gordon process is historically important as the first commercial application of hydrometallurgy to the recovery of nickel from sulfide concentrates.[22] The primary mineral is pentlandite, (Fe, Ni)S, associated with pyrrhotite, $Fe_{11}S_{12}$. The concentrate is produced at Lynn Lake, Manitoba, and is processed in Sherritt Gordon's plant at Fort Saskatchewan, Alberta, Canada. The concentrate contains approximately 10% nickel and is treated by pressure, autoclave leaching using air as the oxidant. Leaching is based on the formation of soluble amine complexes of the form $(Me(NH_3)_x)^{n+}$, where Me may be Ni, Co, Cu, Zn, or Fe(II). The steps in the process are leaching, copper removal, oxidation and hydrolysis of soluble sulfur intermediates, and nickel reduction using gaseous hydrogen. The effluent aqueous stream is treated for cobalt recovery.

Leaching is carried out in two stages with excellent recoveries under the conditions 344–360 K (160–190°F) and 6.9×10^5–1.0×10^6 N/m^2 (100–150 psig). The leaching reactions are

$$NiS + 6NH_3 + O_2 = Ni(NH_3)_6^{2+} + SO_4^{2-} \tag{9.2-38}$$

$$4FeS + 9O_2 + 8NH_3 + 4H_2O = 2Fe_2O_3 + 8NH_4^+ \; 4SO_4^{2-} \tag{9.2-39}$$

All sulfur is shown to convert to sulfate ions in solution. Under the conditions of leaching, much of the sulfur remains in solution as metastable soluble intermediates. Oxidation of sulfur occurs in the sequence: thiosulfate ions, $S_2O_3^{2-}$; thionate ions, $S_nO_6^{2-}$; sulfamate ions, $SO_3NH_2^-$; and sulfate ions, SO_4^{2-}. The partially oxidized sulfur ions in solution must be converted to sulfate ions prior to nickel reduction to maintain nickel purity. One of the important features of the pressure-leaching process is the concurrent

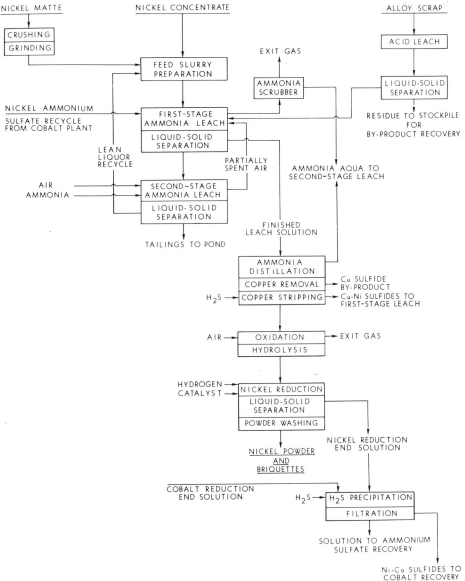

FIGURE 9.2-7 Sherritt Gordon pressure leaching process for recovery of nickel from nickel concentrates.

precipitation of iron as Fe_2O_3 during leaching and subsequent iron removal by thickening and filtration. The leaching circuit is illustrated in Fig. 9.2-7.

Pregnant solution from the leaching circuit contains typically 40–50 g/L nickel, 0.7–1.0 g/L cobalt, 5–10 g/L copper, 120–180 g/L ammonium sulfate, 5–10 g/L sulfur as thiosulfate and polythionates, and 85–100 g/L free ammonia.[1] Copper removal requires initial removal of ammonia followed by precipitation of copper in the form of copper sulfides. The unoxidized sulfur in solution assists in the precipitation process. Removal of ammonia by steam injection releases cupric ions to react with these sulfur compounds, as shown by the reactions:

$$Cu^{2+} + S_2O_3^{2-} + H_2O = CuS + 2H^+ + SO_4^{2-} \qquad (9.2\text{-}40)$$

Similar reactions occur with dithionates and polythionates in solution. Final copper residuals are stripped using H_2S under modest pressure. Aqueous effluent from copper removal is heated to 490 K (425°F) and oxidized in an autoclave using air. The unsaturated sulfur compound is converted to sulfate in solution by combined oxidation and hydrolysis—oxydrolysis—reactions:

$$S_2O_3^{2-} + 2O_2 + 2NH_3 + H_2O = 2NH_4^+ + 2SO_4^{2-} \tag{9.2-41}$$

$$SO_3NH_2^- + H_2O = NH_4^+ + SO_4^{2-} \tag{9.2-42}$$

Effluent from oxydrolysis goes to the nickel reduction autoclave where hydrogen is used as the reducing agent at 477 K (400°F), and a gage pressure of 3.1×10^6 N/m^2 (450 psig). During the previous steps of the process, solution make-up and ammonia removal are balanced so that the reduction feed solution contains NH$_3$ and nickel in the approximate ratio of 2:1. This is necessary to maintain constant pH and nickel ammine stoichiometry during reduction. Reduction occurs in a series of densification steps by the reaction

$$Ni(NH_3)_2^{2+} + H_2 = Ni^0 + 2NH_4^+ \tag{9.2-43}$$

The principal ores of zinc are sulfides containing variable amounts of iron. Most recent plants produce electrolytic zinc by a roast-leach–electrowinning process. Flotation concentrates are roasted and leached using sulfuric acid and spent electrolyte. In recent years, the major advance in the technology has been the introduction of the jarosite process in which iron is precipitated as a sodium or ammonium basic iron sulfate. Alternate processes precipitate iron as geothite or hematite. Excellent reviews of these developments have been presented in recent years.[23,24] Improvements in iron removal have increased zinc recovery because more stringent leaching conditions may be applied when iron is removed by jarosite precipitation. Stronger acids are required to leach zinc ferrite, $ZnO \cdot Fe_2O_3$, which forms in variable amounts during roasting depending on the roasting temperature and the amount of iron present in the feed concentrate. The steps in the process are:

Roasting

$$ZnS + O_2 = ZnO + SO_2 \tag{9.2-44}$$

Leaching

$$ZnO + H_2SO_4 = ZnSO_4 + H_2O \tag{9.2-45}$$

Precipitation

$$3Fe_2(SO_4)_3 + 2(NH_4, Na)OH + 10H_2O = 2(NH_4, Na)Fe_3(SO_4)_2(OH)_6 + 5H_2SO_4 \tag{9.2-46}$$

Electrowinning

$$ZnSO_4 + 2e^- = Zn^0 + SO_4^{2+} \qquad \text{(cathode)} \tag{9.2-47}$$

$$H_2O = 2H^+ + 0.5O_2 + 2e^- \quad \text{(anode)} \tag{9.2-48}$$

A simplified flowsheet of the jarosite process is illustrated in Fig. 9.2-8.

In 1981, Cominco, Trail, British Columbia placed in operation a pressure leaching process for the direct treatment of zinc sulfide concentrates. The process was a joint development of Cominco and Sherritt Gordon[25] and followed 2 years of pilot development.[26] The process is supplementary to the present conventional roast-leach–electrowinning plant and will eventually treat 25% of the zinc production. A simplified flowsheet of the process is illustrated in Fig. 9.2-9. The process includes an oxygen pressure leach of preground concentrate at 417–428 K (145–155°F) under an oxygen partial pressure of 7.5×10^5 N/m^2 (110 psia). The concentrate contains typically 49% Zn, 11% Fe, and 4% Pb. In 100 min retention time, approximately 98% of the zinc is extracted. The process is unique to that molten sulfur is removed in a decantation step following leaching. The slurry is then flashed to reduce pressure and recycle process steam. Flotation is used to recover the 5–10% residual elemental sulfur and all sulfur produced in the process is pressure filtered for cleaning.

During leaching a sequence of reactions occurs which results in a zinc-rich solution and precipitated jarosite salts containing lead and iron. The reactions are:

Initial Leaching

$$ZnS + H_2SO_4 + 0.5O_2 = ZnSO_4 + S^0 + H_2O \tag{9.2-49}$$

$$PbS + H_2SO_4 + 0.5O_2 = PbSO_4 + S^0 + H_2O \tag{9.2-50}$$

$$FeS + H_2SO_4 + 0.5O_2 = FeSO_4 + S^0 + H_2O \tag{9.2-51}$$

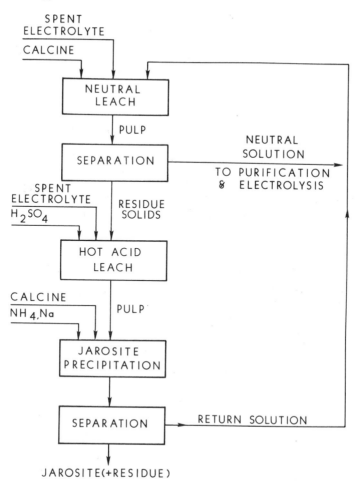

FIGURE 9.2-8 Simplified flowsheet for recovery of zinc from roasted sulfide concentrates. (Reproduced from Gordon and Pickering.[23])

Iron Oxidation

$$2FeSO_4 + H_2SO_4 + 0.5O_2 = Fe_2(SO_4)_3 + H_2O \qquad (9.2\text{-}52)$$

Ferric Ion Leaching

$$Fe_2(SO_4)_3 + ZnS = 2FeSO_4 + ZnSO_4 + S^0 \qquad (9.2\text{-}53)$$

Jarosite Precipitation

$$3Fe_2(SO_4)_3 + PbSO_4 + 12\ H_2O = PbFe_6(SO_4)_4(OH)_{12} + 6H_2SO_4 \qquad (9.2\text{-}54)$$

$$3Fe_2(SO_4)_3 + 14H_2O = (H_3O)Fe_6(SO_4)_4(OH)_{12} + 5H_2SO_4 \qquad (9.2\text{-}55)$$

Clarified zinc sulfate solution goes to electrowinning for zinc recovery.

9.2.4 Solution Mining Systems

Recently, in-place (in situ) leaching of ore deposits has received increased emphasis and appears to be an area in which significant advances will be made in the future. This type of hydrometallurgical operation is often referred to as solution mining. It is useful to consider dump leaching as part of solution mining since the physical and chemical features are similar to the leaching of fragmented or rubblized deposits in place.

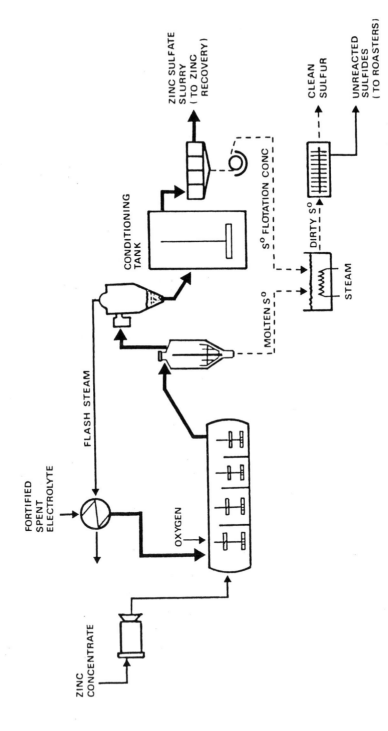

FIGURE 9.2-9 Cominco process for the direct pressure leaching of zinc sulfide concentrates. (Reproduced from Parker et al.[26])

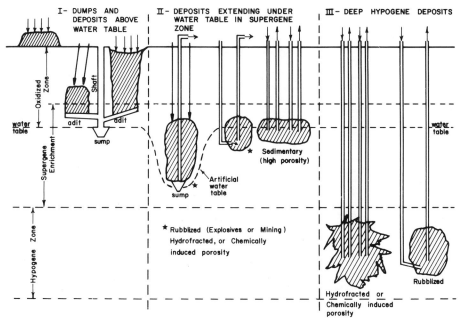

FIGURE 9.2-10 Three generalized conditions for solution mining in situ.

Much of what is to be expected from in situ extraction can be derived from current experience and practice in dump leaching.

Deposits amenable to in situ leaching may be classified into the three general groupings shown in Fig. 9.2-10:

I. Surface dumps or deposits having one or more sides exposed, and deposits within the earth's crust but above the natural water table.

II. Deposits located below the natural water table but accessible by conventional mining or well-flooding techniques.

III. Deposits below the natural water table and too deep for economic mining by conventional methods.

Dump leaching is placed in the first classification. Type II is characteristic of what is to be expected in the near future. Type III is expected to develop more slowly.

Type I would be the leaching of a fractured ore body near the surface above the natural water table in the surrounding area. This would apply to mined out regions of old mines such as a block-caved portion of a copper mine, or regions which have been fractured by hydrofracturing or by the use of explosives. The chemistry and physical requirements would be essentially the same as in dump leaching.

Type II refers to the leaching of deposits that exist at relatively shallow depths, less than approximately 500 ft, and that are under the water table. Such deposits will have to be fractured in place and dewatered so they may be subjected to alternate oxidation and leach cycles or percolation leaching, although the use of special oxidants may eliminate the drainage cycle. This is a special problem requiring a complete knowledge of the hydrology of the region. Water in the deposit, if removed during the oxidation cycle, must be processed, stored, and returned under carefully controlled conditions. An alternate method of leaching would be by flooding as described for Type III below. An important, rapidly developing example of Type II is the application of flooding using wells distributed on a grid such as is currently used for uranium extraction.[27]

The third general type (Type III) of solution mining is represented by deep deposits below the water table and below approximately 500 ft in depth. The ore body is shattered, hydrofractured, or chemically penetrated. Again, the hydrology of the region must be well known for proper containment of solutions. This represents a unique situation in that the hydrostatic head will increase the oxygen solubility to the point that the direct oxygen oxidation of sulfide minerals becomes possible.

9.3 THERMODYNAMICS OF LEACHING

Thermodynamics are important in explaining geological mineralization, corrosion, and dissolution of minerals. The two most important parameters are voltage (or free energy) and pH. Pourbaix[1] has provided a very useful graphic tool for the presentation of thermodynamic data in the form of potential–pH diagrams. These diagrams cover conditions from very oxidizing to very reducing and in effect make it possible in visualize virtually all stable and metastable phases which can exist between the various gas, solid, and aqueous solution phases. To the geochemist and geologist, the diagram represents equilibria between the lithosphere, hydrosphere, and atmosphere. To the metallurgist and physical chemist it provides a useful graphic tool to describe passivity, corrosion, and minerals dissolution. In the context of this discussion, it provides valuable information in describing reaction paths and phases that influence hydrometallurgical processes.

Equilibrium reactions at the solid–aqueous solution interface may be characterized by those in which oxidation or reduction does or does not occur. A reaction without oxidation or reduction may be represented by the reaction

$$a\mathrm{A} + c\mathrm{H_2O} = b\mathrm{B} + m\mathrm{H^+} \tag{9.3-1}$$

The standard free energy is given by

$$\Delta G^\circ = b\mu_\mathrm{B}^\circ + m\mu_\mathrm{H^+}^\circ - (a\mu_\mathrm{A} + c\mu\mathrm{H_2O}) \tag{9.3-2}$$

or in general terms

$$\Delta G^\circ = \Sigma \nu_i \mu_i^\circ \tag{9.3-3}$$

also

$$\ln K = -\frac{\Sigma \nu_i \mu_i^\circ}{RT} \tag{9.3-4}$$

Using log (base 10), Eq. (9.3-4) becomes for room temperature

$$\log K = -\frac{\Sigma \nu_i \mu_i^\circ}{1364} = \log \frac{a_\mathrm{B}^b \, a_\mathrm{H^+}^m}{a_\mathrm{A}^a \, a_\mathrm{H_2O}^c} \tag{9.3-5}$$

Taking $\mu_\mathrm{H^+}^0 = 0$ and $a_\mathrm{H_2O} = 1$, Eq. (9.3-5) may be written in the form

$$\log K = \log \frac{a_B^b}{a_A^a} - m\mathrm{pH} \tag{9.3-6}$$

Whenever electron transfer is not involved, Eq. (9.3-6) is the most useful form for evaluating the thermodynamics of the reaction.

If ΔG is the free energy change for a reaction

$$\Delta G = \Sigma \nu_i \mu_i \tag{9.3-7}$$

where μ_i is defined by Eq. (9.3-6). At equilibrium,

$$\Sigma \nu_i \mu_i = 0 \tag{9.3-8}$$

At concentrations other than equilibrium

$$\Delta G = \Sigma \nu_i \mu_i + RT\Sigma \nu_i \ln a_i \tag{9.3-9}$$

If oxidation–reduction couples exist in the reaction, electrons must be transferred. Electrochemical reactions, in which electrons flow through the solid from anodic to cathodic electrode surface sites, are best represented by anodic and cathodic half-cell reactions:

$$\text{oxidation} \rightarrow e \text{ production of electrons} \quad \text{(anodic)}$$

$$\text{reduction } e \rightarrow \text{ absorption of electrons} \quad \text{(cathodic)}$$

Free energy is related to half-cell potential by the equations

$$\Delta G^\circ = -nFE_0^\circ \quad \text{and} \quad \Delta G = -nFE_0 \qquad (9.3\text{-}10)$$

where n is the number of electrons transferred in the reaction, F is the Faraday number (23,060 cal/mol), and E_0 and E_0° are the half-cell and standard half-cell potentials for the reaction.

9.3-1 Nonoxidative Processes

The dissolution of minerals in which there is no net charge transfer is termed a nonoxidative leaching process. Of greatest importance in leaching are oxidized minerals such as carbonates, oxides, and hydrated oxides. The acid dissolution reactions previously referred to in Eqs. (9.2-11), (9.2-12), and (9.2-13) are examples. In general, for a divalent metal cation, a reaction of this type may be represented by the reaction

$$\text{MeO} + 2\text{H}^+ = \text{Me} + \text{H}_2\text{O} \qquad (9.3\text{-}11)$$

The Gibbs free energy, ΔG, may be expressed by the equation

$$\Delta G = \Delta G^\circ + RT \ln \left(\frac{a_{\text{H}_2\text{O}}}{a_{\text{H}^+}^2} \right) \qquad (9.3\text{-}12)$$

where

$$\Delta G^\circ = \Delta G^\circ(\text{H}_2\text{O}) - \Delta G^\circ(\text{MeO}) \qquad (9.3\text{-}13)$$

The activity of water, $a_{\text{H}_2\text{O}}$, is usually taken as unity, although for a solution of high ionic strength the activity of water may vary significantly from unity and must be evaluated for such cases. Equation (9.3-13) includes the free energy of formation of water (-56.69 kcal/mol) which provides the large driving force for the nonoxidative dissolution of oxidized minerals. The water drive reaction is important in such reactions, making them essentially irreversible.

Nonoxidative leaching of sulfide minerals generally is not favored thermodynamically. Back-reaction with H_2S limits the extent of reaction. An exception is galena (PbS), which dissolves nonoxidatively in chloride solutions, according to the reaction

$$\text{PbS(s)} + 2\text{HCl(aq)} = \text{PbCl}_2(\text{s}) + \text{H}_2\text{S(g)} \qquad (9.3\text{-}14)$$

The standard molar free energy for the equation as written is 1.98 kcal/mol. The reaction does proceed because of the formation of a series of complex lead chloride complexes in solution.

The equilibria between gases and water represent an important aspect of hydrometallurgy since the solubility of the gases often determines both thermodynamic and kinetic features of the system. Gases dissolve in two ways. The first is by reacting with the aqueous phase as in the case of CO_2, where for a certain pH range

$$\text{CO}_2(\text{g}) + \text{H}_2\text{O(l)} = \text{HCO}_3^- + \text{H}^+ \qquad (9.3\text{-}15)$$

Gases that dissolve by reacting with water as shown in Eq. (9.3-15) usually have a greater solubility than in the case of the dissolution of diatomic gases such as O_2, H_2, and N_2, which dissolve without chemical combination with the solvent and remain as diatomic species in solution according to the reaction

$$\text{G}_2(\text{g}) = \text{G}_2(\text{l}) \qquad (9.3\text{-}16)$$

The equilibrium constant for Eq. (9.3-16) is the Henry constant (K_H) which is

$$K_\text{H} = \frac{[\text{G}_2]}{P_{\text{G}_2}} \qquad (9.3\text{-}17)$$

Figure 9.3-1 illustrates the solubility of oxygen and hydrogen in water at various temperatures.[2] The solubility decreases as temperature increases to approximately the normal boiling point of water. As temperature increases further, the solubility then increases. The latter increase is important in autoclave leaching of minerals. The dissolution therefore changes from an exothermic to an endothermic process as temperature increases. At a given temperature the concentration of dissolved gas increases linearly according to the Henry equation, Eq. (9.3-17).

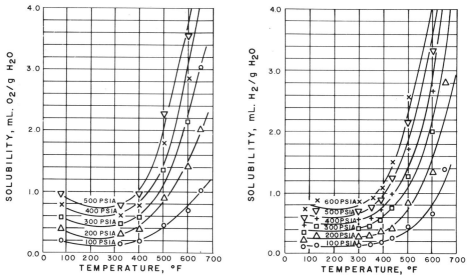

FIGURE 9.3-1 Solubility of oxygen and hydrogen in water at various temperatures and pressures. (Reproduced from Pray et al.[2])

9.3-2 Oxidative Processes

The leaching of naturally occurring minerals often requires oxidation. Many important reactions of this type were presented earlier. Examples are Eqs. (9.2-5)–(9.2-10) applicable to copper dump leaching; oxidative leaching of gold, Eq. (9.2-20); oxidative leaching of NiS, Eq. (9.2-28); and oxidative leaching of uranium minerals, Eqs. (9.2-32) and (9.2-34). If the minerals undergoing leaching are good electrical conductors, as most sulfide minerals are, then the leaching processes may be electrochemical in nature and occur by well-recognized corrosion phenomena. Electrochemical processes in general are extensively involved in oxidative mineral and metal-leaching reactions.[3]

9.3-3 Pourbaix Diagrams

The construction of Pourbaix or potential–pH diagrams has been discussed in detail by Pourbaix[1] and Garrels and Christ.[3] All reactions involving aqueous solution equilibria may be expressed in the form

$$aA + cH_2O + ne^- = bB + mH^+ \tag{9.3-18}$$

If $n = 0$, the reaction is a chemical reaction without oxidation or reduction. For $n \neq 0$, A represents the reactant in the oxidized state while B is in the reduced state. According to the Nernst equation,

$$E_0 = -\frac{\Sigma \nu_i \mu_i^\circ}{nF} - \frac{RT}{nF} \sum_i \nu_i \ln a_i \tag{9.3-19}$$

where $\Sigma \nu_i \mu_i^\circ = \Delta G^\circ$, and μ_i°, ν_i, and a_i refer to the chemical potential, stoichiometry coefficient, and activity of the ith component. At room temperature, Eq. (9.3-19) applied to reaction (9.3-18) becomes

$$E_0 = E_0^\circ - \frac{m}{n} 0.0591 \text{ pH} - \frac{0.0591}{n} \log \frac{a_B^b}{a_A^a} \tag{9.3-20}$$

If voltages are referred to hydrogen half-cell potentials, $E_0 = E_h$. The Pourbaix diagram is a plot of E_h versus pH for solid, gaseous, and dissolved components in equilibrium. It is apparent that if m and n appear on the same side of the equation the ratio m/n will be negative. If $n = 0$, Eq. (9.3-18) represents a vertical line on a potential–pH diagram which is voltage independent. If $m = 0$, the reaction is pH dependent.

The upper and lower limits of water stability are represented by the following equations:

Upper Limit

$$0.5\ O_2(g) + 2H^+(aq) + 2e^- = H_2O(l) \tag{9.3-21}$$

Lower Limit

$$2H^+(aq) + 2e^- = H_2(g) \tag{9.3-22}$$

The corresponding Nernst equations are:

Upper Limit

$$E_0 = 1.228 - 0.0591\ pH + 0.0147\ \log P_{O_2} \tag{9.3-23}$$

Lower Limit

$$E_0 = -0.0591\ pH - 0.0295\ \log P_{H_2} \tag{9.3-24}$$

The upper and lower limits of stability are dependent on the pressure and range from strongly oxidizing to strongly reducing conditions. It is also interesting that large pressure changes affect the voltage (or limits) only slightly.

Figure 9.3-2 illustrates the region of water stability (shaded area) between 1 atm oxygen and 1 atm hydrogen pressure. Increasing pressure to 10^3 atm moves the upper and lower limits to the positions indicated. It is thus clear that the thermodynamic boundaries for water stability are little influenced by pressure, although pressure often has a profound influence on kinetic processes. Virtually all hydrometallurgical reaction interactions can be expected to fall within the shaded region. Also, virtually all dissolution and corrosion final states can be predicted to fall within this region. Stable surface layers and expected surface products can, in many instances, be predicted.

Unstable intermediates also may form, often as kinetic transients. The dotted lines in Fig. 9.3-2 divide it into predominant areas in which the ion or molecule shown is in greatest concentration. The dashed line

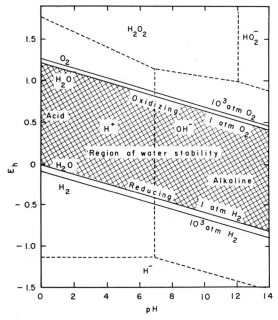

FIGURE 9.3-2 Potential–pH diagram showing region of water stability. (Reproduced from Pourbaix.[1])

in the upper left-hand portion of the diagram represents the condition $H_2O_2/H^+ = 1$. Clearly, hydrogen peroxide, peroxide ion, and hydride ion are unstable under normal conditions. It would require oxygen overpressures in excess of 10^{30} atm to stabilize H_2O_2 at concentrations as low as 10^{-3} M. In spite of this, H_2O_2 is often formed as an intermediate. In the anodic dissolution of gold and silver, in the presence of cyanide, H_2O_2 forms as an intermediate in the cathodic reduction of oxygen according to the reactions

Anodic

$$Au = Au^+ + e^- \qquad (9.3\text{-}25)$$

$$Au + 2CN^- = Au(CN)_2^- \qquad (9.3\text{-}26)$$

Cathodic

$$O_2 + 2H^+ + 2e^- = H_2O_2 \qquad (9.3\text{-}27)$$

The continued discharge of H_2O_2 to water,

$$2H^+ + H_2O_2 + 2e^- = 2H_2O \qquad (9.3\text{-}28)$$

is so slow that measurable concentrations of H_2O_2 appear in solution.

Figure 9.3-3 illustrates the superposition of several metal electrode reactions for various metal ion activities. It is apparent that there are three methods for the reduction of metal ions to metal. By applying an external potential more negative than the half-cell potential, metal reduction occurs, resulting in the deposition of surface layers at the metal–metal ion electrode surface by electrolysis. A second method results when a metal ion in solution, $M_1^{z_1+}$, is contacted by another metal, M_2, whose potential is more negative. This results in the deposition of M_1 on M_2 and is known as contact reduction or cementation. In

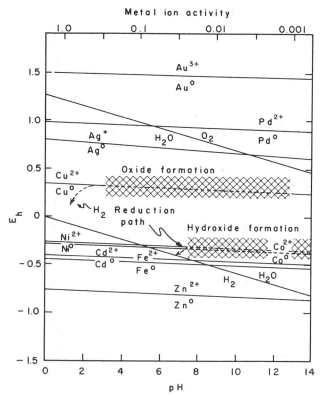

FIGURE 9.3-3 Metal electrode half-cell potentials, for various metal ion activities, superimposed on E_h–pH diagram.

general, contact reduction may be represented by the overall equation

$$M_1^{z_1+} + \frac{z_1}{z_2} M_2 \rightleftarrows M_1 + \frac{z_1}{z_2} M_2^{z_2+} \tag{9.3-29}$$

where $E_{02} < E_{01}$. It is apparent that each metal will reduce metal ions of those metals shown at more positive potentials. Important commercial systems are Cu^{2+}/Fe^0, $Ag(CN_2)^-/Zn^0$, Cu^{2+}/Ni^0, and Cd^{2+}/Zn^0.

A third method for reducing metals in solution is by use of hydrogen as a reductant. Hydrogen is capable of reducing metals having more positive E_h values. At higher pH values difficulty arises due to the formation of passive layers of oxides and hydroxides. Also, kinetics are slow at room temperature and low hydrogen pressure. Thermodynamically, cupric ions should be reduced by bubbling H_2 gas through the solution at room temperature. The kinetics are very slow under ambient conditions and hydrogen reduction must be carried out in autoclaves at elevated pressures and temperatures. The general reaction is

$$M_1^{z_1+} + \frac{z_1}{2} H_2 = M_1 + z_1 H^+ \tag{9.3-30}$$

Since a solid substrate is necessary, metal seed nuclei must be present which then grow as metal is deposited. Consequently the voltage diminishes as the metal ion actively decreases. The pH similarly decreases owing to the generation of H^+ according to Eq. (9.3-30). The dashed line extending from the Cu^{2+}/Cu^0 boundary indicates the course taken for copper during reduction in a batch reactor. When its potential meets the line for the lower limit of water stability the reaction is in equilibrium and reduction terminates. This presents serious problems for Ni^{2+} and Co^{2+} reduction as is evident from Fig. 9.3-3. Equilibrium is attained rapidly and extensive reduction cannot occur. This may be overcome by complexing the cobalt and nickel ammines which results in a reduction couple that can proceed without pH drift according to the reaction

$$Ni(NH_3)_2^{2+} + H_2 = N^0 + 2NH_4^+ \tag{9.3-31}$$

and is the basis for the commercial production of nickel and cobalt in the Sherritt Gordon process.

The ability to dissolve a metal or its oxides may be presented graphically according to the definition of the boundaries used. For equilibria involving a dissolved metal ion or metal ion complex in equilibrium with its oxides, the concentration in solution must be specified. Pourbaix arbitrarily established 10^{-6} as the maximum activity for the dissolved species for regions of passivation or protection due to the formation of surface films. If dissolution is desired, for example, in the hydrometallurgical extraction of metal values, 10^{-3} is a more realistic value. Regions between these two values would represent conditions resulting in corrosion, that is, concentrations that would result in excessive metal corrosion but insufficient for effective extraction by dissolution. Figure 9.3-4 is a potential–pH diagram for the $Cu–O–H_2O$ system showing regions of passivation and dissolution. The log activity values for the soluble species are indicated on the boundaries. Pourbaix[1] has presented detailed diagrams for most of the metal–oxygen systems of importance in determining conditions leading to corrosion and the buildup of surface films resulting in passivation.

Figure 9.3-5 is a Pourbaix, predominant-area, diagram for the $S–O–H_2O$ system for a total activity of all dissolved sulfur species of 10^{-1} ($\Sigma S = 10^{-1}$). The only stable sulfur species are HSO_4^-, SO_4^{2-}, H_2S, HS^-, and elemental sulfur. The formation of elemental sulfur films occurs in acid solutions as indicated. In basic solution, during the oxidation of sulfur-bearing compounds, intermediate metastable sulfur species such as thiosulfate, dithionate, and polythionates form. This is a problem previously described in the Sherritt Gordon process. Under acid conditions, during the dissolution of sulfide minerals, elemental sulfur layers often form but metastable sulfur intermediates such as thiosulfate and sulfite are not observed.

Stability relationships and the sequence of formation of surface reaction layers can be predicted from potential–pH diagrams. The $Cu–O–S–H_2O$ system will be used as an example. Figure 9.3-6 represents the $Cu–O–S–H_2O$ system at $\Sigma S = 10^{-1}$. Stable regions for Cu_2S and CuS indicate that sulfur films will not form adjacent to Cu_2S since the reaction

$$Cu_2S = CuS + Cu^{2+} + 2e^- \tag{9.3-32}$$

occurs with the formation of surface layers of CuS on the Cu_2S substrate. If the oxidant is oxygen, cathodic reduction will consume hydrogen ions and oxygen at the surface according to the reaction

$$0.5 O_2 + 2H^+ + 2e^- = H_2O \tag{9.3-33}$$

If the cathodic couple involves ferric ions, then

$$2Fe^{3+} + 2e^- = 2Fe^{2+} \tag{9.3-34}$$

without consumption of hydrogen ions.

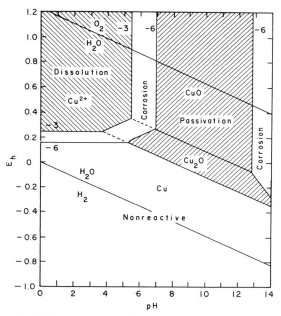

FIGURE 9.3-4 Potential–pH diagram of Cu–O–H$_2$O system showing regions where leaching (dissolution) can occur.

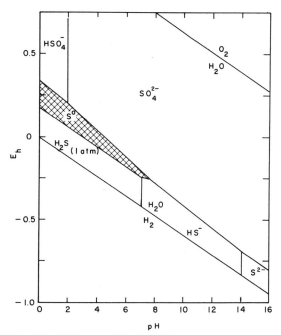

FIGURE 9.3-5 Potential–pH diagram of the S–O–H$_2$O system showing stable forms of sulfur.

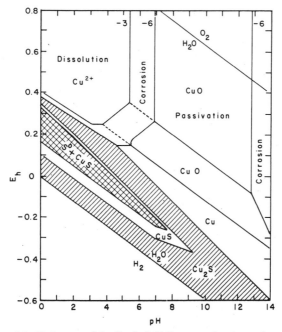

FIGURE 9.3-6 Potential–pH diagram of the Cu–O–S–H₂O system showing regions where leaching (dissolution) can occur.

If the kinetics are rapid enough, as experienced at high oxidation concentration and high temperature, surface polarization may occur sufficient to cause both CuS and sulfur to form. Sulfur formation occurs by the reaction

$$CuS = Cu^{2+} + S^0 + 2e^- \tag{9.3-35}$$

resulting in the formation of surface layers of S^0 on the CuS.

9.4 KINETICS OF LEACHING

9.4-1 Electrochemical Processes

Most metal sulfides and certain metal oxides are electronic conductors and are capable of establishing corrosion and galvanic couples in aqueous solutions. Similarly, metals react by well-established kinetic patterns involving corrosion couples displaying charge transfer or diffusion overvoltages or combinations of these depending on the metal.

Metal sulfides and several important oxides display n-type or p-type semiconducting or metallic properties. As a result of their electronic conductivity, certain minerals can participate in coupled charge transfer processes analogous to a metal corroding in an electrolyte, and the kinetics of leaching can be related to the potential of the solid in contact with the aqueous electrolyte.

Electrochemical processes are unique in that the solid assumes a uniform potential throughout, providing ohmic resistance is negligible. Consequently, the chemical reaction may be influenced over relatively long distances. The potential may directly affect the kinetics of the reaction and may also serve to stabilize intermediate solid phases. Concepts developed in the field of corrosion are transferable to electron-conducting solids in hydrometallurgical systems. The mineral particle behaves as an internally short-circuited (no external applied voltage) system where the solid assumes a mixed potential determined by associated anodic and cathodic processes. At the mixed potential, the sum of all anodic currents, ΣI_a, is such that

$$\Sigma I_a = -\Sigma I_c \quad \text{or} \quad \Sigma i_a A = -\Sigma i_c A \tag{9.4-1}$$

where A is the electrode area and i_a and i_c are the anodic and cathodic current densities. As in corrosion processes, two types of mixed-potential regimes may be operative: corrosion and galvanic couples. The corrosion type involves a single phase having both anodic and cathodic reactions on a single mineral surface. A galvanic couple is operative where two or more solid phases are in electrical contact. In this union, each solid assumes either anodic or cathodic behavior and has its own surface area for reaction. If a metal sulfide, MS, is placed in contact with a solution containing an oxidant N^{n+} (e.g., Fe^{3+} in acid solutions) having a more positive equilibrium potential, a corrosion cell will result in the anodic oxidation of MS and the cathodic reduction of N^{n+}.

The half-cell reactions are

$$MS = M^{m+} + S^0 + me^- \tag{9.4-2}$$

$$N^{n+} + e = N^{(n-1)+} \tag{9.4-3}$$

The mineral particle is essentially a short-circuited electrochemical cell that assumes a mixed potential between the half-cell potentials of reactions [Eqs. (9.4-2) and (9.4-3)]. The mixed potential depends on the properties of the electrode surface. Potentials more positive than the equilibrium potential drive the half-cell reaction in the net anodic direction and potentials more negative than the equilibrium potential drive the reactions in the net cathodic direction.

The net current density (i) for the system is the sum of the partial current densities i_a (anodic) and i_c (cathodic). The mixed potential (E) occurs where the net current density is zero; the values of the partial current densities are therefore equal and of the opposite sign ($i_a = -i_c$).

The kinetics of electrochemical processes involve anodic and cathodic reactions influenced by voltage biases controlled by mixed potential generated during leaching. Recently, Hiskey and Wadsworth[1] reviewed methods for evaluating mixed potentials and the net effect on anodic and cathodic mineral rate processes. Such details are beyond the scope of this discussion but do explain two important kinetic observations in the leaching of electron-conducting phases: (1) the often-observed fractional reaction orders and (2) the formation of solid phases in sequence. The kinetic processes included here are those observed to be operative during the leaching of ore fragments where diffusional mass transfer of lixiviants is rate controlling. While electrochemical processes may be involved at reaction interfaces, mass transfer predominates and is the basis for engineering simulation and design.

9.4-2   Leaching of Matrix Ore Mineral

The treatment presented here relates to the leaching kinetics of an ore fragment with its array of minerals. For purposes of clarity, the ore fragment is referred to as a *matrix block*. In this section the treatment includes primarily the kinetics of single matrix-block reactions exclusive of bulk flow parameters.

Virtually all researchers agree that the leaching of ore fragments involves penetration of solution into the rock pore structure. The kinetics thus involve diffusion of lixiviant into the rock where reaction with individual mineral particles occurs. The kinetics are complicated by changing porosity, pH, and solution concentration. Bartlett[2] applied the continuity equation to a system in which a copper porphyry ore, containing chalcopyrite as the major copper sulfide mineral, is subjected to oxidation under conditions proposed by researchers at the Lawrence Livermore Laboratory (LLL) for deep solution mining.[3] Diffusion of oxygen and subsequent reaction with mineral particles combine to give a non-steady-state concentration gradient within the ore fragment. This treatment provides a basis for modeling of all similar leaching systems. The shrinking-core concept used by several researchers is a special case of this general treatment. Figure 9.4-1 illustrates the mineral fragment, showing different regions at some time t. Oxygen and copper gradients vary markedly over a broad zone in which chalcopyrite (Cp) particles extend from partly to totally reacted. The fragment, considered as spherical, may be viewed in cross section, as divided into annular rings of thickness Δr. If J is the net total flux of dissolved reactant diffusing into and out of this incremental volume, the continuity equation for an individual ore fragment is

$$\epsilon \frac{\partial C}{\partial t} = -\Re + \frac{1}{A} \frac{\partial J}{\partial r} \tag{9.4-4}$$

where ϵ is porosity, C is concentration, t is time, A is the area at radius r, and $\Re$ is the rate of loss of reactant per unit volume because of reaction with mineral particles. Actually, this is a summation of rate processes for all mineral types and sizes, within the incremental volume. The porosity ϵ must be included since only the solution volume is involved. Combining Eq. (9.4-4) with Fick's diffusion equations gives

$$\epsilon \frac{\partial C}{\partial t} = -\Re + D_{\mathrm{eff}} \nabla^2 C \tag{9.4-5}$$

where D_{eff} is the effective diffusivity. The effective diffusivity is related to the intrinsic diffusivity, D, by the equation

$$D_{\mathrm{eff}} = \frac{D\epsilon}{\tau} \tag{9.4-6}$$

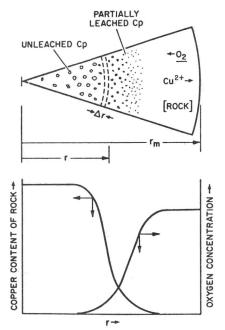

FIGURE 9.4-1 Schematic drawing illustrating the leaching of a matrix mineral fragment containing disseminated copper sulfide particles.

where τ is the tortuosity. The operator ∇^2 depends on the geometry of the ore particles and is defined by the following equation for the case of spherical ore fragments:

$$\nabla^2 = \frac{\partial^2}{\partial r^2} + \frac{2}{r} \frac{\partial}{\partial r} \tag{9.4-7}$$

Normally, spherical fragment geometry may be used, although nonspherical geometry may be required for matrix blocks from layered deposits. For spheres, Eq. (9.4-4) becomes

$$\epsilon \frac{\partial C}{\partial t} = -\sum \Re_k + D_{\text{eff}} \left(\frac{\partial^2 C}{\partial r^2} + \frac{2}{r} \frac{\partial C}{\partial r} \right) \tag{9.4-8}$$

A steady-state approximation of the continuity equation can be applied if the consumption rate of lixiviant, $\sum \Re_k$, is much greater than the accumulation term, $\epsilon \, \partial C / \partial t$.[4]

The assumption of steady state reduces the complex partial differential equation to a simpler ordinary differential expression of the continuity equation for the k mineral types present:

$$\sum \Re_k = D_{\text{eff}} \left(\frac{d^2 C}{dr^2} + \frac{2}{r} \frac{dC}{dr} \right) \tag{9.4-9}$$

Braithewaite[5] and Madsen and Wadsworth[6] used the simplified continuity equation for copper sulfide ores using finite difference approximations. The spherical ore matrix was divided into j concentric shells of thickness Δr, with $j = 1$ corresponding to the center and $j = n$ to the outer edge of the spherical matrix particle. The term $\Re$ may be evaluated using the equation

$$\Re = \sum_k \Re_{jk} \tag{9.4-10}$$

The term $\sum_k \Re_{jk}$ is the rate of consumption of lixiviant at position j for all k minerals in the fragment. Equation (9.4-9) satisfactorily correlated experimental results for various ore types and matrix-block size distributions.

If the rate of reaction of individual mineral particles is fast enough, the region of partially leached

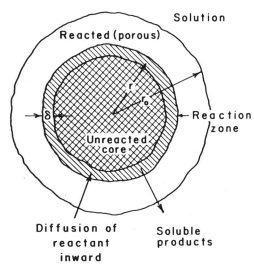

FIGURE 9.4-2 Matrix mineral fragment illustrating reaction zone and reacted and unreacted regions.

minerals (Fig. 9.4-1) is fairly narrow. Such a condition is illustrated in Fig. 9.4-2 where a reaction zone of thickness δ moves topochemically inward during the course of the reaction. According to the reaction zone model, steady-state diffusion occurs through the reacted outer region and is equal to the rate of reaction within the reaction zone itself. The effective area of mineral particles within the moving reaction zone is assumed to be essentially constant and independent of the mineral particle-size distribution since new particles in each size fraction will begin to leach at the leading edge of the reaction zone just as similar particles are completely leached at the tail of the reaction zone. The rate of reaction within the reaction zone may be expressed by the equation

$$\frac{dn}{dt} = -\frac{4\pi r^2 \delta n_p A_p}{\phi} C_s k_s' \tag{9.4-11}$$

where n = moles of leachable minerals
   t = time
   n_p = number of mineral particles per unit volume of rock
   A_p = average area per particle in the reaction zone
   k_s' = rate constant of mineral particle
   C_s = average concentration in the reaction zone
   ϕ = geometric factor accounting for deviation from sphericity

Diffusion through pores to the reaction zone may be expressed by the equation

$$\frac{dn}{dt} = -\frac{4\pi r^2 D_{\text{eff}}}{\phi \sigma} \frac{K_H}{dr} dC \tag{9.4-12}$$

where K_H = Henry constant
   C = bulk solution concentration
   D_{eff} = effective diffusion coefficient
   σ = stoichiometry factor (number of moles of reactant required per mole of metal released).

Equation (9.4-12) may be integrated for steady-state transport and combined with Eq. (9.4-10), giving, for a matrix-block fragment of radius r_i,

$$\frac{dn}{dt} = -\frac{4\pi r_i^2}{\phi_{0i}} \left[\frac{1}{G\beta} + \left(\frac{\sigma}{D_{\text{eff}}}\right)\left(\frac{r_i}{r_{i0}}\right)(r_{i0} - r_i) \right]^{-1} \tag{9.4-13}$$

where G is the grade (weight fraction of copper sulfide mineral), and G and β are given by

$$G = \frac{\delta A_p r_p \rho_p}{3\rho_r} \quad \text{and} \quad \beta = \frac{3\rho_r \delta k'_s}{r_p \rho_p}$$

where r_p = average mineral particle radius

ρ_p = mineral particle density

ρ_r = bulk rock density

Equation (9.4-13) may be integrated numerically and summed over all matrix-block sizes. Braun et al.[3] applied this model to a 5.8×10^6 g sample of primary copper ore with matrix-block sizes between 0.01 and 16 cm. The effect of weathering was accounted for by a systematic correction of the sphericity factor, ϕ_{0i}.

Madsen et al.[7] applied the reaction zone model to the leaching of Butte ore for particles up to 6 in. in diameter. Columns 5 ft in diameter, containing 5 tons of ore, were used. The principal copper mineral was Cu_2S. Equation (9.4-13) in integrated form was used:

$$1 - \frac{2}{3}\alpha_i - (1 - \alpha_i)^{2/3} + \frac{\beta'}{Gr_{0i}}[1 - (1 - \alpha_i)^{1/3}] = \frac{\gamma' t}{Gr_{i0}^2} \tag{9.4-14}$$

where α_i is the fraction reacted for the ith particle size. Also

$$\beta' = \frac{2D_{eff}}{\sigma\beta}$$

and

$$\gamma' = \frac{2MWD_{eff}C}{\rho_r \sigma \phi_{i0}}$$

where MW is the molecular weight of the copper sulfide mineral.

Researchers in general agree that ore matrix-block rate processes in copper dump leaching involve mass transfer of Fe(III) species by diffusion through pores, cracks, and fissures. Ferric ion in solution [Fe(III)] exists in several forms. Thermodynamically, the more important forms are $Fe(SO_4)_2^-$, $FeSO_4^+$, $Fe_2(OH)_2^{4+}$, $FeHSO_4^{2+}$, and $FeOH^{2+}$. As noted earlier, the sulfate complexes are favored over the hydroxyl complexes. Figure 9.4-3 illustrates an ore fragment showing the pore structure and diffusion of ions. A surface film, depicted flowing over the cross section on the right, is exposed under optimum conditions to convective air flow. Oxygen transfer to the liquid film occurs by diffusion. In well-aerated dumps, transfer of oxygen to the liquid phase is not rate limiting because of the large surface areas involved. Ferrous ion in the film is oxidized to the ferric state, a reaction requiring bacterial oxidation to occur at a rate suitable for leaching to proceed. The Fe(III) complex species produced by bacterial oxidation diffuse through the pore structure, reacting with sulfide minerals and releasing cuprous and ferrous ions, which, in turn, must diffuse in directions indicated by the localized concentration gradients. Copper may diffuse to the center of the ore fragment and continue to diffuse if a gradient is established by secondary enrichment reactions. Release of copper and ferrous iron occurs by outward diffusion through the pore structure of the rock fragment joining the flowing film. During rest cycles, leaching continues with a buildup of dissolved metal values in the immobile aqueous phase held by capillary forces. When the rock fragment is first exposed to solution, surface reaction control may occur, followed by increased kinetic dependence on diffusion as the process continues. The rate of extraction may thus be the sum of surface convective transfer, surface chemical reaction, and diffusion through the pore structure.

Roman et al.[8] developed a model for the leaching of copper oxide ores using the shrinking-core model. They found that the surface reaction is fast so that only the diffusion term need be considered. The consumption of acid, A, is related to the copper recovery by the expression

$$-\frac{dA}{dt} = a\frac{dCu}{dt} \tag{9.4-15}$$

where a = (weight of acid)/(weight of Cu) consumed in the reaction. The value of a may be determined experimentally. The diffusion equation becomes accordingly

$$\frac{d\alpha_{ij}}{dt} = \frac{3D_{eff}G_j^0(1 - \alpha_{ij})^{1/3}}{r_{0i}^2 \rho_r a\phi[1 - (1 - \alpha_{ij})^{1/3}]} \tag{9.4-16}$$

where α_{ij} refers to the fraction recovered from the ith particle size in volume increment j, G is the grade,

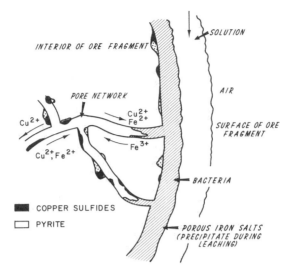

FIGURE 9.4-3 Matrix mineral fragment showing pore network and diffusion paths for solubilized metal ions.

ρ_r is the ore density, ϕ is a geometry factor, and D_{eff} is the effective diffusivity. Equation (9.4-16) may be evaluated numerically and summed over all unit volumes in the flow path. Within one incremental volume,

$$\alpha_j = \sum_j \alpha_{ij}\omega_i$$

where ω_i is the weight fraction of particle size i. Summing over all values of j gives the total fraction recovered:

$$\alpha = \sum_j \alpha_j = \sum_j \sum_i \alpha_{ij}\omega_i$$

In addition to these studies, Cathles and Apps[9] modeled a dump-leaching system including flow, air convection, and heat flow. This analysis was a field study of the Midas Dump at Bingham Canyon, Utah, and contributed significantly to dump-leach modeling in general. Recently, Gao et al.[10,11] developed a detailed model for solution mining of rubblized ore in situ including flow and ore fragment kinetics for single- and two-phase flow.

9.4-3 Aqueous Phase Oxidation–Reduction Processes

The aqueous solutions of importance to leaching technology for most conditions must provide suitable potential–pH conditions to promote oxidative dissolution of metals and metal sulfides. To provide this capability, solutions must contain a lixiviant that can be reduced. Conceptually, the lixiviant half-cell potential must be above that of the corresponding metal or metal sulfide zero-current potential. The greater this difference in voltage the greater the driving force. The important lixiviants to be considered are Fe(II)/Fe(III), oxygen, and hydrogen peroxide. These reactions are not all inclusive but simply represent important reactants currently used in leaching.

Ferric-ion complexes are important in acid-sulfate leaching because ferric ion can be generated from ferrous ion using air or oxygen in situ. The reduction of ferric iron to ferrous occurs as the ferric-ion complex diffuses through fluid-filled pores and channels in the rock matrix and encounters reactive metals or sulfides. In most instances, as already discussed, the rate of ferric ion reduction is a diffusion-limited process. The oxidation of ferrous iron to ferric in aqueous solution becomes of primary importance because of its in situ regeneration capacity under appropriate oxidation potentials.

Burkin[12] has reviewed some of the reaction mechanisms expected to be important in solution extraction systems. The rate of oxidation of ferrous ion in acid sulfate, nitrate, and perchlorate solutions follows the rate expression at low temperature and, in the absence of cupric ions,[13]

$$\frac{d[Fe^{2+}]}{dt} = -k_1' P_{O_2}[Fe^{2+}]^2 \tag{9.4-17}$$

The reaction is third order overall with first-order dependence on oxygen and second-order dependence on ferrous ion. These authors and others[14] have noted the catalytic effect of cupric ions on the kinetics of ferrous ion oxidation, explained by the reaction

$$-\frac{d[Fe^{2+}]}{dt} = k_1' P_{O_2}[Fe^{2+}]^2 + k_2'[Fe^{2+}][Cu^{2+}] \tag{9.4-18}$$

McKay and Halpern[15] developed a somewhat different rate expression for the catalytic contribution, having the form

$$-\frac{d[Fe^{2+}]}{dt} = -k_1' P_{O_2}[Fe^{2+}]^2 + k_2'[Fe^{2+}]^2[Cu^{2+}]^{0.5} \tag{9.4-19}$$

At 30.5°C the values for the constants in Eq. (9.4-18) are $k_1' = 1.5 \times 10^{-5}$ L/m · atm · s and $k_2' = 7.6 \times 10^{-3}$ L/m · s. For small concentrations of Cu^{2+} (e.g., 0.1 g/L) the second term of Eq. (9.4-18) is much larger than the first. The second term of Eq. (9.4-19) includes an oxygen pressure term, which seems to be in accord with observed results in dump leaching where little oxidation occurs in the absence of autotrophic bacteria for the retention time observed for large dumps.

Brierley[16] reviewed research on the biogenic oxidation of iron and cited the work of Lacey and Lawson[17] which indicates that the biogenic oxidation rate is 500,000 times faster than the nonbiogenic oxidation rate. Autotrophic bacteria are thus essential for effective leaching at low oxygen potentials where the activity of oxygen is much lower than that of Fe^{3+} generated biogenically. The rate of biogenic oxidation was proposed to follow the expression

$$\frac{-d[Fe^{2+}]}{dt} = \frac{\mu m[Fe^{2+}]X}{Y(K + [Fe^{2+}])} \tag{9.4-20}$$

where $[Fe^{2+}]$ is in g/L, μm is the maximum specific growth rate of bacteria per hour, Y is the mass of bacteria produced per gram of Fe^{2+} oxidized, K is the saturation constant in g/L Fe^{2+}, and X is the concentration of bacteria (g/L).

Oxygen is an essential intermediate in ferrous iron oxidation. In solution mining systems at elevated pressure, oxygen may become the dominant oxidizing lixiviant in the system. Diffusion of oxygen would thus be rate controlling in matrix-block kinetics. The solubility of oxygen in water–air equilibrium at 30° is 2.28×10^{-4} mol/L. In dump leaching, typically $[Fe^{3+}] = 10^{-2}$ mol/L. Leaching systems using pure oxygen under hydrostatic pressure would accommodate greater solubilities of oxygen. For example, pure oxygen under 300 ft of hydrostatic head would have a solution activity approximately 10^{-2} mol/L Fe^{3+}. For greater depths the relative contribution of oxygen would increase.

Oxygen reduction occurs through the formation of peroxide intermediates. Vetter[18] has reviewed oxygen redox reactions at metal electrode surfaces. On platinum the current for cathodic reduction is

$$i = -k'[O_2] \exp\left(-\frac{EF}{2RT}\right) \tag{9.4-21}$$

and for acid and neutral solutions

$$i = -k'[O_2][H^+] \exp\left(-\frac{3EF}{2RT}\right) \tag{9.4-22}$$

The rate of hydrogen peroxide reduction is important both as a lixiviant and as an intermediate in oxygen reduction. The rate of H_2O_2 reduction is given by the equation[18]

$$i = -k' \frac{C_{H_2O_2}[H^+]}{[H^+] + K} \exp\left[-\frac{EF}{2RT}\right] \tag{9.4-23}$$

where K is the dissociation constant for H_2O_2 and $C_{H_2O_2} = [H_2O_2] + [HO_2^-]$. The use of hydrogen peroxide as an oxidant is limited by its rapid anodic decomposition

$$H_2O_2 = O_2 + 2H^+ + 2e^- \tag{9.4-24}$$

The decomposition is catalyzed by ferrous ion and metallic surfaces. Even in alkaline solutions, such as used in carbonate uranium solution mining, iron present as colloidal ferric oxides (hydroxides) catalyze the decomposition. The half-life of H_2O_2 is approximately 30 min under alkaline solution mining conditions.

NOTATION

a, a_i	activity and activity of ith component
A	area
C	concentration
D	intrinsic diffusivity
D_{eff}	effective diffusivity
E	potential
E_0, E_0°	half-cell and standard half-cell potentials, respectively
F	Faraday number
ΔG°	standard Gibbs free energy of reaction
I	current
i_a, i_c	anodic and cathodic current densities, respectively
J	net total flux of dissolved reactant
k	rate constant
K_H	Henry constant
MW	molecular weight
P	pressure
P_{O2}, P_{H2}	partial pressures of oxygen and hydrogen, respectively
R	ideal gas constant
$\Re$	rate of loss of reactant per unit volume
r	radius
t	time
T	temperature

Greek Letters

ϵ	porosity
ν_i	stoichiometry coefficient
μ_i	chemical potential
ρ	density
τ	tortuosity

REFERENCES

Section 9.1

9.1-1   R. H. Lamborn, *The Metallurgy of Copper*, p. 180, Lockwood and Company, London, 1875.

9.1-2   F. Bushnell, "Precipitating of Copper from Mine Waters," in T. Read (Ed.), *Mining and Scientific Press*, Nov. 1914, editorial comment by W. G. Nash, pp. 329–330.

9.1-3   H. Kondo, *Lives in Science*, Part 4, II, pp. 127–140, Simon and Schuster, New York, 1957.

9.1-4   W. H. Dennis, *A Hundred Years of Metallurgy*, p. 141, Aldine Publishing, Chicago, 1964.

9.1-5   Fathi Habashi, *Principles of Extractive Metallurgy*, Vol. 2, Gordon and Breach, New York, 1970.

9.1-6   David J. I. Evans, Production of Metals by Gaseous Reduction from Solution—Processes and Chemistry, in *Advances in Extractive Metallurgy*, pp. 831–907, The Institute of Mining and Metallurgy, London, 1968.

9.1-7   R. Derry, Pressure Hydrometallurgy: A Review, *Min. Sci. Eng.*, **4**, 2–24 (1972).

9.1-8   Fathi Habashi, Pressure Hydrometallurgy: Key to Better and Non-polluting Process, *Eng. Min. J.*, 88–94 (May 1977).

9.1-9   F. Habashi, Karl Josef Bayer (1847–1904), in F. Habashi (Ed.), *Progress in Extractive Metallurgy*, Vol. 1, pp. 1–16, Gordon and Breach, New York, 1973.

9.1-10 W. H. Dennis, *A Hundred Years of Metallurgy*, pp. 234–239, Aldine Publishing, Chicago, 1964.

9.1-11 G. M. Ritcey and A. W. Ashbrook, *Solvent Extraction Principles and Applications to Process Metallurgy*, Part II, Elsevier, New York, 1979.

9.1-12 K. L. Power, Operation of the First Commercial Copper Liquid Ion Exchange and Electrowinning Plant, in R. P. Ehrlich (Ed.), *Copper Metallurgy*, pp. 1–26, TMS–SME–AIME Symposium, Denver, CO, February 15–19, 1970.

9.1-13 V. Arregui, A. R. Gordon, and G. Steinveit, The Jarosite Process—Past, Present and Future, in J. M. Cigan, T. S. Mackey, and T. J. O'Keefe (Eds.), *Lead–Zinc–Tin '80*, pp. 97–123, The Metallurgical Society of AIME, Warrendale, PA, 1979.

9.1-14 M. E. Wadsworth and C. H. Pitt, ''An Assessment of Energy Requirements in Proven and New Copper Processes,'' December 1980, DOE/CS/40132.

9.1-15 H. H. Kellogg, The State of Nonferrous Metallurgy, *J. of Metals*, **30**, 35–42 (1982).

9.1-16 R. R. Dorfler and J. M. Laferty, Review of Molybdenum Recovery Processes, in H. Y. Sohn, O. N. Carlson, and J. T. Smith (Eds.), *Extractive Metallurgy of Refractory Metals*, pp. 1–17; TMS–AIME, Warrendale, PA, 1980.

9.1-17 M. Shamsuddin and H. Y. Sohn, Extractive Metallurgy of Tungsten, in H. Y. Sohn, O. N. Carlson, and J. T. Smith (Eds.), *Extractive Metallurgy of Refractory Metals*, pp. 205–230; TMS–AIME, Warrendale, PA, 1980.

9.1-18 J. R. Boldt, Jr. and P. Queneau, *The Winning of Nickel*, p. 348, Longmans Canada, Ltd., Toronto, 1967.

Section 9.2

9.2-1 W. J. Schlitt, Current Status of Copper Leaching and Recovery in the U.S. Copper Industry, in *Leaching and Recovering Copper*, Chap. 1, SME–AIME Symposium, Las Vegas, Nevada, February 1979.

9.2-2 D. L. Whiting, ''Groundwater Hydrology of Dump Leaching and In Situ Solution Mining,'' Mackay School of Mines, Groundwater Hydrology and Mining Short Course, University of Nevada, Reno, Nevada, October 15, 1976.

9.2-3 J. S. Jackson and B. P. Ream, Solution Management in Dump Leaching, in *Leaching and Recovering Copper*, Chap. 7, SME–AIME Symposium, Las Vegas, Nevada, February 1979.

9.2-4 R. L. S. Willis, Ferrous–Ferric Redox Reaction in the Presence of Sulfate Ion, *Trans. Faraday Soc.*, **59**, 1315 (1963).

9.2-5 H. J. Heinen, G. E. McClelland, and R. E. Lindstrom, Enhancing Percolation Rates in Heap Leaching of Gold–Silver Ores, *U.S. Bur. Mines Rep. Invest.*, **8388**, 1979.

9.2-6 G. E. McClelland and J. A. Eisele, Improvements in Heap Leaching to Recover Silver and Gold from Low-Grade Resources, *U.S. Bur. Mines Rep. Invest.*, **8612**, 1982.

9.2-7 H. J. Heinen, D. G. Peterson, and R. E. Lindstrom, Processing Gold Ores Using Heap Leach–Carbon Adsorption Methods,'' *U.S. Bur. Mines Inf. Circ.*, **8770**, 1979.

9.2-8 G. J. McDougall, R. D. Hancock, M. J. Nichol, O. L. Wellington, and E. F. Copperthwaite, The Mechanism of the Adsorption of Gold Cyanide on Activated Carbon, *J. S. Afr. Inst. Min. Metall.*, **80**, 344–356 (1980).

9.2-9 L. White, Heap Leaching Will Produce 85000 oz/year of Dore Bullion for Smoky Valley Mining Co., *Eng. Min. J.*, **178**, 70–72 (1977).

9.2-10 R. Pizarro, J. D. McBeth, and G. M. Potter, Heap Leaching Practice at the Carlin Gold Mining Co., in F. F. Aplan, W. A. McKinney, and A. D. Pernichele (Eds.), *Solution Mining Symposium*, SME and TMS–AIME, New York, 1974.

9.2-11 F. W. McQuiston, Jr. and R. S. Shoemaker, *Gold and Silver Cyanidation Plant Practice*, Monograph SME–AIME, Vol. I (1975), Vol. II (1981), Denver, CO.

9.2-12 G. M. Potter and H. P. Salisbury, Innovations in Gold Metallurgy, *Min. Congr. J.*, **60**, 54–57 (1974).

9.2-13 G. Wargalla and W. Brandt, Processing of Diaspore Bauxites, in G. M. Bell (Ed.), *Light Metals 1981*, pp. 83–100, TMS–AIME, Warrendale, PA, 1981.

9.2-14 V. G. Hill and R. J. Robson, The Classification of Bauxites from the Bayer Plant Standpoint, in G. M. Bell (Ed.), *Light Metals 1981*, pp. 15–27, TMS–AIME, Warrendale, PA, 1981.

9.2-15 M. H. Caron, ''Treating Nickel and Cobalt Ores,'' U.S. Patent 1487145 (1924).

9.2-16 J. R. Boldt, Jr. and P. Queneau (Eds.), *The Winning of Nickel*, p. 425, Longmans Canada, Ltd., Toronto, 1967.

9.2-17   L. F. Power and G. H. Geiger, The Application of the Reduction Roast-Ammoniacal Ammonium Carbonate Leach to Nickel Laterites, *Min. Sci. Eng.*, **9**, 32–51 (1977).

9.2-18   N. Calvin and J. W. Gulyas, The Marinduque Surigao Nickel Refinery, in D. J. I. Evans, R. S. Shoemaker, and H. Veltman (Eds.), *International Laterite Symposium*, pp. 346–356, SME–AIME, New York, 1979.

9.2-19   W. P. C. Duyvesteyn, G. R. Wicker, and R. E. Doane, An Omniverous Process for Laterite Deposits, in D. J. I. Evans, R. S. Shoemaker, and H. Veltman (Eds.), *International Laterite Symposium*, pp. 553–570, SME–AIME, New York, 1979.

9.2-20   R. C. Merritt, *The Extractive Metallurgy of Uranium*, Colorado School of Mines Research Institute, Golden, CO, 1977.

9.2-21   E. G. Parker, D. R. McKay, and H. Salomon-de-Freidberg, Zinc Pressure Leaching at Cominco's Trail Operation, in K. Osseo-Asare and J. D. Miller (Eds.), *Hydrometallurgy—Research, Development and Practice*, pp. 927–940, TMS–AIME, Warrendale, PA, 1982.

9.2-22   J. R. Boldt, Jr. and P. Queneau (Eds.), *The Winning of Nickel*, pp. 299–314, Longmans Canada, Ltd., Toronto, 1967.

9.2-23   A. R. Gordon and R. W. Pickering, Improved Leaching Technologies in the Electrolytic Zinc Industry, *Met. Trans. B.*, **6B**, 43 (1975).

9.2-24   V. Arregui, A. R. Gordon, and G. Steintveit, The Jarosite Process—Past, Present and Future, in J. M. Cigan, T. S. Mackey, and T. J. O'Keefe (Eds.), *Lead–Zinc–Tin '80*, pp. 97–123, TMS–AIME, Warrendale, PA, 1979.

9.2-25   V. Arregui, A. R. Gordon, and G. Steinveit, Pilot Plant Demonstration of Zinc Sulfide Pressure Leaching, in J. M. Cigan, T. S. Mackey and T. J. O'Keefe (Eds.), *Lead–Zinc–Tin '80*, pp. 407–425, TMS–AIME, Warrendale, PA, 1979.

9.2-26   E. G. Parker, D. R. McKay, and H. Solomon-de-Friedberg, Zinc Pressure Leaching at Cominco's Trail Operation, in K. Osseo-Asare and J. D. Miller (Eds.), *Hydrometallurgy—Research, Development and Plant Practice*, pp. 927–940, TMS–AIME, Warrendale, PA, 1982.

9.2-27   D. A. Shock and F. R. Conley, Solution Mining—Its Promise and its Problems, in F. F. Aplan, W. A. McKinney, and A. D. Pernichele (Eds.), *Solution Mining Symposium*, pp. 79–97, SME–TMS–AIME, New York, 1974.

Section 9.3

9.3-1   M. Pourbaix, *Atlas of Electrochemical Equilibria in Solutions*, Pergamon Press, New York, 1966.

9.3-2   H. A. Pray, C. E. Schweickert, and B. N. Minnich, *Ind. Eng. Chem.*, **44**, 1146 (1952).

9.3-3   R. M. Garrels and C. L. Christ, *Solutions, Minerals, and Equilibria*, Harper & Row, New York, 1965.

Section 9.4

9.4-1   J. B. Hiskey and M. E. Wadsworth, Electrochemical Processes in the Leaching of Metal Sulfides and Oxides, in M. C. Kuhn (Ed.), *Process and Fundamental Considerations of Selected Hydrometallurgical Systems*, pp. 303–325, SME–AIME, New York, 1981.

9.4-2   R. W. Bartlett, Pore Diffusion—Limited Metallurgical Extraction from Ground Ore Particles, *Met. Trans.*, **3**, pp. 913–977 (1972).

9.4-3   R. L. Braun, A. E. Lewis, and M. E. Wadsworth, In-Place Leaching of Primary Sulfide Ores: Laboratory Leaching Data and Kinetics Model, *Met. Trans.*, **5**, 1717–1726 (1974).

9.4-4   H. Y. Sohn and J. Szekely, A Structural Model for Gas–Solid Reactions with a Moving Boundary—Part III, *Chem. Eng. Sci.*, **27**, 763–778 (1972).

9.4-5   J. W. Braithewaite, "Simulated Deep Solution Mining of Chalcopyrite and Chalcocite," Ph.D. Dissertation, Deptartment of Metallurgical Engineering, University of Utah, 1976.

9.4-6   B. W. Madsen and M. E. Wadsworth, A Mixed Kinetics Dump Leaching Model for Ores Containing a Variety of Copper Sulfide Minerals, *U.S. Bur. Min. Rep. Invest.*, **8547**, 1981.

9.4-7   B. W. Madsen, M. E. Wadsworth, and R. D. Groves, Application of a Mixed Kinetics Model to the Leaching of Low Grade Copper Ores, *Trans. Soc. Min. Eng.*, **258**, 69–74 (1974).

9.4-8   R. J. Roman, B. R. Benner, and G. W. Becker, Diffusion Model for Heap Leaching and Its Application to Scale-Up, *Trans. Soc. Min. Eng.*, **256**, 247–256 (1974).

9.4-9   L. M. Cathles and J. A. Apps, A Model of the Dump Leaching Process that Incorporates Oxygen Balance, Heat Balance and Air Conductivity, *Met. Trans. B*, **6B**, 617–624 (1975).

9.4-10   H. W. Gao, H. Y. Sohn, and M. E. Wadsworth, A Mathematical Model for the Solution Mining of Primary Copper Ore: Part I. Leaching by Oxygen-Saturated Solution Containing No Gas Bubbles, *Met. Trans. B*, **14B**, 541–551 (1983).

9.4-11 H. W. Gao, H. Y. Sohn, and M. E. Wadsworth, A Mathematical Model for the Solution Mining of Primary Copper Ore: Part II. Leaching by Solution Containing Oxygen Bubbles, *Met. Trans. B*, **14B**, 553–558 (1983).

9.4-12 A. R. Burkin, *The Chemistry of Hydrometallurgical Processes*, Van Nostrand, Princeton, NJ, 1966.

9.4-13 R. E. Huffman and N. Davidson, Kinetics of the Ferrous Iron–Oxygen Reaction in Sulfuric Acid Solution, *J. Am. Chem. Soc.*, **78**, 4836 (1956).

9.4-14 P. George, The Oxidation of Ferrous Perchlorate by Molecular Oxygen, *J. Chem. Soc. (London)*, **280**, 4836 (1954).

9.4-15 D. R. McKay and J. Halpern, A Kinetic Study of the Oxidation of Pyrite in Aqueous Suspension, *Trans. AIME*, **212**, 301–308 (1958).

9.4-16 C. L. Brierley, Bacterial Leaching, *Crit. Rev. Microbiol.*, **6**, 207–262 (1978).

9.4-17 D. T. Lacey and F. Lawson, Kinetics of the Liquid Phase Oxidation of Acid Ferrous Sulfate by the Bacterium *Thiobacillus thiooxidans*, *Biotech. Bioeng.*, **12**, 29–50 (1970).

9.4-18 K. J. Vetter, *Electrochemical Kinetics*, Academic Press, New York, 1967.

CHAPTER 10

Leaching—Organic Materials

HENRY G. SCHWARTZBERG

Department of Food Engineering
University of Massachusetts
Amherst, Massachusetts

10.1 DEFINITION OF PROCESS

Leaching is the transfer of solutes from a solid, usually in particulate form, to contiguous liquid, the extract. It almost invariably involves diffusion of solutes in the solid but may also involve washing of solutes or extract off the solid's surfaces, displacement of extract from interparticle pores, and solubilization, reaction-induced creation of solutes from insoluble precursors. The solutes diffuse through occluded solution contained in pores and cells in the solid. Dry solids must first imbibe solvent or extract, which then dissolves their solute content before diffusion occurs. In homogeneous extractions the solvents in the extract and occluded solution are the same. In heterogeneous extractions these solvents differ and are usually immiscible. Heterogeneous extraction is the rarer case, but it can provide greater extraction selectivity.

Generally, the solids are not structurally homogeneous, but the solid and liquid nevertheless will be called phases; and leaching will be treated as a two-phase, mass transfer process. The solid consists of a matrix of insoluble solids, the "marc," and the occluded solution. It may also contain undissolved solute and a nonextractable secondary phase, for example, coffee oil in water-soaked coffee grounds. This secondary phase is treated as part of the marc. Dimensionless parameters that can affect solute transfer include: the solute equilibrium distribution coefficients, m and M; the Fick number, τ; the stripping factor, α; the Biot number, Bi; and the Peclet number, Pe. These parameters are defined more precisely in the Notation section.

10.2 CONTACTING METHODS

The basic solid–liquid contacting methods used for leaching are: (1) batch extraction—batch mixing followed by phase separation; (2) cocurrent slurry extraction—cocurrent two-phase flow followed by phase separation; (3) differential extraction—continuous passage of extract through a well-mixed slurry containing the solids; (4) fixed-bed extraction—passage of extract through a fixed bed of solids; (5) crossflow extraction—transverse passage of extract through a moving bed of solids; (6) countercurrent percolation—continuous countercurrent flow of extract through a moving bed of solids; (7) cocurrent percolation—continuous cocurrent passage of extract through a moving bed of solids; and (8) countercurrent slurry extraction—the conveying of dispersed solids through a counterflowing stream of extract.

The methods used to analyze mass transfer in leaching systems depend on the methods of contacting used and how they are combined in multistage equipment. Appropriate methods of analysis may be the same for two superficially different types of contacting. Thus, crossflow extraction can be analyzed in terms of fixed-bed extraction; and, if the two-phase flow in cocurrent slurry extraction is sufficiently pluglike, it can be analyzed exactly like batch extraction. Moreover, when cocurrent slurry extractors are interconnected to form a multistage countercurrent extraction system, the performance of the system can be analyzed by the methods used for multistage, countercurrent batch extraction. Countercurrent percolation and countercurrent slurry extraction can also be analyzed by identical mathematical methods, provided the same types and extents of backmixing and same contacting imperfections occur in both cases.

10.3 INDUSTRIAL LEACHING EQUIPMENT

Multistage extractors are used frequently for industrial-scale leaching. Batch mixer–separator stages, interconnected to form a multistage batch countercurrent leaching system, are used to produce soluble tea.[1] The solids unloading–loading and solvent feed and extract discharge arrangement used in such systems is shown in Figure 10.3-2 and is known as the Shank's system.[2] Continuous multistage countercurrent leaching is also obtained by interconnecting cocurrent slurry extractors, as in the Silver Scroll and Oliver-Morton systems, and by interconnecting cocurrent moving-bed extractors, as in Silver Chain extractors.[3] Solid and liquid transfers between the stages are arranged so that countercurrent contacting in an overall sense is obtained even though contacting is cocurrent within the individual stages.

10.3-1 Diffusion Batteries

Fixed-bed extractors[4-6] are often interconnected to form diffusion batteries; see Fig. 10.3-1. The individual extractors may be completely enclosed columns that can operate at elevated pressures and temperatures higher than the normal boiling point of the solvent. When operations are carried out at atmospheric pressure, open tanks may be used. In such batteries, there is a continuous flow of extract, usually in the upward direction, in each active bed. Extract drawoff is obtained from the bed containing the freshest solids and fresh solvent is fed into the bed containing the most spent solids. After drawoff, extract flowing out of a bed is fed directly into the opposite end of the next fresher bed. Spent solids are discharged periodically from the system and fresh solids are loaded into the vessel vacated by the spent solids. The drawoff and solvent feed locations also are shifted in a corresponding fashion. The solid unloading–loading and solvent feed and extract discharge sequences, which are shown in Fig. 10.3-2, are similar to those used in the Shank's system.

SOLUBLE COFFEE EXTRACTION BATTERY

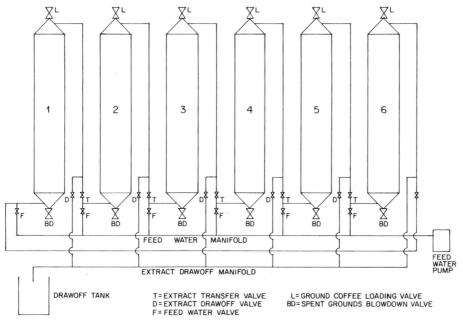

FIGURE 10.3-1 Six-cell diffusion battery. (Reproduced by permission, *Chemical Engineering Progress*, American Institute of Chemical Engineers.)

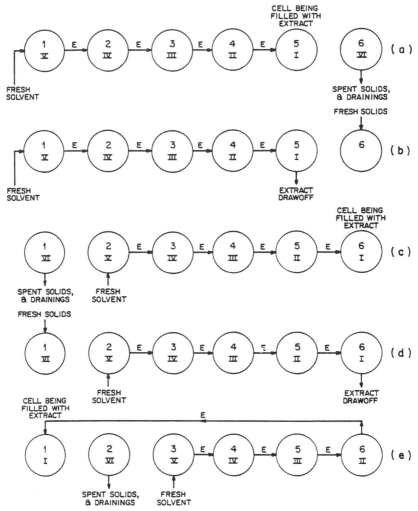

FIGURE 10.3-2 Extract feed and discharge and solids loading and discharge sequences for diffusion battery.

Diffusion batteries are used for extracting soluble coffee, soluble tea, spices, pickling salts, and corn steep solids and formerly were used in very large numbers for extracting beet sugar. Coffee or spice extraction batteries usually contain four to eight columns, typically six; beet sugar batteries contained 10–16 cells, typically 14. Coffee extraction columns usually have diameters ranging between 0.25 and 0.75 m and are 4.5–6.0 m tall.

10.3-2 Multistage Crossflow Extraction

Crossflow extractors are almost always used in multistage countercurrent systems in which a continuous bed of solids, deposited at a solids feed zone, is pushed across or carried through a sequence of stages on a perforated surface and is then discharged at a solids unloading zone. Extract percolates through the bed, drains through the support surface into a series of sumps, and then is resprayed on top of the bed. The sprays and sumps are positioned so that the extract is transferred opposite to the direction of solids movement, thereby providing countercurrent leaching. When drainage is very rapid, extract is recirculated in each stage to keep the bed wetted; overflow from sump to sump provides countercurrent flow. When drainage is slow, recirculation is not used; when drainage is very slow, extract is sprayed on the bed two stages ahead of the sump into which it drains. Conveying may be done by a single belt (see Fig. 10.3-3),

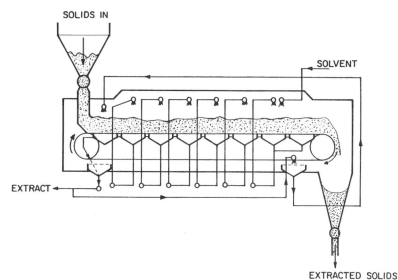

SOLIDS IN

SOLVENT

EXTRACT

EXTRACTED SOLIDS

FIGURE 10.3-3 Belt-type crossflow extractor. (Reproduced by permission, *Chemical Engineering Progress*, American Institute of Chemical Engineers.)

as in the DeSmet, Vatron-Mau, and Amos extractors;[7-9] a double tier of belts in the Lurgi extractor;[10] and a nested cascade of tilted belts resting one upon another in the Van Hengel extractor.[11] In other cases, the solids are pushed down a long perforated trough, as in the BMA-Egyptian extractor,[12] or are conveyed or pushed around a perforated horizontal circular surface, as in the Dravo Rotocel (Fig. 10.3-4), Extraktions-technik Carrousel,[13] Silver Ring,[14] EMI, and Filtrex extractors. Most horizontal basket-conveyor extractors are essentially crossflow extractors.

Crossflow extractors are used to extract many different solutes and can handle a wide variety of solid feeds.[15] Some crossflow extractors are quite large, for example, 11 m wide and 52 m long and can handle 10,000 tons of solid feed per day. They may contain up to 18 stages. The solid beds are usually between 0.5 and 3.0 m deep, although in the Filtrex extactor 0.05 m deep beds are used.

10.3-3 Stationary Basket Extractor

In the French Oil Machinery Co. stationary basket extractor (Fig. 10.3-5), extract percolates through beds of solids contained in a circular array of sector-shaped compartments with perforated bottoms and drains into sumps positioned below the beds. Unlike crossflow extractors, the solids do not move.[16] Instead, the solid feed spout and solids discharge zone rotate about the circle and the extract feed and discharge connections are switched periodically. These extractors are like automated diffusion batteries in which downflow is used; but, because extract backmixes in each sump, the extract concentration leaving a stage is somewhat different from that entering the next stage.

10.3-4 Trommel Extractors

These extractors consist of a large, horizontal rotating tube or trommel containing an Archimedes spiral.[17] The rotation causes the extract and solids to move down the spiral. Perforated plates or baskets cross the trommel. The extract flows through the perforations; but the solids are lifted, and when their angle of repose is exceeded, they slide down chutes or the spiral and are conveyed countercurrent to the direction of extract flow. The solids conveying system used in older models is shown in Fig. 10.3-6. More complex solid and liquid transfer arrangements are used in newer models.[18] These extractors function roughly like a countercurrent array of cocurrent slurry extractors. Trommel extractors are used almost exclusively to extract beet sugar. Their diameters range between 4.7 and 7.0 m and their lengths between 34 and 52 m. They can process up to 10,000 tons of feed per day.

10.3-5 Screw-Conveyer Extractors

In these extractors the solids are conveyed countercurrent to the flow of extract by a perforated or interrupted screw or screws or a screwlike array of paddles or propellers. In slope extractors, such as the DDS extractor

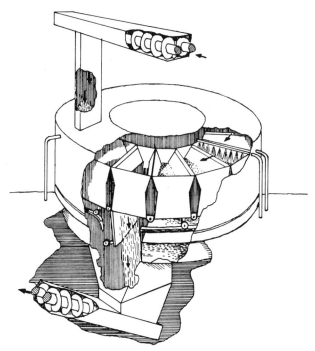

FIGURE 10.3-4 Rotocel crossflow extractor. (Reproduced by permission, *Chemical Engineering Progress*, American Institute of Chemical Engineering.)

(Fig. 10.3-7),[19] a U.S. variant, the Silver DDS extractor,[18] the Amos screw extractor,[20] and the Niro extractor, the solids are conveyed up a slightly tilted trough by intermeshing perforated twin screws. Niro also provides single-screw slope extractors for low-capacity operations. The tilt of the trough causes extract to flow at the requisite rate countercurrent to the flow of solids. Slope extractors are used primarily for beet sugar extraction but have been used to extract many other substances.[22] When DDS extractors are used for cane sugar extraction, one screw is driven slightly faster and then slightly slower than the other for part of each revolution so the solids are squeezed intermittently as they are conveyed.

The Greerco extractor[23] uses a horizontal, interrupted screw made up of a sequence of flat paddles. In contrast to slope extractors, where the solids form a compact bed, the solids are dispersed in the liquid. Therefore, better particle surface exposure is obtained; but, because of the high volume fraction of liquid, extract holdup is relatively long. Liquid flow is caused by the difference in level between the liquid inlet port and the liquid discharge weir, which are positioned at opposite ends of the trough in which the screw is housed. Solids enter the trough at the liquid discharge end and are scooped out at the liquid feed end. The troughs may be connected in series (up to 11 troughs in series have been used). The small available liquid head limits the flow rates that can be used without flooding. This problem may be compounded when series of troughs are used.

In tower extractors, such as the BMA and Buckau-Wolf extractors,[24,25] sugar beet strips (cossettes) are augered upward through a tall tower (Fig. 10.3-8) by screw segments set in a helical pattern with gaps between each segment. Radial baffles placed between these gaps prevent the solid bed from turning en masse. Extract flows down through the moving bed and out through a perforated screen at the base of the tower. Part of the discharged extract is mixed with fresh cossettes. The mixture is heated to denature protein membranes in the beets and then fed back into the tower through a rotating distributor attached to the bottom on the screw shaft. Tower extractors do not work well when the directions of solid and liquid flow are reversed.[26]

Screw conveyor extractors, such as the Hildebrand, Ford, and Detrex extractors,[27] were once used to extract flaked oilseeds. Their use has been abandoned, presumably because the conveying action was too rough for fragile oilseed flakes. The Vulcan-Kennedy extractor used rakelike rotating scoops to transfer solids upward from trough to trough in an ascending sequence of circular troughs. Extract cascaded downward through the troughs in the opposite direction. It is no longer used to any great extent.

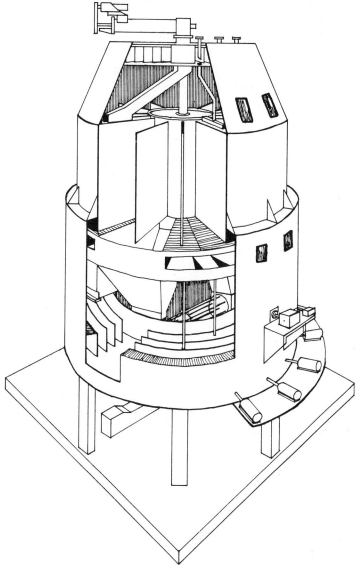

FIGURE 10.3-5 French stationary basket extractor. (Reproduced by permission, French Oil Mill Machinery Co.)

10.3-6 Perforated Plate Extractors

Series of towed, perforated plates have been used to convey solids through closely fitting shells, while liquid flowed through the perforations and solids between the plates. The plates were propelled by chain drives that linked the plates together. Shells in the form of a single vertical loop, as in the Hungarian J extractor,[28] or a serpentine succession of vertical loops (the Olier extractor,[29] the Silver Chain extractor,[30] and Opperman and Diechmann extractor[31]) have been used. These extractors, which were used for beet sugar extraction, have tended to fall into disuse because the chain drives are subject to wear and costly to replace. In the Crown extractor[32] (Fig. 10.3-9), which is widely used for oilseed extraction, the chain and shell form a relatively flat vertical loop. While liquid percolates through the plates and intervening beds in the vertical legs of the loop, in the horizontal sections it percolates across the bed as in crossflow extractors. In the Saturne extractor,[33] which is used to extract cane sugar, the plates are attached to a large vertical ring which is propelled in a vertical circular shell by a hydraulic piston drive.

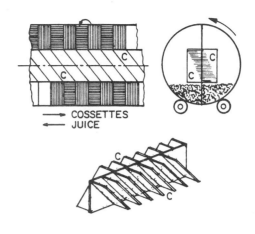

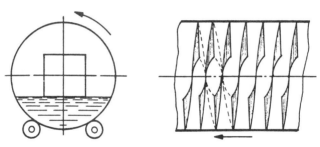

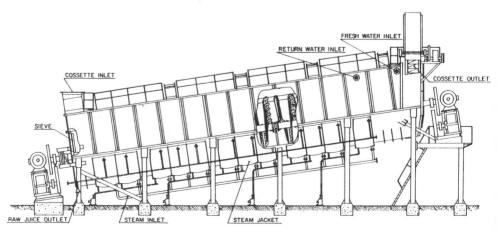

FIGURE 10.3-6 Extract and solids conveying system for RT2 trommel extractor. (Reproduced by permission, Keter Publishing House Jerusalem Ltd.)

FIGURE 10.3-7 DDS Slope extractor. (Reproduced by permission, Danish Sugar Corporation.)

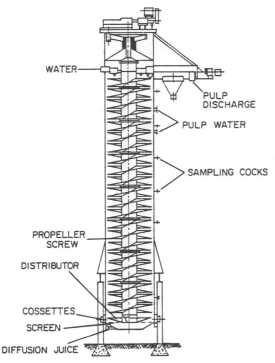

FIGURE 10.3-8 BMA Tower extractor. (Reproduced by permission, Keter Publishing House Jerusalem Ltd.)

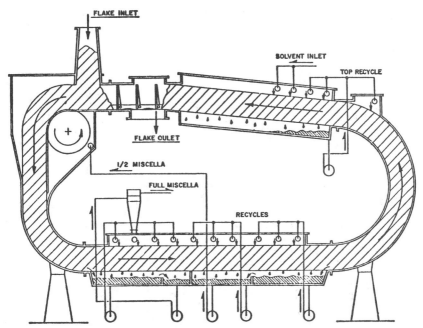

FIGURE 10.3-9 Crown Iron Works extractor. (Reproduced by permission, *Chemical Engineering Progress*, American Institute of Chemical Engineers.)

10.3-7 Vertical Basket Conveyor Extractors

In the Bollman extractor (Fig. 10.3-10),[34] baskets with perforated bottoms are driven around a high vertical loop by a chain drive. Liquid percolates through the solids in the baskets and drips onto the solids in the basket below. Spent solids are dumped out of the baskets by temporarily inverting them at the top of the extractor. The baskets are then reinverted and filled with fresh solids at the top of the leg in which the baskets descend. Solid–liquid flow is concurrent in this leg and countercurrent in the leg where the solids ascend. The extract (half-miscella) that collects in a sump at the bottom of the ascending leg is sprayed onto the contents of the freshly filled baskets at the top of the descending leg. The extract product (full-miscella) is collected at the bottom of the descending leg.

Bollman extractors have largely been superseded by extractors in which the baskets are conveyed around a relatively flat horizontal loop or sequence of loops. These horizontal basket conveyor extractors are usually operated as countercurrent crossflow extractors, but in some cases mixed crossflow countercurrent and upflow–downflow sequences are used.

10.3-8 Gravity Counterflow Extractor

In the extractor shown in Fig. 10.3-11, acid-treated, lime-conditioned pieces of cattle hide form a descending bed through which upflow of acidified hot water occurs.[35] The pieces dissolve progressively as the collagen in the hides is converted into gelatin, which passes into the extract. The extract flows out through screens at the top of the tank. The slight amounts of residual solids that remain are purged out of the bottom of the tank from time to time.

In Allis Chalmers, Anderson, and Bonotto extractors,[36] solids fell through ascending extract and were intermittently deposited on and swept off circular trays. These extractors are no longer used, probably because effective contacting occurred only during the descent of the solids and not when they were on the trays, and because extract turnover and backmixing frequently occurred.

10.3-9 Batch Extractors

Batch extractors are used for single-stage extraction. In most of these the solid–liquid mixture is agitated (e.g., by slowly turning, large-diameter propellers) to suspend the solids in the liquid. In other cases the solids settle and are resuspended by conveying mixers (e.g., Nauta mixers). In other batch extractors a

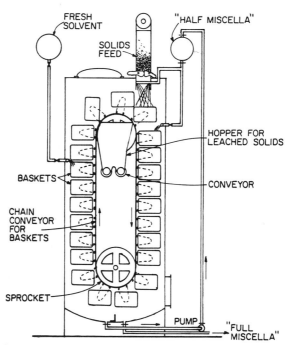

FIGURE 10.3-10 Bollman vertical basket extractor. (Reproduced by permission, *Chemical Engineering Progress*, American Institute of Chemical Engineers.)

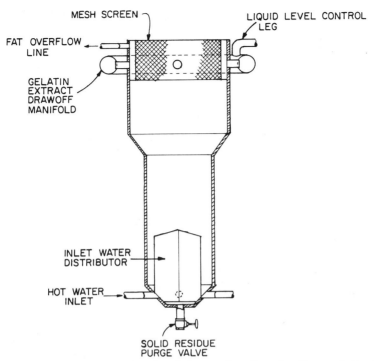

FIGURE 10.3-11 Countercurrent extractor for gelatin extraction. (Reproduced by permission, *Chemical Engineering Progress*, American Institute of Chemical Engineers.)

horizontal rotating drum containing a longitudinal perforated divider is used. The divider intermittently lifts the solids out of and dumps them back into the extract as the drum rotates. Batch extractors which are fitted with screens or sedimentation zones that retain the solids while permitting extract outflow can be used as differential extractors.

10.3-10 Reflux Extractors

In these units the solvent in the extract draining out of the extractor is distilled off, condensed, and recycled to the extractor. The distillation provides both concentrated extract and fresh solvent feed. If the solid is moist and a nonaqueous solvent is used, water will codistill with the solvent. If the water-rich portions of the distillate are replaced with fresh solvent and not recycled to the bed, the solid will dry as leaching proceeds. The amount of solvent replacement required can be minimized if the solvent and water form a heteroazeotrope.

Fixed-bed extraction systems in which supercritical gas is used as the solvent employ various means to remove the extracted solute from the gas before it is repassed through the bed. In terms of solute concentration in the solids these extractors act in the same way as extractors in which solvent distillation is used. Depending on the way solutes are removed from the supercritical gas, concentrated extracts may or may not be obtained.

10.3-11 Continuous Mixer–Separator Systems

Continuous mixer–separator systems may be set up by using solid–liquid separation devices such as continuous vacuum filters, continuous decanting or continuous screening centrifuges, or disk centrifuges and hydroclones that provide a continuous discharge of solids-rich slurry. Surge tanks in which the solid and liquid streams are remixed must be provided between these separators. If the solute being removed is found mainly in liquid cling at the surface of the solids (e.g., the protein-rich liquid clinging to the surface of starch during the wet milling of corn), unbaffled tanks can safely be used. However, if solute is inside the solids and diffusional resistance is appreciable, the surge tanks should be baffled to minimize backmixing and provide uniform, adequately long solids holdup in each stage.

10.4 PRODUCTS, RAW MATERIALS, AND EXTRACTION CONDITIONS

Typical products, solutes, raw materials, solvents, and extraction times for various commercially important leaching processes are listed in Table 10.4-1.

10.5 PHASE EQUILIBRIUM

In homogeneous extractions, $Y_{int} = Y$ at equilibrium. If $Y_{int} < Y_{sat}$ and the solute is not sorbed selectively by the marc, $X = \psi Y$. Consequently, $M = 1/\psi$. M usually remains fairly constant during leaching processes, but m varies somewhat because the density of the occluded and external solutions change as x changes.

In heterogeneous extractions, $Y = M'Y_{int}$, where M' is the solute equilibrium distribution coefficient for the external and internal solvent pair. Hence, for heterogeneous extractions $M = M'/\psi$ and usually remains fairly constant. In all cases, $m = M(v_E/v_R)$.

For homogeneous extractions, triangular diagrams for the system (marc, solute, solvent) can be used to depict phase equilibria. In diagrams such as Fig. 10.5-1, the I, S, and W vertices, respectively, represent 100 wt. % marc, 100 wt. % solute, and 100 wt. % solvent. The saturation line represents compositions of the solid when saturated with occluded solution. If the marc merely encloses the occluded solution and does not sorb selectively the solute or solvent, the tie lines that connect these solid compositions with corresponding equilibrium extract compositions are extensions of straight lines passing through I. Marcs often sorb water selectively. If water is used as the solvent and b, the mass of water sorbed per unit mass of marc, remains constant, the extended tie lines all pass through $Z_W = b/(1 + b)$ on side IW. In practice, b decreases as the solute concentration increases and the tie line extensions terminate in a narrow band on IW. The average value of $m = [(1 - (Z_I)_{sat}(1 + b)]^{-1}$. If $Y_{int} < Y_{sat}$ for all solutes and no solute is sorbed selectively by the marc, then, in terms of phase equilibrium, all solutes may be treated as a single solute.

Because M is nearly constant in most cases, saturation lines are usually fairly straight and nearly parallel to side SW, with the $(Z_I)_{sat}$ tending to decrease slightly as x increases along the line. Typical average $(Z_I)_{sat}$ values are roughly 0.35 for coarse, ground roasted coffee, 0.05 for sugar beets, 0.12 for sugar cane, 0.025 for apples (all in contact with aqueous extracts), and 0.6 for flaked soybeans in contact with hexane extracts. M can be estimated as follows:

$$M = \frac{x\mathbf{v}_S + (Z_W)_{sat}\mathbf{v}_W + (Z_I)_{sat}\mathbf{v}_I}{x\mathbf{v}_S + [(Z_W)_{sat} - b(Z_I)_{sat}]\mathbf{v}_W} \tag{10.6-1}$$

Typical approximate $\mathbf{v}$ values divided by 10^{-3} m³/kg are 1.00 for water, 0.65 for structural carbohydrates, 0.62 for soluble carbohydrates, 0.75 for structural proteins, 1.53 for hexane, and 1.09 for vegetable oils. In cases where the solid volume changes markedly as extraction proceeds, for example, when alcohols are used to extract water from solids with carbohydrate or protein-based marcs, the saturation line is strongly curved.

Triangular diagrams, such as Fig. 10.5-1, can be used to carry out equilibrium-stage leaching calculations in the same way that analogous calculations are carried out for ternary liquid–liquid extractions.[1-3] Other coordinate systems, for example, Janecke coordinates, can also be used for such calculations.

Because of extract cling on the surface of the solid it is difficult to determine the position of the saturation line precisely. The apparent value of $(Z_I)_{sat}$ tends to decrease as the solid particle size decreases and cling consequently increases. If the amount of cling does not change, it can be treated as part of the solid for equilibrium-stage operations, and equilibrium compositions for the solid can be based on the apparent saturation line. For nonequilibrium processes, however, solid composition should be computed on a cling-free basis.

10.6 MULTISTAGE EQUILIBRIUM COUNTERCURRENT EXTRACTION

If the size of the solid and M do not change during an extraction, α, the stripping factor, will also remain constant. Then, for multistage equilibrium countercurrent extractions in which solute-free solvent and saturated solids, respectively, are used as the liquid and solid feeds:

$$\alpha \neq 1 \qquad \eta = \frac{\alpha^{N+1} - \alpha}{\alpha^{N+1} - 1} \tag{10.6-1}$$

$$N = \frac{\ln[(\alpha - 1)/\alpha(1 - \eta)]}{\ln(\alpha)} \tag{10.6-2}$$

$$\alpha \neq 1 \qquad \eta = \frac{N}{N + 1} \tag{10.6-3}$$

TABLE 10.4-1 Materials Involved in Commercial Leaching Processes

Product	Solids	Solute	Solvent	Extraction Time (min)
Anthrocyanins	Chokeberries, grapeskins	Anthrocyanins	Ethanol, water	75–85
Apple juice solutes	Apple chunks	Apple juice solutes	Water	300–360
Apple juice solutes	Pressed apple pomace	Apple juice solutes	Water	
Betanines	Red beets	Betanines	Ethanol, water	
Brewing worts	Malted barley	Sugars, grain solutes	Water	120–300
Butter	Rancid butter	Low molecular weight organic acids	Water	
Carrageenan	Kelp	Carrageenan	Water	
Carotenoid pigments	Leaves	Water first, then pigments	Ethanol, isopropanol	
Cassava	Cyanogenetic glycosides	Manioc	Water	
Citrus molasses	Juice pressing residues	Citrus sugars	Water	
Collagen	Limed hides	CaOH	Water	1400
Cottonseed oil	Cottonseed	Cottonseed oil	Hexane	60–85
Gelatin	Collagen	Gelatin	Water or dilute acid	240 repeated
Cytoplasmic alfalfa protein	Coagulated alfalfa protein	Chlorophyll, chlorogenic acid	Acetone, ethanol, butanol	
Decaffeinated coffee	Green coffee beans	Caffeine[a]	Methylene chloride[b]	480–720
Decaffeinated coffee	Green coffee beans	Caffeine[a]	Supercritical CO_2	
Decaffeinated coffee	Green coffee beans	Caffeine[a]	Caffeine-free green-coffee extract	
Desalted kelp	Giant kelp	Sea salts	Dilute HCL	120–180
Fish oil	Fish scraps	Fish oil	Hexane, CH_2Cl_2, butanol	15–60
Fish protein concentrate	Trash fish	Fish oil	Butanol	
Fruit juice solutes	Sliced fruit or pomace	Fruit juice solutes	Water	
Hop extracts	Hop flowers	Hop solutes	CH_2Cl_2	
Hop extracts	Hop flowers	Hop solutes	Supercritical CO_2	
Hopped worts	Hop flowers	Hop solutes	Water	90–120
Insulin	Beef or pork pancreas	Insulin	Acidic alcohol	
Iodine	Seaweed	Iodine	Aqueous H_2SO_4	
Limed hides	Cattle hides	Nongelatin base proteins, carbohydrates	Aqueous CaOH	40,000–130,000
Liver extract	Mammalian livers	Peptides	Water	
Low-moisture fruit	Moist fruit	Water	50% Aqueous sucrose	480

551

TABLE 10.4-1 (*Continued*)

Product	Solids	Solute	Solvent	Extraction Time (min)
Low-moisture desalted pectin	Alcohol precipitated pectin	NaCl, water	Isopropanol	
Malt extract	Germinated grain	Malt extract	Water	
Methylated pectin	Pectin shreds	Water	Methanol	
Ossein-base collagen	Cattle bones	Ca salts, phosphates	Dilute acid	1400
Pancreatin	Hog pancreas	Pancreatin		
Papain	Papaya latex	Papain	Water	
Pectin	Desugared apple pomace	Pectin	Dilute acid	30–240, twice sometimes
Pectin	Treated citrus peel	Pectin	Dilute acid	
Pepsin	Hog stomachs	Pepsin	Aqueous HCl	
Pickles	Cucumbers	NaCl	Water	7200
Pickle relish	Cucumber bits	NaCl	Water	15
Rennin	Calf stomach lining	Rennin	Aqueous NaCl	
Single-cell protein	Lysed cells	Protein	Water	
Single-cell protein	Intact cells	Nucleic acids	Aqueous NaCl	
Soluble coffee	Ground roasted coffee	Coffee solutes	Water	120–180
Soluble tea	Dry tea leaves	Tea solutes	Water	45–120
Soybean oil	Soybeans	Soybean oil	Hexane	18–45
Soy protein concentrate	Defatted soy flour	Sugars, nonprotein solids	70% Ethanol at isoelectric point	
Soy protein isolate	Defatted soy flour	Protein	pH 9 Aqueous NaOH	
Spices extract	Paprika, cloves, pepper, thyme, marjoram, etc.	Spice solutes	80% Ethanol	
Spice oleoresins	Paprika, etc.	Spice solutes	Methyl ethyl ketone	
Steeped corn	Corn kernels	Corn steep solids[a]	Dilute H_2SO_3	1800–3000
Steroids	Fungi mycelium	Steroids	Acetone and methylene chloride	
Sugar-free pomace	Apple pomace	Sugars	Water	
Sucrose	Sugar beets	Sucrose	Water	20–90
Sucrose	Sugar cane	Sucrose	Water	25–60
Treated citrus peel	Citrus peel	Flavonoids, hesperidin, sugars	Water	
Vanilla	Vanilla beans	Vanilla	65% Ethanol	
Vitamin B_1	Rice polishings	Vitamin B_1	Alcohol-water	
Zein	Corn	Zein	90% Ethanol	10,000

[a] By-product.

[b] Heterogeneous extraction, internal solvent: water.

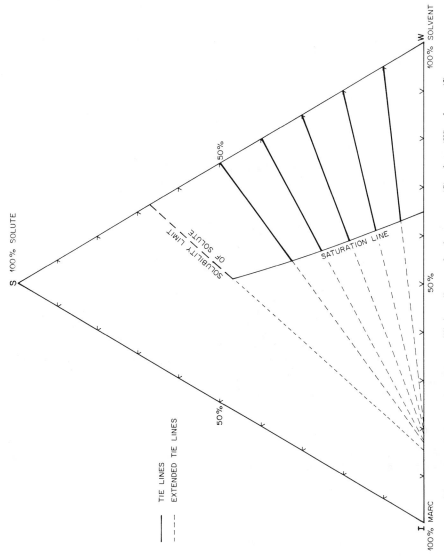

FIGURE 10.5-1 Triangular phase-equilibrium diagram for the system: (*S*) solute, (*W*) solvent, (*I*) marc.

S 100% SOLUTE

W 100% SOLVENT

I 100% MARC

50%

50%

50%

SATURATION LINE

SOLUBILITY LIMIT OF SOLUTE

TIE LINES
EXTENDED TIE LINES

$$N = \frac{\eta}{1 - \eta} \tag{10.6-4}$$

and for all α $$Y_1 = \frac{\eta\, MX_o}{\alpha} \tag{10.6-5}$$

If $\alpha < 1.0$ the maximum attainable yield is α, no matter how many stages are used. When $\alpha = 1.0$,

$$\frac{X_n}{X_o} = \frac{\alpha - \eta}{(\alpha - 1)\alpha^n} - \frac{1 - \eta}{\alpha - 1} \tag{10.6-6}$$

Equation (10.6-1)–(10.6-6) are special forms of the Kremser–Brown–Souders (KBS) equations, which can be expressed in more general forms[1] which are useful when $Y_{N+1} \neq 0$. These more general forms can be used to calculate yields and ideal-stage requirements when occluded liquid is pressed out of spent solids and recycled to the extraction system (as is usually done for beet sugar extraction).

When dry solids are used as a feed and $\alpha \neq 1$,

$$\eta = \frac{(\alpha - f_R m)(\alpha^N - 1)}{(\alpha - f_R m)(\alpha^N - 1) + (\alpha - 1)} \tag{10.6-7}$$

and

$$N = \frac{\ln\left[\dfrac{(\alpha - \eta) - f_R m(1 - \eta)}{(\alpha - f_R m)(1 - \eta)}\right]}{\ln(\alpha)} \tag{10.6-8}$$

In Eq. (10.6-7), α is based on E', the extract mass before imbibition, R', the solid mass after imbibition, and m; N includes the imbibition stage. The yield $\eta = E'_o y_o / F' Z_S$. In continuous flow systems, the corresponding values of E', E'_o, F', and R' would be used.

Although equilibrium is usually not achieved in leaching systems, the above equations are useful for calculating lower bounds on the N required to achieve a given yield at a given α, and upper bounds for η when N and α are given. When there is significant diffusional resistance in the solid, overall stage efficiencies for multistage leaching systems cannot be calculated on the basis of constant Murphree efficiencies.

10.7 FICKS'S LAWS

Diffusive transfer of solutes in solids is governed by Fick's First and Second Laws. Even though biological solids are not structurally homogeneous and diffusion occurs mainly in the fluid occluded within the solid, Fick's Laws will be expressed in terms of X, that is,

$$J_s = -D_s \frac{\partial X}{\partial r} \tag{10.7-1}$$

and the more frequently used Fick's Second Law (FSL) expressed in a form[1] that is valid for infinite slabs, infinite cylinders, and spheres:

$$\frac{\partial X}{\partial t} = \frac{1}{r^{\nu-1}} \frac{\partial}{\partial r}\left(r^{\nu-1} D_s \frac{\partial X}{\partial r}\right) \tag{10.7-2}$$

In solving Eq. (10.7-2) it usually is assumed that $X = X_o$ for all r at $t = 0$, and that $\partial X/\partial r = 0$ at $r = 0$. Furthermore, since Bi usually is quite large except for certain heterogeneous extractions (including supercritical extractions), it is assumed that $X = Y/M$ at $r = a$ for $t > 0$. The most commonly used solutions for Eq. (10.7-2) are exponential series[2] which are expressed in general form by Eq. (10.7-3):

$$\mathbf{X} = \mathbf{Y} = \sum_{i=0}^{\infty} C_i \exp(-q_i^2 \tau) \tag{10.7-3}$$

The q_i are functions of the boundary conditions and the C_i are functions of the boundary conditions and the initial conditions. If Y is maintained constant, $\mathbf{X}$ in Eq. (10.7-3) is replaced by $\mathbf{X}'$ and $\mathbf{Y}$ is not used. When Eq. (10.7-3) is applicable, plots of log $\mathbf{X}$, log $\mathbf{X}'$, or log $\mathbf{Y}$ versus t will resemble Fig. 10.7-1. Values of D_s can be obtained from the slope of such plots at suitably large t.

When flaked oilseeds[3] are extracted, plots of log $\mathbf{X}'$ versus t, as shown in Fig. 10.7-2, do not obey

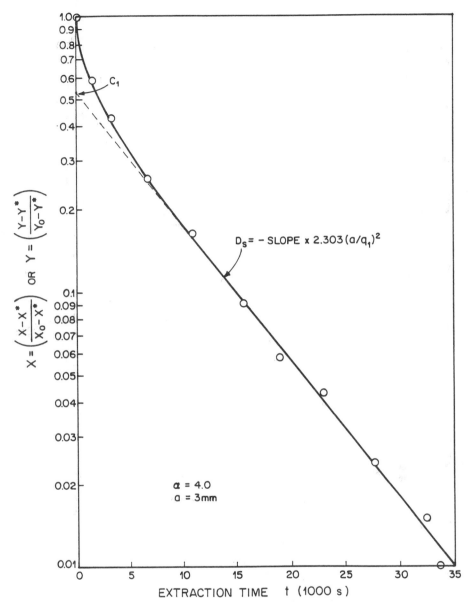

FIGURE 10.7-1 Log $[(Y - Y^*)/(Y_o - Y^*)]$ or log $[(X - X^*)/(X_o - X^*)]$ versus time for the batch extraction of infused sucrose from $-6 + 8$ mesh spent coffee grounds. [Reprinted from *Food Technology*, **36**(2), 73–86 (1982). Copyright © 1982 by Institute of Food Technologists.]

Eq. (10.7-3). The causes for this abnormality are not wholly clear, but solid-structure nonuniformity is probably a major factor. Because of this abnormal behavior it is difficult to carry out valid unsteady-state extraction calculations for oilseeds.

10.8 UNSTEADY-STATE BATCH EXTRACTION

The values of C_i and q_i which can be used for solids with various shapes when Eq. (10.7-3) is used in the case of batch extractions are listed in Table 10.8-1. When α is infinite Y remains constant and the dimensionless concentration used in Eq. (10.7-3) should be X'.

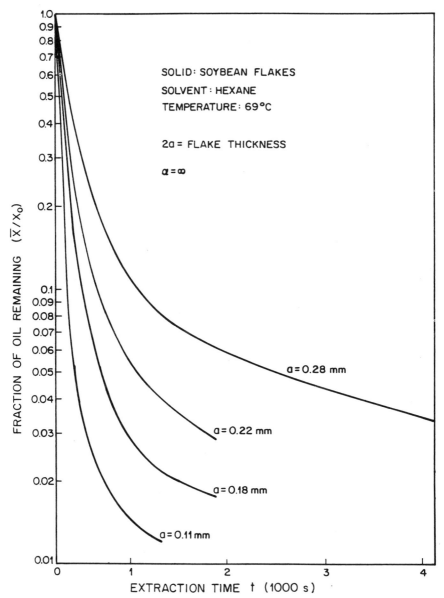

FIGURE 10.7-2 Log (X/X_o) versus time for the extraction of soybean oil from flaked soybeans by oil-free hexane. [Reprinted from *Food Technology*, **36**(2), 73–86 (1982). Copyright © 1982 by Institute of Food Technologists.]

When τ is small[1,2] it is more convenient to use Eqs. (10.8-1) and (10.8-2) than Eq. (10.7-3):

$$(1 - \mathbf{X}) = (1 - \mathbf{Y}) = (1 + \alpha)[1 - A \text{ eerfc}(B\phi) + C \text{ eerfc}(D\phi)] \qquad (10.8\text{-}1)$$

$$(1 - \mathbf{X}) = (1 - \mathbf{Y}) = (1 + \alpha)[F\phi - G\phi^2 + H\phi^3] \qquad (10.8\text{-}2)$$

where $\phi = \sqrt{\tau}/\alpha$ and $\text{eerfc}(x) = \exp(x^2) \text{erfc}(x)$. While Eq. (10.8-2) is more convenient than Eq. (10.8-1), particularly when a hand calculator is used, Eq. (10.8-1) can be used without serious error at slightly larger values of τ. As a rough rule of thumb, in the α range normally encountered in practice, it is more convenient to use Eq. (10.7-3) when $\tau > 0.1$ and Eq. (10.8-2) when $\tau < 0.01$.

TABLE 10.8-1 Coefficients and Eigenvalues for Fick's Second Law Solutions for Batch Extractions in Which α Is Constant and Bi Is Infinite[a]

Solid Shape	α	Equation for q_i	C_i
Infinite slab	Finite	$q_i = -\dfrac{\tan(q_i)}{\alpha}$	$\dfrac{2\alpha(\alpha + 1)}{(\alpha + 1) + (\alpha q_i)^2}$
	Infinite	$q_i = \dfrac{(2i - 1)\pi}{2}$	$\dfrac{2}{q_i^2}$
Infinite cylinder	Finite	$q_i = -\dfrac{2J_1(q_i)}{\alpha J_o(q_i)}$	$\dfrac{4\alpha(\alpha + 1)}{4(\alpha + 1) + (\alpha q_i)^2}$
	Infinite	$J_o(q_i) = 0$	$\dfrac{4}{q_i^2}$
Sphere	Finite	$q_i = \dfrac{(3 + \alpha q_i^2)\tan(q_i)}{3}$	$\dfrac{6\alpha(\alpha + 1)}{9(\alpha + 1) + (\alpha q_i)^2}$
	Infinite	$q_i = i\pi$	$\dfrac{6}{q_i^2}$

[a] The values of C_i and q_i listed in this table are taken from various parts of *Diffusion in Solids* by J. Crank, Oxford University Press, London, 1970.

When Y is constant and τ is small $(1 - X')$ is given by Eq. (10.8-3):

$$(1 - X') = J\sqrt{\tau} - K\tau - L\tau^{3/2} \tag{10.8-3}$$

The coefficients that should be used in Eq. (10.8-1), (10.8-2), and (10.8-3) for various solid shapes are listed in Table 10.8-2. While Eq. (10.7-3) is valid for all values of τ, provided enough terms are used, Eq. (10.8-1), (10.8-2), and (10.8-3) are valid only when τ is small enough. When in doubt, the range of validity of these equations should be checked by comparison with Eq. (10.7-3).

10.9 SOLUTE DIFFUSIVITIES

Schwartzberg[1] has tabulated most of the available D_s data for solutes diffusing in biological solids. The same reference provides D_L data for these solutes and other important solutes. Typical D_s values for sucrose in water-rich cellular vegetable matter at room temperature range between 0.5 and 1.0×10^{-10} m²/s when the cell walls are hard (e.g., sugar cane and roasted coffee) and between 1.5 and 4.5×10^{-10} m²/s when the walls are soft (e.g., sugar beets, potatoes, apples, cucumbers, blanched carrots, celery, onions). D_s values for salts, such as NaCl, KCl, and $K_2C_2O_4$, at the same conditions are four to six times as large. The corresponding NaCl:sucrose D_L ratio is 3–4. Therefore, the cell walls in the solid hinder the diffusion of sucrose more than that of salts which have a lower molecular weight and volume. D_s values for sucrose in materials with hard cell walls are between 0.1 and 0.2 times as large as D_L for sucrose; with soft cell walls the corresponding D_s:D_L ratio is 0.3–0.9.

In a few cases[2] it has been demonstrated that there is a solute molecular weight cutoff or size (MWC) above which a solute will not diffuse into a solid. Therefore, bottlenecks in cell walls may control the rate of diffusion in solids. Solute molecules that cannot pass through these bottlenecks cannot enter or leave the solid (excepting those portions in which the cell walls are ruptured). This suggests that if MWC is known, D_s for a solute i can be estimated from the D_s for a reference solute r and corresponding known values of D_L and MW; that is,

$$\frac{(D_s)_i}{(D_s)_r} = \frac{(D_L)_i}{(D_L)_r} \frac{(\text{MWC}^{1/3} - \text{MW}_i^{1/3})^2}{(\text{MWC}^{1/3} - \text{MW}_r^{1/3})^2} \tag{10.9-1}$$

Furthermore, if pairs of D_s, D_L, and MW values are known, MWC can be estimated and used to estimate other D_s values. In cases where D_s for a low-molecular-weight solute must be estimated and $(D_s)_r$ and MWC are known but both D_L values are not known, $(\text{MW}_i/\text{MW}_r)^{0.6}$ can be substituted for $(D_L)_i/(D_L)_r$ in Eq. (10.9-1).

While Eq. (10.9-1) provides a reasonable basis for estimating D_s in some cases, it should not be applicable when (1) solutes cause changes in the pore structure, (2) nongeometric pore–solute interactions affect the mobility of the solute[3] or its concentration in the pores,[4] (3) particular solutes are sorbed selectively by the marc, or (4) facilitated or activated diffusion occurs. Selective sorption is particularly likely to occur

TABLE 10.8-2 Coefficients for use in Eqs. (10.8-2) and (10.8-3)

Coefficient	Particle Shape		
	Infinite Slab	Infinite Cylinder	Sphere
A	1	$\dfrac{1 + (\alpha + 1)^{1/2}}{2(\alpha + 1)^{1/2}}$	$\dfrac{1 + (1 + 4\alpha/3)^{1/2}}{2(1 + 4\alpha/3)^{1/2}}$
B	1	$1 + (\alpha + 1)^{1/2}$	$1.5\,[1 + (1 + 4\alpha/3)^{1/2}]$
C	0	$\dfrac{1 - (\alpha + 1)^{1/2}}{2(\alpha + 1)^{1/2}}$	$\dfrac{1 - (1 + 4\alpha/3)^{1/2}}{2(1 + 4\alpha/3)^{1/2}}$
D	—	$1 - (\alpha + 1)^{1/2}$	$1.5[1 - (1 + 4\alpha/3)^{1/2}]$
F	$\dfrac{2}{\sqrt{\pi}}$	$\dfrac{4}{\sqrt{\pi}}$	$\dfrac{6}{\sqrt{\pi}}$
G	1	$4 + \alpha$	$3(3 + \alpha)$
H	$\dfrac{4}{3\sqrt{\pi}}$	$\dfrac{32 + 16\alpha - \alpha^2}{3\sqrt{\pi}}$	$\dfrac{12(3 + 2\alpha)}{\sqrt{\pi}}$
J	$\dfrac{2}{\sqrt{\pi}}$	$\dfrac{4}{\sqrt{\pi}}$	$\dfrac{6}{\sqrt{\pi}}$
K	0	1	3
L	0	$\dfrac{1}{3\sqrt{\pi}}$	0

when the marc contains polyelectrolyte macromolecules and the solute dissociates into ions (i.e., when the solid acts like an ion exchanger). This may partly explain why D_s for NaCl in meats[5] and cheeses[6,7] is only 0.25–0.5 times as large as D_s for NaCl in vegetable matter with soft cell walls. On the other hand, D_s values for neutral molecules in both dilute neutral and dilute polyelectrolyte gels[8-10] including cottage cheese[11,12] are roughly 0.7–0.8 times as large as the corresponding D_L values.

As temperature, viscosity, and Y_{int} vary, the change in D_s for a given solute will parallel the change in D_L in the occluded solvent; that is, D_s will be directly proportional to the absolute value of T and inversely proportional to μ.

D_s for soybean oil during the initial period of its extraction from soybeans with hexane is around 1.0 $\times 10^{-10}$ m²/s but drops off to 0.12–0.2 times that value at the end of extraction.[13] Both initial and final D_s values for cottonseed and flaxseed are roughly only 0.1 times as large as the corresponding initial and final values for soybeans; and the extraction times required when these oils are extracted commercially is considerably longer than that for soybeans.

Solute–solute interactions will occur when two solutes codiffuse in a solid. Such interactions may be caused by (1) the effect of the second solute on the viscosity of the occluded solution,[14] (2) partial blockage of pores by large solute molecules, (3) the binding of small molecules to slow-moving large molecules,[15] (4) the formation of other types of complexes between two solutes, (5) osmotic flows and counterflows generated by osmotically active solutes,[16,17] (6) Donnan-membrane effects, and (7) effects that can be interpreted in terms of irreversible thermodynamics.[18,19]

In solids that contain vascular systems, D_s parallel to the vascular bundle may be twice as large as it is at right angles to the bundle. Furthermore, in such solids, D_s as determined from batch leaching tests increases as the particle size increases (below a threshold size of 3–5 mm). In dense cellular materials, on the other hand, ruptured surface cells make up a progressively smaller fraction of the particle volume as size increases; D_s consequently decreases slightly as particle size increases.

10.10 FIXED BEDS

Solutions for Fick's Second Law for leaching, in which extract with a constant inlet concentration Y_{in} passes through a fixed bed of solids initially containing a uniform solid concentration X_0, have been developed by Rosen,[1] Babcock et al.,[2] Neretniecks,[3] and Rasmussen and Neretniecks.[4] Rosen's solutions apply only when D_A is negligibly small; the other solutions take axial dispersion into account. These solutions can be used to determine $S = (Y_{out} - MX_o)/(Y_{in} - MX_o)$ at time t. When $\tau_f > 50$ Rosen indicates that

$$S = 0.5 \{1 + \text{erf}[F(\beta\theta - 1)]\} \tag{10.10-1}$$

where if Bi is finite

$$F = \left[\frac{15 \text{ Bi } \tau_f}{4\beta(\text{Bi} + 5)} \right]^{1/2} \tag{10.10-2}$$

If Bi is very large F reduces to $(15\tau_f/4\beta)^{1/2}$. In most fixed-bed leaching setups τ_f is usually less than 1.0 and frequently less than 0.1.

When $\tau_f < 50$ all the previously cited solutions become quite complicated and extensive computation is required to obtain values for S. Rosen[5] provides tables of values of S versus τ_f and θ for various values of Bi, but relatively few S values are listed in the small τ_f range. The S versus θ curves are usually sigmoidal and spread out more as τ_f decreases. At high values of τ_f, a straight line will be obtained over most of the θ range when S is plotted versus θ on probability paper, and at low values of τ_f, almost straight lines will be obtained over most of the θ range for similar plots on log probability paper. In each case the S is located on the probability coordinate.

Experimental S versus θ curves usually are not sigmoidal and hence do not agree with the solutions provided by the previously cited authors. The lack of agreement is probably largely caused by unstable displacement. The shapes of the experimental curves depend strongly on whether the bed is filled initially with equilibrated extract as well as with solids and on whether upflow or downflow is used.[6] Typical log $(1 - S)$ versus θ curves for these conditions are shown in Figs. 10.10-1, 10.10-2, and 10.10-3. The final linear decrease in log $(1 - S)$ as θ increases apparently characterizes the diffusive extraction process proper; in those cases where the bed is filled initially with equilibrated extract, the initial portion of the curve characterizes the extract displacement process.

If the bed initially does not contain extract and if the exponential decrease in $(1 - S)$ versus θ shown in Fig. 10.10-1 starts at $\theta = 0$ and continues indefinitely,

$$(1 - S) = (1 - S_o) \exp\left[-(1 - S_o)\beta\theta\right] \tag{10.10-3}$$

Furthermore, the maximum value of $(1 - S_o)$ is given by

$$(1 - S_o)_{\max} = 1 - \exp\left[-\frac{s_p(1 - \epsilon)kL}{v_p\epsilon V}\right] \tag{10.10-4}$$

The exponent in (10.10-4) is also equal to $(-v\text{Bi}\tau_f/\beta)$. When backmixing and axial dispersion occur $(1 - S_o)$ will be less than $(1 - S_o)_{\max}$. In the usual Re_p ranges encountered in fixed-bed extractions k can be estimated by using the following equations:[7]

$$k = \frac{1.09 V}{\text{Sc}^{2/3}\text{Re}_p^{2/3}} \qquad 0.0016 < \text{Re}_p < 55 \tag{10.10-5}$$

$$k = \frac{0.250 V}{\text{Sc}^{2/3}\text{Re}_p^{0.31}} \qquad 5 < \text{Re}_p < 1500 \tag{10.10-6}$$

Average values of Y_{out} during specified cycle time intervals will be useful in analyzing crossflow extraction systems. These values are contained in the dimensionless concentrations U_n, which are obtained by integrating S/θ_c with respect to θ between the limits $(n - 1)\theta_c$ and $n\theta_c$. When Bi is very large and Rosen's solution for $\tau_f > 50$ is applicable, this yields

$$U_n = \frac{1}{2F\beta\theta_c}[\text{ierfc } (F(n\beta\theta_c - 1)) - \text{ierfc } (F((n - 1)\beta\theta_c - 1))] \tag{10.10-7}$$

where ierfc $(x) = (1/\sqrt{\pi}) \exp(-x^2) - x \operatorname{erfc}(x)$. When Eq. (10.10-3) is applicable,

$$1 - U_1 = \frac{1 - \exp[-(1 - S_o)\beta\theta_c]}{\beta\theta_c} \tag{10.10-8}$$

and

$$1 - U_n = (1 - U_1) \exp\left[-(1 - S_o)(n - 1)\beta\theta_c\right] \tag{10.10-9}$$

$\Delta U_n = U_n - U_{n+1}$, which will also be useful subsequently, is given by

$$\Delta U_n = (1 - U_1)(U_n - 1)\beta\theta_c \tag{10.10-10}$$

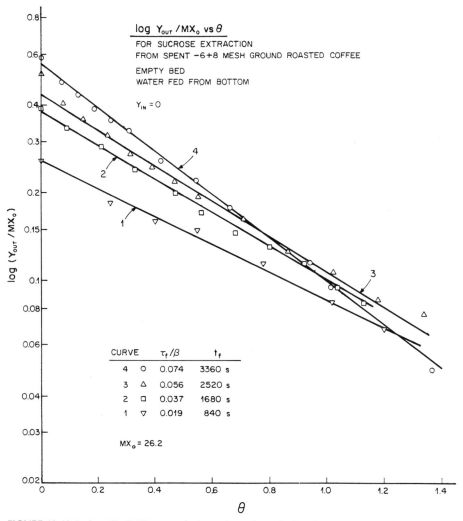

FIGURE 10.10-1 Log (Y_{out}/MX_o) versus θ when using upflow of solute-free water to leach infused sucrose from $-6 + 8$ mesh spent coffee grounds in a bed that initially does not contain liquid in the interparticle pores.

10.11 AXIAL DISPERSION AND FLOW MALDISTRIBUTION

For pluglike extract flow through fixed beds, Y should satisfy the following equation when $t > z/V = t_f$:

$$\frac{\partial Y}{\partial t} = -V \frac{\partial Y}{\partial z} + D_A \frac{\partial^2 Y}{\partial z^2} - \frac{(1 - \epsilon)}{\epsilon} \frac{\partial \overline{X}}{\partial t} \tag{10.11-1}$$

In steady-state continuous countercurrent extraction, when both the solid and extract flow are pluglike,

$$\epsilon A_c D_A \frac{\partial^2 Y}{\partial z^2} = \mathbf{E} \frac{\partial Y}{\partial z} - \mathbf{R} \frac{\partial \overline{X}}{\partial z} \tag{10.11-2}$$

At unsteady-state conditions $A_c \epsilon(\partial Y/\partial t)$ and $A_c(1 - \epsilon)(\partial \overline{X}/\partial t)$ should be added to the right-hand side of (10.12-2). Axial dispersion will reduce extraction efficiency unless D_A is negligibly small, that is, Pe $= Va/D_A$ is very large. Pe is used to characterize the magnitude of axial-dispersion-induced effects.

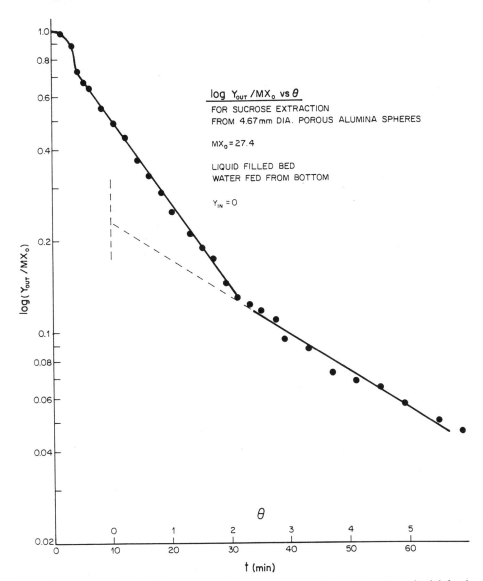

FIGURE 10.10-2 Log (Y_{out}/MX_o) versus t and θ when using upflow of solute-free water to leach infused sucrose from 4.67 mm diameter, porous alumina spheres in a bed initially filled with extract in equilibrium with the spheres.

Marked axial dispersion in both the liquid and solid phases has been observed in continuous countercurrent leaching systems.[1-4] The solid-phase dispersion is probably caused by nonuniform conveying and by backmixing caused by the baffles which are used to prevent solid beds from turning en masse. $V_R/(D_A)_R$ values of 16.1 m^{-1} and 20 m^{-1}, respectively, have been reported for sugar beet extraction in tower and slope extractors. Local flow nonuniformity and larger-scale flow maldistribution are the primary factors that cause axial dispersion in the extract.

Flow maldistribution can be caused by (1) imperfect flow convergence and divergence near extract inlet and outlet ports, (2) radial viscosity gradients induced by radial temperature gradients caused by heated extractor walls, (3) increased porosity near the extractor walls, and (4) unstable displacement. Unstable displacement will occur when V exceeds a critical value, V_c. When downflow is used during extraction and $\partial Y/\partial z$ or $\partial Y/\partial t$ are smooth and continuous,[5] $V_c = Kg(d\rho/d\mu)$. When downflow is used and there are step changes[6] in Y, such as occur during multistage crossflow leaching, $V_c = Kg(\rho_2 - \rho_1)/(\mu_2 - \mu_1)$, where

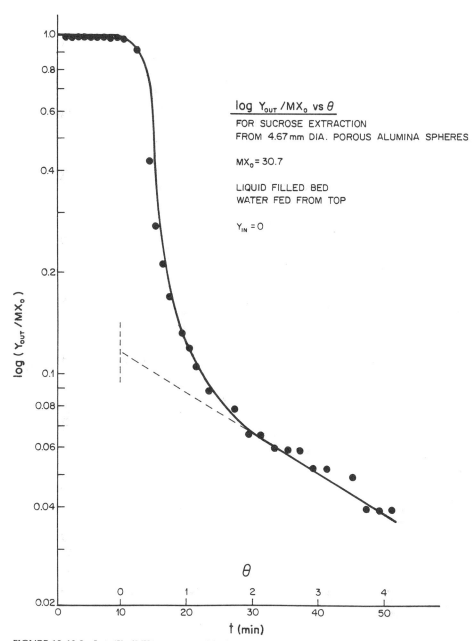

FIGURE 10.10-3 Log (Y_{out}/MX_o) versus t and θ when using downflow of solute-free water to leach infused sucrose from 4.67 mm diameter, porous alumina spheres in a bed initially filled with extract in equilibrium with the spheres.

the subscript 2 refers to the displaced extract and 1 to the entering extract. Unstable displacement almost invariably occurs when upflow is used during leaching. On the other hand, during infusion (the transfer of solute from liquid into a contiguous solid), upflow will be conditionally stable and downflow will be unstable.

Radial dispersion tends to reduce the radial concentration gradients caused by flow maldistribution, thereby producing axial dispersion. Hence, in certain cases the effects of flow maldistribution can be characterized in terms of reductions in Pe. If sharp radial concentration differences persist in spite of radial dispersion, different techniques, for example, parallel extractor models, must be used to account for the effects produced by maldistribution.

D_A and Pe can be determined by measuring the longitudinal spreading of the concentration of a pulse of a noninfusing tracer, such as Dextran Blue, injected into the liquid entering an extractor. The relationship between Pe and the variance of the dye concentration versus time curve for the exiting pulse is[7]

$$\sigma^2 = \frac{2}{Pe'} - \frac{2[1 - \exp(-Pe')]}{(Pe')^2} \qquad (10.11\text{-}3)$$

which for large values of Pe' reduces to

$$\frac{1}{Pe'} = 0.5[1 - (1 - 2\sigma^2)^{1/2}] \qquad (10.11\text{-}4)$$

Therefore, $Pe = Pe'(a/L)$. During upflow in fixed beds, in the absence of extraction, Pe ranged between 0.25 and 0.30. During extraction, Pe slowly increased from roughly 0.1 near the leading edge of the entering liquid and gradually approached 0.25–0.30 as $\partial Y/\partial\theta$ approached zero.[8] Dextran Blue pulses injected at the leading edge of the entering liquid yield an exit Dextran Blue concentration versus θ curve which parallels the $(1 - S)$ versus θ curve for the solute being extracted. This strongly suggests that axial dispersion controls the extraction process.

10.12 SUPERPOSITION—MULTISTAGE COUNTERCURRENT EXTRACTION

Step changes in Y occur when the extract and solids are transferred from stage to stage in multistage extraction systems. Therefore, the BC for Eq. (10.7-2) change when the solids move from stage to stage. Furthermore, $X = X_o$ for all r only when the solids enter the first stage, that is, the uniform X initial condition no longer applies after the first stage. The mathematical difficulties caused by this behavior can be circumvented by using superposition, that is, by applying Duhamel's Theorem.[1-3] If this is done for an N-stage extraction system, a series of N simultaneous linear equations are obtained for the $Q_n = (Y_n - Y_{N+1})/(MX_o - Y_{N+1})$ which characterize the N values of Y_n for the extracts leaving the stages. These equations can be arranged in the following matrix form:

$$\begin{bmatrix} A_1 & B_1 & 0 & 0 & \cdots & 0 \\ A_2 & B_2 & B_1 & 0 & \cdots & 0 \\ A_3 & B_3 & B_2 & B_1 & \cdots & 0 \\ \vdots & \vdots & \vdots & \vdots & & \vdots \\ \vdots & \vdots & \vdots & \vdots & & \vdots \\ A_{N-1} & B_{N-1} & B_{N-2} & B_{N-3} & \cdots & B_1 \\ A_N & B_N & B_{N-1} & B_{N-2} & \cdots & B_2 \end{bmatrix} \times \begin{bmatrix} Q_1 \\ Q_2 \\ Q_3 \\ \vdots \\ \vdots \\ Q_{N-1} \\ Q_N \end{bmatrix} = \begin{bmatrix} C_1 \\ C_2 \\ C_3 \\ \vdots \\ \vdots \\ C_{N-1} \\ C_N \end{bmatrix} \qquad (10.12\text{-}1)$$

Because many of the elements above the principal diagonal are zero, Eq. (10.12-1) can be readily solved by Gauss reduction. The extraction yield is

$$\eta = \alpha Q_1 \qquad (10.12\text{-}2)$$

The values of A_j, B_j, and C_j in Eq. (10.12-1) depend on the type of contacting used in the extraction system.[4,5] These values are listed in Tables 10.12-1 and 10.12-2 for various types of contacting. It can be seen that fairly simple formulas involving the functions X_j and U_j can be used to define most of the elements. If a fraction c of the extract in each stage clings to the solids leaving the stage, the X_j in the A_j, B_j, and C_j listed in Table 10.12-1 and also the right hand side of Eq. (10.12-2) should be multiplied by $(1 - c)$.

TABLE 10.12-1 Elements of Matrix Eq. (10.12-1) When Using Match Mixer–Separator Stages

j	$A_j(\alpha + 1)$	$B_j(\alpha + 1)$	$C_j(\alpha + 1)$
1	$(\alpha + 1)$	$-(\alpha + X_1)$	$(1 - X_1)$
2	$(\alpha + X_1)$	$(1 - X_2)$	$(1 - X_2)$
>2	$(\alpha + X_{j-1})$	$(X_{j-2} - X_j)$	$(1 - X_j)$

TABLE 10.12-2   Elements of Matrix Eq. (10.12-1) When Using Multistage Crossflow Extractors

j	A_j	B_j	C_j
		Without Recirculation	
1	1	$-U_1$	$(1 - U_1)$
2	0	$1 + \Delta U_1$	$(1 - U_2)$
>2	0	ΔU_j	$(1 - U_j)$
		With Recirculation	
1	$1 - fU_1$	$(f - 1)U_1$	$(1 - U_1)$
2	$f \Delta U_1$	$1 + (1 - f)\Delta U_1 - fU_1$	$(1 - U_2)$
>2	$f \Delta U_{j-1}$	$(1 - f)\Delta U_{j-1} + f \Delta U_{j-2}$	$(1 - U_j)$

Additional matrix formulations have been developed for crossflow extractors in which accidental recirculation occurs because the sumps have been incorrectly positioned. When the sumps are placed too far forward and the extract mixes completely on the top of the bed in each stage the matrix is exactly the same as for the recirculation case listed in Table 10.12-2.

10.13   CONTINUOUS COUNTERCURRENT EXTRACTION

FSL solutions have been developed for continuous countercurrent extractions in which the effects of flow maldistribution and axial dispersion are negligibly small.[1-4] These solutions involve the following infinite series:

$$W(t) = \sum_{i=1}^{\infty} C_i \exp \left[-q_i^2 \tau\right] \qquad (10.13\text{-}1)$$

Equation (10.13-1) is similar to Eq. (10.7-3), but the q_i, C_i, and dimensionless concentrations are defined by different equations. When $\alpha < 1.0$, $C_o \exp(q_o^2 \tau)$ should be added to the right hand side of (10.13-1). When $\alpha = 1.0$, $W(t)$ is given by the following equation:

$$W(t) = G\tau + P - \sum_{i=1}^{\infty} C_i \exp(-q_i^2 \tau) \qquad (10.13\text{-}2)$$

Different C_i must be used in (10.13-2), but the q_i remain the same as in (10.13-1). Defining equations for C_o and q_o, and C_i and q_i and values for G and P are presented in Tables 10.13-1 and 10.13-2 for various shapes of solids. These defining equations are valid only when Bi is infinite, which is a good approximation for most continuous leaching systems. Alternative values for these parameters, which are valid when Bi is finite, are available in some cases.[5,6] Extraction yields, $\eta = EY_{out}/RX_o$, are given by the following equations:

$$\alpha \neq 1.0 \qquad \eta = \frac{\alpha[1 - W(t_s)]}{\alpha - W(t_s)} \qquad (10.13\text{-}3)$$

$$\alpha = 1.0 \qquad \eta = \frac{W(t_s)}{1 + W(t_s)} \qquad (10.13\text{-}4)$$

where $W(t_s)$ represents $W(t)$ when $t = t_s$.

TABLE 10.13-1 C_i and q_i for Eq. (10.13-1) When $\alpha \neq 1.0$

Solid Shape	q_i	C_i
Infinite slab	$q_i = \dfrac{\tan(q_i)}{\alpha}$	$\dfrac{2\alpha(\alpha - 1)}{(\alpha q_i)^2 - (\alpha - 1)}$
Infinite cylinder	$q_i = \dfrac{2J_1(q_i)}{\alpha J_o(q_i)}$	$\dfrac{4\alpha(\alpha - 1)}{(\alpha q_i)^2 - 4(\alpha - 1)}$
Sphere	$q_i = \dfrac{(3 - \alpha q_i^2)\tan(q_i)}{3}$	$\dfrac{6\alpha(\alpha - 1)}{(\alpha q_i)^2 - 9(\alpha - 1)}$

TABLE 10.13-2 C_i When $\alpha = 1.0$, G, P, q_o, and C_o

Solid Shape	C_i	G	P	q_o	C_o
Infinite slab	$\dfrac{2}{q_i^2}$	3	$\dfrac{6}{5}$	$q_0 = \dfrac{\tanh(q_o)}{\alpha}$	$\dfrac{2\alpha(1-\alpha)}{(\alpha q_o)^2 - (1-\alpha)}$
Infinite cylinder	$\dfrac{4}{q_i^2}$	8	$\dfrac{4}{3}$	$q_o = \dfrac{2I_1(q_o)}{\alpha I_o(q_o)}$	$\dfrac{4\alpha(1-\alpha)}{(\alpha q_o)^2 - 4(1-\alpha)}$
Sphere	$\dfrac{6}{q_i^2}$	15	$\dfrac{10}{7}$	$q_o = \dfrac{(3 + \alpha q_o^2)\tanh(q_o)}{3}$	$\dfrac{6\alpha(1-\alpha)}{(\alpha q_o)^2 - 9(1-\alpha)}$

Values of $(1 - \eta)$ obtained from Eqs. (10.13-1), (10.13-2), and (10.13-3) are plotted versus τ_s for spherical particles with α as a parameter in Fig. 10.13-1. It can be seen that when $\alpha < 1.0$ the maximum value of η is α. The necessary extractor volume is $t_s R/f_F(1 - \epsilon)$. Typical length to diameter ratios for continuous countercurrent extractors (i.e., tower extractors and slope extractors) range between $2:1$ and $4:1$.

Values of $\overline{X}_t$ and Y_t, the solid and liquid concentrations at positions along the length of the extractor, are provided by the following equations:

$$\alpha \neq 1.0 \qquad \frac{\overline{X}_t}{X_o} = \frac{\alpha W(t) - W(t_s)}{\alpha - W(t_s)} \tag{10.13-5}$$

$$\alpha \neq 1.0 \qquad \frac{Y_t}{MX_o} = \frac{W(t) - W(t_s)}{\alpha - W(t_s)} \tag{10.13-6}$$

$$\alpha = 1.0 \qquad \frac{\overline{X}_t}{X_o} = \frac{1 + W(t_s) - W(t)}{1 + W(t_s)} \tag{10.13-7}$$

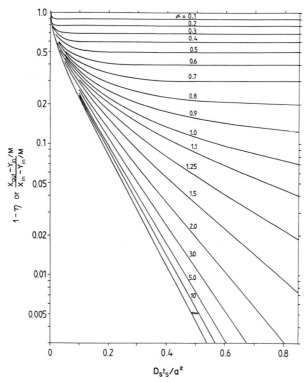

FIGURE 10.13-1 Log $(1 - \eta)$ or log $[(X_{out} - Y_{in}/M)/(X_o - Y_{in}/M)]$ versus $\tau_s = D_s t_s/a^2$ with $\alpha = ME/R$ as a parameter for spherical particles in a continuous countercurrent leaching system.

$$\alpha = 1.0 \qquad \frac{Y_t}{MX_o} = \frac{W(t_s) - W(t)}{1 + W(t_s)} \qquad (10.13\text{-}8)$$

10.13-1 Continuous Extraction Versus Multistage Batch Extraction

Comparisons of yields for mutistage batch and continuous contercurrent leaching systems for equal solids holdup time, that is, for $Nt_c = t_s$, show that η_c is always greater than η_b. However, the difference between η_b and η_c decreases and approaches zero as N increases. When $\alpha > 1.5$ and N is greater than or equal to 3, plots of log $(\eta_c - \eta_b)$ versus log N for equal α and $Nt_c = t_s$ are straight lines. By determining η_c and η_b for $N = 3$ and 4 and using these values to extrapolate to higher values of N on such log–log plots one can obtain η_b for multistage batch systems with large N without having to solve large matrix equations.[7]

10.14 DIFFUSION BATTERIES PERFORMANCE

The performance of diffusion batteries can be characterized in terms of the dimensionless concentration, $S_b = (Y_{out} - MX_o)/(Y_{in,b} - MX_o)$, where $Y_{in,b}$ is Y_{in} for the most spent cell in the battery. $(1 - S_b)$ versus θ curves for cells in diffusion batteries somewhat resemble the $(1 - S)$ versus θ curve shown in Fig. 10.10-1. In $(1 - S)$, Y_{in} is constant, whereas in diffusion batteries, Y_{in} varies with t and $Y_{in,b}$ is usually zero. Ideally, $(1 - S_b)$ versus θ for each cell of the battery should exactly duplicate $(1 - S_b)$ for the next cell, when θ is measured from the start of discharge from the cell in question. After product extract drawoff, the extract discharged from a cell provides the feed to the next cell in the battery. Therefore, after $t = t_c$, Y_{out} for any cell should ideally equal Y_{in} for the same cell t_c earlier. In practice, the $(1 - S_b)$ and Y_{in} and Y_{out} curves exhibit considerable cycle to cycle variation. These variations are probably caused by unstable displacement.

When, as is usually the case, more than four cells are used in a diffusion battery, the performance of the battery on the average tends to resemble that of a continuous countercurrent extractor. Because of this, corrected forms of Eqs. (10.13-3) and (10.13-4) have been used in attempts to predict the performance of diffusion batteries.[1] Finite-difference-based, mass transfer calculations have been used in other attempts[2,3] to predict such performance. While both sets of predictions agree with each other reasonably well, both groups of predicted yields are markedly higher than corresponding experimental yields. On the other hand, solute transfer during infusion tests in diffusion batteries[4] agrees quite well with the transfer predicted by using generalized forms of Eqs. (10.13-3) and (10.13-4). Since displacement instability is suppressed by infusion (when upflow is used), it appears that the poor predictability and impaired mass transfer observed during extraction (with upflow) is due to unstable displacement.

Duhamel's Theorem can be used in conjunction with experimental $(1 - S)$ versus θ correlations for single fixed beds to develop $(1 - S_b)$ versus θ equations which predict the replicating nature of Y_{in} and Y_{out} for the fixed beds in diffusion batteries. These equations may possibly also account for the average effects produced by unstable displacement. If Duhamel's Theorem can be validly applied when such displacement occurs, the equations may also provide an adequate basis for accurately predicting diffusion battery yields.

10.15 DIFFERENTIAL EXTRACTION

FSL solutions, which can be used for differential leaching, can be obtained from analogous heat transfer solutions, which are used to predict solid temperatures in nonadiabatic immersion calorimeters.[1,2] These solutions are similar in form to Eqs. (10.7-3) and (10.13-1), but the dimensionless concentration involved is $(M\overline{X} - Y_{in})/(MX_o - Y_{in})$ and the q_i and C_i depend on both α and an additional dimensionless parameter, $\gamma = EMa^2/D_sR$, where E is the volume flow rate into and out of the extractor, in which the extract volume E remains constant. The q_i and C_i for various solid shapes are defined in Table 10.15-1.

10.16 REFLUX EXTRACTORS

If a deep bed is used in a reflux extractor the extractor can be analyzed like fixed-bed extractors operating with downflow and $Y_{in} = 0$. In such a case $V = E_b/A_c \, \epsilon$. However, if the bed is flooded and consequently part of the reflux bypasses the bed, V should be based on an E_b that barely results in bed saturation rather than the actual E_b. If the refluxed solvent is not passed through a bed but is added to a well-mixed batch of extract and solid and the rate of extract discharge is E_b, the extractor can be treated like a differential extractor. Reflux extractors in which very short beds are used can also be treated like differential extractors as a reasonable approximation.

TABLE 10.15-1 Eigenvalues and Coefficient to Use in FSL Solutions for Differential Extraction

Solid Shape	q_i	C_i
Infinite slab	$q_i = \dfrac{(\gamma - \alpha q_i^2)}{\tan(q_i)}$	$\dfrac{2(\gamma - \alpha q_i^2)^2}{q_i^2[(\gamma - \alpha q_i^2)^2 + q_i^2(\alpha + 1) + \gamma]}$
Infinite cylinder	$q_i = \dfrac{(\gamma - \alpha q_i^2) J_o(q_i)}{J_1(q_i)}$	$\dfrac{4(\gamma - \alpha q_i^2)^2}{q_i^2[(\gamma - \alpha q_i^2)^2 + q_i^2(2\alpha + 1)]}$
Sphere	$q_i = \dfrac{[3(1 - \gamma) + \alpha q_i^2]\tan(q_i)}{3}$	$\dfrac{6(3\gamma - \alpha q_i^2)^2}{q_i^2[(3\gamma - \alpha q_i^2)^2 - 9q_i^2(1 + \alpha) - 9\gamma]}$

10.17 SOLUBILIZATION

Many leaching processes involve "solubilization," the reaction-induced conversion of insoluble precursors into solutes. Examples include protopectin into pectin, collagen into gelatin, coffee bean polysaccharides into low-molecular-weight sugars and oligosaccharides, and malt and brewing grain starches into soluble sugars. Protein dissolution by pH adjustment, for example, the dissolution of keratin during the lime treatment of cattle hides prior to the extraction of gelatin, may also be classified as solubilization. Biological macromolecules must frequently be converted into units with a lower molecular weight to render them soluble or to facilitate diffusion out of the solid matrix in which they are contained. Since the properties of the extracted solute may depend on its molecular weight, this process can be carried too far. Thus, if excessively stringent conditions are used during the extraction of gelatin or pectin, their gelling power will be impaired.

Reactions that convert solutes into insoluble matter also occur during leaching. Coffee solubles are sometimes precipitated during high-temperature coffee extractions; caffeine may be precipitated by forming complexes with tannins. The scalding of sugar beets facilitates sugar extraction by denaturing plasma membranes that hinder diffusion, but it also improves selectivity by coagulating soluble proteins that otherwise would be extracted. In some cases solubilization must be avoided because it creates undesirable products, for example, pectin solubilized during beet sugar or apple juice solutes extraction.

Solubilization rates may control the required duration of leaching processes. Like other reaction rate constants, solubilization rate constants are governed by the Arrhenius equation, and solubilization reactions are greatly speeded up by using higher temperatures. Since such temperatures also accelerate adverse reactions, there frequently is an optimum leaching temperature range. Many solubilization reactions involve hydrolysis. Because water is usually present in large excess, these reactions are frequently pseudo-first-order. Solute degradation reactions that involve the breakdown of macromolecules are also frequently pseudo-first-order. On the other hand, solute precipitation reactions are usually second order. To minimize the extent of such second-order reactions it may be desirable to carry out a leaching using a higher than normal solvent to solid ratio.

If diffusion is very rapid, first-order solubilization reactions can be characterized by the following equation:

$$\ln\left(\frac{Y_{max} - Y}{Y_{max}}\right) = -k_r t \tag{10.17-1}$$

For batch leaching in which both diffusion and first-order solubilization affect the rate of solute release,[1]

$$\frac{Y_{max} - Y}{Y_{max}} = \exp(-k_r t)\left(1 - \sum_{i=1}^{\infty} p_i [1 - \exp(-q_i^2 \tau)]\right) \tag{10.17-2}$$

where $p_i = C_i a^2 k_r / (a^2 k_r - q_i^2 D_s)$, and C_i and q_i are the normal values that apply for batch extraction. Solubilization in continuous and multistage extraction equipment can be analyzed by using superposition, but the infusion of solubles into the solid upstream of the point of solute creation greatly complicates such analysis.

10.18 SOLVENT SELECTION

The Delaney amendment to the Food and Drug Act prohibits the use in food production of materials that exhibit any evidence of carcinogenicity. Therefore, even though great efforts are made to reduce solvent residues in products to extremely low levels, for example, parts per billion, many common solvents, such as benzene, and most chlorinated solvents, with the notable exception of methylene chloride, cannot be used in food processing. On the other hand, for pharmaceuticals, where it is recognized that therapeutic benefits may more than offset solvent-induced risks, wider ranges of solvents can be used. The most commonly used solvents for food processing are water, aqueous solutions of acids and nontoxic salts, commercial hexane, and in some cases other alkanes, ethanol and to a lesser extent the other lower alcohols, methylene chloride, methyl ethyl ketone, and acetone. In certain cases low-molecular-weight aliphatic esters and vegetable oils are also used. When plant pigments have to be extracted it may be necessary to extract the water in the plant material with an alcohol or a ketone, before the pigment can be extracted.

The use of alcohols and alcohol–water mixtures for extracting vegetable oil has attracted attention recently. These solvents can provide greater selectivity than hexane, which is currently used for most vegetable oil extractions. Alcohols and alcohol–water mixtures can also be separated from extracted oil more readily and with less expenditure of energy.

Because it is nontoxic, can be cleanly removed from extracted solutes, and provides good selectivity, there is a great deal of interest in using dense supercritical CO_2 for extracting various food and biological constituents. Because of ease of separation, liquid CO_2 may also be a good solvent for extracting highly volatile flavors and aromas. Although other potential applications have been reported,[1] supercritical CO_2 so far has only been used commercially for extracting caffeine and hop extracts.[2-4] Pressures between 12.5 and 36 MPa gage (120–350 atm) have been used or tested[5,6] in these extractions. Pilot-scale, supercritical CO_2 extractions of vegetable oil[7] have been carried out at pressures between 12.5 and 50 MPa (120–480 atm), and even at pressures as high as 104 MPa (1000 atm).[8] Supercritical hydrocarbons and halocarbons have also been used to extract vegetable oil.[9] These provide much greater extraction capacity at much lower pressures than those used for CO_2-based supercritical extraction. Solutes can be recovered from supercritical gases by either reducing the gas pressure, raising the gas temperature, or contacting the gas with an adsorbent or absorbent.

As is usually the case, it is desirable for solvents to be cheap, noncorrosive, nonflammable, nonexplosive, nontoxic, easily removable, and easily recoverable. In the case of heterogeneous extractions, the external solvent and the occluded solvent must be immiscible. In some cases the solvent should be selective; but when extracts with balanced flavors are desired, selectivity with respect to important flavor components is undesirable. It obviously may be impossible to meet all these objectives.

If possible, flammable solvents should be used at temperatures slightly below their boiling point. This prevents pressure-induced outleakage of solvent vapor; air inleakage yields air–vapor mixtures that are above the upper flammable limit. Automatic flame and explosion arrestor systems should be used whenever highly flammable or explosive solvents are used.

10.19 EQUIPMENT SELECTION

Fixed-bed or crossflow leaching systems should be used whenever the solids to be extracted are very fragile. Fixed-bed systems, including diffusion batteries and crossflow leaching systems, also provide depth filtration and usually yield extracts with good clarity. Particles smaller than 1 mm in diameter are preferably handled in batch mixer–separator systems; if their diameter is smaller than 100 μm, auxiliary separation devices such as centrifuges or rotary vacuum filters should be used. Crossflow percolators, continuous countercurrent extractors (i.e., tower extractors and slope extractors), and trommel extractors all can be advantageously used for reasonably sturdy particles, particularly when very large processing capacity is required. When extraction times in excess of 5000 s are necessary or slow solubilization reactions are involved, batch extractors or diffusion batteries will frequently be the equipment of choice. Diffusion batteries also can be advantageously used when relatively high temperatures and pressures are used or when flow through the solid bed causes large pressure drops. Slurry reactors employing high-pressure slurry feed and discharge pumps or locks may also be used for high-pressure, high-temperature extractions. Batch and crossflow extractors and diffusion batteries are employed when volatile, toxic, flammable, or explosive solvents are used.

10.20 SOLIDS FEED PREPARATION

Solids feed preparation usually plays a crucial role in leaching processes. The use of fine particles usually will speed up diffusive transfer of solutes from the solid to the extract. However, if too fine a size is used, extract may not be able to flow through solid beds as rapidly as desired, flow pressure drop may be excessive and cause undesirable bed compaction, displacement instability will be enhanced, extract cling will increase, solid–liquid separation difficulty may increase, and excessive amounts of intracellular colloidal material may be released. Therefore, the use of a compromise particle size is almost invariably desirable. Except for flaked oilseeds, where flake thicknesses of the order of 0.2 mm are necessary to obtain adequate oil release, particle diameters or thicknesses in the 2–5 mm range usually represent a good choice for industrial scale extractions.

In certain cases, however, the natural or processing-induced form of a material, for example, tea leaves, expeller cakes, and microorganisms, will dictate the use of a size outside the aforementioned range. In such cases it may be necessary to use flocculating agents, pelletizing, filter aids, precoat filters, or centrifuges to facilitate solid–liquid separation or prevent flow difficulties.

Rolling, flaking, shredding, serrating, and maceration may be used in some cases to adequately expose internal pockets of solute or occluded solution.

Size-reduction equipment used to prepare solids for leaching is usually selected so as to provide the desired mean particle size without excessive generation of fines or oversize particles. Therefore, in many instances grinders or cutters employing sharp uniformly spaced knives are used. In certain cases these cutters also provide corrugated particle surfaces that maximize surface exposure to the extract.

The solids moisture content may be adjusted to minimize fines production. When nonaqueous solvents are used, drying a solid before extraction may facilitate extraction. On the other hand, when a heterogeneous extraction is to be carried out and water is the occluded solvent, the solid may have to be steamed or otherwise moistened prior to extraction.

10.21 SPENT SOLIDS TREATMENT

Spent solids are often pressed to recover occluded liquid. The press liquor may be recycled to the leaching system as part of the solvent feed, or it may be concentrated and added back to the pressed solids, which are then dried to produce by-products, such as animal feed. The pressed solids may be combined with fuel and burned. If the pressing removes sufficient water, the solids can provide substantial net fuel value. In some cases spent solids can be chemically hydrolyzed or treated with microorganisms to generate fermentable sugars, single-cell proteins, or other valuable products.

When valuable or hazardous solvents are used, they must be almost completely removed from the spent solids. This is often done by heating the solids on a series of scraped, vapor or steam-heated hearths while superheated steam is passed over them. When the feed value or functional properties of the solids are important, excessively high stripping temperatures should be avoided. However, the temperatures should be high enough to inactivate undesirable enzymes and neutralize nutrient antagonists.

In some cases, for example, spent apple pomace and fermentation mycelia, it is extremely difficult to find uses for spent solids, and they must be disposed of in landfills.

10.22 HYDRODYNAMIC CONSIDERATIONS

The Ergun equation[1] can be used to predict extract-flow pressure drops in beds of solids in leaching systems and in filtration systems used to separate mixed phases discharged from batch extractors. Usually, the viscous term in the equation is dominant; but in regions where flows converge in a bed, the turbulent term may dominate. Particularly large pressure drops can occur near bayonet strainers and other convergence-inducing strainers. Hence, strainers with adequate flow area should be used. Strainer openings should usually be slightly larger than the diameter of the fines in a bed. If the openings are too small, fines will clog the strainer. Fines that pass through strainers are usually trapped in the next bed.

In continuous and fixed-bed extractors there is a critical throughput rate beyond which flooding will occur. The viscosity of extract usually increases as it moves toward the fresh solids end of extraction systems. Hence, in crossflow extractors and stationary-basket extractors, progressive changes in recirculation rates may have to be used to compensate for such changes in viscosity.

In many cases solids progressively soften during extraction. This is particularly true when solids tend to decompose because of solubilization reactions. Flow pressure drop can strongly compact such solids. In such cases, compression permeability measurements should be used to determine the flow resistance characteristics of the solid bed. If flow-induced compaction is excessive it may be necessary to reduce bed depth or use a larger than normal flow cross-sectional area.

Standard, drag-coefficient, drag-force versus gravitational-force balances can be used to predict settling rates when gravitational or centrifugal sedimentation is used to separate solids and liquid discharged from well-mixed extraction stages.

10.23 SOLID–LIQUID HEAT EXCHANGE

Hot solvent is often used to extract cold solids. The solids and extract consequently exchange heat. This exchange can be used to cool down extract immediately prior to discharge, while concurrently heating the solids to temperatures suitable for extraction. The entering solvent can also be partially heated to the required inlet temperature by countercurrent heat exchange with the exiting solids, thereby reducing energy expenditures for leaching. Usually, the entering solvent must be heated to some extent to provide the required inlet temperature; intermediate heat exchangers or heating jackets may be used to modify the liquid and solid temperatures in the leaching system.

The partial differential equations and boundary and initial conditions characterizing heat exchange between solids and extract are formally analogous to FSL and mass transfer boundary conditions and initial conditions. Consequently, analogous differential equation solutions can often be used to predict extract and solid temperatures in leaching systems. These analogues are obtained by substituting dimensionless temperature for dimensionless concentrations and dimensionless heat transfer parameters for corresponding dimensionless mass transfer parameters. However, Bi_h, the heat transfer Biot number, is almost invariably much smaller than Bi, the mass transfer Biot number. As a consequence, the heat transfer analogues must be based on FSL solutions which apply when Bi is small or moderately small.

Finite Bi_h heat transfer solutions corresponding to the infinite Bi mass transfer solutions listed for continuous countercurrent leaching are available. In these solutions α_h corresponds to α, λ to τ, T_E to Y or Y/M, and T_R to X at corresponding locations in the leaching system. When either Bi or Bi_h is finite the equations used to predict q_i and C_i are more complex than those listed in Tables 10.8-1, 10.13-1, and 10.13-2. For example, for spherical particles when $\alpha_h > 1.0$ (the usual case) and Bi_h is finite,

$$q_i = \frac{[3Bi_h + \alpha_h(1 - Bi_h) \, q_i^2] \, \tan{(q_i)}}{3 \, Bi_h + \alpha_h q_i^2} \qquad (10.23\text{-}1)$$

$$C_i = \frac{6 \, Bi_h^2 \alpha_h(\alpha_h - 1)}{(\alpha_h q_i)^2 \, [Bi_h^2 - Bi_h + q_i^2] - 9Bi_h^2(\alpha_h - 1) - 6Bi_h\alpha_h q_i^2} \qquad (10.23\text{-}2)$$

Appropriate values of the surface heat transfer coefficient h, which can be used to determine Bi_h, can be obtained from the heat transfer analogues of Eqs. (10.10-5) and (10.10-6) in which Pr is substituted for Sc, and $V\rho c_E$ is substituted for V. λ will usually be much larger than τ; consequently, heat transfer equilibrium will be approached very closely during batch extraction. Consequently, heat transfer analogues of the Kremser–Brown–Souders equations can be used to predict stage temperatures during multistage countercurrent batch extraction. Furthermore, if flow maldistribution and axial dispersion can be neglected, a heat transfer analogue of Rosen's large τ_f solution can be used to predict extract and solid temperatures for extractions involving fixed beds. In this analogue β_h replaces β, $(T_E)_{out}$ replaces Y_{out}, $(T_E)_{in}$ replaces Y_{in}, $(T_R)_o$ replaces MX_o, and λ_f replaces τ_f.

NOTATION

a	particle radius for spheres and cylinders, half-thickness for slabs (m)
A	coefficient in Eq. (10.8-1)
A_c	cross-sectional area of bed (m²)
A_j	element in jth row of first column of matrix Eq. (10.12-1)
b	mass of bound solution per mass of marc (kg/kg)
$\bar{b}$	average value of b (kg/kg)
B	coefficient in Eq. (10.8-1)
Bi	mass transfer Biot number kMa/D_s
Bi_h	heat transfer Biot number, ha/k_t
B_j	element of jth row of second column of matrix Eq. (10.12-1)
c	(volume of cling/volume of extract) in stage
c_E	extract heat capacity (J/kg·K)
c_R	solid heat capacity (J/kg·K)
C	coefficient, Eq. (10.8-1)
C_i	coefficient of ith term in series solutions to FSL
C_j	element of jth row of right-hand side of matrix Eq. (10.12-1)
C_o	coefficient which occurs in Eq. (10.13-1) when $\alpha < 1$ for continuous countercurrent leaching
D	coefficient, Eq. (10.8-1)
D_A	axial dispersion coefficient for extract (m²/s)
$(D_A)_R$	axial dispersion coefficient for solid (m²/s)
D_L	solute diffusivity in occluded solution (m²/s)
D_s	solute diffusivity in solid (m²/s)
E	extract volume (m³)
E	extract volume flow rate (m³/s)
E'	extract mass (kg)
E'	extract mass flow rate (kg/s)

E_b solvent boilup and reflux rate (m³/s)

E_o' extract mass leaving imbibition stage (kg)

$\mathbf{E}_o'$ extract flow rate leaving imbibition stage (kg/s)

f fraction of extract passing through bed which is recycled to same inlet position at top of bed

f_F fraction of continuous extractor volume occupied by moving solid bed

f_R $(R' - F')/R'$

F coefficient, Eq. (10.8-2)

F term defined by Eq. (10.10-2)

F' mass of dry solid feed (kg)

g gravitational acceleration, 9.806 m/s²

G coefficients, Eqs. (10.8-2) and (10.13-3)

h heat transfer coefficient (W/m²·K)

H coefficient, Eq. (10.8-2)

I_o modified Bessel function of order 0

I_1 modified Bessel function of order 1

J coefficient, Eq. (10.8-3)

J_o Bessel function of order zero

J_1 Bessel function of order one

J_s mass flux of solute (kg/m²·s)

k solute mass transfer coefficient in extract (m/s)

k_r reaction rate constant (s⁻¹) [Eqs. (10.17-1) and (10.17-2)]

k_t thermal conductivity of solid (W/m·K)

K coefficient, Eq. (10.8-3)

K bed permeability (m²)

L coefficient, Eq. (10.8-3)

L depth of bed (m)

m solute equilibrium distribution coefficient, $(y/x)^*$

M solute equilibrium distribution coefficient, $(Y/X)^*$

M' solute distribution coefficient, $(Y/Y_{int})^*$

MW   molecular weight of solute

MWC   molecular weight cutoff

p_i $C_i k_r a^2/(k_r a^2 - q_i^2 D_s)$ in Eq. (10.17-2)

N total number of stages or fixed beds

P constant in Eq. (10.13-2)

Pe   Peclet number, Va/D_A

Pe′   bed-length-based Peclet number, VL/D_A

Pr   Prandtl number, $c_E \mu/k_t$

q_i ith eigenvalue in series solutions to FSL

q_o eigenvalue which occurs in Eq. (10.13-1) when $\alpha < 1$ for continuous countercurrent leaching

Q_n $(Y_n - Y_{N+1})/(MX_0 - Y_{N+1})$

r distance from center, axis, or midplane of solid (m)

R solid volume (m³)

$\mathbf{R}$ solid volume flow rate (m³/s)

R' solid mass (kg)

$\mathbf{R}'$ solid mass flow rate (kg/s)

Re$_p$ particle Reynolds number, $2a\rho V/\mu$

S $(Y_{out} - MX_o)/(Y_{in} - MX_o)$ at θ for a single fixed bed for which Y_{in} is constant

S_b $(Y_{out} - MX_o)/(Y_{in,b} - MX_o)$ at θ for a bed in a diffusion battery

S_o S at $\theta = 0$

Sc   Schmidt number, $\mu/D_L\rho$

t extraction time or solids holdup time (s)

t_c batch extraction time, cycle time, time between an operation in a cell in a diffusion battery and the same operation in the next cell in the battery (s)

t_f extract transit time through bed, L/V or z/V (s)

t_s total holdup time for solids in continuous extraction system, Nt_c in multistage system (s)

T temperature (K)

T_E extract temperature (K)

T_R solid temperature (K)

U_n S/θ_c integrated with respect to θ between $(n-1)\,\theta_c$ and $n\theta_c$

ΔU_n $U_n - U_{n+1}$

v_E specific volume of extract (m³/kg)

v_R specific volume of solid (m³/kg)

$\mathbf{v}_I$ partial volume of marc (m³/kg)

$\mathbf{v}_S$ partial volume of solute (m³/kg)

$\mathbf{v}_W$ partial volume of solvent (m³/kg)

V fluid velocity in bed (m/s)

V_R solid velocity in bed (m/s)

$W(t)$ variable defined by Eq. (10.13-1)

$W(t_s)$ $W(t)$ evaluated at $t = t_s$

x weight fraction of solute in saturated solid (kg/kg)

X solute concentration in saturated solid (kg/m³)

$\overline{X}$ mean value of X for solid (kg/m³)

X_o initial value of X (kg/m³)

$\overline{X}_{\text{out}}$ $\overline{X}$ for solids leaving continuous countercurrent extractor (kg/m³)

$\overline{X}_n$ $\overline{X}$ for solid leaving nth stage (kg/m³)

$\overline{X}_N$ $\overline{X}$ for solid leaving last stage (kg/m³)

$\overline{X}_t$ $\overline{X}$ for solid at contacting time t in continuous extractor

$\mathbf{X}$ $(\overline{X} - X^*)/(X_o - X^*)$

$\mathbf{X}_n$ $\mathbf{X}$ evaluated at $t = nt_c$

$\mathbf{X}'$ $(\overline{X} - Y/M)/(X_o - Y/M)$

X^* equilibrium value of X (kg/m³)

y solute weight fraction in extract (kg/kg)

y_o y for extract leaving imbibition stage (kg/kg)

Y solute concentration in extract (kg/m³)

Y_1 Y for extract leaving stage 1 (kg/m³)

Y_{in} Y for extract entering fixed-bed or continuous countercurrent extractor (kg/m³)

$Y_{\text{in},b}$ Y_{in} for the most spent cell in a diffusion battery (kg/m³)

Y_{N+1} Y for extract or solvent fed into a multistage countercurrent extraction system (kg/m³)

Y_n Y for extract leaving stage n (kg/m³)

Y_o initial value of Y (kg/m³)

$\mathbf{Y}$ $(Y - Y^*)/(Y_o - Y^*)$

Y_{int} Y for occluded solution in solid (kg/m³)

Y_{max} maximum Y achievable by solubilization at fixed E/R (kg/m³)

Y_{out} Y for extract leaving fixed bed of solids, Y for extract leaving continuous extractor (kg/m³)

Y_{sat} the solubility limit of a solute (kg/m³)

Y^* equilibrium value of Y (kg/m³)

z distance from liquid inlet end of a bed (m)

Z_I weight fraction of marc in solid (kg/kg)

$(Z_I)_{\text{sat}}$ value of Z_I along saturation line (kg/kg)

Z_S weight fraction of solute in solid feed (kg/kg)

Z_W weight fraction of solvent in solid (kg/kg)

$(Z_W)_{\text{sat}}$ value of Z_W along saturation line (kg/kg)

Greek Letters

α stripping factor, EM/R, $\mathbf{EM/R}$, $E'm/R'$, or $\mathbf{E}'m/\mathbf{R}'$; note: $EM/R = E'm/R'$ and $\mathbf{EM/R} = \mathbf{E}'m/\mathbf{R}'$

α_h $E'c_E/R'c_R$ or $\mathbf{E}'c_E/\mathbf{R}'c_R$

β	$M\epsilon/(1 - \epsilon)$
β_h	$\epsilon v_R c_E/(1 - \epsilon)v_E c_R$
γ	$EMa^2/D_s R$, parameter used in differential extraction
ϵ	interparticle porosity
η	extraction yield, EY_1/RX_o, $\mathbf{E}Y_1/\mathbf{R}X_o$, $E'y_1/R'x_o$, $\mathbf{E}'y_1/\mathbf{R}'x_o$ or $E'y_o/FZ_S$ when Y_{N+1} or $Y_{in} = 0$; and $(X_o - X_N)/(X_o - Y_{N+1}/M)$ or $(X_o - X_{out})/(X_o - Y_{in}/M)$ when Y_{N+1} or $Y_{in} \neq 0$
η_b	η for multistage batch countercurrent system when $Nt_c = t_s$
η_c	η for continuous countercurrent extractor when $t_s = Nt_c$
θ	$(t - t_f)/t_f$
θ_c	t_c/t_f
λ	Fourier number, $k_t v_R/c_R a^2$
λ_f	$k_t f v_R/c_R a^2$
μ	extract viscosity (kg/m·s)
ν	geometric index, 1 for slabs, 2 for cylinders, 3 for spheres
ρ	extract density (kg/m³)
σ	standard deviation of exit concentration versus t curve for dye pulse
τ	Fick number, $D_s t/a^2$
τ_f	$D_s t_f/a^2$
τ_s	$D_s t_s/a^2$
ϕ	τ/α
ψ	fraction of solid volume occupied by occluded solution

Subscripts

i	ith term in multiterm solution to DPDE, solute i
in	entering fixed bed
in, b	for extract entering diffusion battery
out	leaving fixed-bed or continuous countercurrent extractor
j	jth row in matrix Eq. (10.12-1), solute j
n	nth stage
N	last stage
o	original
r	reference solute
1	entering extract, first stage
2	displaced extract

REFERENCES

Section 10.3

10.3-1 E. Seltzer and F. A. Saporito, "Method of Producing a Tea Extract," U.S. Patent 2,902,368 (Sept. 1, 1959).

10.3-2 W. M. Barger, "Deep Bed Extraction and Desolventizing Systems for Oilseeds," paper presented at AIChE National Meeting, Denver CO, Aug. 28–31, 1983.

10.3-3 R. A. McGinnis, *Beet Sugar Technology*, 3rd ed., pp. 119–153, Beet Sugar Development Foundation, Fort Collins, CO, 1982.

10.3-4 M. Sivetz and N. Desrosier, *Coffee Technology*, pp. 317–372, AVI Publishing, Westport, CT, 1979.

10.3-5 H. G. Schwartzberg, A. Torres, and S. Zaman, Mass-Transfer in Solid–Liquid Extraction Batteries, in H. G. Schwartzberg, D. Lund, and J. Bomben (Eds.), *Food Process Engineering*, AIChE Symposium Series 218, Vol. 78, pp. 90–100, AIChE, New York, 1982.

10.3-6 P. M. Silin, *Technology of Beet Sugar Refining and Production*, Pischepromizdat, Moscow, translation Israel Program for Scientific Translation, Jerusalem, 1964 (original publication, 1958, pp. 114–154).

10.3-7 De Smet, *Solvent Extraction*, Bulletin, Extraction De Smet S. A., Edegem, Antwerp, Belgium (undated).

10.3-8   P. Possman, Green Light for Extraction in Germany, *Flussiges Obst*, **9**, 534–540 (1982).

10.3-9   Vatron-Mau & Cie, 6 Modeles de Diffuseurs Continus, advertisement, *Ind. Aliment. Agric.*, **95**, 1340 (1978).

10.3-10 Lurgi, Continuous Solvent Extraction of Oilseeds and Lecithin Production, *Lurgi Express Information T 1136/11.77*, Lurgi Umwelt und Chemotechnik GmbH, Frankfurt, Federal Republic of Germany, 1977.

10.3-11 J. R. Fitzgerald, G. E. Salt, and A. Van Hengle, The FS Diffuser (Van Hengle System), *Int. Sugar J.*, **80**(949), 3–9 (1978).

10.3-12 E. Hugot, *Handbook of Sugar Cane Engineering*, 2nd ed., pp. 386–388, Elsevier, Amsterdam, 1972.

10.3-13 Extraktionstechnik, *Continuous Equipment for Processing Natural Drugs and Natural Raw Materials for Flavors and Fragrances*, Bulletin, Extraktionstechnik Gesellschaft für Anlagenbau m.b.H., Hamburg, 1979.

10.3-14 C. Iverson and H. G. Schwartzberg, Developments in Beet and Sugar Cane Extraction, *Food Technol.*, **38**(1), 40–44 (1984).

10.3-15 H. G. Schwartzberg, Continuous Countercurrent Extraction in the Food Industry, *Chem. Eng. Prog.*, **76**(4), 67–85 (1980).

10.3-16 W. M. Barger, "Deep Bed Extraction and Desolventizing Systems for Oilseeds," paper presented at the AIChE National Meeting, Denver, CO, Aug. 28–31, 1983.

10.3-17 G. V. Genie, Evolution et Progres des Diffuseurs Continus Rotatifs, *Ind. Aliment. Agri.*, **96**, 763–770 (1979).

10.3-18 C. Pinet, Le Nouveau Diffuseur Continu R.T.4, *Sucr. Belge*, **90**, 111–121 (1971).

10.3-19 H. Bruniche-Olsen, *Solid–Liquid Extraction*, pp. 423–449, NYT Nordisk Forlag, Arnold Busck, Copenhagen, 1962.

10.3-20 R. A. McGinnis, *Beet Sugar Technology*, 3rd ed., pp. 119–153, Beet Sugar Development Foundation, Fort Collins, CO, 1982.

10.3-21 Amos, Amos Extrakteur ET 3035/3055/3010, advertisement, *Flussiges Obst*, **9**, 525 (1982).

10.3-22 H. G. Schwartzberg, Continuous Countercurrent Extraction in the Food Industry, *Chem. Eng. Prog.*, **76**(4), 67–85 (1980).

10.3-23 Greerco, *Counter-Current Contactor*, Bulletin G-102, Greerco Corp, Hudson, NH, undated.

10.3-24 BMA, *BMA Extraction Plants for the Beet Sugar Industry*, Information Leaflet B00000210e, Braunschweigische Maschinenbauanstalt, Braunschweig, Federal Republic of Germany, 1976.

10.3-25 Buckau-Wolf, *Continuous Counter-Current Extraction for Sugar Beet*, Bulletin, Maschinenfabrik Buckau R. Wolf, Grevenbroich, Federal Republic of Germany, 1980.

10.3-26 J. Trimbosch (Manager, Sukkerunie Plant, Roosendaal, The Netherlands), Flow Reversal in Tower Extractors," personal communication, 1980.

10.3-27 G. Karnofsky, The Mechanics of Solvent Extraction, *J. Am. Oil Chem. Soc.*, **26**, 570–574 (1949).

10.3-28 Autorenkollektiv, *Die Zuckerherstellung*, pp. 164–221, VEB Fachbuchverlag, Leipzig, 1980.

10.3-29 P. M. Silin, *Technology of Beet Sugar Refining and Production*, Pischepromizdat, Moscow, translation Israel Program for Scientific Translation, Jerusalem, 1964; original publication, pp. 169–171, 1958.

10.3-30 R. A. McGinnis, *Beet Sugar Technology*, 3rd ed., pp. 119–153, Beet Sugar Development Foundation, Fort Collins, CO, 1982.

10.3-31 F. Schneider, *Technologie des Zuckers*, pp. 175–258, Verlag M. and H. Schaper, Hannover, Federal Republic of Germany, 1968.

10.3-32 Crown, *Crown Solvent Extraction System*, Bulletin, Crown Iron Works, Co., Minneapolis, MN, undated.

10.3-33 SUCATLAN, *Saturne Diffuser*, Bulletin, Societe Sucriere de l'Atlantique, Paris, 1977.

10.3-34 W. Becker, Solvent Extraction of Soybeans, *J. Am. Oil Chem. Soc.*, **55**, 754–761 (1978).

10.3-35 L. E. Garono, F. E. Kramer, and A. E. Steigmann, "Gelatin Extraction Process," U.S. Patent 2,743,265 (Apr. 24, 1956).

10.3-36 G. Karnofsky, The Mechanics of Solvent Extraction, *J. Am. Oil Chem. Soc.*, **26**, 570–574 (1949).

Section 10.5

10.5-1   J. C. Elgin, "Graphical Calculation of Leaching Operations: I. The Carrier Solid Is Insoluble", paper presented at the AIChE National Meeting, Baltimore, Nov. 11–13, 1936, pp. 1–17.

10.5-2   R. E. Treybal, *Mass Transfer Operations*, 3rd ed., pp. 717–765, McGraw-Hill, New York, 1980.

10.5-3   W. Kehse, Solid–Liquid Extraction in the Carousel Extractor, *Chem. Ztg. Chem. Appar. Verfahrenstechnik*, **94**, 56–62 (1970).

Section 10.6

10.6-1   R. E. Treybal, *Mass Transfer Operations*, 3rd ed., pp. 126–130, McGraw-Hill, New York, 1980.

Section 10.7

10.7-1   J. Spaninks, "Design Procedures for Solid–Liquid Extractors," Ph.D. thesis, Agricultural University of the Netherlands, Wageningen, 1979.

10.7-2   J. Crank, *The Mathematics of Diffusion*, Oxford University Press, London, 1970.

10.7-3   G. Karnofsky, The Theory of Solvent Extraction, *J. Am. Oil Chem. Soc.*, **26**, 561–569 (1949).

Section 10.8

10.8-1   J. Crank, *The Mathematics of Diffusion*, pp. 52–56, 70–73, 88–91, Oxford University Press, London, 1970.

10.8-2   P. C. Carman and R. A. Haul, Measurement of Diffusion Coefficients, *Proc. R. Soc. London*, **222-A**, 109–118 (1954).

Section 10.9

10.9-1   H. G. Schwartzberg and R. Y. Chao, Solute Diffusivities in Leaching Processes, *Food Technol.*, **36**(2), 73–86 (1982).

10.9-2   E. Holtz, I. Olson, and H. G. Schwartzberg, "Selective Infusion," paper presented at the IFT National Meeting, Las Vegas, NV, June 22–25, 1982.

10.9-3   C. P. Bean, The Physics of Porous Membranes—Neutral Pores, in G. Eisenman (Ed.), *Membranes*, Vol. 1, *Macroscopic Systems and Models*, pp. 1–54, Marcel Dekker, New York, 1972.

10.9-4   N. Lakshminarayanaiah, *Transport Phenomena in Membranes*, pp. 301–303, Academic Press, New York, 1969.

10.9-5   J. Fox, Diffusion of Chloride, Nitrite, and Nitrate in Beef and Pork, *J. Food Sci.*, **45**, 1740–1744 (1980).

10.9-6   T. J. Geurts, P. Walstra, and H. Mulder, Transport of Salt and Water During Salting of Cheeses. Analysis of the Processes Involved, *Netherlands Milk Dairy J.*, **28**, 102–109 (1974).

10.9-7   T. J. Geurts, P. Walstra, and H. Mulder, Transport of Salt and Water During Salting of Cheeses. 2. Quantities of Salt Taken Up and Water Lost, *Netherlands Milk Dairy J.*, **34**, 229–254 (1980).

10.9.8   L. Friedman and E. O. Kraemer, The Structure of Gelatin Gels from Studies of Diffusion'', *J. Am. Chem. Soc.* **52**, 1295–1304 (1930).

10.9-9   L. Friedman and E. O. Kraemer, Diffusion of Non-Electrolytes in Gelatin Gels, *J. Am. Chem. Soc.*, **52**, 1305–1310 (1930).

10.9-10   L. Friedman and E. O. Kraemer, Structure of Agar Gels from Studies of Diffusion, *J. Am. Chem. Soc.*, **52**, 1311–1314 (1930).

10.9-11   J. A. Bressan, P. A. Carroad, R. L. Merson, and W. L. Dunkley, Temperature Dependence of Effective Diffusion Coefficients for Total Solids During Washing of Cheese Curd, *J. Food Sci.*, **46**, 1958–1959 (1981).

10.9-12   J. A. Bressan, P. A. Carroad, R. L. Merson, and W. L. Dunkley, Modelling of Isothermal Diffusion of Whey Components from Small Curd Cottage Cheese During Washing, *J. Food Sci.*, **47**, 84–88 (1982).

10.9-13   J. O. Osborn and D. L. Katz, Structure as a Variable in the Application of Diffusion Theory to Extraction, *Trans. AIChE*, **40**, 511–531 (1944).

10.9-14   H. Bruniche-Olsen, *Solid–Liquid Extraction*, pp. 163–173, NYT Nordisk Forlag, Arnold Busck, Copenhagen, 1962.

10.9-15   C. J. Geankoplis, E. A. Grulke, and M. R. Okos, Diffusion and Interaction of Sodium Caprylate in Bovine Serum Albumin Solutions, *Ind. Eng. Chem. Fundam.*, **18**, 233–237 (1979).

10.9-16   H. G. Schwartzberg and R. Y. Chao, Solute Diffusivities in Leaching Processes, *Food Technol.*, **36**(2), 73–86 (1982).

10.9-17   J. T. Van Bruggen and W. R. Galey, Solute Interactions and Asymmetric Solute Transfer, in *Mass Transfer in Biological Systems*, AIChE Symposium Series 99, Vol. 66, pp. 53–58, AIChE, New York, 1970.

10.9-18   E. L. Cussler, *Multicomponent Diffusion*, Elsevier, Amsterdam, 1976.

10.9-19   O. Kedem and A. Katchalsky, A Physical Interpretation of the Phenomenological Coefficients of Membrane Permeability, *J. Gen. Physiol.*, **45**, 143–179 (1961).

Section 10.10

10.10-1   J. B. Rosen, Kinetics of a Fixed Bed System for Diffusion into Spherical Particles, *J. Chem. Phys.*, **20**, 387–394 (1952).

10.10-2   R. E. Babcock, D. W. Green, and R. H. Perry, Longitudinal Dispersion Mechanisms in Packed Beds, *AIChE J.*, **12**, 922–926 (1966).

10.10-3   I. Neretnieks, A Simplified Theoretical Comparison of Periodic and Countercurrent Adsorption, *Chem. Ing. Techn.*, **47**, 773 (1975).

10.10-4   A. Rasmussen and I. Neretnieks, Exact Solution of a Model for Diffusion in Particles and Longitudinal Dispersion in Packed Beds, *AIChE J.*, **26**, 686–690 (1980).

10.10-5   J. B. Rosen, General Numerical Solution for Solid Diffusion in Fixed Beds, *Ind. Eng. Chem.*, **46**, 1590–1594 (1954).

10.10-6   H. G. Schwartzberg, S. Flores, and S. Zaman, "Analysis of Multi-Stage Belt Extractors," paper presented at the AIChE National Meeting, Denver CO, Aug. 28–31, 1983.

10.10-7   E. J. Wilson and C. J. Geankoplis, Mass Transfer at Very Low Reynolds Numbers in Packed Beds, *Ind. Eng. Chem. Fundam.*, **5**, 9–14 (1966).

Section 10.11

10.11-1   D. Schliephake and A. Wolf, Die Verweilzeitverteilung von Zuckerruben Schnitzeln in Einer Technischen Extraktions Anlage, *Zucker*, **22**, 493–497 (1969).

10.11-2   D. Schliephake, B. Mechias, and E. Reinefeld, Verweilzeitverteilung und Bewegungsmechanismen in Extraktionsanlagen ohne Schnitzelzwangsfuhrung, in *Proceedings of the 15th Assembly International Technical Commission of Sugar Industry*, Vienna, International Technical Commission of Sugar Industry, Tienen, Belgium, 1976, pp. 31–50.

10.11-3   T. Cronewitz, G. Muller, and H. Schiwek, La Determination de la Repartition des Temps de Sejour des Cossettes et du Jus au Cours de l'extraction et au Cours d'autres Processes Techniques de Sucre, *Ind. Alim. Agric.*, **94**, 771–777 (1978).

10.11-4   H. Bruniche-Olsen, *Solid–Liquid Extraction*, pp. 294–343, NYT Nordisk Forlag, Arnold Busck, Copenhagen, 1962.

10.11-5   J. M. Dumore, Stability Considerations in Downward Miscible Displacement, *Soc. Pet. Eng. AIME Pap.*, **4**(4), 356–362 (Dec. 1964).

10.11-6   M. A. Hill, Channeling in Packed Columns, *Chem. Eng. Sci.*, **1**, 247–253 (1952).

10.11-7   O. Levenspiel and K. J. Bischoff, Patterns of Flow in Chemical Process Vessels, in T. B. Drew, J. W. Hoopes and T. Vermeulen (Eds.), *Advances in Chemical Engineering*, Vol. 4, pp. 95–198, Academic Press, New York, 1963.

10.11-8   S. Zaman, *Single Column Leaching and Axial Dispersion Tests*, Internal Report, Department of Food Engineering, University of Massachusetts, Amherst, MA, 1980.

Section 10.12

10.12-1   H. S. Carslaw and J. C. Jaeger, *Conduction of Heat in Solids*, pp. 30–33, Oxford University Press, London, 1959.

10.12-2   G. E. Meyers, *Analytical Methods in Conduction Heat Transfer*, pp. 151–166, McGraw-Hill, New York, 1971.

10.12-3   H. G. Schwartzberg, Applications of Superposition in Solid–Liquid Extraction," paper presented at the IFT National Meeting, Las Vegas, NV, June 22–25, 1982.

10.12-4   M. Desai, "Analysis of Multistage Nonequilibrium Countercurrent Leaching System," Ph.D. thesis, University of Massachusetts, Amherst, 1977.

10.12-5   H. G. Schwartzberg, S. Flores, and S. Zaman, "Analysis of Multi-Stage Belt Extractors," paper presented at the AIChE National Meeting, Denver, CO, Aug. 28–31, 1983.

Section 10.13

10.13-1   F. P. Plachco and J. H. Krasuk, Solid–Liquid Countercurrent Extractors, *Ind. Eng. Chem. Process Des. Dev.*, **9**, 419–432 (1970).

10.13-2 P. R. Kasten and N. P. Amundson, Analytical Solutions for Simple Systems in Moving Bed Absorbers, *Ind. Eng. Chem.*, **44**(7), 1704–1711 (1952).

10.13-3 K. Tettamanti, J. Manczinger, J. Hunek, and R. Stomfai, Calculation of Countercurrent Solid–Liquid Extraction, *Acta Chim. Acad. Sci. Hung.*, 27–45 (1975).

10.13-4 J. Spaninks, "Design Procedures for Solid–Liquid Extractors," Ph.D. thesis, Agricultural University of the Netherlands, Wageningen, 1979.

10.13-5 H. G. Schwartzberg, "Progress and Problems in Solid–Liquid Extraction," plenary lecture, First Latin American Congress on Heat and Mass Transfer, La Plata Argentina, Nov. 1–4, 1982, *Latin Am. Rev. Heat Mass Trans.*, in press.

10.13-6 J. Spaninks, "Design Procedures for Solid–Liquid Extractors," Ph.D. thesis, Agricultural University of the Netherlands, Wageningen, 1979.

10.13-7 M. Desai and H. Schwartzberg, Mathematical Modelling in Leaching Processes, in Linko et al. (Eds.), *Food Process Engineering*, pp. 86–91, Applied Science Press, London, 1980.

Section 10.14

10.14-1 H. G. Schwartzberg, A. Torres, and S. Zaman, Mass-Transfer in Solid–Liquid Extraction Batteries, in H. G. Schwartzberg, D. Lund, and J. Bomben (Eds.), *Food Process Engineering*, AIChE Symposium Series 218, Vol. 78, pp. 90–100, AIChE, New York, 1982.

10.14-2 J. Spaninks, "Design Procedures for Solid–Liquid Extractors," Ph.D. thesis, Agricultural University of the Netherlands, Wageningen, 1979.

10.14-3 J. Spaninks and S. Bruin, Mathematical Simulation of Solid–Liquid Extractors—I. Diffusion Batteries, *Chem. Eng. Sci.*, **34**, 199–205 (1979).

10.14-4 E. Holtz, I. Olson, and H. G. Schwartzberg, "Selective Infusion," paper presented at the IFT National Meeting, Las Vegas, NV, June 22–25, 1982.

Section 10.15

10.15-1 H. S. Carslaw and J. C. Jaeger, *Conduction of Heat in Solids*, pp. 128–130, 240–241, Oxford University Press, London, 1959.

10.15-2 J. W. Hemmings and J. Kern, The Non-Adiabatic Calorimeter Problem and Its Application to Transfer Processes in Suspensions of Solids, *Int. J. Heat Mass Trans.*, **22**, 99–109 (1979).

Section 10.17

10.17-1 H. G. Schwartzberg, Mathematical Analysis of Solubilization Kinetics and Diffusion in Foods, *J. Food Sci.*, **40**, 211–213 (1975).

Section 10.18

10.18-1 P. Hubert and O. Vitzhum, Fluid Extraction of Hops, Spices, and Tobacco with Supercritical Gases, *Angew. Chem. Int. Ed. Engl.*, **17**, 710–715 (1978).

10.18-2 H. Kurzhals, "Caffeine Extraction," paper presented at the Society of Chemical Industry Food Engineering Panel Symposium, CO_2 in Solvent Extraction, London, Feb. 4, 1982.

10.18-3 R. Vollbrecht, "Industrial Scale Hops Extraction," paper presented at the Society of Chemical Industry Food Engineering Panel Symposium, CO_2 in Solvent Extraction, London, Feb. 4, 1982.

10.18-4 D. S. J. Gardner, "Industrial Scale Hops Extraction," paper presented at the Society of Chemical Industry Food Engineering Panel Symposium, CO_2 in Solvent Extraction, London, Feb. 4, 1982.

10.18-5 W. Roselius, O. Vitzhum, and P. Hubert, "Method for the Production of Caffeine-Free Coffee Extract," U.S. Patent 3,843,824 (Oct. 22, 1974).

10.18-6 K. Zosel, "Process for Recovering Caffeine," U.S. Patent 3,806,619 (Apr. 23, 1974).

10.18-7 J. M. Moses and R. D. deFilippi, "Critical Fluid Extraction of Vegetable Oil from Oilseeds," Paper 39d, presented at the AIChE National Meeting, Denver CO, Aug. 1983.

10.18-8 J. P. Friedrich, G. R. List, and A. J. Heakins, Petroleum-Free Extraction of Oil from Soybeans with Supercritical CO_2, *J. Am. Oil Chem. Soc.*, **59**, 288–292 (1982).

10.18-9 J. M. Moses and R. D. deFilippi, "Critical Fluid Extraction of Vegetable Oil from Oilseeds," Paper 39d, presented at the AIChE National Meeting, Denver, CO, Aug. 1983.

Section 10.22

10.22-1 S. Ergun, Flow Through Packed Columns, *Chem. Eng. Prog.* **48**, 89–94 (1952).

CHAPTER 11

Crystallization Operations

CHARLES G. MOYERS, JR.
Union Carbide Corporation
South Charleston, West Virginia

RONALD W. ROUSSEAU
School of Chemical Engineering
Georgia Institute of Technology
Atlanta, Georgia

11.1 INTRODUCTORY COMMENTS

Crystallization is employed heavily as a separation process in the inorganic chemical industry, particularly where salts are recovered from aqueous media. In production of organic chemicals, crystallization is also used to recover product, to refine intermediate chemicals and to remove undesired salts. The feed to a crystallization system consists of a solution from which solute is crystallized (or precipitated) via one or more of a variety of processes. The solids are normally separated from the crystallizer liquid, washed, and discharged to downstream equipment for additional treatment. High recovery of refined solute is generally the desired design objective, although sometimes the crystalline product is a residue.

The process of forming a solid phase from solution is termed *crystallization*, and the operation occurs in a vessel called a *crystallizer*. A crystallizer provides residence time for the process streams to approach equilibrium, possibly a capability of selectively removing fines or coarse product, a mixing or contacting regime to give uniform crystal growth, and may include provision for addition or removal of heat.

Crystallization is distinguished from other unit operations in that a solid phase is generated. The solid phase is characterized in part by its inherent shape (habit) and size distribution. The natural habit of the solid phase is important since it influences product purity, yield, and capacity of the crystallizer system.

11.1-1 Advantages and Disadvantages

Crystallization offers the following advantages:

1. Pure product (solute) can be recovered in one separation stage. With care in design, product purity greater than 99.0% can be attained in a single stage of crystallization, separation, and washing.
2. A solid phase is formed that is subdivided into discrete particles. Generally, conditions are controlled so that the crystals have the desired physical form for direct packaging and sale.

The major disadvantages of crystallization are:

1. Purification of more than one component is not normally attainable in one stage.

2. The phase behavior of crystallizing systems prohibits full solute recovery in one stage; thus, the use of additional equipment to remove solute completely from the remaining crystallizer solution is necessary.

Since crystallization involves processing and handling of a solid phase, the operation is normally applied when no alternative separation technique is discernible. The choice of crystallization over, say, distillation as the preferred separation technique may hinge on one or more of the following considerations:

1. Solute is heat sensitive and/or a high boiler and decomposes at temperatures required to conduct distillation.
2. Low or nil relative volatility exists between solute and contaminants and/or azeotropes formed between solute and contaminants.
3. Solute (product) is desired in particulate form. For example, if solute can be purified via distillation then it must be solidified subsequently by flaking or prilling and crystallization may be a more convenient scheme to employ in such cases.
4. Comparative economics favor crystallization. If distillation requires high temperatures and energy usage, crystallization may offer economic incentives.

Crystallization is frequently the initial step in a solids processing sequence, similar to that shown in Fig. 11.1-1, that subsequently includes solid–liquid separation and drying equipment. Since product size

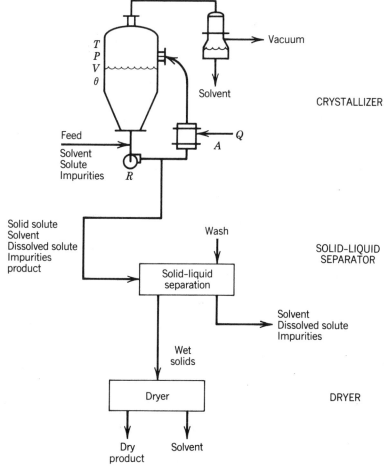

FIGURE 11.1-1 Solids processing sequence.

and suspended solids concentration are controlled, to a large extent, in the crystallizer, predictable and reliable crystallizer performance is essential for smooth operation of the downstream system.

11.1-2 Crystallization Terminology

Binary eutectic. The location on the phase diagram at which simultaneous crystallization of solvent and solute occurs.

Brodie purifier. A specific type of continuous melt crystallizer.

Crystal habit. The external crystal shape that results from different rates of growth of the various crystal faces.

Crystallization kinetics. Expressions that describe crystal growth and nucleation rates from solution.

CSD. Crystal size distribution.

Desupersaturation time. The average time that solution in a crystallizer vessel has to deposit solute before entering the zone where supersaturation is created.

Draft tube. A device inserted in a crystallizer to induce a uniform axial flow pattern inside a crystallizer.

Drawdown time. The time to empty the contents of a crystallizer if the feed is stopped and product is removed at the normal rate.

DTB. Draft tube baffle—a crystallizer that contains a draft tube and an internal baffle to provide a crystal settling zone.

Elutriation leg. A settling leg in a crystallizer that classifies and washes the crystals leaving the vessel by addition of upflow liquid (usually the feed).

Fines removal system. System designed to increase particle size by preferential removal and subsequent destruction of smaller-size crystals.

Forced circulation crystallization. Continuous crystallizer in which agitation inside the vessel is created by external circulation of large quantities of liquid through a heater or cooler.

Heterogeneous nucleation. Nucleation induced by foreign matter in a supersaturated liquid.

Homogeneous nucleation. Spontaneous nucleation caused by supersaturation only.

ICT. Initial crystallization temperature.

Labile zone. Zone on concentration–temperature diagram in which spontaneous homogeneous or heterogeneous nucleation of the solid phase will occur.

Metastable zone. Zone on concentration–temperature diagram in which homogeneous or heterogeneous nucleation will not immediately occur but in which crystal growth will occur.

MSMPR. Mixed-suspension mixed-product removal crystallizer.

Population density. A number density function frequently described as number of crystals per unit volume of clear liquor (or slurry) per size increment (number/$cm^3 \cdot \mu m$).

Product classification. Classification device that removes large product crystals and returns smaller crystals to the crystallizer.

Secondary nucleation. Nucleation of a supersaturated liquid caused directly or indirectly by the presence of crystals of the same species as the solute.

Solid solution. Mixed crystals formed when isomorphous substances crystallize together out of a solution.

Supersaturation. The departure from solution saturation usually caused by cooling of the mixture and/or by evaporating solvent.

Understanding the concept of supersaturation is necessary when discussing crystallization rate processes. Most liquids sustain a certain level of subcooling depending on the rate of cooling, temperature, and degree of agitation. Often a clear solution can be slowly subcooled several degrees below its equilibrium temperature before a profusion of nuclei appear. A diagram describing this phenomena is shown in Fig. 11.1-2. Feed A is cooled to temperature T_D. Thus, feed is subcooled by $(T_B - T_D)$ degrees or has $(T_B - T_D)$ degrees of supersaturation. The temperature where nuclei first appear (T_C), defines the metastable limit. Inside the temperature range from T_C to T_B crystals grow, but spontaneous nucleation will not occur immediately. In the labile zone (temperatures less than T_C) crystal growth and secondary nuclei formation occur simultaneously, and both rate processes are competing for available solute. Crystallizers are designed most commonly to control supersaturation at low levels to minimize nucleation, especially by homogeneous or heterogeneous mechanisms.

11.2 FUNDAMENTALS

Final design of a crystallizer is the culmination of the design strategy depicted in Fig. 11.2-1. In the conceptual design stage, equilibrium data and operating mode (the method by which supersaturation is

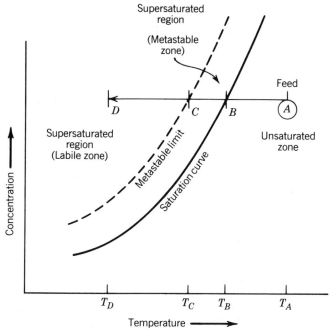

FIGURE 11.1-2 Depiction of supersaturation. The metastable limit is a function of the rate of cooling, temperature, impurities, degree of agitation, and presence of seeds.

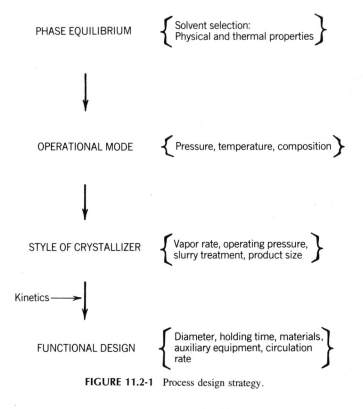

PHASE EQUILIBRIUM $\left\{ \begin{array}{l} \text{Solvent selection:} \\ \text{Physical and thermal properties} \end{array} \right\}$

OPERATIONAL MODE $\left\{ \text{Pressure, temperature, composition} \right\}$

STYLE OF CRYSTALLIZER $\left\{ \begin{array}{l} \text{Vapor rate, operating pressure,} \\ \text{slurry treatment, product size} \end{array} \right\}$

Kinetics ⟶

FUNCTIONAL DESIGN $\left\{ \begin{array}{l} \text{Diameter, holding time, materials,} \\ \text{auxiliary equipment, circulation} \\ \text{rate} \end{array} \right\}$

FIGURE 11.2-1 Process design strategy.

generated) are surveyed. Solvent choice and processing conditions are determined in this step. The type of crystallizer and finally the crystallizer functional design, that is, external and internal construction details, are then established. The discussion in this section is concerned primarily with generation, interpretation, and use of solid–liquid equilibrium data and the selection of operating mode.

11.2-1   Equilibria

Accurate solid–liquid equilibrium data must be obtained to evaluate the process design options for crystallization processes. These data are required in the earliest stages of the conceptual design phase and are necessary for the following reasons:

1. Screening the feasibility of the potential process; that is, determining if pure solute can in fact be crystallized from the feed solution.
2. Determining the best solvent to use in the process if solution crystallization is employed.
3. Establishing the temperature and/or pressure ranges of the crystallizer operation and the composition of the residue liquor exiting the crystallizer.
4. Determining the maximum recovery of solute possible. The feed composition and position of the eutectic fix the maximum attainable solute recovery. Practical limitations such as temperature of cooling medium, slurry concentration that can be pumped, or impurity level that can be tolerated may restrict solute recovery.

Phase diagrams occur in many forms. A common type of diagram is the binary eutectic in which a pure solid component is formed by cooling an unsaturated solution until solids appear. Continued cooling will increase the yield of pure component. At the eutectic temperature both components solidify and additional purification is not normally possible. A typical eutectic-forming system (naphthalene–benzene) is shown in Fig. 11.2-2:

If mixture X is cooled, crystals of benzene will form.

If mixture Y is cooled, naphthalene crystals will result.

Point Z indicates the position of the binary eutectic. Solid mixtures are formed at temperatures below this point.

A few binary systems when cooled do not deposit one of the components in a totally pure state. Instead, behavior resembles that of many vapor–liquid systems and the solid is a true solution. Figure 11.2-3 depicts

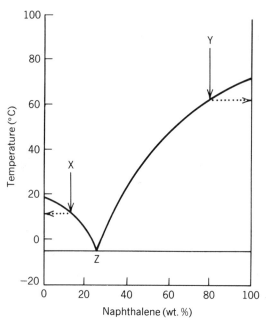

FIGURE 11.2-2 Phase diagram for naphthalene–benzene.

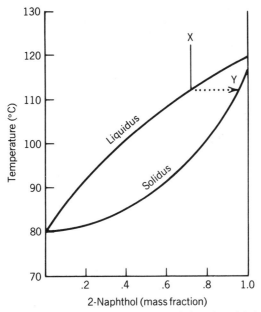

FIGURE 11.2-3   Phase diagram for naphthalene–2-naphthol.

a phase diagram of naphthalene and 2-naphthol, which exhibit solid solution behavior. If the liquid mixture of naphthalene and 2-naphthol of composition X is cooled, a mixed crystal of composition Y is formed rather than a solid containing pure 2-naphthol. This type of behavior is found in only a small fraction of crystallizing systems.

Figure 11.2-4 shows typical solubility diagrams for solutions of various salts in water. The curve for salt (NaCl) is nearly vertical, which indicates little effect of temperature on solubility. The sodium sulfate curve shows reverse solubility as temperature increases; thus, sodium sulfate has a tendency to coat heat-exchanger surfaces where heat is added to saturated solutions of this system.

Often solid–liquid equilibrium data are not available for the system of interest, and experimental determination of the solidus–liquidus curves is required. If the system of interest is simple (i.e., two to three components) and well behaved (ideal), then reliable predictive methods are available. Techniques for predicting nonideal solid–liquid phase behavior and multicomponent equilibria are emerging.

PREDICTING SOLID–LIQUID EQUILIBRIA
The extent to which solids can dissolve in liquids varies enormously; in some cases a solid solute may form a highly concentrated solution and in others the solubility is barely detectable. Some of the principles

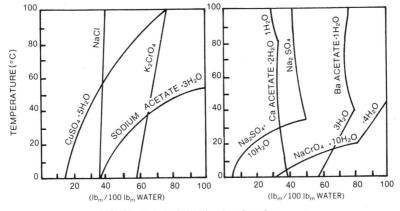

FIGURE 11.2-4   Solubility data for salt systems.

that govern equilibrium between a solid phase and a liquid phase are discussed in this section. In many situations reliable predictive techniques are available to estimate binary and multicomponent solubility behavior, and several of these approaches also are discussed.

FRAMEWORK

The fundamental relationship for equilibrium to exist between two phases (where the solute is designated by subscript 2) is

$$f_2(\text{pure solid}) = f_2(\text{solute in solution}) \tag{11.2-1}$$

or

$$f_2(\text{pure solid}) = \gamma_2 x_2 f_2^{\circ} \tag{11.2-2}$$

where f_2 is the fugacity, x_2 is the solubility (mole fraction) of the solute in the solvent, γ_2 is the liquid phase activity coefficient, and f_2° is the reference-state fugacity to which γ_2 refers. The solubility can then be defined as

$$x_2 = \frac{f_2}{\gamma_2 f_2^{\circ}} \tag{11.2-3}$$

Thus, the solubility depends on the activity coefficient and the ratio of two fugacities. It is convenient to define the reference-state fugacity (of solute) as the fugacity of *pure, subcooled liquid* at the temperature of the solution. As derived elsewhere[1] the pure fugacity ratio can be calculated from

$$\ln\left(\frac{f_2(\text{pure solid})}{f_2^{\circ}(\text{pure subcooled liquid})}\right) = \frac{\Delta h_f}{RT}\left(\frac{T}{T_M} - 1\right) \tag{11.2-4}$$

where Δh_f is the enthalpy of fusion, T_M is the melting temperature of pure substance 2, and T is the system temperature. Correction terms associated with differences between the heat capacities of the liquid and solid are neglected.

If Eq. (11.2-4) is substituted into (11.2-3) the solubility relationship becomes

$$\ln(\gamma_2 x_2) = \frac{\Delta h_f}{RT}\left(\frac{T}{T_M} - 1\right) \tag{11.2-5}$$

If the solution is assumed to be ideal, then $\gamma_2 = 1.0$ and Eq. (11.2-5) becomes

$$\ln x_2 = \frac{\Delta h_f}{RT}\left(\frac{T}{T_M} - 1\right) \tag{11.2-6}$$

Equation (11.2-6) is often referred to as the van't Hoff relationship. To use this equation to compute solute solubility for an ideal solution only the heat of fusion Δh_f, and the pure-component melting temperature T_M are required. The interesting feature of this equation is that the solubility depends only on the properties of the solute and is independent of the nature of the solvent.

Another equation that is employed frequently for ideal systems uses cryoscopic constants that have been obtained empirically for a wide variety of materials in the American Petroleum Institute Research Project No. 44:[2]

$$\ln\frac{1}{x_2} = A(T_m - T)[1 + B(T_m - T)\cdots] \tag{11.2-7}$$

where the cryoscopic constants A and B are defined by

$$A = \frac{\Delta h_f}{RT_m^2} \quad \text{and} \quad B = \frac{1}{T_m} - \frac{\Delta C_p}{2\Delta h_f} \tag{11.2-8}$$

Freezing point curves were computed for *para-* and *ortho*-xylene using Eq. (11.2-6) (the van't Hoff equation) and Eq. (11.2-7) which used cryoscopic constants. Data from Table 11.2-1 were used in the equations and calculated equilibrium compositions are shown as a function of temperature in Table 11.2-2. The computed liquid compositions are nearly identical in the high concentration range and diverge

TABLE 11.2-1 Data Used for Freezing-Point Curve Calculations

Data	p-Xylene	o-Xylene
T_M	286.41 K	247.97 K
Δh_f	4090 cal/mol	3250 cal/mol
A	0.02599 mol frac/K	0.02659 mol frac/K
B	0.0028 mol frac/K	0.0030 mol frac/K

TABLE 11.2-2 Computed Freezing-Point Curves for o- and p-Xylene

	p-Xylene			o-Xylene		
		Mole Fraction in Solution			Mole Fraction in Solution	
Temperature (K)	Eq. (11.2-8)	Eq. (11.2-9)	Temperature (K)	Eq. (11.2-8)	Eq. (11.2-9)
286.41	1.00	1.00	249.97	1.00	1.00
285.0	0.965	0.964	245.0	0.923	0.923
280.0	0.848	0.844	240.0	0.803	0.805
270.0	0.646	0.640	235.0	0.695	0.699
260.0	0.482	0.478			
250.0	0.351	0.352			
240.0	0.249	0.256			
230.0	0.172	0.183			

slightly in the dilute regime. Figure 11.2-5 is a plot of freezing-point depression curves for mixtures of xylene isomers. The binary eutectic temperature for *para* and *ortho* isomers is about 236.3K. This is the temperature at which both components solidify together. Thus, the mole fraction of *ortho*-xylene in solution plus the mole fraction of *para*-xylene must equal 1 at the eutectic ($x_o + x_p = 1.0$). This condition also fixes the eutectic temperature for multicomponent systems. The ternary eutectic temperature for a mixture of xylene isomers is determined by finding the common freezing point at which the sum of the mole fractions of mixed isomers in solution equals 1 ($x_o + x_p + x_m = 1.0$). The eutectic conditions for the three xylene isomer binaries are indicated in Fig. 11.2-5. The ternary eutectic is also indicated.

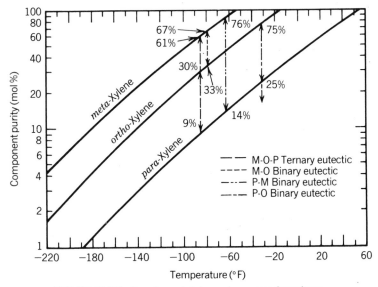

FIGURE 11.2-5 Freezing-point depression curves for xylenes.

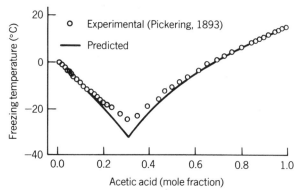

FIGURE 11.2-6 Comparison of predicted and experimental equilibrium curves for acetic acid–water. Excerpted by special permission from *Chemical Engineering*, Feb. 22, copyright © 1982 by McGraw-Hill, Inc., New York, NY.

Determination of eutectic points is important since yield of a desired component is limited by the position of the eutectic. The above procedures give accurate predictions of eutectic points in ideal binary and multicomponent systems and are frequently sufficient for preliminary evaluation of a crystallization system. For the final design of a commercial system, experimental verification of the eutectic is advisable.

For nonideal systems a number of approaches are available. The method to use in a specific case may depend on the availability of thermodynamic parameters necessary to perform the specific calculation. Some of the methods are direct and others require iterative techniques best relegated to computers. Activity coefficients for solid-solubility calculations can often be estimated directly from vapor–liquid equilibrium data obtained at higher temperatures. Experimental vapor–liquid equilibrium data frequently are easier to find than solid–liquid equilibrium data; thus, this approach provides a convenient technique for conducting preliminary design calculations to determine the feasibility of a given process. Data are correlated using a semiempirical equation, such as the Wilson equation, whose characteristic binary parameters are extrapolated with respect to temperature. Muir and Howat[3] have reported the results of using this technique for several ideal and nonideal systems and found good agreement with experimental data. Figures 11.2-6 and 11.2-7 compare predicted and experimental equilibrium data for acetic acid–water and caprolactam–water binaries. Agreement between predicted and experimental freezing points is good.

For nonpolar systems the activity coefficient can be estimated using the Hildebrand–Scatchard[4] theory of regular solutions. To calculate the activity coefficient of a dissolved solute using regular solution theory, solubility parameters must be available for the components. For many materials these parameters can be calculated and/or are available in standard engineering references.

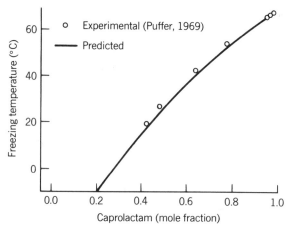

FIGURE 11.2-7 Comparison of predicted and experimental equilibrium curves for caprolactam–water. Excerpted by special permission from *Chemical Engineering*, Feb. 22, copyright © 1982 by McGraw-Hill, Inc., New York, NY.

A group-contribution method, called UNIFAC,[5] has been developed for estimating activity coefficients in nonelectrolyte solutions. By using this approach solubilities for several solutes in a variety of solvents may be calculated. In many cases good agreement is obtained. The activity coefficient consists of two parts, the combinatorial contribution, arising from differences in molecular size and shape, and the residual contribution, arising from differences in intermolecular forces of attraction. Group interaction parameters must be available to use this approach. Parameters are listed by Fredenslund et al.[6] for many systems of interest.

Other approaches to the computation of solid–liquid equilibria are shown in Table 11.2-3. The Soave-Redlich–Kwong equation of state evaluates fugacities to calculate solid–liquid equilibria,[7] while Wenzel and Schmidt[8] developed a modified van der Waals equation of state for the representation of phase equilibria. The Wenzel–Schmidt approach generates fugacities, from which the authors developed a trial-and-error approach to compute solid–liquid equilibrium. Unno et al.[9] recently presented a simplification of the solution of groups model (ASOG) that allows prediction of solution equilibrium from limited vapor–liquid equilibrium data.

Often complex liquid–solid equilibria cannot be predicted reliably, nor are data available in the literature. In this case, experimental determination of equilibrium behavior becomes necessary. Experimental data should be obtained at conditions that allow fitting by a model so that interpolation and extrapolation are possible.

The ASTM apparatus[10] for determining freezing point is shown in Fig. 11.2-8. Freezing point is obtained from freezing or melting temperature versus time curves. Figure 11.2-9 shows a typical temperature–time trace from an ASTM apparatus. The freezing point can be determined from the peak of the temperature rise during slow cooldown or from the break in the temperature–time curve during heating. A DSC (differential scanning calorimeter) can be used for freezing-point determination (as well as heat of fusion) if subcooling is low. Inorganic systems usually are handled well in a DSC apparatus. If many components are present, an experimental freezing-point curve can be determined by fixing the temperature of a well-mixed slurry containing an excess of solids, allowing the system to equilibrate, and analyzing the composition of the liquid phase. By repeating this process at successively lower temperatures a freezing-point curve can be developed.

11.2-2 Choice of Operational Mode

The technique employed to generate supersaturation in a solution is referred to as the mode of operation. The mode chosen by the designer is strongly influenced by the phase-equilibrium characteristics of the system, and it dictates the material and energy balance requirements of the system. The common techniques for producing solids (or generating supersaturation) from a solution include:

1. Lowering the temperature of the feed solution by direct or indirect cooling. If solute solubility is strongly temperature dependent, this is the preferred approach.
2. Adding heat to the system to remove solvent and thus "salt out" the solute. This technique is effective if solubility is insensitive to temperature.
3. Vacuum cooling the feed solution without external heating. If solubility is strongly dependent on temperature, this method is attractive.
4. Combining techniques. Especially common is vacuum cooling supplemented by external heating for systems whose solubility has an intermediate dependence on temperature.
5. Adding nonsolvent. This is a common technique for precipitating solute from solution and is useful as both a laboratory technique and as an industrial process for product recovery.

The methods described above can be employed in single- or multistage crystallization or in batch operations. Multistage operations are employed where evaporative requirements exceed the capabilities of a single vessel and/or energy costs dictate staging of the operation. Sometimes, staging is useful to produce uniform and/or larger crystals. Operation of crystallizers in series generates crystal size distributions having a narrower size spread than the same volume of crystallizers in parallel. Crystal growth kinetics usually are favored at higher temperatures, so if the first stage of a two-stage crystallization system is operated at a higher temperature, the overall CSD is enhanced. Batch crystallizers produce a narrower CSD than continuous well-mixed units. For capacity requirements less than 500 kg/h, batch crystallization is often more economical. If highly uniform crystals are required, which is the case for sugar, then batch operation offers advantages. Table 11.2-4 compares the operational characteristics of several common modes of crystallization. Both advantages and disadvantages are indicated.

11.2-3 Crystallization Kinetics

The kinetic phenomena that influence crystal size distributions are *nucleation* and *growth*. The driving force for both these phenomena is supersaturation, and at some levels of supersaturation, both nucleation and growth occur and compete for available solute. The purpose of the present discussion is to summarize key

TABLE 11.2-3 Predictive Methods for Solid–Liquid Equilibria

System	Approach	Parameters Required	References
1. Ideal	van't Hoff equation	Δh_f, T_M	Prausnitz[1]
2. Ideal	Cryoscopic constants	A, B, T_M	API Project No. 44[2]
3. Nonideal	Compute γ–Wilson equation	Vapor–liquid equilibrium data	Muir and Howat[3]
4. Nonideal	Compute γ–regular solution theory	Solubility parameters	Hildebrand and Scott[4]
5. Nonideal	Compute γ–UNIFAC	Group interaction parameters	Gmehling et al.[5]
6. Nonideal	Compute γ–groups model	Vapor–liquid equilibrium data	Unno et al.[9]
7. Nonideal	Compute fugacity: modified van der Waals equation of state	Equation of state parameters	Wenzel and Schmidt[8]
8. Nonideal	Compute fugacity: Soave–Redlich–Kwong equation of state	Vapor pressure, critical constants, acentric factors	Soave[7]

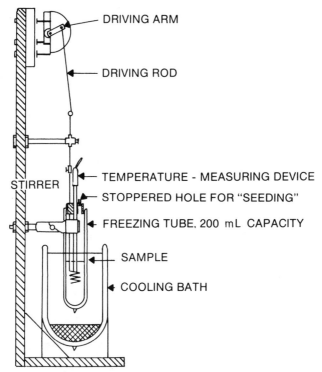

FIGURE 11.2-8 Apparatus for freezing-point determination.

aspects regarding the kinetics of crystal nucleation and growth and to indicate the process variables that are likely to affect each.

It is obvious that supersaturation is a key variable in setting nucleation and growth rates and, as will be shown in the following discussion, the dependence of these rates on supersaturation is affected by the mechanism through which the process is occurring. This is made plain in Fig. 11.2-10, which shows the influence of supersaturation on nucleation and growth. The key aspects in this figure are the qualitative relationships of the two forms of nucleation to growth and to each other: growth rate and secondary nucleation kinetics are low-order (shown as linear) functions of supersaturation, while primary nucleation

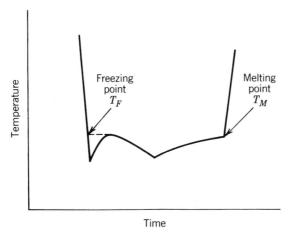

FIGURE 11.2-9 Typical temperature–time trace from freezing-point apparatus.

TABLE 11.2-4 Features of Crystallization Mode

Mode 1. Indirect Cooling Crystallization

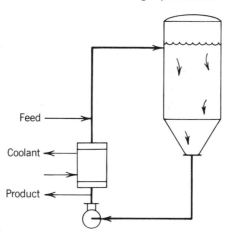

Operational Characteristics: Feed enters at higher saturation temperature than is maintained in the crystallizer and is cooled in shell-and-tube or scraped-surface exchangers to remove sensible heat and heat of crystallization. Solids encrustation problems are generally confined to the cooling surfaces.

Advantages: Operation and control is simple. No vacuum equipment is necessary. Slurry density and product recovery are fixed by the feed composition and by the temperature maintained in the crystallizer body.

Disadvantages: Care must be taken to prevent fouling of the cooling surfaces by maintaining low process-to-coolant temperature differences across cooling surfaces. If severe fouling is anticipated, scraped-surface heat exchangers may be necessary to assure reliable operation.

Mode 2. Evaporative Cooling Crystallization

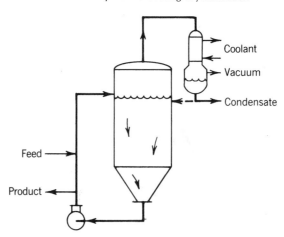

Operational Characteristics: Feed enters at higher saturation temperature than is maintained in the crystallizer body. Crystallizer temperature, product recovery, and slurry density are regulated by vacuum control. Heat of crystallization and the sensible heat of the feed are removed by evaporation and condensation of solvent. The condensate may either be removed or a portion or all returned.

Advantages: No heater is required in this mode of operation, and condensate can be returned to wash down walls to control the formation of encrustations.

TABLE 11.2-4 (*Continued*)

Disadvantages: Slurry concentration and product yield are fixed by material and energy balance constraints.

Mode 3. Cooling and Salting Out Crystallization

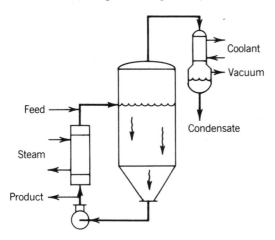

Operational Characteristics: The feed stream saturation temperature is higher than that maintained in the crystallizer body; thus, adiabatic cooling occurs. Also, external heat is added to evaporate solvent and precipitate additional solute.

Advantages: Slurry density and yield can be controlled to some degree by adjustment of external heat input.

Disadvantages: An external (or internal) heat exchanger and a vacuum system are required. Fouling at vapor release surfaces is common.

Mode 4. Salting Out Crystallization (Addition of Nonsolvent)

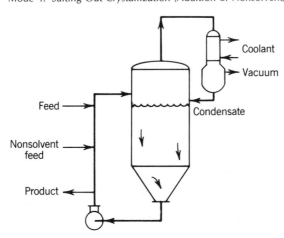

Operational Characteristics: Crystallizer temperature is controlled by vacuum level and ratio of nonsolvent to feed. A reflux condenser normally is used to remove heat by condensing the solvent.

Advantages: System is reliable to operate. Encrustations are minimized since heating and cooling surfaces are eliminated and vaporization is minimal.

Disadvantages: An additional component (nonsolvent) must be separated from all liquid streams and recycled. The technique is somewhat limited since the nonsolvent must be miscible in solvent.

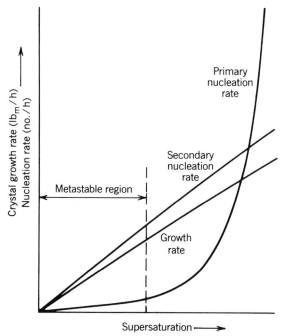

FIGURE 11.2-10   Influence of supersaturation on growth and nucleation rates.

is shown to follow a high-order dependence on supersaturation. Design of a crystallizer to produce a desired crystal size distribution requires kinetic data that quantify the relationships of nucleation and growth to externally controlled variables and to supersaturation. The following discussion necessarily is brief and simply outlines some of the major factors that influence nucleation and growth kinetics. More complete discussions of each phenomenon can be found in the references cited.

As with all kinetic phenomena, the first concept to understand is the driving force for the rate processes. While supersaturation was represented above as the driving force for nucleation and growth, no clear definition of this quantity was given. In fact, there are various ways in which the driving force or supersaturation can be defined:

1. The difference between the solute concentration and the concentration at equilibrium, $C - C^*$.
2. The difference between the system temperature and the temperature at equilibrium, $T - T^*$.
3. The ratio of the solute concentration and the equilibrium concentration, C/C^*.
4. The ratio of the difference between the solute concentration and the equilibrium concentration to the equilibrium concentration, $s = (C - C^*)/C^*$.

Garside[11] cites the analyses and restrictions which have led to the conclusion that the most appropriate driving force for crystal growth is that given by the fourth of the possibilities listed above, $s = (C - C^*)/C^*$. This definition of supersaturation will be used throughout the ensuing discussion. Garside also notes that when dealing with hydrated salts, concentration should not be based on the anhydrous solute.

NUCLEATION

In crystallization, nucleation is the formation of a solid phase from a liquid phase. The process differs from growth in that a new crystal results from the transfer of solute from the liquid to the solid; in growth, solid is deposited on an existing crystal. Because it is the phenomenon of crystal formation, nucleation sets the character of the crystallization process, and it is therefore the most critical component in relating crystallizer design and operation to crystal size distributions.

Mechanisms.   Classical nucleation theory is based on homogeneous and heterogeneous mechanisms, both of which call for the formation of crystals through a process of sequentially combining the constituent units that form a crystal. These mechanisms are referred to as *primary nucleation* because existing crystals play no role in the nucleation. Both homogeneous and heterogeneous mechanisms require relatively high supersaturations in order for them to occur and they exhibit a high-order dependence on supersaturation. These factors often lead to production of excessive fines in systems where primary nucleation mechanisms are important.

Mullin[12] outlines the classical theoretical treatment of primary nucleation, which results in the expression

$$B° = A \exp\left(-\frac{16\pi\sigma^3 v^2}{3k^3 T^3[\ln(s+1)]^2}\right) \tag{11.2-9}$$

where k is the Boltzmann constant, σ is surface energy per unit area, v is molar volume, and A is a constant. Note that this equation can be simplified by recognizing that s is often much less than 1, so that $\ln(s+1)$ approaches s:

$$B° = A \exp\left(-\frac{16\pi\sigma^3 v^2}{3k^3 T^3 s^2}\right) \tag{11.2-9a}$$

The most important variables affecting primary nucleation rates are shown by Eq. (11.2-9) to be interfacial energy σ, temperature T, and supersaturation s. The high-order dependence of nucleation rate on supersaturation is especially important as a small variation in supersaturation may produce an enormous change in nucleation rate. This gives rise to the often observed phenomenon of having a clear liquor transformed into a slurry of very fine crystals with only a slight increase in supersaturation, for example, by decreasing the solution temperature. The catalytic effect of solid particles (as in heterogeneous nucleation) is to reduce the energy barrier to formation of a new phase. This, in effect, can reduce the interfacial energy σ by orders of magnitude.

An empirical approach to modeling primary nucleation requires a knowledge of the metastable limit that a given solution can withstand before spontaneous nucleation occurs. This limit, which was first observed in experiments by Miers and Issac,[13] must be determined through experimentation, and nucleation rate is then correlated with the equation

$$B° = k(C - C_m)^i \qquad C^* < C_m \tag{11.2-10}$$

where C is solute concentration, C_m is the solute concentration at which spontaneous nucleation occurs, and C^* is the solute concentration at saturation. Randolph and Larson[14] indicate that C_m is very close to C^* for many inorganic systems, and they report satisfactory correlations with C^* substituted for C_m in Eq. (11.2-10).

Secondary nucleation is the formation of new crystals as a result of the presence of solute crystals; in other words, crystals of the solute *must* be present for secondary nucleation to occur. Garside and Davey[15] and Larson[16] give excellent reviews of the subject, so only a brief synopsis is provided here.

Several features of secondary nucleation make it more important than primary nucleation in industrial crystallizers. First, continuous crystallizers and seeded batch crystallizers have crystals in the magma that can participate in secondary nucleation mechanisms. Second, the requirements for the mechanisms of secondary nucleation to be operative are fulfilled easily in most industrial crystallizers. Finally, most crystallizers are operated in a low supersaturation regime so that crystal growth is regular and the resulting product is pure and of good habit; low supersaturation can support secondary nucleation but not primary nucleation.

Secondary nucleation can occur as the result of several mechanisms which have been identified and observed in selected systems. These include initial breeding, contact nucleation (also known as collision breeding), and shear breeding. *Initial breeding* results from immersion of seed crystals in a supersaturated solution, and it is thought to be caused by dislodging extremely small crystals that were formed on the surface of the crystals during drying. Although this mechanism is unimportant in continuous and unseeded batch crystallization, it can have a significant impact on the operation of seeded batch crystallizers. Girolami and Rousseau[17] demonstrate the effects of several process variables on nucleation rates caused by initial breeding. *Contact nucleation* results from collisions of crystals with one another, crystallizer internals, or with an impeller of an agitator or circulation pump. It should be recognized that the collision energy for contact nucleation is small and does not necessarily result in the macroscopic degradation of the contacted crystal. Because of the importance of this mechanism it will be discussed in more detail below. *Shear breeding* results when supersaturated solution flows by a crystal surface and carries with it crystal precursors believed formed in the region of the growing crystal surface. Sung et al.[18] observed nucleation of $MgSO_4 \cdot 7H_2O$ by this mechanism and found that very high levels of supersaturation were required for it to produce significant numbers of nuclei.

Process Variables Affecting Contact Nucleation. As stated above, contact nucleation results from a collision of a crystal with an object, such as an impeller or other crystallizer internals, in a supersaturated medium. The pioneering studies that elucidated many factors affecting this nucleation mechanism were carried out by Strickland-Constable and coworkers[19,20] and Clontz and McCabe.[21] These and subsequent works demonstrate that the number of crystals produced by a controlled impact of an object with a seed crystal depends on the energy of impact, supersaturation at impact, supersaturation at which crystals mature, material of the impacting object, area of impact, angle of impact, and system temperature. Although it is impossible to account quantitatively for each of these variables, certain generalizations can be drawn from the research on this nucleation mechanism.

Tai et al.[22] demonstrate that the number of crystals produced under controlled conditions is an exponential function of impact energy E. Furthermore, a threshold or minimum collision energy E_t is required for some crystalline systems. Also, Shah et al.[23] and Evans et al.[24] found that the material of construction of the impacting object greatly affects the number of crystals resulting from the collision. In particular, coating the crystallizer internals or manufacturing them from soft materials reduces contact nucleation rates. This is attributed to a reduction in the energy transmitted from the coated object to the impacted crystal. Based on these two experimental observations, the following expression was proposed for systems at constant supersaturation:[25]

$$B° = k_N \exp (E - E_t) \qquad E > E_t \qquad (11.2\text{-}11)$$

In a crystallizer with a circulation pump or agitator, it has been shown that collisions between such a device and crystals in the circulating magma dominate nucleation resulting from other collisions.[26] Consequently, it might be expected that a relationship of impact energy E to crystallizer variables must include the mass of the impact crystal M_c, and the rotational velocity of the impeller ω. Moreover, the fraction of the available energy actually transmitted to the crystal, ϵ, should also be included in the model:

$$E = f(\epsilon, \omega, M_c) \qquad (11.2\text{-}12)$$

The variables shown in Eq. (11.2-12) can be manipulated to some extent, thereby modifying nucleation rates and the concomitant crystal size distribution. For example, internal classification can be used to keep larger crystals away from energetic collisions with an impeller, but doing so may create other problems with stability of the crystal size distribution. The rotational velocity of an impeller can be changed if there are appropriate controls on the pump or agitator. Caution must be exercised, however, for a reduction in circulation velocity can reduce heat transfer coefficients and increase fouling or encrustation on heat transfer surfaces. Moreover, the crystals in the magma must be kept suspended or else crystal habit and growth rates could be affected adversely. It is suspected that impact energy may have a high-order dependence on rotational velocity, and if that is the case, modest changes in this quantity could alter nucleation rates substantially. As already indicated, the fraction of the impact energy transmitted from an impeller to the crystal can be manipulated by changing the material of construction of the impeller. Tai[27] also showed the hardness of the crystal surface to be important in comparing nucleation characteristics of different crystalline materials. Therefore, the importance of ϵ may vary from one crystalline system to another: those systems in which the crystal face is soft may be more susceptible to nucleation rate changes due to the variables in Eq. (11.2-12) than those crystalline systems where the face is hard.

Correlations of nucleation rate with crystallizer variables have been developed for several systems. For example, Bennett et al.[28] examined the effect of slurry circulation rate on nucleation and developed a correlation based on the tip speed of the impeller. Grootscholten et al.[29] considered the scale-up of nucleation kinetics for sodium chloride crystallization, and they provide an analysis of the role of mixing and mixer characteristics in contact nucleation. Garside and Shah[30] reviewed published kinetic correlations though about 1979.

Supersaturation has been observed to affect contact nucleation, but the mechanism by which this occurs is not clear. There are data[22] that infer a direct relationship between contact nucleation and crystal growth. These data show that the number of nuclei produced by an impact is proportional to the linear growth rate of the impacted face. This is true even for citric acid which exhibits an unusual relationship between growth rate and supersaturation. It has been proposed that the known effect of growth rates, which are determined primarily by supersaturation, on the surface of the growing crystal provides the link between crystal growth and contact nucleation.

A second perspective is that supersaturation influences the fraction of entities formed by a collision that survive to form large crystals. This concept largely has been rejected based on the observation that such entities are much larger than the size of a critical nucleus. Recent research[31-33] used photographic and electronic zone-sensing techniques to examine the size distributions of particles formed by impacts with seed crystals in a supersaturated solution. These experiments on potassium alum show that the particles produced cover a size range from 50 μm down to the smallest size capable of being observed by the instrument. Furthermore, the influence of supersaturation on the number of particles produced in the 20–50 μm range was much greater than it was on the number produced in the smaller size range. Data that will be discussed in the section on crystal growth have shown that dissolution of crystals formed by contact nucleation does not occur, but that a fraction of the crystals grow very slowly and some may not grow at all.

Still another possible role of supersaturation is that it affects the solution structure and perhaps causes the formation of clusters of solute molecules. These clusters may participate in nucleation, although the mechanism by which this would occur is not clear. Mullin and Leci[34] present evidence for the formation of citric acid clusters in aqueous supersaturated solutions, and McMahon et al.[35] have examined cluster formation in solutions of KNO_3 using Raman spectroscopy.

The ease with which nuclei can be produced by contact nucleation is a clear indication that this mechanism is dominant in many industrial crystallization operations. Research on this nucleation mechanism is continuing with the objective of building an understanding of the phenomenon that will allow its successful inclusion in models describing commercial systems.

CRYSTAL GROWTH

Crystal growth rates may be expressed in a variety of ways: as the linear advance rate of an individual crystal face, as the change in a characteristic dimension of a crystal, or as the rate of change in mass of a crystal. These different expressions can be related through an understanding of crystal geometry. However, it is often convenient to use the method of measurement as the basis of the growth rate expression. In certain instances an analysis of the crystallization process will require growth rate to be defined in a specific way; for example, the use of a population balance to describe crystal size distribution requires that growth rate be defined as the rate of change of a characteristic dimension.

Single-crystal studies of growth kinetics often involve the rate of advance of an individual crystal face or the rate of change in crystal size associated with exposure to a supersaturated solution. The latter type of study will be dealt with in discussions of multicrystal magmas. The rate of advance of a single crystal face can be quantified by observation of the face through a calibrated eyepiece of an optical microscope. Using this procedure, it is possible to examine the structure of the advancing crystal face and perhaps to isolate surface reaction kinetics from mass transfer kinetics. These phenomena will be discussed later. An additional advantage of this system is that it is possible to examine crystal growth kinetics without interference from competing processes such as nucleation.

Multicrystal magma studies usually involve examination of the rate of change of a characteristic crystal dimension or the rate of increase in the mass of crystals in a magma. The characteristic dimension in such analyses depends on the method used in the determination of crystal size; for example, the second largest dimension is measured by sieve analyses, while an equivalent spherical diameter is determined by both electronic zone-sensing and laser light-scattering instruments. A relationship between these two measured dimensions and between the measured quantity and the actual crystal dimensions can be derived from appropriate shape factors. Recall that volume and area shape factors are defined by the equations

$$V_c = k_v L^3 \tag{11.2-13}$$

and

$$A_c = k_a L^2 \tag{11.2-14}$$

where V_c and A_c are volume and area of a crystal, k_v and k_a are volume and area shape factors, and L is the characteristic dimension of the crystal. Suppose an equivalent spherical diameter L_s is obtained from an electronic zone-sensing instrument and the actual dimensions of the crystal, which is known to have a cubic habit, are to be calculated. Let L_c be the edge length of the crystal, and k_{vs} and k_{vc} be the volume shape factors for a sphere and a cube, respectively. Since the volume of the crystal is the same, regardless of the characteristic dimension,

$$V_c = k_{vs} L_s^3 = k_{vc} L_c^3 \tag{11.2-15}$$

which means that

$$L_c = \left(\frac{k_{vs}}{k_{vc}}\right)^{1/3} L_s \tag{11.2-16}$$

Since k_{vs} is $\pi/6$ and k_{vc} is 1.0, the numerical relationship between these two dimensions can be determined from Eq. (11.2-16).

If the rate of change of a crystal mass, dM_c/dt, is measured, the quantity can be related to the rate of change in the crystal characteristic dimension by the equation

$$\frac{dM_c}{dt} = \frac{d(\rho_c k_v L^3)}{dt} = 3\rho_c k_v L^2 \frac{dL}{dt} \tag{11.2-17}$$

where ρ_c is crystal density. Since $k_a = A_c/L^2$ and G is defined as dL/dt

$$\frac{dM_c}{dt} = 3\rho_c \left(\frac{k_v}{k_a}\right) A_c G \tag{11.2-18}$$

In multicrystal magma systems, growth kinetics become difficult to isolate from other phenomena, although a combined analysis, such as that described in Section 11.2-4 for the perfectly mixed crystallizer, gives a realistic view of the actual process that occurs in a crystallizer; that is, nucleation and growth kinetics are measured simultaneously and therefore at the same process conditions.

Growth Models. At least two resistances contribute to a determination of growth kinetics. The rate processes to which these resistances apply are (1) integration or incorporation of the crystalline unit (e.g., solute molecules) into the crystal surface (lattice) and (2) molecular or bulk transport of the unit from the surrounding solution to the crystal face. Since bulk transport is addressed elsewhere (see Chapter 2), this discussion will focus only on surface incorporation. Detailed reviews of this topic can be found elsewhere[11,36-39] so this discussion will be brief.

Numerous models have been proposed to describe surface reaction kinetics. Among these are models that assume crystals grow by layers and other models that consider growth to occur by the movement of a continuous step. Each model results in a specific relationship between growth rate and supersaturation, but none can be used for a priori predictions of growth kinetics. Such models do provide insights as to the effects of certain process variables on growth and, with additional research, may lead to predictive capabilities. For these reasons and because of the extensive literature on the subject, all that will be pointed out here are the key aspects of the physical models and the resulting relationship between growth and supersaturation predicted by each theory.

The model used to describe the growth of crystals by layers calls for a two-step process: (1) formation of a two-dimensional nucleus on the surface and (2) spreading of the solute from the two-dimensional nucleus across the surface. The relative rates at which these two steps occur give rise to the mononuclear two-dimensional nucleation theory and the polynuclear two-dimensional nucleation theory. In the mononuclear two-dimensional nucleation theory, the surface nucleation step occurs at a finite rate while the spreading across the surface occurs at an infinite rate. The reverse is true for the polynuclear two-dimensional nucleation theory. From the mononuclear two-dimensional nucleation theory, growth is related to supersaturation by the equation

$$G = C_1 hA[\ln (1 + s)]^{1/2} \exp \left[- \frac{C_2}{T^2 \ln (1 + s)} \right] \tag{11.2-19}$$

where C_1 and C_2 are system-dependent constants, h is the height of the nucleus, A is surface area, and s and T are as defined earlier. The polynuclear two-dimensional theory produces the equation

$$G = \left(\frac{C_3}{T^2[\ln (1 + s)]^{3/2}} \right) \exp \left(- \frac{C_2}{T^2 \ln (1 + s)} \right) \tag{11.2-20}$$

where C_3 is a system-dependent constant. Finally, if both formation of the two-dimensional nucleus and spreading of the surface layer are important in determining growth rate, the following equation can be derived:

$$G = C_4 s^{2/3} [\ln (1 + s)]^{1/6} \exp \left(- \frac{C_2}{3T^2 \ln (1 + s)} \right) \tag{11.2-21}$$

where C_4 is a system-dependent constant.

Equations (11.2-19), (11.2-20), and (11.2-21) can be simplified considerably by recognizing that in many systems the quantity s is much less than 1. In that case, $\ln (1 + s)$ is approximately s. Making this substitution, the growth rate from the mononuclear two-dimensional theory becomes

$$G = C_1 hAs^{1/2} \exp \left(- \frac{C_2}{T^2 s} \right) \tag{11.2-22}$$

For the polynuclear two-dimensional nucleation theory

$$G = \left(\frac{C_3}{T^2 s^{3/2}} \right) \exp \left(- \frac{C_2}{T^2 s} \right) \tag{11.2-23}$$

For both steps occurring at similar rates

$$G = C_4 s^{5/6} \exp \left(- \frac{C_2}{T^2 s} \right) \tag{11.2-24}$$

The screw dislocation theory, which was formulated by Burton, Cabrera, and Frank[40] and is often referred to as the BCF theory, shows that the dependence of growth rate on supersaturation can vary from a parabolic relationship at low supersaturations to a linear relationship at high supersaturations. In the BCF theory, growth rate is given by

$$G = C\left(\frac{\epsilon s^2}{\sigma_1'}\right) \tanh\left(\frac{\sigma_1'}{\epsilon s}\right)$$

(11.2-25)

where ϵ is screw dislocation activity and σ_1' is a system-dependent quantity that is inversely proportional to temperature. The dependence of growth rate on supersaturation is linear if the ratio $\sigma_1'/\epsilon s$ is small, but this dependence becomes parabolic as the ratio becomes large. This is because $\tanh x$ approaches x as x becomes small (supersaturation becomes large), and $\tanh x$ approaches 1.0 as x becomes large (supersaturation becomes small). It is possible then to observe variations in the dependence of growth rate on supersaturation for a given crystal–solvent system.

An empirical approach can also be used to relate growth kinetics to supersaturation. This approach simply fits the data with a power-law function of the form

$$G = k_G s^g$$

(11.2-26)

where k_G and g are constants determined by fitting the equation to growth rate data. Such an approach should be valid over small ranges of supersaturation. In fact, careful analysis of the theories discussed above will show that the more fundamental equations can be fit by Eq. (11.2-26) over limited ranges of supersaturation.

Effects of Impurities and Solvent. The presence of impurities can alter substantially the growth rates of crystalline materials. The alteration that is most common is for the growth rates to be decreased. It is often thought that impurities must be molecularly similar to the solute to have an impact on crystal growth. However, impurities with few similarities to the crystallizing species have been observed to reduce growth rates. This may be because structurally similar complexes have been formed between the impurity and solvent or another species in the solution, but such events are difficult to identify. Because additions of impurities to a crystallizing system most often result in a reduced crystal growth rate, it is critical that feed solutions be as free from contamination as possible. This is one way in which the operation of process units upstream of a crystallizer can affect crystal size distribution. Therefore, it is suggested that a strict protocol be followed in operating units such as reactors or other separation equipment upstream of the crystallizer. Equally important is monitoring the composition of recycle streams so as to detect the accumulation of impurities. Moreover, crystallization kinetics obtained on small-scale equipment for use in design or analysis should be obtained using solutions as similar as possible to that expected in the full-scale process.

Another important effect associated with the presence of impurities is that they may change the crystal habit. Habit alteration is considered to result from unequal changes in the growth rates of different crystal faces. Davey[41] reviews the role of impurities in the general context of habit modification. Surfactants, especially, have been observed to modify growth rates of individual faces and thereby change the habit of a crystal.[42]

The mechanism by which an impurity affects crystal growth rate is considered to involve adsorption of the impurity onto the crystal surface. Once located on the surface, the impurity forms a barrier to solute transfer from the solution to the crystal. This model has been used to relate the growth rate of a crystal to the concentration of impurity in solution through an adsorption isotherm.[43,44] An alternative theory calls for the adsorbed impurity to occupy active growth sites, thereby preventing attachment of solute to the surface. Still another concept of the impact of an impurity on growth rate calls for integration of the impurity into the crystal structure, which leads to two deleterious effects: reduction of growth rate and production of impure product. Few generalities exist in describing the role of impurities in crystallization phenomena, and it is usual to rely on experimental data that are often correlated empirically. Mullin[12] lists several systems in which habit modifiers have been used effectively. An interesting prospect raised by the work of McMahon et al.[35] is that impurities may have a strong effect on the formation of solute clusters and ultimately crystal growth.

The solvent from which a material is crystallized can influence the crystal habit and growth rate. Bourne[45] ascribes the effects of a solvent to two sets of factors: one has to do with the effects of solvent on mass transfer of the solute through adjustments in viscosity, density, and diffusivity; the second is concerned with the structure of the interface between the crystal and solvent. The analysis provided by Bourne concludes that a solute–solvent system having a high solubility is likely to produce a rough interface and concomitantly large crystal growth rates.

Crystal Growth in Mixed Crystallizers. As described earlier, crystal growth rates in multicrystal magmas are defined in terms of the rate of change of a characteristic dimension

$$G = \frac{dL}{dt}$$

(11.2-27)

It will be shown in a later section that the solution of a differential population balance requires a knowledge of the relationship between growth rate and size of the growing crystals. Moreover, this relationship can often be deduced from the form of population density data. A special condition, which simplifies such balances, results when all crystals in the magma grow at the same constant rate. Crystal–solvent systems that show this behavior are said to follow the ΔL Law proposed by McCabe,[46] while systems that do not are said to exhibit anomalous growth.

Anomalous growth is a term used here to indicate that growth rates of crystals in a magma are not identical or that the growth rate of an individual crystal or mass of crystals is not constant. Two theories have been used to explain growth rate anomalies: size-dependent growth and growth rate dispersion. As with systems that follow the ΔL Law, anomalous growth results in characteristic forms of population density data. Unfortunately, such data cannot be used to distinguish between size-dependent growth and growth rate dispersion, because both have the same qualitative effects on population density.

Size-Dependent Crystal Growth. A number of empirical expressions that correlate the apparent effect of crystal size on growth rate are summarized by White et al.[47] However, only a few of these meet the following requirements for use in population balances: (1) the expression must be continuous over the entire size range including $L = 0$; (2) the growth rate of zero-size crystals must not be zero; (3) the growth rate expression must allow convergence of moments of the crystal size distribution. The most often used correlation of size-dependent growth kinetic data is the Abegg–Stevens–Larson (ASL) growth rate expression.[48] This equation uses three empirical parameters to correlate growth rate with crystal size:

$$G = G°(1 + \gamma L)^b \qquad b < 1 \qquad (11.2-28)$$

where $G°$, γ and b are determined from experimental data.

There have been attempts to relate the kinetic parameter b to crystallizer variables; Garside and Jancic[49] showed a qualitative dependence on crystallizer volume, and Rousseau and Parks correlated b with $G°$ and τ. Several theories have been proposed to explain size-dependent growth kinetics, but none has been substantiated by direct observation or used to predict the onset of such behavior. One explanation seems particularly appealing: larger crystals impact impellers and other crystallizer internals with higher frequency and energy than smaller crystals; the larger crystals are therefore recipients of more surface breaks and irregularities that lead to higher growth rates.

At the small end of the size range, crystals in mixed magmas have been observed[33,51,52] to give the appearance of growing at much reduced rates. Theories proposed to explain these observations are as follows: (1) it is assumed that smaller crystals have fewer surface defects to enhance crystal growth; (2) smaller crystals are assumed to have an increased solubility; (3) when small crystals are formed they are more spherical than more-developed crystals and a high surface energy is involved when the crystal shape goes from spherical to that characteristic of the crystal.

Growth Rate Dispersion. The phenomenon of growth rate dispersion is the exhibition of different growth rates by crystals in a magma, even though they have the same size and are exposed to identical conditions. Both zero growth and abrupt changes in growth rates of individual crystals can be exhibited. This dispersion in growth rates can have significant effects on crystal population, leading in general to a broadening of the size distribution. As was discussed under size-dependent growth, growth rate dispersion may lead to population densities with characteristics that are different from those following the ΔL Law. These characteristics will become more apparent in Section 11.2-4. The phenomenological aspects of growth rate dispersion are discussed in the following paragraphs.

White and Wright[53] first characterized the effects of growth rate dispersion on a population of sucrose crystals by correlating the variance of the population about a mean size $\bar{L}$. They correlated the variance with the extent of growth of the crystals in a seeded batch crystallizer. The original crystals were of uniform size and the mean of the crystal size distribution was used as a measure of the extent of growth. The relationship between the variance and extent of growth was linear and had slope p, which was proportional to the rate of spreading of the distribution. Growth rate dispersion for sucrose was found to be significant ($p = 67$ μm), but negligible dispersion ($p = 1$ μm) was observed for aluminum trihydroxide. The slope p did not depend on the initial seed size (10–500 μm) or the stirring rate in the crystallizer; it decreased when growth rate or impurity concentration increased, or when brief intervals of dissolution occurred. Janse and de Jong[54,55] found growth rate dispersion important in the growth of large potassium dichromate and potassium alum crystals and demonstrated that this phenomenon could account for anomalous characteristics in the population density of crystals obtained from continuous, steady-state crystallizers. They concluded that growth rate dispersion could provide an explanation of what had been considered to be size-dependent growth, but they also indicated that both phenomena could occur simultaneously and cause the aforementioned anomalous behavior; in other words, the occurrence of growth rate dispersion does not rule out the possibility that single-crystal growth may be size dependent.

Recent microscopic studies of individual secondary nuclei have observed growth rate dispersion and size-dependent growth directly.[31,56] Garside et al.[32] report that two types of potassium alum secondary nuclei form after gentle contact of a seed crystal. Those less than 4 μm in size were created in larger numbers and were formed even if the solution was undersaturated. The number of such nuclei was relatively independent of supersaturation. These nuclei grew more slowly than larger ones and, in some cases, did not grow at a measurable rate. The number and size of nuclei between 4 and 50 μm in size increased as supersaturation was increased. Large crystals were not formed in saturated or undersaturated solution and in supersaturated solutions tended to grow more rapidly than small crystals.

Girolami and Rousseau[57] analyzed batch data to show that apparent size-dependent growth of potassium alum crystals is in fact a manifestation of growth rate dispersion. Crystals of citric acid monohydrate generated by contact nucleation were found by Berglund and Larson[58] to exhibit growth rate dispersion but not size-dependent growth. Very slow or zero growth rates for small secondary nuclei have been reported for aqueous solutions of pentaerythritol[59] and sodium chloride.[60] Several types of nuclei were distinguished by their size, shape, and individual growth behavior.

Two distinctly different mechanisms that lead to growth rate dispersion have experimental support. The first, which was proposed by Randolph and White,[61] assumes that all crystals have the same time-averaged growth rate but the growth rates of individual crystals fluctuate about some mean value. Direct evidence of random fluctuations in growth rates has been reported for magnesium sulfate heptahydrate[62] and potassium alum.[63] The second proposed mechanism for growth rate dispersion assumes that crystals are born with a characteristic distribution of growth rates but individual crystals retain a constant growth rate throughout their residence in a crystallizer. This mechanism is supported by findings on citric acid,[58] potassium nitrate,[64] and ammonium dihydrogen phosphate.[65,66] All these studies found nuclei to have a distribution of growth rates and individual crystals to have constant growth rates.

The surface integration step for crystal growth is thought to be the primary factor in both mechanisms of growth rate dispersion, and the BCF growth theory can be used to provide a qualitative explanation of the growth rate dispersion phenomenon. From the BCF theory, the growth rate of a crystal face is dependent on the number, sign, and location of screw dislocations on the surface of a growing crystal. Collisions of crystals with each other and crystallizer internals result in changes in the dislocation network of a crystal and lead to the random fluctuations of growth rates. Changes in the dislocation networks also occur simply due to the imperfect growth of crystal faces. A distribution of growth rates is a result of the varying dislocation networks and densities among nuclei and seed crystals.

Although evidence for both mechanisms of growth rate dispersion exists, separate mathematical models have been developed for incorporating the two mechanisms into descriptions of crystal populations. Random growth rate fluctuations were characterized by Randolph and White[61] with a growth rate diffusivity parameter D_G which is representative of the magnitude of the growth rate fluctuations. This model is similiar to that used to describe molecular diffusion and axial dispersion associated with velocity fluctuations. Mathematical models of growth rate dispersion due only to the growth rate distribution mechanism have been developed using different methods by Janse and de Jong[67] and Larson and coworkers.[68-70] These models characterize the growth rate distribution mechanism of growth rate dispersion by a distribution function having a mean growth rate $\overline{G}$ and a growth rate variance σ_G^2. Zumstein and Rousseau[71] have extended such models by showing how both mechanisms of growth rate dispersion can be included and by illustrating the relative effects of the two mechanisms on CSD from batch and continuous crystallizers.

Experimental efforts to determine the importance of the two mechanisms of growth rate dispersion have been involved mainly with the measurement of the spread in crystal size distribution during constant-supersaturation, isothermal batch crystallizations. The model for random growth rate fluctuations predicts the variance of the crystal size distribution will increase linearly with the extent of growth, while the model for a distribution of constant crystal growth rates predicts the variance to increase with the extent of growth squared. The original work by White and Wright on sucrose[53] supports the growth rate fluctuation mechanism, while the mechanism associated with a growth rate distribution is supported by data on contact nuclei of ammonium dihydrogen phosphate[66] and citric acid.[70] Although there is experimental support for both mechanisms, it is still unclear how crystallizing conditions affect the relative importance of the two mechanisms.

11.2-4 MODEL OF A WELL-MIXED CRYSTALLIZER

In this section, the principles involved in using a population balance to describe the crystal size distribution in the product from a well-mixed continuous crystallizer are illustrated. The primary objective of this treatment is to show how nucleation and growth kinetics can be evaluated from data on the crystal size distribution produced in such crystallizers. Detailed development of the theory and extensions of these principles to crystallizers that employ selective removal of crystals from the crystallizer internals or to batch or transient continuous crystallizers is provided by Randolph and Larson.[14]

APPLICATION OF A POPULATION BALANCE
The basis of a population balance is that the number of crystals in a system is a balanceable quantity. Such balances are coupled to the usual mass and energy balances describing any system, and the crystals are

assumed to be sufficiently small and numerous so that the crystal size distribution may be considered to be a continuous function of a variable with which individual crystals are to be characterized. In this discussion, the variable is the characteristic dimension (size) of the crystal.

A balance on the number of crystals in any size range—say L_1 to L_2—must account for crystals that enter and leave that size range by convective flow into and out of the control volume V_T and for crystals that enter and leave the size range by growth. Crystal breakage and agglomeration are ignored in this discussion, and it is assumed that crystals are formed by nucleation at or near size zero. Now define a quantity called *population density n* so that the number of crystals $dN(L)$ in the size range L to $L + dL$ is given by

$$dN(L) = V_T n(L)\, dL \tag{11.2-29}$$

It is important to note that either clear liquor volume or slurry volume can be taken as a basis for the definition of n. For example, if clear liquor volume is taken as the basis, V_T becomes the volume of clear liquor in the control volume. With this definition of the population density, it can be shown that the population balance on a well-mixed continuous crystallizer is given by

$$V_T \frac{\partial(nG)}{\partial L} + Q_o n - Q_i n_i = -\left(V_T \frac{\partial n}{\partial t} + n \frac{\partial V_T}{\partial t} \right) \tag{11.2-30}$$

where Q_i and Q_o are volumetric flow rates of either clear liquor or slurry into and out of the crystallizer, respectively. Equation (11.2-30) can now be simplified as follows: assume that the feed liquor contains no crystals ($n_i = 0$); restrict growth to follow the ΔL Law [$\partial(nG)/\partial L = G \partial n/\partial L$]; specify that the magma volume is constant ($\partial V_T/\partial t = 0$); and operate the crystallizer at steady state ($dn/dt = 0$).

If all four of these restrictions apply, and if a mean residence time τ (often called drawdown time) is defined as V_T/Q_o, the population balance can be written

$$G \frac{\partial n}{\partial L} + \frac{n}{\tau} = 0 \tag{11.2-31}$$

which may be integrated, using the boundary condition $n(0) = n°$, to give

$$n = n° \exp\left(-\frac{L}{G\tau} \right) \tag{11.2-32}$$

If the assumptions that lead to Eq. (11.2-32) are valid, and they often are, a straight line will result if n is plotted versus L on semilog coordinates. Furthermore, the growth rate of crystals can be determined from the slope of the plot, which is $-1/G\tau$. The nucleation rate is related to the intercept of the plot, $n°$, by the expression

$$B° = n° G \tag{11.2-33}$$

Equations (11.2-32) and (11.2-33) can be used to obtain nucleation and growth rates from population density measurements on a crystallizer that conforms to the assumptions stated above. This means that n must be determined as a function of L. There are various techniques for obtaining particle size distributions, and the procedures for determining n from these data varies according to the device used for the analysis. Although complete coverage of the advantages and disadvantages of available particle size analyzers is beyond the scope of this chapter, two common techniques are discussed briefly in the following paragraphs.

Sieves. Dry and/or wet sieve analyses provide differential or cumulative weight distributions. Obviously, crystals must be in the size range appropriate for a given set of sieves in order for this procedure to be used. The processes of sampling, sample treatment, and sieving can modify the crystal size distribution if care is not taken to prevent agglomeration, crystal abrasion, and crystal breakage. In most sieving techniques, obtaining representative material for sieving requires that crystals be filtered carefully from the mother liquor and dried, and prior to drying it may be necessary to wash the crystals with a nonsolvent to prevent agglomeration. When working the first time with a crystalline material, some experimentation may be necessary to develop a good sieving practice and to find an appropriate wash liquor.

Once satisfactory sieving has been accomplished, $n(L)$ is determined from the following relationship derived from the definition of the cumulative weight distribution data:

$$n(L) = \frac{M_T \Delta W(L)}{\Delta L\, \rho_c k_v L^3} \tag{11.2-34}$$

where M_T is magma density or mass of crystals per unit volume of clear liquor or slurry, $\Delta W(L)$ is the fraction of the total mass of crystals retained on the sieve having an average size L, ΔL is the interval in size between the sieve and the next largest sieve in the nest, ρ_c is the density of the crystal, k_v is the volume shape factor, and L is the average of the size of the sieve and the next largest sieve in the nest.

Electronic Particle Size Analyzers. These devices typically measure changes in a property such as light diffraction, light scattering, light blockage, or electrical conductance to determine the size of a particle. Changes in these measured quantities depend on the volume of the particle being sized, and hence this is the property actually determined by the instrument. The characteristic dimension assigned to the particle is that of a sphere having a volume equivalent to the measured volume. The output from these instruments can be in two forms: a particle volume distribution or a number distribution. The volume distribution data can be converted to population densities using Eq. (11.2-34) and assigning the particles the shape factor of a sphere, $\pi/6$. Several devices give population density data directly, and conversions to other forms are unnecessary.

There is considerable technique required in appropriately interpreting the data from electronic particle size analyzers. Vendors can be quite helpful in the development of a measurement procedure, but some experimentation should be expected with each new system.

KINETIC MODELS

Once nucleation and growth kinetics have been obtained using the procedures outlined above, correlations of them as functions of process variables are required to use the population balance model for predictive purposes. As discussed in Section 11.2-3, the important variables affecting nucleation and growth kinetics include temperature, supersaturation, magma density, and external stimuli that can cause nucleation, such as agitation or circulation rate of the magma. In Section 11.2-3 some of the important mechanisms by which nucleation and growth occur were described. These mechanisms should be the basis of models of crystallization kinetics, even though the actual correlations are empirical. Power-law functions are used most frequently in correlating nucleation and growth rates, but the choice of the independent variables can be justified from a mechanistic perspective. For example, systems that are believed to follow secondary nucleation mechanisms should include a variable reflecting the character of crystals in the crystallizer, such as magma density. The most commonly used power-law functions are

$$B^\circ = k_1 s^b M_T^j \tag{11.2-35}$$

and

$$G = k_2 s^g \tag{11.2-36}$$

Several methods for expressing supersaturation were described in Section 11.2-3; it is often very difficult to measure this quantity, especially in systems that have high growth rates. In these systems, supersaturation is often so small that it is difficult to measure accurately, and it can be neglected in writing a solute mass balance. However, it is important in setting nucleation and growth rates. A useful technique in such instances is to substitute growth rate for supersaturation by combining Eqs. (11.2-35) and (11.2-36). This gives

$$B^\circ = k_n G^i M_T^j \tag{11.2-37}$$

The constant k_n depends on process variables other than supersaturation and magma density, such as temperature, rate of agitation or circulation, and presence of impurities. If sufficient data are available, these variables may be separated from the constant by adding more terms in a power-law correlation. However, k_n is essentially a constant specific to the operating equipment and not transferable from one scale of equipment to another. The solute–solvent-specific constants i and j are obtainable from experimental data and may be used in scale-up.

PARAMETER EVALUATION

As shown above, growth rate G can be obtained from the slope of a plot of the log of population density data against crystal size; nucleation rate B° can be obtained from the same data by using the relationship $B^\circ = n^\circ G$, where $\ln n^\circ$ is the intercept of the plot with the axis at $L = 0$. Nucleation rates obtained by these procedures should be checked by comparison with values obtained from a mass balance. It can be shown that the magma density of a slurry in a perfectly mixed crystallizer is given by

$$M_T = 6 \rho_c k_v n^\circ (G\tau)^4 \tag{11.2-38}$$

Rearranging this equation and substituting $n^\circ = B^\circ/G$, we obtain

$$B^\circ = \frac{M_T}{6 \rho_c k_v G^3 \tau^4} \tag{11.2-39}$$

TABLE 11.2-5 Typical Kinetic Orders Reported[72]

System	Mode of Crystallization	Kinetic Order	Comments
Calcium sulfate–phosphoric acid	Precipitation	2.6–2.8	Indication that homogeneous nucleation dominates
Ammonium sulfate–water	Cooling	1.7	No enhancement in crystal size with increased suspension density noted
Ammonium alum–water	Cooling	2.1	Strong secondary nucleation effects were noted
Ammonium alum–water–ethanol	Salting out	1.0	Increased residence time had no effect on crystal size distribution
Ammonium sulfate–water–methanol	Salting out	4.0	Increased residence time had moderate effect on CSD
Sodium chloride–water–ethanol	Salting out	9.0	Increasing residence time strongly affected CSD
Ammonium alum–water–ethanol	Salting out	2.0	Used MSMPR theory to analyze data. Secondary nucleation effects not accounted for
Cyclonite–water–nitric acid	Precipitation	1.0	Used MSMPR theory to obtain kinetic order

Reproduced by permission of the American Institute of Chemical Engineers.

After nucleation and growth rates have been obtained from population density data, the exponents i and j in Eq. (11.2-37) can be determined using graphical techniques or multivariable linear regression. For example, i can be obtained from data on several runs of varying residence times. Each run will yield values of $B°$, $n°$, and G. If slurry concentration (magma density) M_T is the same for all runs, the slope of a logarithmic plot of $n°$ versus G will give i. If slurry concentration varies slightly from run to run, then a logarithmic plot of $B°/M_T$ versus G will smooth the data so that i can be obtained from the slope of the plot. Confidence in the parameters i and j requires that sufficient data be accumulated to make such correlations statistically reliable. Typical values of kinetic orders are given in Table 11.2-5, and a more complete list is given by Garside and Shah.[30]

CHARACTERISTICS OF CRYSTAL SIZE DISTRIBUTIONS FROM MSMPR CRYSTALLIZERS
The preceding discussion centered on the development of expressions for the population density function in terms of nucleation and growth kinetics. It is also possible to express the properties of a crystal size distribution in terms of a mass density function m. The two density functions can be shown to be related by the expression

$$m = \rho_c k_v L^3 n \tag{11.2-40}$$

The perfectly mixed, continuous, steady-state mixed-suspension mixed-product removal (MSMPR) crystallizer is restrictive in the degree to which characteristics of a crystal size distribution can be varied. Indeed, examination of Eqs. (11.2-32) and (11.2-40) shows that once nucleation and growth kinetics are fixed in these systems the crystal size distribution is determined in its entirety. In addition, such distributions have the following characteristics:

1. The mode or maximum of the mass density function occurs at the *dominant crystal size* L_D, which is given by

$$L_D = 3G\tau \tag{11.2-41}$$

2. *Moments* of the population density are given by the expression

$$m_i = i!n°(G\tau)^{i+1} \tag{11.2-42}$$

3. The *coefficient of variation* of the mass density function, which is a measure of the spread of the distribution about the dominant crystal size, is 50%. Such a distribution may be too broad to be acceptable for certain crystalline products, such as sugar.

4.   The magma density M_T (mass of crystals per unit volume of slurry or liquor) may be obtained from the third moment of the population density function. As shown above, this quantity is related to nucleation and growth rates by Eq. (11.2-38). Although magma density is a function of the kinetic parameters $n°$ and G, it can be measured independently of crystal size distribution and, where possible, it should be used, as indicated above, as a constraint in evaluating nucleation and growth rates from measured crystal size distributions.

5.   A pair of kinetic parameters, one for nucleation rate and another for growth rate, describe the crystal size distribution for a given set of crystallizer operating conditions. Furthermore, variation in one of the kinetic parameters without changing the other is not possible.

6.   Properties of the distribution may be evaluated explicitly from the moment equations:

$$N = \text{total number of crystals per unit volume}$$

$$= \int_0^\infty n \, dL = m_0 = n°G\tau \tag{11.2-43a}$$

$$L_T = \text{total length of crystals per unit volume}$$

$$= \int_0^\infty nL \, dL = m_1 = n°(G\tau)^2 \tag{11.2-43b}$$

$$A_T = \text{total area of crystals per unit volume}$$

$$= k_a \int_0^\infty nL^2 \, dL = k_a m_2 = 2k_a n°(G\tau)^3 \tag{11.2-43c}$$

$$V_C = \text{total volume of crystals per unit volume}$$

$$= k_v \int_0^\infty nL^3 \, dL = k_v m_3 = 6k_v n°(G\tau)^4 \tag{11.2-43d}$$

PARAMETERS FROM SYSTEMS EXHIBITING ANOMALOUS GROWTH

Two types of anomalous growth behavior were described in Section 11.2-3: growth rate dispersion and size-dependent growth. As pointed out, the effects of these mechanistically different phenomena on crystal size distributions are the same; both cause curvature in semilog plots of population density against size. Moreover, when size-dependent growth rates increase with crystal size, both cause a broadening of the distribution. In this discussion, methods will be presented that illustrate evaluation of the parameters in systems following size-dependent growth kinetics. Research is in progress on development of similar procedures for systems deemed to follow growth rate dispersion anomalies, and the current literature should be followed for such cases.

A number of empirical size-dependent growth expressions have been developed. Of these, the ASL model given in Eq. (11.2-28) is the most commonly used. Substituting this equation into the differential population balance given by Eq. (11.2-31), the steady-state population density function can be derived as

$$n = n°(1 + \gamma L)^{-b} \exp\left(\frac{1 - (1 + \gamma L)^{1-b}}{G°\tau\gamma(1 - b)}\right) \quad (b < 1) \tag{11.2-44}$$

This equation is often simplified by restricting γ to be equal to $1/G°\tau$, which reduces the number of model parameters by one and often results in a fit to data that is nearly as good as the three-parameter model. The reduced ASL model ($\gamma = 1/G°\tau$) leads to the following equation for population density:

$$n = n°\left(1 + \frac{L}{G°\tau}\right)^{-b} \exp\left(\frac{1 - (1 + L/G°\tau)^{1-b}}{1 - b}\right) \tag{11.2-45}$$

Garside and Jancic[49] noted the relationship of several magma properties to the parameter b in the ASL growth model. O'Dell and Rousseau[73,74] derived formulas for estimating the characteristics of crystal size distributions in terms of parameters in size-dependent growth kinetic models and for crystallizers with selective removal of fines and/or product. For systems whose growth follows the reduced ASL equation, it was shown that total length of crystals per unit volume of magma or liquor, surface area of crystals per unit volume, mass of crystals per unit volume (magma density), mass average size, coefficient of variation, and dominant size could be calculated from the equations

$$L_T = C_1 n°(G°\tau)^2 \tag{11.2-46}$$

$$A_T = k_a C_2 n°(G°\tau)^3 \tag{11.2-47}$$

TABLE 11.2-6 Constants for Evaluating Magma Properties of Systems that Follow Reduced ASL Kinetics

b	C_1	C_2	C_3	C_L	CV	C_D
0.9	3.6602	526.71	5.8409 E06	307680.0	333.40	1681.2
0.8	2.5104	48.373	5205.4	606.09	154.28	57.003
0.7	2.0041	18.058	539.05	78.417	108.47	19.032
0.6	1.7038	9.9331	140.86	28.428	86.786	10.748
0.5	1.5000	6.5000	57.750	15.351	74.203	7.4314
0.4	1.3502	4.6865	29.933	10.064	66.014	5.6882
0.3	1.2344	3.5929	17.862	7.3600	60.256	4.6220
0.2	1.1414	2.8732	11.696	5.7630	55.975	3.9044
0.1	1.0646	2.3964	8.1769	4.7239	52.656	3.3887
0.0	1.0000	2.0000	6.0000	4.0000	50.000	3.0000

$$M_T = k_v C_3 n°(G°\tau)^4 \tag{11.2-48}$$

$$\bar{L} = \left(\frac{C_4}{C_3}\right) G°\tau = C_L G°\tau \tag{11.2-49}$$

$$CV = \left(\frac{C_3 C_5}{C_4^2} - 1\right)^{1/2} \times 100\% \tag{11.2-50}$$

$$L_D = C_D G°\tau \tag{11.2-51}$$

The constants in Eqs. (11.2-46)–(11.2-51) are functions of b only, and they are given in Table 11.2-6. The table can be used to estimate the desired property of a crystal size distribution from the model parameters and Eqs. (11.2-46)–(11.2-51), or it can be used in evaluating the model parameters from experimental data as described below.

It should be clear that the form of the population density function given by Eq. (11.2-32) (systems that follow the ΔL Law) is much simpler than those of Eqs. (11.2-44) and (11.2-45). This increase in complexity for systems that exhibit anomalous growth makes extraction of model parameters more difficult. Not only do anomalous growth models have more parameters, but the curvature in plots of the log of the population density against size means that parameters must be determined by using nonlinear least-squares regressions. For example, fitting the reduced form of the ASL equation requires evaluation of the parameters $n°$, $G°$, and b in Eq. (11.2-45), and the estimate of $n°$ is extremely sensitive to the few data points frequently obtained at very small sizes. Moreover, nonlinearities in the semilogarithmic relationship between population density and size frequently are most severe at small sizes. Consequently, it is very difficult to determine $n°$ with any real accuracy.

Rousseau and Parks[50] simplified this problem by substituting Eq. (11.2-48) into Eq. (11.2-45) to eliminate $n°$

$$n = \left[\frac{M_T}{C_3(b)\,\rho_c k_v(G°\tau)^4}\right]\left(1 + \frac{L}{G°\tau}\right)\exp\left(\frac{1 - (1 + L/G°\tau)^{1-b}}{1 - b}\right) \tag{11.2-52}$$

Experimentally determined population density, measured M_T and the values of C_3 in Table 11.2-6 can then be used to evaluate the two remaining adjustable parameters b and $G°$. Interpolation of C_3 is assisted by a polynomial fit to b

$$\ln C_3(b) = 1.8881 + 1.5829b + 5.5316b^2 + 7.618b^3$$
$$- 32.36b^4 + 34.22b^5 \tag{11.2-53}$$

The effects of the two growth rate dispersion mechanisms on the CSD in a mixed-suspension, mixed-product removal (MSMPR) continuous crystallizer are different. Random growth rate fluctuations do not affect the spread of the CSD from an MSMPR crystallizer and the resulting population density function is indistinguishable from that obtained assuming applicability of the ΔL Law. However, growth rate fluctuations do increase the mean crystal size from that which is expected from the ΔL Law. On the other hand, CSD from a continuous crystallizer is affected noticeably by assigning a distribution of constant crystal growth rates to the nuclei. This mechanism causes the type of curvature in population density plots that typically has been ascribed to size-dependent growth.

A model that combines growth rate dispersion mechanisms has been derived by Zumstein and Rousseau.[74] The expression for population density derived from their model is

$$n(L) = \frac{B^\circ}{\int_0^\infty [Gf^\circ(G)/S]\, dG} \int_0^\infty f^\circ(G) \exp\left(-\frac{SL}{G\tau}\right) dG \qquad (11.2\text{-}54)$$

where $f^\circ(G)$ is the distribution of nuclei growth rates and S, which is between 0 and 1, is given by

$$S = \left(\frac{\tau G^2}{2D_G}\right)\left[\left(1 + \frac{4D_G}{\tau G^2}\right)^{1/2} - 1\right] \qquad (11.2\text{-}55)$$

The other terms are as defined earlier. This equation was used to demonstrate that random growth rate fluctuations may damp out the effect of the growth rate distribution function, thereby causing the effects of growth rate dispersion to be unobservable in a continuous crystallizer.[71] Moreover, it is shown that an understanding of growth rate dispersion may be a key factor in obtaining realistic models for scale-up.[75]

PREFERENTIAL REMOVAL OF CRYSTALS
Crystal size distributions produced in a perfectly mixed crystallizer are highly constrained, and nucleation and growth kinetics are determined exclusively by the crystalline system and the variables that define the state of the crystallizer. Accordingly, specification of the feed variables (flow rate, temperature, pressure, and solute concentration), the crystallizer variables (temperature, pressure in an adiabatic cooling process, agitation or magma circulation rate, and solids fraction in an evaporative crystallizer), and crystallizer geometry determines magma density and other characteristics of the crystal size distribution.

Crystallizers are made more flexible by the introduction of selective removal devices that alter the residence time distribution of materials flowing from the crystallizer. The functions of classified removal are best described in terms of idealized models of clear-liquor advance, classified-fines removal, and classified-product removal. Clear-liquor advance is the removal of mother liquor from the crystallizer without simultaneous removal of crystals. The effects of implementing this removal function are described in Section 11.5-2. The primary objective of fines removal is preferential withdrawal from the crystallizer of crystals whose size is below some specified value; with the withdrawal, dissolution, and return of the solution to the crystallizer, the growth of larger crystals is enhanced. Classified-product removal is carried out to remove preferentially those crystals whose size is larger than some specified value.

The effects of each of the selective removal functions on crystal size distributions can be described in terms of the population density function n. If it is assumed that perfect classification of fines and product crystals is implemented, then the following expression for population density results:

For $L \leq L_F$

$$n = n^\circ \exp\left(-\frac{RL}{G\tau}\right) \qquad (11.2\text{-}56)$$

For $L_F < L < L_C$

$$n = n^\circ \exp\left(-\frac{(R-1)L_F}{G\tau}\right) \exp\left(-\frac{L}{G\tau}\right) \qquad (11.2\text{-}57)$$

For $L \geq L_C$

$$n = n^\circ \exp\left(-\frac{(R-1)L_F}{G\tau}\right) \exp\left(\frac{(Z-1)L_C}{G\tau}\right) \exp\left(-\frac{ZL}{G\tau}\right) \qquad (11.2\text{-}58)$$

where the fines are below size L_F and have a removal rate R times what would exist for perfect mixing, the coarse product is greater than size L_C and has a removal rate Z times that of the perfectly mixed crystallizer. If fines removal is implemented alone, $Z = 1$, and if classified product removal is implemented without fines removal, $R = 1$. Selection of a crystallizer that has both classified-fines and classified-product removal functions usually is done so as to combine the best features of each: increased dominant size and narrower distribution. The model of the crystallizer and selective removal devices that led to Eqs. (11.2-56)–(11.2-58) is referred to as the R-Z crystallizer. It is an obvious idealization of actual crystallizers because of the perfect cuts assumed at L_F and L_C. However, it is a useful approximation to many systems and it allows qualitative analyses of complex operations.

Although many commercial crystallizers operate with some form of selective crystal removal, devices that implement these functions add to the complexity of the operation. In addition, Randolph et al.[76] have established that classified-product removal can lead to cycling of the crystal size distribution. A review of the simulation and control of crystal size distributions has been provided by Randolph.[77] Properties of the crystal size distribution have been given in terms of R and Z and the moments of the crystal size distribution.[74]

BATCH CRYSTALLIZATION

As with continuous crystallizers, the mode by which supersaturation is generated affects the yield and crystal size distribution, but batch crystallizers are strongly influenced by the rate at which supersaturation is generated. For example, in a cooling mode there are several avenues that can be followed in reducing the temperature of the batch system: natural cooling in which a batch is simply allowed to release heat to a constant-temperature heat sink, cooling in which the rate of heat transfer is constant, cooling in which the supersaturation is maintained constant, and size-optimal cooling whose objective is to vary the cooling rate so that the supersaturation in the crystallizer is adjusted to produce an optimal crystal size distribution. The effects of these and other modes of operation on crystal size distribution are discussed by Tavare et al.[78] Other important features of batch crystallizers are reviewed by Wey[79] and Randolph and Larson.[14]

Perhaps the most troublesome aspect of batch crystallizers is the difficulty associated with reproducing crystal size distributions in going from one batch to the next. This may be overcome through seeding[17] and control of mixing conditions. In general, however, the development of methods for design and analysis of batch crystallizers lags those for continuous systems.

11.3 SOLUTION CRYSTALLIZATION

Crystallization is a commonly used industrial separation and purification technique. If the desired product is an evaporate (or filtrate) rather than a crystalline phase, then the process emphasis is primarily that of separation rather than purification. In either case there is a strong interaction and dependence between both degree of separation and purification and the particulate nature of the solid phase produced. Fundamental research on crystallization has focused mainly on understanding the variables that influence the structure and size of the crystalline phase, recognizing that better knowledge and control of this aspect would permit improvement of the unit operation of crystallization, both as a separation and purification technique.

11.3-1 State-of-the-Art

Use of mathematical models for designing industrial crystallizers has lagged other chemical processing operations because of the complexities of rationally describing and interrelating growth and nucleation kinetics with the process configuration and mechanical features of the crystallizer. Crystallizers have traditionally been designed by scale-up of pilot- or bench-scale data, taking care to control supersaturation levels and to maintain similar crystal residence time, vessel configuration, and hydraulic regimes. System kinetics, even though controlling the ultimate crystal size distribution, were thought to be subjects better relegated to the laboratory and were rarely used for crystallizer design.

The design and analysis of crystallization processes of the continuous well-mixed suspension type have developed into formal design algorithms which can now be applied in situations of industrial importance. Specific process configurations that can be modeled rigorously include fines destruction, clear-liquor advance, classified-product removal, vessel staging, and seeding. Such process configurations can be modeled rigorously with a population balance CSD algorithm or, in many cases, with an alternative mass-based formulation. Crystal size distribution transients and stability can also be evaluated as well as the effects of size-dependent growth rate and suspension concentration-dependent secondary nucleation. Simulation capabilities have in fact outstripped the ability to monitor appropriate control variables and to manipulate and control liquid and solids residence-time distributions in a crystallizing environment.

A considerable amount of skepticism naturally is expressed by crystallization practitioners concerning the unbridled use of parameters generated in well-controlled small-scale equipment to predict performance of industrial-scale crystallizers. Many factors contribute to the nonideal behavior of industrial crystallizers, and not all can be easily explained.

The ability to predict crystal behavior in complex systems is ahead of our ability to manipulate crystal and liquid residence times. Generalizations of the MSMPR equations have been made to predict CSD with arbitrary process configurations and kinetics; however, there is no guarantee that an assumed process residence-time distribution can be physically implemented to produce a customized product CSD. Size distributions in cascaded crystallizers, for example, multiple-effect evaporators, can be computed if kinetic parameters are known or assumed. No investigator, however, has reported comparative data showing the applicability of using these kinetic parameters for predicting complex crystallizer behavior. In addition, no definitive technique exists for predicting size-dependent growth rates in systems that exhibit strong diffusion-controlled crystal growth. If the specific size dependency of growth rate is known, the effect on crystal size distribution in a given crystallizer configuration can be computed. Quantifying the effects of secondary nucleation in industrial crystallizers is moving forward rapidly, but much is yet to be learned concerning the complex mechanisms that create nuclei. Moreover, study of the phenomenon of growth rate di. persion is new and little has been done to investigate its role in crystallizer scale-up.

Although numerous design methods and kinetic theories to analyze specific crystallizer config' rations now exist, no clear-cut guidance has been provided by researchers to assist in the choice between crystallizer types and/or modes of operation. Design algorithms for classified and plug-flow crystallizers require considerable strengthening to be of practical value, and documented case studies of the application of mixed-suspension, mixed-product-removal (MSMPR) design methods, especially for systems employing fines removal, product classification, and staging, are desired to establish further the usefulness of MSMPR theory. Suggested fundamental research areas include: (1) interphase kinetic mechanisms, particularly that of secondary nucleation, (2) growth and dissolution (digestion) rates of submicron crystals, (3) formation, migration, and removal of crystal occlusions, and (4) growth rate dispersion. With continued improvement of design techniques and the emergence of realistic kinetic models, the practice of crystallization has the potential of becoming a highly developed science.

11.3-2 Developing Material and Energy Balances

In optimizing any separation process it becomes necessary to compute the product recovery efficiency of the operation as a function of design variables. For crystallization operations the theoretical maximum product recovery or yield from the crystallizer is defined by the following general relationship:

$$\text{Yield (\%)} = \frac{100(F_s R_f - S R_c)}{F_s R_f} \tag{11.3-1}$$

where R_f = mass of dissolved solute in the feed per unit mass of solvent
 R_c = mass of dissolved solute in liquid discharged from the crystallizer per unit mass of discharged solvent
 F_s = mass feed rate of solvent into crystallizer
 S = mass discharge rate of solvent leaving crystallizer

Solvent feed rate, F_s, and the ratio of dissolved solute to solvent in the feed, R_f, are usually fixed by upstream requirements, although the crystallizer designer may have some flexibility in this regard. The composition of dissolved solute in the liquid leaving the crystallizer, R_c, is determined by solubility which is set by the temperature of the operation. Thus, R_c can be manipulated to some extent by controlling temperature or by changing solvent. The other manipulative variable is the solvent discharge rate, S. In evaporative systems, solvent is vaporized by adding heat to the system; thus, S decreases and recovery increases as solvent is evaporated and removed as a separate product stream. In adiabatic cooling the quantity of solvent evaporated is fixed by energy balance constraints. For an evaporative system or a combination adiabatic cooling and evaporation operation, the designer can control pressure to set R_c and supply external heat to control S and maximize yield.

The conclusion of the preceding discussion is that for a given feed concentration and solubility relationship the mode of crystallizer operation governs the maximum recovery of solute. For a cooling crystallization system, where no solvent is removed, the only control that affects maximum solute recovery is crystallizer temperature. For evaporative systems, recovery is influenced by both the system temperature, often set by pressure, and by the quantity of solvent vaporized and removed from the system. Derivation and use of expressions that describe the yield and suspension concentration as functions of feed composition, crystallizer composition, and quantity of solvent evaporated follow in this section.

The overall material balance for a binary system, wherein a pure solid component is crystallized by cooling or evaporation techniques, is derived and rearranged in the following section to provide maximum solute yield and total suspended solids information. The approach described below is also useful for deriving material balance expressions for binary systems that form hydrates and adducts.

In Fig. 11.3-1 the quantities, F_s, R_s, and R_c are described in conjunction with Eq. (11.3-1). R_{se} is the ratio of the mass of solvent evaporated per unit mass of solvent fed, whereas S and C are the solution and crystal removal rate, respectively.

From the overall material balance,

$$F_s + F_s R_f = F_s R_{se} + S + C \tag{11.3-2}$$

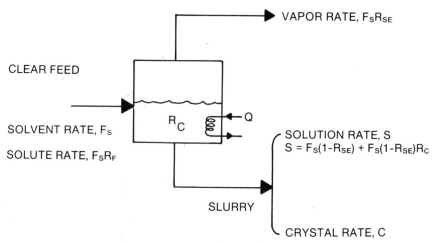

FIGURE 11.3-1 Mass balance for binary system.

the solution removal rate becomes

$$S = F_s(1 - R_{se}) + F_s(1 - R_{se})R_c \qquad (11.3-3)$$

Letting $F_s = 1$ unit of mass and substituting Eq. (11.3-3) into (11.3-2) and solving for C, we obtain

$$C = R_f - R_c(1 - R_{se}) \qquad (11.3-4)$$

The maximum yield is defined by

$$Y_{max} = \frac{C}{R_f} = \frac{R_f - R_c(1 - R_{se})}{R_f} \qquad (11.3-5)$$

Note: Equation (11.3-1) has been recast into Eq. (11.3-5) which is computationally more useful.

The maximum suspended solids (weight fraction) is determined by

$$TS = \frac{C}{1 + R_f - R_{se}} = \frac{R_f - R_c(1 - R_{se})}{1 + R_f - R_{se}} \qquad (11.3-6)$$

Equations (11.3-5) and (11.3-6) are modified to satisfy specific situations. For example, Eq. (11.3-5) reduces to the following equations for the operating modes shown:

Cooling Crystallization Only

$$R_{se} = 0 \Rightarrow Y_{max} = \frac{R_f - R_c}{R_f} \qquad (11.3-7)$$

Evaporation Only

$$R_f = R_c \Rightarrow Y_{max} = \frac{R_f - R_f(1 - R_{se})}{R_f} = R_{se} \qquad (11.3-8)$$

Adiabatic Evaporation

$$R_{se} = R_{ad} \Rightarrow Y_{max} = \frac{R_f - R_c(1 - R_{ad})}{R_f} \quad \text{(solvent evaporation fixed)} \qquad (11.3-9)$$

The preceding expressions can be solved provided the composition of the exit stream is known. In many instances, it is acceptable to assume that the composition corresponds to saturation conditions; systems in which this occurs are said to exhibit fast-growth or Class II behavior. Should growth kinetics be too slow to use essentially all the supersaturation (i.e., the solution concentration is greater than that at saturation), the system is said to exhibit slow-growth behavior and is classified as a Class I system.

EXAMPLE
Determine the maximum yield and slurry concentration obtained if a feed containing 80% naphthalene and 20% benzene is cooled indirectly to 45°C.

Solution. From the phase diagram in Fig. 11.2-2, the mass ratios of solute to solvent are

$$R_f = \frac{80}{20} = 4.0 \quad \text{and} \quad R_c = \frac{60}{40} = 1.5$$

Using Eq. (11.3-7), the maximum yield is

$$Y_{max} = \frac{R_f - R_c}{R_f} = \frac{40.0 - 1.5}{4.0} = 0.625$$

Therefore, 62.5% of naphthalene in the feed is theoretically converted to solid crystals.
From Eq. (11.3-6), the weight fraction of solids in suspension is

$$TS = \frac{R_f - R_c}{R_f + 1} = \frac{4.0 - 1.5}{4.0 + 1.0} = 0.50 \text{ weight fraction}$$

Thus, 1 lb of feed will produce 0.5 lb of crystals and an equal quantity of residual solution.

EXAMPLE
Explore the maximum yield and suspended solids behavior of a binary system having the solubility data shown in Table 11.3-1 for a feed containing 27.5 wt. % A and a feed temperature of 100°C. (Assume that the heat of fusion for the solute is 60 Btu/lb$_m$ and the heat of vaporization for the solvent is 180 Btu/lb$_m$.)

Solution. The yield from a cooling and evaporating crystallizer is given by Eq. (11.3-5):

$$Y_{max} = \frac{R_f - R_c(1 - R_{se})}{R_f}$$

where

$$R_{se} = \frac{\text{solvent evaporated (lb}_m)}{\text{solvent fed (lb}_m)}$$

In the special case of adiabatic vacuum cooling, the quantity of solvent evaporated is fixed by the crystallizer heat balance. Thus, R_{se} is a fixed ratio—referred to as R_{ad}—which can be determined from the following relationship:

$$R_{ad}\lambda_v = \lambda_f(R_f - R_c) + C_p\Delta T_c(1 + R_f) + \lambda_f R_c R_{ad}$$

Heat removed by solvent evaporation	Heat released by cooling crystallization	Heat input from sensible heat of feed	Heat released by "salting out"	(11.3-10)

TABLE 11.3-1 Solubility Data for Binary System

Solute per Solvent (g/g)	Temperature (°C)	Solution Vapor Pressure (mm Hg)
0.100	65	160
0.072	60	135
0.043	50	90
0.028	40	60
0.018	30	40

TABLE 11.3-2 Crystallizer Conditions as Function of Temperature

Pressure (mm Hg)	Temperature (°C)	R_c	Yield (%)	Suspended Solids (wt.%)	R_{ad}
160	65	0.100	83.5	31.5	0.374
135	60	0.072	89.0	35.1	0.418
90	50	0.043	94.4	40.9	0.503
60	40	0.028	96.9	46.1	0.583

The relationship above can be solved for R_{ad}:

$$R_{ad} = \frac{\lambda_f(R_f - R_c) + C_p\Delta T_c(1 + R_f)}{\lambda_v - \lambda_f R_c} \qquad (11.3\text{-}11)$$

The yield and suspended solids percentages as a function of operating pressure were computed for adiabatic vacuum operation as shown in Table 11.3-2.

Figure 11.3-2 indicates the sensitivity of product yield and total suspended solids to the ratio of solvent evaporated to the solvent contained in feed. The adiabatic operating line and constant temperature evaporation curves for 30 and 65°C are indicated. The mass and energy balances will not be satisfied in the region to the left of the adiabatic operating line unless condensed solvent is readmitted to the system. Table 11.3-3 summarizes material balance equations for pure-component forming binary systems.

11.3-3 Process Equipment for Solution Crystallization

This section discusses various features of commercial types of solution crystallizers used in industrial processes. In solution crystallization a diluent solvent is added to the mixture or is already present as a carrier liquid. The solution is then cooled and/or solvent is evaporated to cause crystallization. Another feature of solution crystallization is that the solid phase is formed and maintained below its pure-component freezing temperature.

Selection of a specific crystallizer type is controlled to some extent by choice of the operating mode, three types of crystallizer configurations are commonly used. Variations of these basic types provide special features or circumvent certain difficulties. Table 11.3-4 lists advantages and disadvantages for specific types of crystallizers.

A cross-section of a classified suspension crystallizer is shown in Fig. 11.3-3. The circulating line drawoff is typically in the upper section of the suspension chamber and returns tangentially into the evap-

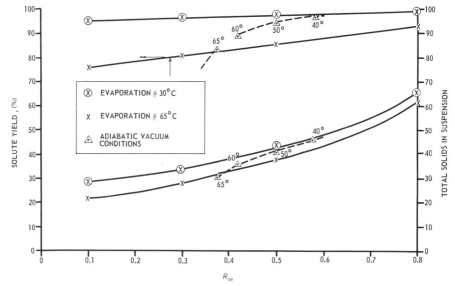

FIGURE 11.3-2 Crystallizer performance as a function of solvent removal rate.

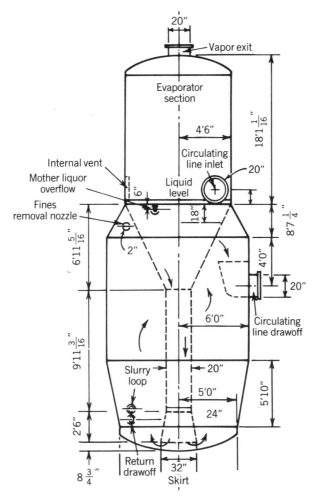

FIGURE 11.3-3 Classifying crystallizer cross section.

orator section. Fines and/or clear liquor drawoff provision is also available in the upper section of the suspension chamber. Product is removed from discharge nozzles in the base or by fitting the base with an elutriation leg. Figure 11.3-4 illustrates closed- and open-type classifying crystallizer designs. The closed type is necessary for systems that process nonaqueous solvents whereas the open type can in fact use an open-top and less expensive suspension chamber. Two modified vapor release arrangements and a technique for providing uniform upward flow in the suspension chamber are depicted on Fig. 11.3-5.

Figures 11.3-6 and 11.3-7 show four types of mixed suspension crystallizers, several of which are fitted with upflow draft tubes and internal annular baffles. An external circulation loop generally emanates from the upper section of the baffled zone and enters near the propeller in the bottom of the draft tube. A stream containing fines and/or a stream of clear liquor are taken from the upper section of the annular baffled zone.

Several forced circulation crystallizer configurations are shown in Fig. 11.3-8. In most of the designs the liquid is well mixed even though the solid phase may or may not be. Many of the designs remove circulating magma from or near the bottom, cycle it through an external heat exchanger, and return it into the evaporation chamber. Hybrid crystallizers that combine other features are illustrated in Fig. 11.3-9.

11.3-4 Using Theory for System Design

This case study involves the use of crystallization kinetics and overall material and energy balances to analyze and design a solution crystallization system. Using kinetic information to guide design adds another

TABLE 11.3-3 Summary of Binary Material Balance Equations

Mode of Operation	Total Solids (lb solids/lb slurry)	Fraction of Solute in Feed Recovered as Solid
Cooling crystallization	$\dfrac{R_f - R_c}{R_f + 1.0}$	$\dfrac{R_f - R_c}{R_f}$
Adiabatic vacuum cooling (with reflux)	$\dfrac{R_f - R_c}{R_f + 1.0}$	$\dfrac{R_f - R_c}{R_f}$
Adiabatic vacuum cooling (without reflux)	$\dfrac{R_f - R_c + R_c\,R_{ad}}{R_f + 1.0 - R_{ad}}$	$\dfrac{R_f - R_c\,(1 - R_{ad})}{R_f}$
Vacuum cooling and evaporation	$\dfrac{R_f - R_c + R_c\,R_{se}}{R_f + 1.0 - R_{se}}$	$\dfrac{R_f - R_c\,(1 - R_{se})}{R_f}$
Evaporation	$\dfrac{R_f\,R_{se}}{R_f + 1.0 - R_{se}}$	R_{se}

where $R_{ad} = \dfrac{\lambda_f (R_f - R_c) + C_p\,\Delta T_c\,(1 + R_f)}{\lambda_v - \lambda_f R_c} = \dfrac{\text{solvent (lb}_m) \text{ evaporated via adiabatic cooling}}{\text{solvent (lb}_m) \text{ in feed}}$

$R_{se} = \dfrac{\text{solvent (lb}_m) \text{ evaporated}}{\text{solvent (lb}_m) \text{ in feed}}$

$\Delta T_c = \text{feed temperature} - \text{crystallizer temperature} = (T_f - T_c)$

$C_p = \text{heat capacity of feed material}$

$\lambda_v = \text{heat of vaporization of solvent}$

$\lambda_f = \text{heat of crystallization of solute}$

TABLE 11.3-4 Advantages and Disadvantages of Specific Types of Crystallizer

Type 1. Classified Suspension Crystallizer

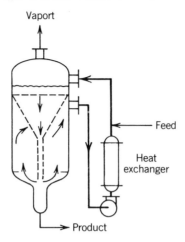

Major Application: General purpose. A fluidized bed of solids is present in the suspension chamber of the crystallizer.

Advantages: Not sensitive to liquid-level fluctuations. Can have an inventory of large ''chunks'' of agglomerated crystalline material.

Disadvantages: Natural solid sump in bottom. Boilout cycles are usually frequent. Internal down-pipe may plug. Difficult to model and scale-up. Flow patterns in suspension chamber not uniform.

Type 2. Draft Tube Crystallizer—Well Mixed

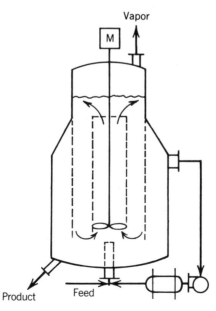

Major Application: Used for adiabatic cooling and medium evaporation loads; also adaptable for cooling crystallization.

Advantages: Uniform liquid flow pattern. Has chunk inventory. Can be scaled up from small-scale experiments.

TABLE 11.3-4 (*Continued*)

Disadvantages: Draft tube subject to fouling if liquid level is too low. Internal baffles, supports, mixer, and so on clutter up internals. Internal agitator requires maintenance.

Type 3. Forced Circulation Crystallizer

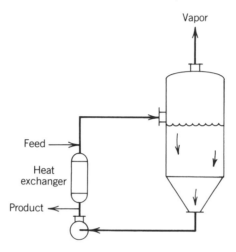

Major Application: General purpose crystallizer. Can be used for any mode of crystallization including indirect cooling.

Advantages: Simple design. Additional loop can be added for large evaporative loads. Operation not sensitive to liquid level.

Disadvantages: No chunk inventory. Internal flow patterns difficult to discern. Scale-up from small-scale tests unreliable.

---

dimension to the design engineer's tools. The sensitivity of crystal size distribution to a number of design variables can be evaluated if kinetics are available. These include the influence of:

- Holding time
- Fines removal
- Product classification
- Slurry density
- Temperature
- Seeding
- Series operation

Direct scale-up of kinetic information from bench-scale tests usually is unreliable since the influence of mixing intensity on secondary nucleation is difficult to assess, as is the simulation of actual plant operating conditions. However, the exponents of the nucleation rate model shown in Eq. (11.3-12) can be evaluated in small-scale equipment in addition to the sensitivity of the nucleation rate coefficient k_N to changes in temperature and/or level of agitation.

$$B° = k_N G^i M_T^j \qquad (11.3\text{-}12)$$

The problem in scale-up from small-scale tests is often the magnitude of the coefficient k_N rather than the exponents i and j. Table 11.3-5 summarizes the useful crystallizer information that is gleaned from tests at various scales of operation.

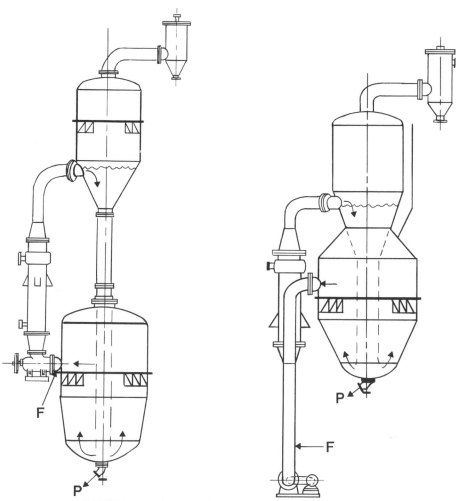

FIGURE 11.3-4 Open and closed types of classifying crystallizer.

Example

The project is to design a crystallization system to produce 1000 lb_m/h of compound A from solution via crystallization. Solubility and vapor pressure data are available as are limited crystal size distribution data from an existing pilot crystallizer.

(a) Performance Criteria. The following performance criteria must be met: Seventy-five percent of the crystals must be greater than 100 μm. The suspended solids content of the product stream is to be 50% by weight and the theoretical yield of product must be greater than 95%.

(b) Data and Operating Conditions
    Feed composition = 0.275 weight fraction A.
    Feed temperature = 100°C.
    Heat of vaporization of solvent = 180 Btu/lb_m.
    Solid specific gravity = 1.23.
    Liquid specific gravity = 0.84.
    Heat of crystallization of solute = 60 Btu/lb_m.
    Feed and slurry heat capacity = 1.0 Btu/lb_m·C.
    Crystal volumetric shape factor = 1.0 (cubes).

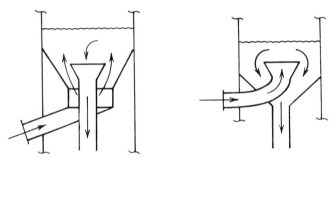

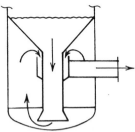

FIGURE 11.3-5 Classifying crystallizer modifications.

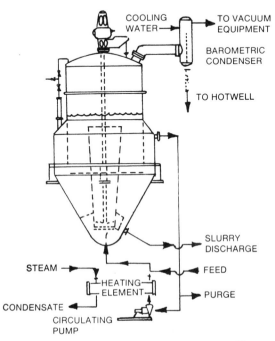

FIGURE 11.3-6 Top-driven mixed suspension crystallizer.

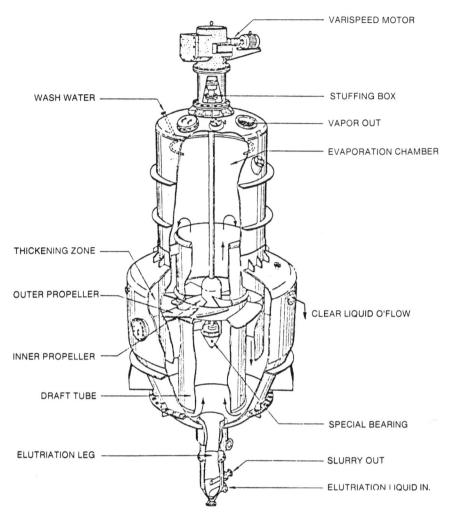

WASH WATER

THICKENING ZONE

OUTER PROPELLER

INNER PROPELLER

DRAFT TUBE

ELUTRIATION LEG

VARISPEED MOTOR

STUFFING BOX

VAPOR OUT

EVAPORATION CHAMBER

CLEAR LIQUID O'FLOW

SPECIAL BEARING

SLURRY OUT

ELUTRIATION LIQUID IN.

FIGURE 11.3-6 (*Continued*) Courtesy of Tsukishima Kihai Co., Ltd.

(c) Design Approach. The following design approach is discussed in this section.

- Develop material and energy balances.
- Obtain kinetics.
- Explore crystal size distribution as a function of solids holding time τ, fines system cut point L_f, and flow index into fines sytem R.
- Finalize system design.
- Size, design, and specify equipment.

(d) Material and Energy Balances. The binary solubility data for the case study example are tabulated in Table 11.3-6. It is evident from the solubility data that high recovery of solute can be achieved by merely cooling the feed. Using the material balance equations presented in Table 11.3-3, we can compute the solute yield and suspended solids concentration for both indirect and adiabatic cooling assuming a Class II system (saturated mother liquor leaving the system). Results are shown in Table 11.3-7.

The material balance information in Table 11.3-7 suggests that cooling crystallization, although effective in obtaining product yield, does not provide the desired suspension density. To meet this condition, external heat must be added to evaporate additional solvent. For practical reasons (such as avoidance of refrigerated coolant in the condenser) the crystallizer operating temperature is selected as 50°C. Thus, the operating absolute pressure will be 90 mm Hg (see Table 11.3-6).

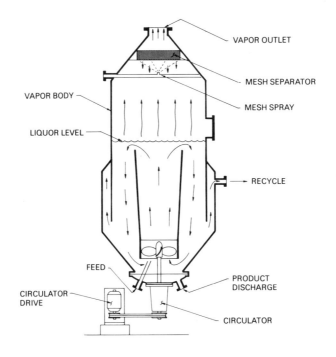

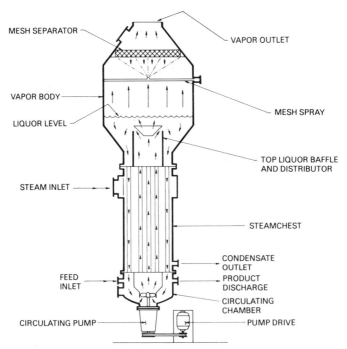

FIGURE 11.3-7 Bottom-driven mixed suspension crystallizers. Courtesy of Evaporator Technology Corp.

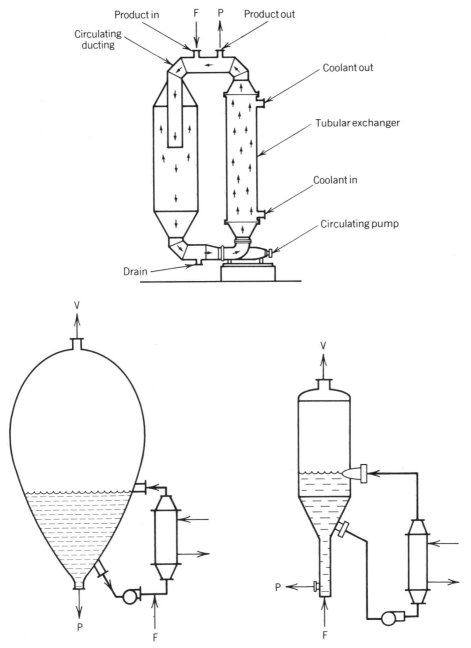

FIGURE 11.3-8 Forced-circulation crystallizers.

The final material balance computations for the case study are based on a crystallizer temperature of 50°C and are shown below:

Feed concentration $(R_f) = \dfrac{0.275}{0.725} = 0.38$ mass of dissolved solute per unit mass of solvent

Crystallizer concentration (R_c) at 50°C = 0.043 mass of dissolved solute per unit mass of solvent

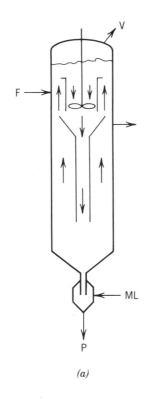

(a)

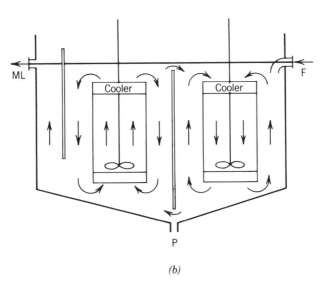

(b)

FIGURE 11.3-9 Hybrid crystallizers.

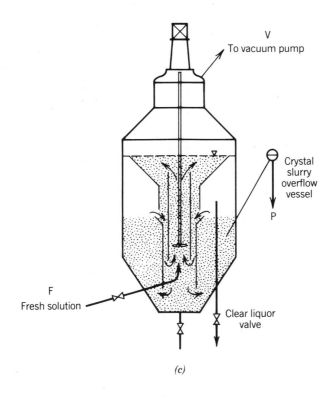

(c)

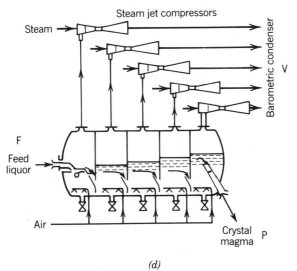

(d)

FIGURE 11.3-9 *(Continued)*

TABLE 11.3-5 Kinetic Data Reliability as a Function of Scale of Operation

Scale of Operation	Comments
Laboratory $V < 10$ L	Good control. Data reliable. Can observe crystal habit and/or agglomeration behavior. Can quantify i and j. Can determine Class I/Class II interface. Can rarely duplicate actual plant conditons. Nucleation rate coefficient k_N not reliable. Usually restricted to atmospheric pressure tests.
Pilot tests $10 < V < 2000$ L	Can usually duplicate actual plant operating conditions Data acquisition is expensive. Data replication is difficult. Determining i and j may not be practical. If i and j are available, k_N can be determined.
Full scale $V > 2000$ L	Actual plant conditions duplicated. If i and j are known, best source of reliable k_N. Determination of i and j often is not practical.

TABLE 11.3-6 Solubility Data for Case Study

Crystallizer Absolute Pressure (mm Hg)	Crystallizer Temperature °C	Equilibrium Solubility Mass Ratio of Dissolved Solute per Unit Mass of Solvent
160	65	0.100
135	60	0.072
90	50	0.043
60	40	0.028
40	30	0.018

TABLE 11.3-7 Case Study-Material Balance

Crystallizer Temperature (°C)	Adiabatic Cooling		Indirect Cooling	
	Yield (%)	Suspended Solids (%)	Yield (%)	Suspended Solids (%)
65	84	31	74	20
60	89	35	81	22
50	94	41	89	24
40	97	46	93	26

The specified total solids in suspension is 50%. From the generalized equation for total solids given in Table 11.3-3 the total solids in suspension (TS) is

$$\text{TS} = \frac{R_f - R_c + R_c R_{se}}{R_f + 1.0 - R_{se}} = 0.5$$

Solving for R_{se} (the mass ratio of solvent evaporated to solvent fed), we obtain

$$R_{se} = 0.65$$

Thus, 65% of solvent entering with the feed must be evaporated and removed from the system to meet the specified suspension concentration. The overall material balance then becomes

Suspended solids rate leaving: 1000 lb$_m$/h.

Solution (mother liquor) rate leaving: 41.2 lb$_m$/h dissolved solute
                                  958.8 lb$_m$/h solvent
                               1000.0 total

Feed contains: 2740.0 lb$_m$/h of dissolved solute
                1041.2 lb$_m$/h of dissolved solute
                3781.2 total

Solvent evaporated overhead: 1781.2 lb$_m$/h.

A final check on the theoretical yield of solids gives us

$$Y = \frac{R_f - R_c(1 - R_{se})}{R_f} = \frac{0.38 - 0.043(1 - 0.65)}{0.38} = 0.96$$

Therefore, two of the three performance criteria are met. The remaining design problem is to determine how to satisfy the crystal size requirement. It is necessary to delve into crystallization kinetics to analyze this aspect of the case study.

The preceding evaluation of the various modes of crystallization indicate that:

Cooling crystallization (to 40°C) will not meet the suspended solids or the minimum yield criteria.

Adiabatic cooling (to 40°C) will not meet the suspended solids criteria.

An evaporative crystallizer operating under vacuum with suitable external heat input will meet the material balance performance objectives.

Figure 11.3-10 summarizes the overall material balance information.

(e) Kinetics. For this example the following model will be employed.

Crystal Growth

$$G = k_g s \tag{11.3-13}$$

Nucleation Rate

$$B° = k_n s^i \tag{11.3-14}$$

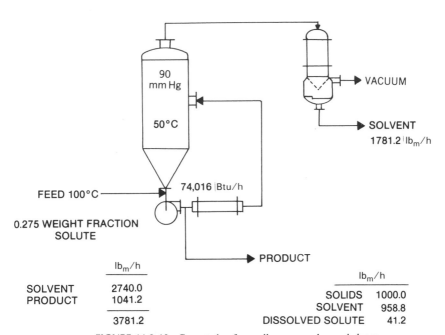

FEED 100°C

0.275 WEIGHT FRACTION SOLUTE

	lb$_m$/h
SOLVENT	2740.0
PRODUCT	1041.2
	3781.2

	lb$_m$/h
SOLIDS	1000.0
SOLVENT	958.8
DISSOLVED SOLUTE	41.2

FIGURE 11.3-10 Case study of overall energy and mass balance.

TABLE 11.3-8 Reduction of Crystal Size Distribution Data

Sieve Mesh	Retained (wt. %)	Cumulative Retained (wt. %)	Mesh Size (μm)	$\bar{L}$ (cm)	$\Delta W/\Delta L$ ($L = \bar{L}$)	$n^a(L)$ (no./cm$^3 \cdot \mu$m)
40	2.9	2.9	425			
		$\Delta W = 0.176$	$\Delta L = 245$	0.030	0.000718	18.3
80	17.6	20.5	180			
		$\Delta W = 0.369$	$\Delta L = 74$	0.0143	0.00499	1175.3
150	36.9	57.4	106			
		$\Delta W = 0.096$	$\Delta L = 31$	0.0091	0.0031	2833.5
200	9.6	67.0	75			
		$\Delta W = 0.133$	$\Delta L = 12$	0.0069	0.011	23064.2
250	13.3	80.3	63			
		$\Delta W = 0.122$	$\Delta L = 18$	0.0054	0.0068	29745.4
325	12.2	92.5	45			
		$\Delta W = 0.075$	$\Delta L = 45$	0.0023	0.00167	94542.3
PAN	7.5	100.0	0			

$^a n(L) = (\Delta W/\Delta L)M_T/\rho_s k_v \bar{L}^3$, where $\rho_s = 1.23$ g/cm^3, $M_T = 0.84$ g/cm^3 of clear liquor, and $k_v = 1.0$ (cubes).

Since s is rarely known, it is eliminated by solving for s in Eq. (11.3-13) and substituting for s in Eq. (11.3-14). Therefore,

$$B° = \frac{k_n}{k_g^i} G^i = k_N G^i \tag{11.3-15}$$

The approach used in this example case study is to obtain the kinetic order i from laboratory data. Since i is "system specific," it should not change with scale. If *one* reliable set of crystal size distribution data at a known residence time is available from plant operation, then $B°$ and G can be extracted after the data are converted to population density. Knowing $B°$, G, and i, k_N is computed easily using Eq. (11.3-15). *Note:* In this development, the solid concentration M_T is assumed constant.

Determination of i usually is conducted in small-scale laboratory equipment. Typically, a 4–8 L MSMPR crystallizer is used for this purpose. Other techniques for obtaining nucleation order include carefully conducted pilot or full-scale tests, mininucleator tests (see Randolph and Larson in the General Bibliography), and reported data in the literature.

For this study, a mininucleator was used and the value of 0.5 was determined for i. Low-order dependence of nucleation rate on supersaturation is somewhat typical of large molecules and indicates that nucleation mechanisms not overly sensitive to supersaturation, particularly secondary nucleation, are dominating.

Sieve analysis of product was obtained from a pilot unit and is shown in Table 11.3-8. Tests were conducted at operating conditions similar to those expected in the plant. A holding time of 10.5 h was employed for the test. The weight distribution data from the plant are converted into population density in Table 11.3-8.

The data from Table 11.3-8 are graphed on a semilog plot as shown in Fig. 11.3-11. The growth rate G and nuclei density, $n°$ are extracted from the slope and intercept, respectively. Eq. (11.3-15) can then be used to obtain the nucleation rate coefficient, k_N.

(f) Control of Crystal Size. Crystal size is sensitive to holding time and to the design of fines removal and dissolving equipment. To explore the influence of holding time on crystal size (cumulative mass percent less than 100 μm), the following equations, which are derived in Randolph and Larson (see General Bibliography) are used.

New Growth Rate

$$G = \left(\frac{M_T}{6\rho \, k_v k_N \tau^4}\right)^{1/i+3} \tag{11.3-16}$$

The cumulative weight fraction less than 100 μm then can be computed from the weight distribution function

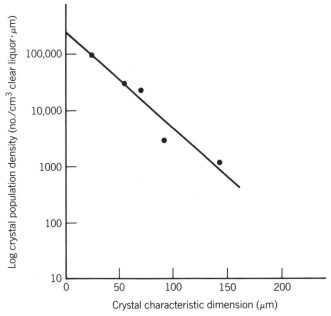

FIGURE 11.3-11   Case study of population balance versus crystal size.

$$W(X) = 1 - \left[\exp\left(-X\right)\left(1 + X + \frac{X^2}{2} + \frac{X^3}{6}\right)\right. \tag{11.3-17}$$

where the dimensionless argument X, is defined as

$$X = \frac{L}{G\tau} \tag{11.3-18}$$

The procedure for using Eqs. (11.3-16)–(11.3-18) is first to compute a new growth rate G for the holding time of interest τ from Eq. (11.3-16). Using the known τ, computed new growth rate G, and crystal size of interest L, compute the dimensionless parameter X from Eq. (11.3-18). Then solve Eq. (11.3-17) for the weight fraction less than size L. For this investigation, L is specified as 100 μm. Table 11.3-9 summarizes the sensitivity of crystal size to holding time.

Two conclusions are apparent from Table 11.3-8. The first is that increasing hold time *decreases* crystal size, and the second is that the influence of this variable is not dramatic. At this point, a holding time τ of 315 min is selected since additional reduction in holding time could result in loss of yield; that is, the system may revert from a Class II system (no residual supersaturation in effluent liquor) to a Class I system (measurable supersaturation in exit liquor).

Exploring the influence of fines removal and dissolution on particle size enhancement is the next topic of interest. Using the kinetic parameters generated from the laboratory tests (i was obtained) and the pilot

TABLE 11.3-9   Sensitivity of Weight Fraction Crystals less than 100 μm to Holding Time

Holding Time τ (min)	Growth Rate G (μm/min)	Dimensionless Parameter X^a	Weight Fraction < 100 μm
315	0.092	3.4	0.45
630	0.042	3.8	0.53
1260	0.019	4.2	0.61

$^a X = L/G\tau$ for $L = 100$ μm.

plant (k_N was extracted), the effect of fines system flow rate and classifier cut size on crystal size distribution is predictable.

The model for describing fines removal is derived in Chapter 8 of *Theory of Particulate Processes* by Randolph and Larson (see General Bibliography). The procedure for deriving and for using the fines removal model equations is the same as for the MSMPR case; however, the equations are more complex. A computer program is useful for quickly solving the equation for the new growth rate although a hand calculator will suffice.

The new growth rate is determined from Eq. (11.3-19):

$$G = \left(\frac{M_T}{6\rho \, k_v K_n T^4 D_p(G)} \right)^{1/i+3} \tag{11.3-19}$$

where

$$D_p(G) = \frac{W(p_1)}{R^4} + \exp\left[-(R-1)X_1\right][1 - W(X_1)] \tag{11.3-20}$$

$W(p_1)$ is evaluated from Eq. (11.3-17) with $X = p_1$ and

$$p_1 = \frac{RL_F}{G_T} \tag{11.3-21}$$

R is the index of fines system flow rate,

$$R = \frac{Q_f + Q_o}{Q_o} \tag{11.3-22}$$

and L_F is the fines cut size for an ideal fines trap. $W(X_1)$ is evaluated from Eq. (11.3-17) with X_1 defined by

$$X_1 = \frac{L_F}{G_T} \tag{11.3-23}$$

Equation (11.3-19) is solved for a new growth rate by a trial-and-error procedure and the new growth rate is then used in the following equation to predict crystal size distribution:

$$W(L) = \frac{W(p)/R^4 + \exp\left[-(R-1)X_1\right][W(X) - W(X_1)]}{W(p_1)/R^4 + \exp\left[-(R-1)X_1\right][1 - W(X_1)]} \tag{11.3-24}$$

where $W(p)$ is evaluated from Eq. (11.3-17) with $p = RL/G_T$ for $L < L_F$.

Figure 11.3-12 indicates the cumulative weight fraction of crystals less than 100 μm as a function of fines cut size L_f and the fines flow index R. The sensitivity of $W(L < 100 \ \mu m)$ to R is pronounced and the larger the fines cut size the more dramatic the change. Table 11.3-10 summarizes the results of fines removal computations.

An R of 1.8 and an L_f of 60 μm are selected for final design. Since only 4 ft^2 of settling area, based on Stokes settling calculations, are required to classify 60 μm crystals, a draft tube baffle (DTB) crystallizer with an internal fines removal bustle is a realistic design choice. Since 0.08 weight fraction of crystals is less than 100 μm, the fines-containing stream will contain no more than 0.04 weight fraction solids. A 10°C increase of the fines stream temperature and a dissolving hold time of several minutes will dissolve the crystals.

A number of options exist for satisfying the fines removal system and the overall energy requirement. Figure 11.3-13 is a conceptual design of a system that satisfies the objectives of the case study.

11.4 MELT CRYSTALLIZATION

Purification of a chemical species by solidification from a liquid mixture can be referred to as either *solution* crystallization or crystallization from the *melt*. The distinction between these two operations is somewhat subtle. The term *melt crystallization* is defined as the separation of components of a mixture without addition of solvent, but this definition is somewhat restrictive. In *solution* crystallization a diluent solvent is added to the mixture; the solution is then directly or indirectly cooled and/or solvent is evaporated to effect crystallization. The solid phase is normally formed and maintained somewhat below its pure-component

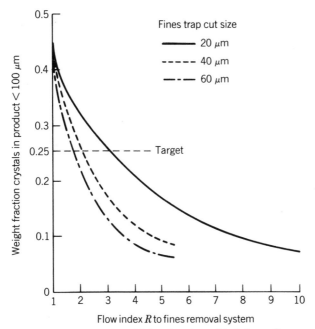

FIGURE 11.3-12 Case study of sensitivity of product critical size to fines trap design.

freezing-point temperature. In *melt crystallization* no diluent solvent is added to the reaction mixture and the solid phase is formed by direct or indirect cooling of the melt. Product is frequently maintained near or above its pure-component freezing-point temperature in the refining section of the apparatus and is removed in a molten liquid state.

11.4-1 State-of-the-Art

A large number of techniques are available for carrying out crystallization from the melt. An abbreviated list includes partial freezing and solids recovery in cooling crystallizer–centrifuge systems; partial melting, for example, sweating; staircase freezing; normal freezing; zone melting; and column crystallization. Description of all these methods is not within the scope of this discussion. Zief and Wilcox[1] have compiled a comprehensive book that describes many of these processes. Current trends in the practice of melt crystallization and ultrahigh purification of organic chemicals are presented by Atwood.[2]

High or ultrahigh product purity is obtained with many of the melt purification processes. Table 11.4-1 compares the product quality and product form that is produced from several of these operations. Zone refining can produce very pure material when operated in a batch mode; however, other techniques also provide high purity and become attractive if continuous high-capacity processing is desired.

Zone melting relies on the distribution of solute between the liquid and solid phases to effect a separation. In this case, however, one or more liquid zones are passed through the ingot. This extremely versatile technique, which was invented by W. G. Pfann, has been used to purify hundreds of materials. Zone melting in its simplest form is illustrated in Fig. 11.4-1. A molten zone can be passed through an ingot

TABLE 11.3-10 Summary of Fines Dissolving Calculations

Fines Cut Size L_f	Fines Flow Index R (min)	Fines Flow (lb$_m$/h)	Fines Trap Area (ft^2)	Weight Fraction Crystals $< L_f$
20	3.1	2100	38.6	0.0024
40	2.1	1100	9.7	0.025
60	1.8	800	4.3	0.08

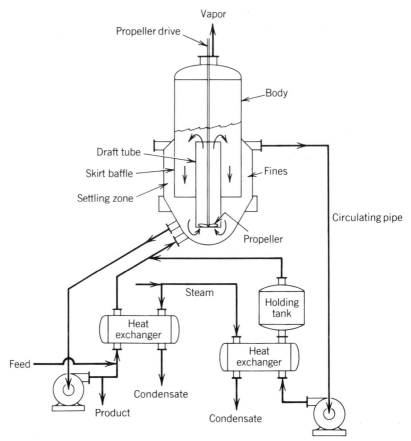

FIGURE 11.3-13 Case study of proposed system configuration.

from one end to the other by either a moving heater or slowly drawing the material to be purified through a stationary heating zone.

Normal freezing or progressive freezing is the slow, directional solidification of a melt. Basically, this involves slow solidification at the bottom or sides of a vessel by indirect cooling. The impurity is rejected into the liquid phase by the advancing solid interface. This technique can be employed to concentrate an impurity or, by repeated solidifications and liquid rejection, to produce a very pure ingot. Figure 11.4-2

TABLE 11.4-1 Comparison of Melt Purification Processes

Processes	Approximate Upper Melting Point (°C)	Materials Tested	Minimum Impurity Level Obtained (ppm, wt.)	Product Form
Normal freezing	1500	All types	1	Ingot
Zone melting				
Batch	3500	All types	0.01	Ingot
Continuous	500	SiI$_4$	100	Melt
Column crystallization				
Continuous end-feed	300	Organic	10	Melt
Continuous center-feed	400	Organic	1	Melt

Source: Adapted from Zief and Wilcox.[1]

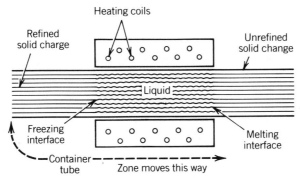

FIGURE 11.4-1 Zone refining principles. From *Perry's Chemical Engineer's Handbook*, 6th ed., McGraw-Hill, Inc., New York, NY.

illustrates a normal-freezing apparatus. The solidification rate and interface position are controlled by the rate of movement of the tube and the temperature of the cooling medium. There are many variations of the apparatus. A commercial process that employs sweating in addition to directional solidification has been described by Saxer and Papp.[3] Operation is sequential and the process steps include (1) partial freezing of a falling film of melt inside 12 m long vertical tubes, (2) partial melting or sweating, and (3) complete melting and recovery of refined product.

For a variety of reasons the concept and practice of conducting crystallization inside a column in a manner somewhat analogous to distillation are of considerable interest. Primary impetus is to attain higher-purity product than can be achieved in a single stage of conventional crystallization; however, much lower energy requirements (compared to distillation) and avoidance of high-speed mechanical separation equipment are additional incentives.

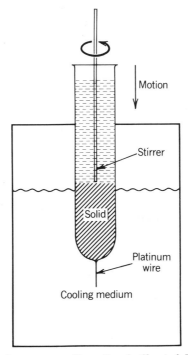

FIGURE 11.4-2 Normal-freezing apparatus. From *Perry's Chemical Engineer's Handbook*, 6th ed., McGraw-Hill, Inc., New York, NY.

The concept of a column crystallizer is to form a crystal phase, internal or external to the column, and to force the solids to flow countercurrently against a stream of enriched reflux liquid formed by melting the crystals. A temperature gradient is thus established in the column through which crystals are forced. In principle, such a device can be operated continuously or batchwise on either eutectic or solid solution systems. A limitation inherent in column crystallization is the difficulty of controlling the solid-phase movement since, unlike most distillation operations, the phases have similar densities, causing gravitational separation to be generally difficult.

Column crystallizers and ancillary equipment are designed in a multitude of ways to achieve controlled solid-phase movement, high product yield, and efficient heat addition and removal. Column crystallizers have been systematized into either end-feed or center-feed types. The Phillips style column[4] design is called an end-feed unit because crystals are formed in external equipment and the slurry is introduced into the column at the top. This type of column has no mechanical internals to transport solids and instead relies on hydraulic force to move the solids phase into the melting zone. Impure liquid is removed through a filter directly above the melter. Another commercial design is the spiral-conveyor Schildknecht column as described by Schildknecht and Maas.[5] In this device, a rotating or oscillating and rotating spiral is used to convey the crystals. This column is classed as a center-feed column crystallizer since liquid feed is introduced between the two enriched streams leaving the column, and the crystals are formed within the unit. A horizontal version of the center-feed crystallizer, which successfully uses spiral conveyors to transport crystals from the cold to hot zones, was reported by Brodie.[6]

A comparison of the process features of the center-feed and end-feed types of column crystallizer is given on Table 11.4-2. Column crystallization is compared with other common separation modes in Table 11.4-3.

11.4-2 Column Crystallization Equipment

Commercial melt crystallizers were introduced by Phillips Petroleum for separation of xylenes. A sketch of this apparatus is shown in Fig. 11.4-3. Crystals are formed by indirect cooling of melt in scraped-surface heat exchangers, and the resultant slurry is introduced into the top of the column. This type of column has no mechanical internals to transport solids and instead relies on an imposed hydraulic gradient to force the solids through the column into the melting zone. Residue liquid is removed through a filter directly above the melter. A pulse piston in the product discharge improves washing efficiency and column reliability.

A continuous horizontal melt crystallizer is available and is named the Brodie Purifier after its inventor. Originally developed for separation of ortho- and para-dichlorobenzene, it is used for purification of a wide variety of materials. Features of the Brodie Purifier are shown in Fig. 11.4-4. Liquid feed enters the column between the hot refining section and the cold freezing or recovery zone. Crystals are formed internally by indirect cooling of the melt through the walls of the refining and recovery zones. Residue liquid that has been depleted of product exits from the coldest section of the column. A spiral conveyor controls the transport of solids through the unit.

Another commercial melt crystallization process is the MWB system depicted schematically in Fig. 11.4-5. Operation is sequential. Steps include partial freezing of a falling film of melt inside vertical tubes,

TABLE 11.4-2 Comparison of Two Types of Column Crystallizer

Center-Feed Column	End-Feed Column
Solid phase is formed internally; thus, only liquid streams enter and exit the column.	Solid phase is formed in external equipment and fed as slurry into the purifier.
Internal reflux can be controlled without affecting product yield.	The maximum internal liquid reflux is fixed by the thermodynamic state of the feed relative to the product stream. Excessive reflux will diminish product yield.
Operation can be continuous or batchwise at total reflux.	Total reflux operation is not feasible.
Center-feed columns can be adapted for both eutectic and solid solution systems.	End-feed columns are inefficient for separation of solid solution systems.
Either low- or high-porosity solid-phase concentrations can be formed in the purification and melting zones.	End-feed units are characterized by low-porosity solids packing in the purification and melting zones.
Scale-up depends on the mechanical complexity of the crystal transport system and techniques for removing heat. Vertical oscillating spiral columns are likely limited to about 0.2 m in diameter whereas horizontal columns of several meters are possible.	Scale-up is limited by design of melter and/or crystal washing section. Vertical or horizontal columns of several meters in diameter are possible.

TABLE 11.4-3 Comparison of Column Crystallization with Standard Separation Techniques[a]

Selection Criteria	Staged Crystallization/Washing	Distillation	Column Crystallization
Field of application: number organic products	Moderate (50)	Very high (1000)	Moderate (50)
Industrial experience	Moderate	Very large	Small
Reliability	Moderate	Good	Good
Relative costs			
Investment	100–120	60–120	100
Energy	Low	High	Very low
Maintenance	High	Small	Small
Personnel	Moderate	Small	Small
Start-up losses	Moderate	None	None
Product purity	Moderate	Moderate–high	Very high
Corrosion	Moderate	Much	Little

[a] Personal communication with the C. W. Nofsinger Company.

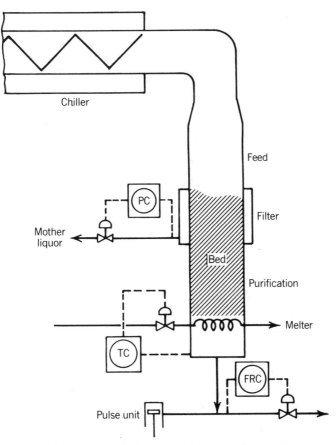

FIGURE 11.4-3 Vertical end-feed column melt crystallizer.

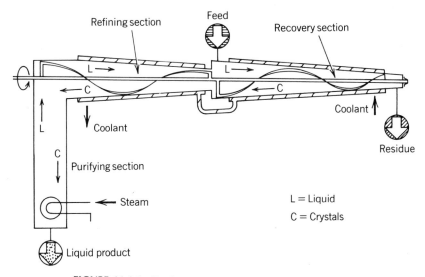

FIGURE 11.4-4 Continuous horizontal Brodie melt crystallizer.

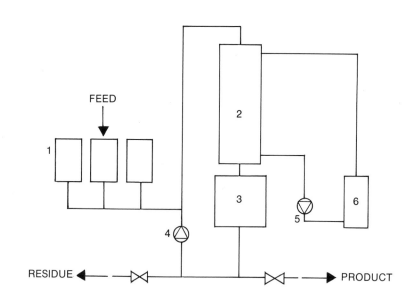

1. STORAGE TANKS
2. CRYSTALLIZER
3. COLLECTING TANK
4. PRODUCT CIRCULATION PUMP
5. CIRCULATION PUMP
6. COOLING–HEATING SYSTEM

FIGURE 11.4-5 MWB batch-automatic melt crystallizer schematic.

followed by slight heating, a "sweating" operation, and complete melting and recovery of the refined product. Separation capacity depends on the number of stages, the reflux ratio, and the distribution coefficient. The device is particularly useful for separation of components that form solid solutions.

11.4-3 Preliminary Design of Brodie Purifier—Case Study

In the early stages of a process it is prudent to develop a conceptual design that provides an estimate of the size and configuration of the equipment involved. These computations are by no means the final answer—since they include certain assumptions—but they provide early insight into the potential process with respect to feasibility and practicality.

The conceptual design of a Brodie Purifier can be accomplished by evaluating key factors that determine the overall configuration. One of the major considerations is the quantity of heat that must be removed to satisfy the material and energy balance. Thus, preliminary equipment design can be established by conducting material and energy balances around the various sections, assuming a reflux ratio, and by approximating the heat transfer rate. The approach used in this section is described in detail by Brodie.[6]

Case Study
The problem is to determine the approximate size of melt crystallizer required to produce 5000 lb_m/h of naphthalene from a feed stream containing 80% by weight naphthalene and 20% benzene. Saturated residue is rejected at 45°C. The overall material balance is computed easily, and is shown in Fig. 11.4-6; the phase diagram is given in Fig. 11.2-2. For evaluation purposes, the continuous melt crystallizer is conveniently divided into three sections: the *recovery* section where solute is crystallized and removed from solution, the *refining* section where reflux melt is recrystallized, and the *purification* zone where crystals are contacted with reflux, are melted, and the melt removed and/or refluxed.

Overall Material and Energy Balance. Assuming specific heats are constant and identical for solid and liquid, the heat of crystallization λ_f is constant, and the reference states are solid naphthalene and liquid benzene at T_B, the spent liquor discharge temperature. We can express the overall enthalpy balance as

$$Q_H + H_F F = H_P P + Q_{ref} + Q_{rec} \tag{11.4-1}$$

If sensible heating is small in comparison to the latent heat required for melting, the heat input to the purification section, Q_H, can be determined from the expression

$$Q_H = (P + RP)\lambda_f \tag{11.4-2}$$

where R is the reflux ratio, that is, the ratio of reflux rate to product removal rate. H_F is the feed enthalpy (referred to liquid at T_B),

$$H_F = a\lambda_f + C_p(T_F - T_B) \tag{11.4-3}$$

where a = fraction of feed crystallized = (P/F)
$\qquad C_p$ = specific heat of crystals and liquid

H_P = product enthalpy (referred to solid at T_B)

$$H_P = \lambda_f + C_p(T_P - T_B) \tag{11.4-4}$$

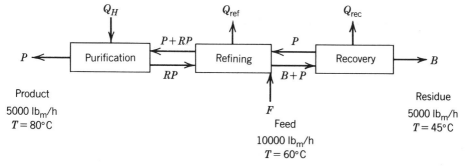

FIGURE 11.4-6 Case study of mass balance for Brodie crystallizer.

Q_{ref} is the heat outflow from the refining section, and Q_{rec} is the heat outflow from the recovery section. Then,

$$Q_{ref} + Q_{rec} = (P + RP)\lambda_f + [a\lambda_f + C_p(T_F - T_B)]F$$
$$- [\lambda_f + C_p(T_p - T_B)]P$$

or

$$Q_{ref} + Q_{rec} = PR\lambda_f + P\lambda_f - P(T_p - T_F)C_p + B(T_F - T_B)C_p \tag{11.4-5}$$

The energy balance for the recovery section is given by the following equations:

$$P\lambda_f + PC_p \, \Delta T_{F-B} + BC_p \, \Delta T_{F-B} = PC_p \, \Delta T_{F-B} + Q_{rec} \tag{11.4-6}$$

Therefore,

$$Q_{rec} = P\lambda_f + B \, \Delta T_{F-B}C_p \quad \text{(design equation)} \tag{11.4-7}$$

The energy balance for the refining section is then determined by

$$Q_{ref} = PR\lambda_f - P \, \Delta T_{P-F}C_p \quad \text{(design equation)} \tag{11.4-8}$$

Note: If there is no refining section ($Q_{ref} = 0$) then

$$PR\lambda_f = P \, \Delta T_{P-F}C_p$$

or

$$R_N = \frac{\Delta T_{P-F}C_p}{\lambda_f} \tag{11.4-9}$$

This is the natural refreezing ratio or maximum product reflux that can occur due to the cooling effect of the crystals. Normally, a reflux higher than R_N is necessary to achieve high product purity.

Using Eqs. (11.4-7) and (11.4-8) to obtain estimates of the heat duties (note that C_p is assumed constant = 0.43 Btu/lb$_m$·°C and λ_f = 70.1 Btu/lb$_m$), the heat removal requirement of the *recovery* section is

$$Q_{rec} = P\lambda_f + B \, \Delta T_{F-B}C_p = (5000)(70.1) + 5000(15)(0.43)$$
$$= 415,000 \text{ Btu/h} \quad (437,000 \text{ kJ/h}) \tag{11.4-10}$$

The heat removal requirement of the *refining* section is

$$Q_{ref} = PR\lambda_f - P \, \Delta T_{P-F}C_p = (5000)(0.5)(70.1) - (5000)(20)(0.43)$$
$$= 132,250 \text{ Btu/h} \quad (139,000 \text{ kJ/h}) \tag{11.4-11}$$

The natural refreezing ratio is

$$R_N = \frac{\Delta T_{P-F}C_p}{\lambda_f} = \frac{(20)(0.43)}{70.1} = 0.12 \tag{11.4-12}$$

A reflux ratio of 0.5 is assumed for the case study. Normally, this value must be determined experimentally.

Design Assumptions

1. Heat tracing of the conveyor shaft and unscraped surfaces is incorporated into the Brodie Purifier to assure that no solute freezing occurs where it may hamper operation. Thus, a considerable quantity of external heat is added intentionally to the system to avoid solid encrustations. In addition, there may be heat gain through insulated sections if the process operating temperature is below ambient. The term "cooling efficiency" is used to quantify the ratio of adiabatic energy removal to that actually required, that is, to correct the energy balance for nonadiabatic energy contributions. Cooling efficiency is a function of scale of operation, operating temperature, and supplier design practice.

For this study the cooling efficiency is assumed to be 0.7.

2. An overall heat transfer coefficient and a temperature difference between the process fluid or slurry

TABLE 11.4-4 Heat Exchanger Tube Sizes

Diameter		Length		Area	
(ft)	(m)	(ft)	(m)	(ft²)	(m²)
5.25	1.6	52.5	16.0	865.9	80.4
3.28	1.0	52.5	16.0	541.0	50.3
1.64	0.5	52.5	16.0	270.5	25.1

and refrigerant must be estimated to establish a heat transfer rate. Again, experimental measurement for the specific system is necessary. For this study, the overall heat transfer coefficient U is 25 Btu/h · ft² °F, and the temperature difference between the process and coolant is 20°F.

Using these assumptions, the jacketed cooling areas required for the recovery and refining sections are

$$A_{rec} = \frac{Q_{rec}}{\text{cooling efficiency} \times U \times \Delta T} = \frac{415,000}{0.7 \times 25 \times 20}$$

$$A_{rec} = 1186 \text{ ft}^2 \ (110 \text{ m}^2)$$

$$A_{ref} = \frac{Q_{ref}}{\text{cooling efficiency} \times U \times \Delta T} = \frac{132,250}{0.7 \times 25 \times 20}$$

$$A_{ref} = 378 \text{ ft}^2 \ (35 \text{ m}^2)$$

Table 11.4-4 summarizes the standard size tubes available. Standard sizes should be used if possible to minimize costs. A preliminary tube arrangement for the melt crystallizer designed in this study is shown in Fig. 11.4-7.

The preliminary design established for this system is based on overall energy and mass balances and estimated heat transfer rate. It is also assumed that the crystals grow sufficiently large and that the density difference between liquid and crystals is sufficient for the crystals to settle rapidly in the purifier section and at the residue end. The settling rate of a thick suspension of pure crystals in their "melt" should be determined early in the exploratory studies to establish the feasibility of the process. There are also retention time considerations and temperature gradient concerns that the supplier may impose based on satisfactory design practice. Thus, the calculations included herein represent a "first pass" estimate and need additional support from laboratory observations and pilot tests.

11.5 General Design and Operational Considerations

In continuous evaporating or cooling crystallizers there are five main functional steps that must be accommodated in the design:

1. Feed liquor is introduced into the system.
2. Mother liquor is circulated, with or without crystal content, through a heat exchanger where heat is added or removed from the system. (Adiabatic vacuum cooling crystallizers do not require this step.)
3. Supersaturation is generated by flashing vapor, by direct or indirect cooling, by evaporative concentration, or by salting out.
4. Supersaturation is discharged by growth on crystals suspended in the liquor and, to a varying extent, generation of new crystals or nuclei.
5. The product crystals and residue liquor (mother liquor) are discharged either together or as separate streams.

There are many variations of how the above functions are incorporated into the design. The following discussion outlines concepts and design considerations that are useful in both design and troubleshooting of continuous crystallizers.

11.5-1 Crystal Habit and Size Control

The "natural" shape or habit of a crystal obtained in the medium from which it is crystallized is of utmost importance in designing an industrial-scale crystallizer. Crystal habit is also of extreme interest to crystallization theorists. Many articles have been written describing variables that affect natural habit, such

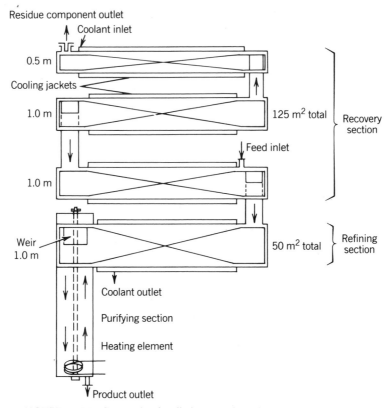

FIGURE 11.4-7 Case study of preliminary configuration of Brodie crystallizer.

as including modifiers that alter the natural crystal habit. However, each compound seems to have its own peculiarities and generalizations are difficult. From an industrial-scale point of view, crystal habit ultimately affects crystal settling velocities, centrifuge drainage rates, and product purity. As a consequence, crystallizer and associated separation equipment are heavily influenced by crystal habit.

Minimizing the formation or birth of new crystals (nuclei) and/or using segregation and destruction of a portion of the fine crystals are techniques employed to control particle size. To prevent formation of nuclei, the following design practices should be employed.

1. Provide even boiling at the vapor-liquid interface.
2. Move crystals to the boiling surface.
3. If cooling crystallization is used, introduce feed into the circulation loop downstream of the cooler.
4. Provide low supersaturation at the boiling interface. This is accomplished by designing for temperature rises (ΔT's) in the circulation loops of only a few degrees. Thus, high recirculation rates are common.
5. Avoid "shock" nucleation, which can occur by a sudden temperature drop at a mix point or by a salting out effect if streams of different compositions are combined.
6. Avoid gross flashing of feed stream.

If prevention of the formation of nuclei is unsuccessful in achieving the desired crystal size distribution, then fines destruction techniques may be employed. Approaches include the following:

1. Provide classifier to segregate fines. Dissolve fines by heating the fines-containing stream, and then pump the solution back to the crystallizer.
2. If the solubility diagram is flat, that is, the concentration of solute in the solvent is not temperature dependent, then fines can be destroyed by dissolving in fresh solvent or an unsaturated stream.
3. If crystal–mother liquor separation is difficult, a nonclassified fines destruction circuit can be con-

sidered. In this mode, the entire slurry is circulated by a pump through a heater, dissolution hold tank, and back to the crystallizer. The stream temperature is raised only by a few degrees.

11.5-2 System Reliability

HEAT EXCHANGE

As in many other separation processes, the selection and design of heat transfer equipment are essential for the success of many crystallizer operations. A few guidelines are presented below:

1. Heat exchanger fouling due to subcooled boiling at the discharge of the exchanger is a common problem in crystallization systems. Extreme care is needed in selection and orientation of the heat exchanger, temperature of the heating medium, submergence of the heat exchanger, and pressure drop in the lines discharging from the heat exchangers.

2. Subcooled boiling inside the heat exchanger and/or boiling at the discharge of the heat exchanger is difficult to control when the vapor pressure rise is greater than 3.4 kPa/K (0.5 psi/°C). Systems that operate under pressure or that employ low boiling solvents present the most difficulties. Sometimes horizontal heat exchangers are used to increase hydrostatic head and minimize boiling.

3. Fluid velocities of 1.5–3.0 m/s (5–10 ft/s) are recommended to minimize plugging of heat-exchanger tubes.

4. Exchanger tube diameters of 2.54 cm (1 in.) are recommended to prevent blockage by small chunks. Spiral heat exchangers also work quite well in slurry service.

5. Cooling crystallization can be accomplished reliably by using scraped-surface exchangers; however, vertical shell and tube exchangers can be used if periodic thawing of solute from the inside of the tubes is not overly disruptive to the process. Heat transfer coefficients of 500–100 kJ/h·m²·°C (25–50 Btu/h·ft²·°F) are typical for scraped-surface units.

CONTROL

The method for computing the quantity of solids contained in the crystallizer liquid is discussed in the section on computation of product recovery. This computation gives the ''natural'' slurry density, which is controlled by temperature and the evaporation rate. If it is desired to control the slurry density at some level other than the natural density, clear liquor advance or accelerated solids removal techniques are employed. Advancing clear liquor increases the slurry density inside the crystallizer and increases the solids residence time. This is accomplished by providing a settling zone inside or outside the crystallizer from which clear liquor can be extracted. Accelerated solids removal by removing solids and returning a portion of the liquid to the crystallizer allows slurry density control *below* the natural slurry density.

By advancing clear liquor and adjusting the feed rate to the centrifuge and returning clear liquor, slurry concentration can be controlled over wide ranges. It should be noted that slurry concentration is difficult to measure, and samples of crystallizer contents usually are extracted manually and analyzed for solids content.

Liquid level in crystallizers usually is controlled by measuring the pressure differential between the vapor space and some point in the crystallizer. Problems are encountered with this approach because of changing liquid properties due to boiling and the presence of solids. Sometimes dual systems are used to increase the reliability of level control. Many crystallizers have viewing ports to observe the liquid level. Closed-circuit television monitors can be used if level control is critical and/or extremely difficult.

ENCRUSTATIONS

The buildup of solids inside a crystallizer is very troublesome. In adiabatic vacuum crystallizers, solvent that is vaporized can be returned and distributed at the vapor–liquid interface to dissolve deposits of solids. In crystallizers where the circulating line exits at the bottom, screens or ''chunk breakers'' have proved effective in protecting downstream equipment and prolonging operating periods between boilouts. From a design standpoint, it is desirable to eliminate natural sumps where large chunks can accumulate and eventually plug the crystallizer. (This can be a serious disadvantage of the suspension type crystallizer.)

11.5-3 Product Quality

In solution crystallization product purity is controlled to a great extent by the efficiency of solid–liquid separation. Impurities dissolved in the crystallizer liquid are normally removed by spinning liquor from the crystals and then washing the resulting cake. The residual impurity level is a function of factors listed in the following equation:

$$\overset{\text{(a)}}{I} = \overset{\text{(a)}}{R_{I/L}} \times \overset{\text{(b)}}{R_{L/S}} \times \overset{\text{(c)}}{\left(1 - \frac{WE}{100}\right)} + \overset{\text{(d)}}{I^\circ} \qquad (11.5\text{-}1)$$

where I = impurity content of product (g impurity/g pure solute)

$I°$ = impurity occluded in solids (g impurity/g pure solute after drying)

$R_{I/L}$ = concentration of impurity in crystallizer liquor (g impurity/g liquid)

$R_{L/S}$ = residual liquid after separation (g liquid/g pure solute)

WE = washing efficiency (% impurity removed by washing)

Each of the factors in Eq. (11.5-1) can be manipulated to a certain degree; however, there are limitations inherent in each. Factor (a) can be minimized by backfeeding crystals to the dilute liquor stage in a multistage system or by contacting the exiting product with feed liquor (via an elutriation leg). Factor (b) is controlled by the type and speed of the liquid–solid separation device and the physical properties of the solids and liquid. Factor (c) is affected by the quantity of wash employed, and factor (d) is controlled by the crystallization environment.

11.5-4   Product Recovery

As shown in Fig. 11.1-1, the solids–liquid separation device is the link between the solids formation equipment and the drying operation. Thus, predictable and reliable operation is essential in this stage of the processing sequence. Continuous or batch-automatic centrifuges and continuous vacuum filters are typically used to recover solid material from continuous crystallization operations. Many factors are considered in the final selection of the device. In any case, an equipment testing program is always essential to determine the capacity and performance of solids separation equipment.

Troubleshooting liquid–solid separation equipment usually involves rerating the separation device for the slurry characteristics actually present in the commercial facility. To evaluate centrifugal filters, material "drain rate" studies are necessary whereas vacuum filters are normally sized by scaling up "formation time" data. Procedures for conducting these tests are best found in equipment supplier literature. The discussion in Chapter 3 also should be consulted.

To correct an unfavorable situation often requires an adjustment to the slurry characteristics rather than modification of the separation device. Increasing solids particle drainage characteristics via size enlargement techniques, concentrating the slurry, heating the slurry, and/or removing "fines" is the usual approach.

NOTATION

A_c	surface area of crystal
b	parameter in ASL model, Eq. (11.2-28)
$B°$	nucleation rate
C	crystal removal rate [Eq. (11.3-2)] or solute concentration
C^*	concentration at equilibrium
C_m	solute concentration at which spontaneous nucleation occurs
CV	coefficient of variation
D_G	parameter describing growth rate dispersion by growth fluctuations
E	impact energy
E_t	threshold or minimum collision energy
f	fugacity
$f°$	reference-state fugacity
$f°(G)$	distribution of nuclei growth rates
F_s	mass feed rate of solvent into crystallizer
G	growth rate
$\overline{G}$	mean growth rate
$G°$	parameter in ASL model, Eq. (11.2-28)
H_F	feed enthalpy
H_P	product enthalpy
Δh_f	enthalpy of fusion
i	exponent in relationship of nucleation and growth kinetics
j	exponent in relationship between nucleation rate and magma density
k_N	nucleation rate constant
k_v, k_a	volume and area shape factors
L	characteristic dimension of crystal
L_C	product cut size

L_D dominant crystal size

L_F fines cut size

M_c mass of crystal

m_i i^{th} moment of population density

M_T magma density (mass of crystals per unit volume of clear liquor or of slurry)

n population density function

Q_H heat inflow

Q_i volumetric flow rate (clear liquor or slurry, depending on definition of n) entering crystallizer

Q_o volumetric flow rate (clear liquor or slurry, depending on definition of n) leaving crystallizer

Q_{rec} heat outflow from recovery section

Q_{ref} heat outflow from refining section

R ideal gas constant, fines flow index, or reflux ratio

R_c mass of dissolved solute in liquid discharged from the crystallizer per unit mass of discharged solvent

R_f mass of dissolved solute in feed per unit mass of solvent

R_N natural refreezing ratio

s supersaturation

S solution removal rate

T temperature

T_B spent liquor discharge temperature

T_M melting temperature

v molar volume

V_c volume of crystal

V_T magma volume (clear liquor or slurry depending on definition of n)

x mole fraction

Y yield

Z index for accelerated course crystal removal

Greek Letters

ϵ fraction of available energy actually transmitted to crystal or screw dislocation activity

γ activity coefficient or parameter in ASL model, Eq. (11.2-28)

λ_f heat of crystallization

λ_v heat of vaporization

ρ_c crystal density

σ surface energy per unit area

σ_G^2 variance in growth rate distribution function

τ solids holding time

ω rotational velocity of impeller

REFERENCES

Section 11.2

11.2-1   J. M. Prausnitz, *Molecular Thermodynamics of Fluid-Phase Equilibria*, Prentice-Hall, Englewood Cliffs, NJ, 1969.

11.2-2   American Petroleum Institute Research Project No. 44, ''Selected Values of Properties of Hydrocarbons.''

11.2-3   R. F. Muir and C. S. Howat, Predicting Solid–Liquid Equilibria from Vapor–Liquid Data, *Chem. Eng.*, 89 (Feb. 22, 1982).

11.2-4   J. H. Hildebrand and R. L. Scott, *Regular Solutions*, Prentice-Hall, Englewood Cliffs, NJ, 1962.

11.2-5   J. G. Gmehling, T. F. Anderson, and J. M. Prausnitz, Solid–Liquid Equilibria Using UNIFAC, *Ind. Eng. Chem. Fundam.*, **17**, 269 (1978).

11.2-6   Aa. Fredenslund, J. Gmehling, M. L. Michelsen, P. Rasmussen, and J. M. Prausnitz, Computerized Design of Multicomponent Distillation Columns Using the UNIFAC Group Contribution

Method for Calculation of Activity Coefficients, *Ind. Eng. Chem. Process Des. Dev.*, **16**, 450 (1977).

11.2-7 G. S. Soave, Application of the Redlich–Kwong–Soave Equation of State to Solid–Liquid Equilibria Calculations, *Chem. Eng. Sci.*, **34**, 225 (1979).

11.2-8 H. Wenzel and H. L. Schmidt, A Modified van der Waals Equation of State for the Representation of Phase Equilibria Between Solids, Liquids, and Gases, *Fluid Phase Equilibria*, **5**, 3 (1980).

11.2-9 Y. Unno, D. Hoshimo, K. Nagahama, and M. Hirata, Prediction of Solid–Liquid Equilibria from Vapor–Liquid Equilibrium Data Using the Solution of Groups Model, *J. Chem. Eng. Jpn.*, **12**(2), 81 (1979).

11.2-10 ANSI/ASTM D1177-65 (Reapproved 1978), Standard Test Method for Freezing Point of Aqueous Engine Coolant Solution.

11.2-11 J. Garside, Advances in the Characterization of Crystal Growth, *AIChE Symp. Ser. No. 240*, **80**, 23 (1984).

11.2-12 J. W. Mullin, *Crystallisation*, 2nd ed., p. 137, CRC Press, Cleveland, 1972.

11.2-13 H. A. Miers and F. Issac, as discussed in H. E. Buckley, *Crystal Growth*, p. 7, Wiley, New York, 1951.

11.2-14 A. D. Randolph and M. A. Larson, *Theory of Particulate Processes*, p. 103, Academic Press, New York, 1971.

11.2-15 J. Garside and Roger J. Davey, Secondary Contact Nucleation: Kinetics, Growth and Scale-up, *Chem. Eng. Commun.*, **4**, 393 (1980).

11.2-16 M. A. Larson, Advances in the Characterization of Crystal Nucleation, *AIChE Symp. Ser. No. 240*, **80**, 39 (1984).

11.2-17 M. W. Girolami and R. W. Rousseau, Initial Breeding in Seeded Batch Crystallizers, *Ind. Eng. Chem. Process Des. Dev.*, **25**, 66 (1986).

11.2-18 C. Y. Sung, J. Estrin, and G. R. Youngquist, Secondary Nucleation of Magnesium Sulfate by Fluid Shear, *AIChE J.*, **19**, 957 (1973).

11.2-19 R. E. A. Mason and R. F. Strickland-Constable, Breeding of Crystal Nuclei, *Trans. Faraday Soc.*, **62**, 2, 455 (1966).

11.2-20 D. P. Lal, R. E. A. Mason, and R. F. Strickland-Constable, Collision Breeding of Crystal Nuclei, *J. Cryst. Growth*, **5**, 1 (1969).

11.2.21 N. A. Clontz and W. L. McCabe, Contact Nucleation of Magnesium Sulfate, *Chem. Eng. Prog. Symp. Ser. No. 110*, **67**, 6 (1971).

11.2-22 C. Y. Tai, W. L. McCabe, and R. W. Rousseau, Contact Nucleation of Various Crystal Types, *AIChE J.*, **21**, 351 (1975).

11.2-23 B. C. Shah, W. L. McCabe, and R. W. Rousseau, Polyethylene vs. Stainless Steel Impellers for Crystallization Processes, *AIChE J.*, **19**, 194 (1973).

11.2-24 T. W. Evans, G. Margolis, and A. F. Sarofim, Mechanisms of Secondary Nucleation in Agitated Crystallizers, *AIChE J.*, **20**, 950 (1974).

11.2-25 R. W. Rousseau and W. L. McCabe, "Nucleation and Growth from Solution," paper presented at the World Congress of Chemical Engineering, Amsterdam, June 1976.

11.2-26 T. W. Evans, A. F. Sarofim, and G. Margolis, Models of Secondary Nucleation Attributable to Crystal–Crystallizer and Crystal–Crystal Collisions, *AIChE J.*, **20**, 959 (1974).

11.2-27 C. Y. Tai, "Contact Nucleation of Various Crystal Types," Ph.D. thesis, North Carolina State University, Raleigh, NC, 1974.

11.2-28 R. C. Bennett, H. Fiedelman, and A. D. Randolph, Crystallizer Influenced Nucleation, *Chem. Eng. Prog.*, **69**(7), 86 (1973).

11.2-29 P. A. M. Grootscholten, L. D. M. v.d. Brekel, and E. J. de Jong, Effect of Scale-up on Secondary Nucleation Kinetics for the Sodium Chloride–Water System, *Chem. Eng. Res. Des.*, **62**, 179 (1984).

11.2-30 J. Garside and M. B. Shah, Crystallization Kinetics from MSMPR Crystallizers, *Ind. Eng. Chem. Process Des. Dev.*, **19**, 509 (1980).

11.2-31 J. Garside and M. A. Larson, Direct Observation of Secondary Nuclei Production, *J. Cryst. Growth*, **43**, 694 (1978).

11.2-32 J. Garside, I. T. Rusli, and M. A. Larson, Origin and Size Distribution of Secondary Nuclei, *AIChE J.*, **25**, 57 (1979).

11.2-33 R. W. Rousseau, S. Craig, and W. L. McCabe, Formation, Survival and Growth of Nuclei from Secondary Nucleation, in E. J. de Jong and S. J. Jančić (Eds.), *Industrial Crystallization '78*, p. 19, North-Holland, Amsterdam, 1979.

11.2-34 J. W. Mullin and C. L. Leci, Evidence of Nucleation Cluster Formation in Supersaturated Solutions, *Philos. Mag.*, **19**, 1075 (1969).

11.2-35 P. M. McMahon, K. A. Berglund, and M. A. Larson, Raman Spectroscopic Studies of the Structure of Supersaturated KNO_3 Solutions, in S. J. Jančić and E. J. de Jong (Eds.), *Industrial Crystallization 84*, p. 229, Elsevier, Amsterdam, 1984.

11.2-36 N. A. Clontz, R. T. Johnson, W. L. McCabe, and R. W. Rousseau, The Growth of Magnesium Sulfate Heptahydrate Crystals from Solution, *Ind. Eng. Chem. Fundam.*, **11**, 368 (1972).

11.2-37 R. F. Strickland-Constable, *Kinetics and Mechanism of Crystallization*, Academic Press, New York, 1968.

11.2-38 J. W. Mullin, *Crystallisation*, 2nd ed., p. 207, CRC Press, Cleveland, 1972.

11.2-39 M. Ohara and R. C. Reid, *Modeling Crystal Growth Rates from Solution*, Prentice-Hall, Englewood Cliffs, NJ, 1973.

11.2-40 W. K. Burton, N. Cabrera, and F. C. Frank, The Growth of Crystals and the Equilibrium Structure of their Surfaces, *Philos. Trans. R. Soc.*, **243**(A866), 299 (1951).

11.2-41 R. J. Davey, The Control of Crystal Habit, in E. J. de Jong and S. J. Jančić (Eds.), *Industrial Crystallization '78*, p. 169, North-Holland, Amsterdam, 1979.

11.2-42 A. S. Michaels and A. R. Colville, The Effect of Surface Active Agents on Crystal Growth Rate and Crystal Habit, *J. Phys. Chem.*, **64**, 13 (1967).

11.2-43 R. W. Rousseau, C. Y. Tai, and W. L. McCabe, The Influence of Quinoline Yellow on Potassium Alum Growth Rates, *J. Cryst. Growth*, **32**, 73 (1976).

11.2-44 R. W. Rousseau and R. Woo, Effects of Operating Variables on Potassium Alum Crystal Size Distribution, *AIChE Symp. Ser. No. 193*, **76**, 27 (1980).

11.2-45 J. R. Bourne, The Influence of Solvent on Crystal Growth Kinetics, *AIChE Symp. Ser. No. 193*, **76**, 59 (1980).

11.2-46 W. L. McCabe, Crystal Growth in Aqueous Solutions, *Ind. Eng. Chem.*, **21**, 30 (1929).

11.2-47 E. T. White, L. L. Bendig, and M. A. Larson, The Effect of Size on the Growth Rate of Potassium Sulfate Crystals, *AIChE Symp. Ser. No. 153*, **72**, 41 (1976).

11.2-48 C. F. Abegg, J. D. Stevens, and M. A. Larson, Crystal Size Distribution in Continuous Crystallizers When Growth Rate Is Size Dependent, *AIChE J.*, **14**, 118 (1968).

11.2-49 J. Garside and S. J. Jančić, Prediction and Measurement of Crystal Size Distributions for Size-Dependent Growth, *Chem. Eng. Sci.*, **33**, 1623 (1978).

11.2-50 R. W. Rousseau and R. M. Parks, Size-Dependent Growth of Magnesium Sulfate Heptahydrate, *Ind. Eng. Chem. Fundam.*, **20**, 71 (1981).

11.2-51 J. Garside and S. F. Jančić, Growth and Dissolution of Potash Alum Crystals in the Subsieve Size Range, *AIChE J.*, **22**, 887 (1976).

11.2-52 J. Garside and S. J. Jančić, Measurement and Scale-up of Secondary Nucleation Kinetics for the Potash Alum–Water System, *AIChE J.*, **25**, 948 (1979).

11.2-53 E. T. White and P. G. Wright, Magnitude of Size Dispersion Effects in Crystallization, *Chem. Eng. Prog. Symp. Ser. No. 110*, **67**, 81 (1971).

11.2-54 A. H. Janse and E. J. de Jong, The Occurrence of Growth Dispersion and Its Consequences, in J. W. Mullin (Ed.), *Industrial Crystallization*, p. 145, Plenum Press, New York, 1976.

11.2-55 A. H. Janse and E. J. de Jong, Growth and Growth Dispersion, in E. J. de Jong and S. J. Jančić (Eds.), *Industrial Crystallization '78*, p. 135, North-Holland, Amsterdam, 1979.

11.2-56 J. Garside, The Growth of Small Crystals, in E. J. de Jong and S. J. Jančić (Eds.), *Industrial Crystallization '78*, p. 143, North-Holland, Amsterdam, 1979.

11.2-57 M. W. Girolami and R. W. Rousseau, Size-Dependent Growth—A Manifestation of Growth Rate Dispersion in the Potassium Alum–Water System, *AIChE J.*, **31**, 1821 (1985).

11.2-58 K. A. Berglund and M. A. Larson, Growth of Contact Nuclei of Citric Acid Monohydrate, *AIChE Symp. Ser. No. 215*, **78**, 9 (1982).

11.2-59 P. D. B. Bujac, Attrition and Secondary Nucleation in Agitated Crystal Slurries, in J. W. Mullin (Ed.), *Industrial Crystallization*, p. 12, Plenum Press, New York, 1976.

11.2-60 C. M. van't Land and B. G. Wienk, Control of Particle Size in Industrial NaCl-Crystallization, in J. W. Mullin (Ed.), *Industrial Crystallization*, p. 51, Plenum Press, New York, 1976.

11.2-61 A. D. Randolph and E. T. White, Modeling Size Dispersion in the Prediction of Crystal Size Distribution, *Chem. Eng. Sci.*, **32**, 1067 (1977).

11.2-62 C. Y. Lui, H. S. Tsuei, and G. R. Youngquist, Crystal Growth from Solution, *CEP Symp. Ser. No. 110*, **67**, 43 (1971).

11.2-63 H. J. Human, W. J. P. van Enckevork, and P. Bennema, Spread in Growth Rates on the (111),

(100), and (110) faces of Potash Alum Growing from Aqueous Solutions, in S. J. Jančić and E. J. de Jong (Eds.), *Industrial Crystallization 81*, p. 387, North-Holland, Amsterdam, 1982.

11.2-64   K. A. Berglund, E. L. Kaufman, and M. A. Larson, Growth of Contact Nuclei of Potassium Nitrate, *AIChE J.*, **25**, 867 (1983).

11.2-65   J. Garside and R. I. Ristic, Growth Rate Dispersion among ADP Crystals Formed by Primary Nucleation, *J. Cryst. Growth*, **61**, 215 (1983).

11.2-66   K. E. Blem and K. A. Ramanarayanan, Generation and Growth of Secondary Ammonium Dihydrogen Phosphate, *AIChE J.*, in press.

11.2-67   A. H. Janse and E. J. de Jong, The Occurrence of Growth Dispersion and Its Consequences, in J. W. Mullin (Ed.), *Industrial Crystallization*, p. 145, Plenum Press, New York, 1976.

11.2-68   M. A. Larson, E. T. White, K. A. Ramanarayanan, and K. A. Berglund, Growth Rate Dispersion in MSMPR Crystallizer, *AIChE J.*, **31**, 90 (1985).

11.2-69   K. A. Berglund and M. A. Larson, Modeling of Growth Rate Dispersion of Citric Acid Monohydrate in Continuous Crystallizers, *AIChE J.*, **30**, 280 (1984).

11.2-70   K. A. Ramanarayanan, K. A. Berglund, and M. A. Larson, Growth Rate Dispersion in Batch Crystallizers, *Chem. Eng. Sci.*, **40**, 1604 (1985).

11.2-71   R. C. Zumstein and R. W. Rousseau, Growth Rate Dispersion by Initial Growth Rate Distributions and Growth Rate Fluctuations, *AIChE J.*, **33**, 121 (1987).

11.2-72   C. G. Moyers and A. D. Randolph, Crystal-Size Distribution and Its Interaction with Crystallizer Design, *AIChE J.*, **23**, 500 (1973).

11.2-73   F. P. O'Dell and R. W. Rousseau, Magma Density and Dominant Size for Size-Dependent Crystal Growth, *AIChE J.*, **24**, 738 (1978).

11.2-74   R. W. Rousseau and F. P. O'Dell, Moments of Crystal Size Distributions in Systems with Selective Crystal Removal or Size-Dependent Crystal Growth, *Chem. Eng. Commun.*, **6**, 293 (1980).

11.2-75   R. C. Zumstein and R. W. Rousseau, "Growth Rate Dispersion in the Crystallization of Copper Sulfate Pentahydrate," paper presented at AIChE National Mtg., New Orleans, 1986.

11.2-76   A. D. Randolph, J. R. Beckman, and Z. I. Kraljevich, Crystal Size Distribution Dynamics in a Classified Crystallizer, *AIChE J.*, **23**, 500 (1977).

11.2-77   A. D. Randolph, Advances in Crystallizer Modeling and CSD Control, *AIChE Symp. Ser. No. 240*, **80**, 14 (1984).

11.2-78   N. S. Tavare, J. Garside, and M. R. Chivate, Analysis of Batch Crystallizers, *Ind. Eng. Chem. Process Des. Dev.* **19**, 653 (1980).

11.2-79   J. S. Wey, Analysis of Batch Crystallization Processes, *Chem. Eng. Commun.* **35**, 231 (1985).

Section 11.4

11.4-1   M. Zief and W. R. Wilcox, *Fractional Solidification*, Vol. 1, p. 24, Marcel Dekker, New York, 1967.

11.4-2   G. R. Atwood, *Recent Advances in Separation Science*, Vol. 1, p. 1, CRC Press, Cleveland, 1972.

11.4-3   K. Saxer and A. Papp, The MWB Crystallization Process, *Chem. Eng. Prog.*, 64 (Apr. 1980).

11.4-4   D. L. McKay and H. W. Goard, Continuous Fractional Crystallization, *Chem. Eng. Progr.*, **61**, 99 (1965).

11.4-5   H. Schildknecht and K. Maas, Kaloonenkristalliseren, *Warme*, **69**, 121 (1963).

11.4-6   J. A. Brodie, *A Continuous Multi-Stage Melt Purification Process*, p. 37, The Mechanical and Chemical Engineering Transactions of the Institution of Engineers, Australia, May 1971.

GENERAL BIBLIOGRAPHY

A. W. Bamforth, *Industrial Crystallization*, Grampion Press, London, 1965. Bamforth emphasizes the factors that influence the design and selection of crystallizers. Flowsheets and industrial data are presented for many systems of commercial interest.

J. W. Mullen, *Crystallization*, Butterworths, London, 1961. Mullen presents a wide spectrum of crystallization fundamentals including concepts of the crystalline state, discussion of phase equilibria, mechanism of crystallization, and selection of industrial equipment.

J. W. Mullen, *Crystallization*, 2nd ed., Butterworths, London, 1972. Mullen presents an updated version of his earlier work incorporating many of the advances made in the last decade.

J. Nyvlt, *Industrial Crystallization from Solutions*, Butterworths, London, 1971. Nyvlt discusses the theoretical foundations of crystallization and presents crystallizer design methods.

J. Nyvlt, *Solid-Liquid Phase Equilibria*, Elsevier–North Holland, New York, 1977. Consolidates various methods for handling and assessing experimental data. Presents correlation methods for multicomponent systems.

J. Nyvlt, *Industrial Crystallization*, Verlag Chemie, Weinheim, NY, 1978. Review of the conclusions of the European Working Party on Crystallization (WPC).

A. D. Randolph and M. A. Larson, *Theory of Particulate Processes*, Academic Press, New York, 1971. This book consolidates the present theories regarding crystal nucleation and growth and presents predictive methods for determining CSD.

J. Wisniak, *Phase Diagrams*, Physical Sciences Data Vol. 10, Elsevier–North Holland, New York, 1981. Extensive compilation of physical property data. Mostly from *Chemical Abstracts*.

CHAPTER 12

Adsorption

GEORGE E. KELLER II
Union Carbide Corporation
South Charleston, West Virginia

RICHARD A. ANDERSON
CARMEN M. YON
Union Carbide Corporation
Tarrytown, New York

12.1 INTRODUCTION

The first use of adsorption is lost in antiquity. Perhaps it was associated with the observation that water tasted differently—presumably better—when it was treated with charred wood. The ability of certain materials to remove color from solutions was known in the fifteenth century, and bone char was used commercially for decolorizing sugar solutions in the late eighteenth century. In the mid nineteenth century wood charcoal was being used in respirators in hospitals for air purification.[1] The first large-scale, bulk gas adsorption processes were commercialized almost simultaneously in the early 1920s—the removal of alcohol and benzene from a gas stream by Bayer AG in Germany[2] and the recovery of ethane and higher hydrocarbons from natural gas by Union Carbide Corporation in the United States.[3]

From such humble and diverse beginnings has grown a separation technology which today, it can be argued, is the most widely used nonvapor–liquid technique for molecular separations in the petroleum, natural gas, petrochemical, and chemical industries. A representative list of separations made by adsorption is given in Table 12.1-1.

Adsorption processes consist of the selective concentration (adsorption) of one or more components (adsorbates) of either a gas or a liquid at the surface of a microporous solid (adsorbent). The attractive forces causing the adsorption are generally weaker than those of chemical bonds, and by increasing the temperature of the adsorbent or reducing an adsorbate's partial pressure (or concentration in a liquid), the adsorbate can be desorbed. The desorption or regeneration step is quite important in the overall process. First, desorption allows recovery of adsorbates in those separations where they are valuable; and second, it permits reuse of the adsorbent for further cycles. In a few cases, desorption is not practical, and the adsorbate must be removed by thermal destruction or another chemical reaction, or the adsorbent is simply discarded.

In the following sections we will discuss a number of important aspects of adsorption technology, including adsorbents, criteria for when to use adsorption, a description of various adsorption process flowsheets, criteria for choosing a process flowsheet, process-design considerations, and speculations on new and expanded uses for adsorption processes.

In addition to the information presented here, the reader should be aware of the other sources of general and specific information on various aspects of adsorption technology.[4–20]

TABLE 12.1-1   Representative Commercial Adsorption Separations

Separation[a]	Adsorbent
Gas Bulk Separations	
Normal paraffins, isoparaffins, aromatics	Zeolite
N_2/O_2	Zeolite
O_2/N_2	Carbon molecular sieve
CO, CH_4, CO_2, N_2, A, NH_3/H_2	Zeolite, activated carbon
Acetone/vent streams	Activated carbon
C_2H_4/vent streams	Activated carbon
H_2O/ethanol	Zeolite
Gas Purifications[c]	
H_2O/olefin-containing cracked gas, natural gas, air, synthesis gas, etc.	Silica, alumina, zeolite
CO_2/C_2H_4, natural gas, etc.	Zeolite
Organics/vent streams	Activated carbon, others
Sulfur compounds/natural gas, hydrogen, liquified petroleum gas (LPG), etc.	Zeolite
Solvents/air	Activated carbon
Odors/air	Activated carbon
NO_x/N_2	Zeolite
SO_2/vent streams	Zeolite
Hg/chlor-alkali cell gas effluent	Zeolite
Liquid Bulk Separations[b]	
Normal paraffins, isoparaffins, aromatics	Zeolite
p-Xylene/*o*-xylene, *m*-xylene	Zeolite
Detergent-range olefins/paraffins	Zeolite
p-Diethyl benzene/isomer mixture	Zeolite
Fructose/glucose	Zeolite
Liquid Purifications[c]	
H_2O/organics, oxygenated organics, chlorinated organics, etc.	Silica, alumina, zeolite
Organics, oxygenated organics chlorinated organics, etc./H_2O	Activated carbon
Odor, taste bodies/drinking H_2O	Activated carbon
Sulfur compounds/organics	Zeolite, others
Various fermentation products/fermentor effluent	Activated carbon
Decolorizing petroleum fractions, sugar syrups, vegetable oils, etc.	Activated carbon

[a] Adsorbates listed first.

[b] Adsorbate concentrations of about 10 wt.% or higher in the feed.

[c] Adsorbate concentrations generally less than about 3 wt.% in the feed.

12.2   ADSORBENTS

Adsorbents have been developed for a wide range of separations. Commercial materials are provided usually as pellets, granules, or beads, although powders are used occasionally. The adsorbent may be used once and discarded, or, as is more common, it is employed on a regenerative basis and used for many, many cycles. Adsorbents are generally used in cylindrical vessels through which the stream to be treated is passed. In the regenerative mode, two or more beds usually are employed with suitable valving to allow for continuous processing. Absorbents are used in applications requiring from a few ounces to over 2 million pounds in one plant.

Commercial adsorbents are divided into four major classes: molecular-sieve zeolites, activated alumina, silica gel, and activated carbon. Since adsorption is a surface-related phenomenon, the useful adsorbents are all characterized by a large surface area per unit of weight (or volume). The typical range of areas covers from about 100 to over 3000 m^2/g. However, the most common commercially useful materials exhibit surface areas ranging from about 300 to 1200 m^2/g.

Estimates of the 1983 world-wide market for adsorbents are as follows:

Activated carbon	$380 million
Molecular-sieve zeolites	100 million
Silica gel	27 million
Activated alumina	26 million

The applications for these adsorbents depend on their particular adsorptive properties. The surface selectivities can be broadly classed as hydrophilic or hydrophobic. For example, activated alumina and a majority of the molecular-sieve zeolites possess hydrophilic surfaces, and as such adsorb water strongly in preference to organic molecules. The surfaces of the vapor-phase-activated carbon products are hydrophobic and prefer organics to water. The surface of silica gel lies between these extremes and has a reasonable affinity for both water and organics. The terms ''organophilic'' and ''organophobic are also used.

12.2-1 Molecular Sieve Zeolites

DESCRIPTION
Molecular-sieve zeolites are crystalline aluminosilicates of group IA and group IIA elements such as sodium, potassium, magnesium and calcium. Chemically, they are represented by the empirical formula:

$$M_{2/n}O \cdot Al_2O_3 \cdot YSiO_2 \cdot wH_2O$$

where Y is 2 or greater, n is the cation valence, and w represents the water contained in the voids of the zeolite. Structurally, zeolites are complex, crystalline inorganic polymers based on an infinitely extending framework of AlO_4 and SiO_4 tetrahedra linked to each other by the sharing of oxygen ions. This framework structure contains channels or interconnected voids that are occupied by the cations and water molecules. The cations are mobile and ordinarily undergo ion exchange. The water may be removed reversibly, generally by the application of heat, which leaves intact a crystalline host structure permeated by micropores which may amount to 50% of the crystals by volume.

The structural formula of a zeolite is based on the crystal unit cell, the smallest unit of structure, represented by

$$M_{x/n}[(AlO_2)_x(SiO_2)_y] \cdot wH_2O$$

where n is the valence of cation M, w is the number of water molecules per unit cell, x and y are the total number of tetrahedra per unit cell, and y/x usually has values of 1–5. Recently, however, high-silica zeolites have been prepared in which y/x is 10–100 or even higher and, in one case, a molecular-sieve silica has been prepared.

Zeolites were first recognized as a new type of mineral in 1756. Studies of the gas-adsorption properties of dehydrated natural zeolite crystals more than 60 years ago led to the discovery of their molecular-sieve behavior. As microporous solids with uniform pore sizes that range from 0.3 to 0.8 nm, these materials can selectively adsorb or reject molecules based on their molecular size. This effect, with obvious commercial overtones leading to novel processes for separation of materials, inspired attempts to duplicate the natural materials by synthesis. Many new crystalline zeolites have been synthesized, and several fulfill important functions in the chemical and petroleum industries. More than 150 synthetic zeolite types and 40 zeolite minerals are known. The most important molecular sieve zeolite adsorbents are the synthetic Type A, Type X, synthetic mordenite, and their ion-exchanged variations, and the mineral zeolites, chabazite and mordenite.

ABSORPTIVE PROPERTIES AND APPLICATIONS
The molecular-sieve zeolites are distinct from other three major adsorbents in that they are crystalline and that adsorption takes place inside the crystals, the access to which is limited by the pore size. Although other microporous solids are used as adsorbents for the separation of vapor or liquid mixtures, the distribution of pore diameters does not enable separations based on the molecular-sieve effect, that is, separations caused by differences in the molecular size of the materials to be separated. The most important molecular-sieve effects are shown by dehydrated crystalline zeolites. Zeolites selectively adsorb or reject molecules based on differences in molecular size, shape, and other properties such as polarity. During the adsorption of various molecules, the micropores fill and empty reversibly. Adsorption in zeolites is a matter of pore filling, and the usual surface-area concepts are not applicable.

The channels in zeolites are only a few molecular diameters in size, and overlapping potential fields from opposite walls result in a flat adsorption isotherm which is characterized by a long horizontal section as the relative partial pressure approaches unity. The adsorption isotherms do not exhibit hysteresis.

To use the adsorption properties of the synthetic crystals (1–5 μm) in processes, the commercial materials are prepared as pelleted agglomerates containing a high percentage of the crystalline zeolite together with an inert binder. The formation of these agglomerates introduces macropores in the pellet which may result in some capillary condensation at high adsorbate concentrations. In commercial materials, the macropores contribute diffusion paths. However, nearly all the adsorption capacity is contained in the voids within the crystals (99%).

Zeolites are high-capacity, selective adsorbents because they separate molecules based on the size and configuration of the molecule relative to the size and geometry of the main apertures of the structures; zeolites adsorb molecules, in particular those with permanent dipole moments which show other interaction effects, with a selectivity that is not found in other solid adsorbents.

Separation may be based on the molecular-sieve effect or may involve the preferential or selective adsorption of one molecular species over another. These separations are governed by several factors.

1. The basic framework structure, or topology, of the zeolite determines the pore size and the void volume.
2. The exchange cations, in terms of their specific location in the structure, their population or density, their charge, and size, affect the molecular-sieve behavior and adsorption selectivity of the zeolite. By changing the cation types and number, one can tailor or modify within certain limits the selectivity of the zeolite in a given separation.
3. The cations, depending on their locations, contribute electric field effects that interact with the adsorbate molecules.
4. The effect of the temperature of the adsorbent is pronounced in cases involving activated diffusion.

Sieving by dehydrated zeolite crystals is based on the size and shape differences between the crystal apertures and the adsorbate molecule. In some instances, the aperture is circular, such as in zeolite A. In others, it may take the form of an ellipse such as in dehydrated chabazite. In this case, subtle differences in the adsorption of various molecules result from this shape factor.

When two or more molecular species involved in a separation are both adsorbed, selectivity effects become important because of interaction between the zeolite and the adsorbate molecules. These interaction energies include dispersion and short-range repulsion energies, polarization energy, and components attributed to electrostatic interactions.

Some typical molecular dimensions are shown in Fig. 12.2-1, based on the Lennard-Jones potential function.

The unique adsorption properties made available by the almost infinite variety of molecular-sieve adsorbent products have led to a multitude of applications. In the area of purification, they are used to dry nearly any fluid stream, liquid or vapor phase. Molecular sieves provide lower dew points than any other commercially available adsorbent desiccant. The use of molecular sieves to dehydrate natural gas prior to cryogenic processing is accepted as the standard. Likewise, the use of the 3-A type, whose pore size allows only the adsorption of water, in the drying of cracked gas before cryogenic recovery of ethylene has become standard. In this case, no coadsorption and subsequent loss of valuable product occur, as they do if activated alumina is employed.

Molecular sieves are used to remove hydrogen sulfide and other sulfur compounds from natural gas and LPG liquids to meet exacting specifications. The removal of carbon dioxide from air and natural gas before cryogenic processing is routine. Molecular sieves are the premium desiccant for insulated windows. Special varieties are used for refrigerant drying, both in the large commercial production plants and in the individual refrigeration units to protect the system for its lifetime.

Broad use of tailored molecular sieves is found in bulk separations. World-scale plants are operating to recover normal paraffins from refinery streams (a molecular-sieving application), paraxylene form mixed aromatic streams, oxygen from air, and fructose from sugar mixes, to name a few.

MANUFACTURE

Zeolites are formed under hydrothermal conditions, defined here in a broad sense to include zeolite crystallization from aqueous systems containing various types of reactant. Most synthetic zeolites are produced under nonequilibrium conditions and must be considered as metastable phases in a thermodynamic sense.

Although more than 150 synthetic zeolites have been reported, many important types have no natural mineral counterpart. Conversely, synthetic counterparts of many zeolite minerals are not yet known. The conditions generally used in synthesis are reactive starting materials such as freshly coprecipitated gels or amorphous solids; relatively high pH introduced in the form of an alkali metal hydroxide or other strong base, including tetraalkylammonium hydroxides; low-temperature hydrothermal conditions with concurrent low autogenous pressure at saturated water vapor pressure; and a high degree of supersaturation of the gel components, leading to the nucleation of a large number of crystals.

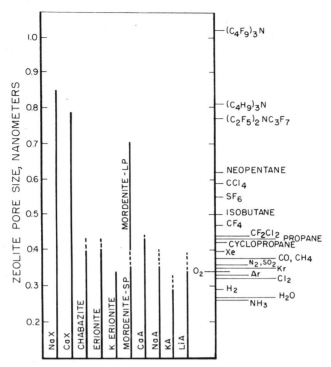

FIGURE 12.2-1   Molecular dimensions and zeolite pore size. Chart shows effective pore sizes of various zeolites over temperatures of 77–420 K. (From Ref. 12.1-8, copyright John Wiley & Sons, Inc., reprinted with permission.)

The gels are crystallized in a closed hydrothermal system at temperatures varying from room temperature to about 200°C. The time required for crystallization varies from a few hours to several days. The alkali metals form soluble hydroxides, aluminates, and silicates. These materials are well suited for the preparation of homogeneous mixtures.

Typical gels are prepared from aqueous solutions of reactants such as sodium aluminate, sodium hydroxide, and sodium silicate; other reactants include alumina trihydrate ($Al_2O_3 \cdot 3H_2O$), colloidial silica, and silicic acid. The temperature strongly influences the crystallization time.

The crystalline products are filtered (cake) from the other liquor. The cake is washed; it can be ion exchanged to a different cationic form, then mixed with suitable binder clays, formed, dried, and fired to provide a final product.

The preparation of zeolite–binder agglomerates as spheres or cylindrical pellets which have high mechanical attrition resistance is not difficult. However, to use the zeolite in a process of adsorption or catalysis, the diffusion characteristics must not be unduly affected. Consequently, the binder system must permit a macroporosity that does not increase unduly the diffusion resistance. Therefore, the problem, is to optimize the zeolite–binder combination to achieve a particle of maximum density (to produce a high volumetric adsorption capacity) with maximum mechanical attrition resistance and minimum diffusion resistance.

Typical properties of commercially available molecular-sieve zeolite adsorbents are presented in Table 12.2-1. Typical single-component adsorption data are presented in Table 12.2-2.

12.2-2   Activated Alumina

DESCRIPTION

Activated alumina adsorbents are amorphous or "transition aluminas" whose chemical composition is Al_2O_3. Their surface area is generated by the removal of water of constitution from hydrated aluminas. They possess reasonably large surface areas. Their fine pore structure and surface chemistry provide the observed adsorption selectivities. The preferred materials have been purified during processing to high Al_2O_3 contents (about 99%), but activated bauxite has also been employed. The product is commonly provided as granules, and shaped forms are also available.

TABLE 12.2-1 Major Commercial Molecular-Sieve Adsorbent Products[a] and Some Physical Properties

Zeolite Type	Designation	Cation	Pore Size (nm)	Bulk Density (kg/m³)
A	3A	K	0.3	670–740
	4A	Na	0.4	660–720
	5A	Ca	0.5	670–720
X	13X	Na	0.8	610–710
Mordenite, small port	AW-300 Zeolon-300	Na + mixed cations	0.3–0.4	720–800
Chabazite	AW-300	Mixed cations	0.4–0.5	640–720

[a] All products available in pellets, beads, and granulated mesh in different size ranges. They are supplied activated with ~1.5% by weight residual water.

ADSORPTIVE PROPERTIES AND APPLICATIONS

The most important industrial application for activated aluminas are in drying processes—both liquids and gases. Activated alumina has a high affinity for water, not as high as that of the molecular-sieve zeolites, but they can produce dried gases to less than 1 ppm moisture content. Among the gases commonly dried by activated alumina are air, argon, helium, hydrogen, methane, chlorine, hydrogen chloride, sulfur dioxide, and the refrigerant fluorocarbons. Liquids are also dried, and the list includes kerosene, aromatics, gasoline fractions, and chlorinated hydrocarbons.

The capacity for water is strongly dependent on the relative humidity of the gas to be dried. A typical isotherm is shown in Fig. 12.2-2 along with water isotherms for a molecular sieve and a silica gel. The capacity for water is also strongly affected by the adsorption temperature. Figure 12.2-3 presents comparative data for the three adsorbents plotted as isobars.

MANUFACTURE

The oldest commercial product, still in use, is made from Bayer α-trihydrate. During precipitation from sodium aluminate solutions, gibbsite accumulates on the vessel walls as a crust. This material is ground to the proper size, usually a few millimeters. Activation is carried out by heating to about 400°C in a current of air to remove the steam formed during the dehydration. The alumina obtained contains less than 1% soda and traces of silica and iron oxide. Typical surface areas range from 200 to 500 m²/g and a pore volume of 0.2–0.3 cm³/g are observed. The predominant pore diameters are in the 2.0–5.0 nm range.

Activation of bauxite yields a similar product except that 10–30% of silica, iron oxide, and titania will be present.

Rapid activation of Bayer hydrate up to 800°C results in a fine particle material nearly free of boehmite. This product can be formed into spheres due to the tendency of the material to rehydrate. The formed product is reactivated at 400°C. The final product exhibits 300–350 m²/g surface area and pore volumes as high as 0.4 cm³/g.

Alumina gels can also be used to prepare activated alumina products, for example, acidified sodium aluminate solutions or mixtures of aluminum sulfate and ammonia. The precipitate is washed and dried, and granules or formed shapes can be prepared.

Typical physical properties and adsorption data are presented in Table 12.2-3.

TABLE 12.2-2 Typical Adsorption Data on Molecular-Sieve Adsorbent Products (Capacity Expressed in Weight Percent on Basis of Activated Pellet)

Designation	Type	O_2 at 100 mm Hg, -183°C	H_2O at 4.6 mm Hg, 25°C	CO_2 at 250 mm Hg, 25°C	n-C_4H_{10} at 250 mm Hg, 25°C
3A	LiA	NA[a]	20	NA	NA
4A	NaA	22	23	13	NA
5A	CaA	22	21	15	10
13X	NaX	24	25	16	12
AW-300	Mordenite	7	9	6	NA
AW-500	Chabazite	17	16	12	NA

[a] NA: Not adsorbed.

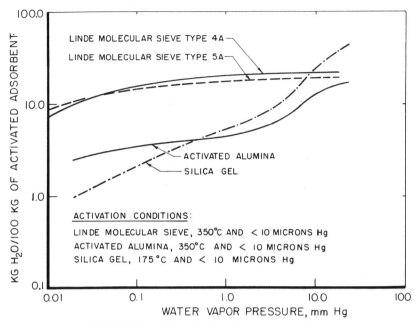

FIGURE 12.2-2 Water isotherms for various adsorbents.

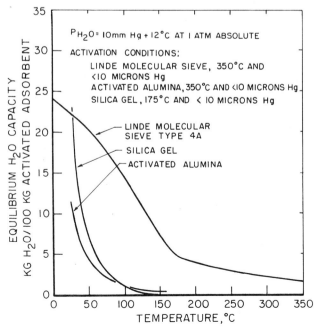

FIGURE 12.2-3 Water isobars for various adsorbents.

TABLE 12.2-3 Typical Properties of Adsorbent-Grade Activated Alumina

	Physical Properties
Surface area (m²/g)	320
Density (kg/m³)	800
Reactivation temperature (°C)	150–315
Pore volume (% of total)	50
Pore size (nm)	1–7.5
Pore volume (cm³/g)	0.40

Adsorption Properties	Percent by Weight
H_2O capacity at 4.6 mm Hg, 25°C	7
H_2O capacity at 17.5 mm Hg, 25°C	16
CO_2 capacity at 250 mm Hg, 25°C	2

12.2-3 Silica Gel

DESCRIPTION

Silica-gel adsorbents are composed of a rigid three-dimensional network of spherical particles of amorphous colloidal silica (SiO_2). They were reported as early as 1640 and commercial production was begun around 1919. The surface area is generated by the very fine size of the colloidal particles. They exhibit surface areas from as little as 100 m²/g for the "aerogels" to over 800 m²/g. The product is provided both in granular and spherical forms.

ADSORPTIVE PROPERTIES AND APPLICATIONS

As mentioned earlier, the silica-gel surface has an affinity for water and organics, although water is preferred. The surface of the silica gel can be in a fully hydroxylated form (Si—O—H) or in a dehydrated siloxane form (Si—O—Si). The former is the result of drying the gel or precipitate below 150°C, and the surface is readily wetted by water. The dehydration of the fully hydroxylated form by heating from 300 to 1000°C results in the siloxane-type surface.

The primary adsorptive applications of silica gel is in the dehydration of gases and liquids. It has been used in drying applications as diverse as gas and liquid drying in fixed beds in a regenerative mode, in the space channels of dual-pane insulated windows to dehydrate the air space to prevent fogging, and to dehydrate flowers. Until the advent of the application of cryogenic technology in the natural gas industry, silica gel was commonly used to recover hydrocarbons from natural gas streams.

MANUFACTURE

The most common preparation is by the acidification of sodium silicate solutions to a pH of less than 10. The precipitate is usually provided in granular form although beaded forms are available. Fine powders are also occasionally used in some adsorptive applications.

Silica gel can also be prepared from a silica hydrosol. In this method, sodium silicate is reacted with a strong mineral acid. A rigid mass will set up which can be granulated and activated.

The most common adsorbent-grade silica gel, termed "regular density," is prepared by gelling in an acid medium. The product is made up of very fine colloidal silica particles and exhibits a 750–800 m²/g surface area. The pores are in the 2.0–4.0 nm range and the pore volume ranges from 0.37 to 0.40 cm³/g. This material has a high capacity for water and polar materials.

An intermediate density silica gel is also available and finds some use as a desiccant. The pore size is much larger (12–16 nm) and the pore volume much higher (0.9–1.1 cm³/g). The surface area is around 300 m²/g. It has a good capacity at high relative humidities and is usually used as a powder.

The low density "aerogels" are used as very fine powders, exhibit 100–200 m²/g surface area, and are rarely used as an adsorbent.

Typical properties of adsorbent grade silica gel are given in Table 12.2-4.

12.2-4 Activated Carbon

DESCRIPTION

Activated carbon is a microcrystalline, nongraphitic form of carbon that has been processed to develop internal porosity. This porosity yields the surface area that provides for the ability to adsorb gases and vapors from gases, and dissolved or dispersed substances from liquids. The range of surface areas is vast, from 300 to 2500 m²/g, although commercially practical materials are usually limited to surface areas of

TABLE 12.2-4 Typical Properties of Adsorbent-Grade Silica Gel

	Physical Properties
Surface area (m^2/g)	830
Density (kg/m^3)	720
Reactivation temperature (°C)	130–280
Pore volume (% of total)	50–55
Pore size (nm)	1–40
Pore volume (cm^3/g)	0.42
Adsorption Properties	Percent by Weight
H$_2$O capacity at 4.6 mm Hg, 25°C	11
H$_2$O capacity at 17.5 mm Hg, 25°C	35
O$_2$ capacity at 100 mm Hg, −183°C	22
CO$_2$ capacity at 250 mm Hg, 25°C	3
n-C$_4$ capacity at 250 mm Hg, 25°C	17

less than 1200 m^2/g. Two distinct types of activated carbon are recognized commercially. Liquid-phase, or decolorizing, carbons are generally light, fluffy powders that exhibit surface areas of about 300 m^2/g. Gas- or vapor-phase carbons are hard granules or formed pellets that exhibit surface areas from 800 to 1200 m^2/g.

The pore size of the liquid-phase materials is usually 3.0 nm or larger, which allows for more rapid diffusion. The pore size of the vapor-phase material is usually less than 3.0 nm. As mentioned before, the activated carbon surface is generally hydrophobic, especially in the vapor-phase grades which are activated at high temperatures. The lower-surface-area liquid-phase materials can be made to be more "wettable" by the proper activation method. This property is important for the liquid-phase products used in decolorizing aqueous solutions.

The liquid-phase materials are usually characterized by sorption tests using phenol, iodine, or "molasses number." The vapor-phase activated carbons are usually characterized by carbon tetrachloride or benzene adsorption tests. The adsorption capacity and the bulk density define the volumetric treating capability of the material.

Another property of concern is the kindling point. This must be high enough to prevent excessive oxidation in the gas phase during regeneration or in the case where a high heat of adsorption is present during adsorption, for example, in the case of ketones. The kindling point in oxygen should be over 370°C.

Three other product considerations are the ash content and ash composition and the pH of the carbon (in contact with water).

Recent usage of adsorbent-grade activated carbons is about 100,000 metric tons annually in the United States. Close to 85% is used in liquid-phase applications, wherein the product is frequently used once and discarded. The liquid-phase is divided at about 60% powder and 40% granulated forms. The vapor-phase product usage is much lower due to the fact that the material is employed in regenerative service and used for many cycles.

ADSORPTIVE PROPERTIES AND APPLICATIONS

Liquid-phase activated carbons are used primarily for decolorizing and deodorizing aqueous solutions. This is due to their affinity for organic compounds in the presence of water. The decolorizing of sugar solutions accounts for about one-third of the consumption. The removal of odor and taste components from drinking water in municipal plants accounts for an additional 25% of the usage. Another major application is in the reclaiming of dry-cleaning solvents. The balance is used in a myriad of industries, including pharmaceutical, food and beverage processing, purifying oils, fats, and waxes, and electroplating baths.

The use of powders is generally in a batch phase. The solution to be treated is mixed with an appropriate quantity of the activated-carbon powder, agitated for less than 1 h, and then separated by settling or filtration. The used carbon may be discarded or "eluted" (regenerated) for further use. The use of particulate forms in liquid treating employs fixed beds through which the solution to be treated is passed. When the beds become saturated, they can be revivified by gas oxidation or solvent extraction.

Although bone char was historically used for sugar decolorization, the use of activated carbon has grown very rapidly. Color improvement is key, but the activated carbon also removes nitrogenous substances and other colloids. This improves both filterability and crystallization rate. In the dextrose industry, granulated carbon is used, and the spent carbon is reactivated with steam and air in rotary multiple-hearth furnaces.

Drinking water has been successfully treated since 1930 with activated-carbon powders. The granular packed-bed method proved to be too expensive and inadequate during periods of high pollution.

The vapor-phase carbons are more highly activated and the particles are stronger, bigger, and denser. They have higher surface areas, usually around 1000 m²/g, and a smaller pore size, as mentioned previously. They are usually employed on a regenerative basis.

The largest application for vapor-phase activated carbons is solvent recovery. Industry abounds with applications that emit a solvent during processing. Ethers, alcohols, ketones, chlorinated hydrocarbons, and hydrocarbon solvents are commonly recovered from vent and air streams. Furthermore, although not considered a regenerative process in the conventional sense, the use of activated carbon in canisters on automobile fuel tanks to reduce hydrocarbon emissions may now be the single largest application. A typical activated-carbon recovery system consists of two beds of activated carbon and suitable valving to direct stream flow. The stream to be processed is passed through one bed until it is exhausted and then switched to the second bed. The spent bed is usually regenerated with low-pressure steam and the steam–solvent mixture condensed. If the solvent is insoluble the mixture can be decanted. If not, the mixture can be distilled. Most units are automated.

Activated-carbon systems are also used for purification of a great variety of gaseous materials. Hydrogen is purified for use in refineries or for the hydrogenation of oils, carbon dioxide is deodorized before dry ice manufacture, and oil vapor is removed from compressed gases. In these applications, the regeneration can be accomplished by a temperature change (thermal purge) or in some cases by a pressure change—with or without a purge gas.

Vapor-phase activated carbons are also used for deodorizing air in ventilation systems, treating exhaust gases from manufacturing plants to avoid contamination of the atmosphere, and of course in gas masks.

MANUFACTURE

Almost any carbonaceous material can be used to manufacture activated carbons, if properly treated. It has been made from the blood and bones of animals, hard and soft woods, rice hulls, nutshells, refinery residuals, peat, lignin, coal, coal tars, pitches, and carbon black. Usually, wood, peat, lignite, and lignin are favored for decolorizing liquid-phase carbons, and nutshells, coal, peat, and petroleum residues are used for gas-phase adsorbent carbons.

There are two basic processes employed to manufacture activated carbons. They are referred to as chemical activation and gas activation. Chemical activation depends on the action of inorganic chemicals on the raw material to dehydrate the organic molecules during carbonization (or calcination). Gas activation is achieved by the oxidation of the carbonaceous matter with air at low temperatures, or steam, flue gas, or carbon dioxide at high temperatures. This step is preceded by a preliminary carbonization of raw material.

Decolorizing carbons are frequently prepared from coal, lignite, sawdust, or peat. These materials are mixed with the oxidizing agent (e.g., alkali metal hydroxides, carbonates, sulfates, and phosphates), heated to 500–900°C, cooled, washed, filtered, and dried.

Some vapor-phase carbons are also prepared by chemical activation. Using sawdust or peat with phosphoric acid or zinc chloride, we obtain a product that can then be further activated with steam to produce a vapor-phase-quality material.

The most common method of manufacturing activated carbons for vapor-phase applications is by gas activation. This method is employed for granular forms of activated carbon. The material is formed and initially calcined at 400–500°C. This material is then selectively oxidized at 800–1000°C to develop porosity and surface activity. The higher temperature in combination with steam, carbon dioxide, or flue gas is preferred since the reaction is less exothermic and easier to control than low-temperature oxidation with air.

Some vapor-phase materials are prepared from hard starting materials. For nutshells, the procedure used is to calcine the shells, granulate to size, and activate. In the case of coal or charcoal, the material is ground to powder, bonded with pitch and formed, calcined at 500–700°C, and activated with steam or flue gas at 500–950°C. The high-temperature furnaces employed include kilns, multiple-hearth, and vertical retort.

Physical properties of typical activated carbons are presented in Table 12.2-5.

12.3 CRITERIA FOR ADSORPTION USE

Adsorption processes have a number of competitors, including primarily distillation, extractive and azeotropic distillation, absorption, solvent extraction, and, more recently, membrane-based processes. Vapor-liquid-based processes and, in particular, distillation are especially formidable competitors because of their relatively simple flowsheets. Thus, if a separation can be performed easily by distillation, it will usually be the process of choice, based on relatively low capital costs and tolerable, if not low, energy costs. In addition, systems of distillation columns can oftentimes be energy integrated to effect lower energy costs per unit of feed processed than for stand-alone columns.

Nevertheless, it is clear that there are separations for which adsorption is the proper choice. Although precise criteria for making this choice are not possible, several rough criteria can be enumerated. In the following list, distillation is assumed to be the chief competitor.

TABLE 12.2-5 Typical Properties of Activated-Carbon Adsorbents

	Liquid-Phase Carbons		Vapor-Phase Carbons	
Physical Properties	Wood Base	Coal Base	Granular Coal	Granular Coal
Mesh size (Tyler)	-100	$-8 + 30$	$-4 + 10$	$-6 + 14$
CCl$_4$ activity (%)	40	50	60	60
Iodine number	700	950	1000	1000
Bulk density, kg/m^3	250	500	500	530
Ash (%)	7	8	8	4

Adsorptive Properties	Vapor-Phase Carbons (wt. %)
H$_2$O capacity at 4.6 mm Hg, 25°C	1
H$_2$O capacity at 250 mm Hg, 25°C	5–7
n-C$_4$ capacity at 250 mm Hg, 25°C	25

1. *Ease of Separability.* If the relative volatility between the key components to be separated is in the range of 1.2–1.5 or less, then both investment and energy costs rise precipitously. This case of course includes azeotrope-forming systems.

2. *Feed Composition.* If the bulk of a liquid feed is a relatively low-value, more-volatile product and the product of interest is in relatively low concentration (10–25 wt. % or less), then a large amount of low-boiling product must be vaporized per unit of valuable product. This situation can lead to excessive energy costs. A variation on this case is one in which a small amount of an undesirable, high-boiling material must be separated from a large amount of low-boiling material; seawater desalination is a notable example of this category.

If a separation must be made between one set of feed components whose boiling range overlaps the boiling range of a chemically dissimilar set, then even though various component relative volatilities may be reasonably large, several columns will be required and investments can become excessive.

3. *Pressure and Temperature.* If dew points and bubble points are such that extreme conditions of temperature (cryogenic conditions or greater than 500–600 K) or pressure (less than 2–50 kPa or greater than 4–6 mPa) are required, then column investments and/or heat source and sink costs can escalate rapidly.

4. *Chemical Reactions.* If thermal damage to the products or rapid column fouling occurs at practical operating conditions, then distillation will not be viable.

When one or more of these criteria pertain to a separation to be done by distillation, then alternatives should be investigated. The most common of these use a mass separating agent to improve the ease of separation. Such processes include extractive and azeotropic distillation, absorption, solvent extraction, and of course adsorption. Criteria for selecting the best of these are not yet available, but adsorption should definitely be considered if a suitable adsorbent exists. In general, a suitable adsorbent is one that shows proper selectivity and capacity, can be regenerated easily, does not foul rapidly, and causes no damage to the products by promoting by-product-forming reactions. Selectivity between key components in general should be greater than 2; capacity criteria will be given for various process embodiments later.

12.4 BASIC ADSORPTION CYCLES

Commercial adsorptions can be divided into bulk separations, in which about 10 wt. % or more of a stream must be adsorbed, and purifications, in which usually considerably less than 10 wt. % of a stream must be adsorbed. Such a differentiation is desirable because in general different process cycles are used for the two categories. (For feeds with adsorbate concentrations from a few up to about 10 wt. %, cycles used in both categories should be investigated.)

As opposed to distillation, adsorption processes come in many different physical embodiments and cycles. Below, four basic cycles and two combinations are described in their simplest forms. Then recent uses and modifications of these cycles, as well as other new process cycles, are described.

12.4-1 Temperature-Swing Cycle

In this cycle, shown in Fig. 12.4-1, a stream containing a small amount of an adsorbate at partial pressure P_1 is passed through the adsorbent bed at temperature T_1. The equilibrium loading on the adsorbent is X_1,

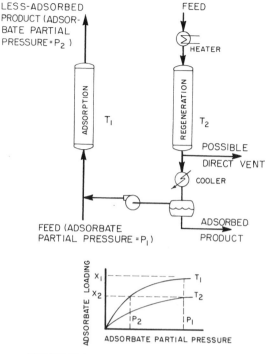

FIGURE 12.4-1 Temperature-swing cycle.

expressed usually in the units of either weight or moles of adsorbate per unit weight of adsorbent. After equilibrium between adsorbate in the feed and on the adsorbent is reached, the bed temperature is raised to T_2, and more feed is passed through the bed. Desorption occurs and a new equilibrium loading, X_2, is established. The net bed removal capacity, called the delta loading, $X_1 - X_2$. (This value is actually an upper limit, since equilibrium loadings in practical operation are not attained, as will be discussed later.) When the bed is subsequently cooled to T_1 and feed is again passed through the bed, purification will occur down to an outlet value of P_2, as shown in Fig. 12.4-1.

Typical delta loadings for temperature-swing cycles are generally in excess of 1 kg per 100 kg of adsorbent and may range up to 10 kg or more per 100 kg of adsorbent in some cases.

The time required to heat, desorb, and cool a bed is usually in the range of a few hours to over a day. Since during this long regeneration time the bed is not productively separating feed, temperature-swing processes are used almost exclusively to remove small concentrations of adsorbates from feeds. Only in such situations can the on-stream time be a significant fraction of the total cycle time of the process.

Temperature-swing cycles can consume a substantial amount of energy per unit of adsorbate. Not only must the heat of desorption be supplied, but also the sensible heat to raise the adsorbent and the adsorber vessel and internals to the desorption temperature. However, if the adsorbate concentration in the feed is only a few percent or less, the energy cost per unit of feed processed can be reasonable.

12.4-2 Inert-Purge Cycle

In this cycle, shown in Fig. 12.4-2, the adsorbate, instead of being removed by temperature increase, is removed by passing a nonadsorbing gas containing very little or no adsorbate through the bed. This has the effect of lowering the partial pressure or concentration of the adsorbate around the particles and desorption occurs. If enough pure purge gas is passed through the bed, the adsorbate will be completely removed, and the maximum delta loading will be X_1. During the subsequent adsorption step, removal of adsorbate from the feed would be essentially complete until the adsorbent becomes nearly fully loaded and breakthrough occurs.

As adsorption occurs, the adsorbent increases in temperature as the heat of adsorption is released. Conversely, cooling occurs during regeneration. Because adsorption capacity is reduced as adsorbent temperature rises, inert-purge processes are usually limited to from 1 to only a very few kg adsorbate per 100 kg adsorbent. Cycle times, in contrast to temperature-swing processes, are only a few minutes and almost always less than 10 min.

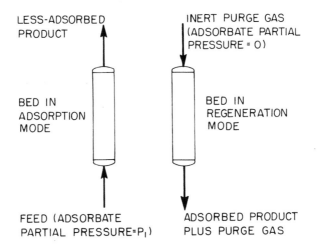

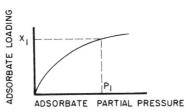

FIGURE 12.4-2 Inert-purge cycle.

12.4-3 Displacement-Purge Cycle

This cycle, though somewhat similar to the inert-purge cycle, differs from it in that a gas or liquid that adsorbs about as strongly as the adsorbate is used to remove the adsorbate (see Fig. 12.4-3). Desorption is thus facilitated both by adsorbate partial-pressure or concentration reduction in the fluid around the particles and by competitive adsorption of the displacement medium. As with the inert-purge cycle, the maximum delta loading is X_1.

The use of the two different types of purge fluid causes two major differences in the processes. First, since the displacement-purge fluid is actually adsorbed on the adsorbent, it is present when the adsorption part of the cycle begins and therefore contaminates the less-adsorbed product. (In the inert-purge cycle this contamination is usually much smaller.) In practical terms this means that the displacement-purge fluid must be recovered from both product streams. Second, since the heat of adsorption of the displacement-purge fluid will be approximately equal to that of the adsorbate, as the two exchange on the adsorbent, the net heat generated (or consumed) is virtually zero, and the adsorbent's temperature remains virtually unchanged throughout the cycle. This fact makes it possible to effect larger delta loadings in this process than in the inert-purge process.

Typical cycle times are usually a few minutes.

12.4-4 Pressure-Swing Cycle

In a gas-adsorption cycle, the partial pressure of an adsorbate can be reduced by reducing the total pressure of the gas. This change can be used to effect a desorption, as shown in Fig. 12.4-4. The lower the total pressure of the regeneration step, the greater the maximum delta loading, $X_1 - X_2$, will be.

The time required to load, depressurize, regenerate, and repressurize a bed is usually a few minutes and can in some cases be only a few seconds. Thus, even though practical delta loadings are almost always less than 1 kg per 100 kg adsorbent to minimize thermal-gradient problems, the very short cycle times make a pressure-swing cycle quite attractive for bulk-gas separations.

12.4-5 Combined Cycles

Often a temperature-swing cycle is combined with an inert purge to further facilitate regeneration. Several possibilities are shown in Fig. 12.4-5. The inert-purge stream can be a fraction of the less-adsorbed product,

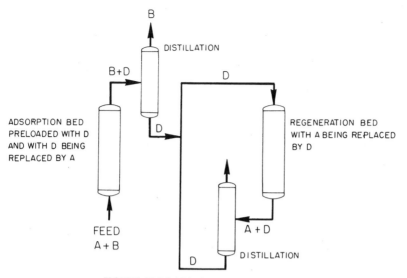

FIGURE 12.4-3 Displacement-purge cycle.

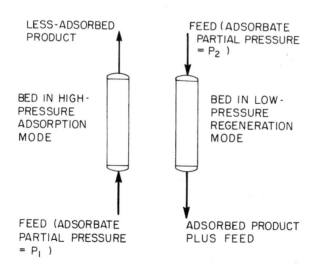

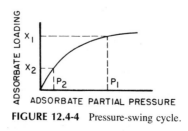

FIGURE 12.4-4 Pressure-swing cycle.

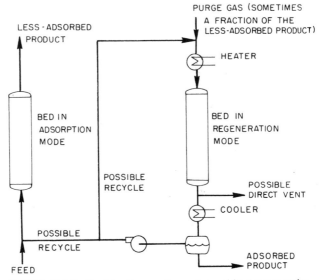

FIGURE 12.4-5   Combined temperature-swing and inert-purge cycles.

or a separate purge stream can be used. If the feed is at super-atmospheric pressure, often the regeneration step will be carried out at atmospheric pressure.

In pressure-swing processes, it is quite common to use a fraction of the less-adsorbed gas product as a low-pressure purge gas, as shown in Fig. 12.4-6. Often, the purge flow is in the opposite direction from the feed flow. The simplest version of this process has been called heatless fractionation or, more commonly, pressure-swing adsorption (PSA). The latter name will be used here. Rules for the minimum fraction of less-adsorbed gas product for displacing the adsorbate have been given by Skarstrom.[1] The larger the high-to-low-pressure ratio, the less the amount of purge required. This fact makes vacuum desorption desirable in some instances. In the limit of very high vacuum, very little if any less-adsorbed gas would be required to complete the desorption, and two virtually pure products would result.

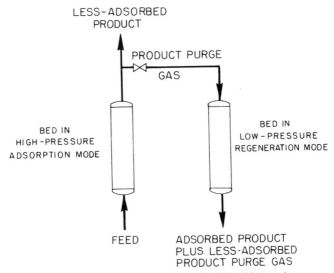

FIGURE 12.4-6   Pressure-swing adsorption (PSA) cycle.

12.5 PROCESS FLOWSHEETS

Adsorption processes have a bewildering variety of flowsheets, brought about by such factors as whether the feed is a liquid or a gas, the feed composition, product recovery and purity, and other factors. In this section we discuss a number of the more common processes, using state of the feed and feed composition as ways of categorizing these processes.

12.5-1 Gas Bulk-Separation Processes

Gas bulk-separation processes consist mostly of PSA variations and displacement-purge and inert-purge processes. These cycles are discussed below. Also, even though chromatography is covered in another chapter, reference is made here and in following sections to several commercial chromatographic processes for completeness of the discussion.

PSA

As previously mentioned, PSA processes use part of the less-adsorbed product to aid in purging the adsorbate from the adsorbent. Thus, an inevitable feature of present-day PSA processes is the loss of that part of the less-adsorbed product to the purge stream. This has the additional implication that the purge stream cannot be of high purity, and as a result PSA is limited at present to those applications where only one pure product is desired: oxygen and nitrogen (but not both simultaneously) from air, hydrogen recovery in cases in which complete recovery is not required, and chemical process purge-stream treatment in which a less-adsorbed product is removed from the process and an adsorbate-containing stream is recycled to the process.

Most often, PSA processes use three or more beds in parallel. The advantages are (1) energy savings, which accrue from using gas from a high-pressure bed to repressurize partially a low-pressure bed following desorption, and (2) higher product recoveries, which accrue from using a combination of purging steps in the cycle. A typical four-bed process for producing 90–95 mol % oxygen from air is shown in Fig. 12.5-1. The steps are sequenced to provide continuous feed and less-adsorbed product flows. Following the adsorption step, rich less-adsorbed gas is removed cocurrently and transferred to another bed to increase its pressure from one intermediate level to another. This is called an equalization step. Next, another volume of less-adsorbed gas is withdrawn cocurrently to purge another low-pressure bed countercurrently. Then a second equalization step occurs, followed by a countercurrent blowdown. Finally, a countercurrent purge completes the pressure-letdown part of the cycle, and the bed is then repressurized by the equalization steps and the adsorption step.

Recently, two widely divergent process variations of PSA have been commercialized. The first of these is called Polybed PSA, which is used for hydrogen recovery.[1-3] Plants with capacities of up to about 1.2 × 10^6 m^3/day of hydrogen have been built. Hydrogen recoveries of 86% versus 70–75% for other PSA processes have been demonstrated, as well as purities of 99.999 mol %.[3] Polybed PSA involves the use of five or more beds, with extensive gas interchanges and pressure equalizations between the beds. One process embodiment[1] is shown in Fig. 12.5-2.

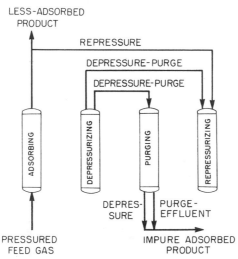

FIGURE 12.5-1 Four-bed PSA process used for producing 90–95 mol % oxygen.

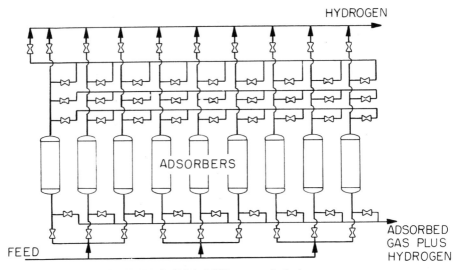

FIGURE 12.5-2 Polybed PSA process for hydrogen recovery.

Thus, Polybed PSA can be thought of as a process that aims for maximum recovery of the less-adsorbed component in very high purity at the expense of a complex flowsheet. Its growing commercial acceptance suggests that it competes quite successfully with cryogenic distillation in many cases.

The second process, called pressure-swing parametric pumping or rapid pressure-swing adsorption (RPSA), was developed to minimize process complexity and investment but at the expense of product recovery. The name parameteric pumping was coined by Wilhelm et al.,[4] who described an adsorption-based separation process involving reversing flows. When the flow is in one direction a parameter, such as temperature, which influences adsorptivity, is at one value, while the parameter is changed to another value when the flow is in the opposite direction. Such a process creates a separation between components with different adsorptivities. Chen[5] has correctly pointed out that PSA processes constitute a subset of parametric pumping, in which pressure is the parameter used to influence adsorptivity.

RPSA exists in single-bed[6,7] and multiple-bed[8] flowsheets (see Fig. 12.5-3). Beds are relatively short (about 0.3–1.3 m), and particle sizes are very small (about 177–420 μm). Overall cycle times, which include adsorption, dead time, countercurrent purge, and sometimes a second dead time, range from a few seconds to about 30 s.

The most unusual feature of RPSA is the continuously changing axial-pressure profile. Whereas all other processes operate with virtually a constant pressure axially through the bed at any given time, the first cycle of RPSA and the flow resistance caused by the use of very small adsorbent particles create substantial pressure drops in the bed, as shown in Fig. 12.5-4. These pressure profiles are highly critical to the performance of the process. For example, the unusual profile during the purge part of the cycle, which shows a maximum pressure somewhere inside the bed, simultaneously permits purging of part of the bed while less-adsorbed products are continually produced at the opposite end of the bed.

RPSA has been commercialized for the production of oxygen and for the recovery and recycle of ethylene and a small of chlorocarbons (the adsorbed stream) to an ethylene-chlorination process while purging nitrogen (the less-adsorbed component) from the process.

As mentioned before, Polybed PSA and RPSA represent quite different approaches to improving the basic PSA process. Table 12.5-1 shows the advantages of these approaches compared to a basic PSA process.

A new PSA process has been developed by Bergbau Forschung GmbH for the production of nitrogen from air.[9] The process uses a unique carbon-based molecular sieve as the adsorbent. The adsorbent microgeometry is such that, even though the adsorption isotherms of oxygen and nitrogen are almost identical, oxygen, because of its approximately 0.02 nm smaller molecular diameter, diffuses many times more rapidly into the pores. This creates an oxygen-depleted gas phase until both gases equilibrate, which requires over an hour. By operating the process with a time cycle considerably less than the equilibration time, the adsorbent becomes oxygen selective and produces a high-nitrogen product. Nitrogen purities of up to 99.9% can be produced. The process, depicted in Fig. 12.5-5, uses a vacuum desorption step to facilitate the production of high-purity nitrogen.

Nitrogen production using carbon molecular sieves is the only known commercial process using differ-

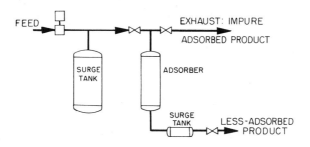

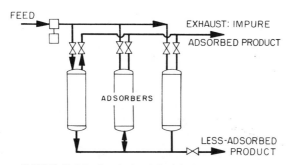

FIGURE 12.5-3   One-bed and three-bed RPSA processes.

ences in intraparticle diffusivity, rather than inherent adsorbent selectivity or selective molecular exclusion, as the basis for the separation.

A second general method for using PSA to obtain nitrogen has been revealed in somewhat different forms by Toray Industries, Inc.[10] and Air Products Corp.[11] In these processes a relatively pure adsorbate (nitrogen) stream is produced, along with a relatively impure oxygen less-adsorbed stream. A diagram of the Toray process is shown in Fig. 12.5-6. Dried air is passed through the adsorber at super-atmospheric pressure, and nitrogen is preferentially adsorbed. Part of the nitrogen product from previous cycles is then passed into the bed to desorb small amounts of oxygen, after which the bed is reduced to atmospheric pressure to desorb nitrogen. Finally, more nitrogen is recovered by vacuum desorption. The adsorbent used in this process is apparently the same—zeolite molecular sieves—as that used in oxygen PSA processes.

This process has been demonstrated in pilot scale. Its commercial status is unknown. The Air Products process has been commercialized.

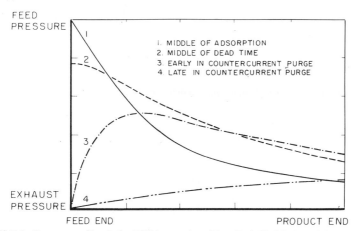

FIGURE 12.5-4   Pressure profiles during RPSA operation. (From Ref. 12.5-7, Copyright American Chemical Society, reprinted with permission.)

TABLE 12.5-1 Comparison of the Performance of Polybed PSA and RPSA with Basic PSA

	Polybed PSA	RPSA
Compression cost	Less	Same or greater
Productivity	About the same	Much greater
Degree of separation	Greater	Same or less
Process complexity	Greater	Less
Adaptability to large flows	Greater	Less
Cycle time	About the same (several minutes)	Much less (several seconds)
Nonadsorbed product pressure	Same	Lower

DISPLACEMENT-PURGE AND INERT-PURGE CYCLES

The most common use for these cycles is the separation of normal and isoparaffins in a variety of petroleum fractions. Distillation is not practical for these separations because the boiling ranges of the two products overlap. The feedstock most often contains several carbon numbers and can range from about C_5 to C_{18}. The adsorbent used is 5A molecular sieve, whose 0.5 nm pores admit normal paraffins and exclude isoparaffins.

Six different processes are listed in a recent review.[12] When higher-molecular-weight fractions are being processed, a displacement-purge cycle is specified because the adsorbates are held so tightly by the molecular sieve that a coadsorbing displacement agent must be used to effect desorption. The presence of the displacement agent, however, means that two distillation columns must be used to recover the displacement agent from the separated streams. When lower-molecular-weight fractions are processed, inert-purge cycles are used in which coolers and vapor–liquid separators are used to separate products and purge gas.

The normal isoparaffin separation process used most often is the Union Carbide IsoSiv process. Units with feed capacities of up to 3600 metric tons/day have been built. In the mid-1970's the Hysomer paraffin-isomerization process, developed by Shell Research B.V., was combined with the IsoSiv process in such a way that normal paraffins are nearly completely isomerized.[13,14] The combination is called the Total Isomerization Process (TIP) and is shown schematically in Fig. 12.5-7. Product from a TIP unit has a Research Octane Number of 88–92, compared to 79–82 for the product from a Hysomer unit alone.

The concept of achieving higher-than-equilibrium conversions by recovering one component of an equilibrium mixture and recycling the other to extinction is certainly not new, but it is probably underused.

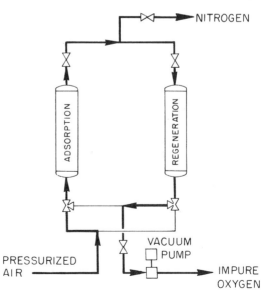

FIGURE 12.5-5 Bergbau Forschung carbon molecular-sieve process for nitrogen production. (From Ref. 12.5-9, copyright McGraw-Hill Inc., reprinted with permission.)

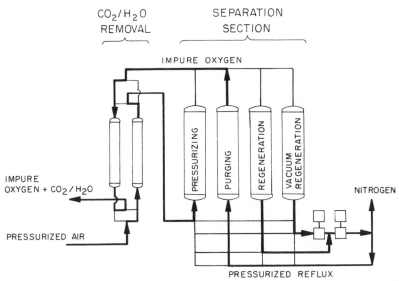

FIGURE 12.5-6 Toray Industries process for nitrogen production. (From Ref. 12.5-10, copyright Chemical Economy Research Institute, reprinted with permission.)

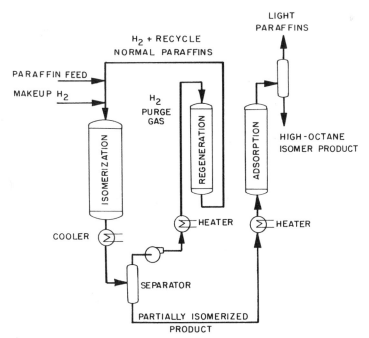

FIGURE 12.5-7 Total isomerization process. (From Ref. 12.5-14, copyright Gulf Publishing Co., reprinted with permission.)

Adsorption, which can sometimes separate materials more easily than distillation, represents an extra degree of freedom in conceiving of such processes.

Recently, a quite different use for an inert-purge cycle has emerged: the removal of large amounts (up to and exceeding 20 wt. %) of water from organic streams.[15] Thus, inert-purge adsorption can now compete directly with azeotropic and extractive distillation and other means for a number of azeotrope-breaking separations, the most common of which is the production of dry ethanol. The process, as depicted in Fig. 12.5-8, operates in a highly nonisothermal manner, with the heat of adsorption of the water stored in the bed as sensible heat. The maximum temperature rise is nearly 200°C. The water is then desorbed by a countercurrent purge of hot gas, and the bed temperatures subside to their original, lower values.

CHROMATOGRAPHY

Gas chromatography has been a common laboratory analytical technique for over a quarter of a century, and, because of the sharp separations that it can often achieve, it has been an appealing model for a commercial process. Only relatively recently, however, has the scale-up effort been successful. Chromatographic theory and practice are covered in another chapter; here, we merely list some of the commercial processes.

The Société Nationale Elf Aquitaine and Société de Recherches Techniques et Industrielles (SRTI) have announced a process for separating mixtures difficult to separate by distillation.[16] Hydrogen is used as the carrier gas. The Elf–SRTI process is presently commercially separating 100 metric tons/year of perfume ingredients. Elf Aquitaine has also announced a demonstration unit for separating 100,000 metric tons/year of C_4–C_{10} normal and isoparaffins.[17] Several beds in parallel are used to effect a constant product flow.

The successful scale-up of a gas-chromatographic process infers that two serious problems—flow maldistribution, which reduces the achievable number of theoretical stages, and capacity-limiting thermal gradients—have been solved.

12.5-2 Gas-Purification Processes

The use of adsorption for gas purification is much more widespread than for bulk separations (see Table 12.1-1 and Refs. 12, 18, and 19). At the same time, the technology is more conventional, for the most part. The most common process cycle is temperature-swing adsorption, which is often combined with inert-purge stripping, using fixed-bed adsorbers. Generally, two or more beds in parallel are employed to allow for continuous feeding while the other bed is being regenerated. These processes, if a thorough regeneration is carried out, can produce extremely high degrees of adsorbate removal. For example, many dehydration processes can produce product gases containing on the order of 1 ppm water.

In situations in which the adsorbate concentration is quite low or the feed flow very low, it may be economically practical simply to discard the adsorbent after it is loaded or to have it regenerated off-site and returned. In fact, a number of companies offer such regeneration services in the United States and other countries. The process savings include eliminating the investment and operating costs associated with the regeneration step. On the other hand, adsorbent on-stream times must be fairly long—several weeks or more—for off-site reprocessing to be economical overall.[20]

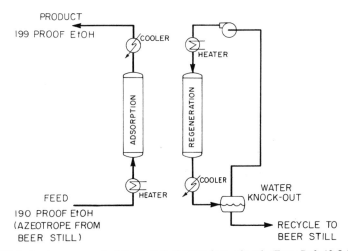

FIGURE 12.5-8 Inert-purge process for removal of water from ethanol. (From Ref. 12.5-15, copyright American Institute of Chemical Engineers, reprinted with permission.)

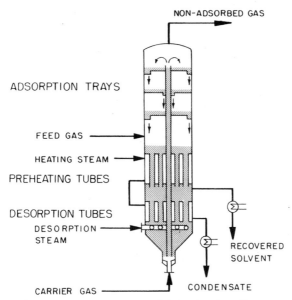

NON-ADSORBED GAS

ADSORPTION TRAYS

FEED GAS

HEATING STEAM

PREHEATING TUBES

DESORPTION TUBES

DESORPTION STEAM

RECOVERED SOLVENT

CARRIER GAS

CONDENSATE

FIGURE 12.5-9 PURASIV HR fluid-bed/moving-bed process.

A fluidized or moving bed, with the adsorbent circulating between adsorbing and regeneration zones, would seem to offer certain process advantages over multiple fixed beds in parallel, and indeed as early as the 1930s moving-bed processes were proposed. About 1950, Union Oil Company commercialized the Hypersorption moving-bed process for separating various light hydrocarbons.[21,22] The processes were plagued by attrition and loss of the activated-carbon adsorbent and were shut down.

More recently, Kureha Chemical Company, Ltd. of Japan has developed a hard, microspherical activated carbon called bead activated carbon (BAC). From this development has come the commercialization of a fluidized-bed/moving-bed process, called GASTAK, by Taiyo Kaken Company, Ltd. (now a part of Kureha).[23] The process is licensed in the United States and Canada through Union Carbide Corporation under the name PURASIV HR.[24] The process is finding applications primarily in the removal of small amounts of solvent vapors (typically 100–10,000 ppm) from air and vent streams. A schematic diagram is shown in Fig. 12.5-9. Feed gas passes upward through fluidized BAC on trays similar to those in a distillation column. The BAC passes downward through downcomers and then drops into a heated, moving-bed regeneration zone where the adsorbate is desorbed and recovered. Stripped BAC is then air-lifted to the top tray in the adsorption section. Removals of adsorbate from the feed gas typically range from 90 to 99%+. Compared to typical fixed-bed processes, the PURASIV HR process is claimed to reduce energy costs by 70–75%, improve the quality of the regenerated adsorbent, and reduce investment costs through greater mechanical simplicity and compact size.

12.5-3 Liquid Bulk-Separation Processes

Bulk liquid-phase adsorptions have been widely commercialized for only about 20 years. A major problem with bulk liquid-phase adsorption is that a certain volume of adsorbent can usually only separate a similar or smaller volume of liquid before needing to be regenerated. Furthermore, the liquid left in the interstices and pores of the bed after the adsorption cycle, even if the liquid were drained from the bed, would still contain certain amounts of the less-adsorbing components, making it quite difficult to obtain a high-purity adsorbate. Both of these facts create severe design problems for bulk liquid-phase adsorption, and their resolution can be obtained only in equipment more complex than that for typical gas-phase adsorption.

These problems can be dealt with in several ways. First, the adsorbent could be moved countercurrently to the feed and then transported to a second zone in which desorption would occur. The adsorbent would thus be moving in a pattern similar to that of solvents in absorption and solvent-extraction processes. Although moving-bed processes of this sort, primarily for ion-exchange separations, have been proposed and built, they have never reached a high degree of popularity. The chief reasons are the mechanical complexity and the cost associated with transporting and contacting solids countercurrently to liquids, and the problem of adsorbent attrition.

A second way of eliminating the above mentioned problems is by the use of chromatography, which is discussed in another chapter. A third way is by the use of parametric pumping. The most successful

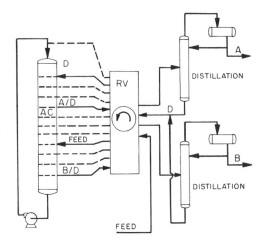

AC - ADSORBENT CHAMBER
RV - ROTARY VALVE
D - DISPLACEMENT AGENT
A - ADSORBATE
B - LESS-ADSORBED COMPONENT

FIGURE 12.5-10 Sorbex simulated moving-bed process.

way, if the extent of commercialization is a criterion, is an elegant variation of the moving-bed concept, and a description of it will be given below. The status of the other two ways will be discussed briefly.

SIMULATED MOVING-BED PROCESSES

The problems associated with moving-bed processes have been solved by the development of the simulated moving bed.[25] This process can also be thought of as a variation of a displacement-purge cycle. In this process, the bed is maintained stationary, and the feed and the displacement-liquid inlets and the two product outlets are moved as a function of time. In addition, the displacement liquid is continuously circulated by a pump from the bottom to the top of the bed. The most widely used version of this scheme is the Sorbex process, developed by Universal Oil Products, Inc. (UOP). A schematic diagram is given in Fig. 12.5-10. The lines shown as open are 2, 5, 9, and 12. After a short period of time each of the first three streams is moved to the next-higher-number point by the rotary valve. The less-adsorbed product line is switched from 12 to 1. During subsequent time periods the four lines are moved in the same manner. A more detailed description is given in Ref. 26.

Other versions of the simulated moving-bed process have been commercialized by Toray Industries, Inc.[27,28] and Mitsubishi Chemical Industries, Ltd.[29] These processes vary from the Sorbex technology in details rather than in their basic concept.

Although simulated moving-bed processes have now reached wide commercial acceptance for a number of separations, the need for a displacement liquid, which must subsequently be distilled from both product streams, constitutes a process complexity and an economic hindrance. For this reason, applications of these processes have been confined to those separations that are difficult or impossible to make by distillation.

The separation of fructose and glucose via simulated moving-bed adsorption is an example, like IsoSiv in the Total Isomerization Process for isoparaffin production, of the close coupling of reaction and separation systems. Glucose, which can be obtained from hydrolysis of starch, is isomerized to a near-equilibrium mixture of fructose and glucose (high-fructose corn syrup, HFCS). The resulting mixture can then be further enriched in fructose while also producing a glucose stream for recycle to isomerization. HFCS, because it rivals sucrose (cane sugar) in sweetness, is finding rapidly growing uses in beverages, candies, baking products, and so on.

CHROMATOGRAPHY

Just as laboratory gas chromatography has been an appealing model for large-scale gas adsorptions, the same has been true in the liquid-separation area. Liquid chromatography has apparently been used to separate glucose from fructose and other sugar isomers, for recovery of nucleic acids, and other uses. A recent patent to Sanmatsu Kogyo Co., Ltd. describes an improved chromatographic process that minimizes many of the problems encountered in other processes.[30] A diagram is given in Fig. 12.5-11, showing a binary separation. Pure-component fractions are recovered while fractions that contain both components

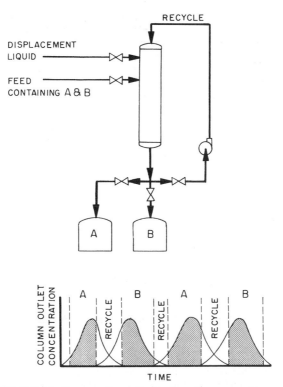

FIGURE 12.5-11   Sanmatsu Kogyo liquid-phase chromatographic process.

are recycled to the feed end of the column. A displacement liquid is also fed to the column. The process is claimed to be fundamentally simpler to build and operate than simulated moving-bed processes; its commercial status is not known, however.

PARAMETRIC PUMPING

Another approach to bulk-liquid separations is parametric pumping. A schematic diagram is shown in Fig. 12.5-12. Liquid is passed back-and-forth through the bed. As explained earlier, a parameter, such as temperature, is varied depending on the direction of the flow. The result is that the more strongly adsorbed components concentrate at that end of the bed toward which the liquid flows when the adsorptivity is low (i.e., temperature is high), and the less strongly adsorbed materials concentrate at the opposite end. High degrees of separation have been demonstrated in a number of cases. A strong positive aspect of thermally driven parametric pumping is that no displacement liquid is needed, eliminating the need for downstream distillation.

Despite a large amount of both theoretical and experimental studies on both simple and more complex versions of thermally driven parametric pumping, no bulk-liquid process seems to have been commercialized. The likely reasons are the mechanical complexity associated with reversing the liquid flow, the inherently long cycle time (hours) for a thermally driven process, and poor thermal efficiency. This poor efficiency derives from three sources. First, the liquid has to be heated and cooled, that is, cycled, several times before it leaves the system. Second, during each heating/cooling cycle the heat of adsorption must be removed by the coolant, and the heat of desorption must be added by the heating medium. And third, the adsorbent and vessel must also be heated and cooled during each cycle. As a result, thermally driven parametric pumping may only be a realistic solution in those situations in which a very low-cost heating medium in the range of 100°C or higher and a low-cost cooling medium in the range of 40°C or lower are available.

Of course, parameters other than temperature can be used in parametric pumping, and some results have been published in which pH variation has been used. No commercial process seems to have resulted.

12.5-4   Liquid-Purification Processes

Liquid purifications by adsorption, like gas purifications, are widely used.[31-33] But also like gas purifications, the technology is rather conventional. Fixed-bed processes and processes using loose, powdered

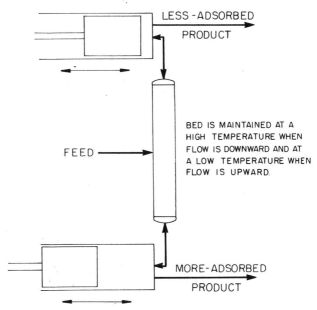

FIGURE 12.5-12   Schematic of parametric pump.

adsorbent (usually activated carbon) are used. Fixed-bed processes usually involve two or more beds in parallel to accommodate both a continuous feed and the need for periodic regenerations. In the powdered-adsorbent processes, the adsorbent and feed are added to an agitated vessel, and the adsorbent is subsequently removed by settling or filtration. In a number of these cases powdered adsorbent loaded with adsorbate is simply discarded instead of being regenerated following use. Obviously, discarding is only economical if the usage is small per unit of feed or if the feed rate itself is very small.

When adsorbents are regenerated, as much liquid as possible is drained or filtered from the adsorbent, and then a thermal-swing cycle is nearly always used. Often a purge gas is used to help remove adsorbates.

12.6   SELECTING A PROCESS

The previous sections present many process types and variations, and a question naturally arises as to how one can choose the most economical process embodiment for a particular separation. The purpose of this section is to present a method by which a reasonable answer can be given to this question. Economics of the processes are chiefly a function of capital (investment-related) costs and energy costs, and to a lesser extent labor and other costs. Since capital and energy costs can vary from site to site, hard-and-fast criteria cannot be given, but the following procedure will seldom mislead.

In Table 12.6-1, nine statements are given that can be used to characterize the separation desired. The numbers of the statements that are true are then used in the matrix given in Table 12.6-2. Encountering a "No" for any true statement under a given process eliminates that process from consideration; a process having all "Yeses" would deserve strong consideration. Entries other than "Yes" or "No" provide a way of ranking processes if more than one process seems to be a possibility. Below are several examples to illustrate the use of the tables.

EXAMPLE

An ambient-temperature vent gas contains about 1 wt.% of a toxic organic, at least 95% of which must be recovered and condensed. A selective adsorbent for the organic has been found. The isotherm changes rapidly as temperature increases from ambient to 150°C. The normal boiling point of the organic is 105°C. In Table 12.6-1, we note that statements 1, 4, and 7 are true for this separation. From Table 12.6-2, we can see that the processes of choice are temperature swing and inert purge. We might further reason that, since refrigeration would be required to condense the organic from the purge gas in a nearly isothermal inert-purge process, it would be reasonable to select the temperature-swing process as the number one choice.

TABLE 12.6-1 Process Descriptors

Nine statements are given below. To find the adsorption process most likely to be the best for a particular separation, first determine which of these statements are true for that separation. Then refer to Table 12.6-2 and read across. If a ''No'' is encountered under a certain process, then that process can effectively be eliminated as an economic possibility. A certain process that has all ''Yeses'' should be investigated.

1. Feed is a vaporized liquid or a gas.
2. Feed is a liquid that can be fully vaporized at less than about 200°C.
3. Feed is a liquid that cannot be fully vaporized at 200°C.
4. Adsorbate concentration in the feed is less than about 3 wt.%.
5. Adsorbate concentration in the feed is between about 3 and 10 wt.%.
6. Adsorbate concentration in the feed is greater than about 10 wt.%.
7. Adsorbate must be recovered in high purity (greater than 90–99% rejection of nonadsorbed material).
8. Adsorbate can only be desorbed by thermal regeneration.
9. Practical purge or displacement agents cannot be easily separated from the adsorbate.

EXAMPLE

A nearly equimolar mixture of ethylene and ethane is to be separated. The stream is available at about 3 atm, and high-purity (greater than 99%) ethylene is desired at greater than 95% recovery. An adsorbent with reasonable selectivity has been found which will desorb ethylene either at pressures well below atmospheric, or in the presence of a C_3 hydrocarbon or heavier, or in the presence of various inert gases. Ethylene in turn cannot displace hydrocarbons with more than four carbons, however. In Table 12.6-1, the true statements are 1, 6, 7, and 9. Statement 9 is considered to be true because cryogenic distillation would be required to separate the ethylene from the inert or purge material. In Table 12.6-2 we see that PSA is the most likely choice, using vacuum desorption.

EXAMPLE

An aqueous stream containing about 5 wt.% of oxygenated organics boiling over a wide range—greater than 200°C—must be purified by removal of the organics. No satisfactory adsorbent for all the organics has been found, and only a water-selective adsorbent has shown any promise. No satisfactory displacement agents for water can be found at the operating temperature of 35°C, and the only practical means for desorbing the water seems to be by temperature swing. From Table 12.6-1, we see that statements 3, 6, and 8 are true. Table 12.6-2 indicates that, for these three statements, there is no process that appears to be suitable. Thus, even though a temperature-swing, liquid-phase adsorption process might be developed, processing costs per unit of adsorbate would in all likelihood be prohibitive, and other means should be sought to accomplish the separation.

12.7 PROCESS-DESIGN CONSIDERATIONS

Compared to distillation, the process-design methodology for adsorption is neither as advanced theoretically nor as all-encompassing in its details. This is because of the inherently more complex nature of adsorption processes, caused by the presence of the adsorbent, and the fact that substantially less design information is available in the literature. Nevertheless, there is a growing body of design criteria, techniques, and practices which one can profitably use, and these are discussed in the following sections.

12.7-1 Adsorbent Considerations

TYPE

The type[1-3] of adsorbent is selected based on considerations of the selectivity, the equilibrium capacity, the dynamic adsorption rate, ease of regeneration, compatibility with the stream, coadsorption effects, type of process cycle, and economics. Many of these have been discussed at some length above.

PARTICLE SIZE

Solid adsorbents are usually offered in particle sizes[4,5] that range from a 100×200 mesh screen analysis to $\frac{1}{4}$ in. (0.64 cm) nominal size. The smaller-size particles are usually in an irregular granular form while the larger sizes are produced in regular shapes such as spheres and cylinders. Shape as a parameter will be discussed in more detail below.

There are two major considerations to be made when selecting the particle size of the adsorbent. They are the effects of size on the mass transfer characteristics and on the pressure drop. Pressure drop through

TABLE 12.6-2 Process-Selection Matrix

Statement Number	Gas- or Vapor-Phase Processes					Liquid-Phase Processes		
	Temperature Swing	Inert Purge	Displacement Purge	PSA	Chromatography	Temperature Swing[a]	Simulated Moving Bed	Chromatography
1	Yes	Yes	Yes	Yes	Yes	No	No	No
2	Not likely	Yes	Yes	Yes	Yes	Yes	Yes	Yes
3	No	No	No	No	No	Yes	Yes	Yes
4	Yes	Yes	Not likely	Not likely	Not likely	Yes	Not likely	Maybe
5	Yes	Yes	Yes	Yes	Yes	No	Yes	Yes
6	No	Yes	Yes	Yes	Yes	No	Yes	Yes
7	Yes	Yes	Yes	Maybe[b]	Yes	Yes[c]	Yes	Yes
8	Yes	No	No	No	No	No	No	No
9	Maybe[d]	Not likely	Not likely	Not applicable	Not likely	Maybe[d]	Not likely	Not likely

[a] Includes powdered, fixed-bed and moving-bed processes.
[b] Very high ratio of feed to desorption pressure (greater than 10:1) will be required. Vacuum desorption will probably be necessary.
[c] If adsorbate concentration in the feed is very low, it may be practical to discard the loaded adsorbent or reprocess off-site.
[d] If it is not necessary to recover the adsorbate, these processes are satisfactory.

packed beds of adsorbent particles is usually predicted by relationships such as those of Ergun[4] and of Leva.[5] Generally, the pressure drop per unit length of packed bed is inversely proportional to the particle size to a power not less then unity. Thus, pressure drop can be reduced by selecting the larger particle size.

The mass transfer rate for adsorption is also inversely proportional to the particle size to a power not less than unity. High mass transfer rates are desirable, since less adsorbent is required for the same separation. Therefore, the size of the packed bed can be reduced by selecting the smaller particle size. Since these two criteria are not compatible, trade-offs must be made in the design.

PARTICLE SHAPE

Adsorbents are commercially available in a variety of geometric shapes:[4-8] beads (spheres), pellets (cylinders), granular (mesh), and extended surface shapes. The shape affects both pressure drop and mass transfer resistance.

Granular materials are irregular in shape and may vary from platelet to spheroid to cubic. They derive their irregularity from the manufacturing processes where the desired-size particles result from crushing larger material. Typical size ranges for granules are 100×200 to 4×8 mesh screen analysis. One measure of the irregularity of particles is the shape factor ϕ_s, which is defined as the ratio of the surface area of a sphere with volume equivalent to the particle divided by the actual surface area of the particle. Published values of ϕ_s for granules range from 0.45 to 0.65.

Beads are also usually denoted by their screen analysis because the manufacturing techniques cannot make a single uniform size. Beaded adsorbents are available in sizes ranging from 16×40 to 4×8 screen size. Although not perfectly spherical, most commercial beads can be taken to have a shape factor of 1.

Pelleted adsorbents are produced by extrusion through dies and therefore have a very uniform diameter but a range of length-to-diameter ratios. Typical commercial-particle sizes range from $\frac{1}{32}$ to $\frac{1}{4}$ in. (0.08–0.64 cm) in diameter and possess a shape factor of about 0.63.

Recently, an extended-surface shape molecular sieve, TRISIV™, has been made available for natural gas drying and for air purification.[7] This trilobal or cloverleaf-shaped particle provides a unique combination of mass transfer and pressure drop performance in one particle.

In Leva's correlation for pressure drop in packed beds of particles, lower pressure drops are predicted for larger shape factors. For particles of the same volume per particle, the pressure drop per unit length is proportional to ϕ_s to the $n - 3$ power where n increases from 1 to 2 as the flow increases from the laminar to the turbulent regime.

The effect of the adsorbent shape on mass transfer is much more complex. There are several mass transfer resistances in series and/or parallel, each one of which may be controlling for a particular set of conditions. However, in general, mass transfer rates will be larger for particles with larger specific surface area.

COMPOUND-SIZE BEDS

There are sometimes design circumstances that warrant the use of compound-size beds (packed beds of two different particle sizes, usually arranged in series). This technique is often used to avoid some of the trade-offs that must be made of pressure drop versus mass transfer rate. For example, a bed is sized to treat a certain volume of fluid using $\frac{1}{16}$ in. (0.16 cm) particles; however, there is not adequate pressure drop to achieve the necessary flow rate in the vessel diameter selected. One solution is to replace some of the $\frac{1}{16}$ in. particles with $\frac{1}{8}$ in. particles. The larger-sized adsorbent would be placed at the adsorption flow inlet where it would have the least detrimental effect on the mass transfer rate, which is important at the effluent end when the adsorption period ends. Although some additional adsorbent may be required, the pressure drop may be low enough to be acceptable.

COMPOUND-TYPE BEDS

Other design conditions may dictate the use of compound-type beds[9] (packed beds containing two or more types of adsorbent). Compound-type beds are used when more than one adsorbate is to be adsorbed from a stream and the adsorbates are of different types, for example, water and hydrocarbons, water and acid gas, carbon dioxide and hydrogen sulfide. When the species to be removed are dissimilar, it is often not possible to select a single adsorbent type that has good selectivity for all compounds. For example, molecular sieves have exceptional selectivity for water but poor selectivity for light hydrocarbons, while activated carbon has poor selectivity for water but outstanding selectivity for hydrocarbons from inert gases. Thus, a packed bed could be compounded of molecular sieve and activated carbon, with each section designed to do only what that type does best. The resulting bed would usually be smaller than one designed completely of either adsorbent. For this kind of compounding, the order in which the beds are compounded is usually not important. However, there are combinations of adsorbates when sequence is important.

12.7-2 Adsorption Design Considerations

TEMPERATURE
The temperature[10] at which the adsorption stroke will be carried out is most commonly selected as that at which the stream exists. If there is some choice of temperature, the coldest value should usually be picked to increase the adsorptive loading as much as possible. There are some applications where other considerations arise. If the feed stream is (or contains) a condensable vapor, the temperature must be maintained sufficiently high to prevent condensation in the system.

PRESSURE
As with temperature, the most convenient pressure for adsorption is that of the feed. When a range of pressure is available, the highest pressure should be chosen so that the adsorptive loading is maximized. The comments above on condensation are applicable to pressure selection also. For pressure-swing cycles, the adsorption pressure must be arrived at from economic trade-offs, since larger swings in pressure yield better adsorbent utilization but require more compression power.

FLOW DIRECTION
In most cases, flow through a packed bed should be in a vertical direction. Because some settling or movement of the bed will always occur, horizontal flow would bypass over the top of the adsorbent unless some complex mechanical seal were employed. In vapor-phase adsorption, the flow direction is usually dictated by the process cycle step with the highest potential for lifting the adsorbent. Since allowable velocities for crushing exceed those for lifting, flow during the limiting step should normally be downward.

OPERATING LOADING
Consider an adsorbent bed with feed entering at one end and product leaving at the other end diminished in the adsorbable component. The graph in Fig. 12.7-1a shows the concentration Y of the adsorbate in the fluid leaving the bed as the ordinate and time Θ as the abscissa. At some time $\Theta = \Theta_b$, "breakthrough" is said to have occurred when the concentration of the adsorbate out of the bed rises to some arbitrarily defined "breakthrough concentration," which is usually a minimum detectable or maximum allowable level of the component to be removed. Θ_b is defined as the "breakthrough time." If the quantity of adsorbable

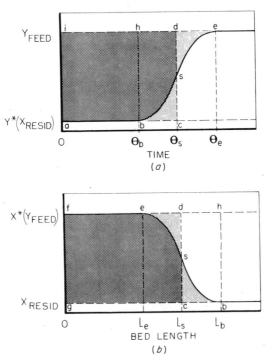

FIGURE 12.7-1 (*a*) Time trace of adsorbate composition in an adsorber effluent during adsorbent. (*b*) Adsorbate loading along the flow axis of an adsorber during adsorption.

species that has flowed into the bed from $\Theta = 0$ to $\Theta = \Theta_b$ is divided by the total bed weight, the result is the "breakthrough loading," or the "differential operating loading," or just the "operating loading." Data exist in the literature as operating loading and some design procedures use such data which usually are only correlated against adsorbate concentration, and possibly against temperature and pressure. However, varying mass transfer resistance causes these data to be strongly dependent on flow conditions, and it is advisable to treat breakthrough data rather carefully.

MASS TRANSFER ZONE

The graph in Fig. 12.7-1b shows the adsorbate loading X as the ordinate and distance L from the inlet end of the bed as the abscissa at some time after flow has been started.[7-9, 11-13] In the region of the bed from $L = 0$ to $L = L_e$ of Fig. 12.7-1a, the loading is the equilibrium loading corresponding to the concentration of the adsorbate in the feed. This portion of the bed is defined as the "equilibrium section." From $L = L_b$ to the end of the bed the adsorbate loading is essentially the same as it was at time zero, that is, at the "residual loading" level. This effluent end of the bed is unused bed. In the intermediate zone between $L = L_e$ and $L = L_b$, the adsorbate loading traces an S-shaped curve as it varies from the equilibrium to the residual loading. It is in this zone that the mass transfer and the dynamic adsorption are occurring, and it is for that reason called the "mass transfer zone" or MTZ. The length of the bed, $L_b - L_e$, is referred to as the "mass transfer zone length" or MTZL. Some adsorption data have been analyzed and reported as MTZ or MTZL, usually correlated to flow rate or flow velocity.[13] However, if an equilibrium zone is added to a mass transfer zone to obtain a design, more bed will be installed than is necessary because some adsorption is actually achieved in the MTZ.

WES AND WUB

There is an alternative way of subdividing an adsorption bed at breakthrough.[14] The length of the MTZ in Fig. 12.7-1b is a function of the mass transfer resistance; as the resistance increases, the length increases. When the mass transfer rate is extremely large, the mass transfer front *esb* reduces to the straight line *dsc*. It is called the "stoichiometric front" since its position at any time is determined strictly by a material balance. Mass transfer fronts can be analyzed with reference to this stoichiometric front. The area *febgf* under the loading curve represents used adsorbent capacity; the area *ehbe* between the equilibrium loading represented by *hb* and the curve *eb* represents unused capacity. Because area *fdcgf* up to the stoichiometric front also represents the used capacity, areas *febgf* and *fdcgf* are equal. This portion of the bed up to the "stoichiometric point" *s* is called the "length (or weight) of equivalent equilibrium section," LES (or WES). The remainder of the bed from the stoichiometric point to the "breakthrough point" is termed the "length (or weight) of unused bed," LUB (or WUB), since it is equivalent to a bed at the residual loading in the stoichiometric interpretation. Now a bed can be designed by adding a WES calculated from equilibrium adsorption loading to a WUB derived from mass transfer.

Since it is difficult to obtain point by point concentration data in adsorption beds, most dynamic adsorption data are obtained in the form of Fig. 12.7-1a. In Fig. 12.7-1a, the area *iebai* represents the uptake of the adsorbate, as does the stoichiometric area *idcai*, and can therefore be used to calculate the equilibrium loading. If a stable mass transfer zone has developed, the breakthrough time Θ_b and the "stoichiometric time" Θ_s can be used to calculate LUB by the equation

$$\text{LUB} = \left(1 - \frac{\Theta_b}{\Theta_s}\right) L_{\text{bed}}$$

In trying to model dynamic adsorption mathematically and predict the shape and size of the MTZ, assumptions must be made about the equilibrium model, molecular diffusion, axial-flow dispersion, and isothermality. The simplest model is isothermal equilibrium operation with infinite diffusion rate and negligible dispersion, that is, a stoichiometric front.

DIFFUSION

The major contributor to the MTZ, especially in vapor-phase adsorption, is diffusion resistance.[15-30] Adsorptive diffusion is the sum of several different mechanisms. There is bulk diffusion through the film around the adsorbent particle. Then the molecules must overcome the macropore diffusion resistance that is usually characterized as Maxwell diffusion. In those pores that have dimensions on the order of molecular dimensions, the transport is by Knudsen diffusion. In these micropores, the diffusion can also occur by surface diffusion. Typically, the entire process is characterized by a single overall diffusion coefficient.

FLOW DISPERSION

Axial dispersion or backmixing[31-36] can make a major contribution to the mass transfer rate in liquid-phase adsorption and cannot be ignored. This is especially true at low Reynolds numbers.

NONISOTHERMAL EFFECTS

For simplicity, many adsorption design approaches assume that adsorption occurs isothermally.[21,37-44] When the adsorbable component concentration is low and/or the heat of adsorption is low, this is a good approximation. There is a simple criterion to measure the nonisothermality of a system. Since most heat is always generated in the MTZ where adsorption is occurring, the rate at which the heat can be carried forward out of this zone compared to the speed at which the MTZ can travel is critical. This can be quantified by the "crossover ratio" R, defined by[40]

$$R = \frac{C_{pg}(X_i - X_{\text{res}})}{C_{ps}(Y_i - Y_o)}$$

where Y is the molar ratio of adsorbate to the carrier (i denotes inlet, o denotes outlet), and the gas and solid head capacities, C_{pg} and C_{ps}, include the effect of the adsorbate. When $R \gg 1$, the heat is easily removed from the MTZ and adsorption can be assumed to be isothermal. As R approaches a value of 1, more and more heat will be retained in the MTZ. An increase in the temperature of the "leading" or breakthrough end of the MTZ will lower the equilibrium loading from the isothermal value and cause the curve to become less favorable relative to the operating line, until ultimately the MTZ has no stable limit but continues to expand as it moves through the bed. When $R = 1$, the heat front is moving through the bed at the same velocity as that of the MTZ, and essentially all the heat of adsorption is found in the MTZ. For cases where the crossover ratio is less than 1, the heat front will lag the adsorption front and heat will be stored in the WES or equilibrium section. Here the temperature rise will cause the equilibrium loading to decrease. Thus, the crossover ratio is an indication of the nonisothermality and also of the extent of the deleterious effects of the temperature rise due to adsorption.

Unstable MTZ
Not all adsorption beds will develop stable MTZs.[41,45-49] One requirement for an MTZ to be stable (i.e., reach a limiting size) is that the equilibrium line must be "favorable;" in the case of a single adsorbate isothermally removed from a nonadsorbable component, the curve of loading as a function of composition must be concave downward in the region of loading below the stoichiometric point. In nonisothermal adsorption it is possible for the temperature effects to cause a favorable isotherm to become an unfavorable equilibrium line.

Short Beds
Another example of unstable MTZ occurs in "short beds," where the length is shorter than an MTZL.[50] As is illustrated in Fig. 12.7-2, the MTZ takes a finite bed length to grow to its ultimate length. However, the previous analysis would add the WES to a stable WUB to arrive at a larger bed than is necessary. Lukchis'[50] interpretation of Rosen's[24] analysis yields the following expression for the breakthrough time Θ_b' in short beds

$$\Theta_b' = \left(\frac{L_{\text{bed}}}{\text{MTZ}}\right)^2 (\Theta_s - \Theta_b)$$

where MTZ, Θ_s, and Θ_b are steady-state values.

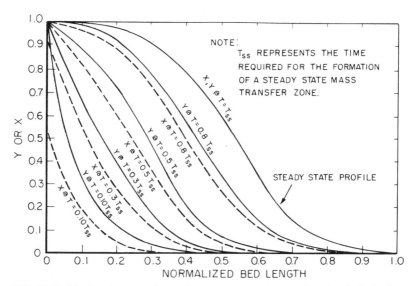

FIGURE 12.7-2 Development of a steady-state mass transfer zone in a short-bed adsorber.

COADSORPTION EFFECTS

When multicomponent mixtures are adsorbed on fixed beds, separate adsorption zones tend to form due to coadsorption.[2,11,29,51] Using the removal of water, hydrogen sulfide, and carbon dioxide from methane by a 5A molecular sieve as an example, the inlet portion of the bed adsorbs the most strongly adsorbed compound (water) but very little of the others, because their equilibrium loading is severely reduced by the strong coadsorption effect of water. In the next zone, the next most strongly adsorbed compound (hydrogen sulfide) is adsorbed easily, since there is little or no water coadsorbed to hinder its adsorption. However, very little carbon dioxide is adsorbed since it is the most weakly adsorbed of the three. A third zone will be required to effect removal of carbon dioxide where there is negligible interference by either of the more strongly adsorbed components. The total amount of adsorbent required for these three WESs to adsorb all the species calculates out to be significantly less using segregated zones than if it were designed large enough to remove all of each species in one zone in equilibrium with the feed.

Detailed design of multicomponent adsorption may also have to take into consideration the "rollup" effect due to coadsorption. The three-adsorbate system used above will serve to illustrate this phenomenon. Consider some time after feed has started through the adsorbent bed and the three distinct zones have formed. The MTZ in which carbon dioxide is adsorbing is moving through the bed at its stoichiometric velocity. However, the MTZ in which the hydrogen sulfide is adsorbing is not as simple, because the more strongly adsorbed H_2S is displacing (or desorbing) a large portion of the CO_2 that had previously adsorbed. Thus, any given molecule of carbon dioxide may be adsorbed and desorbed many times. The desorbed CO_2 causes the concentration of CO_2 to "rollup" to a level above that of the feed. From another viewpoint, if the CO_2 MTZ has already passed out of the bed, the effluent carbon dioxide level will be higher than that of the feed due to the desorption of some CO_2 by the hydrogen sulfide. In the same way, water will rollup the H_2S. These rollup concentrations can all be calculated from the multicomponent equilibrium loadings and the adsorption-front velocities.

LEAD/TRIM OPERATION

When the WUB or MTZ is large relative to the WES, the percent of adsorbent bed fully used is small. One design technique to improve this situation is to use lead/trim adsorption.[9,50,52] Figure 12.7-3 shows the flow scheme for this approach. Two (or more) adsorbent beds are on-stream in series at the same time treating the feed. The feed enters the "lead" bed first and then the "trim" bed. The trim bed is that which has had the least time on-stream since regeneration. The MTZ is allowed to pass all the way through the lead bed(s), without allowing breakthrough of the trim bed. In this way, the lead bed can be almost totally exhausted for adsorption before it is taken off-line to be regenerated. When a lead bed is taken out of service the trim bed is placed in lead position and a new freshly regenerated bed is placed in the trim position.

PARALLEL TRAINS

When very large flows are to be treated, the adsorber vessel size may be so large that it is not practical and ship them. It then becomes advantageous to design the adsorbers in multiple parallel trains.[9] The feed

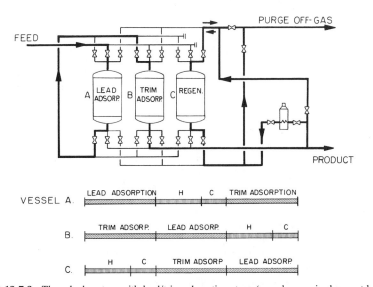

FIGURE 12.7-3 Three-bed system with lead/trim adsorption steps (vessel usage is shown at bottom).

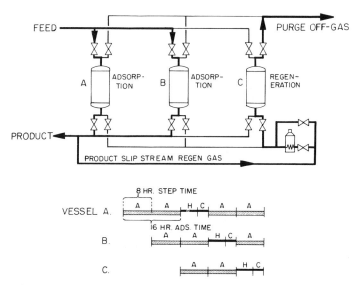

FIGURE 12.7-4 Three-bed system using two ripple beds during adsorption.

flow is split equally between adsorbers and the product streams are rejoined. Because of flow distribution and control considerations, it is usually desirable to supply each of the trains with an independent regeneration system. Such a configuration makes it possible to turn down the regeneration portion of an installation to save on utilities. In general, the adsorption flow can be turned down greatly without detriment to performance.

RIPPLE BEDS
Another design technique to deal with very large flow rates is possible when the regeneration flow rates are only a fraction of the feed flow.[53] All the beds except one are configured in parallel adsorption flow, with the odd bed being in the regeneration mode. This process arrangement is shown in Fig. 12.7-4. When an adsorber has been regenerated, it is returned to adsorption service and the bed that has been adsorbing the longest is removed from service to be regenerated. Thus, each bed "ripples" through the positions.

12.7-3 Temperature-Swing Regeneration: Heating

GENERAL
A temperature-swing regeneration cycle is one in which desorption into a regeneration fluid takes place at a temperature much higher than adsorption. It is most often applied to separations where a contaminant (adsorbate) is present at low concentration such as drying and purification. Such cycles have low residual loadings and high operating loadings. A purge and/or vacuum is nearly always required to remove the thermally desorbed components from the bed, and a cooling step must be added to return the bed to adsorption condition.

STRIPPING VERSUS HEATING LIMITED
In temperature-swing regeneration, there are two steps necessary to accomplish desorption: (1) heat must be added to the adsorbate and adsorbent to raise them to desorption conditions and to provide the endothermic heat of desorption; (2) the desorbed species must be transferred away from the adsorbent. When the former is the limiting step, it is termed heating- (or stoichiometric-) limited regeneration.[2,50,54-56] The latter is referred to as stripping- (or equilibrium-) limited regeneration.

The heating step must provide the heat to:

raise the adsorbate and adsorbent to the temperature at which desorption occurs (not necessarily the same as the ultimate regeneration temperature);

endothermally desorb the adsorbate;

raise the temperature of the adsorption vessel, including piping, valves, internals, and bed supports;

make up for heat losses to the surroundings;

vaporize any liquid not drained when the feed is a liquid; and

raise the adsorbate to the final regeneration temperature.

Like adsorption, the temperature front can be roughly broken down into a stoichiometric part and a transfer part. The stoichiometric portion is determined by an energy balance of the items above. There have been several studies of heat transfer to packed beds of particles that can be used to predict the heat transfer zone that will form.[54-56]

The stripping step becomes the limiting step when the purge gas does not have sufficient capacity to remove the adsorbate as fast as the same gas can provide heat for desorption. The capacity of the purge gas is a function of the equilibrium pressure exerted by the adsorbed component; that is, the maximum molar concentration that can be achieved is the equilibrium pressure divided by the regeneration pressure. High regeneration pressure, low regeneration temperature, and strongly held adsorbates all tend to result in stripping-limited regeneration.

Indirect Versus Direct Heat

Whether the cycle is stripping or heat limited has an influence on the type of bed heating that is selected—indirect or direct.[7-9,50,57] Indirect heating can be considered whenever the regeneration is heating limited. Electric heating elements, heating coils, and panel coils have all been used to heat the adsorbent indirectly. Elements and coils usually have extended fins to provide better heat transfer. Steam, heat transfer fluids, and hot furnace gases can all be used to supply the heat requirement when coils are installed. Coils have the additional advantage that they may be used to introduce indirect cooling. The major disadvantages of indirect heating are that it greatly complicates the mechanical features of the adsorption vessels and may detrimentally affect flow distribution.

When the process is purge-gas limited, direct heating should be the method of choice. Direct heating is accomplished external to the adsorber vessel by passing the purge gas through a heat exchanger or fired heater. This is much simpler, offers better temperature control, and results in a more uniform heating of the adsorbent.

Fluid

Because the desorbed contaminant will be diluted by the regeneration fluid, the fluid must be selected by considering how the contaminant will ultimately be removed.[2,58,59] If the contaminant is easily condensable, such as water or solvents, then a noncondensable fluid such as nitrogen, carbon dioxide, or fuel gas will allow recovery by condensing. If the adsorbate is condensable and immiscible with water, steam can be used for regeneration and both can be condensed and decanted. At other times, the fluid may be selected as one that can easily be distilled from the contaminant. If the adsorbate is to be discarded and can be incinerated, fuel gas may be used as the regenerant and the mixture used in a burner or furnace. When the fluid selected is different than the feed stream, care must be taken to prevent excessive cross-contamination of streams.

Temperature

There are several considerations in selecting the temperature to be used for temperature-swing regeneration. The greater the difference in temperature above that of adsorption, the easier it is to desorb the contaminant and the lower the ultimate residual loading that is achievable. However, higher temperatures can be attained only through the use of a higher-value energy. The thermal stability of the adsorbent must be considered. Even if the adsorbent is stable at a certain temperature, continuous thermal cycling may degrade its performance. The thermal stability of the adsorbate must also be considered, since the products of decomposition or polymerization may very well not be removable from the adsorbent. And lastly, whenever possible the temperature level should be selected from utilities already in place or planned (e.g., existing steam pressures).

Pressure

The pressure used in temperature-swing regeneration is usually that of the adsorption step. This assures that there is negligible pressure drop across closed adsorber valves so that cross-contamination is minimized. It also means there is no need for separate steps to pressurize and depressurize the adsorbers. Occasionally, it may be worthwhile to use a lower pressure for regeneration, either to achieve better desorption or because the fluid is only available at and/or deliverable to that pressure.

Flow Direction

Regeneration flow direction has an impact on product quality, regeneration requirements, and operational simplicity. Figure 12.7-5 illustrates the nature of the residual loading that results from different heating flow directions. Temperature-swing regeneration heating with flow countercurrent to adsorption flow provides the lowest residual loading at the adsorption effluent end. Therefore, any product quality will be easier to attain. Countercurrent flow assures that adsorbate never contacts unused bed. Besides leaving a higher residual at the adsorption effluent end, cocurrent heating requires more total flow since the adsorbate must be desorbed from the entire bed. The reason is that some of the adsorbate readsorbs on the unused portion of the bed and must be redesorbed. Countercurrent heating, on the other hand, uses some heat in the unused portion but needs to desorb no adsorbate.

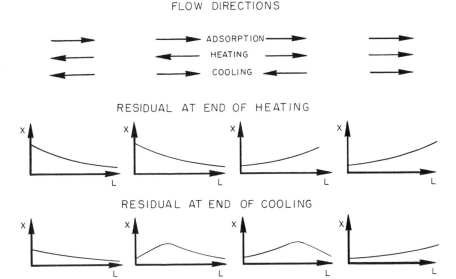

FIGURE 12.7-5 Adsorbate residual loading along the length of an adsorber after heating and cooling in various directions relative to adsorption.

Closed Loop

Closed-loop regeneration is used for three reasons.[51,57] It may be necessary if the amount of fluid available is limited. In other cases, it may be used to provide a higher concentration of the contaminant so that it is easier to reject it. And third, closed-loop regeneration is used when the adsorbate can be condensed away from the regenerant. For the first two cases, there is usually a constant bleed and makeup to the closed loop as shown in Fig. 12.7-6. The major difficulty with all three applications is that the contaminant is at a higher level than in an open loop and product purity will be harder to achieve, and the adsorbate(s) may be spread throughout the bed.

Peak Flare

Peak flaring is a variation of temperature-swing regeneration where a portion of the regeneration effluent is sent to a flare, incinerator, or furnace. The rest of the effluent then must be blended with the feed fluid to be retreated. The utility is that the contaminant can be rejected with a smaller amount of the valuable stream.

Thermal Pulse

One means of conserving thermal energy in heating-limited desorption is to use a thermal-pulse cycle.[52,54-56] When a process is heat limited, it needs only a very small time at temperature to achieve complete regeneration. If the entire bed is brought up to the desired temperature before the cooling step is begun, each portion of the bed will be at temperature for the time it takes to move a temperature front along the length of the bed. In addition, all the heat that is in the bed at the start of cooling may be wasted when it is swept from the bed. Ideally, it is feasible to flow hot purge to the bed until only the stoichiometric heat needed for desorption has been introduced as sensible heat of the purge gas above the adsorption temperature. Then the purge gas can be continued as a cool purge until the bed returns to adsorption temperature. Effectively, a "thermal pulse" has been created which moves through the bed, desorbing the adsorbate until it reaches the outlet, at which time it has exhausted its usable thermal energy. The bed has been returned to its lower temperature and no excess heat has been wasted. Practically, heat fronts do not proceed through packed beds as square waves but have a characteristic "S" shape. But it is still possible to design the regeneration to have only a small thermal pulse remaining at the end of the regeneration step. Figure 12.7-7 demonstrates how such a thermal pulse would proceed through a bed. Because the cooling step is carried out cocurrent to heating, which is usually countercurrent to adsorption, the purge fluid must be low enough in the adsorbate that product purity can be maintained.

Series Cool/Heat

There are other ways in which the heat that is purged from the bed during cooling can be conserved.[51,52] This heat is the sum of all the sensible heat terms that are delineated above. Sometimes the outlet gas is

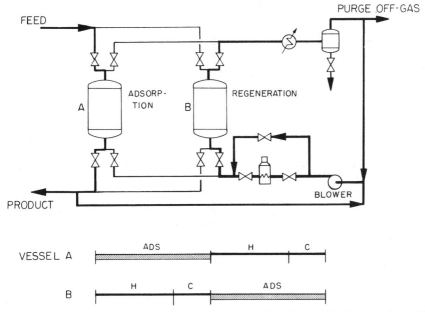

FIGURE 12.7-6   Adsorption using a closed-loop regeneration. Flow bypasses the heater during cooling.

passed to a heat sink where it is stored to be reused to preheat the hot purge. At other times it may be cross-exchanged against the purge gas to recover energy. However, there is also a process cycle that accomplishes the same effect. Three adsorber beds are used with one on adsorption, one on heating, and one on cooling. As Fig. 12.7-8 shows, the purge gas is used in series first to cool the bed just heated and then to heat the bed to be desorbed. Thus, all the heat swept from the bed and vessel can be recovered to reduce the heating requirement. Unlike thermal pulse, this cycle is applicable to heat- or stripping-limited regeneration. The considerations for selection of purge gas are the same as for a thermal pulse.

IN SITU VERSUS EX SITU
Operating conditions used for regeneration are often much more severe than those for adsorption. Therefore, using a single regenerator with materials of construction capable of handling the conditions is more cost effective than constructing all vessels of the expensive material. Hirschoff furnaces used for the rejuvenation of wastewater-treating activated carbons are an example. In other cases, it may be economically attractive to remove the adsorbent from the adsorber when it becomes spent and have an outside contractor regenerate it rather than install regeneration facilities. This is obviously only feasible if the adsorbent can treat feed for weeks or months rather than hours or days. Other opportunities for ex situ regeneration could arise when the solid is to be regenerated chemically.

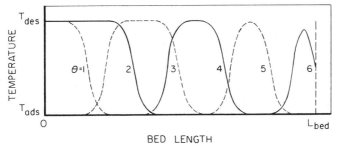

FIGURE 12.7-7   Progression of temperature fronts through an adsorber during a thermal-pulse regeneration.

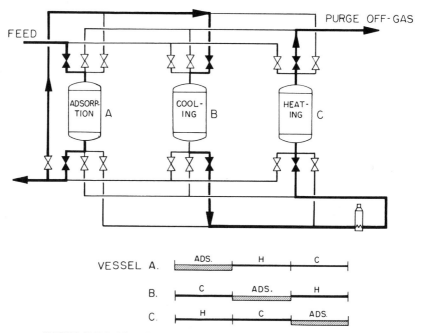

FIGURE 12.7-8 Three-bed system with series cool/heat during regeneration.

12.7-4 Temperature-Swing Regeneration: Cooling

FLUID
The most common fluid used for cooling is that which will be used for heating. However, although feed may be used as purge, it should never be used for cooling countercurrent to adsorption since the effluent end of the bed would become too contaminated to maintain product quality. Rather than use feed to cool cocurrent to adsorption, one would merely start adsorbing onto a hot bed. Therefore, product fluid is usually the best choice for cooling when feed (or product) is being used for regeneration heating. For situations where the feed is too reactive to be used at the elevated temperature of regeneration and another purge fluid is selected, it may be desirable to use product as the cooling fluid.

TEMPERATURE
The temperature of the cooling medium should be at least as cool as the feed. Cooling should proceed until the bed has cooled to within a reasonable range (20°C) of the adsorption conditions.

PRESSURE
Similarly, the most desirable cooling gas pressure is that of the feed. If the purge is at a lower pressure, there will have to be an extra step to pressurize the bed before adsorption can begin. If the pressure is too high, depressurization will be required and compression energy will be wasted.

FLOW DIRECTION
The direction selected for cooling depends on the direction for heating as well as that for adsorption. Figure 12.7-5 presents curves illustrating the shape of the residual loading that results from different combinations of heating and cooling flow directions. When heating is countercurrent to adsorption, cooling in the same direction provides the lowest residual since the cooling fluid continues to strip adsorbate as it passes through the hot portion of the bed. However, if the cooling purge is not free of the adsorbate, this flow scheme can be detrimental to product quality. For contaminated cooling fluid, cooling should be cocurrent to adsorption so that the bed actually cleans up the fluid before it passes through the effluent region of the bed.

CLOSED LOOP
One major reason that closed-loop cooling might be considered, as in the case of closed-loop heating, is that the amount of fluid may be limited.

12.7-5 Pressure-Swing Regeneration

GENERAL

In general, it is not possible to recover the more strongly adsorbed components in concentrated form from a temperature-swing cycle because of dilution by the purge gas used for heating.[61-63] Pressure-swing cycles provide an adsorption process where the adsorbate is recovered at high concentrations, making them attractive for bulk separations. Because of the higher concentration of the adsorbate in bulk separations, much larger beds would be needed than those used for drying and purification. This would be compounded in pressure-swing cycles where the operating loadings tend to be significantly lower. However, packed beds of adsorbent respond much faster to changes in pressure than to changes in temperature and can therefore be cycled in seconds or minutes rather than hours. These cycles operate at close to constant temperature and require no heating or cooling steps. They use compression energy to effect separation rather than the heat energy used in temperature-swing applications. Systems of weakly adsorbed species are especially suited to pressure-swing adsorption.

TEMPERATURE

Since these cycles operate at a single temperature, the available operating loading depends on the slope of the adsorption isotherm in the selected pressure range. If there is not sufficient operating delta loading at existing feed temperatures, it may be necessary to raise the temperature to get satisfactory delta loadings from a moderate pressure swing.

DEPRESSURIZATION

At the end of the adsorption step, the less strongly adsorbed components have been recovered as product. However, besides the more strongly adsorbed species on the adsorbent, there is a significant amount of a mixture of both strongly and weakly adsorbed species held up in the adsorbent and vessel void spaces. Depressurization and blowdown are used to enrich the concentrations of the more strongly adsorbed components. The first method of enrichment is "cocurrent depressurization" where the bed pressure is reduced by allowing flow out of the bed in the adsorption flow direction. This reduces the total quantity of holdup in the voids as well as increases the concentration of the adsorbate due to some desorption. The second type of enrichment is accomplished by purging with a quantity of gas rich in the strongly held species. For either method, the step should be completed with nearly all the weakly held material purged from the bed while most of the strongly held is retained.

BLOWDOWN

The first step in removing the adsorbate from the bed is the blowdown, usually a depressurization countercurrent to adsorption. The final pressure level is often set by the use of the blockdown waste gas, for example, the fuel or vent-header pressure. The extension of the blowdown by vacuum desorption can greatly increase the operating delta loading of the cycle, but the economics of the compression energy must be examined. Using vacuum equipment in pressure-swing cycles, while increasing the operating loading, presents another design difficulty. The duty is far from constant and is more complex than the evacuation of a simple vessel. As the vacuum draws down the pressure, more adsorbate is desorbed from the bed. In addition, the pressure drop across the bed decreases with time as the flow decreases, but increases as the pressure decreases.

PURGE

Final stripping of the adsorbate is carried out by purging countercurrent to adsorption. This should be done with a fluid such as product which is low in the adsorbate concentration to provide a low residual at the effluent end of the bed.

REPRESSURIZATION

Repressurization of the adsorber to feed pressure completes the steps of a pressure-swing cycle. Pressurization can be done with product and/or feed. When product is used, it should be applied countercurrent to adsorption to continue the purging of the effluent end. Pressurization with feed should be carried out cocurrent to adsorption to prevent contamination of the effluent end.

EQUALIZATION

The use of feed or product gas to pressurize the beds consumes gases already at pressure and lowers productivity in the case of product use. Therefore, an efficient cycle will maximize its use of equalization steps. At least portions of the gas released during depressurization, enrichment purge, and blowdown can often be used for some repressurization. The task is one of matching available pressure level and adsorbate concentration to the appropriate portion of the bed at the proper time. The penalty for equalization steps is usually a need for more vessels and the complexity of more piping, valves, and process steps.

FLOW DIRECTIONS

Most of the flow directions in a pressure-swing process are fixed relative to one another. The most common configuration is for adsorption to take place upflow, thus the name "blowdown" for countercurrent depressurization. Therefore, some precautions must be taken to assure that no bed lifting or crushing occurs (see below). Pressure-swing steps are more difficult to assess because many steps such as depressurization, blowdown, pressurization, and equalization begin at a high flow rate and decrease to low or no flow.

12.7-6 Inert-Purge and Displacement-Purge Regeneration

INERT PURGE

The word "inert" in inert-purge stripping refers to the relationship to the adsorbent; that is, the purge stream is not adsorbable at the conditions selected. An inert-purge cycle is one where the driving force for desorption is derived by the reduced partial pressure of adsorbate in the fluid. Inert-purge cycles are most commonly used for bulk separation. Since a pure inert-purge step is carried out with fluid at the same temperature as the feed fluid, the heat of desorption must be provided by the previously released heat of adsorption, which therefore must be kept stored within the bed. Thus, although inert purge appears to be adiabatic and isothermal from an overall balance, the temperatures are cycling within the bed. Desorption by reduction in partial pressure is analogous to the desorption by reduction in system pressure used in pressure-swing cycles. As with pressure-swing regeneration, the operating delta loading will be improved by selecting higher temperature and lower pressure. The purge fluid should be selected with a consideration for ease of separation from the adsorbate.

DISPLACEMENT PURGE

The term "displacement" in this type of regeneration refers to the fact that the purge fluid is adsorbable and has a displacing action on the adsorbent. This effect is in addition to its partial-pressure reduction effect that is identical to that of inert-purge stripping. The relative selectivity of the purge fluid to the adsorbate will determine whether displacement or partial-pressure stripping is more important; high selectivity favors displacement, low selectivity stripping. Unlike inert-purge stripping, displacement purging does not need to depend on stored heat for desorption. Especially for relative selectivities near unity, the heat needed to desorb the adsorbate will be provided by the heat of adsorption of the adsorbing displacement fluid. The displacement fluid must be desorbed before any more adsorbate can be removed from feed. This can be accomplished by another purge fluid from which it must ultimately be separated. Or if the displacement fluid is tolerable in (or easily separated from) the product stream, the bed with displacement fluid adsorbed can be used directly for adsorption where the fluid will be displaced by adsorbate. The major disadvantage is that the purge fluid must be separated from both the product stream and the recovered adsorbate.

12.7-7 Adsorption Vessels

HORIZONTAL VERSUS VERTICAL

The majority of fixed-bed adsorbers are cylindrical, vertical vessels. However, when large volumes of fluid are being treated by small inventories of adsorbent, the pressure drop may become excessive unless the bed height (or depth) is very small. If cylindrical vessels are needed for pressure or ease of fabrication, they have been installed horizontally in such cases with vertical flows. Because of the low pressure drop and the low ratio of bed height to the horizontal dimension, flow distribution is much more critical than in vertical vessels.

BED SUPPORT

There are two types of support systems used to support fixed beds of adsorbent. The first is a series of grids and screens. In this layered support system, each higher layer screen has successively smaller openings to prevent adsorbent from passing through, while each lower layer has greater strength. For example, a series of I beams can be used to support a layer of subway grating which in turn supports several layers of screening. Sometimes a layer of inert ceramic balls or gravel may be used on top of the screening, or in place of the screens. In other cases, special support grills such as Johnson™ screens may rest on the I beams or on clips at the vessel wall and thus directly support the adsorbent. The final support openings in any system should be selected to assure that not only are the original-size particles retained but also that some reasonable size of broken particles created by use are retained. Sufficient care must be taken in installing screens to prevent punctures and overlap.

The second type of support is a graded system of particles such as ceramic balls or gravel. In this case, a typical application may use several inches of 2 in. material, covered by several inches of 1 in. material and succeeding layers of $\frac{1}{2}$ and $\frac{1}{4}$ in. material which supports a $\frac{1}{8}$ in. adsorbent. In water treatment, the support may actually start with filter blocks and have an upper layer of sand. In some particular applications, especially where the adsorbent may have to be removed through a bottom outlet, there may be no support system; however, the fluid flow distributors must be much more complex.

TOP BALLAST AND RETENTION SCREEN

To prevent movement of the adsorbent at the top of the bed, a layer of support balls is often used as a ballast. This will prevent strong local fluid velocities from shifting the particles, which could lead to flow maldistribution or undesirable attrition. The ballast should be denser and of significantly larger particle size than the adsorbent to be effective. Unfortunately, these properties would allow the ballast to migrate easily downward through the adsorbent bed. Therefore, a retention screen must be installed on top of the adsorbent before the balls are added. The screen openings should be small enough to retain the adsorbent, and the screen must be puncture-free. Because the bed tends to be dynamic and shift or settle during operation, the retention screen must be floating rather than tacked to the walls.

FLOW DISTRIBUTORS

If there is poor flow distribution of fluid entering or leaving an adsorber, there will be less than maximum utilization of the adsorbent during adsorption and less than maximum desorption during regeneration. The simplest means of achieving good flow distribution is to employ ample plenum space above and below the fixed bed.[53] A much more cost-effective approach is to use simple baffle plates with symmetrically placed inlet and outlet nozzles. Baffles may be solid or perforated and should be large enough to break the momentum of the incoming fluid and redirect it so that it does not impinge directly on the adsorbent. When ceramic balls or gravel are used in lower heads, the baffles should be surrounded by a screen small enough, and strong enough, to retain the balls; the screen should have considerable open area, such as 50%. An alternative to screening is slotted metal of sufficient strength and open area. When a graded support system is used, the baffle must also be strong enough to support the balls and adsorbent above it.

Some applications may require more sophisticated flow distributors. Shallow horizontal beds often have such large cross-sectional flow area that a single inlet and outlet nozzle is insufficient. For these vessels, carefully designed nozzle headers are needed to balance the flow to each pair of nozzles. In liquid systems, a single inlet may enter the vessel and branch into several pipes that are often perforated along their length. Such ''spiders'' and ''Christmas trees'' may need holes, which are not necessarily uniformly spaced or sized, to provide equal flow per bed area.

INTERNAL VERSUS EXTERNAL INSULATION

Any process that will have temperature-swing regeneration or will be operated significantly above ambient temperature should have insulated adsorber vessels to prevent heat loss and for personnel protection. It is also usually cost effective to insulate all heated lines that are not carrying flow to coolers. The easiest and least expensive insulating approach is the normal external insulation. However, for a temperature-swing cycle all the vessel steel will need to be heated to regeneration temperature and cooled before adsorption. One means of reducing these utilities is to employ internal insulation. If the insulation is of a stand-alone type such as castable, only a thin layer of low-heat-capacity material will be thermally cycled. When a pourable insulation is used with an internal liner, the thin liner heat capacity must be included. However, there are several disadvantages to using internal insulation. If there are condensables in any of the streams, the insulation can become loaded with such liquid and become ineffectual. If internal liners are used, such liquids may actually flood the insulation since the liner (which cannot withstand pressure) must be open to the vessel. Therefore, breathing and drain holes at the bottom, sealed at the top, would be a preferred configuration. Any liner that develops a leak at its sealed end can contribute to significant amounts of flow bypassing the adsorbent. Castable or laid-up insulation can also allow flow bypassing if major cracks develop owing to shock or thermal cycling.

OTHER VESSEL FEATURES

It may become necessary to install intermediate bed supports for applications where a full bed would be subject to crushing.[53] Intermediate supports might also be used for compound-type beds to provide easier adsorbent changeout. For other applications, the available foundation area may be so limited that two adsorbers share the same vertical vessel, with the adsorbers separated by pressure bulkheads. In applications such as wastewater treating where expanded beds and/or backwashing are employed, vessel sizing must take into consideration the necessary extra freeboard.

12.7-8 Special Considerations for Liquids

FLOW DIRECTIONS

Flow direction considerations for liquid systems are somewhat different than those for vapor flow. In liquid or dense-phase flow the buoyancy force of the liquid must be considered as well as the pressure drop (see below). Thus, for upflow adsorption the flow velocity must not cause bed lifting. As the flow rate exceeds lifting velocity, the pressure drop increases only very slowly with increasing velocity. Because of this, sometimes liquid systems are designed with some bed expansion when it is desirable to limit pressure drop. Since too much expansion will cause the adsorbent to become well mixed, with a concomitant drop in removal efficiency, expansion is usually limited to about 10%. Higher velocities also tend to create too much particle turbulence, abrasion, attrition, and erosion. Upflow adsorption is a preferred direction if the liquid contains any suspended solids, since the bed will not act as a filter and become plugged. If all the adsorption steps are carried out with liquids, the remaining design considerations for flow are similar to those for vapor flows.

However, when regeneration is to be carried out in the vapor phase, a fill step and a drain step must be added to the process. The drain step will be downflow. When the adsorption step is downflow, the drained fluid can be collected as part of the treated product since it has exited from the WUB end of the bed. For upflow adsorption, drained fluid must be returned to feed storage. The fill step can be carried out in either direction although upflow is the preferred direction because it is easier to sweep the vapor out and prevent gas pockets that could cause flow maldistribution.

FILL

When an adsorbent bed is to be filled with a liquid, there must be sufficient time for any gas that may be trapped in the pores to outgas. Otherwise the vapor may later contaminate a product or, in the case of upflow adsorption, the effective bulk density may be lowered enough to cause excessive bed lifting or flow channeling.

DRAIN

Draining of the adsorption fluid is accomplished by gravity flow, sometimes assisted by a 10–20 psig pressure pad.[64] The liquid must be given at least about 30 min to drain thoroughly. Even then, there will be significant holdup to account for.

HOLDUP

The holdup of fluid after draining must be minimized because it can adversely affect product streams and regeneration requirements.[53] Even after careful draining, holdup can amount to 40 cm^3 fluid per 100 g of adsorbent. This fluid is retained in the micro- and macropores and bridges between particles. Any liquid that is not drained in a temperature-swing cycle will consume extra thermal energy when it is vaporized from the bed, and the fluid will end up recovered with the adsorbate.

HYDRODYNAMIC STABILITY

In liquid adsorption systems where liquids are also used for purge or displacement, care must be taken to prevent "fingering." Fingering is the displacing of one liquid by another at their interface due to density or viscosity differences. The phenomenon creates columns of the intruding fluid even in uniformly packed beds of adsorbent. It is obvious that a denser fluid above a less dense fluid will cause instability. However, it is also true that when a less viscous fluid is displacing a more viscous one, any bulge in the interface will grow because the resistance to flow is less, and the less viscous fluid will continue to intrude. Whenever the upper fluid is less dense or the more viscous fluid is displacing, the effect will be to correct any flow instabilities that occur.

12.7-9 Pressure Drop

GENERAL

The calculation of pressure drop through packed beds of adsorbent is necessary to assure that it is neither too low nor too high.[4,66,67] When the pressure loss is too low, the flow distribution in the bed will be poor. High pressure drop is an economic detractor. In many commercial applications, the highest pressure drop occurs during regeneration when the fluid is at its highest temperature and/or lowest pressure. However, a pressure drop analysis should be carried out for each of the steps of the cycle.

The most commonly used pressure drop equations for fixed beds of solid adsorbent are those of Brownell et al.,[66] of Ergun,[4] and of Leva.[67] Each of these correlations requires the use of the particle Reynolds number,

$$\text{Re} = \frac{D_p G}{\mu}$$

where D_p is the particle diameter and G is the mass flux. The Reynolds number is used in the specific correlation to calculate a friction factor f, such that the pressure drop per unit length, $\Delta P/L$, is given by

$$\frac{\Delta P}{L} \frac{f G^2}{2 g_c D_p \rho}$$

The Ergun correlation uses a particle diameter defined to be the equivalent diameter of a sphere having the same specific surface (area of particle/volume of particle) as the particle. The Ergun equation for the friction factor is then

$$f = \left[3.5 + 300\left(\frac{1-\epsilon}{Re}\right) \right]\left(\frac{1-\epsilon}{\epsilon^3}\right)$$

where ϵ is the external void fraction of the bed.

The particle diameter for the Leva approach is the equivalent diameter of a sphere with the same volume as the particle. Leva then correlates f by an equation of the form

$$f = \frac{f'(1-\epsilon)^{3-n}}{\phi_s^{3-n}\epsilon^3}$$

where f' is a function of only Re and ϕ_s is the shape factor. n increases from 1 to 2 as the flow increases from the laminar to the turbulent regime.

The Brownell method's particle diameter is taken as the equivalent diameter of a sphere having the same surface area as the particle. In this approach the particle Reynolds number is multiplied by a factor F_{Re} to arrive at a modified Reynolds number Re', given by

$$Re' = Re\ F_{Re}$$

where F_{Re} is a function of the void volume and the shape factor. The resulting modified Reynolds number is used with a standard Fanning friction factor plot to determine a modified friction factor f'. The particle friction factor is then determined from the equation

$$f = f'F_f$$

where F_f is also a function of the void volume and the shape factor.

PRESSURE DROP VERSUS BED DIAMETER
There is an optimum vessel configuration to contain a given quantity of an adsorbent. Figure 12.7-9 shows the equipment cost for a vertical vessel as a function of the diameter that will contain a fixed bed of adsorbent. As shown, there is a minimum cost diameter that is a function of the system pressure, and the optimum diameter increases with decreasing pressure. On the other hand, the pressure drop for a fixed quantity of adsorbent monotonically decreases with increasing bed diameter. Figure 12.7-10 shows the equipment cost of a compressor to overcome the pressure drop as a function of the diameter selected. Therefore, the sum of the equipment costs will indicate an optimum vessel diameter that will be larger than the usual vessel-only optimum. Since larger diameters also reduce the power necessary for compression, inclusion of the investment equivalent of the power will tend to further increase the optimum diameter.

BED LIFTING
The critical parameter for upflow in packed beds of adsorbent at high flow rates is the potential for particle movement or bed lifting.[68] This occurs as the conditions approach those for the onset of fluidization. At that point, the pressure drop is given by

$$\frac{\Delta P}{L} = (1-\epsilon)(\rho_s - \rho_f)\frac{g}{g_c}$$

where ρ_s is the density of the solid. Without a retention screen or other holddown device, the bed will start to expand as flow rate exceeds that for fluidization. However, even with a retention screen, the particles will tend to move and attrition will occur at these velocities.

When testing conditions for possible fluidization, the worst set of conditions should be checked even if they only occur for a small part of a step.

BED CRUSHING
In downflow steps, the possibility of crushing the packed-bed adsorbent must be checked.[69] There will be two forces acting to crush the adsorbent: pressure drop and the weight of the bed. The maximum force will be at the bottom. A conservative design approach is to maintain the sum of these two,

$$\Delta P + (1-\epsilon)\rho_s L$$

at less than the pressure that is known to cause adsorbent crushing.

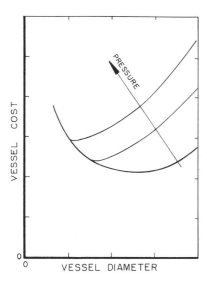

FIGURE 12.7-9 Equipment cost for a pressure vessel to contain a given quantity of adsorbent. Optimum vessel diameter vaires with pressure.

LIQUIDS

In liquid or dense-phase flow, the buoyancy force of the liquid must be considered as well as the pressure drop. Thus, for upflow adsorption the maximum upward flow velocity without bed lifting is that where the sum of the pressure drop and the buoyancy just equals the force due to gravity; that is, the pressure drop equation is the same as that used above to test for bed lifting.

12.7-10   Other Process Considerations

LOW-PRESSURE SYSTEMS

Adsorption in low-pressure gas systems such as vent-stream cleanup and solvent-recovery applications presents some special problems. Because of their low density and the resulting large velocities, such near-atmospheric-pressure streams have much higher pressure drops. This is compounded by the fact that the compression ratio (and thus the power) to overcome any ΔP is much greater than in high-pressure cases. For this reason, low-pressure adsorption is often carried out in very shallow beds, sometimes in horizontal vessels.

However, the use of thin beds and their low aspect ratios result in flow distribution problems. Therefore, some of the pressure drop that is saved by using shallow beds must be sacrificed by adding more sophisticated flow-distribution systems of manifolds, baffles, and screens.

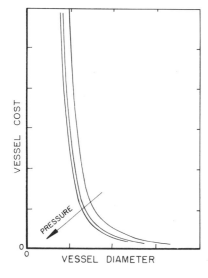

FIGURE 12.7-10 Equipment cost for compression in a system containing a given amount of adsorbent as a function of the vessel diameter.

BATCH OPERATION

Although it is desirable in most applications to treat a continuous stream of fluid, there are some circumstances when a single adsorber bed and batchwise operation are preferable.[57] For applications where a moderate-size vessel can contain enough adsorbent for treatment lasting many hours or even days, the savings associated with one vessel not requiring automatic switching may well justify the inconvenience of interrupting feed periodically for a short regeneration. If the plant operation is on a shift basis that is less than 24 h, it may be possible to treat for one or two shifts and regenerate unattended during the off-shift.

Liquid treating can often lend itself to single-bed adsorption because intermediate holding tankage can be used to provide the surge necessary to allow adsorbent regeneration. The need for draining and filling steps will usually make smooth, continuous treatment difficult anyway.

Nonregenerative systems are applications where a single batch operation should be the choice. The beds are designed to last weeks or even months and only an occasional downtime is needed to recharge the fresh adsorbent.

CONSTANT VERSUS INTERRUPTED FLOW

Fixed-bed adsorption, by its nature, is a cyclic or batch process; for example, a bed is used in the adsorption mode for a period of time followed by a time of heating and desorption followed by a cooldown step and return to the adsorption mode. Since fixed beds usually must interface with continuous processes and often use rotating equipment, multiple beds are used so that a continuous feed can be accepted, continuous product provided, and rotating equipment has a nearly constant load. In spite of this, many streams have cyclic characteristics. Product purity may start at some high level and drop slowly until a predetermined cutoff is reached. In pressure-swing systems, the product pressure may cycle, as well as the vessel pressure into which feed gas flows. The gas leaving a bed during heating and purge will swing between adsorption and desorption temperatures, and the adsorbate concentration of the gas will change by orders of magnitude. Similarly, blowdown and purge gases out of pressure-swing fixed beds have cyclic pressures and concentrations.

REGENERATION BUMPS

The fluid leaving a bed during heating and purge will have adsorbate concentrations that change by orders of magnitude over the cycle. Since adsorption units are often acting as concentrating unit operations rather than true separations, there are difficulties in having such variations in the load to the element that does the ultimate separation or recovery. For example, if water or solvent is to be condensed out of the regeneration loop, the condenser must be sized to handle the peak duty rather than the average.

REGENERATION VERSUS REPLACEMENT

In some applications, it may be advantageous to use the adsorbent on a throw-away basis.[70] Some of the reasons that might precipitate such a choice are the following:

low-cost adsorbent;

difficult or costly regeneration such as chemical extraction, high temperatures, or deep vacuums;

moderate adsorbent inventories that could treat for weeks or months;

chemisorption where reversibility is impractical;

high-value products requiring very high purities.

Magnesium perchlorate and barium oxide are examples of desiccants that have regeneration limitations but excellent drying properties for special applications. Hydrated iron oxide on wood chips (iron sponge) has been used to remove hydrogen sulfide but can be regenerated at best only a few times. Caustic or potassium hydroxide is sometimes used in nonregenerative processes for removal of sulfur compounds or carbon dioxide. When powdered activated carbon (PAC) is used in wastewater treatment to enhance biological treatment, it is not regenerated but remains with the sludge. Other nonregenerative adsorption applications may present more difficult disposal problems; for some, incineration may be possible.

DEACTIVATION

Essentially all solid adsorbents tend to lose adsorption capacity with time, some more rapidly than others.[2,9,71] In thermal cycles, temperature, steam, and other impurities can attack the adsorbent and gradually reduce capacity over long periods of use. In pressure-swing cycles, the rate of degradation is very low. In this case degradation is due to the accumulation of "dirt" and possibly to trace corrosives. This loss of capacity must be factored into the design and the process economics. The adsorption system must be sized so that performance is assured up to the selected changeout interval by adsorbent that has only a percentage of fresh capacity. There are three ways in which adsorbents may be deactivated:

1. A gradual buildup of a substance that is not fully removed during regeneration, thus hindering adsorption.

2. Chemical attack of the adsorbent where the new species is not an effective adsorbent.
3. Physical damage owing to thermal or hydrothermal cycling.

The buildup of foreign compounds can arise from deposition of impurities in the feed or regeneration fluids, or from substances in the stream that react to form nondesorbables. This type of deactivation is due to the unwanted substance either adsorbing on the active adsorption sites or blocking the pore structure, preventing access to the sites. Solids such as dust, fibers, rust, or pipe scale tend to block pores, especially by covering the exterior surface. Carryover of liquid droplets or mist from upstream equipment, such as lubricating oils from compressors and amines or glycol from absorbers, will cause deactivation by internal pore blockage as well as by site coverage when pore size permits access. Components of the feed stream with boiling points that are significantly higher than the key adsorbates will tend to build up on the adsorption sites because they are relatively more difficult to desorb.

Because sites that are active for adsorption also tend to be active catalytic sites, compounds in the streams that would not normally cause deactivation may react to form nondesorbables. Olefins, diolefins, and other unsaturated hydrocarbons are especially difficult since they easily polymerize to long-chain species in the presence of high-surface-area solids. Hydrocarbons in the presence of oxygen can form oxygenated species such as aldehydes and ketones, which can further react by aldol condensation to form heavier components. The presence of oxygen with sulfur compounds can create elemental sulfur.

The second type of deactivation is by chemical attack. Acids or acid gases can attack adsorbents and cause irreversible damage such as loss of crystallinity; this is especially true of adsorbents that are basic in nature such as molecular sieves.

The thermal cycling of adsorbents can cause damage to the internal pore and/or crystal structure due to the constant expansion and contraction. The additional factor of hydrothermal cycling in the presence of water adds the potential for explosive steam release which can physically deteriorate some adsorbents.

REACTIVATION

The process by which an adsorbent is returned to a near fresh state is referred to here as reactivation; more specifically it is the reversal of deactivation.[72-74] This is in contrast to regeneration which removes as much of the key adsorbates as were adsorbed on the previous cycle but often leaves a residual level. In liquid adsorption where solids have deposited, the bed may merely be backwashed to restore capacity. When it is desired to reactivate beds with heavy hydrocarbon buildups, it is usually necessary to use temperatures much higher than normal regeneration as well as controlled low levels of oxygen; such controlled "burn-offs" are performed periodically on Isosiv™ adsorbents. Hirschoff furnaces provide a similar ex situ reactivation of spent activated carbon; steam and carbon monoxide are added rather than oxygen to achieve the proper oxidation state of the adsorption sites. Other high boiling components may be chemically extractable. Some chemical attack may be reversible with suitable reagents. Other deactivation may not be reversible. In other cases, even though an adsorbent may be reactivateable, the cost may be unattractive compared to replacement.

SOLIDS AND MIST REMOVAL

The deposition of solids onto adsorbents not only causes deactivation but can substantially increase the pressure drop. Thus, if there is any possibility of solids or mist, it is good design practice to provide for their removal just upstream of the adsorbers by installing filters and/or mist eliminators.

STEP TIMES

The selection of step times is usually a compromise. Shorter adsorption times allow smaller adsorbent inventories, but the relationship is not linear owing to the mass transfer zone which becomes a larger proportion of the bed at short step times. Therefore, the purge and regeneration flow rates must be increased for shorter steps since the total flow needed per step is normally proportional to the bed weight. Pressure drops are lower, but cycling mechanical equipment such as adsorber valving will have a more severe duty in quick cycles. In temperature-swing regeneration, the adsorbent may deactivate more rapidly with short cycles due to thermal cycling. There may be circumstances where it is advantageous to have the step times correspond to shift lengths or fractions thereof. In addition, holdup becomes a bigger problem for short step times.

SWITCHING VALVES

In fixed-bed adsorption systems, the sequencing of the beds is accomplished with sets of adsorber valves placed as shown in Figs. 12.7-3, 12.7-4, 12.7-6, and 12.7-8. In general, one valve will be needed for each bed at each end for each step that is performed; for example, a three-bed system with an adsorption step plus a heating step plus a cooling step would require $3 \times 2 \times 3 = 18$ valves (see Fig. 12.7-8). If only two beds were used with heating and cooling carried out in the same direction and sequentially in the same step, only $2 \times 2 \times 2 = 8$ valves would be needed (see Fig. 12.7-6). In some cycles such as pressure-swing systems, it may be possible to use valves for more than one function, for example, repressurization with feed gas using the same manifold as adsorption feed. In some simple two-bed systems, it is possible

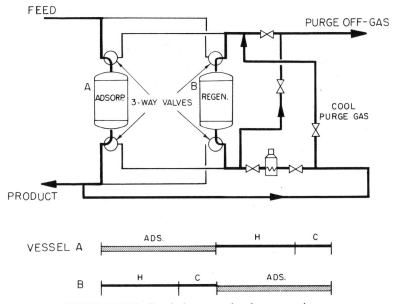

FIGURE 12.7-11 Two-bed system using three-way valves.

to replace several valves by three- or four-way valves such as shown in Figs. 12.7-11 and 12.7-12. For most systems it is mandatory that the opening and closing of the valves be controlled by automatic timers. Only for long step times (8 h or more) would it be advisable to try to operate with manual switching. Another reason for automatic control is that it may be necessary to provide for a short period when flow is going through both the fresh bed and the exhausted bed to assure a smooth transition. At other times a flow such as regeneration gas may need to be by-passed during bed switching to prevent deadheading a compressor. Other special sequencing could be used to prevent cross-contamination of streams. Cross-contamination can also result if the valves are not leak tight, especially in systems with large pressure differentials between streams.

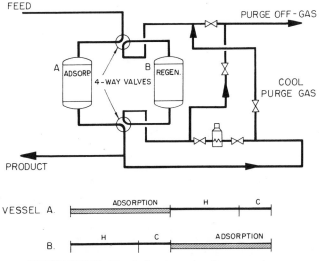

FIGURE 12.7-12 Two-bed system using four-way valves.

12.8 FUTURE DIRECTIONS FOR ADSORPTION TECHNOLOGY AND USES

Distillation and related vapor–liquid processes are by far the most widely used molecular separation processes in the petroleum, natural gas, petrochemical, and chemical industries, as mentioned earlier. It is highly unlikely that adsorption will ever rival distillation in frequency of use, but adsorption will continue to make inroads into its domain. Adsorption's serious competition for the separations for which it is now used would seem to come chiefly from membrane-based processes, and especially fixed-membrane processes. For example, Monsanto's Prism hollow-fiber-based process has been commercialized in a number of hydrogen-upgrading applications, and a growing number of other applications are being pursued.

In the sections that follow we will point out the likely areas of growth for adsorption in existing and new applications. We will also suggest some technological innovations that could stimulate this growth.

12.8-1 Gas Bulk Separations

Until recently, the use of adsorption for oxygen and nitrogen recovery has been limited primarily to applications requiring a few pounds to a few tons per day. Oxygen is being produced via PSA for waste-treatment plants, welding shops, and for in-home care of patients with various types of lung disease. Onboard generation of oxygen for aircraft will also become commercial. But as process improvements continue, it is clear that PSA oxygen will be able to compete economically with cryogenically produced oxygen in much larger plants—100 tons/day or more—in situations in which the higher argon level in the PSA oxygen is not a problem.

Nitrogen production via PSA is growing rapidly, as it is used more extensively for inerting of storage tanks and other vessels and areas, purging of process lines and vessels, creating modified atmospheres to prolong storage life in food-storage areas, and so on. And if grain-storage fumigants such as ethylene dibromide come under increasing environmental pressure, nitrogen inerting may prove to be an ideal replacement for use in enclosed areas.

Hydrogen recovery via PSA will also grow in popularity in comparison to cryogenic methods. It will be possible to process feed streams of 50,000–100,000 m^3/h at pressures up to 7 mPa. PSA will also be favored in cases in which the virtual absence of inert gases in the hydrogen product would minimize purge stream losses in other parts of the process. A related growth area is adjustment of the hydrogen to carbon monoxide ratio in synthesis-gas streams.

The major challenge for PSA is to produce two nearly pure products at once in a simple fashion; such a capability would allow PSA to compete head-on with distillation for an increasing number of separations. Increasing the feed-to-purge pressure ratio is the primary means for improving the purity of the adsorbed product. Desorption under vacuum is one means—although not without added costs—for increasing this ratio. Other PSA cycles not involving the need for high ratios would constitute a true breakthrough.

Inert-purge cycles, given the recently demonstrated success in drying of azeotropes,[1] would seem to be poised for use in several new separations. Prime candidates include those systems now separated by azeotropic and extractive distillation, many of which contain water as one constituent. The use of inert-purge cycles for isomer and other close-boiler separations should also grow for those systems whose components can be easily separated from the purge gas.

Displacement-purge cycles will not find many new uses, primarily because of the inherent complexity of these cycles.

Fixed-bed, temperature-swing processes rarely turn out to be economical for bulk separations. Moving-bed and fluidized-bed processes based on thermal regeneration may prove to be much more economical because of lower heat requirements per unit of feed. The key to the success of these processes lies in the development of highly attrition-resistant adsorbent particles such as Kureha's bead activated carbon.[2]

12.8-2 Gas Purifications

It is not likely that many process innovations will be needed for these separations. Fixed-bed, temperature-swing processes will continue to predominate, although moving-bed and fluidized-bed processes such as PURASIV HR should become more popular. Air-pollution concerns should bring about increased applications for recovery of organics from various process vent streams, storage-tank vents, and air streams from solvent-painting and other operations involving vaporization of organics. Major-use areas such as gas dehydration, removal of sulfur compounds and carbon dioxide, and various specialty separations will grow about in proportion to the growth of those industries in which these separations are found.

12.8-3 Liquid Bulk Separations

As long as displacement liquids are used in liquid bulk-separation processes—be they simulated moving beds, chromatographs, or other configurations—these processes will not be able to compete successfully with distillation for very many of the separations now made by distillation. The problem is one of process complexity, which reflects itself in high investment, and energy usage in recovery of the displacement liquid. Nevertheless, one can speculate that further uses will be found for displacement-purge processes. These could include separation of other saccharides, separation of optically active isomers, purification of agricultural chemicals and drugs, and separation of various azeotropes.

Parametric pumping remains an intriguing process concept because it eliminates the need for a displacement liquid. Thermally driven parametric pumping will likely not be economical, as discussed earlier, and of course pressure is not an effective parameter in liquid systems. Obviously, other parameters should be investigated. pH swing has been proposed, but such a process would require consumption of both acid and base to effect the swing. Recently, a study was done showing the effects on adsorbent capacity of imposing slight voltages on the adsorbent.[3] Whether such a phenomenon can be used in a practical parametric-pumping cycle cannot be said with any assurance yet, but the potentially very low power input certainly suggests that a research effort is warranted.

12.8-4 Liquid Purifications

As with gas purifications, very few process innovations are likely to occur with liquid purifications. Fixed-bed processes and processes in which powdered adsorbent is used will continue to predominate. If process technology is not likely to advance much, uses for the technology are. Perhaps the largest area for expanded use is in municipal and industrial waste treatment. Activated carbon adsorbs a wide spectrum of organics from water and can be useful in improving taste and lowering the concentrations of toxic or other objectionable materials. Also as chemical process effluents are reduced and more streams are recycled, additional adsorption processes will be required to remove traces of contaminants from these recycles.

Applications for liquid purifications by adsorption should grow faster than the industries they serve.

REFERENCES

Section 12.1

12.1-1   C. L. Mantell, *Industrial Carbon. Its Elemental, Adsorptive and Manufactured Forms*, 2nd ed., Van Nostrand, New York, 1946.

12.1-2   W. Kast, Adsorption from the Gas Phase—Fundamentals and Processes, *Ger. Chem. Eng.*, **4**, 265–277 (1981).

12.1-3   A. E. Marcinkowsky and G. E. Keller II, Ethylene and Its Derivatives: Their Chemical Engineering Genesis and Evolution at Union Carbide Corporation, in W. F. Furter (Ed.), *A Century of Chemical Engineering*, pp. 293–352, Plenum Press, New York, 1982.

12.1-4   V. R. Deitz, *Bibliography of Solid Adsorbents*, N.B.S. Circular 566, National Bureau of Standards, Washington, DC, 1956.

12.1-5   D. M. Young and A. D. Crowell, *Physical Adsorption of Gases*, Butterworths, London, 1962.

12.1-6   S. Ross and J. P. Olivier, *On Physical Adsorption*, Wiley-Interscience, New York, 1964.

12.1-7   T. Vermeulen, M. D. LeVan, G. Klein, and N. K. Hiester, Adsorption and Ion Exchange, in R. H. Perry and D. W. Green (Eds.), *Perry's Chemical Engineers' Handbook*, 6th ed., pp. 16-1–16-48, McGraw-Hill, New York, 1984.

12.1-8   D. W. Breck, *Zeolite Molecular Sieves*, Wiley-Interscience, New York, 1974.

12.1-9   V. Ponec, Z. Knor, and S. Cerny, *Adsorption on Solids*, Butterworths, London, 1974.

12.1-10   C. W. Chi and W. P. Cummings, Adsorptive Separation (Gases), in *Kirk-Othmer Encyclopedia of Chemical Technology*, 3rd ed., Vol. 1, pp. 544–563, Wiley-Interscience, New York, 1978.

12.1-11   R. W. Soffel, Carbon (Carbon and Artificial Graphite), in *Kirk-Othmer Encyclopedia of Chemical Technology*, 3rd ed., Vol. 4, pp. 556–631, Wiley-Interscience, New York, 1978.

12.1-12   T. Vermeulen, "Adsorptive Separation," in *Kirk-Othmer Encyclopedia of Chemical Technology*, 3rd. ed., Vol. 1, pp. 531–544, Wiley-Interscience, New York, 1978.

12.1-13   R. A. Hutchins, Activated-Carbon Systems for Separation of Liquids, in P. A. Schweitzer (Ed.), *Handbook of Separation Techniques for Chemical Engineers*, pp. 1-415–1-447, McGraw-Hill, New York, 1979.

12.1-14   J. L. Kovach, Gas-Phase Adsorption, in P. A. Schweitzer (Ed.), *Handbook of Separation Techniques for Chemical Engineers*, pp. 3-3–3-47, McGraw-Hill, New York, 1979.

12.1-15   E. M. Flanigen, Molecular Sieve Zeolite Technology—The First Twenty-Five Years, *Pure Appl. Chem.*, **52**(9), 2191–2211 (1980).

12.1-16   W. H. Flank (Ed.), *Adsorption and Ion Exchange with Synthetic Zeolites*, American Chemical Society Symposium Series 135, American Chemical Society, Washington, DC, 1980.

12.1-17   D. W. Breck and R. A. Anderson, Molecular Sieves, in *Kirk-Othmer Encyclopedia of Chemical Technology*, 3rd ed., Vol. 15, pp. 638–669, Wiley-Interscience, New York, 1981.

12.1-18 G. E. Keller, Separation of Gases and Liquids by Adsorption, in *Proceedings of the Joint Meeting of Chemical Engineering, Chemical Industry and Engineering Society of China, American Institute of Chemical Engineers*, Vol. II, pp. 475–495, Chemical Industry Press, Beijing, China, 1982.

12.1-19 T. E. Whyte, C. M. Yon, and E. H. Wagener (Eds.), *Industrial Gas Separations*, American Chemical Society Symposium Series 223, American Chemical Society, Washington, DC, 1983.

12.1-20 D. M. Ruthven, *Principles of Adsorption and Adsorption Processes*, Wiley, New York, 1984.

Section 12.3

12.3-1 C. J. King, *Separation Processes*, 2nd ed., McGraw-Hill, New York, 1982.

Section 12.4

12.4-1 C. W. Skarstom, Heatless Fractionation of Gases over Solid Adsorbents, in N. N. Li (Ed.), *Recent Developments in Separation Science*, Vol. 2, pp. 95–106, CRC Press, Cleveland, 1972.

Section 12.5

12.5-1 A. Fuderer and E. Rudelstorfer, "Selective Adsorption Process," U.S. Patent 3,986,849 (Oct. 19, 1976).

12.5-2 J. L. Heck and T. Johansen, Process Improves Large Scale Hydrogen Production, *Hydro. Proc.*, **57**(1), 175–177 (1978).

12.5-3 R. T. Cassidy, Polybed Pressure-Swing Adsorption Hydrogen Processing, in W. H. Flank (Ed.), *Adsorption and Ion Exchange with Synthetic Zeolites*, American Chemical Society Symposium Series 135, pp. 247–259, American Chemical Society, Washington, DC, 1980.

12.5-4 R. H. Wilhelm, A. W. Rice, and A. R. Bendelius, Parametric Pumping, A Dynamic Principle for Separating Fluid Mixtures, *Ind. Eng. Chem. Fundam.*, **5**(1), 141–144 (1966).

12.5-5 H. T. Chen, Parametric Pumping, in P. A. Schweitzer (Ed.), *Handbook of Separation Techniques for Chemical Engineers*, pp. 1-467–1-486, McGraw-Hill, New York, 1979.

12.5-6 R. L. Jones, G. E. Keller, and R. C. Wells, "Rapid Pressure-Swing Adsorption with High Enrichment Factor," U.S. Patent 4,194,892 (Mar. 25, 1980).

12.5-7 G. E. Keller and R. L. Jones, A New Process for Adsorption Separation of Gas Streams, in W. H. Flank (Ed.), *Adsorption and Ion Exchange with Synthetic Zeolites*, American Chemical Society Symposium Series 135, pp. 275–286, American Chemical Society, Washington, DC, 1980.

12.5-8 D. E. Earls and G. N. Long, "Multiple-Bed Rapid Pressure-Swing Adsorption for Oxygen," U.S. Patent 4,194,891 (Mar. 25, 1980).

12.5-9 K. Knoblauch, Pressure-Swing Adsorption: Geared for Small-Volume Users, *Chem. Eng.*, **85**(25), 87–89 (1978).

12.5-10 K. Miwa and T. Inoue, Production of Nitrogen Adsorptive Separation, *Chem. Econ. Eng. Rev.*, **12**(11), 40–42 (1980).

12.5-11 S. Sircar and J. W. Zondlo, "Fractionation of Air by Adsorption," U.S. Patent 4,013,429 (Mar. 22, 1977).

12.5-12 C. W. Chi and W. P. Cummings, Adsorptive Separation (Gases), in *Kirk-Othmer Encyclopedia of Chemical Technology*, 3rd ed., Vol. 1, pp. 544–563, Wiley-Interscience, New York, 1978.

12.5-13 M. F. Symoniak, R. A. Reber, and R. M. Victory, Adsorption Makes Isom Better, *Hydro. Proc.*, **52**(5), 101–104 (1973).

12.5-14 M. F. Symoniak, Upgrade Naphtha to Fuels and Feedstocks, *Hydro. Proc.*, **59**(5), 110–114 (1980).

12.5-15 D. R. Garg and J. P. Ausikaitis, Molecular Sieve Dehydration Cycle for High Water Content Streams, *Chem. Eng. Prog.*, **79**(4), 60–65 (1983).

12.5-16 R. G. Bonmati, G. Chapelet-Letourneux, and J. R. Margulis, Gas Chromatography—Analysis to Production, *Chem. Eng.*, **87**(6), 70–72 (1980).

12.5-17 J. R. Bernard, J. P. Gourlia, and M. J. Guttierrez, Separating Paraffin Isomers Using Chromatography, *Chem. Eng.*, **88**(10), 92–95 (1981).

12.5-18 M. N. Y. Lee, Novel Separation with Molecular Sieve Adsorption, in N. N. Li (Ed.), *Recent Developments in Separation Science*, Vol. 1, pp. 75–102, CRC Press, Cleveland, 1972.

12.5-19 R. A. Anderson, Molecular Sieve Adsorbent Applications State of the Art, in J. W. Katzer (Ed.),

Molecular Sieves. II, American Chemical Society Symposium Series 40, pp. 637–649, American Chemical Society, Washington, DC, 1977.

12.5-20 R. C. Bailie, West Virginia University, personal communication, May 1983.

12.5-21 C. Berg, Hypersorption Design. Modern Advancements, *Chem. Eng. Progr.*, **47**(11), 585–591 (1951).

12.5-22 C. Berg, Recent Trends in Refining Processing. Hypersorption in Modern Gas Processing Plants, *Pet. Ref.*, **30**(9), 241–246 (1951).

12.5-23 Y. Sakaguchi, Development of Solvent Recovery Technology Using Activated Carbon, *Chem. Econ. Eng. Rev.*, **8**(12), 36–43 (1976).

12.5-24 Anonymous, Beaded Carbon Ups Solvent Recovery, *Chem. Eng.*, **84**(18), 39–40 (1977).

12.5-25 D. B. Broughton, Adsorptive Separations (Liquids), in *Kirk-Othmer Encyclopedia of Chemical Technology*, 3rd ed., Vol. 1, pp. 563–581, Wiley-Interscience, New York, 1978.

12.5-26 D. B. Broughton, Bulk Separations via Adsorption, *Chem. Eng. Prog.*, **73**(8), 49–51 (1977).

12.5-27 S. Otani, Adsorption Separates Xylenes, *Chem. Eng.*, **80**(21), 106–107 (1973).

12.5-28 H. Odawara, M. Ohno, T. Yamazaki, and M. Kanoka, "Continuous Separation of Fructose from a Mixture of Sugars," U.S. Patent 4,157,267 (June 5, 1979).

12.5-29 H. Ishikawa, H. Tanabe, and K. Usui, "Process of the Operation of a Simulated Moving Bed," U.S. Patent 4,182,633 (Jan. 8, 1980).

12.5-30 K. Yoritomi, T. Kezuka, and M. Moriya, "Method for the Chromatographic Separation of Soluble Components in Feed Solution," U.S. Patent 4,267,054 (May 12, 1981).

12.5-31 T. Vermeulen, Adsorptive Separation, in *Kirk-Othmer Encyclopedia of Chemical Technology*," 3rd ed., Vol. 1, pp. 531–544, Wiley-Interscience, New York, 1978.

12.5-32 R. A. Hutchins, Activated-Carbon Systems for Separation of Liquids, in P. A. Schweitzer (Ed.), *Handbook of Separation Techniques for Chemical Engineers*, pp. 3-3–3-47, McGraw-Hill, New York, 1979.

12.5-33 R. W. Soffel, Carbon (Carbon and Artificial Graphite), in *Kirk-Othmer Encyclopedia of Chemical Technology*, 3rd ed., Vol. 4, pp. 556–631, Wiley-Interscience, New York, 1978.

Section 12.7

12.7-1 D. W. Breck, *Zeolite Molecular Sieves*, pp. 699–718, Wiley, New York, 1974.

12.7-2 H. L. Brooking and D. C. Walton, Less Common Methods of Separation. The Specification of Molecular Sieve Adsorption Systems, *Chem. Eng.*, **257**, 13–17 (1972).

12.7-3 J. W. Carter and D. J. Barret, Comparative Study for Fixed-Bed Adsorption of Water Vapor by Activated Alumina, Silica Gel, and Molecular Sieve Adsorbents, *Trans. Inst. Chem. Eng.*, **51**(2), 75–81 (1973).

12.7-4 S. Ergun, Fluid Flow Through Packed Columns, *Chem. Eng. Progr.*, **48**(2), 89–94 (1952).

12.7-5 M. Leva, Variables in Fixed-Bed Systems. Influence of Particle Characteristics on the Voidage, Pressure Drop, and Flow Resistance in Packed Systems, *Chem. Eng.*, **56**(5), 115–117 (1949).

12.7-6 R. Aris, Shape Factors for Irregular Particles. I. The Steady-State Problem. Diffusion and Reaction, *Chem. Eng. Sci.*, **6**, 262–268 (1957).

12.7-7 J. P. Ausikaitis, "TRISIV Adsorbent—The Optimization of Momentum and Mass Transport Via Adsorbent Particle Shape Modification," paper presented at the International Conference on Fundamentals of Adsorption, Klais, West Germany, May 6–11, 1983.

12.7-8 J. H. Perry, C. H. Chilton, and S. D. Kirkpatrick, *Chemical Engineers' Handbook*, 5th ed., pp. 5-52–5-55, McGraw-Hill, New York, 1973.

12.7-9 H. M. Barry, Fixed-Bed Adsorption, *Chem. Eng.*, **67**(3), 105–120 (1960).

12.7-10 F. A. Dullien, *Porous Media. Fluid Transport and Pore Structure*, pp. 68–72, Academic Press, New York, 1979.

12.7-11 J. M. Campbell, F. E. Ashford, R. B. Needham, and L. S. Reid, More Insight into Adsorption Design, *Hydro. Proc.*, **42**(12), 89–96 (1963).

12.7-12 A. S. Michaels, Simplified Method of Interpreting Kinetic Data in Fixed-Bed Ion Exchange, *Ind. Eng. Chem.*, **44**(8), 1922–1930 (1952).

12.7-13 E. A. Simpson and W. P. Cummings, A Practical Way to Predict Silica Gel Performance, *Chem. Eng. Prog.*, **60**(4), 57–60 (1964).

12.7-14 J. J. Collins, The LUB/Equilibrium Section Concept for Fixed-Bed Adsorption, *Chem. Eng. Progr. Sym. Ser.*, Vol. 63, No. 74, 1967, pp. 31–35.

12.7-15 C. R. Antonson and J. S. Dranoff, "Nonlinear Equilibrium and Particle Shape Effects in Intraparticle Diffusion Controlled Adsorption," *Chem. Eng. Prog. Symp. Ser.*, **65**(49), 20–26 (1969).

12.7-16 C. R. Antonson and J. S. Dranoff, Adsorption of Ethane on Type 4A and 5A Molecular Sieve Particles, *Chem. Eng. Prog. Symp. Ser.*, **65**(49), 27–33 (1969).

12.7-17 J. W. Carter and J. Husain, Carbon Dioxide Adsorption in Fixed Beds of Molecular Sieves, *Trans. Inst. Chem. Eng.*, **50**(1), 69–75 (1972).

12.7-18 D. R. Garg and D. M. Ruthven, The Performance of Molecular Sieve Adsorption Columns: Systems with Micropore Diffusion Control, *Chem. Eng. Sci.*, **29**, 571–581 (1974).

12.7-19 O. A. Hougen and W. K. Marshall, Adsorption from a Fluid Stream Flowing Through a Stationary Granular Bed, *Chem. Eng. Prog.*, **43**(4), 197–208 (1947).

12.7-20 W. S. Kyte, Non-linear Adsorption in Fixed Beds: Freundlich Isotherm, *Chem. Eng. Sci.*, **28**, 1853–1856 (1973).

12.7-21 O. A. Meyer and T. W. Weber, Nonisothermal Adsorption in Fixed Beds, *AIChE J.*, **13**(3), 457–465 (1967).

12.7-22 M. Morbidelli, A. Servida, G. Storti, and S. Carra, Simulation of Multicomponent Adsorption Beds. Model Analysis and Numerical Solution, *Ind. Eng. Chem. Fundam.*, **21**(2), 123–131 (1982).

12.7-23 E. L. Morton and P. W. Murrill, Analysis of Liquid Phase Adsorption Fractionation in Fixed Beds, *AIChE J.*, **13**(5), 965–972 (1967).

12.7-24 J. B. Rosen, General Numerical Solution for Solid Diffusion in Fixed Beds, *Ind. Eng. Chem.*, **46**(8), 1590–1594 (1954).

12.7-25 S. Sircar and R. Kumar, Adiabatic Adsorption of Bulk Binary Gas Mixtures: Analysis by Constant Pattern Model, *Ind. Eng. Chem. Process Des. Dev.*, **22**(2), 271–280 (1983).

12.7-26 C. Tien and G. Thodos, Ion Exchange Kinetics for Systems of Nonlinear Equilibrium Relationships, *AIChE J.*, **13**(5), 373–378 (1959).

12.7-27 R. E. Treybal, *Mass Transfer Operations*, 3rd ed., pp. 585–641, McGraw-Hill, New York, 1980.

12.7-28 S-C. Wang and C. Tien, Further Work on Multicomponent Liquid Phase Adsorption in Fixed Beds, *AIChE J.*, **28**(4), 565–573 (1982).

12.7-29 Y. W. Wong and J. L. Niedzwiecki, A Simplified Model for Multicomponent Fixed Bed Adsorption, *AIChE Symp. Ser.*, **78**(219), 120–127 (1982).

12.7-30 I. Zwiebel, R. L. Gariepy, and J. J. Schnitzer, Fixed Bed Desorption Behavior of Gases with Non-Linear Equilibria: Part I. Dilute, One Component, Isothermal Systems, *AIChE J.*, **18**(6), 1139–1147 (1972).

12.7-31 R. Aris and N. R. Amundsen, Some Remarks on Longitudinal Mixing or Diffusion in Fixed Beds, *AIChE J.*, **3**(2), 280–282 (1957).

12.7-32 L. Z. Balla and T. W. Weber, Axial Dispersion of Gases in Packed Beds, *AIChE J.*, **15**(1), 146–149 (1949).

12.7-33 S. F. Chung and C. Y. Wen, Longitudinal Dispersion of Liquid Flowing Through Fixed and Fluidized Beds, *AIChE J.*, **14**(6), 857–866 (1968).

12.7-34 C. J. Colwell and J. S. Dranoff, Nonlinear Equilibrium and Axial Mixing Effects in Intraparticle Diffusion-Controlled Sorption by Ion Exchange Resin Beds, *Ind. Eng. Chem. Fundam.*, **8**(2), 193–198 (1969).

12.7-35 L. Petrovic and G. Thodos, Mass Transfer in the Flow of Gases Through Packed Beds, *Ind. Eng. Chem. Fundam.*, **7**(2), 274–280 (1968).

12.7-36 N. L. Ricker, F. Nakashio, and C. J. King, *AIChE J.*, **27**(2), 277–284 (1981).

12.7-37 J. W. Carter, Isothermal and Adiabatic Adsorption in Fixed Beds, *Trans. Inst. Chem. Eng.*, **46**(7), T213–221 (1968).

12.7-38 C. W. Chi and D. T. Wassan, Fixed Bed Adsorption Drying, *AIChE J.*, **16**(1), 23–31 (1970).

12.7-39 D. A. Cooney, Numerical Investigation of Adiabatic Fixed-Bed Adsorption, *Ind. Eng. Chem. Process Des. Dev.*, **13**(4), 368–373 (1974).

12.7-40 D. R. Garg and J. P. Ausikaitis, Molecular Sieve Dehydration Cycle for High Water Content Streams, *Chem. Eng. Prog.*, **79**(4), 60–65 (1983).

12.7-41 F. W. Leavitt, Non-isothermal Adsorption in Large Fixed Beds, *Chem. Eng. Prog.*, **58**(8), 54–59 (1962).

12.7-42 H. Lee and W. P. Cummings, A New Design Method for Silica Gel Air Driers under Nonisothermal Conditions, *Chem. Eng. Prog. Symp. Ser. No. 74*, **63**, 42–49 (1967).

12.7-43 R. G. Lee and T. W. Weber, Interpretation of Methane Adsorption on Activated Carbon by Nonisothermal and Isothermal Calculations, *Can. J. Chem. Eng.*, **47**(1), 60–65 (1969).

12.7-44 C. Y. Pan and D. Basmadjian, An Analysis of Adiabatic Sorption of Single Solutes in Fixed Beds: Pure Thermal Wave Formation and Its Practical Implications, *Chem. Eng. Sci.*, **25**, 1653–1664 (1970).

12.7-45 G. Bunke and D. Gelbin, Breakthrough Curves in the Cyclic Steady State for Adsorption Systems with Concave Isotherms, *Chem. Eng. Sci.*, **33**, 101–108 (1978).

12.7-46 D. O. Cooney and E. N. Lightfoot, Multicomponent Fixed-Bed Sorption of Interfering Solutes. System under Asymptotic Conditions, *Ind. Eng. Chem. Process Des. Dev.*, **5**(1), 25–32 (1966).

12.7-47 O. Grubner and W. A. Burgess, Calculation of Adsorption Breakthrough Curves in Air Cleaning and Sampling Devices, *Env. Sci. Tech.*, **15**(11), 1346–1351 (1981).

12.7-48 L. Libreti and R. Passino, Simplified Method for Calculating Cyclic Exhaustion–Regeneration Operations in Fixed-Bed Adsorbers, *Ind. Eng. Chem. Process Des. Dev.*, **21**(2), 197–203 (1982).

12.7-49 C. Y. Pan and D. Basmadjian, Constant Pattern Adiabatic Fixed-Bed Adsorption, *Chem. Eng. Sci.*, **22**, 285–297 (1967).

12.7-50 G. M. Lukchis, Adsorption Systems. Part I: Design by Mass-Transfer-Zone Concept, *Chem. Eng.*, **80**(13), 111–116 (1973); Part II: Equipment Design, *Chem. Eng.*, **80**(16), 83–87 (1973); Part III: Adsorbent Regeneration, *Chem. Eng.*, **80**(18), 83–90 (1973).

12.7-51 J. C. Enneking, How Activated Carbon Recovers Gas Liquids, *Hydro. Proc.*, **45**(10), 189–191 (1966).

12.7-52 W. P. Cummings, Save Energy in Adsorption, *Hydro. Proc.*, **54**(2), 97–98 (1975).

12.7-53 U.S. Environmental Protection Agency, *Process Design Manual for Carbon Adsorption*, 2nd ed., Oct. 1973.

12.7-54 A. R. Balakrishnan and D. C. T. Pei, Heat Transfer in Fixed Beds, *Ind. Eng. Chem. Process Des. Dev.*, **13**(4), 441–446 (1974).

12.7-55 C. C. Furnas, Heat Transfer from a Gas Stream to a Bed of Broken Solids, *Trans. AIChE*, **24**, 142–193 (1930).

12.7-56 G. O. G. Lof and R. W. Hawley, Unsteady-State Heat Transfer between Air and Loose Solids, *Ind. Eng. Chem.*, **40**(6), 1061–1069 (1948).

12.7-57 A. L. Weiner, Drying Gases and Liquids. Dynamic Fluid Drying, *Chem. Eng.*, **81**(19), 92–101 (1974).

12.7-58 C. L. Humphries, Now Predict Recovery from Adsorbents, *Hydro. Proc.*, **45**(12), 88–95 (1966).

12.7-59 R. D. Picht, T. R. Dillman, D. J. Burke, and R. P. deFilippi, Regeneration of Adsorbents by a Supercritical Fluid, *AIChE Symp. Ser.*, **78**(219), 136–149 (1982).

12.7-60 D. K. Friday and M. D. Levan, Solute Condensation in Adsorption Beds During Thermal Regeneration, *AIChE J.*, **28**(1), 86–91 (1982).

12.7-61 Y. N. I. Chan, F. B. Hill, and Y. W. Wong, Equilibrium Theory of a Pressure Swing Adsorption Process, *Chem. Eng. Sci.*, **36**, 243–251 (1981).

12.7-62 L. H. Shendalman and J. E. Mitchell, A Study of Heatless Adsorption in the Model System CO_2 in He, *Chem. Eng. Sci.*, **27**, 1449–1458 (1972).

12.7-63 H. A. Stewart and J. L. Heck, Pressure Swing Adsorption, *Chem. Eng. Prog.*, **65**(9), 78–83 (1969).

12.7-64 M. Shirato et al., Gravitational Drainage of a Packed Bed, *Int. Chem. Eng.*, **21**(2), 294–302 (1981).

12.7-65 S. Hill, Channeling in Packed Beds, *Chem. Eng. Sci.*, **1**, 247–253 (1952).

12.7-66 L. E. Brownell, H. S. Dombrowski, and C. A. Dickey, Pressure Drop Through Porous Media. Part IV—New Data and Revised Correlation, *Chem. Eng. Prog.*, **46**(8), 415–422 (1950).

12.7-67 M. Leva, Fluid Flow Through Packed Beds, *Chem. Eng.*, **64**(8), 263–266 (1957).

12.7-68 M. Leva, *Fluidization*, pp. 45–61, McGraw-Hill, New York, 1959.

12.7-69 E. Ledoux, Avoiding Destructive Velocity Through Adsorbent Beds, *Chem. Eng.*, **55**(3), 118–119 (1948).

12.7-70 D. K. Taylor, Natural-Gas Desulfurization. IV. Iron-Sponge Desulfurization Gains Popularity, *Oil Gas J.*, **54**(84), 147 (1956).

12.7-71 J. W. Carter, Some Aspects of the Prediction of Performance of a Large Air-Dryer, *Trans. Inst. Chem. Eng.*, **46**(7), T222–224 (1968).

12.7-72 L. A. Hernandez and P. Harriott, Regeneration of Powdered Active Carbon in Fluidized Beds, *Env. Sci. Tech.*, **10**(5), 454–456 (1976).

12.7-73 Z. Matsumoto and K. Numasaki, Regenerate Granular Carbon, *Hydro. Proc.*, **55**(5), 157–160 (1976).

12.7-74 A. K. Reed, T. L. Tewksbury, and G. R. Smithson, Jr., Development of a Fluidized-Bed Technique for the Regeneration of Powdered Activated Carbon, *Env. Sci. Tech.*, **4**(5), 432–437 (1970).

Section 12.8

12.8-1 D. R. Garg and J. P. Ausikaitis, Molecular Sieve Dehydration Cycle for High Water Content Streams, *Chem. Eng. Prog.*, **79**(4), 60–65 (1983).

12.8-2 Y. Sakaguchi, Development of Solvent Recovery Technology Using Activated Carbon, *Chem. Econ. Eng. Rev.*, **8**(12), 36–43 (1976).

12.8-3 R. C. Alkire and R. S. Eisinger, Separation by Electrosorption of Organic Compounds in a Flow-Through Porous Electrode, *J. Electrochem. Soc.*, **130**(1), 85–93 (1983).

CHAPTER 13

Ion Exchange

MICHAEL STREAT
Department of Chemical Engineering and Chemical Technology
Imperial College
London, England

FRANCIS LOUIS DIRK CLOETE
Department of Metallurgical Engineering
University of Stellenbosch
Stellenbosch, South Africa

13.1 PRINCIPLES OF ION EXCHANGE

13.1-1 Ion-Exchange Resins

All modern resins are polymeric structures, generally based on either a styrene or an acrylic matrix.

POLYSTYRENE SULFONIC ACID CATION RESINS

Styrene (vinylbenzene) is polymerized readily, using an organic peroxide catalyst, to form linear polystyrene. If divinylbenzene (DVB) (usually about 8% of the total) is mixed with the styrene, a three-dimensional polymer network is formed. The DVB cross-links give a three-dimensional structure that renders the polymer insoluble. The characteristic spherical ion-exchange beads therefore are made by suspension polymerization. The catalyzed monomer mixture is stirred into water under conditions designed to give the desired droplet size which, after several hours heating, will yield solid spherical beads.

At this stage, the beads are hydrophobic but can absorb some organic liquids such as toluene, swelling as a result. The extent of swelling is used in fact as a measure of the crosslinking of the polymer. The beads are converted into an ion-exchange resin, by first swelling the polymer with an inert solvent such as ethylene dichloride. They are then treated with concentrated sulfuric acid, at about 80°C, to make a cation exchanger. The cation exchanger is totally insoluble but is freely permeable by water and contains about 50% H_2O by weight in an apparently dry state. The final material is sulfonated crosslinked polystyrene which is the most widely used cation-exchange resin in commercial use. The total capacity of the resin for sorbing any ion is about 0.00525 equivalent per gram, normally expressed as 5.25 meq/g, calculated on oven-dried resin.

ACRYLIC CATION RESINS

These materials are made usually by copolymerizing acrylic or methacrylic acid with DVB. The total capacity is extremely high, at 13.0 meq/dry gram, corresponding with about 6.5 meq/wet gram, or 4500 meq/L. Polyacrylics, with lower molecular weight and high capacity, have replaced almost entirely the methacrylics. The manufacture is a little more complex, since acrylic acid is fairly soluble in water and cannot be polymerized in suspension. Thus, the procedure is to make crosslinked poly ethyl acrylate, or acrylonitrile, which is then hydrolyzed to give the final product.

POLYSTYRENE ANION EXCHANGERS

The basic polymer is exactly as for the polystyrene cation exchangers, but the introduction of the active group requires two steps. A chloromethyl group is introduced into each benzene ring, after which any amine group can be introduced by a simple addition reaction with chloromethyl ether.

If a tertiary amine such as trimethylamine is used, the product is a strong base quaternary ammonium compound. This resin is the anionic equivalent of the sulfonic cation materials. The capacity of a typical strong base resin is 3.9–4.2 meq/dry gram. The use of a secondary amine, such as dimethylamine, gives a tertiary amine product, which is more weakly basic. However, whereas there are only two acidic groups readily available for cation resins, very large numbers of amines can be used to produce anion-exchange materials for different characteristics. This fact is used to produce special resins for uranium and gold recovery.

ACRYLIC ANION EXCHANGERS

A range of polyacrylic anion-exchange resins has been produced in recent years. The exact structure and method of manufacture have not been published, but it is clear that they contain quaternary ammonium groups (or in the case of the weakly basic materials, amino groups) attached to an acrylic skeleton. They are physically robust materials, with good operating characteristics, and are serious competitors to the polystyrene resins.

MODIFIED RESIN SKELETONS

The polystyrene resins have an extremely irregular structure, owing to the fact that DVB polymerizes with itself more rapidly than with styrene. Consequently, the first polymer formed is highly crosslinked, and the liquid mixture is depleted in DVB. During the progress of polymerization, which takes several hours, the product becomes progressively less crosslinked, and, ultimately, some linear polymer is formed, attached to the copolymer matrix. The final resin, while having an average crosslinking of say 8%, contains tightly knotted regions in which crosslinking may be up to 25%, and the intermolecular distance is correspondingly reduced.

This heterogeneous structure has two results. First, in the case of simple, relatively small, inorganic ions, it is found that the law of mass action is not wholly obeyed, and the defined equilibrium "constants" are not fixed. As an ion enters the resin, the process becomes progressively more difficult as the concentration in the resin approaches the total capacity, and the less accessible active groups come into use.

A second, and more important, problem occurs with large ions such as the natural humic and fulvic acids occurring in the surface-water supplies. These are large polycarboxylic molecules, which diffuse freely through the less crosslinked regions of anion resins but become trapped in the tightly crosslinked regions. As a result, resins can become poisoned, losing capacity and giving inferior treated-water quality. New types of resin, which are referred to as "macroporous" or "macroreticular," overcome this problem. On the molecular scale these resins have a sintered structure that gives free water channels (in contrast to bound gel water) of about 1000 Å diameter. Large molecules travel freely through these channels and have only short distances to travel through the resin itself. Therefore, these materials have a high resistance to organic poisoning. They take up and release natural organic matter freely. Macroporous resins, anion or cation, generally have greater osmotic shock stability than the standard gel resins, but not necessarily greater compressive strength or attrition resistance. Their use in anion form has become very widespread.

Macroporous cation resins are also available but are less widely used than the anion materials. This is because the standard gel cation resins are themselves very strong and also because the poisoning of cation resins by large organic cations is virtually unknown.

INORGANIC ION EXCHANGE MATERIALS

The earliest ion-exchange phenomena were discovered using naturally occurring ion-exchange materials, especially zeolites consisting mainly of an aluminosilicate structure. Interest in the application of inorganic materials for ion exchange has continued with considerable vigor and in some instances these materials are preferred to synthetic polymeric ion-exchange resins.

Important materials that have found widespread application in the process industries fall under the following general headings:

Natural and synthetic zeolites

Insoluble salts, for example, hydrous zirconium phosphate

Heteropolyacids, for example, ammonium phosphomolybdate

Complex ferrocyanides, for example, sodium copper cyanoferrate (II)

Clays, for example, montmorillonite

Zeolites, whose use as molecular sieve adsorbents is discussed in Chapter 12, possess well-defined crystal structures composed of an aluminosilicate lattice containing Si, Al, and O atoms in a cage-type structure. The matrix is capable of exchanging cations, usually alkali or alkaline earth ions in aqueous solution. There is considerable affinity for divalent ions such as Ca^{2+} and Mg^{2+} as well as for Sr^{2+}. The latter renders zeolites of exceptional importance in the removal of strontium (and cesium) from low- and intermediate-level nuclear waste solutions. Naturally occurring zeolites, such as clinoptilolite, have found extensive use in the nuclear industry for this and related applications. Zeolites offer other important advantages over polymeric materials since they can withstand higher operating temperatures, are resistant to ionizing radiation, and are generally more chemically stable. The advantage of naturally occurring materials is their relative cheapness if they are used on a once-through basis. It is common practice in the nuclear industry to extract traces of radioactive strontium and cesium onto zeolites and subsequently immobilize the exhausted ion-exchange material in concrete or cement for permanent storage. Naturally occurring clays, such as montmorillonite, have also been considered for this application but are less favored.

The use of finely divided zeolites in detergents for water softening will lead undoubtedly to large-scale use in the future. Most inorganic ion-exchange materials are cationic, although a few synthetic materials are amphoteric in character. The most interesting salt is hydrous zirconium phosphate, which acts as a cation exchanger in alkaline solution and as an anion exchanger in acid. Other insoluble salts of zirconium can be prepared in crystalline form and exhibit ion-exchange properties, for example, zirconium tungstate, zirconium molybdate, and zirconium arsenate. These compounds hydrolyze above pH 6 and therefore are of more limited use.

Complex transition metal ferrocyanide salts can be produced in crystalline and powdered forms and possess a very high selectivity for cesium in solutions having high salt concentrations. Potassium cobalt cyanoferrate (II) and sodium copper cyanoferrate (II) have been identified as useful reagents for cesium removal from radioactive waste streams and other cyanoferrates are capable of separating Cs from Rb and from a fission product mixture.

13.1-2 Ion-Exchange Selectivity

Ion-exchange equilibria in cation- and anion-exchange resins depend largely on the type of functional group and the degree of crosslinking. The degree of crosslinking determines the tightness of the matrix structure and thus its porosity. It cannot be measured directly and is rarely homogeneous.

Crosslinking varies from the outer shell to the center of a bead and is usually quoted by the manufacturers as the "nominal DVB content." Modern ion-exchange resins contain a more rigid and well-defined structure and are referred to as "macroreticular" or "isoporous" materials. These are composed of highly crosslinked microspheres with a macrospherical structure. However, the laws governing equilibria are essentially the same for these materials.

The important properties of ion exchange materials are briefly given below:

1. To preserve electroneutrality, (a) ion exchange is stoichiometric and (b) capacity is independent of the nature of the counterion.
2. Ion exchange is nearly always a reversible process;
3. Ion exchange is a rate-controlled process, usually governed by diffusion in the bead or the surrounding stagnant liquid film.

SIMPLE BINARY SYSTEMS

The most widely employed method for expression of ion-exchange equilibria has been developed from the law of mass action or the Donnan membrane theory.[1,2] Consider the exchange of cations A and B between a cation-exchange resin and a solution containing no other cations. Assume that the ion exchanger is initially in the B from and that the solution contains ions A. The mass action expression for cation exchange can be written

$$z_A \overline{B^{z_B^+}} + z_B A^{z_A^+} \rightleftharpoons z_B \overline{A^{z_A^+}} + z_A B^{z_B^+} \tag{13.1-1}$$

Here, overbars denote the ionic species in the resin phase, z_A^+ and z_B^+ denote the valency and charge of counterions A and B. We can define the thermodynamic equilibrium constant K_a as follows:

$$K_a = \frac{(\overline{a_A})^{z_B} (a_B)^{z_A}}{(\overline{a_B})^{z_A} (a_A)^{z_B}} \tag{13.1-2}$$

The difficulty in the evaluation of the activity coefficients and therefore activities in the resin phase is great. For most practical applications it is usually satisfactory to assume that the solution-phase activity coefficients are almost unity, which is particularly valid in dilute solutions. Therefore, the resin-phase activity coefficients usually are combined into the equilibrium constant K_a, to provide a new pseudoconstant, that is, the selectivity coefficient K_c. Thus,

$$(K_c)_B^A = \frac{(\overline{C}_A)^{z_B}(C_B)^{z_A}}{(\overline{C}_B)^{z_A}(C_A)^{z_B}} \tag{13.1-3}$$

Often it is desirable to use equivalent ionic fractions to represent concentrations in the solution and resin phases. Thus, in a binary system,

$$(K_c)_B^A \left(\frac{\overline{C}}{C}\right)^{z_A - z_B} = \frac{(y_A)^{z_B}(x_B)^{z_A}}{(y_B)^{z_A}(x_A)^{z_B}} \tag{13.1-4}$$

The distribution coefficient m_A is defined as

$$m_A = \frac{\overline{C}_A}{C_A} = \frac{y_A \overline{C}}{x_A C} \tag{13.1-5}$$

the separation factor α is defined as

$$\alpha_B^A = \frac{y_A/x_A}{y_B/x_B}$$

$$= \frac{y_A x_B}{y_B x_A} \tag{13.1-6}$$

Univalent Exchange. If we consider the exchange of univalent ions, that is, $z_A = z_B = 1$, then the selectivity coefficient becomes

$$(K_c)_B^A = \frac{y_A x_B}{y_B x_A} = \frac{y_A(1-x_A)}{x_A(1-y_A)} = \alpha_B^A \tag{13.1-7}$$

A typical plot of K_c for univalent exchange is given in Fig. 13.1-1. The equilibrium data are plotted with respect to species A and show that if $K_c > 1$, species A is preferred by the exchanger; the preferential attraction becomes more pronounced as K_c increases. If, $K_c < 1$ then species A is less preferred and the ion exchange preferentially sorbs species B. The selectivity of an ion-exchange resin is affected greatly by the degree of crosslinking. Slightly crosslinked and highly swollen resins exhibit reduced selectivity for one small ion over another. With increase in crosslinking, selectivity is increased (see Table 13.1-3). Gregor[3] indicates that swelling of resins during the exchange process is a result of an osmotic process during which the osmotic pressure of the resin gel is opposed by tension in the resin structure. Under

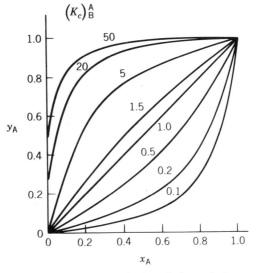

FIGURE 13.1-1 Equilibrium plot for univalent–univalent exchange.

TABLE 13.1-1 Calculated Values of K_c for Sodium-Hydrogen
Exchange for Sulfonated Polystyrene–DVB

DVB (%)	K_c Calculated	K_c Observed
2	1.18	1.06
5	1.49	1.25
10	2.2	1.75
17	3.2	2.6

Source: Reichenberg et al.[4]

TABLE 13.1-2 Calculated Values of K_c for Sodium–Hydrogen Exchange for Varying Degrees of
Saturation

DVB (%)	Fractional Saturation with Sodium	0.1	0.3	0.5	0.7	0.9
2	K_c (calc.)	1.18	1.23	1.27	1.29	1.31
	K_c (obs.)	1.06	1.10	1.11	1.14	1.16
5	K_c (calc.)	1.49	1.59	1.68	1.77	1.86
	K_c (obs.)	1.25	1.40	1.45	1.41	1.29

Source: Gluekauf and Duncan.[5]

TABLE 13.1-3 Selectivity (K_a) Scale for Univalent Ions in Sulfonic Acid
Resins Using the Lithium Ion as Reference

	4% DVB	8% DVB	16% DVB
Li	1.00	1.00	1.00
H	1.32	1.27	1.47
Na	1.58	1.98	2.37
NH_4	1.90	2.55	3.34
K	2.27	2.90	4.50
Rb	2.46	3.16	4.62
Cs	2.67	3.25	4.66
Ag	4.73	8.51	22.9
Tl	6.71	12.4	28.5

Source: Bonner and Smith.[6]

conditions of equal valency, large hydrated ions are less favored since they require expansion of the network. Tables 13.1-1 and 13.1-2 show the predicted values of K_c as a function of degree of crosslinking for the exchange of sodium and hydrogen. Increased selectivity with increasing crosslinking is observed and can be predicted from theory.[4,5]

Bonner and Smith[6] have given the selectivity of univalent ions in sequence with lithium as the reference ion. The selectivity scale (see Table 13.1-3) is based on the parameter K_a and similarly shows the improved selectivity with decrease in hydrated ion size and with increase in the degree of crosslinking.

Divalent–Univalent Exchange.   In the industrially important case of divalent–univalent exchange, that is, $z_A = 2$, $z_B = 1$, the expression for the selectivity coefficient becomes

$$(K_c)_B^A \left(\frac{\overline{C}}{C}\right) = \frac{y_A(1 - x_A)^2}{x_A(1 - y_A)^2} \tag{13.1-8}$$

Here the relationship between y_A and x_A becomes greatly dependent on the value of C, the solution-phase concentration. A typical equilibrium plot is shown in Fig. 13.1-2 and it is obvious that the preference for

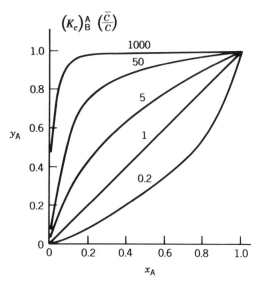

FIGURE 13.1-2 Equilibrium plot for divalent–univalent exchange.

A is enhanced greatly as the solution concentration is decreased. Increasing the concentration results in a lowering of the selectivity of the divalent ion over the univalent ion.

A selectivity scale for divalent cations is given in Table 13.1-4. Again the arbitrary value of unity is given to the lithium ion. Bonner and Smith corrected their value of the equilibrium constant by allowing for activity coefficients by using the method developed by Argersinger and Davidson.[7] Thus,

$$\log K_a = \int_0^1 \log \left(K_c \frac{\gamma_B}{\gamma_A} \right) d\overline{C}_A \tag{13.1-9}$$

which is the result of the Gibbs–Duhem relationship.

Anion Exchange. Extensive values of the selectivity coefficients for strongly basic anion resins of the quaternary ammonium types have been tabulated by Bauman and Wheaton,[8] Kunin and McGarvey,[9] and Gregor et al.[10] Gregor and coworkers have shown two classes of ions: those that exhibit little change in selectivity coefficient with resin composition, and those with composition-dependent selectivity coefficients. The latter presumably form clusters within the resin phase while the former are distributed randomly at equilibrium.

TABLE 13.1-4 Selectivity (K_a) Scale for Divalent Cations in Sulfonic Acid Resins Using the Lithium Ion as Reference

	4% DVB	8% DVB	16% DVB
UO_2	2.36	2.45	3.34
Mg	2.95	3.29	3.51
Zn	3.13	3.47	3.78
Co	3.23	3.74	3.81
Cu	3.29	3.85	4.46
Cd	3.37	3.88	4.95
Ni	3.45	3.93	4.06
Ca	4.15	5.16	7.27
Sr	4.70	6.51	10.1
Pd	6.56	9.91	18.0
Ba	7.47	11.5	20.8

Source: Bonner and Smith.[6]

MULTICOMPONENT ION EXCHANGE

The theoretical treatment of multicomponent ion exchange is quite complex, especially if it involves species of different valency. The treatment of ion-exchange equilibria involving only ions of the same valence is quite straightforward since it can be assumed that the selectivity coefficient or separation factor is constant over the complete range of ionic concentration.

Consider any two species, i and j; the separation factor is defined as

$$\alpha_j^i = \frac{y_i x_j}{x_i y_j} \tag{13.1-10}$$

(even for ions of different valencies, the assumption of constant α's may be a good approximation). If the total number of exchangeable ions is n, there are $n - 1$ independent equations of this type. To find the liquid-phase concentrations from given solid-phase concentrations, the above equations can be combined to yield

$$x_i = \frac{y_i}{\sum_j \alpha_j^i y_j} = \frac{\alpha_i^k y_i}{\sum_j \alpha_j^k y_j} \tag{13.1-11}$$

where the summation is carried out over all the components and k designates an arbitrarily chosen component, α_i^i, α_j^j, and α_k^k are equal to 1. An analogous equation can be obtained for y_i in terms of the x's:

$$y_i = \frac{x_i}{\sum_j \alpha_i^j x_j} = \frac{\alpha_k^i x_i}{\sum_j \alpha_k^i x_j} \tag{13.1-12}$$

If x_i is constant, y_j becomes a linear function of the other y's. Similarly, if y_i is constant, x_j becomes a linear function of the other x's. Consider a ternary system involving three components A, B, and C. Assuming univalent ions, then $(K_c)_B^A = \alpha_B^A$, $(K_c)_C^B = \alpha_C^B$, and $(K_c)_A^C = \alpha_A^C$. It follows that

$$\alpha_B^A \, \alpha_C^B \, \alpha_A^C = 1 \tag{13.1-13}$$

Three-component equilibrium data can be represented on a triangular diagram. The regular grid of the diagram is used to represent the composition of one of the phases (usually the resin phase): the composition of the other phase is given by a set of contour lines, each corresponding to a constant concentration of one of the components. Figure 13.1-3 shows the triangular composition diagram for a hypothetical system having $\alpha_B^A = 2$ and $\alpha_C^B = 2$. A contour line (which is straight in this case) for $x_A = 0.1$ is constructed by connecting its intercepts with the lines $y_B = 0$ and $y_C = 0$. The intercepts are obtained directly from the binary A–C and A–B equilibrium relationships. For the case of a divalent–univalent system, the separation factor (expressed as α_B^A) will vary with composition. Thus, the effect of concentration must be taken into

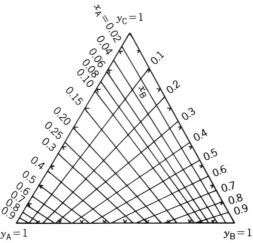

FIGURE 13.1-3 Graphical representation of ternary constant-separation-factor isotherm $\alpha_C^A = 4$, $\alpha_C^B = 2$.

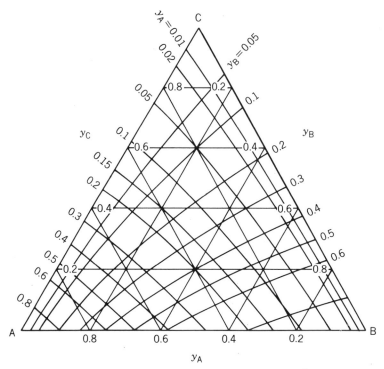

FIGURE 13.1-4 Graphical representation of ternary isotherm $(y_A/x_A)/(x_C/y_C)^2 = 8.06(y_B/x_B)/(x_C/y_C)^2 = 3.87$.

account in presentation of equilibrium data. In such a case, it is still possible to draw the appropriate ternary diagram and an example of variable-separation factor equilibria is shown in Fig. 13.1-4. The contour lines are now curved so that a variable-separation factor system may be viewed as a distorted constant-separation factor system that exhibits identical properties over an infinitesimal composition change. Qualitatively, the behavior over finite composition change is similar to the univalent system.

For ternary systems, the theory is tidy and quite precise, although the representation relies heavily on the constancy of the relevant separation factors. Regrettably, the values of the selectivity coefficients and separation factors are not constant in practice. There is much experimental evidence to show that these parameters are far from constant even in the simplest cases. Figures 13.1-5, 13.1-6, and 13.1-7 show experimentally determined values of the binary separation factors in the K–Na–H system. Note the significant decrease in α_H^K as a function of x_K. The deviation of the experimental points from the derived contour lines is shown in Figs. 13.1-8 and 13.1-9. A method of qualitative representation and interpretation of multicomponent ion-exchange equilibria has been published by Brignal et al.[11] and this paper should be consulted for a rational procedure. A general method for the prediction of multicomponent ion-exchange equilibria from the data of relevant binary systems has emerged in recent years.[12] This requires the calculation of thermodynamic equilibrium constants for the reaction and parameters to calculate activity coefficients of the exchanging species in both phases. Two parameters for each ionic component are required in the solution phase to calculate the activity coefficients by the extended Debye–Hückel equation.[13] Two interaction coefficients are required in the resin phase by analogy with vapor–liquid equilibrium data and the Wilson method has been successfully applied.[14] If a multicomponent system can be resolved into a set of binary equilibria, then it is necessary only to characterize the binary data. If the equilibrium constants constants and the interaction parameters are applied, then the multicomponent data can be obtained by solving the resulting binary-exchange equations simultaneously.

13.1-3   Ion Exchange Kinetics

GENERAL INTRODUCTION
The kinetics of ion exchange may be divided into five steps:

1.   Diffusion of the counterions through the bulk solution to the surface of the ion exchanger.

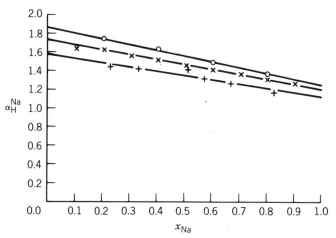

FIGURE 13.1-5 The effect of solution concentration on the separation factor in the Na⁺–H⁺ system. For the 8% DVB resin; ○, 0.01 N Cl⁻; ×, 0.1 N Cl⁻; and +, 1.0 N Cl⁻.

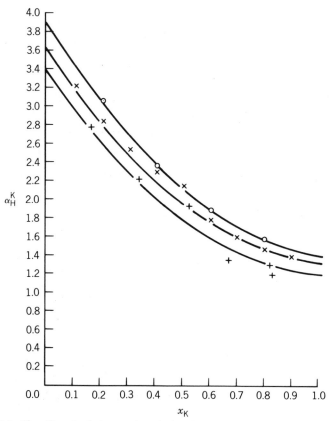

FIGURE 13.1-6 The effect of solution concentration on the separation factor in the K–H system. For the 8% DVB resin: ○, 0.01 N Cl⁻; ×, 0.1 N Cl⁻; and +, 1.0 N Cl⁻.

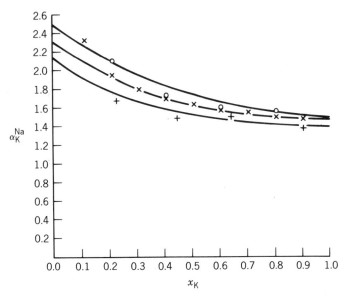

FIGURE 13.1-7 The effect of solution concentration on the separation factor in the K^+–Na^+ system. For the 8% DVB resin: $\circ$, 0.01 N Cl^-; $\times$, 0.1 N Cl^-; $+$, 1.0 N Cl^-.

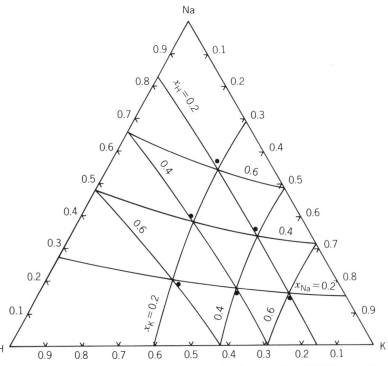

FIGURE 13.1-8 Ternary equilibria for the system H^+–Na^+–K^+: 1.0 N Cl^-, 8% DVB resin. The curves are estimated from binary data and solid circles ($\bullet$) represent experimental points.

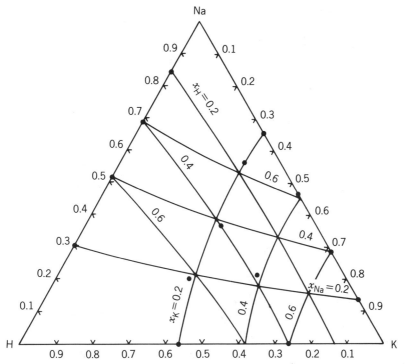

FIGURE 13.1-9 Ternary equilibria for the system H^+–Na^+–K^+: $0.01\ N\ Cl^-$, 8% DVB resin. The curves are estimated from binary data and the solid circles ($\bullet$) represent experimental points.

2. Diffusion of the counterions within the solid phase.
3. Chemical reaction between the counterions and the ion exchanger.
4. Diffusion of the displaced ions out of the ion exchanger.
5. Diffusion of the displaced ions from the exchanger surface into the bulk solution.

Steps 4 and 5 are the reverse processes of steps 2 and 1, respectively. The kinetics of ion exchange are governed by either a diffusion or mass action mechanism, depending on which is the slowest step. In general, the diffusion of ions in the external solution is termed film diffusion control. This is a useful concept but hydrodynamically it is ill defined. The diffusion or transport of ions within the exchanger phase is commonly termed particle diffusion control. Chemical reaction at the exchanger sites can be rate controlling in certain cases.

DIFFUSIONAL PROCESSES
It should be stated that simple diffusion-based kinetic treatments are idealized cases of ion exchange. They are only strictly valid for systems undergoing isotopic redistribution where precise boundary conditions can be described. Usually, it is assumed that all ion-exchange particles are spherical and of uniform size. In this case, diffusional processes are fundamentally described by Fick's First Law:

$$J_i = D \text{ grad } C_i \tag{13.1-14}$$

Here J_i is the flux of the diffusing species i of concentration C_i, and D is the diffusion coefficient. Fundamental treatment of the rate laws of ion exchange is given by Helfferich.[15]

The nature of the rate-determining step can be predicted by use of the simple dimensionless criterion given by Helfferich:[15]

$$\frac{\overline{CD}\delta}{CDr_0}(5 + 2\alpha_B^A) \ll 1 \quad \text{particle diffusion control}$$

$$\frac{\overline{CD}\delta}{CDr_0}(5 + 2\alpha_B^A) \gg 1 \quad \text{film diffusion control}$$
$$\tag{13.1-15}$$

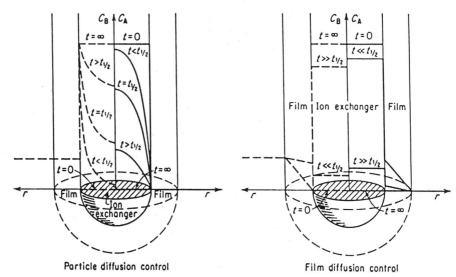

Particle diffusion control   Film diffusion control

FIGURE 13.1-10 Radial concentration profiles for ideal particle diffusion control and ideal film diffusion control (schematic). The right sides of the diagrams show the profiles of species A (initially in the ion exchanger) and the left sides those of species B (initially in the solution). The various curves are for different contact times. Figure reproduced from F. Helfferich, *Ion Exchange,* McGraw-Hill, New York, 1962. Available from University Microfilms International, Ann Arbor, MI.

where the variables are defined in the Notation section. If film diffusion is much faster than diffusion within the ion-exchange particles, then concentration differences in the liquid film level out instantaneously (see Fig. 13.1-10). Concentration gradients exist within the exchanger, and particle diffusion is the rate-controlling step. The quantitative aspects of particle diffusion are more complex since the internal solid phase is only partially available as a diffusion medium. A large fraction of the interior of a bead is occupied by the exchanger matrix and this leads to steric hindrance and tortuous diffusion paths. Furthermore, migration of counterions is confined to the array of fixed ionic groups on the matrix of the exchanger. Thus, steric hindrance and ionic interactions result in slower diffusion kinetics within the particle.

Under the infinite solution volume condition, fractional conversion or attainment of equilibrium is given by the expression

$$X = 1 - \frac{6}{\pi^2} \sum_{n=1}^{\infty} \exp\left(-\frac{\overline{D}\pi^2 n^2 t}{r_0^2}\right) \tag{13.1-16}$$

In actual ion exchange, the counterdiffusing ions possess different mobilities. For a quantitative treatment, Fick's Law has to be replaced by the Nernst–Planck equation which allows both particle and film diffusion kinetics in exchanging ideal systems to be described:

$$J_i = -D_i \left(\text{grad } C_i + \frac{z_i C_i F}{RT} \text{ grad } \phi\right) \tag{13.1-17}$$

where ϕ is the electrical potential. The Nernst–Planck theory predicts that the forward and reverse rates of ion exchange should differ markedly with different mobilities of the counterions concerned (Fig. 13.1-11). A striking feature is the behavior of the concentration profiles within the beads.[16] If the ion initially present in the bead is the much faster one, then a comparatively sharp boundary moves toward the center of the particle (see Fig. 13.1-12). This is a result of the great dependence of the interdiffusion coefficient on ionic composition. Hence, in this case, the concentration of the faster-moving ion is low in the outer shells and high at the center of the bead. Therefore, the interdiffusion coefficient decreases toward the center of the bead, and the edge is converted quickly. This behavior is similar to the proposed shell progressive or ash-layer diffusion model for interacting species.

Chemical Reaction Kinetics. The possibility of a chemical reaction between the counterions and the fixed exchange sites is also recognized as one of the general ion-exchange mechanisms in certain cases.

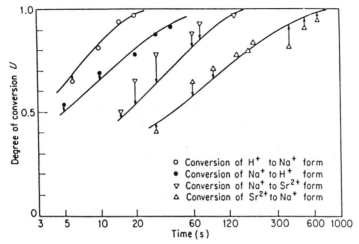

The rate-controlling step is no longer a conventional diffusional process but chemical reaction kinetics are assumed. The rate of ion exchange is governed by the rate constant of the corresponding chemical reaction. Basic laws of chemical kinetics can be used in the mathematical treatment.

Homogeneous Reaction Kinetics. Rate laws that describe bimolecular second-order chemical reactions can be treated as an analogue of the ion-exchange process. Although ion exchange involves a heterogeneous system, homogeneous chemical kinetics may be applied if one assumes that the exchanger phase is a fully dissolved reactant. The rate of ion exchange is now represented by the following rate equation:

$$\frac{dC}{dt} = k_2(\overline{C}_A - C)(C_B - C) \qquad (13.1\text{-}18)$$

which is integrated to give

$$t = \frac{1}{k_2(\overline{C}_A - C_B)} \ln \frac{C_B(\overline{C}_A - C)}{C_A(C_B - C)} \qquad (13.1\text{-}19)$$

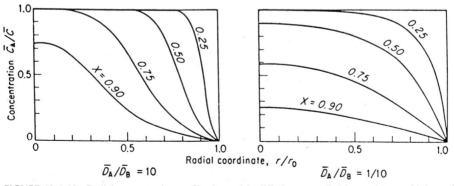

$$\overline{D}_A/\overline{D}_B = 10 \qquad\qquad \overline{D}_A/\overline{D}_B = 1/10$$

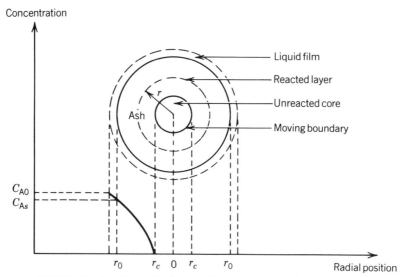

FIGURE 13.1-13 Schematic diagram of a partially reacted ion-exchanger bead.

where $\overline{C}_A$ and C_B are the initial concentrations of counterions in the exchanger and external solution, respectively. The coefficient k_2 is the second-order rate constant and C represents the external ionic concentration of the counterion originally in the solid phase at time t. Equation (13.1-19) can be arranged to a more appropriate form:

$$t = \frac{1}{k_2(\overline{C}_A - C_B)} \ln\left[\frac{C_B(1 - X)}{\overline{C}_A(C_B/\overline{C}_A - X)}\right] \tag{13.1-20}$$

It must be noted that in this case there are no concentration gradients within the ion exchanger and no dependence on particle size.

Heterogeneous Reaction Systems The mathematical treatment is quite different if the exchanger is treated as a solid phase. The conceptual models are then similar to those developed for noncatalytic fluid–solid reactions.[17,18] These kinetic models have been applied successfully to some ion-exchange processes.[19-24]

The "shell progressive" or "shrinking-core" model assumes that a liquid reactant containing counterions A is advancing into a spherical ion-exchange bead (unreacted core); see Fig. 13.1-13. The chemical reaction between the counterions and the exchanger sites progresses in shells or layers and produces a solid product layer that is often called the ash layer. Initially, the reaction takes place at the outer surface of the bead, and as the reaction proceeds, the interface between the reacted and unreacted core moves gradually into the interior of the bead, leaving behind an envelope of reacted material. During this process, a sharp inward-moving reaction boundary is formed, resulting in an unreacted core that shrinks with time.

The shell progressive mechanism can be applied to some ion-exchange processes. Any of the three subsequent steps, that is, film diffusion, ash-layer diffusion, and chemical reaction control, can be rate determining, depending on the prevailing conditions. The following relationships have been obtained for the indicated conditions.

Liquid Film Diffusion Control

$$t = \frac{ar_0 C_{s0}}{3C_{A0}k_{mA}} X \tag{13.1-21}$$

Ash-Layer Diffusion Control

$$t = \frac{ar_0^2 C_{s0}}{D_e C_{A0}}\left[\frac{1}{2} - \frac{1}{2}(1 - X)^{2/3} - \frac{1}{3}X\right] \tag{13.1-22}$$

Chemical Reaction Control

$$t = \frac{r_0}{C_{A0}k_s} [1 - (1 - X)^{1/3}] \tag{13.1-23}$$

It must be noted that the above derivations are confined to cases involving a large excess of liquid reactant. However, other authors have achieved the above relationships with different conceptual approaches.[25,26]

CONCLUSIONS

Conventional ion-exchange reactions often can be explained by homogeneous diffusional kinetics. For example, simple ion-exchange reactions encountered in water treatment, such as Na^+–H^+ of Cl^-–OH^- exchange with polyelectrolyte gels or macroreticular resins, can be fitted by diffusional theory. However, more complex ion-exchange reactions involving complex ions or chelating ion-exchange materials are not satisfied so easily by these relationships. For example, the sorption and desorption or uranium or plutonium from acid solutions or the sorption of copper with an amino-diacetic acid exchanger are better fitted by chemical reaction-based models. Also, sorption into inorganic exchangers can be fitted better by chemical reaction rates. The field of mass transfer and ion-exchange kinetics is in a period of transition owing to the development of new resin products and new industrial applications.

13.2 APPLICATIONS OF ION EXCHANGE

13.2-1 Introduction

The industrial applications of ion exchange are extremely widespread and range from the purification of low-cost commodities such as water to the purification and treatment of high-cost pharmaceutical derivatives as well as precious metals such as gold and platinum. The largest single application, measured in terms of ion-exchange resin usage is water treatment, that is, water softening, water demineralization for high-pressure boilers, and dealkalization. Indeed, enormous advances in ion-exchange technology have occurred because of the relentless requirement for pure and ultrapure water. Other major industrial applications are the processing and decolorization of sugar solutions and the recovery of uranium from relatively low-grade mineral acid leach solutions.

Examples of major ion-exchange applications are listed in Table 13.2-1. The fields of water treatment, effluent treatment, and pollution control are predominant and there have been many recent advances. For example, the partial demineralization of brackish water using the Sirotherm process has been developed in Australia and this is probably one of the most innovative developments in recent years.[1] Important applications in the fields of medicine, pharmacology, chemical processing, catalysis, and analytical techniques are also mentioned. The remainder of this section describes some important applications in detail.

13.2-2 Water Treatment

The removal of hardness from water by ion exchange has been carried out since the beginning of this century and represents the first major application of ion exchange in the water industry. Softening has been used extensively in domestic water supplies in the United States, although the practice is less common in the United Kingdom. In essence, the hardness present in water is removed by passing the water over a cation-exchange column in the sodium form. A typical reaction is as follows:

$$\overline{2R^-Na^+} + Ca^{2+} \rightleftharpoons \overline{R_2^-Ca^{2+}} + 2Na^+$$

Practically all modern cation-exchange materials are effective for this reaction. The equilibrium coefficient of the above reaction favors calcium sorption (or any other divalent ion, such as magnesium) at low concentrations. At total hardness of about 500 ppm, the equilibrium diagram is similar to Fig. 13.1-2. Since the overall selectivity of the cation exchangers for most common divalent ions decreases with concentration, the resin may be regenerated conveniently with concentrated sodium chloride solutions. Because of the change in selectivity with concentration, the efficiency of the cation-exchange softening process decreases with increasing salinity of the water. It is possible nevertheless to soften water up to concentrations of about 5000 ppm and ion exchange has even been suggested for the pretreatment of seawater prior to multistage flash evaporation provided that blowdown can be used as a regenerant.

The removal of salts and other ionic impurities from water by means of ion-exchange resins is based primarily on the exchange of cations (Na^+, K^+, Ca^{2+}, Mg^{2+}, etc.) with the hydrogen form of a cation-exchange resin and the exchange of anions (Cl^-, SO_4^{2-}, NO_3^-, etc.) with the hydroxide form of an anion-exchange resin. Usually, deionization of water is achieved by using a mixed bed of both cation- and anion-exchange resins and this is the classical method of producing ultrapure water for boiler feed, electronics,

TABLE 13.2-1 Examples of Major Ion-Exchange Applications

Water Treatment	*Reagent Purification*
Water softening	Hydrochloric acid
Dealkalization	Formaldehyde
Deionization	Phenol
Fluoride removal	Acrylates
Color removal	
Oxygen removal	*Inorganic Sols (Preparation)*
Iron and manganese removal	SiO_2
Nitrate removal	$Fe(OH)_3$
Ammonia removal	$Al(OH)_3$
	Thoria
Sugars and Polyhydric Alcohols	Zirconia
Purification of cane, corn, and beet sugars	*Catalysis*
Glycerine purification	
Sorbitol	Sucrose inversion
	Esterification
Biological Recovery and Purification	Condensation
Antibiotics	*Medicine*
Vitamins	
Amino acids	Antacids
Proteins	Sodium reduction
Enzymes	Taste masking
Plasma	Sustained release
Blood	Diagnostic
Viruses	Tablet disintegration
	pH control
Hydrometallurgy (Recovery and Purification)	Potassium removal
Uranium	Skin treatment
Thorium	Toxin removal
Rare earths	
Transition metals	*Analysis*
Transuranic elements	Separation
Gold, silver, platinum	Concentration
Chromium	Purification
Solvent Purification	
Alcohols	
Benzene	
Chlorinated hydrocarbons	
Acetone	
Carbon tetrachloride	

and for more general use in the chemical and allied industries. A mixed bed uses a cation-exchange resin in the hydrogen form intimately mixed with an anion-exchange resin in the hydroxide form. The passage of a salt solution, for example, $CaSO_4$, leads to the following irreversible reaction within the bed:

$$\overline{2H^+} + \overline{2OH^-} + Ca^{2+} + SO_4^{2-} \rightarrow \overline{Ca^{2+}} + \overline{SO_4^{2-}} + H_2O$$

Under these conditions the selectivity coefficient for impurities is infinite and therefore the efficiency of sorption is substantially better than if two columns of cation and anion resins are used in series. Regeneration requires separation of the resin prior to treatment with acid and alkali. It is quite difficult to obtain complete regeneration of each resin and therefore some cation and anion contamination of the treated water will occur. Natural groundwaters contain appreciable amounts of long-chain aliphatic acids, for example, humic and fulvic acid, and these are known to foul conventional strong base anion-exchange resins. In recent years, anion-exchange resin synthesis methods have been improved and more modern materials possess a structured matrix rather than a polyelectrolyte gel backbone. The advantage of "macroreticular" and "isoporous" anion resins lies in the relative ease of regeneration of associated organic pollutants.

Ion-exchange is used widely in condensate polishing, and the application of mixed-bed ion-exchange columns in high-pressure boilers (conventional and nuclear) is now widespread. Sorption of impurities at concentrations of 50 ppb and less is customary and this can be achieved with considerable elegance. The use of powdered ion-exchange materials as precoats on conventional candle filters enables condensate treatment to be performed at high flow rate and high efficiency.[2]

Ion-exchange resins also are used widely in effluent treatment and pollution control. Process strategy depends entirely on the waste to be treated, concentration of pollutants, flow rate, and so on, and applications are extremely diverse. The treatment of drainage water from mines, removal of ammonia and nitrates from groundwater, and the treatment of nuclear waste solutions are some typical examples of present-day processes. The efficiency of waste treatment is strongly dependent on the regenerant consumption and thus it is usual to attempt treatment processes with either weak acid or weak base exchangers. Success is likely if the process solutions are either acidic or alkaline. Ion-exchange resins can be used for the removal of noxious gases from gas streams. For example, H_2S and NH_3 have been removed using macroreticular carboxylic acid resins and quaternary ammonium anion-exchange resins, respectively. The selective removal of these two impurities in hydrogen-cycle gas streams from oil refinery processes and their subsequent recovery by thermal elution using an inert gas are of particular value.

13.2-3 Sugar Processing

Practically all sugar that is milled and refined is produced from either sugar cane or sugar beet. The term sugar is used generally as a synonym for the crystalline sucrose derived from either cane or beet. However, other sugars are used in the food and beverage industry and the most widely used alternative is glucose (dextrose), a monosaccharide derived from corn starch.

The principal applications of ion exchange in the purification and treatment of sugar solutions, juices, and syrups are as follows:

1. Softening and demineralization of sugar juices to remove scale-forming elements prior to evaporation.
2. Decolorization using anion-exchange resins.
3. Catalytic inversion of sucrose to fructose and glucose.
4. Glucose–fructose separation.

In general, the largest field of application for ion exchange in sugar processing is in the sugar beet industry. Sugar syrup contains significant amounts of calcium and magnesium salts, and these can be exchanged with a conventional cation-exchange resin in the sodium form. This prevents scale formation in pipelines and evaporators. Deionization of the sugar syrups using both cation- and anion-exchange resins also reduces the molasses, and thus the sugar yield is increased. Ion-exchange resins are used for decolorizing sugar solutions on a wide scale. Generally, an ion-exchange resin column follows an alternative adsorption column, for example, granular carbon, powdered carbon or bone char. Rarely are resins used exclusively for decolorizing due to irreversible fouling, although in recent years, macroreticular and isoporous strong base anion-exchange resins have been used without prefilters. Resins have several advantages over granular carbonaceous adsorbents as decolorizing agents. High capacity for color retention, rapid equilibration, low rinse water requirement, and in situ regeneration are some of the important factors.

The inversion of sucrose can be catalyzed by a cation-exchange resin in the hydrogen form. The resin acts as a heterogeneous catalyst and its effectiveness depends on the chemical nature, degree of crosslinking, and particle size of the ion exchanger, reaction temperature, and contact time. A strong acid cation-exchange resin containing sulfonic acid groups in the hydrogen form and possessing a relatively low degree of crosslinking is a more effective catalyst than, for example, a weak carboxylic acid type exchanger.

13.2-4 Pharmaceutical and Medical Applications

Ion exchange is used extensively in the fields of pharmaceutical manufacture and medicine. Some of the more important applications are (1) processing of pharmaceuticals, (2) use of ion exchangers in pharmaceutical formulations, (3) use of ion exchangers and related materials in artificial organs, and (4) analytical uses in medicine.

The separation of antibiotics from fermentation broths is an interesting application for ion exchange. Streptomycin is a high-molecular-weight, water-soluble organic base containing two guanidine groups and a glucose amine group. The former are strongly basic and the latter is a weakly basic group. Therefore, streptomycin is susceptible to ion exchange with cation-exchange materials and can be sorbed from solution under neutral pH conditions:

$$2 \text{ R·COO}^- \cdot \text{Na}^+ + (\text{streptomycin})^{2+} \rightarrow (\text{R·COO})_2 (\text{streptomycin}) + 2\text{Na}^+$$

It is customary to use a weak acid ion-exchange resin in the sodium form (e.g., a polyacrylic carboxylic resin) and regenerate with hydrochloric acid. Neomycin, a related antibiotic, is also recovered with a carboxylic acid ion exchanger. However, since the basic amine groups in neomycin are considerably weaker than the guanidine groups in streptomycin, it is possible to elute the neomycin with ammonia solutions at pH values high enough to suppress completely the ionization of the neomycin.

Vitamin B_{12} is produced by microbial fermentation and can also be separated from the broth using a carboxylic acid exchanger. It is adsorbed and eluted by a different mechanism, since vitamin B_{12} is essentially a nonionic compound. It does sorb onto the acid form of the ion-exchange resin at pH 3, and after treatment with HCl to remove impurities the vitamin B_{12} product is eluted with an acid–acetone–water solution. Various polymeric adsorbents, that is, macroreticular structured hydrocarbons containing no ionogenic functional groups, have also been found to be effective for the separation of vitamin B_{12} from fermentation broths.

In medicine, ion-exchange materials have found use as preparative media and for various clinical treatments. In the collection of blood, citrate and dextrose solutions are added to prevent coagulation. However, these substances do not influence long-term stability. If the blood is passed over the sodium form of a cation-exchange resin to remove calcium and magnesium, then coagulation can be prevented. Treatment of blood to remove intentional or accidental overdoses of drugs can be effected by hemodialysis. Traditional hemodialysis treatment involves the use of selective membranes, although ion-exchange resins offer the advantage of rapid kinetics and high surface area. It is obvious that the treatment of blood over ion-exchange resins and polymeric adsorbents (hemoperfusion) has considerable advantages. Medical applications are too extensive to be reviewed here and exhaustive detail can be found elsewhere.[3]

13.2-5 Hydrometallurgy

A list of metals that have been recovered and purified commercially by ion exchange is given in Table 13.2-1. In some cases the scale of operation is relatively small, for example, the rare earth elements, the transuranic elements, and the platinum group metals, although the intrinsic value of metal recovered is usually extremely high. Ion exchange is particularly suitable for high-cost, low-throughput purification processes. Separately, the recovery of trace amounts of metal from effluent and waste streams accounts for many applications, for example, chromium from spent metal plating solutions and copper and zinc from wastes arising in the rayon and synthetic fiber industry. However, the largest single application in hydrometallurgy is the recovery and concentration of uranium from naturally occurring mineral ore bodies. At the present time there is a recession in uranium demand and worldwide production is at a low level. About 20,000 tonnes/year of uranium concentrate are produced as yellow cake (ammonium diuranate) in the Western World. Principal producers are the United States, Canada, and South Africa. The most cost-effective recovery processes involve the separation of uranium as a by-product of other minerals or metals. Typical examples are the recovery of by-product uranium from gold and the recovery of uranium during the refining of wet process phosphoric acid. Fixed-bed ion-exchange plants were installed to meet the early demand for uranium in the nuclear industry, although more recently liquid extraction using liquid ion-exchange reagents has found acceptance in some mining locations. Uranium plants are typically capable of processing 100–1000 m^3/h of pregnant solution and it is the uranium industry that has been responsible for many innovations that have occurred in the development of ion-exchange technology in recent years. The foremost equipment development is continuous countercurrent ion exchange in multistage fluidized beds. This novel technology has been advanced most vigorously in South Africa and several large plants have already been erected[4].

Ion exchange is particularly advantageous for the treatment of low-grade uranium ore deposits. A typical example of one such application is the Rossing uranium mine in Namibia. Here, the mineral contains about 0.035 wt.% U_3O_8 on average. The Rossing flowsheet is given in Fig. 13.2-1 and incorporates both ion exchange and liquid extraction and employs the considerable advantages of both techniques. Continuous ion exchange is used to upgrade the low concentration feed (0.15 g/L U_3O_8) and to produce an eluate feed of constant composition containing 3–4 g/L U_3O_8. This eluate is the feed solution to a liquid–liquid extraction plant that performs the effective separation of uranium from trace impurities. The final strip solution contains about 10 g/L U_3O_8 and after precipitation and calcination produces a product containing approximately 97% U_3O_8 by weight.

The combination of ion exchange and liquid extraction is not new in the uranium industry and many early plants in the United States have used this technique. Merritt[5] has written a comprehensive review of the extractive metallurgy of uranium with particular reference to North America. The Eluex plants operated in the United States were essentially similar to the Rossing plant described above, but the ion-exchange equipment was far less elegant.

Continuous ion exchange also facilitates the separation of metals in high-concentration aqueous solutions. A process was developed in South Africa to recover spent hydrochloric acid from pickle liquors. The Metsep process used three continuous countercurrent fluidized-bed columns to separate zinc and iron (Fe^{2+}) in hydrochloric acid prior to recovery in a pyrohydrolysis reactor[6]. The flowsheet for the process is given in Figs. 13.2-2 and 13.2-3. Separation is possible with feed streams containing about 20 g/L zinc, 120 g/L iron, and 30 g/L hydrochloric acid. A special weak base anion-exchange resin was used with a specific gravity of about 1.2 g/cm^3. This was vital, since the feed solution density was about 1.1 g/cm^3. Selective separation of the chloride complexes of Zn^{2+} and Fe^{2+} is relatively easy since the latter does not

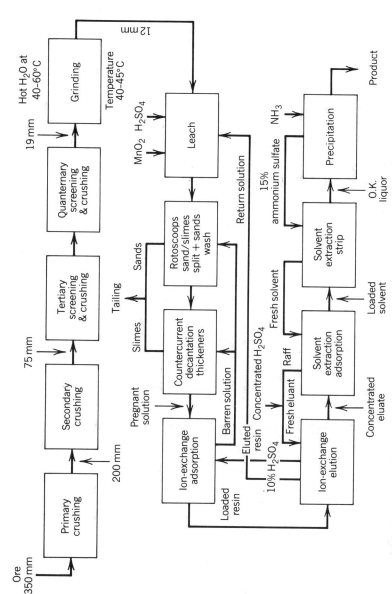

FIGURE 13.2-1 Simplified flowsheet of the uranium recovery circuit at the Rossing Uranium Mine in Namibia.

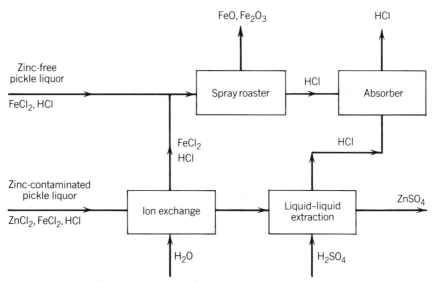

FIGURE 13.2-2 Simplified flowsheet of the Metsep process.

form stable anion complexes at low acid concentration. The removal of zinc from the acid stream is achieved by ion exchange and this can be recovered subsequently by liquid extraction using a cationic extractant (di-2-ethyl hexyl phosphoric acid), thereby converting zinc chloride to zinc sulfate as a fertilizer additive.

It has become common practice to recover gold from low-concentration side-streams with activated carbon. Conventional multistage fluidized-bed columns are used for the sorption process, although the regeneration step is more difficult and requires high-temperature elution and thermal reactivation of the adsorber. Recently, there has been renewed interest in the application of ion-exchange resins for gold recovery from cyanide liquors. Resins are less susceptible to poisoning by calcium or organic impurities and possess higher sorption capacity and selectivity for gold.[7] Pilot-plant work has been performed at MINTEK (Council for Mineral Technology, Randburg, South Africa) using a continuous countercurrent

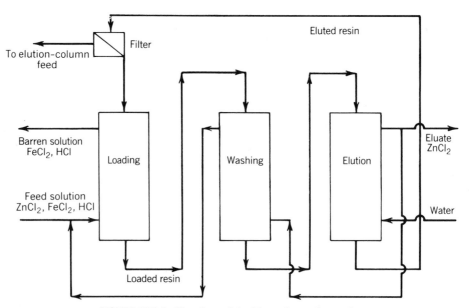

FIGURE 13.2-3 Flowsheet of the Metsep ion-exchange plant.

fluidized-bed column to recover gold from clarified leach liquors with a feed concentration of about 2 ppm and to yield a barren solution containing less than 2 ppb gold. This is achieved with both weak and strong base resins, although the former will probably prove easier to elute in practice and thus provide a cheaper recovery process.

13.3   EQUIPMENT FOR ION EXCHANGE

13.3-1   Principles of Equipment Design

The main variable in the design of an ion-exchange plant is the configuration of the resin while it reacts with the aqueous feed solution. In the past, only fixed beds were considered for most ion-exchange processes, whereas modern technology allows the designer to choose between fixed beds, moving packed beds, fluidized beds, and stirred tanks. The amount of suspended solids in the feed solution determines which of these possibilities are feasible. Packed beds of resin act as efficient filters and rapidly block if suspended solids are present, whereas fluidized beds can handle up to several hundred parts per million of fine solids. Stirred tanks can handle slurries containing up to 20 wt.% solids provided that no particles are larger than the resin beads.

The size of the plant is a further factor that may lead the designer to consider other options besides the conventional fixed bed. A very small plant may be comprised of overdesigned fixed beds to simplify operation and require no instrumentation. On the other hand, the use of continuous ion exchange may significantly reduce the capital cost of a large plant.

The above mentioned factors, combined with the fact that high flow rates obtainable in packed beds give faster reaction rates, have led to the development of present-day plants. In general, clear dilute solutions are treated in fixed or moving packed beds while turbid effluents and ore slurries are handled in fluidized beds or stirred tanks.

PLANT DESIGN PROCEDURE
The basis of ion-exchange plant design is the superficial velocity of liquid feed passing through the vessels, sometimes termed the "empty-bed velocity". Once the type of resin bed—either fixed, fluidized, or stirred—has been selected, a design velocity from the feasible range for that type must be assumed.

The cross-sectional area of the vessels is calculated directly from the specified total throughput and the superficial velocity. If this is larger than current practice, then parallel streams must be considered.

The pressure drop over the resin bed may be an important variable in fixed beds while in other techniques it is not relevant in comparison with pressure losses incurred in pumping liquid through the pipework. The superficial velocity is one of the main variables in the pressure-drop and bed-expansion calculations.

The change in concentration of the solution passing through a resin bed fluidized at the velocity assumed above can be calculated from correlations of reaction rate or mass transfer coefficient. The number of beds and bed depth can be derived.

The type of resin pumping or transfer system to be used is a major decision in ion exchange since resin is both fragile and expensive. Other auxiliary systems such as methods of detecting resin levels and flows, materials of construction and control systems must also be selected.

13.3-2   Flow Through Fixed Beds of Resin

The size of resin beads and the pressure drop generally used in industrial applications result in laminar flow conditions in fixed beds of resin. The characteristic of such flow is that the pressure drop varies linearly with flow rate. Actual measurements of pressure drop over beds of resin include the effect of flow distribution manifolds which may show quadratic dependence of pressure drop against flow rate.

The two most important factors influencing the pressure drop over a bed of resin are the size distribution of the beads and the voidage of the bed. Modern resins are spherical in shape due to their method of manufacture, but older resins were sometimes granular. Many manufacturers publish data on pressure drop as a function of flow rate for their resins at stated water temperatures. These data can also be obtained readily in laboratory experiments or estimated from the particle size distribution of resin and the appropriate solution properties.

The pressure drop for laminar flow through a fixed bed of particles can be predicted by the Carman–Kozeny equation as expressed in the following form:

$$\frac{\Delta P}{L} = \frac{K_1 S^2 \, \mu \, (1 - \epsilon)^2}{\epsilon^3} \, U \qquad (13.3\text{-}1)$$

where U is the superficial or empty-bed velocity and is equal to the volumetric flow rate divided by the cross-sectional area without resin present, and L is the depth of the bed. The fraction of bed volume

TABLE 13.3-1 Typical Data for Pressure Drop over Packed Beds of Resin[a]

Type of Resin	Amberlite IR-120 Cation		Amberlite IRA-420 Anion	
Pressure drop[b] $\Delta P/L$ (kPa/m)	22.62	5.66	22.62	22.62
Flow rate[c] U (m/h)	20.78	24.45	12.23	22.01
Water temperature (°C)	5	100	5	32
Water viscosity μ (Pa·s)	1.5×10^{-3}	0.29×10^{-3}	1.5×10^{-3}	0.76×10^{-3}
Constant[d] K_2 (kPa·h/m^2)	1.09	0.232	1.85	1.03

[a]Resin manufacturers usually give their data as graphs of pressure drop (psi/ft) against flow rate (gal/min·ft^2) for various temperatures. The conversion factors are: (psi/ft) × 22.62 = (kPa/m): kilopascal per meter of bed depth (gal/min·ft^2) × 2.445 = (m/h): superficial bed velocity in meters per hour
[b]Pressure drop is correlated by $\Delta P/L = K_2 U$.
[c]Maximum quoted value for flow rate is about 244 m/h (100 gal/min·ft^2) causing a pressure drop of about 226 kPa/m (10 psi/ft).
[d]The constant K_2 may be estimated at another temperature by multiplication of the ratio of water viscosities at the two temperatures.

occupied by voids between resin particles is given by ϵ and S is the specific surface area of the resin beads and is equal to the total surface of the beads divided by their volume. The dynamic viscosity of the liquid, μ, will change with the temperature and nature of the dissolved solids. The constant K_1 is equal to 5. Equation (13.3-1) reduces to the very simple form below for any given bed of resin and liquid viscosity:

$$\frac{\Delta P}{L} = K_2 U \tag{13.3-2}$$

Therefore, a single measurement of pressure drop ΔP for a bed of depth L with a flow rate U of liquid having a known viscosity enables one to calculate the appropriate value of K_2. The value of K_2 can equally well be read off the graphs or tables of pressure drop given by manufacturers, as shown in Table 13.3-1.

13.3-3 Flow Through Fluidized Beds of Resin

Upward flow of sufficient velocity causes a bed of resin to expand and fluidize unless it is restrained at the upper surface. The resin particles move freely relative to one another during fluidization since they are no longer in contact, and the upper level expands as the liquid velocity increases. Fine particles are carried upward out of the bed. The pressure drop over the fluidized resin is marginally larger than the static head of liquid since the density of resin is only slightly greater than water.

Even flow distribution across the complete cross-section of the bed is the most important aspect of design. This can be achieved by flow distribution using pipe manifolds with carefully sized orifices. A plate with perforations or bubble caps to distribute flow can also be used to support a bed.

The expansion of a fluidized bed of resin is an important variable in determining how much resin will be held in a vessel. This property of a resin is measured very readily in a simple laboratory experiment. For uniform bead size, correlation of data can be done with the Richardson–Zaki equation expressed in the form

$$\frac{U}{U_t} = \epsilon^n \tag{13.3-3}$$

where U_t is the terminal settling velocity of typical resin beads and n is a constant with values between 2.8 and 3.5 for resin beads. The same data can be expressed more conveniently in the form of the ratio of bed depth at one velocity to that at another velocity. The effect of viscosity is very significant since a change of 1°C will change the viscosity by about 3% and noticeably affect the resin inventory of a stage.

Resin manufacturers often express the fluidization properties of their resins in terms of the percentage expansion of the bed related to its packed depth. This expansion cannot be calculated directly from the above equation since a range of bead sizes is present. Typical data are shown in Fig. 13.3-1, taken from a Rohm and Haas pamphlet. These curves can be correlated by an equation giving the expansion as a function of the 1.5 power of the flow rate. An equation of almost identical form can be derived from the Richardson–Zaki equation by expressing the expansion in terms of the void fraction and using appropriate values for n.

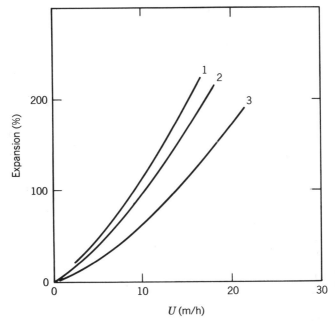

FIGURE 13.3-1 Expansion of resin in a fluidized bed. Type of resin: Rohm and Haas Amberlite IRA 400 UC, sulfate form, density 1130 kg/m³; size range: Minus 0.84 mm, plus 0.42 mm; curve 1: solution density 1037 kg/m³, temperature 15°C; curve 2: solution density 1037 kg/m³, temperature 24°C; curve 3: solution density 1000 kg/m³, temperature 24°C; curves correlated by: bed expansion = $100 (L - L_0)/L_0$ = $100 (k_3 U^{1.5})$, where L_0 is the settled bed depth.

13.3-4 Flow of Resin Slurries

The flow of resin slurries is an extremely important aspect of the design of ion-exchange plants since all movement of resin both in continuous and batch plants is done using some form of hydraulic transport. Resin beads are also comparatively fragile and attrition during transfer operations must be avoided. Resin is crushed by being trapped in closing valves, by passing through positive displacement pumps, by passing through regions of high shear such as the impellers of centrifugal pumps or high-speed agitators, and by water hammer from valves opening or closing too rapidly.

The two common methods of moving resin are by using airlifts between open tanks and by displacing resin hydraulically in dense-phase flow between closed vessels. In both cases the pipeline velocities are relatively low and resin is not broken. An early type of continuous ion-exchange plant, the Dorr Hydro-softener,[1] used liquid jet ejectors to move resin slurries but no data are available on the effect of this operation on resin life. The Wemco recessed impeller pump is used on a very large ion-exchange plant recovering uranium[2] to the satisfaction of the operators. Resin particles do not pass through the impeller.

The variables in the use of hydraulic displacement systems for resin transfer are shown in Fig. 13.3-2. It is obvious that the pump handles only clear liquid which is forced into the top of a pressurized tank to move the settled resin bed as a dense slurry out of a pipe near the base of the tank. Systematic studies done on this phenomenon showed that the slurry moved along the pipe as closely packed as in the settled bed itself. The voidage of the delivered slurry was proved to be constant. This technique can be used for accurate control of resin flow rates since the flow rate of slurry is a constant fraction of the flow rate of displacing liquid.[3]

Hydraulic displacement is used in the Higgins technique, which includes flow through vessels up to 1 m in diameter.[4] Typical line velocities are of the order of 1 m/s. The pressure drop for clean resin and clear water is marginally greater than that for water at the same flow rate but accumulation of tramp rubbish such as wood chips or fine sediment can form blockages at low points in pipelines. Plant practice is to install centrifugal pumps with twice the static head requirements of the delivery line to drive the displacement liquid.[2] Pipelines of lengths up to several hundred meters have been used to carry resins. The installation of pipe tees with valves at low points in the line facilitates clearing out blockages. An important aspect of the flowsheet shown in Fig. 13.3-2 is that resin slurry passes through a number of valves that

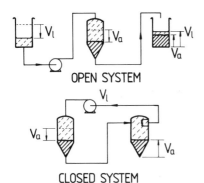

OPEN SYSTEM

CLOSED SYSTEM

FIGURE 13.3-2 Dense-phase resin transport. V_a, bulk volume of resin transported; V_l, volume of transfer water handled by pump; $V_a/V_l = 0.9$ for resins.

open to allow passage. However, control interlocks must prevent these valves from closing until the line has been flushed clear of resin slurry.

Airlifts are simple devices for pumping liquids or slurries through low heads and have been used in hydraulics for many generations. Their throughput is sensitive to the submergence and to the flow rate of air. The output pulsates and cannot be controlled accurately. The air injection pipe must end downward-facing to avoid settlement and must always have a small bleed of air to keep the end open. The formation of bubbles around the air injection point causes severe local abrasion of the pipe in slurry systems which must be reinforced with abrasion-resistant lining.

The performance of an airlift can be derived from hydrostatic principles in terms of the submergence of the air injector below the liquid level in the feed tank (h_s), the delivery height above this level (h_r), and the pressure and density of air at ambient atmospheric pressure (P_a and ρ_a). The atmospheric pressure is expressed as an equivalent height h_a of a liquid of density ρ. The following equation gives an estimate of the mass of air required per unit mass of liquid (or slurry) ignoring frictional losses in the system:

$$\frac{\text{kg air}}{\text{kg liq}} = \frac{\rho h_r g}{P_a \ln \left(\dfrac{h_s + h_a}{h_a} \right)} \tag{13.3-4}$$

13.3-5 Fixed-Bed Equipment

The simplest type of fixed bed is shown in Fig. 13.3-3 which is illustrative of a modern large plant for water purification. The vessel is of a conventional dished-end pressure construction with supports and an access manhole. Construction materials for the vessel are usually mild steel with a lining of rubber or plastic. Stainless steel is sometimes used and smaller vessels are made of reinforced plastic. The depth of settled resin bed is seldom more than about 1 m but the vessel must allow freeboard of about another meter above the upper surface of resin for expansion during fluidization to clean rubbish out of the bed.

The normal direction of flow in such a bed is downward so that the resin acts as a packed bed without any relative movement of particles. The bed of resin can be supported by various means. The simplest is to fill the vessel to cover the lower dished end with coarse sand in which distributor manifold pipes are buried. This can create problems because the volume of liquid trapped in the voids of the sand can cause cross-contamination between cycles of operation. An internal false floor with distributor cups of plastic screwed into place on a grid pattern overcomes this problem. Such construction is more expensive and has to be supported structurally to take the entire pressure thrust on the bed during flow. Some plants use an elaborate manifold of distributor pipes laid against the lower dished end which is itself filled with resin.

Regeneration of resin takes place in the same vessel and can be done downward using the same distributors. However, it is more efficient to regenerate the resin using flow countercurrent to that in normal service. This involves feeding eluant upward through the bed while avoiding fluidizing the resin. Various methods of preventing fluidization include the use of collector pipe manifolds just below the upper surface and placing rubber bellows either above the bed or submerged in the resin to prevent movement during the upward flow of liquid.[5]

The upper surface of the resin bed acts as an efficient filter for any fine particles present in the feed during normal service. It is therefore necessary to backwash or fluidize the resin with clear water to remove these solids to waste at the end of a service cycle. The flow distributor at the base and a collecting manifold at the top of the vessel are used in this operation.

A common type of process in water purification involves the use of a mixed bed of cation- and anion-exchange resins to deionize water, often following separate cation- and anion-exchange beds. The use of such mixed-bed columns enables very low levels of impurities to be attained. Regeneration of such columns

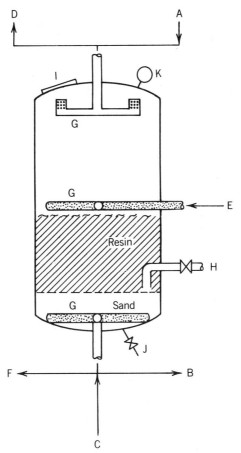

FIGURE 13.3-3 Typical fixed bed. A, feed; B, effluent; C, backwash supply; D, backwash overflow to resin trap; E, eluant supply; F, spent eluant; G, distributor manifolds; H, resin removal line; I, access hole; J, drain; K, pressure gage and vent.

can be done by making use of the different densities of anion- and cation-exchange resins. After fluidizing the bed for some time with water, the anion-exchange resin forms a distinct layer on top of the cation-exchange resin. The two layers are allowed to settle and are regenerated in sequence. First, alkaline solution is fed downward from a distributor just above the top of the bed and removed by a distributor near the interface of the two layers. Second, a short rinse removes most of the alkaline solution; this is followed by acidic regenerant being fed at the interface and taken off below the bed. Third, the bed is remixed with air agitation after the residual acid has been rinsed out.

At least two fixed beds are required to give continuous service of purified water but three are normally used for recovery processes. This ensures that barren effluent is always below the limit and that each bed is fully loaded with product before regeneration takes place. Automatic control of a fixed-bed plant regulates the times of the various cycles and controls feed rates at predetermined values. For fully automatic control of cycles, up to six automatic shut-off valves per vessel are required.

13.3-6 Moving-Bed Equipment

One class of semicontinuous ion-exchange plant is based on moving a packed bed in counterflow to the normal service flow of solution. It is always necessary to shut off normal flow to achieve bed movement. Hydraulic displacement is used to move the bed with either a piston or pump driving the displacement water. A number of different designs have been suggested in the literature but only two have achieved commercial success, that is, the Higgins technique and a design suggested by Porter and developed and marketed by Asahi Chemical Company of Japan.

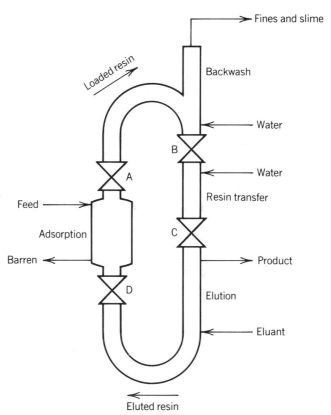

FIGURE 13.3-4   Higgins moving packed bed.

The Higgins loop, developed by I. R. Higgins and Roberts[4] at Oak Ridge in 1954, is illustrated in Fig. 13.3-4 which shows a vertical sketch of a typical plant incorporating an extraction or treating section through which the main liquid feed passes in downflow. This section is connected with a large-diameter pipework as a continuous loop around which resin circulates in increments during resin transfer operations. There is therefore a net flow of resin from extraction to backwash to regeneration to rinse and again to extraction. During normal service resin flow exists as a packed bed in the extraction and regeneration sections which are isolated from one another by valves A, B, C, and D. During resin transfer, external flows of liquid are valved off and the main loop valves A, C, and D are opened to allow resin to be moved around incrementally by the action of displaced water. Valves A, C, and D are closed and the slug of resin is backwashed before opening valve B to let it settle into the pulse section.

The basic idea behind the Asahi design was suggested by R. Porter in 1956.[6] Since then the concept has been developed by the Asahi Chemical Company in Japan to handle rayon processing effluent by recovering copper and the system has also been installed worldwide for standard water treatment operations.[7] The extraction or treating section is comprised of a packed bed with upward flow which pins the bed to the top of the vessel during normal operation, as shown in Fig. 13.3-5. At the end of the service cycle, the feed flow is valved off and the bed falls to the bottom of the vessel, drawing a supply of regenerated resin through a nonreturn valve into the top of the vessel. When normal feed flow is resumed, that part of the bed above the feed distributor is pinned again to the top of the vessel while that below is displaced downward and out of the vessel to the top of a backwash column where it is fluidized to remove fines. Resin settling to the base of the backwash column is fed to a regeneration vessel which operates in a manner similar to the extraction column.

Normal service flow in both these designs takes place through a packed bed of resin which is shallower than conventional fixed beds. Very high flow rates are therefore possible, allowing use of vessels smaller than conventional designs.

13.3-7   Fluidized-Bed Equipment

Fluidized beds of resin have always been used for backwashing in ion-exchange practice. The use of fluidized beds as a means of contacting fluids with particles has been applied since the 1940s and 1950s.

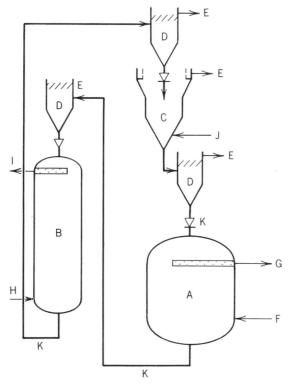

FIGURE 13.3-5 Asahi moving packed bed. A, adsorption section; B, elution section; C, fluidized resin backwash; D, resin collection hoppers with screen top and nonreturn valve outlet for resin; E, transfer and backwash water overflow; F, feed; G, barren effluent; H, eluant; I, eluate product; J, backwash supply; K, resin flow.

Fluidization of particles with density only slightly greater than the liquid does not create the large bubbles characteristic of fluidization with gases; however, other fluidization phenomena such as movement of particles and expansion of the bed take place.

A number of designs using fluidization to promote effective contact between liquid and resin particles have appeared in the literature and have been applied on laboratory or pilot-plant scale. The main problem common to all such ideas is that of transferring resin countercurrent to liquid flow in a manner such that both resin flow rate and resin inventory in each stage are controlled. The advantage of fluidization lies in the fact that the bed does not filter out fine particles but allows them to flow out with the liquid product. Also, evolved gas does not accumulate in a fluidized bed. The other advantage of fluidized over packed beds is negligible pressure drop, but this can be negated by the need to use much lower liquid velocities leading to a larger plant.

The first design based on fluidized beds to contact resin with solution was suggested by the Dorr Company as the Dorrco softener[1]. This used a column comprising a series of trays on which fluidized beds of resin were supported. Liquid flowed continuously upward through the beds while resin was drawn from each bed and fed to the one below using a liquid ejector driven by product water. Transfers of resin from extraction to regeneration columns were also carried out with liquid ejectors.

Another principle of operation has been described by Cloete and Streat[8] and George et al.[9] of the U.S. Bureau of Mines working independently. This techinque relies on periodic cutoff of liquid flow through a series of fluidized beds which allows the beds to settle onto trays. The entire content of the column then flows by gravity or is pumped in a reverse direction for a distance of a few hundred millimeters: this operation causes a part of each bed to be transferred simultaneously to the bed below. Upward normal service flow is then resumed and the series of fluidized beds is re-established. Various related ideas have been published: for example, the suggestion by McNeill et al.[10] of using slow pulsations to cause resin flow, and the idea of stopping liquid flow to allow reverse resin flow by sedimentation as suggested by Levin in a confidential report in 1955 at the National Institute for Metallurgy in Johannesburg, South Africa. The Cloete–Streat principle is illustrated in Fig. 13.3-6 which gives the basic flowsheet for a plant recovering uranium from a dilute leach solution. Liquid feed containing about 200 ppm uranium flows upward through

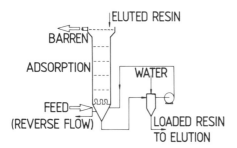

FIGURE 13.3-6   Cloete–Streat fluidized bed.

the 4.85 m diameter column at a flow rate of 332 m^3/h for a forward flow period of 4.5 h. At the end of this period the feed is shut off for a settling period of 5 min and then reversed for 5 min, during which time the top level falls by 400 mm and about 6.3 m^3 of resin is transferred downward between each of the 12 stages. Forward flow is then re-established and the slug of 6.3 m^3 of resin in the conical base of the column is displaced out to the regeneration column by the transfer pump. Thus, the average flow rates are 320 m^3/h of leach solution and 1.35 m^3/h of resin. Each tray contains a fluidized bed of 9.25 m^3 resin expanded to twice its settled level by the flow rate of feed at a superficial velocity of 18 m/h. flow is distributed by 1945 holes of 12 mm diameter in the tray, giving an open area of 1.2% of the tray area.

The first commercial plant of this type was constructed in Germiston, South Africa for separating zinc and iron from spent hydrochloric acid pickle liquor.[11] Subsequently, much larger units have been built in South Africa for recovering uranium at the following mines: Randfontein Estates, Blyvooruitzicht, Chemwes, Harmony, and Vaal Reefs.[2] The same type of equipment based on a 3 m diameter column has been recovering traces of gold from 528 m^3/h of effluent mine water at Welkom gold mine in South Africa using activated charcoal instead of ion-exchange resin. The feed contains 30 ppb gold and about 5 kg of gold per month is recovered.[12]

The studies on the basic principles of engineering and process design were done at Imperial College in England;[13] subsequent demonstration on pilot plants was done by industrial companies in England,[14,15] by the U.S. Bureau of Mines at Salt Lake City,[9] and by the National Institute for Metallurgy (now MINTEK) in South Africa.[16]

A further variation on the use of fluidized beds to contact resin with solution is that adopted by Himsley.[17,18] This is illustrated in Fig. 13.3-7 showing an elevation sketch of a typical plant. Liquid feed flows continuously upward through a series of stages in a column. All except one of the stages contain fluidized beds of resin. Batches of resin are transferred sequentially downward by selectively reversing the flow from stage to stage. The bed selected for transfer is effectively collapsed by causing liquid flow reversal, which then displaces the resin downward into the vacant stage below. The batch of resin in the base of the column is transferred out to the top of the regeneration section by the same method. The design of the trays is unusual since there is only one distributor with the shape indicated in Fig. 13.3-7. The successive stage-by-stage flow reversals require a manifold of valves around the transfer pump whose throughput must be larger than that of the feed pump. Screens or cups at the top of each stage prevent resin being sucked through the transfer pump.

A design was developed by Porter specifically for very large flow rates and outdoor construction based on a cascade of open-topped tanks.[19] Solution was pumped from stage to stage in an early design which was based on rectangular concrete tanks. Modern plants use circular rubber-lined steel tanks and elevate each tank slightly so that gravity flow can take place. Liquid flow is continuous, entering the tanks at a low level through pipe distributor manifolds and overflowing into weirs at the top and finally passing over

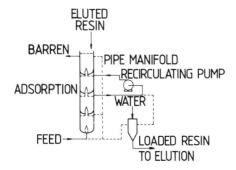

FIGURE 13.3-7   Himsley fluidized bed.

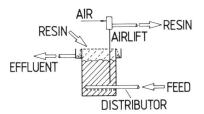

FIGURE 13.3-8 Porter fluidized bed.

screens to collect traces of entrained resin. Resin is transferred in batches from one stage to the next one upstream, which is vacant, by the use of airlifts which draw from near the base of a stage and recycle a slurry of resin and solution into the stage upstream. Fully loaded resin is taken in a batch from the feed stage. A typical stage is illustrated in Fig. 13.3-8.

The largest version of this design was installed at the Rossing Uranium plant in South West Africa (Namibia). The stages are open rectangular tanks made of rubber-lined concrete 3.8 m high with a plan size of 6.2 m. There are four parallel streams handling a total of 3600 m³/h through a cascade of five stages giving a superficial velocity of 23 m/h. Resin inventory per stage is 26 m³, and the total resin inventory in use in the process is about 914 m³. This is the largest uranium plant in the world and probably one of the largest ion-exchange operations of any type.[2]

13.3-8 Stirred-Tank Equipment

The idea of agitating a batch of resin with solution in a beaker to cause reaction is well known for ion-exchange laboratory experiments and is used to study the kinetics of ion-exchange reactions. The use of this concept as a plant technique is limited by the following factors:

1. Agitation must be mild to avoid attrition of resin.
2. A number of tanks in series must be used to effect high conversions.
3. The concentration of resin in suspension is low.

The main application for this technique is in a system where the liquid feed contains fine solids that would block a fixed bed or tend to float resin out of a fluidized bed owing to the high density or viscosity of the feed. The process is used for the recovery of uranium from ore deposits containing a high proportion of clay minerals. Such ores produce a slimy pulp that is expensive to filter and thus there is an incentive to extract dissolved metal directly from the leach pulp. A number of plants were constructed in the 1950s in the United States on this principle—known as resin-in-pulp (RIP) operations. The simplest possible method of agitation, namely, to inject air into the base of the mixing tanks, was used in the early plants.[20]

One of the examples of such a plant is illustrated in Fig. 13.3-9. Air lifts take a slurry of the contents of the mixer to a vibrating screen above the next stage upstream and the airlifts are operated continuously to provide a flow of resin that is countercurrent to the flow of pulp. Very similar equipment is used for the recovery of gold from leached pulps using activated-charcoal granules in the size range 2–4 mm. This is now the preferred technology for gold recovery in new plants in the United States, Canada, Australia, and South Africa.[21]

The flowsheet for gold plants consists of equipment very similar to that used in RIP operations but the equipment differs mainly in the process used for elution of gold under pressure and regeneration of charcoal in a kiln at 600–700°C.

The main problems in design and operation of RIP plants are associated with the screens which are necessary to separate resin (or activated-carbon granules) from the pulp overflowing from each tank and finally to the effluent. These not only block readily with tramp rubbish but they are broken by careless operators and abrade resin if vibrated. Extensive pilot-plant work has been done by Carman[22] and Read[23] to develop a plant with resin floating in the pulp in efforts to develop a screenless RIP recovery process for both uranium and gold.

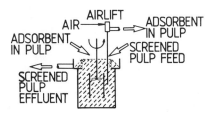

FIGURE 13.3-9 Stirred-tank resin-in-pulp process: similar technology is used for adsorption with activated carbon.

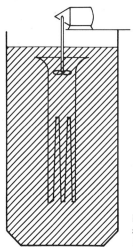

FIGURE 13.3-10 Stirred tank for adsorbent-in-pulp process with resuspension slots in draft tube. Figure reproduced from J. Y. Oldshue, *Fluid Mixing Technology*, McGraw-Hill, New York, 1983.

This far-sighted work was further extended with the concept of an inverted fluidized bed of floating resin in the leach reactors themselves, the resin-in-leach or RELIX process.[24] These ideas have not yet proved successful enough for full-scale plant deployment, so development has concentrated on obtaining high throughput with minimal attrition through screens.

One screen design that has been widely applied in South African ion-exchange plants is the DSM (Dutch State Mines) design based on triangular-sectioned wire mesh. These units require low head, are static, and are strongly built. An alternative screen concept has been suggested by Laxen et al.[25] with somewhat similar proposals from Shoemaker.[26] In both these designs the overflow screen is submerged below the surface of the pulp and pulp is caused to flow tangentially across the screen cloth. Laxen and coworkers achieve this effect by bubbling air across the screen, with a weir behind the screen so that a pulsating flow is caused by the curtain of air bubbles; in Shoemaker's design the surges of pulp resulting from the agitation of the tank cause the same effects. Pulp flow velocities through the screen at a superficial rate of 120 m/h can be achieved in practice with the design of Laxen and coworkers. The use of slow pulsing to clear screens in an RIP plant is described by Arden et al.[27]

The use of air for agitation in pulp systems is very well known in the mineral processing industry. The energy used for mixing can be reduced by the use of mechanical agitation, which is generally more efficient, but at the cost of increased capital. An illustration of the type of plant used in the recovery of gold from leach pulp is given in Fig. 13.3-10 taken from Oldshue.[28] This plant uses draft tube agitators with provision for restarting from shutdown. The slots cut into the draft tube follow a suggestion from Australia. The agitator impeller itself is mounted in the top part of the tube so that it remains above the level of settled pulp.

The design of mechanical agitators is based on predicting the energy required just to suspend all solids off the bottom of the tank. Correlations for predicting this value exist for most types of agitator but cannot be relied on since they often give contradictory results, as pointed out in a recent book by Oldshue.[28] Successful plants have therefore been based on extensive pilot-scale testwork. The effect of the agitator on the attrition rate of resin or carbon must also be considered.

13.4 RECENT DEVELOPMENTS IN ION EXCHANGE

13.4-1 Selective Ion-Exchange Materials

Conventional ion-exchange resins have been discussed in detail in Section 13.1-1 and these materials must represent in excess of 90% of all ion-exchange resin used in traditional applications such as water treatment, sugar processing, and purification of pharmaceuticals. However, there is a growing interest in the development of selective polymers for application in a wide range of specialty operations ranging from the recovery of trace uranium in seawater to the separation and purification of the platinum group metals in hydrometallurgy. Research and development is most exhaustive in hydrometallurgy, especially in the field of precious metals recovery, platinum group metals separation, and base metal recovery. It is interesting to note that separations in the first-row transition metals have been achieved with selective liquid extractants consisting of nitrogen- and oxygen-containing ligands, for example, hydroxyoximes. However, these elements will form weak anionic complexes, especially in chloride media, and this is also true of the second- and third-row transition metals; such characteristics make it possible to carry out separations among the precious metals, that is, gold and silver, and the platinum group metals. There is a considerable economic incentive to develop solid-phase ion-exchange processes for the recovery of precious or alternatively toxic metals, such as mercury, from low-grade dilute solutions. Therefore, the literature concerning the development of selective ion-exchange polymers is expansive and cannot be fully discussed here. Recently, Warshawsky[1] presented a comprehensive review article that discusses the trends in the synthesis of selective chelating polymers containing oxygen, nitrogen, phosphorus, and sulfur ligands.

CHELATING RESINS WITH NITROGEN-CONTAINING PENDANT GROUPS

Tertiary and quaternary polymeric amines are produced in large quantities and are effective for some specialty separation processes. The most notable is the recovery of uranyl sulfate or uranyl carbonate from acid or alkali leach solutions. Some selectivity is also found among the first-row transition-metal chloride complexes. The incorporation of a chelating ethyleneimine group—$(NH-CH_2-CH_2-NH)_n$—increases markedly the stability of the polymeric ligand–metal complex. For example, crosslinked chloromethylated styrene divinylbenzene copolymers containing di, tri, or tetraethylene imine groups show high affinity for Au(III), Hg(II), and Cu(II). Jones and Grinstead[2] have prepared more complex ethylenediamine derivatives and suggested their use for Fe(II) and Cu(II) separation.

CHELATING RESINS WITH NITROGEN- AND OXYGEN-CONTAINING PENDANT GROUPS

Amino-diacetic acid ion-exchange resins have been available for commercial use since the early 1960s. Resins of this type are useful for the separation of the first-row transition metals, although the selectivity is very dependent on pH. More recently, attempts have been made to synthesize more elegant polymer supports and ligands. Many workers have proposed chelating ligands based on poly(vinylimidazole) or poly(vinylimidazole) dicarboxylate groups.[3] It has been suggested that the loading rates for base metals, Cu(II), Ni(II), and Co(II) are enhanced with this type of structural modification.

Many workers have proposed chelating ligands based on poly(vinylimidazole) or poly(vinylimidazole) decarboxylate groups.[3] It has been suggested that the loading rates for base metals, Cu(II), Ni(II), and Co(II) are enhanced with this type of structural modification.

Hydroxyamine polymers have also been explored extensively. Vernon and Eccles[4] have prepared several hydroxyoxime-type polymers and recommended their use for copper separation from iron at low pH value. Amidoxime-type functional groups are found to chelate uranium at low concentrations in seawater and to offer a potential solution to the problem of separating heavy metals from acidic solutions.

CHELATING RESINS WITH SULFUR-CONTAINING PENDANT GROUPS

Sulfur ligands are known to complex or precipitate most of the heavy transition metals. Early developments in this field arose out of pollution control and analytical applications. Typical polymer derivatives are based on macroreticular polymethacrylate beads containing pendant mercapto groups. Such resins are reactive for Ag(I), Hg(II), and Au(III). Thioglycolate resins have been developed and used for the laboratory separation of Ag(I), Bi(III), Sn(IV), Sb(III), Hg(II), Cd(II), Pb(II), and U(VI). Similar resins are dithiocarbamates and their derivatives. Recently, the platinum group metals (PGMs) have been separated in chloride media using a combined ion-exchange and liquid–liquid extraction process.[5] The adsorption step involves a weak base isothiouronium group capable of extracting the chlorocomplexes of the PGMs. Industrial exploitation of this technique has occurred already.

CHELATING RESINS WITH PHOSPHORUS-CONTAINING PENDANT GROUPS

Phosphoric acid, esters, and phosphine oxides are very effective extractants for uranium, gold, and the first-row transition metals. One commercially available chelating resin contains aminophosphonic groups attached to a crosslinked polystyrene matrix. Although developed for the decalcification of brines, this resin shows high selectivity for the separation of trace amounts of uranium in wet process phosphoric acid.[6] A general review of phosphorus containing polymers is given by Efendiev.[7]

CHELATING RESINS WITH OXYGEN-CONTAINING PENDANT GROUPS AND MACROCYCLIC STRUCTURES

Phenolic ion exchangers derived from a phenol–formaldehyde condensation reaction appeared in the first generation of ion-exchange polymers. More recently, styrene–divinylbenzene copolymers incorporating azo-substituted cresol and salicylic acid, catechol, hydroquinone, and benzoquinone have been described. The quinone-type polymers selectively sorb Hg(III) and the catechol resins sorb Cr(VI).

The complexation of metal salts by neutral macrocyclic ligands is well known.[8] Polymeric crown ethers are an expanding group of functional ion exchangers capable of selective sorption of alkali metals such as K, Cs, Na, and Li. The crown ether may be derived from a conventional chlormethylated hydrocarbon backbone which is converted to a polybenzylated catechol. Crown ethers are highly reversible and possess rapid reaction kinetics, thus allowing for an interesting thermal elution procedure whereby a species is sorbed at 20°C and eluted at 60°C.

SOLVENT-IMPREGNATED ION-EXCHANGE RESINS

The idea of developing solvent impregnated ion-exchange resins was to combine the selectivity and specificity of conventional liquid extractants with the advantages of a discrete polymer support material, thus tailoring adsorbents for a specific separation process, usually in the field of hydrometallurgy. Although it is now possible to functionalize polymers as outlined above, it is still difficult to overcome some steric problems and thus it is interesting to consider the potential use of liquid extractants immobilized within a polymer matrix. This can be achieved by physical impregnation of the reagent onto a polymeric or other porous support without chemical binding of any sort. Alternatively, copolymerization of a monomer (e.g., styrene) crosslinking agent (divinylbenzene) in the presence of a reagent (e.g., tri-*n*-butyl phosphate) will produce a polymer "encapsulated" product. Typical of these products are the Levextrel resins developed by Bayer AG.[9] An exhaustive review of extraction with solvent-impregnated resins has been published by Warshawsky.[10] Selective ion-exchange resins have been prepared by either route and applied to the recovery of uranium from radioactive waste, treatment of aqueous wastes containing nonferrous metals, and certain specialty applications. The principal difficulty in the use of these materials is the slow diffusion of the reagent out of the polymer matrix. Although this can be overcome by reimpregnation, the possible environmental implications and cost would make large-scale commercial use unlikely.

13.4-2 New Types of Ion-Exchange Materials

Ion-exchange materials are normally synthesized in granular form and in most cases spherical particles of a precise size distribution are supplied. This facilitates their use in packed columns or in sophisticated continuous countercurrent contactors. For most applications, the particle surface area and therefore the reaction rate are quite adequate and most processes can be performed efficiently and economically. In some applications, for example, the treatment of unclarified liquors, the sorption of slow-diffusing species, and the use of fluidized ion-exchange particles—it is desirable to modify the properties of the ion-exchange materials. Increasing particle size or the relative density will improve the hydraulics in a fluidized bed, whereas a reduction in particle size might improve kinetics for a very slow-diffusing species. Alternatively, ion-exchange fibers and woven fabrics have been prepared for these and similar applications. Sorption of a solute onto fibers is inherently rapid due to the large surface area of reactive sites that can be exposed at any one instant in time. However, supporting the fibers or fabric and contacting the ion exchanger with the liquid phase calls for novel engineering design. The usual idea involves the use of an endless belt and this has been tried for the recovery of copper from a dilute aqueous solution using phosphorylated cotton toweling.[11] A similar idea involved the polymerization of a quaternary ammonium resin onto cotton cloth and this was used to remove chromate ions from a dilute aqueous solution.[12] Separately, Vernon and Shah[13] have synthesized a poly(amidoxime)–poly(hydroxamic acid) fiber and shown that this will sorb significant amounts of uranium from seawater. It is suggested that this fiber could be produced as an endless belt and thus sustain continuous retrieval of uranium by continuous countercurrent operation.

Electrodialysis requires selective ion-exchange membranes. In the simplest form these are synthetic ion exchangers in the form of sheet or film. Heterogeneous membranes are produced from finely milled ion-exchange particles and formed into sheets with an inert binder. Interpolymeric membranes are made from a homogeneous mixture of two polymer solutions, one forming the polyelectrolyte and the other a film-forming material, for example, polystyrene sulfonic acid and polyacrylonitrile. Homogeneous membranes can be produced which contain ionic groups in the film-forming matrix, for example, methacrylic acid. Ion-exchange membranes can carry either cationic or anionic functional groups and can exclude mobile ions of similar charge by the high concentration of fixed groups in the membrane. Thus, for example, a flow of electric current through a cation membrane is carried entirely by the flux of mobile cations. Ion-exchange membranes are capable of selective ion transport. More information on electrodialysis can be found in Chapter 21.

Specialist ion-exchange resins have been developed in recent years in an attempt to overcome typical problems encountered in ion-exchange process technology. Blesing et al.[14] have described the synthesis of novel magnetic microresins suitable for application in desalination, water treatment, and hydrometallurgy. These resins are usually manufactured in the form of beads, typically in the size range of 100–500 μm and containing about 10–15% by volume of a magnetic material such as iron oxide as an inert core. Reactive sites are produced by shell graft polymerization of organic monomers onto the inert core. Since the discrete particles are small, they react faster than conventional ion-exchange resins, but they can be used successfully in fluidized-bed systems at economic flow rates because the microbeads agglomerate magnetically into large flocs when agitation ceases. These flocs have hydraulic properties similar to conventional ion exchangers. The recent status of continuous ion exchange using magnetic microresins has been presented by Swinton et al.[15]

A slightly different approach to the preparation of composite materials has been tried in an attempt to synthesize high-density ion-exchange materials possessing rapid kinetics and thus suitable for fluidized-bed application in hydrometallurgy, especially for uranium recovery from unclarified solutions. Usually, an inert porous core material such as alumina or silica gel is impregnated with a reactive monomer, for example, a substituted vinyl pyridine, and crosslinking agent such as divinylbenzene. The resultant impregnate is reacted to polymerize the organic reagent within the pores of the inorganic matrix. Very-high-density sorbents have been synthesized using a stannic oxide core, although the ion-exchange capacity is relatively low and the material only moderately stable over the entire pH range.[16] An interesting idea is the impregnation of liquid extractants such as Alamine 336 (a long-chain tertiary amine) into an inert porous support such as crushed fireclay or firebrick. It was claimed that materials of this kind were selective for uranium sorption but suffered from instability to the leaching of the reagents and poor hydraulic behavior.[17] None of these or similar ideas have yet proved to be commercially viable.

NOTATION

a stoichiometric coefficient

a_i ionic activity of species i

C concentration of ions in solution phase

$\overline{C}$ resin-phase concentration of ions

C_i solution-phase concentration of species i

$\overline{C}_i$ resin-phase concentration of species i

d_p diameter of resin beads

D diffusion coefficient in solution phase

$\overline{D}$ diffusion coefficient in resin phase

$\overline{D}_e$ effective diffusivity coefficient

F Faraday constant

g gravitational acceleration

h height of liquid vessel

J_i flux of species i in solution

k_2 second-order reaction rate constant

k_s rate constant based on surface area

k_{mi} mass transfer coefficient of species i in the liquid film

K_a thermodynamic equilibrium constant [Eq. (13.1-3)]

K_c selectivity coefficient [Eq. (13.1-3)]

L depth of bed

m_i distribution coefficient

ΔP pressure drop across bed

P_a atmospheric pressure

R gas constant

r_0 radius of ion-exchange bead

S surface area of resin beads per unit volume of beads

T temperature

U superficial velocity of liquid through empty bed

U_t terminal settling velocity of particle through liquid

t time

V solution volume

$\overline{V}$ resin volume

X extent of resin conversion

x equivalent ionic fraction in solution

y equivalent ionic fraction in resin phase

z_i valency of species i

Greek Letters

α separation factor

ϕ electrical potential gradient

δ liquid film thickness

γ_i activity coefficient of species i

ϵ fraction of voids in bed

μ dynamic viscosity of liquid
ρ density of liquid
ρ_s absolute density of resin beads

Subscripts

A   ionic component or species
B   ionic component or species
i refers to either species A or B in a binary ion-exchange system

Note: Overbars refer to species in resin phase.

REFERENCES

Section 13.1

13.1-1   F. G. Donnan, *Z. Elektrochem.*, **17**, 572 (1911); *Z. Phys. Chem.*, **A168**, 369 (1934).

13.1-2   F. G. Donnan and E. A. Guggenheim, *Z. Phys. Chem.*, **A162**, 346 (1932).

13.1-3   H. P. Gregor, *J. Am. Chem. Soc.*, **70**, 1293 (1948); **73**, 642 (1951).

13.1-4   D. Reichenberg, K. W. Pepper, and D. J. McCauley, *J. Chem. Soc. (London)*, 493 (1951).

13.1-5   E. Gluekauf and J. F. Duncan, U.K.A.E.A. Report CR808 (1951).

13.1-6   O. D. Bonner and L. L. Smith, *J. Phys. Chem.*, **61**, 236 (1957).

13.1-7   W. J. Argersinger and A. W. Davidson, *J. Phys. Chem.*, **56**, 92 (1952).

13.1-8   W. C. Bauman and R. M. Wheaton, *Ind. Eng. Chem.*, **43**, 1088 (1951).

13.1-9   R. Kunin and F. X. McGarvey, *Ind. Eng. Chem.*, **41**, 1265 (1949).

13.1-10   H. P. Gregor, J. Belle, and R. A. Martins, *J. Am. Chem. Soc.*, **77**, 2713 (1955).

13.1-11   W. J. Brignal, A. K. Gupta, and M. Streat, "The Theory and Practice of Ion Exchange," Paper 11, Society of Chemical Industry, London, 1976.

13.1-12   R. P. Smith and E. T. Woodburn, *AIChE J.*, **24**, 577, (1978).

13.1-13   R. A. Robinson and R. H. Stokes, *Electrolyte Solutions*, Butterworths, London, 1959.

13.1-14   G. M. Wilson, *J. Am. Chem. Soc.*, **86**, 127 (1964).

13.1-15   F. Helfferich, *Ion Exchange*, McGraw-Hill, New York, 1962.

13.1-16   F. Helfferich and M. S. Plesset, *J. Chem. Phys.*, **28**, 418 (1958).

13.1-17   C. Y. Wen, *Ind. Eng. Chem.*, **60**, 34, (1968).

13.1-18   P. B. Weisz and R. D. Goodwin, *J. Catal.*, **2**, 397 (1963).

13.1-19   M. Nativ, S. Goldstein, and G. Schmuckler, *J. Inorg. Nucl. Chem.*, **37**, 1951 (1975).

13.1-20   R. E. Warner, A. M. Kennedy, and B. A. Bolto, *J. Macromol. Chem.*, **A4**, 1125 (1970).

13.1-21   G. Adams, P. M. Jones, and J. R. Millar, *J. Chem. Soc. A*, 2543 (1969).

13.1-22   P. R. Dava and T. D. Wheelock, *Ind. Eng. Chem. Fundam.*, **13**, 20 (1974).

13.1-23   F. Helfferich, *J. Phys. Chem.*, **69**, 1178, (1965).

13.1-24   W. Holl and H. Sontheimer, *Chem. Eng. Sci.*, **32**, 755 (1977).

13.1-25   O. Levenspiel, *Chemical Reaction Engineering*, 2nd ed., Wiley, New York, 1972.

13.1-26   J. M. Smith, *Chemical Engineering Kinetics*, 3rd ed., McGraw-Hill, New York, 1981.

Section 13.2

13.2-1   B. A. Bolto and D. E. Weiss, The Thermal Regeneration of Ion Exchange Resins in J. A. Marinsky and Y. Marcus (Eds.), *Ion Exchange and Solvent Extraction*, Vol. 7, Marcel Dekker, New York, 1977.

13.2-2   S. B. Appelbaum, *Demineralisation by Ion Exchange*, Academic Press, New York, 1968.

13.2-3   C. Calmon and T. R. E. Kressman, *Ion Exchangers in Organic and Biochemistry*, Interscience, New York, 1957.

13.2-4   D. W. Boydell, Continuous Ion Exchange for Uranium Recovery: An Assessment of the Achievements over the Past Five Years, in *Hydrometallurgy '81*, Society of Chemical Industry, London, 1981.

13.2-5 R. C. Merritt, *The Extraction Metallurgy of Uranium*, Colorado School of Mines Research Institute, 1971.

13.2-6 A. K. Haines, T. H. Tunley, W. A. M. Te Riele, F. L. D. Cloete, and T. D. Sampson, *J. S. Afr. Inst. Min. Metall.*, **74**(4), 149 (1973).

13.2-7 A. Mehmet and W. A. M. Te Riele, The Recovery of Gold from Cyanide Liquors in a Countercurrent Contactor Using Ion Exchange Resin, in D. Naden and M. Streat (Eds.), *Ion Exchange Technology*, Ellis Horwood, London, 1984.

Section 13.3

13.3-1 A. L. Wilcox, et al., U.S. Patent 2, 528, 099 (1950), as described by J. W. Michener and H. E. Lundberg, in F. C. Nachod and J. Schubert (Eds.), *Ion Exchange Technology*, Academic Press, New York, 1956.

13.3-2 F. L. D. Cloete, Comparative Engineering and Process Features of Operating Continuous Ion Exchange Plants in South Africa, in D. Naden and M. Streat (Eds.), *Ion Exchange Technology*, Ellis Horwood, London, 1984.

13.3-3 B. A. Bennett, F. L. D. Cloete, A. I. Miller, and M. Streat, *Chem. Eng.* 412 (Nov. 1969).

13.3-4 I. R. Higgins and J. T. Roberts, *Chem. Eng. Prog. Symp. Ser.*, **50**(14), 87 (1954).

13.3-5 K. D. Dorfner, *Ion Exchangers*, 3rd ed., Ann Arbor Science Publishers, Ann Arbor, MI, 1973.

13.3-6 Editorial article, *Chemical Week*, June 9, 1956.

13.3-7 J. Bouchard, *Ion Exchange in the Process Industries*, p. 91, Society of Chemical Industry, London, 1970.

13.3-8 F. L. D. Cloete and M. Streat, Brit. Patent 1,070,251 (1967) and U.S. Patent 3,551,118 (1970). (Priority date: Oct. 19, 1962).

13.3-9 D. R. George, J. R. Ross, and J. D. Prater, *Min. Eng.* **20**, 1 (1968).

13.3-10 R. McNeill, E. A. Swinton, and D. E. Weiss, *J. Met.* **7**, 912 (1955).

13.3-11 A. K. Haines, T. H. Junley, W. A. M. Te Riele, F. L. D. Cloete, and T. D. Sampson, *J. S. Afr. Inst. Min. Metall.*, **74**(4), 149 (1973).

13.3-12 F. C. Harvey and R. A. V. Smith, Colloquium held at Klerksdorp, South African Institute of Mining and Metallurgy, Johannesburg, 1983.

13.3-13 B. A. Bennett, F. L. D. Cloete, and M. Streat, in *Ion Exchange in the Process Industries*, p. 133, Society of Chemical Industry, London, 1970.

13.3-14 D. G. Stevenson, in *Ion Exchange in the Process Industries*, p. 114, Society of Chemical Industry, London, 1970.

13.3-15 D. Naden and G. Willey, in *Theory and Practice of Ion Exchange*, p. 42, Society of Chemical Industry, London, 1976.

13.3-16 A. K. Haines, *J. S. Afr. Inst. Min. Metall.* **78**, 303 (1978).

13.3-17 A. Himsley and E. J. Farkas, in *Theory and Practice of Ion Exchange*, p. 45, Society of Chemical Industry, London, 1976.

13.3-18 A. Himsley, *Can. Min. Metall. Bull.*, Sept. 1977.

13.3-19 R. Porter, U.S. Patent 3, 879, 287 (1975).

13.3-20 T. Izzo, L. A. Painter, and R. Cheminski, *Eng. Min. J.*, **158**, 90 (1957).

13.3-21 A. S. Dahya and D. J. King, *Can. Min. Metall. Bull.*, **76**, 55 (1983).

13.3-22 E. H. D. Carman, *J. S. Afr. Inst. Min. Metall.*, **60**, 647 (1960).

13.3-23 F. O. Read, *J. S. Inst. Min. Metall.*, **60**, 105 (1959).

13.3-24 F. L. D. Cloete, *J. S. Afr. Inst. Min. Metall.*, **81**, 66 (1981).

13.3-25 P. A. Laxen, G. S. M. Becker, and R. Rubin, *J. S. Afr. Inst. Min. Metall.*, **79**, 315 (1979).

13.3-26 R. A. Shoemaker, U.S. Patent 4,251,352 (1981).

13.3-27 T. V. Arden, E. A. Swinton, and D. E. Weiss, in *Proceedings of the Second United Nations Conference on Peaceful Uses of Atomic Energy, Geneva*, vol. 3, p. 396 (1958).

13.3-28 J. Y. Oldshue, *Fluid Mixing Technology*, McGraw-Hill, New York, 1983.

Section 13.4

13.4-1 A. Warshawsky, Selective Ion Exchange Polymers, *Angew. Makromol. Chem.*, **109–110**, 171 (1982).

13.4-2 K. C. Jones and R. R. Grinstead, *Chem. Ind.*, 637 (1977).

13.4-3   B. R. Green and E. Jaskulla, Poly(vinylimidazole)—A Versatile Matrix for the Preparation of Chelating Resins, in D. Naden and M. Streat (Eds.), *Ion Exchange Technology* Ellis Horwood, London, 1984.

13.4-4   F. Vernon and H. Eccles, *Anal. Chim. Acta*, **77,** 145 (1975).

13.4-5   A. Warshawsky, Hydrometallurgical Processes for the Separation of Platinum Group Metals (PGM) in Chloride Media, in D. Naden and M. Streat (Eds.), *Ion Exchange Technology*, Ellis Horwood, London, 1984.

13.4-6   S. Gonzalez-Luqe and M. Streat *Hydrometallurgy*, **11,** 207, 227 (1983).

13.4-7   A. A. Efendiev, *Issled Obl. Kinet. Model. Optim. Khim. Protsessov,* 247 (1974).

13.4-8   A. Warshawsky, et al., *J. Am. Chem. Soc.*, **101,** 4249 (1979).

13.4-9   H. W. Kauczor and A. Meyer, *Hydrometallurgy*, **3,** 65 (1978).

13.4-10 A. Warshawsky, Extraction with Solvent-Impregnated Resins, in J. A. Marinsky and Y. Marcus (Eds.), *Ion Exchange and Solvent Extraction*, Vol. 8, Marcel Dekker, New York, 1981.

13.4-11 C. H. Muendel and W. A. Selke, *Ind. Eng. Chem.*, **47,** 374 (1955).

13.4-12 D. A. Brown et al., *Appl. Polym. Symp.*, **29,** 189 (1976).

13.4-13 F. Vernon and T. Shah, *Reactive Polym.* **1,** 301 (1983).

13.4-14 N. V. Blesing, B. A. Bolto, D. L. Ford, R. McNeil, A. S. Macpherson, J. D. Melbourne, F. Mort, R. Siudak, E. A. Swinton, D. E. Weiss, and D. Willis, in *Ion Exchange in the Process Industries*, Society of Chemical Industry, London, 1969.

13.4-15 E. A. Swinton, B. A. Bolto, R. J. Eldridge, P. R. Nadebaum, and P. C. Coldrey, The Present Status of Continuous Ion Exchange Using Magnetic Micro Resins in D. Naden and M. Streat (Eds.), *Ion Exchange Technology*, Ellis Horwood, London, 1984.

13.4-16 M. Streat, Brit. Patent, 1,456, 974 Dec. 11, 1973.

13.4-17 P. J. D. Lloyd, S. Afr. Patent, 70/4209.

CHAPTER 14

Large-Scale Chromatography

PHILLIP C. WANKAT
School of Chemical Engineering
Purdue University
West Lafayette, Indiana

Immediately following the development of analytical chromatographic methods there was considerable interest in large-scale applications of these techniques for separation of compounds on a commercial scale. Although there were some reports of large-scale uses, interest decreased when it became clear that large-scale chromatography was not a panacea for all separation problems. Currently, there is a renaissance of interest in large-scale chromatography as shown by an increase in patents, publications, and companies selling large-scale systems.

Large-scale chromatography has been a meeting ground for the ideas of analytical chromatographers trying to scale-up their systems and for chemical engineers trying to apply standard chemical engineering techniques to chromatography. The result has been a proliferation of ideas of how to do chromatographic separations on a large scale. Since methods for doing these separations have not been standardized, this chapter includes a progress report and an estimate of where the field will go.

First, a simple theory of solute movement useful in understanding and comparing the different processes is developed. Next, the scale-up of analytical elution and displacement chromatography is presented. Then countercurrent and simulated countercurrent methods are explained and compared to elution chromatography. After this, hybrid systems which combine aspects of elution chromatography and simulated countercurrent processes are discussed. In Section 14.5 other alternatives including two-dimensional systems, chromatothermography, and cyclic processes are reviewed briefly. Finally, the systems are compared and some general ideas on selecting a process are given.

14.1 THEORY

In this section we are concerned first with the rate of movement of solute and second with the effects that broaden zones in linear systems. More detailed mathematical presentations are available in the references.

14.1-1 Rate of Solute Movement

The equations for rate of solute movement can be derived rigorously,[1,2] but a physical argument is used here. In all types of chromatography, the solute distributes between the stationary phase and the mobile phase. When the solute is in the stationary phase, its velocity with respect to the solid is zero, while when the solute is in the mobile phase, it has the same relative velocity as the moving fluid. Thus, we can calculate the solute velocity from the fraction of time the solute spends in the mobile phase

$$u_{\text{solute},i} = \left(\frac{\text{amount solute in mobile phase}}{\text{total amount of solute in segment}} \right) v \qquad (14.1\text{-}1)$$

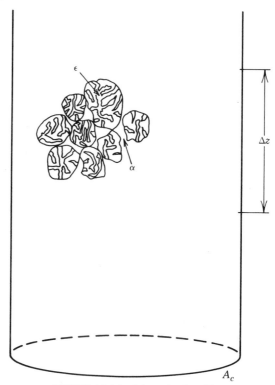

FIGURE 14.1-1 Schematic of packing.

where v is the interstitial velocity of the fluid. The ratio in Eq. (14.1-1) depends on the exact type of chromatography being studied. For the packing shown in Fig. 14.1-1, only the fluid volume outside the particles is moving (external porosity α). The material that is not moving is the fluid volume inside the particles (internal porosity ϵ) plus the solute held up on the solid. The solute may be held on the surface of the solid as in adsorption chromatography or it may be dissolved in a thin layer of liquid coated on the solid as in gas–liquid or liquid–liquid chromatography. Finally, since the pores are small and are not of uniform size, large molecules may be excluded from some fraction of the internal void volume. The fraction of internal void volume which a species i can penetrate is measured as K_{d_i}.

Consider an experiment where an incremental mass of solute is added to the bed segment shown in Fig. 14.1-1. Then the fraction in the mobile phase can be calculated as

fraction i in mobile phase

$$= \frac{(\Delta z \, A_c)\alpha \, \Delta c_i}{(\Delta z \, A_c)\alpha\Delta c_i + (\Delta z \, A_c)(1 - \alpha)\epsilon K_{d_i}\Delta c + (\Delta z \, A_c)(1 - \alpha)(1 - \epsilon)K_{d_i}\rho_s\Delta q_i} \quad (14.1\text{-}2)$$

where Δc_i is the change in solute concentration in the fluid (e.g., in kmol/m^3). In the denominator the first term is the incremental amount of solute in the mobile phase, the second term is the incremental amount of solute held in the fluid inside the pores, and the last term is the amount of solute held by the solid. Δq_i is the change in amount of solute sorbed (e.g., in kmol/kg solid) and ρ_s is the structrural density of the solid (i.e., without any pores) in kg/m^3. For gas or liquid chromatography with small molecules, $K_{d_i} = 1.0$ for all solutes. For size exclusion chromatography (SEC) ideally $\Delta q_i = 0$ for all solutes and K_{d_i} varies from 0 to 1.0. Equations (14.1-1) and (14.1-2) can be combined to give the velocity of this incremental mass of solute. After rearrangement this is

$$u_{\text{solute},i} = \frac{v}{1 + [(1 - \alpha)/\alpha]\epsilon K_{d_i} + [(1 - \alpha)/\alpha](1 - \epsilon)\rho_s K_{d_i}\Delta q_i/\Delta c_i} \quad (14.1\text{-}3)$$

To use Eq. (14.1-3) to predict solute movement in a variety of chromatographic systems, Δq_i must be related to Δc_i. The easiest way to do this is to assume that solid and fluid phases are in equilibrium. If

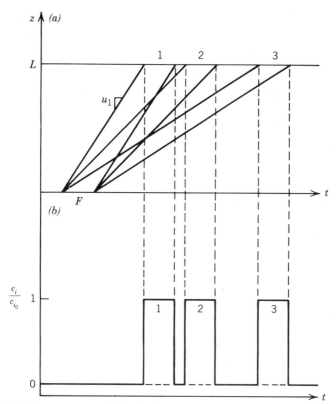

FIGURE 14.1-2 Solute movement solution for three component linear chromatography: (*a*) solute move-ment diagram and (*b*) predicted product concentrations.

this is true, the equilibrium isotherm can be used. At low concentrations all chromatographic systems approach a linear isotherm of the form

$$q_i = k_i c_i \qquad (14.1\text{-}4)$$

At low concentrations the equilibrium constant k_i depends on the temperature (and perhaps pH, ionic strength, or pressure) but not on the concentrations of the other solutes. Then the solute velocity for linear systems is

$$u_{\text{solute},i} = \frac{v}{1 + [(1 - \alpha)/\alpha]\epsilon K_{d_i} + [(1 - \alpha)/\alpha](1 - \epsilon)\rho_s K_{d_i} k_i} \qquad (14.1\text{-}5)$$

If the fluid velocity v is constant and the thermodynamic variables of the system are held constant, then $u_{\text{solute},i}$ is constant for each solute.

Predicting the movement of solute in a column is now easy. This is illustrated in Fig. 14.1-2 for a pulse input of feed containing three solutes. On a plot of axial distance versus time the solutes move at a slope, $\Delta z/\Delta t$, equal to their solute velocities. Solute velocities differ if K_{d_i} and/or k_i are different. In practice, this model and the results shown in Fig. 14.1-2 are greatly oversimplified since mass transfer and dispersion cause a broadening and dilution of the peaks. These processes are discussed in Section 14.1-2. Despite (or perhaps because of) its oversimplification, this model is very useful in comparing different operating methods. That is the major use of the model in this chapter.

The solute velocities are measured easily by analytical chromatography. When a small pulse is injected, the peak exits with a Gaussian profile but the peak maximum moves through the column at a velocity given by Eq. (14.1-5). The velocity is determined easily as

$$u_{\text{solute},i} = \frac{L}{t_{R_i}} \qquad (14.1\text{-}6)$$

where the retention time t_{R_i} is the time the peak maximum exits a bed of length L. The interstitial velocity can be determined from the measured volumetric flow rate or from the retention time of a large, nonretained solute. In the chromatographic literature Eq. (14.1-5) usually is replaced by

$$u_{solute,i} = \frac{u}{1 + k_i'}$$

(14.1-7)

where k_i' is the relative retention of the solute, and u is the velocity of a small, nonretained molecule ($u \neq v$) with a residence time t_0

$$u = \frac{L}{t_0}$$

(14.1-8)

Large-scale chromatography often is operated outside the linear range. If the species interact and the equilibrium amount of species i depends on the concentrations of all species present, the approach presented here is too simple and more complex mathematics are required.[3-5] If the speices do not interact, Eq. (14.1-3) is still valid but a nonlinear isotherm must be used to relate q_i to c_i. The solute velocity now depends on the solute concentration. For adsorption systems the isotherms often have a Langmuir shape. Then the equilibrium expression is of the form

$$q_i = \frac{a_i c_i}{1 + b_i c_i}$$

(14.1-9)

Now as c_i increases q_i increases but at a decreasing rate. Thus, Eq. (14.1-3) predicts that solute moves faster when the fluid is more concentrated. This agrees with experiment. If a concentrated fluid is displaced by a dilute fluid, the concentrated portion moves faster than the dilute portion and the fluid concentration decreases smoothly. In this "diffuse wave" $\Delta q_i/\Delta c_i$ can be determined at any concentration from the derivative of the isotherm, $\partial q_i/\partial c_i$. The result is a different velocity for each concentration and the solute wave spreads out. This is illustrated in Fig. 14.1-3.

If a dilute fluid is displaced by a concentrated fluid, the situation is different. Now there is a discrete change (a "shock wave") in concentration and discrete differences should be retained in Eq. (14.1-3). Now $\Delta q_i/\Delta c_i$ is calculated as the change in fluid and solid concentrations from one side of the shock wave to the other. The shock wave is also shown in Fig. 14.1-3. If one tries to replace $\Delta q_i/\Delta c_i$ by $\partial q_i/\partial c_i$, one finds that the more concentrated fluid moves faster and the solute waves intersect the slower moving dilute waves. This implies that two fluids of different concentration occupy the same location, which is physically impossible. The use of the finite differences, which represent a macroscopic mass balance, avoids this problem. Note that addition of a pulse of feed produces both a shock wave and a diffuse wave. The tailing caused by the diffuse wave may cause the solute band to overlap with the next solute. This effect of the nonlinear isotherms is one of the major causes of decreased resolution during the chromatography of concentrated systems.

Figure 14.1-3 is an oversimplification. In practice, mass transfer and dispersion effects smooth out the shock wave into an S-shaped curve. Equation (14.1-3) with finite differences does give a good prediction of where the center of the S-shaped curve will be. When equilibrium theory predicts a shock wave, a constant pattern is observed and constant pattern solutions can be used.[1] Except at the two corners, the diffuse wave is often a good fit to the experimental data. This allows the calculation of nonlinear equilibrium constants from diffuse waves.[1]

In gas–liquid chromatography Langmuir isotherms are not followed. Instead, the isotherm bends in the opposite direction and more solute is held up by the stationary phase at higher gas concentrations. Diffuse and shock waves still occur, but they are reversed in position in Fig. 14.1-3. Thus, the observed diffuse wave caused by the isotherm in gas–liquid chromatography is different than in adsorption chromatography.

In gas chromatography (both partition and adsorption) there is a second cause of tailing which occurs at high concentrations. This is the "sorption effect." The absorption or adsorption of large amounts of solute from the gas causes a very large volume change, which affects the gas flow. The result is the development of a shock wave at the front of the band and a tail at the back end. The shape is similar to tailing produced by non-linear adsorption (see Fig. 14.1-3). In gas adsorption chromatography the tailing caused by the isotherm and the sorption effects reinforce each other. In gas–liquid (partition) chromatography the two effects, to a large extent, can cancel each other out. This important in the design of large-scale gas systems.

In biospecific affinity chromatography and ion-exchange chromatography of proteins an "on-off" operating produce often is used. During the sorption ("on") step the conditions (pH, ionic strength, buffer, etc.) are set so that once sorbed the protein does not come off. Then q equals the saturation value for all concentrations and a shock wave predicted by Eq. (14.1-3) results. Before breakthrough occurs the feed is stopped and nonsorbed proteins are washed out. Then the conditions are changed so that the desired protein

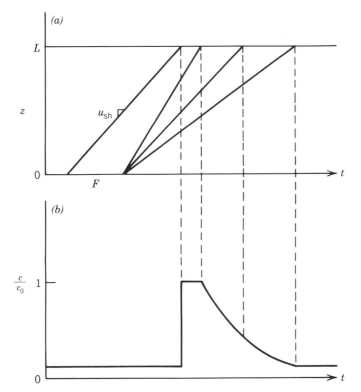

FIGURE 14.1-3 Solute movement solution for pulse of one nonlinear component: (a) solute movement diagram, u_{sh} = shock wave velocity and (b) predicted product concentrations.

is not sorbed or prefers to complex with a species in the liquid ("off" step) and the protein is eluted from the column. The desorption solution is washed out of the column and the cycle is repeated. These methods usually concentrate the protein in addition to purifying it. The solute movement theory can be adapted to this case. This requires determining the velocity of the elution and wash waves from equations similar to (14.1-3) and determining the effect of changing conditions on the protein concentration. Similar calculations have been done for cyclic separation techniques.[2]

14.1-2 Zone Broadening in Linear Systems

The sharp concentration fronts shown in Fig. 14.1-2 and 14.1-3 never occur in practice. The zones are always diluted and broadened by mass transfer and dispersion for both linear and nonlinear isotherms. The complete solution of the equilibrium equations, mass balances, and mass transfer equations for nonlinear systems is a formidable task requiring numerical solutions. For linear systems the task is much easier and very useful solutions have been developed. Even though large-scale chromatography often is operated in the nonlinear range, the linear analyses are valuable since they can provide a qualitative feel (quantitative for linear systems) for band broadening effects.

The analysis of linear chromatography with small pulses usually is based on analysis of a Gaussian profile. The solution can be written

$$c = c_{max} \exp\left(-\frac{x^2}{2\sigma^2}\right) \tag{14.1.10}$$

where x is the distance from the peak maximum and σ is the standard deviation in units of length. Results which reduce to this can be derived for linear isotherms from equilibrium plate theories,[6] random walk or stochastic analysis,[6] or solution of the mass transfer equations including dispersion.[7] The equilibrium plate theories are useful in large-scale studies because they can serve as a base for a numerical method which includes nonlinear isotherms. The random walk analysis is useful because it explains zone spreading. The linear dispersion models are useful because with superposition they can be applied easily to a variety of operating methods.

The standard deviation σ can be related to the plate height H,

$$\sigma = \sqrt{HL} \tag{14.1-11}$$

where L is the column length. Equation (14.1-11) shows that σ in length units and hence the spreading of the solute band is proportional to the square root of the distance traveled by the solute. Since the distance between peak centers is proportional to L, we can obtain in theory a sufficient resolution of two solutes with different k' values by increasing the column length. This is a crude approach which also increases pressure drop and cost. The plate height can be determined from the sum of the contributions due to flow, axial diffusion, and mass transfer. Usually, this is done by the van Deemter equation or one of the modifications of this equation.[6,8-11]

$$H = A + \frac{B}{v} + C_M v + C_{SM} v + C_S v \tag{14.1-12}$$

The A term depends on eddy dispersion and is proportional to the diameter of the particles, d_p. The axial diffusion constant B is proportional to the molecular diffusivity D_M of the solute. The mass transfer effect, the C terms, includes mass transfer outside the particles—C_M proportional to d_p^2/D_M; and mass transfer in the solid or the coated liquid phase—C_S proportional to d_f^2/D_S, where d_f is the film thickness and D_S is the diffusion coefficient in the stationary phase.

Equation (14.1-12) shows that there is a velocity which minimizes H and hence minimizes band spreading, but this velocity usually gives too low a throughput in large-scale separations. In liquid systems the axial diffusion effects are usually small but in gas systems with much higher D_M values the B term can be important. For high molecular weight materials with low D_M values, the C terms are dominant. The A and C terms can be reduced by decreasing the particle diameter. The absolute minimum value of H is about $2\,d_p$. However, since the pressure drop is proportional to $1/d_p^2$, there is an optimum particle diameter for constrained systems.[11] Relatively large-diameter packings are used in large-scale chromatographic systems, although there are theoretical reasons why smaller-diameter particles should be used.[12] Note that column diameter does *not* enter into Eq. (14.1-12) or into the A, B, or C terms. Theoretically, there is *no* reason why properly designed and packed large-scale systems cannot have low H values when operated with small pulses.

The elements of the linear model all come together when we look at how well two solutes can be separated. For Gaussian peaks resolution is defined as

$$R = \frac{2(t_{R2} - t_{R1})}{w_1 + w_2} \tag{14.1-13}$$

where the t_R terms are retention times of the peaks and the w are the bandwidths. An $R = 1$ has the maxima of the two peaks separated by 4σ which gives about 2% overlap between the peaks. The resolution can be predicted from the Gaussian distribution. This "fundamental equation" of chromatography is

$$R = \frac{1}{2} \left(\frac{\alpha - 1}{1 + \alpha} \right) \frac{\bar{k}'}{1 + \bar{k}'} N^{1/2} \tag{14.1-14}$$

where the selectivity $\alpha = k_2'/k_1' = k_2/k_1$, $\bar{k}'$ is the average value of the relative retentions of the two solutes, and the number of plates N is

$$N = \frac{L}{H} \tag{14.1-15}$$

The resolution between two peaks is increased if selectivity increases, $\bar{k}'$ increases, column length increases, or the plate height decreases. The relative retention cannot be increased indefinitely since it will make operating time too long. A value of $\bar{k}'$ from 4 to 6 seems reasonable. The larger the selectivity the higher the resolution or smaller N can be for a given resolution. For α below 1.1 the number of stages required increases drastically.[8] This is just another way of saying that solutes must have different velocities to separate. Above $\alpha = 2.0$ the number of stages required is reasonable and decreases further slowly. Practically, an $\alpha > 2$ or higher is desirable although an $\alpha \approx 1.5$ can be used for large dedicated systems. For medium scale, α values as low as 1.1 can be used. One advantage of high selectivity is that the throughput can be increased. The low stage requirement allows a relatively high H which implies a high velocity [see Eq. (14.1-12)]. In large-scale chromatography one does sloppy separations ($R < 1.0$) and recycles material which is not separated. Equation (14.1-14) shows that this procedure decreases drastically the N required, which again allows significantly higher throughputs.

Large-scale chromatography often operates in the overloaded regime where nonlinear effects are important. In this case the quantitative predictions of this section are no longer valid. However, the qualitative predictions are valid when they are modified with the nonlinear effects discussed in the preceding section.

14.2 SCALE-UP OF ELUTION CHROMATOGRAPHY

Chromatographs first were developed as separating devices and on-line detectors were added later. Because of the success of on-line detection, the detection (or analytical) aspects of chromatography were emphasized. When chromatographers needed to separate more material, they first tried injecting more feed and making the column larger. Up to a point this procedure worked fairly well. However, it soon became evident that uniformly packing larger-diameter columns was not easy. For a long time it was believed that large-diameter columns could not be packed as efficiently (i.e., with as low an H value) as small-diameter columns. This is now known to be false; however, packing large-diameter columns is still difficult.

A schematic of a large-scale chromatography system is shown in Fig. 14.2-1. Feed and recycle material are fed as a pulse through an injector. The carrier gas or solvent is fed continuously and usually is reused in the system since large quantities of pure gas or solvent are too expensive to throw away. The column is packed carefully and probably is well insulated or in an oven. A timer or detector is used for splitting the column outlet stream. Different product streams are sent to different separators where carrier gas or solvent is removed from the product. These separators are probably cold traps for gas systems and some other type of separation device (distillation columns, evaporators followed by crystallizers etc.) for liquid systems. Since column operation is at low resolution ($R < 1.0$), a mixed product fraction is collected and recycled to the column. This recycle stream may not be separated from the solvent or carrier gas. The recovered solvent or carrier gas may require cleanup to remove traces of solute before it is returned to the column. A variety of measurement and control devices for temperature, pressure, and flow rates also are required but are not shown in Fig. 14.2-1. Operation of the system is automatic and controlled by electronic timers or microprocessors. In a large-scale system failure to design any of the components properly may cause the system to perform poorly or fail.

In this section large-scale applications of liquid chromatography including adsorption, ion exchange, affinity chromatography, and size exclusion chromatography and gas chromatography using the direct scale-up approach are discussed.

14.2.1 Liquid Chromatography (LC)

In liquid elution chromatography, separation is based on adsorption on the solid or on partition to a stationary or bonded liquid phase. Distribution coefficients are modest so that solutes migrate through the column as shown in Figs. 14.1-2 and 14.1-3. Gradients can be used but usually are not since the column has to be reequilibrated afterward. Two somewhat different approaches have been taken:

1. Relatively small-scale—say less than 15 cm (6 in.) in diameter—highly efficient columns based on the scale-up of analytical columns. A standard sorbent is used and α may be as low as 1.1. Standard equipment applicable to a variety of separation problems is available commercially.[1-4] In the range up to a few centimeters in diameter these methods have been extensively reviewed.[4-8]
2. Large-scale, custom built systems using a packing with relatively large α (preferably > 2). Equipment is dedicated to one separation. A few applications were reported as early as 1947,[9,10] but interest has increased recently.[11,12] The packing material may be specially developed to be highly selective.[9,11,13]

In both approaches the column has auxiliary equipment as shown in Fig. 14.2-1 and most of the requirements are the same. In both cases a combination of packing and solvent should be selected to give a high selectivity. This decreases the column length required or allows operation at higher velocities. A high capacity is desirable but k' values should be modest ($\sim 1-6$) to keep the column length reasonable. The packing should be rigid, nonfriable, chemically stable, noncatalytic, long life, free from impurities, highly selective, readily available, and cheap. Usually, some compromise among these properties is required. The particle diameter should be significantly less than the column diameter ($30\ d_p <$ column diameter). Since H is proportional to d_p^2 and ΔP to $1/d_p^2$, there is a trade-off on particle size, and the optimum size is still a matter for debate.[31] Very tight sieving is desirable since the largest particles control H and the smallest ΔP. The solvent, in addition to giving a high α, must dissolve the solutes, be easy to separate from the solutes in the downstream separators, and have a low viscosity. The last requirement is important since it leads to lower pressure drops and higher diffusivities. If possible, the solvent also should be noncorrosive, nontoxic, noncarcinogenic, readily available, and cheap. Again some compromise usually is required.

Absolutely even plug flow is desired. Any deviations from plug flow cause zone spreading and decrease resolution. To achieve the desired plug flow the sorbent must be packed carefully and the distribution system must be designed carefully. A low H is desired and external dead volumes must be minimized.

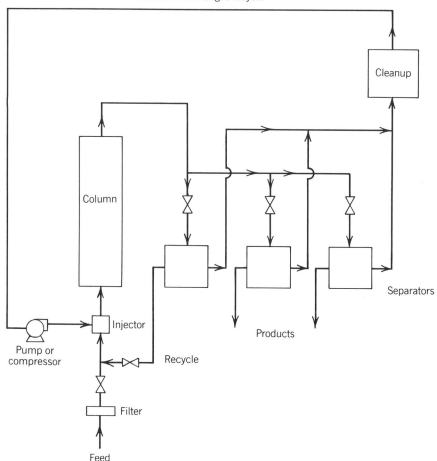

Solvent or carrier gas recycle

FIGURE 14.2-1 General system for large-scale chromatography.

This is where major differences in the different designs occur. For small-scale preparative systems a large number of packing procedures have been developed.[1-8] These include controlled vibrations,[5-8] tamping,[5-8] slurry methods,[5-8] axial compression,[1,4,8] and radial compression.[2,8] The slurry and tamping methods appear to be difficult to scale up while controlled vibrations,[11] axial compression,[1] and radial compression[2] can be scaled up. Whatever packing procedure is used, the goal is to obtain a tight, uniform packing with no voids or channeling. This can be checked with a pulse experiment in which H should be no larger than in an analytical column with the same size packing. The ability to empty and repack the column easily is an advantage.

The distribution system should put in and withdraw liquid without disturbing the plug flow. As the size increases this becomes more important. A combination of a distribution system plus a frit to hold the packing seems to work well. The feed pulses should be input across the entire column diameter in as close to a square wave as possible. Otherwise, the obtained resolution is decreased. The distribution system and all piping and valves should have a minimum volume and distance of travel between the injection device and the downstream valves which isolate the products. This minimizes mixing and makes control easier.

Chromatographic packings act as depth filters if the feed or solvent is dirty. Since this increases pressure drop and decreases capacity, the feed, solvent, and recycle streams all should be filtered. In addition, the liquids need to be deaeraeted to prevent air bubbles in the column.

The operating method used should optimize throughput of feed at the desired purity. This can be done by operating at high flow velocities (well above the velocity that minimizes H), by feeding large feed pulses, by overlapping the outlet peaks and using heart cuts to obtain desired products, and by recycling partially separated materials.[8,14] A single-pass recovery ratio in the range from 0.6 to 0.8[14] appears to be optimum. The feed pulses should be added frequently so that some product always is exiting the column. Very slow solutes greatly increase the cycle time and should be removed either in a precolumn or with segmented columns (see Section 14.4). The concentration of solutes in the feed should be relatively high. This concentration must be below the solubility limits of the solvent, should avoid overflooding the stationary liquid phase (see Section 14.2-3), and should be low enough that the feed pulse is not much more viscous than the solvent. Deviation from the last requirement can lead to viscous fingering. The concentration should be above the linear limit, which is a limit for analytical methods only. Separations where components have a wide range in k' values or where the separation of key components is very difficult might be improved by two-step operation. The operation should be automated on either a time or concentration measurement basis.

Things do go wrong. A routine maintenance procedure should be followed. The filters should be cleaned or replaced periodically. A periodic backflush of the column may help clean frits and remove trapped air bubbles. If the packing becomes fouled a chemical rinse such as a caustic wash may help. The manufacturer's instructions should be followed since packings are stable only within given pH ranges. The ability to remove and repack the column easily is an advantage if the packing becomes fouled. It may be necessary to remove and discard only the top several centimeters of packing. Then the remainder can be removed, mixed with new packing, and the column can be repacked. Modular systems also can be advantageous, particularly if cleaning is frequent.

Examples of fairly small-scale, highly efficient column chromatographic separations are shown in Figs. 14.2-2 and 14.2-3. Figure 14.2-2[3] shows the purification of 2,4,6-tri-*tert*-butylphenol on a 6 in. diameter by 6 ft long column packed with 37–53 μm reverse-phase LRP-2 packing. The column was packed by a modified tap-fill method. Two liters of a 10% solution were injected at a flow rate of 1.8 L/min. The desired product is the last peak while the peak before this is a relatively small amount of a colored impurity which absorbs strongly in the UV. A higher throughput could be achieved by increasing the size of the injection and recycling overlapping material. Figure 14.2-3[15] shows the purification of penicillin V on a 6 in. diameter by 2 ft long column packed with a Prep Pak C_{18} reverse-phase packing. A radial compression packing procedure was used.[2] One liter of solution with a loading of 28 ng/g C_{18} was injected at a flow rate of 3 L/min. The four fractions shown all contain penicillin V with purities of 90, 99, 97, and 99%. The pooled product had a recovery of 95%. Figure 14.2-3 shows that large-scale chromatography often gives chromatograms with a nonclassical shape, but the resulting separation is very useful.

Several large-scale separations have been reported.[9–12] The large-scale separation of sugars is fairly common for enriching fructose and separating cane molasses.[12] Columns range from 2.0 to 12.0 m high and 0.5 to 4.0 m in diameter. Separation of up to 60,000 metric tons/yr of molasses in seven columns is planned at one site. The resin separates the sugars based on ion exclusion. Typical separation results for cane molasses are shown in Fig. 14.2-4.[12] Note that a perfect resolution is not obtained and is not needed.

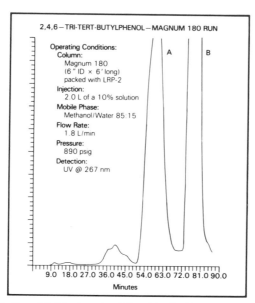

FIGURE 14.2-2 Purification of 2,4,6-tri-*tert*-butylphenol by liquid chromatography.[3] By permission of Whatman Chemical Separation, Inc.

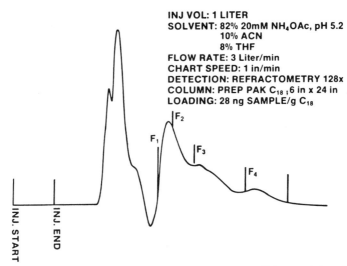

INJ VOL: 1 LITER
SOLVENT: 82% 20mM NH₄OAc, pH 5.2
         10% ACN
          8% THF
FLOW RATE: 3 Liter/min
CHART SPEED: 1 in/min
DETECTION: REFRACTOMETRY 128x
COLUMN: PREP PAK C₁₈ ; 6 in x 24 in
LOADING: 28 ng SAMPLE/g C₁₈

FIGURE 14.2-3 Purification of penicillin V by liquid chromatography.[15] By permission of Waters Associates, subsidiary of Millipore Corp.

14.2-2 Size Exclusion Chromatography (SEC)

Size exclusion chromatography (also known as gel permeation chromatography and gel filtration) is a liquid chromatographic technique, but there are two major differences. First, the mechanism is different. SEC is based on excluding large molecules from some of the pores. Since these molecules see a smaller fraction of the volume, they exit first from the column. In Eq. (14.1-5) K_{d_i} is smaller for the large molecules and hence they have a higher solute velocity. Many secondary effects can occur,[16] but size exclusion is the dominant mechanism. Usually, k_i in Eq. (14.1-5) is close to zero. Since the mechanism is size exclusion, it is difficult to tailor-make a packing for the fractionation of molecules of similar sizes. Thus, the most common large-scale application has been desalting where large molecules such as proteins are separated from small ones such as salts. A second consequence of the size exclusion mechanism is that all solutes (in the ideal case) must exit in the total void volume of the column. This is illustrated in Eq. (14.1-5) since $k_i = 0$ and K_{d_i} varies between zero and one.

The second difference between SEC and LC occurs when proteins or other biological materials are fractionated. To fractionate large molecules the packing must have very large pores. For protein separations the packing must be hydrophilic and have minimal interaction with the solutes. Up to now this combination

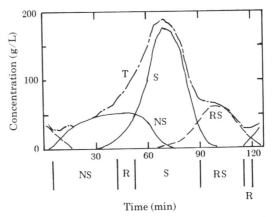

FIGURE 14.2-4 Purification of cane molasses by large-scale liquid chromatography.[12] Reprinted by special permission from *Chemical Engineering* (Jan. 24, 1983). Copyright 1983 by McGraw-Hill, Inc., New York, NY 10020. T = total concentration, S = sucrose, RS = reducing sugars, R = recycle.

of properties has been achieved with crosslinked dextran gels (Sephadex), crosslinked polyacrylamide gels (Biogel P), and various agarose based gels (Biogel A, Sepharose, Ultrogel).[17] Unfortunately, these gels are soft. This places severe flow rate and height limitations on the column and significantly changes the column design for fractionation. For desalting applications the gels are much more rigid and column design is similar to LC.

Since SEC is an essentially linear system, the theories developed in Sections 14.1-1 and 14.1-2 are applicable. A chromatographic separation thus looks like Fig. 14.1-2 with zone spreading added on. Since the mass transfer terms in Eq. (14.1-12) are inversely proportional to the diffusivity, large solutes that penetrate the pores have high H values. Desalting is an easy separation and has a relatively low H since the large molecules do not penetrate the pores.

Commercially, SEC has been used for desalting proteins and as one step in protein purification procedures.[16-19] The equipment for desalting is similar to LC equipment and is described in detail elsewhere.[17] The SEC desalting and buffer exchange of human blood plasma is shown in Fig. 14.2-5.[19] Repeated injections of the feed are illustrated. Size exclusion chromatography is used in blood banks for purification of various factors from human plasma either in conjunction with ion-exchange chromatography (see Section 14.2-4) or after Cohn ethanol fractionation.[18]

Fractionation of protein has not been so successful on a large scale but human serum albumin is separated from both salts and contaminating proteins by SEC as the last step in albumin purification.[18,19] Insulin also is purified on a large scale by SEC.[17] The gels used for fractionation of proteins are soft and cannot be packed in a single tall, large-diameter column. Instead, several short, fat columns ($L/D = 15/37$) are hooked together in series.[17,20] This procedure helps support the gel and solves some of the gel compression problem. The flow rates used are still quite low.[17,20] The use of short, fat columns in series does make distributor design very critical. This has been solved by connecting six pipes in a hexagonal pattern to the top and bottom of each column and with a 1 mm distribution channel above the bed.[17,20]

In the future we can anticipate that more-rigid gels will be available which are compatible with proteins. More-rigid gels are currently available[17,21] but they are not compatible with many biochemicals. Rigid synthetic polymer gels have been used on a preparative scale (up to 10 cm ID) to fractionate polymers.[21]

14.2-3 Gas-Liquid Chromatography (GLC)

There has been considerable interest in large-scale gas–liquid chromatography for many years, and some systems have been commercialized.[22,23] Many of the design principles are similar to the design of liquid chromatographic systems discussed in Section 14.2-1. In this section the major differences are discussed. Complete design details are discussed elsewhere.[14]

In large-scale GLC there is a tremendous difference in volume of sorbed and nonsorbed solutes. Thus the "sorption effect" is important unless the feed pulse is very dilute which would be rare in production systems. The sorption effect causes a sharpening of the front of the peak and a tailing of the rear. In GLC the isotherm effect does the opposite since sorption is favored as more solute dissolves in the stationary liquid phase. The two effects thus cancel to some extent and peaks can be quite sharp if operation is at the

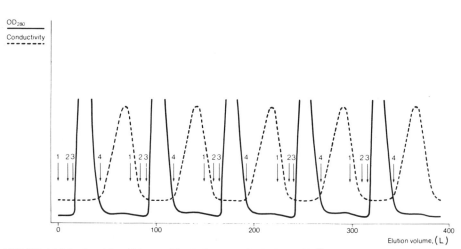

FIGURE 14.2-5 Desalting of human blood plasma on Sephadex H-25.[19] Column Sephamatic Gel filter GF 04-06; $V_t = 75$ L; flow rate 240 L/h; eluent, 0.025 M sodium acetate; sample, 10 L human/plasma. 1. Sample application; 2. 0.025 M sodium acetate; 3. plasma fraction; 4. 0.025 M sodium acetate. By permission of J. M. Curling, J. H. Berglof, S. Erikson, J. M. Cooney, and O. Person.[19]

optimum temperature. This optimum temperature, known as the Valentin temperature, can be calculated from[14,24,25]

$$P_A^0 = P_0 \frac{3}{4} \frac{P^4 - 1}{P^3 - 1} \tag{14.2-1}$$

where P_A^0 = vapor pressure of species A, P_0 is the outlet pressure, and P is the ratio of inlet to outlet pressure. Equation (14.2-1) assumes ideal solutions. The Valentin temperature is typically about 15°C above the boiling point of the main component which may be too high for the stationary liquid phase. Since the stationary phase slowly vaporizes and contaminates the products, a temperature which is at least 50°C below the limit used for analysis is desirable. As a rule of thumb at least half of the stationary phase should remain after two years.[25] The stationary phase should have as high an operating temperature as possible.

The presence of a liquid stationary phase has another effect. When solutes dissolve into the liquid, the liquid must expand. If the concentration of the feed pulse is too high, the stationary phase can overflood the pores of the support and wash down the column. This destroys the column. Thus, long feed pulses with a maximum total concentration between 0.2 and 0.5 are required.[14,24] The support for the stationary phase should be highly porous and inert.

The method of packing the column is absolutely critical and a very sharp cut of packing size is desirable. The best packing methods appear to be proprietary.[26]

Hydrogen or, if safety is a major problem, helium appears to be the best carrier gas. In the usual operating regime the mass transfer terms in Eq. (14.1-12) are controlling. Since these terms are inversely proportional to diffusivity, the column is more efficient when a light carrier gas is used.[24] Alternatively, the column can be operated at a higher velocity and the same H.

The results obtained in a pilot plant for preparation of pure *cis*- and *trans*-1,3-pentadiene are shown in Fig. 14.2-6.[24] The material between cuts 2 and 3 in Fig. 14.2-6b can be recycled. Note that the next

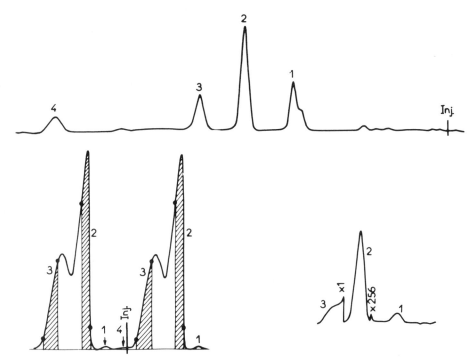

FIGURE 14.2-6 Preparation of pure *cis*- and *trans*-1.3-pentadiene. (*a*) Analysis of the feedstock. Column diameter = $\frac{1}{4}$ in., 6 m long packed with 80–100 mesh Chromosorb P coated with 10% 2.2 oxydipropionitrile; temperature 30°C. (1) isoprene; (2) *trans*-1,3-pentadiene; (3) *cis*-1,3-pentadiene; (4) cyclopentadiene. (b) Preparative chromatogram. Column diameter = 4 cm, 4 m long packed with 20% squalane on 60–80 mesh Chromosorb P; temperature 35°C, helium flow-rate 3 L/min (5 cm/s). Injection 4 mL every 20 min. The isoprene of one injection is eluted at about the same time as the cyclopentadiene of the previous injection. (c) Analysis of the *trans*-1,3-pentadiene prepared (purity 99.8%, cyclopentadiene ~1.5 ppm).[24] Reproduced from the *Journal of Chromatographic Science* by permission of Preston Publications, Inc.

injection is input before the last component, 4, has exited. For more difficult separations two-stage systems can be used.[22-24] Commercial applications include the production of very pure solvents, thiophene, and indole in the Soviet Union[23] and the purification of perfume ingredients and flavor chemicals[22,26] in the United States.

14.2-4 On–Off Chromatography. Biospecific Affinity and Ion-Exchange Chromatography

In on–off chromatography the distribution coefficient for the desired solute is very high at the conditions (temperature, pH, ionic strength, etc.) used for the feed step. Thus, the solute moves until it finds an open site, adsorbs to the site, and then stays there. Desorption is done by changing conditions so that the distribution coefficient is very small and all solute is removed quickly from the bed. Solute is concentrated in the product. Several elution steps can be used to elute different solutes separately. Elution can be done cocurrent or countercurrent to the feed step. The basic operating cycle for on–off chromatography is:

Feed	Solute adsorbs
Wash	Remove non-adsorbed material
Desorption 1	Elute solute 1
Desorption 2	Elute solute 2
Wash	Remove desorption materials

Note that these systems do not require long beds. No zone spreading occurs when the solute is attached to the sorbent. Short beds reduce mixing and zone spreading during the desorption steps. The faster cycles used with short beds increase the productivity of the sorbent.

In biospecific affinity chromatography a ligand which has a specific affinity for the biomolecule is attached to an inert solid support.[27,28] An example would be attaching an inhibitor for an enzyme to agarose. A huge number of these systems have been developed in laboratory studies. The solid support should be chemically stable, have large pores to allow access of the large molecules, have volumetric stability when conditions are changed, be easy to attach ligands to, and be hydrophilic. The most common supports have been crosslinked agaroses[17,27-29] which unfortunately are soft. Thus, the same design precautions discussed in Section 14.2-2 for soft SEC gels must be used.

Large-scale applications of biospecific affinity systems are just starting.[17,29,31] Many reports in the literature are for beds of 1 L or less in volume producing small amounts of enzyme.[17] The packing materials are very expensive since the supports are expensive, the ligands are often expensive, and the chemicals for attachment are expensive. This problem is exacerbated by the additional problem that the packing often has a relatively short life (5–20 cycles), which can increase costs drastically.[29] This occurs due to both loss of ligand from chemical cleavage and fouling of the packing by very tightly adsorbed materials. Careful control of all steps and better attachment chemistry can alleviate the ligand loss problem. Fouling can be controlled by precleaning the feed with precipitation and ion-exchange steps. When these problems have been solved, large-scale biospecific affinity chromatography will become more common.

Ion-exchange chromatography is often an on–off procedure.[17-19,30] A large number of hydrometallurgical solutions and wastes are purified in this way,[30] although the separations usually are not thought of as chromatographic. Since proteins are zwitterions, they can be made to go on or off the ion-exchange resin by changing the pH. Thus, the on–off operatinng cycle is useful for protein purification on a large scale.[17-19] Since one pass will not give a pure protein, a series of chromatographic steps are used. First, the feed is pretreated by centrifugation, precipitation, and so on and then passed through separate cation and anion columns. SEC is used for desalting and final cleanup and the protein is concentrated by ultra-filtration.[17-19] Commercially, human blood proteins, insulin, and egg-white lysozyme are purified in this way.[17] An example is shown in Fig. 14.2-7[19] for the partial purification of human serum albumin. Note that a large amount of nonsorbed material first passes through the column and that successive elution steps are used. The albumin fraction is then sent to a CM-Sepharose column for further purification.[18,19] For proteins, design considerations for ion exchange columns are similar to those for biospecific affinity chromatography and SEC.

14.3 COUNTERCURRENT AND SIMULATED COUNTERCURRENT SYSTEMS

A typical chemical engineering approach to a large-scale separation problem is to try to devise a steady-state, countercurrent system. This has led to methods that are quite different from the scale-up of analytical elution chromatography. First, approaches where the solid is moved are considered and then methods where the solid does not move but movement is simulated are discussed. Finally, the different systems are compared. In all cases the countercurrent or simulated countercurrent system replaces the column in Figure 14.2-1. The equipment for product and solvent recovery still is required.

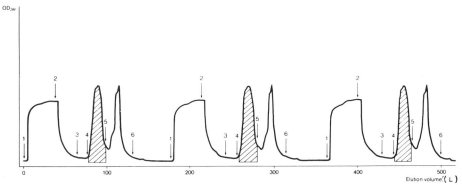

FIGURE 14.2-7 Partial purification of human serum albumin on DEAE Sepharose CL-6B. Column Pharamacia KS 370/15. $V_e = 16$ L; flow rate 24 L/h; eluents are sodium acetate buffers; sample, 42 L containing ~ 500 g albumin; product, 20 L containing ~ 500 g albumin. (1) Sample application; (2) sodium acetate pH 5.2, $I = 0.025$; (3) sodium acetate pH 4.5, $I = 0.025$; (4) elution of albumin fraction; (5) sodium acetate pH 4.0, $I = 0.15$; (6) sodium acetate pH 5.2, $I = 0.025$. Shaded portion is the albumin fraction. By permission of J. M. Curling, J. H. Berglof, S. Eriksson, J. M. Cooney and O. Persson.[19]

14.3-1 Moving-Bed Process

The goal of a countercurrent process is to have all regions of the bed doing useful separative work at all times. This minimizes the amount of sorbent and desorbent or carrier gas required. Minimizing sorbent requirements reduces capital costs for the separator while minimizing desorent or carrier gas reduces both capital and operating costs for the recovery and recycle equipment. A schematic of the basic countercurrent design is shown in Fig. 14.3-1. The system may be built in a single column or a series of columns or vessels may be used.

Like countercurrent distillation, the countercurrent chromatographic systems are binary separators; that is, they can make a cut between two key components. Most industrial applications have separated binary mixtures. Then the purpose of zone I in Fig. 14.3-1 is to remove the more strongly adsorbed solute B from A. Thus, in this zone solute A should go up and solute B should go down. In terms of the average solute velocities,

$$u_{Acc_1} > 0 > u_{Bcc_1} \tag{14.3-1}$$

In zone II solute A is removed from B and the separation condition is the same as in Eq. (14.3-1):

$$u_{Acc_2} > 0 > u_{Bcc_2} \tag{14.3-2}$$

In zone III the solid is regenerated by passing a solvent, desorbent, or carrier gas countercurrently to the solids flow. This zone also may be heated, particularly if a separate vessel is used. In this zone solute B should flow upward:

$$u_{Bcc_3} > 0 \tag{14.3-3}$$

The purpose of zone IV is to recover some of the solvent for recycle. This zone is optional but its use reduces the load on the downstream separator which separates A and S. In zone IV solute A should move downward:

$$u_{Acc_4} < 0 \tag{14.3-4}$$

For nonlinear systems these four equations must be satisfied for the fastest and slowest solute velocities for each solute.

All the solute velocities in a countercurrent system can be determined using the solute movement theory[1] developed in Section 14.1. The solute velocities developed in Eqs. (14.1-1), (14.1-3), and (14.1-5) are all with respect to the solid. Thus, in the countercurrent system the appropriate fluid velocity v is

$$v = \frac{v_{super}}{\alpha} + v_{solid} \tag{14.3-5}$$

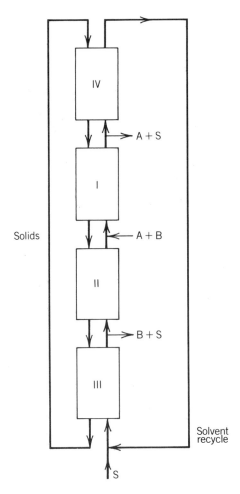

IV

A + S

I

Solids

A + B

II

B + S

III

Solvent
recycle

S

FIGURE 14.3-1 Schematic of equipment for countercurrent and simulated countercurrent separation.

where a perfect plug flow movement of the solids is assumed, and v_{super} is the superficial fluid velocity. The solute velocity u_i with respect to the solid uses this v in Eq. (14.1-3) or (14.1-5). The countercurrent solute velocity seen by a stationary observer is then

$$u_{CCi} = u_i - v_{solid} \qquad (14.3-6)$$

The solute movement diagram is plotted easily.[1]

Equations (14.3-1)–(14.3-4) show that the solute velocities must be varied in the different zones. Since there are input and output ports between the different zones, the fluid velocities must be different in the different zones. This varies v and hence u_{CCi}. This difference in fluid velocity is always helpful. A second way to vary the solute velocities is to change the equilibrium isotherm [k_i in Eq. (14.1-5)] by varying concentrations of desorbent in the zones.

In an ideal separation, the column sections shown in Fig. 14.3-1 could be infinitesimal in size. In actual practice, a finite depth is required because of the usual zone spreading phenomenon and because of mixing when the solids flow countercurrent to the fluid. Ideally, the systems work well, but practical problems have kept moving-bed systems from being a commercial success for chromatographic type separations. Methods for moving the solids have been studied extensively.[1–3] These have included commercial systems such as hypersorption[1,2] (which is now shut down) where the solid actually moved, and systems where the solid is held rigidly and rotates in a direction countercurrent to the fluid flow.[2–4] The systems with solids movement have been plagued with problems of solids attrition and axial mixing. A new development which may overcome these problems is to use magnetically stabilized moving beds.[5] This is an area with considerable recent patent activity. The rotating systems are mechanically very complex and are difficult to scale up to large sizes.

An alternative way of moving the solid is to move intermittent pulses of a slurry. This method is used commercially for ion exchange[6,7] and for activated carbon treatment of waste water.[8] In these applications a fractionation of solutes is not being done. Although axial mixing is undesirable, the requirements are not nearly as stringent as in chromatographic separations. The solute movement theory developed in Section 14.1 can be used easily to analyze the pulsed systems.[1] These pulsed moving-bed systems may be applicable to fractionation by ion exchange or affinity chromatography (see Section 14.2-4).

14.3-2 Simulated Moving-Bed Processes

The difficulties of actually moving the solid in a controlled plug flow led to the development of systems where the solid is stationary but movement is simulated. Simulated countercurrent systems have a long history, going back at least to the 1840s when the Shanks system for leaching soda ash was used in England. The application of simulated countercurrent systems applied to liquid chromatographic separations was first developed by U.O.P.[1,9-12,14] The countercurrent apparatus shown in Fig. 14.3-1 was simulated by subdividing each zone into a series of packed sections with plumbing in between. Every few minutes the positions of all inlet and outlet ports are moved up (in the direction of fluid flow). For an observer fixed at a port it looks like the solid has moved down. Thus, countercurrent motion is simulated. When a port reaches the top of the column, it starts over again at the bottom. In pilot plant studies[9,10] a series of columns arranged in a circle were used while in large-scale plants[9,11,12] a column is used. A complex rotating valve was used[9-13] to control the feed and product lines, but other arrangements can be used.[1] Additional information on these systems can be found in Chapter 12.

The solute movement theory developed in Section 14.1 is applied easily to the simulated moving-bed process.[1] In between shifts in the port location each section is a fixed bed. The solute velocities are then given by Eq. (14.1-3) or (14.1-5). The solute movement for a linear system can be plotted as shown in Fig. 14.3-2. Feed is introduced at all locations labeled A + B, but most of the solute waves are not shown

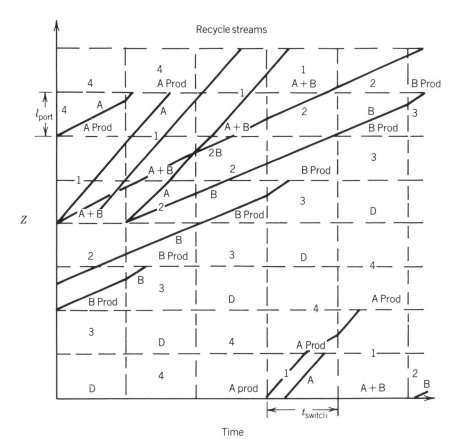

FIGURE 14.3-2 Solute movement theory for simulated countercurrent process.[1] Reprinted by permission of Corn Refiner's Association, Inc.

in the figure. In the simulated countercurrent systems, the solutes always move upward. The average rate of movement of the ports is

$$u_{port} = \frac{l_{port}}{t_{switch}} \tag{14.3-7}$$

where l_{port} is the height of the packing between ports and t_{switch} is the period between switching of the ports. To achieve the desired separation, the velocity of solute A should be greater than u_{port} in zones 1 and 2 and less in zone 4:

$$u_{A_1} > u_{A_2} > u_{port} > u_4 \tag{14.3-8}$$

The condition for solute B is

$$u_{B_3} > u_{port} > u_{B_1} > u_{B_2} \tag{14.3-9}$$

Equations (14.3-8) and (14.3-9) are equivalent to Eqs. (14.3-1)–(14.3-4). Equations (14.3-8) and (14.3-9) are satisfied in Fig. 14.3-2. Note that solute A exits in the A product port and B in the B product port. For nonlinear systems, Eqs. (14.3-8) and (14.3-9) should be satisfied for the fastest (most concentrated for adsorption) and slowest ($c \rightarrow 0$) moving waves for each solute. The larger the value of the selectivity α, the easier it is to satisfy these conditions.

The more sections in each zone and the faster switching is done the closer the simulation is to truly countercurrent operation. Four segments per zone closely approaches truly countercurrent operation.[1] However, this is not necessary since some commercial systems use only one segment (a separate column) per zone.[1]

In actual practice, the solute bands shown in Fig. 14.3-2 are spread by nonlinear isotherms, diffusion, and mass transfer resistances. These effects can be included in an equilibrium staged model of the apparatus.[9,12] In addition, any mixing in the space between the packed sections broadens the solute bands. Thus, in addition to the usual care in packing the column, the distribution and plumbing in the spaces between sections must be designed to minimize mixing. This is probably easiest to do by using a separate column for each section, but this may not be the most economical way in a large-scale system. To simulate countercurrent motion, each packed section or column must be as close to identical as possible. The columns must be packed and tested carefully.

A large number of commercial and experimental results have been reported.[1,9-12] The products from a truly countercurrent process would be constant. In a simulated countercurrent process, the product concentration varies, but the cycle repeats every few minutes when the valves are switched. It is common to pool each product and report the average compositions. Table 14.3-1[12] gives the results for a commercial Parex purification of p-xylene. A specially formulated molecular sieve adsorbent was used and p-diethylbenzene was the desorbent. The p-xylene is the most strongly adsorbed species. A typical plant for producing p-xylene would produce 100,000 metric tons/yr of p-xylene product.

The separations can be custom designed. Figure 14.3-3[10] shows the pilot plant separation of ethylbenzene from a mixed xylenes feed. The adsorbent and desorbent are different than in the Parex system. In this case ethylbenzene is the least adsorbed component (A in Figs. 14.3-1 and 14.3-2). Figure 14.3-3 shows the composition profiles along the column (actually, 24 small columns were used). The profile looks like a countercurrent profile and all parts of the apparatus are doing useful separation. The xylenes are not separated in Fig. 14.3-3. If both pure p-xylene and ethylbenzene were required from a mixed xylenes-ethylbenzene feed, two simulated countercurrent separators would be needed.

The simulated countercurrent method is a general idea that can be applied also to size exclusion

Table 14.3-1 Commercial Parex Operation with p-Diethylbenzene. Separation of p-Xylene from Crystallizer Mother Liquor.[12]

	(wt.%) Feed	(wt%) Extract	Extract After Toluene Removal	(wt%) Raffinate	Product Recovery (%)
Nonaromatics	0.29	0.00	0.00	0.29	
Toluene	0.45	1.54	0.00	0.28	
Ethylbenzene	12.38	0.34	0.35	14.10	
p-Xylene	11.76	97.94	99.48	0.53	96.7
m-Xylene	62.96	0.09	0.09	70.93	
o-Xylene	12.16	0.09	0.08	13.87	

Source: Reprinted with permission of Sijthoff and Noordhoof International Publishers.

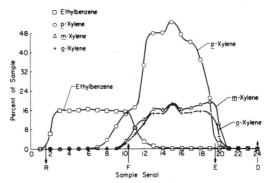

FIGURE 14.3-3 Pilot plant separation of ethylbenzene from a mixed xylenes feed by simulated counter-current system.[10] Reprinted with permission from *Ind. Eng. Chem. Process Des. Dev.*, **15**, 261 (1976). Copyright 1976 American Chemical Society.

separators and to gas chromatography. This has been done experimentally,[1,4,13] but large-scale commercial applications have not been reported. The basic idea for the simulated moving bed is the same, but the details of the separation differ. The fractionation of dextran into two different molecular weight distributions is shown in Fig. 14.3-4. A system with 20 columns packed with 200–400 μm Spherosil XOB075 was used. This system did not use a zone IV. The bottom part of Fig. 14.3-4 shows the variation of total dextran concentration along the apparatus while the top part indicates K_d values which are a measure of molecular weight. The separation of the products is not sharp, but they have clearly different molecular weight distributions. The system can operate with quite high dextran concentrations and successfully treated a feed that was 10% wt.% dextran. Possible commercial applications would include desalting or fractionating proteins and production of polymers with controlled molecular weight distributions.

Extensive experimental studies of simulated countercurrent gas–liquid chromatography have been done.[13] A system with 12 chromatography columns was enclosed in an oven. The system did not use a zone IV. Figure 14.3-5[13] shows one run for the separation of ethylcaprate from ethyl laurate. The columns were packed with 15% OV-275 on chromosorb P and were operated at 160°C with nitrogen as the carrier. This is a fairly difficult separation with a separation factor of 1.44. For a feed flow rate of 75 cm³/h (Fig. 14.3-5) the ethyl caprate product was 94.2% pure and the ethyl laurate was 91.2% pure. When the feed flow rate was lowered to 25 cm³/h, the products were 99.4 and 99.3% pure, respectively. The results taken

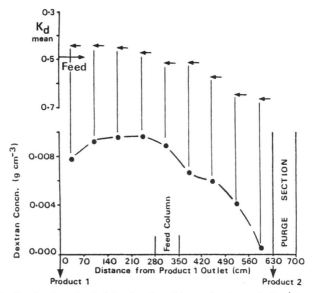

FIGURE 14.3-4 Simulated moving-bed fractionation of dextran by size exclusion.[4] By permission of Ellis Horwood Limited, Publishers, Market Cross House, Cooper Street, Chichester, England.

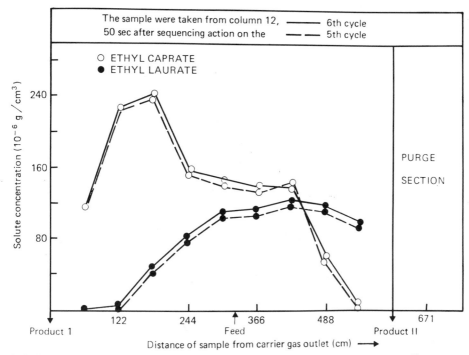

The sample were taken from column 12, —————— 6th cycle
50 sec after sequencing action on the — — — 5th cycle

○ ETHYL CAPRATE
● ETHYL LAURATE

PURGE

SECTION

Solute concentration (10^{-6} g/cm^3)

240

160

80

122   244   ↑ 366   488   671
Product 1   Feed   Product II

Distance of sample from carrier gas outlet (cm) ⟶

FIGURE 14.3-5   Simulated moving-bed separation of ethyl caprate from ethyl laurate by GLC.[13] Reprinted with permission from the *Canadian Journal of Chemical Engineering*, **57**, 42 (1979).

during two different cycles show that a repeating steady state is obtained. Staged models can be used for design or simulation. Larger throughputs probably could be obtained by operating near the Valentin temperature and using helium or hydrogen as the carrier gas.

14.4   HYBRID SYSTEMS

Scaled-up elution chromatography and simulated countercurrent systems both have advantages and disadvantages. It makes sense to try to combine these two methods and develop hybrid systems that have some of the characteristics of both the other types.

14.4-1   Column Switching Methods

When several compounds in the feed are to be purified, the difficulty of the separation is controlled by the two components with the closest selectivities. Thus, the column length and the period of the feed pulse are chosen based on these key components. Other components will come out either much quicker or much slower than the key components.

The time when the next pulse can be input is controlled by the difference between the retention times of the slowest and fastest components. Thus, total throughput is controlled by the easy to separate compounds. The easier the separation of these nonkey components the *lower* the throughput since the retention times of the fastest and slowest components will differ greatly.

To eliminate the throughput problem caused by nonkey components, each nonkey component should be removed as soon as it is purified. This can be accomplished with column switching methods.[1-7] A variety of ways this can be done are shown in Fig. 14.4-1.[1] In Fig. 14.4-1A the fast-moving product is removed at port 1. The key components are sent on to the second column which can use a different chromatographic packing. The slowest-moving component is removed at port 1. The solute movement diagram for this is shown in Fig. 14.4-2.[1] In essence, column A separates component 1 from 2 and 3 and components 2 and 3 from 4; column B separates 2 from 3. Note that the next feed pulse can be input much sooner than would be the case if a column long enough to separate components 2 and 3 had been used. This type of column switching can be extended easily to a series of columns as shown in Fig. 14.4-1B. This technique is called "moving withdrawal chromatography"[1,4,6] and has been simulated with a computer model.[6] Large increases in throughput can be obtained compared to normal operation.

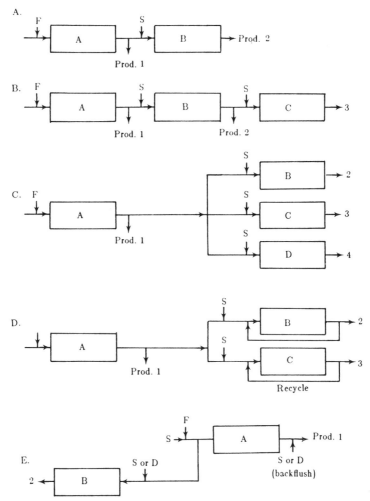

FIGURE 14.4-1   Column switching methods.[1] Reprinted with permission from P. C. Wankat, *Large-Scale Adsorption and Chromatography*, CRC Press, Boca Raton, FL, 1986.

Column switching is also useful instead of recycle for further separation of partially separated peaks. This is shown in Fig. 14.4-1C.[1] Heart cuts which are pure are withdrawn at port 1 while the overlapping tails are sent to other columns for further separation. This is advantageous compared to recycle since the partial separation achieved in column A is retained when transferring columns. This procedure has been used in large-scale systems[5] and can be combined with recycle as shown in Fig. 14.4-1D.[1]

When the slow components are adsorbed very strongly, the arrangement shown in Fig. 14.4-1E.[1] is useful. The faster products are removed at port 1 while slow components are backflushed, often with a desorbent. The backflushed material can be sent to another column for further separation. Backflush is particularly useful when column A is a guard column and product 1 is sent to another column for further separation.

14.4-2   Moving-Port Chromatography

The column switching methods reduce the regions near the product end where no useful separation is being done. The inefficient regions near the feed can be reduced by moving-feed chromatography[1,4,6] where the location the feed is introduced is moved up the column during the period of the feed pulse. Moving feed is most useful when it is operated in conjunction with moving withdrawal. The result has been called moving-port chromatography.[6]

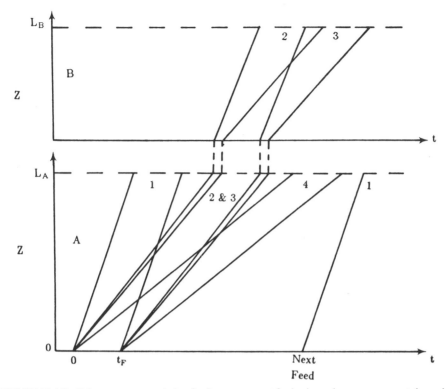

FIGURE 14.4-2 Solute movement solution for four-component feed using column arrangement shown in Figure 14.4-1A.[1] Reprinted with permission from P. C. Wankat, *Large-Scale Adsorption and Chromatography*, CRC Press, Boca Raton, FL, 1986.

A column arrangement similar to Fig. 14.4-3 could be used for moving-port chromatography. After each column section a product can be withdrawn and a feed or solvent stream can be injected or the stream can be sent to the next column. When a product stream is withdrawn at an intermediate point, the entire stream is withdrawn and new solvent or carrier gas is injected into the next column section. This is different than a simulated countercurrent system where most of the stream continues up the column, and new solvent is not injected. Analysis by the theoretical methods of Section 14.1 is straightforward although the graphs can become cluttered.[4] A computer simulation showed large increases in both throughput and resolution when a moving-port system was compared to normal operation.[6]

Moving-port chromatography is more complex than normal preparative chromatography. It is slightly less complex than simulated countercurrent systems. The moving-port system can be converted into a simulated countercurrent system by recycling the top product, continually moving the ports around the column as if it were a donut, withdrawing only part of the stream as products, and adding solvent at only one location in the column. When this is done, the apparatus loses its ability to separate multicomponent mixtures. Whether there will be industrial applications of moving-port chromatography remains to be seen.

14.5 OTHER ALTERNATIVES

A variety of other ingenious schemes have been developed which could be applied to large-scale chromatography. In this section these methods are reviewed briefly.

14.5-1 Two-Dimensional and Rotating Designs

In the search to develop continuous chromatographic processes, a variety of two-dimensional designs have been developed. The most popular design has been the rotating annulus shown in Fig. 14.5-1. An annulus packed with sorbent is rotated slowly past the feed point. Solvent or carrier gas flows continuously downward in the remainder of the annulus. Solutes flow downward with the carrier and are carried around by

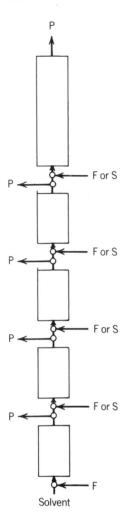

FIGURE 14.4-3 Schematic of apparatus for moving-port chromatography: F = Feed, S = Solvent, P = Products. Reprinted with permission from *IEC Process Des. Dev.*, **23**, 256 (1984). Copyright 1984 American Chemical Society.

the rotation of the annulus. The result is that one solute traces a helical path and is separated from the other solutes since it spends a different amount of time on the solid. Operation is steady state. The second direction θ effectively replaces the time coordinate in one-dimensional elution chromatography. If dispersion effects are ignored this apparatus is exactly analogous to one-dimensional elution chromatography,[1] and the output will look like Fig. 14.1-2 or 14.1-3 with θ/ω replacing t.

The rotating annulus system is probably one of the world's most reinvented ideas. It has been applied to gas chromatography, liquid chromatography, gel permeation chromatography, and paper chromatography.[2,12,13] The annulus has been replaced by paper or by a series of packed columns arranged in a circle like a Gatling gun. Other two-dimensional configurations also have been devised such as feeding radially from the center of the apparatus or using rotating flat plates to develop a continuous two-dimensional thin-layer chromatography.[2,12,13] Two-dimensional devices also have been adapted to utilize changes in thermodynamic variables to force separation (see Section 14.5-2).[1,2,12,13] Related two-dimensional designs have been used for continuous recovery of solvent with activated carbon, heat recovery, and extraction.[1,13]

There is no doubt that the two-dimensional designs can be made to work. However, they are mechanically more complex than the preparative chromatographs they would replace. In addition, the uniform packing of a single, large-diameter column is simpler than uniformly packing an annulus or a series of columns. Despite these difficulties a fairly large-scale pressurized rotating annulus system has been developed successfully.[3] The economics of this system were not discussed. This type of equipment probably will have only limited applications in special circumstances.

A rather different application of rotating equipment is to do normal elution chromatography in a centrifuge (chromatofuge). For example, in a basket type centrifuge first a layer of packing is laid down by

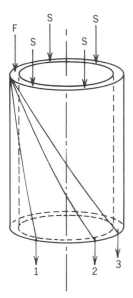

FIGURE 14.5-1 Rotating annulus system for steady-state, two-dimensional chromatography.

feeding in a slurry. Then, while solvent is fed to the centrifuge, a pulse of feed is introduced. The solutes develop radially outward. A second pulse can then be added. This type of device has been studied intermittently for years[2,4] and a rotating disk thin layer chromatograph is now commercially available.[4] The chromatofuge may have applications with very viscous solutions or with unstable compounds where short residence times are desirable. It has been used commercially for gel filtration separation of proteins from lactose.

14.5-2 Methods Using Changes in Thermodynamic Variables

Isotherms change when thermodynamic variables such as temperature, pressure, or pH are varied. When the isotherm changes, the solute velocity much change. For example, in Eqs. (14.1-4) and (14.1-5) k_i is a function of temperature. For adsorption systems, if temperature is increased k_i decreases and hence $u_{\text{solute},i}$ must increase. This phenomenon has been used to develop several different separation methods.

In chromatothermography[5-7] a moving furnace travels along a chromatographic column. A moving temperature gradient is developed ahead of this furnace. If the furnace temperature and velocity are selected appropriately, Eq. (14.1-5) predicts that there is a temperature for each solute where its solute velocity equals the furnace velocity. At temperatures lower than this, the solute moves slower than the furnace and can be overtaken by the moving furnace. At higher temperatures the solute moves faster than the furnace. The result is that each solute concentrates at the moving location where its velocity equals the furnace velocity. Feed of material to the column is continuous and products come out quite concentrated. This procedure works quite well on a small scale and was commercially available as an analytical instrument in the Soviet Union during the 1960s. One alternate which may be easier to scale-up is to have a series of fixed jackets around the column. Each jacket can be heated or cooled to simulate movement of a hot wave. This procedure also works well on a small scale[8] and probably could be scaled up as a shell-and-tube heat exchanger. The analysis methods developed in Section 14.1 can be applied to these problems where energy balances and temperature effects are included.[6,8,12]

A variety of cyclic methods including parametric pumping, cycling zone adsorption, and pressure swing adsorption have been developed which use changes in thermodynamic variables to force separation.[9,12] These methods usually have been applied to removing all solutes from a solvent or carrier gas, but some fractionation or chromatographic methods have be developed.[8-12] Cycling zone adsorption[8-10,12] is quite similar to chromatothermography except fixed heat exchange jackets or changes in the feed temperature are used instead of a moving furnace. Multicomponent separations are obtained again by moving a temperature wave through the column at a velocity between the velocities of all solutes at the highest and lowest temperatures. Large-scale equipment using changes in the feed temperature have been designed. Parametric pumping[9,12] is a similar process except the flow is reversed periodically and reflux is used. This has advantages for removal of one solute but can have disadvantages for multicomponent fractionation. Multicomponent ion-exchange separations[9] and enzyme separations on two columns using affinity resins[11] have been reported. These methods are probably harder to scale up than cycling zone adsorption. The theoretical methods developed in Section 14.1 are adapted easily to these problems.[8-12]

14.6 SYSTEM COMPARISONS

Which operating system should be used for a particular separation problem? Should the system be operated in the gas or liquid phase? Should adsorption, partition, size exclusion, or ion exchange be used? These questions ultimately revert to economics. At the current state of the art, exact answers are not available. However, some qualitative comparisons and general rules can be drawn which should be helpful in selecting the process. The important question of whether a chromatographic or other type of separation should be used is not addressed.

The choice of a gas or liquid system may be very simple. Compounds with very low vapor pressures or which decompose easily require a liquid system. Thus, the recent efforts in biochemical and genetic engineering have focused on liquid systems. Compounds which are a gas at normal temperatures and pressures probably are cheapest to separate in a gas system. For separation problems which can use either gas or liquid systems one needs to consider the availability of stationary phases, the cost of both the chromatographic system and the downstream separators, and the availability of commercial vendors or well-known design techniques. GLC has the advantage that a large number of stationary phases are available and a costly development effort probably is not required to find a system with at least a modest separation factor. However, the possibility for very large α values for similar compounds probably is greater for adsorption systems. This is illustrated by biospecific affinity systems. Packing costs are less for GLC when compared to bonded phase LC packings or affinity packings, but not when compared to commercial adsorbents and ion-exchange resins. However, packing costs per kilogram are reduced greatly when large quantities of a packing are ordered. The downstream separators are quite simple in gas systems, but energy costs for evaporation, refrigeration, and compression may be fairly high. Downstream separation costs for liquid systems are heavily dependent on the desorbent used. Moving-feed and simulated moving-bed systems are commercially available for liquid adsorption and ion-exchange systems. If one of these techniques was used with a gas system, some further development probably would be necessary. The choice of gas or liquid may depend on company experience with the different types of system. Finally, the choice comes down to economics. If a vendor is to be used, quotations with suitable performance guarantees can be obtained and different systems can be compared on this basis.

The choice of operating method also involves multiple decisions. The complexity of the equipment increases in the following order:

1. Scaled-up chromatography
2. Chromatography with multiple elution steps
3. Moving withdrawal
4. Moving-port chromatography
5. Simulated moving-bed
6. Moving bed.

The thermodynamic efficiency of the system generally increases in the same order. Thus, the more complex systems probably produce a more concentrated product and have smaller downstream separators. The very selective bioaffinity methods (and perhaps ion exchange) are an exception. For small-scale systems efficiency is not critical and the simpler systems are preferred. The simpler systems also are preferred when the same unit produces multiple products and frequent changeovers are required. Units dedicated to a single product can be optimized for that product and thus can be more complex. This is modified by the number of components in the feed which are desired as product. (The feed should be preseparated by other separation methods.) Countercurrent and simulated countercurrent systems do a binary split. Thus, if a single separator is to be used the desired product must be either the slowest- or the fastest-moving component. In current applications of simulated moving-bed systems, one component is the product and the other components usually are recycled to a reactor. The countercurrent techniques have a disadvantage when center cuts or several products are desired from a single feed and are unlikely to be used in these cases.

The ready availability of the equipment also needs to be considered. Use of moving-withdrawal or moving-port chromatography currently would require some development work, but the application is fairly straightforward. An economical, reliable, moving-bed system for chromatographic separations has not been demonstrated at this time. Until these systems are demonstrated, which may happen fairly soon, moving-bed systems are not a feasible alternative. When such systems are developed, they will compete with simulated countercurrent separators.

Special situations can modify these suggestions. For on–off chromatography, short fat columns are very useful, particularly if compressible gels must be used. Since these systems are similar to water softening by ion exchange, the pulsed moving-bed methods used there may be applicable (as yet the scale has been fairly small). For viscous solutions the chromatofuge is a viable alternative.

ACKNOWLEDGMENT

The hospitality of the Laboratoire des Sciences du Genie Chimique of the École Nationale Supérieure des Industries Chimiques in Nancy, France, is acknowledged gratefully. This work was supported partially by a NSF/CNRS exchange program.

REFERENCES

14.1-1. T. K. Sherwood, R. L. Pigford, and C. R. Wilke, *Mass Transfer*, Chap. 10, McGraw-Hill, New York, 1975.

14.1-2 P. C. Wankat, Cyclic Separation Techniques, in A. E. Rodrigues and D. Tondeur (Eds.), *Percolation Processes, Theory and Applications*, pp. 443–516, Sijthoff and Noordhoff, Alphen aan den Rijn, Netherlands, 1981.

14.1-3 F. Helfferich and G. Klein, *Multicomponent Chromatography*, Marcel Dekker, New York, 1970.

14.1-4 H.-K. Rhee, Equilibrium Theory of Multicomponent Chromatography, in A. E. Rodrigues and D. Tondeur (Eds.), *Percolation Processes, Theory and Applications*, pp. 285–328, Sijthoff and Noordhoff, Alphen aan den Rijn, Netherlands, 1981.

14.1-5 L. Jacob and G. Guiochon, Theory of Chromatography at Finite Concentrations, *Chromatogr. Rev.*, **14**, 77 (1971).

14.1-6 J. C. Giddings, *Dynamics of Chromatography, Part I. Principles and Theory*, Marcel Dekker, New York, 1965.

14.1-7 E. N. Lightfoot, R. J. Sanchez-Palma, and D. O. Edwards, in H. M. Schoen (Ed.), *New Chemical Engineering Separation Techniques*, p. 125, Interscience, New York, 1962.

14.1-8 H. Engelhardt, *High Performance Liquid Chromatography*, Chap. 2, Springer-Verlag, New York, 1979.

14.1-9 W. W. Yau, J. J. Kirkland, and D. D. Bly, *Modern Size-Exclusion Liquid Chromatography*, Chap. 3, Wiley, New York, 1979.

14.1-10 L. R. Snyder and J. J. Kirkland, *Introduction to Modern Liquid Chromatography*, 2nd ed., Wiley, New York, 1979.

14.1-11 J. H. Knox, Practical Aspects of LC Theory, *J. Chromatogr. Sci.*, **15**, 352 (Sept. 1977).

14.1-12 P. C. Wankat, *Large Scale Adsorption and Chromatography*, CRC Press, Boca Raton, FL, 1986.

Section 14.2

14.2-1 "Series 200 LC. Design, Operation, and Performance of an Industrial Scale High Performance Liquid Chromatography (HPLC) System—Preliminary Technical Notes," Elf Aquitane Development, New York (no date).

14.2-2 "New Waters Kiloprep Process Scale Separation Systems," Waters Division, Millipore, Milford, MA, 1983.

14.2-3 "Whatman Magnum Production LC," Whatman Separation Inc., Clifton, NJ, 1983.

14.2-4 F. W. Karasek, Laboratory Prep-scale Liquid Chromatography, *Res. Dev.*, **28**, 32 (July 1977).

14.2-5 L. Ettre, Preparative Liquid Chromatography: History and Trends: Supplemental Remarks, *Chromatographia*, **12**, 302 (1979).

14.2-6 L. R. Snyder and J. J. Kirkland, *Introduction to Modern Liquid Chromatography*, 2nd ed., Chap. 15, Wiley, New York, 1979.

14.2-7 E. Geeraert and M. Verzele, Preparative Liquid Chromatography: History and Trends, *Chromatographia*, **11**, 640 (1978).

14.2-8 M. Verzele and E. Geeraert, Preparative Liquid Chromatography, *J. Chromatogr. Sci.*, **18**, 559 (1980).

14.2-9 A. J. de Rosset, R. W. Neuzil, and D. B. Broughton, Industrial Applications of Preparative Chromatography, in A. E. Rodrigues and D. Tondeur (Eds.), *Percolation Processes, Theory and Applications*, pp. 249–281, Sijthoff and Noordhoff, Alphen aan den Rijn, Netherlands, 1981.

14.2-10 P. C. Wankat, An Engineer's Perspective—Increasing the Efficiency of Packed Bed Chromatographic Separations, in B. Bidlingmeyer (Ed.), *Solving Problems with Preparative Liquid Chromatography*, Chap. 9, Elsevier, Bronxville, NY, in press.

14.2-11 M. Seko, H. Takeachi, and T. Inada, Scale-up for Chromatographic Separation of *p*-Xylene and Ethylbenzene, *Ind. Eng. Chem. Prod. Res. Dev.*, **21**, 656 (1982).

14.2-12 H. Heikkila, Separating Sugars and Amino Acids with Chromatography, *Chem. Eng.*, 50 (Jan. 24, 1983).

14.2-13 J. L. Glajch and J. J. Kirkland, Optimization of Selectivity in Liquid Chromatography, *Anal. Chem.*, **55**, 319A (Feb. 1983).

14.2-14 J. R. Conder, Design Procedures for Preparative and Production Gas Chromatography, *J. Chromatogr.*, **256**, 381 (1983).

14.2-15 B. A. Bidlingmeyer, Waters Assoc., private communication, 1983.

14.2-16 W. W. Yau, J. J. Kirkland, and D. D. Bly, *Modern Size Exclusion Liquid Chromatography,* Wiley, New York, 1979.

14.2-17 J.-C. Janson and P. Hedman, Large-Scale Chromatography of Proteins, in A. Fiechter (Ed.) *Advances in Biochemical Engineering,* Vol. 25, *Chromatography,* pp. 43–99, Springer-Verlag, Berlin, 1982.

14.2-18 J. M. Curling, (Ed.), *Separation of Plasma Proteins,* Joint Meeting of the 19th Congress of the International Society of Haematology and the 17th Congress of the International Society of Blood Transfusion, Budapest, Aug. 1–7, 1982, Pharmacia Fine Chemicals AB, Uppsala, Sweden, 1983.

14.2-19 J. M. Curling, J. H. Gerglof, S. Eriksson, and J. M. Cooney, "Large Scale Production of Human Albumin by an All-Solution Chromatographic Process," Joint Meeting of the 18th Congress of the International Society of Haematology and the 16th Congress of the International Society of Blood Transfusion, Montreal, Quebec, Canada, Aug. 16–22, 1980.

14.2-20 *Pharmacia Sectional Columns KS 370, the Stack. Instructional Manual,* Pharmacia Fine Chemicals AB, Uppsala, Sweden, 1982.

14.2-21 M. F. Vaughan, and R. Dietz, Preparative Gel Permeation Chromatography (GPC), in R. Epton (Ed.), *Chromatography of Synthetic and Biological Polymers,* Vol. 1, pp. 199–217, Ellis Horwood, Chichester, U.K., 1978.

14.2-22 R. G. Bonmati, C. Chapelet-Letournex, and J. R. Margulis, Gas Chromatography—Analysis to Production, *Chem. Eng.,* 70 (Mar. 24, 1980).

14.2-23 K. I. Sakodynskii, S. A. Volkov, Yu. A. Kovan'ko, V. Yu. Zel'venskii, V I. Rezmkov, and V. A. Averin, Design of and Experience in Operating Technological Preparative Installations, *J. Chromatogr.,* **204,** 167 (1981).

14.2-24 B. Roz, R. Bonmati, G. Hagenbach, P. Valentin, and G. Guiochon, Practical Operation of Prep-Scale Gas Chromatographic Units, *J. Chromatogr. Sci.,* **14,** 367 (1967).

14.2-25 P. Valentin, Design and Optimization of Preparative Chromatographic Separations, in A. E. Rodrigues and D. Tondeur (Eds.), *Percolation Processes, Theory and Applications,* pp. 141–195, Sijthoff and Noordhoff, Alphen aan den Rijn, Netherlands, 1981.

14.2-26 R. Bonmati, J. R. Margulis, and G. Chapelet, "Gas Chromatography—A New Industrial Separation Process." Elf Technologies, New York (no date).

14.2-27 T. C. J. Gribnau, J. Visser, and R. J. F. Nivard (Eds.), *Affinity Chromatography and Related Techniques,* Elsevier, Amsterdam, 1982.

14.2-28 W. Scouten, *Affinity Chromatography: Bioselective Adsorption on Inert Matrices,* Wiley, New York, 1981.

14.2-29 J. C. Janson, Scaling-up of Affinity Chromatography, Technical and Economic Aspects, in T. C. J. Gribnau, J. Visser, and R. J. F. Nivard (Eds.), *Affinity Chromatography and Related Techniques,* pp. 503–512, Elsevier, Amsterdam, 1982.

14.2-30 H. Gold and C. Calmon, Ion Exchange: Present Status, Needs and Trends, *AIChE Symp. Ser.,* 76, 192, 60 (1980).

14.2-31 P. C. Wankat, *Large Scale Adsorption and Chromatography,* CRC Press, Boca Raton, FL, 1986.

Section 14.3

14.3-1 P. C. Wankat, *Large Scale Adsorption and Chromatography,* CRC Press, Boca Raton, FL, 1986.

14.3-2 M. V. Sussman, Continuous Chromatography, *Chemtech,* **6,** 260 (1976).

14.3-3 P. E. Barker, Developments in Continuous Chromatographic Refining, in J. H. Knox (Ed.), *Developments in Chromatography—1,* p. 41, Applied Science Publishers, Barking, Essex, U.K., 1978.

14.3-4 P. E. Barker, F. J. Ellison, and B. W. Hatt, Continuous Chromatography of Macromolecular Solutes, in R. Epton (Ed.), *Chromatography of Synthetic and Biological Polymers,* Vol. 1, pp. 218–239, Ellis Horwood, Chichester, U.K., 1978.

14.3-5 R. E. Rosensweig, Magnetic Stabilization of the State of Uniform Fluidization, *Ind. Eng. Chem. Fundam.,* **18,** 260 (1979).

14.3-6 M. Streat, Recent Developments in Continuous Ion Exchange, *J. Sep. Proc. Technol.,* **1,** 10 (1980).

14.3-7 M. J. Slater, Recent Industrial-Scale Applications of Continuous Resin Ion Exchange Systems, *J. Sep. Proc. Technol.,* **2,** 2 (1981).

14.3-8 T. D. Reynolds, *Unit Operations and Processes in Environmental Engineering,* Chap. 6, Brooks/Cole, Monterey, CA., 1982.

14.3-9 D. B. Broughton, R. W. Neuzil, J. M. Pharis, and C. S. Breasley, The Parex Process for Recovering Paraxylene, *Chem. Eng. Prog.,* **66,** 70 (Sept. 1970).

14.3-10 A. J. de Rosset, R. W. Neuzil, and D. J. Korous, Liquid Column Chromatography as a Predictive Tool for Continuous Counter-Current Adsorptive Separations, *Ind. Eng. Chem. Proc. Des. Dev.*, **15**, 261 (1976).

14.3-11 R. W. Neuzil, D. H. Rosback, R. H. Jensen, J. R. Teague, and A. J. de Rosset, An Energy-Saving Separation Scheme, *Chemtech*, **10**, 498 (1980).

14.3-12 A. J. de Rosset, R. W. Neuzil, and D. B. Broughton, Industrial Applications of Preparative Chromatography, in A. E. Rodrigues and D. Tondeur (Eds.) *Percolation Processes, Theory and Applications*, pp. 249–281, Sijthoff and Noordhoff, Alphen aan den Rijn, Netherlands, 1981.

14.3-13 P. E. Barker, S. E. Liodakis, and M. I. Howari, Separation of Organic Mixtures by Sequential Gas–Liquid Chromatography, *Can. J. Chem. Eng.*, **57**, 42 (1979).

14.3-14 D. Ruthven, *Principles of Adsorption and Adsorption Processes*, Wiley, New York, 1984.

Section 14.4

14.4-1 P. C. Wankat, *Large Scale Adsorption and Chromatography*, CRC Press, Boca Raton, FL, 1986.

14.4-2 L. R. Snyder and J. J. Kirkland, *Introduction to Modern Liquid Chromatography*, 2nd ed., Chap. 16, Wiley, New York, 1979.

14.4-3 K. I. Sakodynskii, S. A. Volkov, Yu. A. Kovan'ko, V. Yu. Zel'venskii, V. I. Rezmkov, and V. A. Averin, Design of and Experience in Operating Technological Preparative Installations, *J. Chromatogr.*, **204**, 167 (1981).

14.4-4 P. C. Wankat, Improved Preparative Chromatography—Moving Port Chromatography, *Ind. Eng. Chem. Fundam.*, **23**, 256 (1984).

14.4-5 M. Seko, H. Takeachi, and T. Inada, Scale-up for Chromatographic Separation of *p*-Xylene and Ethylbenzene, *Ind. Eng. Chem. Prod. Res. Dev.*, **21**, 656 (1982).

14.4-6 G. Miller, and P. C. Wankat, Moving Port Chromatography: A Method of Improving Preparative Chromatography, *Chem. Eng. Commun.*, 1984, **31**, 21.

14.4-7 L. R. Snyder, J. W. Dolan, and Sj. van der Waal, Boxcar Chromatography. A New Approach to Increased Analysis Rate and Very Large Column Plate Numbers, *J. Chromatogr.*, **203**, 3 (1981).

Section 14.5

14.5-1 P. C. Wankat, The Relationship Between One-Dimensional and Two-Dimensional Separation Processes, *AIChE J.*, **23**, 859 (1977).

14.5-2 M. V. Sussman, Continuous Chromatography, *Chemtech*, **6**, 260 (1976).

14.5-3 R. M. Canon, J. M. Begovich, and W. G. Sisson, Pressurized Continuous Chromatography, *Sep. Sci. Technol.*, **15**, 655 (1980).

14.5-4 R. J. Laub and D. L. Zink, Rotating-Disk Thin-Layer Chromatography, *Am. Lab.*. 55 (Jan. 1981).

14.5-5 A. A. Zhukhovitsky, Some Developments in Gas Chromatography in the USSR, in R. P. W. Scott (Ed.), *Gas Chromatography 1960*, pp. 293–300, Butterworths, London, 1960.

14.5-6 A. P. Tudge, Studies in Chromatographic Transport III. Chromathermography, *Can. J. Phys.* **40**, 557 (1962).

14.5-7 M. B. Moshinskaya and M. S. Vigdergauz, The Evolution of the Construction and Manufacturing of Gas Chromatographs in the Soviet Union, *J. Chromatogr. Sci.*, **16**, 351 (1978).

14.5-8 S. C. Foo, K. H. Bergstrom, and P. C. Wankat, Multicomponent Fractionation by Direct, Thermal Mode Cycling Zone Adsorption, *Ind. Eng. Chem. Fundam.*, **19**, 86 (1980).

14.5-9 P. C. Wankat, Cycle Separation Techniques, in A. E. Rodrigues and D. Tondeur (Eds.), *Percolation Processes, Theory and Applications*, pp. 443–516, Sijthoff and Noordhoff, Alphen aan den Rijn, Netherlands, 1981.

14.5-10 P. Jacob and D. Tondeur, Nonisothermal Adsorption: Separation of Gas Mixtures by Modulation of Feed Temperatures, *Sep. Sci. Technol.*, **15**, 1563 (1980).

14.5-11 J. F. Chao, J.-J. Huang, and C. R. Huang, Continuous Multiaffinity Separation of Proteins: Cyclic Processes, *AIChE Symp. Ser.*, **78**, 39, 219 (1982).

14.5-12 P. C. Wankat, *Large Scale Adsorption and Chromatography*, CRC Press, Boca Raton, FL, 1986.

14.5-13 P. C. Wankat, Two-Dimensional Separation Processes, *Sep. Sci. Technol.*, **19**, 801 (1984-85).

CHAPTER **15**

Separation Processes Based on Reversible Chemical Complexation

C. JUDSON KING
Department of Chemical Engineering
University of California
Berkeley, California

SUMMARY

Separation processes based on reversible chemical complexation can have high capacities for dilute solutes and also high selectivities. In these processes the mixture is contacted with a second phase containing a complexing agent that reacts reversibly with the solute(s) of interest. In a second part of the process the reaction is reversed and the solute is recovered. Complexation is used commonly on very large scales for gas absorption and hydrometallurgical refining, and such applications are discussed in Chapters 6 and 8, respectively. This chapter is limited to the application of these chemical complexations to the separation of polar organic solutes from aqueous solutions.

For this purpose, complexation reactions can be implemented in many different ways, including solvent extraction, distillative processes, adsorption, liquid membranes, and foam or bubble fractionation. A solute suitable for separation by complexation usually has Lewis-acid and/or Lewis-base functional groups, low concentration, low volatility relative to water, and a low activity coefficient in water. The complexing agent is chosen so as to give a strong, specific, yet reversible reaction with the solute. Frequently, the complexing agent is dissolved in a diluent. The diluent can improve the equilibrium through solvation of the complex and can affect the process in other ways.

The specific cases of separation of acetic acid, phenol, and ethanol from aqueous solution are considered in some detail. Complexation proves to be very effective for acetic acid, leading to several promising new processing approaches. It is less attractive for phenol, primarily because phenol is extracted readily with conventional solvents; the benefits of complexation are more important for polyhydroxy benzenes. For ethanol, it is difficult to identify extractants giving strong complexation. However, the effects of weaker complexation generate important differences among solvents. Lewis acidity and branching give improved selectivity for ethanol over water at a given solvent capacity.

Complexation also offers attractive possibilities for selective recovery of dicarboxylic acids, hydroxy-carboxylic acids (lactic, citric, etc.), phenolic carboxylic acids (gallic, vanillic, caffeic, etc.), amino acids, quinolines, and alkaloids from aqueous solution.

15.1 INTRODUCTION

The principal separation methods used in classical analytical chemistry are based on chemical reactions. These methods are attractive because many chemical reactions are specific for an individual component of a mixture, and the reactants frequently react completely to the desired products.

Chemical reactions are used far less frequently for separations in the chemical process industries. One reason for this is that processes employing chemical reactions often consume large amounts of expensive reactants, and the reaction product is usually less valuable than the original compound.

These disadvantages can be overcome by using a readily reversible chemical reaction to separate the component of interest from the feed mixture. In a second step of the process, the reaction can be reversed to regenerate the added reactant and recover the desired component in its original form. In order for the reaction to be reversible in an economically attractive process, it must have a relatively low bond energy. Such reactions are known as complexation or association reactions. Figure 15-1.1 shows some reactions of this type along with the bond energies typically involved. Complexation reactions with bond energies less than 10 kJ/mol are similar to ordinary associations by van der Waals forces in the condensed state, and reactions with bond energies above 50 kJ/mol tend to be difficult to reverse without undue expense.

15.1-1 Equilibria

A simple complexation reaction gives an equilibrium of the form:

$$\text{solute} + n \cdot \text{complexing agent} \rightleftharpoons \text{complex}$$

described by an equilibrium constant,

$$K_c = \frac{[\text{complex}]}{[\text{solute}][\text{complexing agent}]^n} \tag{15.1-1}$$

If $n = 1$ and the uncomplexed solute assumes a linear distribution between one phase and a second phase containing the complexing agent, then Eq. (15.1-1) leads to a nonlinear equilibrium of the form shown in Fig. 15.1-2. This relationship shows that a complexation separation can give very high equilibrium distri-

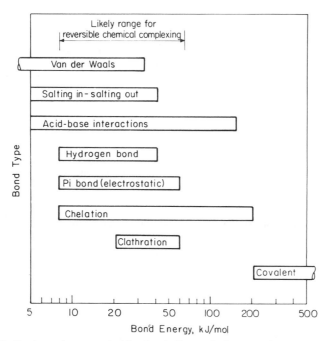

FIGURE 15.1-1 Bond energies most suited for chemically complexing separation processes. (From King;[1] courtesy of Dr. G. E. Keller.)

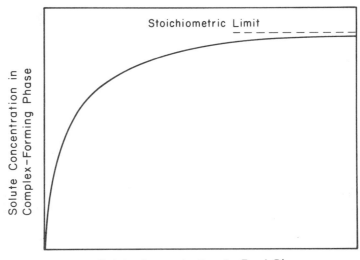

FIGURE 15.1-2 Shape of typical phase-equilibrium relationship for chemical complexation.

bution coefficients at low solute concentrations, but that the distribution becomes less favorable at higher solute concentrations where stoichiometric saturation of the complexing agent is approached. Thus, one important conclusion is that separations based on chemical complexation tend to be *more attractive for cases of relatively low solute concentration*.

In addition to high capacity at low solute concentration, the other major potential advantage of separation processes based on chemical complexation is *selectivity*. Since the complexation reaction can be selective for solutes with particular functional groups, these processes have the potential to separate only certain solutes from a complex mixture. They also can give less coextraction of water than separations with more conventional solvents.

These advantages of chemical complexation for separations are offset by some important potential disadvantages. These separations necessarily employ a mass separating agent rather than an energy separating agent.[1] Therefore, regeneration is required to isolate the product and allow recycle of the separating agent. The expense of regeneration must be taken into account, and the complexation equilibria must have satisfactory properties for the regeneration. Finally, the rate of the complexation reaction may be slow compared to the mass transfer rates in the contactor. This increases the size and cost of the contacting equipment.

15.1-2 Process Characteristics

Chemical complexation already has become well established as a separation method in analytical chemistry and on a large scale in the applications of gas absorption and metals extraction. The ethanolamine and carbonate systems for removal of CO_2 and H_2S from gas streams are classic examples of complexation. Practical aspects of these systems are discussed in Chapter 6 and by Kohl and Riesenfeld,[2] and the more theoretical aspects are discussed by Danckwerts,[3] Danckwerts and Sharma,[4] and Astarita et al.[5] In the metals industry, complexation is used extensively in the hydrometallurgical production of copper and uranium and in the refining of nickel, molybdenum, cobalt, and other metals: for examples see Chapter 8 or the *Proceedings* of the triennial International Conferences on Solvent Extraction.

The topic of this chapter is the application of reversible chemical complexation for removal and recovery of polar organic solutes from aqueous solution. There are some current examples of such separations, and many more are subjects of active research. One of the most important applications is the recovery of products of biochemical synthesis processes, such as fermentation and enzymatically catalyzed reactions. These separation problems range from recovery of commodity chemicals or fuel substances such as acetic acid and ethanol to isolation of much more complex pharmaceuticals and compounds produced by recombinant DNA and other recently developed biological techniques. Another important application is removal and recovery of polar organics from effluent or recycle water streams.

Chemical complexation has been implemented most often in absorption, extraction, and gas and liquid chromatography. A more comprehensive list of alternative processes where chemical complexation could be employed for the recovery of polar compounds from aqueous solution is shown in Table 15.1-1. In

TABLE 15.1-1 Alternatives for Removing Polar-Organic Solutes from Aqueous Solution by Chemical Complexation

Solvent extraction
Extractive distillation
Azeotropic distillation
Adsorption
Solid-infusion processes
Emulsion liquid membranes
Solid-supported liquid membranes
Foam fractionation
Bubble fractionation

extraction, extractive distillation, and azeotropic distillation, the complexing agent is added to modify liquid-phase activity coefficients and thereby establish more favorable phase equilibrium and selectivity. In adsorption, the adsorbent has functional groups on the surface of the solid which form a complex with the solute. An alternative related to adsorption is a solid-infusion process, such as "adsorption" with a polymer gel. In this process the solute can have sufficient mobility in the solid phase to allow transport to complexation sites in the interior of the solid. In an emulsion liquid membrane process, the complexing agent could be present in the outer-emulsion phase to provide facilitated transport of a solute into the inner-emulsion phase, or the complexing agent could be in the inner-emulsion phase where it would have a high capacity for the solute. In the latter case, the complexing agent would have to be regenerated after the emulsion was broken apart. In a solid-supported liquid-membrane process, the complexing agent would be in the liquid impregnated into a polymeric support and would facilitate transport of the solute. In a foam-fractionation[6] or bubble-fractionation[7] process the complexing agent would be surface active or attached to a surfactant and would attract the solute to the gas–liquid interface.

15.1-3 Solute Characteristics

Chemical complexation is most useful for the separation of organic solutes from water when the solute has certain physical properties. Some of the most important criteria favoring the use of complexation are the following:

1. *Lewis-Acid or Lewis-Base Functional Groups.* The solute(s) of interest should have one or more functional groups that can participate in moderately strong complexation. Since most complexation processes involve interactions of Lewis acids with Lewis bases, acidic and basic functional groups are useful.

2. *Low Solute Concentration.* Since complexing agents provide particularly high equilibrium distribution coefficients for low solute concentrations and tend to saturate stoichiometrically at high concentrations (Fig. 15.1-2), complexation is most attractive for relatively dilute aqueous solutions. It is difficult to generalize, but a solute concentration of 5 wt.% is an approximate upper limit beyond which complexation becomes less attractive.

3. *Low Activity Coefficient in Water.* Complexation separation processes are useful for hydrophilic solutes, because these solutes are difficult to recover from aqueous solution by most other means. Complexation is capable of providing a very low organic-phase activity coefficient and hence an acceptable phase-distribution coefficient, even though the activity coefficient in the aqueous phase is low.

4. *Low Solute Volatility.* Solutes that are less volatile than water are good candidates for separation by complexation, because they cannot be separated by steam stripping. Solutes in this category include acetic acid, dicarboxylic acids (succinic, malonic, etc.), glycols, glycol ethers (Cellosolves), lactic acid, and polyhydroxybenzenes (catechol, pyrogallol, etc.), among others. Stripping is also expensive if the solute has a relative volatility close to 1 with respect to water (e.g., ethanol) or if it forms a low-concentration heterogeneous azeotrope with water (e.g., phenol).

15.1-4 Characteristics of the Complexing Agent

There also are a number of criteria by which potential chemical-complexation agents can be evaluated:

1. The *complexation bond energy should be great enough* to give a substantial improvement over conventional mass separating agents, which usually are less expensive.

2. The *complexation bond energy should be low enough* so that the complex can be regenerated easily and completely.

3. *Coextraction of water* can be an important economic disadvantage. The complexing agent should be selected to minimize coextraction of water and/or facilitate subsequent removal of that water.

4. There should be *no side reactions*, and the complexing agent should be *thermally stable*, to avoid irreversible loss.

5. The complexation reaction should have *sufficiently fast kinetics* in both directions so that equipment sizes do not become prohibitively large.

In a number of processes it is either necessary or desirable to use a cosolvent or diluent with the complexing agent. For example, if the complexing agent is a solid—such as trioctyl phosphine oxide—the diluent is needed to form a liquid solution for extraction or liquid-membrane processes. In an extraction process the diluent controls the viscosity, density, and interfacial tension of the mixed solvent. With a relatively nonvolatile complexing agent, the diluent volatility also determines the temperature in the reboiler of a distillation column in a regeneration process. In a liquid-membrane process the diluent may increase the transport rates in the liquid phase.

Two other important functions of a diluent are more subtle:

1. The complexing agent itself may be a poor solvating medium for the complex, in which case an effective diluent can solvate the complex and thereby encourage its formation. Very large changes in equilibrium distribution coefficients can be achieved by changing the diluent.

2. When coextraction of water is an important consideration, the diluent can have a large effect on the amount of water coextracted and also can improve the separation of water in a subsequent distillation step. It can be worthwhile to choose a diluent with a lower equilibrium distribution coefficient for the solute, if the ratio of solute to water extracted is increased.

15.1-5   Regeneration Methods

Any complexation separation process requires a regeneration scheme to recover the complexation agent. In some unusual cases the reaction product can be discarded, but then an inexpensive reactant and an irreversible reaction would be desirable. The following list gives some alternatives for regeneration processes:

1. If the solute is volatile, it can be taken overhead in a distillation or stripping process. The volatility of the solute may be suppressed by the complexation equilibrium. In cases of nonvolatile solutes and sufficiently volatile complexing agents, the complexing agent can be taken overhead.

2. If the equilibrium constant of the complexation reaction is sufficiently sensitive to temperature, back-extraction into water at a different temperature can give an overall concentration of the solute. Even if the resulting solution were not more concentrated, this process can isolate one solute from other solutes.

3. Back-extraction of the solute into a strong base or a strong acid can be used for acidic or basic solutes, respectively. In this case a high degree of concentration may be achieved, but the chemical form of the solute will be changed. A change in the solution pH also may affect the solute in other ways so as to facilitate regeneration. These regeneration methods can consume large quantities of chemicals.

15.2   SPECIFIC EXAMPLES

The recovery of three solutes—acetic acid, phenol, and ethanol—from aqueous solution is considered in detail to illustrate the important factors in these separations. Then separation by chemical complexation is evaluated from some other classes of organic solutes.

15.2-1   Acetic Acid

Acetic acid often is found in dilute solution from processes that use it as a raw material or solvent. Fermentation processes also produce acetic acid in dilute solution. In many of these cases the solution pH is high enough so that the acetic acid is ionized partially or completely. The present discussion is restricted to the recovery of un-ionized acetic acid. At a pH below the pK_a, all the acetic acid can be removed in the un-ionized form. At a higher pH the acetate ion does not distribute into the second phase, and a large distribution coefficient is required for the un-ionized form to shift the equilibrium and obtain an attractive removal capacity.

Acetic acid is less volatile than water, and the relative volatility is close enough to unity so that simple distillation is not the process of choice for recovery of acetic acid, except possibly at extremely high concentrations in aqueous solution.[1] For the past 30 or more years, azeotropic distillation has been the conventional technology for recovering acetic acid from feeds having greater than 35–45% w/w acetic acid in water. For more dilute feeds the favored process has been extraction with solvents such as ethyl acetate, mixtures of ethyl acetate and benzene, and isopropyl acetate.[1,2] These solvents give values of the equilibrium

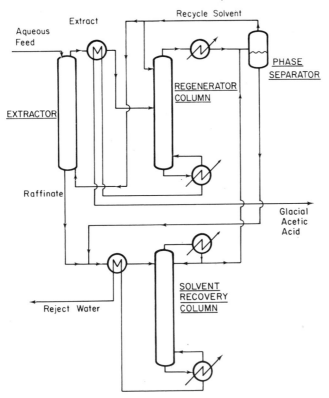

FIGURE 15.2-1 Conventional process for recovery of acetic acid by solvent extraction, followed by azeotropic distillation.[1,2]

distribution coefficient[‡] K_D less than 1.0. An example of such a process is shown in Fig. 15.2-1. An important property of these solvents is their ability to improve the separation of the coextracted water from the acetic acid. The extract is fed to the regeneration column where the solvent modifies the vapor–liquid equilibria and acts as an entrainer for an azeotropic distillation. The distillate, which contains the solvent and the coextracted water, splits into two liquid phases—a solvent-rich stream suitable for recycle to the extractor and a water-rich stream that is combined with the raffinate and sent to a solvent stripping column. The bottom product from the regeneration column is glacial acetic acid.

The lowest feed concentration allowing economical recovery with these conventional processes is 3–5% w/w acetic acid, depending on the value accorded to recovered acetic acid. For more economical processing of dilute feed streams, a solvent giving a higher value of K_D is needed. This leads to chemical complexation. Acetic acid in these aqueous streams meets all the criteria listed above (Lewis-acid functional group, low concentration, low volatility, and low activity coefficient in water) for solutes which are good candidates for separations based on chemical complexation.

Since acetic acid is a Lewis acid, it is appropriate to study solvents that are moderately strong Lewis bases. Complexing agents that have received the most investigation to date incorporate either phosphoryl or amine groups. Tributyl phosphate (TBP) gives a K_D of about 2.3 for acetic acid at high dilution and a somewhat lower value at higher concentrations of acetic acid.[3–5] When TBP is mixed with hydrocarbon diluents, the resulting K_D is approximately a mass-weighted average of the values for the pure solvents.[4] Stronger Lewis bases that have received attention are tertiary amines and trioctyl phosphine oxide (TOPO). All these extractants are used in the hydrometallurgical industry. They are available commercially; therefore, operating experience has been gained with them.

With these more basic extractants the nature of the diluent becomes quite important. Blumberg and Gai[6] have interpreted diluent effects in terms of Lewis acidity and basicity for extraction of mineral acids by tertiary amines. Frolov et al.[7] interpreted diluent effects for extraction of acetic acid by amines in terms

[‡]K_D is the weight fraction solute in organic phase per weight fraction solute in aqueous phase, at equilibrium. All equilibrium distribution coefficients reported here are defined in this way unless otherwise noted.

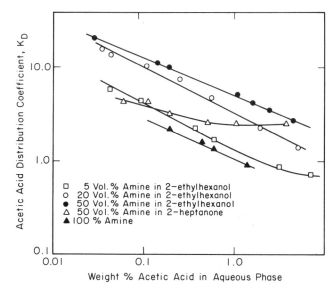

FIGURE 15.2-2 Equilibrium distribution coefficients for solvent extraction of acetic acid from aqueous solution into solvent mixtures containing Alamine 336.[8]

of empirical polarity parameters. Wardell and King,[4] Ricker et al.,[8] and Spala and Ricker[9] have observed very pronounced maxima in K_D at intermediate solvent compositions for extraction of acetic acid by mixtures of tertiary amines or TOPO with diluents such as alcohols, ketones, and chlorinated hydrocarbons. These maxima have been interpreted in terms of the ability of the diluent to solvate the complex formed by acetic acid and the Lewis-base extractant. An example of the behavior of K_D is shown in Fig. 15.2-2, where the intermediate compositions of a mixture of Alamine 336 (a commercial tertiary amine mixture, Henkel Corp.) and 2-ethylhexanol give values of K_D as much as a factor of 5 greater than either the undiluted amine or the alcohol without amine. It is interesting to note in Fig. 15.2-2 that K_D decreases with increasing concentration of acetic acid in the raffinate. This is due to the effect of stoichiometry on the complexation equilibrium [Eq. (15.1-1)]. The ketone diluent, 2-heptanone, also increases K_D as compared to the undiluted amine, but the increase is not very large at low acetic acid concentration. Spala and Ricker[9] interpret that behavior in terms of the lack of an electron acceptor group in the ketone to hydrogen bond with the electronegative carbonyl oxygen in the complex. The ketone diluent becomes more favorable than the alcohol diluent at higher raffinate concentrations of acetic acid because the ketone itself has an appreciable solvent capacity for the uncomplexed carboxylic acid through hydrogen bonding.

Sakai et al.[10] measured degrees of extraction of acetic acid from water using solvent mixtures composed of a secondary amine mixture, Amberlite LA-2 (Rohm & Haas Corp.), with diluents such as hexane, CCl_4, $CHCl_3$, or methyl isobutyl ketone (MIBK). The ketone and $CHCl_3$ diluents gave the highest values of K_D, consistent with the results for the extraction of acetic acid with tertiary amines. A weak effect of temperature on K_D was observed for LA-2 with $CHCl_3$ diluent. Tertiary amines probably are better extractants because primary and secondary amines can react irreversibly when heated with acetic acid to form amides.

Process aspects for extraction of acetic acid with amines are discussed by Ricker et al.[11] Alcohol diluents gave the highest values of K_D, but the alcohols were subject to esterification with acetic acid upon regeneration by distillation. Ketones appeared to be satisfactory diluents from the standpoint of high K_D. Chloroform is a superior diluent because it is a Lewis acid and can interact with the complex.[9] However, chloroform is toxic, and this fact may limit its use.

Process aspects for extraction by TOPO are discussed by Helsel,[12] Ricker et al.,[11] and Golob et al.[13] Among hydrocarbons, aromatics are better diluents since they have higher solvent capacities for TOPO. Alcohol diluents appear to interact preferentially with the phosphoryl group and thereby give values of K_D even lower than found with aromatic hydrocarbon diluents. Ketone diluents cannot interact with the phosphoryl group, and they solvate the acetic acid–TOPO complex to some extent; hence, they give higher values of K_D when used as diluents than do aromatics.[8]

In these extractant–diluent systems for acetic acid, the optimal molecular weight for the extractant reflects a compromise between high K_D (low molecular weight) and low enough solubility of the extractant or the complex in the aqueous phase (high molecular weight). TOPO and Alamine 336 (tri-octyl/decyl amines) appear to be near optimal in this sense. The optimum diluent molecular weight reflects a compromise between high K_D (low molecular weight) and either low solubility in water or low enough volatility

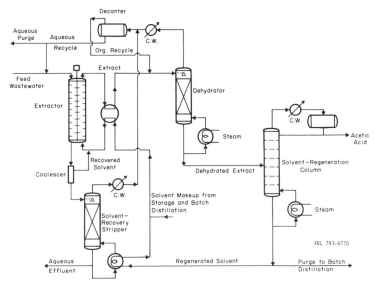

FIGURE 15.2-3 Extraction process for recovery of acetic acid from aqueous solution using a heavy solvent.[11]

relative to acetic acid for solvent regeneration to be carried out in the presence of the extractant (high molecular weight). Among the ketones, diisobutyl ketone (DIBK) or a decanone seem to be an effective compromise in that regard. Values of K_D for extraction of acetic acid (1% w/w in raffinate) are about 2.5 for both 50% Alamine 336 in DIBK and 40% TOPO in 2-heptanone.[8]

TBP, TOPO, and the amines are all higher-boiling solvents than acetic acid. This reduces steam costs for regeneration as compared to low-boiling solvents, but it can lead to higher reboiler temperatures in the regenerator and accumulation of nonvolatile impurities in the solvent. A portion of the solvent might have to be discarded or processed further to remove these impurities. The complexing extractants tend to be more expensive than conventional solvents; TBP and amines are in the range of $2.2–3.3/kg, while TOPO costs about $17/kg. Losses are therefore important. Assuming that extractant losses will be about the same in either case, amines are probably more attractive than TOPO because they are less expensive.

Chemically complexing extractants can be much more selective than conventional solvents and therefore can reduce the amount of coextracted water. On a solvent-free basis, the extract in equilibrium with a 6.6% w/w aqueous acetic acid solution contains about 84% acetic acid[8] for extraction with Alamine 336–DIBK. The selectivity for acetic acid over water is lower for extraction with TBP,[5] but is still substantially higher than with conventional solvents such as acetates or ketones. Coextracted water can be removed in an extractive distillation column located before a solvent regeneration column, as shown in Fig. 15.2-3, in a heavy-solvent analogue of the process shown in Fig. 15.1-2.

Jagirdar and Sharma[14] have employed tertiary amine extractants to recover and fractionate among several carboxylic acids in aqueous solution by means of dissociation extraction, in which a stoichiometrically deficient amount of extractant is used. Comparative equilibria for different carboxylic acids also are given by Niitsu and Sekine[15] for TOPO-based solvents and by Wardell and King[4] for TBP, TOPO, and amines.

Extractive distillation was used in the past in the Suida process for recovery of acetic acid from pyroligneous acid, with recycle wood oils used as the extractive agent.[16] Although this process has long since been discontinued, it is possible that extractive distillation could become attractive again for more concentrated acetic acid feeds if a water-soluble complexing agent were used, such as one of the lower molecular weight phosphates, amines, or phosphine oxides.

Chemical complexing can be used for recovery of acetic acid in process configurations other than extraction, as listed in Table 15.1-1. Smith[17] reports studies using solid-supported liquid membranes, where the impregnating agents are solvents containing amines or TOPO. The acetic acid permeate is taken up by a solution of an aqueous base, such as $Ca(OH)_2$, thereby converting the acetic acid to a salt. TOPO was found to give better sustained performance and a more plasticized polyvinyl chloride membrane than either primary or secondary amines. Kuo and Gregor[18,19] also have studied TOPO-impregnated membranes for removal of acetic acid from dilute aqueous feeds by facilitated transport. They have modeled factors influencing transport rates, explored effects of different diluents for TOPO, and investigated ways of sustaining the membrane strength.

15.2-2 Phenol

Phenol meets several of the criteria for a desirable solute for complexation separations. Phenol is a Lewis acid due to the hydroxyl group. It often is present in dilute aqueous solutions from industrial processes. Although it is more volatile than water in dilute aqueous solutions, it is difficult to separate by stripping because it forms a heterogeneous azeotrope at low concentration (9.2% w/w in water). However, an essential difference in contrast with acetic acid is that phenol has a substantially higher activity coefficient in aqueous solution, and therefore it is extracted much more readily by conventional solvents. For example, diisopropyl ether (DIPE) and methyl isobutyl ketone (MIBK) provide K_D values of about 37 and 100, respectively, for phenol at high dilution.[20] This means that extraction processes with either of these solvents can operate at relatively low solvent-to-feed ratios, and there is therefore less incentive to gain the still higher values of K_D which should be provided by complexing extractants. It might be noted, however, that DIPE and MIBK provide substantially higher values of K_D than do many other solvents of comparable molecular weight. That fact can be attributed to their Lewis basicity, although they are substantially weaker bases than amines or phosphoryl compounds. One disadvantage of conventional solvents is that residual dissolved solvent must be recovered from the aqueous raffinate.

Despite the advantages of conventional solvents, some work has been done with chemically complexing extractants for phenol. Amines have been studied as extractants for phenol by Wolf and Fuertig,[21] Pollio et al.,[22] Pittman,[23] and Inoue et al.[24] These researchers found that diluents such as benzene and 2-ethyl-hexanol provide substantially higher values of K_D than does CCl$_4$, and that CCl$_4$ provides higher K_D values than do alkane diluents. Nonetheless, the values of K_D realized are not high enough to warrant the use of the more expensive extractant rather than DIPE, MIBK, or similar solvents. The amine group, which is a stronger Lewis base, might be expected to give very high K_D values compared to conventional solvents. However, if the complexation occurs through a proton transfer to the amine, it must be recognized that phenol has a relatively high value of pK_a.

Phosphoryl solvents also have been studied for extraction of phenol. Tricresyl phosphate has a K_D of about 72 for phenol.[25] This value is comparable to those for DIPE and MIBK, an important difference being that the phosphate is a high-boiling solvent, whereas DIPE and MIBK are low-boiling solvents. The Lewis basicity increases in the order: phosphates [PO(OR)$_3$] < phosphonates [PO(OR)$_2$R] < phosphinates [PO(OR)R$_2$] < phosphine oxides [POR$_3$]. Complexes formed by phenol with these compounds have been studied extensively by spectroscopic, dielectric, and equilibrium measurements.[26-32] Complexes formed with phosphine oxides are particularly strong, reflecting the hydrogen bond.

Figure 15.2-4 shows measured equilibrium data for extraction of phenol from an aqueous feed containing 5000 ppm phenol at 22.5°C into a solvent composed of 25% w/w TOPO in diisobutyl ketone (DIBK).[33] Here the data are expressed as K_M, the concentration (mass or moles/volume) in the solvent divided by the concentration of phenol in the equilibrium aqueous phase. Predicted curves are drawn for a theoretical model, in which fitted parameters are K_P, the physical (or unreacted) distribution coefficient, and K_R, the equilibrium constant of a complexation reaction with one-to-one stoichiometry. The curves are drawn for

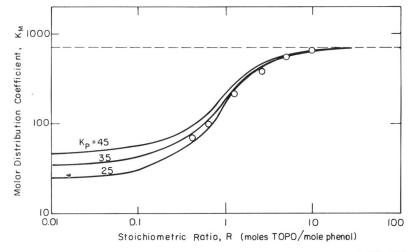

XBL822-5198

FIGURE 15.2-4 Concentration-based equilibrium distribution coefficients for extraction of phenol from dilute aqueous solution into a solvent mixture composed of 25% w/w TOPO in DIBK.[33]

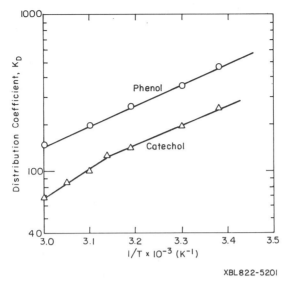

XBL822-5201

FIGURE 15.2-5 Equilibrium distribution coefficients vs. temperature for extraction of phenol and catechol from dilute aqueous solution into a solvent mixture of 25% w/w TOPO in DIBK. Stoichiometric ratio = 2.5 mol TOPO/mol solute.[33]

$K_P = 25$, 35, and 45, and for $K_P K_R T_{io} = 700$, where T_{io} is the initial molar concentration of TOPO in the organic phase. From these results, K_R is in the range of 40–50 L/mol.

One possible regeneration method, mentioned above, is temperature-swing back-extraction. Figure 15.2-5 shows measured values of K_D as a function of temperature for extraction of phenol and catechol individually from water into 25% w/w TOPO in DIBK with feed concentrations of 5000 ppm and a molar TOPO solute ratio of 2.5.[33] There is a substantial change of K_D with temperature but probably not enough to support this method of regeneration.

An alternative is to regenerate by distilling phenol overhead from the solvent, in which case a diluent much heavier than DIBK is needed. Tests by Bixby[34] have shown that the volatility of phenol is reduced by the complexation reaction even at temperatures above 200°C. Even though pure phenol boils at 182°C, it is desirable to use a diluent boiling at 250°C or higher to achieve economical regeneration.

From these results it appears that the more conventional "physical" solvents, such as DIPE or MIBK, have more advantages for phenol removal than the more common Lewis-base complexing extractants. However, complexing extractants may be advantageous for feed solutions containing di- and trihydoxy benzenes. Dihydroxy benzenes have sufficiently low values of K_D into DIPE[20] that solvent-to-water mass phase ratios of about 1.0 would be required. Phase ratios required to remove trihydroxy benzenes would be prohibitively high. MIBK performs significantly better, but TOPO-based extractants give considerably higher values of K_D. A comparison is made in Table 15.2-1 of measured values of K_D for extraction of phenol and di- and trihydroxy benzenes from dilute aqueous solution at ambient temperature using DIPE, MIBK, and 25% w/w TOPO in MIBK (molar ratio of TOPO to solute is about 2.5). All measurements are at low enough pH so that the solutes are not ionized.

15.2-3 Ethanol

Ethanol also has several desirable characteristics for separation by chemical complexation. Ethanol is produced in dilute aqueous solution by fermentation processes. Ethanol has a low activity coefficient in water; therefore, extraction with conventional solvents is difficult. Although ethanol is substantially more volatile than water in dilute solutions, it forms an azeotrope with water at high concentration, making distillation processes complicated and possibly expensive. However, chemically complexing extractants have not been identified that provide high values of K_D for ethanol. Measurements by Roddy,[35] Munson and King,[36] and others have shown that amines and phosphoryl compounds do not provide values of K_D appreciably greater than those provided by conventional solvents. Alcohols have both Lewis-acid and Lewis-base sites. One possible explanation for the low K_D values is that the association of hydroxyl groups with one another in the aqueous phase is significant compared to the association with the extractant in the organic phase. Although solvents with high capacity have not been identified, complexation effects can improve the selectivity of potential solvents.

Table 15.2-1 Equilibrium Distribution Coefficients for Extraction of
Phenol and Higher Phenols from Water with Various Solvents

Solute	Measured Value of K_D for the Following Solvents		
	$DIPE^a$	$MIBK^a$	25% w/w TOPO in $DIBK^b$
Phenol	36.5	(100)	460
Catechol (1,2)	4.9	18.7	200
Resorcinol (1,3)	2.1	17.9	98
Hydroquinone (1,4)	1.03	9.9	35
Pyrogallol (1,2,3)	—	3.6	53
Hydroxyquinol (1,2,4)	0.18	5.0	24
Phloroglucinol (1,3,5)	—	3.9	21

[a] Greminger, et al.[20]
[b] MacGlashan et al.[33]

In the recovery of ethanol and many other organic solutes from aqueous solution, coextraction of water has a large effect on the process economics. Solvents may be compared by plotting the selectivity (α = separation factor between ethanol and water) versus the solvent capacity for ethanol, expressed as K_D. Figure 15.2-6 is such a plot for extraction of ethanol from relatively dilute aqueous solution by many different solvents.[36] This figure includes data from Roddy,[35] Souissi and Thyrion,[37] and Munson and King.[36] It is apparent that the Lewis-acid solvents (alcohols, carboxylic acids, and chlorinated hydrocarbons) provide much better selectivity for a given capacity than do the Lewis-base solvents (ketones, esters, amines, phosphoryls). Furthermore, branching of the solvent molecule is important, as shown in Fig. 15.2-7. The dashed lines relate selectivity to K_D for normal carboxylic acids (solid points) and normal alcohols (open points). Branched carboxylic acids give substantially higher selectivities than do straight-chain acids for a given value of K_D. The same is true for alcohols.

Many investigations of the extraction of ethanol from water have postulated that a very high selectivity is needed to enrich the solvent-free extract to an ethanol content near or above the binary azeotrope with water. However, this degree of enrichment is not necessary. The extraction step can be followed by an extractive-distillation dewatering step similar to the process shown for acetic acid recovery in Fig. 15.2-3.

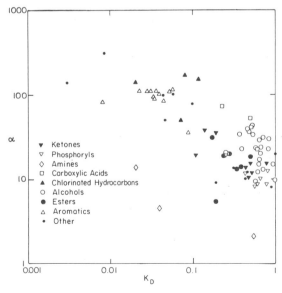

FIGURE 15.2-6   Selectivity (α) for ethanol over water versus K_D for extraction of ethanol from dilute aqueous solution with various common solvents.[36]

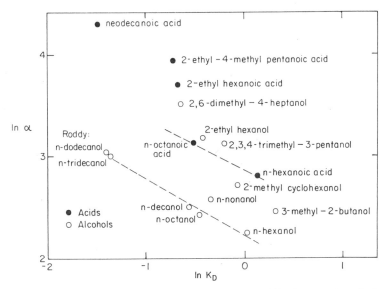

FIGURE 15.2-7 Effect of branching on relationship between α and K_D for extraction of ethanol from dilute aqueous solution by carboxylic acid and alcohol solvents.[36]

15.2-4 Solutes With Multiple Functional Groups

Separations based on reversible chemical complexation may be useful for recovery of many other organic solutes from dilute aqueous solution. Solutes that have multiple Lewis-acid or Lewis-base functional groups are particularly good candidates. Such solutes also have low activity coefficients in water and low relative volatilities with respect to water, thereby satisfying all the desirable criteria identified above. The results shown in Table 15.2-1 demonstrate that separation by complexation with the strong Lewis base TOPO is more attractive for polyhydroxybenzenes than for phenol, especially if the hydroxyl groups are located on adjoining carbon atoms of the aromatic ring.

DICARBOXYLIC ACIDS
Vieux et al.[38,39] have investigated extraction of oxalic, malonic, succinic, and glutaric acids from water, using triisooctyl amine in various diluents. The values of K_D are substantially lower than are found for extraction of acetic acid by similar amines—a result that could be anticipated because the two carboxylic acid groups decrease the solute activity coefficient in water. Vieux and coworkers also found substantial effects of the diluent on K_D, with chloroform and 1,2-dichloroethane giving higher values of K_D than 1,2-dichlorobenzene, which in turn gave higher values of K_D than benzene as diluent. These results are consistent with the results observed for extraction of acetic acid with amines (see above), where the stronger Lewis acids are more effective diluents because they solvate the carbonyl group of the complex better.

HYDROXYCARBOXYLIC ACIDS
Lactic acid is an important article of commerce. It is difficult to separate from aqueous solution because it has a strong affinity for water, resulting from the presence of a hydroxyl group and a carboxylic acid group. Solvent extraction of lactic acid from aqueous solution is discussed in detail by Short and Eaglesfield,[40] who report concentration-based equilibrium distribution coefficients (K_M) for common alcohol, ketone, ether, and ester solvents ranging from 0.04 to 0.82. The stronger Lewis base TBP gives K_M values of 1.3–1.4.[3]

Citric acid contains three carboxylic acid groups and one hydroxyl group and is therefore even more hydrophilic. Common alcohol, ester, ketone, and ether solvents give values of concentration-based distribution coefficients (K_M) ranging from 0.1 to 0.3.[41] K_M ranges from 2.0 to 2.3 for TBP.[3,42] Wennersten[42] also found that K_D increased significantly from tributyl phosphate, to dibutyl phosphonate, to trioctyl phosphine oxide. This reflects increases in the basicity of the phosphoryl group. At 25 and 80°C, respectively, a solvent composed of 50% v/v triisooctyl amine in a hydrocarbon diluent (Shellsol H, 16% aromatics) gave K_D values of 6.5 and 0.33—different by a factor of nearly 20! It has been reported that a recent commercial process for extraction of citric acid uses trilauryl amine as extractant, with regeneration through a temperature change and back-extraction into water.

CHROMATOGRAPHIC MEASUREMENTS

A number of studies of separations by means of liquid chromatography (HPLC), paper chromatography, and related laboratory techniques provide useful information on the utility of various complexing extractants for polyfunctional organic solutes. From such studies it is possible to obtain distribution coefficients, effects of diluents, and information on the complexation stoichiometry and bond strength. An example of such a study is the work of Stuurman et al.,[32] who used HPLC to study complexation of phenol, hydroxybenzoic acids, and other hydroxycarboxylic acids with TOPO in a diluent of n-hexane.

Soczewinski and coworkers have carried out an extensive series of measurements with paper chromatography to observe the characteristics of various complexing systems. These include extraction of various phenolic carboxylic acids (gallic acid, vanillic acid, caffeic acid, etc.) into TBP[43] and of polyhydroxybenzenes and naphthols into both TBP with various diluents and tributyl amine with various diluents.[44] Soczewinski and Rojowska[45] studied effects of pH on chromatographic extraction of several amino acids into di-2-ethyl hexyl phosphoric acid (D2EHPA), with ketones and ethers as diluents, and later extended these measurements to solvents using hexanol as the diluent for D2EHPA.[46] Mixed solvents composed of a Lewis acid (CHCl$_3$) and a Lewis base (phosphate, ketone, or ether) were used for chromatographic extractions of polyfunctional solutes combining hydroxyl, amino, and nitro groups along with phenols, anilines and quinolines.[47,48] In earlier work, oleic acid and D2EHPA in various diluents were used for separations of various quinolines and alkaloids.[49,50]

COMPLEXING ADSORBENTS

Adsorbents, particularly those made from synthetic polymers, can be made to contain specified and controlled functional groups capable of complexation. In some cases the solute-uptake process is one of solid infusion (Table 15.1-1) or bulk absorption. As an example of such a system, Kawabata and Ohira[51] have used a Lewis-base resin made of cross-linked poly(4-vinyl pyridine) to separate phenol from water. Kawabata et al.[52] used this same polymeric material to separate various carboxylic acids from aqueous solution. For a complexing, regenerable absorbent it is important to have a complexation reaction which is strong enough to give a substantial increase in capacity due to complexation, but which is not so strong as to complicate regeneration. In a series of experiments with adipic acid as solute, Kawabata et al.[52] show that poly(4-vinyl pyridine) resin gives substantially greater capacities than the more common styrene–divinylbenzene copolymeric resins, presumably because of the greater basicity of the pyridyl group. A resin containing amine groups in a styrene–divinylbenzene matrix also gave high capacities for adipic acid, comparable to the pyridyl resin, but required more methanol for regeneration. This presumably reflects the much higher basicity of the amine groups compared to the pyridyl groups.

GENERAL REVIEWS

The extraction chemistry of carboxylic acids with both conventional and complexing extractants has been reviewed by Kertes and King.[53] A review of extraction chemistry of the low-molecular-weight chemistry of the low-molecular-weight alcohols has been prepared by the same authors.[54]

REFERENCES

Section 15.1

15.1-1   C. J. King, *Separation Processes*, 2nd ed., McGraw-Hill, New York, 1980.

15.1-2   A. L. Kohl and F. C. Riesenfeld, *Gas Purification*, 4th ed., Gulf Publishing, Houston, 1985.

15.1-3   P. V. Danckwerts, *Gas–Liquid Reactions*, McGraw-Hill, New York, 1970.

15.1-4   P. V. Danckwerts and M. M. Sharma, *The Chemical Engineer*, No. 202, p. CE244 (1966).

15.1-5   G. Astarita, D. W. Savage, and A. Bisio, *Gas Treating with Chemical Solvents*, Wiley, New York, 1983.

15.1-6   R. Lemlich (Ed.), *Adsorptive Bubble Separation Techniques*, Academic Press, New York, 1972.

15.1-7   E. Valdes-Krieg, C. J. King, and H. H. Sephton, *Sep. Purif. Methods*, **6**, 221 (1977).

Section 15.2

15.2-1   C. J. King, Acetic Acid Extraction, in T. C. Lo, M. H. I. Baird, and C. Hanson (Eds.), *Handbook of Solvent Extraction*, Wiley-Interscience, New York, 1983.

15.2-2   P. Eaglesfield, B. K. Kelly, and J. F. Short, *Ind. Chem.*, **29**, 147, 243 (1953).

15.2-3   H. A. Pagel and F. W. McLafferty, *Anal. Chem.*, **20**, 272 (1948).

15.2-4   J. M. Wardell and C. J. King, *J. Chem. Eng. Data*, **23**, 144 (1978).

15.2-5   D. J. Shah and K. K. Tiwari, *J. Sep. Proc. Technol.*, **2**, 1 (1981).

15.2-6   R. Blumberg and J. E. Gai, *Proc. Int. Solvent Extr. Conf. (ISEC'77)*, Can. Inst. Metall., Spec. Vol. 21, pp. 9–16, (1977).

15.2-7   Yu. G. Frolov, A. A. Pushkov, and V. V. Segievsky, *Proc. Int. Solv. Extr. Conf. (ISEC'77)*, Can. Inst. Metall., Spec. Vol. 21, pp. 1236–1242 (1977).

15.2-8   N. L. Ricker, J. N. Michaels, and C. J. King, *J. Sep. Proc. Technol.*, **1**, 36 (1979).

15.2-9   E. E. Spala and N. L. Ricker, *Ind. Eng. Chem. Process Des. Dev.*, **21**, 409 (1982).

15.2-10 W. Sakai, F. Nakashio, T. Tsuneyuki, and K. Inoue, *Kagaku Kōgaku*, **33**, 1221 (1969).

15.2-11 N. L. Ricker, E. F. Pittman, and C. J. King, *J. Sep. Proc. Technol.*, **1**, 23 (1980).

15.2-12 R. W. Helsel, *Chem. Eng. Prog.*, **73**, 55 (1977).

15.2-13 J. Golob, V. Grilc, and B. Zadnik, *Ind. Eng. Chem. Process Des. Dev.*, **20**, 433 (1981).

15.2-14 G. C. Jagirdar and M. M. Sharma, *J. Sep. Proc. Technol.*, **1**, 40 (1980).

15.2-15 M. Niitsu and T. Sekine, *Bull. Chem. Soc. Jpn.*, **51**, 705 (1978).

15.2-16 W. F. Schurig, Acetic Acid, in *Kirk–Othmer Encyclopedia of Chemical Technology*, 1st ed., Vol. 1, pp. 56–78, Interscience, New York, 1947.

15.2-17 B. R. Smith, "Organic Acid Recovery with Coupled-Transport Membranes," paper presented at the Engineering Foundation Conference on Advances in Fermentation Recovery Process Technology, Banff, Alberta, Canada, June 1981. See also Annual Reports for 1980 and subsequent, C.S.I.R.O., Division of Chemical Technology, South Melbourne, Victoria, Australia.

15.2-18 Y. Kuo, D. Eng. Sci. dissertation in Chemical Engineering, Columbia University, New York, 1980.

15.2-19 Y. Kuo and H. P. Gregor, paper presented at the American Chemical Society Meeting, Seattle, WA, March 1983.

15.2-20 D. C. Greminger, G. P. Burns, S. Lynn, D. N. Hanson, and C. J. King, *Ind. Eng. Chem. Process Des. Dev.*, **21**, 51 (1982).

15.2-21 F. Wolf and H. Fuertig, *Chem. Technol.*, **18**, 405 (1966).

15.2-22 F. X. Pollio, R. Kunin, and A. F. Preuss, *Environ. Sci. Technol.*, **1**, 495 (1967).

15.2-23 E. F. Pittman, M.S. Thesis in Chemical Engineering, University of California, Berkeley, 1979.

15.2-24 K. Inoue, K. Tsubonoue, and I. Nakamori, *Sep. Sci. Technol.*, **15**, 1243 (1980).

15.2-25 N. E. Bell, M.S. Thesis in Chemical Engineering, University of California, Berkeley, 1980.

15.2-26 G. Aksnes, *Acta Chem. Scand.*, **14**, 1475 (1960).

15.2-27 G. Aksnes and T. Gramstad, *Acta Chem. Scand.*, **14**, 1485 (1960).

15.2-28 G. Aksnes and P. Albriktsen, *Acta Chem. Scand.*, **22**, 1866 (1968).

15.2-29 D. Hadzi, H. Ratajczak, and L. Sobczyk, *J. Chem. Soc. A*, 48 (1967).

15.2-30 A. A. Shvets, E. G. Amarskii, O. A. Osipov, and L. V. Goncharova, *Zh. Obshch. Khim.*, **46**, 1701 (1976).

15.2-31 N. M. Karayannis, C. M. Mikulski, L. S. Gelfand, and L. L. Pytlewski, *J. Inorg. Nucl. Chem.*, **40**, 1513 (1978).

15.2-32 H. W. Stuurman, K.-G. Wahlund, and G. Schill, *J. Chromatogr.*, **204**, 43 (1981).

15.2-33 J. D. MacGlashan, J. L. Bixby, and C. J. King, *Solvent Extr. Ion Exchange*, 3, 1 (1985); J. D. MacGlashan and C. J. King, Report No. LBL-13963, Lawrence Berkeley Laboratory, Berkeley, CA, March 1982.

15.2-34 J. L. Bixby, M.S. Thesis in Chemical Engineering, University of California, Berkeley, 1983.

15.2-35 J. W. Roddy, *Ind. Eng. Chem. Process Des. Dev.*, **20**, 104 (1981).

15.2-36 C. L. Munson and C. J. King, *Ind. Eng. Chem. Process Des. Dev.*, **23**, 109 (1984).

15.2-37 A. Souissi and F. C. Thyrion, *Proc. 2nd World Congr. Chem. Eng.*, **4**, 443 (1981).

15.2-38 A. S. Vieux, N. Rutagengwa, J. B. Rulinda, and A. Balikungeri, *Anal. Chim. Acta*, **68**, 415 (1974).

15.2-39 A. S. Vieux and N. Rutagengua, *Anal. Chim. Acta*, **91**, 359 (1977).

15.2-40 J. F. Short and P. Eaglesfield, *Trans. Inst. Chem. Eng.*, **30**, 109 (1952).

15.2-41 C. S. Marvel and J. C. Richards, *Anal. Chem.*, **21**, 1480 (1949).

15.2-42 R. Wennersten, *J. Chem. Tech. Biotechnol.*, **33B**, 85 (1983); paper presented at the International Solvent Extraction Conference (ISEC'80), Liege, Belgium, 1980.

15.2-43 G. Matysik and E. Soczewinski, *Sep. Sci.*, **12**, 657 (1977).

15.2-44 E. Soczewinski and G. Matysik, *J. Chromatogr.*, **48**, 57 (1970).

15.2-45 E. Soczewinski and M. Rojowska, *Rocz. Chem.*, **47**, 1025 (1973).

15.2-46 E. Soczewinski and M. Rojowska, *J. Chromatogr.*, **32,** 364 (1978).

15.2-47 E. Soczewinski, G. Matysik, and W. Dumkiewicz, *J. Chromatogr.*, **132,** 379 (1977).

15.2-48 G. Matysik and E. Soczewinski, *J. Chromatogr.*, **160,** 29 (1978).

15.2-49 E. Soczewinski, G. Matysik, and H. Szumilo, *Sep. Sci.*, **2,** 25 (1967).

15.2-50 E. Soczewinski and G. Matysik, *J. Chromatogr.*, **32,** 458 (1968).

15.2-51 N. Kawabata and K. Ohira, *Environ. Sci. Technol.*, **13,** 1396 (1979).

15.2-52 N. Kawabata, J.-I. Yoshida, and Y. Tanigawa, *Ind. Eng. Chem. Prod. Res. Dev.*, **20,** 386 (1981).

15.2-53 A. S. Kertes and C. J. King, *Biotechnol. Bioeng.*, **28,** 269 (1986).

15.2-54 A. S. Kertes and C. J. King, ''Extraction Chemistry of Low Molecular Weight Aliphatic Alcohols,'' Lawrence Berkeley Laboratory Report No. LBL-21210, March 1986.

CHAPTER 16

Bubble and Foam Separations—Ore Flotation

P. SOMASUNDARAN
K. P. ANANTHAPADMANABHAN
Henry Krumb School of Mines
Columbia University
New York, New York

16.1 INTRODUCTION

Flotation processes are useful for the separation of a variety of species ranging from molecular and ionic to microorganisms and mineral fines from one another for the purpose of extraction of valuable products as well as cleaning of wastewaters. They are particularly attractive for separation problems involving very dilute solutions where most other processes usually fail. The success of flotation processes is dependent primarily on the tendency of surface-active species to concentrate at the water–fluid interface and on their capability to make selected non-surface-active materials hydrophobic by means of adsorption on them or association with them. Under practical conditions, the amount of interfacial area available for such concentration is increased by generating air bubbles or oil droplets in the aqueous solution. A classification of flotation processes based on the mechanism of separation and the size of the material that is being separated is given in Table 16.1-1.[1,2] Thus, separation of surface-active species such as detergents from aqueous solution is known as foam fractionation while that of non-surface-active species such as mercury and phosphates that can be complexed with various surfactants is called molecular flotation or ion flotation. The separations of surface-active and non-surface-active subsieve size colloids are known as foam flotation and microflotation, respectively. Froth flotation is used currently for the separation of subsieve size particulates preaggregated by various means to the sieve-size range (Cleveland Cliff Co.). These can be called aggregate flotation. Separation of subsieve-size particulates has been attempted by a number of other techniques using fine bubbles generated by a variety of means or by using oil as the hydrophobic medium. A brief description of various flotation processes is given below.

It is to be noted that, although froth flotation of ores is the only process that has been used industrially on a large scale, other flotation techniques have considerable potential for treating dilute solutions and industrial wastes. Examples of potential areas for large-scale application include treatment of primary and secondary sewage effluents, acid mine drainage, laundry waste, and wastes of textile, paper, leather, dying, printing, and meat processing industries.

16.2 FLOTATION TECHNIQUES

16.2-1 Froth Flotation

In froth flotation, first a pulp of crushed and ground particles in water is conditioned with desired flotation reagents including pH modifiers and surfactants. Then it is agitated in a cell, as shown schematically in

TABLE 16.1-1 Flotation Techniques Classified on the Basis of Mechanism of Separation and Size of Material Separated

	Size Range		
Mechanism	Molecular	Microscopic	Macroscopic
Natural surface activity	Foam fractionation: for example, detergents from aqueous solutions	Foam flotation: for example, microorganisms, proteins	Froth flotation of nonpolar minerals: for example, sulfur
In association with surface-active agents	Ion flotation, molecular flotation, adsorbing colloid flotation: for example, Sr^{2+}, Pb^{2+}, Hg^{2+}, cyanides	Microflotation, colloid flotation, ultraflotation: for example, particulates in wastewater, clay, microorganisms	Froth flotation: for example; minerals such as silica. Precipitate flotation (1st and 2nd kind): for example, ferric hydroxide

Source: Reprinted from *Separation and Purification Methods,* Courtesy of Marcel Dekker, Inc.

Fig. 16.2-1, in the presence of air that is sucked or fed into the impeller zone where the air is well dispersed owing to the intense agitation in that zone. The air bubbles collide with particles and are attached to those that are hydrophobic or have acquired hydrophobicity. The bubble–particle aggregates rise to the top of the cell and are removed by skimming. Various types of machine that are used by the industry have been described in detail by Harris in a recent publication on flotation.[1]

Two cells used in the laboratory for studying the physical chemistry of flotation process are the Hallimond cell and Fuerstenau cell.[2,3] Tests can be conducted in these cells under controlled chemical conditions. Tests in a Hallimond tube cell, shown schematically in Fig. 16.2-2, require only about 1 g of the mineral and do not require the use of a frother. Rigorous control of flotation time, gas flow, and agitation that have been made possible by recent modifications enable one to conduct tests with a reproducibility of $\pm 1\%$. Also, application of the results obtained using the Hallimond tube cell has been demonstrated recently by correlating such results with those obtained using conventional laboratory large-scale cells.[4]

16.2-2　Fine Bubbles Flotation

In contrast to froth flotation, foam separation techniques generally use low aeration rates without any intense agitation. Foam containing the colligend, called foamate, is collected by suction or overflow and then is broken by chemical, mechanical, or thermal methods.[5] A schematic diagram of a typical foam separation unit is given in Fig. 16.2-3. Recovery and grade of the product can be increased by introducing a stripping and enriching mode, respectively. In the stripping mode, the descending feed is introduced into the foam

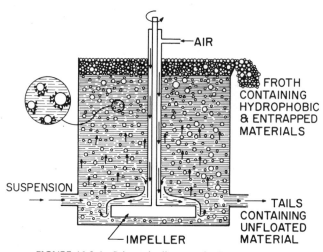

FIGURE 16.2-1　Schematic diagram of a flotation cell.

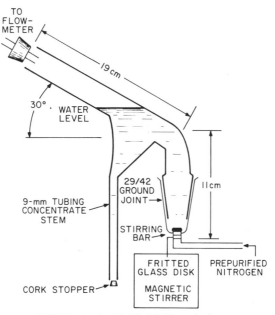

FIGURE 16.2-2   Modified Hallimond tube.

and in the enriching mode part of the foamate is recycled to the top of the flotation column for refluxing action.

16.2-3   Foam Fractionation

Foam fractionation involves the removal of naturally surface-active species by aeration at low flow rates in the absence of any agitation. This technique is particularly useful for the removal of highly surface-active contaminants from surfactants used for basic surfaces and colloid chemistry research work.

16.2-4   Foam Flotation

The above flotation when conducted for microscopic size species that are naturally surface active is called foam flotation. It has been used under laboratory conditions for the removal of microorganisms, dyes, and so on.

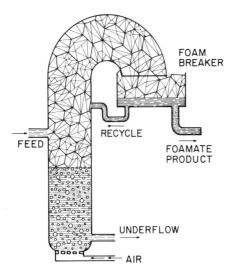

FIGURE 16.2-3   Schematic diagram of a foam separation unit.

16.2-5   Ion Flotation

Ion flotation involves separation of ions capable of association with a surfactant from other ions, molecular matter, or waste material from aqueous solutions. In this case, equimolar amounts of surface-active agents often are needed, making it less attractive as a process for recovering valuable products.

16.2-6   Precipitate Flotation

Ions also can be removed first by precipitating them by changing the pH or by bubbling, for example, hydrogen sulfide in the case of copper, and then by providing appropriate surface-active agents that can adsorb selectively on the surface of the precipitate to make it hydrophobic. This process, known as precipitate flotation, has to be conducted under nonturbulent conditions, because the precipitates usually are colloidal and bulky in nature. An interesting variation of the technique, called precipitate flotation of the second kind,[6] involves precipitation of the species with an organic reagent so that the resulting precipitate is naturally hydrophobic and can be floated without the help of any additional reagents. Examples of this include flotation of nickel with dimethylglyoxime.[7,8]

16.2-7   Microflotation

Flotation of colloidal-size colligends with the aid of surfactants under mild agitation and aeration conditions is called microflotation. This technique has been used recently under laboratory conditions for removal of clays and other colloidal matter from wastewater effluents. It is to be noted that the term microflotation also is used by those working on mineral flotation chemistry to froth flotation conducted in the laboratory with 1–10 g of mineral feed.

16.2-8   Pressure Release and Vacuum Flotation

Conventional froth flotation using cells such as that shown in Fig. 16.2-1 usually fail in processing micron-size particles. A basic handicap of the conventional operation is its inability to control the size of bubbles. Examples of techniques in which fine bubbles are generated include pressure release flotation and vacuum flotation. Pressure release flotation consists of release of gas predissolved in the pulp under pressure, whereas vacuum flotation involves release of gas normally present in the pulp by application of vacuum. In either case, numerous microbubbles are generated on the hydrophobic particles causing their levitation. Generation of bubbles preferentially on hydrophobic sites can produce enhanced selectivity. The air pockets in crevices and pores also can act as nucleation sites for the bubbles; this of course can be detrimental.

16.2-9   Electroflotation

Bubbles that are extremely fine and homogeneous in size can be produced by electrolysis of water using electrodes of a given design. Flotation using such bubbles has been used reportedly in Russia in various industries. An attractive feature of this technique is that the bubbles resist coalescence, possibly because of similar charges on the bubble. It has potential for operation in combination with the conventional flotation (where external air is used) for treating ores containing particles in all size ranges.

   Vacuum, pressure release, and electroflotation can be used to remove a variety of materials such as oils, fats, heavy metals, and other suspended solids from municipal or industrial waste.[9]

16.2-10   Oil Flotation

The flotation processes using oil–water interface for collection of particles are emulsion flotation and liquid–liquid flotation. In the former the reagentized particles are collected by oil–water emulsion droplets and by aeration of the system, whereas in the latter removal of the particles collected at the interface is achieved mostly by phase separation. The only commercial use of emulsion flotation, to our knowledge, is that of the separation of apatite from iron ore at LKAB, Malmberget, Sweden.

16.2-11   Aggregate Flotation

Conventional flotation processes can be made applicable to the treatment of fines simply by preaggregating them among themselves or with another carrier material. Techniques in this category include floccflotation, carrier flotation (ultraflotation), and spherical agglomeration.

FLOCCFLOTATION

A technology with enormous potential in the mineral processing area is selective flocculation accompanied by flotation. Such a process already has become commercial for the separation of iron minerals from low-grade iron ore.[10] In this case starch is used as flocculant for iron oxide and quartz is floated using amine as the collector. Flocculation also can be achieved by the adsorption of polyelectrolytes or ionic species. Past laboratory work on selective flocculation deals mostly with binary mineral systems in which the valuable mineral was a metal sulfide (galena, pyrite, or sphalerite)[11-15] or a metal or its oxide (hematite, chromite, iron, and titanium)[15-18] and the other component was a gangue mineral. Reports of separation by selective flocculation on multicomponent natural ores itself are scant. One noteworthy attempt in this regard is that by Carta et al.[19] for the beneficiation of ultrafine fluorite from latium.

In this technique, known also as piggyback flotation, a carrier material is used for floating the fine particles. For example, anatase is removed on a commercial scale from clay for use in the paper industry by using calcite as the carrier. While anatase does not float by itself, it is cofloated with a coarse auxiliary mineral such as calcite.

An analogous process is one called adsorbing colloid flotation in which the colligend is adsorbed on a colloid that can be floated using various microflotation techniques.[20]

SPHERICAL AGGLOMERATION

Fines are tumbled in this case in an aqueous solution containing an immiscible liquid which forms capillary bridges between reagentized particles and causes their aggregation. Since Stock's original observation of this phenomenon in 1952 with barium sulfate precipitates in benzene, containing a small amount of water, it has been examined mainly by Puddington and coworkers for agglomeration of graphite, chalk, zinc sulfide, coal, iron ore, and tin ore suspensions in aqueous solutions.[21] Also, Farnard et al.[22] have claimed good separation of each component from a mixture of zinc sulfide, calcium carbonate, and graphite in water with nitrobenzene as binding liquid by stepwise agglomeration.

Physicochemical principles governing the various flotation processes are essentially identical, even though there can be significant differences in the actual mechanics used in their application. Basic principles involved in flotation are discussed below with appropriate examples.

16.3 PHYSICOCHEMICAL PRINCIPLES

The success of selective flotation depends primarily on the differences in the hydrophobicity of the species or particles that are to be floated. Except for a small fraction, colligends are generally hydrophilic and therefore, to impart hydrophobicity, surfactants that selectively will associate with or adsorb on them are added to the system. These surfactants, generally called *collectors*, have at least one polar head and one hydrophobic tail in their molecular structure. Collectors adsorb on minerals with their hydrophobic tail turned toward the bulk solution, thereby making the minerals hydrophobic. Typical examples of collectors used in practice include long-chain amines for quartz, potash, and anionic complexes such as ferrocyanide and short-chain xanthates for base metal sulfides.

In the recent past, a number of excellent reviews and books have appeared on the physicochemical aspects of flotation.[1-6] Only a brief overview of the mechanistic aspects are included here and for more details readers should consult the above references.

The association or adsorption of surfactants with the colligend species occurs due to various interactive forces, operating individually or in combination with each other. Major forces that can contribute to the adsorption arise from electrostatic attraction, covalent bonding, hydrogen bonding, van der Waals cohesive interaction among the adsorbate species, and solvation or desolvation of adsorbate or adsorbent species in the interfacial region. The concentration c_s, in kmol/m^3, of counterions in the interfacial region can be given on the basis of the Boltzmann distribution function as

$$c_s = c_b \exp\left(\frac{-\Delta G^\circ_{B \to S}}{RT}\right) \tag{16.3-1}$$

when c_b is the concentration in bulk and $\Delta G^\circ_{B \to S}$ is the free-energy change involved in the transfer of surfactants from the bulk to the surface of the colligend. Equation (16.3-1) can be rewritten in terms of adsorption density Γ_s by multiplying the right-hand side by the thickness τ of the adsorbed layer:

$$\Gamma_s = \tau c_b \exp\left(\frac{-\Delta G^\circ_{ads}}{RT}\right) \tag{16.3-2}$$

In the case of an inorganic species, τ can be assumed to be equal to its diameter in the hydrated or dehydrated form as the case may be. Even though a similar assumption has been made in the past with respect to surfactants, it must be noted that in this case τ can vary anywhere from the cross-sectional diameters of the molecule (for flat adsorption on particles) to the diameter of a partially coiled molecule or even that of a fully extended molecule (for surface micellization). ΔG°_{ads}, the driving force for adsorption, will be the sum of a number of contributing forces as shown below:

$$\Delta G_{ads}^{\circ} = \Delta G_{elec}^{\circ} + \Delta G_{cov}^{\circ} + \Delta G_{c\text{-}c}^{\circ} + \Delta G_{c\text{-}s}^{\circ} + \Delta G_{H}^{\circ} + \Delta G_{solv}^{\circ} \qquad (16.3\text{-}3)$$

ΔG_{elec}° is the term that arises from the electrostatic interaction between ionic species and the charged colligend; similarly, ΔG_{cov}° is due to any covalent bonding that leads to chemisorption; $\Delta G_{c\text{-}c}^{\circ}$ is due to the cohesive chain–chain interaction between surfactant species upon adsorption; $\Delta G_{c\text{-}s}^{\circ}$ is the nonpolar interaction between the chain and the solid substrate; ΔG_{H}° is the term due to hydrogen bonding; and ΔG_{solv}° is the result of solvation or desolvation of any species owing to the adsorption process. For each system, one or more of the above terms can be contributing, depending on the type of the colligend, surfactant, and other chemical species in the system, concentration of the surfactant, pH, temperature, ionic strength, and so on. Thus, for adsorption of alkyl sulfates on nonmetallic minerals such as quartz, electrostatic and lateral chain–chain interaction forces are considered to play a governing role, whereas for adsorption of xanthates on sulfides, the covalent forces are considered to be predominant.

16.3-1　Electrostatic Forces

Electrostatic properties of the solid surfaces generally result either from the preferential dissolution of the lattice ions, as in the case of silver iodide, or from the hydrolysis of the surfaces followed by the pH-dependent dissociation of the surface hydroxyls as in the case of silica:[7]

$$-M(H_2O)_{surf}^{+} \underset{H^{+}}{\overset{OH^{-}}{\rightleftharpoons}} -MOH_{surf} \rightleftharpoons -MO_{surf}^{-} + H_2O$$

The sign and magnitude of the electrical field is determined primarily by the concentration of positive and negative (surface) potential-determining ions. Lattice ions are considered to be potential-determining ions for AgI-type solids and H^{+} and OH^{-} are the corresponding ions for oxide minerals. For salt-type minerals such as calcite and apatite, both of the above mechanisms can be operative since their lattice ions can undergo preferential dissolution as well as hydrolysis reactions with H^{+} and OH^{-}. In such cases, H^{+}, OH^{-}, and all charged complexes that are the result of the hydrolysis reactions can play a major role in determining the surface potential. Even for the oxide minerals such as silica, it will be more accurate to consider dissolved hydrolyzed species as potential determining, since these minerals do have finite solubilities that can amount to significant levels. Silicate minerals, with layered structures, possess a net negative charge under most natural conditions due to substitutions, for example, Al^{3+} for Si^{4+} and Mg^{2+} for Al^{3+} in the structure.

The surface potential Ψ_0 for the above minerals is given by

$$\Psi_0 = \frac{RT}{Z_+ F} \ln\left(\frac{a_+}{a_+^{PZC}}\right) = \frac{RT}{Z_- F} \ln\left(\frac{a_-}{a_-^{PZC}}\right) \qquad (16.3\text{-}4)$$

where F is the Faraday constant and a_+ and a_- are activities of the positive and negative potential-determining ions with valencies Z_+ and Z_- (inclusive of sign); a_+^{PZC} and a_-^{PZC} are activities under conditions of zero charge of the particle surface. Such a condition of zero charge is called the point of zero charge (PZC). Particles will carry a positive charge below the PZC, represented in terms of the negative of the logarithm of the positive potential-determining ions, and negatively charged above it. Since the system as a whole must be electrically neutral, there should be an equivalent amount of ions in the interfacial region, called counterions, with charge opposite to that of the particle surface. A schematic diagram of the resultant diffuse soluble layer is given in Fig. 16.3-1.

For oxides and salt-type materials, adsorption of both organic and inorganic flotation reagents are often the result of electrostatic attraction between the solid and the reagent. The PZC of the solid is an important characteristic property in such cases and can be determined easily by experiment. Typical PZC values of some common minerals are given in Table 16.3-1. It is to be noted that the PZC of the minerals has been shown, using zeta potential (potential of the shear region) measurements, to be affected significantly by various factors such as pretreatment of the solid, extent of aging, storing, as well as the pH and even the ionic strength of the solution in which it is stored.[45,68-70]

It has been shown recently that the commonly used cleaning procedures such as leaching in acidic and hot solutions can affect drastically both the sign and magnitude of the experimentally measured parameters such as the zeta potential.[71,72] In addition to the above mentioned variables, surface chemical heterogeneity of the particles also can contribute significantly to the range of PZC values that can be obtained for a given solid.[73]

An interesting study, in this regard, by Kulkarni and Somasundaran[73] involved the analysis of various spots on a typical hematite particle using scanning electron microscopy, energy dispersive X-ray analysis, and Auger spectroscopic techniques. Figure 16.3-2 shows the electron micrograph of a hematite particle. Spots E, G, and H shown in the figure were analyzed and the results showed a very high percentage of silica at spots F and G whereas spot H showed almost pure hematite. For this particular sample, whereas the bulk analysis indicated a silica content of 4%, surface analysis using the Auger technique gave a value

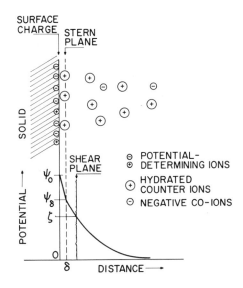

FIGURE 16.3-1 Schematic diagram of the diffuse double layer—Stern's model.

as high as 50% for the silica content. The presence of such large amounts of silica on the surface will decrease the PZC of the hematite to lower pH values. It is to be noted that, because of the presence of positive hematite regions on the mineral, the adsorption of anionic surfactants still can take place above the net PZC of the sample. In fact, results from the literature (see Fig. 16.3-3) indicate the adsorption of anionic surfactants above the net PZC of hematite.[74,75] In such cases, the observed adsorption could be due to the surface chemical heterogeneities rather than any chemisorption of the collector as speculated in the past.[74,75] In addition, chemical heterogeneity of the particles can contribute significantly to the range of PZC values that can be obtained for a given solid.

In the case of sulfide minerals, oxidation of the surface also can affect the PZC considerably. As the pH increases, the surface may get oxidized and the potential obtained at any particular pH may be the net value of the oxide and the sulfide. In fact, it may be possible to obtain two PZC values for such minerals, one corresponding to that of the sulfide at lower pH and the other corresponding to that of the oxide at higher pH values. Some of the wide range of values shown in Table 16.3-1, for example, chalcocite, chalcopyrite, pentlandite, and sphalerite, could have resulted partly from such surface oxidation. The role of the electrical nature of the interface in determining adsorption can be seen for the case of adsorption of dodecylsulfonate on alumina[76] (see Fig. 16.3-4). It can be seen that only below the point of zero charge of alumina, when the solid is positively charged, is there measurable adsorption of the anionic sulfonate. This effect is shown more clearly for the case of calcite, for which adsorption and resultant flotation with cationic amine is significant only above the point of zero charge (see Fig. 16.3-5). Indeed, change in electrical characteristics due to adsorption of inorganic species can affect significantly the flotation response of the particulates, as will be seen later. Thus, flotation can be depressed by adding electrolytes that will compete with the flotation reagents for adsorption in the interfacial region or enhanced by adding those electrolytes that can adsorb specifically and change the charge in the desired direction. Depression of amine flotation of quartz using monovalent and divalent inorganic cations has been analyzed recently on the basis of the double-layer model and its compression.[77] Toward this purpose we rewrite Eq. (16.3-1) with the ΔG°_{ads} consisting of the electrical term and a term to account for the specific adsorption of the bivalent ion

$$\Gamma_s = \tau c_b \exp\left(\frac{-(ZF\psi^{\delta} + \phi)}{RT}\right) \qquad (16.3\text{-}5)$$

where ϕ is the specific adsorption energy. Assuming that ψ^{δ} is equal to the zeta potential, that the adsorption density of the collector ions at the solid–liquid interface is constant for a given amount of flotation and that the addition of electrolytes has no effect on the specific adsorption potential of the collector ions, one can write the following equation for the ratio c_{Na}/c_{Ba} for equivalent flotation:

$$\frac{\Gamma_s^{Na}}{\Gamma_s^{Ba}} \simeq \frac{r^{Na}}{r^{Ba}} \frac{c_b^{Na}}{c_b^{Ba}} \exp\left(\frac{\phi^{Ba} + 2F\zeta^{Ba} - F\zeta^{Na}}{RT}\right) \qquad (16.3\text{-}6)$$

Γ_s^{Na} and Γ_s^{Ba} are, respectively, the adsorption densities of sodium and barium counterions under conditions

TABLE 16.3-1 Typical Point of Zero Charge (PZC) Values of Common Minerals

Oxides	IEP/PZC[a] pH	Reference
Anatase	5.9	8
Cassiterite	4.5	9
Chromite	5.6–7.2	10, 11
Corundum	9–9.4	12, 13
Cuprite	7–9.5	14, 15
Goethite	6.7	16
Hematite (natural)	4.8–6.7	17
Hematite (synthetic)	8.6	17
Magnesia	12.0	18
Magnetite	6.5	19
Pyrolusite	5.6, 7.4	20
Rutile	6.0	21
Tenorite	9.5	22

Salt Type	IEP/PZC[a] pH	Reference
Aragonite	5.0	23
Barite, pBa 3.7–7.0	3.4	24
Calcite, pCa 3.5, pCO$_3$ 3.0	8–10.8	25, 26
Celestite	2.3	23
Dolomite	7.0	27
Eggonite	4.0	28
Fluorapatite	4.6	29, 30
Fluorapatite (synthetic), pCA 4.4, pF 4.6	5.2	31
Francolite, pHPO$_4$	3.8–4.9	32
Magnesite	6–6.5	27
Monazite	3.4	24
Scheelite, pCA 4.8		28
Silver, pAg 4.1–4.6		28
Silver iodide, pAg 5.6		33
Silver sulfide, pAg 10.2		34
Strengite	2.8	28

Silicates	IEP/PZC[a] pH	Reference
Andalusite	5.2–7.5	35, 36
Augite	2.7	37
Bentonite	3.0	38
Beryl	3.1–4.4	39, 40, 41
Biotite	0.4	41
Chrysocolla	2.0	42
Garnet	4.4	41
Kaolinite	3.4	38
Kyanite	6.2–7.9	35, 36, 43
Muscovite	0.95	44
Quartz	2.3–3.7	45, 46, 47
Rhodonite	2.8	42
Spodumene	2.5–4	48, 49
Talc	3.5	51
Tourmaline	4.0	52
Zircon	5.8	41

Sulfides	IEP/PZC pH	Ref.	Comments
Antimonite	2–3.0	53	Conditioning time not specified
Chalcocite	3–10	54	
Chalcopyrite	2.0–3.0	55, 56	Conditioning time not specified
Chalcopyrite	5.25, 8.0	57	
Cinnabar	3.0–4.0	58	
Galena	2.0–4.0	59	

TABLE 16.3-1 (*Continued*)

Sulfides	IEP/PZC pH	Ref.	Comments
Molybdenite	3.0	60, 61	
Nickel sulfate	2.5–3.0	62	
Pentlandite	11.5	63	
Pyrite	6.2–6.9	64	Short conditioning time
Pyrrhotite	3.0	55	Conditioning time not specified
Sphalerite	2–7.5	51, 65 66, 67	Conditioning time not specified

[a] Any condition corresponding to zero electrokinetic potential is referred to as IEP. In the absence of specific adsorption PZC = IEP

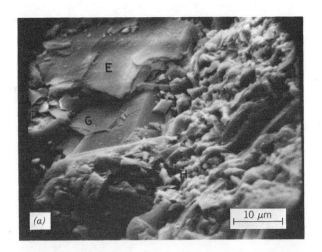

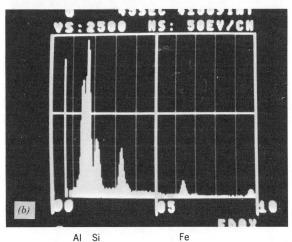

Al   Si   Fe

FIGURE 16.3-2 (*a*) Scanning electron micrograph of a hematite sample on which spot analysis was conducted. (*b*) EDAX analysis of spot G shown in part (*a*). Analysis of spot E was similar to that obtained for spot G. (*c*) EDAX analysis of spot H as shown in part (*a*). (After Kulkarni and Somasundaran;[73] courtesy of Elsevier Seuqoia S.A., Lausanne, Switzerland.)

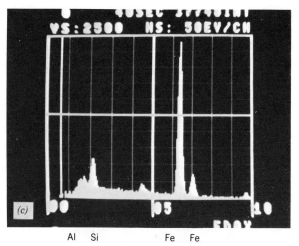

Al   Si   Fe   Fe

FIGURE 16.3-2 (*Continued*)

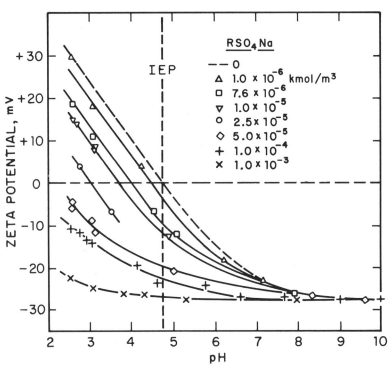

FIGURE 16.3-3 Zeta potential of natural hematite in sodium dodecylsulfate solution. (Data from Shergold and Mellgren.[74,75])

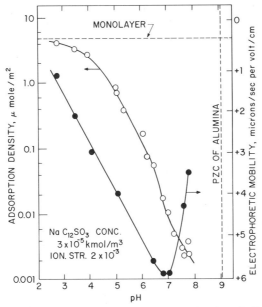

FIGURE 16.3-4 Adsorption of dodecylsulfonate on alumina as a function of pH. (After Somasundaran and Fuerstenau;[76] courtesy of the American Chemical Society.)

of equivalent flotation and therefore, by assumption, under constant concentration of the collector. c_b^{Na} and c_b^{Ba} are the corresponding bulk concentrations of sodium and barium ions and ζ^{Na} and ζ^{Ba} are the corresponding zeta potentials. r^{Na} and r^{Ba} are considered in this case to be the radii of the hydrated sodium and unhydrated barium ions, respectively, and ϕ^{Ba} is the specific adsorption energy of 1 mol of barium ions. The adsorption of hydrated sodium ion on oxide minerals is nonspecific, since its presence is not known to change the point of zero charge of these minerals. Using a value of 2 for Γ^{Na}/Γ^{Ba} and $3RT$ for ϕ^{Ba}, we obtain values on the basis of Eq. (16.3-6) for the ratio of $c_b^{Na} c_b^{Ba}$ for constant flotation. These theoretical values are seen in Table 16.3-2 to be in fair agreement with experimental values. Considering the complex

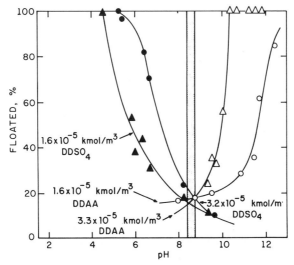

FIGURE 16.3-5 Flotation of calcite with dodecylammonium acetate (DDAA) and sodium dodecyl sulfate (DDSO$_4$) solutions. (After Somasundaran and Agar;[25] courtesy of Academic Press.)

TABLE 16.3-2 Calculated and Experimental Values for the Relative Effectiveness of Sodium Ions and Barium Ions in Depressing Quartz Flotation at Natural pH Using 10^{-4} kmol/m^3 per Dodecylammonium Acetate

Floated (%)	c^{Na} (kmol/m^3)	c^{Ba} (kmol/m^3)	ζ^{Na} (mV)	ζ^{Ba} (mV)	Present Calculated c^{Na}/c^{Ba}	Experimental c^{Na}/c^{Ba}
90	1×10^{-3}	2×10^{-5}	-63	-50	64	50
70	6×10^{-3}	1.5×10^{-4}	-49	-35	33	40
50	1.5×10^{-2}	8×10^{-4}	-40	-22	18	20
30	5×10^{-2}	5×10^{-3}	-32	-10	10	10

nature of the flotation process, this agreement must be considered to provide an excellent support for the role of the electrostatic interactions in determining the flotation of such materials.

16.3-2 Chain–Chain Interactions

The zeta potential plot for alumina given in Fig. 16.3-4 as a function of pH of dodecylsulfonate solutions shows a reversal of slope below about pH 7, suggesting increased adsorption below this pH involving forces in addition to electrical attraction. The adsorption isotherm obtained for dodecylsulfonate on alumina in fact shows a marked change in slope at a particular surfactant adsorption supporting such an interpretation. Based on the experimental observations that a number of other interfacial properties such as flotation, contact angle, and suspension settling rate undergo a marked change in a given surfactant concentration range, it was proposed that at low concentrations of the surfactant, the surfactant ions are adsorbed on the mineral due to electrostatic forces, while at high concentrations adsorption is assisted further by forces arising from lateral associative interactions of the adsorbed surfactant species (see Figs. 16.3-3–16.3-7). The concentration at which such two-dimensional lateral interactions begin has been shown to depend on pH, temperature, and the chemical state and the structure of the surfactant.

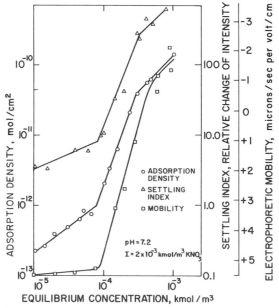

FIGURE 16.3-6 Adsorption density of dodecylsulfonate, electrophoretic mobility, and settling rate of alumina–sodium dodecylsulfonate system as a function of the concentration of sodium dodecylsulfonate.[78] (Courtesy of Academic Press.)

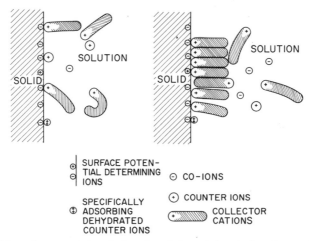

SURFACE POTEN-
TIAL DETERMINING
IONS

⊖ CO-IONS

⊙ COUNTER IONS

⊕ SPECIFICALLY
ADSORBING
DEHYDRATED
COUNTER IONS

COLLECTOR
CATIONS

FIGURE 16.3-7   Schematic representation of adsorption of a long-chain anionic surfactant: (*a*) individually at low concentrations and (*b*) with lateral association between chains at higher concentrations.

16.3-3   Covalent Bonding

In contrast to that of adsorption of sulfonates on alumina where electrostatic forces are primarily responsible, the adsorption of fatty acids on solids such as calcite, fluorite, barite, and apatite has been considered to be the result of covalent bonding of surfactant species to the solid surface species. Such adsorption, usually referred to as chemisorption, first was proposed by French et al.[79] for the adsorption of oleate on fluorite on the basis of results obtained from infrared analysis of the solid surface contacted with the oleate. Subsequently, Peck and Wadsworth[80] reported chemisorption of oleate on calcite and barite also. The chemisorption has been proposed to take place by an ion-exchange mechanism in which the surfactant anions replace an equivalent amount of lattice ions such as F^- to form a surface layer of alkaline earth oleate. The ion-exchange mechanism has been confirmed by Bahr et al.[81,82] and Bisling[83] who obtained stoichiometric release of fluoride ions into the system during the adsorption of oleate on fluorite. Similar mechanisms resulting in the formation of salts have been proposed by Shergold[84] for the adsorption of dodecylsulfate on fluorite, Cumming and Schulman[85] for that of dodecylsulfate on barite, and Fuerstenau and Miller[86] for that of alkylsulfonates on calcite.

As indicated above, a technique used for studying adsorption in such systems is the infrared analysis of the surface contacted with the surfactant. In general, use of such spectroscopic techniques and interpretation of such results have to be attempted with caution. The preparatory steps for the spectroscopic analysis such as washing, drying, or any aging conceivably can alter the nature of the surface complexes. In fact, complexes observed during the analysis could have been generated during such preparatory steps.

Also, oleate has been suggested to adsorb on calcite due to electrostatic bonding below its point of zero charge and covalent bonding above it. Results of oleate adsorption and zeta potential changes supporting such a mechanism are given in Figs. 16.3-8 and 16.3-9. The sharp increase in the slope of the adsorption isotherm is attributed to the precipitation of calcium oleate since, with calcium present at a concentration of 1.5×10^{-4} kmol/m^3, the solubility limit is exceeded for calcium oleate above about 5×10^{-5} kmol/m^3 of oleate. Oleate adsorption at concentrations below 10^{-6} kmol/m^3 oleate and at pH values below the PZC of calcite is considered to be due to electrostatic attraction between the negative oleate ions and positive surface sites, since under these conditions there is very little change in the zeta potential.

16.3-4   Hydrogen Bonding

Surfactants containing phenolic, hydroxyl, carboxyl, and amine groups have been considered to form hydrogen bonds in many systems. For example, adsorption of oleic acid on fluorite and beryl pretreated with hydrofluoric acid has been proposed to involve hydrogen bonding.[88] According to Parks' review of Giles' work, adsorption of phenols on alumina and silica and some textile substrates also involves hydrogen bonding. In this regard, the mechanism suggested by Sorensen,[89] based on the structural similarity between the crystal and the surfactant, is noteworthy.

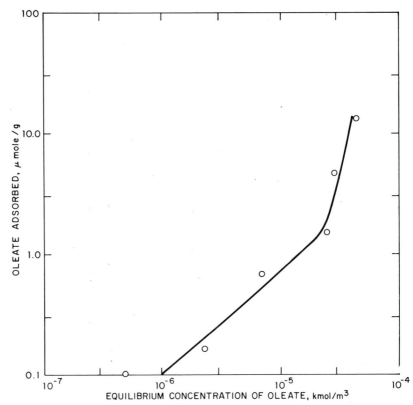

FIGURE 16.3-8 Adsorption isotherm of potassium oleate on calcite at a natural pH of 9.6. (After Somasundaran,[87] courtesy of Academic Press.)

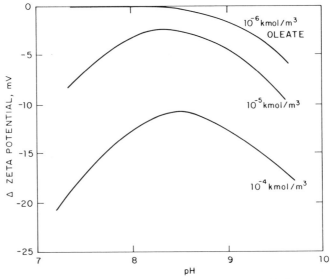

FIGURE 16.3-9 Change in zeta potential of calcite particles as a function of pH at constant ionic strength 10^{-3} kmol/m³ KNO₃). (After Somasundaran;[87] courtesy of Academic Press.)

16.3-5 Structural Compatibility

For the case of anionic flotation of simple salts such as fluorite, it has been suggested by Sorensen that hydrogen bonding between the oxygen of the collector and the fluoride species is active and that it is assisted by the electron resonance of the polar groups, the structure of which must be compatible with the geometry of the mineral crystal. The role of structural compatibility also has been examined for the case of soluble salts. For example, Fuerstenau and Fuerstenau[90] proposed that the surfactant adsorption on soluble salts is governed by a matching of size of the functional group of the surfactant with that of similarly charged lattice ions of the solid. Thus, aminium ions adsorb on sylvite (KCl) but not on halite (NaCl), owing to the comparable size of the aminium ion and the potassium ion. It is to be noted, however, that this theory fails to explain why a long-chain anionic sulfate will adsorb only on KCl but no on NaCl.

An alternate mechanism in terms of hydration properties of the solid has been put forward by Rogers and Schulman[91] for the adsorption of surfactants on soluble salts.

16.3-6 Hydration Factors

According to Rogers and Schulman,[91] adsorption on soluble salts is governed by their solvation properties, the ones with the largest negative heat of solution being a better adsorbent than the others. However, this theory provides no adequate explanation for several adsorption systems involving soluble salts.

16.3-7 Precipitation

An interesting consideration by du Rietz involves a condition of precipitation of the surfactant–lattice ion complex for incipient flotation.[92] This mechanism has been examined subsequently in detail by a number of investigators.

16.3-8 Adsorption at Liquid–Air Interface

Even though the flotation process involves three phases and three interfaces, most research work has been solely on the behavior of the solid–liquid interface. This is in spite of the fact that adsorption at the solid–liquid interface, as shown in Fig. 16.3-10, is of a considerably smaller magnitude than that at the solid–gas or liquid–gas interface.[93] It is to be noted that excellent correlation has been obtained recently between surfactant adsorption at the liquid–gas interface and flotation for the hematite–oleate system (see Fig. 16.3-11). It is also important to note that the migration of the surfactant at the liquid–gas interface is faster than its diffusion from bulk to the interface, as least for this system.[94] Such a migration at the interface can help toward faster attainment of required surfactant adsorption density at the solid–gas interface upon the contact of the bubble with the particle.

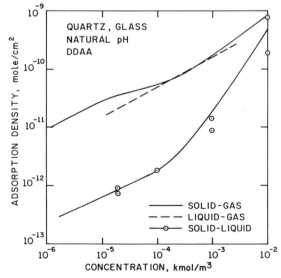

FIGURE 16.3-10 Comparison of adsorption of dodecylammonium acetate at different interfaces. (After Somasundaran;[93] courtesy of American Institute of Metallurgical Engineers.)

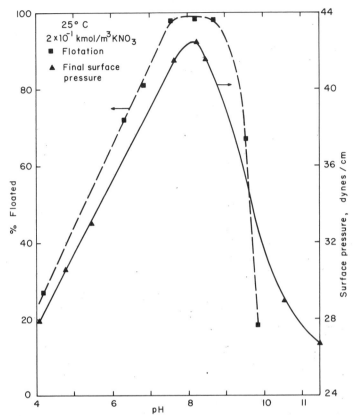

FIGURE 16.3-11 Comparison of flotation properties of 3×10^{-5} kmol/m^3 potassium oleate solutions with final surface pressure at 25°C. (After Kulkarni and Somasundaran;[94] courtesy of American Institute of Chemical Engineers.)

16.3-9   Role of Ionomolecular Complexes

Studies of the liquid–gas interfacial properties have provided a new insight into the flotation mechanisms by revealing the prominent role of the complexes formed between different surfactant species in flotation. Surfactants such as fatty acids and amines will undergo hydrolysis in water and produce various complexes depending on the pH of the solution. Thus, oleate will exist in the ionic form in the alkaline pH range, in the molecular form in the acidic range, and in the ionomolecular complex in the intermediate pH range.

$$\underset{\text{oleic acid}}{RH} \;=\; \underset{\text{oleate}}{R^-} + H^+$$

$$\underset{\text{oleate}}{R^-} + \underset{\text{oleic acid}}{RH} \;=\; \underset{\text{ionomolecular complex}}{RRH^-}$$

The role of such ionomolecular complexes has been shown to be a potentially important factor in flotation. In fact, the pH of maximum flotation recovery for the hematite–oleate system is found to correspond with the pH where maximum complex formation is expected.[94] Evidence for the formation of the highly surface-active complex was obtained using surface-tension measurements shown in Fig. 16.3-11. Similarly, the pH of maximum flotation of quartz using amine has been shown to coincide with the pH at which maximum lowering of the adhesive tension of the system and of the surface tension of amine solutions occurs (see Figs. 16.3-12–16.3-14). This is also the pH region in which stable amine–aminium ion complexes can be expected. The thermogravimetric analysis results of Kung and Goddard[96] have provided some evidence for the existence of such ionomolecular complexes in the bulk phase also. From these results, the formation of complexes between neutral molecules and ions appears to play an important part in their enhanced adsorption and resultant flotation using them. It must be pointed out that such correlation has not been obtained during studies using tertiary amines.

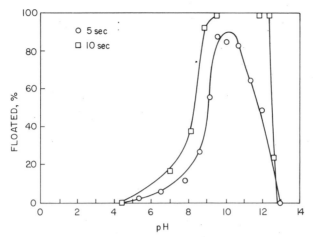

FIGURE 16.3-12 Flotation recovery of quartz as a function of solution pH using dodecylammonium acetate as collector for flotation duration of 5 and 10 s. (After Somasundaran;[95] courtesy of Elsevier S.A., Lausanne, Switzerland.)

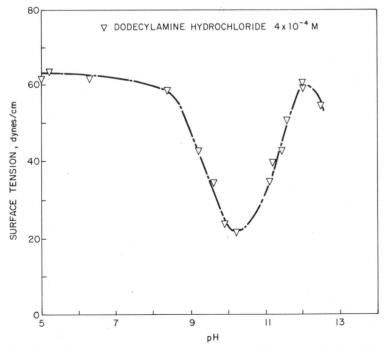

FIGURE 16.3-13 Adhesion tension of dodecylammonium chloride solution of various concentrations as a function of solution pH. (After Somasundaran;[95] courtesy of Elsevier S.A., Lausanne, Switzerland.)

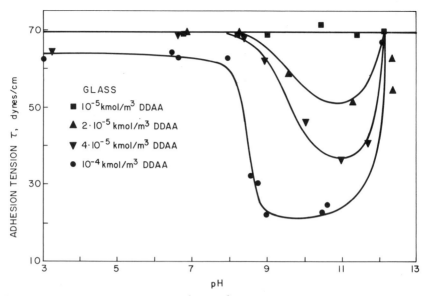

FIGURE 16.3-14 Surface tension of 4×10^{-4} kmol/m^3 dodecylamine hydrochloride solution as a function of pH determined by pendant drop method, measured 15 s after forming drop.[95] (After R. W. Smith, personal communication, 1967.)

The mechanism of adsorption of complexes on minerals has not been investigated. It can be noted, however, that the electrostatic factor can be important in this case also, since the complexes like the monomer ions are charged. In addition, increase in the effective size of the species due to complex formation can be expected to make the surfactant less soluble in water and hence more active.

16.4 FLOTAIDS

In addition to collectors, a number of other chemical additives are used in flotation to aid separation by this process. They include frothers, activators, depressants, deactivators, flocculants, and dispersants.

16.4-1 Frothers

To produce the desired froth stability, nonionic surfactants, such as the sparingly soluble monohydroxylated cresols, usually are added, particularly when the collector used is of the short-chain type. The optimum concentration of the frother in the system is approximately that at which there is a significant change in surface tension with surfactant addition (Fig. 16.4-1). It is possible, even though not proved, that the restoring force that becomes available upon any distention of the bubble to prevent its rupture might contribute toward the required froth stability in the flotation cell. Indeed, this is applicable only if the diffusion of the surfactant to the locally extended surface region is not fast enough to reduce the surface pressure difference between this region and the surrounding surface, before the distention is repaired by such pressure difference.

In addition to inducing froth stability, frother species can take part in the overall process of adsorption on the mineral surface. Like the collector species, the frother species also can be expected to migrate to the particle–gas interface during the time of contact and assist in establishing the attachment of the bubble to the particle. Coadsorption of frother along with the collector species[2-4] can be favorable for flotation, possibly because the neutral molecules adsorbed between charged collector ions can reduce the repulsion between the latter species and thereby enhance the overall surfactant adsorption.

16.4-2 Activators

Activators are used for enhancing flotation of the minerals that may not possess any flotability in their absence. Flotation of quartz using calcium salts and of sphalerite using copper sulfate (see Fig. 16.4-2) are typical examples of activation. In the case of oleate flotation of quartz in the presence of calcium, activation can be attributed to electrostatic adsorption of the calcium ions on the negatively charged quartz

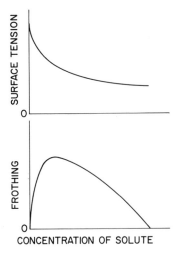

SURFACE TENSION

FROTHING

CONCENTRATION OF SOLUTE

FIGURE 16.4-1 Diagram illustrating the correlation between froth stability and surface tension lowering due to the addition of a surfactant. (After Cooke;[1] courtesy of John Wiley & Sons.)

and thereby providing sites for adsorption of the oleate collector species. Bivalent ions, upon adsorption, can reverse the sign of the stern potential and thus cause adsorption and flotation with collectors that have a charge of the same sign as that of the mineral (see Fig. 16.4-3).

However, sphalerite activation by copper sulfate is the result of adsorption of copper ions on the surface of this mineral, due to ion-exchange processes. The effect of activators also can be due to their reactions with the collectors to form compounds of low-solubility product.[7]

Another typical example of activation is that of the oxide or carbonate minerals by sodium sulfide. For example, in the case of cerrusite,[8] the following reactions can produce a surface layer of sulfide:

$$Na_2S + H_2O \rightleftharpoons NaSH + NaOH$$

$$PbCO_3 + 3NaOH \rightleftharpoons H_2O + Na_2CO_3 + NaHPbO_2$$

$$NaSH + NaHPbO_2 \rightleftharpoons 2NaOH + PbS$$

$$Na_2S + PbCO_3 \rightleftharpoons Na_2CO_3 + PbS$$

The cerrusite surface, which is altered in the above manner, can be floated with xanthate collectors. In certain cases, on the other hand, it is necessary to remove altered surfaces using acids to obtain flotation (see Fig. 16.4-4). Acids can also enhance flotation possibly by generating microbubbles on the mineral surface as has been suggested in the case of calcite.[9]

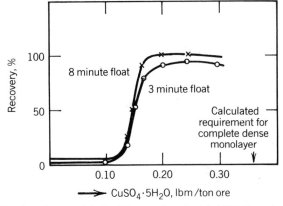

Recovery, %

100

8 minute float

3 minute float

50

Calculated requirement for complete dense monolayer

0

0.10   0.20   0.30

CuSO₄·5H₂O, lbm/ton ore

FIGURE 16.4-2 Flotation of pure sphalerite. The granular mineral, 100/150 mesh was floated with xanthate 0.10 lb/ton, terpineol, 0.20 lb/ton, sodium carbonate, 2.00 lb/ton, and copper sulfate, as shown. (From Ref. 5; courtesy of McGraw-Hill, New York.)

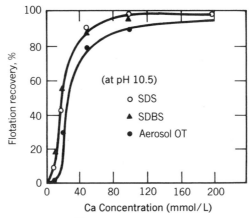

FIGURE 16.4-3 Effect of amount of Ca^{2+} in the flotation of quartz by different surfactants (From Ref. 6.)

16.4-3 Depressants

Depressants are organic or inorganic reagents that prevent the collector adsorption on the mineral by interacting with the mineral or the collector. Silicates, phosphates, aluminium salts, chromates, and dichromates are typical inorganic salts used as depressants. The action of sodium silicate in flotation is considered to be due usually to its depressing effect on quartz present in the pulp and to its ability to control the dispersion of the slimes that are present in the pulp. The effectiveness of silicates has been related to the degree of polymerization that it can undergo under the given conditions.[10] The effect of polyvalent cations in the case of flotation of phosphate or calcite, on the other hand, is attributed primarily to collector precipitation in the presence of these ions.[11,12] Polyvalent ions also can prevent collector adsorption by causing charge reversal of minerals. For example, addition of soluble phosphate can depress flotation of apatite using oleate (see Fig. 16.4-5).

Organic reagents used as depressants include starch, tannin, quebracho, and dextrin.[13] Except for some short-chain organic acids, these reagents are characterized by their relatively high molecular weight (10^5–10^6) and the presence of strongly hydrated polar groups such as $-OH$, $-COOH$, $-NH_2$, $-SO_3H$, and

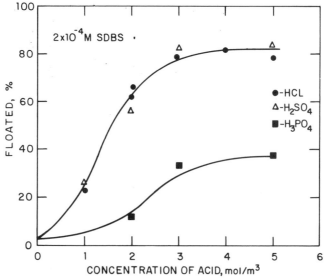

FIGURE 16.4-4 Effect of the type and concentration of mineral acids on the flotation of phosphate-depressed calcite. (After Saleeb and Hanna.[6])

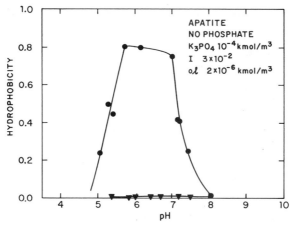

FIGURE 16.4-5 Effect of phosphate species on the flotation of apatite[125]. K-oleate 2×10^{-6} kmol/m^3.

—COH. These reagents can act by mere adsorption on the minerals as well as by causing flocculation of the slime that is responsible for excessive collector consumption. Even though such reagents have been used extensively as modifiers, the actual mechanisms by which they act are not understood totally. Schultz and Cooke[14] and Balajee and Iwasaki[15] have shown that the adsorption of starch on iron oxide materials depends on the type of starch, its preparation, the extent of branching, and so on. Experiments done recently on the adsorption of starch and oleate in the presence of each other have provided some insight into the mechanisms involved here (Ref. 16.3-8). It was found that, in this case, the starch does not reduce flotation by inhibiting the adsorption of surfactant on calcite particles. In fact, the adsorption of oleate on calcite was found to be higher in the presence of starch than otherwise (see Figs. 16.4-6–16.4-8). Similarly, the adsorption of starch also was enhanced by oleate. Depression of mineral flotation obtained under these conditions suggested that even though the mineral has adsorbed more oleate, it has remained hydrophilic. This unusual phenomenon was ascribed to the formation of the helical-type structure that starch assumes in the presence of hydrophobic materials or in alkaline solutions and to the fact that the interior of this helix is hydrophobic and the exterior is hydrophilic. It was suggested that the adsorbed oleate is wrapped inside the starch helices. Interactions between the nonpolar surfactant chain and the hydrophobic starch interior can be expected to produce mutual enhancement of adsorption. The hydrophilic nature of calcite in the presence of starch and oleate results from the fact that the adsorbed oleate is obscured from the bulk

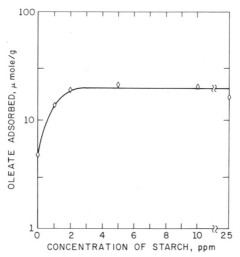

FIGURE 16.4-6 Adsorption density of oleate on calcite at natural pH 9.6–9.8 as a function of starch added prior to the oleate addition. (After Somasundaran, Ref. 16.3-87; courtesy of Academic Press.)

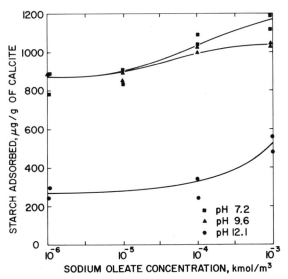

FIGURE 16.4-7 Effect of oleate addition on the adsorption of starch on calcite at various pH values. (After Somasundaran, Ref. 16.3-87; courtesy of Academic Press.)

solution by starch helices with a hydrophilic exterior and from possible masking of the collector adsorbed on the mineral in the normal manner by massive starch species.

Simple organic compounds such as citric acid, tartaric acid, oxalic acid, and EDTA often are used as modifiers.[16] Some of these reagents react by complexing various interfering ions such as calcium that are usually present in the solution.

16.4-4 Deactivators

Deactivators interact with activators to form an inert species and thus prevent activation. An example is the deactivation of copper in the xanthate flotation of sulfides by cyanides.

16.5 VARIABLES IN FLOTATION

A number of variables are encountered in flotation due to the variations in the raw materials, methods of their preparation, reagentizing, and the actual flotation process itself. An understanding of the effect of the

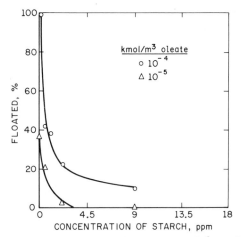

FIGURE 16.4-8 Percent of calcite floated as a function of starch in 10^{-4} and 10^{-5} kmol/m³ sodium oleate solutions. (After Somasundaran, Ref. 16.3-87; courtesy of Academic Press.)

variables can help toward proper control of the flotation for optimum performance. The effects of major physical and chemical variables have been reviewed elsewhere.[1,2]

The chemical variables that can play a major role include chemical state and structure of the surfactant, including its chain length, concentrations of surfactant, complexing ions, flocculants, and dispersants, pH, ionic strength, and the temperature of the solution. As mentioned earlier, flotation can be expected to be maximum if the collector is present as ionomolecular complexes. Substitution of hydrogen in the $-CH_2-$ groups of the long chain with fluorine has been found to produce a significant increase in flotation.[3] Of course, flotation is strongly dependent on collector concentration. Rubin et al.[4,5] among others have studied the dependence of precipitate flotations and ion flotations on the concentration of the collector. While a collector to colligend ratio of 0.2 : 1 normally is necessary for achieving good flotation of precipitates, the ion flotation was found to need much larger quantities. For foam separation techniques, the most suitable surfactant concentration is the lowest one that provides the desired foaming properties.[6-8] Transiency of the foam was found desirable for extraction in solutions of low collector concentrations. In froth flotation also, an excess of collector has sometimes been found to produce reduced extraction.[9]

Flocculants and polymers can cause an increase or a decrease in flotation depending on the properties of the colligens. Thus, while flotation of B cereus and illite with sodium lauryl sulfate can be enhanced by adding alum,[10,11] that of calcite and apatite using sodium oleate can be depressed totally using starch.

The pH of the solution is an important variable controlling flotation since the pH can affect the electrical properties of the particle surfaces and its solubility as well as the chemical state of the surfactant. The effect of pH due to its role in determining the surface charge of the particles is illustrated in Fig. 16.3-5 where only the collector that is charged oppositely to the mineral surface is capable of producing significant flotation. The role of pH in determining the chemical state of oleate and amine and thereby flotation using them has been discussed earlier. Maximum flotation was obtained in both cases under pH conditions that generate ionomolecular complexes of the collector.

Ionic strength has a significant role to play in determining the adsorption of collector on the mineral as well as on the bubble due to both the increased electrical double-layer compression and the increased salting out of the collector from the aqueous solution with increase in ionic strength. While the effect on the double layer will cause a decrease in flotation for systems where electrostatic attraction is a major factor, the salting unit effect will produce an increase in flotation. When the increase in ionic strength is the result of a salt containing a bivalent counterion, the depression of flotation is even larger. This larger effect results from the tendency of the bivalent ions to adsorb strongly and compete with the collector more than the monovalent ions. This effect also can be used to activate the flotation of a particle that has a charge similar to that of a collector (see Fig. 16.4-3). Enough bivalent ions are introduced in this case to cause a particle charge reversal, thereby making the collector adsorption possible.

The effect of variation in the temperature of the pulp or solution or flotation has not been studied in detail. Elevation in temperature is expected to decrease adsorption of collectors on minerals if the adsorption is due to physical forces, and to increase adsorption if it is due to chemical forces. An interesting observation in this regard has been the results obtained for the flotation of hematite using oleate under various ionic

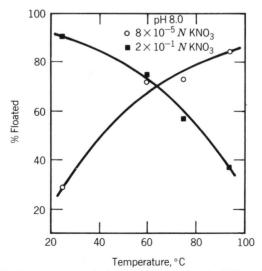

FIGURE 16.5-1 Conditioning temperature–ionic strength interactions in Hallimond cell flotation of hematite. (After Kulkarni and Somasundaran.[12])

TABLE 16.5-1 Industrial Flotation Systems

Minerals	Gangue	Collector	Modifiers	Remarks
Coal	Pyrite, clay, kaoline, shales, quartz, calcite, feldspar	Kerosene, fuel oil, alcohols	Sodium silicate, NaCN, $SnCl_2$, $KMnO_4$, starch, dextrin, lime	Coal is naturally flotable; the modifiers are for depressing the gangue flotation
Graphite	Calcite, clay, quartz	Kerosene	Sodium silicate, water glass	Naturally flotable; modifiers are for gangue depression; graphite, being soft, can coat other minerals and thus cause difficulties
Copper sulfides, chalcopyrite, chalcocite, bornite, covellite	Pyrite, pyrrhotite, other sulfide minerals, quartz	Xanthates, dithiophosphates	Cyanide, caustic soda, lime	Separation of nonsulfide minerals from sulfide minerals is relatively easy; separating sulfides from one another needs careful control of conditions such as pH and depressant concentration
Cu–Zn sulfides	Pyrite, other sulfides	Xanthates	Cyanide, Na_2S, $ZnSO_4$, lime	Separating finely disseminated Cu–Zn sulfide is a problem
Pb–Zn sulfides	Pyrite	Xanthates	Cyanide, $CuSO_4$	PbS is floated first by xanthate by depressing ZnS by cyanide; later ZnS is activated by Cu^{2+}
Pentlandite	Pyrrhotite, chalcopyrite, other sulfides	Xanthates	Cyanide, lime	Flotation properties of pentlandite are not well understood; flotability is in between chalcopyrite and pyrrhotite
Molybdenum sulfide	Pyrite, chalcopyrite galena, etc.	Kerosene, fuel oil	Lime, cyanide, Na_2S	Naturally flotable; since Mo_2S is soft, it can coat other minerals and thus enhance gangue flotation
Quartz	Usually quartz is the gangue in other minerals	Amines, fatty acids (for activated quartz)		Quartz is the gangue for other minerals; it can be floated easily with amines during the flotation of minerals; sodium silicate can be used to depress quartz

Mineral	Gangue	Collector	Depressant/Modifier	Remarks
Feldspar	Quartz, mica	Amines	Lignin sulfate, Na_2CO_3, sodium silicate	Mica is floated by depressing quartz and feldspar; during second stage feldspar is floated by depressing quartz with HF
Hematite	Quartz	Fatty acids, amines	Starch, dextrin	Flotation of hematite using fatty acids around neutral pH and reverse flotation of quartz using amines
Calcite		Fatty acids, amines		Easily flotable in the acidic range; calcite is usually present as the gangue in other salt-type minerals
Fluorite	Calcite	Oleic acid	Water glass, aluminum salts	Water glass depresses calcite flotation and aluminum salts increase the selectivity
Apatite	Calcite, quartz, dolomite	Oleic acid	Starch, phosphates, inorganic acids	Separation of calcareous and dolomitic apatite is still a difficult problem
Scheelite	Quartz, calcite	Fatty acids	Sodium silicate, starch, dextrin	
Barite	Calcite	Fatty acids	Water glass	
Halite	Clays, sylvite	Carboxylic acids	Pb or Bi salts	Floated in saturated medium; flotation is closely dependent on the composition of the medium
Sylvite	Clays, halite	Sodium dodecyl sulfate		Same as for halite

Flotation is a powerful mineral beneficiation tool that has wide potential in effluent treatments in the following industries:[22] oil industry, engineering industry wastes, dairy wastes, food industry, textile fiber wastes, cellulose fibers, rubber wastes, asbestos wastes, polymeric wastes, paper industry wastes, dyes, electroplating industry, vegetable wastes, poultry processing wastes strength conditions. In this case an increase in temperature enhanced flotation but only under low ionic strength conditions (Fig. 16.5-1). Above an ionic strength of 2×10^{-3} kmol/m^3, flotation was found to decrease with an increase in temperature. These temperature–ionic strength interactions are attributed to be the effect of adsorption of the oleate on the mineral and the salting out of the oleate from the solution. It is to be noted that the alterations in temperature also can affect the performance of a foam separation technique due to its effects on foam drainage, transiency, and adsorption on the bubble surface.

Physical hydrodynamic variables that can affect the flotation, even though not to as great an extent as the chemical variables listed above, include gas flow rate, bubble size distribution, agitation, feed rate, foam height, and reagent addition modes.

A detailed list of foam separation studies is available in a number of reviews Refs. 16.1-1, 16.1-2, 16.3-6, 16.5-13–16.5.15). A compilation of major minerals and other particulate matter that have been separated by froth flotation are given in Table 16.5.1.

REFERENCES

Section 16.1

16.1-1　P. Somasundaran, Foam Separation Methods, in E. S. Perry and C. J. Vann Oss (Eds.), *Separation and Purification Methods*, Vol. 1, p. 117, Marcel Dekker, New York, 1972.

16.1-2　P. Somasundaran, Separation Using Foaming Techniques, *Sep. Sci.*, **10**, 93 (1975).

Section 16.2

16.2-1　C. C. Harris, Flotation Machines in M. C. Fuerstenau (Ed.), *Flotation*, Vol. 1, p. 753, A. M. Gaudin Memorial International Flotation Symposium. AIME, New York, 1976.

16.2-2　D. W. Fuerstenau, P. H. Metzger, and G. D. Seele, *Eng. Min. J.*, **158,** 93 (1957).

16.2-3　M. C. Fuerstenau, *Eng. Min. J.*, **165,** 108 (1964).

16.2-4　R. D. Kulkarni, "Flotation Properties of Hematite–Oleate System and Their Dependence on the Interfacial Adsorption," D. Eng. Sci., Dissertation, Columbia University, New York, 1976.

16.2-5　M. Goldberg and E. Rubin, *Ind. Eng. Chem. Proc. Des. Dev.*, **6**, 195 (1967).

16.2-6　T. A. Pinfold, Adsorptive Bubble Separation Methods, *Sep. Sci.*, **5**, 379 (1970).

16.2-7　E. J. Mahne and T. A. Pinfold, Precipitate Flotation, *J. Appl. Chem.*, **18**, 52 (1968).

16.2-8　E. J. Mahne and T. A. Pinfold, Selective Precipitate Flotation, *Chem. Ind.*, 1299 (1966).

16.2-9　D. B. Chambers and W. R. T. Cottrell, Flotation: Two Fresh Ways to Treat Effluents, *Chem. Eng.*, **83**(16), 95 (Aug. 2, 1976).

16.2-10　R. Sizzelman, Cleveland–Cliffs Takes the Wraps Off Revolutionary New Tilden Ore Process, *Eng. Min. J.* **10,** 79 (1975).

16.2-11　D. N. Collins and A. D. Read, The Treatment of Slimes, *Miner. Sci. Eng.*, **3**, 19 (1971).

16.2-12　S. K. Kuzkin and V. P. Nebera, "Synthetic Flocculants in Dewatering Processes," Trans. J. E. Baker, National Lending Library, Boston Spa 278, 1966.

16.2-13　O. Griot and J. A. Kitchener, Role Surface Silanol Groups in the Flocculation of Silica Suspensions by Polyacrylamides, *Trans. Faraday Soc.*, **61**, 1026 (1965).

16.2-14　A. M. Gaudin and P. Malozemoff, Recovery by Flotation of Mineral Particles of Colloidal Size, *J. Phys. Chem.*, **37**, 597 (1932).

16.2-15　A. D. Read, "Selective Flocculation of Fine Mineral Suspensions," Stevenage, Warren Spring Lab. Report LR88 (MST), 1969.

16.2-16　A. D. Read, Selective Flocculation Separation Involving Hematite, *Trans. IMM (London)*, **80**, C24 (1971).

16.2-17　I. Iwasaki, W. J. Carlson, Jr., and S. M. Parmerter, The Use of Starches and Starch Derivatives as Depressants, and Flocculants in Iron Ore Beneficiation, *Trans. AIME*, **244**, 88 (1969).

16.2-18　A. D. Read and A. Whitehead, Treatment of Mineral Combinations by Selective Flocculation, *Proc. X Int. Min. Proc. Cong.*, *IMM (London)*, 949 (1974).

16.2-19　M. Carta, G. B. Alsano, C. Deal Fa, M. Ghiani, P. Massacci, and F. Satta, "Investigations on Beneficiation of Ultrafine Fluorite from Latium," *Proc. XI Int. Min. Proc. Cong.*, Paper 41, 1975.

16.2-20　Y. S. Kim and H. Zeitlin, *Anal. Chem. Acta*, **46**, 1 (1969).

16.2-21　H. M. Smith and I. E. Puddington, Spherical Agglomeration of BaSO$_4$, *Can. J. Chem.*, **38,** 1911 (1960).

16.2-22 J. R. Farnard, F. W. Meadow, P. Tymchuk, and I. E. Puddington, Application of Spherical Agglomeration to the Fractionation of Sn-Containing Ore, *Can. Metall. Q.*, **3**, 123 (1964).

Section 16.3

16.3-1 M. C. Fuerstenau, *Flotation*, A. M. Gaudin Memorial Vols. 1 and 2, AIME, New York, 1976.

16.3-2 J. Leja, *Surface Chemistry of Flotation*, Plenum, New York, 1982.

16.3-3 R. P. King, *Principles of Flotation*, South African Institute of Mining and Metallurgy, Johannesburg, 1982.

16.3-4 D. W. Fuerstenau, *Froth Flotation*, 50th Ann. Vol. AIME, New York, 1962.

16.3-5 P. Somasundaran and R. B. Grieves, *Interfacial Phenomena of Particulate/Solution/Gas Systems, Applications to Flotation Research*, AIChE Symp. Ser. Vol. 71, No. 150, AIChE, New York, 1975.

16.3-6 A. N. Clarke and D. J. Wilson, *Foam Flotation, Theory and Applications*, Marcel Dekker, New York, 1983.

16.3-7 P. Somasundaran, T. W. Healy, and D. W. Fuerstenau, Surfactant Adsorption at the S/L Interface—Dependence of Mechanism on Chain Length, *J. Phys. Chem.*, **68**, 3562 (1964).

16.3-8 Y. G. Berube and P. L. de Bruyn, Adsorption at the Rutile–Solution Interface, *J. Colloid Interface Sci.*, **27**, 305 (1968).

16.3-9 P. G. Johansen and A. S. Buchanan, An Application of the Micro-electrophoresis Method to the Study of Surface Properties of Insoluble Oxides, *Austr. J. Chem.*, **10**, 398 (1957).

16.3-10 B. R. Palmer, M. C. Fuerstenau, and F. F. Aplan, Mechanisms Involved in the Flotation of Oxides and Silicates with Anionic Collectors: Part II, *Trans. AIME*, **258**, 261 (1975).

16.3-11 J. Laskowski and S. Sobieraj, Zero Points of Charge of Spinel Minerals, *Inst. Min. Met.*, **28**, C163 (1969).

16.3-12 M. C. Fuerstenau, D. A. Elgillani, and J. D. Miller, *Trans. AIME*, **247**, 11 (1970).

16.3-13 D. W. Fuerstenau and H. J. Modi, Streaming of Potentials of Corundum in Aqueous Organic Electrolyte Solutions, *J. Electrochem. Soc.*, **106**, 336 (1959).

16.3-14 G. A. Parks, Isoelectric Points of Solid Oxides, Solid Hydroxides and Aqueous Hydroxo-complex Systems, *Chem. Rev.*, **65**, 177 (1965).

16.3-15 D. R. Nagaraj and P. Somasundaran, unpublished results.

16.3-16 I. Iwasaki, S. R. B. Cooke, and A. F. Colombo, Flotation Characteristics of Goethite, *U.S. Bur. Mines, Rep. Invest.*, No. 5593 (1960).

16.3-17 D. W. Fuerstenau, Interfacial Processes in Mineral/Water Systems, *Pure Appl. Chem.*, **24**, 135 (1970).

16.3-18 M. Robinson, J. A. Pask, and D. W. Fuerstenau, Surface Charge of Al_2O_3 and MgO in Aqueous Media, *J. Am. Chem. Soc.*, **47**, 516 (1964).

16.3-19 I. Iwasaki, S. R. B. Cooke, and Y. S. Kim, Surface Properties and Flotation Characteristics of magnetite, *Trans. AIME*, **223**, 113 (1962).

16.3-20 M. C. Fuerstenau and D. A. Rice, *Trans. AIME*, **241**, 453 (1968).

16.3-21 Y. G. Berube and P. L. Bruyn, *Electroanal. Chem. Interface Electrochem.*, **37**, 99 (1972).

16.3-22 W. Stumm and J. J. Morgan, *Aquatic Chemistry*, Wiley, New York, 1970.

16.3-23 P. Ney, Zeta Potential and Flotability of Minerals, in V. D. Frechette et al. (Eds.), *Applied Mineralogy*, Vol. 6, Springer Verlag, Wien, 1973 (German text).

16.3-24 G. A. Parks, Aqueous Surface Chemistry of Oxides and Complex Oxide Minerals, *Adv. Chem. Ser.* **6**, 121 (1967).

16.3-25 P. Somasundaran and G. E. Agar, The Zero Point of Charge of Calcite, *J. Colloid Interface Sci.*, **24**, 433 (1967).

16.3-26 M. C. Fuerstenau, G. Gutierrez, and D. A. Elgillani, The Influence of Sodium Silicate in Nonmetallic Flotation Systems, *Trans. AIME*, **241**, 319 (1968).

16.3-27 J. J. Predali and J. M. Cases, Zeta Potential of Magnesium Carbonate in Inorganic Electrolytes, *J. Colloid Interface Sci.*, **45**, 449 (1973).

16.3-28 G. A. Parks, Adsorption in the Marine Environment, in J. P. Riley and G. Skirrow (Eds.), *Chemical Oceanography*, 2nd ed., p. 241, Academic, New York, 1975.

16.3-29 P. Somasundaran, Zeta Potential of Apatite in Aqueous Solutions and Its Change During Equilibration, *J. Colloid Interface Sci.*, **27**, 659 (1968).

16.3-30 P. Somasundaran and G. E. Agar, Further Streaming Potential Studies on Apatite in Inorganic Electrolytes, *Trans. SME/AIME*, **252**, 348 (1972).

16.3-31 F. Z. Saleeb and P. L. Bruyn, Surface Properties of Alkaline Earth Apatites, *Electroanal. Chem. Interface Electrochem.*, **37**, 99 (1972).

16.3-32 M. S. Smani, J. M. Cases, and P. Blazy, Beneficiation of Sedimentary Moroccan Phosphate Ores—Part I: Electrochemical Properties of Some Minerals of the Apatite Group; Part II: Electrochemical Phenomena at the Calcite/Aqueous Interface, *Trans. SME/AIME*, **258**, 168 (1975).

16.3-33 J. Th. G. Overbeek, in H. R. Kruft (Ed.) *Electrokinetic Phenomena in Colloid Science*, Vol. 1, Elsevier, New York, 1952.

16.3-34 W. L. Freyberger and P. L. de Bruyn, Electrochemical Double Layer on Ag_2S, *J. Phys. Chem.*, **63**, 1475 (1957).

16.3-35 H. S. Choi and J. H. Oh, Surface Properties and Flotability of Kyanite and Andalusite, *J. Inst. Min. Metall., Jpn.*, **81**, 614 (1965).

16.3-36 T. J. Smolik, X. Harman, and D. W. Fuerstenau, Surface Characteristics and Flotation Behavior of Aluminosilicates, *Trans. AIME*, **235**, 367 (1966).

16.3-37 M. C. Fuerstenau, B. R. Palmer, and G. B. Gutierrez, Mechanisms of Flotation of Selected Iron Bearing Silicates, *Trans. SME/AIME*, to appear.

16.3-38 I. Iwasaki, S. R. B. Cooke, D. H. Harraway, and H. S. Choi, Fe-Wash Ore Slimes—Mineralogical and Flotation Characteristics, Trans. AIME,, **223**, 97 (1962).

16.3-39 R. W. Smith and N. Trivedi, Variation of PZC of Oxide Minerals as a Function of Aging Time in Water, *Trans. AIME*, **255**, 73 (1974).

16.3-40 M. C. Fuerstenau, D. A. Rice, P. Somasundaran, and D. W. Fuerstenau, Metal Ion Hydrolysis and Surface Charge in Beryl Flotation, *Bull. Inst. Min. Met. London*, No. 701, 381 (1965).

16.3-41 J. M. Cases, Normal Interaction Between Adsorbed Species and Adsorbing Surface, *Trans. AIME*, **247**, 123 (1970).

16.3-42 B. R. Palmer, G. B. Gutierrez, and M. C. Fuerstenau, Mechanisms Involved in the Flotation of Oxides and Silicates with Anionic Collectors: Part I, *Trans. AIME*, **258**, 257 (1975).

16.3-43 J. M. Cases, Zero Point of Charge and Structure of Silicates, *J. Chim. Phys. Phys. Chim. Biol.*, **66**, 1602 (1969).

16.3-45 R. D. Kulkarni and P. Somasundaran, Effect of Aging on the Electrokinetic Properties of Quartz in Aqueous Solutions, in *Oxide Electrolyte Interfaces*, p. 31, American Electrochemical Society, 1972.

16.3-46 A. M. Gaudin and D. W. Fuerstenau, Quartz Flotation with Anionic Collectors, *Trans. AIME*, **202**, 66 (1955).

16.3-47 I. Iwasaki, S. R. B. Cooke, and H. S. Choi, Flotation Characteristics of Hematite, Geothite and Activated Quartz with C-18 Aliphatic Acids and Related Compounds, *Trans. AIME*, **220**, 394 (1961).

16.3-48 R. A. Deju and R. B. Bhappu, A Chemical Interpretation of Surface Phenomena in Silicate Minerals, *Trans. AIME*, **235**, 329 (1966).

16.3-49 A. N. Dolzhenkova, Determination of the Electrokinetic Potential Applicable to Flotation Processes, *Obogashch Rud*, **12**, 59 (1967).

16.3-50 Y. Y. A. Azim, The Effect of Fluoride on the Flotation of Non-sulfide Minerals, Part II, Zeta Potential Measurements, Stevenage: Warren Spring Lab., ref. R.R./MP/137, 9, 1964.

16.3-51 O. Huber and J. Weigl, Influence of the Electrokinetic Charge of Inorganic Fillers on Different Processes of Paper Making, *Wochenbl. Papierfabr.*, **97**, 359 (1969).

16.3-52 D. A. Rice, D. Sci. Thesis, Colorado School of Mines, 1968.

16.3-53 B. V. Derjaguin and N. D. Shukakidse, Dependence of the Flotability of Antimonite on the Value of the Zeta Potential, *Trans. IMM*, **70**, 564 (1960-61).

16.3-54 C. A. Ostreicher and D. W. McGlashan, "Surface Oxidation of Chalcocite," Paper presented at the Annual AIME Meeting, San Francisco, 1972.

16.3-55 P. Ney, *Zetapotential und Flotierbarkeit von Mineralen*, Springer-Verlag, Vienna, 1973.

16.3-56 D. McGlashan, A. Rovig, and D. Podobnik, "Assessment of Interfacial Reactions of Chalcopyrite, *Trans. AIME*, **244**, 446 (1969).

16.3-57 G. C. Sresty and P. Somasundaran, unpublished results.

16.3-58 R. O. James and G. A. Parks, Adsorption of Zn (II) at the Hgs (Cinnabar)–Water Interface, *AIChE Symp. Ser. No. 150*, **71**, 157 (1975).

16.3-59 P. C. Neville and R. J. Hunter, "The Control of Slime Coatings in Mineral Processing," paper presented at the 4th RACL Electrochemistry Conference Adelaide, 1976.

16.3-60 S. Chander and D. W. Fuerstenau, On the Natural Flotability of Molybdenite, *Trans. AIME*, **252**, 62 (1972).

16.3-61 J. M. Wie and D. W. Fuerstenau, The Effect of Dextrin on Surface Properties and the Flotations of Molybdenite, *Int. J. Miner. Proc.*, **1**, 17 (1974).

16.3-62 M. S. Moignard, D. R. Dixon, and T. W. Healy, Electrokinetic Properties of the Zinc and Nickel Sulfide–Water Interfaces, *Proc. Austr. IMM*, **263**, 29 (1977).

16.3-63 S. Ramachandran and P. Somasundaran, unpublished results.

16.3-64 M. C. Fuerstenau, M. C. Kuhan and D. Elgillani, Role of Dixanthogen in Xanthate Flotation of Pyrite, *Trans. AIME*, **241**, 148 (1968).

16.3-65 H. A. Bull, B. S. Ellegson, and N. W. Taylor, Electrokinetic Potentials and Mineral Flotation, *J. Phys. Chem.*, **38**, 401 (1934).

16.3-66 D. McGlashan, A. Rovig, and D. Podobnik, "Assessment of Interfacial Reactions of Sulfide Copper, Lead and Zinc Minerals," paper presented at the AIME Annual Meeting, New York, 1968.

16.3-67 M. Bender and H. Mouquin, Brownian Motion and Electrical Effects, *J. Phys. Chem.*, **56**, 272 (1952).

16.3-68 P. Somasundaran, Pre-treatment of Mineral Surfaces and Its Effect on Their Properties, in *Clean Surfaces: Their Characterization for Interfacial Studies*, p. 285, Marcel Dekker, New York, 1970.

16.3-69 R. W. Smith and N. Trivedi, Variation of Zero Charge of Oxide Mineral as a Function of Aging Time in Water, *Trans. AIME*, **256**, 69 (1974).

16.3-70 G. A. Parks, *Am. Mineralogist*, **57**, 1163 (1972).

16.3-71 R. D. Kulkarni and P. Somasundaran, paper presented at 164th meeting of American Chemical Society, New York, 1972.

16.3-72 P. Somasundaran and R. D. Kulkarni, New Streaming Potential Apparatus and Study of Temperature Using It, *J. Colloid Interface Sci.*, **45**, 591 (1973).

16.3-73 R. D. Kulkarni and P. Somasundaran, Mineralogical Heterogeneity of Ore Particles and Its Effect on Their Interfacial Characteristics, *Powder Technol.*, **14**, 279 (1976).

16.3-74 H. L. Shergold and O. Mellgren, Concentration of Minerals at the Oil–Water Interfaces: Hematite–Iso-octane–Water Systems in the Presence of Sodium Sulfate, *Trans. IMM*, **78**, C121 (1969).

16.3-75 D. W. Fuerstenau, The Adsorption of Surfactants at Solid–Water Interfaces, in M. L. Hair (Ed.), *The Chemistry of Biosurfaces*, Vol. 1, p. 143, Marcel Dekker, New York, 1971.

16.3-76 P. Somasundaran and D. W. Fuerstenau, Mechanisms of Alkyl Sulfonate Adsorption at the Alumina–Water Interface, *J. Phys. Chem.*, **70**, 90 (1966).

16.3-77 P. Somasundaran, Cationic Depression of Amine Flotation of Quartz, *Trans. AIME*, **255**, 64 (1974).

16.3-78 P. Somasundaran, Ph.D. Thesis, University of California, Berkeley, 1964.

16.3-79 R. O. French, M. E. Wadsworth, M. A. Cook, and I. P. Cutler, Applications of Infrared Spectroscopy to Studies in Surface Chemistry, *J. Phys. Chem.*, **58**, 805 (1954).

16.3-80 A. S. Peck and M. E. Wadsworth, Infrared Studies of Oleic Acid and Sodium Oleate Adsorption on Fluorite, Barite, and Calcite, *U.S. Bur. Mines, Rep. Invest.*, No. 6202, 188 (1963).

16.3-81 A. Bahr, "The Influence of Some Electrolytes on the Flotation of Fluorite by Na–Oleate," Dissertation, T. H. Clausthal, Clausthal, 1966.

16.3-82 A. Bahr, M. Clement, and H. Surmatz, On the Effect of Inorganic and Organic Substances on the Flotation of Some Non-sulphide Minerals by Using Fatty Acid Type Collectors, *Proc. 8th Int. Min. Proc. Cong.*, **S-11**, 12 (1968).

16.3-83 U. Bisling, "The Mutual Interaction of the Minerals During Flotation, for Example, the Flotation of CaF_2 and Ba SO_4," Dissertation Bergakademie Freiberg, 1969 (German text).

16.3-84 H. L. Shergold, Infrared Study of Adsorption of Sodium Dodecylsulfate by CaF_2, *Trans. Inst. Min. Met. Sec. C*, **81**, 148 (1972).

16.3-85 B. D. Cumming and J. H. Schulman, Two Layer Adsorption of Dodecylsulfate on Barium Sulfate, *Aust. J. Chem.*, **12**, 413 (1959).

16.3-86 M. C. Fuerstenau and D. J. Miller, The Role of Hydrocarbon Chain in Anionic Flotation of Calcite, *Trans. AIME*, **238**, 153 (1967).

16.3-87 P. Somasundaran, Adsorption of Starch and Oleate and Interaction Between Them on Calcite in Aqueous Solutions, *J. Colloid Interface Sci.*, **31**, 557 (1969).

16.3-88 A. S. Peck and M. E. Wadsworth, Infrared Studies of the Effect of Fluorite, Sulfate and Chloride on Chemisorption of Oleate on Fluorite and Barite, in N. Arbiter (Ed.), *Proceedings of the 7th International Mining Procedures Congress*, p. 259, Gordon and Beach, New York, 1965.

16.3-89   E. Sorensen, On the Adsorption of Some Anionic Collectors on Fluoride Minerals, *J. Colloid Interface Sci.*, **45**, 601 (1973).

16.3-90   D. W. Fuerstenau and M. C. Fuerstenau, Ionic Size in Flotation Collection of Alkali Halides, *Trans. AIME*, **204**, 302 (1956).

16.3-91   J. Rogers and J. H. Schulman, Mechanism of the Selective Flotation of Soluble Salts in Saturated Solutions, in *Proceedings of the 2nd International Congress on Surface Activity*, p. 243, Vol. III, Butterworths, London, 1957.

16.3-92   C. du Rietz, Fatty Acids in Flotation, progress in Mineral Dressing, in *Transactions of the 4th International Mineral Dressing Congress*, p. 417, Almqvist & Wiksells, Stockholm, 1958.

16.3-93   P. Somasundaran, The Relationship Between Adsorption at Different Interfaces and Flotation Behavior, *Trans. SME/AIME*, **241**, 105 (1968).

16.3-94   R. D. Kulkarni and P. Somasundaran, Kinetics of Oleate Adsorption at the Liquid/Air Interface and Its Role in Hematite Flotation, in P. Somasundaran and R. B. Grieves (Eds.), *Advances in Interfacial Phenomena on Particulate Solution, Gas Systems, Applications to Flotation Research*, AIChE Symposium, Series No. 150, p. 124, AIChE, New York, 1975.

16.3-95   P. Somasundaran, The Role of Ionomolecular Surfactant Complexes in Flotation, *Int. J. Mineral. Proc.*, **3**, 35 (1976).

16.3-96   H. C. Kung and E. D. Goddard, Interaction of Amines and Amine Hydrochlorides, *Kolloid Z. Z. Polym.*, **232**, 812 (1969).

Section 16.4

16.4-1   S. R. B. Cooke, in H. Mark and E. J. W. Verwey (Eds.), *Advances in Colloid Sci.*, Vol. III, Interscience, New York, 1950.

16.4-2   J. H. Schulman and J. Leja, Molecular Interactions at the Solid–Liquid Interface with Special Reference to Flotation at Solid–Particle Stabilized Emulsions, *Kolloid-Z.*, **136**, 107 (1954).

16.4-3   J. Leja, *Proc. 2nd Int. Conf. Surface Activity, London*, **3**, 273 (1957).

16.4-4   J. H. Schulman and J. Leja, in J. F. Danielli et al. (Eds.), *Surface Phenomena in Chemistry and Biology*, p. 236, Pergamon, New York, 1958.

16.4-5   A. M. Gaudin, *Flotation*, 2nd ed., pp. 310–314, McGraw-Hill, New York, 1957.

16.4-6   F. Z. Saleeb and H. S. Hanna, Flotation of Calcite and Quartz with Anionic Collectors, the Depressing of Activating Action of Polyvalent Ions, *J. Chem. U.A.R. (Egypt)*, **12**, 237 (1969).

16.4-7   M. C. Fuerstenau, The Role of Metal Ion Hydrolysis in Oxide and Silicate Flotation Systems, *AIChE Symp. Ser.*, No. 150, 71 (1975).

16.4-8   V. A. Glembotskii, V. I. Klassen, and I. N. Plaskin, *Flotation*, translated by R. E. Hammond, Primary sources, New York, 1972.

16.4-9   A. K. Biswas, Role of CO_2 in the Flotation of Carbonate Minerals, *Indian J. Tech.*, **5**, 187 (1967).

16.4-10   A. S. Joy and A. J. Robinson, in J. F. Danielli, K. G. A. Pankhurst, and A. C. Riddiford (Eds.), *Flotation, Recent Progress in Surface Science*, Vol. 2, p. 169, Academic, New York, 1964.

16.4-11   S. C. Sun, R. E. Snow, and W. I. Purcell, Flotation Characteristics of Florida Leached Zone Phosphates Ore with Fatty Acids, *Trans. AIME*, **208**, 70 (1957).

16.4-12   K. P. Ananthapadmanabhan, "Effects of Dissolved Species on the Flotation Properties of Calcite and Apatite," M.S. Thesis, Columbia University, New York, 1976.

16.4-13   H. S. Hanna and P. Somasundaran, in M. C. Fuerstenau (Ed.), *Flotation of Salt Type Minerals in Flotation*, A. M. Gaudin Memorial Volume, AIME, New York, 1976.

16.4-14   N. F. Schultz and S. R. B. Cooke, Froth Flotation of Iron Ores, *Ind. Eng. Chem.*, **45**, 2767 (1953).

16.4-15   S. R. Balajee and I. Iwasaki, Adsorption Mechanisms of Starches in Flotation and Flocculation of Iron Ores, *Trans. AIME*, **244**, 401 (1969).

16.4-16   F. F. Aplan and D. W. Fuerstenau, Principles of Non-Metallic Mineral Flotation, in D. W. Fuerstenau (Ed.), *Froth Flotation*, 50th Ann. Vol., AIME, New York, 1962.

Section 16.5

16.5-1   P. Somasundaran, Interfacial Chemistry of Particulate Flotation, in P. Somasundaran and R. B. Grieves (Eds.), *Advances in Interfacial Phenomena on Particulate Solution/Gas Systems, Applications to Flotation Research*, AIChE Symposium Series No. 150, p. 1, AIChE, New York, 1975.

16.5-2 M. C. Fuerstenau and B. R. Palmer, Anionic Flotation of Oxides and Silicates, in M. C. Fuerstenau (Ed.), *Flotation*, A. M. Gaudin Memorial Volume, AIME, New York, 1976.

16.5-3 P. Somasundaran and R. D. Kulkarni, Effect of Chainlength of Perfluoro Surfactants as Collectors, *Trans. IMM*, **82**, c164 (1973).

16.5-4 A. J. Rubin, J. D. Johnson and C. Lamb, Comparison of Variables in Ion and Precipitate Flotation, *Ind. Eng. Chem. Proc. Des. Dev.*, **5**, 368 (1966).

16.5-5 A. J. Rubin and W. L. Lapp, Foam Fractionation and Precipitate Flotation, *Sep. Sci.*, **6**, 357 (1971).

16.5-6 R. W. Schneff, E. L. Gaden, E. Microcznik, and E. Schonfeld, Foam Fractionation, *Chem. Eng. Prog.*, **55**, 42 (1959).

16.5-7 H. M. Schoen, E. Rubin, and D. Ghosh, Radium Removal from Uranium Mill Waste Water, *J. Water Pollut. Contr. Fed.*, **34**, 1026 (1962).

16.5-8 R. B. Grieves, Foam Separation for Industrial Wastes: Process Selection, *J. Water Pollution Control Federation* **42**, R336, 1970.

16.5-9 P. Somasundaran and B. M. Moudgil, The Effect of Dissolved Hydrocarbon Gases in Surfactant Solutions on Froth Flotation of Minerals, *J. Colloid Interface Sci.*, **47**, 290 (1974).

16.5-10 A. J. Rubin, in *Adsorptive Bubble Separation Techniques*, p. 216, Academic, New York, 1972.

16.5-11 R. B. Grieves and D. Bhattacharya, Foam Separation of Complexed Cyanide, *J. Appl. Chem.*, **19**, 115 (1969).

16.5-12 R. D. Kulkarni and P. Somasundaran, Effects of Reagentizing Temperature and Ionic Strength and Their Interactions in Hematite Flotation, *Trans. SME*, **262**, 120–125 (1977).

16.5-13 E. Rubin and E. J. Gaden, Jr., Foam Separation, in H. M. Schoen (Ed.), *New Chemical Engineering Separation and Techniques*, Interscience, New York, 1962.

16.5-14 R. Lemlich, Principles of Foam Fractionation, in E. S. Perry (Ed.), *Progress in Separation and Purification*, Vol. 1, p. 1, Interscience, New York, 1968.

16.5-15 R. B. Grieves, Foam Separations: A Review, *Chem. Eng. J.*, **9**, 93 (1975).

16.5-16 A. T. Kuhn, The Electrochemical Treatment of Aqueous Effluent Streams, in J. O'M. Bockris (Ed.), *Electrochemistry of a Cleaner Environment*, Plenum, New York, 1972.

16.5-17 S. W. Reed and F. F. Woodland, Dissolved Air Flotation of Poultry Processing Waste, *J. WPCF*, **48**, 107 (1976).

16.5-18 L. Logue and E. A. Hassan, Jr., "Peeling of Wheat by Flotation," Denver Bull., F10-B16.

16.5-19 N. H. Grace, G. J. Klassen, and R. W. Watson, Denver Bull., F10-B21.

16.5-20 C. A. Roe, "Froth Flotation, Industrial and Chemical Application," Denver Bull., F10-B46.

16.5-21 J. W. Jelks, "Flotation Opens New Horizons in the Paper Industry," Denver Bull., F10-B64.

16.5-22 S. H. Hopper and M. C. McCowen, "A Flotation Process for Water Purification," Denver Bull., F10-B71.

CHAPTER 17

Bubble and Foam Separations— Waste Treatment

DAVID J. WILSON
Departments of Chemistry and Environmental Engineering
Vanderbilt University
Nashville, Tennessee
ANN N. CLARKE
AWARE, Incorporated
Nashville, Tennessee

17.1 BACKGROUND

In today's world of waste treatment, bubble and foam separations remain relatively little used techniques. Dissolved or induced air flotation is the most commonly used variation, having been employed for many years in the treatment of wastewaters for the separation of suspended soils, oils, greases, fibers, and other low-density solids as well as for the thickening of activated sludge and flocculated chemical sludges.[1] The most active commercial use was and is in ore flotation—as discussed in Chapter 16. The potential for the use of these separation techniques, however, is very high in both the areas of traditional and hazardous waste management.

This chapter provides an overview of the various bubble and foam separation techniques. The overview includes not only the results of laboratory and larger-scale studies but also the mathematical models that can be employed to optimize removal efficiencies and predict removal behavior.

17.1-1 Historical Perspective

Ore flotation developed and grew in the three decades beginning in 1915.[2,3] Our areas of interest—ion, molecular, precipitate, and adsorbing colloid flotation—are relatively new subjects. While Adamson's original text[4] on surface chemistry provided an excellent introduction to basic principles, Sebba's book[5] in 1962 first discussed ion flotation and solvent sublation in detail. Bikerman's book[6] provides detailed information on foam characteristics.

As will be seen in the later sections of this chapter, research in bubble and foam separations has continued to be performed throughout the world. Major centers of effort have developed in the United States, France, Israel, Italy, and Japan, as well as Poland and the Soviet Union. Indeed, a very large number of foam separation techniques have been developed in recent years. The most commonly employed nomenclature was recommended in 1967 by Karger et al.[7] This discussion also uses that nomenclature.

17.1-2   Principles of Operation

The basis for the separation by bubbles and foam (adsorptive bubble separation) is the difference in the surface activities of the various materials present in the solution or the suspension of interest. The material may be cellular or colloidal substances, crystals, minerals, ionic or molecular compounds, precipitates, proteins, or bacteria, but in any case it must be surface active at the air–liquid interface (Fig. 17.1-1). These surface-active materials tend to attach preferentially to the air–liquid interfaces of the bubbles or foams. As the bubbles or foams rise through the column or pool of liquid, the attached material is removed. When this combination reaches the surface, the material can be removed in the relatively small volume of collapsed foam or surface "scum."

When it is necessary to remove a substance that is not surface active, it can be adsorbed onto a surface-active "collector." This combination when correctly selected is surface active at the air–liquid interface and can be removed as described above. The substance of interest which is removed is called the "colligend."

The Gibbs equation [Eq. (17.1-1)] forms the basis of adsorptive bubble separations of molecular and ionic species:

$$\Gamma = -\frac{a}{RT}\frac{d\gamma}{da} = \frac{-1}{RT}\frac{d\gamma}{d\ln a} \qquad (17.1\text{-}1)$$

where Γ = surface excess of solute
a = activity of the solute
γ = surface tension

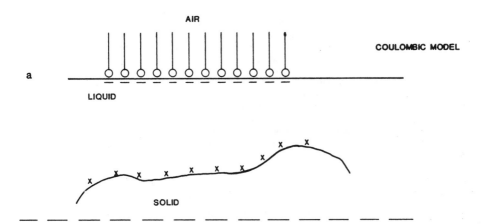

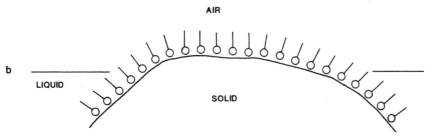

FIGURE 17.1-1   Schematic representation of the two basic models used in bubble and foam separation technology.

TABLE 17.1-1   Classification and Principles of Major Separation Techniques

I. Nonfoaming adsorptive bubble separations
    A. Solvent sublation
    B. Bubble fractionation

II. Form separations
    A. Foam fractionation (surface-active material removed at gas–solvent interface)
    B. Microgas dispersion (extremely small-sized bubbles)
    C. (Froth) Flotation

    1. Ore flotation
    2. Precipitate flotation (formation in situ—the flotation of insoluble precipitates)
    3. Adsorbing colloid flotation (adsorption onto or coprecipitation with a carrier floc which is floated)
    4. Ion flotation (reaction with now surface-active material with surface-active collector-surfactant to produce surface-active precipitate which is then foamed)
    5. Molecular flotation (same principle as ion flotation)

This equation simply says that substances which decrease the surface tension of the solvent tend to concentrate at the air–solvent interface. The Gibbs and other basic equations are discussed more fully in Section 17.3.

TYPES OF SEPARATIONS

The separation techniques are categorized initially as foaming or nonfoaming. Nonfoaming techniques include solvent sublation and bubble fractionation. Solvent sublation is a comparatively little studied technique that originated with Sebba.[5,8] Material adsorbs onto a bubble's surface and is collected into an immiscible liquid layer on top of the liquid layer (frequently water) that is being treated. In bubble fractionation, the surface-active substance is adsorbed onto bubbles and carried to the liquid's surface. At the top of the column of liquid, the bubble bursts and the transported material can be removed from the surface.

There are several subcategories for foam separations, as summarized in Table 17.1-1. These include froth flotation,[‡] foam fractionation, and microgas dispersion. Froth flotation is further subdivided. Foam fractionation and ion flotation require large quantities of surfactant—stoichiometric or greater amounts. The cost of surfactants usually precludes use of these techniques on a large scale. Removal rates also are rather slow for these two techniques. By comparison, precipitate and adsorbing colloid flotation require relatively little surfactant—only enough for foam formation and sufficient surfactant to make the material hydrophobic. In addition, removal rates for these techniques are generally rapid. This permits a high feed rate and therefore increased potential for full-scale use.

MODES OF OPERATION

Separation by foam flotation techniques can be performed in several different modes of operation. These are summarized in Table 17.1-2. Lemlich's book,[9] *Adsorptive Bubble Separation Techniques*, provides a discussion of the advantages and disadvantages of these various modes of operation.

17.1-3   Literature Review

As the field of bubble and foam separations grew and the literature proliferated, both general and specialized reviews appeared periodically to help one keep track of current developments. A summary of the major review articles (not including ore flotation reviews) is provided in Table 17.1-3.

TABLE 17.1-2   Modes of Foam Flotation Operation

Options		
1. Batch	or	Continuous
2. Reflux of collapsed foamate	or	No reflux of collapsed foamate
3. Multiple staging	or	No multiple staging
4. Feed to pool of bottom of column	or	Feed directly into rising foam

[‡]Froth flotation commonly is referred to as "flotation."

TABLE 17.1-3 Summary of Major Literature Reviews

	Date	Author	Comments	Reference
Specialized	1972	Grieves	Micro Organism	9
	1973	Lemlich	Heavy Metals—Trace Levels	10
	1974	Izumi	Heavy Metals	11
	1975	Grieves	Oxyanions	12
	1977	Eklund	Wastewater Treatment	13
	1977	Hyde	Wastewater Treatment	14
	1977	Marks	Wastewater Treatment	15
	1977	Thomas	Enzymes and Protein	16
	1977	Loftus	Cell Component	17
General	1957	Cassidy		18
	1962	Rubin		19
	1963	Eldib		20
	1967	Rubin	(through 1965)	21
	1968	Grieves	Wastewater	22
	1968	Karger	Principles/Application	23
	1968	Lemlich		24
	1972	Somasundaran		25
	1973	Ho		26
	1975	Grieves		27
	1975	Ahmed		28
	1975	Somasundaran		29
	1976	Bahr		30
	1976	Panov		31
	1977	Richmond		32
	1977	Balcerzak		33
	1978	Clarke	286 references	34
	1982	Grieves	353 references	35
	1983	Clarke	292 references	36

17.2 THEORY OF SEPARATION

As discussed earlier, the flotation techniques most promising for adaptation to full scale are precipitate and adsorbing colloid flotation. To assist in this development, there has been considerable effort expended in modeling these techniques at the molecular level. Both techniques are sensitive to interfering factors at the molecular level: these include pH, ionic strength, surfactant concentration, and specifically interfering ions. However, before going into a discussion of the molecular level mathematical models, we wish to address the role of fluid dynamics in particle flotation.

17.2-1 Role of Fluid Dynamics

The particle is bound to the air–water interface of bubbles or foam by electrical energy. This energy is much larger than random thermal energies (kT, approximately 4×10^{-14} ergs at 298 K). There are, however, viscous drag forces that must be considered. If the drag is too great a particle may become detached and removal efficiencies decreased. Another fluid mechanical consideration is the collision cross section of the bubble–floc encounter.

VISCOUS DRAG FORCES

Viscous drag forces result from the counterflow of liquid downward between the rising water–air interfaces (see Fig. 17.2-1). Generally, it can be shown that the magnitude of these forces is not sufficient to detach the adsorbed floc. Calculations result in a viscous force of an order of magnitude of 10^{-8} dyn while the binding force is 10^{-5} dyn for particles of radius 10^{-5} cm (using spherical floc particles and bubbles). However, since the binding force increases with the radius and the viscous drag increases with the square of the radius, when the particle reaches a size of 10^{-2} cm or larger interference with removal may occur [see Eqs. (17.2-1) and (17.2-2)].

$$\text{Viscous force } F_v = 6\pi\eta\, rv \quad \text{(Stokes Law)} \tag{17.2-1}$$

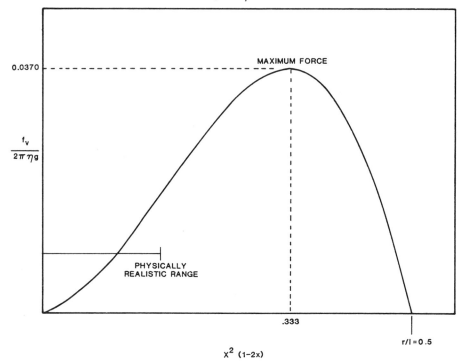

FIGURE 17.2-1 Viscous drag force for counterflow of liquid downward between rising water–air interfaces.

where η = viscosity (in poise)
     g = gravitational constant
     r = radius at particle
     v = velocity in downward directions

But

$$v = \left(\frac{\rho g}{2\eta}\right)(2\,lr - 4r^2)$$

where l = film thickness
     ρ = liquid density

$$\text{Binding force } F_b = \frac{|\Delta G|}{r} \qquad\qquad (17.2\text{-}2)$$

where ΔG = Gibbs free energy (ergs)
     r = radius of particle (cm)

    To further approximate the real world conditions, the buoyancy and rise of the bubble–floc system must be included. This has been done[1-3] in varying degrees of quantitation and computational difficulty. However, it can be concluded that optimum sizes exist for optimum removal efficiencies for a given system. Similarly, size limits exist beyond which floc particles will detach. It should be noted in review of Eq. (17.2-1) that for very wet films (a large l value) interference also might occur.

COLLISION CROSS SECTION

It is assumed that there exists a minimal distance within which a bubble and particle must pass to attach. We wish to calculate the rate at which these attachments occur.

If γ is the distance across which coulombic attraction could pull a particle to the bubble surface in the available time of passage, it can be shown that the particle must be within a distance of $(\frac{3}{2})^{1/2}\gamma$ of the path of the center of the bubble in order for it to attach to the bubble as it passed (see Fig. 17.2-2). This "capture" volume is $\frac{3}{2}\pi\gamma^2 h$, where h is the distance the bubble rises. This analysis is based on small bubbles which are within the viscous creeping flow regime.

For larger bubbles inertial effects become more important. In this case, an ideal inviscid liquid is assumed (i.e., no turbulent wake as the bubble passes). Similar calculations provide a volume of liquid from which particles are captured by these larger bubbles: $V = 3\pi\gamma ah$, where γ and h are defined as before and a is the bubble radius. This is a considerably larger volume than in the first case (viscous flows).

This information enables us to determine the removal rate of the flotation process based on attachment to the bubble in a system of known floc concentration and volumetric flow rate. For a continuous flow, single-stage apparatus in steady state, the effluent floc concentration (C_e) is given by Eq. (17.2-3):

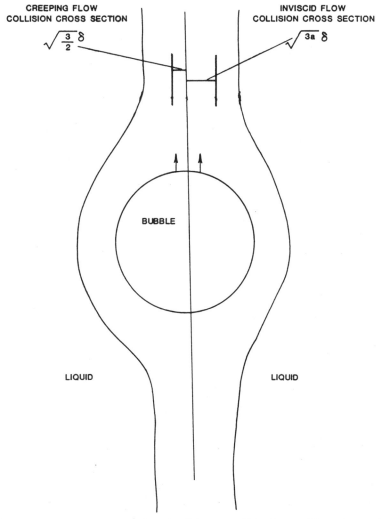

FIGURE 17.2-2 Schematic representation of collision cross section for two flow regimes—creeping and inviscid.

$$C_e = \frac{C_o}{1 + 9\,\gamma h Q_a/4a^2 Q_l}$$ (17.2-3)

where C_o = influent floc concentration
           Q_a = volumetric air flow rate
           Q_l = volumetric liquid flow rate
      γ, h, a = defined as before

In concentrated slurries, in which the bubbles may become fully loaded, the maximum number M_{max} of floc particles which a bubble can carry is

$$M_{max} = \frac{4a^2}{b^2}$$ (17.2-4)

where b is the effective radius of the floc (much less than a, the bubble radius).

Imposing this limit [Eq. (17.2-4)] on the effluent floc calculation in Eq. (17.2-3), we find that the particle removal rate in a continuous flow system for concentrated slurries in which bubbles are saturated with floc is simply

$$C_e - C_o - \frac{3Q_a}{\pi a b^2 Q_l}$$ (17.2-5)

17.2-2 Mathematical Models

Detailed discussion of the mathematical models[4,5] for precipitate and adsorption colloid flotation alone could fill a complete text. The modeling is generally complex and requires varying amounts of computer time to complete. Therefore, in this chapter we present the overall approach and some resulting major mathematical relationships that can help maximize operating efficiency.

A nonlinear Poisson–Boltzmann equation,

$$\frac{d^2\psi}{dx^2} = \frac{A\,\sinh\,(3z\psi/kT)}{1 + B\,\cosh\,(ez\psi/kT)}$$ (17.2-6)

where $A = 8\pi\,zec_\infty/(1 - 2C_\infty/C_{max})D$
         $B = 2\,C_\infty/(C_{max} - 2C_\infty)$
         $z = 1$ (|charge| of ions in the electrolyte solution)
         e = electronic charge
        C_∞ = electrolyte concentration (molecules/cm^3)
    C_{max} = maximum possible electrolyte concentration
        D = dielectric constant of the solution

forms the basis for the calculation of the electric potential in the vicinity of a floc–water interface. The finite volumes of the ions responsible for the space charge in the solution are taken into account in this equation. The solution of this Poisson–Boltzmann equation is outlined for the geometry that will be employed. This yields the electric potential distributions between a solid–liquid interface and bulk solution; between a solid–liquid interface and a liquid–air interface; and between two solid–liquid interfaces. These electric potentials then are used to calculate the free energies of interaction between a solid–liquid interface and a liquid–air interface and between two solid–liquid interfaces.

These results then provide the foundation for a coulombic model of particle flotation, one in which a charged solid particle is attracted to an air–water interface oppositely charged by the presence of ionic surfactant. These results and the techniques of statistical mechanics are employed to calculate the adsorption isotherms of particles on air–water interfaces within the framework of this model.

Next, a noncoulombic model for the attachment of a floc particle to a bubble is developed. In this model, it is assumed that the adsorption of surfactant on the particle leads to the formation of a hydrophobic surface on which the air–water contact angle is greater than zero, which permits bubble attachment. The magnitude of the floc–bubble binding energy for this model is estimated. Then this is compared with estimates of the viscous drag forces exerted on a floc particle attached to an air–water interface in a wet foam or on a rising bubble.

Using the noncoulombic model, one can investigate the details of surfactant adsorption at the solid–water interface. The van der Waals interactions of the hydrocarbon tails of the surfactant ions lead to the abrupt condensation of the surfactant on the surface as the concentration of surfactant in solution increases. The formation of such a condensed film (called a hemimicelle) appears to be necessary in the development of a hydrophobic solid surface (Fig. 17.2-3).

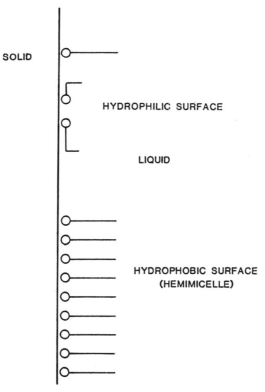

FIGURE 17.2-3   Schematic representation of the hemimicelle–hydrophobic surface.

This surface phase change can be studied by statistical mechanics.[2] It is affected by changes in ionic strength and surface charge density. An analogous approach can be employed to study the formation of condensed surfactant monolayers on air–water interfaces. This provides a mechanism for the establishment of a surface charge density at this interface in the coulombic model.

The information resulting from the mathematical modeling helps address some practical problems on flotation. For example, two practical problems in precipitate and adsorbing colloids are:

1. The recovery for recycle of surfactant adsorbed on the sludge separated from the collapsed foamate.
2. Interference by foreign ions with the formation of a condensed hemimicelle of surfactant on the solid–water interface.

Information is obtained by modeling the competition between surfactant ions and other ions for adsorption sites on the floc surface. A statistical mechanical approach, as previously employed for cooperative surface phenomena (e.g., hemimicelle formation), allows one to observe that excessive concentrations of surfactant interfere with the flotation of particulate material. This apparently results from the formation of a second hemimicelle of condensed surfactant on top of the first, with the surfactant polar or ionic heads presented to the water; this results in a hydrophilic surface (Fig. 17.2-4). The effects of ionic strength and surfactant hydrocarbon chain length on this behavior have been modeled mathematically.

The results of the coulombic and noncoulombic modeling generally agree. Both predict the formation of hemimicelles prior to particle flotation and similar effects on this resulting from varying parameters such as surfactant chain length and ionic strength.

The resulting relationships between adsorption and several major parameters are summarized below. These relationships were developed mathematically from the models and a number have been verified in laboratory studies.

1. Increasing ionic strength: decreases adsorption, decreases thickness of double layer, decreases strength of attractive force between bubble (surfactant coated and floc particle), decreases effective size of floc particles, and reduces flotation efficiency.
2. Increasing temperature: decreases adsorption—small effect.

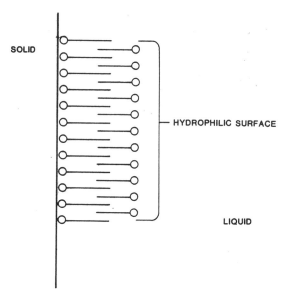

FIGURE 17.2-4   Schematic representation of the micellar–hydrophilic surface.

3. Increasing surface potential for air–water interface: increases adsorption of oppositely charged particles.
4. Increasing magnitude of ionic atmosphere ions: reduces flotation efficiency through increased screening of electrical interactions.
5. Increased particle size: increased flotation efficiency.
6. Increasing hydrocarbon chain length of surfactant: reduces concentration of surfactant needed to form hemimicelles (onset of particle flotation) and micellar layers (flotation inhibition).

These relationships can be quite helpful in selecting reagent systems for a given task and in optimizing operating conditions for a foam flotation apparatus. For example, industrial wastewaters are prime targets for treatment by foam flotation. However, their multicomponent nature frequently causes interference with the removal process. Indeed, only small amounts of such common ions as phosphates can block completely the normally excellent flotation of hydroxide flocs with the surfactant sodium lauryl sulfate (NLS). The models describe this phenomenon as the displacement of NLS from the adsorption sites on the floc surface by the more strongly adsorbed phosphate ions which do not create a hydrophobic surface. This suggests either the tying up of phosphate or the use of a more strongly binding surfactant as a solution to the problem.

Similarly, another practical problem in the commercial use of foam flotation treatment is the recovery and reuse of the surfactant. The concept of competitive adsorption also affords hope in this area; one displaces the surfactant from the sludge in the collapsed foamate by addition of a competing ion such as OH^- or CO_3^{2-} so that the surfactant could be recovered from the sludge and recycled.

17.3   LABORATORY STUDIES

As evidenced by the frequency and size of the literature reviews, there continues to be extensive laboratory work in the area of bubble and foam separations. This work provides the technical community with an ever increasing base of removal/treatment data—both in coordination with and independent of the mathematical models.

In an effort to organize this extensive literature, this section provides the results of laboratory studies on precipitate and adsorbing colloid flotation, which, as described previously, are most promising for large-scale waste treatment, and on solvent sublation, which is promising in the area of hazardous waste management. The former is divided further by the type of removal performed—that is, metals, anions, or organic compounds. Solvent sublation addresses organic removal to trace level residuals.

17.3-1   Metals Removal Studies

When using precipitate or adsorbing colloid flotation techniques for the removal of metals,[3] one finds that many combinations will work—some better than others. When removal efficiencies are comparable, other considerations must come into play prior to the selection among alternatives. These considerations include safety, question of ultimate disposal, and costs.

The majority of laboratory experiments have dealt with the removal of metals. The nine most important metals from wastewater treatment considerations are addressed in this section. These are cadmium, chromium, copper, iron, lead, manganese, mercury, nickel and zinc. In consideration of space constraints only the studies producing the most efficient and potentially applicable results are summarized.

Two other areas also must be included in any discussion of metals removal by flotation. These are (1) the concentration of trace level elements from seawater—studied extensively by Zeitlin and coworkers at the University of Hawaii—and (2) applications in radioactive waste treatment.

CADMIUM

Precipitate flotation was shown to remove cadmium (as the sulfide) using hexadecyltrimethylammonium bromide (HTA) at a pH of 5-9.[1] The residual cadmium concentration was approximately 0.5 mg/L. Using adsorbing colloid flotation with ferrous sulfide and HTA at pH 9 resulted in residual cadmium concentrations of 0.003 mg/L. In an effort to eliminate the use of sulfide, cadmium hydroxide was floated with ferric hydroxide and HTA. The lowest residual was 0.010 mg/L at pH 12. This system, however, was very sensitive to ionic strength variations.[2] Other studies removed cadmium precipitates with various systems over large ranges of pH by varying the coprecipitates and surfactants.[3,4]

CHROMIUM

Bhattacharyya et al.[5] have developed a process to remove Cr^{6+} by reducing it to Cr^{3+} with bisulfite followed by precipitation as the hydroxide with NaOH and precipitate flotations with sodium dodecylsulfate (SDS).‡ A 99% removal rate was achieved in the pH range 7.0-8.8. This method was modified later for use in a small, continuous flow pilot plant.[6]

Several other studies were performed using both precipitate and adsorbing colloid flotation which used various surfactants and pH values.[7] It was found that a critical parameter was the pH at which the chromium was precipitated. If the pH were too high, the flotation resulted in poor removal.

COPPER

The removal of copper is one of the most studied topics in foam separation. Rubin showed that copper could be removed by ion or precipitate flotation, with the latter being the most effective technique. Precipitate flotation was as cupric hydroxide and used SDS. Beitelshees and coworkers[8] flotated copper precipitated as the sulfide with a cationic surfactant for removals of 90% or more for initial concentrations of 10–1000 mg/L.

Copper also was removed by adsorbing colloid flotation using ferric hydroxide floc and SDS surfactant. Optimum removal occurred in the pH range 6–8, depending on ionic strength. The 99.98% removal was achieved at pH 7.5 with an initial copper concentration of 50 mg/L.[9]

Many industrial wastewaters contain copper as well as other heavy metals. Table 17.3-1 provides the results of the adsorbing colloid flotation of such a mixture of metals.[10] The floc was ferric hydroxide and the surfactant was SDS. Removals are generally good (based on initial concentrations of 20 mg/L for each metal).

TABLE 17.3-1 Removal of Metals in Mixtures by Adsorbing Colloid Flotation

pH	Add Fe(III) (mol/L)	Added NaNO₃ (mg/L)	Residual Metal[a] Concentration (mg/L) Cu (II)	Pb (II)	Zn (II)
6.1	100	0.02	0.30	0.20	3.2
6.5	100	0.02	0.25	0.15	2.1
7.02	100	0.02	0.20	0.20	1.8
6.05	100	0.05	0.27	0.15	2.9
6.55	100	0.05	0.26	0.15	2.9
7.05	100	0.05	0.30	0.30	2.2
6.05	100	0.10	0.35	0.4	3.2
6.56	100	0.10	0.26	0.30	2.8
7.02	100	0.10	1.25	1.2	2.9
7.02	100	0.075	7.75	>10	3.3
8.05	150	0.075	>8.0	>10	3.5
8.05	200	0.075	>8.0	>10	3.5

[a]Original concentration of each metal was 20 mg/L.

‡Same as NLS.

IRON

The flotation of iron as ferric hydroxide has been studied extensively.[11-14] Flotation was effective over a pH range of 4.5–8.5 using anionic surfactants (e.g., SDS), and above a pH of 9 using cationic surfactants (e.g., HTA). While SDS is a good choice in terms of cost and toxicity considerations, it is displaced rather easily from precipitates by competing anions. The anions most effective in blocking flotation were phosphate, hexaphosphate, arsenate, EDTA, and oxalate.

LEAD

At pH values below 8, lead can be removed by ion flotation. For pH values of 8–9, precipitate flotation as the hydroxide with SDS is successful. Removal is poor at pH values above 9.[15] Lead sulfide was removed rapidly with HTA over a pH range of 4.0–9.5 with best removal occurring at pH 8–9. The removal was improved tenfold (residual lead concentrations of 0.01–0.02 mg/L) by coprecipitation with ferrous sulfide in the same range of pH (8–9).

However, the use of sulfides (and HTA) must be considered in terms of toxicity and cost. The use of ferric hydroxide and SDS did provide relatively good removal combination at pH 6.5—7.5, to below detection limits (about 0.05 mg/L) from an initial concentration of 50 mg/L Pb.[16]

MANGANESE

Manganese dioxide was floated with gelatin at pH values of 2.3, 4.0, 7.0, and 10.0.[17] Separation was improved if 0.3 mg/L of methanol were used as a frother. The precipitate and adsorbing colloid flotation of manganese was also studied.[18] Since manganous hydroxide is relatively soluble, the preliminary runs floating $Mn(OH_2)$ coprecipitated with ferric or aluminum hydroxides and with SDS as a surfactant were not promising. However, extended aeration at pH 10 to oxidize the manganese to MnOOH or MnO_2, followed by flotation with ferric hydroxide and SDS at pH 5.5, yielded residual concentrations of about 1 mg/L. Flotation of $MnCO_3$ with HTA at pH 9, flotation of $MnCO_3$ with SDS and $Fe(OH)_3$ with HTA at pH 9, flotation of $MnCO_3$ with SDS and $Fe(OH)_3$ at pH 5–6, and flotation of MnS with FeS and HTA are not promising. Flotation of MnS with CuS and HTA at pH 9–10 yielded residual manganese concentrations as low as 0.15 mg/L, but the separation was not reliable.

MERCURY

HgS was not removed by flotation with SDS at any pH but was removed readily with HTA at acid pH values with a residual mercury concentration of 0.1 mg/L at pH 0.8.[2,19] Adsorbing colloid flotation of mercury with aluminum or ferric hydroxide and SDS was also ineffective. Flotation of mercury with HTA and FeS at pH 9 yielded residual concentrations of about 0.05–0.07 mg/L (the separation is not affected strongly by ionic strength). The most effective separation appeared to be adsorbing colloid flotation with copper sulfide and HTA. This method is effective at ionic strengths as high as 2.0, is insensitive to large pH variations, and can yield residual mercury concentrations of 0.005 mg/L.

It also was found that mercury was removed by coprecipitation with ferric hydroxide and flotation with sodium oleate and pine oil. However, it was necessary to add sodium sulfide.[3]

NICKEL

The flotation of nickel with SDS and ferric or aluminum hydroxide was carried out.[18] Residual nickel concentrations of about 1 mg/L were obtained for pH values of 8–9.5 with aluminum hydroxide and SDS for ionic strength less than 0.1. Ferric hydroxide performed somewhat less satisfactorily. Nickel concentrations of 420 mg/L from plating waste were precipitated with lime slurry at pH 9.5 and floated using Duomac "T" and Dowfroth 250. Residual nickel concentrations were generally less than 3 mg/L.[20]

ZINC

For pH values above 8, precipitate flotation was more efficient in removing zinc than foam separation of soluble zinc species.[15] The optimum pH was 9.2. The surfactant was SDS. The adsorbing colloid flotation of zinc used aluminum hydroxide and NLS in the pH range 8.0–8.6; ferric hydroxide flocs were not effective. Removal was 99.8% or higher.[16] Table 17.3-1 shows the removal of zinc in association with other metals.

SEAWATER

Adsorbing colloid flotation is used to concentrate trace level components in seawater. Zeitlin and coworkers from the University of Hawaii have studied seawater extensively in this manner. Table 17.3-2 summarizes their results.[2] Supplementing this work is a paper by Matsuzaki and Zeitlin[21] which studies nine surfactants and their performance with six collectors [$Fe(OH)_3$, $Th(OH)_4$, $Al(OH)_3$, HgS, CdS and MnO_2].

RADIOACTIVE WASTES

The United States, the Soviet Union, and Japan lead in the study of flotation techniques to treat radioactive wastes.[2,7] The metals of interest include cobalt, strontium, zirconium, niobium, ruthenium, cesium, and cerium, as well as rhodium and yttrium. The studies, all of which employed precipitate or adsorbing colloid flotation, indicated that an average of over 90% removal is achievable. Some removals approached 100%.

TABLE 17.3-2 Summary of the Adsorbing Colloid Flotation of Sea Water

Metal	Floc	Surfactant	Optimum pH	Approximate Recovery
Zinc	$Fe(OH)_3$	Dodecylamine	7.6	94
Copper	$Fe(OH)_3$	Dodecylamine	7.6	95
Silver	PbS	Stearylamine	2	92
Mercury	CdS	Octadecyltrimethyl-ammonium chloride	1	"Quantitative"
Molybdenum (as molybdate)	$Fe(OH)_3$	SDS	4	95
Uranium [as $UO_2(CO_3)_3^{-4}$]	$Fe(OH)_3$	SDS	5.7	82
	$Th(OH)_4$	Sodium dodecanoate	5.7	90
	TiO_2	SDS	6.4–6.8	91
Vanadium	$Fe(OH)_3$	SDS	5	86
Selenium	$Fe(OH)_3$	SDS	3.5–5.3	100
Phosphorus (as phospate)	$Fe(OH)_3$	SDS	4	91
Arsenic (as arsenate)	$Fe(OH)_3$	SDS	4	91

In the United States, most of the work was carried out at Radiation Applications, Inc. and at Oak Ridge National Laboratory. Most of the Japanese work was performed by Koyanaka and coworkers. In the Soviet Union, Pushkarev and associated developed much of the methodology.

17.3-2 Anion Removal Studies

The work on anion removal by flotation techniques is much less extensive than that on metals removed. We shall address the five major anions studied: arsenate, cyanide, fluoride, phosphate, and sulfite. Sometimes the removal of anions is studied in combination since they frequently occur together in a given wastestream.

ARSENATE
Two methods of arsenate removal are reported which reduce initial concentrations in the range of 10–20 mg/L to 0.1–0.2 mg/L. Adsorbing colloid flotation uses either $Fe(OH)_3$ with sodium dodecyl sulfate (SDS) at pH 4–5 or MnO_2 with SDS at pH 8.5.[2,31]

CYANIDE
There are two major studies in the removal of cyanides. First, Grieves and Bhattacharyya used ethylhexadecyldimethylammonium bromide to float colloidal or precipitated foams with iron at acidic pH values.[22-24] Clarke used precipitate and adsorbing colloid flotations to remove free cyanide as ferric ferrocyanide on ferric hydroxide floc with SDS at pH 5.[25] Performance was good (over 90% removal from 50 mg/L solution) for chromium, cobalt, copper, and nickel cyanides but not for zinc.

FLUORIDE
Clarke and coworkers[25] demonstrated fluoride removal was performed easily at neutral pH 7.3–7.8 using adsorbing colloid flotation with $Al(OH)_3$ and SDS. Residual fluoride was less than detection limits with an Al:F ratio of 2.54 or more.

Bhattacharyya and coworkers[32] demonstrated fluoride removal in conjunction with *ortho*-phosphate removal by precipitation of the anions with lanthanum(III) and floating with SDS. The operating pH range was 3.6–6.0 and removal efficiency was 95%.

SULFITE
Sulfite is removed by precipitate flotation with sodium dodecylbenzenesulfonate at an optimum pH of 8.6. Removal was 93% for precipitate ($CaSO_3$) sulfite and 85% for total sulfite.[27,28] The simultaneous removal of sulfite and carbonate was studied using precipitate flotation as the calcium salts. Removal was 95% of precipitated sulfite and 97% of the carbonate.

17.3-3 Organic Compound Removal Studies

The removal of oils by flotation techniques comprises the largest segment of organic compound studies. For example, Skrylev and coworkers have studied the removal of emulsified vegetable oils (sunflower, corn, coconut) at low concentrations with a gelatin collector and mixed cationic surfactants.[29,30] Best performance occurred at pH 2–3; worst performance occurred at pH 10–11. Oils in wastewaters have been reduced to 10–30 mg/L using $FeSO_4$ and 1–5 mg/L using alkyl pyridinium chloride.

However, laboratory studies in the foam flotation of common wastewater organic compounds such as phenols have not led to high-efficiency separations so far. One promising lead is suggested by the work of Grieves and Chouinard on the flotation of phenol-charged powdered activated carbon.[33]

17.3-4 Solvent Sublation

Solvent sublation is a nonfoaming separation technique in which surface-active substances are transferred from an aqueous phase into an immiscible supernatant organic phase of much smaller volume by aeration. The technique originally was proposed by Sebba,[34] and the early work is reviewed in Lemlich's book.[35] We also have reviewed the literature and discussed solvent sublation column modeling.[5] Recently, there has been a resurgence of interest in the technique as a method for removing low concentrations of toxic hydrophobic organics from wastewaters. A recent government report[36] presents a complete review through 1983 as well as the experimental and theoretical results summarized here.

Laboratory-scale studies on o- and p-dichlorobenzene, Aroclor 1254 (a PCB mixture), lindane, endrin, CH_3CCl_3, naphthalene, and phenanthrene suggest the feasibility of using these techniques for removing hydrophobic organics from aqueous systems. Methylene blue, methyl orange, and two nitrophenols were removed as ion complexes with ionic surfactants. Preliminary batch pilot-scale studies with napthalene indicated the importance of small bubble size in improving removal efficiency. The presence of added salts appears to increase slightly the removal rates of hydrophobic organics, but to decrease markedly the removal rates of ionic complexes. The presence of organic solvents decreases removal rates somewhat.

This report also includes theoretical work on the estimation of adsorption isotherm parameters and mass transfer rate coefficients, and the mathematical modeling of equilibrium-controlled and mass-transfer-controlled solvent sublation column operation in batch and continuous flow modes.

Additional recent work includes the solvent sublation of aldrin and of alkylphthalates.[37]

17.4 ROLE OF COLUMN DESIGN

Before reviewing larger-scale studies and applications, we briefly discuss the modes of operation possible in foam flotation and the corresponding details of column design to optimize results. Although mathematical models currently provide excellent simulations of column design and operation,[1] that discussion is beyond the scope of this chapter. The discussion herein is qualitative.

Column operation is either batch or continuous mode. Flow can be simple or countercurrent. For countercurrent systems, there are three major operational modes: stripping, enriching, and a combination (see Fig. 17.4-1).

In the stripping mode, influent liquid is introduced into the foam part of the way up the column and flows countercurrently to the rising foam; the section of column above the point at which the influent is introduced permits foam drainage. In stripping columns, the objective is to reduce the effluent solute concentration to the lowest feasible level. Sometimes partial recycling of effluent is used to improve solute removals.

In enriching columns, some of the collapsed foamate is recycled back to the upper section of the column to increase the solute concentration in the foamate as much as possible.

Another option is the use of baffles. The use of baffles[2] in continuous flow foam columns improves removal efficiency and increases maximum hydraulic loading rates by reducing channeling in the foam and other types of axial dispersion.

Both experimental and modeling results show that the effectiveness of separation increases with increasing gas flow rate and decreases with increasing cross-sectional area of a column, axial dispersion, and bubble size.[3,4] Other design considerations include perfecting the horizontal and vertical alignments and uniformly distributing the influent across the column cross section. Misaligned systems reduce removal efficiency by increasing channeling in the foam and foam overturn.[5] Adequate flocculation times and fairly close pH control are also necessary.

When dealing with foam generation techniques, it is important to break the foam efficiently after the separation has been effected. Several foam breaking techniques exist. These are fairly descriptive and include thermal, sonic, chemical, liquid spray, orifice, whirling paddle, and high-speed spinning disks. The last technique is very effective at relatively low energy demands.[6]

Figure 17.4-2 is a generalized schematic of a pilot-plant sized unit.

17.5 LARGER-SCALE STUDIES

After combining the information obtained from mathematical modeling and laboratory experiments in maximizing operating parameters, the next step is the development of pilot-plant studies. The objectives

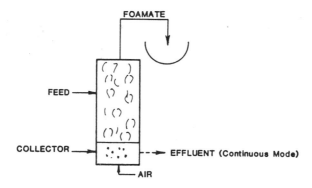

I. STRIPPING MODE

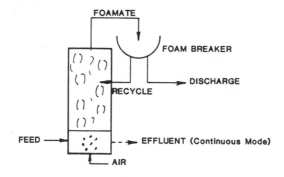

II. ENRICHING (REFLUX) MODE

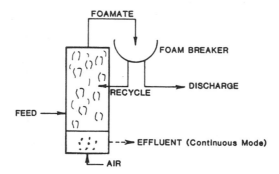

III. COMBINATION (I & II)

FIGURE 17.4-1 Schematic representation of the three major operational modes for foam separation columns.

of these larger-scale studies was to test the feasibility of the foam separations on full-scale waste treatment problems.

Both adsorbing colloid and precipitate flotation have been studied in larger columns. Results for lead, copper, chromium, and magnesium indicate drinking level residuals are possible after treatment at optimum conditions. The operating conditions and results for several studies are provided in Table 17.5-1.

Problems encountered in larger-scale studies included channeling of the foam and foam overturn. Vertical alignment of the column is an important factor in reducing channeling. The horizontal alignment of baffles is important in the elimination of foam overturn. Increasing the number of baffles decreases axial dispersion at higher hydraulic loading rates.

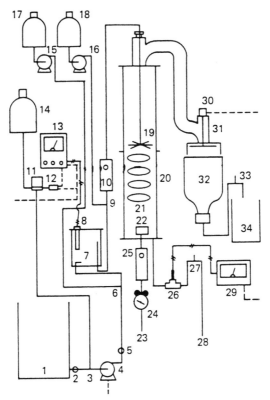

FIGURE 17.4-2 Generalized schematic of a pilot-plant-sized foam separation system. (After *Foam Flotation of Heavy Metals and Fluoride—Bearing Industrial Wastewaters*, USEPA, 600/2-77-072.)

1. Waste Tank	18. NLS Tank
2. Waste Tank Valve	19. Flow Dispersion Head
3. Sodium Hydroxide Injection Tee	20. Column
4. Main Pump	21. Baffle
5. Flow Control Valve	22. Air Diffuser
6. Ferric Chloride Injection Tee	23. Air Supply Line
7. Mixing Chamber	24. Air Pressure Regulator
8. Control pH Electrode	25. Air Flow Rotometer
9. NLS Injection Tee	26. Monitoring pH Electrode
10. Waste Flow Rotometer	27. Column Liquid Level Contrc
11. NaOH Solenoid Valve	28. Effluent Line
12. Electrical Junction Box	29. Monitoring pH Meter
13. Control pH Meter	30. Foam Breaker Motor
14. NaOH Tank	31. Foam Breaker
15. Ferric Chloride Feed Pump	32. Clarifier
16. NLS Feed Pump	33. Clarifier Liquid Level Contrc
17. Ferric Chloride Tank	34. Broke Foam Container

When "real world" industrial effluents are treated on this or any scaled system, two major problems could exist for any given waste. One is solids content, which can be corrected easily, if necessary, by filtering or settling as pretreatment. The other is ionic strength (dissolved solids). Most flotation systems are very sensitive to ionic strength—with efficiency decreasing rapidly with increasing ionic strength. Some systems do not exhibit as sensitive a relationship. The question of a waste's ionic strength and its ultimate compatibility must be addressed on an individual basis. Approaches therefore include pretreatment (e.g., precipitation) and/or developing a set of flotation operating parameters less sensitive to ionic strength. It is possible that ionic strength more than any other single wastewater characteristic could limit foam flotation as a treatment option.

TABLE 17.5-1 Summary of Larger-Scale Studies

Parameters	_____ Metals _____					
	Pb	Pb	Pb	Cu^a	Mg	Cr
Initial concentration (mg/l)	60	70	20	100	b	100^c
Column diameter (inside) (cm)	10	30	30^d	7.6	12.7	9.7
Column length (cm)	244	244	244	120	45.7	89.5
Hydraulic loading rate (m³/m²·day)	≤118	≤150–180	326	246		30
Air flow rate (m³/m²·min)	0.24	0.4–0.5	0.21	1100^e	270^e	1300^e
Surfactant	SDS	SDS	SDS	Mixed	SDS	SDS
Concentration of surfactant (mg/L)	35	35–40	25	45–60	f	10
Floc	Fe^{3+}	Fe^{3+}	Fe^{3+}	—		
Concentration of floc (mg/L)	100–150	100–150	90	—		
pH	5.0–6.5	6.0–7.0	6.9–7.1	—	11.75	6.5–7.5
Ion stength (mol/L)	0.1	0.1	0.1	—		
Residual concentrations (mg/L)	0.15	0.5	0.1	0.05	$N.D.^g$	5^c
Reference	1	2	3	4	5	6

^{a}Precipitate flotation as CuS.
^{b}0.35–5.85 $\times$ 10^{-3} g mol/L.
^{c}Two-stage.
^{d}Increased baffles.
^{e}mL/min.
^{f}1.5 $\times$ 10^{-5} g · L/mol.
^{g}N.D. = no data given.

17.6 APPLICATIONS

As indicated previously, froth flotation has found its major commercial use in the mining industry. However, investigations are continuing at both laboratory and pilot-scale levels to adapt bubble and foam separation technology to other useful commercial applications. It should be remembered that flotation offers several important features. These include high removal efficiency, low energy requirements, reasonable capital investment, and comparatively low levels of maintenance and operational requirements. In addition, there is the possibility in many instances of recovery and reuse, not only of the separated substances but also of the somewhat costly surfactants.

17.6-1 Industrial Waste Treatment

One of the major areas of interest, of course, is the full-scale application of flotation techniques to the treatment of industrial wastewaters. Some pilot-scale studies have been performed on the removal of metals from actual industrial wastes. These wastes included electroplating wastewaters,[1] textile wastewaters,[2] wood preservation industry,[3] tanneries,[4] and, importantly, laundry wastewaters.[5] This last category is an interesting approach in that the surfactants could be reclaimed for reuse. This in turn would give it high applicability for mobile units in field hospitals, military camps, and so on.

Perhaps the most cost effective use of flotation techniques would lie in recovery of valuable metals from the wastewaters generated in the photographic, electroplating, and jewelry manufacturing industries.

Another use of flotation techniques would be their incorporation prior to biological treatment. This would expedite the secondary treatment by eliminating solids which can be problems in biological treatment units. Another aspect to this pretreatment application of flotation technology is the neutralization of alkaline wastes by the use of CO_2 rather than air as the gas phase. This also aids in biological treatment.

Another interesting side effect is exemplified in the removal of cullet glass from solid wastestreams originating from urban refuse using flotation techniques. The glass, once separated, was cleaned by additional reflotation so that the ultimate product was deslimed to approximately 90%.[6]

17.6-2 Domestic Sewage

Foam flotation techniques can serve a dual role when used in domestic sewage treatment. Not only can contaminants be removed by foaming but there is the input of additional oxygen prior to discharge. Foam flotation could be the means for the removal of refractory and nonbiodegradable constituents.

As discussed earlier, simple flotation techniques have been used for the removal of oils from water and sewage. This is important since the oils tend to coat the biofloc, reducing its efficiency in degrading the waste. The introduction of air also can be used to remove dissolved detergents common in municipal wastes and thereby improve the efficiency in the overall system.

17.6-3 Energy Oriented Uses

As discussed earlier, foam flotation techniques were used in the removal of metals from wastewaters generated in nuclear power plants. One problem which occurred was the coating of the column walls by the removed floc. This could be quite a major problem on a large scale. The problem, however, could be reduced or eliminated by proper coating of the column walls.

Other areas related to energy can use foam flotation techniques. For example, this technology provides a means of separating the oils in sand tars and on water resulting from spills. A mobile unit was tested for field applicability of this idea by the U.S. EPA.[7] The unit would arrive at the spill and clean up oil-contaminated beaches, then relocate as necessary.

Dissolved air flotation or induced air flotation are commonly used techniques in the oil refining industry for the removal of both oils and suspended solids.[8] The coal industry has been investigating foam flotation technology as well. One use would include cleaning the wet scrubber air pollution control devices used in coal-burning power plants.[9]

17.7 CONCLUSIONS

In the last 20 years, the field of adsorptive bubble separations has developed greatly in areas other than ore flotation. Development included theoretical analysis and mathematical modeling as well as practical pilot-scale testing of real wastewaters. Ingenuity has been applied to the use of these techniques as discussed in the previous section. As the problems of wastewater and its treatment stay with us, one could anticipate that the future will bring even more extensive research in the application of these techniques to the practical problems faced by the scientists and engineers of today. Primary areas for increased research would include solvent sublation and microgas dispersion (the very small bubbles rise quite slowly and provide large interfacial areas for adsorption, both of which favor efficient mass transfer).

Another area of research would be improving column design to maximum hydraulic loading rates so that high flow rates can be accommodated without sacrificing the quality of the effluent produced. Additional work is needed on the collapsing of foams and recycling of surfactants as well as solutions to the problems of the precipitates and adsorbed colloidal flocs adhering to the column walls.

These goals should be accomplished by the interaction of the chemist and chemical engineer, with close attention paid to environmental regulations and economics. As we hope it has been made clear in this brief chapter, the basis does exist for these and other advances. It is hoped that new orientations and ideas will be brought into the field, in an effort to have these techniques maintain their vigorous growth.

REFERENCES

Section 17.1

17.1-1   Adams, Ford, and Eckenfelder, *Development of Design and Operational Criteria for Wastewater Treatment*, CBI Publishing, Boston, 1981.

17.1-2   A. M. Gaudin, *Flotation*, McGraw-Hill, New York, 1957.

17.1-3   M. C. Fuerstenau, *Flotation, A. M. Gaudin Memorial Volume*, American Institute of Mining, Metallurgical and Petroleum Engineers, New York, 1976.

17.1-4   A. W. Adamson, *Physics and Chemistry of Surfaces*, Wiley-Interscience, New York, 1960.

17.1-5   F. Sebba, *Ion Flotation*, American Elsevier, New York, 1962.

17.1-6   J. J. Bikerman, *Foams*, Springer-Verlag, New York, 1973.

17.1-7   B. L. Karger, R. B. Grieves, R. Lemlich, A. J. Rubin, and F. Sebba, *Sep. Sci.*, **2**, 401 (1967).

17.1-8   F. Sebba, *AIChE Inst. Chem. Eng. Symp. Ser.*, **1**, 14 (1965).

17.1-9   R. Lemlich, *Adsorptive Bubble Separation Techniques*, Academic, New York, 1972.

17.1-10   R. Lemlich, *Traces Heavy Met. Water Removal Proc. Monit., Proc. Symp.*, 211 (1973).

17.1-11   G. Izumi, *Sekiyu Gakkai Shi*, **17**, 404 (1974).

17.1-12   R. B. Grieves, *AIChE Symp. Ser. No. 150, 71*, 143 (1975).

17.1-13   L. Eklund and C. Andersson, *Ind.-Anz.*, **99**, 875 (1977).

17.1-14   R. A. Hyde, D. G. Miller, R. F. Packham, and W. N. Richards, *J. Am. Water Works Assoc.*, **69**, 369 (1977).

17.1-15   R. H. Marks and R. J. Thurston, *Environ. Pollut. Manage.*, **7**, 94 (1977).

17.1-16   A. Thomas and M. A. Winkler, *Top. Enzyme Ferment. Biotechnol.*, **1**, 43 (1977).

17.1-17   S. C. Loftus, J. C. Yeung, and D. R. Christman, Report 1977, ORNL/MIT-248, 1977.

17.1-18   H. G. Cassidy, *Techniques of Organic Chemistry*, Vol. 10, Wiley-Interscience, New York, 1957.

17.1-19   E. Rubin and E. L. Gaden, Jr., in H. M. Schoen (Ed.), *New Chemical Engineering Separation Techniques*, Chap. 5, Wiley-Interscience, New York, 1962.

17.1-20   I. A. Eldib, in K. A. Kobe and J. F. McKette (Eds.), *Advances in Petroleum Chemical Refining*, Vol. 7, p. 98, Wiley-Interscience, New York, 1963.

17.1-21   A. J. Rubin, "Foam Separation of Microcontaminants by Low-Flow-Rate Methods," Ph.D. dissertation, University of North Carolina, 1966; University Microfilms, Inc., Ann Arbor, MI, No. 67-1045, 1967.

17.1-22   R. B. Grieves, *Br. Chem. Eng.*, **13**, 77 (1968).

17.1-23   B. L. Karger and D. G. DeVivo, *Sep. Sci.*, **3**, 393 (1968).

17.1-24   R. Lemlich, *Ind. Eng. Chem.*, **60**, 16 (1968).

17.1-25   P. Somasundaran, *Sep. Purif. Methods*, **1**, 117 (1972).

17.1-26   G. E. Ho, *SNIC Bull.*, **2**, 29 (1973).

17.1-27   R. B. Grieves, *Chem. Eng. J.*, **9**, 93 (1975).

17.1-28   S. I. Ahmed, *Sep. Sci.*, **10**, 649 (1975).

17.1-29   P. Somasundaran, *Sep. Sci.*, **10**, 93 (1975).

17.1-30   A. Bahr and J. Hense, *Fortschr. Verfahrenstech. Abt. F*, **14**, 503 (1976).

17.1-31   G. Panov, *Rapp. Tech., Cent. Belge Etude Corros.*, **129** (RT. 234), 153 (1976).

17.1-32   P. Richmond, *Chem. Ind. (London)*, **19**, 792 (1977).

17.1-33   W. Balcerzak, *Gaz Woda Tech. Sanit.*, **51**, 131 (1977).

17.1-34   A. N. Clarke and D. J. Wilson, *Sep. Purif. Methods*, **7**, 55 (1978).

17.1-35   R. B. Grieves, in P. J. Elving (Ed.), *Treatise on Analytical Chemistry*, 2nd ed., Pt. 1, Vol. 5, p. 371, Wiley, New York, 1982.

17.1-36 A. N. Clarke and D. J. Wilson (Eds.), *Foam Flotation, Theory and Application*, Marcel Dekker, New York, 1983.

Section 17.2

17.2-1 D. J. Wilson, *Sep. Sci. Technol.*, **13**, 107 (1978).

17.2-2 B. L. Currin, F. J. Potter, D. J. Wilson, and R. H. French, *Sep. Sci. Technol.*, **13**, 285 (1978).

17.2-3 R. H. French and D. J. Wilson, *Sep. Sci. Technol.*, **15**, 1213 (1980).

17.2-4 R. M. Kennedy and D. J. Wilson, *Sep. Sci. Technol.*, **15**, 1239 (1980).

17.2-5 A. N. Clarke and D. J. Wilson, *Foam Flotation, Theory and Application*, Marcel Dekker, New York, 1983.

Section 17.3

17.3-1 B. B. Ferguson, C. Hinkle, and D. J. Wilson, *Sep. Sci.*, **9**, 125 (1974).

17.3-2 D. J. Wilson, *Foam Flotation Treatment of Heavy Metals and Fluoride Bearing Industrial Wastewaters*, U.S. EPA, 600/2-77-072, Washington, DC, 1977.

17.3-3 S. Mukai, T. Wakamatsu, and Y. Nakahiro, *Rec. Dev. Sep. Sci.*, **5**, 67 (1979).

17.3-4 J. H. Oh, H. S. Lee, I. Z. You, and D. S. Cho, *Tachan Kawangsan Hakkoe Chi*, **15**, 144 (1978).

17.3-5 D. Bhattacharyya, J. A. Carlton, and R. B. Grieves, *AIChE J.*, **17**, 419 (1971).

17.3-6 R. B. Grieves and R. W. Lee, *Ind. Eng. Chem. Process Des. Dev.*, **10**, 390 (1971).

17.3-7 A. N. Clarke and D. J. Wilson (Eds.), *Foam Flotation, Theory and Practices*, Marcel Dekker, New York, 1983.

17.3-8 C. P. Beitelshees, C. J. King, and H. G. Sephton, *Rec. Dev. Sep. Sci.*, **5**, 43 (1979).

17.3-9 T. E. Chatman, S.-D. Huang, and D. J. Wilson, *Sep. Sci.*, **12**, 461 (1977).

17.3-10 B. L. Currin, R. M. Kennedy, A. N. Clarke, and D. J. Wilson, *Sep. Sci. Technol.*, **14**, 669 (1979).

17.3-11 R. B. Grieves and D. Bhattacharyya, *Can. J. Chem. Eng.*, **43**, 286 (1965).

17.3-12 R. B. Grieves and D. Bhattacharyya, *J. Am. Oil Chem. Soc.*, **44**, 498 (1967).

17.3-13 R. B. Grieves, D. Bhattacharyya, and C. J. Crandall, *J. Appl. Chem.*, **17**, 163 (1967).

17.3-14 R. B. Grieves and D. Bhattacharyya, in A. N. Clarke and D. J. Wilson (Eds.), *Foam Flotation, Theory and Practices*, Chaps. 11 and 12, Marcel Dekker, New York, 1983.

17.3-15 A. J. Rubin and W. L. Lapp, *Anal. Chem.*, **41**, 1133 (1969).

17.3-16 R. P. Robertson, D. J. Wilson, and C. S. Wilson, *Sep. Sci.*, **11**, 569 (1976).

17.3-17 V. V. Pushkarev and E. A. Budnekov, *Kolloid Zh.*, **25**, 589 (1963).

17.3-18 J. C. Barnes, J. M. Brown, N. A. K. Mumallah, and D. J. Wilson, *Sep. Sci. Technol.*, **14**, 777 (1975).

17.3-19 S. D. Huang and D. J. Wilson, *Sep. Sci.*, **11**, 215 (1976).

17.3-20 D. Pearson and J. M. Shirley, *J. Appl. Chem. Biotechnol.* **23**, 101 (1973).

17.3-21 C. Matsuzaki and M. Zietlin, *Sep. Sci.*, **8**, 185 (1973).

17.3-22 R. B. Grieves and D. Bhattacharyya, *Sep. Sci.*, **3**, 185 (1968).

17.3-23 R. B. Grieves and D. Bhattacharyya, *J. Appl. Chem.*, **19**, 114 (1969).

17.3-24 R. B. Grieves and D. Bhattacharyya, *Sep. Sci.*, **4**, 301 (1969).

17.3-25 A. N. Clarke, B. L. Currin, and D. J. Wilson, *Sep. Sci. Technol.*, **14**, 141 (1979).

17.3-26 D. Bhattacharyya, J. D. Romans, and R. B. Grieves, *AIChE J.*, **18**, 1024 (1972).

17.3-27 R. B. Grieves, P. M. Schwartz, and D. Bhattacharyya, *Sep. Sci.*, **10**, 777 (1975).

17.3-28 R. B. Grieves, D. Bhattachrayya, and W. T. Strange, *AIChE Symp. Ser. No. 150*, **71**, 40 (1975).

17.3-29 L. D. Skrylev and V. K. Ososkov, *Ottrytiya Izobret. Prom. Obraztsy Tovarnye Znaki*, **54**, 68 (1977).

17.3-30 L. D. Skrylev, R. E. Savina, and V. V. Sviridov, *Khim. Technol. (Kiev)*, **1**, 57 (1978).

17.3-31 F. E. Chaine and H. Zeitlin, *Sep. Sci.*, **9**, 1 (1974).

17.3-32 D. Bhattacharyya, J. D. Romans, and R. B. Grieves, *AIChE J.*, **18**, 1024 (1972).

17.3-33 R. B. Grieves and E. F. Chouinard, *J. Appl. Chem.*, **19**, 60 (1969).

17.3-34 F. Sebba, *Ion Flotation*, Elsevier, Amsterdam, 1962.

17.3-35 B. L. Karger, in R. Lemlich (Ed.), *Adsorptive Bubble Separation Techniques*, Chap. 8, Academic, New York, 1972.

17.3-36 D. J. Wilson, *Solvent Sublation of Organic Contaminants for Water Reclamation*, RU-83/6, Bureau of Reclamation, U.S. Department of the Interior, 1984.

17.3-37 D. J. Wilson, unpublished work, 1983.

Section 17.4

17.4-1 A. N. Clarke and D. J. Wilson (Eds.), *Foam Flotation, Theory and Applications*, Marcel Dekker, New York, 1983.

17.4-2 M. A. Slapik, E. L. Thackston, and D. J. Wilson, *J. Water Pollut. Control. Fed.*, **54**, 238 (1982).

17.4-3 L. K. Wang, M. L. Granstrom, and B. T. Kown, *Environ. Lett.*, **3**, 251 (1972).

17.4-4 L. K. Wang, M. L. Granstrom, E. L. Boursdeimas, and B. T. Kown, *Environ. Lett.*, **4**, 233 (1973).

17.4-5 K. S. Kalman and G. A. Ratcliff, *Can. J. Chem. Eng.*, **49**, 626 (1971).

17.4-6 M. Goldberg and E. Rubin, *Ind. Eng. Chem., Proc. Des. Rers.*, **6**, 195 (1967).

Section 17.5

17.5-1 D. J. Wilson, *Foam Flotation of Heavy Metals and Fluoride—Bearing Industrial Wastewaters*, Environmental Protection Agency EPA-600/2-77-072, U.S. Government Printing Office, Washington, DC, 1977.

17.5-2 J. S. Hanson, M.S. thesis, Vanderbilt University, Nashville, TN, 1976.

17.5-3 E. L. Thackston, J. S. Hanson, D. L. Miller, Jr., and D. J. Wilson, *J. Water Pollut. Control Fed.*, **52**, 317 (1980).

17.5-4 D. L. Miller, Jr., M.S. Research Report, Vanderbilt University, Nashville, TN, 1977.

17.5-5 K. S. Kalman and G. A. Ratcliff, *Can. J. Chem. Eng.*, **49**, 626 (1971).

17.5-6 R. B. Grieves and R. W. Lee, *Ind. Eng. Chem. Proc. Des. Dev.*, **10**, 390 (1970).

Section 17.6

17.6-1 D. Pearson and N. McPhater, Chem. Eng. Hostile World Conf. Paper, Eurochem. Conf. No. 39, 1977.

17.6-2 Ribot, *Rev. Quim. Text.*, **42**, 41 (1977).

17.6-3 Oikawa, Shoji, Sato, and Izumi, *Tohoku Kogyo Gijutsu Shikensho Hokoku*, **7**, 46 (1976).

17.6-4 Rogov, Flipchuk, Anopol'skii, and Shamt'ko, *Elecktron. Obnab. Mater.*, **6**, 80 (1978).

17.6-5 R. B. Grieves and Bewley, *JWPCF*, **45**, 470 (1973).

17.6-6 Heginbotham, *U. S. Bur. Mines Rep. Invest.*, RI8327 (1978).

17.6-7 M. J. Sittig, *Oil Spill Prevention and Removal Handbook*, Noyes Data Corp., Park Ridge, NJ 1974.

17.6-8 American Petroleum Institute, *Land Treatment Practices in the Petroleum Industry*, 1983.

17.6-9 R. B. Grieves, P. M. Schwartz, and D. Bhattacharyya, *Sep. Sci.*, **10**, 777 (1975).

CHAPTER 18

Ultrafiltration and Reverse Osmosis

WILLIAM EYKAMP
JONATHAN STEEN
Koch Membrane Systems, Inc.
Wilmington, Massachusetts

18.1 ULTRAFILTRATION

18.1-1 Introduction and Definitions

Ultrafiltration (UF) is a membrane process capable of retaining solutes as small as 1000 daltons (1 dalton is $\frac{1}{16}$ the mass of an oxygen atom), while passing solvent and smaller solutes. By convention, UF is distinguished from *reverse osmosis* in that UF does not retain species for which *bulk solution osmotic pressure* is significant, and distinguished from *microfiltration* in that UF does exhibit some retention for soluble macromolecules regardless of pore size.

UF usually is applied to aqueous streams which may contain soluble macromolecules, colloids, salts, sugars, and so on. It is used to *concentrate* or *fractionate*, often simultaneously.

Flux (J) is the measure of a membrane's productivity. SI units for flux, μm/s, are convertible into more conventional units by multiplying by 3.60 to obtain L/m^2·h or by 2.12 to obtain U.S. gal/ft^2·day.

Permeate is the material that has passed through the membrane.

Retention, also called rejection or reflection, is defined as

$$1 - \frac{\text{permeate concentration}}{\text{feed concentration}}$$

Retention ignores the phenomenon of *concentration polarization*, which generally increases significantly the true concentration of the retained species at the membrane surface (Fig. 18.1-1).

UF is practiced in the laboratory and on a large scale industrially. The approach can differ radically. Industrially, UF always involves *crossflow*, with conversion per pass generally quite low. Laboratory applications often use stirred cells, which give an approach to crossflow. Scale-up of laboratory rate data is difficult unless the laboratory-scale equipment is designed carefully and the engineer is experienced. Occasionally, when dilute solutions are processed in the lab, UF is run as a conventional unstirred normal flow filter. This application is similar to ordinary filtration and is not treated here.

The UF membrane can be pictured as a sieve on a molecular-dimensional scale. It is usually polymeric and asymmetric, designed for high productivity and resistance to plugging. Compared to crystalline molecular sieves, it is imprecise because pore size is not controllable within narrow limits. Adsorption effects are usually significant, altering both the rate and retention characteristics of UF. The UF membrane

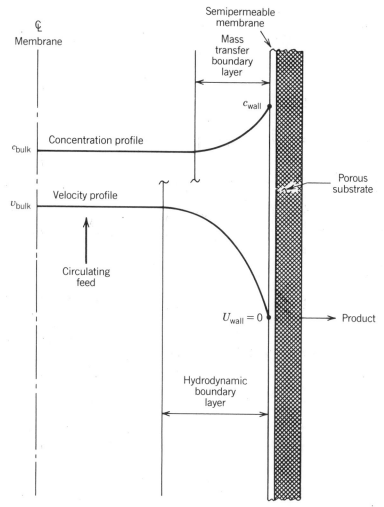

FIGURE 18.1-1 Concentration polarization and velocity profiles in semipermeable membranes.

functions both as a concentrating device (removal of solvent) and as a fractionating device. It is useful to consider that the concentration of unrejected small solutes in solvent is largely unchanged across the membrane, while the *retained* species rise in concentration *and in relation to unretained solutes* as ultrafiltration proceeds.

Diafiltration is a washing process analogous to filter cake washing. Solvent is added to the concentrate to reduce the concentration of all species, after which UF selectively increases the concentration of retained species.

Pressure is the driving force for the flow of mass through a membrane and, in the absence of macrosolutes, flow is proportional to the pressure difference across the membrane and inversely proportional to viscosity. This obvious case is unusual.

Concentration polarization is defined next. Virtually all commercial applications for UF are flux limited by the fact that macrosolutes carried toward the membrane with the permeating solvent build up near or on the membrane. The redistribution of this polarized material becomes the rate-limiting step and is far more important to rate than is pressure. A consequence of concentration polarization is the anomalous *independence* of flux and pressure (above a threshold value) in most operating ultrafilters. Figure 18.1-2 shows the typical dependence of flux-J on pressure difference ΔP. At low enough pressures, all ultrafiltrations are linearly dependent on ΔP. As the retained species become concentrated at the membrane, pressure independence generally becomes the rule (Region III), and the rate is controlled by mass transfer, not pressure. Region II results from contributions by both effects and dominance by neither: it is particularly

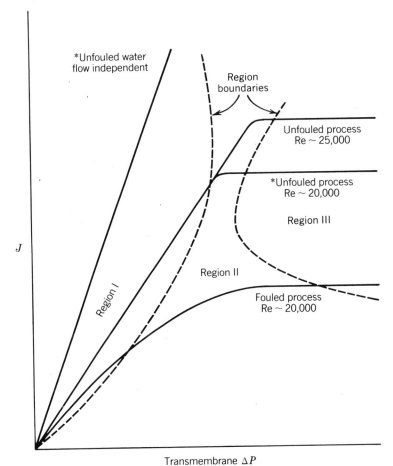

FIGURE 18.1-2 Flux versus transmembrane pressure drop. Region I—pressure dominated; region II—mixed; region III—mass transfer dominated. * Lab data.

evident with partially fouled membranes. In turbulent, fully polarized (pressure-independent) UF, a useful guide for predicting the dependence of flux on flow rate is derivable from a modified form of the Sherwood number for turbulent flow

$$\text{Sh} = \frac{kd}{D} = A\text{Sc}^{1/3}\text{Re}^{m} \qquad (18.1\text{-}1)$$

where k = mass transfer coefficient
        d = hydraulic diameter
        D = Diffusivity
        Sc = Schmidt number, ν/D
        Sh = Sherwood number, kd/D
        Re = Reynolds number, $d\nu/\nu$
        ν = velocity
        ν = kinematic viscosity
        A = system-specific constant
        m = experimental constant depending on the system.

While this expression is not useful for predicting fluxes *a priori*, it is very important in correlating and understanding experimental data. Assume the geometry and the solution properties remain constant. Flux is proportional to k, the mass transfer coefficient, since redistribution of the material polarized at the

membrane is rate determining. The expression simplifies to $J \sim Q^m$ or v^m, Q being volumetric flow rate.
Thus,

$$\log J = m \log Q \tag{18.1-2}$$

Equation (18.1-2) is very useful for experimental data reduction and for equipment design. The value
and constancy of the slope m is also a particularly sensitive indicator of fouling.

If membrane flux is limited by the back diffusion of retained solutes, another useful expression may be
derived assuming 100% membrane retention and other simplifications:[1]

$$J = k \ln \frac{c_m}{c_b} \tag{18.1-3}$$

If a gel layer is formed at the membrane,

$$c_m = c_g = \text{constant}$$

The useful expression

$$dJ = -kd \ln c_b \tag{18.1-4}$$

results, and plots of J versus $\ln c_b$ are predicted to be linear.

The assumptions regarding constancy of c_m are rarely true; nonetheless, the expression is very useful
as a predictor of flux in streams operating in Region III of Fig. 18.1-2 when they are being concentrated.
By extrapolating the semilog plot to $J = 0$, the value of retentate concentration at the membrane can be
estimated. This value is the maximum theoretical concentration, since $C_b = C_m$ at this point. Semilog
plots for many soluble macromolecules extrapolate to zero flux between 25–35% concentration: the value
for colloids is usually around 75% by volume. Experimental values taken at high concentrations often
deviate significantly from prediction, because the assumptions underlying the expression are invalid near the
limit.[2–4]

18.1-2 Common Applications

In the laboratory, UF finds many clinical and biological uses for isolating proteins and other macromolecules
from salts, metabolites, and so on, and for solution concentration. Table 18.1-1 gives a sample of a large
number of industrial applications by industry.

18.1-3 Membranes

CHARACTERISTICS
Ultrafiltration membranes are commonly asymmetric (skinned) polymeric membranes prepared by the phase
inversion process. Materials commercially made into membranes include cellulose nitrate, cellulose acetate,
polysulfone, aramids, polyvinylidene fluoride, and acrylonitrile polymers and copolymers. Inorganic membranes of hydrous zirconium oxide deposited on a tubular carbon backing are also commercially available.

Membrane manufacturers seek a variety of pore sizes and inherent flux characteristics to meet customer
requirements. Resistance to temperature, pH, solvents, and aggressive chemicals has been a major goal.
Because of its high residual stress and specific surface, a membrane is rarely as chemically resistant as is
its parent polymer, and some caution is appropriate.

Nonetheless, both organic and inorganic membranes exhibit acceptable lives at 80°C, pH 2–12, in the
presence of solvents, and on exposure to oxidizing agents, albeit not simultaneously.

PROPERTIES
Molecular weight cutoff is a convenient fiction that provides a rough guide to a membrane's pore size.
Most sources define the cutoff as the molecular weight of the substance to which the membrane shows
90% retention in very dilute solution. Molecular weight is but one of many factors influencing retention.
Shape near the membrane and affinity for the membrane are the determining factors in whether a molecule
is retained. When a membrane is operating either with concentration polarization-dependent behavior or
when it is severely fouled, the polarized or fouling layer itself may become a significant filter. Furthermore,
pressure changes often affect the retention of a polarized layer. Taking account of all the factors that
contribute to a membrane's retentiveness in an operating process is not yet possible, and "cutoff" is useful
only in the most general sense, although measurements run on very dilute solute homologues are helpful
in ranking relative properties of membranes in a family. Dextrans, Carbowax fractions, polymeric surfactants, and proteins commonly are used for the purpose.

The expression "molecular weight cutoff" has fostered a misconception that membranes can be used

TABLE 18.1-1 Applications of UF by Industry

Feed Stream	Purpose
Food	
Blood	Fractionation, concentration
Egg	Concentration
Fruit juices	Clarification
Gelatin	Concentration, fractionation
Soy whey	Fractionation, concentration
Starch	Concentration, desalting
Vinegar	Clarification
Wastes	Concentration for anaerobic digestion
Wine	Clarification, stabilization, protein removal
Pharmaceutical	
Fermentation broths, especially enzymes	Fractionation, concentration, desalting
Dairy	
Milk	Fractionation, preparation of cheese precursors, concentration
Whey	Fractionation
Industrial	
Electrophoretic paint	Bath control, production of rinse water
Flexographic ink washwater	Concentration, pollution abatement
Kraft mill effluent	Decolorization
Latex (PVC, SBR, etc.)	Concentration
Leather tanning	Bath recycle, product recovery
Lignosulfonate	Fractionation, concentration
Oil–water emulsions	Concentration, pollution abatement
Paper machine whitewater	Recycle
Textile size (polyvinyl alcohol)	Concentration for recycle
Water	Colloid removal
Wool scouring	Lanolin recovery

to separate, for example, a 100,000 dalton solute from a 30,000 dalton solute by using a 50,000 MW cutoff membrane. Because of molecular interactions in the boundary layer, which are complicated by size and shape considerations, success in such a separation would be rare.

MEMBRANE LIFE

Modern membranes have long economic lives when run at moderate temperatures and pH and in the absence of chemically aggressive materials. Generally, life is limited by cleaning, intractable fouling, or mechanical failure. Of these, cleaning is the dominant consideration. Food-related streams require daily cleaning for hygienic reasons. Most membranes can withstand the aggressive cleanings used to remove proteins and polysaccharides for 2 yr at best. Failure usually occurs by a slow decline in the retentiveness of the membrane or by a tendency to foul faster and clean up slower, reducing the membrane's throughput. Cleaning occasionally leads to mechanical failure due to environmental stress cracking or seal failure. Membranes always are used in conjunction with other materials (backings, seals, sealants, etc.) whose failure may precede that of the membrane. While some cleaning is benign, membrane lifetime is usually inversely proportional to cleaning.

Membranes are intolerant of freezing, drying, and mechanical abuse; therefore, they usually are impregnated with glycerol during shipping and handling.

UF membranes are made commercially in sheet, capillary, and tubular forms. Sheet membrane is used in plate-and-frame devices, stack devices, and spiral elements. Capillary membranes are generally 0.4–1 mm diameter, are self-supporting, and are pressurized from inside. Usually, they are made with a skin on both inside and outside surfaces so that they can be pressure reversed for cleaning. Tubular devices are cast on supports and are generally in the 10–25 mm diameter range.

All these types have commercial niches in which they dominate, but generally there is overlap in the utility of the various configurations. Large volumes of feed are processed in either thin-channel devices (capillary, spiral, plate-and-frame) or open-channel devices (tubes). By comparison, open-channel devices are robust, tolerant of debris, expensive, and big; thin-channel devices are intolerant of debris (especially fibers), more compact, and less expensive. Laboratory-scale equipment is usually stirred cell or mini-thin channel.

18.1-4   Ultrafiltration

YIELD EQUATIONS
Two useful yield equations for use on a *batch process* are

$$Y = \left(\frac{V_0}{V}\right)^{R_i - 1} \tag{18.1-5}$$

$$= \left(\frac{c}{c_0}\right)_{ib}^{(R_i - 1)/R_i} \tag{18.1-6}$$

$$R_i = \left(\frac{c_b - c_p}{c_b}\right)_i \tag{18.1-7}$$

where c = solute concentration (c_0, initial; c_b, bulk; c_p, permeate)

V = volume (V_0, initial)

CONVERSION
There are important applications for which conversion must be low. A good example is the UF of electrophoretic paint. The desired product is permeate, which is required for rinsing newly painted objects for recovery of excess paint and for removal of electrolytes and contaminants. Since flux, paint stability, and membrane cleanliness are affected adversely by rising concentration, the UF is run at as low a paint concentration as feasible, that is, at low conversion. In such a situation, yield is not a factor, and the absence of leakage is the dominant consideration.

At the other extreme, some waste emulsions are concentrated 100-fold. Permeates containing even a few ppm of emulsified material are unacceptable, so both conversion and yield must be very high; yields less than ~100% mean that the membrane process will not be employed.

Many applications are controlled by economic yield criteria, where there is a valuable component to be recovered and its loss has only economic consequences. Two examples illustrate two aspects of the problem.

1. *Polyvinyl Alcohol Size Recovery.*   A textile mill washes the PVA size from woven fabric in a countercurrent washer giving 2% PVA in hot water. PVA is required upstream at 14% total solids to treat warp thread before weaving. An ultrafilter is employed to recover both the PVA concentrate and the hot water (Fig. 18.1-3)

PVA accompanying the permeate is lost permanently, since it leaves with the fabric, and reduces the quality of the removal step. Yield is based on membrane retention and concentration factor. In practice, the process is run in continuous stages, so yield must be calculated from a stage-by-stage computation, but the batch equations (18.1-6) and (18.1-7) give a reasonable approximation.

Retention is *assumed* constant, a good first approximation. In fact, retention is generally a function of concentration, being lower for more dilute streams. Conceptually, this finding is consistent with the discussion of concentration polarization (above).

2. *Apple Juice Clarification.*   After being pressed from the fruit, apple juice undergoes filtration and clarification. Enzymes normally are employed to reduce levels of pectin and other components that may induce haze in storage. UF provides an excellent means for removing colloidal material and soluble macromolecules. Conversion must be very high, since the UF is economically justified on its ability to recover >99% of the fruit sugar fed. Yield is measured both in terms of the recovery of undiluted juice and as the measure of overall sugar recovery. Dilution water is used to wash residual sugar out of the concentrate, with obvious economic disadvantages. In process design, the dilution phenomenon and yield generally are considered together in the economic evaluation (Fig. 18.1-4)

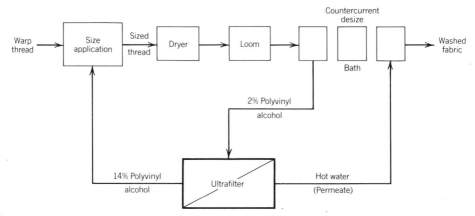

FIGURE 18.1-3 Schematic diagram of process using UF for recovery of polyvinyl alcohol size.

FLUX

The *value* of an ultrafilter is determined by yield, the qualities of the two product streams, capacity, and the cost of alternative means of producing an acceptable product. The *cost* is strongly dependent on the ultrafiltration membrane area. Area is dictated by flux, the amount of permeate per area, and time.

Flux Estimation. Flux is determined by careful experiment. Many factors influence flux. Already mentioned are concentration and velocity, which is in turn a function of configuration. Also important are fouling, pretreatment, and feed consistency. Natural streams often exhibit seasonal variation. Nonetheless, thousands of ultrafilters operate on large-scale applications worldwide. Table 18.1-2 provides industrial data on diverse applications, intended to give the reader a general idea of the magnitudes of process variables.

Flux Dependence on Concentration. Equation (18.1-3) provides general guidance that is useful for rough design until experimental data become available on the system of interest. Figure 18.1-5 gives two examples of experimental data.

Staging. When operating an ultrafilter as a concentrator, the dependence of flux on concentration rewards the designer for operating at the lowest concentration possible. Batch operation requires minimum membrane area, but it is used rarely because of inconvenience, inappropriateness, spoilage due to long residence time, or tankage requirements. *Semibatch* operation is more popular, since it lowers residence time and tankage requirements. Feed is added to a batch tank as processing is carried out, with feed being discontinued at an appropriate time, following which the batch tank is concentrated to the final requirement.

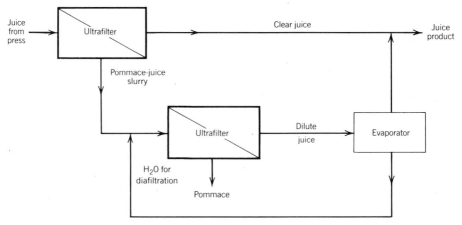

FIGURE 18.1-4 Schematic diagram of process using UF for apple juice clarification.

TABLE 18.1-2

Process Stream	T(°C)/pH	Weight Percent (%) or Volumetric Concentration Ratio (X)	Configuration	Re[a]	J[b] (μm/s)	m [Eq. (18.1-2)]	Cleaning Interval
Electrophoretic paint Cationic Uniprime 3042 (PPG)	35/6.2	21%	Tubular	34,000	6.0	1.25	Months
			Spiral	700[a]	3.8		
			Capillary	2,600	8.6		[c]
Electrophoretic paint Anionic Electrolure 106 (Glidden)	24/9	10%	Tubular	30,000	7.1	1.25	Years
			Spiral	700[a]	5.2		
Acid whey	50/4.5	5X	Spiral	1,000[a]	9.7		Daily
		25X	Spiral	300[a]	2.9		
Sweet whey	50/5.6	5X	Capillary	1,660	5.1		Daily
		5X	Spiral	1,000[a]	11.1		
		.25X	Spiral	300[a]	3.3		
Skim milk (cow)	50/6.3	1X	Spiral	1,100[a]	9.8		Daily
		6X	Spiral	50[a]	1.8		
PVC latex	46/5	30%	Tubular	20,000	22.4	1.7	Days
		55%	Tubular	9,100	3.3		
Polyvinyl alcohol[d] Dupont T-66	80/7	3%	Spiral	200[a]	5.4	0.7	Months
		10%	Spiral	10[a]	1.0	0.5	
Oil–water emulsion cutting oil waste	35/8	1.7%	Tubular	115,000	16.7	1.25	Weekly
Water (15 MΩ)	25/6.0		Spiral	[e]	47.0[e]	0[e]	Months
Gelatin (low Bloom)	43/2.5	3%	Spiral	600[a]	7.8	0.7	Daily
		18%	Spiral	110[a]	1.3	0.4	

[a] For spirals, Re is calculated as if it were a parallel-plate device: the presence of the feed channel spacer is ignored.
[b] SI units for flux, μm/s are convertible into more conventional units by multiplying by 3.60 to obtain L/m²h or by 2.12 to obtain U.S. gal/ft²·day.
[c] Backflushed hourly.
[d] Laminar flow
[e] Pressure dependent; flux at 250 kPa.

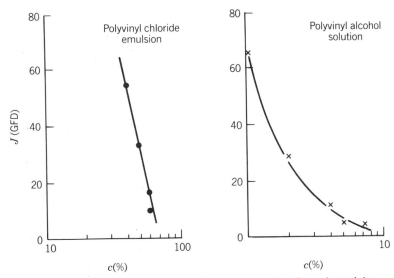

FIGURE 18.1-5 Flux J versus concentration c: two examples of experimental data.

Stages in series is the dominant industrial process design. Two- and three-stage designs are rare. Above five stages, the incremental flux benefit diminishes, and larger number of stages are used primarily when required by limitations on stage size. Stages usually are of equal size, although smaller diafiltration stages and special high-concentration stages are used. Flux and stage concentrations are determined from the flux versus log concentration curve by trial-and-error calculations. As a rule, only the concentration in the final stage is known, since the final stage concentration is the product concentration. Therefore, the flux of the final stage is also known. After making an overall estimate of the area required, calculate backward to the first stage (where concentration is higher than feed but is *not* known) and check for closure. Solutions converge rapidly.

FOULING

Fouling[5] causes the time-dependent irreversibility in flux through a membrane when operating under constant conditions. So viewed, fouling includes *membrane compaction* (see Section 18.2-4) but, as a practical matter, UF compaction is far less important than true fouling. Transient flux changes at start-up are not fouling; they are concentration related, not time related, and flux may be regained without cleaning.

It is a fact that membranes foul. Even carefully prefiltered water produces irreversible flux decline.

UF membranes are believed to have a significant distribution of pore sizes. Since volumetric flow rate in cylindrical pores varies with r^4, it is always true that a minority of the pores transport a majority of the flow. Any plugging or restriction in these large pores has significant impact on flux. Generally, a membrane's retention increases as it fouls, suggesting that it is the larger pores that are obstructed. Rough estimates of fouling tendency can be inferred from the "cleaning frequency" column in Table 18.1-2.

Installations producing edible products are cleaned daily or more often for hygienic reasons; otherwise, cleaning frequency is to maintain 80% of postcleaning flux. The fouling tendency of some streams is related inversely to velocity: paints and oily emulsions are examples. Oily emulsions are intolerant of stagnation points at the membrane and usually are processed in tubes or capillaries. If not stabilized by surfactant, the oil preferentially wets the membrane, upsetting the separation. Oily emulsions represent an important class of membrane applications, the metastable system. In spite of the latent instability, properly designed membrane systems function well on a wide variety of metastable systems.

18.1-5 Experimental Design

Since UF is so dependent on experimental data, some suggestions are offered on obtaining meaningful data.

1. Be sure the feed is representative and does not change during the test. Normal bulk measurements such as pH, temperature, viscosity, and concentration may not be sufficient, since the membrane may interact with the stream in unforeseen ways. Periodically returning to the starting conditions with a clean membrane to see that flux, retention, and response to flow are unchanged is a wise precaution. Save the permeate and use it to redilute a concentrated sample.

2. Be very alert for time-dependent changes. Run flux and flow experiments in a loop, watching for time-related changes.

3. Test for fouling. A good experimental practice to use for long-term testing is to pick a point just inside region III (Fig. 18.1-2) at high Re; then, holding pressure constant, reduce the flow rate substantially for the life test. When flow is restored to the higher value, fouling tendencies should be easy to observe. Check the slope of log flux versus log flow before and after this test.

4. Run flux versus concentration experiments at more than one flow rate.

Meaningful experiments can be very difficult. For example, consider ultrafiltration of whey. There are a number of proteins and microorganisms that may foul the membrane and the presence of these materials is dependent on the cheese-making process upstream. There also are inorganic salts present at or above their solubility. A change in the concentration of the organic solutes influences the tendency of the inorganic materials to precipitate. Two streams that are "the same" may behave completely differently. Means have been found to pretreat many problem feeds; running others in certain flow regimes increases success.

18.1-6 Economics

CREDITS
The partial list presented below may be evaluated and supplemented by the user.

1. Enhanced value of the primary product, including yield increase.
2. Coproduct credits (water, heat recovered, etc.).
3. Downstream savings in hauling, sewage, or waste disposal.
4. Savings in materials; diatomaceous earth, filters, disposables, and so on are not required by UF.
5. Process change savings: elimination of other unit operations and attendant labor.

COSTS
The following list gives an idea of the cost involved in using UF membranes.

1. Capital charges.
2. Power and cooling water.
3. Cleaning chemicals and cleaning water.
4. Membrane replacement.
5. Maintenance.
6. Labor.
7. Taxes, insurance, and space costs.

Capital Charges. Ultrafiltration membranes are critically dependent on hydrodynamics, and industrial membrane equipment usually is sold as a skid-mounted package complete with pumps, valves, and instruments. Units requiring daily cleaning usually include automatic cleaning processes. Longer-cleaning-interval equipment generally includes cleaning pumps and solution-mixing tanks. Cost estimates assume 304 SS construction, cleaning capability, and totally enclosed electrical equipment. Costs scale at about the 0.6 power of size. Costs vary widely among configurations, but so do performance and suitability. Costs assume skid-mounted equipment, without tanks, and a 100 m^2 area.

Type of Membrane	1986 $/m^2	Design Energy Density (W/m^2)
Tubular	900–2000	300–120
Spiral	600–1500	120–25
Capillary	600–1600	280–100
Plate and frame	800–1600	280–180

Power and Cooling. Power costs per unit permeated can be estimated by dividing energy density by the flux. Typical values found *in practice* are as follows:

Feed	Energy (W·h/L) Range	Energy (W·h/L) Most Common Value
Paint	3–12	6
Milk	3–13	4
Oil–water	2–4	2
Polyvinyl alcohol	0.8–2.1	1

Cooling water is required to dissipate essentially all the energy consumed. Cooling water quality requirements are standard.

Cleaning Chemicals and Cleaning Water. Cleaning costs vary considerably but they do not dominate the overall cost picture. A normal cleaning cycle is a water rinse, a chemical cleaning cycle, water rinse, sanitizing treatment if appropriate, and (rarely) a proprietary posttreatment. Plants cleaned frequently usually employ inexpensive cleaners such as NaOH, NaOCl, H_2O_2, H_3PO_4, and detergents.

Membrane Replacement. Membranes generally lose flux or rejection so slowly that their replacement represents a balance between yield or productivity loss and replacement cost. However, cleaning frequency generally is correlated inversely with membrane life; daily cleaning usually wears out membranes in 12–24 months. For infrequently cleaned applications, replacement in 30–42 months is reasonable. Membranes are subject to operating limits on pressure, temperature, dryout, and freezing. Losses to these causes are infrequent.

Maintenance. Maintenance is generally very low, with pump seals, instruments, and valves the principal equipment requiring attention.

Labor. Most medium-sized ultrafilters run unattended except for labor required to mix and supervise cleaning operations. Rarely if ever is labor a significant cost.

Taxes, Insurance, and Space Costs. Membrane systems are compact but space availability and cost are so variable that no generalization can be made. Taxes and insurance are generally negligible.

18.1-7   Conclusion

UF processes are well established industrially, but predictive tools are still quite primitive. For many applications, the economic benefits from UF are so dramatic as to make a favorable feasibility analysis possible with no more than what is given here. As the technology and the science progress, quantitative predictive methods can be expected to appear.

1. Enhanced value of the primary product, including yield increase.
2. Coproduct credits (water, heat recovered, etc.).
3. Downstream savings in hauling, sewage, or waste disposal.
4. Savings in materials; diatomaceous earth, filters, disposables, and so on are not required by UF.
5. Process change savings: elimination of other unit operations and attendant labor.

18.2   REVERSE OSMOSIS

18.2-1   Introduction and Definitions

Reverse osmosis (RO), also called *hyperfiltration*, is a membrane process capable of separating a solvent from a solution by forcing the solvent through a semipermeable membrane by applying a pressure greater than the osmotic pressure of the solute. RO usually is applied to aqueous streams at ambient temperatures. Unlike *freezing* and *evaporation*, reverse osmosis requires no phase change. *Electrodialysis* is conceptually similar, being driven by electromotive force rather than pressure.

Usually the solution-diffusion model is used to explain the flow of a solvent through an RO membrane.[1] This explanation holds that solvent dissolves in the membrane material then diffuses across it in response to a chemical potential gradient. Solute is presumed to pass through the membrane by diffusion driven by the solute concentration difference across the membrane. The explanation implies that solute retention is proportional to flux, in the absence of other effects.

Osmotic pressure poses a practical limit to the use of RO; a difference in system pressure and osmotic pressure $(p - \pi)$ of approximately 1 MPa is required for RO membranes. Rarely are useful membranes operated above a system pressure of 5.5 MPa, giving a practical upper limit to osmotic pressure of 4.5 MPa. This osmotic pressure is approximately that of 1 M NaCl and 35 wt.% sucrose solutions.

For many common solutes at low concentrations, the van't Hoff equation

$$\pi = cRT \qquad (18.2\text{-}1)$$

is sufficiently accurate for the estimation of osmotic pressure. In Eq. (18.2-1) c is molar concentration, which must include the effects of solute dissociation, R is the gas constant in appropriate units, and T is absolute temperature. Either experimental values of π or a more sophisticated model should be used at higher concentrations.

18.2-2 Concentration Polarization

The tendency of the retained solutes to concentrate at the membrane, that is, concentration polarization, is of less importance than it is in UF.[2]

18.2-3 Membranes

RO membranes have undergone dramatic improvements in the past decade, and there is every reason to believe that significant changes will continue. RO first became practical with the invention of the Loeb–Sourirajan membrane,[3] an asymmetric membrane based on cellulose acetate. Membranes of this type are still important in commerce. The critical discovery was a method for producing membranes by the *phase inversion process*. Membranes thus produced have an extremely thin active layer, permitting high flux, and have integrity, permitting excellent salt retention. Because the thin barrier is integral with a thick, porous support, the membranes are robust, stable, and commercially useful, especially when manufactured into spiral elements. Many other membrane materials have been tried, and among the early successes were polyaramids based on polymers of diamino benzenes and benzene dicarboxylic acids. Asymmetric membranes based on various polymers of these monomers were commercialized as fine hollow fibers (84 μm OD by 42 μm ID) and have been offered subsequently as flat sheet membranes in spiral modules.

The next major commercial success was the family of *composite membranes*. They feature a very thin RO membrane on a suitable substrate, usually a UF membrane. Most of the RO composite membranes are polyamide coatings in which the separating layer is produced by interfacial polymerization of a diamine and a multibasic acid chloride. The most successful recent membrane is based on an interfacial polymer of 1,3-diamino benzene and 1,3,5-benzene tricarboxylic acid chloride coated on a polysulfone membrane substrate.

Many other membranes have been tried, and some have found commercial niches. Properties most sought after, in addition to excellent flux and salt retention, are hydrolytic stability, resistance to compaction under pressure, resistance to biological and chemical assault, and tolerance of chlorine disinfectants.

Membrane retentions to inorganic ions span the entire spectrum. Sodium chloride is used universally as a test solute, and membranes are available commercially with <10% to >99% sodium chloride retention. Most other ions have higher retentions on a given membrane than does sodium chloride, and calcium, magnesium, sulfate, and other divalent ions always are retained to a much higher degree than is sodium chloride.

Retention of organic species is very specific, with a few materials passing through the membrane faster than water (therefore having negative rejection). Phenol passage through cellulose acetate is a classic example. Groves[4] lists retentions for 11 organic solutes (ethanol, propanol, phenol, acetic acid, oxalic acid, citric acid, urea, ethylene glycol, ethylene diamine, methylethyl ketone, and ethyl acetate) by several commercial membranes (Filmtec FT30, Toray PEC1000, Teijin PBIL, and UOP PA300). Reportedly, however, the membranes discussed by Groves have been surpassed by newer devices.

18.2-4 Compaction

It is customary in RO to combine membrane creep and fouling into one term referred to as compaction. This quantity is defined in terms of the flux J_t, which is determined at some time t after installation, and the flux J_0, which is determined at some reference time t_0. The phenomenon is quantified in terms of the compaction slope, which is defined by the equation

$$\text{compaction slope} = s_c = \frac{\log(J_t/J_o)}{\log(t/t_0)}$$

A value of $s_c = -0.03$ corresponds to a flux decline of 17% over a 1 yr period. Current state-of-the-art in membrane manufacture gives a value of s_c of about -0.01, which corresponds to a 5 yr life of the membrane.

18.2-5 Applications

RO membranes are employed most frequently with the desalination of brackish water and, to a lesser extent, for seawater. Membranes also find process applications: for example, the preparation of high-purity water (as a successor to distillation) and as a precursor for ion exchange in ultrapure water systems; in the food industry for concentrating whey, milk, and maple sap; for recovery of salts, such as the reconcentration of rinse waters containing $NiSO_4$; for concentration and partial fractionation of lignosulfonates in wood pulping; and for the recovery and concentration of many industrial effluents.

18.2-6   Process Considerations: Membrane Modules

There are four basic types of RO module currently manufactured:

1. *Plate and Frame.* Conceptually very simple but relatively rare.

2. *Hollow Fiber.* All commercial designs to date use shell-side pressurization and tube-side permeate collection. The dominant design has very fine fibers, but larger fiber devices also are found. Fine hollow fiber devices are intolerant of colloidal matter in the feed and require care in feed pretreatment.

3. *Tubular.* Relatively uncommon in RO and used where particulates require the open channel or where hygienic requirements are paramount.

4. *Spiral Wound.* This device, pictured in Fig. 18.2-1, consists of a membrane envelope containing a permeate-conducting channel, usually a woven fabric. Between membrane envelopes, a feed spacer conducts the feed across the membrane face. There are several variants, but the usual device has the permeate channel connected to an axial product tube. The spiral has been very successful commercially, as an inexpensive device for commercializing flat sheet membranes. Spirals are made commonly in 100 and 200 mm diameters, but new desalination plants are using elements 300 mm in diameter by 1.5 m long.

18.2-7   Process Design

Fundamental design of a RO plant is quite simple. The feed is prepared, usually by filtration and, where appropriate, biological stabilization, then fed to a high-pressure pump. The outlet pressure of the pump must be high enough to keep the feed at the design difference between system and osmotic pressures ($p > \pi$) in the last membrane stage. As the feedstream passes through membrane stages, it loses pressure because of frictional losses and gains osmotic pressure because the solvent is permeated selectively. Thus, a design "pinch" occurs in the final membrane stage. Booster pumps sometimes are employed, and forced recirculation is employed when the presence of macrosolutes creates an abnormally high concentration polarization problem. Excessive feed pressures are not acceptable because they impose unacceptable stresses on the earlier membrane stages, giving rise to creep or compaction, and for energy reasons.

The dominant energy consumer is the high-pressure pump. Depending on the application, conversion can be low; seawater plants run at 30–40% conversion because both the osmotic pressure of higher salt concentrations and the limits on membrane retention and thus the quality of permeate produced make operation at higher conversions impractical. On large plants, energy recovery is practiced, but it is not economically viable for most applications.

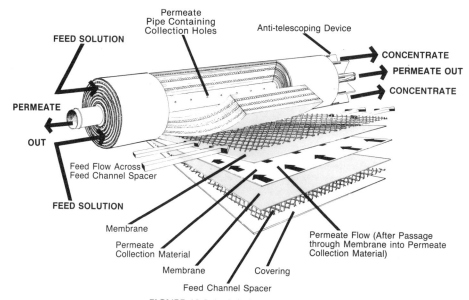

FIGURE 18.2-1 Spiral-wound module.

18.2-8 Process Economics

In a parametric study on seawater RO, Soo-Hoo et al.[5] estimated the capital cost for a stand-alone seawater RO plant to produce 400 kL/day of potable water from seawater to be $1,100,000 (1981). Energy requirements were between 9 W·h/L for a high-pressure plant with no energy recovery, and 5 W·h/L for a lower pressure plant with energy recovery. Membrane replacement costs were approximately $40,000/year.

A fully automated, tubular RO plant for removing 200 kL/day of water from a food stream costs about $400,000, excluding buildings, pretreatment utilities, and installation, which were included in the Soo-Hoo study. Energy requirements are about 3 W·h/L, and membrane replacement costs are $80,000/year. Cleaning frequency dictates annual membrane replacement in this process application.

While the plant designs for these two applications are totally different, the capital costs are in the same general range.

REFERENCES

Section 18.1

18.1-1 M. C. Porter, *Am. Inst. Chem. Eng. Symp. Ser. 171*, **73**, 83–103 (1977).

18.1-2 J. G. Wijmans, S. Nakao, and C. A. Smolders, *J. Memb. Sci.*, **20**, 115–124 (1984); this reference gives a comparison of the gel layer and osmotic pressure as a cause of this phenomenon.

18.1-3 A. G. Fane, *J. Sep. Proc. Technol.*, **4**, 15–23 (1983).

18.1-4 V. L. Vilker, C. K. Colton, K. A. Smith, and D. L. Green, *J. Memb. Sci.*, **20**, 63–77 (1984).

18.1-5 For excellent general reviews, see D. E. Potts, R. C. Ahlert, and S. S. Wang, *Desalination*, **36**, 235–264 (1981); G. Belfort and F. W. Altena, *Desalination*, **47**, 105–127 (1983). Both articles incorporate excellent bibliographies.

Section 18.2

18.2-1 H. K. Lonsdale, U. Merten, and R. L. Riley, *J. Appl. Polym. Sci.*, **9**, 1341–1362 (1965).

18.2-2 P. L. T. Brian, in U. Merten (Ed.), *Desalination by RO*, pp. 161–202, M.I.T. Press, Cambridge, MA, 1966.

18.2-3 S. Loeb and S. Sourirajan, *Adv. Chem. Ser.*, **38**, 117 (1962).

18.2-4 G. R. Groves, *Desalination*, **47**, 277–284 (1983).

18.2-5 R. Soo-Hoo, Sh. May, and L. Awerbuch, *Desalination*, **46**, 3–15 (1983).

BIBLIOGRAPHY

References cited were selected to add depth and scope to the item referred to. General references include:

A. G. Fane, *J. Sep. Proc. Technol.*, **4**, 15–23 (1983).

H. K. Londsdale, *J. Membr. Sci.*, **10**, 81–181 (1982).

S. Sourirajan and T. Matsuura, Eds., *Reverse Osmosis and Ultrafiltration*, ACS Symp. Ser. No. 281, American Chemical Society, Washington, DC, 1985.

Many excellent articles appear in *Journal of Membrane Science* and *Desalination*.

CHAPTER 19

Recent Advances in Liquid Membrane Technology

J. W. FRANKENFELD

Exxon Research and Engineering Company
Linden, New Jersey

NORMAN N. LI

Allied-Signal, Inc.
Des Plaines, Illinois

19.1 INTRODUCTION

The separation of mixtures using semipermeable membranes has been the subject of continuing interest. However, this concept has not met with much success in industrial applications since the polymeric membranes in common use have suffered from low flux rates and low selectivities in general. Thus, area and staging requirements become too great for any large-scale process.[1,2] An alternative is to use liquid films for membranes. In general, such liquid films possess much higher selectivities than polymeric membranes, thereby reducing staging requirements markedly. However, membrane systems based on thin liquid films have not been able to overcome the costs associated with achieving sufficient area to make a significant impact in the separation area. Liquid membranes,[3] invented in 1968, overcome this liability by generating the necessary surface area without the need for mechanical support. Thus, these devices offer a more effective means for the separation of mixtures in an efficient manner. The large number of membranes with different formulations renders this an extremely versatile process useful for a variety of applications, including hydrocarbon separations,[1,3-5] minerals recovery,[6-15] wastewater treatment,[1,8,16-21] and a number of biochemical and biomedical applications.[2,22-26] Progress in all these fields up to 1978 is covered in four previous reviews (Refs. 2 and 26 for biochemical and biomedical applications and Refs. 8 and 20 for industrial separation and wastewater treatment). This chapter reviews work that has been reported after 1978 as well as material not reviewed previously. Three separate areas are covered: (1) developments in the theory of membrane transport, (2) liquid membranes in industrial reactions and separations, and (3) biochemical and biomedical applications. In general, only unsupported liquid membranes are considered in this chapter. The data on the so-called "supported liquid membranes," or porous polymer membranes with liquid filling the pores, are not reviewed specifically, although the separation principles are the same.

19.2 GENERAL DESCRIPTION OF LIQUID MEMBRANES

Liquid membranes are made by forming an emulsion of two immiscible phases and then dispersing it in a third phase (the continuous or donor phase). Frequently, the encapsulated (or receiving) phase and the continuous phase are miscible. The membrane phase must not be miscible with either if it is to remain

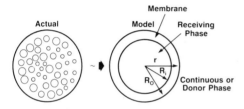

FIGURE 19.2-1   Liquid-membrane model.

stable. Therefore, the emulsion is of the oil-in-water type if the continuous phase is oil and of the water-in-oil type if the continuous phase is water. To maintain the integrity of the emulsion during the separation process the membrane phase usually contains certain surfactants and other additives as stabilizing agents and a base material which is a solvent for all the other ingredients. For a given application, liquid membranes usually are tailor-made to meet all the specific requirements. Liquid membranes sometimes are called emulsion membranes, liquid surfactant membranes, or double emulsions. The term "double emulsions" is not really suitable because the third phase is never emulsified.

When the emulsion is dispersed by agitation in a continuous phase (the third or donor phase), many small globules of emulsion are formed. Their size depends strongly on the emulsion viscosity, the mode and intensity of mixing, and the nature and concentration of the surfactants in the emulsion. In general, the globule size is controlled in the range of 0.1–2.0 mm in diameter. Thus, a very large number of globules of emulsion can be formed easily to produce a very large membrane surface area for rapid mass transfer from either the continuous phase to the encapsulated phase or vice versa. It should be noted that many such smaller droplets, typically 1 μm in diameter, are encapsulated within each globule. Figure 19.2-1 illustrates an actual liquid-membrane globule and a simplified model which is useful for discussing separation principles.[1-5] For a detailed account of general emulsion technology and the effects of formulation, papers by Cahn et al.[4] and Marr and Kopp[6] are recommended.

19.3   PRINCIPLES OF SEPARATION

19.3-1   Extraction Mechanisms

Liquid membranes are versatile reagents that can be used to perform separations by a variety of mechanisms. Three of these are shown in Fig. 19.3-1. The simplest of these is that of selective permeation (Fig. 19.3-1a) wherein mixtures can be separated by taking advantage of their different rates of diffusion through the liquid membrane. This type of mechanism has been used largely for the separation of hydrocarbons.[1-3]

More complicated separations can be achieved by using one or more of the "facilitated transport" mechanisms that can be built in a liquid-membrane system.[1,2] These mechanisms are:

Type 1 (Fig. 19.3-1b).   These minimize the concentration of the diffusing species in the receiving phase. This can be done by reacting the diffusing species with a reagent in the receiving phase to form a product (or products) which may or may not be able to diffuse back through the membrane. Separations of ammonia,[4-6] hydrogen sulfide,[7] and phenols[8-11] are good examples.

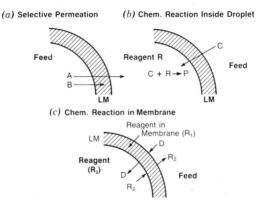

FIGURE 19.3-1   Mechanisms of liquid-membrane transport.

Type 2 (Fig. 19.3-1c). These carry the diffusing species across the membrane by incorporating "carrier" or chelating compounds in the membrane. This kind of carrier-mediated transport can be illustrated by the separation of various metal ions, such as cadmium, chromium, copper, and mercury, from their aqueous solutions by the use of oil-type liquid membranes containing oil-soluble liquid ion-exchange agents.[4,12-16] These mechanisms have been described in detail elsewhere[1,2,17] and are not repeated here.

19.3-2 Factors Influencing Transport Through Liquid Membranes

There are a number of parameters that can affect the transport of various substances across liquid-membrane barriers. Identifying these and assessing their relative importance has been the subject of much recent study, especially as the treatment of liquid-membrane phenomena becomes more sophisticated. Cahn et al.[12] have described the effects of several of these parameters on both the extraction of ammonia (type 1 facilitated transport) and copper ions (type 2 facilitated transport). The chemistry of the copper extraction is described below. In the ammonia case, where simple diffusion is rate determining, the two most important parameters are solubility of NH_3 in the oil phase and the viscosity of the membrane. Similar observations were made by Yang and Rhodes[18] and Frankenfeld et al.[19] who studied the uptake of organic acids by membrane systems quite similar to Cahn's. Some of Cahn's results are given in Table 19.3-1. In such cases the product of the viscosity (μ) and the extraction rate constant (D) is a constant (Table 19.3-1). In the case of copper extraction, where the diffusing species is a complex of copper and a carrier, the product μD is no longer a constant, although viscosity certainly plays an important role in determining the permeation rate constant. The copper case is more complicated as is pointed out in Section 19.4-1.

The data in Table 19.3-1 suggest that liquid membranes should be as nonviscous as possible for optimum extraction rates. Up to a point this is true. However, membrane stability generally is reduced as the viscosity decreases and the tendency to rupture or "leak" the internal phase becomes greater, especially at long contact times. This effect is illustrated by the curves in Fig. 19.3-2. Initially, the extraction rate is more rapid with the less viscous membrane. However, at longer contact times the weaker formulation shows a tendency to "leak" the internal phase at a rate exceeding the membrane's ability to reabsorb it.[12] Membrane stability is an important consideration which is discussed in more detail below.

Cahn and coworkers also studied the effects of the concentration of complexing agent (copper carrier) in the membrane phase, concentration of the extracted species in the internal phase, and treat ratio (volume of external phase/volume of emulsion) on the rate of copper extraction (type 2 facilitated transport). They found that, within rather wide limits, none of these had major effects on the rate of copper uptake. Increasing the concentration of complexing agent, LIX 64N‡ from 0.5 to 2.0% resulted in a small rate enhancement. Further increases showed no effect. It is a characteristic of liquid membranes that very high loadings of the internal phase can be achieved before any reduction in extraction rate is observed; that is, the extracted species can be "pumped" against a strong concentration gradient. Cahn et al.[12] found that emulsions preloaded to 30 g/L of copper were still effective in extracting copper from synthetic mine water containing as little as a few hundred ppm of dissolved copper. In addition, feed ratios of at least 10:1 were found to be feasible, an important economic consideration.

One parameter in addition to membrane viscosity that was important in determining the rate of copper uptake was the size of the internal microdroplets. This can be controlled by the way in which emulsions are made. Cahn et al.[12] found that a 30% increase in rate accrued from reducing the average size of the microdroplet from 14 to 2 μm. This is illustrated by the extraction curves in Fig. 19.3-3. Membrane

TABLE 19.3-1 Effects of Membrane Viscosity on Transport Rates

Extracted Species	Viscosity μ of Oil Phase (C_p at 100°F)[a]	Permeation Constant D' (min^{-1})[b]	$\mu D'$
Copper	3.75	4.0	15.0
6.83	2.5	17.1	—
7.13	3.1	22.3	—
26.3	1.5	39.5	—
Ammonia	6.0	0.75	4.5
11	0.48	5.3	—
18	0.32	5.8	—
24	0.23	5.5	—

[a] Controlled by adjusting the amount of nonviscous solvent in the oil phase of the liquid membrane.
[b] Average value over several time intervals.
Source: From Cahn et al.[12] with permission.

‡LIX is a trademark of the General Mills Corp.

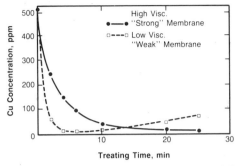

FIGURE 19.3-2 Effect of membrane viscosity on copper extraction. From Cahn et al.[12] with permission.

stability also is enhanced by small microdroplet size and this also contributes to an apparent rate increase by reducing leakage.[12] Similar results were reported by Yang and Rhodes[18] on the extraction of organic acids (type 1 facilitated transport).

Halwachs et al.[20] studied the permeation of hydrocarbons through oil–water–oil systems. They found that the most important variables governing the permeation rates were membrane viscosity, type and concentration of surfactant, and ionic strength of the medium. However, the permeation rates of hydrocarbons through aqueous membranes were several orders of magnitude slower than the permeation rates of phenol through hydrocarbon membranes.[20]

The parameters discussed above have direct effects on extraction rates. In addition, several variables exist which, since they influence membrane stability, have secondary effects on rates. Unstable membranes which liberate some of their internal phase to the external phase, either by "leakage" or rupture, show a significantly lower "apparent" extraction rate. The most important of these parameters are the viscosity of the emulsion, the amount and type of surfactant used in emulsion preparation, the size of the internal micro-subdroplets, and the operating conditions of the mixer.[2,12,21,22] Membranes of very low viscosity or with inadequate surfactant levels are unstable.[12,14,22] Hochhauser and Cussler[14] observed that a 50% increase in membrane rupture was caused by either a 1500% reduction in viscosity, a 0.2 wt.% lowering of surfactant concentration, or a 25% reduction in the volume fraction of the internal micro-subdroplets. The last-mentioned finding is in apparent disagreement with reported results of Cahn et al.,[12] Yang and Rhodes,[18] and Martin and Davies,[23] all of whom found that a decrease in the size of the subdroplets—and thereby an increase in the percentage of the encapsulated phase—resulted in more stable emulsions. At any rate, it is clear that for each individual membrane formulation work is required to strike a balance between rapid extraction and diffusion and membrane stability.

The mixing rate during contact with the external phase is another parameter that can have opposing effects on overall extraction rates. Obviously, rapid mixing increases uptake of the extracted species. However, membrane rupture also increases at high stirring rates.[1,12,22,23] Recently, Stroeve et al.[22,24,25] studied the hydrodynamic stability of liquid-membrane systems in both well-stirred and poorly stirred reactors. Stroeve and coworkers point out that, under well-stirred conditions, mass transfer limitations in the external phase are negligible (see also Ho et al.[26]). However, in some instances rapid stirring may result in intolerably high breakage rates and reduced agitation may be required. Under such conditions, external phase resistance could become important.

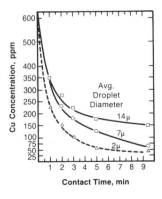

FIGURE 19.3-3 Effects of internal droplet size on copper extraction. From Cahn et al.[12] with permission.

In their studies of the effect of shear rate, Stroeve and coworkers established that the breakup of liquid-membrane droplets can be correlated with the ratio of the apparent viscosity of the disperse (membrane) phase to that of the continuous (external or feed) phase. They also observed internal circulation of sub-droplets within the internal phase of the liquid-membrane droplets[22,24] (Fig. 19.2-1). The effects of this circulation on extraction rates and membrane stability are unknown as yet.

19.3-3 Mathematical Modeling of Liquid-Membrane Transport

Continuing progress is being made in the mathematical modeling of mass transport in liquid-membrane systems. This should contribute to a better understanding of the parameters that control the rate and efficiency of liquid-membrane processes.

The simple rate equations for the various liquid-membrane separation processes in use prior to 1978 were reviewed previously.[2] Since that time, much has been done to improve them, especially by considering the influence of the internal microdroplets and the effects of leakage of the internal reagent which was previously ignored.[12,22,24] Cahn et al.,[12] studying copper extraction, have taken membrane instability into account by modifying the basic permeation rate equation[2,8] to include a leakage term. This modified equation has the form

$$\frac{dC}{d\theta} = -Pt\,\Delta C + l \tag{19.3-1}$$

where C = concentration of permeating species in the external phase
  t = treat ratio, (volume emulsion)/(volume external phase)
  l = leakage rate
  P = revised permeation constant
  θ = contacting time

Solving Eq. (19.3-1) for P, we obtain

$$P = \frac{t}{\theta t} \ln \left(\frac{C_1 - l/Pt}{C_2 - l/Pt} \right) \tag{19.3-2}$$

The leakage rate l can be determined experimentally for a given emulsion formulation. With this, P can be determined by following the concentration C of the permeating species as a function of contact time θ. Cahn et al.[12] found a good agreement with measured and predicted P values when the correction was applied.

Rhodes and coworkers[18,19,27–29] have studied the extraction of various organic acids by liquid membranes (type 1 facilitated transport). They developed a two-phase kinetic model for the transport of a substance from the aqueous donor phase (continuous phase) across the liquid membrane into the internal phase. This mode is described by Eqs. (19.3-3) and (19.3-4),[18,28]

$$C_e \underset{k_{21}}{\overset{k_{12}}{\rightleftarrows}} C_{\mathrm{LM}} \underset{k_{32}}{\overset{k_{23}}{\rightleftarrows}} C_i \tag{19.3-3}$$

$$C_e^t = Ae^{-\alpha t} + Be^{-\beta t} + C_e^s \tag{19.3-4}$$

where C_e, C_{LM}, and C_i represent the concentrations of solute in the external aqueous donor phase, the liquid membrane, and the internal aqueous phase, respectively; k_{12}, k_{21}, k_{23}, and k_{32} are the first-order microrate constants; α and β represent the complex first-order macrorate constants; A and B are preexponential terms; C_e^t represents the concentration of solute in the external aqueous phase at any time t; and C_e^s represents the concentration of solute in the external aqueous phase when the steady state is reached.

All four rate constants have been approximated and have been used to rationalize transport in several liquid-membrane systems. The α constant was assigned to the transport of solute into the membrane and the β constant to the transport of solute from the membrane into the internal (or receiving) phase.[18] If this is true one would expect changes in treat ratio and oil viscosity to have a greater effect on α than β while variation in internal microdroplet size would influence β but not α. In general, this is what was observed (Fig. 19.3-4), although the size of the microdroplets did affect the α constant somewhat, perhaps due to a secondary effect such as leakage.[18] The Rhodes model may be oversimplified but it does provide a good starting point for further development work.

Ho et al.[26,30] and Stroeve and Varanasi[22] have developed sophisticated mathematical models of liquid-membrane transport. These models are of useful theoretical interest and provide some important insights into the most important parameters affecting the rate and efficiency of liquid–membrane extraction process.

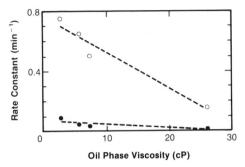

FIGURE 19.3-4 Effect of changes in oil viscosity on kinetics of phenobarbital transport: $\bigcirc$, α rate constant; $\bullet$, β rate constant. From Yang and Rhodes[18] with permission.

19.4 PRACTICAL APPLICATIONS FOR LIQUID MEMBRANES

This discussion is presented in two sections: industrial applications and applications in the biochemical and biomedical areas.

19.4-1 Industrial Applications

As pointed out above, liquid membranes possess a number of advantages for use in industrial separations. These include versatility (by proper membrane formulation a great variety of separations can be achieved, sometimes several different ones at the same time), nondependence on equilibrium considerations (substances can be extracted against large concentration gradients); and relative low cost compared to other membrane systems. Disadvantages include the fragility of some liquid membranes, which leads to swell and leakage problems, and the power required to disperse them in dilute aqueous continuous phases. Although no commercial processes are yet in operation, interest in liquid membranes is high[1] and, as the market situations for metals and wastewater treatment improve and some of the stability problems are solved, commercial processes should be forthcoming. Since our last reviews,[2,3] a number of new developments have been reported in the areas of wastewater cleanup, hydrometallurgy and well control fluids, and the use of liquid membranes as chemical "reactors."

LIQUID MEMBRANES IN THE TREATMENT OF WASTEWATER
Liquid membranes are ideally suited for this application since they have the potential for removing toxic substances from wastewater down to very low levels. Both molecular and ionic species (anions, cations, and complex ions) have been extracted successfully by properly constituted liquid membranes. A partial list of processes reported prior to 1978 is given in Table 19.4-1. These have all been reviewed.[2,3] Recent improvements in these separations as well as new methods reported recently are summarized in this section.

TABLE 19.4-1 Wastewater Cleanup With Liquid Membranes

Substance Removed	Transport Mechanism[a]	Carrier	Stripping Agent	References
Phenols and acids	Type 1 facilitated	None	Base	4–6
Ammonia	Type 1 facilitated	None	Acid	4, 7–9
Phosphates	Type 2 facilitated	Amines	Ca^{2+}	2–4, 6
Chromates	Type 2 facilitated	Amines or salts	Acid or base	2, 3
Copper[b]	Type 2 facilitated	LIX[c]	Acid	2, 8, 10
Mercury[d]	Type 2 facilitated	LIX[c]	Acid	2, 3, 8
Cadmium[d]	Type 2 facilitated	LIX[c]	EDTA	2, 8
Nitrates and nitrites	Type 2 facilitated	Amines	Biological	2, 11, 12

[a] See Section 19.3-1.
[b] As the cation.
[c] LIX is the General Mills trademark for various oil-soluble, liquid ion-exchange agents.
[d] As complex anions.

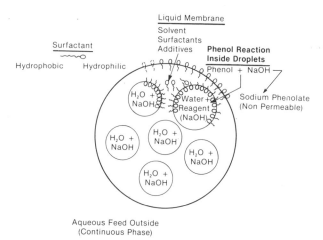

FIGURE 19.4-1   Schematic of liquid membrane for phenol and acid extraction.

Much of the recent work on wastewater treatment has been concerned with the removal of phenols and organic acids. Terry et al.,[13] following the earlier work by Li and Shrier[6] and Cahn and Li,[4] studied various aspects of the simultaneous removal of phenol and organic acids from wastewater in order to improve the efficiency of the process. The schematic diagram in Fig. 19.4-1 illustrates the system that they employed in their studies. This process, which is a good example of type 1 facilitated transport, involves unassisted permeation of the acidic compound through the membrane phase to the internal phase where it is trapped in the form of its oil-insoluble anion by aqueous NaOH. Terry et al.[13] investigated a number of process variables including membrane composition, mixing speed, and treat ratio (ratio of continuous phase to emulsion) on the efficiency of phenol extraction. As expected, they found that the membrane viscosity had a large effect on extraction efficiency with the less viscous formulation having significantly faster extraction rates. Rapid extraction rates where observed with all formulations, however; the contact time for $>98\%$ extraction ranged from 1 to 5 min. Increasing the mixing speed increased the extraction rate slightly but also had a deleterious effect on membrane stability. Increased breakage was related directly to mixing rates and the difference increased steadily with contact time. Also, it was found that reducing the treat ratio from 5:1 to 2.5:1 increased the initial extraction rate significantly. In addition, leakage of the internal phase was higher at the higher treat ratio.

Terry et al.[13] also observed that membrane swell (the increase in the volume of the emulsion phase during a run) was much greater with the less viscous formulations. This suggests that swell is due at least partly to osmotic pressure. Swell is a serious problem because it increases the volume of the emulsion phase that must be handled and decreases the concentration of the extracted species, thereby making recovery more difficult and expensive. These observations illustrate the need for trade-offs between extraction rate and emulsion stability in nearly all liquid-membrane processes.

Terry et al.[13] studied multicompound systems and found that the relative extraction rates paralleled the partition coefficients of the substances between water and oil. Thus, phenols were removed more rapidly than acids, and propionic acid more rapidly than acetic acid. However, the extraction rate for each component of a mixture was the same whether present as mixtures or as individual compounds. The carboxylic acids were extracted at much slower rates than phenols. However, if the caustic concentration in the internal phase of the liquid membrane was very low, the acids were extracted preferentially. The pH of the external phase was also important. The acids were extractable only at low pH while phenols were extracted at pH 7.0.

An independent study with a closely related system has been reported by Shu et al.[14] These workers obtained a phenol removal rate of $>92.3\%$ in a single-stage treatment and $>99\%$ in a two-stage process, with the phenol concentration in the wastewater reduced to <7 ppm.

The extraction of phenol has been used by Schlosser and Kossaczky[15] and Halwachs et al.[16] as a model for studying liquid-membrane transport. The former group compared the liquid-membrane method to a conventional liquid–liquid double extraction process. They concluded that liquid membranes have distinct economic advantages in cases where type 1 or type 2 facilitation methods can be used (e.g., phenol or metal ion extractions). However, where one must rely on simple differential permeation (hydrocarbon separations) the advantage is less clear-cut.

Similarly, Halwachs et al.[16] compared the extraction of phenol to that of benzene in liquid-membrane systems. The mass transfer coefficient in the former case was estimated to be some two orders of magnitude

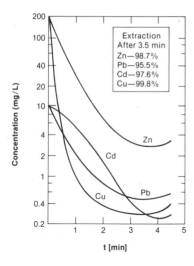

FIGURE 19.4-2 Kinetic curves for the removal of zinc, lead, cadmium, and copper ions in a stirred-batch reactor. From Boyadzhiev and G. Kyuchoukov[17] with permission.

higher than for benzene extraction. They conclude that the liquid-membrane technique is very promising for such applications as phenol removal.

Boyadzhiev and Kyuchoukov[17] have shown recently that mineral acids as well as organic acids may be removed from wastewater using liquid membranes. They report the successful extraction of HNO_3 using a tributylphosphate carrier. These same workers used fatty acids as carriers to scavenge zinc, lead, cadmium, and copper from aqueous solutions. Stripping in the internal phase was accomplished with $2\ N\ H_2SO_4$. Some of their results are shown in Fig. 19.4-2. In every case, >95% removal was achieved in short periods of time. These results are comparable to those reported previously by Kitagawa et al.[8] using liquid ion-exchange agents as carriers. Fatty acids also were employed by Boyadzhiev and Bezenshek for removal of mercury from wastewater.[18]

The simultaneous extraction of trivalent chromium, hexavalent chromium, and zinc from cooling tower blowdown water was studied in detail by Fuller and Li.[19] They investigated Alamine 336, an oil-soluble tertiary amine, and Aliquat 336-S, an oil-soluble quaternary amine salt, both manufactured by General Mills, as carriers in various liquid-membrane systems. Alamine was found to be an effective extractant for Cr^{3+}, Cr^{6+}, and Zn. However, for Cr^{3+} and Cr^{6+}, a pH of 3.0 or less in the external (feed) phase was required while Zn could be extracted only at a pH > 7. Thus, a combined process could not be developed. Aliquat 336-S was more promising. The chemistry of extraction with this carrier involves the following steps:[2,3,19]

Extraction

$$
\overset{\underset{\displaystyle |}{CH_3}}{2R_3NX} + Cr_2O_4^{2-} \rightleftarrows \overset{\underset{\displaystyle |}{CH_3}}{(R_3N)_2}\ Cr_2O_4 + 2X^-
\tag{19.4-1}
$$

Oil   Feed   Oil   Feed

Stripping

$$
\overset{\underset{\displaystyle |}{CH_3}}{(R_3N)_2}\ Cr_2O_4 + 2\ NaX \rightleftarrows \overset{\underset{\displaystyle |}{CH_3}}{2R_3NX} + Na_2Cr_2O_7
\tag{19.4-2}
$$

Oil   Internal   Oil   Internal
          aqueous   aqueous

where $X = Cl^-$ or OH^-. The best stripping agent was found to be a mixture of NaOH and NaCl. Fuller and Li[19] found that, under special conditions, all three ions could be removed successfully (Table 19.4-2).

Unfortunately, the process fails when a large excess of chloride is present in the feed, conditions that normally exist in cooling tower blowdown waters. The effect of excess Cl^- on the extraction of Cr^{6+} is shown is Fig. 19.4-3. This is caused by a competition between Cl^- and $Cr_2O_7^{2-}$ for the carrier. Fuller

TABLE 19.4-2   Liquid-Membrane Extraction of Cr and Zn From Cooling Tower Blowdown

	Blowdown Feed (ppm)	Product	
		Target (ppm)[a]	Achieved (ppm)
Total Cr	18.65	0.25	0.030
Cr^{6+}	14	0.05	0.030
Zn	4.63	1	0.005
pH	6.5	6-9	11.3

[a] EPA standards.

Source: From Fuller and Li[19] with permission.

and Li[19] also observed a deleterious interaction between the surfactant used to strengthen the membrane and the Aliquat. Thus, for this process to be feasible a new carrier must be found.

Liquid membranes are potentially useful for the removal of alkali metal and alkaline earth cations from wastewater streams and in biological systems. The most useful carriers for this purpose are the so-called "crown ethers" which can be highly selective for various cationic species.[20-22] Most of the work in this area has involved supported liquid membranes. Recently, however, efforts have been made to extend this to emulsion-membrane systems.[21,23] Hopfenberg et al.[21] have studied the transport of K^- ions through liquid membranes using dicyclohexyl-18-crown-6 ether (DCH-18C6) as the carrier. Initial efforts were disappointing because of the tendency of the DCH-18C6 to partition toward the aqueous phase of the membrane system instead of the oil. Addition of methylene chloride, a good solvent for DCH-18C6, to the membrane greatly increased the efficiency of K^+ extraction but the resulting liquid membrane was unstable.[21] Success in this area requires the development of stable membranes that are also good solvents for the crown ethers.

USE OF LIQUID MEMBRANES IN HYDROMETALLURGY

Liquid membranes possess a number of advantages over solvent extraction methods for the recovery of valuable metals from dilute aqueous solutions. These include the following:

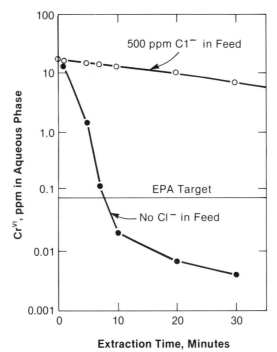

FIGURE 19.4-3   Effect of chloride ion on extraction of chromium from cooling tower blowdown. From Fuller and Li[19] with permission.

1. Simultaneous extraction and stripping in a single operation; this circumvents the usual equilibrium limitations of solvent extractions.
2. Very high loading of the receiving phase; this, of course, is an outgrow of 1.
3. Because of 2, a lower reagent requirement than conventional solvent extraction.
4. Reduced dependence on reaction temperature, again an outgrowth of 1.

As a result, liquid-membrane processes have received a great deal of attention for the potential recovery of copper,[2,16,19,24-32] uranium,[1,33,34] and nickel.[20,35,36]

The extraction of copper from acid leachates has been studied most extensively, especially by Cahn et al. and Igawa et al.[23,24,31] The chemistry of copper extraction is illustrated below:[2]

Extraction

$$2RH + Cu^{2+} \rightleftarrows R_2 Cu + 2H^+ \tag{19.4-3}$$

$$\text{Oil} \quad \text{Feed} \quad \text{Oil} \quad \text{Feed}$$

Stripping

$$R_2Cu + 2H^+ \rightleftarrows 2RH + Cu^{2+} \tag{19.4-4}$$

$$\text{Oil} \quad \underset{\text{aqueous}}{\text{Internal}} \quad \text{Oil} \quad \underset{\text{aqueous}}{\text{Internal}}$$

where RH represents an oil-soluble, liquid ion-exchange reagent. The most common of these for copper is the General Mills LIX 64N. Extraction occurs at the membrane-external aqueous phase interface, while stripping occurs at the membrane-internal aqueous phase interface. The overall reaction represents an exchange of a copper ion for two hydrogen ions. The copper is trapped effectively in the interior of the liquid membrane by the large excess of hydrogen ions. LIX 64N is a mixture of LIX 63 and LIX 65N as shown below:

Chelating Extractant **Structure**

LIX 63 (General Mills)

LIX 64N/Mixture of 63 and 65N (General Mills)

LIX 65N (anti form) (General Mills)

SME 529 (Shell)

Some of the results from Cahn and coworkers are shown in Table 19.4-3. Ninety nine percent of the copper was extracted in 10 min with a fresh emulsion and 98% with an emulsion preloaded to >30 g/L of copper. Cahn et al.[23,24] studied a number of the parameters which affect the efficiency of the process. These were described in Section 19.3-1. A 9 day continuous test of the process was conducted[24] in which excellent extraction efficiency was maintained throughout. A preliminary economic estimate, based on this continuous run, suggested that a 40% investment savings over solvent extraction could be attained. The operating costs for the two processes were comparable.

Similar results were obtained by Volkel et al.[29] who also used LIX 64N as the carrier. Their extraction rates were somewhat slower than those reported by Cahn and coworkers, probably because of greater membrane viscosity. Volkel and coworkers also developed a fairly complicated mathematical model for membrane transport from which they determined mass transfer parameters. These ranged from 1×10^{-3} to 1×10^{-4} s^{-1}. These are comparable to the values obtained for the extraction of phenol (type 1 facilitated transfer in which no carrier is used) by the same group.[16]

Cussler and Evans[20] and Lee et al.[32] used the extraction of copper with LIX 64N to compare supported

TABLE 19.4-3 Extraction of Copper From Synthetic Mine Water by Liquid Membranes

| Contact Time (min) | Concentration of Cu^{2+} | | Extracted (%) |
	External (Feed) Phase[a] (g/L)	Internal Phase (g/L)	
	With Fresh Emulsion		
0	2.0	b	—
2	0.61	9.3	70
4	0.30	11.4	85
6	0.20	12.1	90
10	0.03	13.2	99
	With Preloaded Emulsion		
0	0.50	30.0	—
2	0.07	33.6[c]	85
4	0.02	34.0	96
6	0.01	34.1	98

[a]Present as $CuSO_4$ in synthetic mine water.
[b]Based on the external phase to internal phase ratio of 6.7:1.
[c]Based on the external phase to internal phase ratio of 8.3:1.
Source: From Cahn et al.[24] with permission.

and unsupported liquid membranes. The extraction rates in both cases were high and of about the same magnitude. Some difficulty was encountered, however, in making stable, reproducible unsupported membranes.[32] Copper extraction also was employed by Schlosser and Kossaczky[15] in comparing liquid-membrane "pertraction" to double liquid–liquid extraction. As discussed in the previous subsection, they found liquid membranes to be superior for this type of separation. They point out that liquid membranes require much lower carrier concentrations than in liquid–liquid extraction processes. This would result in both reduced investment and operating costs. Although LIX 64N appears to be the reagent of choice for copper extraction, a number of other carriers have been used or proposed for this purpose. These include Shell SME 529,[37] stearic acid,[37] and benzoylacetone.[26]

LIX 64N has been used to separate nickel ions as well as copper.[20,22] The chemistry of extraction in the two cases is the same [see Eqs. (19.4-3) and (19.4-4)]. However, data on transport through supported liquid membranes suggest that the copper flux across the membrane is four times that of nickel and selective extraction of copper can be achieved.[22] The compound di-(2-ethylhexyl) phosphoric acid also has been proposed for the extraction of nickel II using liquid membranes.[35]

An especially intriguing use of liquid-membrane technology, reported for the first time recently by Fox[1] and Hayworth et al.,[33] is the recovery of uranium from wet process phosphoric acid (WPPA). In the manufacture of fertilizer from phosphate rock, the acid is solubilized by treatment with sulfuric acid. In addition to H_3PO_4, this crude leachate can contain up to 0.18 g/L of uranium.[33,38] Under oxidizing conditions this occurs as the uranate ion, UO_2^{2+}. Solvent extraction (SX) processes have been developed to recover the uranium values. One of the most common of these uses a mixture of di-(2-ethylhexyl) phosphoric acid (D2EHPA) and trioctylphosphine oxide (TOPO) in kerosene solution. The chemistry of solvent extraction is as follows:[38]

Step 1: Extraction

$$UO_2^{2+} + 2(RO)_2PO_2H \overset{TOPO}{\rightleftarrows} UO_2[(RO)_2PO_2]_2 + 2H^+ \qquad (19.4\text{-}5)$$

$\quad$ Aqueous $\qquad$ Oil $\qquad\qquad$ Oil $\qquad\qquad$ Aqueous

Step 2: Reductive Stripping

$$UO_2[(RO)_2PO_2]_2 + 2Fe^{2+} + 6H^+ \rightleftarrows U^{4+} + 2Fe^{3+} + 2H_2O + 2(RO)_2PO_2H \qquad (19.4\text{-}6)$$

$\quad$ Oil $\qquad\qquad$ Aqueous $\qquad\qquad$ Aqueous $\qquad\qquad$ Oil

where

$$R = CH_3(CH_2)_2\overset{\displaystyle CH_3}{\overset{|}{C}}HCH-$$

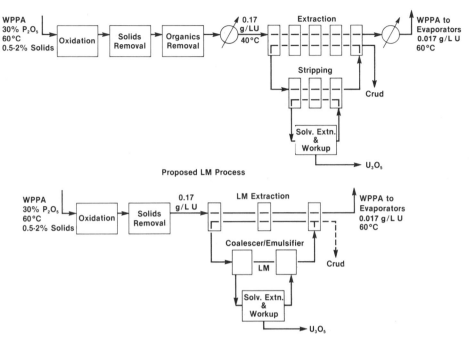

FIGURE 19.4-4 Comparison between the LM and SX processes for uranium recovery from wet process phosphoric acid. Redrawn from data in Hayworth et al.[33]

The uranium is extracted into the oil phase as shown in Eq. (19.4-5). The function of TOPO is not known precisely but it is essential to the process. In a second operation, the uranium first is stripped into strong aqueous acid by reversing Eq. (19.4-5) and then reduced to U^{4+} by ferrous ion [Eq. (19.4-6)]. This is to prevent reextraction by the D2EHPA-TOPO. The process then proceeds to a finishing step wherein the U^{4+} is reoxidized, extracted, concentrated, and precipitated as $(NH_4)_4UO_2CO_3$ (AUT or "yellow cake"). The finishing step is the same for both the solvent extraction and liquid-membrane processes.

The solvent extraction method presents several processing problems:

1. The equilibrium in Eq. (19.4-5) is not very favorable. This requires a high organic to feed ratio (1:1) and entails large reagent losses. In addition, at least eight stages of extraction and stripping are required.
2. The equilibrium in Eq. (19.4-5) is especially unfavorable at high temperatures, requiring an expensive cooling step before uranium extraction, followed by reheating before the phosphoric acid can be concentrated to levels required for fertilizer manufacture.[33,38]

The liquid-membrane (LM) procedure employs essentially the same extraction chemistry. However, by simultaneous extracting and stripping, liquid membranes largely overcome the limitations listed above. At most, three stages are required so that the treat ratio (feed/organic) is estimated at 18:1 instead of 1:1. This results in a 10-fold reduction in reagent losses.[33] In addition, the liquid-membrane process is actually more efficient at higher temperatures.[33] These differences are illustrated by a comparison of flow plans in Fig. 19.4-4. The potential advantages are obvious. More details on the economics of the liquid-membrane process are given later.

Hayworth et al.[33] and Bock et al.[39] have studied the effects of some process parameters on both batch and continuous extractions of uranium. Extraction rates under batch conditions were good; about 13 min contact time was required for 90% extraction. The most important factors were mixing rate and temperature. The extraction rate was not affected significantly by either the concentration of D2EHPA-TOPO or the acid strength of the feed. This presents a distinct improvement over solvent extraction which is limited by equilibrium considerations.[33,38]

The effect of temperature deserves special mention since it represents a major advantage for the liquid-membrane process.[33,38] Figure 19.4-5 compares the influence of temperature on extraction of uranium by the SX and LM techniques. In the case of liquid membranes, extraction efficiency increases with temperature

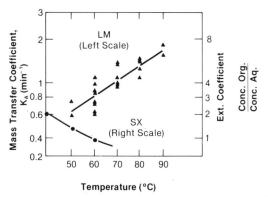

FIGURE 19.4-5 Effect of temperature on mass transfer rate in uranium extraction from wet process phosphoric acid. Redrawn from data in Hayworth et al.[33] and Hurst et al.[38]

while the opposite is true for solvent extraction. It is interesting to note that the complexation of uranium by D2EHPA-TOPO is favored by low temperature, whereas decomplexation or stripping is favored by high temperature. A solvent extraction process may involve cooling of the acid feed from 60 to 40°C for extraction and heating back to 60°C for stripping. A liquid-membrane process can do extraction and stripping simultaneously and effectively at 60°C or higher.[33] This implies that diffusion of the loaded complex through the membrane and stripping are the rate-controlling steps.

LIQUID MEMBRANES AS WELL CONTROL AGENTS
Liquid membranes have been proposed recently as useful reagents for a variety of oil well control problems.[36,40] The injection of viscous emulsions into oil or gas wells to achieve hydraulic fracturing is well known. Alternately, acidic emulsions may be injected into wells to dissolve formation rock (e.g., carbonates) to cause fracturing. This is called acid fracturing.[36,40] In some cases, well control involves just the opposite treatment, that is, the injection of substances that will close or plug fractures. Liquid membranes have distinct advantages for achieving all these objectives. Viscous liquid-membrane emulsions, for example, can be made with a much higher water-to-oil ratio than conventional emulsions.[36] This creates a viscous emulsion with a saving of expensive oil components. Liquid-membrane formulation also can be used to encapsulate plugging agents which are released in situ, thus achieving very efficient well control with minimum loss of expensive components.[36] Liquid-membrane encapsulation of acids is also a good way to obtain delayed acidization. Again the active acidic component is released at the desired site rather than all along the well bore. Finally, the use of multiple liquid-membrane emulsions permits the simultaneous injection of reactive components which must be kept separate until they reach the subterranean formation where fracturing is needed. The membranes, which rupture or invert, release the reactants which then combine to form the active reagent. One example is the encapsulation of ammonium bifluoride in a water-in-oil emulsion and dispersing this emulsion in aqueous hydrochloric acid. This mixture tends to invert underground, releasing the ammonium bifluoride which reacts with hydrochloric acid to form highly reactive hydrofluoric acid in situ.[40] Well fracturing appears to be an area of considerable promise for liquid-membrane technology. A well control fluid using liquid membrane emulsion has been commercialized.[41,42]

CHEMICAL REACTIONS IN LIQUID MEMBRANES
The use of liquid membranes for controlling chemical reactions such as that just discussed has been proposed for a number of other systems.[12,20,43,44] This type of application, in which liquid membranes are used as heterogeneous catalysts or as reaction moderators, is an area that deserves more study. Ollis et al.[43] and Wolynic and Ollis[44] studied liquid membranes as heterogeneous catalyst systems using the catalytic oxidation of ethylene to acetaldehyde (Wacker process) as a model. This process entails the following three steps:[44]

$$PdCl_2 + C_2H_4 + H_2O \rightarrow CH_3CHO + Pd^0 + 2HCl \qquad (19.4\text{-}7)$$

$$Pd^0 + 2\,CuCl_2 \rightarrow PdCl_2 + 2\,CuCl \qquad (19.4\text{-}8)$$

$$2\,CuCl + \tfrac{1}{2}\,O_2 + 2HCl \rightarrow CuCl_2 + H_2O \qquad (19.4\text{-}9)$$

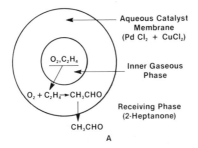

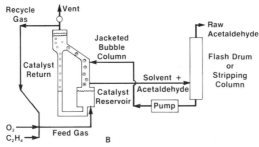

FIGURE 19.4-6 Use of a liquid-membrane system for the catalytic production of acetaldehyde from CO and C_2H_4 (Wacker process): (a) schematic diagram; (b) flow diagram. From Ollis et al.[41] with permission.

Overall

$$C_2H_4 + \tfrac{1}{2}O_2 \rightarrow CH_3CHO \tag{19.4-10}$$

Figure 19.4-6A shows a schematic of the liquid membrane as a chemical "reactor." The reactants, gaseous O_2 and C_2H_4, are encapsulated in an aqueous glycerol membrane which contains the catalyst system $PdCl_2$ + $CuCl_2$. As the gases diffuse through the membrane, the reaction occurs and the product, acetaldehyde, diffuses out into the external, receiving solvent (3-heptanone was used in this case). The apparatus used is shown in Fig. 19.4-6B. It is quite similar to that used by Li and Asher[45] for encapsulating oxygen for blood oxygenation.[11] The reactants are introduced at the bottom of the catalyst chamber where they are encapsulated. The resulting capsules rise through the solvent where the reaction takes place with the product, CH_3CHO being collected in the solvent. At the top the capsules collapse releasing gas and catalyst, which are recycled separately to the bottom of the reactor.[43] Ollis et al.[43] claim this scheme has a number of advantages over the conventional reaction for this process. These include the following:

1. A single vessel which is both chemical reactor and extraction stage.
2. Use of arbitrarily small feed droplets that yield a large liquid–liquid surface area for mass transfer to and from the membrane.
3. Thin catalytic membranes can be used to make the liquid-phase diffusional resistance for dissolved gas transfer small.
4. Conversion of volatile reactants to less volatile products, which permits rapid reactant–product separation.
5. Since the reaction product is extracted into the solvent as formed, high conversions are possible even with reactions having small equilibrium constants.
6. With series reactions, high selectivity to an intermediate product might be achieved by rapid transfer of this product into the solvent phase.

Liquid membranes as reactors are especially promising for biological systems.[11,46] For example, sensitive enzymes may be encapsulated to protect them from deactivating substances while maintaining free access to the substrate ("immobilization"). The substrate may be encapsulated with the enzyme or it may diffuse from the external phase into the internal phase where the reaction takes place. In either case, it is

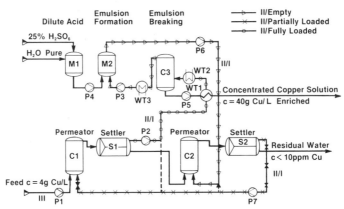

FIGURE 19.4-7 Flow diagram for copper recovery using liquid membranes. From Marrand and Kopp[37] with permission.

usually desirable for the product to diffuse back out to the external phase. These types of "immobilized" enzymes were reviewed in some detail previously.[11,46]

ECONOMIC CONSIDERATIONS

Several papers have been published that discuss the economic comparisons between liquid membranes and competitive processes. A typical liquid membrane "flow plan," in this case for copper extraction as discussed by Marr and Kopp,[37] is shown in Fig. 19.4-7. Other flow plans, in varying detail, are shown or discussed in Refs. 8, 15, 24, and 33. Most liquid-membrane processes, as shown in Fig. 19.4-7, consist of four separate steps:

1. Emulsion makeup.
2. Permeation or contacting with feed.
3. Settling and phase separation.
4. Emulsion breakup and product recovery.

These do not include product concentration and finishing steps which are common to all competitive processes. Major savings in liquid-membrane processes normally accrue from smaller plant size (due to fewer extraction stages) and reduced solvent and carrier inventories.[24,33] Thus, the advantage of liquid membranes over competitive processes depends on the relative importance of these costs. Of course, this can be estimated only on an individual basis. Examples of such estimates which have appeared in the open literature are summarized in Table 19.4-4. It would appear from these estimates that liquid-membrane processes are at least competitive and often can show some distinct economic as well as technical advantages.

19.4-2 Biochemical and Biomedical Applications

Liquid membranes are adapted especially well for biochemical and biomedical applications. This is partly because of their great versatility which permits removal of toxin from biological systems, slow release of enzymes and drugs, and use of the membrane as an immobilization technique or biochemcial "reactor." Ease of preparation, good shelf life, and lack of toxicity are other advantages. In previous reviews[11,46] the use of liquid membranes for blood oxygenation, the treatment of chronic uremia, emergency treatment of drug overdose, and the encapsulation of enzymes were described. In this section, recent developments in these areas as well as applications not previously reviewed are discussed.

BLOOD OXYGENATION AND ARTIFICIAL RED BLOOD CELLS

Frankenfeld, Li, and Asher[11,45,46] have described an apparatus quite similar to that shown in Fig. 19.4-6B for the extracorporeal oxygenation of blood. Bubbles of gaseous O_2, encapsulated in a fluorocarbon membrane, rise through oxygen-depleted blood. As they do so oxygen diffuses from the membrane into the blood while CO_2 diffuses in the opposite direction and is swept out. This process was reviewed earlier.[11,46] The key to its potential success is the use of fluorocarbons in the membrane phase.[11,45,46] Fluorocarbons are uniquely compatible with human blood and circumvent the damage to blood cells which is encountered with conventional devices.

TABLE 19.4-4 Economic Comparisons of Liquid-Membrane and Competitive Processes[a]

| Process | Cost Bases | Liquid Membrane | Competitive Process | | |
			Type	Cost	Reference
Wastewater cleanup	$/10^4 gal	5.5	—	b	8
Phenol removal	$/10^4 gal	1.6	Biological	1.6–2.6	45
			Solvent extraction	5.2	45
Copper recovery[c]	Investment $ × 10^6	8.0	Solvent extraction	13.0	24,40
	Operating cost ¢/1b	1.7	Solvent extraction	1.7	24,46
Uranium recovery[d]	Investment $ × 10^6	27.4	Solvent extraction	28–33.4	33
	Operating cost ¢/1b	15.0	Solvent extraction	20.6–20.9	33

[a]Rough estimates only; comparisons based on cost estimates for year published (see reference given).
[b]Not given but said to be "comparable."
[c]Based on 36,000 tons/yr of copper.
[d]Based on 400,000 tons/yr P$_2$O$_5$ plant (350,000 lb/yr U$_3$O$_8$).

Davis, Asher, and Wallace [49,50] recently described the use of liquid membranes in a novel process for preparing artificial blood cells. In this technique the liquid membranes serve as a template for formation of the artificial cells. The procedure consists of the following steps:[50]

1. Hemoglobin solution, isolated from human cells, is emulsified in oil to form a water-in-oil emulsion such as that shown in Fig. 19.2-1. The emulsion is prepared in a blender at 20,000 rpm so that stable microdroplets of hemoglobin ranging in size from 1 to 4 μm are formed.

2. The emulsion is suspended in an external aqueous phase containing glutaraldehyde, a water-soluble crosslinking agent with slight, but significant, oil solubility.

3. The glutaraldehyde diffuses slowly through the oil to the surface of the microdroplets where the hemoglobin is crosslinked to form a new membrane.

4. A surfactant is added to the liquid-membrane phase causing the liquid membrane to invert and the internal microdroplets, with their newly formed membranes to be ejected.

5. The new cells are washed free of oil and suspended in saline solution.

Both in vitro and in vivo tests with rats demonstrated the artificial cells were effective blood oxygenators and tolerated by the animals. The efficiency of oxygenation was somewhat lower than human blood (Table 19.4-5). However, Davis, Asher, and Wallace feel this can be improved by refining the crosslinking technique.

REMOVAL OF TOXINS FROM BLOOD
Halwachs et al.[51] have described a liquid-membrane system for removing phenolic toxins from blood during liver failure. The technique is similar to the removal of phenols from wastewater except that the internal trapping agent consists of an aqueous solution of the enzymes, uridinediphosphoglucuronic acid (UDPGA) and uridinediphosphoglucuronyl-transferase (UDPGT). The UDPGA serves to link the phenolic compounds to UDPGA to form an oil-insoluble complex which is trapped in the membrane.[51] The extraction is slow compared to most liquid-membrane processes (75–80% removal in 200 min) but Halwachs and coworkers

TABLE 19.4-5 Hemoglobin Activity at Various Stages of Cell Preparation

	P_{50} (torr)	O$_2$ Capacity (mL/g)
Human blood	25	1.34
Hb solution	16	0.96
Hb removed from emulsion	10	0.90
Artificial red cells	9	0.55

Source: From Davis et al.[48] with permission.

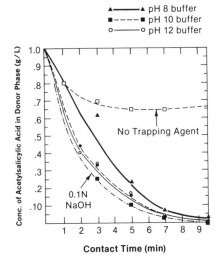

FIGURE 19.4-8 Extraction of acetylsalicylic acid from Ph2 donor by various liquid-membrane formulations. From Frankenfeld et al.[68] with permission.

showed that the process is equally applicable to plasma and whole blood. In a follow-up report[52] these same workers used a mixture of UDPGT and glycuronic acid to complex the phenols.

EMERGENCY TREATMENT OF DRUG OVERDOSE

Liquid-membrane formulations have been proposed by Rhodes et al.[10, 53-56] as agents for the emergency treatment of drug overdose. Liquid membranes have a number of potential advantages over such techniques as peritoneal dialysis or the use of emetics for this purpose. These include ease of preparation and administration, potentially good efficacy, and minimal patient resistance. Preliminary experiments in this area were described in previous reviews.[11, 46] Basically, the technique is similar to that used in extraction of phenols and acids from wastewater (Fig. 19.4-1), since many drugs are organic acids that are not too different in behavior from phenols. Un-ionized drugs, such as aspirin (acetylsalicylic acid) or barbiturates, diffuse through liquid membranes and are trapped in the internal phase by basic buffers in the same way as phenols. The liquid-membrane emulsion, which has the consistency and appearance of a milk shake can be administered orally, even to comatose patients.[53] Some typical results, illustrating the speed with which liquid membranes can remove such species are given in Fig. 19.4-8. The curves in Fig. 19.4-8 also illustrate the efficacy of a trapping agent (various buffered basic solutions in this case) in enhancing the uptake of the drug. Where no trapping agent was present, only about 30% of the drug could be removed. Similar results have been reported by Frankenfeld et al.[11, 46] and Chang et al.[56] for a variety of barbiturates (Table 19.4-6). Recently, these same investigators[10, 54, 55] have studied the transport of drugs across liquid membranes in order to develop a mathematical model. This work is described in Section 19.3-2.

TABLE 19.4-6 Apparant Partition Coefficients, Dissociation Constants, and Transport Constants for Six Barbiturates

	Dissociation Constant pK_a	Apparent Partition Coefficient[a]	Transport Rate Constant[b] (min^{-1})
Barbital	7.86	0.00	0.037
Phenobarbital	7.37	0.21	0.116
Butabarbital	8.01	0.07	0.188
Pentobarbital	8.03	0.50	0.401
Amobarbital	7.87	0.36	0.662
Secobarbital	7.90	0.83	0.727

[a]Obtained from equilibrium partitioning experiments of the drugs between liquid membranes with no internal trapping agents and the aqueous donor phase.
[b]Obtained from first-order plots with least-squares fitting program; regression coefficients > 0.95.

Source: From Chang et al.[54] with permission.

SLOW RELEASE OF ENZYMES AND DRUGS

Although most liquid-membrane separations involve the movement of substances from the external or donor phase into the encapsulated phase of the liquid-membrane system, the reverse is equally feasible. For many applications this controlled-release mechanism is quite useful. For example, in developing an artificial kidney system, Frankenfeld et al.[11,46] and Asher et al.[57-59] demonstrated the slow release of urease in the gastrointestinal tract from liquid membranes. This was required to supplement the naturally occurring urease in the tract in order to get rapid conversion of urea to ammonia. The ammonia is removed by a separate liquid-membrane system. In developing an effective liquid-membrane formulation, Asher et al.[57-59] made use of the fact that the bile salts and pancreatin present in the small intestine have destabilizing effects on certain membranes. One formulation which was relatively stable in the absence of bile salts and pancreatin released the enzyme at a rate almost four times as great in the presence of these substances as in their absence. Thus, the liquid membrane successfully protects the urease from the deactivating stomach acids but allows LTS slow release in the small intestine, the desired site for urease activity. Early work on these formulations was conducted in vitro[11,46,57] where it was shown that the liquid membrane was stable in the presence of stomach acids but broke down slowly in the presence of bile acids and pancreatin to provide an effective slow release of urease. Recently, successful in vivo experiments[58,59] were carried out which demonstrated the efficacy of the membrane in releasing urease in the intestines of the animals.

Asher et al.[60] also studied the slow release of other medicinals through liquid membranes. The data shown in Table 19.4-7 demonstrates the selective release of up to 64% of encapsulated sodium salicylate from a liquid membrane while 80% of the sucrose solution, encapsulated in the same membrane, was retained.

Brodin et al.[62,63] have studied drug release from oil–water–oil emulsion systems, using naltrexone and thymol as model drugs. They discovered a two-phase release mechanism somewhat similar to that reported by Yang and Rhodes[54] for the *uptake* of drugs by liquid membranes. The use of liquid-membrane systems for the slow release of drugs would seem to be an application deserving of a great deal more attention than it has yet received.

LIQUID MEMBRANES IN THE TREATMENT OF CHRONIC UREMIA

Continued progress is reported in the development of an effective artificial kidney using liquid membranes.[58-60,64,65] The process, conceived at Exxon Research and Engineering Company,[11,46,57-60,64-67] consists of a liquid-membrane system which is administered orally to patients. One emulsion contains encapsulated urease which is released slowly in the gastrointestinal tract where it breaks down urea to ammonia; the second emulsion contains acid and serves as a trap for the ammonia. Both emulsions are eliminated in the stool after use. The process has been reviewed elsewhere.[11,46] Encouraging results both in vitro and in vivo have been obtained recently. The in vivo experiments, using uremic dogs, demonstrated both the palatability and efficacy of the liquid-membrane systems. The concept of the double-emulsion system also was confirmed in that both urease release and ammonia clearance could be demonstrated in a single in vivo experiment.[64,65]

At the present time the outlook for the artificial kidney is as an adjunct to dialysis rather than as a replacement for dialysis altogether. Based on demonstrated ammonia removal in dogs,[65] dialysis time with current formulations could be reduced more than 50%. Work is continuing to develop a better formulation in the hope that dialysis time can be reduced still further. The technology has been licensed by Exxon

TABLE 19.4-7 Controlled Release of Medicinals (Sodium Salicylate) by Diffusion in Liquid Membrane[a]

Time (h)	Concentration in External Phase (%)[b]		Percentage of Maximum Equilibrium Concentration in Outer Phase	
	Sodium Salicylate	Sucrose	Salicylate	Sucrose
Experiment 1				
0	0.0	0.005	0.0	0.5
80	0.83	0.010	64.0	1.0
Experiment 2				
0	0.0	0.008	0.0	1.0
80	0.59	0.013	45.0	1.6

[a]Data from Asher et al.[58]
[b]See Matulevicius and Li.[59]

Research to a Japanese firm for further development.[1] It may become the first commercialized liquid-membrane process.

19.5 CONCLUSIONS

Liquid membranes are effective and versatile tools for performing a large variety of separations. They are particularly attractive for separations where equilibrium considerations make solvent extraction methods difficult or where slow flux rates render polymeric membranes too inefficient. Such applications include wastewater treatment and the recovery of metal ions from dilute solutions. In addition, liquid membranes are ideally suited for a variety of biochemical, biomedical, and oil production applications.

REFERENCES

Section 19.1

19.1-1   E. C. Matulevicius and N. N. Li, *Sep. Purif. Methods*, **4**, 73–96 (1975).

19.1-2   J. W. Frankenfeld and N. N. Li, in P. L. Elving (Ed.), *Treatise on Analytical Chemistry*, 2nd ed. Pt. I, Vol. 5, p. 251, Wiley, New York, 1982.

19.1-3   N. N. Li, U.S. Patent 3,410,794 (1968).

19.1-4   R. P. Cahn and N. N. Li, in P. Mears (Ed.), *Membrane Separation Processes*, Chap. 9, Elsevier, New York, 1976.

19.1-5   N. N. Li, *Ind. Eng. Chem. Proc. Res. Dev.*, **10**, 215 (1971).

19.1-6   R. P. Cahn, J. W. Frankenfeld, N. N. Li, D. Naden, and K. N. Subramanian, in N. N. Li (Ed.), *Recent Developments in Separation Science*, Vol. VI, p. 51, CRC Press, Boca Raton, FL, 1981.

19.1-7   H. C. Hayworth, W. S. Ho, W. A. Burns, Jr., and N. N. Li, *Sep. Sci. Technol.*, **18**, 493 (1983).

19.1-8   J. W. Frankenfeld and N. N. Li, in N. N. Li (Ed.), *Recent Developments in Separation Science*, Vol. II, p. 285, CRC Press, Boca Raton, FL, 1977.

19.1-9   A. M. Hochhauser and E. L. Cussler, *AIChE Symp. Ser.*, **71**, 136 (1975).

19.1-10   K. Kondo, K. Kita, I. Koida, J. Irie, and E. Nakashio, *J. Chem. Eng. Jpn.*, **12**, 203 (1979).

19.1-11   N. N. Li, R. P. Cahn, D. Naden, and R. W. Lai, *Hydrometallurgy*, **9**, 277–305 (1983).

19.1-12   J. Strzelbicki and W. Charewicz, *Hydrometallurgy*, **5**, 243 (1980).

19.1-13   W. Vokel, W. Halwacks, and K. Schugerl, *J. Membr. Sci.*, **6**, 19 (1980).

19.1-14   T. P. Martin and G. A. Davies, *Hydrometallurgy*, **2**, 315 (1977).

19.1-15   E. J. Fuller and N. N. Li, "Metal Extraction by Liquid Membranes," *J. Membr. Sci.*, **18**, 251 (1984).

19.1-16   R. P. Cahn, N. N. Li, and R. M. Minday, *Environ. Sci. Technol.*, **12**, 1051 (1978).

19.1-17   T. Kitagawa, Y. Nishikawa, J. W. Frankenfeld, and N. N. Li, *Environ. Sci. Technol.*, **11**, 602 (1977).

19.1-18   N. N. Li, R. P. Cahn, and A. L. Shrier, U.S. Patent 3,779,907, Dec. 18, 1973.

19.1-19   R. P. Cahn and N. N. Li, *Sep. Sci.*, **9**, 505 (1974).

19.1-20   J. W. Frankenfeld and N. N. Li, in C. Calmon, H. Gold, and R. Prober (Eds.), *Ion Exchange for Pollution Control*, Vol. II, CRC Press, Boca Raton, FL, 1978.

19.1-21   H. H. Downs and N. N. Li, *J. Sep. Proc. Technol. (U.K.)*, **2**, 19–24 (1981).

19.1-22   W. J. Asher, K. C. Bovee, J. W. Frankenfeld, R. W. Hamilton, J. W. Henderson, P. G. Holtzapple, and N. N. Li, *Kidney Int.*, **10**, S-254 (1976).

19.1-23   W. J. Asher, H. W. Wallace, M. T. Zabrow, T. P. Stein, and H. Brooks, "Fluorocarbon Liquid Membranes for Blood Oxygenation," Final Report for Contract No. NIH-NILI-71-2369-D, National Heart and Lung Institute, NIH, 1973.

19.1-24   S. W. May and N. N. Li, *Biochem. Biophys. Res. Commun.*, **47**, 1179 (1972).

19.1-25   C. T. Rhodes, J. W. Frankenfeld, and G. C. Fuller, "Use of Liquid Membrane Technology in the Oral Treatment of Drug Overdoses," Symposium on Separation and Encapsulation by Liquid Membranes, American Chemical Society Centennial Meeting, New York, Apr. 6, 1976.

19.1-26   J. W. Frankenfeld, W. J. Asher, and N. N. Li, in N. N. Li (Ed.), *Recent Developments in Separation Science*, Vol. IV, p. 39, CRC Press, Boca Raton, FL, 1978.

Section 19.2

19.2-1 E. C. Matulevicius and N. N. Li, *Sep. Purif. Methods*, **4**, 73–96 (1975).

19.2-2 J. W. Frankenfeld and N. N. Li, in P. L. Elving (Ed.), *Treatise on Analytical Chemistry*, 2nd ed., Pt. 1, Vol. 5, p. 251, Wiley, New York, 1982.

19.2-3 N. N. Li, U.S. Patent 3,410,794 (1968).

19.2-4 R. P. Cahn, J. W. Frankenfeld, N. N. Li, D. Naden, and K. N. Subramanian, in N. N. Li (Ed.), *Recent Developments in Separation Science*, Vol. VI, p. 51, CRC Press, Boca Raton, FL, 1981.

19.2-5 W. S. Ho, T. A. Hatton, E. N. Lightfoot, and N. N. Li, *AIChE J.*, **28**, 602 (1982).

19.2-6 R. Marr and A. Kopp, *Int. Chem. Eng.*, **22**, 44 (1982).

Section 19.3

19.3-1 E. C. Matulevicius and N. N. Li, *Sep. Purif. Methods*, **4**, 73–96 (1975).

19.3-2 J. W. Frankenfeld and N. N. Li, in P. L. Elving (Ed.), *Treatise on Analytical Chemistry*, 2nd ed., Pt. 1, Vol. 5, p. 251, Wiley, New York, 1982.

19.3-3 N. N. Li, *Ind. Eng. Chem. Prod. Res. Dev.*, **10**, 215 (1971).

19.3-4 T. Kitagawa, Y. Nishikawa, J. W. Frankenfeld, and N. N. Li, *Environ. Sci. Technol.*, **11**, 602 (1977).

19.3-5 W. J. Asher, K. C. Bovee, J. W. Frankenfeld, R. W. Hamilton, J. W. Henderson, P. G. Holtzapple, and N. N. Li, *Kidney Int.*, **10**, S-254 (1976).

19.3-6 H. H. Downs and N. N. Li, *J. Sep. Proc. Technol. (U.K.)*, **2**, 19–24 (1981).

19.3-7 R. P. Cahn, N. N. Li, and R. M. Minday, *Environ. Sci. Technol.*, **12**, 1051 (1978).

19.3-8 R. P. Cahn and N. N. Li, *Sep. Sci.*, **9**, 505 (1974).

19.3-9 R. P. Cahn and N. N. Li, *Sep. Purif. Methods*, **4**, 73 (1975).

19.3-10 N. N. Li and A. L. Shrier, in N. N. Li (Ed.), *Recent Developments in Separation Science*, Vol. I, p. 163, CRC Press, Boca Raton, FL, 1972.

19.3-11 R. E. Terry, N. N. Li, and W. S. Ho, *J. Membr. Sci.*, **10**, 305 (1982).

19.3-12 R. P. Cahn, J. W. Frankenfeld, N. N. Li, D. Naden, and K. N. Subramanian, in N. N. Li (Ed.), *Recent Developments in Separation Science*, Vol. VI, p. 51, CRC Press, Boca Raton, FL, 1981.

19.3-13 J. W. Frankenfeld and N. N. Li, in N. N. Li (Ed.), *Recent Developments in Separation Science*, Vol. II, p. 285, CRC Press, Boca Raton, FL, 1977.

19.3-14 A. M. Hochhauser and E. L. Cussler, *AIChE Symp. Ser.*, **71**, 136 (1975).

19.3-15 N. N. Li, R. P. Cahn, D. Naden, and R. W. Lai, *Hydrometallurgy*, **9**, 277–305 (1983).

19.3-16 J. W. Frankenfeld and N. N. Li, in C. Calmon, H. Gold, and R. Prober (Eds.), *Ion Exchange for Pollution Control*, Vol. II, Chap. 18, CRC Press, Boca Raton, FL, 1978.

19.3-17 R. Marr and A. Kopp, *Int. Chem. Eng.*, **22**, 44 (1982).

19.3-18 T. T. Yang and C. T. Rhodes, *J. Appl. Biochem.*, **2**, 7 (1980).

19.3-19 J. W. Frankenfeld, G. C. Fuller, and C. T. Rhodes, *Drug Dev. Commun.*, **2**, 405 (1976).

19.3-20 W. Halwachs, E. Flaschel, and K. Schügerl, *J. Membr. Sci.*, **6**, 33 (1980).

19.3-21 S. W. May and N. N. Li, *Biochem. Biophys. Res. Commun.*, **47**, 1179 (1972).

19.3-22 P. Stroeve and P. P. Varanasi, *Sep. Purif. Methods*, **11**, 29 (1982).

19.3-23 T. P. Martin and G. A. Davies, *Hydrometallurgy*, **2**, 315 (1977).

19.3-24 P. Stroeve, P. V. Prabodh, T. Elias, and J. S. Ulbrecht, "Stability of Double Emulsion Droplets in Shear Flow," 53rd Annual Meeting of the Society of Rheology, Louisville, KY, Oct. 11–15, 1981.

19.3-25 P. P. Varanasi, P. Stroeve, and J. J. Ulbrecht, "Behavior of Double Emulsions in Shear Flow," First Conference of European Rheologists, Graz, Austria, Apr. 14–16, 1982.

19.3-26 W. S. Ho, T. A. Hatton, E. N. Lightfoot, and N. N. Li, *AIChE J.*, **28**, 602 (1982).

19.3-27 C. T. Rhodes, J. W. Frankenfeld, and G. C. Fuller, "Use of Liquid Membrane Technology in the Oral Treatment of Drug Overdoses," Symposium on Separation and Encapsulation by Liquid Membranes, American Chemical Society Centennial Meeting, New York, Apr. 6, 1976.

19.3-28 R. N. Chilamkurti and C. T. Rhodes, *J. Appl. Biochem.*, **2**, 17 (1980).

19.3-29 A. Panaggio, "An Investigation of Some Factors Controlling Solute Transport Across Liquid Membranes," Ph.D. Thesis, University of Rhode Island, Kingston, 1982.

19.3-30 W. S. Ho and N. N. Li, Modeling of Liquid Membrane Extraction Processes, in R. G. Bautista (Ed.), *Hydrometallurgical Process Fundamentals*, Plenum, New York, in press.

Section 19.4

19.4-1   J. L. Fox, *Chem. Eng. News*, **60**, 7 (Nov. 8, 1982).

19.4-2   J. W. Frankenfeld and N. N. Li, in N. N. Li (Ed.), *Recent Developments in Separation Science*, Vol. II, p. 285, CRC Press, Boca Raton, FL, 1977.

19.4-3   J. W. Frankenfeld and N. N. Li, in C. Calmon, H. Gold, and R. Prober (Eds.), *Ion Exchange for Pollution Control*, Vol. II, Chap. 18, CRC Press, Boca Raton, FL, 1978.

19.4-4   R. P. Cahn and N. N. Li, *Sep. Sci.*, **9**, 505 (1974).

19.4-5   R. P. Cahn and N. N. Li, *Sep. Purif. Methods*, **4**, 73 (1975).

19.4-6   N. N. Li and A. L. Shrier, in N. N. Li (Ed.), *Recent Developments in Separation Science*, Vol. I, p. 163, CRC Press, Boca Raton, FL, 1972.

19.4-7   R. P. Cahn, N. N. Li, and R. M. Minday, *Environ. Sci. Technol.*, **12**, 1051 (1978).

19.4-8   T. Kitagawa, Y. Nishikawa, J. W. Frankenfeld, and N. N. Li, *Environ. Sci. Technol.*, **11**, 602 (1977).

19.4-9   N. N. Li, R. P. Cahn, and A. L. Shrier, U.S. Patent 3,779,907, Dec. 18, 1973.

19.4-10 R. N. Chilamkurti and C. T. Rhodes, *J. Appl. Biochem.*, **2**, 17 (1980).

19.4-11 J. W. Frankenfeld and N. N. Li, in P. L. Elving (Ed.), *Treatise on Analytical Chemistry*, 2nd ed., Pt. 1, Vol. 5, p. 251, Wiley, New York, 1982.

19.4-12 R. R. Mohan and N. N. Li, *Biotechnol. Bioeng.*, **16**, 513 (1974).

19.4-13 R. E. Terry, N. N. Li, and W. S. Ho, *J. Membr. Sci.*, **10**, 305 (1982).

19.4-14 R.-S. Shu, Y.-C. Chang, Y. Liu, C.-Y. Tao, and H.-C. Lu, *Huan Ching Ko Hsueh*, **1**, 4 (1980); *CA*, **94**, 7299d (1981).

19.4-15 S. Schlosser and E. Kossaczky, *J. Membr. Sci.*, **6**, 83 (1980).

19.4-16 W. Halwachs, E. Flaschel, and K. Schügerl, *J. Membr. Sci.*, **6**, 33 (1980).

19.4-17 L. Boyadzhiev and G. Kyuchoukov, *J. Membr. Sci.*, **6**, 107 (1980).

19.4-18 L. Boyadzhiev and E. Bezenshek, *J. Membr. Sci.*, **14**, 13 (1983).

19.4-19 E. J. Fuller and N. N. Li, "Metal Extraction by Liquid Membranes," presented at AIChE National Meeting, Orlando, FL, Feb. 28–Mar. 4, 1982. *J. Membr. Sci.* (to appear).

19.4-20 E. L. Cussler and D. F. Evans, *J. Membr. Sci.*, **6**, 113 (1980).

19.4-21 H. B. Hopfenberg, M. Z. Ward, R. D. Pierson, Jr., and W. J. Koros, *J. Membr. Sci.*, **8**, 91 (1981).

19.4-22 C. F. Reusch and E. L. Cussler, *AIChE J.*, **19**, 736 (1973).

19.4-23 M. Igawa, K. Matsumura, M. Tanaka, and T. Yamabe, *Nippan Kaysku Karshi*, 625 (1981); *CA*, **95**, 49869r (1981).

19.4-24 R. P. Cahn, J. W. Frankenfeld, N. N. Li, D. Naden, and K. N. Subramanian, in N. N. Li (Ed.), *Recent Developments in Separation Science*, Vol. VI, p. 51, CRC Press, Boca Raton, FL, 1981.

19.4-25 A. M. Hochhauser and E. L. Cussler, *AIChE Symp. Ser.*, **71**, 136 (1975).

19.4-26 K. Kondo, K. Kita, I. Koida, J. Irie, and E. Nakashio, *J. Chem. Eng. Jpn.*, **12**, 203 (1979).

19.4-27 N. N. Li, R. P. Cahn, D. Naden, and R. W. Lai, *Hydrometallurgy*, **9**, 277–305 (1983).

19.4-28 J. Strzelbicki and W. Charewicz, *Hydrometallurgy*, **5**, 243 (1980).

19.4-29 W. Vokel, W. Halwachs, and K. Schügerl, *J. Membr. Sci.*, **6**, 19 (1980).

19.4-30 T. P. Martin and G. A. Davies, *Hydrometallurgy*, **2**, 315 (1977).

19.4-31 R. P. Cahn and N. N. Li, U.S. Patent 4,086,163 (1978).

19.4-32 K.-H. Lee, D. F. Evans, and E. L. Cussler, *AIChE J.*, **24**, 860 (1978).

19.4-33 H. C. Hayworth, W. S. Ho, W. A. Burns, Jr., and N. N. Li, *Sep. Sci. Technol.*, **18**, 493 (1983).

19.4-34 W. S. Ho and N. N. Li, Modeling of Liquid Membrane Extraction Processes, in R. G. Bautista (Ed.), *Hydrometallurgical Process Fundamentals*, Plenum, New York, in press.

19.4-35 J. Strzelbicki and W. Charewicz, *Chem. Stosow.*, **24**, 239 (1980); *CA*, **94**, 37209s (1981).

19.4-36 W. M. Salathiel, T. W. Muecki, C. E. Cooke, Jr., and N. N. Li, U.S. Patent 4,233,265, Nov. 11, 1980.

19.4-37 R. Marr and A. Kopp, *Int. Chem. Eng.*, **22**, 44 (1982).

19.4-38 F. J. Hurst, D. J. Crouse, and K. B. Brown, *Ind. Eng. Chem. Prod. Res. Dev.*, **11**, 123 (1972).

19.4-39 J. Bock, R. R. Klein, R. L. Valint, and W. S. Ho, Presentation at AIChE Annual Meeting, New Orleans, LA, Nov. 8–12, 1981.

19.4-40 W. M. Salathiel, T. W. Muecki, C. E. Cooke, Jr., and N. N. Li, U.S. Patent 4,359,391, Nov. 16, 1982.

19.4-41 C. R. Dawson and N. N. Li, U.S. Patent 4,397,354 (1983).

19.4-42 C. R. Dawson and N. N. Li, U.S. Patent 4,568,392 (1986).

19.4-43 D. F. Ollis, J. B. Thompson, and E. T. Wolynic, *AIChE J.*, **18**, 457 (1972).

19.4-44 E. T. Wolynic and D. F. Ollis, *Chemtech*, **111** (Feb. 1974).

19.4-45 N. N. Li and W. J. Asher, Blood Oxygenation by Liquid Membrane Permeation, in *Chemical Engineering in Medicine* p. 1, American Chemical Society, Washington, DC, 1973.

19.4-46 J. W. Frankenfeld, W. J. Asher, and N. N. Li, in N. N. Li (Ed.), *Recent Developments in Separation Science*, Vol. IV, p. 39, CRC Press, Boca Raton, FL, 1978.

19.4-47 T. H. Maugh II, *Science*, **193**, 134 (1976).

19.4-48 J. W. Frankenfeld, R. P. Cahn, and N. N. Li, *Sep. Sci. Technol.*, **16**, 385 (1981).

19.4-49 T. A. Davis and W. J. Asher, U.S. Patent 4,376,059 (1981).

19.4-50 T. A. Davis, W. J. Asher, and H. W. Wallace, *Trans. Am. Soc. Artif. Intern. Organs*, **28**, 404 (1982).

19.4-51 W. Halwachs, W. Vokel, and K. Schügerl, *Proc. Int. Solv. Ext. Conf.*, paper 80-88 Liege, Belgium (1980).

19.4-52 W. Halwachs, W. Vokel, G. Brunner, and K. Schügerl, *Proc. Int. Symp.*, *Artif. Liver Support*, 219 (1981); *CA*, **95**, 138561b (1981).

19.4-53 C. T. Rhodes, J. W. Frankenfeld, and G. C. Fuller, "Use of Liquid Membrane Technology in the Oral Treatment of Drug Overdoses," Symposium on Separation and Encapsulation by Liquid Membranes, American Chemical Society Centennial Meeting, New York, Apr. 6, 1976.

19.4-54 T. T. Yang and C. T. Rhodes, *J. Appl. Biochem.*, **2**, 7 (1980).

19.4-55 A. Panaggio, "An Investigation of Some Factors Controlling Solute Transport Across Liquid Membranes," Ph.D. Thesis, University of Rhode Island, Kingston, 1982.

19.4-56 C. W. Chang, G. C. Fuller, J. W. Frankenfeld, and C. T. Rhodes, *J. Pharm. Sci.*, **67**, 63 (1978).

19.4-57 W. J. Asher, D. V. M. Bover, and R. W. Hamilton, "Liquid Membrane Capsule System for the Treatment of Chronic Uremia," Final Report for Contract No. 1-AM-3-2224, NIH, Feb. 1978.

19.4-58 W. J. Asher, K. C. Bovee, T. C. Vogler, R. W. Hamilton, and P. G. Holtzapple, *Trans. Am. Soc. Artif. Intern. Organs*, **26**, 120 (1980).

19.4-59 W. J. Asher and T. C. Vogler, U.S. Patent 4,183,960 (1980).

19.4-60 W. J. Asher, N. N. Li, and A. J., Shrier, U.S. Patent, 4,183,918 (1980).

19.4-61 E. C. Matulevicius and N. N. Li, *Sep. Purif. Methods*, **4**, 73–96 (1975).

19.4-62 A. F. Brodin, S. G. Frank, and D. R. Kavaliunas, *Acta Pharm. Suec.*, **15**, 1 (1978).

19.4-63 A. F. Brodin and S. G. Frank, *Acta Pharm. Suec.*, **15**, 111 (1978).

19.4-64 W. J. Asher, T. C. Vogler, K. C. Bovee, P. G. Holtzapple, and R. W. Hamilton, *Trans. Am. Soc. Artif. Intern. Organs*, **23**, 73 (1977).

19.4-65 W. J. Asher, K. C. Bovee, T. C. Vogler, R. W. Hamilton, and P. G. Holtzapple, *Clin. Nephrology*, **11**, 92 (1979).

19.4-66 W. J. Asher, K. C. Bovee, J. W. Frankenfeld, R. W. Hamilton, J. W. Henderson, P. G. Holtzapple, and N. N. Li, *Kidney Int.*, **10**, S-254 (1976).

19.4-67 W. J. Asher, H. W. Wallace, M. T. Zabrow, T. P. Stein, and H. Brooks, "Fluorocarbon Liquid Membranes for Blood Oxygenation," Final Report for Contract No. NIH-NILI-71-2369-D, National Heart and Lung Institute, NIH, 1973.

19.4-68 J. W. Frankenfeld, G. C. Fuller, and C. T. Rhodes, *Drug. Dev. Commun.*, **2**, 405 (1976).

CHAPTER 20

Separation of Gaseous Mixtures Using Polymer Membranes

WILLIAM J. KOROS
Department of Chemical Engineering
The University of Texas
Austin, Texas

REY T. CHERN
Department of Chemical Engineering
North Carolina State University
Raleigh, North Carolina

20.1 INTRODUCTION

20.1-1 General Overview

The energy efficiency and simplicity of membrane separation devices make them extremely attractive for solution of fluid-phase separation problems. The ability of ideal membrane processes to pass selectively one component in a mixture, while rejecting others in a continuous steady-state manner, defines the perfect separation device. A substantial literature related to gas separation indicates, however, that numerous factors must be considered in achieving a commercially successful membrane-based process. Nevertheless, recently membranes have been shown to act as remarkably effective separating devices.[1-12] By clever engineering, the productivities of these devices have been increased dramatically, and they are strongly competitive with—if not superior to—more traditional chemical engineering separation approaches for several applications involving gas treatment.

The historical development of membrane-based gas separation is traced briefly, and a discussion of current membrane types is presented along with a projection of requirements for the next generation of membrane materials. An overview of the fundamental principles governing the operation of gas separation membranes also is presented. Finally, a review of the state of knowledge of membrane module construction and operating configurations is presented in conjunction with a guide to the mathematical modeling of these devices. The coverage offered here should be useful to both the general reader interested in separation processes and to practicing engineers interested in understanding and evaluating membrane processes for potential application in new situations. The present discussion does not treat the topic of liquid membranes that have potential for gas separations, since this topic is treated in Chapter 19.

20.1-2 Historical Background

In 1831, Mitchell reported that india rubber membranes passed carbon dioxide substantially faster than hydrogen under equivalent conditions.[13] Mitchell's work marked the first known report of gas permselectivity of a membrane. The understanding of mass diffusion as a process driven by a concentration gradient gave rise to Eq. (20.1-1), Fick's First Law for the diffusive flux N:[14]

$$N = -D \nabla C \qquad (20.1\text{-}1)$$

where D is the diffusion coefficient, analogous to the thermal conductivity in Fourier's Law of heat conduction, and C refers to the local penetrant concentration. For simple, one-dimensional diffusion through a flat membrane, Eq. (20.1-1) can be written as

$$N = -D \frac{\partial C}{\partial x} \qquad (20.1\text{-}2)$$

In 1866, Graham made the next major step in understanding the permeation process.[15] He postulated that the permeation process in polymers involved a solution-diffusion mechanism by which the penetrant first was dissolved in the membrane and then transported through it by the same process as that occurring in the diffusion of liquids. He devised a membrane-testing device and demonstrated that atmospheric air could be enriched from 21 to 41% oxygen using a natural rubber membrane. Moreover, he showed that increasing the thickness of a pinhole-free membrane reduces the rate of permeation of both components through the membrane, but does not affect its ability to act as a permselective separator for the two components.[15] Exner showed that the permeation rate of a penetrant through a soap film is proportional to the product of the solubility and diffusivity of the penetrant in the film.[16] von Wroblewski provided a mathematical analysis of the process of permeation through polymer films that incorporated the observations of these earlier workers.[17] He defined a coefficient, the permeability, equal to the observed steady-state flux divided by the driving pressure, $\Delta p = p_2 - p_1$, across the membrane normalized by the membrane thickness l; that is,

$$P \equiv \frac{N}{\Delta p / l} \qquad (20.1\text{-}3)$$

As shown in Fig. 20.1-1, the upstream condition at $x = 0$ is distinguished by a subscript "2" and the downstream condition at $x = l$ is distinguished by a subscript "1" in all cases in the following discussion. This convention ensures that the flux occurs in the positive x direction.

If the dissolved permeant concentration C obeys Henry's Law [Eq. (20.1-4)], there is a simple constant solubility coefficient S relating C to the external penetrant pressure p; that is,

$$C = Sp \qquad (20.1\text{-}4)$$

Furthermore, if Fick's Law [Eq. (20.1-2)] with a constant diffusion coefficient applies to the diffusion process, von Wroblewski showed that the permeability P is equal to the product of the solubility and diffusivity coefficients; that is,

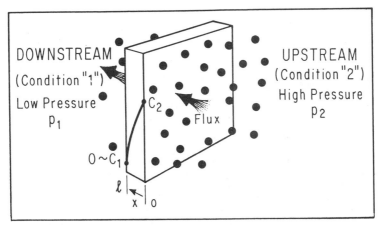

FIGURE 20.1-1 Terminology and coordinate system convention for one-dimensional permeation through a membrane of thickness l.

$$P = \frac{N}{\Delta p/l} = -\frac{D(dC/dx)}{\Delta p/l} = \frac{D(C_2 - C_1)}{\Delta p/l} = DS \qquad (20.1\text{-}5)$$

An important parameter, referred to as the separation factor of the membrane for component A relative to B is defined by Eq. (20.1-6):[18]

$$\alpha_{AB} = \frac{[Y_{1A}/Y_{1B}]}{[Y_{2A}/Y_{2B}]} \qquad (20.1\text{-}6)$$

where the Y_1's and Y_2's refer to the mole fractions of components A and B in the downstream (low-pressure) product and the upstream (high-pressure) feed streams, respectively. When the downstream pressure is negligible compared to that upstream, one can substitute Eqs. (20.1-2) and (20.1-3) into Eq. (20.1-6) to give Eq. (20.1-7):[18]

$$\alpha_{AB}^* = \frac{P_A}{P_B} \qquad (20.1\text{-}7)$$

The above ratio of permeabilities is referred to as the ideal separation factor and provides a useful measure of the intrinsic ability of the membrane to separate a given gas mixture of A and B into its components.

Shakespear demonstrated that the permeability of a gas is independent of the presence of other permeating gases for rubbery polymers at low pressures.[19] This result has been verified by Yi-Yan et al.[20] The assumption of strict noninteraction between permeating components does not hold in general, however, for rubbery polymers at high pressures[21] or for glassy polymer membranes even at relatively low penetrant pressures.[22-29] While the deviations from such an assumption generally tend to be small, as low as a 15 mm Hg partial pressure of water vapor has been reported to cause nearly a 60% reduction in flux of both members of an H_2–CH_4 mixture.[27] Similar but less dramatic results, in the range of 20% flux depression, also have been observed because of the presence of hydrocarbons in feed streams.[22]

Conversely, at higher relative humidities or partial pressures, Stern and coworkers have shown that plasticization of rubbery polymers can produce significant increases in the permeability of mixture components.[30] A recent report also suggests that methane solubility (and perhaps permeability) can be increased owing to the presence of a second component such as carbon dioxide.[31] One therefore might be surprised to find Shakespear's observation supported in actual practice. Because of the current uncertainty in predicting the exact magnitude of such interactive effects, however, a retreat to Shakespear's constant permeability assumption often is made. This issue is treated in more detail in a later section on fundamentals. Cussler has considered an even more subtle general situation in which coupling of the fluxes of the first component to that of the second occurs.[23] These effects, while obviously correct and significant for liquid systems, usually are neglected for gases in polymers.

Two serious limitations faced by pioneers in gas separation were the low selectivities and the rather low permeation fluxes observed for most membranes. The low-flux problem arose because membranes had to be thick (at least 1 mil) to avoid pinholes, which destroyed selectivity as a result of almost indiscriminate passage of all the feed components by Knudsen or viscous flow.

Discovery of the extraordinarily high permeability of silicone polymers (see Table 20.1-1 for comparisons to other typical rubbery and glassy polymers) spurred a renewed interest in gas separation aimed at O_2–N_2 separation from air.[33, 34] Unfortunately, the O_2–N_2 selectivity of silicones is rather low. Moreover, even with silicone polymers 1 mil thick, very large membrane areas were required. Stern showed that membrane separations were not competitive with cryogenic processes using 1 mil silicone rubber membranes even if the feed air was compressed considerably to reduce installed membrane cost.[36]

The development of casting techniques for ultrathin membranes comprised of 150 Å silicone rubber-polycarbonate copolymers permitted formation of pinhole-free selective membranes with thicknesses of around 1000 Å by laminating multiple layers.[37] Such membranes offered flux increases of as much as 50- to 100-fold compared to their 1 mil silicone rubber counterparts. The membranes were supported typically in a plate-and-frame fashion on a porous substrate, so that accommodation of a large membrane area in a small module was not feasible.

To overcome this problem, the so-called hollow-fiber and spiral-wound membrane configurations described below have been adopted in the gas separation field (as in the reverse osmosis and ultrafiltration fields). Permeation area per unit of separator volume, for example, can be increased roughly 30-fold simply by replacing plate-and-frame flat membranes with 200 μm outside diameter (OD) fibers packed at a 50% void factor. The flat membrane configuration typically provides areas of about 300 ft^2/ft^3 of module volume and the 200 μm diameter fibers give more than 10,000 ft^2/ft^3 of module volume.[37] Clearly, even higher areas per unit volume can be achieved using smaller diameter fibers and higher packing densities. Studies have been reported using low-flux polyester fibers of 36 μm inside diameter (ID) fibers.[28, 38] For the high-flux fiber currently in use, however, dimensions in the range of 100–600 μm ID appear to be favored on the basis of the information available in the open literature.[39-43] Consideration of optimum fiber diameter choice is dealt with in greater depth in the section on design case studies.

TABLE 20.1-1 Comparison of Permeabilities of Important Product Gases in Typical Rubbery and Glassy Polymers at 25°–30°C and 1 atm

Polymer	Repeat Structure	$P^a_{H_2}$	$P^a_{CO_2}$	$P^a_{O_2}$
	Rubbery Polymers			
Silicone rubber	$\begin{array}{c} CH_3 \\ \mid \\ -Si-O- \\ \mid \\ CH_3 \end{array}$	550	2700	500
Natural rubber	$\begin{array}{c} -CH_2-C=CH-CH_2- \\ \mid \\ CH_3 \end{array}$	49	131	24
Polychloroprene (Neoprene®)	$\begin{array}{c} -CH_2-C=CH-CH-CH_2- \\ \mid \\ Cl \end{array}$	20	22	4
	Glassy Polymers			
Poly(ethylene terephthalate) (Mylar®)	$\begin{array}{c} O \quad\quad O \\ \parallel \quad\quad \parallel \\ -C-\bigcirc-C-O-CH_2-CH_2-O- \end{array}$	0.6	0.10	0.03
Aromatic polyether diimide (Kapton®)	$-N\overset{\overset{O}{\parallel}}{\underset{\underset{O}{\parallel}}{\overset{C}{\underset{C}{\bigcirc}}}}N-\bigcirc-O-\bigcirc-$	1.5	0.27	0.15
Polycarbonate (Lexan®)	$\begin{array}{c} CH_3 \quad\quad O \\ \mid \quad\quad\quad \parallel \\ -O-\bigcirc-C-\bigcirc-O-C- \\ \mid \\ CH_3 \end{array}$	12	5.6	1.4

[a] Units of permeability are Barrers. 1 Barrer = 10^{-10} (cm³(STP)·cm)/(cm²·s·cm Hg).
Source: Stannett et al.[35]

    Hollow-fiber technology as a means of producing large amounts of permeation area grew out of a coupled program dealing with gas separation, artificial kidney, and desalination work at the Dow Chemical Company in the 1960s.[44] The first mention of hollow-fiber membranes appears to have occurred in a series of patents in 1966.[45] The patents describe cylindrical modules with inlet feed on the side of the module as shown in Fig. 20.1-2. Small hollow-fiber bundles were sealed with an adhesive into header plates at opposite ends of the module in such a way that a large number of fibers could be included. Only the exterior of the fibers was epoxied at the two ends, so both open fiber ends communicated with the two collection chambers at the opposite ends of the module. A feed stream entering at the side moved axially down past the fiber bundles. The residual stream was removed from the exhaust port at the opposite end from the inlet port. The permeate diffused through the fiber walls and moved along the inside of the fiber bores to be removed from the collection chambers at the two ends of the module as shown in Fig. 20.1-2. The current Dow gas separation module undoubtedly has evolved considerably since this early design but it still is based on a hollow-fiber concept.

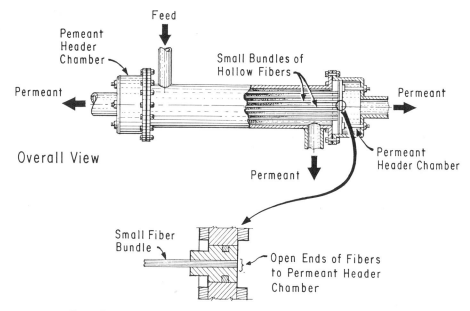

End Seal Unit for Typical Small Hollow Fiber Bundles

FIGURE 20.1-2 Early Dow hollow-fiber permeation device. Note that permeant is withdrawn from both ends.

A second hollow-fiber patent, issued in 1970 to the E. I. Du Pont de Nemours & Co. for a reverse osmosis module, employed a U-shaped loop arrangement of the fiber bundle as shown in Fig. 20.1-3.[46] Several small changes in the design have occurred since then. Both ends of the fibers terminate in the header plate separating the sample collection chamber from the main body of the permeator. The looped end of the fiber bundle is held in an epoxy "deflector block." A cylindrical flow screen confines the fiber bundle and forms an annular region between the outer shell and the sleeve for flow of rejected brine. Fluid feed enters at the center through a perforated tube, where it then passes through numerous perforations in the tube, and flows essentially radially outward. The nonpermeant is collected in the annular channel between the flow screen and flows to the brine header. Permeant diffuses through the fiber walls and is transferred down the length of the bore to the collection chamber. The crossflow effect, arising from the flow arrangement, causes movement perpendicular to the axis of the hollow fiber to minimize blocked flow paths, thereby maximizing the use of the fiber surface. This basic module design also should be useful for gas separation applications based on high-flux membranes.

The Monsanto hollow-fiber module for gas separation appears to be a compromise between the two preceding reverse osmosis module designs.[47] As shown in Fig. 20.1-4, the hollow fibers are closed at one end with an epoxy plug or similar device, and feed enters at the side at one end. The feed then flows axially along the fibers, and the nonpermeant exits at the opposite end from which the feed enters. Permeant diffusing through the hollow-fiber wall into the bore moves countercurrently to the shell-side flow and ultimately is collected in the chamber where all the open ends of the fibers terminate.

The so-called spiral-wound module currently used by Separex, Delta Engineering, and Envirogenics dates back to 1968, when a reverse osmosis module was patented by Gulf General Atomics.[48–50] The essence of the spiral-wound concept is illustrated in Fig. 20.1-5. A sandwich consisting of a porous backing material (e.g., Dacron® felt) placed between the two halves of a folded continuous membrane is assembled with a separator grid (e.g., polypropylene screen) on the top membrane half. The porous backing material is wrapped around the perforated mandrel to provide an easy pathway for gas to the mandrel holes. Then the sandwich is sealed along its edges with a suitable epoxy at three sides. The fourth side terminates at the central perforated mandrel. Finally, the assembly is wound spirally and loaded into a suitable cylindrical chamber such as that shown in Fig. 20.1-5. A feed stream enters at one end and moves through the passageways provided by the polypropylene grid separator. The nonpermeant then exits at the opposite end of the module from where it entered. In the separator, permeant diffuses across the two membranes in the spirally wound sandwich and flows in the porous backing material inwardly toward the open end of the sandwich terminating at the perforated mandrel. The permeant then accumulates in the central collection tube and is delivered as product. As a means of incorporating more area in a compact volume, multileaved cartridges can be used without requiring excessively long flow paths of permeant to the collection tube.[51]

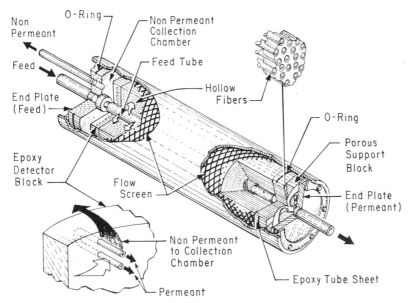

FIGURE 20.1-3 Sketch of a Permasep® hollow-fiber reverse osmosis module. Courtesy of E. I. Du Pont Co.

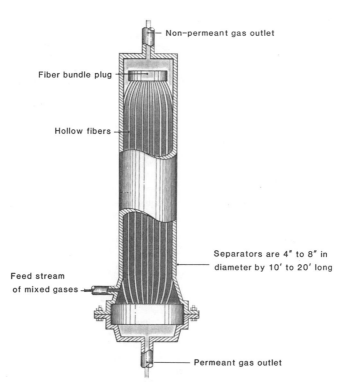

FIGURE 20.1-4 Sketch of Monsanto's Prism® hollow-fiber separator module.

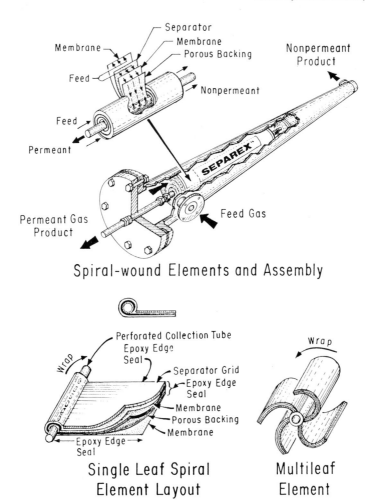

Spiral-wound Elements and Assembly

Single Leaf Spiral Element Layout

Multileaf Element

FIGURE 20.1-5 Spiral-wound elements and assembly. The use of multileaf element avoids long pathways for permeant through the porous backing material to minimize pressure buildup in the permeant channel.

20.1-3 Asymmetric Membranes

The development of asymmetric membranes marked another breakthrough in the drive to increase module productivity. These membranes permitted the formulation of dry, nearly pore-free membranes with selective layers that were only 1000 Å, rather than 1 mil, thereby increasing flux by a factor of more than 200. Coupling the higher permeation area density with the much smaller membrane resistances gives the possibility of flux densities that are on the order of 10,000 times higher than those available in the early days of gas separation in the 1960s.

Typical asymmetric membrane morphologies are shown schematically in Fig. 20.1-6. The dense skin on the outside of the membrane is supported by the porous substructure. The same general form can be obtained for flat and hollow-fiber asymmetric membranes. Although the technology of producing asymmetric structures is reasonably well developed, there is still considerable discussion about the detailed fundamental processes involved. The work of Loeb, Sourirajan, Kesting, Strathmann, Cabasso, Smolders, and others has been important in establishing and interpreting protocols for the formation of asymmetric membranes.[52-62]

Generally, it is agreed that two fundamentally different mechanisms of phase separation can occur, as shown in Fig. 20.1-7. The so-called binodal regions in the two-phase envelope mark metastable regimes in which a single phase can exist until a nucleus (homogeneous or heterogeneous) presents itself and initiates a precipitation process.[59] This process is similar in many respects to classical crystallization from

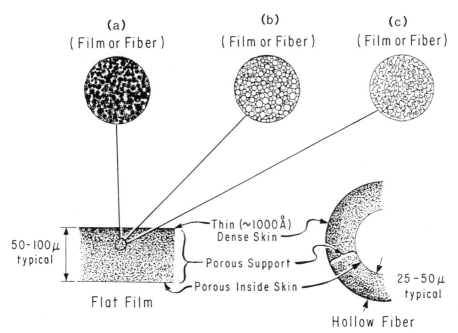

FIGURE 20.1-6 Schematic representation of an asymmetric membrane in flat-film and hollow-fiber form. The dark regions signify polymeric domains and the white regions signify nonpolymer-containing void regions. The designations (a), (b), and (c) refer to the regions shown in Fig. 20.1-7.

solution. The second regime is characterized by absolute instability, and systems lying in this so-called spinodal region undergo spontaneous phase separation without the need for the presence of a nucleus. The process is termed spinodal decomposition.[63] The different properties of the two regimes are summarized below based on the current literature.

Two types of membrane morphology can be envisioned to arise from phase separation due to *nucleation behavior* [regions (a) and (b) in Fig. 20.1-7]. In some cases, precipitation conditions will create a local effective concentration at a point in the coalescing membrane that falls in region (a) in Fig. 20.1-7. This situation produces a concentrated polymer solution, nucleating and growing as the dispersed phase in this local region of the developing membrane. In such cases, a latexlike suspension of polymer spheroids will form.[59,64] Upon further coalescence, the spherical structures will tend to meld, leaving interstitial pores between the basically spherical structures such as those shown in Fig. 20.1-6a. In such a case, the dense skin presumably corresponds to a more coalesced form of these small particles resulting from their more intense contact with the external coagulating nonsolvent bath due to the absence of diffusional limitations at the outside surface. Hoehn and coworkers suggest that the morphology in such cases may not in fact comprise a simple, well-defined surface skin supported by a highly open porous substructure. Instead, it is possible that a steady gradation exists with an open-cell support structure yielding gradually to an essentially pore-free surface region.[64]

On the contrary, in the second binodal region (b) in Fig. 20.1-7, where the local mixing point finds the polymer-rich solution as the continuous phase, dispersed spheroids of nearly polymer-free fluid are nucleated. These domains then coalesce to produce a foam structure whose walls are composed of the solidified dispersed polymer phase. To obtain an open-cell foam with low resistance to flow, defects clearly must occur in the walls of the cells.[59] Such a structure is shown in Fig. 20.1-6b. The dense film on the surface can be promoted by a brief exposure of the cast or spun nascent membrane to air to obtain a more concentrated region at the surface prior to immersion in the nonsolvent precipitation bath, which then sets the dense layer in place and proceeds to nucleate the substructure as described above. This evaporation step, however, is not required in all cases to produce acceptable skins.[56,65,66]

The third regime of precipitation occurs in the central area (c) of Fig. 20.1-7 and is more difficult to achieve. The tendency for nucleation and growth mechanisms to occur while one is attempting to establish a local mixing point in the spinodal region by diffusive exchange of solvent and nonsolvent is a practical problem that must be acknowledged.[59] Nevertheless, it has been suggested that such processes can produce excellent uniformity, highly open-cell foam support structures, which are the most suitable of the three shown (see Fig. 20.1-6c) for asymmetric membranes.[59,65]

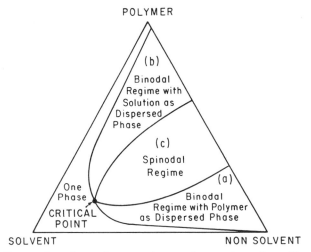

FIGURE 20.1-7 Ternary-phase diagram between polymer, solvent, and nonsolvent. The three regions (*a*), (*b*), and (*c*) refer to the two-phase regimes for which precipitation of polymer occurs with the potential for membrane formation. If the local concentration conditions at a point in a precipitating membrane correspond to one of these domains, the resultant local morphology in the membrane at that point will resemble sketches (*a*), (*b*), or (*c*) in Fig. 20.1-6, respectively.

Whichever membrane morphology is desired, it is clear that the science of how to achieve the morphology trails the technology in the field. As a result, methodical trial-and-error optimization of solvent, nonsolvent, special additives, and spin dope conditions must be performed to produce desirable membrane morphology. Actual asymmetric membrane structures tend not to have all one type or the other of the morphologies shown in Fig. 20.1-6. Instead, depending on the conditions persisting at each point in the membrane at the instant when precipitation occurs, a complex graded structure with aspects of all the forms shown in Fig. 20.1-6 may be observed. Posttreatments by heating can help heal large pores in some cases, but the ideal situation is clearly to produce, by means of manipulation of spinning conditions, an open-cell support structure with an *essentially* pore-free skin.[67] The use of selected additives such as poly(vinylpyrrolidone) in the case of polysulfone spinning was found to be useful in producing good fiber skins by causing subtle alterations in the spin dope viscosity and phase relationship between polysulfone and the solvent DMF.[68]

20.1-4 Actual Membrane Properties

The words "essentially pore free" are extremely restrictive. Pores of even 10–50 Å present at an area fraction of even 10 ppm can destroy the permselectivity of a membrane which must discriminate gas molecules with characteristic dimensions of 3–5 Å. A significant advance in technology occurred when Monsanto introduced a process to posttreat asymmetric fibers with a minuscule amount of highly permeable "stop leak" material such as silicone rubber. This treatment eliminates most of the Knudsen and viscous contributions to flow through the membrane and produces a high-flux membrane with selectivity approaching that of true dense films.[69] As demonstrated in the Monsanto patent on this matter, however, the coating appears to be less than perfect, since dense-film selectivities typically are not achieved but may be approached quite closely. Data from applications of this approach are shown in Table 20.1-2 for coated and uncoated fibers.[70]

A simple analysis, known as the resistance model, has been suggested as a framework for interpreting the various flux-reducing impedances that gas molecules encounter in moving across a composite asymmetric membrane.[69] The treatment is couched in terms of an electrical resistance analogue. The series resistances to flow arising from permeation through the elastomeric coating, permeation through the effective dense selective layer, and ultimately through the open-cell foam support are taken into account. While highly idealized, the model is useful for semiquantitative analysis of such composite membranes. A practical difficulty in application of this approach lies in a priori assignment of the effective layer thicknesses of the elastomeric and dense selective resistances. The Monsanto composite-membrane approach is marketed commercially under the name Prism®, using an asymmetric polysulfone membrane treated with silicone rubber. The modules have proved effective and reliable in both H_2 and CO_2 separation applications.[2, 7]

Cellulose acetate and modified cellulose acetates of various degrees of acetylation comprise a second

TABLE 20.1-2   Extent of Achievement of Dense-Film Ideal Separation Factors Using Coated High-Flux Asymmetric Polysulfone Fibers

Membrane Type	Ideal Separation Factor	
	P_{O_2}/P_{N_2}	P_{H_2}/P_{CO}
Dense	6	40
Coated asymmetric	4.1	31
Achievement of dense film separation factor using coated asymmetric film	68%	78%

Source: Henis and Tripodi.[70]

major commercial polymer used by a number of companies in the field. Dow, Separex, Delta Engineering, and Envirogenics are involved in either hollow-fiber (Dow) or spiral-wound module approaches (Separex, Delta, and Envirogenics) to membrane packaging using these polymers.[4-6,71] Little detailed information is available concerning the performance of cellulose acetate (CA) and cellulose acetate–cellulose triacetate (CTA) blends in actual gas separation applications, but it appears to be roughly equivalent to polysulfone on the basis of the available literature.[9-12] Recent reports suggest a greater tendency toward plasticization of cellulose acetate materials by CO_2 than for polysulfone under equivalent temperature and CO_2 partial pressure conditions.[71] The detailed membrane formation and posttreatments for CA are proprietary. It is believed, however, that the posttreatments are similar to those employed by Gantzel and Merten[72] and Lonsdale.[73] Gantzel and Merten used surfactants to reduce the interfacial tension between the pore walls of the membrane and the water in the pores to permit evaporation of the water from the wet membrane without having the pores collapse. Stagewise solvent exchange to alter gradually the interfacial tension environment in the pores was used by Gantzel and Merten to produce a cellulose acetate membrane with a good ability to separate nitrogen and hydrogen at high flux rates of hydrogen.[72,73]

Du Pont, a leader in reverse osmosis technology built around a unique class of tailored aromatic polyamides, was also an early leader in the gas separation field.[27,28,74,75] Molecularly engineered aromatic polyimides were found by Du Pont to provide extraordinarily good flux and selectivity properties for hydrogen separations.[27] Posttreatment processes for these membranes were not reported.

Many other national and international companies, some with extensive expertise in liquid separations as well as others with primarily chemical and petrochemical histories, are either closely monitoring the field or actively engaged in research aimed at tapping its exciting potential for growth. An interesting alternative approach to membrane formation has been the basis for reverse osmosis membranes marketed by Film Tech, UOP, and Albany International. These composite membranes are essentially the mirror images of the Monsanto composites.

In these alternative cases the membrane support is highly porous, and a thin, selective coating is formed on the porous support by interfacial polymerization.[35,73,76] The patented interfacial polymerization process involves the three steps illustrated in Fig. 20.1-8: (1) imbibing a low-molecular-weight prepolymer or a monomer of type A (e.g., trimesyl chloride) into the pores of a highly porous membrane support such as polysulfone, (2) contacting only the outer film surface with monomer B (e.g., piperazine or an aromatic diamine), and (3) follow with a final heat-curing step. As the thin layer of dense, high-polymer film (~ 500–2500 Å) forms, it restricts further invasion of monomer B, thereby limiting the coating thickness. The technique tends to guarantee a pore-free membrane, because monomer B is convected through any pores and reacts with monomer A to seal the orifice. As noted above, this approach is the reverse of the Monsanto "stop leak" technique since the coating in the interfacial polymerized case forms the actual separating layer, and the highly porous sublayer acts simply as a support.

The interfacial polymerization approach is clearly attractive, since a selective coating can be formed from a wide range of monomers displaying a tremendous spectrum of physical and chemical properties. Currently, no commercial gas separation membranes are available that use this technique. The durability of such composite membranes is reported to be quite good in reverse osmosis applications where they first were developed.[77] Applications in gas separation service involve rather moderate sorption levels, so swelling stresses likely to dislodge such coatings should be low; however, in high-pressure CO_2 service, some difficulties may be encountered. Nevertheless, this interesting approach is likely to see more emphasis in the future.

Membrane processes, like other unit operations, should be approached with a systematic design procedure supported by a solid data base. Because of their novelty, the data base for membrane processes is still much smaller than that for corresponding older unit operations such as distillation. The current lack of information often necessitates the estimation and use of constant values for component permeabilities although these coefficients are known to be functions of pressure and gas composition in many cases.[28,29]

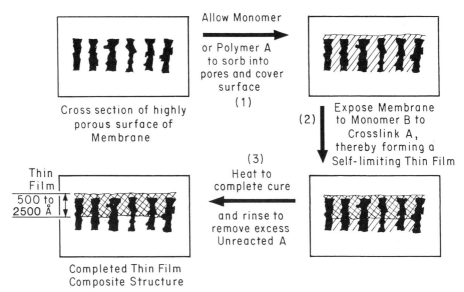

FIGURE 20.1-8 Schematic representation of thin-film composite-type films pioneered by Cadotte for reverse osmosis application. Typical monomers might include for A, mesyl chloride; for B, piperazine or 1,4 phenyldiamine.

As information and working experience with gas separation membranes broaden, analysis undoubtedly will mature also. The incorporation of pressure and composition dependency of permeabilities in permeator modeling is somewhat analogous to the refinements achieved through the use of activity coefficients in traditional vapor–liquid and liquid–liquid operations. These effects can be treated either "phenomenologically" (i.e., empirically), or more fundamentally in terms of physical explanations for the nonideality.

Perhaps the most important by-product benefit from fundamental understanding of the processes involved in membrane systems lies in the area of new membrane development. Analogous to solvent selection in absorption and extraction processes, membrane selection lies at the heart of permselection processes. Therefore, discussions and interpretations of the forms and interpretations of sorption, diffusion, and permeability coefficients are offered in subsequent sections. First, however, to demonstrate the importance of these sections, an overview of general design procedures and applications is given. The final topic to be discussed involves the detailed engineering and mathematical analysis required to arrive at an actual module design. The immediately following general overview section also provides a somewhat qualitative preview of this involved topic.

20.2 GENERAL DESIGN PROCEDURES AND APPLICATION EXAMPLES

20.2-1 Procedures

The following phases of the design process typically are observed in the development of a suitable membrane-based gas separation system. The general elements are similar to those for other more traditional separation processes.

1. *Prepare Flow Diagrams.*   A preliminary schematic of the proposed process option being considered is prepared. As many variables as possible are specified on the diagram (temperatures, pressures, flows, etc.). Required outlet purities and key component recoveries are used to establish material balance constraints where possible.

2. *Acquire Basic Data.*   Because of the novelty of membrane-based gas separations, a scarcity of important design data often is encountered. Some tabulations do exist, for example, Table 20.1-1 and Tables 20.4-1 and 20.4-2, which are discussed later. In addition, membrane manufacturers can be of help for common systems. If one is considering unusual systems or novel membranes, some experimental determination of permeabilities is likely to be necessary. Procedures and equipment for such measurements are described in later sections.

3. *Perform Detailed Design Calculations.*   This subject has been dealt with in varying degrees of complexity by several authors. The first treatment of the problem was by Weller and Steiner in 1950.[1] A

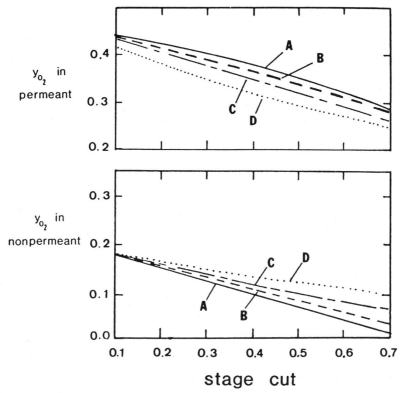

FIGURE 20.2-1 Effect of flow pattern on degree of separation achievable in a single permeation stage.[2] The figure refers to the separation of a mixture of 20.9 mol % O_2 and 79.1 mol % N_2 and shows the oxygen concentration (in mole fraction) in the permeated and unpermeated streams leaving the stage as a function of stage-cut. Conditions: $p_2 = 380$ cm Hg; $p_1 = 76$ cm Hg; $\alpha^* = 5$; $r = 5$; $P_{O_2} = 5 \times 10^{-8}$ [cm^3 (STP) · cm]/cm^3 · s · cm Hg). A: Countercurrent flow with no mixing. B: Crossflow with no mixing. C: Cocurrent flow with no mixing. D: Perfect mixing on both sides of membrane.

simple case dealt with by these authors assumes that both the permeant and the nonpermeant streams are well mixed and a negligible recovery of feed occurs. Clearly, for practical modern modules, this assumption is not valid; nevertheless, one can use the convenient results for this case even in real separators under conditions of low permeant recovery, since compositional changes are small in these cases.

Equation (20.2-1) is the expression for Y_{1A}, the permeant mole fraction of the fast gas (A) in an A–B binary mixture when the total upstream pressure is p_2 and the downstream pressure is p_1:

$$Y_{1A} = \frac{-\psi + [\psi^2 - 4(\alpha^*_{AB} - 1)\,\alpha^*_{AB}(p_2/p_1)Y_{2A}]^{1/2}}{2(1 - \alpha^*_{AB})} \tag{20.2-1}$$

where ψ equals $(\alpha^*_{AB} - 1)(p_2/p_1)\,Y_{2A} + (p_2/p_1) + (\alpha^*_{AB} - 1)$, and α^*_{AB} equals the ratio of permeabilities of component A to component B discussed in Eq. (20.1-7). The parameter Y_{2A}, defined in Eq. (20.1-6), corresponds to the upstream (feed) gas mole fraction. Equation 20.2-1 shows the dependence of permeant composition on the ratio of upstream to downstream pressure. The assumption of perfect mixing can lead to substantial errors for higher product recoveries, which are typically of practical interest. More detailed treatments of module operation require consideration of a variety of permeant–nonpermeant flow patterns. As an example, the effects of crossflow, countercurrent, and cocurrent operation are shown in Fig. 20.2-1 for the O_2–N_2 separation using a hypothetical membrane with $\alpha^*_{O_2/N_2}$ equal to 5.[2] The ratio of upstream to downstream pressure is also equal to 5 in this example, which illustrates the substantial difference in behavior at high stage-cut between the various modes of operation. The stage-cut is simply the fraction of the feed that is collected as permeant.

4. *Modify Preliminary Flow Diagrams.* On the basis of the results of the single-module process, it may be found that target product compositions and recoveries cannot be achieved without recycle strategies.

A considerable literature exists dealing with the potential benefits and energy requirements associated with these approaches. These techniques typically are applied in cases where high recoveries are desired. Therefore, the use of Eq. (20.2-1) to establish compositions for material balances is compromised somewhat, and rather tedious calculations are required. Examples of the applications of these techniques are summarized in the section entitled "Modeling and Design Considerations."

5. *Perform Economic Evaluation of Chosen Designs.* This part of the project is the same as for any other separation operation. After all flows, compositions, and equipment ratings are known, capital, energy, and other operating costs can be assessed by a standard formula.

20.2-2 Application Examples

Applications of membrane-based gas separation technology tend to fall into three major categories:

1. Hydrogen separation from a wide variety of slower permeating supercritical components such as CO, CH_4, and N_2.[3-7]
2. Acid gas (CO_2 and H_2S) and water separations from natural gas.[3-5, 8-15]
3. Oxygen or nitrogen enrichment of air.[15-17]

The order of the various types of applications given above provides a qualitative ranking of the relative ease of performing the three types of separation. The extraordinarily small molecular size of H_2 makes it extremely permeable and easily collected as a permeant compared to the other more bulky gases.[18] Perhaps surprisingly, it is difficult to separate H_2 from CO_2 and H_2S, although these latter gases are clearly much larger.

This observation can be understood from consideration of Eq. (20.1-5). Although H_2 has a high diffusivity, because of its low condensibility, it has a very low solubility in membranes.[19,20] Therefore, a more soluble, lower diffusivity gas such as CO_2 may have a steady-state permeability comparable to that of H_2, since the product of solubility and diffusivity determines the permeability of a component through the membrane. The reasonably high solubilities of CO_2, H_2S, and H_2O in membranes at low partial pressures, coupled with the relatively low solubility and diffusivity of the bulky methane molecule, have made possible the second type of separations. The primary difficulties one encounters in this class of systems arise from possible loss in permselectivity of the dense selective layer and compaction of the open-cell support structure shown in Fig. 20.1-6 at high pressures due to plasticization by the permeating component.

By far, the most difficult of the three types of separation shown is the last one, involving O_2 and N_2. The potential market for O_2-enriched air for medical and furnace applications is considerable. Moreover, N_2-enriched air for blanketing of fuels and stored foods to provide nontoxic, nonresidual protection from fire and oxygen-breathing pests is an interesting possibility. Unfortunately, currently available polymer membranes have only moderate selectivities to separate oxygen and nitrogen.[21] One can understand this situation by considering again Eq. (20.1-5). The size and shape (and hence diffusivity) of O_2 and N_2 are quite similar; moreover, the solubility of the pair in most membranes is similar. Since the factors entering into Eq. (20.1-5) are similar for the two components, the ratio of the permeabilities of the two components tends to be similar, and hence the ideal separation factor [Eq. (20.1-7)] is low. Nevertheless, due to the importance of the problem, processes have been designed that are able to produce economical supplies of O_2- and N_2-enriched air for commercial applications. Recently, Monsanto announced plans to use their Prism® modules, based on polysulfone hollow fibers, to provide N_2 as a blanketing agent for fuels in tanker shipments.[17] Examples of all three of the above types of membrane application are reviewed in the following sections.

HYDROGEN SEPARATION

The following discussion reviews an actual case involving an H_2 recovery permeation unit in an operating 100×10^6 gal/y low-pressure methanol plant owned by Monsanto at Texas City, Texas.[6] The example reemphasizes the fact that H_2 and CO_2 are not separated easily. In this particular case, however, the relative inseparability of the pair turns out to be an *advantage* since both components must be split from N_2 in a purge stream for recycle to the synthesis gas reactor as shown in Fig. 20.2-2. The methanol synthesis loop is based on reactions (I) and (II), which upon combination yield reaction (III), indicating a stoichiometric ratio of three parts hydrogen to one part carbon dioxide entering the reactor loop:

Synthesis $CO + 2H_2 = MeOH$ .... (I)

Water gas shift $\underline{CO_2 + H_2 = CO + H_2O}$ .... (II)

Combined $CO_2 + 3H_2 = MeOH + H_2O$ .... (III)

The detailed separator flow diagram with process conditions indicated is shown in Fig. 20.2-3. The

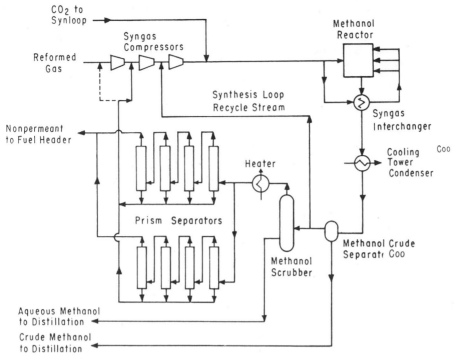

FIGURE 20.2-2 Schematic representation of oxo-alcohol process showing the two banks of 8 in. diameter, 10 ft long Prism® separators.[7]

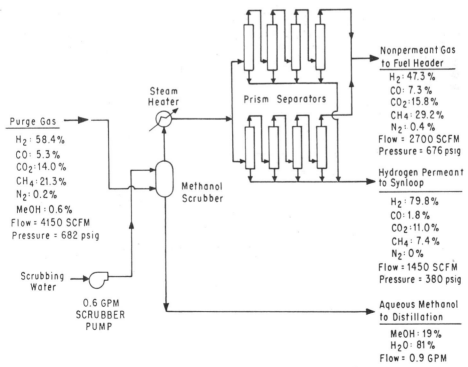

FIGURE 20.2-3 Detailed flow, pressure, and composition diagram for the Prism® separator system shown in Fig. 20.2-2.[7]

separator train involves two parallel banks of four 10 ft long, 8 in. diameter modules arranged in series. The modules contain polysulfone fibers coated with silicone rubber to prevent leaks. Although specific information about the 8 in. diameter modules is proprietary, if one assumes a standard 200 μm OD fiber and a 50% packing factor, roughly 80,000 ft^2 of total membrane area can be estimated to be present in the compact separator train to treat the feed gas required for the process.

Approximately 200 ppm residual methanol leaves the scrubber and enters the separator train. As indicated in Fig. 20.2-3, the major feed components consist of H_2, CO_2, CO, CH_4, and N_2 with a saturated vapor pressure of water pesent from the scrubbing operation. The stream is preheated prior to entering the separator train to prevent depression of the product gas permeation rate. This preheating prevents complications arising from water condensation and other more subtle permeation effects discussed later in the fundamentals section. Then the feed is split and fed in parallel to the series of separators shown in Fig. 20.2-3.

The permeant gas, consisting largely of H_2 and CO_2 for reasons noted earlier, is sent back to the synthesis reactor, while the nonpermeant gas continues through the series of permeators. The permeant from each permeator is recycled to the synthesis reactor, and the nonpermeant from the last separator is sent to the fuel header to be burned. As shown in Fig. 20.2-3, for a feed-side pressure of 682 psig and a bore-side pressure of 380 psig, the purge gas is reconstituted to almost 80% H_2 and depleted somewhat in CO_2 because the permeabilities of H_2 and CO_2 are not identical.

The separators are reported to recover approximately half of the available H_2 and CO_2, which previously had been sent to the fuel header. Reportedly, an impressive net 2.4% increased production rate of methanol can be achieved based simply on elimination of this raw material waste. As noted by the Monsanto authors, the slightly slower permeation rate of CO_2 relative to H_2 requires the addition of a small amount of CO_2 to the recycle stream to maintain the 3-to-1 stoichiometry of H_2 to CO_2 required by reaction (III). Accounting for this supplement CO_2 and a small debit for the fact that the entire purge stream can no longer provide fuel value, an estimate of the net advantage of the permeator installation was reached by the Monsanto engineers. Even with the above debits, the permeator installation permitted a 13% reduction in incremental costs to produce each additional gallon of methanol from the purge stream components that had been burned previously.

ENHANCED OIL RECOVERY, CARBON DIOXIDE RECLAMATION

Enhanced oil recovery (EOR) operations often involve injection of high-pressure (2000 psia) CO_2 into a reservoir. Subsequent removal of CO_2 occurs along with oil and light gases (primarily methane) from wells drilled at a distance from the primary injection point (see Fig. 20.2-4). As shown in Fig. 20.2-5, taken from a recent experimental and modeling study by researchers at Shell, cooling of the gas streams passing through a separator occurs due to a Joule–Thompson-like effect.[12] The 25°F drop in temperature results when the nonideal CO_2 is reduced in pressure as it permeates across the membrane. Removal of a substantial fraction of the CO_2 also tends to increase the mole fraction of condensable components in the residual gas. This effect, producing a dew point above the operating temperature of the module, will cause undesirable condensation of hydrocarbons if precautions are not taken. Preheating the feed is an adequate precaution

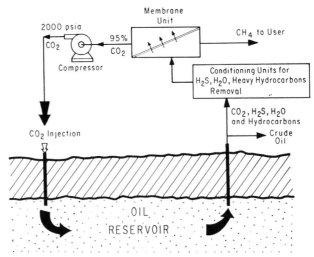

FIGURE 20.2-4 Schematic representation of enhanced oil recovery application of membranes for CO_2–CH_4 separation from gas recovery from a production well after injection of CO_2.

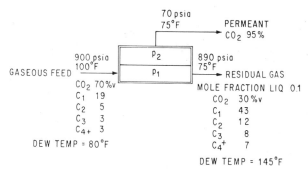

70 psia
75°F
PERMEANT
CO_2 95%

900 psia
100°F
GASEOUS FEED
CO_2 70%v
C_1 19
C_2 5
C_3 3
C_{4+} 3
DEW TEMP = 80°F

p_2

p_1

890 psia
75°F
RESIDUAL GAS
MOLE FRACTION LIQ 0.1
CO_2 30%v
C_1 43
C_2 12
C_3 8
C_4^+ 7
DEW TEMP = 145°F

FIGURE 20.2-5 Representation of the temperature and dew point change resulting from permseparation of a typical enhanced oil recovery gas which contains heavier hydrocarbons in addition to CH_4.[12]

for streams containing low levels of heavier hydrocarbons; however, a condensation step shown in Fig. 20.2-4 may be required in conjunction with preheating to mitigate this problem. Dehydration of the streams with a glycol pretreatment step is sometimes used prior to the membrane unit.[8]

The gaseous products can be separated into a CO_2-rich stream (95% CO_2) used for reinjection and a CH_4-rich stream (98.5% CH_4) used for pipeline gas. Alternatively, the stream can be conditioned to only 90% CH_4 and used as a field fuel. The study by researchers at Shell cited above considers the merits of membrane technology for these applications relative to other competitive technologies such as physical and chemical solvent scrubbing and cryogenic treatment.[9] The Shell study was based on the permeability and selectivity properties of commercial asymmetric cellulose acetate membranes. The permeability and selectivity properties of the membranes were assumed to be independent of gas feed pressure and composition as a first approximation. Considerable evidence indicates that at high CO_2 partial pressures, this assumption may break down for pure CO_2 feeds due to plasticization effects noted earlier in the background section. It has been reported, however, that the plasticization effects appear to be less severe in actual mixed-gas situations in the presence of substantial methane partial pressures.[12]

In view of the ambiguity of published data on this topic, the neglect of the plasticization effect by the Shell group seems reasonable at the present time. The behaviors of polysulfone and cellulose acetate (in the absence of the ill-defined plasticization issue) are similar to a first approximation. Therefore, the results of the study are reasonably representative of both types of existing membrane system. The study permits insights into the strengths and weaknesses of the current generation of membranes compared to chemical and physical solvent treatment systems for these gas purification applications. The detailed feed and product conditions considered by Youn et al.[12] are summarized in Tables 20.2-1 and 20.2-2. In addition, the process flow sheets for the membrane units are shown in Fig. 20.2-6. The flow diagram for case 4 is essentially the same as for case 2 except that the feed pressure is much lower, thereby necessitating more membrane area in case 4.

Cost data used by the Shell group are reported in Table 20.2-3. The data apply to the 1982–1983 period. Clearly, some of the numbers are dependent on the module type employed; however, the values are generally adequate for illustrative purposes. This study indicates operating regimes in which membranes are clearly superior to other technologies. It also identifies regimes where membranes have problems competing with traditional technology due to insufficient selectivity and permeability of current membrane materials. This factor clearly emphasizes the need for the continued development of membrane materials. A point related to the costs of membrane area given below should be made. The prices in Table 20.2-3 apply primarily to spiral-wound units, and even then may be a bit high. For more compact hollow-fiber module configurations in which less steel shell and flanging are required to accommodate the same amount of membrane area, the indicated prices per square foot of membrane area may be somewhat high. Obviously, lower specific costs for new and replacement membrane areas will make the conclusions of the study in some cases even more decisive in favor of membranes. Moreover, it could swing the balance more in favor of membrane processes in other cases in Table 20.2-3.

A computer model of a spiral-wound permeator similar to the Separex design shown in Fig. 20.2-6 was developed by the Shell group to predict performance of the module. Field tests of actual modules were found to agree well with simulation results for conditions where concentration polarization and nonideal flow problems were not encountered. For commercially important flow rates, these conditions were found to be always well satisfied, and the model performed well for all flow rates above 1000 SCFH to the 8 in. diameter spiral-wound modules containing about 1200 ft² of area. A basis of 30×10^6 SCFD of feed gas was taken in all the evaluations, and the system performances were reported on the basis of $/10^3$ SCF of treated feed gas. Credit was taken for heavy hydrocarbons that are retained in the residual gas by membrane processes and that typically are lost along with the CO_2 in physical solvent systems. This benefit can be substantial in some cases for these valuable components.

TABLE 20.2-1 Description of Case Studies Considered in SHELL Evaluations

Case Number	Description of Pertinent Variables
1	Feed gas containing 70% (molar) CO_2/30% CH_4 and other light hydrocarbons at 900 psig Product gases containing less than 1.5% CO_2 (molar) at 850 psig as a nonpermeant and a permeant of 95% CO_2 recompressed to 2000 psia suitable for reinjection as a flood gas
2	Same feed conditions and permeant conditions as in case 1 Nonpermeant stream may contain up to 10% CO_2 and will be used directly as a field fuel
3	Feed gas is 5% CO_2 at 900 psig Product gas contains less than 1.5% CO_2 at 874 psia as a nonpermeant and the low-pressure, CO_2-rich stream is vented to a field heater
4	Feed gas containing 70% CO_2 at 65 psia Product gases containing up to 10% CO_2 at 50 psia as a nonpermeant to be used directly as a field fuel with a permeant of 95% CO_2 at atmospheric pressure and subsequently compressed to 2000 psia; suitable for reinjection as a flood gas
5	Same product stream property requirements as in case 4; however, the feed gas contains 30% CO_2 at 65 psia

Although the details of the physical and chemical solvent systems were not published, it is likely that the physical solvent system was modeled after a material like *N*-methyl-2-pyrrolidone and the chemical solvent was modeled after the characteristics of tertiary amines.

The results of the study are presented in Table 20.2-4. The conclusions of the Shell researchers were:

The membrane processes are competitive for separation of gases containing high CO_2 concentrations into residual gases containing substantial CO_2, say 10% volume. Membrane processes appear to have a greater sensitivity to the CO_2 partial pressure ratio (inlet CO_2 partial pressure/nonpermeant partial pressure) compared to physical solvent processes. However, further optimization of the membrane processes would reduce this difference. Membrane processes are not competitive with the chemical solvent processes for gases with low CO_2 concentrations. This shows that the membrane applications for CO_2 recovery from flue gases are not viable with today's technology.

Clearly, more selective membranes with higher permeabilities could alter the last conclusion as the active research programs in this area yield second-generation materials superior to cellulose acetate and polysulfone. Applications in which membranes are used as "topping units" for high-CO_2-content streams are extremely attractive even with current membrane materials. Another useful discussion of EOR applications of membranes has been presented by Schendel et al.[8] These authors make comparisons with the Ryan–Holmes low-temperature distillation process. They found slight cost advantages associated with the membrane approach. More significant, however, was the important operating flexibility of membrane systems. They noted the following:

The real advantage with membrane use is flexibility and modular design. A reliable forecast of ultimate gas production is not required, and will not significantly affect the economics of the plant. With a pure distillation scheme, a plant must be built to accommodate forecasted gas production. If the plant is oversized, this will result in extra capital being spent for unused capacity; if the plant is undersized, inability to process gas may limit oil production.

Because of extremes in gas volume processed (turn-down ratio), the design basis in this case represents only half the anticipated peak volume. A second distillation plant will be required. With membranes, however, additional membrane capacity at the front end can be added as needed without additional equipment required in the distillation section. Analyses of several associated gas forecasts indicate the exit gas from the first stage of membranes to be remarkably stable in composition and flow rate. This flexibility is not shown in a single case economic comparison.[8]

OXYGEN–NITROGEN SEPARATIONS

Significant advantages in burner efficiency in high-temperature flame applications are realized from even moderate increases in O_2/N_2 ratios. Surprisingly, this market and a corresponding one in N_2-enriched air

TABLE 20.2-2 Detailed Feed and Product Stream Variables in SHELL Case Study Evaluations

Case Number	Feed (psia)	Pressure (psia)		Mole Percent			End Use	
		Nonpermeant	Permeant	CO_2 Feed	Nonpermeant	Permeant	Nonpermeant	Permeant
1	900	850	2000	70	1.5	95	Pipeline	CO_2 Flood
2	900	840	2000	70	10	95	Field fuel	CO_2 Flood
3	900	875	15	5	1.5	40	Pipeline	Vent
4	65	50	2000	70	10	95	Field fuel	CO_2 Flood
5	65	50	2000	30	10	95	Field fuel	CO_2 Flood

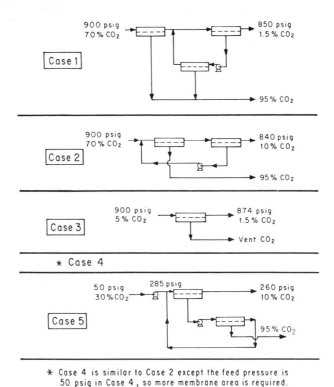

* Case 4 is similar to Case 2 except the feed pressure is
50 psig in Case 4, so more membrane area is required.

FIGURE 20.2-6 System configurations used in Shell case studies of membrane applications in CO_2–CH_4 recovery.[12]

TABLE 20.2-3 Cost Data Used in SHELL Evaluation

Capital Costs to Include

1. Processing facilities
2. Compressors (gas engine drivers)
3. Initial membrane (or solvent charge for solvent stripping alternatives)
4. Off-site facilities (roads, etc.)

Major Component Costs

1. Compressors
   (a) $1400/hp installed with intercoolers and separators
   (b) Overall fuel usage 10 SCF/hp·h
2. Membranes
   (a) $35/ft^2 installed
   (b) $15/ft^2 replacement cost
   (c) 3-year life expectancy anticipated

Operating Costs to Include

1. Capital charge: 25% per year
2. Fixed operating cost: 10% of capital per year

Utility and Product Schedules

1. Fuel gas: $5/10^6 Btu
2. Electrical power: $0.07/kW·h
3. Carbon dioxide: $0.05/10^3 SCF uncompressed

TABLE 20.2-4 Summary of Results for SHELL Case Study

Case Number[a]	Process Alternatives	Capital Cost $ × 10^6	Compression (brake hp)	Treatment Cost ($/10^3 SCF)[b]
1	1. Chemical solvent	25	8,200	1.50
	2. Physical solvent	14	5,200	1.00
	3. Membrane	25	5,200	1.00
	4. Membrane–chemical solvent	18	4,900	0.97
2	1. Chemical solvent	23	7,800	1.40
	2. Physical solvent	12	4,000	0.90
	3. Membrane	15	4,400	0.66
3	1. Chemical solvent	4	0	0.17
	2. Physical solvent	5	100	0.18
	3. Membrane	5	300	0.25
4	1. Chemical solvent	27	8,200	1.60
	2. Physical solvent after compression to 300 psia	26	9,900	1.60
	3. Membrane after compression to 300 psia	34	11,000	1.60
	4. Membrane after compression to 900 psia	24	9,200	1.20
	5. Cryogenic[c]	28	11,000	<1.60>
5	1. Chemical solvent	14	3,500	0.81
	2. Physical solvent after compression to 300 psia	20	6,200	1.80
	3. Membrane after compression to 300 psia	36	6,400	1.50

[a] See Table 20.2-2 for description of details of feed and product stream characteristics.
[b] Accounts for the fact that feed gas contains some heavy hydrocarbons which membranes retain as a small extra benefit.
[c] Process details not specified; standard cryogenic unit presumably considered.

TABLE 20.2-5   System Parameters to Make 10^7 SCF/day of 30% P-11 Membrane

Membrane area	78,000 ft^2
Feed gas pressure	1 atm absolute
Pressure of permeant gas	0.35 atm absolute
Temperature	25°C
Power required to deliver permeant gas at 1 atm	0.15 kW·h/100 ft^3
O$_2$ required to make mixture	47.3 tons
Power required if O$_2$ is supplied from low-temperature air separation plant (basis: 1.4 kW·h/100 ft^3 pure O$_2$)	0.16 kW·h/100 ft^3

Source: Ward et al.[16]

have not driven commercial development of membrane processes in the United States until Monsanto's recent announcement of the N$_2$ blanketing of tanker fuels.[17] A program at General Electric in the 1960s demonstrated the technical feasibility of such processes using flat, low-surface-area, plate-and-frame membranes composed of ultrathin silicone rubber–polycarbonate copolymers. This project, however, has been terminated, and the technology currently is being used only to supply O$_2$-enriched air for personal medical applications.[15]

The most probable reason for the slow development of the O$_2$–N$_2$ membrane systems is the strong competition encountered from existing, highly optimized cryogenic processes. For large-scale operations, cryogenic processes become progressively more economical, while each additional unit of capacity requires an exactly corresponding additional unit of membrane area to perform the separation. Thus, membranes are most competitive in small- and intermediate-scale O$_2$–N$_2$ operations in which flexibility of operation is valuable.

Table 20.2-5 presents the system parameters determined by Ward et al.[16] for producing 10×10^6 SCFD of 30% oxygen with single-stage 1000 Å ultrathin silicone–polycarbonate copolymer membranes. In addition to their thinness, these membranes had high permeabilities due to their silicone rubber component (57% on a mole basis) (see Fig. 20.2-7). The separation factor for the family of silicone–polycarbonate materials shown in Fig. 20.2-8 increases as the fraction of flexibilizing silicone decreases. If one considers the ratio of permeabilities at 0% silicone and at 57% silicone, it is clear that an approximately 100-fold increase in permeation area is required to achieve the same oxygen productivity for pure polycarbonate membranes as for the 57% copolymer. Thus, roughly 7.8×10^6 ft^2 of membrane area with a 1000 Å thick separating layer would be required to supply the same absolute amount of oxygen in the product gas for the polycarbonate case as compared to 78,000 ft^2 for the copolymer case.

The two cases are not entirely equivalent, since the higher selectivity of the polycarbonate fibers will reduce permeation of N$_2$ relative to O$_2$ by roughly a factor of 2 (see Fig. 20.2-8). If a large flow rate of feed air passes through the module, so that feed concentration depletion is not a serious problem, the Weller–Steiner expression in Eq. (20.2-1) can be used for calculating the oxygen composition in the limit of negligible feed recovery as suggested by Ward. The ratio of total downstream to total upstream pressure, p_1/p_2, was equal to 0.35 in the example by Ward summarized in Table 20.2-5, and the mole fraction of

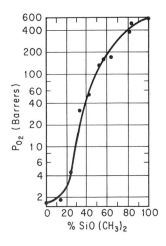

FIGURE 20.2-7   Oxygen permeability of a range of copolymers from pure polycarbonate to pure silicone rubber. The weight percent of silicone rubber in the copolymer is shown as the ordinate.[16]

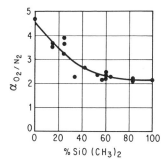

FIGURE 20.2-8 Ideal separation factor for O_2–N_2 versus percent silicone rubber in the copolymer.[16]

O_2 in the feed air is equal to 0.21. Using the ideal separation factor for polycarbonate from Fig. 20.2-8 with a value of $p_1/p_2 = 0.35$, a product gas composition of $Y_{O_2} = 0.38$ results rather than the $Y_{O_2} = 0.30$ obtained with the copolymer membrane used by Ward. Clearly, since the oxygen production has been assumed to be the same, the total product flow rate will amount to only $(0.30/0.38)\ 10 \times 10^6$ SCFD or 7.9×10^6 SCFD of 38% O_2 product gas. This higher purity product is more valuable than the 30% product achievable with the silicone–polycarbonate membranes at the same value of p_1/p_2.

To compare equivalent product qualities, a material balance shows that 8.92×10^6 SCFD of standard air (21% O_2) can be added to dilute the 38% stream to 30%. Alternatively, one could adjust the value of the pressure ratio (p_1/p_2) to make the Y_{O_2} calculated from Eq. (20.2-1) equal to 0.30. Since the primary discussion here relates to membrane area, the first approach is assumed to keep other parameters essentially constant. Therefore, 16.83×10^6 SCFD of 30% air would result if the full 7.8×10^6 ft^2 of membrane area were used. Therefore, to produce 10×10^6 SCFD of 30% air, one would require 7.8×10^6 $(10/16.83) = 4.63 \times 10^6$ ft^2 of membrane area. This area could be accommodated in roughly 450 low-pressure, 8 in. diameter, 10 ft long modules. Given their compact size, this number of modules is not unreasonably large. Blowers, vacuum pumps, and compressors would be the only additional capital costs involved in the system. As mentioned earlier, the membrane system also would provide easy turndown simply by eliminating whatever fraction of the module permeation capacity is not required at the time. The number of modules required could be reduced greatly if the feed gas were compressed to a substantial pressure (say 5–10 atm). In such a case, more expensive module shells and compressor costs would need to be balanced against the reduction in the membrane area requirements.

20.3 MEASUREMENT AND PHENOMENOLOGICAL DESCRIPTION OF GAS SORPTION AND TRANSPORT IN POLYMERS

20.3-1 General Discussion

The detailed engineering design for gas permeators generally involves a rather complex computer simulation of the coupled momentum and mass transfer phenomena occurring in the module. Advantages of recycle and various flow options also must be evaluated to arrive at the best overall design. These considerations, introduced in the previous section, are dealt with in detail in the section entitled ''Modeling and Design Considerations.'' The key to success of the process, however, rests on the optimum choice of membrane properties, just as solvent selection lies at the heart of solvent extraction processes. The present section therefore focuses on basic polymer properties of importance to membrane selection. Emphasis is placed on factors affecting the sorption and transport properties of dense films because most preliminary screening work on novel membrane materials deals with such samples.

Temperature is an extremely important variable that directly affects physical properties of the membrane and hence its permeation and selectivity properties. As an example, if one considers the low-pressure limit, under which condition Eq. (20.1-5) applies, the magnitude of temperature effects can be illustrated for a typical group of gases. Permeabilities, ideal separation factors, diffusivities, and solubility coefficients of gases in natural rubber are plotted in Fig. 20.3-1 in semilogarithmic form as a function of $1/T$.[1,2] The slopes of the solubility and diffusivity curves can be interpreted directly in terms of enthalpies of sorption and activation energies of diffusion, respectively.[1] The temperature dependence of the permeabilities and separation factors is composed of contributions from both sorption enthalpies and activation energies, so that their theoretical interpretation is more complex than that of the simple diffusion and sorption coefficients.

The plots in Fig. 20.3-1 show clearly that the strongest effect of temperature is on the diffusion coefficient. Changes of 200–300% occur over the temperature range from 25 to 50°C, depending on the gas considered. The smallest molecule, H_2, shows the smallest increase with increasing temperature. This

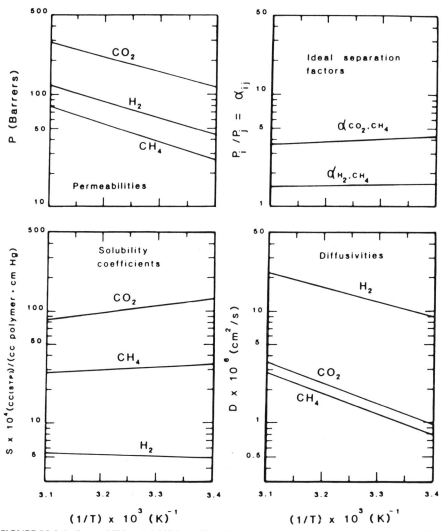

FIGURE 20.3-1 Permeabilities, solubilities, diffusivities, and separation factors for CO_2, H_2, and CH_4 in natural rubber over the temperature range from 25 to 50°C.

trend often is observed; increases in temperature speed up the diffusion processes of the most bulky penetrants the most. The solubility coefficients of the various gases change less than 30% even for CO_2 over this same temperature range. The permeability—comprising the product of the solubility and mobility coefficients—tends to be dominated by the diffusivity and increases by roughly 120–160% between 25 and 50°C. The ideal separation factor, composed of a ratio of permeabilities, is moderated in its temperature dependence because the temperature dependencies of the permeabilities of the various gases are rather similar.

The above trends often are observed even for glassy polymers at low pressures; however, the separation factors tend to be much higher in glasses. Moreover, precipitous drops in separation factors can occur as one approaches the glass transition (softening point) of a glassy polymer. In such cases, marked deviations from linear semilog behavior such as that shown in Fig. 20.3-1 occur.

Most separators tend to be operated over a relatively narrow range of temperatures; however, a wide range of feed pressures and compositions can be encountered. Therefore, it is appropriate to consider in more detail the pressure or concentration dependence of the permeation, sorption, and diffusion coefficients that control module operation.

20.3-2   Concentration Dependence of Sorption and Diffusion Coefficients

Accurate engineering modeling of membrane-based gas permeators can be performed if the composition and pressure dependence of the various component permeabilities are known at the temperature of interest. Permeabilities of gases and vapors have been reported to display behaviors represented by Figure 20.3-2.[3-6] For gases at low and intermediate pressures, behaviors shown in Fig. 20.3-2a-c are most common. Response (b) is characteristic of a simple plasticizing response.[4] Interestingly, response (d) in this figure is simply a combination of responses (b) and (c). At higher pressures, even in plasticization-resistant materials, one can expect the response shown in Fig. 20.3-2c to evolve into a form such as the one shown in Fig. 20.3-2d, since sorption levels will be high under these conditions. The entire response is apparent in Fig. 20.3-2d because of the high solubility of the penetrant in the polymer even at low partial pressures.

The magnitude of the pressure dependence of the various responses illustrated differs substantially. Assumption of a constant value for these coefficients in design calculations is tempting for responses (a), (c), and (d), where only about a 20–30% change occurs over the pressure range shown. In the case of response (b), however, the strong dependence makes it ambiguous as to what average value is reasonable to use in modeling since large pressure differences typically exist across separating membranes. Even in the less extreme cases, however, rational design would be aided by using the correct functional form for the transport coefficients.

As noted earlier, when both the diffusion and solubility coefficients are constants, the simple form shown in Fig. 20.3-2a results. In all other cases, nonconstancy of either the sorption or diffusion coefficients can cause nonconstancy of the observed permeability and selectivity. As illustrated, separation of the two component contributions is straightforward if both equilibrium solubility and steady-state permeation data are available.

Beginning with the definition of permeability in Eq. (20.1-3), one can proceed without the simplifying assumptions in Eq. (20.1-5) that the diffusion and sorption coefficients are constants. In this case we find

$$P = \frac{N}{(p_2 - p_1)/l} = -\frac{D(C)\,dC/dx}{(p_2 - p_1)/l} \tag{20.3-1}$$

where $D(C)$ is the local concentration-dependent diffusion coefficient of the penetrant at an arbitrary point

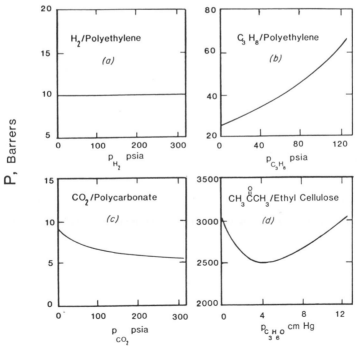

FIGURE 20.3-2   The pressure dependency of various penetrant–polymer systems: (a) at 30°C,[3] (b) at 20°C,[4] (c) at 35°C,[6] and (d) at 40°C.[5]

between the upstream and downstream surface and dC/dx is the local concentration gradient at this same point in the membrane. As before, N is the steady-state flux, and p_2 and p_1 refer to the upstream and downstream penetrant partial pressures, respectively. Although both $D(C)$ and dC/dx may be position dependent, their product is constant at each point in a flat membrane,‡ since the steady-state permeability is a constant for fixed upstream and downstream conditions. Therefore, we see from integration of Eq. (20.3-1) with respect to x that

$$\frac{P}{l} \int_0^l dx = P = - \int_{C_2}^{C_1} D(C)\, dC \qquad (20.3\text{-}2)$$

By the chain rule we find Eq. (20.3-3):

$$P = \frac{1}{p_2 - p_1} \int_{p_1}^{p_2} \underbrace{D(C)}\ \underbrace{\frac{dC}{dp}}\, dp \qquad (20.3\text{-}3)$$

$$\text{kinetic factor} \qquad \text{thermodynamic factor}$$

Clearly, the observed permeability is a normalized integration of the product of a kinetic factor and a thermodynamic factor across the entire membrane from upstream ("condition 2") to downstream ("condition 1"). The thermodynamic term dC/dp is determined by polymer–penetrant interactions. The kinetic term $D(C)$ is determined by polymer–penetrant dynamics and requires localization of a sufficiently large transient packet of volume contiguous to the penetrant to permit it to execute a so-called diffusive jump.

The ideal separation factor, equal to the ratio of the permeabilities of the two components, is also interpretable as a product of two factors: a "solubility selectivity" and a "mobility selectivity." These two selectivity contributions, consisting of the ratios of the respective component solubilities and diffusivities, indicate the relative importance of thermodynamic and kinetic factors in the permselection process. Unfortunately, optimization of product permeability and membrane selectivity is often difficult, and trade-offs in the two parameters may be necessary on economic grounds. A brief discussion of characterization methods and typical forms of sorption isotherms and local diffusion coefficients for gases and vapors in polymers is presented below. This discussion serves as a background for rationalizing pressure dependencies of permeabilities and selectivities.

Sorption Isotherm Forms

As shown in Fig. 20.3-3, a variety of isotherm shapes are observed for gases and vapors in polymers. Gas sorption in rubbery polymers and low-concentration gas sorption in glassy polymers follow an apparent Henry's Law form represented by Fig. 20.3-3a. Simple homogeneous swelling of rubbery polymers by compatible penetrants tends to produce isotherms like that shown in Fig. 20.3-3b. Interestingly, rather complex polar and hydrogen-bonding penetrants, which tend to cluster together in nonpolar rubbery polymers, also tend to exhibit isotherms like that shown in Fig. 20.3-3b. More complicated isotherms involving an inflection, such as that shown in Fig. 20.3-3d, often are observed for vapor and even highly sorbing gases such as CO_2 in glassy polymers under certain conditions.

The isotherm presented in Fig. 20.2-3c is observed commonly for gases up to moderate pressures (20–30 atm) in glassy polymers. It corresponds to the low-activity portion of the curve shown in Fig. 20.3-3d. The term *dual-mode* sorption applied to Fig. 20.3-3c derives from a specific physical model of the glassy state.[6] The model suggests that two different environments exist in nonequilibrium glasses, resulting in two idealized sorbed populations that exist at local equilibrium with each other because of rapid exchange between the two environments. This issue is pursued in greater depth in the section entitled "Fundamentals of Sorption and Transport Processes in Polymers." For present purposes, such considerations are not necessary, and a simple polynomial fit of C versus p to any of the data forms shown in Fig. 20.3-3 would suffice to evaluate dC/dp for use in Eq. (20.3-3). Such an approach is termed *phenomenological* in that it simply describes the phenomenon in question without considering its physical bases.

Although indirect techniques exist for generation of data such as shown in Fig. 20.3-3, the most reliable approaches involve direct determination of the amount of penetrant sorbed into the polymer at a given external pressure. At low pressures with highly sorbing vapors, gravimetric techniques involving weight gain monitoring by quartz spring or electrobalance are effective. For gases at high pressures, however, pressure decay during sorption monitored by accurate transducers with calibrated reservoir and receiving volumes has proved to be the most effective method. Typical data collected using such a cell are shown in Fig. 20.3-4 for a variety of gases in polycarbonate.[7] Design and operation of a pure gas cell (Fig. 20.3-5a) and a multicomponent cell (Fig. 20.3-5b), which requires a gas chromatograph, have been described elsewhere in detail.[8,9]

---

‡Curvature effects are neglected here since most screening studies of candidate membranes are conducted on flat dense films.

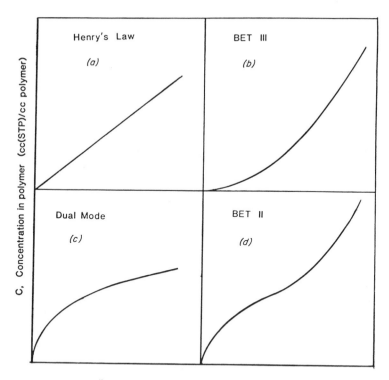

P , partial pressure of penetrant, atm

FIGURE 20.3-3   Schematic representation of typical sorption isotherm forms.

DIFFUSION COEFFICIENT FORMS

Determination of concentration-dependent diffusion coefficient data usually is done indirectly. Typically, one determines $D(C)$ from the observed pressure dependence of the permeability in conjunction with the independently determined dC/dp data discussed in the preceding section.[6] When significant concentration dependence is observed for the local diffusion coefficients, the form generally resembles one of the curves shown in Fig. 20.3-6. Of course, for low-sorbing gases such as H_2 and N_2 in rubbery polymer, both D and dC/dp are essentially constant resulting in adherence to the simple situation indicated by Eq. (20.1-5).

Transport in hydrocarbon polymers exposed to organic vapors or hydrophilic polymers exposed to water vapor typically is characterized by diffusion coefficients that increase exponentially with the concentration of plasticizing penetrant.[10-13] Such a relationship is represented graphically in Fig. 20.3-6a. At relatively low concentrations of a plasticizing penetrant, the linear relationship between $D(C)$ and C shown in Fig.

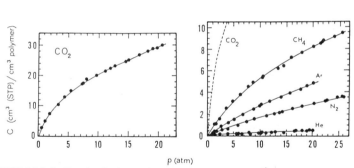

FIGURE 20.3-4   Sorption isotherms for various gases in Lexan® polycarbonate at 35°C.

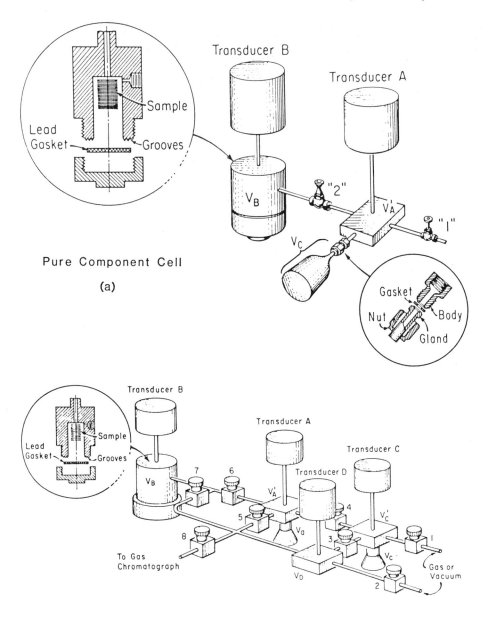

FIGURE 20.3-5 Sketches of (*a*) pure component and (*b*) multicomponent high-pressure sorption cells for collection of data such as those shown in Figs. 20.3-3 and 20.3-4.

20.3-6*b* often is observed. Clustering of penetrant can reduce the effective diffusion coefficient, since in simple terms the size of the mobile penetrant effectively is increased.[14] Such behavior is observed most commonly for the diffusion of hydrogen-bonding penetrants in relatively nonpolar environments. Since clustering becomes more pronounced at higher concentrations, the effective diffusion coefficient decreases with concentration, as shown in Fig. 20.3-6*d*. The monotonically increasing, albeit inflecting, response of $D(C)$ versus C presented in Fig. 20.3-6*c* is characteristic of diffusion coefficients for many penetrants at relatively low concentrations in glassy polymers.[6] This form of the relationship is consistent with the so-

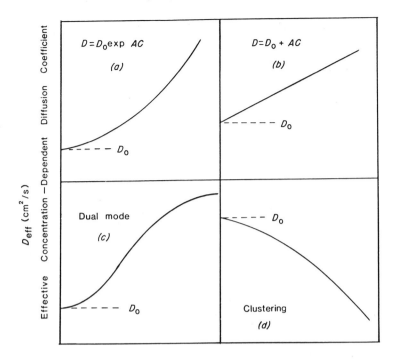

C, Local Penetrant Concentration in Polymer

FIGURE 20.3-6 Schematic representation of typical forms for concentration-dependent diffusion coefficients.

called dual-mode sorption and transport theory which was mentioned earlier in connection with the interpretation of Fig. 20.3-3. This theory is discussed in more detail in the section dealing with the properties of the glassy state.

Three types of laboratory apparatus are in common use for the measurement of gas permeation rates through polymers to obtain data for analyses such as those discussed previously:[15]

Manometric cells
Isobaric cells
Carrier-gas cells

The manometric and isobaric cells are simple and differ primarily in the mode of sensing gas transmission through the polymer. As shown in Fig. 20.3-7a, the manometric approach involves monitoring gas permeation into an accurately calibrated and thermostated, evacuated volume by recording the pressure-rise rate in the volume. Several designs for this type of cell have been reported.[3, 16, 17] The most recent cells generally rely on sensitive transducers such as the MKS-Baratron rather than a mercury McLeod gage, which was used in early designs.

The isobaric cell, shown in Fig. 20.3-7b, generally operates with atmospheric pressure downstream. Gas permeation in this case is monitored by recording the change in position of a mercury or silicone oil plug or a soap bubble moving along a calibrated capillary. As with the manometric cell, several designs have been reported for the isobaric appartus. Stern and coworkers have developed the most advanced form of these cells.[18]

The so-called carrier-gas method is the most complicated approach to gas permeation monitoring.[19] Either pure-component or multicomponent permeation rates through films can be monitored using this device, shown in Fig. 20.3-8. A carrier gas such as helium, containing a desired partial pressure of the desired component or components, flows past the upstream face of the membrane. A downstream sweep gas picks up the permeated components and routes them to a gas chromatograph for analysis of the fluxes of each penetrant. An excellent discussion of such a system has been offered by Pye et al.,[3] and comparisons were made with manometric cells for pure gas permeation to prove that the results for both cells are essentially identical if care is taken in operation.

As an example of analysis of permeation data to evaluate the concentration-dependent diffusion coeffi-

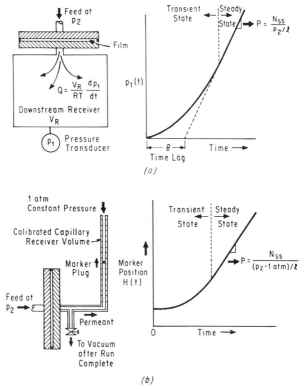

FIGURE 20.3-7 (*a*) Manometric (variable downstream pressure) and (*b*) isobaric (variable downstream volume) permeation cells for determination of gas transport rates.

cients, consider a case in which the observed permeability continuously inflects upward (e.g., Fig. 20.3-2*b*) and the isotherm displays a BET III form (e.g., Fig. 20.3-3*b*). The local value of $D(C)$ can be evaluated readily by differentiation of polynomial fits to both the permeation and sorption data. Equation (20.3-6) is derived by application of the Liebnitz rule for differentiation under an integral sign to Eq. (20.3-2). For present purposes, the downstream pressure (p_1) of the diffusing component in the derivation is assumed to be negligible, since a vacuum or carrier gas generally is present in most testing situations. Thus, Eq. (20.3-2) becomes

$$p_2 P = - \int_{C_2}^{C_1 \nearrow^0} D(C)\, dC \tag{20.3-4}$$

Therefore, applying the Liebnitz rule

$$D(C_2) = p_2 \left.\frac{dP}{dC}\right|_{C_2} + P \left.\frac{dp}{dC}\right|_{C_2} \tag{20.3-5}$$

or

$$D(C_2) = \left.\frac{dp}{dC}\right|_{p_2} \left(p_2 \left.\frac{dP}{dp}\right|_{p_2} + P|_{p_2} \right) \tag{20.3-6}$$

Using Eq. (20.3-6) one can determine $D(C_2)$ versus C_2 for any chosen pressure condition for which both sorption and permeation data exist. If one applies Eq. (20.3-6) over the entire pressure range of interest to curves such as those shown in Fig. 20.3-2*b* and 20.3-3*a* or 20.3-3*b*, a constantly upwardly inflecting $D(C)$ versus C relationship similar to that shown in Fig. 20.3-6*a* will result.

One must be a bit cautious, however, about drawing sweeping conclusions from the general shape of permeation and sorption isotherms. If the slope of the sorption isotherm, dC/dp, increases more rapidly

CARRIER-GAS PERMEATION SYSTEM

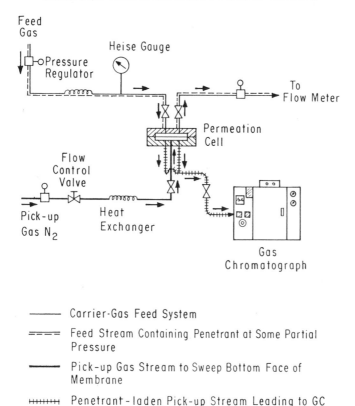

——— Carrier-Gas Feed System

==== Feed Stream Containing Penetrant at Some Partial Pressure

▬▬▬ Pick-up Gas Stream to Sweep Bottom Face of Membrane

++++++ Penetrant-laden Pick-up Stream Leading to GC

FIGURE 20.3-8 Carrier-gas permeation system for determination of gas transport rates. The equipment is especially useful for work with multicomponent mixtures.

than the permeability increases, the effective diffusivity actually may be found to decrease with increasing values of C_2. Such behavior is shown in Fig. 20.3-6c and is observed for hydrogen-bonding penetrants such as water and alcohols in nonpolar polymers like silicone rubber. The concentration dependence of a penetrant's diffusion coefficient therefore provides insight into the nature of polymer–penetrant interactions.

20.3-3 Partitioning Permeability into Thermodynamic and Kinetic Constituents

In cases where either or both of the coefficients appearing in Eq. (20.3-3) are not constants, it is useful to know explicitly which of the two factors is primarily responsible for nonconstancy in the observed permeability. As noted above, this information cannot be known simply by measuring the steady-state permeability. Demonstration of the existence of a continuously upwardly inflecting exponential curve for the $D(C)$ versus C response (Fig. 20.3-6) suggests that the membrane is being plasticized strongly by the penetrant. This information would warn one to expect rather poor selectivity due to loss in ability of the polymers to discriminate between penetrants of different sizes and shapes. On the other hand, concentration dependence identified as "dual mode" in Fig. 20.3-6c is not associated with plasticization in general. This behavior is observed for even low-sorbing gases such as N_2, as well as for more strongly sorbing materials like CO_2 in glassy materials that maintain reasonable selectivity up to the point at which strong plasticization is indicated by a sudden upturn in the diffusivity relationship.[6]

Unfortunately, except for a few gases in a few glassy polymers, high-pressure sorption and permeation data are not available to pursue the concept of decoupling the thermodynamic and kinetic contributions to the ideal separation factor. The technique, however, can be illustrated for the important case of CO_2 and CH_4, for which a fair number of data exist at pressures up to 20 atm. Carbon dioxide permeabilities at 35°C in a variety of glassy polymers at upstream pressures up to 20 atm (with a negligible downstream pressure) are shown in Fig. 20.3-9. Besides the dramatic range of permeabilities represented (more than a

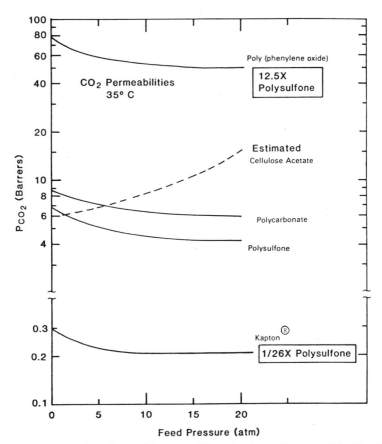

FIGURE 20.3-9 Pressure dependence of the permeability of various polymers to CO_2. The data were measured with a vacuum downstream except for the cellulose acetate (CA), which was estimated from a variety of sources.[21-24]

300-fold difference among polymers), the surprising difference in the reported response of cellulose acetate to increasing CO_2 pressure compared to the other polymers is obvious.

The difference in the responses can be considered in terms of the composite nature of the permeability suggested by Eq. (20.3-3). If $D(C)$ in this equation is not a very strong function of concentration, the strong reduction in dC/dp due to the concave-shaped, dual-mode-type isotherms for such systems (Fig. 20.3-3c) causes the observed permeability to decrease at high pressures. In the case of cellulose acetate, although the dual-mode isotherm shape still is observed,[20] apparently the polymer exhibits a strong plasticization response (like Fig. 20.3-6a rather than like Fig. 20.3-6c). This sharp increase in penetrant diffusivity swamps the effect of the decreasing dC/dp term, thereby causing the dramatic upswing in permeability.

One may pursue the issue of the apparent special affinity of CO_2 for cellulose acetate in an additional fashion by considering the CO_2–CH_4 separation factor for the polymers presented in Fig. 20.3-9. Considering Eq. (20.3-2) for component i for the case where the downstream partial pressure, p_{1i} is negligible, we see that

$$P_i = \int_0^{C_{2i}} \frac{D(C_i)\, dC_i}{p_{2i}} \tag{20.3-7}$$

Multiplying the numerator and denominator of Eq. (20.3-7) by C_{2i}, we obtain the sorbed concentration of component i in the film at the upstream partial pressure p_{2i}:

$$P_i = \left(\int_0^{C_{2i}} \frac{D(C_i)\, dC_i}{C_{2i}} \right) \left(\frac{C_{2i}}{p_{2i}} \right) = \overline{D}_i\, \overline{S}_i \tag{20.3-8}$$

where $\overline{D}_i$ and $\overline{S}_i$ are the *average* diffusivity and *apparent* solubility of component i defined by Eqs. (20.3-9) and (20.3-10)

$$\overline{D}_i = \int_0^{C_{2i}} D(C_i)\, dC_i \bigg/ \int_0^{C_{2i}} dC_i \tag{20.3-9}$$

$$\overline{S}_i = \frac{C_{2i}}{p_{2i}} \tag{20.3-10}$$

Therefore, one may, express the ideal separation factor for the chosen upstream partial pressure, p_{A2} and p_{B2}, as shown in Eqs. (20.3-11) and (20.3-12):

$$\alpha_{A,B}^* = \frac{P_A}{P_B} = \left(\frac{\overline{D}_A}{\overline{D}_B}\right) \left(\frac{\overline{S}_A}{\overline{S}_B}\right) \tag{20.3-11}$$

$$\alpha_{A,B}^* = \left(\begin{array}{c}\text{mobility} \\ \text{selectivity}\end{array}\right) \left(\begin{array}{c}\text{solubility} \\ \text{selectivity}\end{array}\right) \tag{20.3-12}$$

The ideal separation factors and the mobility and solubility contributions comprising this parameter for the various polymers shown in Fig. 20.3-9 are presented in Table 20.3-1 for a case corresponding to a 600 psia total feed pressure with 300 psia of CO_2 upstream and very low downstream pressures of both components. The data for preparation of Table 20.3-1 were estimated from a number of sources, and the absolute value of the entries for cellulose acetate may be in error up to 10–15%.[2,21-24] The data for selectivities reported in the table are higher than those normally reported for asymmetric polysulfone and cellulose acetate membranes since the dense-film forms of the samples tend to eliminate flow through pores that are present in asymmetric membranes and decrease their selectivity (see Table 20.1-2).

It appears significant that the majority of the polymers at 20 atm CO_2 pressure are selectively sorptive for CO_2 relative to CH_4 by a factor of only 2–4. Cellulose acetate, on the other hand, appears to be substantially more sorptive for CO_2 relative to CH_4.[25] Conversely, the mobility selectivities of all the polymers in Fig. 20.3-9 that do not show a plasticization response are more favorable than that of cellulose acetate. This observation assumes that plasticization of cellulose acetate causes no further loss in its ability to distinguish between CO_2 and CH_4 compared to the pure component permeability ratios.

If this assumption is seriously in error, the actual mixed gas mobility selectivity of cellulose acetate may be even lower than indicated by the diffusivity ratios in Table 20.3-1. Polymers such as cellulose acetate which are "solubility selectors" may tend to display plasticization-type responses in the permeability versus pressure plots such as that shown in Fig. 20.3-2b. More detailed sorption and diffusion data on a single, well-characterized film sample for this interesting system are needed badly to investigate these effects further. Understanding the principles at play in the case of cellulose acetate may permit expansion of the ranks of such "solubility selecting" materials for possible use as thin-film composite membranes or in blending with other, more plasticization-resistant membrane materials.

The preceding analysis indicates the value of considering the separate solubility and mobility constituent contributions to the permeability and selectivity of a given membrane. These data also suggest approaches for modifying existing membrane materials at the two extremes of permeability shown in Fig. 20.3-9. The mose selective film (Kapton®) represented in Table 20.3-1 is also the least permeable, and the most permeable [poly(phenylene) oxide] is also the least selective: a rather unfortunate statement of Murphy's Law.

The interesting empirical correlation in Fig. 20.3-10 provides an impressive relationship between the

TABLE 20.3-1 Partitioning of the Ideal Separation Factor for CO_2–CH_4 into Its Kinetic and Thermodynamic Contributions at 35°C and 40 atm for a 50–50 Molar Feed Mixture

Polymer	P_{CO_2}/P_{CH_4}	$\overline{D}_{CO_2}/\overline{D}_{CH_4}$	$\overline{S}_{CO_2}/\overline{S}_{CH_4}$
Poly(phenylene oxide)	15.1	6.88	2.2
Polycarbonate	24.4	6.81	3.6
Polysulfone	28.3	8.85	3.2
Cellulose acetate	30.8	4.21	7.3
Kapton®	63.6	15.38	4.1

Note: Assumes equal percent increase in CO_2 and CH_4 permeability due to any plasticization of cellulose acetate by CO_2.

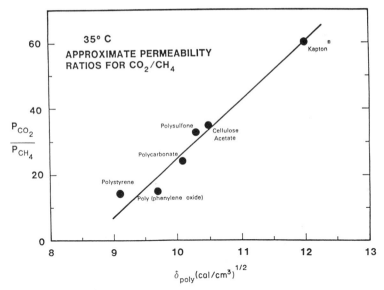

FIGURE 20.3-10 Ideal separation factors for CO_2–CH_4 with several polymers based on pure component permeability ratios. The solubility parameter of the polymer is a successful correlating parameter for the ideal separation factor for the polymers shown here.[21]

solubility parameter of the membrane polymer and its ideal separation factor for CO_2 relative to CH_4.[21-24] This plot suggests that if one were constrained to deal within the limits of the existing polymers in Fig. 20.3-9, modifications should be carried out to increase the cohesive energy density of PPO (e.g., by substitution of a polar or hydrogen-bonding group on the phenylene ring) and to decrease the cohesive energy density of Kapton® [e.g., by introduction of structural irregularities, perhaps using a mixture of aromatic diamines besides bis-4(aminophenyl) ether to make ordering of the imides more difficult, thereby to "open" the structure]. The logical conclusion of this concept indicates that it might be possible to open the structure enough to allow almost unrestricted movement of the streamlined CO_2 while still retarding the bulky CH_4 to a greater extent.

These concepts are largely derivative of the ideas of Hoehn of Du Pont, who developed high-permeability, high-selectivity membranes for hydrogen applications in the 1970s.[3] A systematic study of the effects of structural modifications on the solubility and mobility contributions to the observed permeabilities and selectivities of both the PPO and Kapton® films would be useful. Such an approach, based on the most permeable and least permeable candidate polymers, effectively would "surround" the problem and allow systematic closure on the best membrane in terms of combined permeability and selectivity. Clearly, one should *not* be constrained to deal within the limits of the existing polymers shown in Fig. 20.3-9. The best membrane materials may not be the best engineering plastics, since the requirements are different for the two classes of materials. Stiff-chained, bulky, repeat-unit polymers with purposely introduced structural irregularities are interesting candidates for future study.

A critical need exists for construction of a systematic, reliable data base for a wide number of gases in a wide number of polymers. Until this need is recognized and addressed in the same way that collection of vapor–liquid equilibrium data is accepted, membrane development will be reduced largely to an "Edisonian" approach.

20.3-4 Transient Permeation Analyses to Check for Data Consistency

Most of the measurements to be performed in a membrane-testing program involve asymmetric membranes after the first phase of work to screen candidate polymers. Nevertheless, it is wise to perform a benchmark study using dense films of the polymer of interest. This approach permits an unambiguous determination of the sorption and transport properties of the inherent polymer, unobscured by minuscule, but possibly significant, porous defects in the asymmetric structure. Moreover, if such data are available, drifts in the permeability and selectivity of asymmetric membranes composed of the polymer can be understood better. By making comparisons between the asymmetric membrane results and those of the dense film, which are free of time-dependent support effects, drifts due solely to support aging can be quantified.

Whichever of the methods depicted in Figs. 20.3-7 and 20.3-8 for monitoring permeation is used, a transient period of permeation typically is observed when films of more than a few microns are studied.

Upon analysis of this transient period, either as the zero-pressure intercept on the time axis for constant volume systems (Fig. 20.3-7a) or as the corrected zero-moment of the detector response for carrier-gas systems (Fig. 20.3-8), one can derive a kinetic parameter θ, referred to as the "time lag".[15,26,27]

In most time-lag experiments, the initial downstream penetrant concentration, C_1, is zero and the upstream concentration C_2 is assumed to be established instantaneously at its equilibrium value. Although θ is an additional parameter not required to evaluate permeability and solubility data, it is wise, when possible, to verify that the time lags observed experimentally are consistent with the predictions of Fick's Law. To perform this test, one needs only to fit an empirical equation to the $D(C)$ versus C data shown in Fig. 20.3-6 and to perform the integration indicated:[26,27]

$$\theta = \frac{l^2 \int_0^{C_2} wD(w) \int_w^{C_2} D(u)\, du\, dw}{\left(\int_0^{C_2} D(u)\, du\right)^3} \tag{20.3-13}$$

where l is the membrane thickness and $D(u)$ and $D(w)$ refer to the empirically determined form that best describes the local diffusion coefficient data evaluated according to Eq. (20.3-6) and illustrated schematically in Fig. 20.3-6.

As an example, if the local diffusion coefficient varies exponentially with local sorbed concentration, for example, $D = D_0 \exp(AC)$, we find after integration[27]

$$\theta = \frac{l^2}{4D_0} \left(\frac{4 \exp(AC_2) - 1 + \exp(2AC_2)(2AC_2 - 3)}{[\exp(AC_2) - 1]^3}\right) \tag{20.3.14}$$

If the sorbed concentration at the upstream face obeys Henry's Law as shown in Fig. 20.3-3a, then $C_2 = Sp_2$, and we can express the time lag directly in terms of pressure. If a more complex polynomial fit to the sorption isotherms shown in Fig. 20.3-3a is necessary, it is clear that C_2 versus p_2 could still be substituted into Eq. (20.3-14) to obtain the pressure-explicit prediction of θ versus pressure for comparison to the experimentally measured data.

The time-lag evaluation therefore provides a consistency check to verify that the data collected are behaving in accordance with the standard Fickian transport model. Significant deviations (> 10–15%) may signal either experimental problems in the data acquisition or long-term drifts in the polymer structure. Such long-term drifts could be associated with true glassy-state drifts or with a failure to remove traces of casting solvent. This latter problem could be verified and eliminated by replication of the measurements with a more thoroughly heat-treated film. Failure to remove small traces of solvent residuals can lead to a variety of peculiar anomalies.[28–30]

For cases in which the diffusion coefficient is a constant, even for complex isotherm shapes, Eq. (20.3-14) reduces to a simple relationship:

$$\theta = \frac{l^2}{6D} \tag{20.3-15}$$

In such cases, and only in such cases, it is correct to refer to the diffusion coefficient as

$$D = \frac{l^2}{6\theta} \tag{20.3-16}$$

Unfortunately, the above equation often is used to calculate a parameter sometimes referred to as D_{app} (apparent diffusivity). This parameter does not have a simple physical meaning and is a complicated average of the true local diffusion coefficient at the upstream and downstream conditions, as can be seen from the expression for θ in Eq. (20.3-13). This parameter tends to show less dependence on the upstream pressure or concentration than the true local diffusion coefficient does when evaluated at the upstream condition.

Consider a hypothetical plot of $D(C)$ versus C, such as that shown by the solid line in Fig. 20.3-11. In a time-lag experiment with $C_1 = 0$ downstream and C_2 fixed by the upstream pressure, the parameter D_{app} has a numerical value between the two limits of $D(C_2)$ and $D(0) = D_0$. This is illustrated by the dashed line in Fig. 20.3-11, which is much less concentration dependent than the true $D(C)$. The ambiguity arising from application of Eq. (20.3-16) to calculate a coefficient to characterize the local penetrant mobility at a given upstream pressure or concentration becomes larger at high concentrations. The discrepancies between the true mobility and the value of D_{app} tend to disappear in the limit as results are extrapolated to the zero-pressure limit. Unfortunately, this extrapolation does not permit one to analyze the high-pressure range, where most gas separation operations are likely to occur, so that the use of Eq. (20.3-16) as a fundamental measure of mobility is not desirable in general.

When the local diffusion coefficient evaluated from Eq. (20.3-6) is not too strongly concentration dependent (10–20% variation over the range of interest), the error associated with the use of Eq.

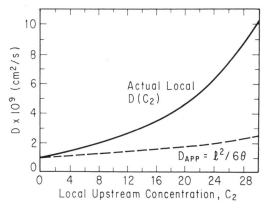

FIGURE 20.3-11 Demonstration of the marked difference in concentration dependence of the true local diffusion coefficient (solid line) and the apparent diffusion coefficient (dashed line) for a hypothetical exponentially varying case such as shown in Fig. 20.3-6a.

(20.3-16) is clearly not very serious. Several rather extensive examples of comparisons of measured and calculated time lags for CO_2 and other gases in glassy polymers have been reported recently. In most cases up to the 20 atm limit which have been investigated, Fick's Law appeared to be obeyed well.

20.4 FUNDAMENTALS OF SORPTION AND TRANSPORT PROCESSES IN POLYMERS

20.4-1 General Discussion

The following two subsections provide physical interpretations of the forms of sorption isotherms and concentration-dependent diffusion coefficients observed for rubbery and glassy polymers, respectively. These sections are not required for simulation of module operations if a complete set of empirical pressure, temperature, and composition-dependent permeation data are available. A simple polynomial fit of permeability data as a function of all operating variables would suffice for design and simulation.

Generally, such detailed data are not available, and a theoretical framework is valuable to aid in the physical interpretation of new data. Moreover, in some cases an appreciation of fundamentals can explain observations that otherwise might be termed "anomalous" and cast doubt on the validity of a perfectly good particular set of data. Examples of this sort of situation can be cited for the case of glassy polymers in which the following consistent, but surprising, observations were reported by two separate investigators regarding H_2 permeation in the presence of a slower permeating gas (either CO or CH_4):

There is some interaction between components of the binary mixture during permeation which effectively decreases the permeability of H_2 relative to CO.[1]

The fast gas was slowed down by the slow gas and the slow gas was speeded up by the fast gas.[2]

Such behavior is not consistent with plasticization, since the slightly sorbing H_2 would not be expected to plasticize the polymer while the more condensable (and hence more strongly sorbing) CO or CH_4 acts as an antiplasticizer. As is discussed in the section on glassy polymers, such behavior as described here is not anomalous and can be shown to be consistent with inherent properties of glassy polymers under certain conditions.

The major emphasis of the following discussion is on glassy polymer membranes since the more rigid size- and shape-selecting glassy materials tend to have substantially higher selectivities than do flexible rubbers, which are essentially high-molecular-weight liquids. This effect can be seen clearly from comparison of the ideal separation factor data in Tables 20.4-1, 20.4-2, and 20.4-3 for rubbery and glassy polymers.[3] However, consistent with the preceding discussion of the two-part nature (solubility and mobility) of both permeability and selectivity, one might envision a rubbery polymer with sufficiently favorable solubility for one of the components in the mixture to make it useful as a thin film coating on a composite membrane. In such a case, plasticization of the separating membrane could be a less serious problem than for the materials whose permselection action relies on molecular sieving action based on penetrant size and shape. Therefore, treatment of transport in rubbers is reviewed prior to consideration of the more important glassy polymer case.

TABLE 20.4-1 Permeabilities and Separation Factors for Typical Rubbery Polymers at Low Pressures

Rubber	T (°C)	P_{He} (Barrer)[a]	P_{H_2} (Barrer)[a]	P_{CO_2} (Barrer)[a]	P_{CH_4} (Barrer)[a]	P_{O_2} (Barrer)[a]	P_{N_2} (Barrer)[a]	P_{He}/P_{CH_4}	P_{O_2}/P_{N_2}	P_{H_2}/P_{CH_4}	P_{CO_2}/P_{CH_4}
Silicone rubber	25	300	550	2,700	800	500	250	0.38	2.0	0.69	3.38
Silicone–polycarbonate block copolymer	25	—	210	970	—	160	70	—	2.29	—	—
Silicone–nitrite copolymer	25	86	123	670	—	85	33	—	2.58	—	—
Natural rubber	25	31	49	131	30	24	8.1	1.03	2.96	1.63	4.4
SBR	25	23.1	40.1	124	21	17	6.3	1.09	2.71	1.91	5.9
Acrylonitrile–butadiene copolymer (27% AN)	25	12.2	15.9	30.9	—	3.84	1.1	—	3.63	—	—
Acrylonitrile–isoprene (26% AN)	25	7.8	7.5	4.3	—	0.85	0.18	—	4.72	—	—
Butyl rubber	25	8.42	7.2	5.18	0.78	1.30	0.33	10.67	3.96	9.17	6.6
Polychloroprene	23	13	20	22	2.6	4.0	1.1	5.0	3.64	7.69	8.5
Polyethylene $\rho = 0.914$	25	4.9	—	12.6	2.9	2.9	0.97	1.69	2.99	—	4.3
Polyethylene $\rho = 0.964$	25	1.14	—	1.7	0.39	0.41	0.14	2.90	2.93	—	4.4
FEP copolymer[d]	30	62	14.1	—	1.4–23[b]	5.9	2.2	44–2.7[b]	2.68	10.1–0.6[b]	—
Teflon[d] 75% crystalline	30	11	12[c]	2.6	—	—	0.79	—	—	—	—
Polyethylene–vinyl acetate copolymer	25	30	—	25	30	—	—	1.00	—	—	0.8
Chlorosulfonated polyethylene	23	7.2	10.8	15.8	1.7	2.1	0.92	4.23	2.28	6.36	9.3
Nylon 66	25	1.0	—	0.17	—	0.034	0.008	—	4.25	—	—

[a] 1 Barrer = 10^{-10} [cm³ (STP)·cm]/[cm²·sec·cm Hg].

[b] Highly pressure dependent at low temperature. Above 60°C, pressure dependence diminishes greatly.

[c] Temperature = 33°C.

[d] These fluorocarbon polymers are difficult to describe unambiguously as rubber or glassy because of multiple thermal transitions.

Source: Stannett et al.[3]

TABLE 2.4-2 Helium and CO$_2$ Permeabilities and Separation Factors Relative to CH$_4$ in a Variety of Glassy Polymers

Polymer Structure	P_{He} (Barrer)^a	P_{CO_2}^{b} (Barrer)^a	P_{He}/P_{CH_4}^{b}	P_{CO_2}/P_{CH_4}
Poly(tetramethyl bis L terephthalate)	148	83.0	16	34
Poly(tetraisopropyl bis A-sulfone)	110	98.0	25	15
Poly(2-6-dimethyl-1,4-phenylene oxide)	60.0	50.0	27	18
Poly(tetramethylcyclo-butane diol carbonate)	75.0	32.0	38	14
Poly(tetramethyl bis A carbonate)	43.0	29.0	49	33
Poly(tetramethyl bis L sulfone)	48.3	65.0	28	24
Poly(tetramethyl bis A isophthalate)	31.0	—	60	—
Poly(tetramethyl bis A sulfone)	38.4	21.0	67	68
Poly(bis A carbonate) Lexan®	15.0	10.0	35	20
Poly(bis A-4,4'-N-methyl phenyl sulfone urethane)	6.05	—	93	—
Poly(ethylene terephthalate)	1.30	—	57	—

^{a} 1 Barrer $= 10^{-10}$ [cm^3 (STP)·cm]/[cm^2·sec·cm Hg].
^{b} Likely to be somewhat pressure dependent.

   Source: Stannett et al.[3] No temperature was reported in the reference; presumably, these data apply at ~30°C.

20.4-2 Rubbery Polymers

"Rubbery polymers" refers to amorphous polymeric materials that are above their softening or glass transition temperatures under the conditions of use.[4] Regions of crystallinity in rubbery matrices act as fillers, restricting segmental motion, and also as tortuous blocks to facile movement of penetrants. Reduction in the sorption level of a desirable permeating component also obviously results from reduction in the volume fraction of material available for accommodating permeant.

   As shown in Fig. 20.4-1, the solubility coefficient [C/p in Eq. (20.3-8)] of a large number of penetrants in polyethylene is essentially directly proportional to the volume fraction of amorphous material in the polymer.[5] Therefore, except to improve mechanical properties and resistance to dissolution by highly sorbing vapors, crystallinity generally is not too desirable since it can reduce productivity greatly. However, Michaels and Bixler[6] and Bixler and Sweeting[5] have discussed cases in which, besides reducing productivity, the introduction of crystallinity can impart an ability to discriminate among molecules of increasing size. The diffusivity of a penetrant in a semicrystalline polymer decreases relative to a totally amorphous sample. This reduction occurs because of the combined effects of added tortuosity and the restriction of amorphous chain mobility necessary for the opening of a transient molecular-scale passageway into which the penetrant can jump.

   In the work of Bixler and Sweeting[5] it was assumed that chain restriction effects are negligible for the compact helium molecule, so that only tortuosity (added path length) contributes to the reduction of the diffusion coefficient for this penetrant. Furthermore, for the 50% crystalline polyethylene sample studied, these authors suggest that the tortuosity effect is essentially nondiscriminating and causes about a 65% reduction in mobility compared to the hypothetical totally amorphous case. The transport properties of a hypothetical totally amorphous branched polyethylene were assumed by these authors to be equivalent to those of unvulcanized natural rubber.

   The ratios of diffusivities for a series of penetrants in the 50% crystalline polyethylene ($D_{0.5}$) and in the hypothetical amorphous sample ($D_{1.0}$) are shown in Fig. 20.4-2. In addition to the substantial nondiscriminating reductions in mobility, there is an additional size-dependent reduction in mobility due to chain restriction effects. Note that since the solubilities of all the components are proportional to the amorphous fraction (Fig. 20.4-1), selectivity enhancements resulting from the introduction of crystallinity rely solely on the chain restriction effect.

TABLE 2.4-3 Hydrogen Permeabilities and Separation Factors Relative to CH₄ and CO for a Variety of Glassy Polymers

Material	P_{H_2} (Barrer)[a]	P_{H_2}/P_{CH_4}[b]	P_{H_2}/P_{CO}[b]
3,5 DBA-6F[c]	33.6	340.6	—
1,5 ND-6F[d]	78.1	108.5	—
PPD-6F[e]	48.2	152.5	—
Polystyrene	11.0	—	12
PVC	8.0	—	12.9
Parylene® N[f]	2.8	—	25.4
Cellulose acetate[g]	13.0	—	37.2
Sulfone[h]	14.0	—	37.8
Polyvinyl fluoride[i]	0.6	—	66.7
Mylar® type S[j]	1.4	—	74.0
Polyimide[k]	2.0	—	74.0
Parylene® C[l]	1.4	—	110.0
Polycaprolactam[m]	1.5	—	115.0

[a] 1 Barrer $= 10^{-10}$ [cm³(STP)·cm]/[cm²·s·cm Hg].
[b] Possibly dependent on driving pressure.
[c] Soluble polyimide prepared by reaction of 4,4'-hexafluorisopropylidene diphthalic anhydride (6F) with 3,5-diaminobenzoic acid.
[d] Soluble polyimide prepared by reaction of (6F) with 1,5-diaminonaphthalene.
[e] Soluble polyimide prepared by reaction of (6F) with 1,4-diaminobenzene.
[f] Poly(para-xylylene) Union Carbide registered trademark.
[g] du Pont 100 CA-43.
[h] Union Carbide Sulfone 47.
[i] du Pont Tedlar®.
[j] du Pont semicrystalline poly(ethylene terephthalate) film.
[k] du Pont Kapton® polyimide.
[l] Poly(monochloro-paraxylylene) Union Carbide registered trademark.
[m] Allied Chemical, amorphous nylon film.

Source: Stannett et al.[3]

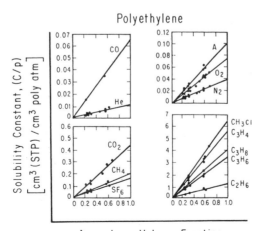

FIGURE 20.4-1 Linear relationship between the infinite dilution solubility coefficient and the amorphous fraction in the polymer. The data are for polyethylene at 25°C.[5]

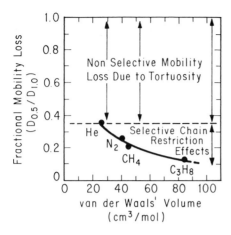

FIGURE 20.4-2 Ratio of diffusivities in a 50% crystalline and a 0% crystalline polyethylene as a function of the van der Waals volume of the penetrant.[8] The diffusion coefficients for amorphous polyethylene were taken to be equivalent to those of unvulcanized natural rubber.[5]

As shown in Fig. 20.4-3, this added effect can improve helium selectivity relative to methane by a factor of 1.75. At the same time, a factor of 2.0 loss in helium solubility and a factor of 2.8 reduction in helium diffusivity due to the crystallinity produces a 5.6-fold reduction in helium productivity relative to the hypothetical amorphous film case. Interestingly, the effects of chain restriction on different sizes of penetrants were found by Michaels to be effective only for rubbery materials, not glassy ones. A logical explanation for this fact was advanced in terms of the already low mobility of chain segments in the glass compared to the rubbery state:

> In glassy amorphous or crystalline polyethylene terephthalate, a diffusing molecule must pass between chain segments which have little rotational mobility. Physically, the concept of chain immobilization by crystallites loses its significance, since the rigidity of the polymer backbone itself apparently outweighs any additional restriction on mobility imposed by crystallinity. A consequence of the absence of this effect in polyethylene terephthalate is the virtual equality, within precision limits, of apparent activation energies in the glassy amorphous and crystalline polymers.[7]

Kreituss and Frisch[9] also have treated the effect of crystallinity on transport in rubbery polymers using the so-called free-volume approach, which is discussed below. In essence, their ideas suggest that introduction of crystallinity reduces the freedom of motion of amorphous chain segments between crystals. By

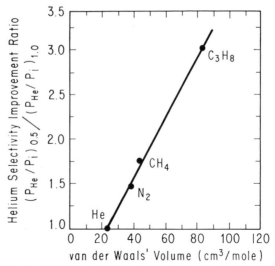

FIGURE 20.4-3 Ratio of selectivities of helium relative to component i (N_2, CH_4, or C_3H_8) for 50% crystalline and amorphous polyethylene as a function of the van der Waals volume of penetrant i[8].

expressing these concepts in the terminology of the free-volume theory, a different analytical form of the diffusion coefficient results from that suggested by Michaels and coworkers. The physical ideas, however, are consistent with those discussed by Bixler and Sweeting[5] and Michaels and Bixler.[6]

Several versions of the free-volume theory of diffusion have been promulgated following its introduction for polymeric systems in the work of Fujita.[10] Other more fundamentally based molecular approaches have been attempted at times based on actual molecular properties of the polymer and penetrant; however, their greater complexity has discouraged widespread application. The free-volume theory is simple and has evolved through the work of Vrentas and Duda[11,12] and Stern, Frisch, and coworkers[13,14] to a readily useful form. Its primary drawback lies in the difficulty in providing a precise physical definition for the parameter defined as the free volume. Excellent discussions of the theory have been offered, and the reader is referred to them for greater detail than that provided here.[12,14]

In general, D can be a function of local concentration as discussed in the context of Fig. 20.3-6. As noted by Stern, a convenient form of Fick's Law can be written as shown in Eq. (20-4.1) in terms of the local volume fraction of the penetrant in the polymer, v, and the mutual diffusion coefficient D.[13]

$$N = -\frac{D}{1-v}\frac{dv}{dx}$$ (20.4-1)

The volume fraction of penetrant is related easily to penetrant sorption concentration if the partial molar volume of the penetrant can be approximated by its molar volume as a hypothetical liquid. Values of the infinite dilution partial molar volume of various gases in a series of low-molecular-weight liquids are shown in Table 20.4-4.[15] Since rubbery polymers are essentially high-molecular-weight liquids, the average values shown in the table for the various penetrants would be reasonable estimates of the partial molar volume in polymeric media also.

The so-called thermodynamic diffusion coefficient D_T is related to D by Eq. (20.4-2):

$$D_T = \frac{D/(1-v)}{[(\partial \ln a)/(\partial \ln v)]_T}$$ (20.4-2)

For gases, because of their relatively low solubilities, $[(\partial \ln a)/(\partial \ln v)]_T$ may be taken as a constant close to 1.0,[13] so that Eq. (20.4-2) becomes

$$N = -D_T \frac{dv}{dx}$$ (20.4-3)

The diffusion coefficient appearing in Eq. (20.4-3) is a true measure of the molecular mobility of the penetrant in question. Intuitively, the free-volume theory proponents argue that a penetrant can execute a diffusive jump when a free-volume element greater than or equal to a critical size presents itself to a penetrant. The native polymer, totally devoid of penetrant, still possesses a certain amount of free-volume packets of distributed size which wander spontaneously and randomly through the rubbery matrix. In fact, when a packet of sufficient size presents itself to a polymer segment, the polymer may execute a self-diffusive motion, and this is the action that causes slow interdiffusion of polymer chains.

Kreituss and Frisch have modified Fujita's ideas by arguing that introduction of crystalline material reduces the free volume in the composite semicrystalline structure in direct proportion to the amount of crystalline material present.[9] Thus,

$$D_T = RTA_d \exp\left(\frac{-B_d}{\phi_a V_f}\right)$$ (20.4-4)

TABLE 20.4-4 Partial Molal Volumes $\overline{V}_\infty$ of Gases in Liquid Solution at 25°C

	H_2	N_2	CO	O_2	CH_4	C_2H_2	C_2H_4	C_2H_6	CO_2	SO_2
Ethyl ether	50	66	62	56	58	—	—	—	—	—
Acetone	38	55	53	48	55	49	58	64	—	68
Methyl acetate	38	54	53	48	53	49	62	69	—	47
Carbon tetrachloride	38	53	53	45	52	54	61	67	—	54
Benzene	36	53	52	46	52	51	61	67	—	48
Methanol	35	52	51	45	52	—	—	—	43	—
Chlorobenzene	34	50	46	43	49	50	58	64	—	48

Source: Reid et al.[15]

where A_d and B_d are free-volume parameters that must be evaluated empirically. The parameter ϕ_a is the amorphous volume fraction, and V_f is the so-called free-volume fraction of a totally amorphous sample in the presence of penetrant at the system hydrostatic pressure and temperature.[9]

Increases in system temperature give rise to volume dilation, resulting in increased amounts of free volume. This explains the strong tendency for the diffusion coefficient to increase with temperature. The free-volume fraction in Eq. (20.4-4) may be represented as a linear addition of the several variables that affect its value:

$$V_f = V_{fs}^\circ + \alpha(T - T_s) - \beta(p - p_s) + \gamma v \tag{20.4-5}$$

where V_{fs}° is the fractional free volume of the pure, penetrant-free, amorphous, rubbery polymer at some reference temperature T_s (usually the glass transition) and reference pressure p_s (usually 1 atm). The coefficients α, β, and γ are positive constants whose values are evaluated empirically. These coefficients characterize the effectiveness of temperature, pressure, and penetrant concentration for increasing (or in the case of pressure, decreasing) the fractional free volume in the amorphous phase that is accessible to penetrant movement.

In principle, one can evaluate each of these coefficients somewhat independently by observing the effects on D_T over a sufficiently wide range of temperatures, external hydrostatic pressures, and sorbed concentrations. The values of D_T for analysis in this manner can be determined by the techniques described earlier using Eq. (20.3-6) under different temperature and hydrostatic pressure conditions. The temperature, pressure, and penetrant volume fraction effects on the observed values of D_T then can be fit by nonlinear least squares to the expression for D_T given by substitution of Eq. (20.4-5) into Eq. (20.4-4):

$$D_T = RTA_d \exp\left(\frac{-B_d}{\phi_a\left[V_{fs}^\circ + \alpha(T - T_s) - \beta(p - p_s) + \gamma v\right]}\right) \tag{20.4-6}$$

One could decouple hydrostatic effects from penetrant sorptive effects by using an increasingly high pressure of an effectively nonsorbing pressurizing medium such as helium in the presence of a fixed partial pressure of the penetrant of interest. Hydrostatic pressure would be expected to have a rather small negative effect on D_T (and V_f) since solid polymers are slightly compressible. On the other hand, increases in temperature and sorbed penetrant concentrations may cause large increases in D_T (and V_f). Because of the low solubility of supercritical gases, Henry's Law typically is obeyed up to high pressures for the solubility relationship, and dC/dp is a constant in Eq. (20.3-3). Because of the relatively low volume fractions of sorbed material present in such cases, strong plasticization generally is not observed for supercritical components.

For more strongly sorbing penetrants, such as CO_2 and C_2 and C_3 hydrocarbons, BET III isotherms indicative of Flory–Huggins swelling behavior are common. The Flory–Huggins model, shown below, relates the penetrant activity (a_i) to the sorbed volume fraction v of the component. Typically, the activity is replaced by the ratio of the penetrant partial pressure p and the component vapor pressure p_0:

$$\ln\left(\frac{p}{p_0}\right) = \ln(v) + (1 - v) + \chi(1 - v)^2 \tag{20.4-7}$$

where χ is the Flory–Huggins parameter that characterizes the polymer–penetrant interactions in the dissolved state. For systems that show a pronounced deviation from Henry's Law, one can anticipate a corresponding tendency to exhibit plasticization in the $D(C)$ versus C response, such as shown in Fig. 20.3-6a.

Stern and coworkers have shown that deviations from constant diffusion coefficient behavior generally occur in rubbery polymers before the onset of deviations from Henry's Law solubility behavior.[16] In either case, substitution of Eq. (20.4-6) along with the appropriate isotherm form into Eq. (20.3-2) permits graphical or analytical integration to obtain the permeability in terms of the upstream and downstream penetrant volume fractions or concentrations. Even for the case in which Henry's Law applies, the resulting expression is rather complicated and is not reproduced here in its general form.

To obtain an expression for permeability in terms of the actual upstream and downstream pressures, one must employ the appropriate isotherm equation. In the case of Henry's Law behavior, $v = Kp$, the following expression can be used to a good approximation for the usual case in which the downstream pressure is negligible relative to the upstream pressure, that is, $p_1 \ll p_2$:[13]

$$\ln P = \ln(RTA_d K) - \frac{B_d}{\phi_a V_f^*} + \frac{B_d/\phi_a}{(V_f^*)^2}\left[-\beta + \frac{\gamma K}{2}\left(1 + \frac{2\beta p_2}{V_f^*}\right)\right]p_2 \tag{20.4-8}$$

where V_f^* is the fractional free volume of the penetrant-free polymer at $p = 0$ and the temperature of the system. All other parameters were defined earlier. If the reference states mentioned earlier are used (1 atm and the glass transition of the polymer), V_f^* can be determined simply in terms of previously specified parameters:

$$V_f^* = V_{fs}^\circ + \alpha(T - T_g) + \beta p_s \qquad (20.4\text{-}9)$$

Therefore, although rather cumbersome, the free-volume theory permits one to prepare theoretical plots of permeability as a function of temperature, penetrant pressure, and amorphous volume fraction in the rubbery polymer.

Vrentas and Duda also have approached the problem of penetrant sorption and diffusion in rubbery polymers in terms of the free-volume theory. Most of their work, however, focuses on the problem of larger penetrant molecules such as benzene and the like. Their approach permits remarkably accurate descriptions of diffusion coefficient behavior over a wide range of temperatures and penetrant activities using a relatively small amount of information, such as the viscosity of the penetrant-free polymer and the effective glass transition temperature of the penetrant. This approach is especially useful in dealing with highly swelling vapors and liquids, but it is beyond the scope of the current discussion dealing with small penetrants. The interested reader is directed to several references describing their work.[11,12,17]

20.4-3 Glassy Polymers

Glassy polymers are inherently more size and shape selective than rubbery materials. This additional "mobility selectivity" arises from the restrictive nature of transport in glassy environments, which are characterized by extremely limited segmental motions. Because of their large potential for use as monolithic membranes and selective coatings on composite membranes, a fairly detailed discussion of the sorption and transport behavior of glassy polymers is presented in the following section.

As noted earlier in the context of Tables 20.4-1, 20.4-2, and 20.4-3, the ratio of pure component permeabilities at an arbitrary pressure is a useful indication of the selectivity of a candidate membrane for the chosen gas pair. Nevertheless, because of a variety of factors, the actual selectivity and productivity observed in the mixed gas case may be somewhat different than predicted on the basis of the pure component data. One of these factors, illustrated in Fig. 20.4-4, is important for low- and intermediate-pressure applications and is believed to be due to competition by mixture components for unrelaxed molecular-scale gaps between glassy-polymer chain segments. As shown in the left-hand side of the figure, the presence of a second component B can depress the observed permeability of a component A relative to its pure component value at a given upstream driving pressure of component A. Hoehn and coworkers note that the permeability of a membrane to a component A may be reduced due to the sorption of a second component B in the polymer which ". . . effectively reduces the microvoid content of the film and the available diffusion paths for the nonreactive gases."[18]

At higher pressures, where microvoid saturation processes are essentially complete, additional sorption simply plasticizes the membrane, making it more rubberlike. As shown in Fig. 20.4-4, this higher activity range corresponds to upswings in the observed permeability due to free-volume effects like those discussed previously for simple rubbery polymers and illustrated in Fig. 20.3-2d. Stern and Saxena have treated these effects theoretically for glassy polymers.[19] Paulson et al.[20] have reported such plasticizing effects for the CO_2–CH_4 system at high relative humidities for a variety of polymers. The effects become especially noticeable at higher temperatures.

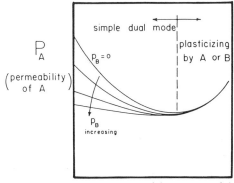

upstream partial pressure of A

FIGURE 20.4-4 Permeability of a glassy polymer to penetrant A in the presence of varying partial pressures of penetrant B.

The two preceding effects are due to true polymer-phase sorption and transport phenomena. At very high pressures, an additional complexity related to nonideal gas-phase effects may arise and cause potentially incorrect conclusions about the transport phenomena involved. This confusion can be avoided by defining an alternative permeability $P^{\#}$ in terms of the fugacity difference rather than the partial pressure difference driving diffusion across the membrane. The benefit of using this "thermodynamic permeability" is illustrated for the CO_2–CH_4 system at elevated pressures.

The pressure level at which the various nonidealities discussed above become important is dependent primarily on the condensability of the penetrant. The most convenient measure of condensability is the penetrant vapor pressure. Even for slightly supercritical components, a hypothetical vapor pressure can be obtained by extrapolation of a semilog plot of vapor pressure versus the inverse absolute temperature up to the system temperature. Water has a low vapor pressure at ambient temperatures and is quite condensable. Only a few mm Hg of water vapor may cause the permeability depression effect shown in the left-hand side of Fig. 20.4-4 for typical less condensable penetrants such as CH_4 and H_2. Conversely, high partial pressures of supercritical components such as N_2, CH_4, CO, and H_2 would be required to cause noticeable depression of each other's permeabilities. Nevertheless, as noted earlier, measurable examples of this effect have been observed even in the H_2–CO and H_2–CH_4 cases where the faster permeating H_2 appeared to have been "slowed down" by the slower CO and CH_4.

Many of the important applications for the separation of CO_2 and CH_4 correspond to CO_2 partial pressures well above 20 atm (e.g., cases 1–3 in the previous discussion of enhanced oil recovery applications). In such cases, competition for unrelaxed volume should be of second-order importance, since the "condensable" nature of CO_2 causes it to dominate the microvoid environments in the polymer even in the face of substantial methane partial pressures. At high CO_2 partial pressures, plasticizing behavior such as that illustrated in Fig. 20.4-4 may be observed. On the other hand, at relatively low CO_2 pressures (20 atm) (cases 4 and 5 in the discussion of enhanced oil recovery applications), competition effects would be anticipated for most glassy polymers. These effects could be masked in the case of cellulose acetate because of its strong tendency to interact with CO_2 and exhibit plasticization at low partial pressures.

The magnitude of the partial pressure of a given penetrant required to cause strong plasticization, illustrated on the right-hand side of Fig. 20.4-4, depends on the magnitude of specific polymer–penetrant interactions and is not currently predictable from first principles. Fortunately, the onset of nonideal gas-phase effects and their effects on the driving force for component permeation can be predicted from existing thermodynamic expressions such as the Soave modification of the Redlich–Kwong equation.[15] This information permits one to decouple and analyze the polymer-related transport and sorption nonidealities without confusion from simple gas-phase nonidealities.

UNRELAXED VOLUME AND ITS RELATION TO SORPTION AND TRANSPORT PROPERTIES OF GLASSY POLYMERS

The concept of "unrelaxed volume", $V_g - V_l$, illustrated in Fig. 20.4-5, may be used to interpret sorption and transport data in glassy polymers exposed to pure and mixed penetrants. The extraordinarily long relaxation times for segmental motion in the glassy state lead to trapping of nonequilibrium chain conformations in quenched glasses, thereby permitting minuscule gaps to exist between chain segments. These gaps can be redistributed by penetrant in so-called conditioning treatments during the initial exposure of the polymer to elevated pressures of the penetrant.[21]

Following such initial exposure to penetrant, settled isotherms characterized by concavity to the pressure axis at low pressures and a tendency to approach linear high-pressure limits are observed as shown in Figs. 20.4-6 and 20.3-3c. Reference to penetrant-induced conditioning effects have been made continuously since the very earliest investigations of penetrant–glassy polymer sorption behavior.[21–23] If one "overswells" the polymer with a highly sorbing penetrant, such as a high-activity vapor, and then removes the penetrant, the excess volume which is introduced will tend to relax quickly at first, followed by a very slow, long-term approach toward equilibrium.[24]

At extremely high gas conditioning pressures, substantial swelling of the polymer sample can occur, with a resultant increase in the value of $V_g - V_l$.[23] After conditioning with gases such as CO_2 at pressures less than 30 atm, however, consolidation in the absence of penetrant is typically unmeasurable, since penetrant-induced swelling is not very extensive (less than 4–5% uptake by weight).[23,24] As a result, in such cases *redistribution* of the originally present intersegmental gaps may be the primary process occurring during the first exposure of the polymer to high pressures of penetrant as shown in Fig. 20.4-6.[21] At high pressures, sorption isotherms such as that shown in Fig. 20.3-3d are anticipated since Flory–Huggins-type behavior typical of Fig. 20.3-3b is superimposed on the simple dual-mode isotherm shape shown in Fig. 20.3-3c. This rather complex topic is beyond the scope of the present discussion.

As noted previously, an interpretation of conditioning that occurs during the primary penetrant exposure in the *absence* of large swelling effects may be couched in terms of coalescence of packets of the original intersegmental gaps. Redistribution of chain conformations, consistent with optimal accommodation of the penetrant in the unrelaxed volume between chain segments, may permit this process during the conditioning treatment. This rearrangement would tend to produce a more or less densified matrix with a small volume fraction of essentially uniformly distributed molecular-scale gaps or "holes" throughout the matrix. In such

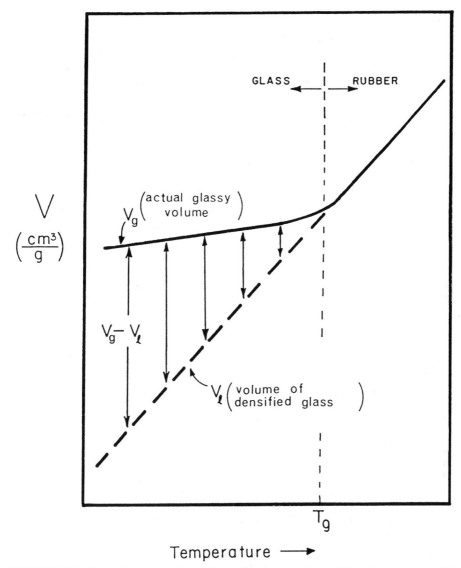

FIGURE 20.4-5 Schematic representation of the unrelaxed volume $V_g - V_l$ in a glassy polymer. Note that the unrelaxed volume disappears at the glass transition temperature T_g.

a situation, one can appreciate the meaning of two slightly different molecular environments in the glass in which sorption of gas may occur.

Consider first the limiting case in which a highly annealed, truly equilibrium densified glass characterized by V_l in Fig. 20.4-5 is exposed to a given pressure of a penetrant. In this hypothetical case, all gaps are missing, but clearly there still will be a certain characteristic sorption concentration (C_D) typical of true molecular dissolution in the densified glass. This sorption environment is believed to be similar to that observed in low-molecular-weight liquids or rubbers (above T_g).

Next, consider a corresponding conditioned *nonequilibrium* glass (illustrated by V_g in Fig. 20.4-5) containing unrelaxed volume in which the surrounding matrix has been densified more or less by the coalescence of gaps to form molecular-scale berths for penetrant. A local equilibrium requirement leads to an average local concentration of penetrant held in the uniformly distributed molecular-scale gaps (C_H) in equilibrium with the "dissolved" concentration (C_D) at any given external penetrant pressure or activity.

This simple physical model can be described analytically up to reasonably high pressures (generally for

Dual-Mode Sorption

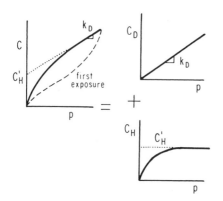

FIGURE 20.4-6 Schematic representation of the dual-mode concept. The sorption observed during the first exposure to the penetrant is often considerably different from that observed after the isotherm has "settled" at the highest pressure of measurement. The conditioning process associated with the first exposure to the penetrant is believed to be associated with redistribution of unrelaxed volume elements in the polymer.

pressures less than or equal to the maximum conditioning pressure employed[21]) in terms of the sum of Henry's Law for C_D and a Langmuir isotherm for C_H:

$$C = C_D + C_H \qquad (20.4\text{-}10)$$

$$C = k_D p + \frac{C_H' b p}{1 + b p} \qquad (20.4\text{-}11)$$

where k_D is the coefficient, analogous to S in Eq. (20.1-4), that characterizes sorption of penetrant in the densified regions that comprise most of the interchain gap in the polymer. The parameter C_H' is the Langmuir sorption capacity of the glassy matrix and can be interpreted directly in terms of Fig. 20.4-5, in which the unrelaxed volume, $V_g - V_l$, corresponds to the summation of all of the molecular-scale gaps in the glass. As shown in Fig. 20.4-7, the Langmuir capacity of glassy polymers tends to approach zero in the same way that $V_g - V_l$ approaches zero at the glass transition temperature. Since C_H', b, and k_D all tend to decrease with increasing temperature, the sorption uptake at a given pressure tends to decrease with increasing temperature.

The preceding qualitative observations about the temperature dependence of C_H' and $V_g - V_l$ can be extended to a quantitative statement in cases for which the effective molecular volume of the penetrant in the sorbed state can be estimated. As a first approximation, one may assume that the effective molecular volume of a sorbed CO_2 molecule is 80 Å^3 in the range of temperatures from 25 to 85°C. This molecular volume corresponds to an effective molar volume of 49 cm^3/mol of CO_2 molecules and is similar to the partial molar volume of CO_2 in various solvents, in several zeolite environments, and even as a pure subcritical liquid (see Tables 20.4-4 and 20.4-5).[21,25] The implication here is *not* that more than one CO_2 molecule exists in each molecular-scale gap, but rather that the effective volume occupied by a CO_2 molecule is roughly the same in the polymer sorbed state, in a saturated zeolite sorbed state, and even in a dissolved or liquidlike state since all these volume estimates tend to be similar for materials that are not too much above their critical temperatures. With the above approximation, the predictive expression given below for C_H' can be compared to independently measured values for this parameter from sorption measurements:

$$C_H' = \frac{V_g - V_l}{V_g} \rho^* \qquad (20.4\text{-}12)$$

where ρ^* is the equivalent density of CO_2 ($\frac{1}{49}$ mol/cm^3) discussed above. The comparison of measured and predicted C_H' calculated from Eq. (20.4-12) using reported dilatometric parameters for various glassy polymers is shown in Fig. 20.4-8.[21,28,29] The correlation is clearly impressive.

Application of Eq. (20.4-12) to highly supercritical gases is somewhat ambiguous since the effective molecular volume of sorbed gases under these conditions is not estimated easily. A similar problem exists in a priori estimates of partial molar volumes of supercritical components even in low-molecular-weight liquids.[30] The principle on which Eq. (20.4-12) is based remains valid, however, and while the total amount of unrelaxed volume may be available for a penetrant, the magnitude of C_H' depends strongly on how condensable the penetrant is, since this factor determines the relative efficiency with which the component can use the available volume.

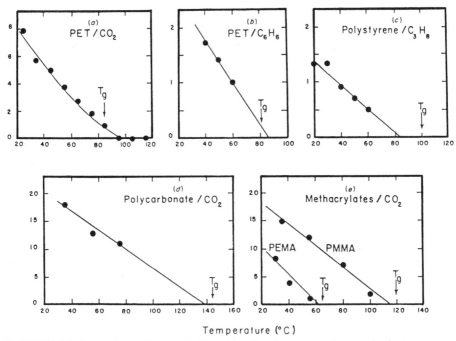

Temperature (°C)

FIGURE 20.4-7 Langmuir sorption capacity C_H as a function of temperature for several polymer–penetrant systems. Note that C_H disappears near T_g in the same fashion as $V_g - V_l$ does in Fig. 20.4-5.[26]

TRANSPORT

A companion transport model that also acknowledges the fact that penetrant may execute diffusive jumps into and out of the two sorption environments expresses the local flux N at any point in the polymer in terms of a two-part contribution:[32-34]

$$N = -D_D \frac{\partial C_D}{\partial x} - D_H \frac{\partial C_H}{\partial x} \tag{20.4-13}$$

where D_D and D_H refer to the mobility of the dissolved and Langmuir sorbed components, respectively. It is typically found that D_D is considerably larger than D_H except for noncondensable gases such as helium.[32,35,36]

The two transport coefficients obey Arrhenius expressions with the activation energy for D_H tending to be slightly larger than for D_D.[35] The above expression also can be written in terms of Fick's Law with an effective diffusion coefficient $D_{eff}(C)$ that is dependent on local concentration:

TABLE 20.4-5 Effective Molecular Size of CO_2 in Various Dissolved and Zeolitic Environments

Environment	Effective Volume at 25°C per CO_2 Molecule ($\overset{\circ}{A}{}^3$)
Carbon tetrachloride	80.0
Chlorobenzene	74.1
Benzene	79.5
Acetone	74.2
Methyl acetate	73.9
4 or 5 Å Zeolites at saturation of sorption capacity	86.9

Note: Average of above values = 78.1 $\overset{\circ}{A}{}^3$.
  Source: Barrer and Barrie[22] and Chan and Paul.[27]

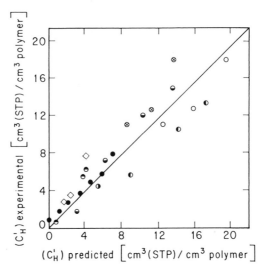

FIGURE 20.4-8 Quantitative comparison of experimentally measured values of C'_H for CO_2 in various polymers with the predictions of Eq. (20.4-12): $\bullet$ = poly(ethylene terephthalate) at 25, 35, 45, 55, 65, 75, and 85°C. $\mathbb{O}$ = poly(benzyl methacrylate) at 30°C. $\mathbb{C}$ = poly(phenyl methacrylate) at 35, 50, and 75°C. $\diamond$ = poly(acrylonitrile) at 35, 55, and 65°C. $\bigcirc$ $\otimes$ = polycarbonate at 35, 55, and 75°C: the $\bigcirc$ and $\otimes$ points refer to predictions using dilatometric coefficients from different sources.[29,31] $\ominus$ = poly(ethyl methacrylate) at 30, 40, and 55°C. $\ominus$ = poly(methyl methacrylate) at 35, 55, 80, and 100°C.

$$N = -D_{\text{eff}}(C) \frac{\partial C}{\partial x} \qquad (20.4\text{-}14)$$

The dual-mobility model expresses the concentration dependency of $D_{\text{eff}}(C)$ in terms of the local concentration of dissolved penetrant, C_D, as shown in Eq. (20.4-15):

$$D_{\text{eff}}(C) = D_D \left[\frac{1 + FK/(1 + \alpha C_D)^2}{1 + K/(1 + \alpha C_D)^2} \right] \qquad (20.4\text{-}15)$$

where $F \equiv D_H/D_D$, $K \equiv C'_H b/k_D$, and $\alpha \equiv b/k_D$. This model explains concentration dependency of the local diffusion coefficient such as that shown for CO_2 in poly(ethylene terephthalate) (PET) in Figs. 20.3-6c and 20.4-9 in terms of a progressive increase in the concentration fraction that is present in the higher mobility Henry's Law environment as the local Langmuir capacity saturates at higher pressures.

The points in Fig. 20.4-9 were evaluated from the permeability and sorption concentration data using Eq. (20.3-6), which do not depend on the dual-mode model in any way. The line through the data points corresponds to the predictions of $D_{\text{eff}}(C)$ using Eq. (20.4-15) along with the independently determined dual-mode parameters for this system.[35] It is also important to note that the form of data in this plot which exhibit a tendency to form an asymptote at high pressures is not typical of plasticization.

Finally, if one considers the low-concentration region of Fig. 20.4-9, it is clear that the diffusion coefficient is surprisingly concentration dependent. For example, $D_{\text{eff}}(C)$ increases by more than 35% as the local sorbed concentration rises from 0.142 cm³ (STP)/cm³ polymer (or 0.00020 wt. fraction) at 50 mm Hg to 1.7 cm³ (STP)/cm³ polymer (or 0.0025 wt. fraction) at 760 mm Hg CO_2 pressure. The above concentrations are extremely dilute, corresponding to less than 1 CO_2/1200 PET repeat units and 1 CO_2/98 repeat units, respectively.[35]

An even more dramatic case corresponds to CO_2 in PVC in which $D_{\text{eff}}(C)$ rises by 86% as the local concentration increases from 1 CO_2/4400 repeat units at 100 mm Hg to 1 CO_2/1300 repeat units at 500 mm Hg at 40°C. The above values of CO_2 solubility are based on measurements by Toi and by Tikhomorov and coworkers, which are in good agreement.[38,39]

At such extraordinarily low penetrant concentrations, plasticization of the overall matrix is not anticipated. As noted by Michaels in the previous discussion concerning crystallinity, motions involving relatively few repeat units are believed to give rise to most short-term glassy-state properties such as diffusion. In rubbery polymers, on the other hand, longer-chain concerted motions occur over relatively short time scales, and one expects plasticization to be easier to induce in these materials. Interestingly, no known

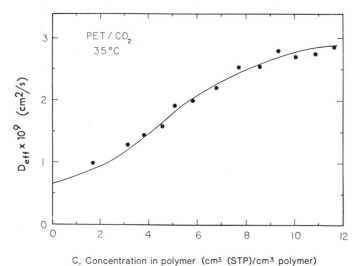

C, Concentration in polymer (cm³ (STP)/cm³ polymer)

FIGURE 20.4-9   Local concentration-dependent diffusion coefficient for CO_2 in poly(ethylene terephthalate).

transport studies in rubbers have indicated plasticization at the low sorption levels noted above for PVC and PET.

The discussion directly following Eq. (20.4-15) provides a simple, physically reasonable explanation for the preceding observations of the marked concentration dependence of $D_{eff}(C)$ at relatively low concentrations. Clearly, at some point, the assumption of concentration independence of D_D and D_H in Eq. (20.4-15) will fail; however, for work with most "conditioned" glassy polymers at CO_2 pressures below 300 psi, these plasticization effects are small. Even at a CO_2 pressure of 10 atm, there still will be less than 1 CO_2 molecule/20 PET repeat units at 35°C.

In the case of cellulose acetate, however, it does appear that breakdown of the assumption of constant values of D_H and D_D occurs. Presumably, this hypersensitivity of cellulose acetate may derive from its apparent special affinity for CO_2 discussed in the context of Table 20.3-1. Stern and Saxena have described a generalized form of the dual-mode transport model that permits handling situations in which nonconstancy of D_D and D_H manifest themselves.[19] It is reasonable to assume that the next generation of gas separation membrane polymers will be even more resistant to plasticization than polysulfone and cellulose acetate, so that the assumption of constancy of these transport parameters should be well justified.

Although not necessary in terms of phenomenological applications, it is interesting to consider possible molecular meanings of the coefficients D_D and D_H. If two penetrants exist in a polymer in the two respective modes designated by D and H to indicate the "dissolved" (Henry's Law) and the "hole" (Langmuir) environments, the molecules can execute diffusive movements within their respective modes, or they may execute intermode jumps.

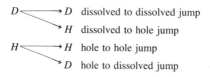

Clearly, the true character (activation energy, entropy, and jump length) of the phenomenologically observed D_D will be a weighted average of the relative frequency of $D \rightarrow D$ and $D \rightarrow H$ jumps, and likewise for D_H in terms of $H \rightarrow H$ and $H \rightarrow D$ jumps. Given the relatively dilute overall volume fraction associated with the nonequilibrium gaps that comprise the H environment (less than 4–5% on a volume basis), one may as a first approximation assume that most diffusive jumps of a penetrant from a D environment result in movement to another D environment, and most diffusive jumps from H environments result in movement to a D environment. The observed activation energies, entropies, and jump lengths therefore have fairly well defined meanings on a molecular scale.

The appropriate equation derived from the dual-mode sorption and transport models for the steady-state permeability of a pure component in a glassy polymer is given by Eq. (20.4-16) when the downstream

Permeability x 10, (Barrers)

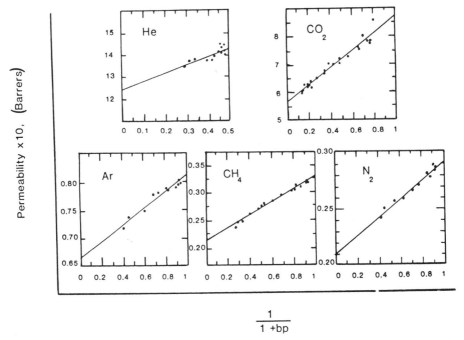

$$\frac{1}{1+bp}$$

FIGURE 20.4-10 Permeability of various gases in polycarbonate at 35°C as a function of the linearizing variable, $1/[1 + bp]$, in Eq. (20.4-16).[32] Similar good linear fits are observed for many other systems.

receiving pressure is effectively zero and the upstream driving pressure is p.[35]

$$P = k_D D_D \left(1 + \frac{FK}{1 + bp}\right) \tag{20.4-16}$$

The first term in Eq. (20.4-16) describes transport through the Henry's Law environment, while the second term is related to the Langmuir environment. The tendency shown in Fig. 30.3-2c for the permeability to approach a limiting value equal to $k_D D_D$ at high pressures is clear from this equation. Physically, the asymptote is approached because after saturation of the upstream Langmuir capacity at high pressures, additional pressure increases result in additional flux contributions *only* from the term related to Henry's Law, which continues to increase as upstream pressure increases.

The remarkable efficacy of the dual-mode sorption and transport model for description of pure component data has been illustrated by plots of the linearized forms of Eq. (20.4-16) for a wide number of polymer–penetrant systems.[21,23,29,34,37,40,41] Typical examples of such data are shown in Fig. 20.4-10 for various gases in polycarbonate.[32] These linearized plots are stringent tests of the ability of the proposed functional forms to describe the phenomenological data. Assink also has investigated the dual-mode model using a pulsed NMR technique and concluded the following:

> We have been able to demonstrate the basic validity of the assumptions on which the dual mode model is based and we have shown the usefulness of NMR relaxation techniques in the study of this model.[42]

MIXED COMPONENT SORPTION AND TRANSPORT

Arguments similar to those presented above for pure components have been extended to generalize the expressions given in Eq. (20.4-11) and Eq. (20.4-16) to account for the case of mixed penetrants.[43,44] The appropriate expressions are given below:

$$C_A = k_{DA} p_A + \frac{C'_{HA} b_A p_A}{1 + b_A p_A + b_B p_B} \tag{2.4-17}$$

$$P_A = k_{DA} D_{DA} \left(1 + \frac{F_A K_A}{1 + b_A p_A + b_B p_B}\right) \tag{20.4-18}$$

In the above expressions, A refers to the component of primary interest while B refers to a second, "competing" component.

As noted in the context of Eq. (20.3-8), the permeability of a polymer to a penetrant depends on the multiplicative contribution of a solubility and a mobility term. These two factors may be functions of local penetrant concentration in the general case as indicated by the dual-mode model. Robeson has presented data for CO_2 permeation in polycarbonate in which both solubility and diffusivity are reduced as a result of antiplasticization caused by the presence of the strongly interacting 4,4'-dichloro diphenyl sulfone.[45] On the other hand, sorption of a less-interacting penetrant, such as a hydrocarbon, may affect primarily only the solubility factor, without significantly changing the inherent mobility of the penetrant in either of the two modes. Flux reduction in this latter context occurs simply because the concentration driving force of penetrant A is reduced. This results from the exclusion of A by component B from Langmuir sorption sites that previously were available to penetrant A in the absence of penetrant B.

Consistent with the preceding discussion concerning sorption and flux reductions by relatively noninteracting penetrants, the data shown in Fig. 20.4-11 clearly illustrate the progressive exclusion of CO_2 from Langmuir sorption sites in poly(methyl methacrylate) (PMMA) as ethylene partial pressure (p_B) is increased in the presence of an essentially constant CO_2 partial pressure of $p_A = 5.05 \mp 0.13$ atm.[46] The tendency of the CO_2 sorption shown in Fig. 20.4-11 to decrease monotonically with ethylene pressure provides impressive support for the "competition" concept on which Eqs. (20.4-17) and (20.4-18) are based. Permeation data are not available for this system to determine if changes in the values of D_D and D_H occur in the mixed gas situation. If offsetting increases in these transport coefficients do not occur in the presence of ethylene, the CO_2 permeability will be depressed in the mixed gas permeation situations.

The data shown in Fig. 20.4-12 illustrate the reduction in permeability of polycarbonate to CO_2 caused by competition between isopentane and CO_2 for Langmuir sorption sites.[47] The flux depression shown in Fig. 20.4-12 was found to be reversible, with the permeability returning to the pure CO_2 level after sufficient evacuation of the isopentane-contaminated membrane. Even at the low isopentane partial pressure considered (117 mm Hg), the tendency is clear for the CO_2 permeability to be depressed from its pure component level toward the limiting value ($P_A = k_{DA}D_{DA}$) corresponding to complete exclusion of CO_2 from the Langmuir environment. This indicates, for example, that the effective CO_2 permeability at a CO_2 partial pressure of 2 atm could be reduced by almost 40% as a result of small amounts of such hydrocarbons in the feed streams. Further increases in the isopentane partial pressure in the feed eventually should complete the depression of CO_2 permeability to its limiting value of 4.57 barrers and ultimately produce plasticizing effects such as indicated in Fig. 20.4-4 at the higher isopentane levels.

The lines drawn through the data in Fig. 20.4-12 were calculated from Eqs. (20.4-16) and (20.4-18) for the pure and mixed penetrant feed situations, respectively, using the *same* CO_2 model parameters in both cases. The affinity constant of isopentane was estimated to be 13.8 atm^{-1}.[47] The excellent fit of the data suggests that D_D and D_H for CO_2 in the mixed feed case are not affected measurably by the presence of the relatively noninteracting isopentane.

One can see that if the upstream pressure in Eq. (20.4-16) is sufficiently large for the pure component, the permeability will tend to form an asymptote to the value of $k_D D_D$. In such a case, further increases in the denominator of Eq. (20.4-18) for the mixed gase case, because of contributions from the second component $b_B p_B$ term, will not depress the magnitude of P_A below the pure component $k_D D_D$ value. This tendency to form an asymptote is clear in Fig. 20.4-12 for CO_2 in polycarbonate and explains why the

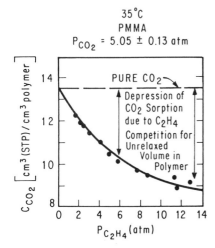

FIGURE 20.4-11 Demonstration of the depression of CO_2 sorption below the pure component CO_2 sorption level as a result of competition of C_2H_4 for unrelaxed volume sorption sites in the polymer at increasing C_2H_4 partial pressures.[46]

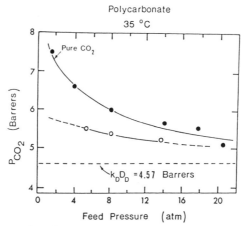

FIGURE 20.4-12 Demonstration of the flux-depressing effects of a small (117 mm Hg) partial pressure of isopentane in the upstream feedstream for CO_2 permeating through polycarbonate at 35°C.[47]

competition effect is more apparent at low values of CO_2 partial pressure, where the permeability is still considerably greater than $k_D D_D$ and has "room" to be depressed. The magnitude of the competition phenomenon in a given application appears to be dependent largely upon the fraction of the unrelaxed volume $(V_g - V_l)/V_g$ in Eq. (20.4-12) that a pure component occupies prior to introduction of the second competing component.

An additional factor, concerning the effectiveness of the second component in competing for the unrelaxed volume, is determined largely by the condensability of the second component. A highly condensable material with a large affinity constant, such as water, is effective in saturation of the microvoid phase at a much lower partial pressure than a supercritical, relatively noncondensable component. Because of their highly condensable nature, such components tend to sorb to substantial extents at low partial pressures in the vapor phase. Therefore, at some point, such materials will begin to act as a flux enhancer rather than a flux depressor. This effect is complex and clearly dependent on polymer–penetrant interactions as well as on simple volume filling, as is the case in the simple dual-mode regime of Fig. 20.4-4. The correlation shown in Fig. 20.4-13 suggests that components below their critical temperatures, that is, vapors, have markedly larger affinities for the microvoid environments in the polymer than corresponding supercritical gaseous components.[46] Carbon dioxide, which has a critical temperature of 303 K, is only slightly supercritical ($T/T_c = 1.01$) at the 308 K measurement temperature shown in Fig. 20.4-13.

On the basis of the preceding discussion, fairly high system pressures are required to observe competition effects for highly supercritical components such as H_2 and CH_4. Pure component permeabilities for such penetrants also appear to be essentially independent of pressure because of the small value of the product bp appearing in the denominator of Eq. (20.4-16) for these low-affinity penetrants. The addition of a small partial pressure of water to the feed, however, may cause dramatic reductions in the accessibility of the unrelaxed pathways to both H_2 and CH_4 because of the large value of b_{H_2O}.

This finding is consistent with the earlier cited quote by Pye et al.[18] It can be understood from Eq. (20.4-18), in which the second term effectively is suppressed, causing the permeability to approach the limiting value of $k_D D_D$. If the polymer is water sensitive, of course, plasticization may occur as the relative humidity is increased further, thereby causing increases in the value of D_D with a resulting upswing in the observed permeability as shown in Fig. 20.4-4.[19] Data for the series of superselective high-flux aromatic polyimides studied by Pye and coworkers are shown in Table 20.4-6.[18] The table compares "dry" (0% RH) behavior to "wet" (exposed to only 15 mm Hg of water vapor at 30°C [50% RH]).

As illustrated in this table, exclusion of both H_2 and CH_4 by the water vapor produces a dramatic (16–59%) reduction in the flux of both components. Only a moderate reduction (3–9%) in selectivity was observed, since the dissolved-mode contribution to the permeability ($k_{Di} D_{Di}$) of each component still experiences the influence of the rigid size- and shape-selecting matrix. The Du Pont workers demonstrated a similar, but less dramatic, effect with n-C_5, consistent with the lower affinity constant of pentane compared to water (see Fig. 20.4-13). In both the water and n-C_5 cases the effects were reversible after protracted degassing followed by a return to contaminant-free feed.

A similar study of the effects of relative humidity on the CO_2 permeability of Kapton® polyimide showed only a 5% reduction compared to the pure component value at 250 psia. This finding is consistent with the previous discussion, since the pure component CO_2 permeability shown in Fig. 20.4-14 is approaching its limiting value of $k_D D_D$ at a CO_2 pressure of 250 psia. This rapid saturation arises as a result

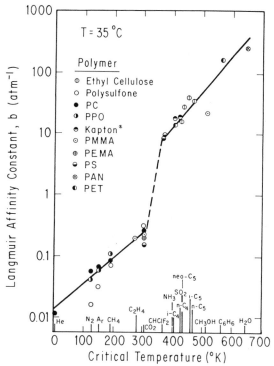

FIGURE 20.4-13   Relationship between observed affinity constants b for a series of penetrants in a variety of polymers at 308 K. Note that the penetrants that are substantially subcritical ($308/T_c < 1.0$) tend to have markedly larger affinity constants because of their higher condensability.[48]

of the high affinity constant of CO_2 compared to the supercritical CH_4 and H_2 (see Fig. 20.4-13). Introduction of the high-affinity water simply completes the small remaining depression of the permeability of the limiting value of $k_D D_D$ shown in Fig. 20.4-14. Hence, the observed flux depressions are rather undramatic compared to the H_2 and CH_4 cases discussed previously.

NONIDEAL GAS-PHASE EFFECTS
Manifestations of external nonideal gas-phase effects on the apparent permeabilities and selectivities for a Kapton® membrane are given in Tables 20.4-7 and 20.4-8 for the CO_2–CH_4 system at 60°C. A CO_2

TABLE 20.4-6   Permeabilities[a] and Permeability Rates at 30°C for Various Stiff-Chain Polyimides

Polymer	P_{H_2} Dry[b]	P_{H_2} Wet[c]	Reduction Due to H_2O Contaminant	P_{CH_4} Dry	P_{CH_4} Wet	Reduction Due to H_2O Contaminant	P_{H_2}/P_{CH_4} Dry	P_{H_2}/P_{CH_4} Wet	Reduction Due to H_2O Contaminant
A[d]	3557	2817	21%	13.2	11.1	16%	270	254	6%
B[e]	7637	6160	19%	68.2	57.0	16%	112	108	3%
C[f]	3168	1273	59%	7.5	3.3	56%	422	386	9%

[a] $P = 10^{-12}$ cm³ (STP)·cm/cm²·s·cmHg.
[b] Dry—The situation in which the feed is truly free of water.
[c] Wet—The situation in which the feed contains 15 mm Hg which corresponds roughly to 50% RH.
[d] A—Prepared by reaction of 4,4'-hexafluoroisopropylidenediphthalic anhydride with 1,3-diamino benzene.
[e] B—Prepared by reaction of 4,4'-hexafluoroisopropylidenediphthalic anhydride with 1,5-diaminonaphthalene.
[f] C—Prepared by reaction of 4,4'-hexafluoroisopropylidenediphthalic anhydride with 3,5-diamino benzoic acid.

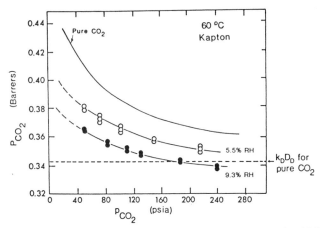

FIGURE 20.4-14 Demonstration of the flux-depressing effect of low relative humidities on the CO_2 permeability of Kapton® polyimide.[48] Paulson et al.[20] has shown that at high relative humidities, the observed flux of CO_2 in a number of polymers is higher than the pure component level. At much higher relative humidities for Kapton®, it may be that this effect also will be observed, consistent with the right-hand side of Fig. 20.4-4.

pressure of 180 psia and a CH_4 pressure of 380 psia were used in both the pure and mixed gas situations.[48] The nonconventional permeabilities $P^\#$ given in Table 20.4-7 are calculated by normalizing the observed steady-state component fluxes with the actual thermodynamic fugacity driving force propelling transport across the membrane rather than with the component partial pressure differences indicated in Eq. (20.1-3). The true driving force for permeation, of course, is the fugacity difference rather than the partial pressure difference that usually is used for convenience. The component fugacities in this case were calculated using the Soave modification of the Redlich–Kwong equation.

Comparison of the differences between pure and mixed component cases for the permeabilities of each component calculated in the standard fashion and in the "thermodynamically normalized" fashion ($P^\#$) is revealing. The difference between the pure and mixed gas cases for the P columns of each component corresponds to the apparent total depression in flux resulting from both true dual-mode competition effects *and* nonideal gas-phase effects. The differences between the pure and mixed gas cases for the $P^\#$ columns are free of complications arising from nonideal gas-phase effects. Therefore, the differences between these columns are manifestations of the rather small competition effect due to dual-mode sorption under these conditions.

These two effects are summarized in Table 20.4-8. Moreover, the percentage reduction in flux for each component due to nonideal gas-phase effects is listed in Table 20.4-8. It is clear that the reduction in fugacity driving force is very small (0.5%) for methane, which tends to be an ideal gas under these conditions. For CO_2, however, the effects amount to roughly 7.5% and account for a roughly 6.9% reduction in the ideal separation factor. To reemphasize, these effects would be observed even if the polymer-phase sorption and transport behavior did not show dual-mode effects and were perfectly ideal.

TABLE 20.4-7 Magnitude of the Nonideal Gas-Phase Effects for Both Pure and Mixed Gas Situations with Kapton® at 60°C for the CO_2–CH_4 System

	Pure Component Data		Actual Mixed Gas Case (32.2% CO_2–67.8% CH_4)	
	CO_2 Permeability at 180 psia CO_2 Pressure (Barrers)	CH_4 Permeability at 380 psia CH_4 Pressure (Barrers)	CO_2 Permeability at 180 psia CO_2 Partial Pressure (Barrers)	CH_4 Permeability at 380 psia CH_4 Partial Pressure (Barrers)
Standard pressure-based normalization P_i	0.382	0.00996	0.342	0.0097
Fugacity-based normalization $P_i^\#$	0.400	0.0102	0.388	0.0100

TABLE 20.4-8 Magnitude of Apparent Permeability Reduction in the Mixtures of CO_2 and CH_4 in a Plasticization-Resistant Polymer, Kapton®, Under Conditions Discussed in TABLE 20.4-6

Component	Total Reduction (%) in Permeability in Mixture Compared to Pure Gas Case A	Reduction (%) in Permeability Owing to Dual-Mode Effects B	Reduction (%) in Permeability Owing to Fugacity Effects C	Total Reduction (%) in Ideal Separation Factor in Mixture Compared to Pure Gas Case D	Reduction (%) in Ideal Separation Factor Owing to Dual-Mode Effects E	Reduction (%) in Ideal Separation Factor Owing to Fugacity Effects F
CO_2	10.5	3	7.5	7.9	1	6.9
CH_4	2.6	2	0.6			

Note: Column A = Column B + C and Column D = Column E + F. The reductions (%) in the table refer to changes relative to the values tabulated under the pure component permeability columns in Table 20.4-6. Separation factor reductions refer to changes relative to the pure component permeability ratios of CO_2 and CH_4 tabulated in the pure component columns in Table 20.4-6. As noted earlier, for plasticization-prone polymers such as cellulose acetate, massive flux *increases* can be observed in the presence of a strongly interacting gas such as CO_2 in mixed gas situations. Such plasticization is generally undesirable because it produces losses in selectivity.

These fugacity effects simply correspond to gas-phase driving force reductions resulting from nonideal gas-phase behavior. Such effects clearly will become more substantial as higher total system pressures are considered in CO_2-containing systems. Conversely, such effects will be relatively unimportant for mixtures containing only supercritical, relatively ideal components.

Fortunately, these phenomena can be predicted readily by using standard thermodynamic equations of state to calculate the fugacity of each component in the upstream and downstream gas phases. For plasticization-prone polymers such as cellulose acetate, the depression of CO_2 fugacity in high CH_4 pressure situations may suppress large upswings in permeability noticed for pure CO_2 (as in Fig. 20.3-9). The area of nonideal gas-phase effects clearly requires considerably more careful investigation.

20.5 CHARACTERIZATION OF ASYMMETRIC MEMBRANES

Although models and techniques such as those described in preceding sections permit characterization of dense-film transport properties, a need exists for improved characterization of the transport behavior of asymmetric membrane properties. Currently, performance tests such as gas permeability and selectivity using standardized feed streams provide useful tools for quality control but are ambiguous for fundamental characterizations of the type possible for dense membranes. The following discussion presents information pertinent to the characterization of porosity and morphology of asymmetric membranes and to analysis of their flux decline as a function of time in service. These are two of the most important additional characteristics that asymmetric structures introduce into the description of transport behavior of the separator module.

20.5-1 Porosity and Morphology Characterizations Void Fraction Determinations

Estimates of the overall void volume in an asymmetric membrane can be made by equilibrating a known dry weight of sample, w_1, in a suitable liquid which will penetrate into the membrane pores but will not swell the polymer substrate. After centrifugation to remove superficial fluid and fluid held in the bore of hollow-fiber membranes, determination of the weight gain of the sample provides an estimate of the void fraction present in the membrane using Eq. (20.5-1):[1]

$$\text{Void volume (\%)} = \frac{(w_2/\rho_2) \times 100}{w_2/\rho_2 + w_1/\rho_1} \tag{20.5-1}$$

where ρ_1 refers to the density of the dry polymer and w_2 and ρ_2, respectively, refer to the weight and density of the liquid imbibed by capillary forces into the membrane pores.

This technique clearly assumes that the polymer is not swollen appreciably by the liquid. Care should be exercised to verify this fact using a dense thin-film sample of the polymer immersed in the candidate liquid. In addition, the apparent value of w_2, determined roughly as the difference between the centrifuged sample weight and the dry weight, should be corrected to account for actual molecular sorption into the polymer mass. This sorption, which occurs in addition to the capillary uptake, can produce an overestimation of the void volume percent in the fiber by as much as 5% if unaccounted for. The simplest means of accounting for such sorption involves adjustment of the value of w_2 using sorption data for the liquid in the dense film for which capillary effects are not present.

SCANNING ELECTRON MICROSCOPY

Scanning electron microscopy (SEM) characterizations of the surface of asymmetric membranes and support structure morphologies are invaluable for general qualitative monitoring of changes in the membrane structure arising from changes in casting or spinning conditions. As a means of characterizing surface porosity, however, the technique is rather poor, since gas molecules can "see" pores as small as 5–10 Å, which are on the borderline of resolution of most SEM equipment. Moreover, SEM often suffers from the problem that variations along the membrane may be sufficiently great to make local characterizations of properties insufficient to provide a representative picture of the overall membrane nature.

As a result of these shortcomings, porosity characterizations generally involve both gas permeability measurements to provide estimates of the average pore size and liquid displacement to measure maximum pore size in the membrane surface. The techniques rely on a number of assumptions for their results to have physical meaning. Depending on the structure of the asymmetric membrane, some of the assumptions may be satisfied only marginally. In such a case, the characterizations are primarily useful as empirical indices for comparison of different samples, and the fundamental meaning of the numbers derived from such analyses is questionable.

AVERAGE PORE SIZE DETERMINATION

Characterization of the average pore size in a porous solid can be performed using a technique described by Yasuda and Tsai.[2] The expression for the steady-state mass or molar flux of a gas through a porous solid of thickness l in the presence of a gas pressure difference, $\Delta p = p_2 - p_1$, between the upstream and downstream points is given by Eq. (20.5-2):[2,3]

$$\text{Flux} = \frac{K \, \Delta p}{l} \tag{20.5-2}$$

where K is the permeability coefficient for the gas.

When the porosity properties of the medium are reasonably uniform, the permeability coefficient calculated using observed fluxes by rearrangement of Eq. (20.5-2) has a well-defined meaning. For media with fine pores, slip flow in which gas molecules experience wall interactions frequently as well as gas-gas viscous interactions must be considered. In such cases, the observed permeability coefficient calculated from Eq. (20.5-2) depends on the average pressure $[p' = (p_1 + p_2)/2]$ and may be represented by Eq. (20.5-3):[3]

$$K = K_0 + \frac{B_0}{\eta} p' \tag{20.5-3}$$

where K_0 is the Knudsen permeability, B_0 is a term characteristic of the solid porosity, and η is the gas viscosity.

Clearly, plots of K versus p' permit evaluation of K_0 and B_0 in such cases. Moreover, Yasuda and Tsai have shown that the average pore size m_x may be calculated as shown in Eq. (20.5-4):

$$m_x = \frac{16}{3} \frac{B_0}{K_0} \left(\frac{2RT}{\pi M} \right)^{1/2} \tag{20.5-4}$$

where R is the gas constant, T is the absolute temperature, and M is the probe gas molecular weight. Cabasso and coworkers have noted that when nitrogen is used as the probe at 25°C, a measure of the "effective porosity" can be derived as given below:[1]

$$\frac{\epsilon}{q^2} = 2.5B \tag{20.5-5}$$

The parameter ϵ is the actual porosity and q is a measure of the tortuosity of the pores for complex pore structures. For the case of straight cylindrical pores, q is simply equal to unity.

Clearly, although one can obtain parameter values from such analyses in the case of nonhomogeneous porous materials, the parameters of K_0, B_0, and ϵ/q^2 have complex physical meanings that are averages of all the local values extending from the dense skin through the support structure. While the above problem exists in a general sense, if the support structure is known to be sufficiently open to provide an order of magnitude or less resistance to flow than the dense surface layer, the parameters discussed above can be given physical meaning in terms of properties of the skin.

In particular, if the pores in the surface are modeled reasonably as cylinders, the average pore diameter and fraction of surface area that is porous can be obtained from m_x and ϵ/q^2, respectively. Clearly, if tortuosity is significant, the above technique will underestimate the actual surface porosity, and one will obtain an effective number of such idealized straight pores. Moreover, in cases where a gradation in dense skin porosity is apparent, one would again obtain a gross average of the pores in this thin dense layer. Nevertheless, the technique is useful for characterizing the average properties of a representative bundle of fibers or quantity of flat asymmetric membranes. The detailed procedures for carrying out the experiments, including designs for test cells, are discussed by Yasuda and Tsai[2] and Cabasso et al.[1]

MAXIMUM PORE SIZE DETERMINATION

Determination of the maximum pore size in a porous solid can be performed using the Kelvin equation:[1]

$$p^* = \frac{2\gamma}{m_e} \tag{20.5-6}$$

where γ is the surface tension of a liquid filling the pores of the solid, and p^* is the pressure that must be exerted to overcome the surface tension holding the liquid in a pore of diameter m_e.

As indicated by Eq. (20.5-6), if there is a distribution of pore sizes, the largest pore will be the first to surrender its imbibed liquid as a steadily increasing gas pressure is applied to one side of the pores. The fiber bundle or flat membrane first is allowed to equilibrate in a nonswelling liquid that wets the pores of the polymer, and then one applies a gradually increasing gas pressure to the porous side of the membrane or the bores of a fiber bundle. The pressure corresponding to the appearance of the first significant number

TABLE 20.5-1 Typical Uncoated Porous Polysulfone Membrane Properties

Sample Number	Void Volume %	Total Wall Thickness (μm)	Calculated Mean Pore Size (μm)	Maximum Pore Size (μm)	Porosity Parameter[a] $(\epsilon/q^2) \times 10^2$
1	75.6	143	0.038	0.09	4.6
2	74.2	116	0.040	0.11	1.5
3	70.6	100	0.040	0.08	1.1
4	73.2	90	0.040	0.11	2.8

[a]For straight pores, the tortuosity weighting parameter (q^2) is unity, and ϵ/q^2 provides an estimate of the equivalent area fraction of pores (ϵ) in the membranes with a length equal to the membrane thickness. For tortuous pores, it is not generally possible to separate decreases in porosity (ϵ) from increases in tortuosity (q), so that the composite parameter given in this table is reported.

  Source: Cabasso et al.[1]

of bubbles on the membrane surface marks the value of p^* and permits calculation of the maximum pore size from Eq. (20.5-6). In cases where the pores are not reasonably cylindrical or where the imbibed liquid swells the membrane surface appreciably, the measurement is compromised as a fundamental characterization tool. As in the previous case for average pore size, even if the parameter estimated is not perfectly meaningful as a well-defined physical property, it is useful for quality control and for comparisons among test lots of different membranes.

Typical data for asymmetric fibers for reverse osmosis applications are reported in Table 20.5-1.[1] The ranges of these variables for as-spun and posttreated cellulose acetate and polysulfone membranes currently used in gas separation are proprietary. Nevertheless, the surface porosity for such membranes is undoubtedly lower than for those described in Table 20.5-1, since, as indicated in Table 20.1-2, in their posttreated forms such membranes have selectivities approaching the values of dense films. Porosities as high as those shown in Table 20.5-1 would produce unacceptably low selectivities as a result of nondiscriminant pore flow.

20.5-2 Permeability Creep Characterization and Analysis

Time-dependent drifts in permeability, such as those shown in Fig. 20.5-1 for cellulose acetate, are important practical features of asymmetric fibers that generally result in a loss in productivity over a period of time.[4] When fouling is known not to be a factor, the so-called permeability creep phenomenon is believed to arise from one or a combination of the following factors:

1. Compaction of the open-cell support structure to a point where the previously negligible resistance to flow of permeated product through the support is no longer insignificant.

2. True glassy-state drifts in the dense selective membrane's properties resulting from a slow volumetric consolidation or densification of this layer of the membrane.

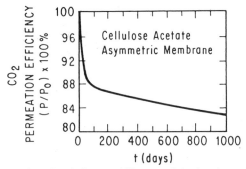

FIGURE 20.5-1 Demonstration of typical permeability creep behavior observed for asymmetric membranes. Both cellulose acetate and polysulfone display this behavior, which is believed to be related to densification of the dense separating layer or compaction of the support foam structure shown in Fig. 20.1-6.[4]

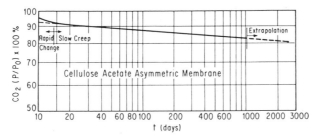

FIGURE 20.5-2 Demonstration of the usefulness of a log–log plot of the permeability creep data in Fig. 20.5-1 to predict behavior over extended periods based on relatively short time behavior (20–60 days after start-up).

Plots of the product permeability versus time on a log–log coordinate system are often linear over relatively long time periods, as shown in Fig. 20.5-2.[4,5] Similar behavior is observed in asymmetric reverse osmosis membranes. The log–log plotting approach provides a simple and reasonably satisfactory means of predicting the performance change of fibers under long-term operation by extrapolation of short-term data. Mechanical creep and volume recovery in glassy polymers after an initial perturbation also are known to be reasonably represented on such log–log plots.[6]

Since the two explanations for permeability drifts are related, respectively, to creep and volume recovery following initial membrane fabrication, it is not surprising that the log–log permeability plots are reasonably linear also. Both creep and volume recovery behavior are expected to be functions of temperature and the composition of the pressurizing medium, since a strongly sorbing permeant such as CO_2 may affect both the dense layer and the support structure properties at high pressures. Unfortunately, definitive data to permit comparisons of flux declines in the presence of CO_2 and inert atmospheres are not available at present.

If the permeability creep occurs primarily as a result of support compaction, one may question the validity of the simple model generally offered for an asymmetric membrane. Specifically, the assumption that a dense surface layer or a composite surface layer constitutes the only significant resistance to flow in a dense-skinned asymmetric membrane under pressure may be incorrect. Instead, upon loading, the effective value of K in Eq. (20.5-2) for the porous substructure may decrease significantly as a result of constriction of flow paths and then slowly continue to decrease as time passes.

Evidence for clarifying whether the observed productivity declines are related to the support compaction effect could be obtained by considering the membrane selectivity as a function of time during the permeability creep process. Compression of the support would cause the generation of a pressure drop through the porous support in order to move permeated material through the compressed porous structure. Depending on the magnitude of the effect, a significant surface concentration of the fast gas could result at the interface just below the dense film at the porous support. This effect would produce a reduction in the concentration difference driving diffusion across the dense membrane and lead to a reduction in flux of the fast gas. Moreover, since the slow gas permeates less rapidly, restrictions to its removal from the downstream face of the dense surface layer should be relatively less serious in the face of support porosity losses, and less polarization would be anticipated. Thus, the depression in the flux of the fast gas, coupled with a smaller depression in the slow-gas flux upon compaction of the support foam, would be characterized by a reduction in selectivity as well as a drop in desired product permeability.

On the contrary, if the permeability creep of the fast gas is associated with the second factor introduced above, that is, a true glassy-state consolidation of the dense selecting membrane, a different selectivity response is anticipated. Although permeation of both species will be restricted by the consolidation, the bulkier slow gas will tend to be retarded *more* because of its greater relative difficulty in moving through the more restrictive densified glass compared to the more streamlined fast gas. In this event, although one would still expect a drop in fast-gas permeability, a corresponding rise in selectivity also would be anticipated.

Of course, these two contributions to permeability creep may occur simultaneously. To aid in determining the relative magnitudes of the two contributions, it is wise to consider the time-dependent behavior of totally dense membranes with a posttreatment history similar to that of the asymmetric membrane. This approach, of course, eliminates the effects of the porous substructure. For example, the time dependence of dense-membrane permeabilities following a conditioning treatment has been observed. As shown in Fig. 20.5-3, the permeability of polycarbonate to pure methane is a decreasing function of time after exposure of both sides of the membrane to 300 psia of CO_2 pressure for 24 h at 35°C. The slow relaxation in permeability shown in Fig. 20.5-3 was still underway after 15 days, although it had slowed noticeably from its original rate. The dissolved-mode concentration C_D of CO_2 in polycarbonate under the 300 psia conditions amounts to almost 3% by weight. Rapid removal of the CO_2 presumably is followed by a time-

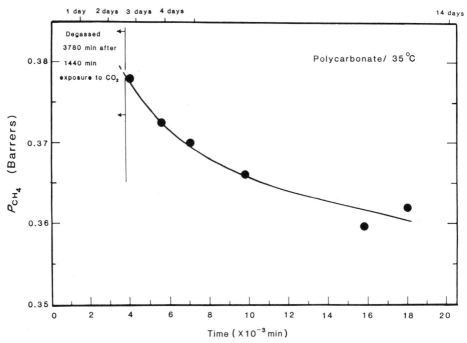

FIGURE 20.5-3 Demonstration of time-dependent response of pure component CH_4 permeation after prior exposure to uniform CO_2 conditioning for 1440 min, followed by prolonged degassing (3780 min) of the 5 mil film. Within 200 min of degassing, the CO_2 had been removed effectively completely (less than 0.01% of the originally present CO_2 remained). The relaxation is related apparently to redistribution of polymer segments in the absence of CO_2.

dependent recovery of the slightly swollen polymer to its unswollen, but still nonequilibrium, state. In the absence of CO_2 this relaxing excess volume can serve as extra sorption and transport pathways for CH_4.

A similar type of relaxation might be anticipated for a membrane which had been cast from a solvent and the solvent quickly removed. Heat treatment at a slightly higher temperature would be expected to speed the relaxation process considerably. In the context of the dual-mode model, one might explain such effects, and even some of the longer term drifts in permeability, in terms of steady reductions and/or redistributions of the excess volume in the glass, $V_g - V_l$. Of course, data such as those in Fig. 20.5-3 do not prove that all or even most of the permeability creep behavior is associated with the dense layer. No selectivity data are available for the above system to verify if the selectivity of the membrane rises as the structure "tightened" during the relaxation of the excess volume. The subject of permeability creep requires considerably more work to permit an unambiguous resolution of its origin and prediction of its magnitude in practical situations involving mixtures of penetrants.

20.6 MODELING AND DESIGN CONSIDERATIONS

20.6-1 General Discussion

As emphasized in the preceding sections, the successful application of membranes for separation purposes depends primarily on discovery of economically competitive membranes with high permselectivities and permeabilities. However, the ensuing considerations, such as membrane configurations and flow patterns of the feed and the permeant streams, are also very important in determining the performance of the final separator system.

The three major membrane configurations, flat sheet, spiral wound, and hollow fiber, each having advantages and limitations, will be reviewed briefly prior to considering the detailed analysis of these devices. Some of these issues include relative magnitudes of active membrane area per unit separator volume, minimization of pressure buildup in the permeant stream, membrane integrity, and the ease of module manufacture and membrane replacement.

The permeability and selectivity of an asymmetric membrane are determined by its complex morphology ranging from the relatively dense top layer to the porous sublayer. As pointed out in Section 20.4, the permeability of even a dense film is subject to complications caused by gas-phase pressure and composition variations, to which an asymmetric membrane is not expected to be immune. Additional time-dependent behavior, presumably associated with compaction and consolidation of the asymmetric morphology, also has been observed routinely for asymmetric structures.[1] Moreover, well-characterized permeability data regarding the effects of membrane asymmetry generally are lacking, although a simplified mathematical description has been performed by Sirkar.[2] In the following model development, the apparent permeability (P_i) of penetrant i will be either represented by a constant or assumed to obey the dual-mode model, and the effects of membrane asymmetry will be disregarded.

Due to the many varieties of separator designs, a comprehensive inclusion of all possible arrangements will not be attempted. Modeling procedures for a few simplified systems will be discussed. In particular, computational results for a single hollow-fiber separator will be presented in more detail. Extension of the single-stage analyses to multiple-stage designs also will be discussed briefly.

MODULE TYPES

Flat membranes are used today mainly for characterizing the permeabilities of membranes. However, commercial flat membrane modules for the production of oxygen-enriched air are sold for individual medical applications by the Oxygen Enrichment Company.[3] The fabrication of such modules is relatively easy, particularly those of smaller sizes. Membrane mounting and supporting are straightforward and permeation areas are well defined. Earlier large-scale industrial modules resembled the plate-and-frame press filter and had a complex layout.[4] The major drawbacks of this type of arrangement are the small membrane area which can be accommodated in a separator of reasonable size and the cumbersome plate-and-frame structure required.

The spiral-wound membrane configuration was adopted to increase the membrane area per unit separator volume while maintaining the ease of fabricating flat membranes. As depicted in Fig. 20.1-5, the spiral-wound module is comprised of essentially flat membranes with the plates and frames replaced by spacers and basestrips which support and protect the membrane from rupture. The feed stream flows in the axial direction of the spiral and the permeant stream flows roughly perpendicularly to the feed stream toward the collecting tube at the center of the separator. This is illustrated in Fig. 20.6-1. The width of the membrane leaf (axial direction of the spiral) is usually less than 5 ft (1.52 m) and the space between the membranes on the feed side is expected to be in the range of 1 mm. As a result, the pressure drop of the feed stream is not very significant at moderate flow rates.[5] The length of the leaf is usually less than 8 ft (2.44 m) and the thickness of the highly porous spacer between the membranes on the permeant side is expected to be in the range of 0.2 mm, so the pressure drop on the permeant side is also relatively small. In practice, pressure buildup in the permeant stream can be alleviated by using multileaf elements with shorter leaves such as those shown in Fig. 20.1-5. The ease of flat membrane preparation, low pressure buildup of the permeant stream, and low pressure loss of the feed stream promote the popularity of spiral-wound membranes in current separator designs.

If the membrane material is spun into hollow fibers with asymmetric walls, the active area per unit separator volume can be increased from less than 100 ft^2/ft^3 (328 m^2/m^3) to more than a few thousand ft^2/ft^3 (greater than 10,000 m^2/m^3).[6] In contrast to flat membranes, hollow fibers are self-supporting. The maximum external or internal pressures at which they can be operated are determined by the modulus of the membrane material, the ratio of fiber OD and ID, and the detailed structure of the asymmetric membrane. A rigorous mechanical strength analysis is not possible due to the ill-defined porous structure of the hollow-fiber wall. An approximate analysis can be made by assuming the hollow fiber to be a homogeneous and isotropic elastic tube as suggested by Stern et al.,[7] Varga,[8] and Thorman and Hwang.[9]

If the hollow fiber is made of glassy polymers, only small fiber deformation is tolerable over the possible operating pressure range to avoid rupture of the fiber wall. Since the selective dense layer on the outside region of the hollow fiber is normally very thin (less than 1500 Å), it is understandable that the shell-feed mode, where the fiber is pressurized externally, is of most commercial interest since the dense layer may be ruptured by dilation if the fiber is pressurized internally.

The inside and outside diameters of the fibers are in the ranges of 100–500 μm and 500–1000 μm, respectively, and the length has been reported to be up to 16 ft.[6,10] When operated at high permeation rates, pressure buildup in the fiber bore can be substantial for small fibers (e.g., less than 100 μm). The increase in bore-side pressure will lower the overall production rate and undermine selectivities compared to cases with milder pressure profiles. Nevertheless, due to its large membrane area per separator volume and self-supporting feature, the hollow fiber is a very desirable configuration as far as the overall performance and economical feasibility of membrane processes are concerned. Optimization of fiber dimensions (diameter and length) becomes particularly important for these systems.

IDEALIZED FLOW PATTERNS

The combination of membrane configurations and input–output port arrangements results in many possible feed and permeant flow patterns. For simplicity, several idealized flow patterns shown in Fig. 20.6-2 have been used to model these different situations. In Fig. 20.6-2a–c, both the feed and permeant streams are

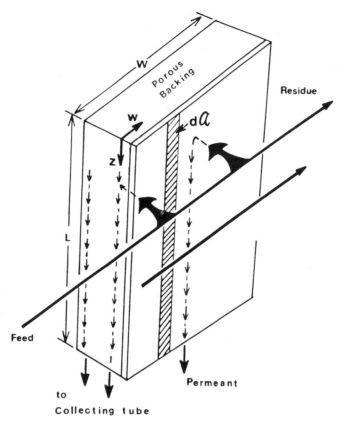

FIGURE 20.6-1 Local gas-flow configuration in a spiral-wound separator.

assumed to be plug flow, that is, uniform concentration across the flow paths, and the compositions change gradually along their respective flow directions. In Fig. 20.6-2e, complete mixing is assumed for both the feed and the permeant chambers with the residue and the permeant product compositions equal to their respective uniform compositions in the chambers. Figure 20.6-2d corresponds to an ideal crossflow pattern where the feed stream is assumed to be plug flow while the permeant flows away from the membrane in the normal direction without mixing. Consequently, for this latter flow pattern, the local mole fraction ratio for the penetrants (Y_{1i}/Y_{1j}) in the permeant stream at any point over the membrane is equal to the ratio of their respective permeation fluxes (N_i) at that point, that is,

$$\frac{Y_{1A}}{Y_{1B}} = \frac{N_A}{N_B} \tag{20.6-1}$$

As recognized by Breuer and Kammermeyer,[11] the crossflow assumption implies that the membrane is situated sufficiently far away from the bulk permeant stream so that the gas composition next to the membrane is not affected by the bulk stream. This flow pattern approximately resembles that in an actual spiral-wound separator with a high-flux asymmetric membrane.[12] An idealized sketch of a small region of the membrane resting on the porous felt backing support is shown in Fig. 20.6-3. Gas exiting the bottom face of the dense skin at l enters pores running from l to l'. If the flux N_i is sufficient to overcome back diffusion from the external permeate stream Q^*, there will be no concentration gradient in the pores, assuming they do not provide any resistance to flow. The composition of this injecting stream is then *completely* independent of the bulk stream Q^*.

All five ideal flow patterns in Fig. 20.6-2 have been studied thoroughly for describing the performance of a single-stage membrane separator; notably, Weller and Steiner,[13] Naylor and Backer,[14] Breuer and Kammermeyer,[11] Oishi et al.,[15] Walawender and Stern,[16] Pan and Habgood,[17] Blaisdell and Kammermeyer,[18] Stern and Wang,[19] and Hwang and Kammermeyer[20] have addressed this subject. In practice, one is not likely to realize the flow patterns in Fig. 20.6-2a–c for a spiral-wound module since the bulk permeant

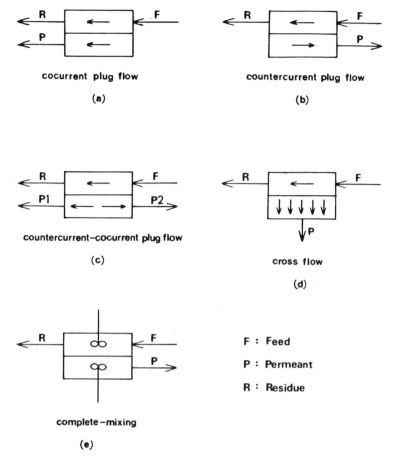

cocurrent plug flow

(a)

countercurrent plug flow

(b)

countercurrent–cocurrent plug flow

(c)

cross flow

(d)

complete–mixing

(e)

F : Feed

P : Permeant

R : Residue

FIGURE 20.6-2 Idealized flow patterns in a membrane gas separator.

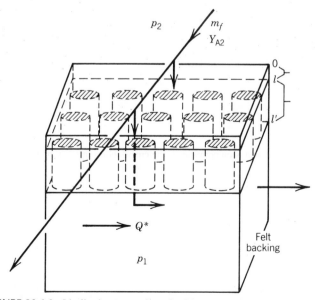

p_2 m_f
         Y_{A2}

Q^*

Felt backing

p_1

FIGURE 20.6-3 Idealized cross section of a felt-supported asymmetric membrane.

and feed streams flow essentially perpendicularly to each other. Moreover, it rarely occurs that the complete mixing model (Fig. 20.6-2d) can be used to describe a hollow-fiber separator due to pressure buildup in the fiber bore and the inevitable composition variation along the fiber when the separator is operated at a reasonable stage-cut. For conditions in which recoveries are small, however, the predictions based on this simple limiting case discussed with respect to Eq. (20.2-1) apply.

INFORMATION FLOW DIAGRAM

To help analyze the characteristics of a single-stage separator, information flow diagrams were prepared for four commonly encountered situations (Fig. 20.6-4). Notations used in the following discussion are summarized at the end of this section. Following the principles of the description rule,[21] there are $(N - 1) + 4$ degrees of freedom for an N-component system whenever the membrane is chosen and the operation is assumed to be isothermal.

In the information flow diagram, the solid arrows represent independent variables, the open arrows represent dependent variables, and the hatched arrows represent externally fixed variables. Figure 20.6-4a corresponds to a typical performance-evaluation problem, where the membrane area ($\mathcal{C}$) and the operating and feed conditions ($p_{2,f}, p_{1,L}, m_{f,f}, Y_{2A,f}, \ldots$) are fixed, while the efficiency of the separator is reflected by the values of the dependent variables ($Y_{1A,L}, m_{p,L}, m_{f,0}, Y_{2A,0}, \ldots$). Figure 20.6-4b corresponds to a design problem where the required membrane area is to be determined for a given membrane and operating conditions with a target mole fraction of component A in the residual stream. For multistage cascade systems, sometimes the situation in Fig. 20.6-4c is encountered. The task is to determine the feed rate a given separator module can process under specified operating conditions with a target mole fraction of component A in the residual stream. Figure 20.6-4d corresponds to the case where a target overall stage-cut ($m_{p,L}/m_{f,f}$) is fixed by process requirements.

20.6-2   Single-Stage Modules

NEGLIGIBLE PERMEANT-PRESSURE VARIATION

The following discussion is most suitable for flat and spiral-wound separators. Assumptions made in the model development are summarized below.

1. Operation is isothermal; that is, the temperature effect associated with transmembrane expansion of the penetrants from the high feed-side pressure to the low permeant-side pressure is neglected (this may introduce some error, for example, when a large amount of CO_2 is recovered at large feed–permeant pressure ratios).

2. There is negligible pressure drop in both the feed and the permeant streams.

3. There are constant "effective" permeabilities for the various gaseous components; that is, either the resistances of the porous sublayer and the support are neglected or a lumped value based on the overall thickness is used without any pressure and composition dependencies.

4. There are negligible gas-phase concentration gradients in the permeation direction.

5. There is insignificant dispersion along the bulk flow direction of the feed stream.

With these assumptions, governing equations for the various ideal flow patterns can be derived from material balances. Note that different information flow arrangements such as those shown in Fig. 20.6-4 can be assigned to each ideal flow pattern shown in Fig. 20.6-2. Solution procedures developed for a particular combination may not be convenient to use for other cases; this is why there are many different modeling "methods" in the literature.[22]

A comprehensive list of the solutions for all possible combinations will not be included, but at least one example will be given for each ideal flow pattern shown in Fig. 20.6-2, and references will be provided for other cases.

Complete Mixing.   This corresponds to the ideal flow pattern in Fig. 20.6-2e. Practically, this flow pattern rarely is realized except in a flat or spiral-wound separator operated at low cuts and low feed rates. Nevertheless, it is useful mainly due to the algebraic simplicity of the corresponding model equations. Stern and Walawender have reviewed three different methods for modeling this flow pattern.[23] The equations derived by Weller and Steiner for a binary-component feed are summarized below.[13]

$$Y_{1A} = \frac{\alpha_A^* - (1 - p_r)/\beta}{\alpha_A^* - 1} \tag{20.6-2}$$

$$Y_{2A} = (\beta + p_r)\left[\frac{\alpha_A^* - (1 - p_r)/\beta}{\alpha_A^* - 1}\right] = (\beta + p_r)Y_{1A} \tag{20.6-3}$$

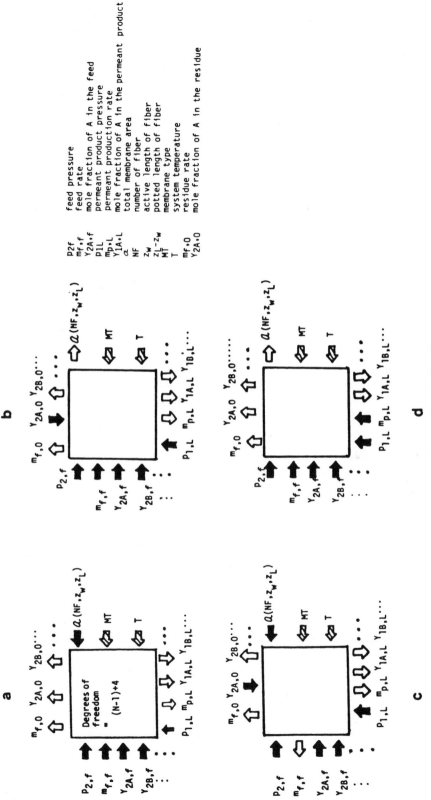

P_{2f}	feed pressure
$m_{f,f}$	feed rate
$Y_{2A,f}$	mole fraction of A in the feed
P_{1L}	permeant product pressure
$m_{p,L}$	permeant production rate
$Y_{1A,L}$	mole fraction of A in the permeant product
α	total membrane area
NF	number of fiber
z_w	active length of fiber
$z_L - z_w$	potted length of fiber
MT	membrane type
T	system temperature
$m_{f,0}$	residue rate
$Y_{2A,0}$	mole fraction of A in the residue

FIGURE 20.6-4 Various information flow diagrams.

925

$$Y_{2A.f} = \frac{[\theta + (1 - \theta)(\beta + p_r)][\alpha_A^* - (1 - p_r)/\beta]}{\alpha_A^* - 1} \tag{20.6-4}$$

$$= \theta Y_{1A} + (1 - \theta)Y_{2A}$$

$$\theta = \text{stage-cut} = \frac{m_p}{m_{f.f}} \tag{20.6-5}$$

$$p_r = \frac{p_1}{p_2} \tag{20.6-6}$$

$$\alpha_A^* = \frac{P_A}{P_B} \quad \text{[equivalent to } \alpha_{A.B}^* \text{ in Eq. (20.1-6)]} \tag{20.6-7}$$

$$\beta = \frac{m_p}{\alpha P_A p_2 / l} \tag{20.6-8}$$

where Y_{2A}, Y_{1A} = mole fractions of component A in the feed and permeant streams, respectively
 p_2, p_1 = feed and permeant pressures respectively
subscripts 0 and L = the residue and feed ends of the separator, respectively
 P_A, P_B = "effective" permeabilities of A and B, respectively
 l = "effective" membrane thickness
 α = active membrane area
subscript f indicates the variable is a feed property.
The calculation procedures are as follows:

(A) For the case in Fig. 20.6-4a where the feed conditions, feed rate, membrane area, and permeant pressure are given (i.e., p_2, $m_{f.f}$, $Y_{2A.f}$, l, p_1, α, P_A and P_B are known):

1. Calculate m_p from Eqs. (20.6-4), (20.6-5), and (20.6-8).
2. Substituting β and other known parameters into Eqs. (20.6-2) and (20.6-3) gives the values for Y_{1A} and Y_{2A}.

(B) For the case in Fig. 20.6-4b where the feed rate, feed conditions, desired residue composition, and permeant pressure are given (i.e., p_2, $m_{f.f}$, $Y_{2A.f}$, p_1, Y_{2A}, P_A, and P_B are known):

1. Substitute Y_{2A}, p_r, and α_A^* into Eq. (20.6-3) and solve for β which corresponds to the positive root.
2. Using the β value calculated above and Eq. (20.6-2), calculate Y_{1A}.
3. From Eqs. (20.6-4) and (20.6-5), obtain θ and m_p.
4. Calculate the required membrane area from Eq. (20.6-8).

If Eqs. (20.6-2)–(20.6-8) are to be used for the case in Fig. 20.6-4d, an iterative procedure has to be adopted since it involves simultaneous solution of Eqs. (20.6-2)–(20.6-4). Stern and coworkers have extended the above treatment and reported an iteration method for multicomponent mixtures.[24] This iteration method was developed for solving the arrangement in Fig. 20.6-4d.

Crossflow. The practical significance and characteristics of the crossflow pattern were discussed earlier. The model equations can be derived by considering a material balance over a small section of the membrane area shown in Fig. 20.6-5. At steady state, decrease in the amount of i in the feed stream, $d(m_f Y_{2i})$, is equal to the amount of i leaving the permeant side of the membrane which is $Y_{1i}(dm_f)$. The model equations for this case are as follows:

Feed-Side Material Balance

$$\frac{d(m_f Y_{2i})}{d\alpha} = \left(\frac{P_i}{l}\right)(p_2 Y_{2i} - p_1 Y_{1i}) \quad i = \text{A, B, C, \ldots, N} \tag{20.6-9}$$

Crossflow Characteristic Equation

$$\frac{d(m_f Y_{2i})}{dm_f} = Y_{1i} \tag{20.6-10}$$

$$\sum Y_{1i} = 1 \tag{20.6-11}$$

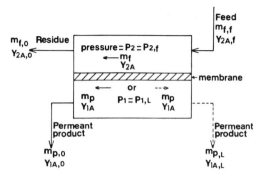

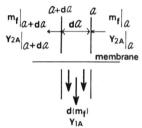

FIGURE 20.6-5   Schematic diagram used for cross flow model development.

$$\Sigma \ Y_{2i} = 1 \qquad (20.6\text{-}12)$$

Weller and Steiner[13] solved these equations and provided the following expressions for a binary component system:

$$\theta^* = 1 - \frac{m_f}{m_{f,f}}$$

$$= 1 - \left(\frac{1 - Y_{2A,f}}{1 - Y_{2A}}\right) \left(\frac{t_f - B'/A'}{t - B'/A'}\right)^{R'} \left(\frac{t_f - \alpha_A^* + C'}{t - \alpha_A^* + C'}\right)^{U'} \left(\frac{t_f - C'}{t - C'}\right)^{T'}$$

$$= \text{stage-cut at the residue product exit} \qquad (20.6\text{-}13)$$

The required membrane area $\mathcal{C}$ is given by

$$\mathcal{C} = \left(\frac{lm_{f,f}}{P_A P_2}\right) \alpha_A^* \int_X^{X_f} \frac{(1 - Y_{2A})(1 - \theta^*)}{[f(X) - X]\{1/(1 + X) - p_r/[f(X) + 1]\}} \, dX \qquad (20.6\text{-}14)$$

where

$$A' = \frac{(1 - \alpha_A^*) \, p_r + \alpha_A^*}{2} \qquad (20.6\text{-}15)$$

$$C' = -\frac{[(1 - \alpha_A^*) \, p_r - 1]}{2} \qquad (20.6\text{-}16)$$

$$B' = -A'C' + \frac{\alpha_A^*}{2} \qquad (20.6\text{-}17)$$

$$R' = \frac{1}{2A' - 1} \qquad (20.6\text{-}18)$$

$$U' = \frac{\alpha_A^*(A' - 1) + C'}{(2A' - 1)(\alpha_A^*/2 - C')} \qquad (20.6\text{-}19)$$

$$T' = \frac{1}{1 - A' - B'/C'} \tag{20.6-20}$$

$$X = \frac{Y_{2A}}{1 - Y_{2A}} \tag{20.6-21}$$

$$t = -A'X + [A'^2X^2 + 2B'X + C'^2]^{1/2} \tag{20.6-22}$$

$$f(X) = A'X - C' + [A'^2X^2 + 2B'X + C'^2]^{1/2} \tag{20.6-23}$$

Undoubtedly, the solution for this system is more complicated than that of the complete mixing case since the concentration of the feed stream is allowed to change along the membrane.

The calculation procedures are as follows:

(A) For the case in Fig. 20.6-4b (i.e., $p_{2,f}$, $m_{f,f}$, $Y_{2A,f}$, p_1, $Y_{2A,0}$, P_A, and P_B are known):

1. Calculate A', B', C', R', U', and T' from their respective definitions.
2. Since $Y_{2A,0}$ is known, X and t can be calculated from their definitions. Substitution of these variables into Eq. (20.6-13) gives the stage-cut θ^*.
3. To calculate the required membrane area, a series of values of Y_{2A} (greater than $Y_{2A,0}$ and less than $Y_{2A,f}$) are substituted into Eq. (20.6-13), giving a series of θ^* values. These values then are used to integrate Eq. (20.6-14) numerically from X ($Y_{2A} = Y_{2A,0}$) to X_f ($Y_{2A} = Y_{2A,f}$).

(B) For the case in Fig. 20.6-4d where the stage-cut θ is given, numerical iterations are required to solve Eq. (20.6-13) for Y_{2A}.

(C) Equations (20.6-13)–(20.6-23) are inconvenient to use for the case in Fig. 20.6-4a where the membrane area is given, but the stage-cut is unknown. Direct numerical integration of the mass balance equations is preferred for this arrangement, which will be shown later for a hollow-fiber separator.

Pan and Habgood have derived an alternative set of analytic expressions that relates the stage-cut to the permeant composition instead of the feed-side composition [cf. Eq. (20.6-13)].[25]

For multicomponent systems, Stern and coworkers have derived a numerical iteration algorithm for solving the case in Fig. 20.6-4d.[24] Pan and Habgood, on the other hand, have proposed a noniterative procedure for solving the case with the arrangement in Fig. 20.6-4b.[17]

Countercurrent and Cocurrent Plug Flows. The model equations for these flow patterns cannot be solved analytically. Oishi and coworkers first derived the general model equations for a binary-component system with porous media.[15] Walawender and Stern,[16] Blaisdell and Kammermeyer,[18] and Pan and Habgood[17] later reported solutions for similar membrane separators. The cocurrent–countercurrent combination flow pattern also has been studied by Pan and Habgood.[17]

Cocurrent Flow. According to the schematic diagram in Fig. 20.6-6a, the following mass balance equations can be derived:

$$m_f + m_p = m_{f,f} \tag{20.6-24}$$

$$m_f Y_{2A} + m_p Y_{1A} = m_{f,f} Y_{2A,f} \tag{20.6-25}$$

$$\frac{d(m_p Y_{1A})}{d\alpha} = \left(\frac{P_A}{l}\right)(p_2 Y_{2A} - p_1 Y_{1A}) \tag{20.6-26}$$

$$\frac{dm_p(1 - Y_{1A})}{d\alpha} = \left(\frac{P_B}{l}\right)[p_2(1 - Y_{2A}) - p_1(1 - Y_{1A})] \tag{20.6-27}$$

Equations (20.6-24)–(20.6-27) can be combined and simplified to

$$\frac{dY_{1A}}{dY_{2A}} = \left(\frac{Y_{1A} - Y_{2A,f}}{Y_{2A} - Y_{2A,f}}\right)\left(\frac{\alpha_A^*(1 - Y_{1A})(Y_{2A} - p_r Y_{1A}) - Y_{1A}[(1 - Y_{2A}) - p_r(1 - Y_{1A})]}{\alpha_A^*(1 - Y_{2A})(Y_{2A} - p_r Y_{1A}) - Y_{2A}[(1 - Y_{2A}) - p_r(1 - Y_{1A})]}\right) \tag{20.6-28}$$

and

$$\frac{dR^f}{dY_{2A}} = \frac{Y_{1A} - Y_{2A,f}}{(Y_{2A} - Y_{1A})\{\alpha_A^*(1 - Y_{2A})(Y_{2A} - p_r Y_{1A}) - Y_{2A}[(1 - Y_{2A}) - p_r(1 - Y_{1A})]\}} \tag{20.6-29}$$

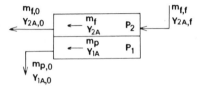

(a) cocurrent

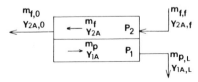

(b) coutercurrent

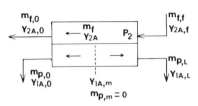

$m_{p,m} = 0$

(c) countercurrent /cocurrent

FIGURE 20.6-6 Schematic diagram for the plug-flow model development.

where

$$R^f = \left(\frac{P_B}{l}\right) \frac{p_2 \mathcal{Q}}{m_{f,f}} \qquad (20.6\text{-}30)$$

To integrate Eq. (20.6-28), it is necessary to evaluate the value of dY_{1A}/dY_{2A} at $Y_{2A} = Y_{2A,f}$ using an equation other than Eq. (20.6-28) since direct substitution of $Y_{2A,f}$ gives an indeterminate value. This can be done by differentiating both the numerator and the denominator of Eq. (20.6-28) with respect to Y_{2A} and letting $Y_{2A} = Y_{2A,f}$ and $Y_{1A} = Y_{1A,L}$:

$$\left(\frac{dY_{1A}}{dY_{2A}}\right)_{Y_{2A,f}} = (Y_{1A,L} - Y_{2A,f}) [\alpha_A^* - (\alpha_A^* - 1) Y_{1A,L}]$$

$$\div \{\alpha_A^*(1 - Y_{2A,f}) (Y_{2A,f} - p_r Y_{1A,L}) - Y_{2A,f}[(1 - Y_{2A,f}) - p_r(1 - Y_{1A,L})]$$

$$- (Y_{1A,L} - Y_{2A,f}) [(\alpha_A^* - 1) (2 p_r Y_{1A,L} - Y_{2A,f} - p_r) - 1]\} \qquad (20.6\text{-}31)$$

where $Y_{1A,L}$ is determined from

$$\frac{Y_{1A,L}}{1 - Y_{1A,L}} = \frac{\alpha_A^*(Y_{2A,f} - p_r Y_{1A,L})}{1 - Y_{2A,f} - p_r(1 - Y_{1A,L})} \qquad (20.6\text{-}32)$$

which is derived from taking the ratio of Eqs. (20.6-26) and (20.6-27).

The calculation procedures are as follows: For the case in Fig. 20.6-4b, $Y_{1A,L}$ is evaluated from Eq. (20.6-32), which then is used to initiate the numerical integration of Eqs. (20.6-28) and (20.6-31) to establish the relationship between Y_{1A} and Y_{2A}. Substituting the Y_{1A} versus Y_{2A} relationship into Eq. (20.6-29) and then integrating from $Y_{2A,f}$ to $Y_{2A,0}$ gives R^f and consequently the required membrane area. Equations (20.6-24) and (20.6-25) are used for calculating $m_{p,0}$ and $m_{f,0}$ and consequently the stage-cut.

Countercurrent Flow. As indicated in Fig. 20.6-6b,

$$m_f = m_{f,0} + m_p \qquad (20.6\text{-}33)$$

$$m_f Y_{2A} = m_{f,0} Y_{2A,0} + m_p Y_{1A} \qquad (20.6\text{-}34)$$

Equations (20.6-26)–(20.6-30) also are applicable for this case, except that Eq. (20.6-32) is replaced by Eq. (20.6-35):

$$\frac{Y_{1A,0}}{1 - Y_{1A,0}} = \frac{\alpha_A^*(Y_{2A,0} - p_r Y_{1A,0})}{1 - Y_{2A,0} - p_r(1 - Y_{1A,0})} \qquad (20.6\text{-}35)$$

Similarly, Eq. (20.6-31) is replaced by Eq. (20.6-36):

$$\left(\frac{dY_{1A}}{dY_{2A}}\right)_{Y_{2A,0}} = (Y_{1A,0} - Y_{2A,0})\,[\alpha_A^* - (\alpha_A^* - 1)Y_{1A,0}]$$

$$\div \{\alpha_A^*(1 - Y_{2A,0})(Y_{2A,0} - p_r Y_{1A,0}) - Y_{2A,0}[(1 - Y_{2A,0}) - p_r(1 - Y_{1A,0})]$$

$$-(Y_{1A,0} - Y_{2A,0})\,[(\alpha_A^* - 1)(2p_r Y_{1A,0} - Y_{2A,0} - p_r) - 1]\} \qquad (20.6\text{-}36)$$

The calculation procedures are as follows: For the case in Fig. 20.6-4b, the calculation procedure is similar to the cocurrent case. Equation (20.6-28) is integrated to give the Y_{1A} versus Y_{2A} relationship from the residual end where $Y_{2A,0}$ is given. $Y_{1A,0}$ and dY_{1A}/dY_{2A} at $Y_{2A} = Y_{2A,0}$ are evaluated via Eqs. (20.6-35) and (20.6-36), respectively. The required membrane area and stage-cut are evaluated from Eqs. (20.6-29), (20.6-33), and (20.6-34) in a similar manner as for the cocurrent-flow case.

Cocurrent–Countercurrent Combination Flow. In this hybrid flow pattern, both the feed and the permeant streams are assumed to be plug flow, but the permeant product is withdrawn at both ends of the separator as shown in Fig. 20.6-6c. In principle, there will be a position in the permeant side where the permeant flow rate is zero, which also divides the separator into a countercurrent region on one side and a cocurrent region on the other side. The calculation procedures discussed previously for the cocurrent and the countercurrent flow patterns can be used for these two sections, respectively.

The zero-permeant-flow position is determined by the permeant pressure drop under the given operating conditions. Pressure drop, however, has not been taken into account in the discussion so far, which makes the zero-permeant-flow position indeterminate. To overcome this difficulty, one must take an approximate measure by adding an additional imaginary degree of freedom which is not rigorously correct. This may be done by either fixing the position of the zero-permeant-flow or fixing the relative flow rate of the two permeant product streams. Pan and Habgood suggest that equal product flows may be a reasonable assumption.[17]

For the case in Fig. 20.6-4b where the residue concentration $Y_{2A,0}$ is given, calculation is started by assuming a value for Y_{2A} at the zero-permeant-flow position. One then proceeds with the calculation following the steps given above for countercurrent flow pattern and $m_{f,m}$; the resulting value of the feed-side flow rate at the location where $m_{p,m} = 0$ is used to initiate calculation for the cocurrent flow section following the procedures given. Trial and error continues until the ratio of calculated $m_{p,0}$ and $m_{p,L}$ satisfy the specified value (e.g., 1). Similar procedures, in principle, can be used for systems containing more than three components; however, computation time may be tremendous.

Comparison of Flow Patterns. Detailed parametric studies have been reported by several authors for a binary-component system.[16,17,25] Comparisons among the four flow patterns, that is, complete mixing, crossflow, cocurrent, and countercurrent were also well documented. For example, one may plot resulting permeant product composition Y_{1A} versus membrane area for a given feed flow rate, feed composition, feed pressure, permeant product pressure, and another design variable. Generally, this latter design variable is either the stage-cut (amount of feed that ends up in the permeant product) or the residue composition $Y_{2A,0}$.[17] Alternatively, if the stage-cut is fixed, one may plot both the permeant product composition of Y_{1A} and the residue composition $Y_{2A,0}$ as functions of stage-cut and prepare a separate figure relating the required membrane to the desired stage-cut.[22]

In general, it has been concluded that countercurrent flow is the most efficient flow pattern, requiring the lowest membrane area and producing the highest degree of separation, at the same operating conditions. The order of efficiency for the other three flow patterns is crossflow > cocurrent flow > perfect mixing.[17,22]

PERMEANT PRESSURE VARIATION IN HOLLOW FIBERS

The model equations and solution procedures discussed so far are derived for systems with negligible pressure variations in both the feed and the permeant streams. These conditions may not be satisfied fully in a commercial membrane separator, particularly when hollow fibers are used. The extent of pressure buildup in the fiber bore will increase with decreasing fiber inside diameter and increasing permeant flow rate. However, pressure variation in the shell side generally can be neglected safely.

With these considerations in mind, an additional equation is introduced to account for the momentum loss caused by viscous friction. Because of the large number of possible arrangements, the following discussion is confined to a countercurrent shell-feed hollow-fiber separator, which is probably the most common hollow-fiber separator arrangement to date. However, only slight modification is required to extend the following treatment to bore-feed or cocurrent flow operations.

Major assumptions made in the development of the following model equations are:

1. Isothermal operation is maintained with negligible pressure drop in the shell side.

2. The differential form of the Hagen–Poiseuille equation is valid for describing the permeant stream pressure buildup inside the fiber bore. This is based on Berman's study of incompressible laminar flow in permeable tubes with constant injection at the tube wall.[26] Pan and Habgood suggested that as long as the parameter (tube ID) (local permeant flux)/(viscosity) is smaller than 1, it is a reasonable approximation.[17] The same approach was adopted by Antonson et al.[27] and Chern.[28] On the other hand, Thorman and Hwang have combined expressions for compressible flow in an impermeable tube and incompressible flow in a permeable tube using a first-order perturbation technique, to describe the momentum balance.[9] Direct experimental verification of the above two approaches has not been reported.[29,30]

3. The permeability of the selective layer to each penetrant obeys the dual-mode transport model.

4. The porous sublayer has a negligible resistance to permeation.

5. Negligible gas-phase concentration gradients exist in the permeation direction, that is, no concentration polarization. The gases are assumed to be ideal.

6. Both the feed and the permeant streams are plug flow, and rapid radial mixing occurs such that the gas-phase composition next to the permselective membrane is equal to that in the bulk stream. There are some doubts regarding the validity of assumption 6 when high-flux asymmetric membranes are used,[12] as explained previously. Slight modifications in the following model equations will be required if assumption 6 is in question, since the crossflow pattern rather than the countercurrent pattern would need to be used.

Antonson et al.[27] and Pan and Habgood[31] have developed model equations for constant-permeability systems and solved them as an initial-value problem for design calculations where the *length* of the separator was to be determined (a special case of Fig. 20.6-4a or b). However, the length of fibers normally will be predetermined by other design considerations such as permeant pressure buildup and module fabrication cost. Consequently, the advantage of being able to treat the fiber length as a dependent variable for determining the overall membrane area requirement is largely diminished since modules tend to be produced in a number of discrete lengths. Practically, membrane area for a given separation task will be provided by controlling the total number of fibers of predetermined length in a separator module and the number of modules to be used.

Since the material and momentum balances for the countercurrent arrangement are intrinsically a boundary-value problem, Chern[28] has proposed an alternative approach using a general boundary-value algorithm developed by Newman.[32] This latter method can be extended readily to multicomponent systems without rearranging the original material and momentum balances (cf. Ref. 25). According to the schematic diagram shown in Fig. 20.6-7, the following relationships can be established for a binary-component feed:

Bore-Side Material Balance

$$\frac{dm_{pA}}{dz} = N_A \pi \mathrm{ID} \left[1 - H(z - z_w) \right] \tag{20.6-37}$$

$$\frac{dm_{pB}}{dz} = N_B \pi \mathrm{ID} \left[1 - H(z - z_w) \right] \tag{20.6-38}$$

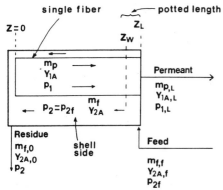

FIGURE 20.6-7 Schematic diagram of a hollow-fiber separator.

where $H(z - z_w)$ is the unit step function which is needed since there is no permeation in the potted section of the fiber; N_A and N_B are as given in Eqs. (20.6-39) and (20.6-40) depending on whether or not competition effect is considered.

Without Competition

$$N_A = \frac{D_{DA}}{(ID/2) \ln [OD/(OD - 2l)]} \left[k_{DA}(p_2 Y_{2A} - p_1 Y_{1A}) + F_A \left(\frac{C'_{HA} b_A p_2 Y_{2A}}{1 + b_A p_2 Y_{2A}} - \frac{C'_{HA} b_A p_1 Y_{1A}}{1 + b_A p_1 Y_{1A}} \right) \right]$$

(20.6-39)

With Competition

$$N_A = \frac{D_{DA}}{(ID/2) \ln [OD/(OD - 2l)]} \left[k_{DA}(p_2 Y_{2A} - p_1 Y_{1A}) + F_A \left(\frac{C'_{HA} b_A p_2 Y_{2A}}{1 + b_A p_2 Y_{2A} + b_B p_2 Y_{2B}} \right. \right.$$
$$\left. \left. - \frac{C'_{HA} b_A p_1 Y_{1A}}{1 + b_A p_1 Y_{1A} + b_B p_1 Y_{1B}} \right) \right]$$

(20.6-40)

Bore-Side Momentum Balance (Hagan–Poiseuille)

$$\frac{dp_1^2}{dz} = -\frac{256 \, RT m_p \mu}{\pi \, ID^4}$$

(20.6-41)

The local molar flow rates of component i in the permeant and in the shell side per fiber are represented by m_{pi} and m_{fi}, respectively; m_p is the total molar flow rate in the fiber; R is the gas constant; and μ is the viscosity of the gas mixture, which is calculated according to the expressions suggested by Reid et al.[33]
The boundary conditions for Eqs. (20.6-37) and (20.6-38) are set at $z = 0$, yet that for Eq. (20.6-41) is set at $z = z_L$:

$$z = 0 \qquad m_{pA} = m_{pB} = 0$$

(20.6-42)

$$z = z_L \qquad p_1 = p_{1,L}$$

(20.6-43)

The overall material balances for components A and B at any location z in the fiber are the remaining balance equations required:

$$m_{fA} = m_{fA,0} + m_{pA}$$

(20.6-44)

$$m_{fB} = m_{fB,0} + m_{pB}$$

(20.6-45)

Equations (20.6-44) and (20.6-45) are used to determine the local shell-side partial pressures of A and B for use in N_i since $Y_{2i} = m_{fi}/(m_{fi} + m_{fj})$.
Equations (20.6-37), (20.6-38), and (20.6-42) with Eqs. (20.6-41) and (20.6-43) constitute a boundary-value problem. However, since the equations are coupled and since the shell-side molar flow rates are known only at z_L, two additional trivial, ordinary differential equations are introduced to use Newman's procedure.[32]

$$\frac{d(m_{fA,0})}{dz} = 0$$

(20.6-46)

$$\frac{d(m_{fB,0})}{dz} = 0$$

(20.6-47)

which are subject to the boundary condition at z_L, respectively:

$$m_{fA,0} + m_{pA,0} = m_{fA,f}$$

(20.6-48)

$$m_{fB,0} + m_{pB,0} = m_{fB,f}$$

(20.6-49)

Equations (20.6-48) and (20.6-49) along with Eqs. (20.6-37), (20.6-38), and (20.6-41) are solved as a coupled boundary-value ordinary differential equation system of five unknowns after nondimensionalization and linearization.
The linearized Eqs. (20.6-37)–(20.6-49) together with the boundary conditions listed above can be used directly for solving case (*a*) in Fig. 20.6-4. For the other arrangements, one needs to include either an

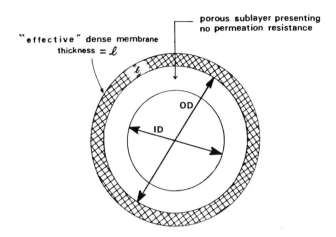

porous sublayer presenting
no permeation resistance

"effective" dense membrane
thickness = ℓ

OD

ID

cross section of an asymmetric hollow fiber

FIGURE 20.6-8 Highly simplified drawing of the permeation resistance of a hollow fiber.

interpolation or an optimization scheme in the main program since a systematic variation of the coefficients in these equations will be required.

Note that the plotted length, $z_L - z_w$, does not contribute to the separation but instead provides extra length for permeant pressure buildup. Clearly, it should be made as short as allowed by module integrity and leak-free considerations. The highly idealized picture in accord with assumptions 1, 2, and 3 is depicted in Fig. 20.6-8. This idealization will allow an independent investigation of the effects of fiber on separator performance, free from complications due to changes in the permeation resistance of the fiber wall.

Useful information regarding the characteristics of a particular separator arrangement often can be obtained through parametric studies. Separator specifications used in the following calculation are shown in Table 20.6-1 unless stated otherwise. Experimental permeability parameters, according to the dual-mode model, obtained from dense film–pure gas measurements[34–36] are summarized in Table 20.6-2. For simplicity, only the results for a binary-component feed will be discussed.

An informative presentation of the performance of a given type of separator is to plot the fast-gas (the higher permeability penetrant) concentration in the permeant product and the slow-gas concentration in the residue product as functions of the fast-gas fractional recovery into the permeant product. These curves will be called the "characteristic curves" and will be used to indicate the effects of varying one of the several important design and operating variables.

Fiber Dimensions. As is apparent from Eq. (20.6-41), friction loss in the fiber bore can be significant if the bore diameter is small and the permeant flow rate (m_p) is high. The significance of permeant pressure buildup is shown in Fig. 20.6-9. Permeability data for polysulfone are used in the calculation. Except for fiber ID, all the other operating variables are the same for the four cases. In Fig. 20.6-9a, the "driving"

TABLE 20.6-1 Separator Dimension and Operation Condition Ranges

Polymer	Polysulfone or polycarbonate
Temperature	35°C
Feed	Pressure, 600 psia
	Composition, CO_2–CH_4 mixture; $Y_{2CO_2,f} = 0.45$
Permeant	Pressure, 44.1 psia
Fiber dimensions	ID, 50, 100, 150 μm
	OD, 300 μm
	z_L, 152.4 cm
	$z_L - z_w$, 15.2 cm
	Number of fibers (NF), 9×10^5
	Effective membrane thickness $l = 1000$ Å

TABLE 20.6-2 Dual-Mode Parameters of Pure Gases at 35°C

Polymer	Penetrant	k_D [cm³ (STP)/cm³ · atm]	C'_H [cm³ (STP)/cm³]	b (atm⁻¹)	D_D (10⁸ cm²/s)	F
Polysulfone	CO_2	0.664	17.90	0.326	4.40	0.105
	CH_4	0.161	9.86	0.070	0.444	0.349
Polycarbonate	He	0.0145	0.313	0.0121	550	1.33
	CH_4	0.147	8.38	0.0841	1.09	0.115

Note: According to the dual-mode model,[37] the diffusive flux $N = -D_D \, dC_D/dx - D_H \, dC_H/dx$. The total sorption concentration $C = C_D + C_H$, and $C_D = k_D p$, $C_H = C'_H bp/(1 + bp)$, and $F = D_H/D_D$, where p is the penetrant pressure in the gas phase.

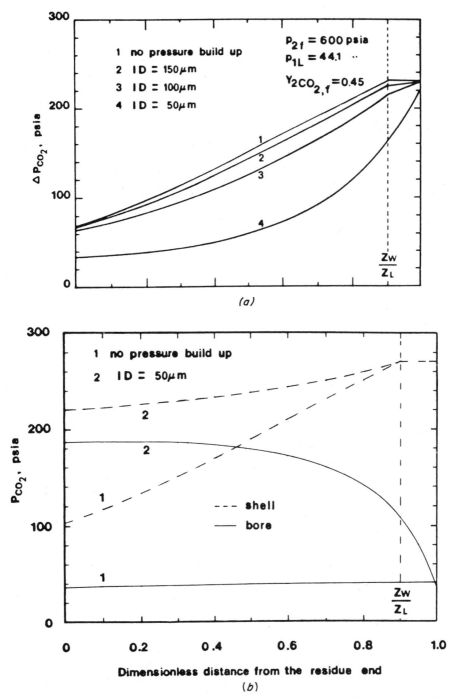

FIGURE 20.6-9 Effects of fiber ID on CO_2 (the fast gas): (*a*) partial pressure difference across the fiber wall; (*b*) shell- and tube-side partial pressure. Fiber OD = 300 μm, NF = 9 $\times$ 10^5, feed rate = 126,374 SCFH.

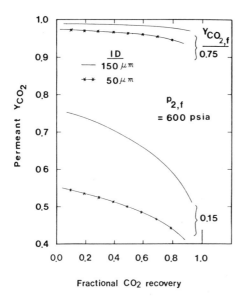

FIGURE 20.6-10 Effects of fiber ID on the characteristic curves. Fiber OD = 300 μm, NF = 9 $\times$ 10^5, feed rate = 126,374 SCFH, p_{1L} = 44.1 psia.

pressure ($\Delta p_i = p_{2i} - p_{1i}$) profiles of CO_2 for three different fiber bore sizes (ID = 50, 100, and 150 μm) are compared with the case where permeant pressure is constant. The constant-permeant-pressure result can be obtained easily by letting Eq. (20.6-41) equal zero while keeping all the other equations intact; it corresponds to the most favorable driving pressure profile at the given operating conditions. The loss of CO_2 driving pressure toward the residue end for this latter case is due solely to depletion of CO_2 from the feed stream as more and more CO_2 molecules permeate to the bore side of the fiber. Partial pressure profiles of CO_2 in the shell and the tube sides are contrasted in Fig. 20.6-9b for two extreme cases; one with significant bore-side pressure buildup, the other with negligible pressure buildup.

The deleterious effects of the diminished CO_2 partial pressure difference are reflected in the characteristic curves shown in Fig. 20.6-10. The characteristics of the two otherwise similar separators are drastically different when the feed gas is lean in the fast gas. When operated at the same fast-gas fractional recovery, the one with ID = 150 μm produces much higher purity permeant product. For a feed stream rich in CO_2, the CO_2 partial pressure difference across the fiber wall will remain relatively high even in the presence of permeant pressure buildup. This accounts for the difference between the dilute and the concentrated cases in Fig. 20.6-10.

The membrane area required to effect the same fast-gas recovery from a fixed feed rate is increased significantly when a smaller fiber ID is used, as shown in Fig. 20.6-11. Due to diminished driving force, larger permeant pressure buildup invariably results in larger membrane requirement even if the feed is rich in the fast gas. For the particular system under consideration, more than a twofold decrease in the permeant production rate results from a change in ID from 150 to 50 μm while the permeant and the residue purities are lowered by 5.7 and 27%, respectively. This reduction in permeant production capacity is contributed mainly by the dramatic decrease in the CO_2 recovery shown in Table 20.6-3. There are also dramatic increases in the CH_4 driving pressure (Δp_{CH_4}) across the fiber wall when larger fiber IDs are used as shown in Fig. 20.6-12. However, due to the much lower permeability of CH_4, the ratio of the fractional CH_4 recovery to the fractional CO_2 recovery actually is *decreased* with increasing ID values, even if the fractional CH_4 recovery itself increases from 3.8 to 5.2% as indicated in the last column of Table 20.6-3. Figures 20.6-10 and 20.6-11 clearly demonstrate the importance of choosing a suitable ID to maintain the separator performance close to that of the case where pressure buildup is negligible.

Smaller fiber outside diameters with a fixed ID will lower the permeant flow rate per fiber roughly proportionally, with a concomitant decrease in permeant pressure buildup if the characteristic membrane thickness remains the same. But it will increase simultaneously more than proportionally the number of fibers which can be accommodated in a separator module of a given diameter. The net result is higher permeant production capacity. Clearly, in principle, fibers with small OD and large ID should be used. The absolute values of these two dimensions will depend on the magnitude of the expected permeant flow rate per fiber and certain fabrication technology and mechanical strength considerations.

Whenever the permeant pressure buildup is expected to be nontrivial, care should be taken not to use fibers of excessive length. This is shown in Table 20.6-4 where a small fiber with ID = 50 μm is used. Due to the excessive permeant pressure buildup, doubling the total membrane area by increasing the fiber length offers no clear improvement in performance, even in the permeant production rate. On the other

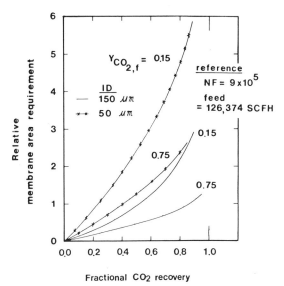

FIGURE 20.6-11 Effects of fiber ID on membrane area requirement. Fiber OD $= 300 \ \mu m$, NF $= 9 \times 10^5$, feed rate $= 126,374$ SCFH, $p_{1L} = 44.1$ psia, $p_{2f} = 600$ psia.

hand, for the case with no permeant pressure buildup, a two fold increase in fiber length results in roughly 60% more permeant production rate with a slight deterioration in its purity as also shown in Table 20.6-4.

Feed Composition. The effects of feed composition on separator characteristics are illustrated in Fig. 20.6-13 which is generated by varying either the feed flow rate (if the separator size is fixed) or the number of fibers (if the feed rate is given) for several feed compositions. The permeant product purity is sensitive to the initial fast-gas mole fraction particularly when the feed is relatively lean in the fast gas as indicated by the significant change from curve 1 to curve 2 in Fig. 20.6-13. The residue characteristic curves converge toward unity as CO_2 fractional recovery approaches unity while the permeant characteristic curves asymptote to their respective limiting values at zero CO_2 recovery. This limiting permeant product purity is determined by the feed composition, operating conditions, separator dimensions, and membrane selectivity.

Membrane area required to effect the same fractional fast-gas recovery is also sensitive to feed stream composition as shown in Fig. 20.6-14. There is little difference in area requirement for feeds with fast-gas mole fraction greater than 0.45 when the separator is operated at smaller than 0.3 fractional fast-gas recovery. However, the difference increases at larger recoveries because Δp_{CO_2} along the fiber length is maintained higher for a more concentrated feed and, consequently, less membrane area is needed to recover the same fraction of fast gas. Together with higher attainable permeant product purity discussed earlier, this explains largely why available reports indicate that the membrane process is more competitive with conventional processes when the gas to be treated has a higher fast-gas concentration, assuming the permeant is the primary product.[5,38]

Feed Pressure. The characteristic curves also can be used to indicate the effects of feed pressure on the performance of a separator as shown in Fig. 20.6-15. Raising the feed pressure results in a higher-purity permeant product when the separator is operated at the same fractional CO_2 recovery, although the extent of the increase diminishes at higher feed pressures. On the other hand, the residue characteristic curve of CH_4 is rather insensitive to the feed pressure at fixed CO_2 recovery. Due to the low permeability of polysulfone to CH_4 relative to CO_2, the absolute amount of CH_4 lost to the permeant is small enough such that an increase in the feed pressure does not change significantly the residue composition for a given CO_2 recovery. Undoubtedly, an increase in the feed pressure will lower the membrane area needed for a given fast gas recovery. This is shown in Fig. 20.6-16 where the relative membrane area requirement for different feed pressures is compared.

Composition-Dependent Permeability. There are some experimental data[28,39] indicating the existence of competition between penetrants in accordance with the postulations of the generalized dual-mode transport model [Eq. (20.6-40)]. It is of interest to examine the magnitude of this competition effect on the overall performance of an actual separator. This is illustrated in Tables 20.6-5 and 20.6-6. Equation

TABLE 20.6-3 Effects of Permeant Pressure Buildup on the Performance of a Hollow-Fiber Separator

	Permeant CO_2 Mole Fraction	Residue CH_4 Mole Fraction	Permeant Production Rate (SCFH)	Overall Stage-Cut (%)	CO_2 Recovery (%)	CH_4 Recovery (%)
No pressure buildup	0.924	0.827	46,766	37.0	75.8	5.2
Fiber ID						
150 μm	0.922	0.806	44,594	35.3	72.2	5.0
100 μm	0.916	0.762	39,622	31.4	63.7	4.8
50 μm	0.872	0.633	20,918	16.6	32.0	3.8

Note: Feed rate = 126,374 SCFH
p_{2f} = 600 psia
Y_{CO_2} = 0.45
Permeant $p_{1,L}$ = 44.1 psia
NF = 9×10^5
ID = 150 μm
OD = 300 μm
z_L = 5 ft
$z_L - z_w$ = 0.5 ft
Polysulfone, 35°C

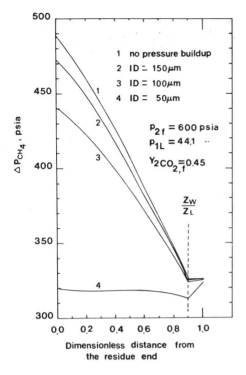

FIGURE 20.6-12 Profile of CH_4 (slow gas) partial pressure difference across the fiber wall. Fiber OD $= 300\ \mu m$, NF $= 9 \times 10^5$, feed rate $= 126,374$ SCFH.

Labels within figure:
1 no pressure buildup
2 ID $\doteq$ 150 μm
3 ID $=$ 100 μm
4 ID $=$ 50 μm

$P_{2,f} = 600$ psia
$P_{1L} = 44.1$
$Y_{2CO_2,f} = 0.45$

$\dfrac{Z_W}{Z_L}$

ΔP_{CH_4}, psia (y-axis)
Dimensionless distance from the residue end (x-axis)

(20.6-39) is used for the no-competition case while Eq. (20.6-40) is used for the competition case calculations. As expected, competition of penetrants for the available sorption sites in the polymer results in a lower permeant production rate.

In Table 20.6-5, competition actually results in higher permeant purity, which also is indicated by the local CO_2-CH_4 wall-flux ratio. The changes in values of stage-cut, permeant purity, residue purity, and permeant production rate, however, are relatively small (less than 5%) for one CO_2-CH_4 mixture/polysulfone system. The effects are somewhat more significant for the He-CH_4 mixture/polycarbonate system. Roughly, a 14% decrease in the permeant purity results from competition when the feed is dilute in He such as in the case of many natural gases. For a feed containing a relatively higher concentration of He, the effects of competition return to the 5-6% level as shown in Table 20.6-6.

20.6-3 Multistage Separators

Typical of all rate-determined unit operations, membrane separation suffers the depletion of separating driving force (i.e., partial pressure difference) along the unit. The characteristic curves in Figs. 20.6-10,

TABLE 20.6-4 Effect of Fiber Length on Separator Performance

	Fiber Length (cm)	Fractional CO_2 Recovery	Permeant Product $Y_{CO_2,L}$	Permeant Product $Y_{CH_4,0}$	Total Stage-Cut
Fiber ID 50 μm	304.8 (10 ft)	0.27	0.86	0.62	0.14
	152.4 (5 ft)	0.22	0.90	0.61	0.11
No permeant pressure buildup	304.8	0.778	0.920	0.838	0.381
	152.4	0.490	0.946	0.700	0.233

Note: At the same operating conditions as shown in Table 20.6-3.

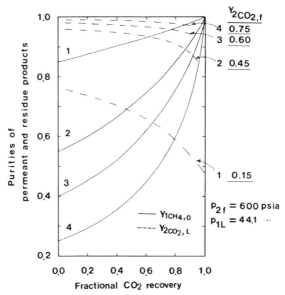

FIGURE 20.6-13 Effects of feed composition on the characteristic curves. Fiber ID = 150 μm, OD = 300 μm, NF = 9 × 10^5.

20.6-13, and 20.6-15 indicate that even with a fairly permselective membrane, high-purity permeant product cannot be produced unless the fast-gas partial pressure in the feed stream is maintained sufficiently high. For cases such as helium recovery from natural gas, where the initial concentration of helium is normally less than 1%, a single-stage module is unlikely to produce high-purity helium with currently available membranes. Sometimes, production of not only high-purity permeant but also high-purity residue product is required. More then one stage of modules would be needed for these applications. The "continuous" membrane column originated by Pfefferle[40] and refined by Hwang and Thorman[41-44] is an exception (where each unit includes one stripping module and one enriching module with a very large permeant product recycle).

Systems with more than three compression stages are not reported to be very attractive with respect to both capital and operating cost.[5,38,45] But even constrained within only three stages, there are many multiple-separator arrangements that, in principle, can meet the same production requirements. Clearly, detailed

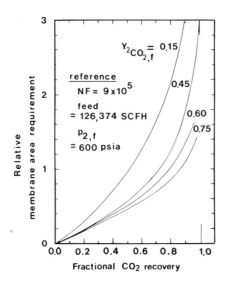

FIGURE 20.6-14 Effects of feed composition on membrane area requirement. Fiber ID = 150 μm, OD = 300 μm, NF = 9 × 10^5, p_{1L} = 44.1 psia.

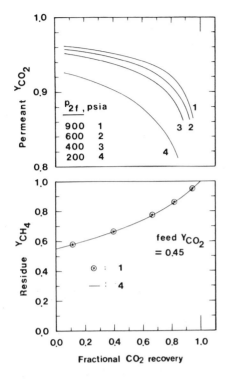

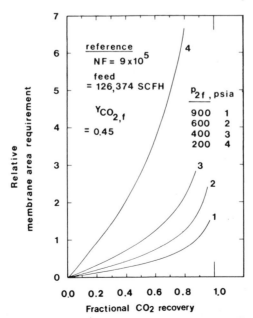

FIGURE 20.6-15 Effects of feed pressure on the characteristic curves. Fiber ID = 150 μm, OD = 300 μm, NF = 9×10^5, p_{1L} = 44.1 psia.

FIGURE 20.6-16 Effects of feed pressure on membrane area requirement. Fiber ID = 150 μm, OD = 300 μm, NF = 9×10^5, p_{1L} = 44.1 psia.

TABLE 20.6-5   Effects of Dual-Mode Competition on CO_2–CH_4 Separation[a]

CO_2–CH_4 Mixture, $Y_{CO_2,f} = 0.45$, Polysulfone at 35°C

	Competition	No competition
CO_2 recovery to the permeant	0.722	0.750
CH_4 recovery to the permeant	0.050	0.062
Permeant product, Y_{CO_2}	0.922	0.908
Residue product, Y_{CH_4}	0.806	0.820
Permeant production rate (SCFH)	44,594	47,023

	Z^b	Competition	No Competition
Local CO_2–CH_4 flux ratio	0.2	6.83	5.99
	0.4	10.27	8.73
	0.6	15.06	12.32
	0.8	21.50	16.85

[a] At the same operating conditions as shown in Table 20.6-3.
[b] Z is the fractional fiber length from the closed end.

engineering calculations and cost analyses are needed before deciding which arrangement is most suitable. Some considerations include questions such as: Are both high-purity permeant and residue products to be produced? Is it cost effective to add an additional compression stage such that one can operate the previous stage at very high recovery to reclaim as much fast gas as possible? Should higher feed pressure be used to save membrane cost? Is it more economical to recycle part of the final permeant product to the feed stream of the last enriching stage, than to add one more compression stage?

Figure 20.6-17 is one of the many viable arrangements for producing greater than 98 mol % CH_4 and greater than 95 mol % CO_2 products from a 12 mol % CO_2–88 mol % CH_4 mixture. Separator dimensions are shown in the figure. Permeability data of dense polysulfone at 35°C are used in the calculation. Other pertinent data are indicated in the figure and Tables 20.6-1 and 20.6-2. The two parallel modules in the feed stage are used to strip as much CO_2 as possible from the high-pressure side to the permeant while producing a permeant product of slightly more than 60% CO_2. This target CO_2 mole fraction is based on the characteristic curve (curve 3) in Fig. 20.6-13. Clearly, at the given conditions, a feed stream of at least 60% CO_2 is needed to produce a final permeant product of greater than 95% CO_2 purity from the second compression stage at a reasonably high fractional CO_2 recovery. On the other hand, no further compression is required in the stripping section where it is relatively easy to produce a residue product of greater than 98% CH_4. In determining the operating conditions for this stripping stage, again we resort to the characteristic curves and choose the minimum fractional CO_2 recovery required to minimize CH_4 loss to the permeant stream while keeping the residue purity above 98% CH_4.

TABLE 20.6-6   Effects of Dual-Mode Competition on He–CH_4 Separation

He–CH_4 Mixture, Lexan Polycarbonate, 35°C

	$Y_{He} = 0.01$		$Y_{He} = 0.1$	
Feed	Competition	No Competition	Competition	No Competition
He recovery to the permeant	0.579	0.669	0.733	0.817
CH_4 recovery to the permeant	0.069	0.069	0.075	0.075
Permeant product, Y_{He}	0.078	0.089	0.522	0.547
Permeant production rate, SCFH	10,572	10,709	20,004	21,275

Note: At the same operating conditions as shown in Table 20.6-3.

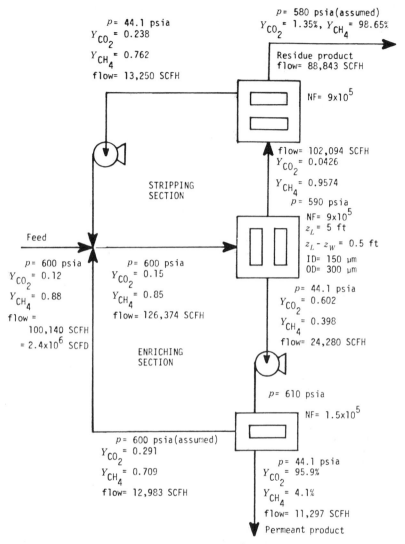

p= 580 psia(assumed)
Y_{CO_2} = 1.35%, Y_{CH_4} = 98.65%

Residue product
flow= 88,843 SCFH

NF= 9×10^5

p= 44.1 psia
Y_{CO_2} = 0.238
Y_{CH_4} = 0.762
flow= 13,250 SCFH

STRIPPING
SECTION

flow= 102,094 SCFH
Y_{CO_2} = 0.0426
Y_{CH_4} = 0.9574
p= 590 psia
NF= 9×10^5
z_L = 5 ft
$z_L - z_W$ = 0.5 ft
ID= 150 μm
OD= 300 μm

Feed

p= 600 psia
Y_{CO_2} = 0.12
Y_{CH_4} = 0.88
flow =
100,140 SCFH
= 2.4×10^6 SCFD

p= 600 psia
Y_{CO_2} = 0.15
Y_{CH_4} = 0.85
flow= 126,374 SCFH

ENRICHING
SECTION

p= 44.1 psia
Y_{CO_2} = 0.602
Y_{CH_4} = 0.398
flow= 24,280 SCFH

p= 610 psia
NF= 1.5×10^5

p= 600 psia(assumed)
Y_{CO_2} = 0.291
Y_{CH_4} = 0.709
flow= 12,983 SCFH

p= 44.1 psia
Y_{CO_2} = 95.9%
Y_{CH_4} = 4.1%
flow= 11,297 SCFH
Permeant product

FIGURE 20.6-17 Example of a two-compression-stage separator system.

Feed pressures and the sizes of the various stages are determined according to characteristic curves similar to those shown in Fig. 20.6-15 for different feed compositions. For the enriching section, the objective is to find the minimum feed pressure needed to produce permeant product of desired purity when the separator is operated at a reasonable fast-gas recovery given the anticipated range of feed rates. As indicated in Fig. 20.6-15, except for low feed pressure ranges (or more precisely, the ratio p_{2f}/p_{1L}) and very high fast-gas recoveries, the permeant product purity is not very sensitive to the feed pressure. Consequently, it is not difficult to choose this near-optimal feed pressure and membrane area size for processing a given feed with fibers of given dimensions.

For the *stripping* section, the objective is to minimize area requirement and CH_4 loss to the permeant stream. As shown in Fig. 20.6-15 and 20.6-16, in each stage, higher feed pressure lowers the membrane area requirement while affecting only slightly the residue purity. The residue purity is dependent primarily on the feed composition as indicated in Fig. 20.6-13. On the other hand, higher feed pressure will hasten the depletion of CO_2 and thus decrease the loss of CH_4 to the permeant stream for a target residue purity. The feed pressures of the stripping stages are determined from a balance between costs associated with membrane area, compression, and CH_4 losses.

In general, the feed stage in a cascade should be operated at maximum recovery of the most permeable component(s) while maintaining a high enough purity such that few enriching stages will be required to satisfy the final permeant purity. This would maximize the fast-gas production (overall recovery into the final permeant product) and minimize waste of membrane area and recompression of the recycled permeant stream(s) in the stripping-stage section.

To ensure reasonable process flexibility for handling feed rate and composition fluctuations, several optional designs may be included. For example, (a) feed pressures to the various stages may be allowed to change over a range of values, (b) part of the final permeant product may be recycled to the feed port of the last enriching stage to boost its feed concentration,[25,31,46] and (c) some limited independent manipulation of stage-cut may be provided by using several parallel smaller modules in each stage which can be turned on or off according to process requirements. The last option may not be very desirable because of increased module cost. Option (b) is a viable means of increasing the purity of the final permeant product but it also causes significant reductions in the overall permeant production rate and increases in the compression load. Option (a) appears to be the most effective approach for providing process flexibility. However, a combination of these options may be needed if the feed rate and composition fluctuate significantly.

In summary, the various model equations are strictly applicable only when the validities of the corresponding assumptions are justified. The generality of the above engineering considerations however, are, believed to be reasonable and have covered the major characteristics of membrane gas separators.

NOTATION

$\mathcal{Q}$	total membrane area in a separator module
D_{DA}, k_{DA}, F_A C'_{HA}, b_A	dual-mode parameters of A, see Table 20.6-2
P	permeability = (diffusive flux)(film thickness)/(partial pressure difference)
ID, OD	inside and outside diameters of the hollow fiber
$H(z - z_w)$	the unit step function
l	the "effective" thickness of the asymmetric fiber wall
$m_{fA,f}$	initial molar flow rate of A in the feed per fiber
m_{pA}	local molar flow rate of A in the fiber bore per fiber
m_p	local total permeant molar flow rate per fiber
N_A	local permeation flux of component A
NF	total number of fibers in the module
p_1	local permeant total pressure in the fiber bore
p_{1L}	total permeant pressure at the open end of the fiber
p_2	feed-side total pressure
R	universal gas constant
Y_{1A}	local mole fraction of A in the permeant stream
Y_{2A}	local mole fraction of A in the shell stream (feed side)
z	distance from the closed end of the fiber
z_w	effective length of the fiber
z_L	total length of the fiber
μ	viscosity

Subscripts

0	indicates the residue end of the module
L	indicates the permeate exit end of the module
1	the permeant side
2	the shell side (feed side)
A or B	indicates component A or B
α_A^*	permeability ratio of penetrant A to a reference penetrant, for example, penetrant B

REFERENCES

Section 20.1

20.1-1   R. J. Gardner, R. A. Crane, and J. F. Hannan, Hollow Fiber Permeator for Separating Gases, *Chem. Eng. Prog.*, **73**, 76 (1977).

20.1-2   W. A. Bollinger, D. L. MacLean, and R. S. Narayan, Separation Systems for Oil Refining and Production, *Chem. Eng. Prog.*, **78**, 27 (1982).

20.1-3   W. J. Schell and C. D. Houston, Process Gas with Selective Membranes, *Chem. Eng. Prog.*, **78**, 33 (1982); *Hydrocarbon Process.*, **61**, 249 (1982); and paper No. 83C presented at AIChE Symposium, Houston, TX, Mar. 28–31, 1983.

20.1-4   W. H. Mazur and M. C. Chan, Membranes for Natural Gas Sweetening and CO_2 Enrichment, *Chem. Eng. Prog.*, **78**, 38 (1982).

20.1-5   A. B. Coady and J. A. Davis, CO_2 Recovery by Gas Permeation, *Chem. Eng. Prog.*, **78**, 45 (1982).

20.1-6   K. C. Youn, G. C. Blytas, and H. H. Wall, "Role of Membrane Technology in the Recovery of Carbon Dioxide: Economics and Future Prospects," paper presented at Gas Processors Association Regional Meeting, Houston, TX, Nov. 1983.

20.1-7   B. M. Burmaster, and D. C. Carter, "Increased Methanol Production Using Prism Separators," paper No. 83A presented at AIChE Symposium, Houston, TX, Mar. 28–31, 1983.

20.1-8   W. J. Schell, Membrane Use/Technology Growing, *Hydrocarbon Process.*, **62**, 43 (1983).

20.1-9   F. G. Russell, Operating Permeation Systems, *Hydrocarbon Process.*, **62**, 55 (1983).

20.1-10   V. O. Lane, Plant Permeation Experience, *Hydrocarbon Process.*, **62**, 56 (1983).

20.1-11   R. L. Schendel, C. L. Mariz, and J. Y. Mak, Is Permeation Competitive?, *Hydrocarbon Process.*, **62**, 58 (1983).

20.1-12   D. L. MacLean, D. J. Stookey, and T. R. Metzger, Fundamentals of Gas Permeation, *Hydrocarbon Process.*, **62**, 47 (1983).

20.1-13   J. K. Mitchell, *Philadelphia J. Med. Sci.*, **13**, 36 (1931).

20.1-14   A. Fick, Über Diffusion, *Pogg. Ann.*, **94**, 59 (1855).

20.1-15   T. Graham, *Philos. Mag.*, **32**, 401 (1866).

20.1-16   F. Exner, *J. Chem. Soc.*, **155**, 321, 443 (1875).

20.1-17   S. von Wroblewski, *Wied Ann.*, **8**, 29 (1879).

20.1-18   D. R. Paul and G. Morel, Membrane Technology, reprinted from *Kirk–Othmer Encyclopedia of Chemical Technology* 3rd ed., Vol. 15, p. 92, Wiley, New York, 1981.

20.1-19   G. A. Shakespear, *Reports of the Advisory Committee on Aeronautics, T. 1164* (1918).

20.1-20   N. Yi-Yan, R. M. Felder, and W. J. Koros, Selective Permeation of Hydrocarbon Gases in Poly(tetrafluoro ethylene) and Poly(fluoroethylene/propylene Copolymer), *J. Appl. Polym. Sci.*, **25**, 1755 (1980).

20.1-21   S. M. Fang, S. A. Stern, and H. L. Frisch, A Free Volume Model of Permeation of Gas and Liquid Mixtures Through Polymeric Membranes, *Chem. Eng. Sci.*, **30**, 773 (1975).

20.1-22   R. T. Chern, W. J. Koros, H. B. Hopfenberg, and V. T. Stannett, Reversible Isopentane-Induced Depression of Carbon Dioxide Permeation Through Polycarbonate, *J. Polym. Sci. Polym. Phys. Ed.*, **21**, 753 (1983).

20.1-23   R. T. Chern, W. J. Koros, E. S. Sanders, and R. E. Yui, Second Component Effects in Sorption and Permeation of Gases in Glassy Polymers, *J. Membr. Sci.*, **15**, 157 (1983).

20.1-24   R. T. Chern, "Measurements and Modeling of Mixed Gas Permeation through Glassy Polymers," Ph.D. dissertation, North Carolina State University, Raleigh, 1983.

20.1-25   R. T. Chern, W. J. Koros, H. B. Hopfenberg, and V. T. Stannett, Selected Permeation of CO_2 and CH_4 Through Kapton® Polyimide: Effects of Penetrant Competition and Gas Phase Non-idealities, *J. Polym. Sci. Polym. Phys. Ed.*, **22**, 1061 (1984).

20.1-26   L. M. Robeson, The Effect of Antiplasticization on Secondary Loss Transitions and Permeability of Polymers, *Polym. Eng. Sci.*, **9**, 277 (1969).

20.1-27   D. G. Pye, H. H. Hoehn, and M. Panar, Measurement of Gas Permeability of Polymers. I. Permeabilities in Constant Volume/Variable Pressure Apparatus, *J. Appl. Polym. Sci.*, **20**, 1921 (1976); Measurement of Gas Permeability of Polymers. II. Apparatus for Determination of Mixed Gases and Vapors, *J. Appl. Polym. Sci.*, **20**, 287 (1976).

20.1-28   G. R. Antonson, R. J. Gardner, C. F. King, and D. Y. Ko, Analysis of Gas Separation by Permeation in Hollow Fibers, *Ind. Eng. Chem. Proc. Des. Dev.*, **16**, 463 (1977).

20.1-29   F. P. McCandless, Separation of Binary Mixtures of CO and H_2 by Permeation Through Polymeric Films, *Ind. Eng. Chem. Proc. Des. Dev.*, **11**, 470 (1972).

20.1-30   S. A. Stern, G. R. Mauze, and H. L. Frisch, Test of a Free Volume Model of Gas Permeation Through Polymer Membranes. I. Pure CO_2, CH_4, C_2H_4, and C_3H_8 in Polyethylene, *J. Polym. Sci. Polym. Phys. Ed.*, **21**, 467, 1275 (1983).

20.1-31   S. A. Stern and S. S. Kulkarni, Solubility of Methane in Cellulose Acetate—Conditioning Effect of Carbon Dioxide, *J. Membr. Sci.*, **10**, 235 (1982).

20.1-32 E. L. Cussler, *Multicomponent Diffusion*, Elsevier, New York, 1976.

20.1-33 K. Kammermeyer, Silicone Rubber as a Selective Barrier, *Ind. Eng. Chem.*, **49**, 1685 (1957).

20.1-34 W. L. Robb, Thin Silicon Membranes: Their Permeation Properties and Some Applications, *Ann. N.Y. Acad. Sci.*, **146**, 119 (1967).

20.1-35 V. T. Stannett, W. J. Koros, D. R. Paul, H. K. Lonsdale, and R. W. Baker, Recent Advances in Membrane Science and Technology, *Adv. Polym. Sci.*, **32**, 69 (1979).

20.1-36 S. A. Stern, Gas Permeation Processes, R. Lacey and S. Loeb (Eds.), in *Industrial Processing with Membranes*, Wiley-Interscience, New York, 1972.

20.1-37 W. J. Ward, W. R. Browal, and R. M. Salemme, Ultrathin Silicone/Polycarbonate Membranes for Gas Separation Processes, *J. Membr. Sci.*, **1**, 99 (1976).

20.1-38 J. A. Gerow, "Permeation Separation Device," U.S. Patent 3,832,830, Sept. 3, 1973, E.I. Du Pont.

20.1-39 Composite Membrane Key to Gas Separators, *Chem. Eng. News*, 57 (May 19, 1980).

20.1-40 J. M. S. Henis and M. K. Tripodi, The Developing Technology of Gas Separating Membranes, *Science*, **220**, 11 (1983).

20.1-41 D. L. MacLean, C. E. Prince, Y. C. Chae, Energy Saving Modifications in Ammonia Plants, *Chem. Eng. Prog.*, **76**, 98 (1980).

20.1-42 R. R. Ward, R. C. Chang, J. C. Danos, and J. A. Carden, Jr., "Coating the Exteriors of Hollow Fibers," U.S. Patent 4,214,020, July 22, 1980, assigned to Monsanto Co.

20.1-43 E. C. Makin and K. K. Okamoto, "Process for Methanol Production," U.S. Patent 4,181,675, Jan. 1, 1980, assigned to Monsanto Co.

20.1-44 R. Heitz, "Hollow-Fiber Membrane Technology," paper presented at the 1983 International Symposium on Milestones and Trends in Polymer Science and Technology: A Tribute to Turner Alfrey, MMI, Midland, Michigan, June 6–9, 1983.

20.1-45 H. I. Mahon, "Permeability Separatory Apparatus and Process Using Hollow Fibers," U.S. Patent 3,228,877, and "Permeability Separatory Apparatus Membrane Element, Method of Making the Same and Process Utilizing the Same," U.S. Patent 3,228,876, Jan. 1966.

20.1-46 W. G. Smith, "Permeation Separation Device for Separating Fluids and Process Relating Thereto," U.S. Patent 3,526,001, Aug. 1970.

20.1-47 Monsato Prism® Separator Product Information Bulletin, 1981.

20.1-48 J. Westmoreland, "Spirally Wrapped Reverse Osmosis Membrane Cell," U.S. Patent 3,367,504, Feb. 1968.

20.1-49 D. T. Bray, "Reverse Osmosis Purification Apparatus," U.S. Patent 3,417,870, Dec. 1968.

20.1-50 U. Merten, "Reverse Osmosis Membrane Module," U.S. Patent 3,386,583, June 1968.

20.1-51 W. King, Separex, Inc., personal communication, 1983.

20.1-52 S. Loeb and S. Sourirajan, Sea Water Demineralization by Means of an Osmotic Membrane, *Adv. Chem. Ser.*, **38**, 117 (1962).

20.1-53 H. K. Lonsdale, U. Merten, and R. L. Riley, Transport Properties of Cellulose Acetate Osmotic Membranes, *J. Appl. Polym. Sci.*, **9**, 1341 (1965).

20.1-54 S. Loeb and S. Sourirajan, "High Flow Porous Membranes for Separating Water from Saline Solutions," U.S. Patent 3,133,132, May 12, 1964.

20.1-55 R. E. Kesting, *Synthetic Polymeric Membranes*, McGraw-Hill, New York, 1971.

20.1-56 H. Strathmann, K. Kock, P. Amar, and R. W. Baker, The Formation Mechanism of Asymmetric Membranes, *Desalination*, **16**, 179 (1975).

20.1-57 H. Strathmann, P. Scheible, and R. W. Baker, A Rationale for the Preparation of Loeb–Sourirajan-Type Cellulose Acetate Membranes, *J. Appl. Polym. Sci.*, **15**, 811 (1971).

20.1-58 I. Cabasso, E. Klein, and J. K. Smith, Polysulfone Hollow Fibers. I. Spinning and Properties, *J. Appl. Polym. Sci.*, **20**, 2377 (1976).

20.1-59 D. M. Koenhen, M. H. V. Mulder and C. A. Smolders, Phase Separation Phenomena During the Formation of Asymmetric Membranes, *J. Appl. Polym. Sci.*, **21**, 199 (1977).

20.1-60 H. Strathmann and K. Kock, The Formation Mechanism of Phase Inversion Membranes, *Desalination*, **21**, 241 (1977).

20.1-61 C. Cohen, G. B. Tanny, and S. Prager, Diffusion-Controlled Formation of Porous Structures in Ternary Systems, *J. Polym. Sci. Polym. Phys. Ed.*, **17**, 477 (1979).

20.1-62 G. B. Tanny, The Surface Tension of Polymer Solutions and Asymmetric Membrane Formation, *J. Appl. Polym. Sci.*, **18**, 2149 (1974).

20.1-63 J. W. Cahn, Spinodal Decomposition, *Trans. Metall. Soc. AIME*, **242**, 166 (1968).

20.1-64   C. W. Allegranti, D. G. Pye, H. H. Hoehn, and M. Panar, The Morphology of Asymmetric Separation Membranes, *J. Appl. Polym. Sci.*, **19**, 1475 (1975).

20.1-65   I. Cabasso, personal communication, 1983.

20.1-66   C. A. Smolders, personal communication, 1983.

20.1-67   I. Cabasso, E. Klein, and J. K. Smith, *Research and Development of NS-1 and Related Polysulfone Hollow Fibers for Reverse Osmosis Desalination of Seawater*, report prepared for Office of Water Research and Technology, Gulf South Research Institute, July 1975.

20.1-68   I. Cabasso, E. Klein, and J. K. Smith, Polysulfone Hollow Fibers. II. Morphology, *J. Apl. Polym. Sci.*, **31**, 165 (1977).

20.1-69   J. M. S. Henis and M. K. Tripodi, A Novel Approach to Gas Separations Using Composite Hollow-Fiber Membranes, *Sep. Sci. Technol.*, **15**, 1059 (1980).

20.1-70   J. M. S. Henis and M. K. Tripodi, "Multicomponent Membranes for Gas Separation," U.S. Patent 4,230,463, Oct. 1980.

20.1-71   C. G. Wensley and S. Z. Zakabhazy, "High Performance Gas Separation Membranes," paper presented at AIChE Winter National Meeting, Atlanta, GA, May 1984.

20.1-72   P. K. Gantzel and U. Merten, Gas Separations with High Flux Cellulose Acetate Membranes, *Ind. Eng. Chem. Proc. Des. Dev.*, **9**, 331 (1970).

20.1-73   H. K. Lonsdale, The Growth of Membrane Technology, *J. Membr. Sci.*, **10**, 81 (1982).

20.1-74   H. Hoehn, "Heat Treatment of Membranes of Selected Polyimides, Polyesters, and Polyamides," U.S. Patent 3,822,202, July 1974.

20.1-75   H. Hoehn and J. W. Richter, "Aromatic Polyimide, Polyester, and Polyamide Separation Membranes," U.S. Patent Reissue 30,351, July 1980.

20.1-76   J. E. Cadotte, R. S. King, R. J. Majerle, and R. J. Petersen, Interfacial Synthesis in the Preparation of Reverse Osmosis Membranes, *J. Macromol. Sci. Chem.*, **A15**, 725 (1981).

20.1-77   R. L. Riley, P. A. Case, A. L. Lloyd, C. E. Milstead, and M. Tagami, "Recent Developments in Thin-Film Composite Reverse Osmosis Membrane Systems," paper presented at Joint Symposium on Water Filtration and Purification, AIChE and Filtration Society, Philadelphia, PA, June 1980.

Section 20.2

20.2-1   S. Weller and W. A. Steiner, Separation of Gases by Fractional Permeation Through Membranes, *J. Appl. Phys.*, **21**, 279 (1950).

20.2-2   W. P. Walawender and S. A. Stern, Analysis of Membrane Separation Parameters. II. Countercurrent and Cocurrent Flow in a Single Permeation Stage, *Sep. Sci.*, **7**, 553 (1972).

20.2-3   R. J. Gardner, R. A. Crane, and J. F. Hannan, Hollow Fiber Permeator for Separating Gases, *Chem. Eng. Prog.*, **73**, 76 (1977).

20.2-4   W. A. Bollinger, D. L. MacLean, and R. S. Narayan, Separation Systems for Oil Refining and Production, *Chem. Eng. Prog.*, **78**, 27 (1982).

20.2-5   W. J. Schell and C. D. Houston, Process Gas with Selective Membranes, *Chem. Eng. Prog.*, **78**, 33 (1982); *Hydrocarbon Process.*, **61**, 249 (1982); and paper No. 83C presented at AIChE Symposium, Houston, TX, Mar. 28–31, 1983.

20.2-6   B. M. Burmaster and D. C. Carter, "Increased Methanol Production Using Prism Separators," paper No. 83A presented at AIChE Symposium, Houston, TX, Mar. 28–31, 1983.

20.2-7   V. O. Lane, Plant Permeation Experience, *Hydrocarbon Process*, **62**, 56 (1983).

20.2-8   R. L. Schendel, C. L. Mariz, and J. Y. Mak, Is Permeation Competitive?, *Hydrocarbon Process.*, **62**, 58 (1983).

20.2-9   D. L. MacLean and T. E. Graham, Hollow Fibers Recover Hydrogen, *Chem. Eng.*, **87**, 54 (1980).

20.2-10   W. H. Mazur and M. C. Chan, Membranes for Natural Gas Sweetening and CO_2 Enrichment, *Chem. Eng. Prog.*, **78**, 38 (1982).

20.2-11   A. B. Coady and J. A. Davis, CO_2 Recovery by Gas Permeation, *Chem. Eng. Prog.*, **78**, 45 (1982).

20.2-12   K. C. Youn, G. C. Blytas, and H. H. Wall, "Role of Membrane Technology in the Recovery of Carbon Dioxide: Economics and Future Prospects," paper presented at Gas Processors Association Regional Meeting, Houston, TX, Nov. 1983.

20.2-13   F. G. Russell, Operating Permeation Systems, *Hydrocarbon Process.*, **62**, 55 (1983).

20.2-14   D. L. MacLean, D. J. Stookey, and T. R. Metzger, Fundamentals of Gas Permeation, *Hydrocarbon Process.*, **62**, 47 (1983).

20.2-15 H. K. Lonsdale, The Growth of Membrane Technology, *J. Membr. Sci.*, **10**, 81 (1982).

20.2-16 W. J. Ward, W. R. Browal, and R. M. Salemme, Ultrathin Silicon/Polycarbonate Membranes for Gas Separation Processes, *J. Membr. Sci.*, **1**, 99 (1976).

20.2-17 Susan C. Kelly, News Release, "Separation Technology Used for Atmospheric Blanketing," Monsanto Public Relations Dept., Oct. 1983.

20.2-18 V. T. Stannett, in J. Crank and G. S. Park (Eds.), *Diffusion in Polymers*, Chap. 2, Academic, New York, 1968.

20.2-19 H. J. Bixler and O. J. Sweeting, in O. J. Sweeting (Ed.), *The Science and Technology of Polymer Films*, Vol. II, p. 2, Wiley, New York, 1971.

20.2-20 G. J. Van Amerongen, Diffusion in Elastomers, *Rubber Chem. Technol.*, **37**, 1065 (1964).

20.2-21 V. T. Stannett, W. J. Koros, D. R. Paul, H. K. Lonsdale, and R. W. Baker, Recent Advances in Membrane Science and Technology, *Adv. Polym. Sci.*, **32**, 69 (1979).

Section 20.3

20.3-1 G. J. Van Amerongen, Diffusion in Elastomers, *Rubber Chem. Technol.*, **37**, 1065 (1964).

20.3-2 H. J. Bixler and O. J. Sweeting, in O. J. Sweeting (Ed.), *The Science and Technology of Polymer Films*, Vol. II, p. 2, Wiley, New York, 1971.

20.3-3 D. G. Pye, H. H. Hoehn, and M. Panar, Measurement of Gas Permeability of Polymers. I. Permeabilities in Constant Volume/Variable Pressure Apparatus, *J. Appl. Polym. Sci.*, **20**, 1921 (1976); Measurement of Gas Permeability of Polymers. II. Apparatus for Determination of Mixed Gases and Vapors, *J. Appl. Polym. Sci.*, **20**, 287 (1976).

20.3-4 S. A. Stern, G. R. Mauze, H. L. Frisch, Test of a Free Volume Model of Gas Permeation Through Polymer Membranes. I. Pure CO_2, CH_4, C_2H_4, and C_3H_8 in Polyethylene, *J. Polym. Sci. Polym. Phys. Ed.*, **21**, 467, 1275 (1983).

20.3-5 R. M. Barrer, J. A. Barrie, and J. Slater, Sorption and Diffusion in Ethyl Cellulose. I. History—Dependence of Sorption Isotherms and Permeation Rates, *J. Polym. Sci.*, **23**, 315 (1957).

20.3-6 R. T. Chern, W. J. Koros, E. S. Sanders, S. H. Chen, and H. B. Hopfenberg, Implications of the Dual Mode Sorption and Transport Models for Mixed Gas Permeation, ACS Symposium Series No. 233, *Industrial Gas Separations*, T. E. Whyte, C. M. Yon, and E. H. Wagener (Eds.), American Chemical Society, Washington, DC, 1983.

20.3-7 W. J. Koros, A. H. Chan, and D. R. Paul, Sorption and Transport of Various Gases in Polycarbonate, *J. Membr. Sci.*, **2**, 165 (1977).

20.3-8 W. J. Koros and D. R. Paul, Design Considerations for Measurement of Gas Sorption in Polymers by Pressure Decay, *J. Polym. Sci. Polym. Phys. Ed.*, **14**, 1903 (1976).

20.3-9 E. S. Sanders, W. J. Koros, H. B. Hopfenberg, and V. T. Stannett, Pure and Mixed Gas Sorption of Carbon Dioxide and Ethylene in Poly(methyl methacrylate), *J. Membr. Sci.*, **13**, 161 (1983).

20.3-10 C. E. Rogers, in D. Fox, M. M. Labes, and A. Weissberger (Eds.), *Physics and Chemistry of the Organic Solid State*, Vol. II, Chap. 6, Wiley-Interscience, New York, 1965.

20.3-11 H. Fujita, in J. Crank and G. S. Park (Eds.), *Diffusion in Polymers*, Chap. 3, Academic, New York, 1968.

20.3-12 J. L. Vrentas and J. S. Duda, Diffusion in Polymer–Solvent Systems. I. Reexamination of the Free Volume Theory, *J. Polym. Sci. Polym. Phys. Ed.*, **15**, 403 (1977); Diffusion in Polymer–Solvent Systems. II. A Predictive Theory for the Dependence of Diffusion Coefficients on Temperature, Concentration, and Molecular Weight, *J. Polym Sci. Polym. Phys. Ed.*, **15**, 417 (1977).

20.3-13 S. A. Stern and H. L. Frisch, The Selective Permeation of Gases Through Polymers, *Ann. Rev. Mater. Sci.*, **11**, 523 (1981).

20.3-14 J. A. Barrie, in J. Crank and G. S. Park (Eds.), *Diffusion in Polymers*, Chap. 8, Academic, New York, 1968.

20.3-15 R. M. Felder and G. S. Huvard, Permeation, Diffusion, and Sorption of Gases and Vapors, *Methods Exp. Phys.*, **16C**, 315 (1980).

20.3-16 W. J. Koros, D. R. Paul, and A. A. Rocha, Carbon Dioxide Sorption and Transport in Polycarbonate, *J. Polym. Sci. Polym. Phys. Ed.*, **14**, 687 (1976).

20.3-17 G. S. Huvard, V. T. Stannett, W. J. Koros, and H. B. Hopfenberg, The Pressure Dependence of CO_2 Sorption and Permeation in Poly(acrylonitrile), *J. Membr. Sci.*, **6**, 185 (1980).

20.3-18 S. A. Stern, P. J. Gareis, T. F. Sinclair, and P. H. Mohr. Performance of a Versatile Variable Volume Permeability Cell—Comparison of Gas Permeability Measurements by the Variable Volume and Variable Pressure Methods, *J. Appl. Polym. Sci.*, **7**, 2035 (1963).

20.3-19 R. A. Pasternak, J. F. Schimscheimer, and J. Heller, A Dynamic Approach to Diffusion and Permeation Measurements, *J. Polym. Sci. A-2*, **8**, 467 (1970).

20.3-20 S. A. Stern and A. H. DeMeringo, Solubility of Carbon Dioxide in Cellulose Acetate at Elevated Pressures, *J. Polym. Sci. Polym. Phys. Ed.*, **16**, 735 (1978).

20.3-21 W. J. Koros, "Membrane-Based Gas Separations: Research Directions for the Eighties," paper presented at Sunriver Membrane Conference, Redmond, OR, Sept. 1983.

20.3-22 S. T. Hwang, C. K. Choi, K. Kammermeyer, Gaseous Transfer Coefficients in Membranes, *Sep. Sci.*, **3**, 461 (1974).

20.3-23 L. Pilato, L. Litz, B. Hargitay, R. C. Osborne, A. Farnham, J. Kawakami, P. Fritze, and J. McGrath, Polymers for Permselective Membrane Gas Separations, *ACS Preprints*, **16**, 42 (1975).

20.3-24 R. T. Chern, W. J. Koros, H. B. Hopfenberg, and V. T. Stannett, Material Selection for Membrane-Based Gas Separations, ACS Symposium Series 269, *Materials Science of Synthetic Membranes*, D. R. Lloyd (Ed.), American Chemical Society, Washington, DC, 1985.

20.3-25 S. A. Stern and S. S. Kulkarni, Solubility of Methane in Cellulose Acetate—Conditioning Effect of Carbon Dioxide, *J. Membr. Sci.*, **10**, 235 (1982).

20.3-26 H. L. Frisch, The Time Lag in Diffusion, *J. Phys. Chem.*, **61**, 93 (1957).

20.3-27 J. Crank, *The Mathematics of Diffusion*, 2nd ed., Clarendon, Oxford 1975.

20.3-28 V. Saxena and S. A. Stern, Concentration-Dependent Transport of Gases and Vapors in Glassy Polymers, *J. Membr. Sci.*, **12**, 65 (1982).

20.3-29 W. J. Koros, G. N. Smith, and V. T. Stannett, High Pressure Sorption of Carbon Dioxide in Solvent Cast Poly(methyl methacrylate) and Poly(ethyl methacrylate) Films, *J. Appl. Polym. Sci.*, **26**, 159 (1981).

20.3-30 W. J. Koros, Model for Sorption of Mixed Gases in Glassy Polymers, *J. Polym. Sci. Polym. Phys. Ed.*, **18**, 981 (1980).

Section 20.4

20.4-1 F. P. McCandless, Separation of Binary Mixtures of CO and H_2 by Permeation Through Polymeric Films, *Ind. Eng. Chem. Proc. Des. Dev.*, **11**, 470 (1972).

20.4-2 G. R. Antonson, R. J. Gardner, C. F. King, and D. Y. Ko, Analysis of Gas Separation by Permeation in Hollow Fibers, *Ind. Eng. Chem. Proc. Des. Dev.*, **16**, 463 (1977).

20.4-3 V. T. Stannett, W. J. Koros, D. R. Paul, H. K. Lonsdale, and R. W. Baker, Recent Advances in Membrane Science and Technology, *Adv. Polym. Sci.*, **32**, 69 (1979).

20.4-4 F. W. Billmeyer, *Textbook of Polymer Science*, 2nd ed., Wiley-Interscience, New York, 1971.

20.4-5 H. J. Bixler and O. J. Sweeting, in O. J. Sweeting (Ed.), *The Science and Technology of Polymer Films*, Vol. II, p. 2, Wiley, New York, 1971.

20.4-6 A. S. Michaels and H. J. Bixler, Flow of Gases Through Polyethylene, *J. Polym. Sci.*, **50**, 413 (1961).

20.4-7 A. S. Michaels, W. R. Vieth, and J. A. Barrie, Diffusion and Solution of Gases in Poly(ethylene terephthalate), *J. Appl. Phys.*, **34**, 1, 13 (1963).

20.4-8 A. R. Berens, The Diffusion of Gases and Vapors in Rigid PVC, *J. Vinyl Technol.*, **1**, 8 (1979).

20.4-9 A. Kreituss and H. L. Frisch, Free Volume Estimates in Heterogeneous Polymer Systems. I. Diffusion in Crystalline Ethylene–Propylene Copolymers, *J. Polym. Sci. Polym. Phys. Ed.*, **19**, 889 (1981).

20.4-10 H. Fujita, in J. Crank and G. S. Park (Eds.), *Diffusion in Polymers*, Chap. 3, Academic, New York, 1968.

20.4-11 J. S. Vrentas and J. L. Duda, Solvent and Temperature Effects on Diffusion in Polymer–Solvent Systems, *J. Appl. Polym. Sci.*, **21**, 1715 (1977).

20.4-12 J. S. Vrentas and J. L. Duda, Diffusion of Small Molecules in Amorphous Polymers, *Macromolecules*, **9**, 785 (1976).

20.4-13 S. A. Stern, S. S. Kulkarni, and H. L. Frisch, Test of a Free Volume Model of Gas Permeation Through Polymer Membranes. I. Pure CO_2, CH_4, C_2H_4, and C_3H_8 in Polyethylene, *J. Polym. Sci. Polym. Phys. Ed.*, **21**, 467 (1983).

20.4-14 H. L. Frisch, D. Klempner, and T. Kwei, Modified Free Volume Theory of Penetrant Diffusion in Polymers, *Macromolecules*, **4**, 237 (1971).

20.4-15 R. C. Reid, J. M. Prausnitz, and T. K. Sherwood, *The Properties of Gases and Liquids*, 3rd ed., McGraw-Hill, New York, 1977.

20.4-16 S. A. Stern, J. T. Mullhaupt, and P. J. Gareis, The Effect of Pressure on the Permeation of Gases and Vapors Through Polyethylene. Usefulness of the Corresponding States Principle, *AIChE J.*, **15**, 64 (1969).

20.4-17 J. L. Vrentas and J. S. Duda, Diffusion in Polymer–Solvent Systems. I. Reexamination of the

Free Volume Theory, *J. Polym Sci. Polym. Phys. Ed.*, **15**, 403 (1977); Diffusion in Polymer–Solvent Systems. II. A Predictive Theory for the Dependence of Diffusion Coefficients on Temperature, Concentration, and Molecular Weight, *J. Polym Sci. Polym. Phys. Ed.*, **15**, 417 (1977).

20.4-18 D. G. Pye, H. H. Hoehn, and M. Panar, Measurement of Gas Permeability of Polymers. I. Permeabilities in Constant Volume/Variable Pressure Apparatus, *J. Appl. Polym. Sci.*, **20**, 1921 (1976); Measurement of Gas permeability of Polymers. II. Apparatus for Determination of Mixed Gases and Vapors, *J. Appl. Polym. Sci.*, **20**, 287 (1976).

20.4-19 S. A. Stern and V. Saxena, Concentration-Dependent Transport of Gases and Vapors in Glassy Polymers, *J. Membr. Sci.*, **7**, 47 (1980).

20.4-20 G. T. Paulson, A. B. Clinch, and F. P. McCandless, The Effects of Water Vapor on the Separation of Methane and Carbon Dioxide by Gas Permeation Through Polymeric Membranes, *J. Membr. Sci.*, **14**, 129 (1983).

20.4-21 W. J. Koros and D. R. Paul, CO_2 Sorption in Poly(ethylene terephthalate) Above and Below the Glass Transition, *J. Polym. Sci. Polym. Phys. Ed.*, **16**, 1947 (1978).

20.4-22 R. M. Barrer, J. A. Barrie, and J. Slater, Sorption and Diffusion in Ethyl Cellulose. I. History—Dependence of Sorption Isotherms and Permeation Rates, *J. Polym. Sci.*, **23**, 315 (1957).

20.4-23 A. G. Wonders and D. R. Paul, Effects of CO_2 Exposure History on Sorption and Transport in Polycarbonate, *J. Membr. Sci.*, **5**, 63 (1978).

20.4-24 J. M. Fechter, H. B. Hopfenberg, and W. J. Koros, Characterization of Glassy State Relaxations by Low Pressure Carbon Dioxide Sorption in Poly(methyl methacrylate), *Polym. Eng. Sci.*, **21**, 925 (1981).

20.4-25 J. Horiuti, *Sci. Papers Inst. Phy. Chem. Res. (Tokyo)*, **17**, 126 (1931).

20.4-26 W. J. Koros and D. R. Paul, Observations Concerning the Temperature Dependence of the Langmuir Sorption Capacity of Glassy Polymers, *J. Polym. Sci. Polym. Phys. Ed.*, **19**, 1655 (1981).

20.4-27 A. H. Chan and D. R. Paul, Effect of Sub-T_g Annealing on CO_2 Sorption in Polycarbonate, *Polym. Eng. Sci.*, **20**, 87 (1980).

20.4-28 R. T. Chern, W. J. Koros, E. S. Sanders, and R. E. Yui, Second Component Effects in Sorption and Permeation of Gases in Glassy Polymers, *J. Membr. Sci.*, **15**, 157 (1983).

20.4-29 S. H. Chen, ''High-Pressure Carbon Dioxide Sorption in a Series of Methacrylate Polymers and Kapton® Polyimide,'' M.S. thesis, North Carolina State University, Raleigh, 1982.

20.4-30 Stanley I. Sandler, *Chemical and Engineering Thermodynamics*, p. 435, Wiley, New York, 1977.

20.4-31 O. G. Lewis, *Physical Constants of Linear Homopolymers*, Springer-Verlag, New York, 1968.

20.4-32 W. J. Koros, A. H. Chan, and D. R. Paul, Sorption and Transport of Various Gases in Polycarbonate, *J. Membr. Sci.*, **2**, 165 (1977).

20.4-33 D. R. Paul and W. J. Koros, Effect of Partially Immobilizing Sorption on Permeability and Diffusion Time Lag, *J. Polym. Sci. Polym. Phys. Ed.*, **14**, 675 (1976).

20.4-34 A. H. Chan, W. J. Koros, and D. R. Paul, Analysis of Hydrocarbon Gas Sorption and Transport in Ethyl Cellulose Using the Dual Mode Sorption/Partial Immobilization Models, *J. Membr. Sci.*, **3**, 117 (1978).

20.4-35 W. J. Koros and D. R. Paul, Transient and Steady State Permeation in Poly(ethylene terephthalate) Above and Below the Glass Transition, *J. Polym. Sci. Polym. Phys. Ed.*, **16**, 2171 (1978).

20.4-36 J. A. Barrie, K. Munday, and M. Williams, Sorption and Diffusion of Hydrocarbon Vapors in Glassy Polymers, *Polym. Eng. Sci.*, **20**, 20 (1980).

20.4-37 R. T. Chern, W. J. Koros, E. S. Sanders, S. H. Chen, and H. B. Hopfenberg, Implications of the Dual Mode Sorption and Transport Models for Mixed Gas Permeation, ACS Symposium Series No. 233, *Industrial Gas Separations*, T. E. Whyte, C. M. Yon, and E. H. Wagner (Eds.), American Chemical Society, Washington, DC, 1983.

20.4-38 K. Toi, Pressure Dependence of Diffusion Coefficient for CO_2 in Glassy Polymers, *Polym. Eng. Sci.*, **20**, 30 (1980).

20.4-39 B. P. Tikhomorov, H. B. Hopfenberg, V. T. Stannett, and J. L. Williams, Permeation, Diffusion, and Solution of Gases and Water Vapor in Unplasticized Poly(vinyl chloride), *Macromol. Chem.*, **118**, 117 (1968).

20.4-40 G. S. Huvard, V. T. Stannett, W. J. Koros, and H. B. Hopfenberg, The Pressure Dependence of CO_2 Sorption and Permeation in Poly(acrylonitrile), *J. Membr. Sci.*, **6**, 185 (1980).

20.4-41 W. J. Koros, D. R. Paul, and A. A. Rocha, Carbon Dioxide Sorption and Transport in Polycarbonate, *J. Polym Sci. Polym. Phys. Ed.*, **14**, 687 (1976).

20.4-42 R. A. Assink, Investigation of the Dual Mode Sorption of Ammonia in Polystyrene by NMR, *J. Polym. Sci. Polym. Phys. Ed.*, **13**, 1665 (1975).

20.4-43 W. J. Koros, Model for Sorption of Mixed Gases in Glassy Polymers, *J. Polym. Sci. Polym. Phys. Ed.*, **18**, 981 (1980).

20.4-44 W. J. Koros, R. T. Chern, V. T. Stannett, and H. B. Hopfenberg, A Model for Permeation of Mixed Gases and Vapors in Glassy Polymers, *J. Polym. Sci. Polym. Phys. Ed.*, **19**, 1513 (1981).

20.4-45 L. M. Robeson, The Effect of Antiplasticization on Secondary Loss Transitions and Permeability of Polymers, *Polym. Eng. Sci.*, **9**, 277 (1969).

20.4-46 E. S. Sanders, "Pure and Mixed Gas Sorption in Glassy Polymers," Ph.D. dissertation, North Carolina State Unversity, Raleigh, NC, 1983.

20.4-47 R. T. Chern, W. J. Koros, H. B. Hopfenberg, and V. T. Stannett, Reversible Isopentane-Induced Depression of Carbon Dioxide Permeation Through Polycarbonate, *J. Polym. Sci. Polym. Phys. Ed.*, **21**, 753 (1983).

20.4-48 R. T. Chern, W. J. Koros, H. B. Hopfenberg, and V. T. Stannett, Selective Permeation of CO_2 and CH_4 Through Kapton® Polyimide: Effects of Penetrant Competition and Gas Phase Non-idealities, *J. Polym. Sci. Polym. Phys. Ed.*, **22**, 1061 (1984).

Section 20.5

20.5-1 I. Cabasso, E. Klein, and J. K. Smith, *Research and Development of NS-1 and Related Polysulfone Hollow Fibers for Reverse Osmosis Desalination of Seawater*, report prepared for Office of Water Research and Technology, Gulf South Research Institute, July 1975.

20.5-2 H. Yasuda and J. T. Tsai, Pore Size of Microporous Polymer Membranes, *J. Appl. Polym. Sci.*, **18**, 805 (1974).

20.5-3 P. C. Carmen, *Flow of Gases Through Porous Media*, Academic, New York, 1956.

20.5-4 W. H. Mazur and M. C. Chan, Membranes for Natural Gas Sweetening and CO_2 Enrichment, *Chem. Eng. Prog.*, **78**, 38 (1982).

20.5-5 A. B. Coady and J. A. Davis, CO_2 Recovery by Gas Permeation, *Chem. Eng. Prog.*, **78**, 45 (1982).

20.5-6 G. Rehage and W. Borchard, The Thermodynamics of the Glassy State, in R. N. Haward (Ed.), *The Physics of Glassy Polymers*, p. 88, Applied Science Publishers, London, 1973.

Section 20.6

20.6-1 A. B. Coady and J. A. Davis, CO_2 Recovery by Gas Permeation, *Chem. Eng. Prog.*, **78**, 45 (1982).

20.6-2 K. K. Sirkar, Separation of Gaseous Mixtures with Asymmetric Dense Polymeric Membranes, *Chem. Eng. Sci.*, **32**, 1137 (1977).

20.6-3 Product bulletin, *OECD Membrane Type Oxygen Enrichers*, Oxygen Enrichment Co., Schenectady, NY, Dec. 1980.

20.6-4 S. A. Stern, Gas Permeation Processes, in R. Lacey and S. Loeb (eds.), *Industrial Processing with Membranes*, Wiley-Interscience, New York, 1972.

20.6-5 K. C. Youn, G. C. Blytas, and H. H. Wall, "Role of Membrane Technology in the Recovery of Carbon Dioxide Economics and Future Prospects," paper presented at Gas Processors Association Regional Meeting, Houston, TX, Nov. 1983.

20.6-6 J. A. Gerow, "Permeation Separation Device," U.S. Patent 3,832,830, Sept. 3, 1973, E. I. Du Pont.

20.6-7 S. A. Stern, F. J. Onorato, and C. Libove, "The Permeation of Gases through Hollow Silicone Rubber Fibers: Effects of Fiber Elasticity on Gas Permeability, *AIChE J.*, **23**, 567 (1977).

20.6-8 O. H. Varga, *Stress–Strain Behavior of Elastic Materials*, Interscience, New York, 1966.

20.6-9 J. M. Thorman and S. T. Hwang, Compressible Flow in Permeable Capillaries Under Deformation, *Chem. Eng. Sci.*, **33**, 15 (1978).

20.6-10 E. Klein, J. K. Smith, and F. C. Morton, U.S. Patent 4,051,300, Sept. 1977.

20.6-11 M. E. Breuer and K. Kammermeyer, Effects of Concentration Gradients in Barrier Separation Cells, *Sep. Sci.*, **2**, 319 (1967).

20.6-12 C. Y. Pan, Gas Permeation by Permeators with High-Flux Asymmetric Membranes, *AIChE J.*, **29**, 545 (1983).

20.6-13 S. Weller and W. A. Steiner, Engineering Aspects of Separation of Gases: Fractional Permeation Through Membranes, *Chem. Eng. Prog.*, **46**, 585 (1950).

20.6-14 R. W. Naylor and P. O. Backer, Enrichment Calculations in Gaseous Diffusion: Large Separation Factor, *AIChE J.*, **1**, 95 (1955).

20.6-15   J. Oishi, Y. Matsumura, K. Higashi, and C. Ika, Analysis of a Gaseous Diffusion Separation
           Unit, *J. Atomic Energy Soc. (Jpn).*, **3**, 923 (1961).

20.6-16   W. P. Walawender and S. A. Stern, Analysis of Membrane Separation Parameters. II. Counter-
           current and Cocurrent Flow in a Single Permeation Process, *Sep. Sci.*, **7**, 553 (1972).

20.6-17   C. Y. Pan and H. W. Habgood, An Analysis of the Single-Stage Gaseous Permeation Process,
           Ind. Eng. Chem. Fundam., **13**, 323 (1974).

20.6-18   C. T. Blaisdell and K. Kammermeyer, Countercurrent and Cocurrent Gas Separation, *Chem.
           Eng. Sci.*, **28**, 1249 (1973).

20.6-19   S. A. Stern and S. C. Wang, Countercurrent and Cocurrent Gas Separation in a Permeation Stage.
           Comparison of Computation Methods, *J. Membr. Sci.*, **4**, 141 (1978).

20.6-20   S. T. Hwang and K. Kammermeyer, *Membranes in Separations*, Wiley-Interscience, New York,
           1975.

20.6-21   D. N. Hanson, J. H. Duffin, and G. F. Somerville, in *Computation of Multistage Separation
           Processes*, Chap. 1, Reinhold, New York, 1962.

20.6-22   S. A. Stern, The Separation of Gases by Selective Permeation, in P. Mears (Ed.), *Membrane
           Separation Processes*, Chap. 8, Elsevier Scientific, New York, 1976.

20.6-23   S. A. Stern and W. P. Walawender, Analysis of Membrane Separation Parameters, *Sep. Sci.*, **4**,
           129 (1969).

20.6-24   S. A. Stern, T. F. Sinclair, P. G. Gareis, N. P. Vahldieck, and P. H. Mohr, Helium Recovery
           by Permeation, *Ind. Eng. Chem.*, **57**, 49 (1965).

20.6-25   C. Y. Pan and H. W. Habgood, Gas Separation by Permeation. Part I. Calculation Methods and
           Parametric Analysis, *Can. J. Chem. Eng.*, **56**, 197 (1978).

20.6-26   A. S. Berman, Laminar Flow in Channels with Porous Walls, *J. Appl. Phys.*, **24**, 1232 (1953).

20.6-27   G. R. Antonson, R. J. Gardner, C. F. King, and D. Y. Ko, Analysis of Gas Separation by
           Permeation in Hollow Fibers, *Ind. Eng. Chem. Proc. Des. Dev.*, **16**, 463 (1977).

20.6-28   R. T. Chern, "Measurement and Modeling of Mixed Gas Permeation Through Glassy Poly-
           mers," Ph.D. dissertation, North Carolina State University, Raleigh, 1983.

20.6-29   R. M. Terrill and P. W. Thomas, On Laminar Flow Through Uniformly Porous Pipe, *Appl. Sci.
           Rev.*, **21**, 37 (1969).

20.6-30   J. P. Quaile and E. K. Levy, Laminar Flow in a Porous Tube with Suction, *J. Heat Transfer*,
           67 (Feb. 1975).

20.6-31   C. Y. Pan and H. W. Habgood, Gas Separation by Permeation, Part II. Effect of Permeate
           Pressure Drop and Choice of Permeate Pressure, *Can J. Chem. Eng.*, **56**, 210 (1978).

20.6-32   J. S. Newman, in *Electrochemical Systems*, p. 414, Prentice-Hall, Englewood Cliffs, NJ, 1973.

20.6-33   R. C. Reid, J. M. Prausnitz, and T. K. Sherwood, *The Properties of Gases and Liquids*, 3rd
           ed., McGraw-Hill, New York, 1977.

20.6-34   W. J. Koros, A. H. Chan, and D. R. Paul, Sorption and Transport of Various Gases in Poly-
           carbonate, *J. Membr. Sci.*, **2**, 165 (1977).

20.6-35   K. Toi, G. Morel, and D. R. Paul, Gas Sorption and Transport in Poly(phenylene oxide) and
           Comparisons with Other Glassy Polymers, *J. Appl. Polym. Sci.*, **27**, 2997 (1982).

20.6-36   A. J. Erb and D. R. Paul, Gas Sorption and Transport in Polysulfone, *J. Membr. Sci.*, **8**, 11
           (1981).

20.6-37   R. T. Chern, W. J. Koros, E. S. Sanders, S. H. Chen, and H. B. Hopfenberg, Implications of
           the Dual Mode Sorption and Transport Models for Mixed Gas Permeation, ACS Symposium
           Series No. 233, *Industrial Gas Separations*, T. E. Whyte, C. M. Yon, and E. H. Wagener
           (Eds.), American Chemical Society, Washington, DC, 1983.

20.6-38   C. S. Goddin, Pick Treatment for High CO_2 Removal, *Hydrocarbon Process.*, 125 (May 1982).

20.6-39   R. T. Chern, W. J. Koros, H. B. Hopfenberg, and V. T. Stannett, Selective Permeation of CO_2
           and CH_4 Through Kapton® Polyimide: Effects of Penetrant Competition and Gas Phase Non-
           idealities, *J. Polym Sci. Polym. Phys. Ed.*, **22**, 1061 (1984).

20.6-40   W. C. Pfefferle, U.S. Patent 3,144,313 (1964).

20.6-41   S. T. Hwang, J. M. Thorman, K. M. Yuen, Gas Separation by a Continuous Membrane Column,
           Sep. Sci. Technol., **15**, 1069 (1980).

20.6-42   S. T. Hwang and J. M. Thorman, The Continuous Membrane Column, *AIChE J.*, **26**, 558
           (1980).

20.6-43   J. M. Thorman, "Engineering Aspects of Capillary Gas Permeators and the Continuous Mem-
           brane Column," Ph.D. dissertation, University of Iowa, Iowa City, 1979.

20.6-44 S. T. Hwang and S. Ghalchi, Membrane Separation by a Continuous Membrane Column, *J. Membr. Sci.*, **11,** 187 (1982).

20.6-45 S. L. Matson, J. Lopez, and J. A. Quinn, Separation of Gases with Synthetic Membranes, *Chem. Eng. Sci.*, **38,** 503 (1983).

20.6-46 S. Teslik and K. K. Sirkar, paper presented at the 182nd ACS National Meeting, New York, Aug. 23–28, 1981.

CHAPTER 21

Membrane Processes—Dialysis and Electrodialysis

ELIAS KLEIN‡
RICHARD A. WARD‡
Division of Nephrology
School of Medicine
University of Louisville
Louisville, Kentucky

ROBERT E. LACEY§
Southern Research Institute
Birmingham, Alabama

21.1 DIALYSIS

Dialysis first was reported in 1861 by Graham,[1] who used parchment paper as a membrane. His experiments were based on the observations of a school teacher, W. G. Schmidt, that animal membranes were less permeable to colloids than to sugar or salt.[2] Over the next 100 years, dialysis became widely used as a laboratory technique for the purification of small quantities of solutes but, with minor exceptions, it realized no large-scale industrial applications. In the last 20 years, development of dialysis for the treatment of kidney failure has brought about a resurgence of interest in dialysis for a wide range of separations.

Dialysis is a diffusion-based separation process that uses a semipermeable membrane to separate species by virtue of their different mobilities in the membrane. A feed solution, containing the solutes to be separated, flows on one side of the membrane while a solvent stream, the dialysate, flows on the other side (Fig. 21.1-1). Solute transport across the membrane occurs by diffusion driven by the difference in solute chemical potential between the two membrane–solution interfaces. In practical dialysis devices, an obligatory transmembrane hydraulic pressure may add an additional component of convective transport. Convective transport also may occur if one stream, usually the feed, is highly concentrated, thus giving rise to a transmembrane osmotic gradient down which solvent will flow. In such circumstances, the description of solute transport becomes more complex since it must incorporate some function of the transmembrane fluid velocity.

The relative transfer of two solutes across a dialysis membrane is a function of both their diffusivities in the membrane and their driving forces. Separations will be efficient only for species that differ significantly in diffusion coefficient. Since diffusion coefficients are a relatively weak function of molecular size,

‡Authors of Section 21.1.
§Author of Section 21.2.

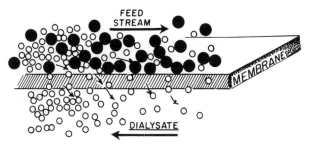

FIGURE 21.1-1 Schematic representation of dialysis. Small, highly mobile species (○) diffuse across the membrane, from feed to dialysate, while larger molecules (●), to which the membrane is relatively impermeable, remain in the feed stream.

dialysis is limited in practice to separating species that differ significantly in molecular size, for example, separation of crystalloids from polymers or colloids. In addition to this limitation, dialysis is a useful technique only when the solute(s) to be separated is present in high concentration. This is because solute fluxes in dialysis are directly dependent on the transmembrane concentration gradient, an intrinsic property of the feed and dialysate streams. (In other membrane separation processes, such as reverse osmosis and ultrafiltration, transmembrane fluxes depend on an applied pressure which is independent of the properties of the process stream.) Thus, if the transmembrane concentration gradient is low, practical solute recoveries can be obtained only by increasing membrane area, which may compromise the economics of the process. Because of these considerations, dialysis is characterized by low flux rates in comparison to other membrane separation techniques. However, the passive nature of dialysis may be an advantage in those circumstances where the species to be separated are sensitive to mechanical degradation by high pressures or high shear rates. Dialytic separations can be enhanced under certain circumstances by taking advantage of charge repulsion effects between a solute and the membrane[3,4] (Donnan dialysis), by complexing one or more of the species to be separated[5,6] (liquid membranes), by chemical conversion of the permeating solute in the dialysate,[7,8] or by staging dialytic cells.[9]

21.1-1 Applications of Dialysis

Until about 1960, only a few industrial uses of dialysis had reached large-scale application. The most prominent of these was dialysis of concentrated sodium hydroxide solutions containing hemicellulose, a by-product of viscose rayon production.[10] A plate-and-frame dialyzer, containing parchmentized cotton cloth membranes, was used to recover sodium hydroxide from the hemicellulose-contaminated solution. In the 1950s, the development of synthetic polymer membranes with greater chemical resistance[11] led to the application of dialysis for recovering nickel in the electrolytic refining of copper.[12] However, the evolution of other membrane processes, in particular ultrafiltration, together with general advances in chemical technology resulted in dialysis remaining a relatively unimportant industrial separation process.

Beginning in the 1960s, dialysis found what has become its major application, the treatment of end-stage renal disease.[13] In a process known as hemodialysis, small-molecular-weight metabolic waste products are removed from a patient's blood by dialytic transfer across a membrane that is impermeable to normal blood proteins. The dialysate is formulated to normalize blood electrolyte concentrations and acid–base balance by dialytic exchange between the blood and dialysate. Fluid balance is restored by superimposing a transmembrane hydrostatic pressure gradient. In 1982, the lives of upward of 57,000 people in the United States alone were sustained by hemodialysis.

The growth of hemodialysis has resulted in the development of new membranes and more efficient dialysis devices. In turn, the availability of such devices has generated new interest in dialysis for a variety of applications. One area where dialysis may find a role as a separation process is in the biotechnology industry, where products must be separated from fragile (e.g., shear-sensitive) or heat-sensitive solutions. In 1969, Schultz and Gerhardt[14] reviewed the use of dialysis as a means of controlling bacterial cultures for a variety of applications. By circulating the culture solution from a fermenter through a dialyzer, the yield of the fermenter can be enhanced either by the dialytic addition of nutrient or removal of products that may inhibit bacterial growth and metabolism. Since then, several groups have conducted laboratory- and pilot-scale studies of a number of microbiological systems. These include the recovery of enzymes and proteins from *Staphylococcus aureus* cultures[15,16] and the growth of yeast cultures using dialytic transfer of lactose from whey as substrate.[17] Other applications involving sensitive solutions include the desalting of cheese whey solids[18] and reduction of the alcohol content in beer.[19]

Generally, dialysis has been restricted to use with aqueous solutions, although there is no fundamental reason why it cannot be used with organic solvents, given the development of suitable membranes. An

TABLE 21.1-1 Dialysis Membrane Materials

Membrane Material	Manufacturer
Regenerated cellulose	Asahi Medical Co., Tokyo, Japan Enka AG, Wuppertal, West Germany Terumo Corporation, Tokyo, Japan
Cellulose acetate	Asahi Medical Co., Tokyo, Japan CD Medical, Miami, Florida Nipro Medical Industries, Tokyo, Japan
Ethylene–Polyvinyl alcohol	Kuraray Co., Osaka, Japan
Polyacrylonitrile	Hospal, Lyon, France
Polycarbonate	Gambro Ab, Lund, Sweden
Polymethylmethacrylate	Toray Industries Inc., Tokyo, Japan
Polyperfluoro (ethylene-co-ethylene sulfonic acid)	E. I. Du Pont de Nemours, Wilmington, Delaware
Polysulfone	Fresenius AG, Bad Homburg, West Germany

example of dialysis in nonaqueous systems has been described by Benkler and Reineccius,[20] who used dialysis to separate volatile flavor components from higher-molecular-weight oils. The membrane used was a perfluorosulfonic acid copolymer. There also has been a continuing effort to use dialysis in the separation and concentration of ionic species. Using the Donnan effect, a number of potential applications have been studied, including the removal of contaminating cations from dilute wastewater solutions,[3,21] softening of feedwater to reverse osmosis desalination systems,[22] and acid recovery in aluminum anodizing plants.[23]

However, none of these new applications of dialysis, with the possible exception of acid recovery in aluminum anodizing, appears to have progressed beyond the pilot-plant scale, and today the only major application of dialysis remains hemodialysis for the treatment of renal failure.

21.1-2 Membrane Materials and Configurations

The materials commonly used for dialysis membranes are listed in Table 21.1-1. The choice of a membrane for a particular application is dependent on both the separation required and the environment in which the membrane is to operate.

The principal membrane material used in medical applications is regenerated cellulose, derived either from hydrolysis of cellulose acetate or by the cuprammonium process.[24] Cellulose membranes are hydrogels that contain up to 50% water when equilibrated with aqueous solutions. The permeability of hydrogels increases with the degree of hydration.[25] To maintain the expanded solvation state in the dry membrane, and thus its permeability, during extended storage, cellulose membranes are produced containing 5–40% by weight of glycerol. Once wetted, subsequent drying of the membrane in the absence of a plasticizer results in an irreversible collapse of the membrane structure and loss of permeability. This property may limit the usefulness of cellulose membranes in some applications.

Dialysis membranes fabricated from glassy polymers, such as methacrylates, polysulfones, polycarbonates, and polyacrylonitriles, are not hydrogels and do not undergo irreversible alteration on drying. However, they are generally hydrophobic and their permeability may decrease after drying due to trapping of air in the membrane pores. This can be reversed by first wetting the membrane with a low-surface-tension solvent, such as alcohol, before exposure to aqueous solutions. Fabrication of membranes from glassy polymers is complicated by the tendency for membranes with adequate diffusive permeabilities to have high hydraulic permeabilities, which limits their use in applications where volume control is important. Methacrylate, polycarbonate, and polysulfone membranes which circumvent this problem have been developed for hemodialysis. The hydrophobic nature of most glassy polymers causes them to adsorb proteins, if present, from the solutions in contact with the membrane. Such adsorption of protein can cause a decrease in membrane permeability[26] and this may limit the role of such membranes for some applications in biotechnology.

The choice of a membrane for a particular separation is governed also by considerations of temperature and pH. Membranes of glassy polymers generally are restricted to use at temperatures below 80°C, whereas cellulosic membranes will function satisfactorily at higher temperatures. Although most membranes fabricated from glassy polymers will operate over a wide range of pH, polycarbonate membranes will deteriorate rapidly at pH values greater than 7.5. Cellulosic membranes hydrolyze slowly at low pH (≤ 2) and swell extensively in highly alkaline solutions (4% sodium hydroxide).

Dialysis membranes are available in three basic configurations: flat sheet, tubular, and hollow fiber (Fig. 21.1-2). Continuously cast flat sheet membranes are used in plate-and-frame devices. Most of the early industrial applications of dialysis were based on dialyzers of the plate-and-frame type. The availability of cellophane tubing, used as sausage casing, led to the development of coil dialyzers for hemodialysis. The dry tubular membrane was laid flat and wound into a coil with a mesh spacer to provide a channel for dialysate flow. Coil configurations provided a relatively large membrane surface area in a compact form; however, their use has been limited entirely to hemodialysis. More recently, hollow-fiber dialysis membranes have been developed[27] that enable fabrication of compact devices with large surface areas. In addition, their structure allows them to be operated without the need for a membrane support structure. These two properties of hollow fibers have made them the dominant membrane form in use today. An industrial hollow-fiber dialysis unit is shown in Fig. 21.1-3.

21.1-3 Membrane Transport

For homogeneous membranes, the flux J_s of a solute across the membrane at any point is given by

$$J_s = -\mathcal{D}_M \frac{dC}{dx} + J_v(1 - \sigma)C \tag{21.1-1}$$

where $\mathcal{D}_M$ is the diffusivity of the solute in the membrane, dC/dx is the solute concentration gradient at the external surface of the membrane, J_v is the transmembrane solvent velocity, σ is the reflection coefficient, and C is the local solute concentration within the membrane.[28] The reflection coefficient is a measure of a solute's ability to enter the pores of the membrane. For solutes able to pass the membrane freely, $\sigma = 0$, while for solutes completely excluded by the membrane, $\sigma = 1$. Integration of Eq. (21.1-1), at steady state, results in the following expression:

$$J_s = P_m \Delta C + J_v(1 - \sigma)\overline{C} \tag{21.1-2}$$

where P_M is the membrane permeability, ΔC is the concentration difference across the membrane, and $\overline{C}$ is the average concentration in the membrane. When the transmembrane Peclet number, Pe $= J_v(1 - \sigma)/P_M$, is less than 3.0, $\overline{C}$ is given by

$$\overline{C} = C_W - \frac{\Delta C}{3} \tag{21.1-3}$$

where C_W is the feed-side concentration at the membrane wall.[28] Use of Eqs. (21.1-2) and (21.1-3) requires knowledge of the solute concentrations at the feed- and dialysate-side membrane surfaces. As a first approximation, the solute concentration at the dialysate-side membrane surface can be equated to the bulk dialysate concentration. The concentration at the feed-side membrane surface, C_W, can be related to the bulk feed-side concentration, C_F, using thin-film boundary-layer theory.[29] For dialysis, Villarroel et al.[28] have shown that the ratio of wall concentration to bulk concentration is given by

$$\frac{C_W}{C_F} = \frac{1 + \psi + (1 - \sigma)\xi C_D/C_F}{(1 - \sigma)(1 + \xi) + \psi} \tag{21.1-4}$$

where $\xi = [\exp(\text{Pe}) - 1]^{-1}$ and $\psi = [\exp(\theta) - 1]^{-1}$; for hollow-fiber membranes, the parameter θ is given by $\theta = 0.709 J_V(xd^3N/Q_F\mathcal{D}^2)^{0.333}$, where x is the axial distance from the beginning of the membrane to the point at which C_W is being evaluated.[29] In dialytic processes, the transmembrane solvent velocity J_v is generally less than 3×10^{-5} cm/s and permeability coefficients are of the order of 10^{-4} cm/s, so that the Peclet number is usually less than 0.1. When high transmembrane pressures are obligatory because of flow constraints, the value of J_v may increase, and convection may become an important contribution to overall mass transfer.

A number of attempts have been made to correlate P_M with membrane structure. The concept generally accepted by membrane chemists for diffusion in very highly swollen membranes, such as those used in dialysis, is that solute diffusion occurs through solvent held in the membrane structure. The membrane structure may be flexible, as in cellulose hydrogels, or it may be glassy, as in porous polysulfone. In either case, the diffusion path is thought to consist of channels with a distribution of sizes and whose lengths are greater than the membrane thickness by virtue of their tortuous paths through the membrane.

Yasuda et al.[25] have developed the concept of a homogeneous solvent-swollen membrane in which thermally induced movement of segments of randomly coiled polymer molecules leaves an interstitial free volume available for solute transport. They concluded that the permeability characteristics of highly swollen systems cannot be represented by a single coefficient. Values of solute and solvent permeabilities depend on the conditions of measurement, in particular, the magnitude of diffusive flux relative to convective flux.

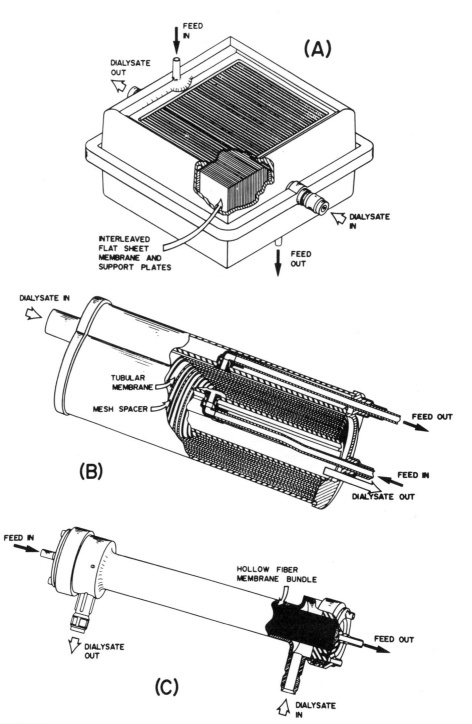

FIGURE 21.1-2 Basic dialyzer configurations: (A) plate-and-frame, (B) coil and (C) hollow fiber. (Courtesy of Travenol Laboratories, Deerfield, IL.)

FIGURE 21.1-3 Industrial hollow-fiber dialysis unit. (Courtesy of Enka AG, Wuppertal, West Germany.)

Yasuda and coworkers found the following relationships between diffusive permeability and membrane structure:

1. The permeability of solutes, whose size is small compared to the membrane pore size, is proportional to the degree of membrane hydration.

2. Membrane permeability decreases exponentially with increasing molecular size, where the latter is expressed in terms of molecular cross-sectional area.

3. Solute reflection coefficients change markedly when solute size approaches the average pore size of the membrane.

A different approach was used by Klein et al.[30] for a range of dialytic membranes that included not only the homogeneous, swollen gels studied by Yasuda and coworkers, but also glassy polymers having porous structures. Their model was based on hypothetical pore structures whose dimensions are hydrodynamically equivalent to cylindrical pores.[31,32] Since neither the fractional cross-sectional area of the pore openings A_P nor the actual pore path length l_P can be measured independently, an experimental method is used to derive their ratio. From the Poiseuille relationship, the hydraulic permeability $L_\wp$ is related to the ratio A_P/l_P and the hydrodynamic pore radius r_P by

$$L_\wp = \left(\frac{A_P}{l_P}\right)\left(\frac{r_P^2}{8\mu}\right) \tag{21.1-5}$$

where μ is the viscosity of the solution. The value of $L_\wp$ is determined by measuring the volume flux of water across the membrane in response to an applied pressure gradient. Water also can diffuse through the membrane in response to a concentration gradient. Using a diffusional model,[31,32] we can relate the diffusive permeability coefficient of the membrane P_M to the ratio A_P/l_P by

$$P_M = \frac{(A_P/l_P)\mathfrak{D}(1-q)^2}{K_1} = \frac{\mathfrak{D}_M}{l_P} \tag{21.1-6}$$

where q is the ratio of solute to pore diameter, $\mathfrak{D}$ is the solute diffusion coefficient in the solvent, K_1 is a power series in q, and $\mathfrak{D}_M$ is the effective solute diffusivity in the membrane.

For a solute, such as tritiated water, whose radius is small compared to the pore radius, q is approximately equal to zero, $K_1 = 1$, and Eqs. (21.1-5) and (21.1-6) can be solved simultaneously to yield[30]

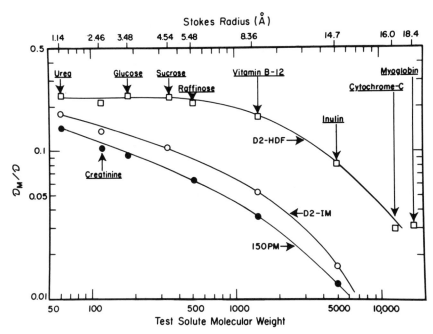

FIGURE 21.1-4 Ratio of solute diffusivity in the membrane ($\mathcal{D}_M$) to solute diffusivity in solution ($\mathcal{D}$) as a function of solute molecular weight for three cellulosic dialysis membranes.

$$r_P^2 = \frac{8\mu\mathcal{D}L_{\wp}}{P_M^{\text{HTO}}} \tag{21.1-7}$$

Once the average pore size of the membrane has been determined using Eq. (21.1-7), the permeability of any other solute of known radius through the same membrane can be calculated from simultaneous solution of Eqs. (21.1-5) and (21.1-6). In the absence of specific solute–membrane interactions, such as charge or hydrophobic bonding, this model is useful for predicting solute permeability coefficients through a characterized membrane, for values of q less than 0.6.

On the basis of either model, it is clear that dialytic transport will decrease with increasing solute molecular size, not only because of the smaller solution diffusivity of a large molecule but also because of increasing values of q. This effect is seen in Fig. 21.1-4, where the ratio of diffusivity in the membrane ($\mathcal{D}_M$) to diffusivity in solution ($\mathcal{D}$) is plotted as a function of solute Stokes radii for several dialysis membranes.

The membrane models described above relate solute and solvent fluxes to concentration differences at the dialyzer's two membrane–solution interfaces, values that cannot be determined experimentally. Since, in practice, dialysis always involves the movement of solute from one bulk phase to another, some means of expressing the fluxes in terms of bulk concentrations is needed.

As solute is removed from the feed-side membrane–solution interface by dialysis, the layer is depleted and its concentration must be restored from the bulk solution. In laminar flow, which is usual in small-bore hollow-fiber and thin-film plate-and-frame devices, there is no convection and repletion of the interfacial layer is solely by diffusion from the bulk solution. As such diffusion occurs, the concentration gradient from the bulk solution to the interfacial layer decreases. Thus, the rate of restoration of the interfacial solute concentration is a function of the solute size, the transmembrane flux, and the rate of solute supply, that is, the axial feed flow rate.

As solute crosses the membrane, it must be transported away from the dialysate-side interfacial layer to maintain the transmembrane concentration difference. The rate at which this occurs is a function of the concentration gradient between the membrane–dialysate interface and the bulk dialysate. Thus, mass transfer on the dialysate side of the membrane is a function of the transmembrane flux, the bulk dialysate concentration, which in turn is dependent on the dialysate velocity, and, in the case of laminar flow, the solute diffusivity.

Thus, the resistance to mass transfer from the bulk feed stream to the bulk dialysate stream can be considered the sum of three terms: concentration polarization on the feed and dialysate sides of the membrane and the resistance to diffusion in the membrane itself. For laminar flow devices, these resistances are combined to give the following expression for the overall mass transfer coefficient k_{OV}:

$$\frac{1}{k_{OV}} = \frac{1}{k_D} + \frac{1}{P_M} + \frac{1}{k_F} \tag{21.1-8}$$

where k_D and k_F are the mass transfer coefficients for the concentration polarization layers on the dialysate and feed sides of the membrane, respectively. Both k_F and k_D can be considered in terms of the ratio of the diffusivity of the solute in a hypothetical stagnant layer to the thickness of the layer. Figure 21.1-4 shows that as solute size increases diffusivity in the membrane decreases more rapidly than diffusivity in the solution. Translating this observation to Eq. (21.1-8), we see that membrane resistance, which is directly proportional to solute diffusivity in the membrane [Eq. (21.1-6)], will become the dominant factor contributing to k_{OV} as solute molecular size increases. However, for small solutes, it is clear from Eq. (21.1-8) that membrane resistance will limit mass transfer only when k_D and k_F are very large relative to P_M. Therefore, dialyzer design and fluid dynamics, in the first instance, can limit transport if they permit excessive concentration polarization layers to develop. When concentration polarization has been minimized, further improvements in the rate of mass transfer depend on increased membrane permeability, which in turn will generate further demands on the design of the fluid pathways to minimize concentration polarization.

21.1-4 Overall Mass Transfer in Dialyzers

In the previous section, mass transfer was discussed for a single point on the membrane. For practical applications, this analysis must be extended to describe mass transfer for the dialyzer as a whole. Neglecting convective contributions, we can write the following equation for the mass of solute $d\dot{m}$ transferred across an element of membrane of length dx and area dA_M per unit time (Fig. 21.1-5):

$$d\dot{m} = k_{OV}(C_F - C_D)\, dA_M \tag{21.1-9}$$

where C_F and C_D are the solute concentrations in the feed and dialysate, respectively, in element dx. Equation (21.1-9) can be integrated along the length of the membrane to give the following equation for overall mass transfer:

$$\dot{m} = k_{OV}A_M \frac{\Delta C_{x=l} - \Delta C_{x=0}}{\ln(\Delta C_{x=l}/\Delta C_{x=0})} \tag{21.1-10}$$

where $\Delta C_{x=0}$ and $\Delta C_{x=l}$ are the transmembrane concentration differences at either end of the device. For countercurrent flow, $\Delta C_{x=0} = C_{Fi} - C_{Do}$ and $\Delta C_{x=l} = C_{Fo} - C_{Di}$. This leads to the following equation for overall mass transfer in a dialyzer with this flow configuration:

$$\dot{m} = k_{OV}A_M \frac{(C_{Fo} - C_{Di}) - (C_{Fi} - C_{Do})}{\ln[(C_{Fo} - C_{Di})/(C_{Fi} - C_{Do})]} \tag{21.1-11}$$

A similar expression can be derived for cocurrent flow by redefining $\Delta C_{x=0}$ and $\Delta C_{x=l}$.

If ultrafiltration is negligible, inlet and outlet flows are equal for both the feed and dialysate streams. Using this assumption, the following overall mass balances may be written for each stream:

$$\dot{m} = Q_F(C_{Fi} - C_{Fo}) = Q_D(C_{Do} - C_{Di}) \tag{21.1-12}$$

Finally, the performance of a dialyzer can be described in terms of a "dialysance" D, which is defined as the rate of mass transfer divided by the concentration difference between inlet feed and inlet dialysate; that is,

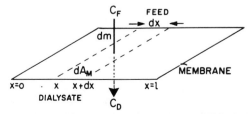

FIGURE 21.1-5 Diffusive mass transfer across an incremental length of membrane dx.

$$D = \frac{\dot{m}}{C_{Fi} - C_{Di}} \tag{21.1-13}$$

Combining Eqs. (21.1-11)–(21.1-13), we arrive at the following expression for the dialysance of a countercurrent dialyzer in terms of membrane properties (k_{OV}, A_M) and operating conditions (Q_F, Q_D):

$$D = Q_F \left\{ \frac{\exp\left[k_{OV}A_M\,(1 - Q_F/Q_D)/Q_F\right] - 1}{\exp\left[k_{OV}A_M\,(1 - Q_F/Q_D)/Q_F\right] - Q_F/Q_D} \right\} \tag{21.1-14}$$

Note that if the feed and dialysate flows are essentially equal ($Q_F = Q_D$), then

$$D = Q_F \left\{ \frac{k_{OV}A_M}{Q_F + k_{OV}A_M} \right\} \tag{21.1-15}$$

Similar expressions can be derived for other flow configurations.[33]

The foregoing assumes that the solute is distributed freely in the solvent phase. This assumption may not always be valid; for example, in protein solutions, small solutes may bind to the protein molecules and exist in equilibrium between the free and bound states. In such circumstances, modifications of Eq. (21.1-14) must be used to determine dialysance.[34]

Equation (21.1-14) can be used to estimate the degree of separation of two solutes by a given dialyzer. For any solute, the fractional extraction into the dialysate, E, of solute from the feed stream is given by

$$E = \frac{Q_F C_{Fi} - (Q_F - Q_{UF})C_{Fo}}{Q_F C_{Fi}} \tag{21.1-16}$$

For negligible ultrafiltration, this reduces to

$$E = \frac{C_{Fi} - C_{Fo}}{C_{Fi}} \tag{21.1-17}$$

If the inlet concentration of solute in the dialysate is zero ($C_{Di} = 0$), Eq. (21.1-17) further reduces to

$$E = \frac{D}{Q_F} \tag{21.1-18}$$

By combining Eqs. (21.1-14) and (21.1-18), fractional extraction can be obtained as a function of $k_{OV}A_M/Q_F$ for different ratios of feed to dialysate flow rate. These relationships are shown in Fig. 21.1-6. The separation factor α_{jk} in a dialytic process can be considered to be the ratio of the fractional masses of two solutes removed from their common feed stream, under a given set of operating conditions. By applying Eq. (21.1-16) to each solute, we can see that α_{jk} is equal to the ratio of the fractional extractions of the two solutes. Thus, Fig. 21.1-6 can be used to estimate the relative separation of two solutes from a knowledge of their overall mass transfer coefficients, membrane area, and feed and dialysate flow rates.

21.1-5 Contributions of Convection to Mass Transfer

The above derivations are based on the assumption that convective mass transfer is negligible, that is, significant ultrafiltration is not occurring. If significant ultrafiltration does occur, mass transfer will be enhanced by the addition of a convective component to the diffusive mass transfer described above. Ultrafiltration will occur from the feed to the dialysate in response to a pressure gradient which may be either applied, in order to concentrate the feed, or obligatory, as a consequence of the geometry of the device and the desired feed flow rate. Less commonly, if the feed is highly concentrated and the dialysate dilute, ultrafiltration may occur from the dialysate to the feed in response to an osmotic gradient. In the first case, ultrafiltration will enhance solute transport while in the second it will impede it.

As discussed earlier, at a steady state the overall solute flux J_s at a point on the membrane can be considered to consist of the sum of a diffusive and a convective component:

$$J_s = P_M\,\Delta C + J_v\,(1 - \sigma)\,\overline{C} \tag{21.1-2}$$

where the first term, $P_M\,\Delta C$, represents diffusive transfer and the second term, $J_v\,(1 - \sigma)\,\overline{C}$, convective transfer. The relative magnitudes of diffusive and convective transfer are functions of the membrane permeability P_M and the reflection coefficient σ. For species of small molecular size, the resistance to diffusive transfer through the membrane is low, that is, P_M is large, and diffusive transfer always significantly exceeds

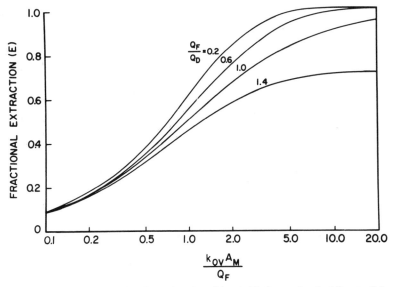

FIGURE 21.1-6   Fractional extraction E as a function of $k_{OV}A_M/Q_F$ for varying feed flow to dialysate flow ratios (Q_F/Q_D).

convective transfer at low Peclet numbers. However, as solute molecular size increases, membrane permeability decreases logarithmically,[35] whereas reflection coefficients increase at a much lesser rate and the relative importance of convection to overall mass transfer increases. This effect is illustrated in the following example.

Figure 21.1-7 shows membrane permeability and reflection coefficient as a function of molecular weight for a cellulosic dialysis membrane.[36] For two solutes, A and B, having molecular weights of 200 and 2000 daltons, respectively, feed-side concentrations at a point on the membrane surface of 0.1 g/cm³, and negligible dialysate-side concentrations, $\Delta C = 0.1$ g/cm³ and $\overline{C} = 0.067$ g/cm³. For solute A, $P_M = 3.6 \times 10^{-4}$ cm/s and $1 - \sigma = 0.8$. Substituting these values into Eq. (21.1-2), we obtain

$$J_s = 3.6 \times 10^{-5} + 0.054 \, J_v \tag{21.1-19}$$

As the ultrafiltration rate increases from 0 to 0.5×10^{-4} cm/s, J_s increases from 3.6×10^{-5} g/s·cm² to 3.9×10^{-5} g/s·cm²; that is, convection increases overall solute transfer by 8%. For solute B, $P_M = 0.40 \times 10^{-4}$ cm/s and $1 - \sigma = 0.6$. Under these conditions, Eq. (21.1-2) reduces to

$$J_s = 0.40 \times 10^{-5} + 0.04 \, J_v \tag{21.1-20}$$

Now, for an increase in ultrafiltration rate from 0 to 0.5×10^{-4} cm/s, J_s increases from 0.40×10^{-5} to 0.60×10^{-5} g/s·cm², an increase in overall mass transfer of 50%.

The above considerations deal with a point on the membrane. Expansion of these concepts to the dialyzer as a whole requires that the feed- and dialysate-side solute concentrations at a point on the membrane be expressed in terms of known concentrations. Such an approach has been developed by Schindhelm et al.,[37] who, by making the assumptions of a zero inlet dialysate concentration and a linear dialysate concentration profile, developed the following expression for the dialysance of solutes greater than 300 daltons molecular weight in a countercurrent dialyzer in the presence of simultaneous diffusion and convection. [Note that the equations for X and Y given in Ref. 37 contain misprints; the correct expressions are as given in Eqs. (21.1-22) and (21.1-23).]

$$D = \frac{X}{1 + Y} \tag{21.1-21}$$

where

$$X = -\frac{[k_{OV}A_M + Q_{UF} (1 - \sigma)]}{Q_{Fi} - Q_{UF}} \left(\frac{Q_{Fi}}{Z}\right) [1 - \exp(Z)] \tag{21.1-22}$$

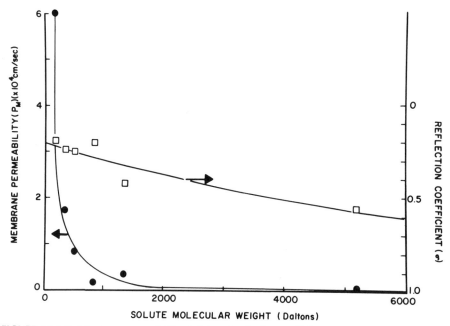

FIGURE 21.1-7 Membrane permeability P ($\bullet$) and reflection coefficient ($\square$) versus solute molecular weight for a regenerated cellulose membrane (Cuprophan 150 PM, Enka AG, Wuppertal, West Germany). Data from Ref. 36.

$$Y = -\left(k_{\text{OV}}A_M + Q_{\text{UF}}(1 - \sigma)\right)\left(\frac{k_{\text{OV}}A_M}{Q_{Fi}Q_{Do}Z^2}\right)\left(1 + \frac{Z}{2} + \frac{1}{Z}[1 - \exp(Z)]\right)$$

$$+ \frac{X}{Q_{Fi}} + \frac{k_{\text{OV}}A_M}{ZQ_{Do}} \tag{21.1-23}$$

$$Z = \frac{k_{\text{OV}}A_M - Q_{\text{UF}}\sigma}{Q_{Fi}} \tag{21.1-24}$$

A similar analysis, but one that requires a numerical methods solution, has been described by Jaffrin et al.[38]

The contribution of convection to overall mass transfer, illustrated previously for a point on the membrane, can now be estimated for the dialyzer as a whole. Referring to the previous example, let us consider the contribution of convection to dialysance for the two solutes A and B in a dialyzer containing 10^4 cm^2 of cellulosic membrane operating in a countercurrent configuration with feed and dialysate flow rates of 8 and 16 cm^3/s, respectively. For such a membrane, the overall mass transfer coefficients of solutes A and B are on the order of 3.23×10^{-4} and 0.37×10^{-4} cm/s, respectively, while the reflection coefficients are on the order of 0.2 and 0.4, respectively. Using these data in Eqs. (21.1-21)–(21.1-24), we can calculate that the dialysance of solute A will increase by 12% (from 2.47 to 2.78 cm^3/s) as the ultrafiltration rate increases from 0 to 0.5 cm^3/s, while the same increase in ultrafiltration rate will result in an increase of 84% (from 0.36 to 0.66 cm^3/s) in the dialysance of solute B. This example further demonstrates the important contribution of convection to overall mass transfer for larger-molecular-weight species.

21.1-6 Estimation of Overall Mass Transfer Coefficient

Use of Eqs. (21.1-21)–(21.1-24) for calculation of dialysance requires the knowledge of a dialyzer's overall mass transfer coefficient k_{OV}. An experimentally determined value of k_{OV} for the solute, membrane, and dialyzer configuration of interest is to be preferred. Estimates of k_{OV} can be made from a knowledge of the geometry and fluid dynamics of the dialyzer and the nature of the solute. Such methods are described below; however, they should be used only to provide a rough guide since they involve numerous assumptions, such as well-defined geometries and Newtonian fluid dynamics. In addition, it can be difficult to determine membrane area accurately in some dialyzers, such as plate-and-frame devices, where membrane masking by the support plates may occur. In such circumstances, experimental determination of the product of mass transfer coefficient and effective membrane area ($k_{OV}A_M$), using Eqs. (21.1-11) and (21.1-12), is the only practical means of obtaining useful design data. Regardless of whether or not an experimental measurement or a calculated estimate of k_{OV} is used, it is important to make the determination at the projected operating temperature because of the strong temperature dependence of solute diffusivity.

As discussed earlier, the overall mass transfer resistance ($R_{OV} = 1/k_{OV}$) can be considered the sum of the individual boundary-layer and membrane resistances; that is,

$$R_{OV} = R_F + R_M + R_D \qquad (21.1\text{-}25)$$

If the individual resistances can be estimated, then a value for k_{OV} can be derived from Eq. (21.1-25). Estimation of the membrane resistance ($R_M = 1/P_M$) at a point has been discussed already, and this value can be assumed to be invariate along the length of the dialyzer. Boundary-layer resistances generally do vary along the length of the dialyzer; however, in practice, average resistances can be used in Eq. (21.1-25). Estimates of these can be obtained for simple geometries such as parallel plates or hollow fibers by making use of the analogy between heat and mass transfer.

Colton et al.[39] used a Graetz-type solution to calculate a log-mean feed-side Sherwood number ($\overline{Sh_F} = k_F h/\mathfrak{D}$, where h is the channel height) for laminar flow in a flat duct with permeable walls as a function of the wall Sherwood number ($Sh_W = P_M h/\mathfrak{D}$) and the dimensionless length of the dialyzer. This analysis assumes a constant dialysate-side resistance which is combined with the membrane resistance in Sh_W; an ultrafiltration rate of zero also is assumed. Subsequently, Cooney et al.[40] and Jagannathan and Shettigar[41] included a variable dialysate-side resistance in reformulating the analysis as a conjugated boundary-value problem for parallel-plate and hollow-fiber configurations, respectively. These latter approaches require solution by numerical methods. The analysis of Jagannathan and Shettigar[41] for hollow fibers, which also considers the effects of ultrafiltration, predicts that, for constant membrane area, fiber diameter and length have very little effect on overall mass transfer for low-permeability membranes. However, for high-permeability membranes, overall mass transfer decreases with increasing membrane diameter.

Unlike the feed side, flow on the dialysate side of the membrane is often turbulent. For turbulent flow, the dialysate-side mass transfer coefficient can be described in terms of a Sherwood number which can be related to two other dimensionless groups, the Reynolds number (Re) and the Schmidt number (Sc), by a correlation of the form:

$$Sh = a(Re)^b(Sc)^c \qquad (21.1\text{-}26)$$

where a, b, and c are constants. Using a stirred-batch dialyzer, Kaufmann and Leonard[42] found values of 0.68 and 0.38 for b and c, respectively, for Re > 20,000, while Marangozis and Johnson[43] determined values of 0.65–0.70 and 0.33 for a variety of published data. Values of the remaining constant, a, vary and may depend on the geometry of the system.[44]

21.1-7   Fluid Dynamics

The contribution of concentration polarization to overall dialytic mass transfer resistance suggests that dialyzers should be operated with flows in the turbulent flow region to minimize boundary-layer formation. While this is typically the case on the dialysate side of the membrane, dialyzers usually are operated with laminar flow on the feed side. In hemodialysis, this is partly to avoid undue mechanical stress on the blood cells which may result in their destruction. However, a consideration of fluid dynamics dictates that feed-side flow be laminar in nearly all applications.

For laminar flow and in the absence of ultrafiltration, the pressure drop (ΔP) on the feed side of a dialyzer is given by the Hagen–Poiseuille equation for hollow fibers,

$$\Delta P = \frac{128\mu Q_F l}{\pi N d^4} \qquad (21.1\text{-}27a)$$

or parallel plates,

$$\Delta P = \frac{12\mu Q_F l}{N w h^3} \qquad (21.1\text{-}27b)$$

where μ is the viscosity of a fluid flowing at a rate Q_F through a dialyzer containing N channels of diameter d, or height h and width w, and length l. In the case of a hollow-fiber dialyzer, the Reynolds number is given by

$$Re = \frac{4\rho Q_F}{\pi N d \mu} \qquad (21.1\text{-}28)$$

where the variables are as defined for Eq. (21.1-27a) and ρ is the fluid density. Typically, hollow-fiber dialysis membranes have an internal diameter of 0.025 cm, so that a dialyzer of 2.5×10^4 cm^2 surface area might contain 13,000 fibers, each 20 cm in length. For a solution with a density of 1.0 g/cm^3 and a viscosity of 0.05 dyn·s/cm^2 flowing through the dialyzer at 200 cm^3/s, the Reynolds number is 16 [Eq. (21.1-28)], well within the laminar flow region, while the pressure drop is 1.61×10^6 dyn/cm^2 [Eq. (21.1-27a)]. Such a pressure drop mandates a high feed-side pressure which leads to difficulties in controlling ultrafiltration, particularly for highly permeable membranes and, in addition, may place an undue mechanical stress on the membrane. For these reasons, fluid flow on the feed side of dialyzers remains well inside the laminar flow region.

In an attempt to decrease boundry-layer resistance, while at the same time maintaining laminar flow, Abel et al.[45] have designed a parallel-plate dialyzer that uses vortex mixing to disrupt the boundary layers. Through this mechanism, they were able to reduce overall mass transfer resistance by a factor of 2–3 compared with conventional parallel-plate dialyzers containing the same membrane.

21.1-8   Dialysis as a Unit Operation

Dialysis can be used as a unit operation in two basic configurations—as a batch process or in a continuous process stream. Figure 21.1-8 depicts the flow diagram for batch dialysis. A well-mixed reservoir of solution volume V and solute concentration C_F is dialyzed continuously against a dialysate with an inlet solute concentration C_{Di}. The feed and dialysate streams enter the dialyzer with flow rates Q_F and Q_D, respectively. The feed stream exiting the dialyzer returns to the feed reservoir while the spent dialysate is discarded. Such a configuration could be used for stripping low-molecular-weight contaminants from a high-molecular-weight product. It represents, in essence, the practice of hemodialysis with the reservoir representing the body solute pool.

A mass balance for the reservoir may be written

$$\frac{d(VC_F)}{dt} = -D(C_F - C_{Di})$$

(21.1-29)

where the dialysance D is given by Eq. (21.1-14) or (21.1-21). A volume balance also may be written for the reservoir:

$$V = V^0 - Q_{UF}t$$

(21.1-30)

where V^0 is the initial reservoir volume and Q_{UF} is the rate of ultrafiltration from feed to dialysate. Substitution for V in Eq. (21.1-29) leads to

$$(V^0 - Q_{UF}t)\frac{dC_F}{dt} - Q_{UF}C_F = -D(C_F - C_{Di})$$

(21.1-31)

Rearrangement of Eq. (21.1-31) and integration lead to the following expression for the concentration of solute in the reservoir as a function of time:

$$C_F^t = \frac{DC_{Di}}{D - Q_{UF}}\left[1 - \left(\frac{V^0 - Q_{UF}t}{V^0}\right)^\beta\right] + C_F^0\left(\frac{V^0 - Q_{UF}t}{V^0}\right)^\beta$$

(21.1-32)

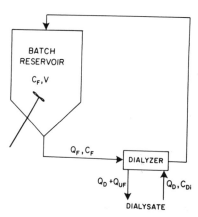

FIGURE 21.1-8   Schematic representation of batch dialysis.

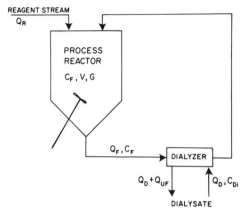

FIGURE 21.1-9 Schematic representation of dialysis in a continuous process.

where β is given by

$$\beta = \frac{D - Q_{\text{UF}}}{Q_{\text{UF}}} \qquad (21.1\text{-}33)$$

If ultrafiltration is negligible, that is, $Q_{\text{UF}} = 0$, then the volume of the reservoir is constant and Eq. (21.1-31) reduces to

$$V \frac{dC_F}{dt} = -D(C_F - C_{Di}) \qquad (21.1\text{-}34)$$

In this instance, the solute concentration of the reservoir is given by

$$C_F^t = C_F^0 \exp\left(-\frac{Dt}{V}\right) + C_{Di}\left[1 - \exp\left(-\frac{Dt}{V}\right)\right] \qquad (21.1\text{-}35)$$

Dialysis may be operated in a variety of configurations as part of a continuous process. Figure 21.1-9 illustrates a typical example. Here, a well-mixed reservoir of volume V represents a reaction vessel where a high-molecular-weight substrate, introduced into the reservoir in a reagent stream of flow rate Q_R, is processed to yield a small-molecular-weight product with a concentration of C_F in the reservoir. The reaction volume is maintained by removing solvent by ultrafiltration through the dialyzer. An example of such a process might be the enzymatic hydrolysis of corn starch to yield low-molecular-weight saccharides. It is assumed that the enzymatic reaction rate G is constant. The reaction mix passes through a dialyzer where product is removed while the substrate returns to the reservoir; the high-molecular-weight enzyme also would be retained in the reservoir by virtue of its negligible dialysance.

Again, a mass balance for the reaction product may be written

$$\frac{d(VC_F)}{dt} = -D(C_F - C_{Di}) + G \qquad (21.1\text{-}36)$$

and a volume balance can be written

$$V = V^0 - Q_{\text{UF}}t + Q_R t \qquad (21.1\text{-}37)$$

Solution of Eqs. (21.1-36) and (21.1-37) leads to the following equation for C_F as a function of time:

$$C_F^t = \frac{DC_{Di} + G}{D + Q_R - Q_F}\left[1 - \left(\frac{V^0 - Q_{\text{UF}}t}{V^0}\right)^{\gamma}\right] + C_F^0\left(\frac{V^0 - Q_{\text{UF}}t}{V^0}\right)^{\gamma} \qquad (21.1\text{-}38)$$

where γ is given by

$$\gamma = \frac{D + Q_R - Q_{\text{UF}}}{Q_{\text{UF}} - Q_R} \qquad (21.1\text{-}39)$$

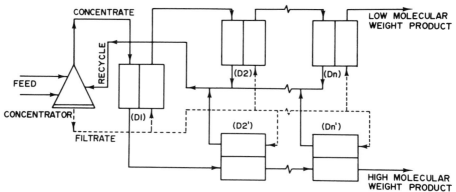

FIGURE 21.1-10 Schematic representation of a multistage dialysis process for continuous fractionation of solutes. The process combines a nonselective solvent stripper (concentrator) with two crosscurrent dialysis cascades (D1–Dn and D1–Dn′). (Reprinted by permission of the American Institute of Chemical Engineers.)

Dialysis also may be used in cascade configurations to give multistage separations. Series staging of dialyzers serves to increase the area available for dialysis and the equations developed in the preceding sections are applicable. Noda and Gryte[9] have described a multistaged system in which two crosscurrent dialyzer cascades are combined with a nonselective solvent stripper, such as a reverse osmosis unit, to increase the slope of the fractional extraction versus $k_{OV}A_M/Q_F$ curves shown in Fig. 21.1-6 and thus the relative separation of two solutes. Their process configuration is shown in Fig. 21.1-10. Assuming identical dialyzers, equal feed and dialysate flow rates throughout the system, and negligible ultrafiltration, they have shown that the fractional extraction of the system as a whole is given by

$$E = \frac{(k_{OV}A_M/Q_F)^n}{1 + (k_{OV}A_M/Q_F)^n} \tag{21.1-40}$$

where n is the number of dialyzers in each cascade. Remembering that the separation factor α_{jk} is equal to the ratio of the fractional extractions of components j and k, we can use Eq. (21.1-40) to calculate the improvement in separation brought about by using the system shown in Fig. 21.1-10 rather than a single dialyzer. For example, if the values of $k_{OV}A_M/Q_F$ for two solutes in a given dialyzer with equal feed and dialysate flow rates are 0.5 and 1.5, respectively, the fractional extractions of the two solutes are 0.33 and 0.60 and their separation factor is 1.82. If the same dialyzer then is used in the system shown in Fig. 21.1-10, with three dialyzers in each cascade ($n = 3$), Eq. (21.1-40) predicts fractional extractions of 0.11 and 0.77 for the two solutes and a separation factor of 7.00. Although multistaging of dialyzers clearly improves separations, fluid flow considerations place a practical limitation on the number of dialytic stages which can be incorporated.

21.2 ELECTRODIALYSIS

In electrodialysis, electrolytes are transferred through solutions and membranes by an electrical driving force. Electrodialysis is used to change the concentration or composition of solutions, or both. The process usually involves multiple, thin compartments of solutions separated by membranes that allow passage of either positive ions (cations) or negative ions (anions) and block the passage of the oppositely charged ions. But the process may be operated with only one membrane that separates two electrode-rinse solutions to purify one of the solutions (e.g., production of salt-free sodium hydroxide).

As in any membrane process, selective membranes are critically important in electrodialysis. Therefore, the nature of the ion-exchange membranes used in electrodialysis will be discussed first.

21.2-1 Ion-Exchange Membranes

Seemingly solid membranes are actually a jumble of intertwined polymer chains. In most ion-exchange membranes, the polymer chains are attached to each other at points by crosslinking. Transferring ions pass through spaces between the intertangled polymer chains. Commercially available ion-exchange membranes usually have high crosslinking densities (i.e., the polymer chains are attached to each other at many points). The segments between points of crosslinking are short enough to be relatively inflexible. The short, inflexible polymer segments and small spaces between segments result in high resistances to transfer because of interactions between the moving ions and the polymer segments.

Cation-exchange membranes are permeable to cations but not anions. Anion-exchange membranes are permeable to anions but not cations. The source of this ability to discriminate between cations and anions is discussed with the aid of Fig. 21.2-1.

In Fig. 21.2-1, polymer chains are shown that have negatively charged groups chemically attached to them. The polymer chains are intertwined and also crosslinked at various points. Positive ions are shown dispersed in the voids between polymer chains but near a negative fixed charge. The fixed negative charges on the chains repulse negative ions that try to enter the membrane and exclude them. Thus, negative ions cannot permeate the membrane, but positive ones can (i.e., the membrane is a cation-exchange membrane). If positive fixed charges are attached to the polymer chains instead of negative fixed charges, positive ions cannot permeate the membrane, but negative ions can; and the membrane is an anion-exchange membrane. This exclusion as a result of electrostatic repulsion is termed Donnan exclusion.

For an ion-exchange membrane to be practical for low-cost processing, it must be low in resistance to ion transfer as well as being ion selective. To decrease resistance, a membrane may be made thinner, the crosslinking density may be decreased, or the fixed-charge density may be increased. Decreases in crosslinking density tend to increase the length and flexibility of polymer segments between points of cross-

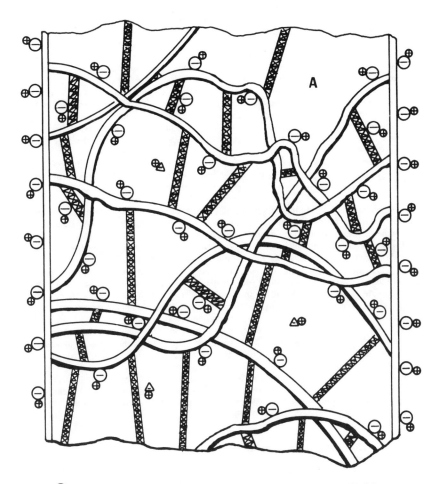

$\ominus$ FIXED NEGATIVELY CHARGED EXCHANGE SITE; I.E., SO_3^-
$\oplus$ MOBILE POSITIVELY CHARGED EXCHANGEABLE CATION; I.E., Na^+
⎯ POLYSTYRENE CHAIN
ⵉⵉⵉ DIVINYLBENZENE CROSSLINK

FIGURE 21.2-1 Representation of an ion-exchange membrane.

linking. This increases the size of the void spaces between segments, which makes it easier for ions to move through the membrane. Increases in fixed-charge density also tend to increase the size of the void spaces because of the repulsion of like fixed charges. But, if the void spaces between polymer chains become too large due to either decreased crosslinking density or increased fixed-charge density, there can be volumes in the center of the voids that are not affected by the fixed charges on the chains (as at point A in Fig. 21.2-1) because the repulsion of fixed charges falls off rapidly with the distance away from the charges. Such volumes that are unaffected by fixed charges result in ineffective repulsion of the undesired ions and lowered selectively. Usually, a compromise between selectivity and low resistance must be made. In membranes now available, it has been possible to combine excellent selectivity with low resistance, high physical strength, and long lifetimes.

Heterogeneous ion-exchange membranes have been made by incorporating ion-exchange particles in solutions of film-forming polymers and allowing the solvents to evaporate. They also have been made by dispersing ion-exchange materials in partially polymerized film-forming polymers, casting films of the dispersion, and completing the polymerization.

Homogeneous ion-exchange membranes have been made by graft polymerization of anionic or cationic moieties into preformed films, by casting films from a solution of a linear film-forming polymer and a linear polyelectrolyte and allowing the solvent to evaporate, or by polymerization of mixtures of reactants that can undergo polymerization by either addition or condensation reactions. In the last method, one of the reactants must be capable of being converted to either anionic or cationic moieties.

The resistances of commercially available ion-exchange membranes range from 2 to 20 ohm·cm^2, depending on the types of membrane and solution being treated. The current efficiencies attainable with commercially available membranes range from 85 to 95% for most operations. Names and addresses of suppliers of ion-exchange membranes are given in Table 21.2-1.

With membranes made by the methods described above, the maximum concentrations of the solutions that can be processed efficiently are limited because Donnan exclusion decreases with increasing solution concentration. With decreases in Donnan exclusion, the current efficiency decreases. Cation-exchange membranes based on fluorocarbon polymers have been developed with which very concentrated solutions can be treated with high current efficiencies. The efficiency achievable with these membranes has been shown to be due to a phenomenon termed "ion clustering" within the membranes.[1-5] Because of the internal structure of such membranes, both high current efficiencies and low resistances can be achieved in concentrated solutions.

These membranes are based on perfluorinated polymers that will withstand oxidative conditions and high temperatures that would be destructive to hydrocarbon-based membranes. Sulfonic or carboxylic acid groups, or both, are affixed chemically to the perfluorinated polymers to impart cation-exchange characteristics.

Some of the membranes are composite structures. These have a thin "tight" layer that provides high current efficiencies (i.e., high transport numbers for cations) backed by a thicker layer of "looser" polymer to provide low resistance combined with strength and convenience of handling. Typical resistances range from 2 to 4 ohm·cm^2 at room temperature depending on the solutions in which they are used. Typical transport numbers are 0.94–0.97, which provide current efficiencies of 94–97%.

The perfluorinated membranes have excellent dimensional stability and long lifetimes. For example, they change in dimensions only 4–6% after being treated in 2% sodium hydroxide solutions and then in 24% sodium chloride/1% sodium hydroxide in preparation for use in chlor-alkali cells. (By assembling the membranes after treatment in such solutions, wrinkle-free surfaces are obtained.)

The major use of perfluorinated membranes, at present, is as separators in chlor-alkali cells. The combination of low resistances, high current efficiencies at high solution concentrations, and high temperatures that can be achieved results in 20–30% lower energy requirements than those achieved with diaphragm or mercury cells. The long membrane lifetime (typically 2 years) results in low cost for membrane replacement. (Asbestos diaphragms usually last for only a year or less.)

The caustic soda produced contains very low levels of salt (less than 60 ppm at 50% caustic strength). In addition, the environmental problems encountered with mercury cells and the potential for health hazards from handling asbestos are eliminated. The chlorine gas that is coproduced is also a highly pure product.

TABLE 21.2-1 Suppliers of Membranes

Asahi Chemical Industry Co., Ltd., 8345 Yako-cho, Daishigawara, Kawasaki, Japan
Asahi Glass Co., Ltd., 14,2-chome, Marunouchi, Chikoda-ku, Tokyo, Japan
E. I. Du Pont de Nemours, Wilmington, Delaware
Ionac Chemical Co., Birmingham, New Jersey
Ionics, Inc., 65 Grove Street, Watertown, Massachusetts
Tokuyama Soda Co., Ltd., Tokuyama, Japan
Zerolit, Ltd., 632-52 London Road, Isleworth, Middlesex, England

For new plants, the capital investment for membrane cells is lower than that for comparable diaphragm or mercury cells. The high conversion of salt reduces the cost of brine handling facilities below that needed for mercury cells. The high concentrations of caustic from membrane cells reduces the cost of evaporators below that needed for diaphragm cells. Moreover, existing mercury cells or diaphragm cells can be replaced by membrane cells at lower costs than required for building new plants. Additional information about perfluorinated membranes can be obtained from Asahi Chemical Industry Co., Asahi Glass Co., E. I. Du Pont Company, and Tokuyama Soda Co. (see Table 21.2-1).

A number of other uses for these membranes have been or are being studied. Some of these uses are:

1. In Donnan dialysis for decontamination of radioactive wastes from nuclear power plants.
2. For removal of zinc from textile wastes.
3. For recovering copper from industrial wastes or from low-grade ores.
4. For recovery of nickel, cadmium, and chromium from electroplating wastes.
5. For stripping common pollutants, such as mercury, cadmium, or lead, from industrial wastes to meet environmental requirements.
6. For production of certain high-purity materials, such as potassium stannate.

21.2-2 Electrodialysis Process and Electrodialysis Stacks

In electrodialysis, cation-exchange membranes are alternated with anion-exchange membranes in a parallel array to form thin solution compartments (0.5–1.0 mm thick). The entire assembly of membranes is held between two electrodes to form an electrodialysis stack as shown in Fig. 21.2-2. A solution to be treated is circulated through the solution compartments. With the application of a dc electrical potential to the electrodes, all cations transfer toward the cathode (negative electrode) and all anions move toward the anode (positive electrode). The ions in the even-numbered compartments can transfer through the first membranes they encounter (cations through cation-exchange membranes, anions through anion-exchange membranes), but they are blocked by the next membranes they encounter as indicated by the arrows in the diagram. Ions in the odd-numbered compartments are blocked in both directions. Ions are removed from the solution circulating through one set of compartments (even) and transferred to the other set of com-

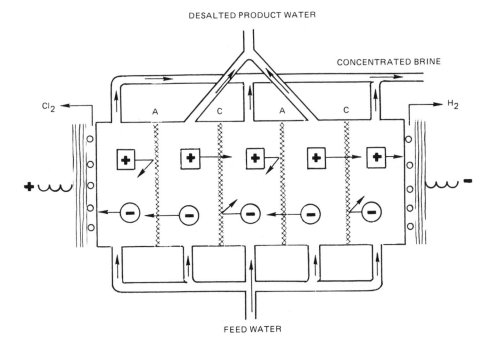

A = ANION-PERMEABLE MEMBRANE
C = CATION-PERMEABLE MEMBRANE

FIGURE 21.2-2 Simplified representation of the electrodialysis process.

partments (odd). Ion-depletion is accomplished for one solution, and ion concentration is accomplished for the second solution.

Figure 21.2-3 is an exploded view of part of an electrodialysis stack that shows the main components. Component 1 in Fig. 21.2-3 is one of the two end frames, each of which has provisions for holding an electrode and introducing and withdrawing the depleting, the concentrating, and the electrode-rinse solutions. The end frames are made thick and rigid to resist bending when pressure is applied to hold the stack components together. The inside surfaces of the electrodes may be recessed, as shown, to form an electrode-rinse compartment when an ion-exchange membrane, component 2, is clamped in place. Components 3 and 5 are spacer frames. Spacer frames have gaskets at the edges and ends so that solution compartments are formed when ion-exchange membranes and spacer frames are clamped together.

Usually, the supply ducts for the various solutions are formed by matching holes in the spacer frames, membranes, gaskets, and end frames. Each spacer frame is provided with solution channels (at point E in Fig. 21.2-3) that connect the solution-supply ducts with the solution compartments. The spacer frames have mesh spacers or other devices in the compartment spaces to support the ion-exchange membranes and so to prevent collapse when there is a differential pressure between two compartments. These mesh spacers are selected or designed to aid in lateral mixing, which decreases the thickness of boundary layers.

An electromembrane stack usually has many repeated sections, each consisting of components 2, 3, 4, and 5, with a second end frame at the end. These repeating units are termed cell pairs.

There are three basic types of electrodialysis stacks—tortuous-path, sheet-flow, and unit-cell stacks. In the tortuous-path stack, the solution flow path is a long, narrow channel which makes several 180° bends between the entrance and exit ports of a compartment as illustrated in Fig. 21.2-4. The bottom half of the spacer gasket in Fig. 21.2-4 shows the individual narrow solution channels and the cross straps used to promote mixing, whereas the cross straps have been omitted in the top half of the figure so that the flow path could be better depicted. The ratio of channel length to width is high, usually greater than 100:1. Spacer screens to support the membranes may or may not be used in tortuous-path stacks.

In sheet-flow stacks, spacer screens almost always are needed because the width between gasketing devices is much greater than that in the usual tortuous-path stacks. The solution flow in sheet-flow stacks is in approximately a straight path from one or more entrance ports to an equal number of exit ports as indicated in Fig. 21.2-3. As the solution flows in and around the filaments of the spacer screens, a mixing action is imparted to the solutions to aid in reducing the thicknesses of diffusional boundary layers at the surfaces of the membranes.

Solution velocities in sheet-flow stacks are typically in the range of 5–14 cm/s, whereas the velocities in tortuous-path stacks usually are much higher, 30–100 cm/s. The drop in hydraulic pressure through a sheet-flow stack, and thus the pumping power, is normally lower than that through a tortuous-path stack because of the lower velocities and shorter path lengths.

Unit-cell stacks were developed specifically for concentrating solutions. Readers desiring information about unit-cell stacks are referred to Ref. 6.

21.2-3 Concentration Polarization

The detailed nature of ion transfer through a system of solutions and ion-exchange membranes warrants thorough discussion because it is not only the source of the depletion and concentration that occurs in electrodialysis but also the source of excessive concentration polarization, which is responsible for most of the difficulties encountered in electrodialytic processing.

Figure 21.2-5 shows a cation- and an anion-exchange membrane mounted between two electrodes. A solution of an electrolyte flows through the compartments formed by the membranes and electrodes. With the passage of an electrical current through the system of membranes and solutions, anions transfer toward the anode, and cations transfer toward the cathode. These ion transfers are the way in which electrical current is carried in an electrolytic medium. The fraction of the current carried by an ionic species is termed its transference number (t^+ or t^-).

The hydraulic characteristics of an electrodialysis solution compartment and the boundary layers of nearly static solution at the surfaces of the membranes can have a controlling effect on the current densities that can be used in electrodialysis and therefore a controlling influence on the rate of demineralization. Because of the flow of solution through the center compartment formed by the two membranes in Fig. 21.2-5, there is a zone of relatively well-mixed solution near the center of the compartment. The velocity of the solution and thus the degree of mixing diminish as the surfaces of the membranes are approached. In Fig. 21.2-5, an idealization is used such that there is a completely mixed zone in the center of the compartment and completely static zones of solution in boundary layers adjacent to the membranes. In the static boundary layers, ions are transferred only by electrical transfer and diffusion, but in the mixed zone ions are transferred electrically, by diffusion, and by physical mixing.

If the transference number of anions through the anion-exchange membranes (t_m^-) is 1.0 and that of anions through the solution (t_s^-) is 0.5, only half as many ions will be transferred electrically through the solution in the static boundary layer on the side of the membrane that anions enter as will be transferred through the membranes. The solution at the membrane interface will be depleted of ions. For the same

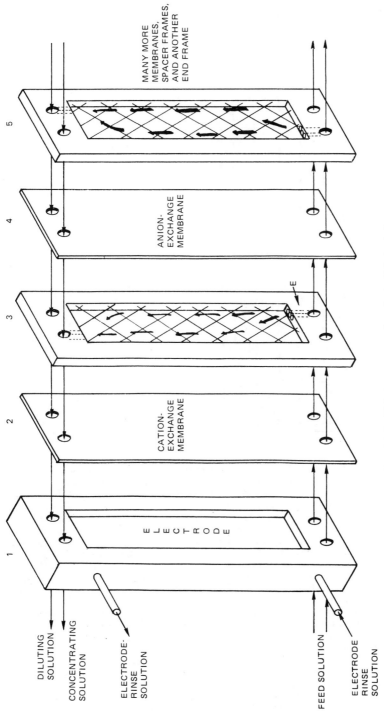

FIGURE 21.2-3 Main components of an electrodialysis stack.

DILUTING SOLUTION

CONCENTRATING SOLUTION

ELECTRODE-RINSE SOLUTION

FEED SOLUTION

ELECTRODE RINSE SOLUTION

ELECTRODE

CATION-EXCHANGE MEMBRANE

ANION-EXCHANGE MEMBRANE

MANY MORE MEMBRANES, SPACER FRAMES, AND ANOTHER END FRAME

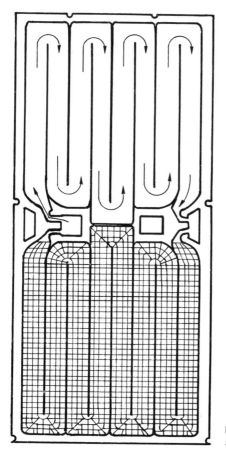

FIGURE 21.2-4 Diagram of a tortuous-path spacer for an electrodialysis stack.

reasons, the static boundary layer on the other side of the membrane will accumulate ions. The interfacial concentrations will change until concentration gradients are established in the boundary layers which are sufficiently large to transfer by diffusion the ions not transferred through the boundary layers electrically, as indicated by the dashed lines in Fig. 21.2-5.

If the current density through the system is increased, the rate of electrical transfer of ions increases, and the diffusional transfer through the boundary layers must increase to supply the additional ions transferred electrically. With any given thickness of boundary layer, a current density can be reached at which the concentrations of electrolytes at the membrane interfaces on the depleting sides will approach zero. At this current density, called the limiting current density, H^+ and OH^- ions from ionization of water will begin to be transferred through the membranes. This continuous ionization of water is termed water splitting. The water splitting caused by exceeding the limiting current density has detrimental effects on the operation of electrodialysis, as discussed below.

Scaling of membranes by pH-sensitive electrolytes occurs at anion-exchange membranes because OH^- ions transfer through the membranes when water splitting occurs. The OH^- ions increase the pH within the membrane and at the interface on the concentrating side so that pH-sensitive substances, such as calcium carbonate, precipitate.

Operation at or above the limiting current density causes the stack to have resistances higher than normal for the following reasons. When water splitting occurs, the rate of dissociation of water is increased because the H^+ and OH^- ions are transferred continuously away from the membrane interfaces, which are the locations of dissociation. An increased voltage is necessary to induce this continuous dissociation of water. Also, a thin film of highly depleted solution forms at the depleting sides of the membranes. These films have high specific resistances, which are in series with the other resistances in an electrodialysis stack. Both the increased voltages needed for continuous ionization of water and the high specific resistances of depleted films contribute to increased stack resistance.

Fouling of ion-exchange membranes by organic materials also can occur when concentration polarization occurs. Fouling of anion-exchange membranes has been shown to be caused by large organic ions becoming attached to charged groups on membranes having the opposite charge.[7-9]

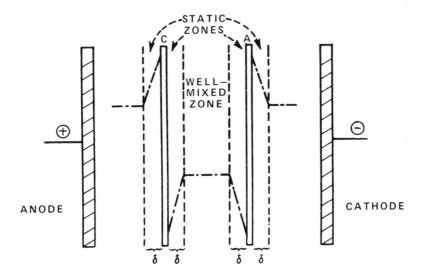

δ = STATIC BOUNDARY LAYERS
C = CATION−EXCHANGE MEMBRANE
A = ANION−EXCHANGE MEMBRANE

FIGURE 21.2-5 Idealized representation of concentration gradients in electrodialysis.

Because of these detrimental consequences of excessive concentration polarization, the minimization of polarization is an important factor in the design of electrodialysis stacks.

21.2-4 Minimizing Concentration Polarization

As was implied in the discussion of Fig. 21.2-5, the changes in electrolyte concentration in the solutions at the membrane–solution interfaces, which are responsible for the deleterious consequences of polarization, can be minimized by decreasing the thickness of the nearly static boundary layers.

With the idealization of completely static and completely mixed zones used in Fig. 21.2-5, Eq. (21.2-1) has been derived to determine the thickness of boundary layers:[10,11]

$$\delta = \frac{DFN}{i_{\lim}(t_m - t_s)} \tag{21.2-1}$$

where δ is the thickness (cm) of the boundary layers in depleting compartments, D is the diffusivity (cm/s) of the salt being treated, F is the Faraday number, 96,500 C/eq, N is the concentration (eq/cm^3) of the salt in the well-mixed zone, $i_{\lim}$ is the limiting current density (mA/cm^2)—that is, the current density when the interfacial concentration becomes zero, t_m is the transference number of the counterion in the membrane, and t_s is the transference number of the counterion in solution.

Equation (21.2-1) has been rearranged to a form in which the polarization parameter $i_{\lim}/N$ is expressed in terms of the above variables:

$$\frac{i_{\lim}}{N} = \frac{DF}{t_m - t_s} \tag{21.2-2}$$

The polarization parameter is more convenient to use in design and scale-up of electrodialytic equipment than the thickness of the boundary layer. It can be measured easily in small-scale stacks with a given value of bulk concentration N and used to predict limiting current densities in larger stacks at other values of N.

Although boundary-layer thickness at any point in a cell compartment depends on several factors,[12] the average boundary-layer thickness in Eq. (21.2-2) is primarily a function of the solution velocity in the depleting compartments and the mixing characteristics of spacer materials. Because the boundary-layer thickness varies with solution velocity, the polarization parameter does also. The relationships between solution velocity, the polarization parameter, and the nature of ion-exchange membrane have been studied by different methods,[13] including determinations of changes in pH and stack resistances which indicate the limiting current density has been reached.

Values of i_{lim}/N may be measured in laboratory stacks assembled with the membranes to be tested. The solution of interest is circulated through the depleting compartments, and the stack voltage is adjusted to give a desired stack current. The flow rate and the influent and effluent concentrations of the solution being depleted are measured. The current is increased in increments, and, at each increment, measurements are made of the stack voltage E_s and stack current I_s. Values of E_s/I_s at each increment are plotted as a function of current density. At the limiting current density (i_{lim}), there will be a change in the slope of the curve. If plots of E_s/I_s versus the reciprocal of current density are used, the break in the curve is more pronounced so that i_{lim} can be defined more closely. Then the polarization parameter is calculated as i_{lim} divided by the log-mean concentration N.

Because of the ill effects of exceeding the limiting current density and the fact that there can be stagnant or semistagnant areas in even well-designed electrodialysis stacks, most process designers use operating values of i/N that are 50–70% of the limiting values.

21.2-5 Electrodes

The most common reaction at the cathode of an electrodialysis stack is the decomposition of water

$$2\,H_2O + 2e^- \rightarrow H_{2(g)} + 2\,OH^-$$

This reaction gives rise to both gas (H_2) and high values of pH. The gas can blanket the electrode and increase the resistance of the stack. High values of pH can cause precipitation of any pH-sensitive materials (e.g., calcium carbonate) in the electrode-rinse stream, which can raise resistance and even make the stack inoperative.

Accumulation of basic precipitates in the catholyte compartment can be prevented by acidification or softening of the electrode-rinse solution. The OH^- ions generated at the cathode can transfer into the multiple compartments and cause precipitates to form. Therefore, isolating compartments that are flushed with acidified or softened catholyte solutions often are used. A high solution velocity is used in the catholyte compartment to sweep out the gas as it is formed.

The choice of cathode materials is not too critical because the cathode is not subjected to severe corrosion. Relatively low-cost materials, such as stainless steel or graphite, are used.

A variety of electrode reactions[14] can occur at the anode. Some of them result in the formation of gases, corrosive attacks on most electrode materials, or oxidative attacks on the membrane in contact with the anolyte solution. Platinum is too costly for use as an anode material, but titanium or tantalum coated with extremely thin (microinches) layers of platinum has been used. Lead dioxide deposited on and within graphite has been used also. Oxidation-resistant membranes (fluorocarbon-based) often are used adjacent to the anolyte solution even though they are expensive.

21.2-6 Design Considerations

In electrodialysis, there are three categories of costs: those that increase with current density (the cost of energy); those that decrease with current density (the cost of membrane replacement and amortization of capital investment); and those that are essentially invariant with current density (cost of chemicals, maintenance, and labor). Because two of these categories of costs vary in opposite directions with current density, there is an economically optimum current density. However, for most applications of electrodialysis, the limiting current density is lower than the optimum current density. Therefore, determination of the limiting current density i_{lim} by the previously described methods is usually the first step in design.

The polarization parameter i_{lim}/N can be calculated next. During the experiments to determine i_{lim}, data may be taken on voltage, current, membrane area, and number of cell pairs in the experimental stack. The cell-pair voltage E_{cp} can be determined for various values of concentration N, and current density i, and curves of E_{cp} versus i for various values of N can be prepared. The current efficiency to be expected in large-scale use can be calculated from the experiments to determine i_{lim}:

$$\eta_c = \frac{QF(N_f - N_e)}{S_{cp}I_s} \tag{21.2-3}$$

where η_c is the current efficiency, Q is the volumetric flow rate (cm³/s) through the stack, F is the Faraday number, 96,500 C/eq, N_f and N_e are the feed and effluent concentration (eq/cm³), respectively, S_{cp} is the number of cell pairs in the stack (exclusive of isolating compartments), and I_s is the current (A) through the stack.

As stated previously, the operating current density i_o is usually 50–70% of the limiting value. Vendors of electrodialysis equipment will recommend the percentage to be used in their equipment.

For a given demineralization problem, the feed concentration N_f and the desired final concentration usually are known. The number of stages of electrodialysis can be calculated by calculating the effluent concentration N_e for each stage:

$$N_e = N_f - \frac{i_o a \eta_c}{F} \qquad (21.2\text{-}4)$$

where N_e and N_f are the effluent and feed concentrations (eq/cm³), respectively, i_o is the operating current density (A/cm²), a is the effective area (cm²) of membrane, η_c is the current efficiency, and F is the Faraday number, 96,500 C/eq.

The concentration of the effluent from the first stage is the feed concentration for the second stage. The calculation of N_e is continued until the effluent concentration from a stage is equal to the desired final concentration, and the number of stages needed is found.

The cell-pair voltage E_{cp} can be read from the curves of E_{cp} versus i for the average concentrations in each stage. The cell-pair voltages multiplied by the number of cell pairs per stack will provide an approximation of the voltage per stack for each stage.

The operating current densities for each stage multiplied by the effective membrane area gives the stack currents for each stage.

The product of stack voltage and stack current will give the dc-power requirements for each stage.

The number of lines of stacks to be arranged in parallel to provide a given throughput can be determined by dividing the desired throughput by the throughput in the depleting compartments of a stack, which can be obtained from the vendor.

Some water will be transferred out of the depleting compartments by electroosmosis. The rate of electroosmotic transfer is determined by the nature of the membranes. Estimates of the decrease in the amount of demineralized product because of this water transfer can be obtained from the supplier of the ion-exchange membranes.

With the information described above, predesign estimates of capital and operating costs can be prepared. Capital costs include costs for pretreatment equipment (if needed), electrodialysis stacks, storage tanks, and pumps. These costs may be estimated by standard predesign estimating methods.

Operating costs include costs for electrical energy (for stacks, pumping, lighting, and auxiliaries), membrane replacement, chemicals (e.g., acids for electrode-rinse streams), maintenance supplies and labor, operating labor, overhead items, amortization of investment, taxes, and insurance.

NOTATION FOR SECTION 21.1

Nomenclature

a	constant defined in Eq. (21.1-26) (dimensionless)
A_M	membrane area (cm²)
A_P	fractional cross-sectional area of membrane pore openings (dimensionless)
b	constant defined in Eq. (21.1-26) (dimensionless)
c	constant defined in Eq. (21.1-26) (dimensionless)
C	solute concentration (g/cm³)
ΔC	solute concentration difference (g/cm³)
d	diameter (cm)
D	dialysance (cm³/s)
$\mathcal{D}$	diffusivity in bulk solution (cm²/s)
$\mathcal{D}_M$	diffusivity in the membrane (cm²/s)
E	fractional extraction (dimensionless)
G	reaction rate (g/s)
h	channel height (cm)
J_s	transmembrane solute flux (g/cm² · s)
J_v	transmembrane solvent velocity (cm/s)
k	mass transfer coefficient (cm/s)
K_1	power series in q defined in Eq. (21.1-6) (dimensionless)
l	length (cm)
l_P	pore path length (cm)
L_φ	hydraulic permeability (cm³/dyn · s)

m mass of solute (g)

N number of channels or fibers (dimensionless)

ΔP pressure difference (dyn/cm^2)

P_M membrane permeability (cm/s)

P_M^{HTO} membrane permeability of tritiated water (cm/s)

Pe   transmembrane Peclet number (dimensionless)

q ratio of solute to pore diameter (dimensionless)

Q volumetric flow rate (cm^3/s)

r_P hydrodynamic pore radius (cm)

R mass transfer resistance (s/cm)

Re   Reynolds number (dimensionless)

Sc   Schmidt number (dimensionless)

Sh   Sherwood number (dimensionless)

t time (s)

V volume (cm^3)

w channel width (cm)

x length (cm)

X variable defined in Eq. (21.1-22) (cm^3/s)

Y variable defined in Eq. (21.1-23) (dimensionless)

Z variable defined in Eq. (21.1-24) (dimensionless)

Greek Symbols

α_{jk} separation factor (dimensionless)

β variable defined in Eq. (21.1-33) (dimensionless)

γ variable used in Eq. (21.1-38) (dimensionless)

θ variable used in Eq. (21.1-4) (dimensionless)

μ viscosity (dyn · s/cm^2)

ξ variable used in Eq. (21.1-4) (dimensionless)

ρ density (g/cm^3)

σ reflection coefficient (dimensionless)

ψ variable used in Eq. (21.1-4) (dimensionless)

Subscripts

D dialysate

F feed

i inlet

o outlet

OV   overall

R reagent

UF   ultrafiltrate

W membrane wall

Superscripts

0   initial value (time = 0)

t value at time = t

$-$ average value

REFERENCES

Section 21.1

21.1-1   T. Graham, Liquid Diffusion Applied to Analysis, *Philos. Trans. R. Soc. London*, **151**, 183–224 (1861).

21.1-2   W. G. Schmidt, Versuche uber filtrationsgeschwindigkeit verschiedener. Flussigkeiten durch thierische membran. *Pogg. Annal. Phys. Chem.*, **99**, 337–388 (1856).

21.1-3   R. M. Wallace, Concentration and Separation of Ions by Donnan Membrane Equilibrium, *Ind. Eng. Chem. Proc. Des. Dev.*, **6**, 423–431 (1967).

21.1-4   M. A. Lake and S. S. Melsheimer, Mass Transfer Characterization of Donnan Dialysis. *AIChE J.*, **24**, 130–137 (1978).

21.1-5   R. W. Baker, M. E. Tuttle, D. J. Kelly, and H. K. Lonsdale, Coupled Transport Membranes. I. Copper Separations. *J. Membr. Sci.*, **2**, 213–233 (1977).

21.1-6   E. L. Cussler and D. F. Evans, Liquid Membranes for Separations and Reactions, *J. Membr. Sci.*, **6**, 113–121 (1980).

21.1-7   E. Klein, J. K. Smith, R. P. Wendt, and S. V. Desai, Solute Separations from Water by Dialysis. I. The Separation of Aniline, *Sep. Sci.*, **7**, 285–292 (1972).

21.1-8   A. K. Fritzsche, Permeation of Phenol Through Ethylene Copolymer Hollow Fibers, *Sep. Sci. Technol.*, **15**, 1323–1337 (1980).

21.1-9   I. Noda and C. C. Gryte, Multistage Membrane Separation Processes for the Continuous Fractionation of Solutes Having Similar Permeabilities, *AIChE J.*, **27**, 904–912 (1981).

21.1-10   H. D. Vollrath, Applying Dialysis to Colloid–Crystalloid Separations, *Chem. Metall. Eng.*, **43**, 303–306 (1936).

21.1-11   N. S. Chamberlin and B. H. Vromen, Make Dialysis Part of Your Unit Operations with New Acid Resistant Membranes, *Chem. Eng.*, **66**, 117–122 (1959).

21.1-12   S. B. Tuwiner, Dialysis in Electrolytic Copper Refining, in *Diffusion and Membrane Technology*, pp. 324–332, ACS Monograph Series No. 156, Reinhold, New York, 1962.

21.1-13   W. Drukker, F. M. Parsons, and J. F. Maher. (Eds.), *Replacement of Renal Function by Dialysis*, 2nd ed., Martinus Nijhoff, Boston, 1983.

21.1-14   J. S. Schultz and P. Gerhardt, Dialysis Culture of Microorganisms, Design, Theory, and Results, *Bacteriol. Rev.*, **33**, 1–47 (1969).

21.1-15   T. Wadstrom and O. Vesterberg, Continuous Dialysis of Protein Solutions on a Large Scale. II. Dialysis of Extracellular Proteins from *Staphylococcus Aureus* Followed by Adsorption on Ion Exchangers and Other Methods, *Sep. Sci.*, **5**, 91–98 (1970).

21.1-16   P. Landwall, Dialysis Culture for the Production of Extracellular Protein A from *Staphylococcus Aureus* A676, *J. Appl. Bacteriol.*, **44**, 151–158 (1978).

21.1-17   A. G. Lane, Production of Food Yeast from Whey Ultrafiltrate by Dialysis Culture, *J. Appl. Chem. Biotechnol.*, **27**, 165–169 (1977).

21.1-18   W. F. Blatt, L. Nelsen, E. M. Zipilivan, and M. C. Porter, Rapid Salt Exchange by Coupled Ultrafiltration and Dialysis in Anisotropic Hollow Fibers, *Sep. Sci.*, **7**, 271–284 (1972).

21.1-19   H. Moonen and H. J. Niefind, Alcohol Reduction in Beer by Means of Dialysis, *Desalination*, **41**, 327–335 (1982).

21.1-20   K. F. Benkler and G. A. Reineccius, Separation of Flavor Compounds from Lipids in a Model System by Means of Membrane Dialysis, *J. Food Sci.*, **44**, 1525–1529 (1979).

21.1-21   T. A. Davis, J. S. Wu, and B. L. Baker, Use of the Donnan Equilibrium Principle to Concentrate Uranyl Ions by an Ion-Exchange Membrane Process, *AIChE J.*, **17**, 1006–1008 (1971).

21.1-22   J. L. Eisenmann and J. D. Smith, "Donnan Softening as a Pretreatment to Desalination Processes," U.S. Department of the Interior, Office of Saline Water, Research and Development Progress Rept. No. 506, 1970.

21.1-23   I. Nakamura, T. Kawahara, and T. Utsunomiya, Use of Dialysis Technique in Metal Finishing Process, *New Mater. New Process.*, **2**, 166–175 (1983).

21.1-24   B. L. Browning, L. O. Sell, and W. Abel, Cellulose Solvents for Viscosity Measurements. The Effect of Copper and Base Concentrations in Cuprammonium and Cupriethylenediamine Solutions, *TAPPI*, **37**, 273–283 (1954).

21.1-25   H. Yasuda, C. E. Lamaze, and L. D. Ikenberry, Permeability of Solutes Through Hydrated Polymer Membranes. I. Diffusion of Sodium Chloride, *Makromol. Chem.*, **118**, 19–35 (1968).

21.1-26   P. Feldhoff, T. Turnham, and E. Klein, Effect of Plasma Proteins on the Sieving Spectra of Hemofilters, *Artif. Organs*, **8**, 186–192 (1984).

21.1-27   H. I. Mahon (assigned to Dow Chemical Corp.), "Permeability Separatory Apparatus and Process Utilizing Hollow Fibers," U.S. Patent 3,228,877 (1966).

21.1-28   F. Villarroel, E. Klein, and F. Holland, Solute Flux in Hemodialysis and Hemofiltration Membranes, *Trans. Am. Soc. Artif. Intern. Organs*, **23**, 225–233 (1977).

21.1-29   L. J. Derzansky and W. N. Gill, Mechanisms of Brine-side Mass Transfer in a Horizontal Reverse Osmosis Tubular Membrane, *AIChE J.*, **20**, 751–761 (1974).

21.1-30   E. Klein, F. F. Holland, and K. Eberle, Comparison of Experimental and Calculated Permeability and Rejection Coefficients for Hemodialysis Membranes, *J. Membr. Sci.*, **5**, 173–188 (1979).

21.1-31   R. E. Beck and J. S. Schultz, Hindrance of Solute Diffusion Within Membranes as Measured with Microporous Membranes of Known Pore Geometry, *Biochim. Biophys. Acta*, **255**, 273–303 (1972).

21.1-32   A. Verniory, R. DuBois, P. Decoodt, J. P. Gassee, and P. P. Lambert, Measurement of the Permeability of Biological Membranes. Application to the Glomerular Wall, *J. Gen. Physiol.*, **62**, 489–507 (1973).

21.1-33   E. Klein et al. (Eds.), *Evaluation of Hemodialyzers and Dialysis Membranes*, U.S. Department of Health, Education and Welfare, National Institute of Health, Rept. No. (NIH) 77-1294, 1977.

21.1-34   P. C. Farrell, J. W. Eschbach, J. E. Vizzo, and A. L. Babb, Hemodialyzer Reuse: Estimation of Area Loss from Clearance Data, *Kidney Int.*, **5**, 446–450 (1974).

21.1-35   P. C. Farrell and A. L. Babb, Estimation of the Permeability of Cellulosic Membranes from Solute Dimensions and Diffusivities, *J. Biomed. Mater. Res.*, **7**, 275–300 (1973).

21.1-36   R. P. Wendt, E. Klein, E. H. Bresler, F. F. Holland, R. M. Serino, and H. Villa, Sieving Properties of Hemodialysis Membranes, *J. Membr. Sci.*, **5**, 23–49 (1979).

21.1-37   K. Schindhelm, P. C. Farrell, and J. H. Stewart, Convective Mass Transfer in the Artificial Kidney, Second Australasian Conference on Heat and Mass Transfer, University of Sydney, 1977, pp. 189–197.

21.1-38   M. Y. Jaffrin, B. B. Gupta, and J. M. Malbrancq, A One-Dimensional Model of Simultaneous Hemodialysis and Ultrafiltration with Highly Permeable Membranes, *J. Biomech. Eng.*, **103**, 261–266 (1981).

21.1-39   C. K. Colton, K. A. Smith, P. Stroeve, and E. W. Merrill, Laminar Flow Mass Transfer in a Flat Duct with Permeable Walls, *AIChE J.* **17**, 773–780 (1971).

21.1-40   D. O. Cooney, E. J. Davis, and S-S. Kim, Mass Transfer in Parallel-Plate Dialyzers—A Conjugated Boundary Value Problem, *Chem. Eng. J.*, **8**, 213–222 (1974).

21.1-41   R. Jagannathan and U. R. Shettigar, Analysis of a Tubular Hemodialyser—Effect of Ultrafiltration and Dialysate Concentration, *Med. Biol. Eng. Comput.*, **15**, 134–139 (1977).

21.1-42   T. G. Kaufmann and E. F. Leonard, Mechanism of Interfacial Mass Transfer in Membrane Transport, *AIChE J.*, **14**, 421–425 (1968).

21.1-43   J. Marangozis and A. J. Johnson, A Correlation of Mass Transfer Data of Solid–Liquid Systems in Agitated Vessels, *Can. J. Chem. Eng.*, **40**, 231–237 (1962).

21.1-44   K. A. Smith, C. K. Colton, E. W. Merrill, and L. B. Evans, Convective Transport in a Batch Dialyzer: Determination of True Membrane Permeability from a Single Measurement, *Chem. Eng. Prog. Symp. Ser. No. 84*, **64**, 45–58 (1968).

21.1-45   K. Abel, M. A. Jeffree, B. J. Bellhouse, E. L. Bellhouse, and W. S. Haworth, A Practical Secondary-Flow Hemodialyzer, *Trans. Am. Soc. Artif. Intern. Organs*, **27**, 639–643, (1981).

Section 21.2

21.2-1   T. Gierke, Ionic Clustering in Nafion Perfluorosulfonic Acid Membranes and Its Relationship to Hydroxyl Rejection and Chloro-alkali Current Efficiency, E. I. Du Pont de Nemours and Co., Inc., Wilmington, Delaware, 1977.

21.2-2   S. G. Cutler, Transport Phenomena and Morphology Changes Associated with Nafion 390 Cation-Exchange Membranes, in *Ions in Polymers*, American Chemical Society Symposium Series 187, pp. 145–154, American Chemical Society, Washington, DC, 1980.

21.2-3   K. A. Mauritz, C. J. Hora, and A. J. Hopfinger, Theoretical Model for the Structures of Ionomers: Application to Nafion Materials, in *Ions in Polymers*, American Chemical Society Symposium Series 187, pp. 123–144, American Chemical Society, Washington, DC, 1980.

21.2-4   K. A. Mauritz and S. R. Lowry, The Fourier Transform Infrared Characterization and Theory of Counterion Associations in Nafion Membranes, *Am. Chem. Soc. Div. Polymer. Chem. Polym. Prep.*, **19(2)**, 336–340 (1978).

21.2-5   S. C. Yeo and A. Eisenberg, Physical Properties and Supermolecular Structure of Perfluorinated Ion-Containing Polymers, *J. Appl. Polym. Sci.*, **21**, 875–898 (1977).

21.2-6   T. A. Davis and G. F. Brockman, Physiochemical Aspects of Electromembrane Processes, in R. E. Lacey and S. Loeb (Eds.), *Industrial Processing with Membranes*, pp. 21–38, Wiley, New York, 1972.

21.2-7   E. Korngold, F. de Korosy, R. Rakov, and M. Taboch, Fouling of Anion-Selective Membranes in Electrodialysis, *Desalination*, **8**, 195–200 (1970).

21.2-8   G. Grossman and A. Sonin, *Experimental Study of the Effects of Hydrodynamics and Membrane Fouling*, Office of Saline Water Research and Development, Report No. 742, Superintendent of Documents, Catalog No. 11 88 742, Washington, DC, 1971.

21.2-9   J. R. Wilson, *Demineralization by Electrodialysis Butterworth*, London, 1960.

21.2-10   A. Sonin and R. F. Probestein, Hydrodynamics—Theory of Desalination by Electrodialysis, *Desalination*, **5,** 293–308 (1968).

21.2-11   H. Belfort and G. Guter, *Hydrodynamic Studies for Electrodialysis*, Office of Saline Water Research and Development, Report No. 459, National Technical Information Service, PB 203270, Springfield, VA, 1969.

21.2-12   H. Gregor and R. Petersen, Concentration Polarization in Membrane Processes, *J. Phys. Chem.*, **68,** 2201–2208 (1964).

21.2-13   B. A. Cooke, Concentration Polarization in Electrodialysis, *Electrochim. Acta*, **3,** 307–316 (1960); **4,** 179–188 (1961).

21.2-14   N. Lakshminarayanaiah, *Transport Phenomena in Membranes*, Academic, New York, 1969.

21.2-15   R. E. Lacey, *Transport of Electrolytes Through Membrane Systems*, Office of Saline Water Research and Development, Report No. 343, National Technology Information Service, Springfield, VA, 1965.

CHAPTER 22

Selection of a Separation Process

HAROLD R. NULL
Monsanto Company
St. Louis, Missouri

22.1 INTRODUCTION

There is no rote procedure by which separation processes are chosen. However, there are certain guidelines that usually are followed in arriving at the separation process sequence to be adopted in the plant design. The degree to which the optimum separation scheme is approached depends on the time and money allotted to development and analysis and on the skill of the process design engineer in applying guidelines to the selection of sequences for detailed evaluation.

It is safe to say that the ultimate criterion is economics. However, the economic criterion is subject to a number of intangible constraints. These constraints may include corporate attitude toward market strategy and timing, reliability, risks associated with innovation, and capital allocation. In this regard, separation processes are no different than any other type of process. To illustrate the influence of the intangible constraints, we might consider two extreme cases.

1. The product is of high unit value with a short market life expectancy.
2. The product is a high-volume chemical with many producers in a highly competitive market.

In Case 1 the procedure would probably be to choose the first successful separation method found. The ultimate overall economics of this situation are determined by getting into the market ahead of the competition and selling as much of the product as possible while the market lasts. However, if the market turns out to have a sustained life, it would be foolish in the extreme to continue to follow this philosophy on second- and third-generation plants. As time passes and competitors move in, products ultimately take on the nature of Case 2.

Even in Case 2, there are time and money constraints on the process development and design teams. However, the ultimate economic viability of the plant dictates that, within those constraints, the team do the most thorough job possible in the development and evaluation of various process schemes to make the closest approach to the economic optimum design possible. In its broadest meaning, this development and evaluation activity constitutes process synthesis. In a narrower context, process synthesis is a rigorous methodology for choosing the optimum flowsheet. In still narrower context process synthesis refers to the use of a computer program to develop the flowsheet.

22.2 INITIAL SCREENING

The first step in the selection of a separation process is to define the problem. Usually, this entails establishing product purity and recovery specifications. Product purity specifications are established ultimately by customers, although they may not be stated explicitly by the customer. Recovery specifications are set to assure an economic process. Ideally, the recovery specification should be a variable to be optimized by the process design. In practice, there is usually insufficient time to consider recovery as a variable, and it is specified either arbitrarily or by consensus.

22.2-1 Property Differences

Having established the products to be separated and the purity and recovery desired for each, the next essential step is to determine which separation methods are capable of accomplishing the separation. In order for two components to be separable, there must be some difference in properties between them. The objective of design of the separation process is to exploit the property differences in the most economical manner to accomplish the separation. The following properties are used as bases for separation processes:

Equilibrium properties
  Vapor pressure
  Solubility
  Distribution between immiscible liquid phases
  Melting point
  Chemical reaction equilibrium
  Electric charge (isoelectric point)
  Surface sorption
Rate properties
  Diffusivity
  Ionic mobility
Molecular size
Molecular shape

22.2-2 Classification of Unit Operations

While separation unit operations often depend on several property differences for their overall success, there is usually one property difference that forms the primary basis for separation. The commonly used methods of separation and the primary basis for separation of each follows:

Distillation—vapor pressure
Extraction—distribution between immiscible liquid phases
Crystallization—melting point or solubility
Adsorption—surface sorption
Reverse osmosis—diffusivity and solubility
Membrane gas separation—diffusivity and solubility
Ultrafiltration—molecular size
Ion exchange—chemical reaction equilibrium
Dialysis—diffusivity
Electrodialysis—electric charge and ionic mobility
Liquid membranes—diffusivity and reaction equilibrium
Electrophoresis—electric charge and ionic mobility
Chromatographic separations—depends on type of stationary phase
Gel filtration—molecular size and shape

22.2-3 Scale of Operation

The scale of operation often is the determining economic factor in the selection between alternative separation processes. For example, air separations on a very large scale are accomplished most economically by cryogenic distillation processes. Small-scale air separations, however, often are accomplished more economically by other means such as pressure swing adsorption or hollow-fiber gas separation membranes. The scale of operation at which the choice switches is case dependent and includes such factors as location and oxygen and nitrogen purity specifications. However, cryogenic distillation rarely is challenged by other methods for high-purity separations at air feed rates exceeding 100,000 standard cubic feet per hour (2832 m^3/h).

Another example is in desalination where reverse osmosis tends to be more economical than multistage flash or evaporation at feed rates less than 20×10^6 gal/day (80×10^6 L/day), while the flash or evaporation methods are more economical in very large installations.

TABLE 22.2-1　Approximate Single-Line Maximum Capacities for Selected Separation Processes

Separation method	Upper commercial scale for single-line operation
Distillation	No limit
Extraction	No limit (certain types of column have been used only up to 6 ft (1.83 m) diameter.
Crystallization	$25–150 \times 10^6$ lb$_m$/yr ($10–70 \times 10^6$ kg/yr)
Adsorption	No limit
Reverse osmosis	1×10^6 lb$_m$/yr (0.45×10^6 kg/yr) water flow per module, with many modules per unit
Membrane gas separation	2×10^6 lb$_m$/yr (0.9×10^6 kg/yr) gas permeated per module, with many modules per unit
Ultrafiltration	1×10^6 lb$_m$/yr (0.45×10^6 kg/yr) water flow per module, with many modules per unit
Ion exchange	1×10^9 lb$_m$/yr (450×10^6 kg/yr) water flow
Electrodialysis	1×10^6 lb$_m$/yr (0.45×10^6 kg/yr) water flow
Electrophoresis	1000 kg/yr carrier-free basis
Chromatographic separations	0.5×10^6 lb$_m$/yr (0.24×10^6 kg/yr) gas, 0.2×10^6 lb$_m$/yr (0.09×10^6 kg/yr) liquid
Gel filtration	100 kg/yr carrier-free basis

Any separation process chosen must be compatible with the scale of operation of the commercial plant. While it is common to see two or three parallel lines of operation in a commercial installation, plant operation becomes unwieldy when more than three parallel lines must be operated. In the case of very-high value products, perhaps as many as ten parallel lines might be considered; however, it is extremely rare that more than ten parallel lines would be considered. One must be careful how one interprets the term "parallel lines." For example, a membrane separation process may require many identical "modules" piped together in a single unit with common feed and product headers and a single instrumentation package which monitors the entire package. In such a case, the entire installation acts as a single unit; operators and control systems treat it as a single unit or line, regardless of the number of individual modules that may be included in the unit. On the other hand, a number of identical chromatographic columns operating in parallel will have to be monitored individually by the operators or have individual monitoring and control instrumentation. In this case, each column is effectively a separate line.

Many separation operations have an upper limit on the capacity which can be handled by a single unit, and this in turn limits the production scale for which that operation will be considered. In some cases, the upper limit represents a true limitation imposed by a physical phenomenon, while in other cases, the upper limit is simply the maximum scale on which the commercial equipment presently is manufactured. In either case, future developments are apt to make any tabulation of "single-line" limits subject to future obsolescence.

At the risk of some entries being obsolete by the time of printing, a list of single-line upper limits for the various types of common separation process is given in Table 22.2-1. An entry of "no limit" in the table implies that some other equipment, either upstream or downstream of the separation, is almost certain to determine the single-line bottleneck.

22.2-4　Design Reliability

On all factors influencing the decision to choose one process in preference to another, design reliability is the most important. Regardless of any other considerations, the plant, when constructed, must work properly to produce an acceptable product that can be sold at a profit. The economic consequences of a design failure are too dire to accept an unworkable separation process design. Design reliability is not really definable in quantitative terms because it actually relates to the amount of testing and demonstration that must be done before a suitable commercial scale design is produced. A general statement regarding design reliability for several separation processes follows.

DISTILLATION
Reliable design methods for distillation have been developed over many years of industrial experience and extensive testing of commercial-scale equipment by Fractionation Research, Inc. A competent engineering firm usually can design a distillation process given a knowledge of the pertinent physical properties of the components and vapor–liquid equilibrium information on the significant binary-pair components of the mixture to be separated. Occasionally, some small-scale testing is required, but scale-up methods for distillation are the most reliable of all the separation methods.

EXTRACTION

The design of extraction processes requires a knowledge of the phase-equilibrium relationships between the components to be separated and the extraction solvents to be considered. With this information and a knowledge of the pertinent physical properties (densities, interfacial tension, viscosities, etc.), one usually can project a preliminary design that is reliable enough to determine the choice of solvents and operating conditions to be used and the type of extraction equipment to use. However, the designs generated in this manner are not nearly so reliable as those for distillation. Consequently, small-scale testing on the type of equipment to be used in the plant is required. Given such tests, the vendors of proprietary extraction equipment can provide a reliable scale-up to the commercial-size equipment. For nonproprietary equipment, reliable scale-up can be done for mixer–settlers, sieve tray columns, and columns packed with Raschig rings. Vendors of proprietary packings also claim to be able to scale-up reliably, but there is little if any public data available to assess the reliability of scale-up methods.

CRYSTALLIZATION

The design of crystallization equipment is quite difficult. Knowledge of the phase equilibrium and physical properties does not allow one to predict the behavior of the crystallization process. Bench-scale experiments on the actual stream always are required, and equipment vendors usually require a pilot-plant test in small commercial equipment before designing the equipment. Even with such testing, operating adjustments usually must be made on the actual commercial installation before the equipment will operate to yield an acceptable product. Even with such testing, there are occasional failures at the commercial scale. However, since crystallization often can offer the potential for the purest product attainable by any separation process at considerably less energy consumption than other methods, we can expect to see its use increase substantially in the future. Many of the design reliability deficiencies are likely to diminish as more widespread use and large-scale testing occur.

ADSORPTION

Adsorption equipment and operating cycles usually can be designed reliably on the basis of measured adsorption isotherm data and a reasonable number of relatively small-scale experiments to determine the mass transfer characteristics of the stream to be treated in the chosen adsorbant bed. If the stream treated contains several adsorbed components it is generally necessary to run experiments on the actual mixture to be treated, since multicomponent isotherm behavior cannot be predicted in general from the individual isotherms.

REVERSE OSMOSIS

The design of reverse osmosis units usually requires that a single module or a smaller-scale test unit be operated using the actual feed stream to be treated. The tests are conducted not only to determine the operating characteristics of the membrane, but also to determine what pretreatment of the stream may be necessary to protect the membrane from fouling and structural damage. Once the appropriate tests have been made, however, the larger-scale design can proceed with confidence since most large-scale reverse osmosis installations consist of a number of modular units operating in parallel.

MEMBRANE GAS SEPARATION

The comments on reverse osmosis apply equally well to gas separation with selective membranes.

ULTRAFILTRATION

The evolution of a design for an ultrafiltration process requires extensive testing on the bench scale and normally requires pilot-plant operation. Not only must the equipment design be determined by testing, but also the operating cycle and recirculation flow rates must be resolved to prevent excessive fouling or "concentration polarization." There are a number of ultrafiltration configurations from which to choose (e.g., hollow fiber, plate-and-frame, and spiral-wound). If an optimum choice is to be made, all the configurations to be considered must be tested, since it is not possible to predict the characteristics of one configuration or vendor source from the data obtained on another. One also must determine by test the passage and retention of the components to be filtered and how this changes with time and operating conditions. The rated molecular weight cutoff associated with a particular membrane is only a nominal rating, and the actual fractional retention may vary considerably between components of the same molecular weight.

ION EXCHANGE

Although ion-exchange systems can be designed with the aid of only a few laboratory tests, a pilot-plant operation of substantial duration generally is recommended. Ion exchangers are subject to fouling and loss of capacity in operation which cannot be detected readily by laboratory tests. Therefore, it is desirable to operate a pilot plant throughout a number of cycles using the commercial plant regeneration procedures and, if at all possible, the actual stream to be treated in the commercial installation.

DIALYSIS AND ELECTRODIALYSIS

The comments on reverse osmosis apply equally well to dialysis and electrodialysis with selective membranes.

ELECTROPHORESIS

Electrophoresis, at this time, is an inherently small-scale process. While there is considerable development underway to increase the capacity of electrophoresis equipment, the only way to determine the separation characteristics at present is to test the stream to be treated in the actual equipment to be used in the commercial installation. Thus, the only reliable scale-up possible consists of using multiple parallel units of the largest type tested.

CHROMATOGRAPHIC SEPARATIONS

Scale-up of chromatographic separations usually is done by testing at several sizes following a sequence such as: analytical column, 1 in. (245 mm) diameter preparative column, 6 in. (15 cm) diameter column, 3 ft (1 m) diameter column, with optimum injection–elution policy determined at each step. Each scale of operation has its own distribution and packing method problems associated with it, which can be solved only partially by testing at a smaller scale.

GEL FILTRATION

Gel filtration is a chromatographic operation, and all the comments regarding chromatography also apply to gel filtration. There are additional problems associated with gel filtration, however, because often the packing consists of a swollen gel with little structural strength. Thus, the maximum packing height is an additional factor to be determined at each scale of operation. Since such gels depend largely on wall friction for their support, the maximum height of packing decreases with increasing column diameter.

22.2-5 Necessity for Pilot Plant

The need for pilot-plant operation to test separation process design is discussed under the various unit operations in the section on design reliability. Aside from the considerations of separation design reliability, it may be necessary to pilot plant a process to test the integrated operation of all the units. If pilot plant operation is needed for that purpose, then all the units must be included in the pilot plant whether or not their individual testing is required for design reliability.

22.2-6 Number of Steps Required

A strong variable in the cost of separation processes is the number of processing steps required to accomplish the desired separation. However, the number of steps is not by any means an absolute determinant of the cost of a separation process. There are many instances in which it is more economical to install a separation sequence incorporating two or more steps when it would be possible to accomplish the separation with a single-step process. A specific example is heavy water separation, which can be accomplished in a single distillation column with very many stages and very high reflux but is accomplished more economically in a multicolumn chemical-exchange process involving hydrogen sulfide. Nevertheless, in the majority of cases, a single-step process will prove more economical than a multistep process. A number of separation methods are discussed with regard to their ability to accomplish essentially complete separations in a single step.

DISTILLATION

Unless there is an azeotrope formed between the components to be separated, it is possible to separate completely a binary mixture into its components in a single distillation column. In some cases it is even possible, with the use of sidestreams, to obtain more than two pure components from a single distillation column. In general, a multicomponent mixture in which there are no azeotropes and no sidestreams withdrawn can be separated completely in a distillation train using one column less than the number of product streams required.

EXTRACTION

Except in the rare case in which the desired product form is a solution of the extracted product in the extraction solvent and the solvent does not contaminate the raffinate, complete component separation cannot be accomplished by extraction. In fact, extraction is a selective transfer process rather than a separation process. To obtain the components of the original mixture in the specified purity, it is necessary to separate the extract stream by recovering the extracted component from the extracting solvent. Usually, it is also necessary to recover the small amount of solvent dissolved in the raffinate stream to prevent either excessive solvent make-up costs or contamination of the raffinate product. Frequently, these auxiliary separations are accomplished by distillation. Therefore, an extraction process virtually always requires at least two steps to separate a feed into two acceptable product streams.

CRYSTALLIZATION

In those cases in which complete solid solution with no eutectic formation occurs, it is possible to separate completely a binary mixture into its components via countercurrent multistage melt crystallization. However, such systems are the rare exception, whereas eutectic formation is the general rule in solid–liquid equilibria. Therefore, in the overwhelming majority of cases, the degree of separation attainable by crystallization is limited by the eutectic composition. To accomplish a complete separation, it is necessary to couple an auxiliary separation step to the crystallization process to break the eutectic.

In the case of solution crystallization, one product is usually the pure crystal with occluded mother liquor and the other is either the mother liquor (a saturated solution of the crystalline material in the solvent) or a vapor or condensate of the solvent. The solid product stream must be subjected to a subsequent wash and filtration step (often with centrifugation) to recover the crystalline material in pure form.

ADSORPTION

Adsorption processes deposit one or more components selectively onto a granular solid material. If the adsorbing material is to be reused or the adsorbed material is to be recovered as a product, the adsorption process must be coupled with a regeneration step. Thus, adsorption is inherently a two-step process.

REVERSE OSMOSIS

The products from a reverse osmosis separation step consist of a purified solvent stream and a solution somewhat more concentrated than the feed stream. Therefore, it is not a complete separation process and must be accompanied by additional processing steps if complete separation or high recovery of solvent is desired. The degree of recovery or extent of concentration of the reject stream is limited by the osmotic pressure of the concentrated solution. The membrane must maintain its integrity and permit reasonable flux of solvent with a pressure difference across the membrane exceeding the osmotic pressure. In general, excessive osmotic pressures are encountered when the solute mole fraction in the concentrated solution exceeds 0.03–0.06.

MEMBRANE GAS SEPARATION

It is theoretically possible to accomplish a complete separation of a pair of gases by membrane separation, either by very highly selective membranes combined with very low partial pressure of the permeating component on the permeate side of the membrane, or by multistage cascades. However, it usually is not economical to accomplish high purity and high recovery of both components of a binary by membrane separation alone. Therefore, it usually is necessary to couple membrane gas separation with other processing steps when nearly complete separation is required.

ULTRAFILTRATION

Theoretically, ultrafiltration can separate two high-molecular-weight solutes from each other completely if the difference in molecular size is sufficiently large (differing by a factor of 10). When the molecular weight–size difference is not quite so great, it still may be possible to accomplish the complete separation by cascading. In any case, the two product streams from ultrafiltration consist of very dilute solutions and additional processing steps will be required if pure products are required. If the desired product form is a more concentrated solution, ultrafiltration may be used to accomplish the concentration.

ION EXCHANGE

Ion exchange is similar to adsorption in that a regeneration step must be included to recover the product and/or reuse the bed. In addition, the product streams from ion exchange are usually aqueous solutions and additional processing steps must be included if the products are desired in pure form. In the case of water purification, probably the most widely used application of ion exchange, the deionized water effluent from the ion exchanger is the desired product and further processing is unnecessary.

DIALYSIS AND ELECTRODIALYSIS

Dialysis processes, like extraction, are selective transfer processes rather than complete separation processes. Therefore, if the transferred components are desired in pure form, additional processing steps are required. Unlike extraction, the solvent does not contaminate the retained stream since the solvent on both sides of the membrane is the same.

ELECTROPHORESIS, CHROMATOGRAPHIC SEPARATION, AND GEL FILTRATION

Electrophoresis, chromatography, and gel filtration all separate components in dilute solution. They must be followed by concentration steps if a dissolved product is satisfactory, and additional processing steps if pure products are desired.

22.2-7 Capital and Capital Related Costs

It is not appropriate to attempt to give a comprehensive treatment of methods of capital estimation here. There are a number of good texts and references available for that purpose. It is appropriate, however, to discuss the various factors in the selection of a separation process and, qualitatively, their effect on the capital required. When we compare two alternative separation processes, we choose the one with greater capital only if the revenue generated (savings in expense) is great enough to justify the expenditure of the difference in capital requirements between the alternatives. It is pertinent to recognize that there are certain expenses generated by or correlated to the capital itself. Such capital related expenses (e.g., depreciation and maintenance) generate annual expenses in the range of 20–25% of the capital expenditure. Among the factors that affect capital are the following.

SCALE OF OPERATION
A widely used rule for estimating the effect of scale of operation on capital requirement is to make capital proportional to the 0.6 power of capacity. While this usually works quite well when applied to the total cost of large plants and to individual operations that can be carried out in a single unit regardless of the capacity required, it must be modified considerably for small-scale operations and for operations that must be carried out in a number of parallel lines. For very small-scale operations, the required capacity may be below that of the minimum feasible size of certain critical equipment items, in which case the capital cost is virtually independent of the scale of operation. An example of this is a situation in which instrumentation is the dominant cost and the same instrumentation can be used regardless of the scale of operation. On the other hand, when multiple parallel lines must be used because the capacity of the plant is greater than the maximum single line size for the method of separation chosen, the capacity exponent tends to be between 0.9 and 1, rather than the conventional 0.6. For those separation methods using a single line of operation composed of multiple modules, the exponent tends to lie between 0.6 and 0.9. The scale of operation applicable to a number of separation methods is discussed in Section 22.2-3.

DESIGN RELIABILITY
Capital requirements are affected indirectly by design reliability in that larger safety factors are applied when design methods are less reliable. There is also a tendency to install more extensive control systems and more spare backup equipment when there is uncertainty in the reliability of the design methods. Such practices significantly increase the capital requirement for those methods with low design reliability. The design reliability of the various separation processes is discussed in Section 22.2-4.

NECESSITY OF A PILOT PLANT
Pilot plants normally are considered part of the research and development expense and usually do not represent a part of the capital of a specific plant. In some cases, especially where design reliability is low, the pilot plant may continue to be maintained as part of the plant and used for trouble-shooting or to test proposed plant modifications. In such cases, the cost of the pilot plant becomes a component of the plant capital. Section 22.2-5 discussese the need for pilot plants for the various methods of separation.

NUMBER OF STEPS REQUIRED
A crude method of very preliminary capital estimation is to count the number of processing steps required and multiply by the capacity and a cost factor. While such a method is not recommended for anything other than very preliminary qualitative thinking, it does point out the fact that capital is related directly to the number of steps in the complete process. A method of separation that can accomplish a complete separation in a single step is usually a lower capital process than one that requires a number of steps. The number of steps associated with the various separation methods is discussed in Section 22.2-6.

22.2-8 Energy Requirements

One of the most important considerations in choosing among separation processes is the energy requirement of the process. Energy cost is often the dominant operating expense, and energy requirements also have a direct influence on a portion of the capital in the form of heat exchangers, compressors, pumps, and so on. "Energy requirement" can be a deceptive term, however, because our real concern regarding energy costs relates not to the quantity of energy that flows through the process but rather to the amount of fuel that must be burned to induce that flow of energy. If we had only the first law of thermodynamics to consider, these two quantities would be in direct proportion. Second law effects distort the relationship between energy flow and fuel burned, thereby forcing us to consider not only the quantity of energy flow but also the quality of the energy that flows through a process. As always, the ultimate choice is based on an economic comparison of the specific cases considered rather than generalizations. However, it is possible to draw some general conclusions regarding comparisons between separation processes which allow one to reduce the number of alternatives to be given the full economic analysis. The discussion that follows is based on the paper by Null.[1]

Typically, a large plant will have associated with it a steam system that supplies both power and heat requirements to the various processes in the plant. Steam is generated at a fairly high temperature and pressure and delivers work by expansion through a turbine. The maximum work that can be extracted from expansion is obtained when the steam is expanded to a low pressure and condensed at a temperature slightly above the ambient air temperature. Generally, the heating value of the fuel burned to deliver this work will be three to four times the quantity of work delivered via this route, depending on the efficiency of the utility steam system. To supply the heating requirements of the plant, steam must be diverted from this expansion path at a sufficiently high temperature that it can be transferred into the process at the point of use. As a result, the amount of work delivered by that steam will be reduced by the following amount:

$$w_{eq} = \frac{qe_p(T_s - T_0)}{T_s} \tag{22.2-1}$$

Since the diverted power must be replaced by additional steam expanded to the ultimate condensing pressure, the fuel heating value required to deliver this heat is given by

$$q_f = \frac{w_{eq}/e_p}{e_c} = \frac{q(T_s - T_0)/T_s}{e_c} \tag{22.2-2}$$

When refrigeration is required to move heat from a lower temperature to a higher temperature, work must be supplied by burning fuel, giving

$$w = \frac{q(T_0 - T_r)/T_0}{e_r} \tag{22.2-3}$$

$$q_f = \frac{q(T_0 - T_r)/T_0}{e_r e_p e_c} \tag{22.2-4}$$

where

$$e_c = \frac{e_b(T_{s,\max} - T_0)}{T_{s,\max}}$$

Energy requirements are based on the ultimate fuel required as given by Eqs. (22.2-2) and (22.2-4) with the recognition that the temperatures T_s and T_r do not represent a continuum of temperatures, but rather only the discrete temperatures at which the steam and refrigeration utilities are available.

We can make generalizations only if we make extensive approximations. While not all methods of separation can be readily generalized, some are summarized in the following discussions. To compare the fuel efficiencies of different separation processes, we derive the approximate equation for fuel equivalent for each and compare the values.

DISTILLATION
Since distillation historically has been the separation method of choice whenever it could be used, it represents a suitable base case against which to compare the energy economy of other separation processes. We can approximate the heat requirement of distillation processes by

$$q_d = D\,\Delta H_v(1 + R_d) \tag{22.2-5}$$

with approximately the same quantity of cooling required for condensation. The fuel requirements for the reboiler are

$$q_{fr} = \frac{D\,\Delta H_v(1 + R_d)(T_s - T_0)/T_s}{e_c} \tag{22.2-6}$$

If the condenser is cooled by air or water, the fuel requirement is negligible compared to the reboiler, but if refrigeration is required a fuel penalty accrues to the condenser as well as the reboiler:

$$q_{fc} = \frac{D\,\Delta H_v(1 + R_d)(T_0 - T_r)/T_0}{e_p e_r e_c} \tag{22.2-7}$$

On the other hand, if useful heat is recovered from the condenser, a fuel credit accrues to the condenser:

$$q_{fc} = \frac{-D\,\Delta H_v(1 + R_d)(T_{s'} - T_0)/T_{s'}}{e_c} \tag{22.2-8}$$

The net fuel penalty for a distillation column is given by Eq. (22.2-6) alone or modified by adding either Eq. (22.2-7) or (22.2-8) if appropriate.

EXTRACTION

If we examine only the typical extraction process, the only direct requirement appears to be the pumping energy to bring the feed and solvent together and the power for internal agitation if internal agitation is used. These energy requirements are certainly negligible in comparison to the energy requirements for distillation. However, the extraction column does not accomplish the desired separation alone; it merely selectively transfers one or more components into the solvent. To complete the separation it is necessary to separate the extracted product from the solvent. In most cases it is also necessary to remove the small amount of dissolved solvent from the raffinate. These downstream separations are accomplished most frequently by distillation, and the energy requirements of the downstream separations determine the true energy requirements associated with the extraction process.

If the solvent is less volatile than the components of the feed stream, the fuel equivalent for the recovery of the extracted product from the solvent is approximated by

$$q_{f,\text{ext}} = \frac{E \, \Delta H_v (1 + R_{\text{ext}})(T_{s,\text{ext}} - T_0)/T_{s,\text{ext}}}{e_c} \tag{22.2-9}$$

and the fuel equivalent for recovery of solvent from the raffinate is given by

$$q_{f,\text{raf}} = \frac{R \, \Delta H_v (1 + R_{\text{raf}})(T_{s,\text{raf}} - T_0)/T_{s,\text{raf}}}{e_c} \tag{22.2-10}$$

If the solvent is more volatile than the feed components, the fuel equivalent for the recovery of extract product is given by

$$q_{f,\text{ext}} = \frac{S \Delta H_{vs} (1 + R_{\text{ext}})(T_{s,\text{ext}} - T_0)/T_{s,\text{ext}}}{e_c} \tag{22.2-11}$$

and the fuel equivalent for solvent recovery from raffinate is given by

$$q_{f,\text{raf}} = \frac{R x_s \, \Delta H_{vs}(1 + R_{\text{raf}})(T_{s,\text{raf}} - T_0)/T_{s,\text{raf}}}{e_c} \tag{22.2-12}$$

It is possible, of course, that the solvent might be less volatile than the raffinate and more volatile than the extract, or vice versa. In any event, the total fuel equivalent is determined by adding the appropriate subset of Eqs. (22.2-9)–(22.2-12). These equations assume that the overhead condensers in the recovery distillations accrue neither penalty (refrigeration) nor credit (heat recovery). If this is not the case, the appropriate adjustments may be made in the manner described in the paragraphs on distillation.

While we cannot make generalizations to cover every case, we can draw some conclusions regarding the energy efficiency of extraction compared to a distillation process for accomplishing the same separation. For example, Eqs. (22.2-9) and (22.2-10) compared with Eq. (23.2-6) lead to the conclusion that extraction is more efficient for the nonvolatile solvent whenever

$$1 + R_d > \left(\frac{T_s}{T_s - T_0}\right)\left(\frac{F}{D}\right)\left[\left(\frac{E}{F}\right)\left(\frac{(1 + R_{\text{ext}})(T_{s,\text{ext}} - T_0)}{T_{s,\text{ext}}}\right)\right.$$
$$\left. + \left(\frac{R}{F}\right)\left(\frac{(1 + R_{\text{raf}})(T_{s,\text{raf}} - T_0)}{T_{s,\text{raf}}}\right)\right]$$

For the case of the volatile solvent, a similar comparison gives

$$1 + R_d > \left(\frac{T_s}{T_s - T_0}\right)\left(\frac{F}{D}\right)\left[\left(\frac{S}{F}\right)\left(\frac{(1 + R_{\text{ext}})(T_{s,\text{ext}} - T_0)}{T_{s,\text{ext}}}\right)\right.$$
$$\left. + \left(\frac{R x_s}{F}\right)\left(\frac{(1 + R_{\text{raf}})(T_{s,\text{raf}} - T_0)}{T_{s,\text{raf}}}\right)\right]\left(\frac{\Delta H_{vs}}{\Delta H_v}\right)$$

The qualitative observations which can be drawn from the above criteria are that extraction is likely to be more energy efficient than distillation in those cases where a very large distillation reflux ratio is required, and a large fraction of the feed must be taken overhead in distillation. On the other hand, distillation will be favored if the recovery steps of the extraction process require significant reflux, the distillation reflux ratio is small, or if only a relatively small fraction of the feed must be taken overhead in distillation.

CRYSTALLIZATION

The heat requirement for continuous countercurrent crystallization can be approximated in a manner similar to that used for distillation:

$$q_c = B_c(1 + R_c)\Delta H_f \tag{22.2-13}$$

with about the same amount of cooling required at the cold end of the cascade. The equivalent fuel associated with the melting is given by

$$q_{fc,h} = \frac{B_c(1 + R_c)\Delta H_f(T_{sc} - T_0)/T_{sc}}{e_c} \tag{22.2-14}$$

The freezing or crystallization end of the cascade usually requires some degree of refrigeration with the associated fuel equivalent given by

$$q_{fc,ref} = \frac{B_c(1 + R_c)\Delta H_f(T_0 - T_{rc})/T_0}{e_r e_p e_c} \tag{22.2-15}$$

Comparison of the above equations with Eq. (22.2-6) yields the criterion that crystallization is more energy efficient than distillation when

$$\frac{1 + R_d}{1 + R_c} > \left(\frac{\Delta H_f}{\Delta H_v}\right)\left(\frac{B_c}{D}\right)\left\{\left(\frac{T_{sc} - T_0}{T_{sd} - T_0}\right)\left(\frac{T_{sd}}{T_{sc}}\right) + \left(\frac{1}{e_p e_r}\right)\left(\frac{T_0 - T_{rc}}{T_{sd} - T_0}\right)\left(\frac{T_{sd}}{T_0}\right)\right\}$$

Since the value of $\Delta H_f/\Delta H_v$ is typically about 0.2 and the required reflux $R_c \ll R_d$, the above criterion usually shows a potential for a substantial energy advantage for melt crystallization whenever it can be used to accomplish the same separation as distillation. This conclusion holds true even when refrigeration is required to a fairly deep level. Discussions in previous sections regarding design reliability, scale of operation, and number of steps required indicate why crystallization is not used more widely in spite of its inherent energy advantages.

GAS ADSORPTION

The energy requirement associated with gas adsorption is the energy needed to regenerate the spent bed. This is accomplished most frequently by heating the entire bed to such a temperature that the adsorbed material is desorbed and leaves as a vapor. The heat of regeneration includes the sensible heat of the entire bed and the vessel containing it plus the heat of adsorption of the adsorbed material. The heat of adsorption is greater than but of the same order of magnitude as the heat of vaporization for physically bound adsorbates, and much greater than the heat of vaporization for chemisorption. A similar analysis to that used above shows that adsorption is more energy efficient than distillation whenever

$$1 + R_d > \left(\frac{N_{ads}}{D}\right)\left(\frac{\Delta H_{reg}}{\Delta H_v}\right)\left(\frac{T_{sd}}{T_{s,reg}}\right)\left(\frac{T_{s,reg} - T_0}{T_{sd} - T_0}\right)$$

Since the least-volatile component is usually the material most strongly adsorbed, in cases in which adsorption or distillation could be used, $B = N_{ads}$. Thus, the criterion above indicates that gas adsorption is preferable to distillation in those cases in which trace quantities of nonvolatile material are to be removed from the bulk of the feed stream. These are the instances in which adsorption normally is considered regardless of energy requirements.

REVERSE OSMOSIS

Reverse osmosis uses a membrane to separate a virtually pure solvent from a solution while simultaneously concentrating the solutes in the solution. The motive force for the separation is the pressure drop across the membrane, which must be considerably greater than the osmotic pressure of the solvent in the concentrated effluent to generate an acceptable flux through the membrane. The fuel equivalent is related to the pumping work required by the expression

$$q_{f,ro} = \frac{w_{ro}}{e_c} = \frac{-F(P/\pi)R_g T_{ro} \ln(\gamma x_{rej})}{e_c} \tag{22.2-16}$$

Reverse osmosis will have a lower energy requirement than most competing methods of separation, since its work requirement differs from the minimum work of separation only by the ratio P/π. However, it is limited to modest concentration of very dilute solutions because the osmotic pressure becomes quite large at solute concentrations exceeding 5 mol %. Thus, the application of reverse osmosis is limited to those cases in which membranes have been developed with the desired selectivity and with sufficient structural integrity to withstand the required pressure drop.

Other Methods of Separation

The energy requirements for other methods of separation are not reduced so readily to simple approximations. For example, energy requirements for gas separation by membranes are strongly dependent on the pressure at which feed is available and the required pressures of the permeate and permeant products, as well as the number of cascade stages required for the separation. Ion-exchange energy requirements are indirect inasmuch as they are associated with the energy required to produce the chemicals used for regeneration. Separation methods that require dilute liquid solutions, such as chromatographic separations of all kinds and electrophoresis, have energy requirements related to the handling and purification of the large quantities of water or carrier gases inherently associated with the process. While the direct energy requirements may be small at times, the indirect energy requirements are usually very large when compared to more conventional bulk separation methods such as distillation and extraction.

22.3 CHOOSING THE BASE CASE

A great deal of research has been reported in the synthesis of process flowsheets in recent years, and attempts have been made to develop systematic algorithms whereby the best separation sequence can be synthesized with a minimum of input to a computer program. In practice, separation processes rarely are chosen in this manner. Neither the computational algorithms nor the quantitative descriptions of all the separation methods is developed sufficiently well at this time to allow such automatic choice between separation methods. However, current developments in process synthesis can be of value in choosing a near-optimum sequence of steps all involving a single type of separation process, such as distillation.

Current practice is to make a qualitative judgment, based on the several criteria described in Section 22.2 of the methods or combination of methods which have the greatest probability of providing a process to accomplish the desired separation. Out of such a list one is chosen as a base case. This choice will represent the qualitative judgment of the process designer as to which of the processes listed is likely to be the most economical method of separation that can be developed within the time and money constraints of the project. *If distillation is among the list of possible methods, it almost always will be the first base case chosen.* A rereading of Sections 22.2-1–22.2-6 will make it quite evident why distillation is favored when it can accomplish the desired separation. Following distillation, the next most likely base case would use extraction when possible. Crystallization, adsorption, and ion exchange would be the next most likely base case choices, with all the others following in no particular order.

Once the base case has been chosen, process designers will propose a design and determine capital and operating costs in as much detail as the current knowledge of the properties of the feed stream mixture and its components and operating characteristics of the chosen unit operation warrants. At this stage, assumptions are made and noted when needed data are not available. Determination of these needed data becomes the subject of subsequent research and development. After the base case has been evaluated, a judgment is made regarding the next most likely economic competitor process. The challenger process then is evaluated with regard to capital and operating costs with assumptions noted in the same manner as the base case. If the challenger process has lower capital and operating costs, it then becomes the new base case. In case the process with the lower operating cost requires the greatest capital, the saving in operating cost is compared to the difference in capital to determine whether the return generated by the savings meets the corporate criterion for return on capital. If it does, the lower operating cost process becomes the base case. If not, the lower capital process becomes the base case.

This sequential procedure is continued until all the candidate processes have been evaluated. However, the final choice has not been made necessarily at this point. In the early stages of process development, many assumptions have to be made regarding unavailable data. The assumptions made in the latest base case must be either verified or modified as additional experimental data are obtained. As additional information is obtained, both capital and operating costs usually rise compared to earlier estimates. This may force a reevaluation of the base case and some of the closest alternatives. The reevaluation may dictate a change in the base case choice. In addition, initial evaluations are done on process schemes that have not been optimized. If the evaluations indicated a large difference between alternatives, optimization of the processes usually will not change the choice. However, in cases where the economic differences between alternatives is not large, optimization of the processes can change the choice of a separation process.

When the continuing activities of evaluation, obtaining data, reevaluation, optimization, and demonstration are completed, the final base case will be the actual process installed.

22.4 PROCESS SIMULATION

In recent years, process simulation has evolved to a highly sophisticated stage. Many large corporations and a number of computer service organizations have their own proprietary versions of simulation programs for complete flowsheets as well as individual unit operations. Not all separation methods are covered in these software packages, but most include at least distillation, absorption, and extraction. In cases where subroutines are not available for the specific separation method of interest, there is usually a "generic separation" subroutine available that at least will allow the computation of the overall material and energy balance of the operation and its effect on the complete flowsheet. Some simulation programs also have cost estimation capability built into the program.

Process simulation programs represent a very powerful tool for the process engineer in the selection, optimization, and design of separation processes. If the procedures suggested in this chapter are to be carried to fulfillment, many evaluations will have to be made before a final choice is possible. When properly used (as a powerful tool and not as a substitute for engineering judgment), process simulation can decrease greatly the time and personnel required to reach a decision among separation processes. It even can be used in the early stages of process development to guide the course of the research.

There is often a hesitancy to use simulation programs in the early stages of process development because many of the parameters required as input to the program are not well known. In such cases, it is proper to use estimated values that bracket the range of probable expected values and to test the simulation program's sensitivity to parameters. In this way, we can decide which properties must be determined experimentally in order for a valid choice to be made. It is also appropriate, in the early stages of process selection and development, to reduce the computation time required in the simulations by using short-cut methods in the unit operation subroutines *if reasonably accurate short-cut subroutines are available*. In so doing, however, the engineer should be aware of the assumptions implicit in the short-cut computations and should use rigorous simulations where the short-cut assumptions can result in serious errors. The process designer also should recognize that the early simulation results are tentative and as the development or design proceeds toward the final selection or design, the simulations will be repeated a number of times with a shift from assumed ranges of parameter values and short-cut methods to firm values of parameters and rigorous simulations.

22.5 PROCESS SYNTHESIS

In recent years a considerable amount of literature has appeared in chemical engineering journals on the subject of "process synthesis." Nishida et al.[2] reviewed the current state of the art as it applies to chemical processes. If process synthesis methodology were developed fully, one could invoke a computer program which, with minimal input, would select not only the separation method of choice but also would determine how the various unit operations should be connected together to form the entire flowsheet. At the present time, however, process synthesis methodology for separation processes is in the early stages of research.

Very little process synthesis literature has dealt successfully with the problem of selection between alternative methods of separation. Indeed, most process synthesis literature has dealt with the selection of the flowsheet sequences for a single separation method with only simple, sharp separations between components of adjacent selectivity, without recycle. Furthermore, most of this literature has used only distillation in illustrative applications. In practice, much use is made of recycle, nonsharp splits, and complex columns (in distillation); and, of course, distillation is not always the separation method of choice.

While process synthesis research has not yet produced the solution to the problem of generation of the best flowsheet, a number of heuristic rules have been stated by various process synthesis researchers which can be of considerable aid in guiding the selection of the near optimum or base case and for systematically choosing those alternate sequences to be investigated. Some of the heuristic rules developed will be competing and in some cases even contradictory, but nevertheless they do serve to reduce the enormous amount of work that would be involved in evaluating all possible separation sequences.

Among the more widely accepted heuristics for selection of separation processes are the following:

1. Favor distillation unless relative volatilities are less than 1.05.
2. Do the easiest separations first.
3. When selectivities do not vary widely, remove the component having the highest mole fraction first.
4. Remove the component with the highest volatility first when distillation is feasible.
5. When mass separating agents are used (i.e., extractive distillation, or extraction), perform the recovery steps for the mass separating agent and/or dissolved products in the immediately following step.
6. Do no use a second mass separating agent to remove or recover a mass separating agent.

While these rules sometimes do not yield the optimum flowsheet, when applied with reasonable judgment they can lead to an acceptable base case against which alternative flowsheets can be prepared.

In addition to attempting to provide heuristic rules for choosing the base case, process synthesis research attempts also to develop systematic methods for the evolution of the flowsheet from the base case to the optimum case. Such methods are still under development and no generally accepted strategy has emerged yet. However, two useful methods which have been proposed are (1) to challenge one by one the heuristics which led to the base case selection and (2) some form of branch and bound technique. The many branch and bound strategies that have been proposed use the common principle that, once a base case has been established, an alternate sequence need be evaluated only through the step in which its cost or other objective function exceeds the base case.

Process simulation is *now* a powerful tool for process design. As process synthesis methodology matures, it is reasonable to expect process simulation to be a powerful subset of process synthesis. Before that becomes a reality, however, much must be done to currently available process simulation packages. More separation operations must be incorporated into the simulation packages with their associated cost calculations. The most serious deficiency in the simulation software, however, is the absence of short-cut computations of sufficient accuracy to lead to valid conclusions.

In any process synthesis algorithm, simulations must be done so many times that only very rapid simulation computations are possible. Usually, this means short-cut computations even with high-speed computers. Unfortunately, short-cut calculations available today have implicit in them assumptions that can lead to completely erroneous conclusions. Distillation calculations provide a ready example. Usually, process synthesis investigations of distillation sequences have made use of Fenski–Underwood equations to simulate the individual columns. These equations carry the assumption of constant relative volatility of all the components. In many cases, this assumption merely affects the magnitude of the results but does not alter the conclusions reached. In other very commonly occurring cases, an unworkable flowsheet will result. Any computation method that assumes constant relative volatility cannot simulate an azeotrope, a very common occurrence in petrochemical processes. The constant relative volatility assumption will lead to the conclusion that the key components can be separated in a single distillation column to the degree of purity and recovery specified, although in some cases an unacceptable number of stages may be required. In actual practice an azeotropic system will require a minimum of two distillation columns although the number of total stages might be modest. Unfortunately, at the present time, only rigorous methods have the capability to simulate the azeotropic system sufficiently well to produce results of the required quality. It is currently more fashionable to spend research effort on the structure of the process synthesis methodology than on the development of high-quality, rapid computation algorithms for the individual unit operations. However, the ultimate utility of the process synthesis software to the process designer will depend on the availability of the rapid, high-quality simulation of the individual separation operations.

Process synthesis, at the present time, falls far short of its goal of providing a systematic procedure that assures us of the selection of the optimum methods of separation and the optimum sequence of those methods. Therefore, for some time to come, we will continue to have to rely on the more subjective and qualitative methods described in previous sections. However, we can begin to adapt some of the methodology being generated by process synthesis research to some phases of the selection of separation processes. We also can continue to encourage the development of process synthesis methodology. In the author's view, however, this development will be slow to come regarding selection between methods of separation. We should expect more rapid progress in the optimization of the separation sequence for a given method of separation such as distillation.

NOTATION

B_c	molten product molar flow rate from high-temperature end of a melt crystallizer
D	molar flow rate of distillation overhead product
E	molar flow rate of extract stream from extractor
e_b	combustion efficiency of the steam plant, that is, ratio of heat actually transferred to steam cycle to the heating value of the fuel
e_c	combined combustion and Carnot efficiency of the steam plant operating across its maximum temperature span
e_p	power cycle efficiency, that is, ratio of the actual power delivered to that of a Carnot engine across the same temperature span
e_r	refrigeration cycle efficiency, that is, ratio of power required by a reverse Carnot cycle to the actual power required by the refrigeration cycle across the same temperature span
F	molar feed rate to a reverse osmosis unit
ΔH_f	heat of fusion of crystallizer product
ΔH_{reg}	heat of regeneration per mole of adsorbed material in a gas adsorption process
ΔH_v	heat of vaporization of distillation overhead
ΔH_{vs}	heat of vaporization of extraction solvent
N_{ads}	average molar rate of adsorption of adsorbed material
P	pressure on retained fluid side of reverse osmosis membrane

q	rate of heat transfer
q_c	heat requirement in a melt crystallizer
q_d	heat requirement for distillation
q_f	fuel equivalent, that is, heating value of fuel required to deliver a power or heat requirement
q_{fc}	fuel equivalent for distillation condenser
$q_{fc,\,h}$	fuel equivalent for heating in melt crystallizer
$q_{fc,\,ref}$	fuel equivalent for refrigeration in melt crystallizer
$q_{f,\,ext}$	fuel equivalent for recovery of extract product in an extraction process
q_{fr}	fuel equivalent for distillation reboiler
$q_{f,\,raf}$	fuel equivalent for solvent recovery from the raffinate in an extraction process
$q_{f,\,ro}$	fuel equivalent of a reverse osmosis process
R	molar flow rate of raffinate from extractor
R_c	reflux ratio of melt crystallization process
R_d	reflux ratio of distillation process
R_{ext}	reflux ratio required for extract product recovery
R_g	gas law constant
R_{raf}	reflux ratio required for solvent recovery from raffinate
S	molar flow rate of extraction solvent
T_0	heat rejection temperature (absolute) of power cycle
T_r	temperature (absolute) of refrigeration utility requirement
T_{rc}	temperature (absolute) of refrigeration utility for distillation condenser
T_{ro}	temperature (absolute) of reverse osmosis process
T_s	temperature (absolute) of utility steam required for distillation reboiler
$T_{s'}$	temperature (absolute) of utility steam replaced by heat recovered from high-temperature distillation condenser
T_{sc}	temperature (absolute) of utility steam required for melt crystallization
$T_{s,\,ext}$	temperature (absolute) of utility steam required for recovery of extract product from solvent
$T_{s,\,max}$	maximum temperature (absolute) of steam in the power cycle
$T_{s,\,raf}$	temperature (absolute) of utility steam required for recovery of solvent from raffinate
$T_{s,\,reg}$	temperature (absolute) of utility steam required for regeneration of adsorption bed
w	work
w_{eq}	power cycle work equivalent to a heat requirement
w_{ro}	work required for reverse osmosis process
x_{rej}	mole fraction of solvent in the retentate from a reverse osmosis process
x_s	Mole fraction of extraction solvent in the raffinate from an extractor
γ	activity coefficient of solvent in the retentate from a reverse osmosis process
π	osmotic pressure of the solvent in the retentate from a reverse osmosis process

REFERENCES

22.2-1 H. R. Null, Energy Economy in Separation Processes, *Chem. Eng. Prog.*, **76(8)**, 42–49 (1980).

22.2-2 N. Nishida, G. Stephanopoulos, and A. W. Westerberg, A Review of Process Synthesis, *AIChE J.*, **27**, 321–351 (1981).

Index